Soil Constraints and Productivity

Identifying, interpreting, and managing soil constraints are major challenges, especially when multiple constraints occur in the same soil at various depth zones. Although amelioration tools and strategies are available to manage some of these constraints, field adoption of these technologies is a major challenge to the farming community. *Soil Constraints and Productivity* helps in identifying and understanding soil constraints, focusing on management practices to alleviate problems associated with these restrictions and their impacts on crop productivity.

Soil Constraints and Productivity aims to:

- Describe various strategies suitable for mitigating soil constraints
- Provide data on cost-benefit analysis of managing soil constraints
- Provide case studies of managing soil constraints to increase productivity

Soil is essential for the doubling of major grain production proposed to be necessary to avoid major food security collapses in the future. This book will be a key resource for soil and environmental scientists, farmers, students majoring in agricultural and environmental sciences, and crop consultants.

Soil Constraints and Productivity

Edited by

N. S. Bolan and M. B. Kirkham

CRC Press
Taylor & Francis Group
Boca Raton London New York

CRC Press is an imprint of the
Taylor & Francis Group, an **informa** business

The cover photo was taken by Professor DeAnn Presley at the Kansas Agricultural Watershed located near Manhattan, Kansas, USA. The metal ring, which has an inside diameter of 5 cm, was being used by Prof. Presley to collect surface (0 to 5 cm depth) bulk density measurements just after soybeans were planted in 2022. She noticed a lot of crusting in the non-cover-cropped plots. (Her study compared cover-cropped and non-cover-cropped plots and what effect the cover crops had on water quality measurements.) The soil texture is a silty clay loam. The photo was taken close to the ground. (Photo courtesy of Prof. Presley, Department of Agronomy, Kansas State University.) Please refer to Chapter 9 in this book for crusting.

The Cooperative Research Centre for High Performance Soils (the Soil CRC) brings together scientists, industry and farmers to find practical solutions for Australia's underperforming soils and it is undertaking research relating to soil constraints and productivity. More information can be found at www.soilcrc.com.au.

First edition published 2023
by CRC Press
6000 Broken Sound Parkway NW, Suite 300, Boca Raton, FL 33487-2742

and by CRC Press
4 Park Square, Milton Park, Abingdon, Oxon, OX14 4RN

CRC Press is an imprint of Taylor & Francis Group, LLC

ISBN: 9780367554392 (hbk)
ISBN: 9781032472027 (pbk)
ISBN: 9781003093565 (ebk)

DOI: 10.1201/9781003093565

Typeset in Times
by codeMantra

Contents

Foreword

Soil is responsible for 95% of the food we eat.
Soil provides ecosystem services critical for life.
Soil is our life support system.
Soil helps to regulate the Earth's climate.
Soil stores more carbon than all of the world's forests combined.
It all starts with the soil.

These statements, singularly and collectively, help to describe the critical importance of soil to our planet, to our environment, to our economy, to our society, and to our survival. Readers of this book will already have a broad appreciation of the essential contribution that soil makes and will also know that its roles and functions extend way beyond the high-level statements in the opening paragraph.

Hence, it logically follows that we must do all that we can to allow soil to function to its fullest potential, whilst preserving and maintaining its ability to do so for generations to come. However, we face some major global challenges, which will have an impact on the way in which we manage soil now and into the future.

The first is global food security. The total food demand of the global population will double over the next 40 years as the population increases, and many countries will increase their per capita food demand as individual incomes increase and diets become more affluent. This places increased pressure on the soil as agriculture extracts an increasing toll on our soil resources.

The second is climate change. In many parts of the world, climate change manifests itself not just in increased average temperatures, but in increased frequency and severity of extreme weather events – droughts, floods, storms – which increase the likelihood of soil degradation events, especially erosion and soil loss. The impact on soil can be severe and sudden.

A third challenge is nutrient supply and soil fertility. As our rate of soil utilization and nutrient extraction increases with increasing agricultural intensity, the need to replace those nutrients also increases. For the greater part of the past century, this has been through the use of synthetic fertilizers. However, there is now increased pressure on fertilizer supply and use, as energy-intensive fertilizer production becomes more expensive, or reserves of natural resources become depleted, and the off-site impacts of excessive fertilizer use are better understood.

Individually, these global challenges are significant in their own right. Collectively, and through the interactions of each of the challenges on each other, they present a call to arms for soil scientists across the world to be devoting their efforts to research that provides farmers, industry, and governments with the tools and knowledge they need to better manage their soils and ensure that they are used to their full potential now and into the future.

The research and knowledge collected and presented in this book represent some of those efforts. Being able to define and identify the scope and scale of various constraints, and then identify possible strategies to address, ameliorate, or overcome those constraints, will hopefully provide the means by which we can better manage our soils to address these global challenges.

In so many areas of the world, land (and in turn soil) is a highly limited resource. The competition for land is such that the opportunity for expanding in a lateral direction is often restricted or highly regulated. However, there is almost always an opportunity for expanding vertically, or in the third dimension. That is, to increase the depth of effective rooting, from say 30 cm to 60 cm, or from 50 cm to 100 cm, by removing constraints to rooting depth.

That is one of the biggest benefits of identifying and addressing soil constraints – the ability to effectively acquire a greater amount of productive soil without having to purchase or lease more land, just by increasing the productive depth of the soil profile.

In Australia, the country of my nationality, there is a major research focus on identifying and addressing soil constraints, as many of the chapters in this book testify. For it to have maximum impact, we have come to understand that the research is most effective when it is end-user driven. To that effect, in 2017, the Australian Government provided funding for 10 years for the Cooperative Research Centre for High Performance Soils, better known as the Soil CRC (www.soilcrc.com.au). The Soil CRC, which I lead, is the largest collaborative soil research effort in Australia's history and involves scientists, industry and farmers working together to develop tools and knowledge to help farmers better manage their soils and increase productivity and profitability.

The Soil CRC brings together 39 participant organizations across Australia, including 8 universities, 4 state government agencies, 7 industry groups, and 20 farmer groups. The involvement and engagement of the farmer groups in the Soil CRC ensure that farmers are involved in the identification of priorities, the development of proposals, the implementation of projects, the interpretation of results, and the communication of findings. This ultimately leads to greater adoption and implementation of the research outputs.

The chapters of this book rightly reflect the diversity of soil constraints and productivity challenges across the world. However, what all authors have in common is a desire to ensure that rigorous scientific theory and practice are brought to bear to ensure that our best efforts are being made in identifying and addressing soil constraints and increasing productivity. So long as 95% of the food we eat has its genesis in the soil, it is only through the application of this science to soil management that we will be able to achieve sustained global field security in a changing climate.

I congratulate the editors, authors, and publishers of this book and hope that the research presented within informs and inspires soil scientists and soil managers at all levels around the world to better manage our soils – for our generation and for those to come.

Dr Michael C. Crawford
Chief Executive Officer,
Cooperative Research Centre for High Performance Soils
Callaghan, NSW, Australia

Preface

It is important to recognize soil constraints that limit crop productivity, farm sustainability, and environmental quality. For example, in Australia, ~75% of agricultural soils have constraints that limit farm productivity. Soil constraints include any physical, chemical, or biological characteristics that limit or inhibit plant access to resources, such as moisture and nutrients, thereby impacting farm productivity. Some examples of soil constraints include the following: low water availability and low nutrient holding capacity, acidity, alkalinity, dispersion, salinity, compaction, erosion, non-wetting soils, and hard-setting soils.

Soil constraints have been shown to occur at various depth zones in soil profiles including surface, subsurface (i.e., subsoil), and throughout the profile. Identifying, interpreting, and managing soil constraints are major challenges, especially when multiple constraints occur in the same soil at various depth zones. The climate-driven genetic yield potential of most agricultural crops can be limited by many of these soil constraints. Management practices can be adjusted to alleviate the problems associated with soil constraints. Although a number of amelioration tools and strategies are available to manage some of these constraints, field adoption of these technologies is a major challenge to the farming community.

This book aims to:

- Cover various types of soil constraints and their impacts on crop productivity
- Discuss various techniques for managing soil constraints
- Describe various strategies suitable for mitigating soil constraints
- Provide data on cost-benefit analysis of managing soil constraints
- Provide case studies of managing soil constraints to increase productivity

This book covers five major themes in soil constraints that include: (i) Dynamics and processes of soil constraints that cover physical, chemical, and biological constraints; (ii) Soil constraints and productivity; (iii) Management of soil constraints; (iv) Soil-constraints mapping; and (v) National case studies. The chapters in these five themes are covered under two major sections: Section I: General Topics covering Salinity, Remediation, Trace elements, Plastics, Fertilizers, Crusts, Biochar, Nutrients, Biology (microbial aspects), Dispersive soils, Sandy soils, Acidity, Mapping, and Language of constraints; and Section II: Country-wide Case Studies from Canada, China, India, the Sub-Sahara, Taiwan, and the USA.

The key features of this book include:

- Fundamental understanding of the various physical, chemical, and biological soil constraints in agriculture
- Various sustainable practices to mitigate and manage these soil constraints
- Case studies of soil constraints and management practices in various regions
- Various approaches for the cost-effective management of soil constraints

Although a large number of books have been published covering particular soil properties and processes, those books tend to focus on managing these properties, whereas this book covers various soil constraints and the sustainable practices to manage these constraints. This book is suitable for undergraduate and postgraduate students majoring in agriculture, environmental, earth, and soil sciences; environmental and soil scientists; and field agronomists.

N.S. Bolan and M.B. Kirkham

Editors

Professor N.S. Bolan is currently serving as the Professor of Soil Science at the School of Agriculture and Environment, The University of Western Australia. He also served as the Program Leader, Cooperative Centre for High Performance Soils (www.soilcrc.com.au), Chair in Environmental Science (2007–2015) and Dean of Graduate Studies (2008–2010) at the University of South Australia (2007–2015), Professor of Soil Science at Massey University, New Zealand (1984–2006).

Professor N.S. Bolan's teaching and research interests include agronomic value of manures, fertilizers and soil amendments, soil acidification, nutrient and carbon cycling, pesticide and metal pollutants interactions in soils, greenhouse gas emission, soil remediation, mine site revegetation, and waste management. Nanthi is a Fellow of American Soil Science Society, American Society of Agronomy and New Zealand Soil Science Society and was awarded the Communicator of the Year Award by the New Zealand Institute of Agricultural Sciences. He has supervised more than 50 postgraduate students and was awarded the Massey University Research Medal for excellence in postgraduate students' supervision. He has published more than 400 book chapters and journal papers with an H index of 112 and total citations of 51,830 (Google Scholar) and is one of the Web of Science Globally Highly Cited Researchers for 2018–2022. He has served as Associate Editor of *Critical Reviews in Environmental Science and Technology*, and Technical Editor of *Journal of Environmental Quality*.

M.B. Kirkham is University Distinguished Professor in the Department of Agronomy at Kansas State University and a graduate of Wellesley College (B.A.) and the University of Wisconsin, Madison (M.S. and Ph.D.). Professor Kirkham's research deals with water movement in the soil-plant-atmosphere continuum and uptake of heavy metals by plants grown on contaminated soil. Professor Kirkham has written three textbooks dealing with soil-plant-water relations and teaches a graduate-level class on soil-plant-water relations. In addition, Professor Kirkham is the author or coauthor of over 300 contributions to scientific publications and is on the editorial boards of 16 journals. Professor Kirkham has received recognitions, including being named fellow of five societies and elected Honorary Member of the International Union of Soil Sciences.

Prologue

This book presents a timely overview of a topic that is of great significance to agricultural productivity and environmental health. It addresses the importance of soil in the broadest sense, as well as more specifically in terms of agricultural production. It highlights issues that can be common across soil types, ecosystems, and locations, as well as those that are specific to particular soil types, land uses, climate, and environmental conditions.

Some soil constraints are specific to local conditions, such as salinity and acidity, but they can be exacerbated at a catchment level or in association with some land use practices. In other situations, contamination and pollution result in persistent constraints to agricultural productivity that may be difficult to address over short time frames. Constraints can be related to intrinsic soil characteristics or imposed by human activities. Consequently, knowledge of soil and the impacts of land use on soil conditions are essential for selecting practices that minimize land degradation and all forms of constraints to productivity, including contamination. While it is generally accepted that this already underpins all land management decisions (including fertilizer use), in practice, there are two key situations that contradict this. The first is that knowledge of soil biological processes is not usually fully integrated in land management decision-making, especially in agriculture. The second is that traditionally there has been a greater focus on research to solve the adverse outcomes of land use practices, which result in soil constraints, than on preventing constraints in the first place.

While the relevance of soil biological processes has always been acknowledged in a general way, incorporation of this information into decision-making is often not seriously addressed in practice. This is primarily because soil biological processes are diverse, complex, less precise, and more variable over

Example of innovative pasture management on the property of Rob and Caroline Rex near Wagin, Western Australia. A mixture of perennial and annual pasture species was introduced across the farm to overcome soil constraints and improve grazing opportunities for sheep. A publication on the impacts on soil carbon at this site was published based on a collaborative project funded by the Australian Government Department of Agriculture and Water Resources. [Pauli N, Abbott, LK, Rex, R, Rex C, Solaiman ZM (2018) A farmer-scientist investigation of soil carbon sequestration potential in a chronosequence establishment of perennial pastures. Land Degradation and Development 29: 4301–4312].

time than chemical and physical processes, which underpin fertilizer use and other soil management practices. There are no simple biological indicators of soil health. Nevertheless, land managers are increasingly searching for an understanding that includes how to manage soil biological processes, which form key components of soil chemical fertility (including processes involved in nutrient cycling) and soil physical fertility (including processes involved in soil aggregate stability). Furthermore, the search for increased understanding of soil health by land managers has potential to re-focus attention from problem-solving about soil constraints to avoidance or minimization of soil constraints that limit productivity.

Best-practice management practices are commonly derived locally according to soil type, topography, climate, and environmental conditions. There are basic principles underlying soil security and agricultural productivity that are common across ecosystems, environmental conditions, and land uses, and they can be applied to specific local circumstances. Unfortunately, financial considerations may introduce an overriding limitation to the selection of the most appropriate management practices at any point in time, and this could restrict the management options that are available to landholders from year to year. Consequently, an important component of selection of locally adapted land use practices needs to be derived from the experience of the land managers themselves. This could involve identifying what is practical by gathering information from their own on-farm trials and interpreting the impacts of practices in consultation with soil scientists and agronomists. Inevitably, a focus on predicting potential constraints should also help avoid longer-term problems from arising, especially where gradual deterioration might otherwise be ignored and lead to low-level constraints becoming more recalcitrant and difficult to overcome.

As demonstrated by contributions to this extensive volume, there already exists wide-ranging understanding of constraints to soil and productivity. Therefore, the capacity to predict and avoid potential constraints is also available. In most cases, it should be preferable to pre-empt and implement practices that limit the development of constraints, rather than wait and attempt to ameliorate moderate or severe damage to soil and land conditions. Modeling scenarios and best-practice guides that take account of pre-existing relevant information as well as predictions are essential.

Emerita Professor Lynette K. Abbott
School of Agriculture and Environment and UWA Institute of Agriculture
The University of Western Australia, Perth, WA, Australia

1 Overview of Salinity in Arable Soils

Global Perspectives

Viraj Gunarathne[1,2], Meththika Vithanage[1], and Jörg Rinklebe[2]
[1]Ecosphere Resilience Research Centre, Faculty of Applied Sciences, University of Sri Jayewardenepura, Nugegoda, Sri Lanka
[2]University of Wuppertal, School of Architecture and Civil Engineering, Institute of Foundation Engineering, Water- and Waste-Management, Laboratory of Soil- and Groundwater-Management, Wuppertal, Germany

CONTENTS

1.1 SOIL SALINITY: A GLOBAL THREAT TO FOOD SECURITY

The world population is rising precipitously, increasing the demand for food. However, the accessibility of arable lands has declined approximately by twofold within the past 50 years threatening global food security (Liu et al., 2020). As a consequence, hunger prevails, especially in many developing countries. Therefore, United Nations (UN) stated that zero-hunger should be a goal, and it

DOI: 10.1201/9781003093565-1

is the second goal of their 17 sustainable development goals for 2030. One of the major focuses of the zero-hunger goal is food security, which is defined as sufficient physical, economic, and social access to food by all people in a community at all times to fulfill their dietary requirements for a healthy and active life (Verburg et al., 2013; UN General Assembly, 2015; Van den Broeck and Maertens, 2016; Zhang et al., 2018). Soil salinity is becoming a fundamental challenge to achieving the food security and zero-hunger goal, because it restricts the use of a vast area of arable land worldwide, limiting annual agricultural production.

It has been estimated that the extent of salinity-affected soil is 20% of the total agricultural land, and it affects 33% of irrigated farming lands throughout the world (Shrivastava and Kumar, 2015). Approximately 1125 million hectares of agricultural land in 13 countries have been degraded by soil salinity affecting 4.03 billion people (Liu et al., 2020). Soil salinization is a continuous process that results from a number of climatic and anthropogenic factors, and it is expected to increase exponentially by the year 2050, impacting on 50% of all arable lands and 5.02 billion people (Wang et al., 2003; CSSRI, 2014; Liu et al., 2020). Agricultural lands in arid and semi-arid regions of the world face the highest threat from soil salinization (Halima et al., 2019).

Soil salinization affects food security in two ways: (i) limiting the land area suitable for agriculture and (ii) depleting the quality and quantity of yield. Further, the magnitude of salinity in soil restricts the number of crop cultivars suitable to grow in the salt-affected area. Major crop types considered to provide staple foods across the globe are highly susceptible to soil salinity. For instance, the world's highly cultivated two crops, *Triticum aestivum* (wheat) and *Oryza sativa* (rice) reduce their mean yield by approximately 40% and 70%, respectively, when exposed to soil salinity (Shabala and Munns, 2017). Expansion of salt-affected land area in many parts of the world is observed, thus it is important to act on mitigating soil salinization to ensure food security. Therefore, this chapter aims to describe the current status of global arable soil salinization, the factors influencing it, and the effects of salinization on soil properties and crop production.

1.2 OVERVIEW OF SOIL SALINITY

Keller et al. (1986) described soil salinity as an excessive accumulation of ions in the soil, including sodium (Na^+), calcium (Ca^{2+}), magnesium (Mg^{2+}), chloride (Cl^-), and sulfate (SO_4^{2-}), which hinder the growth and function of plants. The relative concentrations of the salinity-inducing ions vary in different types of saline soils (Bernstein, 1975). Usually, saline soils are demarcated by measuring their electrical conductivity (EC_e), exchangeable sodium percentage (ESP), and sodium adsorption ratio (SAR) (Allison and Richards, 1954). The EC_e is correlated linearly with total dissolved ions, while ESP and SAR directly depend upon the Na^+ ions (Suarez, 2005). Table 1.1 indicates the classification of salt-affected soils based on the magnitude of the parameters mentioned above.

TABLE 1.1
Classification for Salt-Affected Soils

Salt-affected soil type	EC_e (dSm^{-1})	pH	$SAR_{1:5}$	SAR_e	ESP (%)
Normal soil	<4.0	6–8	<5	<3	<5
Saline soil	>4.0	6–8	<5	<13	<15
Saline-sodic	>4.0	>8	>5	>13	>15
Sodic soil	<4.0	>8	>5	>13	>15

Source: Data from Rengasamy (2002) and Richards (1954).
SAR_e, sodium adsorption ratio in saturated extract; $SAR_{1:5}$, sodium adsorption ratios in 1:5 soil/water extract.

1.2.1 Factors Influencing Salinity in Soil

Soil salinization occurs through two main routes: primary salinization and secondary salinization. Deposition of soluble salts in soils by natural processes, such as capillary action from natural sources including saline groundwater and saline minerals, is the cause for primary or dryland salinity. Secondary salinity, or irrigation salinity, is generated by undesirable anthropogenic activities, such as irrigation of lands with poor quality water that ultimately lead for excessive accumulation of salts in the soil (Allison and Richards, 1954; Rengasamy, 2010) (Figure 1.1). The combination of both primary and secondary salinization processes accelerates soil salinization, especially in regions with high water scarcity (Ondrasek, 2020; Pereira et al., 2020; van de Craats et al., 2020). This section describes each natural and anthropogenic factor influencing salinity in the soil.

1.2.1.1 Natural Factors

1.2.1.1.1 Deposition of Weathered Minerals

Parent rocks bearing primary minerals contain salts including chlorides, sulfates, or carbonates of Na, Ca, and Mg (Rengasamy, 2002; Pessarakli, 2016). Weathering of salt-enriched rock minerals through chemical, physical, or biological factors releases soluble salts. These weathered minerals and associated salts can be transported away from their original location with surface or groundwater flow and deposited further away. Evaporation of water, especially in arid and semi-arid regions, causes the concentration and precipitation of salts in soil (Kumar and Sharma, 2020).

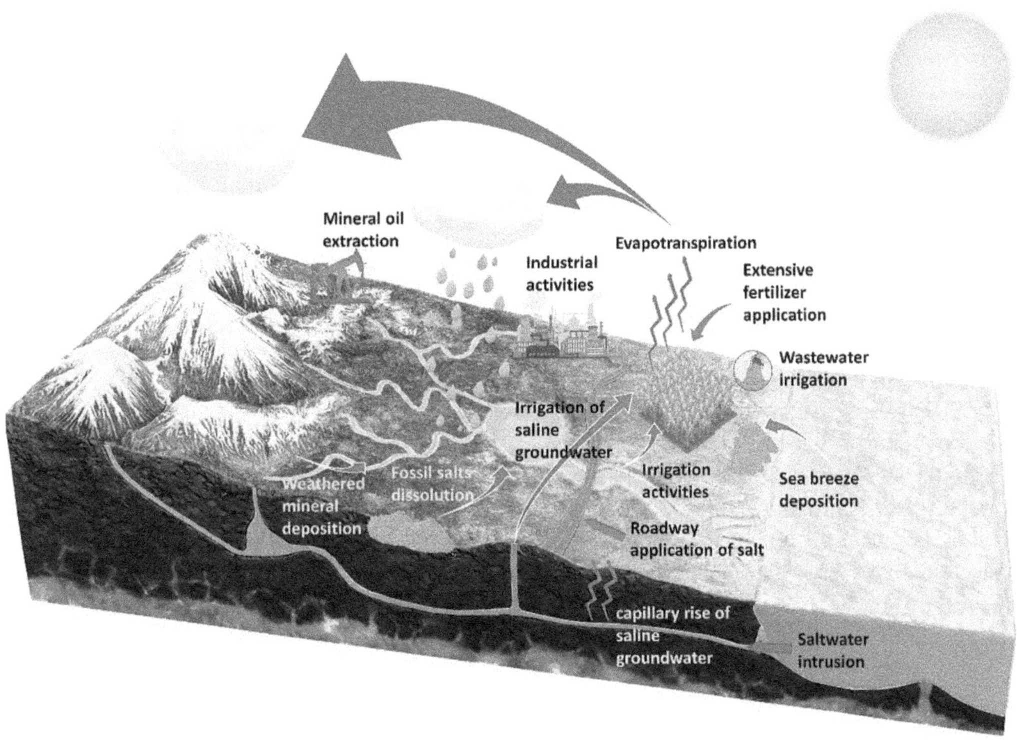

FIGURE 1.1 Factors involved in soil salinization.

1.2.1.1.2 *Fossil Salts*

Fossil salts are salts related to former salt depositions or salts entrapped in concentrated solutions of saline water that have originated from sedimentation in ancient marine or lake systems. They are among the most prominent salts in arid regions. These lacustrine and marine deposits of fossil salts contribute to soil salinity because they are released through natural or anthropogenic activities such as storage or transmission of water (Bressler et al., 1982; Gunarathne et al., 2020b).

1.2.1.1.3 *Sea Breeze Deposition*

Sea breeze carries ample amounts of saltwater droplets with salt concentrations as high as 14.2 $\mu g/m^3$ into inland areas causing soil salinization (Prospero, 1979). Salinization from these events can be seen in areas even more than 80 km away from the sea (Kumar and Sharma, 2020).

1.2.1.1.4 *Saltwater Intrusion and Capillary Rise of Saline Groundwater*

Coastal areas are often associated with shallow, saline groundwater tables, and regular saltwater inundation occurs in low-lying areas along the coasts. Overexploitation of groundwater eventually facilitates saltwater intrusion from the ocean and brackish water estuaries resulting in groundwater salinization. Salinity and the depth of the groundwater table are crucial factors to be considered in determining soil salinization in low-lying lands located in semi-arid and arid regions (Fan et al., 2012). Groundwater is considered to be the major geological force for the accumulation, transmission, and discharging of salt (Salama et al., 1999). Transportation of saline groundwater to the soil surface via capillary rise throughout dry seasons causes surface soil salinization (Bressler et al., 1982). Tsunamis and storm surges also play a role in being natural factors for soil salinization (Vithanage et al., 2012; Capooci et al., 2019).

1.2.1.2 Anthropogenic Factors

1.2.1.2.1 *Irrigation Activities*

Irrigation is a major factor needed for cultivation in arid and semi-arid regions. Therefore, irrigation-driven soil salinization is well understood. Even slight concentrations of salts in irrigation water build up soil salinity over a long period of application (Greene et al., 2016). This process is prominent in agricultural systems where irrigation water is applied in limited quantities to reach only the root zone of crops, which restricts the leaching of salts deep into the soil profile. However, the application of excessive amounts of irrigation water can result in a rising saline groundwater table contributing to excessive salinity in the topsoil (Litalien and Zeeb, 2020).

Wastewater irrigation is an important factor that affects the salinization of arable lands. Even treated wastewater contains total solids ranging between 200 and 3,000 mg/L, or more (Feigin et al., 2012). Therefore, irrigation using wastewater may provide potentially large amounts of salts in agricultural soils. Bond (1998) estimated that the field application of 1,000 mm of low salinity water (*i.e.*, 500 mg/L TDS) annually provides 5 t/ha salts. The salts provided by irrigation with wastewater continues to become concentrated and accumulates in the soil through evapotranspiration, making the soil more saline. Phogat et al. (2020) predicted the outcome of long-term irrigation of recycled wastewater from 2018 to 2050. Their model estimated the ESP and SAR increase beyond threshold levels at the end of the simulation period creating unfavorable conditions for the growth of most crop species.

1.2.1.2.2 *Industrial Activities*

Industrial activities, such as steel manufacturing, cement production, textile manufacturing, mining, power generation, paper production, food production, and aquaculture, generate significant amounts of saline wastes (Muller et al., 2009; Litalien and Zeeb, 2020). The land disposal of these waste materials can contribute to soil and surface water salinization within the proximity of the disposal site.

1.2.1.2.3 *Mineral Oil Extraction*

Saline water is a waste component in crude oil because many of the mineral oil depositions originated in marine environments, and the salt content in oil can occur up to 200,000 ppm (Mohamed et al., 2003; Akinwumi, 2014). Saline water present as small droplets in crude oil is removed to increase oil quality, and subsequent land disposal of extracted salts involves soil salinization around the disposal site (Al-Haddabi and Ahmed, 2007).

1.2.1.2.4 *Roadway Application of Salt*

Since the 1940s, the demand for road salts has increased with the development of the automobile industry and roadways (Daley et al., 2009). Road salt is primarily composed of NaCl, and its application to roadways during the winter season helps to prevent the accumulation of ice by increasing the freezing point of water. This process avoids accidental collisions of automobiles. The application of road salts is associated with increasing salinity in soil and surface water bodies in cold climate regions (Chernousenko et al., 2003; Daley et al., 2009).

Extensive chemical fertilizer usage, deforestation, and overgrazing contribute to soil salinization on a minor scale (Endo et al., 2011; Pessarakli, 2019; Litalien and Zeeb, 2020).

1.3 SALINITY IN ARABLE SOILS: GLOBAL DISTRIBUTION

The global distribution of soil salinity shows a wide variation among different continents (Figure 1.2). Soil salinity is prominent in arid and semi-arid regions of the world. Salt-affected soils are distributed among ~10% of terrestrial lands in over 100 countries, and they cover about 1,128 Mha of land. They are increasing at a rate of more than 1.5 Mha annually (Eynard et al., 2006; Wicke et al., 2011; Ondrasek and Rengel, 2020; Pereira et al., 2020). Asia, South America, and Africa account for the continents with the highest soil salinization, and their contribution to total global salt-affected areas represents 33.97%, 13.84%, and 8.63%, respectively (Mukhopadhyay et al., 2020).

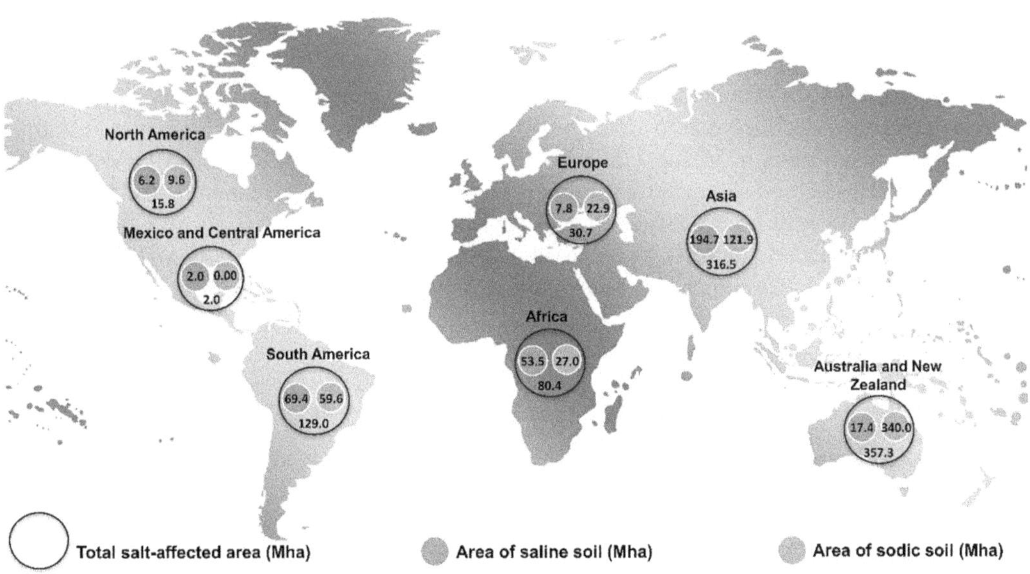

FIGURE 1.2 Global distribution of salt-affected soil. (Data from Mukhopadhyay et al. (2020).)

1.3.1 Asia

Studies in different regions of Central Asia (*i.e.*, Kazakhstan and Uzbekistan) indicate that secondary salinization of soil is prominent, and the major causes are geochemical and edaphic factors (Funakawa and Kosaki, 2007). Other Asian countries that face salinization of arable soils include Bangladesh, China, India, Pakistan, Russia, and Sri Lanka (Hasanuzzaman et al., 2014; Gopalakrishnan and Kumar, 2020).

In Bangladesh, 1.056 Mha out of 1.689 Mha of total coastal land has been affected by various levels of salinity (Szabo et al., 2016). These coastal regions of Bangladesh are highly affected by rising sea levels and saltwater intrusion (Corwin, 2021). Dasgupta et al. (2018) estimated that rice production in coastal areas of Bangladesh will be affected by soil salinization, and if current trends continue, by 2050 there will be a 15.6% decline in the production of high-yielding rice varieties. About 4.88% (3.6×10^7ha) of the total area of China is affected by soil salinity (Wang et al., 2013). The Chinese government started to foresee that soil salinity would become a problem, and for the past 70 years, there has been a comprehensive salinity management program and far-reaching desalinization efforts (Li et al., 2014). Soil salinization affects 7 Mha of land in India (Tripathi et al., 2015). India will need 350 million tons of food grain by 2050 to feed the continuously growing population. Studies show that each year 10% of new land is salinized, and by 2050, more than 50% of the arable land in India will be salt-affected, which will affect the food security of the country (Kumar and Sharma, 2020). About 6.67 Mha of arable land is salinized in Pakistan (Khan, 1998), and 80% of it is in Punjab followed by Sindh and Balochistan (GoP, 2007). Soil salinity in Pakistan is related to adverse economic issues, such as poverty since salinity is negatively associated with wheat production in the country (Makhdum and Ashfaq, 2008). Soil salinity is of concern in the southern part of Russia, which has major agricultural lands; however, most saline soil mappings are based on data from the 1980s to 1990s (Pankova, 2015).

Soil salinization is a major threat in Sri Lanka, and it affects 11,200 ha of the coastal zone resulting in 50% abandonment of paddy-cultivated lands (Perera et al., 2018). Gopalakrishnan and Kumar (2020) estimated that 45% of paddy-cultivated land and 32.8% of the total land in the Jaffna peninsula, a semi-arid region in Sri Lanka, have been affected by salinity. Extensive usage of groundwater, saltwater intrusion, high evaporation rates, and low precipitation are the major causes governing soil salinization in the Jaffna peninsula (Wijeyaratne and Subanky, 2017; Gopalakrishnan and Kumar, 2020). In dry zones in north-central province, 66% of well water irrigated cultivated land has been affected by salinity, while rain-fed lands in the same region, exhibit less salinity (Perera, 2016).

Soil salinization is an ancient phenomenon in Middle Eastern history. Early civilizations such as Mesopotamia (which includes Iraq and parts of surrounding countries in the Middle East) first flourished and then declined due to soil salinity caused by mismanagement. Shahid et al. (2018) indicated that salinization in Mesopotamia continued to increase from 2400 BC to 1200 AD until civilization collapsed. Soil salinization still is a major source of soil degradation in Middle Eastern countries and is widely observed in Iran, Iraq, and Syria (Hasanuzzaman et al., 2014). Consequently, studies have been carried out to assess the feasibility of using fodder grass in salt-affected areas in these countries to reduce salinity, and *Panicum turgidum* is proposed to be a potential species for remediation (El-Keblawy, 2004).

1.3.2 Africa

Rice-cultivated lands in Africa affected by salinity/sodicity are around 0.173 Mha, which is 2% of the total rice-cultivated area of the continent (Van Oort, 2018). Rice production in Tanzania (Kashenge-Killenga et al., 2016), Egypt (El-Shahway et al., 2016), Nigeria (Akinbile et al., 2016), Guinea (Wolanski and Cassagne, 2000), Senegal (Barbiero et al., 2001), and Mozambique (Menete et al., 2008) is affected by soil salinity. Furthermore, Madagascar, Chad, and northwest Tanzania have soil salinization in rice fields (Van Oort, 2018). Evaluation of saline soil in coastal regions of Africa poses numerous challenges, mainly due to strong seasonal dynamics such as saltwater intrusion during certain seasons (Wolanski and Cassagne, 2000). In the 1990s, 10% of Africa's soils were affected by secondary salinization (Thomas and Middleton, 1993). About 25% of South Africa's soils are degraded, and soil salinization is one major reason for such degradation. Salt-induced soil salinization from sodium results in low water infiltration and increased water runoff, leading to water erosion in African soils (Du Preez and Van Huyssteen, 2020).

1.3.3 South America

Data on the distribution of salt-affected soils in Latin America (South America, Mexico, Central America, and the islands of the Caribbean) are scarce, yet there are clear indications that soil salinity and sodicity, under dryland and irrigated conditions, are rapidly increasing due to human-induced processes such as irrigation and drainage practices (Sentís, 2021). Soil salinization in South America has adverse effects on agriculture and human well-being and is observed in various parts of Argentina, Chile, Colombia, and Paraguay (Taleisnik, 2020; Viglizzo and Ricard, 2021). Both natural and anthropogenic source-related soil salinity reduces the soil carbon content of 18.4% of South American soils, and the soil organic carbon turnover in these saline soils indicates the necessity of considering the impact of salinity not only on decomposition rates but also on plant input and yield (Setia et al., 2013).

1.3.4 North America

The Canadian prairies in North America typically suffer from soil salinization. Eilers et al. (1995) stated that ~38% of arable land in the Canadian prairies has been influenced by salt accumulations that extend from moderate to severe levels. In this region, areas demarcated with elevated amounts of soluble salts can be found as surface deposits or as discharges of salinized groundwater in flat and weakly drained fields along with non-percolative water systems (Florinsky et al., 2000). Topography in this region is the most prominent factor that affects the spatial distribution of soil salinization.

In Cuba, ~1 million hectares of land of over 7.08 Mha of total agricultural land have been affected by salinity, while 1.5 Mha of arable land suffers from salinity-related problems creating a severe problem for agricultural production in the country (González, 2000). Furthermore, about 42% of cattle land, situated predominantly in eastern areas of Cuba, is affected by soil salinity (Ramírez et al., 1999; Mesa, 2003).

1.3.5 Europe

Soil salinization is one of the major causes of soil degradation and desertification of European soils (Montanarella, 2007; Tóth et al., 2008). It is estimated that salt-affected soils in Europe contribute to 3.3% of the total global saline and sodic soils (Rengasamy, 2006). Salinity endangers 1–3 Mha of land in Mediterranean countries (European Soil Bureau, 2014). Salt-affected soils in the European part of Russia and neighboring Ukraine, Moldova, Azerbaijan, Armenia, and Georgia are estimated to be 65 Mha (Khitrov et al., 2009).

Soil salinity poses a great threat to agricultural systems around the North Sea region, which include the United Kingdom, Belgium, Germany, Netherlands, Denmark, Sweden, and Norway (De Waegemaeker, 2019). The prime reasons for the salinization around the North Sea are said to be irrigation practices, aerosol salinization due to high winds, floods, and seepage salinization, which is the movement of saline or brackish water to shallow groundwater reserves or the root zone of plants (Gustafsson and Franzén, 1996; Raats, 2015; Gould et al., 2020; Gould et al., 2021). Therefore, the European Commission in its EIP-AGRI program has established a focus group to address the question "How to maintain agricultural productivity by preventing, reducing or adapting to soil salinity?" (EIP-AGRI, 2019). Most agricultural farmlands in the UK are affected by saltwater intrusion, mainly due to coastal flooding (Gould et al., 2020). In the Netherlands, saltwater intrusion through groundwater seepage affects agricultural lands (De Louw et al., 2010). The East Frisian Islands and coastal areas of Northern Germany are vulnerable to soil salinization (Daliakopoulos et al., 2016; Seibert et al., 2018). Soil salinity has been observed in the coastal inner sea regions of Denmark, including Zealand and small islands (Kristiansen et al., 2011). Saltwater intrusion into agricultural land has extensively increased soil salinization in some parts of Greece. Crop rotation from cotton to corn in Southern Greece implies impacts from soil salinity, to change from sensitive to moderately tolerant crops (Pisinaras et al., 2010). In Turkey, the area affected by saline soil extends up to 19.6% of the total land, while about 60% of the total land is slightly saline (Senol and Bayramin, 2013). Road salting over the years has had a great effect on the salinization of soil and surface water bodies in European countries. For example, Lake Svinsjøen in southeastern Norway was subjected to extensive salinization by road salts during the years 1947–1995 (Kjensmo, 1997).

1.3.6 Australia

Sodium salt-related soil salinization affects about 30% of the land area in Australia, whereas 16% of the agricultural lands are exposed to groundwater-associated salinity and irrigation salinity. Studies further suggest that 67% of Australian agricultural lands have the potential for transient salinity or non-groundwater-associated salinity (Rengasamy, 2006). The extent of salinized land may rise from 1.2 to 15 Mha. New South Wales has the highest probability of salinization up to 5 Mha (Greiner and Cacho, 2001). Moreover, the Murray Darling river basin, which contributes to over 40% of the total agricultural production of Australia, has been threatened by soil salinity that has resulted from water scarcity due to hindered rainfall associated with global climate change patterns (Corwin, 2021). Furthermore, forestry is considered an option to reduce soil acidification and minimize off-site impacts in Australia (Scott et al., 2000).

1.4 EFFECT OF SALINITY ON SOIL HEALTH

Soil salinity negatively influences soil health by altering the chemical, physical, and biological properties of soil. The modified soil properties affect average plant growth which results in reworked natural vegetation patterns and changes in agricultural practices. Figure 1.3 summarizes the effect of salinity on soil properties and crop growth.

1.4.1 Chemical Properties

Salinity changes soil chemical properties, including ESP, pH, soil organic carbon content, and cation exchange capacity, and it changes the osmotic potential of the soil solution (Wang et al., 2014). Salinity causes significant soil organic carbon loss from soil aggregates via the impact of sodicity and induces organic matter solubilization, which changes carbon stocks and fluxes in the soil (Munny et al., 2021). Zhu et al. (2020) found that high dissolved organic matter content in coastal wetland soil was positively correlated with high salinity.

Salt-affected soils suffer from nutrient deficiencies. Therefore, the intense use of chemical fertilizers might be required. Macronutrients including N, P, and K in salt-affected soils occur at depleted concentrations. Yang and Sun (2020) saw a reduction in soil total carbon and total nitrogen in the Yellow River Delta, China, with an increasing soil salinity gradient. Zhang et al. (2019b) observed decreased exchangeable potassium and soil organic matter in coastal saline soil. Deficiencies of micronutrients, which are an indirect complication of soil salinization, develop from ion competition and alkalinization (Grattan and Grieve, 1998). The high pH, which is correlated with saline-sodic soils, further causes reduced availability of specific micronutrients including Mn, Cu, Zn, Al,

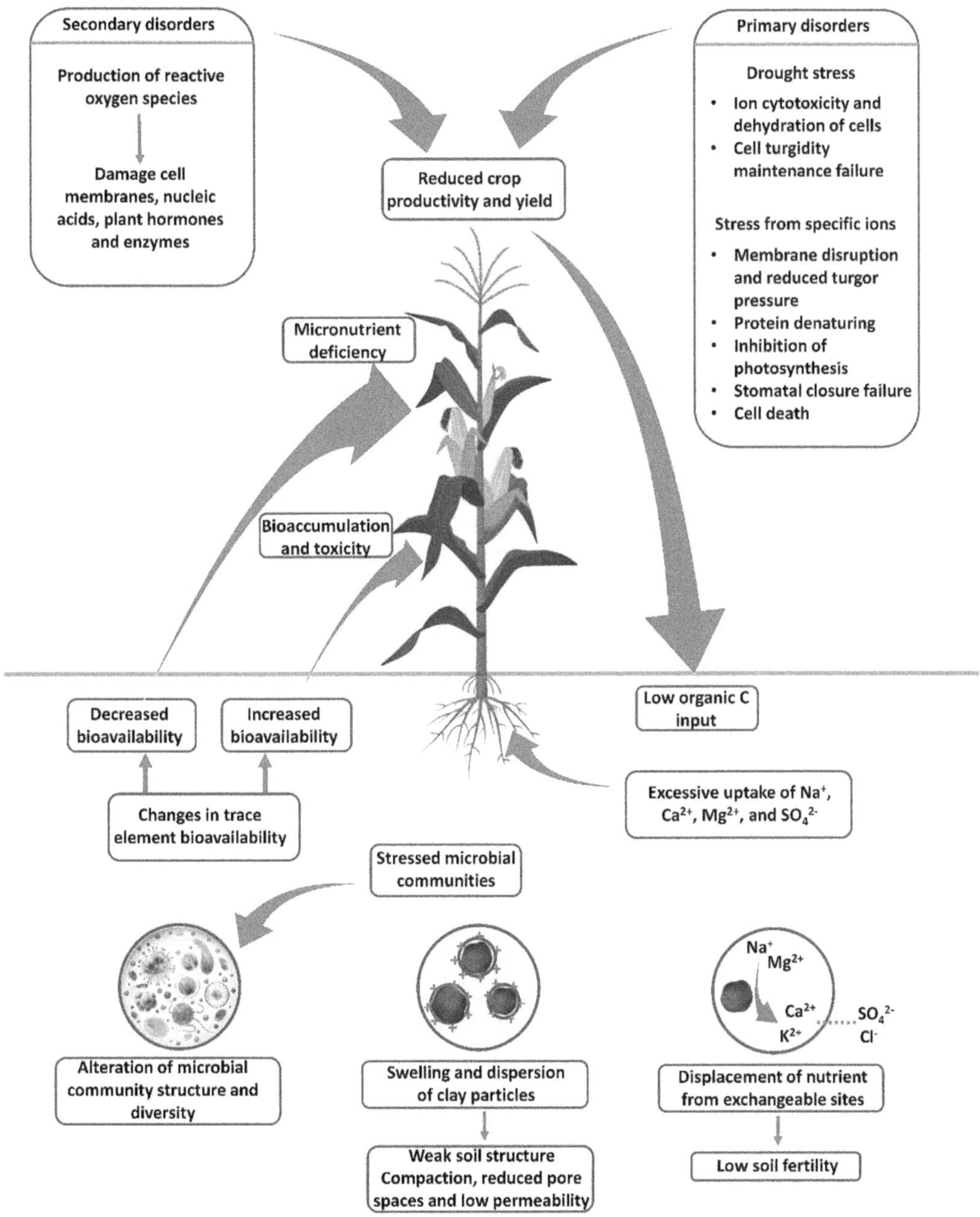

FIGURE 1.3 Adverse effects of salinity on soil properties and crop growth.

and Fe (Pessarakli and Szabolcs, 1999; Lakhdar et al., 2009). In contrast, Zhu et al. (2020) found that increased amounts of dissolved organic carbon in high salinity areas promoted Cr(VI) adsorption by soil particles and reduced Cr(II) adsorption, which creates an ecological risk from Cr(II) toxicity. Low-carbon input into the soil from sparse vegetation cover, which results from undesirable edaphic conditions such as unstable soil structure, salt toxicity, and increased osmotic potential, further degrades the chemical and physical properties of salt-affected soils (Wong et al., 2009).

1.4.2 Physical Properties

Elevated concentrations of different ions that contribute to soil salinity often cause structural degradation of salt-affected soils. Saline soils contain habitually high concentrations of exchangeable and soluble Na^+ that cause swelling and dispersion, resulting in a weak soil structure with low permeability to soil water infiltration (Gharaibeh et al., 2009; Yu et al., 2010). Swelling and clay dispersion increase with increasing ESP and decreasing soil solution cationic concentrations (Cass and Sumner, 1982).

Studies have suggested that Mg^{2+} induces indirect structural degradation in Mg-rich saline soils (Zhang and Norton, 2002). Bronick and Lal (2005) proposed that the soil degradation caused by Mg^{2+} could be analogous to the Na^+-driven particle dispersion in sodic soils. Similarly, SO_4^{2-} ions facilitate structural degradation in sulfate-rich saline soils. Springer et al. (1999) stated that SO_4^{2-} enriched soil can induce Na^+-governed soil particle dispersion by ion paring and bond formation of SO_4^{2-} with Mg^{2+} and Ca^{2+} allowing Na^+ ions to bind with cation exchange sites vacated by Mg^{2+} and Ca^{2+} ions. The magnitude of this effect is higher in SO_4^{2-} ion enriched soils compared to Cl^- ion enriched soils.

There is an increase in dry density of soil and a decrease in soil water content with elevating salinity levels, especially in soils containing a high fraction of clay and swelling minerals. Soils having a low fraction of clay shows neutral or an opposite behavior (Abood et al., 2007; Ajalloeian et al., 2013; Liu and Zhang, 2014; Ying et al., 2021). Therefore, soil salinity affects soil compaction differently depending on soil type, which varies with mineral composition and clay fractions.

1.4.3 Biological Properties

Soil salinity has been recognized as the most crucial factor affecting the global distribution of soil microorganisms (Lozupone and Knight, 2007; Auguet et al., 2010). It is one of the main factors that contribute to shaping of soil bacterial composition and diversity in natural habitats (Campbell and Kirchman, 2013; Zhao et al., 2020). Zhang et al. (2019a) identified a decline in the alpha diversity of bacteria in a desert ecosystem with an increasing salinity gradient, while Rath et al. (2019) detected an alteration of bacterial communities in a salty lakeshore with an increasing salinity gradient. Moreover, Guan et al. (2021b) observed a decline in bacterial richness and altered bacterial community composition in salinity-sodicity-affected soils. Yang and Sun (2020) measured a decline of fungus diversity in saline-alkali soil of the Yellow River Delta, China, with an increasing soil salinity gradient. They found fewer litter saprophytes and plant saprophytes at soil salinity extremes than in non-saline soils.

Wong et al. (2008) and Rietz and Haynes (2003) showed that the metabolic quotient (*i.e.*, respiration per unit biomass) was boosted with increasing salinity and sodicity, indicating the occurrence of a stressed microbial community. Similarly, Ghollarata and Raiesi (2007) found an increase in the metabolic quotient with increasing soil salinity. Yuan et al. (2007) detected an alteration in soil microbial communities with increasing salinity, which can be explained as an adaptive mechanism to mitigate salt stress by incorporating microorganisms with a lower metabolism. Salinity and sodicity substantially decrease soil microbial biomass and enzyme activities (Garcia and Hernandez, 1996; Tripathi et al., 2006). The reduction of microbial community composition and diversity can be explained by salinity-induced osmotic stress and by specific ion stress on microorganisms that cause cell damage, alteration of metabolic activities, and reduced growth rates. Moreover, the degradation of soil physical properties, such as the breakdown of soil aggregates, increased bulk density, reduced porosity, and declined soil water holding capacity caused by sodicity, influences the growth and distribution of microorganisms (Qadir et al., 2007).

1.5 EFFECT OF SALINITY ON PLANT GROWTH, YIELD, AND FOOD PRODUCTION

The majority of cultivated crop varieties are glycophytes, which are known as plants with little ability to survive under high levels of EC_e (electrical conductivity of the soil solution) or EC_w (electrical conductivity of the irrigation water) (Chinnusamy et al., 2005). Soil salinity affects root uptake of water and mineral nutrients that disrupts vital physiological functions of plants resulting in decelerated growth and development (Munns, 2002). Glycophytes show two types of disorders under salinity stress, namely primary and secondary disorders (Figure 1.3), which are based on the salinity level, period of exposure, and other factors, such as plant water status, growth stage, temperature, and solar radiation (Ondrasek and Rengel, 2020).

1.5.1 Effect of Soil Salinity on Plant Growth

1.5.1.1 Primary Disorders

High salinity levels induce two types of primary detrimental effects on plants, from osmotic stress and from impacts of specific ions, which directly affect the physiological functioning of plants, ultimately reducing crop productivity.

1.5.1.1.1 Drought Stress

Excessive concentrations of soluble salts in the soil solution in the root zone reduce the osmotic potential creating osmotic stress for crops (Maas and Nieman, 1978). Due to osmotic stress, root water uptake through cell membranes decelerates creating physiological drought stress for crops (de Oliveira et al., 2013). Osmotic-induced physiological stress affects crops in two ways: induction of ion cytotoxicity and cell dehydration (Munns, 2002). Osmotic stress inhibits K^+ ions transport into cells in exchange for Na^+ affecting cell turgidity maintenance that eventually decreases metabolic processes and crop growth (Horie et al., 2012; Gupta and Huang, 2014). Salt-stressed plants usually regulate the uptake of water by avoiding 97%–98% of Na^+ and Cl^- ions to prevent toxicity. However, this mechanism is not sufficient for glycophytes to cope with elevated concentrations of salts (Munns et al., 2020).

1.5.1.1.2 Stress from Specific Ions

Each type of ion (*i.e.*, Na^+, Ca^{2+}, Mg^{2+}, Cl^-, and SO_4^{2-}) contributes to the buildup of soil salinity. Some of the ions serve as plant nutrients at certain concentrations. However, excessive uptake and accumulation of them in plant tissues generate toxic symptoms. The toxicity of different ions in plant cells or tissues differs. Membrane disruption is one of the common symptoms resulting from excessive accumulation of all of the ions involved with salinity (Volkmar et al., 1998; Butcher et al., 2016). Disrupted membranes lead to reduced turgor pressure in cells, failure of stomatal closure, protein denaturation, inhibition of photosynthesis, and cell death, which cause reduced plant growth and yield (Aslamsup et al., 2011).

However, Na^+ causes the most detrimental effects on plant growth, because it replaces Ca^{2+} ions in cell membranes resulting in protein denaturation (Bernstein, 1975; Cramer et al., 1985). Long-term exposure of plants to Ca^{2+} at elevated concentrations produces stress situations (Parida and Das, 2005). Plant exposure to high concentrations of Cl^- causes toxic effects, including degradation of chlorophyll pigments and reduction in uptake of NO_3^- (Grattan and Grieve, 1998; Tavakkoli et al., 2010). Studies focused on specific toxic effects of excessive Mg^{2+} and SO_4^{2-} are rare. Thus, their toxic effects on plant physiological functioning, growth, and development are less evident (Butcher et al., 2016).

Excessive salinity in soil influences the biogeochemistry of trace elements, increasing their bioavailability, increasing phytoaccumulation and toxicity, or limiting bioavailability, which creates deficiency symptoms in plants. Several studies have shown salinity-induced enhancement of uptake of trace elements, such as Cd, Cu, Zn, and Mn, by cultivated glycophyte crop varieties (Weggler et al., 2004; Khoshgoftarmanesh et al., 2006; He et al., 2017; Guo et al., 2018). Increased uptake and bioaccumulation of toxic trace elements in plant tissues reduce plant growth and yield and decrease nutritional quality, and, at the same time, may cause potential toxicity for humans.

1.5.1.2 Secondary Disorders

Primary stress factors typically induce secondary stresses by producing excessive concentrations of reactive oxygen species inside plant tissues. These reactive oxygen species, such as H_2O_2, O_2^-, and OH, induce oxidative damage to essential cell components, including cell membranes, nucleic acids, plant hormones, and enzymes, creating a diverse range of detrimental effects for plants (Gomez et al., 2004).

However, not only the soil salinity itself, but also several other factors, including crop variety and growth stage, soil water status, nutrient status, and stress induced by pests, ultimately contribute to the extent of salinity stress and, thus, crop productivity (Maas, 1993; Volkmar et al., 1998; Rengasamy, 2010; Butcher et al., 2016).

1.5.2 Soil Salinity on Yield and Food Production

Increasing soil salinity is a major global problem that leads to the loss of productivity in cultivated soil and badly affects both the quality and quantity of yield of many crops in various ways (Machado and Serralheiro, 2017). High soil salinity creates adverse effects on crops in three ways: by creating specific ionic toxicity, causing an ionic imbalance in the soil solution reducing nutrient uptake, and increasing the osmotic potential of the soil water solution, which reduces soil water extraction by plants (Ayers and Westcot, 1985; Ruiz et al., 1997; Fisarakis et al., 2001; Shani and Ben-Gal, 2005). High salinity may also lead to the accumulation of extensively high internal ion concentrations in plants and thus cause growth reduction (Greenway and Munns, 1980). Salts in the soil lead to increased soil osmotic pressure, directly affecting plant growth by interfering with plant nutrition. The osmotic effect of salinity encourages metabolic changes in the plant leading to a condition similar to water stress-induced wilting (Munns et al., 2002).

Visual symptoms of salt injury in plants appear gradually. The first signs of salt stress may include wilting, yellowed leaves, and stunted growth. In later stages, the damage results in chlorosis of green parts, leaf tip burning, necrosis of leaves, and scorching of oldest leaves (Machado and Serralheiro, 2017). Under marginal salinity conditions, the growth of many vegetable crops is inhibited without showing other noticeable injury symptoms (Hayward, 1955). However, some vegetables show well-demarcated injury symptoms in the later stages of their growth due to high salinity levels. For instance, beans show a marginal chlorosis-necrosis with upward cupping of the leaves while onions also show leaf necrosis (Bernstein and Ayers, 1951, 1953). In addition to retarding growth, some vegetables exhibit nutritional deficiency or symptoms of imbalance due to high salinity (Läuchli and Epstein, 1990; Francois and Maas, 1999). Some lettuce cultivars exhibit deficiency symptoms of calcium when they grow in soils with high sulfate levels (Francois and Maas, 1999). Bernstein (1953) suggested that reduced yields of beans and carrots are due to excessive soil calcium levels, which may restrict potassium uptake. High levels of exchangeable Na^+ restrict the vegetative growth of many crops, because of unfavorable soil physical conditions in sodic soils (Francois and Maas, 1999). Some crops, like beans growing in sodic soils, are sensitive to nutritional factors and may be severely affected even before the soil's physical condition is impaired (Francois and Maas, 1999).

High levels of soil salinity affect photosynthesis by lowering the availability of CO_2 due to diffusion limitations and reduction of photosynthetic pigments (Delfine et al., 1999; Flexas et al., 2007; Ashraf and Harris, 2013). Salt accumulation in spinach inhibits photosynthesis, mainly by decreasing stomatal and mesophyll conductance of CO_2 and reducing chlorophyll content (Delfine et al., 1998; Delfine et al., 1999; Di Martino et al., 1999; Alvino et al., 2000). About 80% of the growth reduction of radish growing under saline conditions is attributed to reduced leaf area and reduced light interception (Marcelis and Van Hooijdonk, 1999).

Most vegetable crops growing in saline soils lack market quality due to various reasons. In tomatoes and peppers, there is a smaller fruit size; in tomatoes, there is a decrease in fruit number; in celery, there is a reduction in petiole length; in broccoli and cauliflower, there is a reduction of leaf area (Bernstein, 1959; Jobes et al., 1981; Francois et al., 1982; Giuffrida et al., 2017; Di Gioia et al., 2018). However, in some instances, high salinity positively affects the quality of vegetable crops. For example, tomatoes grown under high salinity have an increased carotenoid content and antioxidant activity (De Pascale et al., 2015). Similarly, high salinity increased the carotenoid content in romaine lettuce (Kim et al., 2008). Under moderate salinity conditions, the nutritional quality of the edible florets of broccoli was increased (Lopez-Berenguer et al., 2009). Moreover, spinach grown under high salinity showed increased polyphenol levels and decreased levels of oxalic acid (Shimomachi et al., 2008). Although salt stress enhances the flavor of carrots and asparagus by increasing sugar contents when grown under saline conditions, the gain in quality is more than offset by lower yields (Bernstein, 1953; Francois, 1987).

Unlike vegetable crops, most of the major cereal crops can withstand high levels of soil salinity. However, because the seedling stage or early vegetative stages are more sensitive, high soil salinity reduces the number of leaves per culm, spikelets per spike, and the number of tillers per plant (Maas, 1990; Maas and Grieve, 1990). The effects of salinity on spikelet and tiller number, established in the early stages of crop development, have a greater influence on the reduction of harvest than salinity that occurs at the final stages (Pearson et al., 1966; Maas et al., 1986).

Some fruit trees, such as stone fruits, citrus, and avocado, are sensitive to salinity and accumulate Cl^- and Na^+ in foliage (Francois and Maas, 1999). The accumulation of these ions to detrimental levels and general osmotic growth inhibition lead to a reduction in plant growth and fruit yield. Chlorine toxicity is caused for more severe stress than Na^+ toxicity in woody crop fruits and leads to leaf tip necrosis, which later develops into marginal necrosis (Francois and Maas, 1999). For instance, strawberry plants exhibit typical necrosis in older leaves due to the accumulation of Cl^- (Ferreira et al., 2019). The further accumulation of Cl^- increases toxicity in the plant body and ultimately results in premature leaf drop, complete defoliation, twig dieback, and sometimes death (Bernstein, 1980; Hayward, 1981). Injury by Na^+ may happen even at small concentrations in the soil solution. Characteristic injury symptoms, such as tip, marginal, and/or interveinal necrosis, may not occur until a substantial time after exposure to salinity. Initially, Na^+ is retained in the sapwood of the plant. During the conversion of sapwood to heartwood, Na^+ is released, translocated in the leaves, and accumulates leading to leaf burn (Abd-El-Rahman et al., 1972).

1.6 ENVIRONMENTAL ISSUES OF SOIL SALINITY

Soil salinization is one of the major environmental issues that cause land degradation, biodiversity loss, and crop yield reduction throughout the world (Rath and Rousk, 2015). Constant fluctuations in the water table caused by salinity in irrigated and non-irrigated areas have adverse environmental effects (Jalili et al., 2011). One of the most significant off-site impacts of salinity is the salinization of rivers (Halse et al., 2003). This affects water quality (drinking and irrigation), with major economic, social, and environmental implications (Rosegrant and Ringler, 2000; Khatri and Tyagi, 2015). High salt content changes the taste of drinking water (Dietrich and Devesa, 2019). Chloride has a low taste threshold level, and the laxative effect of magnesium sulfate and sodium in drinking water makes it unsuitable for grazing animals (Burlingame et al., 2007; Valtorta and Gallardo, 2012). Salt interacts with in-stream fauna and flora and alters the ecological health of streams and estuaries (Cañedo-Argüelles et al., 2013).

One of the biggest salinity issues is the threat imposed on biodiversity through habitat loss (Hart et al., 2003). Riparian zones (the interface between land and water bodies) are severely affected by saline groundwater (McEwan et al., 2006). Suspended particulate materials tend to flocculate in the presence of salt, which allows excessive light penetration into water bodies and may lead to algal blooms and eutrophication (Paul, 2008; Mietta et al., 2009). Even low salinity concentrations in soils can act as an abiotic stress factor that affects vegetation health and distribution (Bui, 2013). The vegetation of salt-affected landscapes quickly gets damaged, posing severe threats to terrestrial biodiversity (Briggs and Taws, 2003). Soil salinity tends to change natural landscapes and causes fragmentation of habitats (Taddese, 2001). Dryland salinity is a reason for soil degradation, including soil erosion (Eswaran et al., 2019). Salinity is closely linked to prolonged wetness, and sparse and reduced surface vegetation cover, which, thus, increases the vulnerability of soils to erosion (Chakraborty et al., 2018). Saline shallow water tables may result in flooding, and the absorption capacity of soil is reduced with high runoff rates. Soils in this situation have limited capacity to absorb rainfall, thereby inducing flooding (Wale and Dejenie, 2013). Saline creeps are created due to excessive leaching and increase the chance of saltwater intrusion (Singh, 2009).

Soil salinity is known to be a crucial factor affecting the global distribution of soil microorganisms (Auguet et al., 2010). The abundance, composition, and distribution of soil bacterial communities are altered with increased soil salinity, and it acts as a critical inhibitor in soil biogeochemical processes, especially in wetland ecosystems (Zhang et al., 2021). Furthermore, soil salinity is a primary factor that shapes soil fungal communities. This eventually changes the composition of plant and litter saprotrophs (Yang and Sun, 2020).

Wetlands are often characterized by unique biodiversity, contributing significantly to the regional diversity of flora and fauna (Waterkeyn et al., 2008). In the presence of sulfate salts, there is a high risk of acid sulfate soil formation in wetlands, which eventually leads to soil degradation and biodiversity loss (Powell and Martens, 2005). Studies showed that changes in the depth distribution of soil organic carbon and total nitrogen in coastal wetlands occur with changing salinity (Guan et al., 2021a). The potential ecological risk of heavy metals (e.g., Pb, Cd) is higher under increased salinity in wetland ecosystems (Zhu et al., 2020). The combined and interacting effects of soil salinization reduce the diversity of aquatic invertebrates, aquatic plants, and plankton communities, resulting in the loss of wetland biodiversity (Nielsen et al., 2003; Waterkeyn et al., 2008).

1.7 FUTURE PERSPECTIVES

Salinization of lands is a continuous process that is accelerated by unfavorable climatic conditions and anthropogenic activities. This process directly impacts global food security by limiting arable lands for sufficient crop production to feed the increasing human population. Therefore, establishing soil salinity mitigation and management strategies at global and regional levels is crucial. However, prior to integrating mitigation and management practices for soil salinity, continuous and accurate monitoring is essential to determine soil salinity distribution on a regional and global scale and ascertain the magnitude of salinization. Three major methods developed over time to determine and monitor soil salinity at the field scale and larger geographical scale are the following: visual crop observation; geospatial measurement of the electrical conductivity of the soil; and multi- and hyperspectral imagery techniques (Corwin, 2021). First, technological advancement of these methods and the establishment of new, cost-effective, and efficient methods could improve real-time monitoring of salinization, offering timely management practices for affected areas.

Secondly, the search for cost-effective, efficient, environmentally friendly, and sustainable techniques for the reclamation of salt-affected soils is vital to re-utilize affected arable areas. The use of biochar and other organic amendments is recognized as a sustainable, cost-effective approach and has proven to be effective for the reclamation of salt-affected soils (Gunarathne et al., 2020a). However, further investigation with different types of biochar and organic amendments is important to decide which amendment has the best performance for a particular region. Moreover, identifying and breeding plant species to use as phytoremediation agents to minimize salinity in soil and the root zone soil solution is crucial. Such plant species may further be a useful source of food, firewood, medicine, or provide aesthetic value. Further, mechanisms and functions of salt tolerance of soil microbial communities in saline-alkaline environments need to be assessed in soil remediation initiatives (Arora et al., 2014).

Thirdly, prevention of further spreading of salinization is necessary to maximize the use of unaffected and partially salt-affected arable lands. For that, careful monitoring of salinization and alteration of existing land management and irrigation methods should be carried out. For instance, drip irrigation systems are known to be highly effective compared to surface irrigation systems for maximum water use efficiency while reducing the salinization of lands. Policymaking and execution of policies for sustainable land use and irrigation by relevant authorities at the regional and global levels are critical in preventing soil salinization and its further spreading.

Finally, cost-effective technologies to maintain sustainable agricultural production in salt-affected environments need to be developed. There is an urgent need to shift traditional farming into salinity management-based agricultural systems. Further studies to find salt-tolerant crop species and varieties to cultivate in salt-affected arable areas are required. The introduction of salt-tolerant improved crop species or varieties can be done using novel plant breeding techniques, such as using transgenic plant technology. However, the utilization of transgenic plants should be carried out after a complete study to determine their safety for use in natural ecosystems and their possible hazards in food products consumed by humans, because the safety of their use as crop plants is still in question (Gunarathne et al., 2019). Furthermore, time-tested and proven salinity reduction technologies need to be incorporated into diverse crop production systems. However, most importantly, soil salinity issues need to be approached in a holistic way by involving all possible mitigation methods and stakeholders to sustain salt-affected soil and water resources.

REFERENCES

Abd-El-Rahman, A.A., El-Shourbagy, M.N., Shalaby, A.F., El-Monayeri, M., 1972. Salinity effects on and development of the Elberta peach on Shalil and Lovell rootstocks. USDA Technical Bulletins, 922.

Abood, T.T., Kasa, A.B., Chik, Z.B., 2007. Stabilisation of silty clay soil using chloride compounds. *Journal of Engineering Science and Technology* 2, 102–110.

Ajalloeian, R., Mansouri, H., Sadeghpour, A.H., 2013. Effect of saline water on geotechnical properties of fine-grained soil. *EJGE* 18, 1419–1435.

Akinbile, C.O., Adegbola, O.A., Akande, S.O., 2016. Determining the effect of salt-induced soil on paddy rice development and yield using GIS mapping. *Paddy and Water Environment* 14, 313–323. doi: 10.1007/s10333-015-0501-3.

Akinwumi, A., 2014. A controlled experiment to determine the effect of high formation water salinity on crude oil in hydrocarbon bearing rock. *International Journal of Science and Research* 3, 2955–2958.

Al-Haddabi, M., Ahmed, M., 2007. Land disposal of treated saline oil production water: Impacts on soil properties. *Desalination* 212, 54–61. doi: 10.1016/j.desal.2006.09.019.

Allison, L.E., Richards, L.A., 1954. *Diagnosis and Improvement of Saline and Alkali Soils.* Scienctific Publishers: India.

Alvino, A., D'Andria, R., Delfine, S., Lavini, A., Zanetti, P., 2000. Effect of water and salinity stress on radiation absorption and efficiency in sunflower. *Italian Journal of Agronomy* 4, 53–60.

Arora, S., Vanza, M.J., Mehta, R., Bhuva, C., Patel, P.N., 2014. Halophilic microbes for bio-remediation of salt affected soils. *African Journal of Microbiology Research* 8, 3070–3078. doi: 10.5897/AJMR2014.6960.

Ashraf, M., Harris, P.J.C., 2013. Photosynthesis under stressful environments: An overview. *Photosynthetica* 51, 163–190. doi: 10.1007/s11099-013-0021-6.

Aslamsup, R., Bostansup, N., Mariasup, M., Safdar, W., 2011. A critical review on halophytes: Salt tolerant plants. *Journal of Medicinal Plants Research* 5, 7108–7118. doi: 10.5897/JMPRX11.009.

Auguet, J.-C., Barberan, A., Casamayor, E.O., 2010. Global ecological patterns in uncultured Archaea. *The ISME journal* 4, 182–190. doi: 10.1038/ismej.2009.109.

Ayers, R.S., Westcot, D.W., 1985. *Water Quality for Agriculture.* Food and Agriculture Organization of the United Nations: Rome.

Barbiero, L., Cunnac, S., Mané, L., Laperrousaz, C., Hammecker, C., Maeght, J.L., 2001. Salt distribution in the Senegal middle valley: Analysis of a saline structure on planned irrigation schemes from N'Galenka creek. *Agricultural Water Management* 46, 201–213. doi: 10.1016/S0378-3774(00)00088-3.

Bernstein, L., 1953. Salt tolerance of five varieties of carrots. *Proceedings of the American Society for Horticultural Science* 61, 360–366.

Bernstein, L., 1959. *Salt Tolerance of Vegetable Crops in the West*. Agricultural Research Service, US Department of Agriculture: Riverdale, MD.

Bernstein, L., 1975. Effects of salinity and sodicity on plant growth. *Annual Review of Phytopathology* 13, 295–312. doi: 10.1146/ANNUREV.PY.13.090175.001455.

Bernstein, L., 1980. Salt tolerance of fruit crops, USDA Agricultural Information Bulletin 292, p. 8.

Bernstein, L., Ayers, A.D., 1951. Salt tolerance of six varieties of green beans. *American Society for Horticultural Science* 57, 243–248.

Bernstein, L., Ayers, A.D., 1953. Salt tolerance of five varieties of onions. *American Society Horticultural Science* 61, 367–370.

Bond, W.J., 1998. Effluent irrigation: An environmental challenge for soil science. *Soil Research* 36, 543–556. doi: 10.1071/S98017.

Bressler, E., Carter, D.L., McNeal, B.L., 1982. *Saline and Sodic Soils: Principles, Dynamics, Modeling*. Springer-Verlag: Berlin Heidelberg, 227.

Briggs, S.V., Taws, N., 2003. Impacts of salinity on biodiversity: Clear understanding or muddy confusion? *Australian Journal of Botany* 51, 609–617. doi: 10.1071/BT02114.

Bronick, C.J., Lal, R., 2005. Soil structure and management: A review. *Geoderma* 124, 3–22. doi: 10.1016/j.geoderma.2004.03.005.

Bui, E.N., 2013. Soil salinity: A neglected factor in plant ecology and biogeography. *Journal of Arid Environments* 92, 14–25. doi: 10.1016/j.jaridenv.2012.12.014.

Burlingame, G.A., Dietrich, A.M., Whelton, A.J., 2007. Understanding the basics of tap water taste. *Journal-American Water Works Association* 99, 100–111. doi: 10.1002/j.1551-8833.2007.tb07930.x.

Butcher, K., Wick, A.F., DeSutter, T., Chatterjee, A., Harmon, J., 2016. Soil salinity: A threat to global food security. *Agronomy Journal* 108, 2189–2200. doi: 10.2134/agronj2016.06.0368.

Campbell, B.J., Kirchman, D.L., 2013. Bacterial diversity, community structure and potential growth rates along an estuarine salinity gradient. *The ISME journal* 7, 210–220. doi: 10.1038/ismej.2012.93.

Cañedo-Argüelles, M., Kefford, B.J., Piscart, C., Prat, N., Schäfer, R.B., Schulz, C.-J., 2013. Salinisation of rivers: An urgent ecological issue. *Environmental Pollution* 173, 157–167. doi: 10.1016/j.envpol.2012.10.011.

Capooci, M., Barba, J., Seyfferth, A.L., Vargas, R., 2019. Experimental influence of storm-surge salinity on soil greenhouse gas emissions from a tidal salt marsh. *Science of the Total Environment* 686, 1164–1172. doi: 10.1016/j.scitotenv.2019.06.032.

Cass, A., Sumner, M.E., 1982. Soil pore structural stability and irrigation water quality: I. Empirical sodium stability model. *Soil Science Society of America Journal* 46, 503–506. doi: 10.2136/sssaj1982.0361599 5004600030011x.

Chakraborty, K., Basak, N., Bhaduri, D., Ray, S., Vijayan, J., Chattopadhyay, K., Sarkar, R.K., 2018. Ionic basis of salt tolerance in plants: Nutrient homeostasis and oxidative stress tolerance. In: Hasanuzzaman, M. et al. (Eds.), *Plant Nutrients and Abiotic Stress Tolerance*. Springer: Berlin/Heidelberg, Germany, pp. 325–362.

Chernousenko, G.I., Yamnova, I.A., Skripnikova, M.I., 2003. Anthropogenic salinization of soils in Moscow. *Eurasian Soil Science C/C Of Pochvovedenie* 36, 92–100.

Chinnusamy, V., Jagendorf, A., Zhu, J.K., 2005. Understanding and improving salt tolerance in plants. *Crop Science* 45, 437–448. doi: 10.2135/cropsci2005.0437.

Corwin, D.L., 2021. Climate change impacts on soil salinity in agricultural areas. *European Journal of Soil Science* 72, 842–862. doi: 10.1111/ejss.13010.

Cramer, G.R., Läuchli, A., Polito, V.S., 1985. Displacement of Ca^{2+} by Na^+ from the plasmalemma of root cells: A primary response to salt stress? *Plant Physiology* 79, 207–211. doi: 10.1104/pp.79.1.207.

CSSRI, 2014. Vision 2050. Pragmatic assessment of the agricultural production and food demand scenario of India by the Year 2050. Central Soil Salinity Research Institute, Karnal, India.

Daley, M.L., Potter, J.D., McDowell, W.H., 2009. Salinization of urbanizing New Hampshire streams and groundwater: Effects of road salt and hydrologic variability. *Journal of the North American Benthological Society* 28, 929–940.

Daliakopoulos, I.N., Tsanis, I.K., Koutroulis, A., Kourgialas, N.N., Varouchakis, A.E., Karatzas, G.P., Ritsema, C.J., 2016. The threat of soil salinity: A European scale review. *Science of the Total Environment* 573, 727–739. doi: 10.1016/j.scitotenv.2016.08.177.

Dasgupta, S., Hossain, M.M., Huq, M., Wheeler, D., 2018. Climate change, salinization and high-yield rice production in coastal Bangladesh. *Agricultural and Resource Economics Review* 47, 66–89. doi: 10.1017/age.2017.14.

De Louw, P.G.B., Essink, G.H.P.O., Stuyfzand, P.J., Van der Zee, S., 2010. Upward groundwater flow in boils as the dominant mechanism of salinization in deep polders, the Netherlands. *Journal of Hydrology* 394, 494–506. doi: 10.1016/j.jhydrol.2010.10.009.

De Oliveira, A.B., Alencar, N.L.M., Gomes-Filho, E., 2013. Comparison between the water and salt stress effects on plant growth and development. In: Akinci, S. (Ed.), *Responses of Organisms to Water Stress*. INTECH Open Access Publisher: Croatia, pp. 67–94.

De Pascale, S., Maggio, A., Orsini, F., Stanghellini, C., Heuvelink, E., 2015. Growth response and radiation use efficiency in tomato exposed to short-term and long-term salinized soils. *Scientia Horticulturae* 189, 139–149. doi: 10.1016/j.scienta.2015.03.042.

De Waegemaeker, J., 2019. SalFar framework on salinization processes: A comparison of salinization processes across the North Sea Region.

Delfine, S., Alvino, A., Zacchini, M., Loreto, F., 1998. Consequences of salt stress on conductance to CO_2 diffusion, Rubisco characteristics and anatomy of spinach leaves. *Functional Plant Biology* 25, 395–402. doi: 10.1071/PP97161.

Delfine, S., Alvino, A., Villani, M.C., Loreto, F., 1999. Restrictions to carbon dioxide conductance and photosynthesis in spinach leaves recovering from salt stress. *Plant Physiology* 119, 1101–1106. doi: 10.1104/pp.119.3.1101.

Di Gioia, F., Rosskopf, E.N., Leonardi, C., Giuffrida, F., 2018. Effects of application timing of saline irrigation water on broccoli production and quality. *Agricultural Water Management* 203, 97–104. doi: 10.1016/j.agwat.2018.01.004.

Di Martino, C., Delfine, S., Alvino, A., Loreto, F., 1999. Photorespiration rate in spinach leaves under moderate NaCl stress. *Photosynthetica* 36, 233–242. doi: 10.1023/A:1007099627285.

Dietrich, A.M., Devesa, R., 2019. Characterization and removal of minerals that cause taste. In: Lin, T.F. et al. (Eds.), *Taste and Odour in Source and Drinking Water: Causes, Controls, and Consequences*. IWA Publishing: London, England, pp. 113–142.

Du Preez, C.C., Van Huyssteen, C.W., 2020. Threats to soil and water resources in South Africa. *Environmental Research* 183, 109015. doi: 10.1016/j.envres.2019.109015.

Eilers, R.G., Eilers, W.D., Pettapiece, W.W., Lelyk, G., 1995. Salinization of soil. In: Acton, D.F., Gregorich, L.J. (Eds.), *The Health of Our Soils: Toward Sustainable Agriculture in Canada*. Centre for Land and Biological Resources Research, Agriculture and Agri-Food Canada: Ottawa, 75-82.

EIP-AGRI, 2019. European Innovation Partnership 'Agricultural Productivity and Sustainability' EIP-AGRI – European Commission.

El-Keblawy, A., 2004. Salinity effects on seed germination of the common desert range grass, Panicum turgidum. *Seed Science and Technology* 32, 873–878. doi: 10.15258/sst.2004.32.3.24.

El-Shahway, A.S., Mahmoud, M.M.A., Udeigwe, T.K., 2016. Alterations in soil chemical properties induced by continuous rice cultivation: A study on the arid Nile delta soils of Egypt. *Land Degradation & Development* 27, 231–238. doi: 10.1002/ldr.2409.

Endo, T., Yamamoto, S., Larrinaga, J.A., Fujiyama, H., Honna, T., 2011. Status and causes of soil salinization of irrigated agricultural lands in southern Baja California, Mexico. *Applied and Environmental Soil Science* 2011. doi: 10.1155/2011/873625.

Eswaran, H., Lal, R., Reich, P.F., 2019. Land degradation: An overview. In: Bridges, E.M. (Ed.), *Response to Land Degradation*. CRC Press: Boca Raton, FL, pp. 20–35.

European Soil Bureau. 2014. Soil themes. Soil salinisation. Joint Research Centre, Ispra (http://eusoils.jrc.ec.europa.eu/library/themes/Salinization/ accessed: 20th February 2020)

Eynard, A., Lal, R., Wiebe, K.D., 2006. Salt-affected soils. *Encyclopedia of Soil Science* 2, 1538–1541.

Fan, X., Pedroli, B., Liu, G., Liu, Q., Liu, H., Shu, L., 2012. Soil salinity development in the yellow river delta in relation to groundwater dynamics. *Land Degradation & Development* 23, 175–189. doi: 10.1002/ldr.1071.

Feigin, A., Ravina, I., Shalhevet, J., 2012. *Irrigation with Treated Sewage Effluent: Management for Environmental Protection*. Springer Science & Business Media: Berlin/Heidelberg, Germany.

Ferreira, J.F.S., Liu, X., Suarez, D.L., 2019. Fruit yield and survival of five commercial strawberry cultivars under field cultivation and salinity stress. *Scientia Horticulturae* 243, 401–410. doi: 10.1016/j.scienta.2018.07.016.

Fisarakis, I., Chartzoulakis, K., Stavrakas, D., 2001. Response of Sultana vines (V. vinifera L.) on six rootstocks to NaCl salinity exposure and recovery. *Agricultural Water Management* 51, 13–27. doi: 10.1016/S0378-3774(01)00115-9.

Flexas, J., Diaz-Espejo, A., Galmes, J., Kaldenhoff, R., Medrano, H., Ribas-Carbo, M., 2007. Rapid variations of mesophyll conductance in response to changes in CO_2 concentration around leaves. *Plant, Cell & Environment* 30, 1284–1298. doi: 10.1111/j.1365-3040.2007.01700.x.

Florinsky, I.V., Eilers, R.G., Lelyk, G.W., 2000. Prediction of soil salinity risk by digital terrain modeling in the Canadian prairies. *Canadian Journal of Soil Science* 80, 455–463. doi: 10.4141/S99-093.

Francois, L.E., 1987. Salinity effects on asparagus yield and vegetative growth. *Journal of the American Society for Horticultural Science (USA)* 112(3), 432–436.

Francois, L.E., Maas, E.V., 1999. Crop response and management of salt-affected soils. In: Pessarakli, M. (Ed.), *Handbook of Plant and Crop Stress*. Marcel Dekker Press Inc.: New York, pp. 169–201.

Francois, L.E., Le, F., Dw, W., 1982. Reduction in yield and market quality of celery caused by soil salinity. *Journal of the American Society for Horticultural Science (USA)* 107(6), 952–954.

Funakawa, S., Kosaki, T., 2007. Potential risk of soil salinization in different regions of Central Asia with special reference to salt reserves in deep layers of soils. *Soil Science & Plant Nutrition* 53, 634–649. doi: 10.1111/j.1747-0765.2007.00186.x.

Garcia, C., Hernandez, T., 1996. Influence of salinity on the biological and biochemical activity of a calciorthird soil. *Plant and Soil* 178, 255–263. doi: 10.1007/BF00011591.

Gharaibeh, M.A., Eltaif, N.I., Shunnar, O.F., 2009. Leaching and reclamation of calcareous saline-sodic soil by moderately saline and moderate-SAR water using gypsum and calcium chloride. *Journal of Plant Nutrition and Soil Science* 172, 713–719. doi: 10.1002/jpln.200700327.

Ghollarata, M., Raiesi, F., 2007. The adverse effects of soil salinization on the growth of *Trifolium alexandrinum* L. and associated microbial and biochemical properties in a soil from Iran. *Soil Biology and Biochemistry* 39, 1699–1702. doi: 10.1016/j.soilbio.2007.01.024.

Giuffrida, F., Cassaniti, C., Malvuccio, A., Leonardi, C., 2017. Effects of salt stress imposed during two growth phases on cauliflower production and quality. *Journal of the Science of Food and Agriculture* 97, 1552–1560. doi: 10.1002/jsfa.7900.

Gomez, J.M., Jimenez, A., Olmos, E., Sevilla, F., 2004. Location and effects of long-term NaCl stress on superoxide dismutase and ascorbate peroxidase isoenzymes of pea (Pisum sativum cv. Puget) chloroplasts. *Journal of Experimental Botany* 55, 119–130. doi: 10.1093/jxb/erh013.

González, L.M., 2000. Extent, cause and management of salt affected soils in Cuba. *FAO. Newsletter on Sustainable Productive Use of Salt Affected Habitats* 4, 8–11.

GoP. 2007. "Economic Survey of Pakistan, 2006–07," Planning Commission, Government of Pakistan, Islamabad.

Gopalakrishnan, T., Kumar, L., 2020. Modeling and mapping of soil salinity and its impact on paddy lands in Jaffna Peninsula, Sri Lanka. *Sustainability* 12, 8317. doi: 10.3390/su12208317.

Gould, I.J., Wright, I., Collison, M., Ruto, E., Bosworth, G., Pearson, S., 2020. The impact of coastal flooding on agriculture: A case-study of Lincolnshire, United Kingdom. *Land Degradation & Development* 31, 1545–1559. doi: 10.1002/ldr.3551.

Gould, I., De Waegemaeker, J., Tzemi, D., Wright, I., Pearson, S., Ruto, E., Karrasch, L., Christensen, L.S., Aronsson, H., Eich-Greatorex, S., 2021. Salinization threats to agriculture across the North Sea Region. In: Negacz, K. et al. (Eds.), *Future of Sustainable Agriculture in Saline Environments*. CRC Press, Boca Raton, FL, pp. 71–92.

Grattan, S., Grieve, C., 1998. Salinity–mineral nutrient relations in horticultural crops. *Scientia Horticulturae* 78, 127–157. doi: 10.1016/S0304-4238(98)00192-7.

Greene, R., Timms, W., Rengasamy, P., Arshad, M., Cresswell, R., 2016. Soil and aquifer salinization: toward an integrated approach for salinity management of groundwater. In: Jakeman, A.J., et al. (Eds.), *Integrated Groundwater Management*. Springer, Cham, pp. 377–412.

Greenway, H., Munns, R., 1980. Mechanisms of salt tolerance in nonhalophytes. *Annual Review of Plant Physiology* 31, 149–190.

Greiner, R., Cacho, O., 2001. On the efficient use of a catchment's land and water resources: Dryland salinization in Australia. *Ecological Economics* 38, 441–458. doi: 10.1016/S0921-8009(01)00192-6.

Guan, Y., Bai, J., Wang, J., Wang, W., Wang, X., Zhang, L., Li, X., Liu, X., 2021a. Effects of groundwater tables and salinity levels on soil organic carbon and total nitrogen accumulation in coastal wetlands with different plant cover types in a Chinese estuary. *Ecological Indicators* 121, 106969. doi: 10.1016/j.ecolind.2020.106969.

Guan, Y., Jiang, N., Wu, Y., Yang, Z., Bello, A., Yang, W., 2021b. Disentangling the role of salinity-sodicity in shaping soil microbiome along a natural saline-sodic gradient. *Science of the Total Environment* 765, 142738. doi: 10.1016/j.scitotenv.2020.142738.

Gunarathne, V., Mayakaduwa, S., Ashiq, A., Weerakoon, S.R., Biswas, J.K., Vithanage, M., 2019. Transgenic plants: Benefits, applications, and potential risks in phytoremediation. In: Prasad, M.N.V. (Ed.), *Transgenic Plant Technology for Remediation of Toxic Metals and Metalloids*. Academic Press: Cambridge, MA, pp. 89–102.

Gunarathne, V., Senadeera, A., Gunarathne, U., Biswas, J.K., Almaroai, Y.A., Vithanage, M., 2020a. Potential of biochar and organic amendments for reclamation of coastal acidic-salt affected soil. *Biochar*, 1–14. doi: 10.1007/s42773-020-00036-4.

Gunarathne, V., Senadheera, J.A.I., Gunarathne, U., Almaroai, Y.A., Vithanage, M., 2020b. Reclamation of salt-affected soils. In: Sik Ok, Y. (Ed.), *Soil and Groundwater Remediation Technologies*. CRC Press: Boca Raton, FL, pp. 183–199.

Guo, S.-H., Hu, N., Li, Q.-S., Yang, P., Wang, L.-L., Xu, Z.-M., Chen, H.-J., He, B.-Y., Zeng, E.Y., 2018. Response of edible amaranth cultivar to salt stress led to Cd mobilization in rhizosphere soil: A metabolomic analysis. *Environmental Pollution* 241, 422–431. doi: 10.1016/j.envpol.2018.05.018.

Gupta, B., Huang, B., 2014. Mechanism of salinity tolerance in plants: Physiological, biochemical, and molecular characterization. *International Journal of Genomics*. doi: 10.1155/2014/701596.

Gustafsson, M.E.R., Franzén, L.G., 1996. Dry deposition and concentration of marine aerosols in a coastal area, SW Sweden. *Atmospheric Environment* 30, 977–989. doi: 10.1016/1352-2310(95)00355-X.

Halima, O.I., Azzouzi, M.E., Douaik, A., Azim, K., Zouahri, A., 2019. Organic and inorganic remediation of soils affected by salinity in the Sebkha of Sed El Mesjoune–Marrakech (Morocco). *Soil and Tillage Research* 193, 153–160. doi: 10.1016/j.still.2019.06.003.

Halse, S.A., Ruprecht, J.K., Pinder, A.M., 2003. Salinisation and prospects for biodiversity in rivers and wetlands of south-west Western Australia. *Australian Journal of Botany* 51, 673–688. doi: 10.1071/BT02113.

Hart, B.T., Lake, P.S., Webb, J.A., Grace, M.R., 2003. Ecological risk to aquatic systems from salinity increases. *Australian Journal of Botany* 51, 689–702. doi: 10.1071/BT02111.

Hasanuzzaman, M., Nahar, K., Alam, M., Bhowmik, P.C., Hossain, M., Rahman, M.M., Prasad, M.N.V., Ozturk, M., Fujita, M., 2014. Potential use of halophytes to remediate saline soils. *BioMed Research International* 2014. doi: 10.1155/2014/589341.

Hayward, H.E., 1955. Factors affecting the salt tolerance of horticultural crops. *Report XIV International Horticultural Congress,* Wageningen, pp. 385–399.

Hayward, H.E., 1981. Effect of chloride and sulfate salts on the growth and development of the Elberta peach on Shalil and Lovell rootstocks.

He, B.-Y., Yu, D.-P., Chen, Y., Shi, J.-L., Xia, Y., Li, Q.-S., Wang, L.-L., Ling, L., Zeng, E.Y., 2017. Use of low-calcium cultivars to reduce cadmium uptake and accumulation in edible amaranth (*Amaranthus mangostanus* L.). *Chemosphere* 171, 588–594. doi: 10.1016/j.chemosphere.2016.12.085.

Horie, T., Karahara, I., Katsuhara, M., 2012. Salinity tolerance mechanisms in glycophytes: An overview with the central focus on rice plants. *Rice* 5, 1–18. doi: 10.1186/1939-8433-5-11.

Jalili, S., Moazed, H., Boroomand, N.S., Naseri, A.A., 2011. Assessment of evaporation and salt accumulation in bare soil: Constant shallow water table depth with saline ground water. *Scientific Research and Essays* 6, 6068–6074. doi: 10.5897/SRE11.509.

Jobes, J.A., Hoffman, G.J., Wood, J.D., 1981. Leaching requirement for salinity control. II. Oat, tomato, and cauliflower. *Agricultural Water Management* 4, 393–407.

Kashenge-Killenga, S., Meliyo, J., Urassa, G., Kongo, V., 2016. Extent of salt-affected soils and their effects in irrigated and lowland rain-fed rice growing areas of southwestern Tanzania. In: Lal, R. et al. (Eds.), *Climate Change and Multi-Dimensional Sustainability in African Agriculture*. Springer, Berlin/Heidelberg, Germany, pp. 97–126.

Keller, L.P., McCarthy, G.J., Richardson, J.L., 1986. Mineralogy and stability of soil evaporites in North Dakota. *Soil Science Society of America Journal* 50, 1069–1071. doi: 10.2136/sssaj1986.0361599500 5000040047x.

Khan, G.S., 1998. Soil salinity/sodicity status in Pakistan. Soil Survey of Pakistan, Lahore, 59.

Khatri, N., Tyagi, S., 2015. Influences of natural and anthropogenic factors on surface and groundwater quality in rural and urban areas. *Frontiers in Life Science* 8, 23–39. doi: 10.1080/21553769.2014.933716.

Khitrov, N.B., Rukhovivh, D.I., Kalinina, N.V., Novikova, A.F., Pankova, E.I., Chernousenko, G.I., 2009. Estimation of the areas of salt-affected soils in the European part of Russia on the basis of a digital map of soil salinization on a scale of 1: 2.5 M. *Eurasian Soil Science* 42, 581–590. doi: 10.1134/S1064229309060015.

Khoshgoftarmanesh, A.H., Shariatmadari, H., Karimian, N., Kalbasi, M., Van der Zee, S., 2006. Cadmium and zinc in saline soil solutions and their concentrations in wheat. *Soil Science Society of America Journal* 70, 582–589. doi: 10.2136/sssaj2005.0136.

Kim, H.-J., Fonseca, J.M., Choi, J.-H., Kubota, C., Kwon, D.Y., 2008. Salt in irrigation water affects the nutritional and visual properties of romaine lettuce (*Lactuca sativa* L.). *Journal of Agricultural and Food Chemistry* 56, 3772–3776. doi: 10.1021/jf0733719.

Kjensmo, J., 1997. The influence of road salts on the salinity and the meromictic stability of Lake Svinsjøen, southeastern Norway. *Hydrobiologia* 347, 151–159. doi: 10.1023/A:1003035705729.

Kristiansen, S.M., Christensen, F.D., Hansen, B., 2011. Does road salt affect groundwater in Denmark? *Geological Survey of Denmark and Greenland (GEUS) Bulletin* 23, 45–48.
Kumar, P., Sharma, P.K., 2020. Soil salinity and food security in India. *Frontiers in Sustainable Food Systems* 4: 533781. doi: 10.3389/fsufs.2020.533781.
Lakhdar, A., Rabhi, M., Ghnaya, T., Montemurro, F., Jedidi, N., Abdelly, C., 2009. Effectiveness of compost use in salt-affected soil. *Journal of Hazardous Materials* 171, 29–37. doi: 10.1016/j.jhazmat.2009.05.132.
Läuchli, A., Epstein, E., 1990. Plant responses to saline and sodic conditions. *Agricultural Salinity Assessment and Management* 71, 113–137.
Li, J., Pu, L., Han, M., Zhu, M., Zhang, R., Xiang, Y., 2014. Soil salinization research in China: Advances and prospects. *Journal of Geographical Sciences* 24, 943–960. doi: 10.1007/s11442-014-1130-2.
Litalien, A., Zeeb, B., 2020. Curing the earth: A review of anthropogenic soil salinization and plant-based strategies for sustainable mitigation. *Science of the Total Environment* 698, 134235. doi: 10.1016/j.scitotenv.2019.134235.
Liu, J., Zhang, L., 2014. The microstructure characters of saline soil in Qarhan Salt Lake area and its behaviors of mechanics and compressive strength. *Arabian Journal for Science and Engineering* 39, 8649–8658. doi: 10.1007/s13369-014-1410-2.
Liu, M., Pan, T., Allakhverdiev, S.I., Yu, M., Shabala, S., 2020. Crop halophytism: An environmentally sustainable solution for global food security. *Trends in Plant Science* 25, 630–634. doi: 10.1016/j.tplants.2020.04.008.
Lopez-Berenguer, C., Martínez-Ballesta, M.A.D.C., Moreno, D.A., Carvajal, M., Garcia-Viguera, C., 2009. Growing hardier crops for better health: Salinity tolerance and the nutritional value of broccoli. *Journal of Agricultural and Food Chemistry* 57, 572–578. doi: 10.1021/jf802994p.
Lozupone, C.A., Knight, R., 2007. Global patterns in bacterial diversity. *Proceedings of the National Academy of Sciences* 104, 11436–11440. doi: 10.1073/pnas.0611525104.
Maas, E.V., 1990. Crop salt tolerance. In: Tanji, K.K. (Ed.), *Agricultural Salinity Assessment and Management*, ASCE Manual Reports on Engineering Practices, Vol. 71, ASCE: New York, pp. 262–304.
Maas, E.V., 1993. Testing crops for salinity tolerance. In: Maranville, J.W., Baligar, B.V., Duncan, R.R., Yohe, J.M. (Eds.), *Proceedings of the Workshop on Adaptation of Plants to Soil Stresses*. Citeseer: Princeton, NJ, pp. 234–247.0
Maas, E.V., Grieve, C.M., 1990. Spike and leaf development of sal-stressed wheat. *Crop Science* 30, 1309–1313. doi: 10.2135/cropsci1990.0011183X003000060031x.
Maas, E.V., Nieman, R.H., 1978. Physiology of plant tolerance to salinity. *Crop Tolerance to Suboptimal Land Conditions* 32, 277–299. doi: 10.2134/asaspecpub32.c13.
Maas, E.V., Poss, J.A., Hoffman, G.J., 1986. Salinity sensitivity of sorghum at three growth stages. *Irrigation Science* 7, 1–11.
Machado, R.M.A., Serralheiro, R.P., 2017. Soil salinity: Effect on vegetable crop growth: Management practices to prevent and mitigate soil salinization. *Horticulturae* 3, 30. doi: 10.3390/horticulturae3020030.
Makhdum, M.S.A., Ashfaq, M., 2008. An economic evaluation of negative impact of water-logging and salinity on wheat productivity. *Soil and Environment*,27(1), 105–108.
Marcelis, L.F.M., Van Hooijdonk, J., 1999. Effect of salinity on growth, water use and nutrient use in radish (*Raphanus sativus* L.). *Plant and Soil* 215, 57–64. doi: 10.1023/A:1004742713538.
McEwan, K.L., Holland, K., Jolly, I.D., 2006. A review of groundwater-surface water interactions in arid/semi-arid wetlands and the consequences of salinity for wetland ecology. *Ecohydrology* 1, 43–58.
Menete, M.Z.L., Van Es, H.M., Brito, R.M.L., DeGloria, S.D., Famba, S., 2008. Evaluation of system of rice intensification (SRI) component practices and their synergies on salt-affected soils. *Field Crops Research* 109, 34–44. doi: 10.1016/j.fcr.2008.06.003.
Mesa, D., 2003. Obtainment of plants resistant to salinity for Cuban saline soils. *Cuban Journal of Agricultural Science* 37, 215–224.
Mietta, F., Chassagne, C., Manning, A.J., Winterwerp, J.C., 2009. Influence of shear rate, organic matter content, pH and salinity on mud flocculation. *Ocean Dynamics* 59, 751–763. doi: 10.1007/s10236-009-0231-4.
Mohamed, A.-M.O., El Gamal, M., Zekri, A.Y., 2003. Effect of salinity and temperature on water cut determination in oil reservoirs. *Journal of Petroleum Science and Engineering* 40, 177–188. doi: 10.1016/S0920-4105(03)00137-2.
Montanarella, L., 2007. Trends in land degradation in Europe. In: Sivakumar, M.V.K., Ndiangui, N. (Eds.), *Climate and Land Degradation*. Springer: Berlin/Heidelberg, Germany, pp. 83–104.
Mukhopadhyay, R., Sarkar, B., Jat, H.S., Sharma, P.C., Bolan, N.S., 2020. Soil salinity under climate change: Challenges for sustainable agriculture and food security. *Journal of Environmental Management*, 111736. doi: 10.1016/j.jenvman.2020.111736.

Muller, M., Schreiner, B., Smith, L., van Koppen, B., Sally, H., Aliber, M., Cousins, B., Tapela, B., Van der Merwe-Botha, M., Karar, E., 2009. Water security in South Africa. Development Planning Division. Working Paper Series 12.

Munns, R., 2002. Comparative physiology of salt and water stress. *Plant, Cell & Environment* 25, 239–250. doi: 10.1046/j.0016-8025.2001.00808.x.

Munns, R., Husain, S., Rivelli, A.R., James, R.A., Condon, A.G.T., Lindsay, M.P., Lagudah, E.S., Schachtman, D.P., Hare, R.A., 2002. Avenues for increasing salt tolerance of crops, and the role of physiologically based selection traits. In: Horst, W. et al. (Eds.), *Plant Nutrition: Plenary Lectures of the XIV International Plant Nutrition Colloquium.* Springer, Dordrecht, pp. 93–105.

Munns, R., Passioura, J.B., Colmer, T.D., Byrt, C.S., 2020. Osmotic adjustment and energy limitations to plant growth in saline soil. *New Phytologist* 225, 1091–1096. doi: 10.1111/nph.15862.

Munny, N.N., Khan, M.Z., Hanif, M.D., 2021. Vertical distribution of soil properties and organic carbon under different land use systems in saline soils of Bangladesh. *Environmental Challenges* 4, 100097. doi: 10.1016/j.envc.2021.100097.

Nielsen, D.L., Brock, M.A., Crosslé, K., Harris, K., Healey, M., Jarosinski, I., 2003. The effects of salinity on aquatic plant germination and zooplankton hatching from two wetland sediments. *Freshwater Biology* 48, 2214–2223. doi: 10.1046/j.1365-2427.2003.01146.x.

Ondrasek, G., 2020. *Drought: Detection and Solutions.* BoD–Books on Demand: Norderstedt, Germany.

Ondrasek, G., Rengel, Z., 2020. Environmental salinization processes: Detection, implications & solutions. *Science of the Total Environment*, 142432. doi: 10.1016/j.scitotenv.2020.142432.

Pankova, E.I., 2015. Salt-affected soils of Russia: Solved and unsolved problems. *Eurasian Soil Science* 48, 115–127. doi: 10.1134/S106422931502009X.

Parida, A.K., Das, A.B., 2005. Salt tolerance and salinity effects on plants: A review. *Ecotoxicology and Environmental Safety* 60, 324–349. doi: 10.1016/j.ecoenv.2004.06.010.

Paul, V.J., 2008. Global warming and cyanobacterial harmful algal blooms. *Cyanobacterial Harmful Algal Blooms: State of the Science and Research Needs*, 239–257. doi: 10.1007/978-0-387-75865-7_11.

Pearson, G.A., Ayers, A.D., Eberhard, D.L., 1966. Relative salt tolerance of rice during germination and early seedling development. *Soil Science* 102, 151–156.

Perera, M.P., 2016. Development of agro-well lands and its impact on soil salinity in the North Central Dry Zone of Sri Lanka. *International Journal of Science and Research (IJSR)* 5, 497–501. doi: 10.21275/v5i7.ART2016217.

Perera, M., Ranasinghe, T., Piyadasa, R.U.K., Jayasinghe, G.Y., 2018. Risk of seawater intrusion on coastal community of Bentota river basin Sri Lanka. *Procedia Engineering* 212, 699–706. doi: 10.1016/j.proeng.2018.01.090.

Pereira, P., Barceló, D., Panagos, P., 2020. Soil and water threats in a changing environment. *Environmental Research* 186, 109501–109501. doi: 10.1016/j.envres.2020.109501.

Pessarakli, M., 2016. *Handbook of Cucurbits: Growth, Cultural Practices, and Physiology.* CRC Press: Boca Raton, FL.

Pessarakli, M., 2019. *Handbook of Plant and Crop Stress.* CRc Press: Boca Raton, FL.

Pessarakli, M., Szabolcs, I., 1999. Soil salinity and sodicity as particular plant/crop stress factors. In: Pessarakli, M. (Ed.), *Handbook of Plant and Crop Stress.* CRC Press: Boca Raton, FL, p. 2.

Phogat, V., Mallants, D., Cox, J.W., Šimůnek, J., Oliver, D.P., Pitt, T., Petrie, P.R., 2020. Impact of long-term recycled water irrigation on crop yield and soil chemical properties. *Agricultural Water Management* 237, 106167. doi: 10.1016/j.agwat.2020.106167.

Pisinaras, V., Tsihrintzis, V.A., Petalas, C., Ouzounis, K., 2010. Soil salinization in the agricultural lands of Rhodope District, northeastern Greece. *Environmental Monitoring and Assessment* 166, 79–94. doi: 10.1007/s10661-009-0986-6.

Powell, B., Martens, M., 2005. A review of acid sulfate soil impacts, actions and policies that impact on water quality in Great Barrier Reef catchments, including a case study on remediation at East Trinity. *Marine Pollution Bulletin* 51, 149–164. doi: 10.1016/j.marpolbul.2004.10.047.

Prospero, J.M., 1979. Mineral and sea salt aerosol concentrations in various ocean regions. *Journal of Geophysical Research: Oceans* 84, 725–731. doi: 10.1029/JC084iC02p00725.

Qadir, M., Oster, J.D., Schubert, S., Noble, A.D., Sahrawat, K.L., 2007. Phytoremediation of sodic and saline-sodic soils. *Advances in Agronomy* 96, 197–247. doi: 10.1016/S0065-2113(07)96006-X.

Raats, P.A.C., 2015. Salinity management in the coastal region of the Netherlands: A historical perspective. *Agricultural Water Management* 157, 12–30. doi: 10.1016/j.agwat.2014.08.022.

Ramírez, R., González, L.M., López, R., 1999. Afectaciones por salinidad en plántulas de Centrosema pubescens y Stylosanthes guianensis. *Pastos y Forrajes* 22, 115–121.

Rath, K.M., Rousk, J., 2015. Salt effects on the soil microbial decomposer community and their role in organic carbon cycling: A review. *Soil Biology and Biochemistry* 81, 108–123. doi: 10.1016/j.soilbio.2014.11.001.
Rath, K.M., Fierer, N., Murphy, D.V., Rousk, J., 2019. Linking bacterial community composition to soil salinity along environmental gradients. *The ISME Journal* 13, 836–846. doi: 10.1038/s41396-018-0313-8.
Rengasamy, P., 2002. Transient salinity and subsoil constraints to dryland farming in Australian sodic soils: An overview. *Australian Journal of Experimental Agriculture* 42, 351–361. doi: 10.1071/EA01111.
Rengasamy, P., 2006. World salinization with emphasis on Australia. *Journal of Experimental Botany* 57, 1017–1023. doi: 10.1093/jxb/erj108.
Rengasamy, P., 2010. Soil processes affecting crop production in salt-affected soils. *Functional Plant Biology* 37, 613–620. doi: 10.1071/FP09249.
Richards, L.A., 1954. *Diagnosis and Improvement of Saline and Alkali Soils*. Handbook 60. US Department of Agriculture, Washington, DC.
Rietz, D., Haynes, R., 2003. Effects of irrigation-induced salinity and sodicity on soil microbial activity. *Soil Biology and Biochemistry* 35, 845–854. doi: 10.1016/S0038-0717(03)00125-1.
Rosegrant, M.W., Ringler, C., 2000. Impact on food security and rural development of transferring water out of agriculture. *Water Policy* 1, 567–586. doi: 10.1016/S1366-7017(99)00018-5.
Ruiz, D., Martínez, V., Cerdá, A., 1997. Citrus response to salinity: Growth and nutrient uptake. *Tree Physiology* 17, 141–150.
Salama, R.B., Otto, C.J., Fitzpatrick, R.W., 1999. Contributions of groundwater conditions to soil and water salinization. *Hydrogeology Journal* 7, 46–64.
Scott, B.J., Ridley, A.M., Conyers, M.K., 2000. Management of soil acidity in long-term pastures of south-eastern Australia: A review. *Australian Journal of Experimental Agriculture* 40, 1173–1198. doi: 10.1071/EA00014.
Seibert, S.L., Holt, T., Reckhardt, A., Ahrens, J., Beck, M., Pollmann, T., Giani, L., Waska, H., Böttcher, M.E., Greskowiak, J., 2018. Hydrochemical evolution of a freshwater lens below a barrier island (Spiekeroog, Germany): The role of carbonate mineral reactions, cation exchange and redox processes. *Applied Geochemistry* 92, 196–208. doi: 10.1016/j.apgeochem.2018.03.001.
Senol, S., Bayramin, I., 2013. Soil resources of Turkey. In: Yigini, Y., Panagos, P., Montanarella, L. (Eds.), *Soil Resources of Mediterranean and Caucasus Countries*. The European Soil Bureau, Joint Research Centre, Italy, pp. 225–237.
Sentís, I.P., 2021. Overview of salt-affected areas in Latin America: Physical, social and economic perspectives. In: Taleisnik, E., Lavado, R.S. (Eds.), *Saline and Alkaline Soils in Latin America*. Springer: Berlin/Heidelberg, Germany, pp. 3–36.
Setia, R., Gottschalk, P., Smith, P., Marschner, P., Baldock, J., Setia, D., Smith, J., 2013. Soil salinity decreases global soil organic carbon stocks. *Science of the Total Environment* 465, 267–272. doi: 10.1016/j.scitotenv.2012.08.028.
Shabala, S., Munns, R., 2017. *Salinity Stress: Physiological Constraints and Adaptive Mechanisms,* 2nd ed. CAB International: Wallingford.
Shahid, S.A., Zaman, M., Heng, L., 2018. Soil salinity: Historical perspectives and a world overview of the problem. In: Zaman, M., Shahid, S.A., Heng, L. (Eds.), *Guideline for Salinity Assessment, Mitigation and Adaptation Using Nuclear and Related Techniques*. Springer,: Berlin/Heidelberg, Germany, pp. 43–53.
Shani, U., Ben-Gal, A., 2005. Long-term response of grapevines to salinity: Osmotic effects and ion toxicity. *American Journal of Enology and Viticulture* 56, 148–154.
Shimomachi, T., Kawahara, Y., Kobashigawa, C., Omoda, E., Hamabe, K., Tamaya, K., 2008. Effect of residual salinity on spinach growth and nutrient contents in polder soil. *Acta Horticulturae* 797, 419–424.
Shrivastava, P., Kumar, R., 2015. Soil salinity: A serious environmental issue and plant growth promoting bacteria as one of the tools for its alleviation. *Saudi Journal of Biological Sciences* 22, 123–131. doi: 10.1016/j.sjbs.2014.12.001.
Singh, G., 2009. Salinity-related desertification and management strategies: Indian experience. *Land Degradation & Development* 20, 367–385. doi: 10.1002/ldr.933.
Springer, G., Wienhold, B.J., Richardson, J.L., Disrud, L.A., 1999. Salinity and sodicity induced changes in dispersible clay and saturated hydraulic conductivity in sulfatic soils. *Communications in Soil Science and Plant Analysis* 30, 2211–2220. doi: 10.1080/00103629909370366.
Suarez, D.L., 2005. Chemistry of salt-affected soils. *Chemical Processes in Soils* 8, 689–705. doi: 10.2136/sssabookser8.c15.
Szabo, S., Hossain, M.S., Adger, W.N., Matthews, Z., Ahmed, S., Lázár, A.N., Ahmad, S., 2016. Soil salinity, household wealth and food insecurity in tropical deltas: Evidence from south-west coast of Bangladesh. *Sustainability Science* 11, 411–421. doi: 10.1007/s11625-015-0337-1.

Taddese, G., 2001. Land degradation: A challenge to Ethiopia. *Environmental Management* 27, 815–824. doi: 10.1007/s002670010190.

Taleisnik, E., 2020. *Saline and Alkaline Soils in Latin America: Natural Resources, Management and Productive Alternatives*. Springer Nature: London.

Tavakkoli, E., Rengasamy, P., McDonald, G.K., 2010. High concentrations of Na+ and Cl–ions in soil solution have simultaneous detrimental effects on growth of faba bean under salinity stress. *Journal of Experimental Botany* 61, 4449–4459. doi: 10.1093/jxb/erq251.

Thomas, D.S.G., Middleton, N.J., 1993. Salinization: New perspectives on a major desertification issue. *Journal of Arid Environments* 24, 95–105. doi: 10.1006/jare.1993.1008.

Tóth, G., Adhikari, K., Várallyay, G., Tóth, T., Bódis, K., Stolbovoy, V., 2008. Updated map of salt affected soils in the European Union. Threats to Soil Quality in Europe EUR 23438 EN, 65–77.

Tripathi, S., Kumari, S., Chakraborty, A., Gupta, A., Chakrabarti, K., Bandyapadhyay, B.K., 2006. Microbial biomass and its activities in salt-affected coastal soils. *Biology and Fertility of Soils* 42, 273–277. doi: 10.1007/s00374-005-0037-6.

Tripathi, R., Nayak, A.K., Shahid, M., Raja, R., Panda, B.B., Mohanty, S., Kumar, A., Lal, B., Gautam, P., Sahoo, R.N., 2015. Characterizing spatial variability of soil properties in salt affected coastal India using geostatistics and kriging. *Arabian Journal of Geosciences* 8, 10693–10703. doi: 10.1007/s12517-015-2003-4.

United Nations General Assembly. 2015. Transforming our world: The 2030 Agenda for Sustainable Development, A/RES/70/1, 21 October 2015, Available: https://www.refworld.org/docid/57b6e3e44.html, Accessed on: 20 May 2021.

Valtorta, S.E., Gallardo, M.R., 2012. Water salinity under heat stress in grazing conditions. In: N. Chaiyabutr (Ed.), *Milk Production-An Up-to-Date Overview of Animal Nutrition, Management and Health*. IntechOpen: London, UK, pp. 217–234.

van de Craats, D., van der Zee, S.E., Sui, C., van Asten, P.J.A., Cornelissen, P., Leijnse, A., 2020. Soil sodicity originating from marginal groundwater. *Vadose Zone Journal* 19, e20010.

Van den Broeck, G., Maertens, M., 2016. Horticultural exports and food security in developing countries. *Global Food Security* 10, 11–20. doi: 10.1016/j.gfs.2016.07.007.

Van Oort, P.A.J., 2018. Mapping abiotic stresses for rice in Africa: Drought, cold, iron toxicity, salinity and sodicity. *Field Crops Research* 219, 55–75. doi: 10.1016/j.fcr.2018.01.016.

Verburg, P.H., Mertz, O., Erb, K.-H., Haberl, H., Wu, W., 2013. Land system change and food security: Towards multi-scale land system solutions. *Current Opinion in Environmental Sustainability* 5, 494–502. doi: 10.1016/j.cosust.2013.07.003.

Viglizzo, E.F., Ricard, M.F., 2021. Climate change and salinity-vulnerable ecosystems in Latin America. Taleisnik, E., Lavado, R.S. (Eds.), *Saline and Alkaline Soils in Latin America*. Springer: Berlin/Heidelberg, Germany, pp. 449–456.

Vithanage, M., Engesgaard, P., Villholth, K.G., Jensen, K.H.J.G., 2012. The effects of the 2004 tsunami on a coastal aquifer in Sri Lanka. *Groundwater* 50, 704–714.

Volkmar, K.M., Hu, Y., Steppuhn, H., 1998. Response physiologique des plantes a la salinite: Mise au point bibliographique. *Canadian Journal of Plant Science* 78, 19–27. doi: 10.4141/P97-020.

Wale, H.A., Dejenie, T., 2013. Dryland ecosystems: Their features, constraints, potentials and managements. *Research Journal of Agricultural and Environmental Management* 2, 277–288.

Wang, W., Vinocur, B., Altman, A., 2003. Plant responses to drought, salinity and extreme temperatures: Towards genetic engineering for stress tolerance. *Planta* 218, 1–14. doi: 10.1007/s00425-003-1105-5.

Wang, F., Chen, X., Luo, G., Ding, J., Chen, X., 2013. Detecting soil salinity with arid fraction integrated index and salinity index in feature space using Landsat TM imagery. *Journal of Arid Land* 5, 340–353. doi: 10.1007/s40333-013-0183-x.

Wang, L., Sun, X., Li, S., Zhang, T., Zhang, W., Zhai, P., 2014. Application of organic amendments to a coastal saline soil in north China: effects on soil physical and chemical properties and tree growth. *PLoS One* 9, e89185. doi: 10.1371/journal.pone.0089185.

Waterkeyn, A., Grillas, P., Vanschoenwinkel, B., Brendonck, L.U.C., 2008. Invertebrate community patterns in Mediterranean temporary wetlands along hydroperiod and salinity gradients. *Freshwater Biology* 53, 1808–1822. doi: 10.1111/j.1365-2427.2008.02005.x.

Weggler, K., McLaughlin, M.J., Graham, R.D., 2004. Effect of chloride in soil solution on the plant availability of biosolid-borne cadmium. *Journal of Environmental Quality* 33, 496–504. doi: 10.2134/jeq2004.4960.

Wicke, B., Smeets, E., Dornburg, V., Vashev, B., Gaiser, T., Turkenburg, W., Faaij, A., 2011. The global technical and economic potential of bioenergy from salt-affected soils. *Energy & Environmental Science* 4, 2669–2681.

Wijeyaratne, W.M., Subanky, S., 2017. Assessment of the efficacy of home remedial methods to improve drinking water quality in two major aquifer systems in Jaffna Peninsula, Sri Lanka. *Scientifica* 2017. doi: 10.1155/2017/9478589.

Wolanski, E., Cassagne, B., 2000. Salinity intrusion and rice farming in the mangrove-fringed Konkoure River delta, Guinea. *Wetlands Ecology and Management* 8, 29–36. doi: 10.1023/A:1008470005880.

Wong, V.N., Dalal, R.C., Greene, R.S., 2008. Salinity and sodicity effects on respiration and microbial biomass of soil. *Biology and Fertility of Soils* 44, 943–953. doi: 10.1007/s00374-008-0279-1.

Wong, V.N., Dalal, R.C., Greene, R.S., 2009. Carbon dynamics of sodic and saline soils following gypsum and organic material additions: A laboratory incubation. *Applied Soil Ecology* 41, 29–40. doi: 10.1016/j.apsoil.2008.08.006.

Yang, C., Sun, J., 2020. Soil salinity drives the distribution patterns and ecological functions of fungi in saline-alkali land in the Yellow River Delta, China. *Frontiers in Microbiology* 11. doi: 10.3389/fmicb.2020.594284.

Ying, Z., Cui, Y.-J., Benahmed, N., Duc, M., 2021. Salinity effect on the compaction behaviour, matric suction, stiffness and microstructure of a silty soil. *Journal of Rock Mechanics and Geotechnical Engineering*. doi: 10.1016/j.jrmge.2021.01.002.

Yu, J., Wang, Z., Meixner, F.X., Yang, F., Wu, H., Chen, X., 2010. Biogeochemical characterizations and reclamation strategies of saline sodic soil in northeastern China. *Clean–Soil, Air, Water* 38, 1010–1016. doi: 10.1002/clen.201000276.

Yuan, B.-C., Li, Z.-Z., Liu, H., Gao, M., Zhang, Y.-Y., 2007. Microbial biomass and activity in salt affected soils under arid conditions. *Applied Soil Ecology* 35, 319–328. doi: 10.1016/j.apsoil.2006.07.004.

Zhang, X.C., Norton, L.D., 2002. Effect of exchangeable Mg on saturated hydraulic conductivity, disaggregation and clay dispersion of disturbed soils. *Journal of Hydrology* 260, 194–205. doi: 10.1016/S0022-1694(01)00612-6.

Zhang, J., He, C., Chen, L., Cao, S., 2018. Improving food security in China by taking advantage of marginal and degraded lands. *Journal of Cleaner Production* 171, 1020–1030. doi: 10.1016/j.jclepro.2017.10.110.

Zhang, K., Shi, Y., Cui, X., Yue, P., Li, K., Liu, X., Tripathi, B.M., Chu, H., 2019a. Salinity is a key determinant for soil microbial communities in a desert ecosystem. *MSystems* 4, e00225–00218. doi: 10.1128/mSystems.00225-18.

Zhang, W.-W., Chong, W., Rui, X.U.E., Wang, L.-J., 2019b. Effects of salinity on the soil microbial community and soil fertility. *Journal of Integrative Agriculture* 18, 1360–1368.

Zhang, G., Bai, J., Tebbe, C.C., Zhao, Q., Jia, J., Wang, W., Wang, X., Yu, L., 2021. Salinity controls soil microbial community structure and function in coastal estuarine wetlands. *Environmental Microbiology* 23, 1020–1037. doi: 10.1111/1462-2920.15281.

Zhao, Q., Bai, J., Gao, Y., Zhao, H., Zhang, G., Cui, B., 2020. Shifts in the soil bacterial community along a salinity gradient in the Yellow River Delta. *Land Degradation & Development* 31, 2255–2267. doi: 10.1002/ldr.3594.

Zhu, M., Kong, F., Li, Y., Li, M., Zhang, J., Xi, M., 2020. Effects of moisture and salinity on soil dissolved organic matter and ecological risk of coastal wetland. *Environmental Research* 187, 109659. doi: 10.1016/j.envres.2020.109659.

2 Plant-Based Strategies for the Sustainable Management of Soil Contamination

Liuwei Wang and Deyi Hou
School of Environment, Tsinghua University, Beijing, P.R. China

CONTENTS

2.1 INTRODUCTION

Soil is a valuable, non-renewable resource that should be preserved with care (Hou, 2022a; Jia et al., 2021; Jin et al., 2021). However, various physical, chemical, and biological soil constraints have made the soil become "unhealthy," causing it not to be able to function as a living system (FAO, 2011; FAO, 2015; FAO, 2018). In particular, soil contaminants, including toxic metals and metalloids (Gao et al., 2022; Hou et al., 2022; Wang et al., 2021a), organic contaminants (Bolan et al., 2021; Haritash and Kaushik, 2009; Qi et al., 2020), and emerging contaminants such as microplastics and nanoplastics (Bolan et al., 2020; Wang et al., 2022a; Wang et al., 2021b), have resulted in global attention due to their high toxicity to human beings and the entire ecosystem.

Phytoremediation is the utilization of plants to extract, degrade, volatize, or stabilize soil contaminants (Salt et al., 1998). Compared with conventional remediation processes, such as soil washing, thermal treatment, dig and haul, and ex situ solidification/stabilization, this approach is usually considered to be a "gentle remediation option" with low disturbance to the soil (Cundy et al., 2013; Kidd et al., 2015; Lacalle et al., 2020). With the emergence and the rapid development of the "green and sustainable" remediation movement, these plant-based remediation strategies with low environmental, social, and economic impacts have become important topics for environmental research (Hou, 2021b; O'Connor et al., 2019b; Song et al., 2018). This chapter provides an overview of these

DOI: 10.1201/9781003093565-2

plant-based remediation processes and a critical assessment of key influencing factors. Furthermore, emerging phytomanagement practices will be discussed with lessons from the field.

2.2 PLANT-BASED REMEDIATION STRATEGIES

2.2.1 Phytomining and Phytoextraction

Phytomining refers to the process where a "crop" of metal is produced as a result of hyperaccumulation (Brooks et al., 1998; Sheoran et al., 2009). This technique enables the recovery of valuable metals from contaminated lands in a green and sustainable manner (Hou, 2020; Wang et al., 2020c). Typically, a plant with the traits of hyperaccumulation (termed hyperaccumulator) could accumulate metals 100 times higher than those that occur in other plants growing in the same substrate (Brooks et al., 1998). Hyperaccumulators could extract metals from contaminated soils, including farmlands, tailings, and contaminated industrial sites, and further translocate metals in their aboveground tissues (Suman et al., 2018). However, there exists a threshold value for the hyperaccumulation ability. Brooks et al. (1998) suggested that the threshold for most metals was 1,000 mg/kg, while plants can accumulate Zn as high as 10,000 mg/kg, Cd 100 mg/kg, and Au 1 mg/kg. Phytomining is theoretically practical, but not economic in most cases. For instance, assuming that biomass yield reaches as high as 10 t/ha, plants should accumulate extremely high concentrations of 3,965 mg/kg Ni, 9,016 mg/kg Cu, 19,230 mg/kg Cd, and 28,011 mg/kg Pb in tissues, far exceeding the threshold value (Figure 2.1). In contrast, metals that are suitable for phytomining include Pd, Au, Pt, and Ag (Figure 2.1).

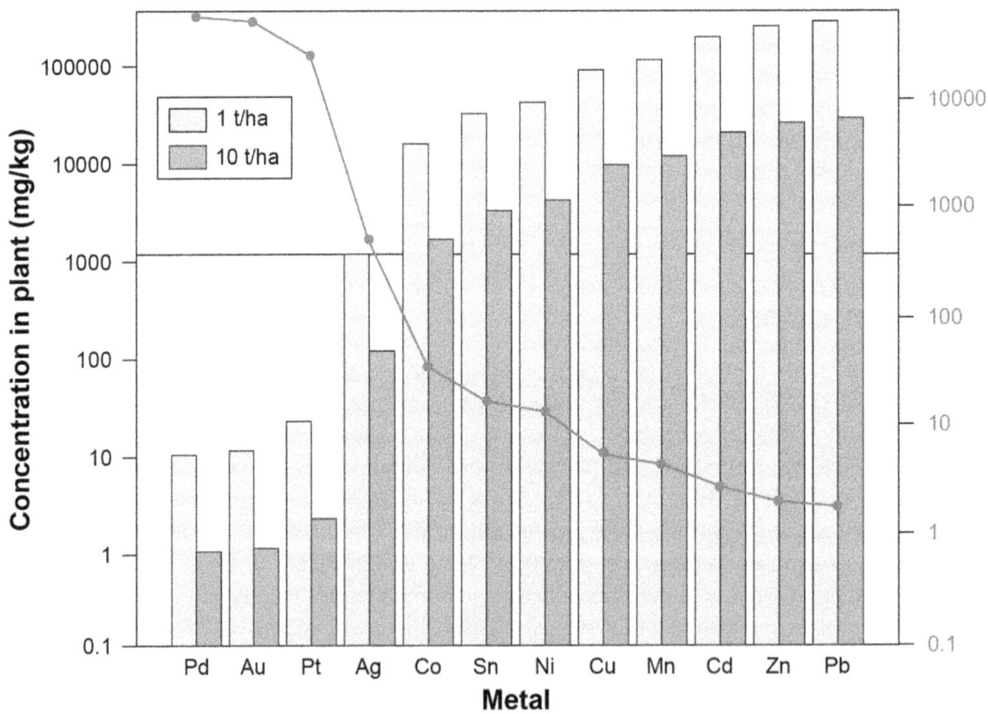

FIGURE 2.1 Concentrations of metals in plant tissues that would be needed to achieve a gross value of $500/ha. Prices of metal (in 2019) were obtained from the London Metal Exchange (https://www.lme.com/) and KITCO (https://www.kitco.com/). For hyperaccumulators, the threshold concentration is regarded as 1,000 mg/kg, except for Au (1 mg/kg), Cd (100 mg/kg), and Zn (10,000 mg/kg). (Reproduced from Wang et al., 2021c.)

The hyperaccumulation traits of certain plants are more widely used for remediation purposes rather than resource recovery. Phytoextraction makes use of hyperaccumulators to remove metals from contaminated soils (Ghori et al., 2015; Li et al., 2012). Evidence from the field is mounting that hyperaccumulators, such as *Pteris vittata* (As) (Wei et al., 2006; Yang et al., 2020), *Sedum plumbizincicola* (Cd & Zn) (Deng et al., 2016; Hu et al., 2019a), *Phytolacca americana* (Mn) (Chen et al., 2015; Yang et al., 2013), and *Alyssum murale* (Ni) (Bani et al., 2007; Bani et al., 2015), have proven effective for soil metal extraction. Apart from hyperaccumulators, non-hyperaccumulators with high biomass yield also can be used for phytoextraction (Liang et al., 2009; Ran et al., 2020). Considering that soil metals should be extracted from the contaminated soil, metal bioavailability is the key factor affecting the extraction (Section 2.3.1). For a comprehensive overview of phytoextraction, readers are referred to Suman et al. (2018).

2.2.2 Phytovolatilization

Volatile organic compounds (VOCs) (Legault et al., 2017; Zhang et al., 2020), or certain forms of metal(loid)s with high volatility such as trimethylarsine (Jia et al., 2012) and elemental mercury (Hussein et al., 2007a), can be volatilized from the stems and leaves of plants (Figure 2.2). Although contaminants would be transported to the air, this technique is usually considered beneficial because the volatized contaminants would be substantially diluted and photochemically transformed in the atmosphere (Limmer and Burken, 2016). Trees, such as willows (Yang et al., 2015) and poplars (Legault et al., 2017), are suitable candidates for phytovolatilization. Ideally, roots penetrate the vadose zone and reach the groundwater aquifer; while growing downward, they take up substantial amounts of VOCs, which are transferred from roots to branches and leaves, enabling the volatilization of contaminants and release to the atmosphere (Arnold et al., 2007; Ferro et al., 2013; Ferro and Tammi, 2009).

Apart from the aforementioned direct translocation and volatilization of contaminants from the plant itself, Limmer and Burken (2016) suggested that indirect phytovolatilization from the soil surface, via flux from the soil to the atmosphere due to root activities, must not be neglected (Figure 2.2). The fluctuation of the groundwater table as a result of root activities, contaminant advection, an increase in permeability of the soil, hydraulic redistribution-induced chemical transport, advection with water, and interception of rainfall are all possible mechanisms contributing to indirect phytovolatilization (Limmer and Burken, 2016).

2.2.3 Phytotransformation

Direct detoxification of organic contaminants by plants, which change them into more "environmentally acceptable" products without the assistance of soil microorganisms, is termed phytotransformation or phytodegradation. Phytotransformation can occur either inside the plant due to metabolic processes or within the rhizosphere where enzymes are produced and released by plants (Newman and Reynolds, 2004; US EPA, 2001). Enzymes play vital roles in phytotransformation through the breakdown of organic contaminants into simpler molecular forms, which could be incorporated into plant tissues. Several successful attempts of contaminant phytotransformation in soil and water include the transformation of 2,4,6-trinitrotoluene (TNT) by *Myriophyllum aquaticum* (Wang et al., 2003); degradation of phorate by *Brassica juncea* (Rani et al., 2012); remediation of organophosphorus (OP) pesticides by aquatic plants including *Myriophyllum aquaticum*, *Spirodela oligorrhiza* L., and *Elodea Canadensis* (Gao et al., 2000); degradation of textile dyes by *Ipomoea hederifolia* L. (Rane et al., 2014); and degradation of methylene blue by *Azolla pinnata* (Al-Baldawi et al., 2018).

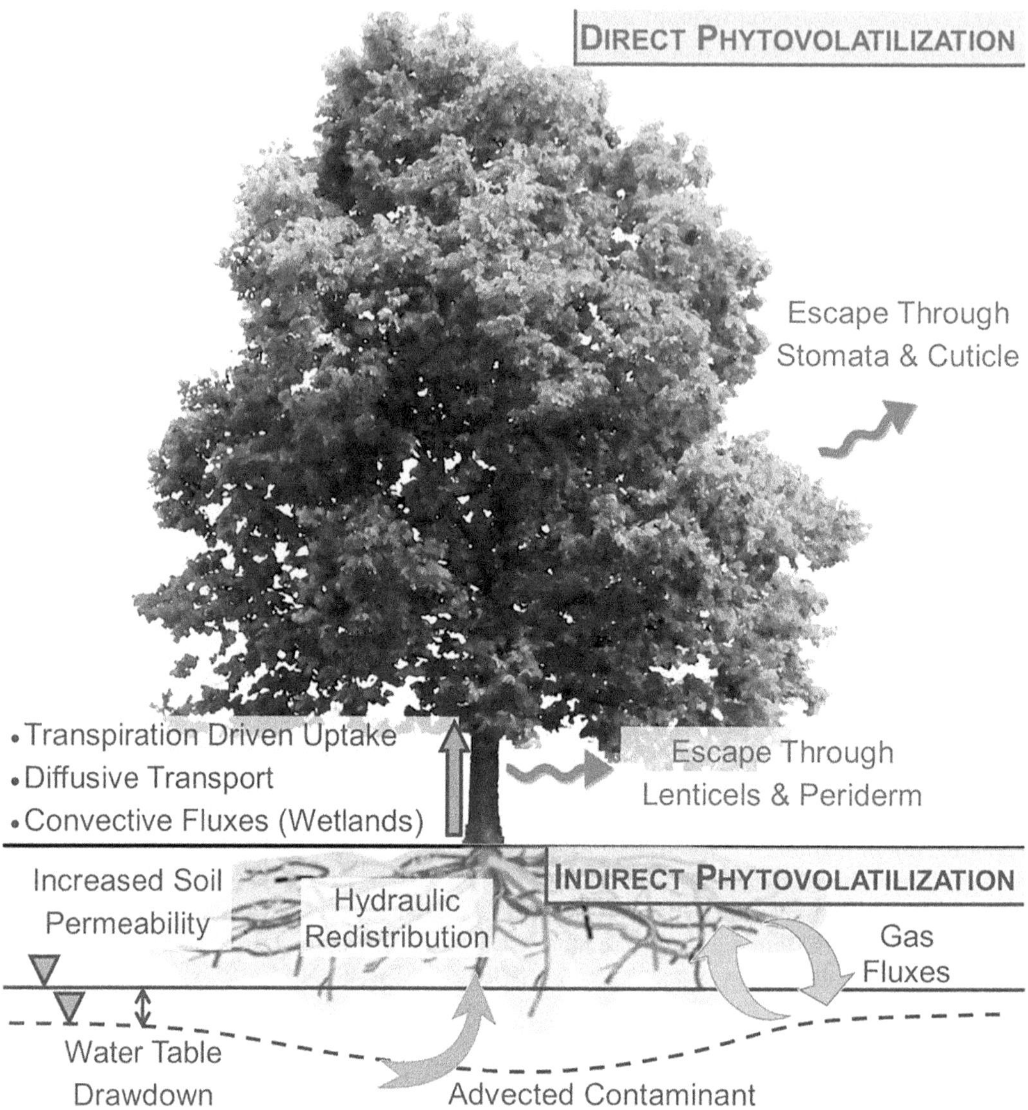

FIGURE 2.2 Direct and indirect phytovolatilization processes. (Reproduced with permission from Limmer and Burken, 2016. Copyright 2016 American Chemical Society.)

2.2.4 Rhizoremediation

The degradation of contaminants by microorganisms within the rhizosphere is termed rhizoremediation (Kuiper et al., 2004). The "rhizo" represents the root zone, while "remediation" refers to the degradation of organic compounds by microorganisms in this zone. Root exudates play vital roles in the stimulation of microbial activity. They are a complex mixture of low-molecular-weight organic acids (e.g., malic acid, citric acid), enzymes (e.g., amylase, phosphatase), growth factors (e.g., auxins, choline), and sugars (e.g., glucose, maltose) and serve as nutritional substances for microorganisms (Hillel and Hatfield, 2005; Saravanan et al., 2020), inducing a mutualism effect that defends against pathogen attack (Hu et al., 2018; Lambers et al., 2009; Raynaud et al., 2006). The microorganisms may either enter the plant tissues as endophytes or be present near the plant root

(Praveen et al., 2018; Saravanan et al., 2020). Successful examples include the rhizoremediation of polycyclic aromatic hydrocarbons (PAHs) (Kotoky et al., 2018; Techer et al., 2012), total petroleum hydrocarbons (TPHs) (Gaskin and Bentham, 2010; Hoang et al., 2021), polychlorinated biphenyls (PCBs) (Tu et al., 2017; Vergani et al., 2019), and pesticides (Abhilash and Singh, 2010; Dams et al., 2007).

2.2.5 Phytostabilization

Phytostabilization aims at reducing the mobility of contaminants within the rhizosphere as a result of root accumulation or immobilization via precipitation or complexation (Bolan et al., 2011; Radziemska et al., 2019; Zhan et al., 2020). Apart from immobilization of contaminants in the vadose zone, the plant layer also serves as a green barrier over the soil diminishing exposure to wind, rainfall, and sunlight irradiation, and cutting off potential contact with human beings (Bolan et al., 2011; Li and Huang, 2015). Phytostabilization is usually enhanced by various soil immobilization agents, including lime (Verdugo et al., 2011; Wu et al., 2011), phosphate (Hashimoto et al., 2011; Yang et al., 2019), biochar (Lebrun et al., 2018; Visconti et al., 2020), and inorganic industrial wastes (Kumpiene et al., 2012; Labidi et al., 2017). During this process, plants have a secondary role in assisting the soil amendments to stabilize toxic metals in the rhizosphere. Sometimes plant-growth-promoting bacteria also can be applied to change root morphology and nutrient bioavailability (Bolan et al., 2011; Grandlic et al., 2009; Saran et al., 2020).

2.3 KEY INFLUENCING FACTORS

2.3.1 Contaminant Bioavailability

The bioavailability of soil contaminants is the key influencing factor determining whether plants can succeed in extraction, volatilization, or stabilization. For toxic metal(loid)s, the bioavailability is usually determined by extraction (leaching) tests. One-step leaching methods, including the use of the Toxicity Characteristic Leaching Procedure (TCLP) or extracting with DTPA, $CaCl_2$, NH_4NO_3, are the most widely used approaches to assess leachability and bioavailability (Cappuyns and Swennen, 2008; Wang et al., 2020a; Zhang et al., 2006a). Sequential extraction procedures, such as the Tessier method and the Commission of the European Community Bureau of Reference (BCR) method, classify the metal fractions in a more detailed manner (Quevauviller et al., 1993; Tessier et al., 1979). Labile fractions, including exchangeable and carbonate-bound metals, can be easily taken up by plants, while less labile ones, such as oxides-bound, organic-bound, and the residual fractions, are considered to be much more difficult, or even impossible, to be bioaccumulated. To enhance phytoextraction, chelating agents can be applied to mobilize soil metals. In contrast, to promote phytostabilization, soil immobilization agents are frequently used to immobilize metals.

The key factor determining whether organic contaminants can be translocated by plants is the molecular structure. Translocatable compounds typically possess a molecular mass below 350 Da, with log octanol-water partitioning coefficients $\log K_{ow} < 5$ (Limmer and Burken, 2014). The performance of plants in phytovolatilization is determined by the partition coefficient of VOCs. The most studied mechanism involved in the transport of VOCs is the removal of subsurface water by plants. This results in a lower water table, thus increasing the thickness of vadose zone. Given that VOCs diffuse much more quickly in the air than in water, the VOC flux increases in turn (Guo et al., 2019; Limmer and Burken, 2016). Limmer and Burken (2016) found that, among the organic contaminants with low octanol-water partitioning coefficients ($\log K_{ow} < 5$) that can be possibly translocated to plant tissues, those with octanol-air partitioning coefficients $\log K_{oa} < 5$ would be phytovolatilized much more easily (Figure 2.3). With a decrease of the K_{oa} value, diffusive transport tends to play a more significant role in VOC migration and volatilization.

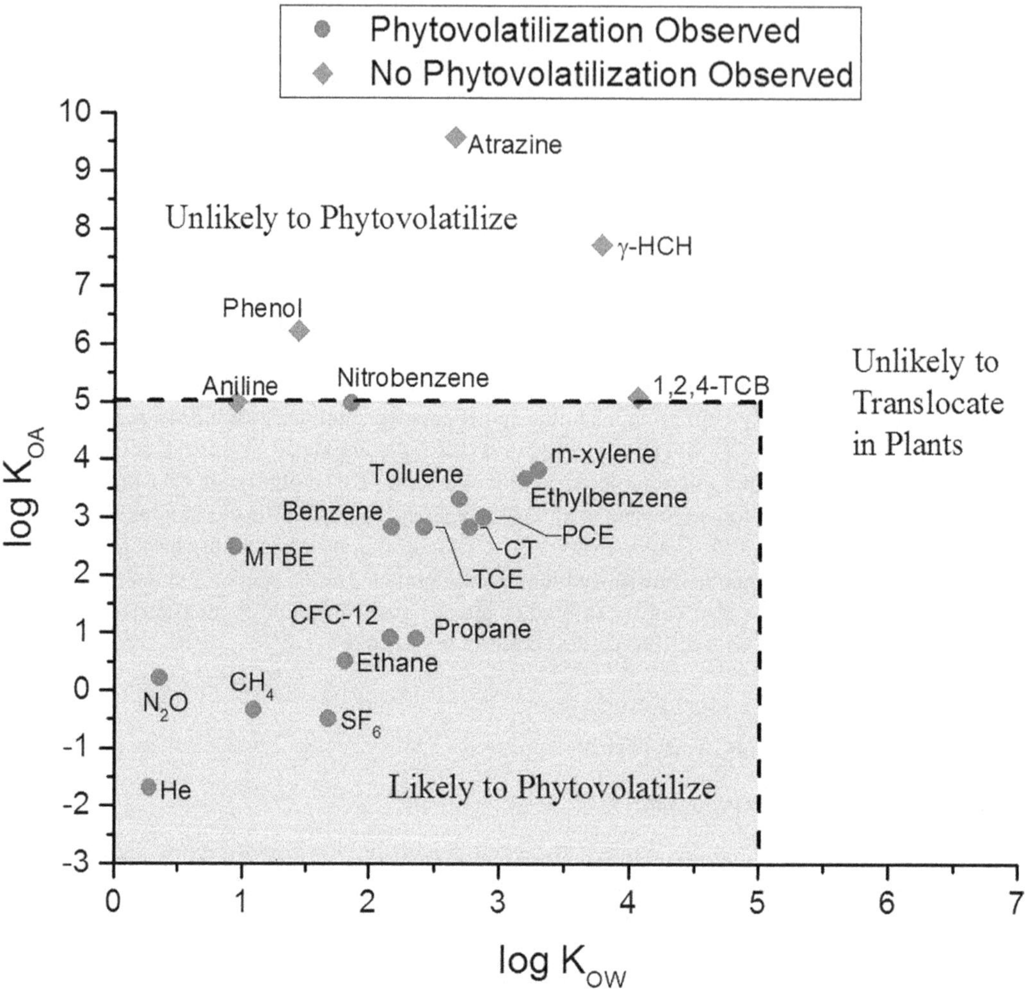

FIGURE 2.3 Physicochemical domains favoring phytovolatilization. (Reproduced with permission from Limmer and Burken, 2016. Copyright 2016 American Chemical Society.)

2.3.2 Plant Characteristics

Hyperaccumulators and non-hyperaccumulators with high biomass yield are the most widely used plant species used for phytoremediation. However, to enhance phytoremediation, genetic modification has been studied in numerous investigations with multiple aims (Table 2.1). During phytomining and phytoextraction, the major goal is to extract as much contaminant as possible. Therefore, genetic modification can be used to enhance the plant tolerance to toxicity, promote translocation and bioaccumulation, and increase the biomass yield (Bañuelos et al., 2005; Nahar et al., 2017). For phytovolatilization, plants have been genetically modified to introduce the *mer* operon for mercury volatilization (Haque et al., 2010; Hussein et al., 2007b). To enhance phytotransformation and rhizoremediation, certain genes can be introduced for organic contaminant degradation (Kabouw et al., 2012; Zhang et al., 2019). For phytostabilization, a genetically modified plant should possess high tolerance to contamination as well as the ability to restrict translocation from roots to aboveground tissues (Perez-Palacios et al., 2017; Shim et al., 2013). However, the application of genetically modified plants is sometimes restricted due to regulations. A mutant may also help with phytoremediation (Kolbas et al., 2018; Thijs et al., 2018).

TABLE 2.1
Targets of Genetic Modification in Phytoremediation (Wang et al., 2020c)

Aim	References
To enhance resistance to toxicity	Bañuelos et al. (2005); Fan et al. (2018); Kebeish et al. (2014); Liu et al. (2015); Lyyra et al. (2007); Raldugina et al. (2018); Zhang et al. (2019); Zhang et al. (2006b); Zhao et al. (2019)
To degrade organic contaminants	Zhang et al. (2019); Zhu et al. (2018); Zhu et al. (2012)
To enhance contaminant translocation of plants	Matsui et al. (2013); Wang et al. (2018)
To enhance contaminant accumulation in plants	Fan et al. (2018); Liu et al. (2015); Nahar et al. (2017); Ruiz et al. (2011); Ruiz and Daniell (2009); Sun et al. (2018)
To enhance rhizosphere microbial activity	Kabouw et al. (2012)
To enhance plant growth rate and biomass production	Shim et al. (2013)
To enhance volatilization of contaminant (Hg)	Haque et al. (2010); Hussein et al. (2007b)
To enhance phytostabilization of contaminants	Barabasz et al. (2011); Perez-Palacios et al. (2017); Shim et al. (2013)

2.3.3 Planting Methods and Agronomic Practices

Planting density determines the biomass yield directly. The self-thinning effect must not be overlooked when selecting a suitable planting density (Mackenzie et al., 2001; Yoda et al., 1963). Plants cannot avoid competition through movement, so, if the density is too high, the extensive competition results in a decrease of biomass yield, which can be described as Yoda's law (Yoda et al., 1963) (Equations 2.1 and 2.2):

$$W = C \times D^n \tag{2.1}$$

$$\lg W = \lg C + n \lg D \tag{2.2}$$

where W is the average individual biomass, D is the plant density, C is a constant, and n is another constant dependent on plant species that varies from –1 to –2 (Begon et al., 1996; Roderick and Barnes, 2004; Yoda et al., 1963). In most cases, the relationship has a slope of –3/2 on a log-log plot, so this relationship is also known as Yoda's –3/2 law (Mackenzie et al., 2001; Yoda et al., 1963).

To overcome this effect, intercropping seems to be a feasible way. In an intercropped system, the biomass production rate of different plants will peak at different times, diminishing the competition for nutrients (Bânuelos and Dhillon, 2011; Carvalho et al., 2019; Viana et al., 2021).

Both transplantation and direct sowing have been used for planting. Transplantation usually achieves higher phytoextraction as compared with sowing but can be costly for field application of phytoremediation in large areas (Jacobs et al., 2018; Papazoglou and Fernando, 2017). Double cropping can sometimes be used to overcome the obstacle of short growth duration of certain species. Double harvesting is another feasible way to extend the effective remediation duration. Compared with double cropping where biomass is removed thoroughly during harvest, in double harvesting plants remain in contaminated soils and they keep growing until harvested (Ji et al., 2011; Qu et al., 2013).

2.3.4 Additive Assistance

Soil amendments, including biochar, compost, and inorganic industrial wastes, decrease the bioavailability of metals directly, assisting plants in phytostabilization (Figure 2.4). As a representative

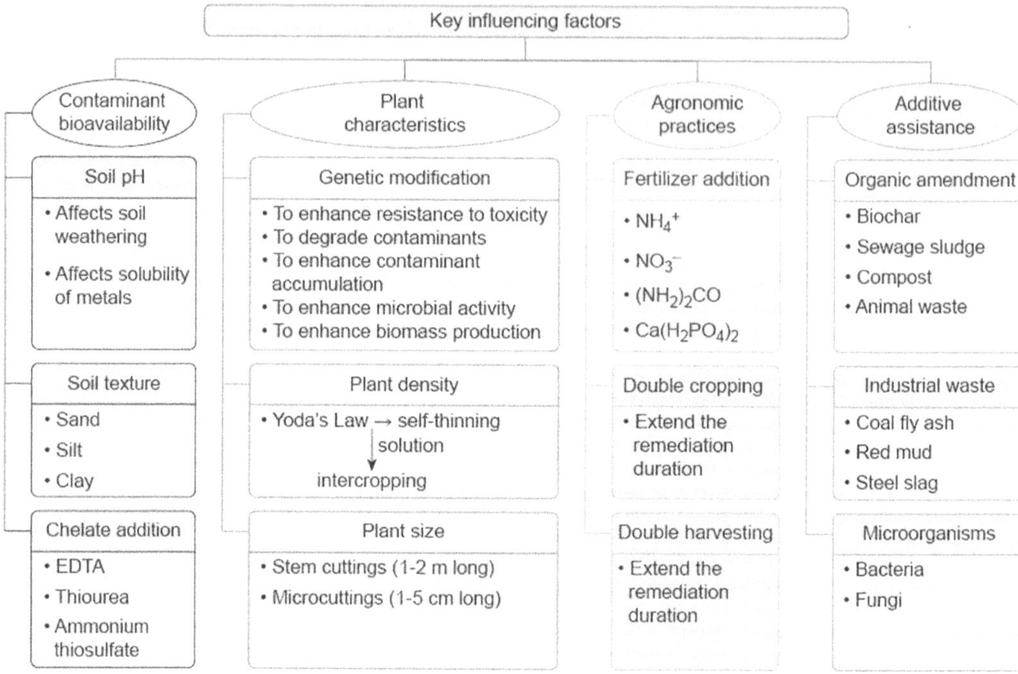

FIGURE 2.4 Key influencing factors determining the performance of phytoremediation strategies (Wang et al., 2020c). Yoda's law is a numerical description of the self-thinning phenomenon among plants, that is, sessile organisms like plants cannot avoid competition by movement, and the competition result can be concluded as the increase of plant density will lead to the decrease of individual biomass (Yoda et al., 1963).

of organic amendments, biochar has received much attention in soil remediation due to excellent metal stabilization (Hou, 2021a; Hou et al., 2021; Yang et al., 2022). Precipitation, surface complexation, and enhancement of electrostatic interactions between soil colloids and metals as a result of soil pH increase (i.e., the liming effect) are possible metal immobilization mechanisms (He et al., 2019; Wang et al., 2022b; Wang et al., 2022c). Many successful attempts have shown that biochar can assist in phytostabilization (Edenborn et al., 2015; Kowitwiwat and Sampanpanish, 2020; Lebrun et al., 2018; Visconti et al., 2020). Other organic amendments, including compost (Touceda-González et al., 2017), sewage sludge (Grobelak et al., 2017), and animal waste (Galende et al., 2014), also can decrease the bioavailability of metals, thus favoring metal stabilization within the rhizosphere. Coal fly ash, steel slag, and red mud consisting of oxides such as Al_2O_3, Fe_2O_3, SiO_2, and CaO, could also elevate soil pH for metal stabilization, or immobilize metals directly via surface complexation and precipitation (Blissett and Rowson, 2012; Lehoux et al., 2013; Leon-Romero et al., 2018). However, these materials may contain metals themselves that can be released in the long run, posing a threat to the environment (Ma and Feng, 2011; Panda et al., 2018).

Microorganisms, including bacteria and fungi, are often used to increase the biomass yield, thus favoring phytoextraction. Evidence from the field has shown that bioaccumulation of Cu, Zn, and Cd can be enhanced by a variety of microorganisms, including plant growth-promoting bacteria (Pérez Rodríguez et al., 2014) and arbuscular mycorrhizal fungi (Cicatelli et al., 2014). Inoculation with endophytes favors phytodegradation or organic contaminants. For instance, a 3-year field study showed that inoculation of the natural trichloroethylene (TCE)-biodegrading strain *Enterobacter* sp. strain PDN3 resulted in successful phytodegradation, because TCE concentrations in the groundwater plume were reduced from 300 to 5 μg/L after passing through the phytoremediation zone (Doty et al., 2017).

2.4 PHYTOMANAGEMENT: AN EMERGING CONCEPT FOR THE SUSTAINABLE MANAGEMENT OF CONTAMINATED AGRICULTURAL SOILS

Phytomanagement (also known as phytoattenuation) can be defined as the use of plants to contain the risks caused by soil contamination, while simultaneously achieving a sustainable and profitable use of soil via producing biomass that is marketable (Shrinkhal, 2018). Therefore, phytomanagement has two purposes: utilization and remediation. Conventional phytoremediation processes are often criticized for the long remediation duration and the handling of hyperaccumulator biomass with high concentrations of contaminants (Figure 2.5). In comparison, phytomanagement provides wider site benefits through the combination of remediation with sustainable site management. A variety of benefits, including economic ones such as land value recovery and renewable energy regeneration, environmental ones such as greenhouse gas mitigation and water management, and social ones including aesthetic and cultural values, renders this management strategy a sustainable one (Figure 2.5) (Cundy et al., 2016; Evangelou et al., 2015; Hou, 2022b).

Evidence from long-term field studies has shown that phytomanagement ensures the sustainable use and management of contaminated lands through energy production (Table 2.2). For instance, instead of using hyperaccumulators to extract Zn, Cd, and Pb from a contaminated soil affected by a smelting industry, Meers et al. (2010) grew the bio-energy crop *Zea mays* (maize) for energy production while slowly bioaccumulating toxic metals. It was calculated that growing maize would result in 33,000–46,000 kW·h of electrical and thermal energy per year per hectare, offsetting approximately 21 tons of CO_2 emissions. Another field study by Cheng et al. (2015 suggested that

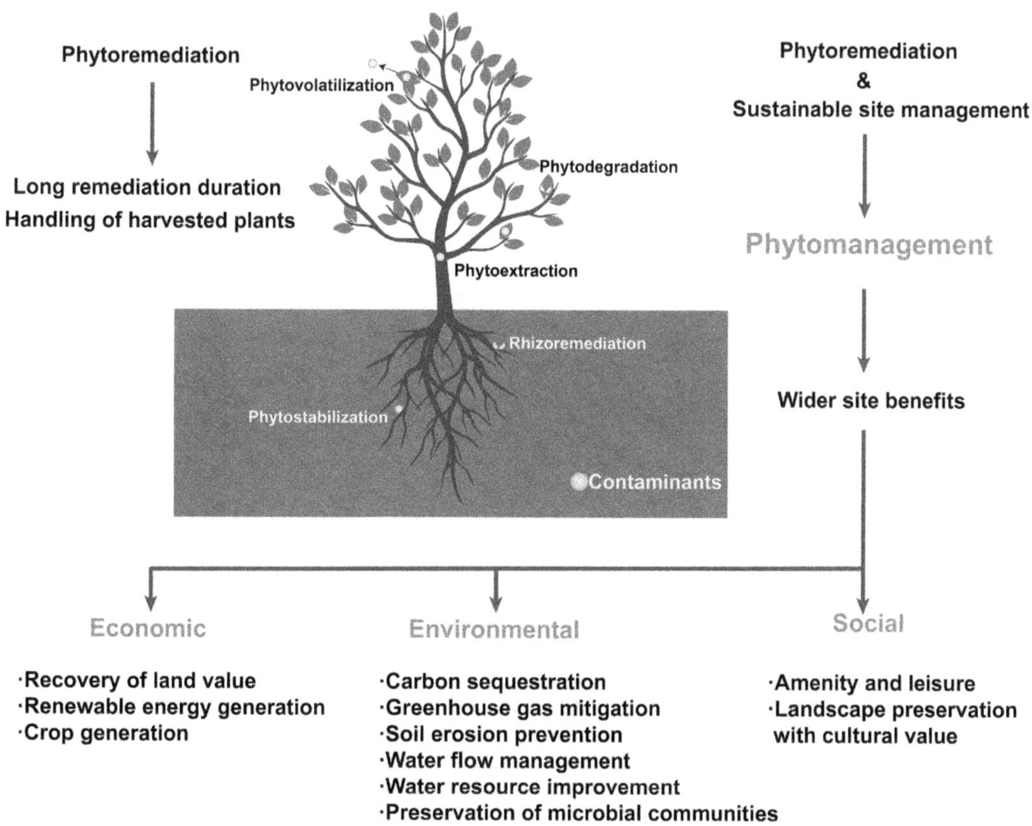

FIGURE 2.5 Economic, environmental, and social benefits of phytomanagement. (Modified with permission from Burges et al., 2018. Copyright 2018 Taylor & Francis.)

TABLE 2.2
Phytomanagement Strategies for the Sustainable Remediation of Contaminated Soils

Contaminant Concentrations in Soil (mean, mg/kg)	Plant	Study Duration	Remediation Performances	Key Findings	Sustainability Considerations	Reference
Se (12.1)[a]	Rapeseed, arhar	11 years	Se removal rate 25.6%	Se removal by rapeseed was 6–7 times higher than that of arhar	The harvested biomass can be used as a blend in animal feed	Bânuelos and Dhillon (2011)
Cd, Cu, Ni, Mn, Zn	Poplar	10 years	Removal rate ~ 30% –50% for Cd, Cu, Ni, Zn; > 80% for Mn	Heavy metal concentrations decreased, while macronutrient concentrations increased	Stem (wood only) harvesting could reduce the risks encountered with contaminated biomass in valorization	Chalot et al. (2020)
Pb (5600)	Peanut	1 year	Removal rate 17.6 kg/(ha yr)	A period of 137 years will be required to completely remove Pb	The harvested biomass can produce energy (654.5 GJ/(ha yr)), peanut kernel can be used as animal feed	Cheng et al. (2015)
Cd (1.37), Cu (536), Zn (684)	Maize, sunflower, tobacco (rotation, with elemental sulfur addition)	6 years	Maximum removal rate 1,392 g Zn/(ha yr) (maize + sulfur); 7.95 g Cd/(ha yr) (tobacco + sulfur)	It will take over 360 years to remove heavy metals from this soil	Animal feeding, biofuel production, or other economic management strategies assure sustainability despite the long remediation duration	Fässler et al. (2010)
Cd, Ni, Pb, Zn	Poplar	25 years	–	Phytomanagement for 25 years increased the bioavailability of Cd, Ni, Pb, and Zn as a result of soil acidification	–	Hu et al. (2019b)
Cd (15), Cu (79), Pb (541), Zn (1506)	Stinging nettle	2 years	–	The tensile properties of the fibers produced from the contaminated biomass were equal to those of industrial hemp and flax	The harvested plant can be used for fiber production	Jeannin et al. (2020)

(Continued)

TABLE 2.2 (*Continued*)
Phytomanagement Strategies for the Sustainable Remediation of Contaminated Soils

Contaminant Concentrations in Soil (mean, mg/kg)	Plant	Study Duration	Remediation Performances	Key Findings	Sustainability Considerations	Reference
Cd (8.9), Pb (189), Zn (398)	Maize	5 months	Very slow (statistically insignificant)	The cultivation of maize can result in 33,000–46,000 kW h of renewable energy per hectare per year, implying a reduction of 21 t CO_2 if used to substitute a coal fired power plant	The harvested biomass can generate renewable energy	Meers et al. (2010)
Cu (760)	Tobacco, sunflower (rotation, with compost and limestone addition)	9 years	<20%	Compost promoted plant growth, favoring long-term phytoextraction	The harvested plant can be used by various sectors processing non-food crops (e.g., biorefineries, co-firing with coal, organic fertilizers in Cu-deficient soils)	Mench et al. (2018)
Cd (5.3), Zn (289)	Maize	5 months	Removal rate 19 g Cd/(ha yr), 4,300 g Zn/(ha yr)	The economic outlook for farmers who intend to grow energy maize on contaminated soils is positive in this region	The harvested plant can be used for animal feeding and anaerobic digestion as per legal threshold values	Van Slycken et al. (2013a)
Cd (6.5), Zn (377)	Willow	4 years	Removal rate 72 g Cd/(ha yr), 2,000 g Zn/(ha yr)	Removal of willow leaves could enhance phytoextraction	The willow wood is suitable for energy production	van Slycken et al. (2013b)
Pb (324), Zn (323), As (29)	Mulberry	Not mentioned	–	Mulberry tree plantation can yield $25,675 for every 1 ha soil	The mulberry leaves could feed the silkworms for silk production	Wan et al. (2017)

[a] kg/ha

using peanut for phytomanagement of Pb-contaminated soils would generate a heat value of 654 GJ per hectare per year, which was comparable with that produced by burning 31 tons of coals.

If plants used for phytomanagement do not bioaccumulate high concentrations of toxic metals, the harvested biomass during phytomanagement can be further used as the animal feed. Incorporation of Se-rich rapeseed biomass as animal feed would supplement the animal's need for this element (Bânuelos and Dhillon, 2011). *Zea mays* grown in an agricultural field with Cd (5.3 mg/kg) only accumulated 0.96 mg/kg Cd, which did not exceed the threshold value for animal feed of 1.14 mg/kg, suggesting that the biomass can be safely reused in that way (Van Slycken et al., 2013a). Mulberry leaves collected from a phytomanagement site contaminated with 28.9 mg/kg As and 324 mg/kg Pb only slightly accumulated As (0.34 mg/kg) and Pb (11.87 mg/kg), and the values did not exceed the allowed concentration according to the Hygienical Standard for Feeds, making it possible for silkworm feeding (Wan et al., 2017).

Harvested biomass with high marketing value obtained from phytomanagement also can be used for other purposes. For instance, Jeannin et al. (2020) used *Urtica dioica* L. (stinging nettle) for phytomanagement of an industrial site contaminated with Cu, Zn, Pb, and Cd. Although fiber yield was lower than that obtained from uncontaminated agricultural lands, the tensile properties of fibers were comparable with the best-performing industrial fibers. However, not all phytomanagement practices achieve promising results. A 25-year field study by Hu et al. (2019b) found that phytomanagement of poplar decreased soil organic matter content, while simultaneously mobilizing soil metals due to acidification. Phytomanagement faces obstacles similar to those of phytoremediation, including the need for long-term monitoring and uncertainty in reproducibility as a result of soil heterogeneity, climate, and other possible causes (Burges et al., 2018; Evangelou et al., 2015). This gentle remediation option is often considered to be applicable only for remote and large areas with low land value (Bardos et al., 2020; Burges et al., 2018; Cundy et al., 2016). It's only applicable when concentrations of contaminants are not too high (Meers et al., 2010). It may serve as a "polishing" step for the cleaning up of marginally contaminated lands (Wang et al., 2021c). More evidence from the field must be collected to further examine the environmental, social, and economic sustainability of phytomanagement.

2.5 FINAL CONSIDERATIONS

Plant-based remediation strategies, including phytomining, phytoextraction, phytovolatilization, phytotransformation, rhizoremediation, and phytostabilization, have attracted much attention over the last few decades. The ability of plants to accumulate, volatize, degrade, or stabilize toxic metals and organic contaminants is appealing to researchers due to low disturbance to the environment and aesthetic and cultural values. However, the majority of phytoremediation studies have confirmed that these phytoremediation methods are feasible only under laboratory or greenhouse conditions. Limited evidence from the field has shown high uncertainty due to various factors that cannot be controlled. The commercialization of phytoremediation has not taken off yet. According to the Superfund Remedy Report, only in 1% of 188 contaminated-site remediation projects has phytoremediation been applied (USEPA, 2017).

Major obstacles exist in all phytoremediation approaches that must be overcome or at least be diminished prior to practical application. Phytomining is theoretically practical, but not economic for nearly all toxic metals. According to the results from a number of studies (Gerhardt et al., 2017; Niazi et al., 2012; Robinson et al., 2015; Tangahu et al., 2011), extracting metals at heavily contaminated lands to an acceptable level would take such a long time that it is often unacceptable. Phytovolatilization removes contaminants from the soil but releases them into the atmosphere. For certain contaminants, such as Hg, the emission of Hg(0) into the air would lead to long-range transport (O'Connor et al., 2019a; Wang et al., 2020b). In most cases, organic contaminants are not mineralized during phytotransformation and rhizoremediation. The degradation products of these processes may not be "environmentally acceptable." Considering that contaminants might not be removed from soil, long-term monitoring is required for phytostabilization processes.

In most cases, phytoremediation is regarded as an approach that is green and sustainable (Hou, 2020; Song et al., 2019). Yet the low remediation efficiency, long duration, need for long-term monitoring, and secondary pollution that comes from handling of contaminated biomass after harvest may result in a low "net environmental benefit" (Wang et al., 2020c). Compared with other remediation approaches such as soil washing and thermal desorption, phytoremediation is a "gentle remediation option" with more flexible removal rates of contaminants. Phytomanagement uses plants to mitigate the risks caused by soil contamination, while simultaneously achieving a sustainable and profitable use of soil. However, more field evidence is needed to examine the environmental, social, and economic sustainability of this strategy.

REFERENCES

Abhilash PC, Singh N. Withania somnifera Dunal-mediated dissipation of lindane from simulated soil: Implications for rhizoremediation of contaminated soil. *Journal of Soils and Sediments* 2010; 10: 272–282.

Al-Baldawi IA, Abdullah SRS, Anuar N, Hasan HA. Phytotransformation of methylene blue from water using aquatic plant (Azolla pinnata). *Environmental Technology and Innovation* 2018; 11: 15–22.

Arnold CW, Parfitt DG, Kaltreider M. Phytovolatilization of oxygenated gasoline-impacted groundwater at an underground storage tank site via conifers. *International Journal of Phytoremediation* 2007; 9: 53–69.

Bani A, Echevarria G, Sulçe S, Morel JL, Mullai A. In-situ phytoextraction of Ni by a native population of Alyssum murale on an ultramafic site (Albania). *Plant and Soil* 2007; 293: 79–89.

Bani A, Echevarria G, Zhang X, Benizri E, Laubie B, Morel JL, et al. The effect of plant density in nickel-phytomining field experiments with Alyssum murale in Albania. *Australian Journal of Botany* 2015; 63: 72–77.

Bañuelos G, Terry N, Leduc DL, Pilon-Smits EAH, Mackey B. Field trial of transgenic Indian mustard plants shows enhanced phytoremediation of selenium-contaminated sediment. *Environmental Science and Technology* 2005; 39: 1771–1777.

Bânuelos GS, Dhillon KS. Developing a sustainable phytomanagement strategy for excessive selenium in Western United States and India. *International Journal of Phytoremediation* 2011; 13: 208–228.

Barabasz A, Mills RF, Trojanowska E, Williams LE, Antosiewicz DM. Expression of AtECA3 in tobacco modifies its responses to manganese, zinc and calcium. *Environmental and Experimental Botany* 2011; 72: 202–209.

Bardos P, Spencer KL, Ward RD, Maco BH, Cundy AB. Integrated and sustainable management of post-industrial coasts. *Frontiers in Environmental Science* 2020; 8: 86.

Begon M, Harper JL, Townsend CR. *Ecology: Individuals, Populations, and Communities* (3rd ed.): Oxford, UK; Cambridge, MA: Blackwell Science, 1996.

Blissett RS, Rowson NA. A review of the multi-component utilisation of coal fly ash. *Fuel* 2012; 97: 1–23.

Bolan NS, Kirkham M, Halsband C, Nugegoda D, Ok YS. *Particulate Plastics in Terrestrial and Aquatic Environments*: Boca Raton, FL: CRC Press, 2020.

Bolan NS, Park JH, Robinson B, Naidu R, Huh KY. Phytostabilization. A green approach to contaminant containment. *Advances in Agronomy* 2011; 112: 145–204.

Bolan NS, Sarkar B, Yan Y, Li Q, Wijesekara H, Kannan K, et al. Remediation of poly- and perfluoroalkyl substances (PFAS) contaminated soils – To mobilize or to immobilize or to degrade? *Journal of Hazardous Materials* 2021; 401: 123892.

Brooks RR, Chambers MF, Nicks LJ, Robinson BH. Phytomining. *Trends in Plant Science* 1998; 3: 359–362.

Burges A, Alkorta I, Epelde L, Garbisu C. From phytoremediation of soil contaminants to phytomanagement of ecosystem services in metal contaminated sites. *International Journal of Phytoremediation* 2018; 20: 384–397.

Cappuyns V, Swennen R. The use of leaching tests to study the potential mobilization of heavy metals from soils and sediments: A comparison. *Water, Air, and Soil Pollution* 2008; 191: 95–111.

Carvalho CFMD, Viana DG, Pires FR, Egreja Filho FB, Bonomo R, Martins LF, et al. Phytoremediation of barium-affected flooded soils using single and intercropping cultivation of aquatic macrophytes. *Chemosphere* 2019; 214: 10–16.

Chalot M, Girardclos O, Ciadamidaro L, Zappelini C, Yung L, Durand A, et al. Poplar rotation coppice at a trace element-contaminated phytomanagement site: A 10-year study revealing biomass production, element export and impact on extractable elements. *Science of the Total Environment* 2020; 699: 134260.

Chen C, Zhang H, Wang A, Lu M, Shen Z, Lian C. Phenotypic plasticity accounts for most of the variation in leaf manganese concentrations in Phytolacca americana growing in manganese-contaminated environments. *Plant and Soil* 2015; 396: 215–227.

Cheng SF, Huang CY, Lin SC, Chen KL, Lin YC. Feasibility of using peanut (Arachis hypogaea L.) for phytoattenuation on lead-contaminated agricultural land-an in situ study. *Agriculture, Ecosystems and Environment* 2015; 202: 25–30.

Cicatelli A, Torrigiani P, Todeschini V, Biondi S, Castiglione S, Lingua G. Arbuscular mycorrhizal fungi as a tool to ameliorate the phytoremediation potential of poplar: Biochemical and molecular aspects. *IForest* 2014; 7: 333–341.

Cundy A, Bardos R, Church A, Puschenreiter M, Friesl-Hanl W, Müller I, et al. Developing principles of sustainability and stakeholder engagement for "gentle" remediation approaches: The European context. *Journal of Environmental Management* 2013; 129: 283–291.

Cundy AB, Bardos RP, Puschenreiter M, Mench M, Bert V, Friesl-Hanl W, et al. Brownfields to green fields: Realising wider benefits from practical contaminant phytomanagement strategies. *Journal of Environmental Management* 2016; 184: 67–77.

Dams RI, Paton GI, Killham K. Rhizoremediation of pentachlorophenol by Sphingobium chlorophenolicum ATCC 39723. *Chemosphere* 2007; 68: 864–870.

Deng L, Li Z, Wang J, Liu H, Li N, Wu L, et al. Long-term field phytoextraction of zinc/cadmium contaminated soil by Sedum plumbizincicola under different agronomic strategies. *International Journal of Phytoremediation* 2016; 18: 134–140.

Doty SL, Freeman JL, Cohu CM, Burken JG, Firrincieli A, Simon A, et al. Enhanced degradation of TCE on a superfund site using endophyte-assisted poplar tree phytoremediation. *Environmental Science and Technology* 2017; 51: 10050–10058.

Edenborn SL, Edenborn HM, Krynock RM, Haug KLZ. Influence of biochar application methods on the phytostabilization of a hydrophobic soil contaminated with lead and acid tar. *Journal of Environmental Management* 2015; 150: 226–234.

Evangelou MWH, Papazoglou EG, Robinson BH, Schulin R. Phytomanagement: Phytoremediation and the production of biomass for economic revenue on contaminated land. *Phytoremediation: Management of Environmental Contaminants* 2015; 1: 115–132.

Fan W, Guo Q, Liu CY, Liu XQ, Zhang M, Long DP, et al. Two mulberry phytochelatin synthase genes confer zinc/cadmium tolerance and accumulation in transgenic Arabidopsis and tobacco. *Gene* 2018; 645: 95–104.

FAO. Save and Grow: A Policymaker's guide to the sustainable intensification of smallholder crop production. Food and Agriculture Organization of the United Nations, 2011.

FAO. Status of the World's Soil Resources. Food and Agriculture Organization of the United Nations, 2015.

FAO. Soil Pollution: A Hidden Reality. Food and Agriculture Organization of the United Nations, 2018.

Fässler E, Robinson BH, Stauffer W, Gupta SK, Papritz A, Schulin R. Phytomanagement of metal-contaminated agricultural land using sunflower, maize and tobacco. *Agriculture, Ecosystems and Environment* 2010; 136: 49–58.

Ferro AM, Kennedy J, LaRue JC. Phytoremediation of 1,4-dioxane-containing recovered groundwater. *International Journal of Phytoremediation* 2013; 15: 911–923.

Ferro AM, Tammi CE. Field note: Irrigation of tree stands with groundwater containing 1,4-dioxane. *International Journal of Phytoremediation* 2009; 11: 425–440.

Galende MA, Becerril JM, Barrutia O, Artetxe U, Garbisu C, Hernández A. Field assessment of the effectiveness of organic amendments for aided phytostabilization of a Pb-Zn contaminated mine soil. *Journal of Geochemical Exploration* 2014; 145: 181–189.

Gao J, Garrison AW, Hoehamer C, Mazur CS, Wolfe NL. Uptake and phytotransformation of organophosphorus pesticides by axenically cultivated aquatic plants. *Journal of Agricultural and Food Chemistry* 2000; 48: 6114–6120.

Gao J, Wang L, Ok YS, Bank MS, Luo J, Wu W-M, et al. Nanoplastic stimulates metalloid leaching from historically contaminated soil via indirect displacement. *Water Research* 2022; 218: 118468.

Gaskin SE, Bentham RH. Rhizoremediation of hydrocarbon contaminated soil using Australian native grasses. *Science of the Total Environment* 2010; 408: 3683–3688.

Gerhardt KE, Gerwing PD, Greenberg BM. Opinion: Taking phytoremediation from proven technology to accepted practice. *Plant Science* 2017; 256: 170–185.

Ghori Z, Iftikhar H, Bhatti MF, Nasar Um M, Sharma I, Kazi AG, et al. Chapter 15: Phytoextraction: The use of plants to remove heavy metals from soil. *Plant Metal Interaction: Emerging Remediation Techniques* 2015: 361–384.

Grandlic CJ, Palmer MW, Maier RM. Optimization of plant growth-promoting bacteria-assisted phytostabilization of mine tailings. *Soil Biology and Biochemistry* 2009; 41: 1734–1740.

Grobelak A, Placek A, Grosser A, Singh BR, Almås ÅR, Napora A, et al. Effects of single sewage sludge application on soil phytoremediation. *Journal of Cleaner Production* 2017; 155: 189–197.

Guo Y, Holton C, Luo H, Dahlen P, Johnson PC. Influence of fluctuating groundwater table on volatile organic chemical emission flux at a dissolved chlorinated-solvent plume site. *Groundwater Monitoring and Remediation* 2019; 39: 43–52.

Haque S, Zeyaullah M, Nabi G, Srivastava PS, Ali A. Transgenic tobacco plant expressing environmental E. coli merA gene for enhanced volatilization of ionic mercury. *Journal of Microbiology and Biotechnology* 2010; 20: 917–924.

Haritash AK, Kaushik CP. Biodegradation aspects of Polycyclic Aromatic Hydrocarbons (PAHs): A review. *Journal of Hazardous Materials* 2009; 169: 1–15.

Hashimoto Y, Takaoka M, Shiota K. Enhanced transformation of lead speciation in rhizosphere soils using phosphorus amendments and phytostabilization: An x-ray absorption fine structure spectroscopy investigation. *Journal of Environmental Quality* 2011; 40: 696–703.

He L, Zhong H, Liu G, Dai Z, Brookes PC, Xu J. Remediation of heavy metal contaminated soils by biochar: Mechanisms, potential risks and applications in China. *Environmental Pollution* 2019; 252: 846–855.

Hillel D, Hatfield JL. *Encyclopedia of Soils in the Environment.* Vol 3: Amsterdam: Elsevier, 2005.

Hoang SA, Lamb D, Seshadri B, Sarkar B, Choppala G, Kirkham MB, et al. Rhizoremediation as a green technology for the remediation of petroleum hydrocarbon-contaminated soils. *Journal of Hazardous Materials* 2021; 401: 123282.

Hou D. *Sustainable Remediation of Contaminated Soil and Groundwater: Materials, Processes, and Assessment*: Oxford: Elsevier, 2020.

Hou D. Biochar for sustainable soil management. *Soil Use and Management* 2021a; 37: 2–6.

Hou D. Sustainable remediation in China: Elimination, immobilization, or dilution. *Environmental Science & Technology* 2021b; 55: 15572–15574.

Hou D. China: Protect black soil for biodiversity. *Nature* 2022a; 604: 40–40.

Hou D. Expediting climate-smart soils management. *Soil Use and Management* 2022b; 38: 1–6.

Hou R, Wang L, O'Connor D, Rinklebe J, Hou D. Natural field freeze-thaw process leads to different performances of soil amendments towards Cd immobilization and enrichment. *Science of the Total Environment* 2022; 831: 154880.

Hou R, Wang L, Shen Z, Alessi DS, Hou D. Simultaneous reduction and immobilization of Cr(VI) in seasonally frozen areas: Remediation mechanisms and the role of ageing. *Journal of Hazardous Materials* 2021; 415: 125650.

Hu L, Robert CAM, Cadot S, Zhang X, Ye M, Li B, et al. Root exudate metabolites drive plant-soil feedbacks on growth and defense by shaping the rhizosphere microbiota. *Nature Communications* 2018; 9: 1–3.

Hu P, Zhang Y, Dong B, Gao W, Cheng C, Luo Y, et al. Assessment of phytoextraction using Sedum plumbizincicola and rice production in Cd-polluted acid paddy soils of south China: A field study. *Agriculture, Ecosystems and Environment* 2019a; 286: 106651.

Hu Y, Gao Z, Huang Y, Chen S, Yang X, Su J, et al. Impact of poplar-based phytomanagement on metal bioavailability in low-phosphorus calcareous soil with multi-metal contamination. *Science of the Total Environment* 2019b; 686: 848–855.

Hussein HS, Ruiz ON, Terry N, Daniell H. Phytoremediation of mercury and organomercurials in chloroplast transgenic plants: Enhanced root uptake, translocation to shoots, and volatilization. *Environmental Science and Technology* 2007a; 41: 8439–8446.

Hussein S, Ruiz ON, Terry N, Daniell H. Phytoremediation of mercury and organomercurials in chloroplast transgenic plants: Enhanced root uptake, translocation to shoots, and volatilization. *Environmental Science & Technology* 2007b; 41: 8439–8446.

Jacobs A, Drouet T, Noret N. Field evaluation of cultural cycles for improved cadmium and zinc phytoextraction with Noccaea caerulescens. *Plant and Soil* 2018; 430: 381–394.

Jeannin T, Yung L, Evon P, Labonne L, Ouagne P, Lecourt M, et al. Native stinging nettle (Urtica dioica L.) growing spontaneously under short rotation coppice for phytomanagement of trace element contaminated soils: Fibre yield, processability and quality. *Industrial Crops and Products* 2020; 145: 111997.

Ji P, Sun T, Song Y, Ackland ML, Liu Y. Strategies for enhancing the phytoremediation of cadmium-contaminated agricultural soils by Solanum nigrum L. *Environmental Pollution* 2011; 159: 762–768.

Jia X, Cao Y, O'Connor D, Zhu J, Tsang DCW, Zou B, et al. Mapping soil pollution by using drone image recognition and machine learning at an arsenic-contaminated agricultural field. *Environmental Pollution* 2021; 270: 116281.

Jia Y, Huang H, Sun GX, Zhao FJ, Zhu YG. Pathways and relative contributions to arsenic volatilization from rice plants and paddy soil. *Environmental Science and Technology* 2012; 46: 8090–8096.

Jin Y, Wang L, Song Y, Zhu J, Qin M, Wu L, et al. Integrated life cycle assessment for sustainable remediation of contaminated agricultural soil in China. *Environmental Science & Technology* 2021; 55: 12032–12042.

Kabouw P, van Dam NM, van der Putten WH, Biere A. How genetic modification of roots affects rhizosphere processes and plant performance. *Journal of Experimental Botany* 2012; 63: 3475–3483.

Kebeish R, Azab E, Peterhaensel C, El-Basheer R. Engineering the metabolism of the phenylurea herbicide chlortoluron in genetically modified Arabidopsis thaliana plants expressing the mammalian cytochrome P450 enzyme CYP1A2. Environmental Science and Pollution Research 2014; 21: 8224–8232.

Kidd P, Mench M, Álvarez-López V, Bert V, Dimitriou I, Friesl-Hanl W, et al. Agronomic practices for improving gentle remediation of trace element-contaminated soils. *International Journal of Phytoremediation* 2015; 17: 1005–1037.

Kolbas A, Kolbas N, Marchand L, Herzig R, Mench M. Morphological and functional responses of a metal-tolerant sunflower mutant line to a copper-contaminated soil series. *Environmental Science and Pollution Research* 2018; 25: 16686–16701.

Kotoky R, Rajkumari J, Pandey P. The rhizosphere microbiome: Significance in rhizoremediation of polyaromatic hydrocarbon contaminated soil. *Journal of Environmental Management* 2018; 217: 858–870.

Kowitwiwat A, Sampanpanish P. Phytostabilization of arsenic and manganese in mine tailings using Pennisetum purpureum cv. Mott supplemented with cow manure and acacia wood-derived biochar. *Heliyon* 2020; 6: e04552.

Kuiper I, Lagendijk EL, Bloemberg GV, Lugtenberg BJJ. Rhizoremediation: A beneficial plant-microbe interaction. *Molecular Plant-Microbe Interactions* 2004; 17: 6–15.

Kumpiene J, Fitts JP, Mench M. Arsenic fractionation in mine spoils 10 years after aided phytostabilization. *Environmental Pollution* 2012; 166: 82–88.

Labidi S, Firmin S, Verdin A, Bidar G, Laruelle F, Douay F, et al. Nature of fly ash amendments differently influences oxidative stress alleviation in four forest tree species and metal trace element phytostabilization in aged contaminated soil: A long-term field experiment. *Ecotoxicology and Environmental Safety* 2017; 138: 190–198.

Lacalle RG, Aparicio JD, Artetxe U, Urionabarrenetxea E, Polti MA, Soto M, et al. Gentle remediation options for soil with mixed chromium (VI) and lindane pollution: Biostimulation, bioaugmentation, phytoremediation and vermiremediation. *Heliyon* 2020; 6: e04550.

Lambers H, Mougel C, Jaillard B, Hinsinger P. Plant-microbe-soil interactions in the rhizosphere: An evolutionary perspective. *Plant and Soil* 2009; 321: 83–115.

Lebrun M, Miard F, Nandillon R, Léger JC, Hattab-Hambli N, Scippa GS, et al. Assisted phytostabilization of a multicontaminated mine technosol using biochar amendment: Early stage evaluation of biochar feedstock and particle size effects on As and Pb accumulation of two Salicaceae species (Salix viminalis and Populus euramericana). *Chemosphere* 2018; 194: 316–326.

Legault EK, James CA, Stewart K, Muiznieks I, Doty SL, Strand SE. A field trial of TCE phytoremediation by genetically modified poplars expressing cytochrome P450 2E1. *Environmental Science and Technology* 2017; 51: 6090–6099.

Lehoux AP, Lockwood CL, Mayes WM, Stewart DI, Mortimer RJG, Gruiz K, et al. Gypsum addition to soils contaminated by red mud: Implications for aluminium, arsenic, molybdenum and vanadium solubility. *Environmental Geochemistry and Health* 2013; 35: 643–656.

Leon-Romero MA, Soto-Rios PC, Nomura M, Nishimura O. Effect of steel slag to improve soil quality of tsunami-impacted land while reducing the risk of heavy metal bioaccumulation. *Water Air and Soil Pollution* 2018; 229: 14.

Li JT, Baker AJM, Ye ZH, Wang HB, Shu WS. Phytoextraction of Cd-contaminated soils: Current status and future challenges. *Critical Reviews in Environmental Science and Technology* 2012; 42: 2113–2152.

Li X, Huang L. Toward a new paradigm for tailings phytostabilization - Nature of the substrates, amendment options, and anthropogenic pedogenesis. *Critical Reviews in Environmental Science and Technology* 2015; 45: 813–839.

Liang HM, Lin TH, Chiou JM, Yeh KC. Model evaluation of the phytoextraction potential of heavy metal hyperaccumulators and non-hyperaccumulators. *Environmental Pollution* 2009; 157: 1945–1952.

Limmer M, Burken J. Phytovolatilization of organic contaminants. *Environmental Science and Technology* 2016; 50: 6632–6643.

Limmer MA, Burken JG. Plant translocation of organic compounds: Molecular and physicochemical predictors. *Environmental Science and Technology Letters* 2014; 1: 156–161.

Liu DL, An ZG, Mao ZJ, Ma LB, Lu ZQ. Enhanced heavy metal tolerance and accumulation by transgenic sugar beets expressing Streptococcus thermophilus StGCS-GS in the presence of Cd, Zn and Cu alone or in combination. *PLoS One* 2015; 10: 15.

Lyyra S, Meagher RB, Kim T, Heaton A, Montello P, Balish RS, et al. Coupling two mercury resistance genes in Eastern cottonwood enhances the processing of organomercury. *Plant Biotechnology Journal* 2007; 5: 254–262.

Ma Y, Feng X. The application of red mud on restoring heavy metal contaminated water and soil. *Advanced Materials Research* 2011; 225–226: 1262–1265.

Mackenzie A, Ball AS, Virdee SR, Mackenzie A, Ball AS, Virdee SR. Instant Notes Series. *Ecology*. Second Edition: Oxford, UK; New York, NY: BIOS Scientific Publishers Ltd., 2001.

Matsui K, Togami J, Mason JG, Chandler SF, Tanaka Y. Enhancement of phosphate absorption by garden plants by genetic engineering: A new tool for phytoremediation. *Biomed Research International* 2013; 7: 182032.

Meers E, Van Slycken S, Adriaensen K, Ruttens A, Vangronsveld J, Du Laing G, et al. The use of bio-energy crops (Zea mays) for 'phytoattenuation' of heavy metals on moderately contaminated soils: A field experiment. *Chemosphere* 2010; 78: 35–41.

Mench MJ, Dellise M, Bes CM, Marchand L, Kolbas A, Coustumer PL, et al. Phytomanagement and remediation of cu-contaminated soils by high yielding crops at a former wood preservation site: Sunflower biomass and ionome. *Frontiers in Ecology and Evolution* 2018; 6: 123.

Nahar N, Rahman A, Nawani NN, Ghosh S, Mandal A. Phytoremediation of arsenic from the contaminated soil using transgenic tobacco plants expressing ACR2 gene of Arabidopsis thaliana. *Journal of Plant Physiology* 2017; 218: 121–126.

Newman LA, Reynolds CM. Phytodegradation of organic compounds. *Current Opinion in Biotechnology* 2004; 15: 225–230.

Niazi NK, Singh B, Van Zwieten L, Kachenko AG. Phytoremediation of an arsenic-contaminated site using Pteris vittata L. and Pityrogramma calomelanos var. austroamericana: A long-term study. *Environmental Science and Pollution Research* 2012; 19: 3506–3515.

O'Connor D, Hou DY, Ok YS, Mulder J, Duan L, Wu QR, et al. Mercury speciation, transformation, and transportation in soils, atmospheric flux, and implications for risk management: A critical review. *Environment International* 2019a; 126: 747–761.

O'Connor D, Zheng X, Hou D, Shen Z, Li G, Miao G, et al. Phytoremediation: Climate change resilience and sustainability assessment at a coastal brownfield redevelopment. *Environment International* 2019b; 130: 104945.

Panda D, Panda D, Padhan B, Biswas M. Growth and physiological response of lemongrass (Cymbopogon citratus (DC) Stapf.) under different levels of fly ash-amended soil. *International Journal of Phytoremediation* 2018; 20: 538–544.

Papazoglou EG, Fernando AL. Preliminary studies on the growth, tolerance and phytoremediation ability of sugarbeet (Beta vulgaris L.) grown on heavy metal contaminated soil. *Industrial Crops and Products* 2017; 107: 463–471.

Perez-Palacios P, Romero-Aguilar A, Delgadillo J, Doukkali B, Caviedes MA, Rodriguez-Llorente ID, et al. Double genetically modified symbiotic system for improved Cu phytostabilization in legume roots. *Environmental Science and Pollution Research* 2017; 24: 14910–14923.

Pérez Rodríguez N, Langella F, Rodushkin I, Engström E, Kothe E, Alakangas L, et al. The role of bacterial consortium and organic amendment in Cu and Fe isotope fractionation in plants on a polluted mine site. *Environmental Science and Pollution Research* 2014; 21: 6836–6844.

Praveen A, Pandey VC, Marwa N, Singh DP. Rhizoremediation of polluted sites: Harnessing plant-microbe interactions. In: *Phytomanagement of Polluted Sites: Market Opportunities in Sustainable Phytoremediation*, Vimal Chandra Pandey and Kuldeep Bauddh (eds.), 2018, Amsterdam: Elsevier, pp. 389–407.

Qi S, Luo J, O'Connor D, Cao X, Hou D. Influence of groundwater table fluctuation on the non-equilibrium transport of volatile organic contaminants in the vadose zone. *Journal of Hydrology* 2020; 580: 124353.

Qu GZ, Tong YA, Gao PC, Zhao ZP, Song XY, Ji PH. Phytoremediation Potential of Solanum nigrum L. under different cultivation protocols. *Bulletin of Environmental Contamination and Toxicology* 2013; 91: 306–309.

Quevauviller P, Griepink B, Ure A, Muntaui H. Improvement of analytical measurements within the bcr-programme: Single and sequential extraction procedures applied to soil and sediment analysis. *International Journal of Environmental Analytical Chemistry* 1993; 51: 129–134.

Radziemska M, Koda E, Vaverkovà MD, Gusiatin ZM, Cerdà A, Brtnický M, et al. Soils from an iron and steel scrap storage yard remediated with aided phytostabilization. *Land Degradation and Development* 2019; 30: 202–211.

Raldugina GN, Maree M, Mattana M, Shumkova G, Mapelli S, Kholodova VP, et al. Expression of rice OsMyb4 transcription factor improves tolerance to copper or zinc in canola plants. *Biologia Plantarum* 2018; 62: 511–520.

Ran J, Zheng W, Wang H, Wang H, Li Q. Indole-3-acetic acid promotes cadmium (Cd) accumulation in a Cd hyperaccumulator and a non-hyperaccumulator by different physiological responses. *Ecotoxicology and Environmental Safety* 2020; 191: 110213.

Rane NR, Chandanshive VV, Khandare RV, Gholave AR, Yadav SR, Govindwar SP. Green remediation of textile dyes containing wastewater by Ipomoea hederifolia L. *RSC Advances* 2014; 4: 36623–36632.

Rani R, Padole P, Juwarkar A, Chakrabarti T. Phytotransformation of Phorate by Brassica juncea (Indian Mustard). *Water, Air, and Soil Pollution* 2012; 223: 1383–1392.

Raynaud X, Lata JC, Leadley PW. Soil microbial loop and nutrient uptake by plants: A test using a coupled C:N model of plant-microbial interactions. *Plant and Soil* 2006; 287: 95–116.

Robinson BH, Anderson CWN, Dickinson NM. Phytoextraction: Where's the action? *Journal of Geochemical Exploration* 2015; 151: 34–40.

Roderick ML, Barnes B. Self-thinning of plant populations from a dynamic viewpoint. *Functional Ecology* 2004; 18: 197–203.

Ruiz ON, Alvarez D, Torres C, Roman L, Daniell H. Metallothionein expression in chloroplasts enhances mercury accumulation and phytoremediation capability. *Plant Biotechnology Journal* 2011; 9: 609–617.

Ruiz ON, Daniell H. Genetic engineering to enhance mercury phytoremediation. *Current Opinion in Biotechnology* 2009; 20: 213–219.

Salt DE, Smith R, Raskin I. Phytoremediation. *Annual Review of Plant Biology* 1998; 49: 643–668.

Saran A, Imperato V, Fernandez L, Gkorezis P, D'Haen J, Merini LJ, et al. Phytostabilization of polluted military soil supported by bioaugmentation with PGP-trace element tolerant bacteria isolated from helianthus petiolaris. *Agronomy* 2020; 10: 204.

Saravanan A, Jeevanantham S, Narayanan VA, Kumar PS, Yaashikaa PR, Muthu CMM. Rhizoremediation - A promising tool for the removal of soil contaminants: A review. *Journal of Environmental Chemical Engineering* 2020; 8: 103543.

Sheoran V, Sheoran AS, Poonia P. Phytomining: A review. *Minerals Engineering* 2009; 22: 1007–1019.

Shim D, Kim S, Choi YI, Song WY, Park J, Youk ES, et al. Transgenic poplar trees expressing yeast cadmium factor 1 exhibit the characteristics necessary for the phytoremediation of mine tailing soil. *Chemosphere* 2013; 90: 1478–1486.

Shrinkhal R. Economics, technology, and environmental protection: A critical analysis of phytomanagement. *Phytomanagement of Polluted Sites: Market Opportunities in Sustainable Phytoremediation*, Vimal Chandra Pandey and Kuldeep Bauddh (eds.), 2018, Amsterdam: Elsevier, pp. 569–580.

Song Y, Hou D, Zhang J, O'Connor D, Li G, Gu Q, et al. Environmental and socio-economic sustainability appraisal of contaminated land remediation strategies: A case study at a mega-site in China. *Science of the Total Environment* 2018; 610–611: 391–401.

Song Y, Kirkwood N, Maksimović Č, Zhen X, O'Connor D, Jin Y, et al. Nature based solutions for contaminated land remediation and brownfield redevelopment in cities: A review. *Science of the Total Environment* 2019; 663: 568–579.

Suman J, Uhlik O, Viktorova J, Macek T. Phytoextraction of heavy metals: A promising tool for clean-up of polluted environment? *Frontiers in Plant Science* 2018; 9: 871.

Sun LP, Ma YF, Wang HH, Huang WP, Wang XZ, Han L, et al. Overexpression of PtABCC1 contributes to mercury tolerance and accumulation in Arabidopsis and poplar. *Biochemical and Biophysical Research Communications* 2018; 497: 997–1002.

Tangahu BV, Sheikh Abdullah SR, Basri H, Idris M, Anuar N, Mukhlisin M. A review on heavy metals (As, Pb, and Hg) uptake by plants through phytoremediation. *International Journal of Chemical Engineering* 2011; 2011. Article number 939161.

Techer D, Martinez-Chois C, Laval-Gilly P, Henry S, Bennasroune A, D'Innocenzo M, et al. Assessment of Miscanthus×giganteus for rhizoremediation of long term PAH contaminated soils. *Applied Soil Ecology* 2012; 62: 42–49.

Tessier A, Campbell PGC, Bisson M. Sequential extraction procedure for the speciation of particulate trace-metals. *Analytical Chemistry* 1979; 51: 844–851.

Thijs S, Witters N, Janssen J, Ruttens A, Weyens N, Herzig R, et al. Tobacco, sunflower and high biomass src clones show potential for trace metal phytoextraction on a moderately contaminated field site in Belgium. *Frontiers in Plant Science* 2018; 871.

Touceda-González M, Álvarez-López V, Prieto-Fernández Á, Rodríguez-Garrido B, Trasar-Cepeda C, Mench M, et al. Aided phytostabilisation reduces metal toxicity, improves soil fertility and enhances microbial activity in Cu-rich mine tailings. *Journal of Environmental Management* 2017; 186: 301–313.

Tu C, Ma L, Guo P, Song F, Teng Y, Zhang H, et al. Rhizoremediation of a dioxin-like PCB polluted soil by alfalfa: Dynamic characterization at temporal and spatial scale. *Chemosphere* 2017; 189: 517–524.

US EPA. Phytotransformation: The Ultimate Cleaner-Upper. United States Environmental Protection Agency, 2001.

USEPA. Superfund Remedy Report 15th Edition EPA-542-R-17-001, Washington DC, 2017.

Van Slycken S, Witters N, Meers E, Peene A, Michels E, Adriaensen K, et al. Safe use of metal-contaminated agricultural land by cultivation of energy maize (Zea mays). *Environmental Pollution* 2013a; 178: 375–380.

van Slycken S, Witters N, Meiresonne L, Meers E, Ruttens A, van Peteghem P, et al. Field evaluation of willow under short rotation coppice for phytomanagement of metal-polluted agricultural soils. *International Journal of Phytoremediation* 2013b; 15: 677–689.

Verdugo C, Sánchez P, Santibáñez C, Urrestarazu P, Bustamante E, Silva Y, et al. Efficacy of lime, biosolids, and mycorrhiza for the phytostabilization of sulfidic copper tailings in chile: A greenhouse experiment. *International Journal of Phytoremediation* 2011; 13: 107–125.

Vergani L, Mapelli F, Suman J, Cajthaml T, Uhlik O, Borin S. Novel PCB-degrading Rhodococcus strains able to promote plant growth for assisted rhizoremediation of historically polluted soils. *PLoS One* 2019; 14: e0221253.

Viana DG, Pires FR, Ferreira AD, Egreja Filho FB, Carvalho CFMD, Bonomo R, et al. Effect of planting density of the macrophyte consortium of Typha domingensis and Eleocharis acutangula on phytoremediation of barium from a flooded contaminated soil. *Chemosphere* 2021; 262: 127869.

Visconti D, Álvarez-Robles MJ, Fiorentino N, Fagnano M, Clemente R. Use of Brassica juncea and Dactylis glomerata for the phytostabilization of mine soils amended with compost or biochar. *Chemosphere* 2020; 260: 127661.

Wan X, Lei M, Chen T, Tan Y, Yang J. Safe utilization of heavy-metal-contaminated farmland by mulberry tree cultivation and silk production. *Science of the Total Environment* 2017; 599–600: 1867–1873.

Wang C, Lyon DY, Hughes JB, Bennett GN. Role of hydroxylamine intermediates in the phytotransformation of 2,4,6-trinitrotoluene by Myriophyllum aquaticum. *Environmental Science and Technology* 2003; 37: 3595–3600.

Wang C, Wang L, Ok YS, Tsang DC, Hou D. Soil plastisphere: Exploration methods, influencing factors, and ecological insights. *Journal of Hazardous Materials* 2022a; 430: 128503.

Wang L, Bolan NS, Tsang DCW, Hou D. Green immobilization of toxic metals using alkaline enhanced rice husk biochar: Effects of pyrolysis temperature and KOH concentration. *Science of the Total Environment* 2020a; 720: 137584.

Wang L, Hou D, Cao Y, Ok YS, Tack FMG, Rinklebe J, et al. Remediation of mercury contaminated soil, water, and air: A review of emerging materials and innovative technologies. *Environment International* 2020b; 134: 105281.

Wang L, Huang J, Li G, Luo J, Bolan NS, Hou D. Long-term immobilization of soil metalloids under simulated aging: Experimental and modeling approach. *Science of the Total Environment* 2022b; 806: 150501.

Wang L, Jin Y, Weiss DJ, Schleicher NJ, Wilcke W, Wu L, et al. Possible application of stable isotope compositions for the identification of metal sources in soil. *Journal of Hazardous Materials* 2021a; 407: 124812.

Wang L, Li P, Zhang Q, Wu W-M, Luo J, Hou D. Modeling the conditional fragmentation-induced microplastic distribution. *Environmental Science & Technology* 2021b; 55: 6012–6021.

Wang L, Ok YS, Tsang DCW, Alessi DS, Rinklebe J, Mašek O, et al. Biochar composites: Emerging trends, field successes and sustainability implications. *Soil Use and Management* 2022c; 38: 14–38.

Wang L, Rinklebe J, Tack FMG, Hou D. A review of green remediation strategies for heavy metal contaminated soil. *Soil Use and Management* 2021c; 37: 936–963.

Wang LW, Hou DY, Shen ZT, Zhu J, Jia XY, Ok YS, et al. Field trials of phytomining and phytoremediation: A critical review of influencing factors and effects of additives. *Critical Reviews in Environmental Science and Technology* 2020c; 50: 2724–2774.

Wang XT, Zhi JK, Liu XR, Zhang H, Liu HB, Xu JC. Transgenic tobacco plants expressing a P1B-ATPase gene from Populus tomentosa Carr. (PtoHMA5) demonstrate improved cadmium transport. *International Journal of Biological Macromolecules* 2018; 113: 655–661.

Wei CY, Sun X, Wang C, Wang WY. Factors influencing arsenic accumulation by Pteris vittata: A comparative field study at two sites. *Environmental Pollution* 2006; 141: 488–493.

Wu Q, Wang S, Thangavel P, Li Q, Zheng H, Bai J, et al. Phytostabilization potential of Jatropha curcas l. in polymetallic acid mine tailings. *International Journal of Phytoremediation* 2011; 13: 788–804.

Yang C, Ho YN, Inoue C, Chien MF. Long-term effectiveness of microbe-assisted arsenic phytoremediation by Pteris vittata in field trials. *Science of the Total Environment* 2020; 740: 140137.

Yang QW, Zeng Q, Xiao F, Liu XL, Pan J, He JF, et al. Investigation of manganese tolerance and accumulation of two Mn hyperaccumulators Phytolacca americana L. and Polygonum hydropiper L. in the real Mn-contaminated soils near a manganese mine. *Environmental Earth Sciences* 2013; 68: 1127–1134.

Yang SX, Liao B, Xiao RB, Li JT. Effects of amendments on soil microbial diversity, enzyme activity and nutrient accumulation after assisted phytostabilization of an extremely acidic metalliferous mine soil. *Applied Sciences (Switzerland)* 2019; 9: 1552.

Yang T, He C, Luo W, Wang T, Sheng L. Progress in study of Salix characteristics and its application in sediment phytoremediation. *Chinese Journal of Applied and Environmental Biology* 2015; 21: 837–841.

Yang X, Wang L, Guo J, Wang H, Mašek O, Wang H, et al. Aging features of metal(loid)s in biochar-amended soil: Effects of biochar type and aging method. *Science of the Total Environment* 2022; 815: 152922.

Yoda K, Kira T, Ogawa H, Hozumi K. Self-thinning in overcrowded pure stands under cultivated and natural conditions. *Journal of Biology, Osaka City University* 1963; 14: 107–129.

Zhan J, Huang H, Yu H, Zhang X, Wang Y, Li T. Characterization of dissolved organic matter in the rhizosphere of phytostabilizer Athyrium wardii (Hook.) involved in enhanced metal accumulation when exposed to Cd and Pb co-contamination. *Environmental Pollution* 2020; 266.

Zhang L, Rylott EL, Bruce NC, Strand SE. Genetic modification of western wheatgrass (Pascopyrum smithii) for the phytoremediation of RDX and TNT. *Planta* 2019; 249: 1007–1015.

Zhang M, Fang L, Zhou C. Evaluation of heavy metals bioavailability and mobility in polluted soils: A comparison of four methods. *Chinese Journal of Applied Ecology* 2006a; 17: 1501–1504.

Zhang Q, Kong W, Wei L, Wang Y, Luo Y, Wang P, et al. Uptake, phytovolatilization, and interconversion of 2,4-dibromophenol and 2,4-dibromoanisole in rice plants. *Environment International* 2020; 142: 105888.

Zhang RQ, Tang CF, Wen SZ, Liu YG, Li KL. Advances in research on genetically engineered plants for metal resistance. *Journal of Integrative Plant Biology* 2006b; 48: 1257–1265.

Zhao H, Wei Y, Wang J, Chai T. Isolation and expression analysis of cadmium-induced genes from Cd/Mn hyperaccumulator Phytolacca americana in response to high Cd exposure. *Plant Biology* 2019; 21: 15–24.

Zhu B, Han HJ, Fu XY, Li ZJ, Gao JJ, Yao QH. Degradation of trinitrotoluene by transgenic nitroreductase in Arabidopsis plants. *Plant Soil and Environment* 2018; 64: 379–385.

Zhu B, Peng RH, Fu XY, Jin XF, Zhao W, Xu J, et al. Enhanced transformation of TNT by arabidopsis plants expressing an old yellow enzyme. *PLoS One* 2012; 7: 7.

3 Green and Sustainable Remediation as an Option to Manage and Recover Soil Polluted by Anthropogenic Activities

Claudia Rocco[1,2], *Diana Agrelli*[3], *and Paola Adamo*[3]
[1]Department of Earth Science and Engineering, Imperial College London, London, United Kingdom
[2]Faculty of Natural Sciences, Department of Life Sciences, MRC Centre for Molecular Bacteriology and Infection, Imperial College London, London, United Kingdom
[3]Department of Agricultural Sciences, University of Naples Federico II, Portici, Naples, Italy

CONTENTS

3.1 SOIL CONTAMINATION EMERGENCY

Soil is a thin layer where various amounts of organic matter, inorganic materials, water, and air are mixed and cover the whole Earth's surface. The organic components of the soil, deriving from plant and animal biomass, are concentrated in the topsoil (the upper part of the soil layer where plants usually grow), while the inorganic materials comprise rock components formed over thousands of years by physical and chemical weathering of parent rock material (Shankar, 2017). Soil plays a crucial role in the regulation of natural and socioeconomic processes for sustaining human survival. The World Soil Information network (ISRIC, 2016) defined soil as "our life support system." Indeed, soil performs many vital functions and activities useful for survival of humans and

DOI: 10.1201/9781003093565-3

ecosystems. For example, soil provides food, biomass, and raw materials; it is a platform for human activities, an archive of geological and archaeological heritage, and it is also a biodiversity reserve where different habitats coexist, such as flora and fauna species, and genes. It also performs several activities of storage, filtration, and transformation of energy and other substances, including water, nutrients, and carbon, and constitutes the major carbon store in the terrestrial ecosystems of the planet. At the same time, soil is a central element of the problem of environmental sustainability, and it links climate change, decline in biodiversity, water, energy, and food security (European Commission, 2006; Hatfield et al., 2017; Fiorentino et al., 2018).

In the last decades, great attention was given to soil degradation factors. Land degradation may be defined as a long-term loss of ecosystem function and productivity caused by disturbances from which land cannot recover unaided (Bai et al., 2008). The Global Assessment of Human-Induced Soil Degradation (GLASOD) project indicated that 15% of the agricultural land is degraded, and the highest proportions of soil degradation were reported for Europe (25%), Asia (18%), and Africa (16%) (Bai et al., 2008). If we consider that soil is responsible for 99% of the world's food production, then the importance of soil becomes evident (Pimentel and Pimentel, 2000). Indeed, productive soils are necessary for good agricultural productivity (Hatfield et al., 2017).

To date, it has been estimated that the world has lost a third of its arable land mainly due to erosion or pollution in the past 40 years, with potentially disastrous consequences on food quality and production (The Guardian, 2015) and negative effects on socioeconomic conditions of the affected areas (Hatfield et al., 2017). Due to environmental pollution, the soil, the basis of agriculture-producing crops for human food and animal feed is under tremendous pressure. For all these reasons, there is urgency in controlling soil pollution in order to preserve soil fertility and productivity, which will improve the health of all living beings (Ashraf et al., 2014).

Pollution may be defined as "an unsolicited alteration in the physicochemical and biological properties of environment and its components causing threat to lives of humans, useful living plants and animals, industrial progress, living conditions and cultural assets" (Shankar, 2017).

A pollutant "is something which adversely interferes with health, comfort, property or environment of the people" (Ashraf et al., 2014).

Soil is getting polluted in a number of ways. Most soil pollution is due to human activities. Usually, contamination can be correlated with the degree of urbanization, industrialization, and intensity of chemical usage, and it is diffused worldwide (Bini, 2008).

Generally, pollutants are introduced into the environment by domestic livestock and municipal wastes and sewage; common agricultural practices (use of pesticides, herbicides, fertilizers); accidental discharge of petroleum-derived products or the by-products/residues of other production activities such as industrial activities including mining, smelting, and manufacturing (FAO, 2018). Another source of soil contamination is atmospheric deposition caused by industrial emissions, motor vehicles, and acid rain (Bini, 2008).

Pollutants may be distinguished, according to their composition and nature, in two categories with different diffusion and health hazards: organics and inorganics. The main organic contaminants are (i) mineral oil (fossil fuel, gasoline, diesel, lubricants), (ii) aromatic compounds and combustion products (PAH, PCB, dioxins), (iii) agrochemicals and pharmaceuticals, and (iv) biological pollutants. Inorganic contaminants are (i) potentially toxic elements (PTEs), also known as potentially toxic metals (PTMs) or heavy metal(loid)s (e.g., Cd, Cr, Cu, Zn, Pb, Al, Tl); (ii) volatile elements (As, Hg, Se); (iii) radionuclides (Cs, U, Ra); (iv) and anions (nitrates, nitrites, phosphates).

It is reported that >10 million contaminated sites exist worldwide, with >50% of the sites contaminated with PTEs (He et al., 2015). The majority of these contaminated sites exist in developed countries, i.e., the United States of America (USA), Australia, Germany, Sweden, and China owing to their increased use in industrial processes (Foucault et al., 2013; Goix et al., 2014). In the USA, around 600,000 ha (especially brownfield sites) have been contaminated with metal(loid)s. The US EPA has designated >50,000 priority metal(loid) polluted sites awaiting urgent remediation in the USA alone (Raskin and Ensley, 2000). Gorospe (2012) analyzed 16 different PTEs in soil samples collected from 91 vegetable gardens in San Francisco, USA. They observed that the majority (>75%) of the gardens surpassed the California Human Health Screening Level for Cd (84%), As (74%), and Pb (62%). Similarly, several metal(loid) contaminated agricultural sites situated near mining areas have been reported in different countries of Europe (Foucault et al., 2013; Goix et al., 2014). According to the European Environmental Agency (EEA, 2007), there exist about 3 million potentially polluting sites in Europe and about 250,000 metal(loid)s polluted sites in the EEA member countries.

Contaminants introduced into the environment can be more or less harmful for plants, animals, humans, and the environment depending on many factors including chemical properties and form, bioavailability, toxicity, solubility, and mobility. Remediation actions for the recovery of polluted sites are important for securing them, to return them to agriculture and to restore all the ecosystem services.

3.2 GENERAL CONSIDERATION ON REMEDIATION SUSTAINABILITY

The concept of remediation of polluted sites evolved over the past few decades. In the 1970–1980s, the goal was to reach the complete removal of all pollutants from soil, while in the 1990–2000s, the goal was to reach acceptable risk levels of soil contamination. This modification in the proposed goals is mainly linked to the very high cost of removal of pollutants that would outweigh the benefits to the society. To solve this problem, a risk-based remediation was considered; from 2010 to date, the need to minimize the impact of the remediation actions themselves, and therefore, the identification of the optimal means for risk management, led to the birth of green and sustainable remediation (USEPA, 2008; Bardos et al., 2013; Martino et al., 2016; Hou and O'Connor, 2019).

There are various adverse effects associated with most traditional remediation operations such as greenhouse gas emissions, global warming, depletion of natural resources, acidification, eutrophication, ozone depletion, ecological toxicity, waste production, and loss of land use and soil ecosystem services (Suèr et al., 2004; Hou and Al-Tabbaa, 2014); in addition, traditional remediation methods can cause damage to the physical structure of soil and microbial communities, with severe consequences on soil fertility and productivity after contaminant-removal operations (Gerhardt et al., 2017). Furthermore, there is increasing attention globally for soil protection, and the preservation of soil quality and functionality must be taken into account in the selection of a remediation strategy (Kibblewhite et al., 2012; Petruzzelli et al., 2016; Ronchi et al., 2019).

To date, "sustainable remediation" is applied in Europe, and it takes into account environmental, social, and economic impacts of the remediation activity (Figure 3.1). The United States performs "green remediation," which considers only environmental concerns because US EPA (United States Environmental Protection Agency) is only allowed authority in the environmental area but not in social and economic issues (Bardos et al., 2013; Hou and O'Connor, 2019). The term "environmental sustainability" mainly refers to reducing the risk derived from contamination and minimizing secondary adverse effects associated with remediation operations; "economic sustainability" considers the cost of the remediation operations and the impact of site restoration on the broader economy; "social impact" takes into account the workers' safety, the impact on the community, public perception and participation, and social inclusion (Hou and Al-Tabbaa, 2014; Hou and O'Connor, 2019).

For the evaluation of sustainability, some consideration should be given during the remediation planning. Hou and O'Connor (2019) list the sustainability considerations by UK and US remediation practitioners. These considerations are related to the environmental issues for the US and environmental, economic, and social impacts for the UK.

Nowadays, green and sustainable remediations are intensively being developed all over the world, involving public and private organizations in the development of projects to improve remediation techniques and make them sustainable. Bardos et al. (2013) list several initiatives taking place with the wider concept of "sustainable regeneration" related to brownfield rehabilitation. In this chapter, green and sustainable remediation are used synonymously with green remediation.

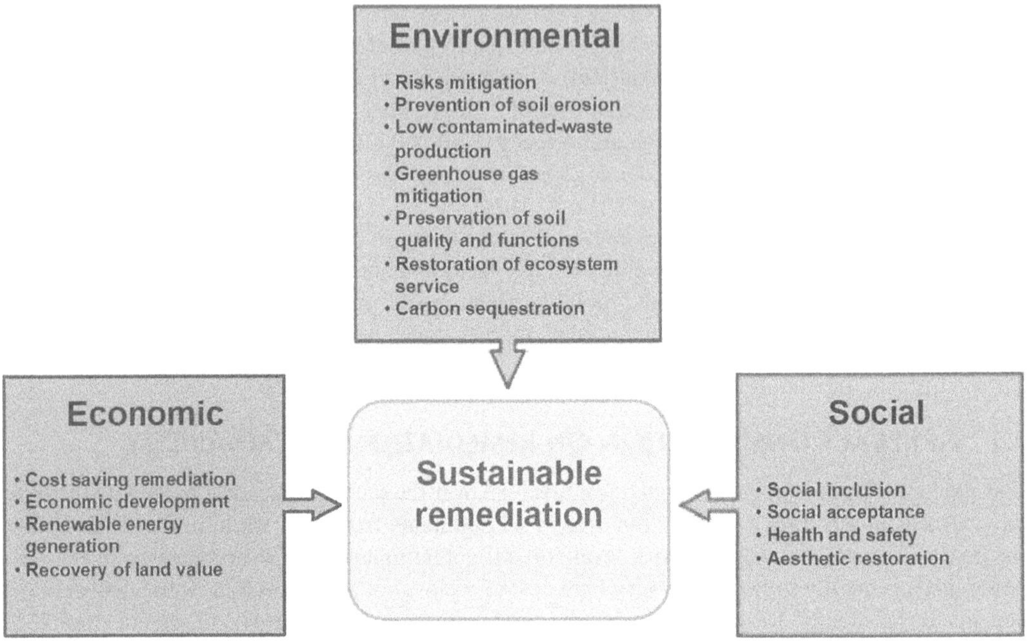

FIGURE 3.1 Main environmental, social, and economic issues of the sustainable remediation.

The various secondary impacts of remediation actions should be quantified by using a life cycle assessment (LCA) (Suèr et al., 2004; Hou and Al-Tabbaa, 2014). LCA methods quantitatively assess the total impact of a remediation plan, taking into account environmental and economic impacts and the time and spatial scales. It is applied before the remediation for choosing the best sustainable actions between various alternatives, and after the remediation to better assess the remediation impact with the collected data during the remediation treatment. The choice of remediation technology to apply, as well as the assessment of its impact, is always site-specific because it must consider the characteristics of both contaminants and site to be remediated (Suèr et al., 2004; Petruzzelli et al., 2016). Also, through this analysis, a sustainable remediation plan for a specific situation should demonstrate its advantage over classical remediation methods.

Green remediation has mainly focused on the achievement of the remediation objectives by low-energy remediation systems that preserve natural resources and minimize degradation of the site (USEPA, 2008). In situ bioremediation and phytoremediation are the most used technologies. They are cost-effective alternatives that allow treatment of a wide range of contaminants in different environmental compartments, with additional benefits of having low environmental impacts and preserving soil ecosystem services and having low and reusable waste production and high public acceptance (Petruzzelli et al., 2016; Wan et al., 2016; Gerhardt et al., 2017; Fiorentino et al., 2018; Hou and O'Connor, 2019). Plants are a "solar-driven, pumping and filtering system that has measurable loading, degradative, and fouling capacity." Roots are "exploratory, liquid-phase extractors that can find, alter and/or translocate elements and compounds against large chemical gradients" (Cunningham and Berti, 1993).

The use of plants to remediate polluted soil can have several positive effects. For example, some plants are used, alone or in combination with other plants, as cover crops. A cover crop is a plant (such as grasses, legumes, or forbs), which grows fast and remains unharvested in order to provide ground cover. Plant cover, beyond the functions of phytoremediation (phytoextraction or phytostabilization, described later in this chapter), can also secure a polluted site and limit the resuspension of

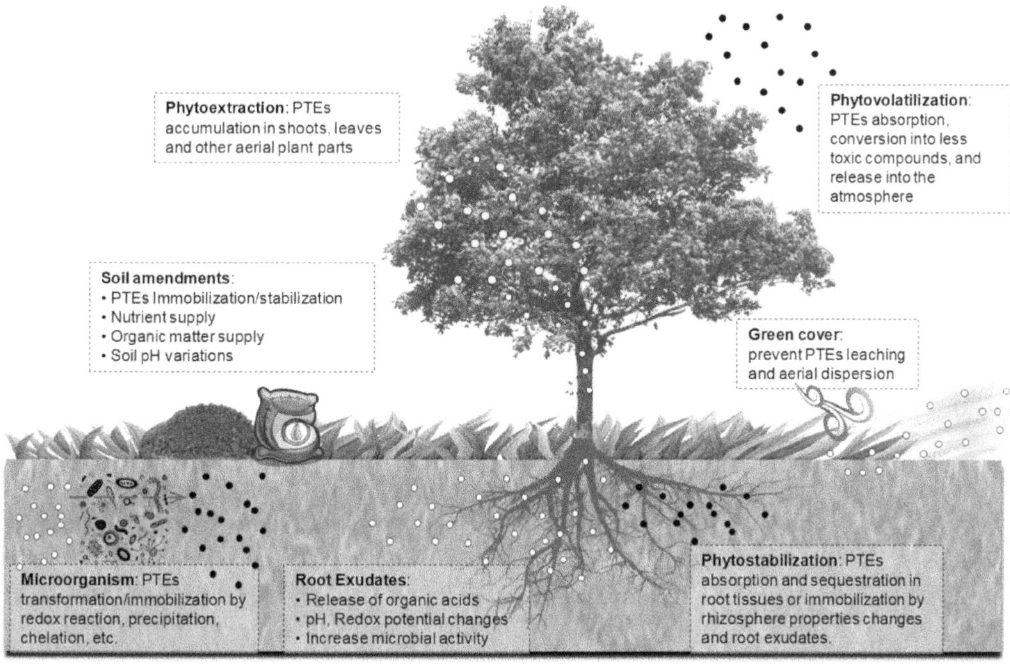

FIGURE 3.2 Main methods of green remediation to PTEs mobility and availability. The white dots represent soil contaminants; the black dots represent the transformation of PTEs into less toxic or less mobile forms.

metal(loid)s present on the soil surface in the form of dust and prevent wind dispersion of contaminants with consequent ingestion/inhalation by living organisms (Figure 3.2). It has a positive impact on the aesthetic aspect of the landscape and on soil fertility, increases organic matter, and improves soil structure, preventing soil erosion and leaching (Grifoni et al., 2017; Fiorentino et al., 2018).

However, if phytoremediation and other green remediation methods are considered as promising strategies, they are still not widely used because of some limitations, which are discussed in Section 3.4.

3.3 PRINCIPAL METHODS OF SOIL GREEN REMEDIATION

Soil is a non-renewable resource, and its new formation and development are closely related to very slow soil genesis factors (Washa et al., 2014). This means its degradation is not recoverable within a human lifespan, and for that reason, it is considered as non-renewable resource (FAO, 2015). Therefore, it is clear that the soil needs to be protected, and there is urgent need to tackle the problem of contaminated-soil recovery.

The soil acts as a long-term sink for PTEs released into the environment by anthropogenic activities, and, unlike organic contaminants, most metals do not undergo under microbial or chemical degradation. As a consequence, their total concentration in soil persists for a long time (from hundreds to thousands of years depending on the element and the properties of the soil) after their introduction (Megharaj et al., 2011; Bolan et al., 2014). Furthermore, changes in the chemical forms (speciation) and bioavailability of these elements are possible, and their presence in soil can also severely inhibit the biodegradation of organic contaminants (Wuana and Okieimen, 2011).

Nowadays, soil contamination by PTEs is a considerable environmental concern. Immobilizing or reducing the mobility and bioavailability of PTEs in soil is one way to reduce the damage of contamination. Immobilizing PTEs refers to reducing their mobility with the aim to minimize the accumulation of metal(loid)s in the food chain and reducing their leaching to groundwater (Bolan et al., 2014; Megharaj and Naidu, 2017). This can be achieved with techniques such as use of amendments or phytostabilization. Another way to reduce damage from contamination can be the removal of PTEs from the soil, and this can be achieved with techniques such as phytoextraction.

The above-mentioned techniques fall into the green-remediation concept, which can generally be described as the practice of implementing remedial actions in an environmentally friendly way, such as utilizing renewable energy sources wherever possible (Hadley and Harclerode, 2015).

In this chapter, ecofriendly, cost-effective approaches to remediate soils from environmental pollutants are discussed.

3.3.1 Chemical Remediation: Element Immobilization by Soil Amendments

The stabilization process involves chemical reactions that immobilize contaminants. The word "immobilization" is used to refer to the decrease in mobility, bioavailability, and bioaccessibility of PTEs in soil due to the redistribution of contaminants from the solution phase to the solid phase resulting from the addition of immobilizing agents (such as amendments) to the polluted soils (Bolan et al., 2014).

The amendments are able to reduce the bioavailability of PTEs in soils through several mechanisms, such as complexation, adsorption, precipitation, redox reactions, and electrostatic interaction. Nevertheless, also soil properties (i.e., soil pH, clay, oxides, and organic matter content) and processes (i.e., sorption/desorption and redox reactions) are strictly involved in the efficacy of amendments for the immobilization of PTEs in soils. The choice of the appropriate immobilizing agents can lead to a cost-effective remediation technique and meet green and sustainable remediation principles (Palansooriya et al., 2020; Bolan et al., 2014).

A wide range of novel stabilization materials (both organic and inorganic), such as biochar, ferrous sulfate, apatite, clay minerals, and fly ash, are currently being researched for this purpose (Sun et al., 2016; Rocco et al., 2018; Shen et al., 2019). Moreover, the addition of stabilizing agents (e.g., biochar) may also have beneficial effects on soil health, by improving soil structure, adding nutrients (N, P, and K), and increasing the capacity of the soil to hold water. In addition, these materials could also release reactants or microbes for enhanced remediation in the long term (Shen et al., 2019).

As a consequence of addition of amendments, the element solubility equilibria in soil are altered and metal(loid)s can be immobilized by complexation, precipitation, or adsorption reactions forming stable complexes (Sabir et al., 2015; Adamo et al., 2018). These processes cause the redistribution of elements from soil solution to solid particles limiting their mobility and bioavailability (Ashraf et al., 2014; Adamo et al., 2018), as shown in Figures 3.3 and 3.4. The application of soil amendments has proven effective at immobilizing PTEs in soils, based on physicochemical changes that occurred in the soil after application. This section will provide information on soil amendments and their potential for use in immobilizing PTEs in contaminated soils.

3.3.1.1 Organic Amendments

Biomaterials have been greatly used in recent years to immobilize metal(loid)s in soil due to their easy availability and low cost. It is well known that organic soil amendments improve soil structure and increase soil organic-matter content enhancing the ability of soil to retain nutrients and moisture, with consequent positive effects on plant growth and health. In addition, organic soil amendments, like biochar, compost, and animal wastes, have been identified as immobilizing agents for PTEs, because of their abilities to restrain PTEs through different mechanisms. However, over time, the effect of organic amendments on element bioavailability varies in relation to organic matter decomposition (Park et al., 2011; Ahmad et al., 2017; Shaheen et al., 2017a).

Some of the most common amendments used for the immobilization of PTEs in soil are biosolids, animal manures, and biochar (Houben et al., 2012; Palansooriya et al., 2020).

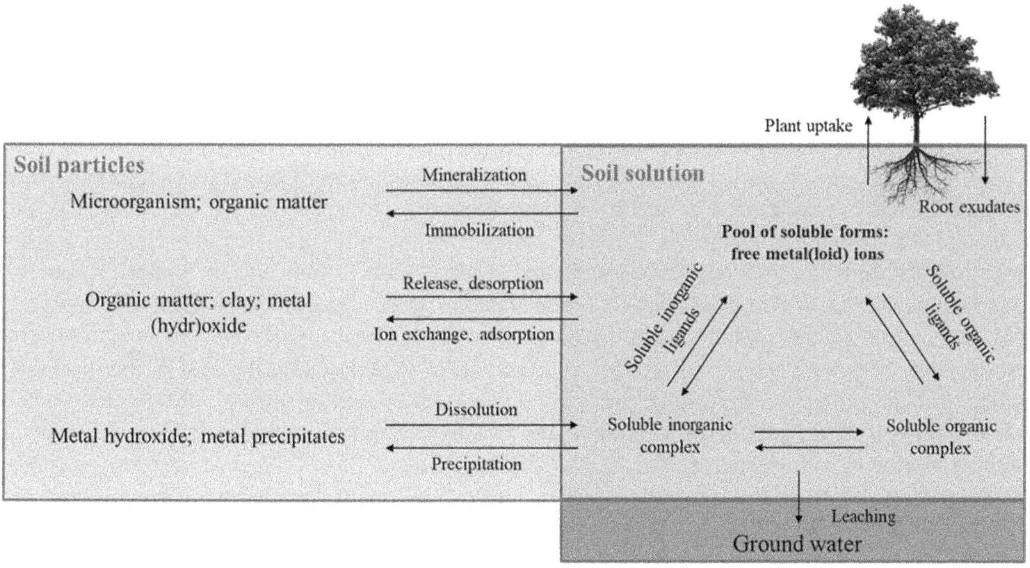

FIGURE 3.3 Illustrated diagram showing the main physical, chemical, and biological reactions of metal(loid)s in soil. (Modified from Seshadri et al., 2015.)

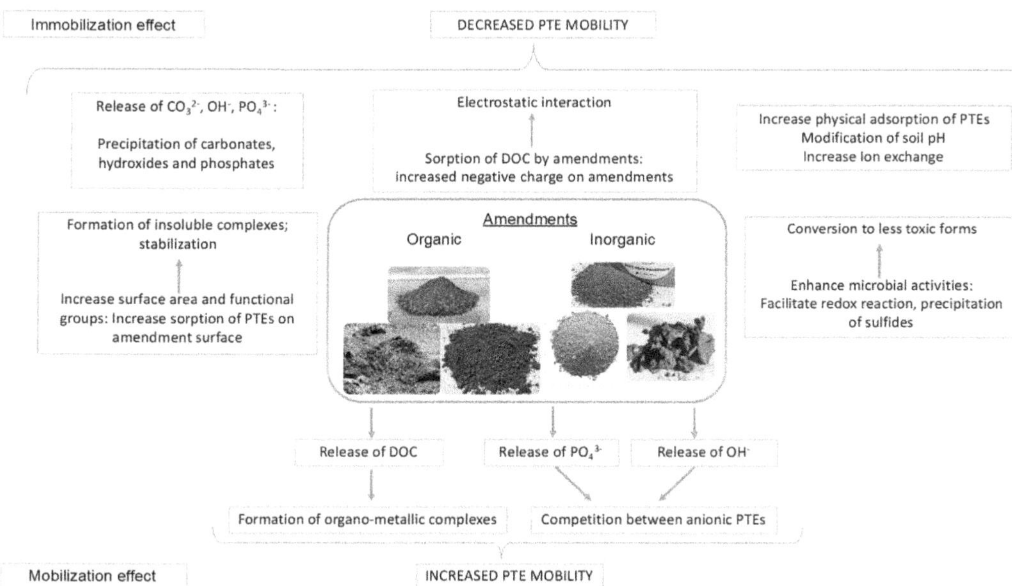

FIGURE 3.4 Effects of organic and inorganic soil amendments on potentially toxic element (PTE) (im) mobilitization. (Modified from Palansooriya et al., 2020.)

Biosolids (treated and stabilized before the use) are derived from wastewater and include a mix of water and organic materials that are a by-product of the sewage treatment processes (Girovich, 1996). Shaheen et al. (2017b) indicated that biosolids could be used to immobilize Cd, Cu, Pb, and Zn in soils with high metal content. Nevertheless, it must be considered that the application of biosolids to soils with a low-pollution level could increase metal desorption and consequently increase environmental risks. The prolonged use of sludges can cause metal accumulation in soils, even after the short-time application of biosolids (Cele and Maboeta, 2016; López-Mosquera et al., 2000). Oliveira and Mattiazzo (2001) observed increases in Cu, Cr, Ni, and Zn concentrations in soils amended for two years with biosolids. The accumulated metals from biosolid sources can also contaminate the food chain and reduce crop yields (Wang et al., 2003). Nevertheless, Mossa et al. (2020) evaluated the dynamics of PTEs in a long-term application of biosolids to an arable soil. Despite the large concentration of metals, measurement of isotopic exchangeability of Ni, Cu, Zn, Cd, and Pb demonstrated low levels of lability, supporting the hypothesis that biosolid constituents help immobilize PTEs. Indeed, when metals are added to soils as biosolids, a "plateau response" occurs in plant uptake. This plateau effect is connected to the presence in the biosolids of adsorptive materials, such as organic matter and amorphous Fe-oxides. These compounds are very stable and exhibit high adsorbing affinity to metals. The fraction of organic matter present in the biosolids, which provides protection against the uptake of PTEs, can resist decomposition (McBride, 1995).

Also, organic fertilizers, such as poultry and livestock manures, are nowadays used for metal immobilization in soils (Venegas et al., 2015, Shi et al., 2018).

Animal wastes from the dairy, beef cattle, swine, and poultry industries (e.g., cow and pig manure, chicken manures/litter) are commonly used by farmers as fertilizers, because of their high nutrient content. Although the application of animal-originated wastes is aimed at improving soil fertility, some of these amendments have been found to be suitable for reducing the mobility of PTEs in soil (Bolan et al., 2004; Rocco et al., 2018).

Some studies reported that the application of cow manure to contaminated soil significantly reduced leaching and phytoavailability of Cd and Zn in soil (Houben et al., 2012). Leaching of Cd was reduced by 63.1% with cow manure. Similarly, cow manure reduced Zn-soil-leachable concentrations by 89%. This could be explained by the high organic-matter content in the cow manure, which can bind PTEs via chemisorption and reduce their mobility (Zhao et al., 2014; Shaheen et al., 2017a). Khan et al. (2015) reported that the use of farmyard manure was highly effective in immobilizing Fe, Cr, Ni, Mn, and Pb. Moreover, organic amendments increase the organic content of soil (Bolan et al., 2014) and, with it, the occurrence of biomolecules such as cellulose, hemicellulose, lignin, proteins, lipids, starches, sugars, and other compounds containing several functional groups (i.e., carbonyl, phenolic, acetamido groups, amido, amino, structural polysaccharides, and esters) (Niazi et al., 2016), which are characterized by high affinity for metal complexation. Organic constituents of soil have a high affinity for metal(loid)s due to high-binding constants of metal ions with phenolic, hydroxyl, and carboxylic (–COOH) groups of dissolved organic matter (Khalid et al., 2016).

Immobilization of metal(loid)s by organic amendments can be related to the increase in soil pH by preventing sulfide oxidation/hydrolysis (Walker et al., 2004). Moreover, organic amendments can reduce PTE bioavailability as consequence of the formation of stable complexes with metal ions and increasing in surface charge (Khalid et al., 2016).

Nevertheless, PTEs such as As, Cu, and Zn are commonly added to animal feed as supplementary nutrients and growth promoters. As consequence, some organic fertilizers are a source of PTEs in soils (Niu et al., 2013). For example, the annual Zn input in China from livestock manures reached 96,000 t (about half of total Zn inputs to soil) (Luo et al., 2009). For that reason, livestock manures have become the most significant inputs of Cu and Zn into soils in Europe and China (Nicholson et al., 2003).

Among biomaterials, biochar has attracted great interest over the past few years because of its promising surface properties. Biochar is a carbon-rich, porous, purpose-produced charcoal manufactured during the pyrolysis of organic residues such as municipal wastes, animal wastes, wood, crop residues, and biosolids. It is a stable compound with high pH and high adsorption capacity, which results in a prominent increase in cation exchange processes (Lu et al., 2012). Several studies revealed that the addition of biochar to soil enhanced the sorption of some metal(loid)s and significantly reduced their mobility and phytoavailability (Al-Wabel et al., 2015; Rocco et al., 2018). For example, Park et al. (2011) reported that the application of biochar reduced the soil readily soluble concentrations of Cd, Cu, and Pb extracted by NH_4NO_3. Fellet et al. (2011) found also that applying biochar reduced the soil availability and mobility of Cd, Cr, Pb, and Zn. Biochar might immobilize metals by transforming the readily available fractions to geochemically more stable residual fractions, resulting in reduced mobility and bioavailability of PTEs (Al-Wabel et al., 2015).

Chen et al. (2019) showed that biochar application reduced plant uptake of Cd from soils. The maximum mobility reduction of Cd (92.7%) was observed in 2% pig biochar-treated soil. The plant responses to biochar applications indicated that less Cd was taken up by plants. Beesley et al. (2010) observed that, although As solubility and mobility were increased after biochar addition, plant uptake and toxicity risk were reduced.

Moreover, biochar addition to soil changes chemical, physical, and biological properties of soil (Figure 3.4). Changes in the soil properties, especially the increase of pH, can cause the precipitation of PTEs and thereby result in their immobilization in soils. Integrated application of appropriate amendments is also recommended to maximize their use efficiency (Palansooriya et al., 2020).

In general, the addition of organic amendments to soils polluted by Cr can be useful to maintain a reducing environment, thus favoring the shift of the Cr(III)-Cr(VI) equilibrium toward the reduced trivalent form, which is much less mobile and dangerous, compared to chromium in the hexavalent form (Agrelli et al., 2020a).

3.3.1.2 Inorganic Amendments

Many different inorganic soil amendments have been used to immobilize PTEs in contaminated soils. Some of the most commonly used in soil remediation are clay minerals, phosphates, and coal fly ash.

Clay minerals, relatively abundant and low-cost materials, are hydrous aluminosilicates that make up the colloidal fraction of soils and sediments (Xu et al., 2017). The addition of clays has been classified as an effective remediation method for contaminated soil, water, and other media (Otunola and Ololade, 2020), and they play an important role in the environment by acting as a natural scavenger of pollutants through the retention of cations and anions (Xu et al., 2017). Indeed, clay minerals, such as kaolinite, illite, zeolite, sepiolite, bentonite, and vermiculite, can act as sinks for PTEs through mechanisms such as ion exchange and specific adsorption (Palansooriya et al., 2020). Isomorphous substitution leads some phyllosilicate minerals, such as smectite and vermiculite, to have permanent negative charges on their surfaces, whereas some clay minerals have charges that are variable or pH-dependent (Bolan et al., 1999). Also, the different origin of the clay surface charge alters the bonding between various metal(loid)s and clay mineral surfaces. Moreover, metal sorption on a given soil component is influenced by the co-existence of other phases, which may change the surface properties and the metal sorption (Violante and Pigna, 2002).

Clay minerals have been widely used for in situ immobilization of PTEs in soil, especially in paddy soil polluted by Cd (Xu et al., 2017). Several authors (Li and Xu 2015; Sun et al., 2015; Liang et al., 2016; Sun et al., 2016) found a significant reduction of Cd uptake in both soils and plants after clay mineral (such as sepiolite and bentonite) application. Clay minerals can increase soil pH, decrease the chemical-extractable fractions and bioavailability of elements in soils, and reduce the content of metal(loid)s in the edible parts of the plant.

Coal fly ash (CFA) is a by-product of coal combustion. It is a low-cost alkaline or acidic material often considered as a problematic solid waste, mainly due to its content of PTEs such as Cd, Cr, Ni, and Pb (Khalid et al., 2016). Nevertheless, the favorable physicochemical properties of fly ash, such as pH, high water-holding capacity, low bulk density, and the presence of substantial plant nutrients (e.g., Ca, Mg, K, P, S, B, Fe, Cu, and Zn), make it a potential soil ameliorant (Pandey and Singh, 2010), and it has also been widely used in the remediation of soil contaminated by PTEs (Wang et al., 2020). However, the effects of CFA on the stability of PTEs in contaminated farmland soil still have not been clearly evaluated (Wang et al., 2020).

Several authors (Janoš et al., 2010; Patra et al., 2012; Wang et al., 2020) reported that the application of alkaline CFA, with or without addition of other amendments, can help to remediate

PTE-contaminated soils by immobilization of mobile metal forms. For example, a greenhouse pot experiment demonstrated that applying CFA reduced soluble and exchangeable Pb (14.7%) in contaminated soil, leading to a reduction (66.1%) of Pb in rapeseed plant tissue (Shaheen and Rinklebe, 2015). Houben et al. (2012) observed that CFA reduced Cd, Zn, and Pb leaching by 32.9%, 41.2%, and 25%, respectively, mainly through increased soil pH. A review conducted by Shaheen et al. (2014) found that the presence of alumina, calcium, magnesium and ferric oxide, silica, and carbon in fly ash may be involved in the retention of PTEs, as those constituents are strong metal sorbents. However, given the high variability in the nature and composition of CFA (e.g., pH, content of macro- and micro-nutrients) and soil (pH, texture, and fertility), the choice of CFA and its application rate need to consider the properties and problems of the soil (Shaheen et al., 2014).

Phosphate amendments (e.g., apatites, hydroxyl apatites, diammonium phosphate, phosphate-based salts, phosphoric acid, phosphate rocks), commonly used to improve soil fertility, are well known to be able to effectively immobilize some metal contaminants in soil. Phosphates can reduce the bioavailability of metals by altering soil properties such as pH and surface charge, or by reacting directly with metal ions, thus increasing metal sorption and precipitation and forming insoluble minerals. Lead, zinc, copper, and cadmium are the most studied PTEs immobilized by phosphates (Yan et al., 2015; Palansooriya et al., 2020). Lead in presence of soluble phosphate ions precipitates as pyromorphite, a very stable compound over a wide soil pH range. Sanderson et al. (2018) cited several studies that examined the effect of various phosphate treatments for the immobilization of Pb, in particular in shooting-range contaminated soils, and they also highlighting some criticisms such as the leaching of phosphate ions, dangerous for the environment, or the influence of environmental conditions and the type of phosphorous amendments on the success of the treatment.

3.3.2 Biological Remediation

Bioremediation is an alternative to traditional methods of removal of contaminants using natural processes. It is one of the most feasible options to remediate and re-establish the natural condition of contaminated soil that has been considered to be damaging to environmental and human health. "Bioremediation" is defined as the process of making use of various biological agents, microorganisms, or plants, to degrade environmental contaminants into less toxic forms or of removing or stabilizing PTEs from soil (Van Dillewijn et al., 2007; Khalid et al., 2016). It is a widely accepted method, mostly carried out in situ; hence, it is suitable for the reestablishment of crops on treated soils (Chibuike and Obiora, 2014). This approach acts as an option for making a clean and safe environment and its resources by moving away, destroying, or mitigating the dangerousness of various contaminants using natural biological activity. Indeed, metal(loid)s can be naturally transformed from one organic complex or oxidation state to another or can be phytoextracted or phytostabilized. For example, due to a change in their oxidation state, PTEs can be transformed into nontoxic or less toxic, easily volatilized, and less water-soluble (which allows them to precipitate and become easily removed from the environment) forms or less bioavailable (Garbisu and Alkorta, 2003) forms. For all these reasons, bioremediation is considered as being a cleaner, noninvasive, cost-effective, and environmental friendly technology that has a high public acceptance. It can be carried out at any site and provides a permanent solution (Sharma and Pandey, 2014). However, success of this approach largely depends on the species of organisms involved in the process. Although bioremediation is a nondisruptive method of soil remediation, it is usually time-consuming, and its use for the treatment of polluted soils is sometimes affected by the climatic and geological conditions of the site to be remediated (Schmoger et al., 2000).

Biological remediation includes several techniques, such as bioremediation, phytoremediation, bioventing, bioleaching, land forming, bioreactors, composting, bioaugmentation, and biostimulation. Among these methods, bioremediation and phytoremediation are the most useful techniques, because they preserve natural soil properties and depend on solar energy (Beskoski et al., 2011; Ullah et al., 2015).

3.3.2.1 Microbial-Based Bioremediation

The activities of microbes can naturally detoxify or "transform" contaminants and consequently remediate soil through different mechanisms, such as redox reactions, bioprecipitation, extracellular complexation, and intracellular accumulation (Yao et al., 2012; Sharma et al., 2019). The ability of microorganisms to break down hazardous contaminants into nontoxic or harmless forms mainly depends on the availability of nutrients, electron donors, and acceptors. Nitrogen and phosphorus are the most important nutrients for microbial growth, and they are usually provided as ammonia and orthophosphate. Oxygen is the electron acceptor par excellence in bioremediation. For example, under aerobic conditions, organic contaminants are transformed into CO_2, H_2O, and microbial cell mass. In the absence of O_2, many microorganisms utilize alternative compounds as electron acceptors, such as nitrate, sulfate, iron, manganese, and CO_2 (Megharaj et al., 2014). Microbial interaction with the PTEs in soil is driven by several factors, such as type and nature of the elements, their concentration, and availability; type of microbial species and their modification in microbial communities causing metabolic interferences (like changes in active conformations of biological molecules); inhibition by blocking essential functional groups and displacement of essential metal ions (Dixit et al., 2015; Sharma et al., 2019).

Transformation of metals, as a consequence of changes in their oxidation state, promotes remediation by modifying their toxicity and bioavailability, making elements more soluble (removal through bioleaching or phytoextraction) or less soluble (removal through bioprecipitation) in water and easy to volatilize (Garbisu and Alkorta, 2003; Chibuike and Obiora, 2014). For example, the solubility of metals like Fe and Mn increases after oxidation-reduction reactions aided by microorganisms, due to the reduction of Fe(III) to Fe(II) and Mn(IV) to Mn(II), respectively. Microbial reduction of soluble mercuric ion Hg(II) to volatile elemental mercury Hg(0) by bacteria and fungi is a very important in situ remediation strategy for Hg contaminated soil (Lovley and Lloyd, 2000).

Bioremediating metals from the soil needs solubilization of insoluble metal compounds (such as phosphates, oxides, sulfides) or their mobilization through chelation, chemical transformation, or protonation (Chibuike and Obiora, 2014). Microorganisms like *Bacillus* (*B. cereus* and *B. thuringiensis*), *Pseudomonas* (*P. aeruginosa* and *P. putida*), and *Azotobacter* can produce siderophores, a complexing molecule with high affinity for iron and other metals, which facilitate the extraction of elements like Cd, Zn, and Pb from soil (Tripathi et al., 2005; Ajaz et al., 2010; Naik and Dubey, 2011). For example, Ajaz et al. (2010) found that *B. cereus* and *B. thuringiensis* increased the phytoextraction of Cd and Zn from soil polluted with effluent from a metal industry. It is assumed that the production of siderophores by bacteria may have facilitated the extraction of these metals from the soil. Several other microorganisms, especially bacteria (i.e., *Bacillus subtilis*, *Bacillus circulans*, *Pseudomonas putida*, and *Enterobacter cloacae*) have been successfully used for the reduction of Cr(VI) to the less toxic Cr(III) (Kanmani et al., 2012; Chibuike and Obiora, 2014).

Biomethylation is another effective process carried out by a range of microorganisms under both aerobic and anaerobic conditions. This mechanism involves the enzymatic transfer of methyl groups to metal(loid)s, like As, Pb, Hg, Se, Sn, and Te, considerably modifying their solubility, volatility, and toxicity (Sharma et al., 2019). Majumder et al. (2013) showed that some isolated bacterial strains, comprising As volatilizing indigenous bacteria, volatilized 37% of As(III) (under aerobic conditions) and 30% As(V) (under anaerobic conditions) in 3 days.

In order to make the soil favorable for soil microorganisms' growth and turnover, it is common to add to the soil C and nutrients in the form of manure or other organic amendments. This process is known as biostimulation. Indeed, the addition of nutrients increases the growth and activities of microorganisms involved in the remediation process, and, thus, this increases the efficiency of bioremediation of both organic and inorganic soil pollutants (Chibuike and Obiora, 2014; Megharaj et al., 2014).

3.3.2.2 Phytoremediation

Phytoremediation, also known as vegetative remediation or green remediation, is one of the safer, most promising, noninvasive, economical, environment-friendly, attractive, and aesthetically pleasing in situ technologies where plants are used to remediate, recover, and revegetate soils polluted by organic and inorganic compounds (Bolan et al., 2011; Sabir et al., 2015; Yan et al., 2020). Plants reduce the mobility and transport of pollutants in the environment, either by uptake or immobilization processes (Figure 3.2) (Bolan et al., 2011). The efficiency of phytoremediation depends on numerous factors, such as the physicochemical properties of the soil, bioavailability of metals in soil, microbial and plant exudates, and the capacity of living organisms to take up, accumulate, translocate, and detoxify PTEs (Khalid et al., 2016; Adamo et al., 2018). Plants that have evolved mechanisms to tolerate high concentrations of metal(loid)s, named "metallophytes," are considered an optimal choice for phytoremediation (Boularbah et al., 2006). Metallophytes can be classified as: (i) accumulators: if they show an active metal uptake and translocation to aerial parts; (ii) indicators: if they regulate metal uptake; this means that the internal concentrations reflect external soil concentrations; and (iii) excluders: if they limit the entry of metals into the root and/or their transport to the shoots (Barrutia et al., 2011). Some metallophytes, known as hyperaccumulators, have specialized mechanisms that enable them to accumulate metals at a concentration of over 1% of their dry weight, and, in some cases, it reaches up to 10% (Burges et al., 2018).

The term phytoremediation includes several processes or mechanisms where plants can immobilize, remove, or degrade metals in soil. Based on different uptake mechanisms, phytoremediation is generally categorized into phytostabilization (the use of tolerant plants to reduce metal bioavailability), phytoextraction (the use of plants to extract metals from soil), and phytovolatilization (the use of plants to take up metals from soil, their conversion to a volatile form, and release into the atmosphere) (Ali et al., 2013; Burges et al., 2018; Yan et al., 2020).

3.3.2.2.1 Phytostabilization

Phytostabilization (Figure 3.2) is the use of plants to immobilize metal(loid)s belowground and/or decrease their bioavailability and mobility in soils (Sylvain et al., 2016), preventing their migration into the ecosystem and reducing the likelihood of metals entering the food chain (Marques et al., 2009). Phytostabilization of PTEs does not reduce the concentration of metal(loid)s in soil but prohibits their movement to the surrounding areas. The vegetation prevents the dispersion of metal-containing soil particles by wind and promotes their accumulation in the vadose zone of plants through accumulation (by roots) or precipitation (in the rhizosphere) (Bolan et al., 2011; Abbas et al., 2016). Phytostabilization can occur through precipitation of PTEs due to binding by organic compounds, metal valence reduction in the rhizosphere, absorption and sequestration within root tissues, or adsorption onto root cell walls (Wuana and Okieimen, 2011; Kumpiene et al., 2012; Gerhardt et al., 2017). For example, metals of different valences differ in toxicity. Plant roots, by excreting redox enzymes, can convert hazardous metal(loid)s to a less toxic state, decreasing possible stress and damage caused by PTEs. Indeed, plant-root reduction of Cr(VI) to Cr(III) is widely studied, because the latter is less mobile and less toxic than Cr(VI) (Wu et al., 2010).

A crucial aspect to have effective phytostabilization is the choice of the most appropriate plant species. To this aim, plants (i) should be tolerant to the metal(loid) conditions; (ii) should have dense rooting systems (plant roots play a pivotal role to immobilize metals, stabilize soil structure, and prevent soil erosion); and (iii) should be able to produce a large amount of biomass and grow fast to establish in a timely manner a vegetative cover at a specific site. In addition, the plant should be adapted to the local environmental conditions and easy to maintain under field conditions (Marques et al., 2009).

The most effective plants for phytostabilization are species able to accumulate high levels of PTEs in their belowground parts, with limited transport to their aerial parts, which inhibits the release of PTEs into food (Ali et al., 2013). During the last few decades, several plant species, particularly the agricultural grasses like *Agrostis* spp. and *Festuca* spp., have been successfully used in the phytostabilization of soils in Europe, China, and America polluted with Cu, Zn, and Pb (Dickinson et al., 2009; Galende et al., 2014). Conesa et al. (2007) demonstrated the possibility of successfully using *Piptatherum miliaceum*, *Zygophyllum fabago*, and *Lygeum spartum* for green cover over a neutral mining sludge containing up to 5,200 mg/kg Pb and 9,100 mg/kg Zn (southern Spain). Also, plants like willows (*Salix* spp.) have been considered effective for both phytostabilization and phytoextraction owing to their elevated tolerance to metal(loid)s, high territorial expansion, and intense evapotranspiration (Sylvain et al., 2016).

Plants can immobilize PTEs in soils also by affecting changes in the rhizosphere (the few millimeters of soil surrounding plant roots), which has proper, distinct physical, chemical, and biological conditions compared to the bulk soil. The rhizosphere is a highly dynamic microenvironment characterized by the interactions between root processes, soil characteristics, and the dynamics of the associated microbial populations (Bolan et al., 2011). Root exudates stimulate microbial activity and biochemical transformations in the rhizosphere (Paterson, 2003). The chemical and biological reactions occurring in the rhizosphere play an important role in the bioavailability of PTEs. Plant roots change the physico-bio-chemical conditions of the soil in the rhizosphere, which is enriched with organic substances from plant and microbial populations, including organic acids, sugars, amino acids, lipids, coumarins, flavonoids, proteins, enzymes, aliphatics, aromatics, and carbohydrates (Shukla et al., 2011). Common organic acids in the rhizosphere are acetic, butyric, citric, fumaric, lactic, malic, malonic, oxalic, propionic, tartaric, and succinic acids. Organic acids in the rhizosphere affect the dynamics of PTEs in soils via their effect on acidification, chelation and complexation of metal(loid)s, precipitation, redox reactions, microbial activity, rhizosphere physical properties, and root morphology (Bolan et al., 2011). Plants secrete compounds able to immobilize contaminants using redox processes or to precipitate insoluble compounds in the rhizosphere. For example, Pb can be precipitated as phosphate, and Cd forms complexes with sulfides, in the roots and the rhizosphere of *Agrostis capillaris* and *Silene vulgaris*, respectively (Bolan et al., 2011).

In order to enhance the immobilization of PTEs in soil, phytostabilization can be assisted by other remediation techniques, such as the use of soil microorganisms, like bacteria and mycorrhiza, and the addition of amendments. Rajkumar et al. (2013) reported that soil microorganisms can reduce the mobility of PTEs, both increasing the content of root metal(loid)s due to an increased growth as well as immobilizing the metal(loid)s in soil. In addition, they can also increase plant root surface and depth to facilitate phytostabilization, and they even serve as a filtration barrier against metal ion translocation from roots to shoots (Yan et al., 2020). One of the advantages of phytostabilization is that disposal of hazardous biomass is not required when compared with phytoextraction (Wuana and Okieimen, 2011).

Moreover, organic and inorganic amendments facilitate plant-induced immobilization of metal(loid)s in soil promoting the precipitation and the sorption of PTEs (Parra et al., 2016). These soil amendments can reduce the solubility and bioavailability of metal(loid)s by changing the pH value and redox status of the soil (Alvarenga et al., 2009; Burges et al., 2018). Moreover, the application of amendments can increase soil organic matter and content of nutrients and improve physicochemical and biological properties, which can benefit plant colonization and improve water-holding capacity.

Phytostabilization is generally used for soils where phytoextraction is not desirable or possible. It can also be used at sites where technical or regulatory limitations do not allow other traditional remediation techniques. Phytostabilization does not produce contaminated secondary waste that needs further management, and it is helpful in achieving ecosystem restoration and in increasing soil fertility. However, because the metal(loid)s are stabilized within the soil, the site needs regular monitoring to make sure that the optimal stabilizing conditions are ensured and retained over the time (Bolan et al., 2011).

3.3.2.2.2 Phytovolatilization

Phytovolatilization or phytoevaporation (Figure 3.2) refers to plant uptake of pollutants from soil, their conversion to less toxic and volatile forms in the plant, and subsequent release into the atmosphere. In the phytovolatilization process, the pollutants are not completely fixed but only transferred from soil to the atmosphere from where they can be redeposited (Mahar et al., 2016). This method is mainly applied to organic pollutants and, in minor part, to some metal(loid)s. In detail, during phytovolatilization, inorganic PTEs are assimilated into volatile organic compounds (volatile hydride and methyl compounds), which are then released into the atmosphere as biomolecules (Marques et al., 2009). Metal(loid)s, such as Se, Hg, and As, may exist as gaseous species in the environment. Some plant species, such as *Arabidopsis thaliana*, *Brassica juncea*, and *Chara canescens*, can take up and volatize metal(loid)s, releasing them into the atmosphere, and effectively cleaning up the polluted soil (Padmavathiamma and Li, 2007; Verbruggen et al., 2009). For example, As was effectively volatilized in the form of As compounds (arsenite and arsenate) by *Pteris vittate* (Sakakibara et al., 2010). Ghosh and Singh (2005) showed that phytovolatilization was successfully used for removing Hg by converting it into gaseous Hg, a less toxic form. *Brassica juncea* can grow under high Se levels and is capable of volatilizing it (Padmavathiamma and Li, 2007). The conversion of metal(loid)s into gaseous and volatilized forms inside plants involves a specific mechanism controlled by some specific enzymes or genes. There are very few plant species that naturally convert PTEs into volatilized forms. For that reason, the phytovolatilization technique is generally performed using genetically modified plants to enhance the ability of plants to volatilize PTEs. For example, *Arabidopsis thaliana* and *Nicotiana tabacum* have been genetically engineered with mercuric reductase, a gene that can volatilize Hg, by introducing Hg reductase genes from bacteria (merA and merB) (Meagher, 2000; Khalid et al., 2016). One drawback of volatilization is that there is no control on the destination of the contaminants (Palansooriya et al., 2020).

3.3.2.2.3 Phytoextraction

Phytoextraction, also known as phytoaccumulation, phytoabsorption, or phytosequestration, is the clean-up of PTEs present in soil by means of plant uptake. This technique is based on the ability of plant roots to take up, translocate, and concentrate or accumulate metal(loid)s from soil to the aboveground parts of the plant without phytotoxicity symptoms, followed by harvesting and then disposal of the plant biomass (Rafati et al., 2011; Wang et al., 2019a) or reuse of the harvested biomass for energy production (Sharma, 2018). The adsorbed metal(loid)s are not destroyed by plants, but they are accumulated in shoots, leaves, and other plant parts (Rashid et al., 2014), as illustrated in Figure 3.2. For effective phytoextraction, translocation of metal(loid)s to shoots is a crucial and desirable process, because plant biomass is very easy to remove, recycle, dispose, treat, or oxidize compared to harvesting the root biomass (often not feasible) and to soil removal and disposal (Tangahu et al., 2011; Khalid et al., 2016). Phytoextraction is the main and most useful phytoremediation technique for the decrease and permanent removal of PTEs from polluted sites (Khalid et al., 2016). However, this technique is preferentially, but not exclusively, suitable to sites with low to moderate levels of metal(loid)s, because most plant species are not able to tolerate heavily polluted sites (Sabir et al., 2015). Phytoextraction is widely utilized for metallic (i.e., Pb, Zn, Cu, As, Ni, Cr, Cd) and radionuclide wastes (Muthusaravanan et al., 2018; Wang et al., 2019b).

The efficiency of phytoextraction depends on many factors, like bioavailability of the PTEs in soil, soil properties, and the plant species involved (Tangahu et al., 2011). Plants suitable for phytoextraction should have some specific characteristics, such as the following (Ali et al., 2013; Yan et al., 2020):

a. high growth rate;
b. high aboveground biomass production;
c. profuse root system;
d. tolerance to the toxic effect of the target PTEs;

e. high accumulation rate of PTEs in the foliar part;
f. high translocation rate of the PTEs from roots to shoots;
g. easy cultivation and harvest and good adaptation to the environmental and climatic conditions;
h. resistance to pathogens and pests;
i. repulsion to herbivores to avoid food chain contamination.

Two different approaches have been tested for phytoextraction: (i) the use of hyperaccumulators, which produce less aboveground biomass but accumulate PTEs to a greater extent; or (ii) the application of other plants, such as *Brassica juncea* and *Brassica napus*, which accumulate target metals to a lesser extent but produce more aboveground biomass so that overall accumulation is comparable to that of hyperaccumulators due to production of more biomass (Rizwan et al., 2018). However, hyperaccumulation and hypertolerance are more important in phytoremediation than high biomass. The use of hyperaccumulators will yield a metal-rich, low-volume biomass, which is economical and easy to handle for both metal recovery and safe disposal (Ali et al., 2013).

According to Yan et al. (2020), hyperaccumulator plants should meet the following criteria: (i) the shoot-to-root ratio of PTE concentration is greater than 1, which is a sign of efficient ability to transport metals from roots to shoots; (ii) the shoot-to-soil ratio of PTE concentration is greater than 1, indicating a higher capability to take up metals from soil; and (iii) the concentration of the metal in the shoot is higher than 10 mg/kg for Hg, 100 mg/kg for Cd and Se, 1,000 mg/kg for Co, Cu, Cr, Ni, and Pb, and 10,000 mg/kg for Zn and Mn. Searching for effective hyperaccumulators is a key, and the most straightforward strategy, for successful phytoremediation of metals. Currently, more than 450 plant species from at least 45 angiosperm families have been identified as metal(loid) hyperaccumulators (Suman et al., 2018), ranging from annual herbs to perennial shrubs and trees, such as Brassicaceae, Fabaceae, Euphorbiaceae, Asterraceae, Lamiaceae, and Scrophulariaceae (Yan et al., 2020). Some species can even accumulate more than two elements, such as *Sedum alfredii*, which can hyperaccumulate Zn, Pb, and Cd (Yang et al., 2004). Plants that offer multiple harvests in a single growth period (like *Trifolium* spp.) can have a great potential for phytoextraction of PTEs (i.e., Cd, Pb, Cu, and Zn) (Ali et al., 2012).

Nevertheless, high biomass producing crops, such as *Helianthus annuus*, *Cannabis sativa*, *Nicotiana tabacum*, and *Zea mays*, have been reported to effectively remove metals from contaminated soil through phytoextraction (Meers et al., 2005; Herzig et al., 2014; Rocco et al., 2018; Yan et al., 2020). Grasses are preferable for phytoextraction compared to shrubs or trees, because of their high growth rate, more adaptability to stressed environments, high tolerance to abiotic stresses, and high biomass (Malik et al., 2010). However, using edible crops (i.e., maize or barley) for phytoremediation should be avoided as metals can accumulate in edible parts of the plant and thus enter the food chain by human or animal consumption, raising concerns for human health (Ali et al., 2013; Chen and Li, 2018). Hence, selection of the non-edible hyperaccumulators is a key for efficient and safe phytoremediation of PTEs (Yan et al., 2020). Grasses employ various mechanisms to take up metal(loid)s from soil. For example, wheat roots produce phytosiderophores, biomolecules able to enhance Fe mobilization in soil and consequently the plant acquisition of Fe. Phytosiderophores also form chelates with Zn, Cu, and Mn and facilitate the solubilization of these metals in calcareous soils (Bolan et al., 2011; Oburger et al., 2014). Therefore, nonessential metal(loid) ions with a similar size to nutrients may be taken up into the symplast and ultimately be translocated to the shoots. Moreover, plants take up arsenate (AsO_4^{3-}) using the same physiological mechanism as phosphate (PO_4^{3-}); Tl^+ may enter via the K^+ ion channel; Cd^{2+} may enter via either the Ca^{2+} or Zn^{2+} transport system; and Ni^{2+} may enter along with Mg^{2+} (Bolan et al., 2011). Moreover, root secretion of H^+ ions can acidify the rhizosphere increasing metal dissolution. Indeed, H^+ ions can displace metal cations adsorbed to soil particles (Alford et al., 2010). Root exudates can lower the rhizosphere soil pH generally by one or two units over that in the bulk soil. Lower soil pH increases the concentration of metals in solution by promoting their desorption (Ali et al., 2013). Furthermore, rhizosphere microorganisms (mainly bacteria and mycorrhizal fungi) may significantly increase the bioavailability of metal(loid)s in soil (Sheoran et al., 2011). Interactions of microbial siderophores can increase labile metal pools and uptake by roots (Mench et al., 2009).

The metal(loid) removal efficiency of phytoextraction can be calculated using Equation 3.1 (Chen and Li, 2018). The phytoextraction efficiency varies with different metal elements, plant species, soil properties, and amendment applications (Tangahu et al., 2011; Mohd et al., 2013).

$$\mathrm{Eff}_{\mathrm{phyto}} = \frac{M_{\mathrm{plant}} \times C_{\mathrm{plant}} \times Y}{M_{\mathrm{soil}} \times C_{o,\,\mathrm{soil}}} \times 100\% \tag{3.1}$$

where $\mathrm{Eff}_{\mathrm{phyto}}$ is the metal(loid) removal efficiency; M_{plant} is the average harvestable biomass of plant used for phytoextraction; C_{plant} is the average metal(loid) concentration in the harvestable part of plant; Y is the treatment period of phytoextraction, generally in years; M_{soil} is the mass of contaminated soil; $C_{o,\,soil}$ is the original metal(loid) concentration in the polluted soil.

3.3.2.2.4 Phytoremediation by-Product Management

As previously described, phytoextraction is the uptake of PTEs from soil by plant roots and their translocation and accumulation in shoots or leaves. This method may involve repeated cropping of plants in polluted soil until the concentration of PTEs drops to tolerable levels. After each cropping, the entire plant or the aboveground part is removed from the site; this leads to accumulation of a huge amount of hazardous biomass. Usually, the biomass can be stored or disposed appropriately, so that it does not pose any risk to the environment. However, handling of a huge amount of this type of waste is a problem, and hence, volume reduction is needed (Blaylock and Huang, 2000). During the last decades, composting and compaction have been proposed as post-harvest biomass treatment by some authors (Garbisu and Alkorta, 2001; Ghosh and Singh, 2005; Sharma, 2018). Studies carried out by Hetland et al. (2001) showed that composting can significantly reduce the volume of harvested biomass; however, metal-contaminated plant biomass would still require treatment prior to disposal. Total dry weight loss of contaminated plant biomass by compaction is advantageous, because it will lower the cost of transportation to a hazardous waste disposal facility. Compaction of harvested plant material was proposed by Blaylock and Huang (2000) for processing metal-rich residue from phytoextraction. One of the ordinary and promising courses, to use biomass produced by phytoremediation in an integrated way, is through a thermochemical conversion process. If a phytoextraction technique is combined with biomass generation, it also can lead to a profitable energy source, and the leftover ash can be used as bio-ore (Brooks et al., 1998). Nicks and Chambers (1994) reported a second potential use for hyperaccumulator plants for economic gain in the mining industry. This operation, named phytomining, includes the generation of revenue by extracting saleable PTEs from the plant biomass ash, also known as bio-ore. Combustion and gasification are the most important subroutes for organized generation of electrical and thermal energy. Recovery of this energy from biomass by burning or gasification could help make phytoextraction more cost-effective, and save energy, if one is considering a global balance of the remediation plan. Thermochemical energy conversion best suits the biomass residue from phytoextraction, because it cannot be utilized in any other way as fodder or fertilizer. Combustion is a crude method of burning the biomass, but it should be under controlled conditions, whereby volume is reduced to 2%–5% and the ash can be disposed properly (Raskin et al., 1997; Bridgwater et al., 1999). It will not be favorable to burn the metal-bearing hazardous waste in the open, because the gases and particulates released in the environment may be detrimental; only the volume is reduced and the heat produced in the process is wasted. Gasification is the process through which biomass material can be subjected to series of chemical changes to yield clean and combustive gases at high thermal efficiencies. This mixture of gases, called producer gas and/or pyro-gas, can be combusted for generating thermal and electrical energy. The process of gasification of biomass in a gasifier is a complex phenomenon; it involves drying, heating, thermal decomposition (pyrolysis) and gasification, and combustion chemical reactions, which occur simultaneously (Iyer et al., 2002). Future experiments should concentrate on the development of combustion systems and methods to recycle different metals from ash. The liberated metals remain in the slag. Modern flue gas cleaning technology assures effective capture of the metal-containing dust. Considering the other technologies for disposal, this method is environment-friendly (Ghosh and Singh, 2005; Sharma, 2018). Pyrolysis is one of the promising methods of municipal waste treatment that might also be used for contaminated plant material (Bridgwater et al., 1999). Pyrolysis decomposes material under anaerobic conditions; there is no emission to the air. The final products of pyrolysis are pyrolytic fluid oil and coke. The metal(loid)s will remain in the coke and could be used in smelter. Koppolu et al. (2003) reported that the metal(loid) recovery was about 99% in the char that was formed when the synthetic hyperaccumulator biomass was pyrolyzed in a pilot-scale reactor. The metal component was concentrated in char by 3.2–6 times as compared to the feed. Helsen et al. (1997) conducted low-temperature pyrolytic experiments with chromium-, copper-, and arsenate-treated wood and concluded that most of the metal was retained in the pyrolytic residue. The large expense of establishing a facility to treat an element is prohibitive if it utilized exclusively for plant transfer. To stay away from this, plant material can be prepared in existing establishments together with metropolitan waste (Singh and Ghosh, 2003).

3.3.3 Assisted Phytoremediation

Phytoremediation can be used effectively in combination with other traditional remediation techniques (Khalid et al, 2016). As mentioned before, in natural phytoextraction, plants remove PTEs from soil under natural conditions, i.e., no soil amendment is added, but phytoextraction can be limited by low availability of the metal and, consequently, low uptake and translocation. To cope with this limitation, an induced or chelate-assisted phytoextraction can be applied. In order to increase the bioavailability of PTEs in soil for plant uptake, several chelating agents, such as EDTA (ethylenediaminetetraacetic acid), EDDS (ethylenediamine-N, N'-disuccinic acid), DTPA (diethylenetriaminepentaacetic acid), EDDHA (ethylenediamine di(o-hyroxyphenylacetic acid)), citric acid, or ammonium sulfate, are added to soil (Sun et al., 2011; Ali et al., 2013). For example, chelating agents like EDTA have been successfully applied to polluted soil. Farid et al. (2013) demonstrated that the addition of EDTA to soil increased plant growth, dry matter, the stress tolerance index, and accumulation of different metals, such as Cd, Zn, and Pb. Similar results using citric acid to enhance uranium uptake have been documented. Results have indicated that chelating agents enhanced or facilitated Pb transport via the xylem and increased Pb translocation from roots to shoots (Huang et al., 1997; Ghosh and Singh, 2005). The fact that chelates are able to increase the availability of PTEs is due to their ability to form water-soluble complexes with metal(loid)s in soil increasing their desorption from soil particles (Adamo et al., 2018). However, these treatments can lead to secondary pollution problems. For example, EDTA is a nonbiodegradable chelate and can leach into groundwater making an additional environmental hazard. In addition, high concentrations of synthetic chelating agents can also cause toxicity problem to plants. For that reason, proper care should be taken in using this method to induce phytoextraction (Song et al., 2012; Ali et al., 2013).

Another approach to increase the bioavailability of PTEs is the combination of both microorganisms and plants to ensure a more efficient clean-up of metal-polluted soils. Microbial-assisted phytoextraction uses microorganisms to induce the absorption, precipitation, oxidation, and reduction of PTEs in soil and helps plants to grow well even under metal(loid)-stress conditions. This remediation approach takes advantage of the synergic actions of a few beneficial, select genera of microorganisms in plant-bacterial communities (Rajkumar et al., 2012; Waigi et al., 2017). Microorganism-assisted phytoremediation uses several mechanisms, such as biosorption, intracellular accumulation, enzyme-catalyzed transformation, bioleaching and biomineralization, and redox reactions (Khalid et al., 2016). Among the beneficial rhizobiota and endomicrobiota, which subdivide into relatively few bacterial classes, are the sphingomonads, a group in the Alphaproteobacteria comprising crucial microorganisms (Waigi et al., 2017). Soil bacteria can enhance the mobility and bioavailability of PTEs in soil through several ways, such as lowering soil pH, producing plant growth-promoting and metal-chelating compounds, such as siderophores, organic acids, and biosurfactants, and by altering soil redox conditions (Ahemad, 2014; Ullah et al., 2015). For example, Yang et al. (2016) demonstrated that the inoculation of *Burkholderia* sp. (Z-90) enhanced the removal efficiency of metal(loid)s in soil by 31% for Pb, 37% for Cd, 32% for As, 52% for Mn, 44% for Zn, and 24% for Cu. However, the effect of soil microorganisms on metal solubility and uptake by plants is not always the same (Ma et al., 2015). Indeed, certain microbial processes can increase metal solubility and bioaccumulation, whereas other processes may cause metal immobilization, with a consequent decrease of uptake of PTEs (Gadd, 2010; Ma et al., 2015).

3.4 ADVANTAGES AND LIMITATIONS OF GREEN REMEDIATION

As described in Section 3.2, one of the greatest advantages of phytoremediation is its low cost compared to conventional clean-up technologies. Moreover, phytoremediation has several other significant advantages, such as (i) a variety of organic and inorganic compounds can be removed in the phytoremediation process; (ii) it is capable of permanently treating a wide range of contaminants in a wide range of environments; (iii) it can be used either as an in situ or ex situ application. In situ applications are frequently considered, because they minimize disturbance of the soil and the surrounding environment and reduce the spread of contamination *via* air and water; (iv) it is both environmentally friendly and aesthetically pleasing to the public; (v) it does not require expensive equipment or highly specialized personnel, and it is relatively easy to implement.

Although phytoremediation is a promising technology, it is not devoid of drawbacks. The major weakness is that it is a time-consuming process. Indeed, compared with other traditional methodology that can take weeks to months to clean up sites polluted with PTEs, phytoremediation may need several years to achieve the same result, and the contamination may still not be fully remediated (Megharaj and Naidu, 2017). Another limitation is that phytoremediation is restricted to the rooting depth of remediative plants selected for the remediation, which makes this technology effective only for sites with shallow contamination. Also, the presence of very high levels of PTEs in soil can induce phytotoxicity, so the level of contaminants plays a significant role in the applicability of this technology. Moreover, the use of invasive, non-native species can affect biodiversity and also plant and microorganism development. Phytoremediation is strongly dependent upon climatic conditions, e.g., unfavorable climate can limit plant growth and phytomass production, thus decreasing the process efficiency (Agrelli et al., 2020b; Ekta and Modi, 2018; Petruzzelli et al., 2016). Other limiting factors are related to the mobility and bioavailability of PTEs in soil. Indeed, low pollutant bioavailability can make phytoremediation ineffective, or, over time, immobilized pollutants can be re-released because the plant itself can induce changes (e.g., pH and organic matter content) in the bioavailability of PTEs (Adamo et al., 2018). For all these reasons, more large-scale in-situ studies are required and long-term monitoring of a site is recommended.

3.5 CONCLUSIONS

Contamination by metal(loid)s is a serious and widespread environmental problem highly related to various anthropogenic activities, which impacts both human and ecological health. Several remediation technologies have been developed over the years to deal with this problem. The most traditional technologies, both physical and chemical remediation, are highly expensive and often may generate a significant number of secondary pollutants. During the last few decades, the focus has shifted to developing "greener" and ecofriendly approaches, such as bioremediation and phytoremediation. These green technologies can be applied to remediate polluted soils without creating any destructive effect on soil structure. Several candidate plants, from grasses to woody species, suitable for phytoremediation, have been used with promising results. Fast-growing plants with high biomass and good ability to take up PTEs are preferred. Nevertheless, several factors must be considered to achieve a high performance of remediation, such as soil properties, climate, concentrations of target elements, and selection of proper plant species. Chelating agents and microorganisms also have been considered to enhance element mobilization and uptake. In addition, several methods of plant disposal have been considered, but data regarding these methods are still scarce. Composting and compaction can be applied for volume reduction, but care should be taken to collect leachate resulting from compaction. To date, to reduce the contaminated biomass, incineration seems to be least time-consuming and environmentally sound compared to direct burning or ashing. Further research is needed to better understand rhizosphere interactions with microorganisms and to minimize all the limitations in order to apply these techniques effectively.

REFERENCES

Abbas, G., Saqib, M., Akhtar, J., Murtaza, G., Shahid, M., Hussain, A. 2016. Relationship between rhizosphere acidification and phytoremediation in two acacia species. *J. Soils Sediments* 16:1392–1399.

Adamo, P., Agrelli, D., Zampella, M. 2018. Chemical speciation to assess bioavailability, bioaccessibility and geochemical forms of potentially toxic metals (PTMs) in polluted soils. In *Environmental Geochemistry: Site Characterization, Data Analysis and Case Histories*, 2nd ed. B. De Vivo, H. Belkin, A. Lima, Eds. Amsterdam, The Netherlands: Elsevier, pp. 153–194.

Agrelli, D., Caporale, A. G., Adamo, P. 2020a. Assessment of the bioavailability and speciation of heavy metal(loid)s and hydrocarbons for risk-based soil remediation. *Agronomy* 10(9):1440.

Agrelli, D., Duri, L. G., Fiorentino, N., Cozzolino, E., Fagnano, M., Adamo, P. 2020b. Potentially toxic element availability and risk assessment of cadmium dietary exposure after repeated croppings of Brassica juncea in a contaminated agricultural soil. *Agronomy* 10(6):880.

Ahmad, M., Lee, S. S., Lee, S. E., Al-Wabel, M. I., Tsang, D. C., Ok, Y. S. 2017. Biochar-induced changes in soil properties affected immobilization/mobilization of metals/metalloids in contaminated soils. *J. Soils Sediments* 17(3):717–730.

Ahemad, M. 2014. Remediation of metalliferous soils through the heavy metal resistant plant growth promoting bacteria: Paradigms and prospects. *Arab. J. Chem.* 12(7):1365–1377.

Ajaz, H. M., Arasuc, R. T., Narayananb, V. K. R., Zahir, H. M. I. 2010. Bioremediation of heavy metal contaminated soil by the exigobacterium and accumulation of Cd, Ni, Zn and Cu from soil environment. *Inter. J. Biol. Technol.* 1(2):94–101.

Alford, E. R., Pilon-Smits, E. A. H., Paschke, M. W. 2010. Metallophytes – a view from the rhizosphere. *Plant Soil* 337:33–50.

Ali, H., Khan, E., Sajad, M. A. 2013. Phytoremediation of heavy metals - Concepts and applications. *Chemosphere* 91(7):869–881.

Ali, H., Naseer, M., Sajad, M. A., 2012. Phytoremediation of heavy metals by Trifolium alexandrinum. *Int. J. Environ. Sci.* 2:1459–1469.

Alvarenga P., Gonçalves A. P., Fernandes R. M., de Varennes A., Vallini G., Duarte E., Cunha-Queda A. C. 2009. Organic residues as immobilizing agents in aided phytostabilization: (I) effects on soil chemical characteristics. *Chemosphere* 74:1292–1300

Al-Wabel, M. I., Usman, A. R. A., El-Naggar, A. H., Aly, A. A., Ibrahim, H. M., Elmaghraby, S., Al-Omran, A. 2015. Conocarpus biochar as a soil amendment for reducing heavy metal availability and uptake by maize plants. *Saudi J. Biol. Sci.* 22:503–511.

Ashraf, M. A., Maah, M. J., Yusoff, I. 2014. Chapter 1. Soil contamination, risk assessment and remediation. In *Environmental Risk Assessment of Soil Contamination*, M. C. Hernandez Soriano, Ed. IntechOpen, London pp. 3–56.

Bai, Z. G., Dent, K. L., Olsson, L., Schaepman, M. E. 2008. Proxy global assessment of land degradation. *Soil Use and Manag.* 24:223–234.

Bardos, P., Bakker, L., Darmendrail, D., Harries, N., Holland, K., MacKay, S., Pachon, C., Slenders, H., Smith, G., Smith, J., Wiltshire, L. 2013. Sustainable and green remediation-global update, ThS E3 sustainable use of the subsurface. *Proceedings Aquacon Soil*, April 2013, Barcelona, Spain. http://dx.doi.org/10.13140/2.1.1812.2887.

Barrutia, O., Artetxe, U., Hernández, A., Olano, J. M., García-Plazaola, J. I., Garbisu, C., Becerril, J. M. 2011. Native plant communities in an abandoned Pb-Zn mining area of northern Spain: Implications for phytoremediation and germplasm preservation. *Int. J. Phytoremediation* 13:256–270.

Beesley, L., Moreno-Jiménez, E., Gomez-Eyles, J. L. 2010. Effects of biochar and greenwaste compost amendments on mobility, bioavailability and toxicity of inorganic and organic contaminants in a multi-element polluted soil. *Environ. Pollut.* 158(6):2282–2287.

Beskoski, V. P., Gojgic-Cvijovic, G., Milic, J., Ilic, M., Miletic, S., Solevic, T., Vrvic, M. M. 2011. Ex situ bioremediation of a soil contaminated by mazut (heavy residual fuel oil)–A field experiment. *Chemosphere* 83:34–40.

Bini, C., 2008. Fate of trace elements in the Venice lagoon watershed and conterminous areas (Italy). In *Causes and Effects of Heavy Metal Pollution*, M. L. Sanchez, Ed. New York: Nova Science Publishers Inc., pp. 137–171.

Blaylock, M. J., Huang, J. W. 2000. Phytoextraction of metals. In *Phytoremediation of Toxic Metals: Using Plants to Clean up the Environment*, I. Raskin, and B. D. Ensley, Eds. New York: John Wiley and Sons, pp. 53–70.

Bolan, N. S, Adriano, D., Mahimairaja, S. 2004. Distribution and bioavailability of trace elements in livestock and poultry manure by-products. *Crit. Rev. Environ. Sci. Technol.* 34(3):291–338.

Bolan, N. S, Kunhikrishnan, A., Thangarajan, R., Kumpiene, J., Park, J., Makino, T., Kirkham, M. B., Scheckel, K. 2014. Remediation of heavy metal(loid)s contaminated soils -To mobilize or to immobilize? *J. Hazard. Mater.* 266:141–166.

Bolan, N. S., Naidu, R., Syers, J. K., Tillman, R. W. 1999. Surface charge and solute interactions in soils. *Adv. Agron.* 67:87–140.

Bolan, N. S., Park, J. H., Robinson, B., Naidu, R., Huh, K. Y. 2011. Phytostabilization. A green approach to contaminant containment. *Adv. Agron.* 112:145–204.

Boularbah, A., Schwartz, C., Bitton, G., Aboudrar, W., Ouhammou, A., Morel, J. L. 2006. Heavy metal contamination from mining sites in South Morocco: 2. Assessment of metal accumulation and toxicity in plants. *Chemosphere* 63:811–817.

Bridgwater, A. V., Meier, D., Radlein, D. 1999. An overview of fast pyrolysis of biomass. – *Org. Geochem.* 30:1479–1493.

Brooks, R. R., Chambers M. F., Nicks L. J., Robinson B. H. 1998. Phytomining. *Trends Plant Sci.* 3:359–62.

Burges, A., Alkorta, I., Epelde, L., Garbisu, C. 2018. From phytoremediation of soil contaminants to phytomanagement of ecosystem services in metal contaminated sites. *Int. J. Phytoremediation* 20(4):384–397.

Cele, E. N., Maboeta, M. 2016. A greenhouse trial to investigate the ameliorative properties of biosolids and plants on physicochemical conditions of iron ore tailings: Implications for an iron ore mine site remediation. *J. Environ. Manage.* 165:167–174.

Chen, H., Yang, X., Gielen, G., Mandal, S., Xu, S., Guo, J., Shaheen, S. M., Rinklebe, J., Che, L., Wang, H. 2019. Effect of biochars on the bioavailability of cadmium and di-(2-ethylhexyl) phthalate to Brassica chinensis L. in contaminated soils. *Sci. Total Environ.* 678:43–52.

Chen, W., Li, H. 2018. Cost-Effectiveness Analysis for Soil Heavy Metal Contamination Treatments. *Water Air Soil Pollut.* 229(4):1–13.

Chibuike, G. U., Obiora, S. C. 2014. Heavy metal polluted soils: Effect on plants and bioremediation methods. *Appl. Environ. Soil Sci.* 2014:1–12.

Conesa, H. M., Schulin, R., Nowack, B. 2007. A laboratory study on revegetation and metal uptake in native plant species from neutral mine tailings. *Water Air Soil Pollut.* 183:201–212.

Cunningham, S. D., Berti, W. R. 1993. Remediation of contaminated soils with green plants: An overview. *In Vitro Cell. Dev. Biol. Plant* 29(4):207–212.

Dickinson, N. M., Baker, A. J. M., Doronila, A., Laidlaw, S., Reeves, R. D. 2009. Phytoremediation of inorganics: Realism and synergies. *Int. J. Phytoremediation* 11:97–114.

Dixit, R., Malaviya, D., Pandiyan, K., Singh, U. B., Sahu, A., Shukla, R., Singh, B. P., Rai, J. P., Sharma, P. K., Lade, H., Paul, D. 2015. Bioremediation of heavy metals from soil and aquatic environment: An overview of principles and criteria of fundamental processes. *Sustainability* 7(2):2189–2212.

Ekta, P., Modi, N. R. 2018. A review of phytoremediation. *J. Pharmacogn. Phytochem.* 7(4):1485–1489.

European Environmental Agency (EEA), 2007. http://ec.europa.eu/environment/emas/pdf/es_library/99_dk_european_environment_agency_08.pdf.

European Commission, 2006. Impact assessment of the thematic strategy on soil protection. *Commission staff working document.* SEC (2006), 620(22.9), 2006.

FAO, 2015. Soil is a non-renewable resource. Its preservation is essential for food security and our sustainable future. http://www.fao.org/fileadmin/user_upload/soils-2015/docs/EN/IYS_fact_sheets_preservation_en_PRINT.pdf.

FAO, 2018. *Be the Solution to Soil Pollution. Outcome Document of the Global Symposium on Soil Pollution.* Rome, Italy: Food and Agriculture Organization of the United Nations, p. 32.
Farid, M., Ali, S., Shakoor, M. B., Bharwana, S. A., Rizvi, H., Ehsan, S., Tauqeer, H. M., Iftikhar, U., Hannan, F. 2013. EDTA assisted phytoremediation of cadmium, lead and zinc. *Int. J. Agron. Plant Prod.* 4(11):2833–2846.
Fellet, G., Marchiol, L., Delle Vedove, G., Peressotti, A. 2011. Application of biochar on mine tailings: Effects and perspectives for land reclamation. *Chemosphere* 83:1262–1267.
Fiorentino, N., Mori, M., Cenvinzo, V., Duri, L. G., Gioia, L., Visconti, D., Fagnano, M. 2018. Assisted phytoremediation for restoring soil fertility in contaminated and degraded land. *Ital. J. Agron.* 13(s1):34–44.
Foucault, Y., Lévêque, T., Xiong, T., Schreck, E., Austruy, A., Shahid, M., Dumat, C. 2013. Green manure plants for remediation of soils polluted by metals and metalloids: Ecotoxicity and human bioavailability assessment. *Chemosphere* 93:1430–1435.
Gadd, G. M. 2010. Metals, minerals and microbes: Geomicrobiology and bioremediation. *Microbiology* 156:609–643.
Galende, M. A., Becerril, J. M., Barrutia, O., Artetxe, U., Garbisu, C., Hernández, A. 2014. Field assessment of the effectiveness of organic amendments for aided phytostabilization of a Pb-Zn contaminated mine soil. *J. Geochem. Explor.* 145:181–189.
Garbisu, C., Alkorta, I. 2001. Phytoextraction: A cost-effective plant-based technology for the removal of metals from the environment. *Bioresour. Technol.* 77:229–36.
Garbisu, C., Alkorta, I. 2003. Basic concepts on heavy metal soil bioremediation. *Eur. J. Miner. Process. Environ. Prot.* 3(1):58–66.
Gerhardt, K. E., Gerwing, P. D., Greenberg, B. M. 2017. Opinion: Taking phytoremediation from proven technology to accepted practice. *Plant Sci.* 256:170–185.
Ghosh, M., Singh, S. P. 2005. A review on phytoremediation of heavy metals and utilization of its byproducts. *Appl. Ecol. Environ. Res.* 6:214–231.
Girovich, M. J. 1996. *Biosolids Treatment and Management: Processes for Beneficial Use.* Boca Raton, FL: CRC Press.
Goix, S., Leveque, T., Xiong, T. T., Schreck, E., Baeza-Squiban, A., Geret, F., Uzu, G., Austruy, A., Dumat, C. 2014. Environmental and health impacts of fine and ultrafine metallic particles: Assessment of threat scores. *Environ. Res.* 133:185–194.
Gorospe, J. 2012. Growing Greens and Soiled Soil: Trends in Heavy Metal Contamination in Vegetable Gardens of San Francisco. Master's Theses. 4131. https://doi.org/10.31979/etd.emgr-wkv4; https://scholarworks.sjsu.edu/etd_theses/4131.
Grifoni, M., Petruzzelli, G., Barbafieri, M., Rosellini, I., Pezzarossa, B., Pedron, F. 2017. Plants as alternative covers at contaminated sites. *Int. J. Environ. Ecol. Eng.* 11(10):729–733.
Hadley, P. W., Harclerode, M. A., 2015. Green remediation or sustainable remediation: Moving from dialogue to common practice. *Remed. J.* 25(2):95–115.
Hatfield, J. L., Sauer, T. J., Cruse, R. M. 2017. Soil: The forgotten piece of the water, food, energy nexus. *Adv. Agron.* 143:1–46.
He, Z., Shentu, J., Yang, X., Baligar, V. C., Zhang, T., Stoffella, P. J. 2015. Heavy metal contamination of soils: Sources, indicators, and assessment. *J. Environ. Indic.* 9:17–18.
Helsen, L., Bulck, E. V. D., Broeck, K. V. D., Vandecasteele, C. 1997. Low-temperature pyrolysis of CCA-treated wood waste: Chemical determination and statistical analysis of metal input and output; mass balances. *Waste Manage.* 17(1):79–86.
Herzig, R., Nehnevajova, E., Pfistner, C., Schwitzguebel, J. P., Ricci, A., Keller, C. 2014. Feasibility of labile Zn phytoextraction using enhanced tobacco and sunflower: Results of five-and one-year field-scale experiments in Switzerland. *Int. J. Phytoremediation* 16:735–754.
Hetland, M. D., Gallagher, J. R., Daly, D. J., Hassett, D. J., Heebink, L. V. 2001. Processing of plants used to phytoremediate lead-contaminated sites. In *Sixth International In Situ and On Site Bioremediation Symposium*, San Diego, California, June 4-7, 2001, pp. 129–136.
Hou, D., Al-Tabbaa, A. 2014. Sustainability: A new imperative in contaminated land remediation. *Environ. Sci. Policy* 39:25–34.
Hou, D., O'Connor, D. 2019. Green and sustainable remediation: past, present, and future developments. In *Sustainable Remediation of Contaminated Soil and Groundwater: Materials, Processes, and Assessment*, D. Hou, Ed. Butterworth-Heinemann, An imprint of Elsevier, headquartered in Amsterdam, pp. 19–42.
Houben, D., Pircar, J., Sonnet, P. 2012. Heavy metal immobilization by cost-effective amendments in a contaminated soil: Effects on metal leaching and phytoavailability. *J. Geochem. Explor.* 123:87–94.

Huang, J. W., Chen, J., Berti, W. R., Cunningham, S. D. 1997. Phytoremediation of lead contaminated soils-Role of synthetic chelates in lead phytoextraction. *Environ. Sci. Technol.* 31:800–806.
ISRIC, 2016. https://www.isric.org.
Iyer, P. V. R., Rao, T. R., Grover, P. D. 2002. *Biomass Thermochemical Characterization*, Third Edition, New Delhi Indian Institute of Technology, MNES, pp. 38.
Janoš, P., Vávrová, J., Herzogová, L., Pilařová, V. 2010. Effects of inorganic and organic amendments on the mobility (leachability) of heavy metals in contaminated soil: A sequential extraction study. *Geoderma* 159(3–4):335–341.
Kanmani, P., Aravind, J., Preston, D., 2012. Remediation of chromium contaminants using bacteria. *Int. J. Environ. Sci. Technol.* 9:183–193.
Khalid, S., Shahid, M., Niazi, N. K., Murtaza, B., Bibi, I., Dumate, C. 2016. A comparison of technologies for remediation of heavy metal contaminated soils. *J. Geochem. Explor.* 182:247–268.
Khan, M. A., Chattha, M. R., Farooq, K., Jawed, M. A., Farooq, M., Imran, M., Iftkhar, M., Kasana, M. I. 2015. Effect of farmyard manure levels and NPK applications on the pea plant growth, pod yield and quality. *Life Sci. Int. J.* 9:3178–3181.
Kibblewhite, M. G., Miko, L., Montanarella, L. 2012. Legal frameworks for soil protection: Current development and technical information requirements. *Curr. Opin. Environ. Sustain.* 4(5):573–577.
Koppolu, L., Agblover, F. A., Clements, L. D., 2003. Pyrolysis as a technique for separating heavy metals from hyperaccumulators. Part II: Lab-scale pyrolysis of synthetic hyperaccumulator biomass. *Biomass Bioenergy* 25:651–663.
Kumpiene, J., Fitts, J. P., Mench, M. 2012. Arsenic fractionation in minespoils 10 years after aided phytostabilization. *Environ. Pollut.* 166:82–88.
Li, J., Xu, Y. 2015. Immobilization of Cd in a paddy soil using moisture management and amendment. *Chemosphere* 122:131–136.
Liang, X., Xu, Y., Xu, Y., Wang, P., Wang, L., Sun, Y., Huang, Q., Huang, R. 2016. Two-year stability of immobilization effect of sepiolite on Cd contaminants in paddy soil. *Environ. Sci. Pollut. Res.* 23:12922–12931.
López-Mosquera, M. E., Moirón, C., Carral, E. 2000. Use of dairy-industry sludge as fertilizer for grasslands in northwest Spain: Heavy metal level in the soil and plant. *Resour. Conserv. Recycl.* 30:95–109.
Lovley, D. R., Lloyd, J. R., 2000. Microbes with a mettle for bioremediation. *Nat. Biotechnol.* 18:600–601.
Lu, H., Zhang, Y. Y., Huang, S., Qiu, R. 2012. Relative distribution of Pb^{2+} sorption mechanisms by sludge-derived biochar. *Water Res.* 46:854–862.
Luo, L., Ma, Y., Zhang, S., Wei, D., Zhu, Y. G. 2009. An inventory of trace element inputs to agricultural soils in China. *J. Environ. Manage.* 90(8):2524–2530.
Ma, Y., Oliveira, R. S., Nai, F., Rajkumar, M., Luo, Y., Rocha, I., Freitas, H. 2015. The hyperaccumulator sedum plumbizincicola harbors metal-resistant endophytic bacteria that improve its phytoextraction capacity in multi-metal contaminated soil. *J. Environ. Manage.* 156:62–69.
Mahar, A., Wang, P., Ali, A., Awasthi, M. K., Lahori, A. H., Wang, Q., Li R., Zhang, Z. 2016. Challenges and opportunities in the phytoremediation of heavy metals contaminated soils: A review. *Ecotoxicol. Environ. Safety* 126:111–121.
Majumder, A., Bhattacharyya, K., Kole, S. C., Ghosh, S. 2013. Efficacy of indigenous soil microbes in arsenic mitigation from contaminated alluvial soil of India. *Environ. Sci. Pollut. Res.* 20(8):5645–5653.
Malik, R. N., Husain, S. Z., Nazir, I. 2010. Heavy metal contamination and accumulation in soil and wild plant species from industrial area of Islamabad, Pakistan. *Pak. J. Bot.* 42:291–301.
Marques, A. P., Rangel, A. O., Castro, P. M. 2009. Remediation of heavy metal contaminated soils: Phytoremediation as a potentially promising clean-up technology. *Crit. Rev. Environ. Sci. Technol.* 39:622–654.
Martino, L. E., Dona, C. L., Dicerbo, J., Hawkins, A., Moore, B., Horner, R. 2016. Green and sustainable remediation practices in Federal Agency cleanup programs. *Environ. Earth Sci.* 75(21):1–13.
McBride, M. B. 1995. Toxic metal accumulation from agricultural use of sludge: Are USEPA regulations protective? *J. Environ. Qual.* 24:5–18.
Meagher, R. B. 2000. Phytoremediation of toxic elemental and organic pollutants. *Curr. Opin. Plant Biol.* 3:153–162.
Meers, E., Ruttens, A., Hopgood, M., Lesage, E., Tack, F. M. G. 2005. Potential of Brassica rapa, Cannabis sativa, Helianthus annuus and Zea mays for phytoextraction of heavy metals from calcareous dredged sediment derived soils. *Chemosphere* 61(4):561–572.
Megharaj, M., Naidu, R. 2017. Soil and brownfield bioremediation. *Microb. Biotechnol.* 10:1244–1249.
Megharaj, M., Ramakrishnan, B., Venkateswarlu, K., Sethunathan, N., Naidu, R. 2011. Bioremediation approaches for organic pollutants: A critical perspective. *Environ. Int.* 37:1362–1375.

Megharaj, M., Venkateswarlu, K., Naidu, R. 2014. Chapter 3: Bioremediation. In *Encyclopedia of Toxicology*, Vol. 1. Elsevier. http://dx.doi.org/10.1016/B978-0-12-386454-3.01001-0.
Mench, M., Schwitzguebel, J. P., Schroeder, P., Bert, V., Gawronski, S., Gupta, S. 2009. Assessment of successful experiments and limitations of phytotechnologies: Contaminant uptake, detoxification and sequestration, and consequences for food safety. *Environ. Sci. Pollut. Res.* 16:876–900.
Mohd, S. N., Majid, N. M., Shazili, N. A. M., Abdu, A., 2013. Growth performance, biomass and phytoextraction efficiency of Acacia mangium and Melaleuca cajuputi in remediating heavy metal contaminated soil. *Am. J. Environ. Sci.* 9:310.
Mossa, A. W., Bailey, E. H., Usman, A., Young, S. D., Crout, N. M. 2020. The impact of long-term biosolids application (> 100 years) on soil metal dynamics. *Sci. Total Environ.* 720:137441.
Muthusaravanan, S., Sivarajasekar, N., Vivek, J. S., Paramasivan, T., Naushad, M., Prakashmaran, J., Gayathri, V., Al-Duaij, O. K. 2018. Phytoremediation of heavy metals: Mechanisms, methods and enhancements. *Environ. Chem. Lett.* 16(4):1339–1359.
Naik, M. M., Dubey, S. K. 2011. Lead-enhanced siderophore production and alteration in cell morphology in a Pb-resistant Pseudomonas aeruginosa strain 4EA. *Curr. Microbiol.* 62(2):409–414.
Niazi, N. K., Bashir, S., Bibi, I., Murtaza, B., Shahid, M., Javed, M. T., Shakoor, M. B., Saqib, Z. A., Nawaz, M. F., Aslam, Z., Wang, H., Murtaza, G. 2016. Phytoremediation of arsenic-contaminated soils using arsenic hyperaccumulating ferns. In *Phytoremediation*, Ansari, et al., Eds. Cham: Springer, pp. 521–545.
Nicholson, F. A., Smith, S. R., Alloway, B. J., Carlton-Smith, C., Chambers, B. J. 2003. An inventory of heavy metals inputs to agricultural soils in England and Wales. *Sci. Total Environ.* 311(1–3):205–219.
Nicks, L. J. Chambers, M. F. 1994. Nickel farming, *Discover Magazine*. 19 September.
Niu, L., Yang, F., Xu, C., Yang, H., Liu, W. 2013. Status of metal accumulation in farmland soils across China: From distribution to risk assessment. *Environ. Pollut.* 176:55–62.
Oburger, E., Gruber, B., Schindlegger, Y., Schenkeveld, W. D., Hann, S., Kraemer, S. M., Wenzel, W.W., Pushenreiter, M. 2014. Root exudation of phytosiderophores from soil-grown wheat. *New Phytol.* 203:1161–1174.
Oliveira, F.C., Mattiazzo, M.E. 2001. Metais pesados em Latossolo tratado com lodo de esgoto e em plantas de cana-de-açúcar. *Sci. Agric.* 58:581–593.
Otunola, B.O., Ololade, O.O. 2020. A review on the application of clay minerals as heavy metal adsorbents for remediation purposes. *Environ. Technol. Innov.* 18:100692.
Padmavathiamma, P. K., Li, L.Y. 2007. Phytoremediation technology: Hyperaccumulation metals in plants. *Water Air Soil Pollut.* 184:105–126.
Palansooriya, K. N., Shaheen, S. M., Chen, S. S., Tsang, D. C. W., Hashimoto, Y., Hou, D., Bolan, N. S., Rinklebe, J., Ok, Y. S. 2020. Soil amendments for immobilization of potentially toxic elements in contaminated soils: A critical review. *Environ. Int.* 134:105046.
Pandey, V. C., Singh, N. 2010. Impact of fly ash incorporation in soil systems. *Agric. Ecosyst. Environ.* 136(1–2):16–27.
Park, J. H., Choppala, G. H., Bolan, N. S., Chung, J. W., Chuasavathi, T. 2011. Biochar reduces the bioavailability and phytotoxicity of heavy metals. *Plant Soil* 348:439–451.
Parra, A., Zornoza, R., Conesa, E., Gómez-López, M. D., Faz, A. 2016. Evaluation of the suitability of three Mediterranean shrub species for phytostabilization of pyritic mine soils. *Catena* 136:59–65.
Paterson, E. 2003. Importance of rhizodeposition in the coupling of plant and microbial productivity. *Eur. J. Soil Sci.* 54:741–750.
Patra, K. C., Rautray, T. R., Nayak, P. 2012. Analysis of grains grown on fly ash treated soils. *Appl. Radiat. Isot.* 70:1797–1802.
Petruzzelli, G., Pedron, F., Grifoni, M., Barbafieri, M., Rosellini, I., Pezzarossa, B. 2016. Soil remediation technologies towards green remediation strategies. *Int. J. Geol. Environ. Eng.* 10(6):654–658.
Pimentel, D., Pimentel, M. 2000. Feeding the world's population. *Bioscience* 50:387
Rafati, M., Khorasani, N., Moattar, F., Shirvany, A., Moraghebi, F., Hosseinzadeh, S. 2011. Phytoremediation potential of Populus alba and Morus alba for cadmium, chromuim and nickel absorption from polluted soil. *Int. J. Environ. Res.* 5:961–970.
Rajkumar, M., Prasad, M. N. V., Swaminathan, S., Freitas, H. 2013. Climate change drivenplant-metal-microbe interactions. *Environ. Int.* 53:74–86.
Rajkumar, M., Sandhya, S., Prasad, M. N. V., Freitas, H. 2012. Perspectives of plant-associated microbes in heavy metal phytoremediation. *Biotechnol. Adv.* 30:1562–1574.
Rashid, A, Mahmood T, Mehmood, F, Khalid, A., Saba, B., Batool, A., Riaz, A., 2014. Phytoaccumulation, competitive adsorption and evaluation of chelators-metal interaction in lettuce plant. *Environ. Eng. Manage. J.* 13:2583–2592.

Raskin, I., Ensley, B. D. 2000. *Phytoremediation of Toxic Metals*. Using Plants to Clean Up the Environment. New York: John Wiley and Sons, Inc.

Raskin, I., Smith, R. D., Salt, D. E. 1997. Phytoremediation of metals: Using plants to remove pollutants from the environment. *Curr. Opin. Biotechnol.* 8(2):221–226.

Rizwan, M., Ali, S., ur Rehman, M. Z., Rinklebe, J., Tsang, D. C. W., Bashir, A., Maqbool, A., Tack, F. M. G., Ok, Y. S. 2018. Cadmium phytoremediation potential of Brassica crop species: A review. *Sci. Total Environ.* 631:1175–1191.

Rocco, C., Seshadri, B., Adamo, P., Bolan, N. S., Mbene, K., Naidu, R. 2018. Impact of waste-derived organic and inorganic amendments on the mobility and bioavailability of arsenic and cadmium in alkaline and acid soils. *Environ. Sci. Pollut. Res.* 25(26):25896–25905.

Ronchi, S., Salata, S., Arcidiacono, A., Piroli, E., Montanarella, L. 2019. Policy instruments for soil protection among the EU member states: A comparative analysis. *Land Use Policy* 82:763–780.

Sabir, M., Waraich, E. A., Hakeem, K. R., Öztürk, M., Ahmad, H. R., Shahid, M. 2015. Chapter 4, Phytoremediation: mechanisms and adaptations. In *Phytoremediation, Soil Remediation and Plants*. Elsevier Inc., pp. 85–105. http://dx.doi.org/10.1016/B978-0-12-799937-1.00004-8.

Sakakibara, M., Watanabe, A., Inoue, M., Sano, S., Kaise, T. 2010. Phytoextraction and phytovolatilization of arsenic from As-contaminated soils by Pteris vittata. *Proceedings of the Annual International Conference on Soils, Sediments, Water and Energy*, 18-21 October 2010, Amherst, Massachusetts, USA. Vol. 12, p. 26.

Sanderson, P., Qi, F., Seshadri, B., Wijayawardena, A., Naidu, R. 2018. Contamination, fate and management of metals in shooting range soils - A review. *Curr. Pollut. Rep.* 4(2):175–187.

Schmoger, M. E. V., Oven, M., Grill, E. 2000. Detoxification of arsenic by phytochelatins in plants. *Plant Physiol.* 122(3):793–801.

Seshadri, B., Bolan, N.S., Naidu, R. 2015. Rhizosphere-induced heavy metal(loid) transformation in relation to bioavailability and remediation. *J. Soil Sci. Plant Nutr.* 15(2): 524–548.

Shaheen, S. M., Antoniadis, V., Kwon, E. E., Biswas, J. K., Wang, H., Ok, Y. S., Rinklebe, J. 2017a. Biosolids application affects the competitive sorption and lability of cadmium, copper, nickel, lead, and zinc in fluvial and calcareous soils. *Environ. Geochem. Health* 39(6):1365–1379.

Shaheen, S. M., Hooda, P. S., Tsadilasc, C. D. 2014. Opportunities and challenges in the use of coal fly ash for soil improvements – A review. *J. Environ. Manage.* 145:249–267.

Shaheen, S. M., Rinklebe, J. 2015. Impact of emerging and low cost alternative amendments on the (im)mobilization and phytoavailability of Cd and Pb in a contaminated flood plain soil. *Ecol. Eng.* 74:319–326.

Shaheen, S. M., Shams, M. S., Khalifa, M. R., Mohamed, A., Rinklebe, J. 2017b. Various soil amendments and environmental wastes affect the (im)mobilization and phytoavailability of potentially toxic elements in a sewage effluent irrigated sandy soil. *Ecotoxicol. Environ. Saf.* 142:375–387.

Shankar, S. 2017. Management and remediation of problem soils, solid waste and soil pollution. In *Principles and Applications of Environmental Biotechnology for a Sustainable Future*, R. L. Singh, Ed. Singapore: Springer, pp. 143–171.

Sharma, B., Singh, P., Chauhan, P. S., Singh, S., Singh, R. P. 2019. Microbes-assisted remediation of metal polluted soils. In *New and Future Developments in Microbial Biotechnology and Bioengineering*, J. S. Singh and D.P. Singh, Eds. headquartered in Amsterdam: Elsevier, pp. 223–232.

Sharma, J. 2018. Utilization of phytoremediation by-product. *Recent Adv. Biol. Med.* 4:47–50.

Sharma, P., Pandey, S. 2014. Status of phytoremediation in world scenario. *Int. J. Environ. Bioremediat. Biodegrad.* 2(4):v178–191.

Shen, Z., Jin, F., O'Connor, D., Hou, D. 2019. Solidification/stabilization for soil remediation: An old technology with new vitality. *Environ. Sci. Technol.* 53:11615–11617.

Sheoran, V., Sheoran, A., Poonia, P. 2011. Role of hyperaccumulators in phytoextraction of metals from contaminated mining sites: A review. *Crit. Rev. Environ. Sci. Technol.* 41:168–214.

Shi, T., Ma, J., Wu, X., Ju, T., Lin, X., Zhang, Y., Li, X., Gong, Y., Hou, H., Zhao, L., Wu, F. 2018. Inventories of heavy metal inputs and outputs to and from agricultural soils: A review. *Ecotoxicol. Environ. Saf.* 164:118–124.

Shukla, K. P., Sharma, S., Singh, N. K., Singh, V., Tiwari, K., Singh, S. 2011. Nature and role of root exudates: Efficacy in bioremediation. *Afr. J. Biotechnol.* 10(48):9717–9724.

Singh, S.P., Ghosh, M. A. 2003. Comparative study on effect of cadmium, chromium and lead on seed germination of weed and accumulator plant species. *Indian J. Environ. Prot.* 23(5):513–518.

Song, X., Hu, X., Ji, P., Li, Y., Chi, G., Song, Y. 2012. Phytoremediation of cadmium-contaminated farmland soil by the hyperaccumulator Beta vulgaris L. var. cicla. *Bull. Environ. Contamin. Toxicol.* 88:623–626.

Suèr, P., Nilsson-Påledal, S., Norrman, J. 2004. LCA for site remediation: A literature review. *Soil Sediment Contam.* 13(4):415–425.

Suman, J., Uhlik, O., Viktorova, J., Macek, T. 2018. Phytoextraction of heavy metals: A promising tool for clean-up of polluted environment? *Front. Plant Sci.* 9:1476.

Sun, Y., Li, Y., Xu, Y., Liang, X., Wang, L. 2015. In situ stabilization remediation of cadmium (Cd) and lead (Pb) co-contaminated paddy soil using bentonite. *Appl. Clay Sci.* 105:200–206.

Sun, Y., Sun, G., Xu, Y., Liu, W., Liang, X., Wang, L. 2016. Evaluation of the effectiveness of sepiolite, bentonite, and phosphate amendments on the stabilization remediation of cadmium-contaminated soils. *J. Environ. Manage.* 166:204–210.

Sun, Y. B., Sun, G. H., Zhou, Q. X., Xu, Y. M., Wang, L., Liang, X. F., Sun, Y., Qing, X. 2011. Induced-phytoextraction of heavy metals from contaminated soil irrigated by industrial wastewater with Marvel of Peru (Mirabilis jalapa L.). *Plant Soil Environ.* 57:364–371.

Sylvain, B, Mikael, M. H., Florie, M., Emmanuel, J., Marilyne, S., Sylvain, B., Domenico, M. 2016. Phytostabilization of As, Sband Pb by two willow species (S. viminalis and S. purpurea) on former mine technosols. *Catena* 136:44–52.

Tangahu, B. V., Abdullah, S. R. S., Basri, H., Idris, M., Anuar, N., Mukhlisin, M. 2011. A review on heavy metals (As, Pb, and Hg) uptake by plants through phytoremediation. *Int. J. Chem. Eng.* http://dx.doi.org/10.1155/2011/939161.

The Guardian, 2015. https://www.theguardian.com/environment/2015/dec/02/arable-land-soil-food-security-shortage.

Tripathi, M., Munot, H. P., Shouche, Y., Meyer, J. M., Goel, R. 2005. Isolation and functional characterization of siderophore-producing lead and cadmium-resistant Pseudomonas putida KNP9. *Curr. Microbiol.* 50(5):233–237.

Ullah, A., Heng, S., Munis, M. F. H., Fahad, S., Yang, X. 2015. Phytoremediation of heavy metals assisted by plant growth promoting (PGP) bacteria: A review. *Environ. Exp. Bot.* 117:28–40.

USEPA, 2008. Green remediation: Incorporating Sustainable Environmental Practices into Remediation of Contaminated Sites; Office of Solid Waste and Emergency Response, Washington. EPA 542-R-08-002.

Van Dillewijn, P., Caballero, A., Paz, J. A., Gonzalez-Perez, M. M., Oliva, J. M., Ramos, J. L. 2007. Bioremediation of 2,4,6 trinitrotoluene under field conditions. *Environ. Sci. Technol.* 41:1378–1383.

Venegas, A., Rigol, A., Vidal, M. 2015. Viability of organic wastes and biochars as amendments for the remediation of heavy metal-contaminated soils. *Chemosphere* 119:190–198.

Verbruggen, N., Hermans, C., Schat, H. 2009. Molecular mechanisms of metal hyperaccumulation in plants. *New Phytol.* 181:759–776.

Violante, A., Pigna, M., 2002. Competitive sorption of arsenate and phosphate on different clay minerals and soils. *Soil Sci. Soc. Am. J.* 66:1788–1796.

Waigi, M. G., Sun, K., Gao, Y. 2017. Sphingomonads in microbe-assisted phytoremediation: Tackling soil pollution. *Trends Biotechnol.* 35(9):883–899.

Walker, D. J., Clemente, R., Bernal, M. P. 2004. Contrasting effects of manure and compost on soil pH, heavy metal availability and growth of Chenopodium album L. in a soil contaminated by pyritic mine waste. *Chemosphere* 57:215–224.

Wan, X., Lei, M., Chen, T. 2016. Cost–benefit calculation of phytoremediation technology for heavy-metal-contaminated soil. *Sci. Total Environ.* 563:796–802.

Wang, J., Shaheen, S. M., Swertz, A. C., Rennert, T., Feng, X., Rinklebe, J. 2019a. Sulfur-modified organoclay promotes plant uptake and affects geochemical fractionation of mercury in a polluted floodplain soil. *J. Hazard. Mater.* 371:687–693.

Wang, P., Li, R., Guo, D., Guo, Z., Mahar, A., Du, J., Zhang, Z. 2020. The influences of fly ash on stabilization for Cd in contaminated soils. *Environ. Sci. Pollut. Res.* 27(35): 43505–43513.

Wang, Q. R., Cui, Y. S., Liu, X. M., Dong, Y. T., Christie, P. 2003. Soil contamination and plant uptake of heavy metals at polluted sites in China. *J. Environ. Sci. Health A Tox. Hazard Subst. Environ. Eng.* 38:823–838.

Wang, Y., Meng, D., Fei, L., Dong, Q., Wang, Z. 2019b. A novel phytoextraction strategy based on harvesting the dead leaves: Cadmium distribution and chelator regulations among leaves of tall fescue. *Sci. Total Environ.* 650(2): 3041–3047.

Washa, M., Bini, C., Zilioli, D., Spiandorello, M., Gallo, M. 2014. Potentially harmful elements in terraced agroecosystems of NE Italy: Geogenic *vs.* anthropogenic enrichment. *J. Geochem. Explor.* 144:355–362.

Wu, G., Kang, H., Zhang, X., Shao, H., Chu, L., Ruan, C. 2010. A critical review on the bio-removal of hazardous heavy metals from contaminated soils: Issues, progress, eco-environmental concerns and opportunities. *J. Hazard. Mater.* 174:1–8.

Wuana, R. A., Okieimen, F. E. 2011. Heavy metals in contaminated soils: A review of sources, chemistry, risks and best available strategies for remediation. *ISRN Ecology* 2011:1–20.

Xu, Y., Liang, X., Xu, Y., Qin, X., Huang, Q., Wang, L., Sun, Y. 2017. Remediation of heavy metal-polluted agricultural soils using clay minerals: A review. *Pedosphere* 27(2):193–204.

Yan, A., Wang, Y., Tan, S. N., Mohd Yusof, M. L., Ghosh, S., Chen, Z. 2020. Phytoremediation: A promising approach for revegetation of heavy metal-polluted land. *Front. Plant Sci.* 11: 359.

Yan, Y., Zhou, Y. Q., Liang, C. H. 2015. Evaluation of phosphate fertilizers for the immobilization of Cd in contaminated soils. *PLoS One* 10(4):e0124022.

Yang, X. E., Long, X. X., Ye, H. B., He, Z. L., Calvert, D., Stoffella, P. 2004. Cadmium tolerance and hyperaccumulation in a new Zn-hyperaccumulating plant species (Sedum alfredii Hance). *Plant Soil* 259:181–189.

Yang, Z., Zhang, Z., Chai, L., Wang, Y., Liu, Y., Xiao, R. 2016. Bioleaching remediation of heavy metal-contaminated soils using Burkholderia sp. Z-90. *J. Hazard. Mater.* 301:145–152.

Yao, Z., Li, J., Xie, H., Yu, C. 2012. Review on remediation technologies of soil contaminated by heavy metals. *Procedia Environ. Sci.* 16:722–729.

Zhao, Y., Yan, Z., Qin, J., Xiao, Z. 2014. Effects of long-term cattle manure application on soil properties and soil heavy metals in corn seed production in Northwest China. *Environ. Sci. Pollut. Res.* 21(12):7586–7595.

4 Trace Elements and the Sustainability of Agricultural Systems

Brett Robinson
School of Physical and Chemical Sciences, University of Canterbury, Christchurch, New Zealand

CONTENTS

4.1 INTRODUCTION

Trace elements (TEs) normally occur at concentrations < 1,000 mg/kg in agricultural plants or livestock and include all the chemical elements that are not macronutrients. TEs may be essential for plant or animal metabolism, or they may be non-essential contaminants that either reduce agricultural production or reduce food quality, potentially endangering human health (Robinson et al., 2009). Most essential TEs are cofactors required for enzyme activation, rather than part of the structural components of the organism. As such, TE deficiencies can result in profound metabolic perturbations that reduce growth or functioning and may even lead to death. Essential TEs that occur in excess of their metabolic requirements, or abnormally high concentrations of non-essential TEs, may inhibit metabolic processes through oxidative stress, enzyme inactivation, or inhibition of the uptake of essential nutrients (Jaishankar et al., 2014). Therefore, while present in organisms at only trace concentrations, TEs have a disproportionately large effect on the quantity and quality of agricultural produce.

Surveys in countries such as India have shown that up to half of agricultural soils provide insufficient TEs to maximize crop production (Shukla et al., 2021). Animal production is similarly affected by TE deficiencies: In New Zealand where most livestock graze pasture year round, inadequate intake of TEs limit production and require expensive supplementation (Grace and Knowles, 2012). Soil is the ultimate source of many TEs that humans ingest through agricultural produce. The diets of one-half to two-thirds of humanity are deficient in one or more TEs (Pfeiffer and McClafferty, 2007; White and Broadley, 2009).

Anthropogenic activity has resulted in elevated concentrations (i.e., above natural "background" concentrations) of one or more TEs in most, if not all, agricultural soils (Mills et al., 2005). While such increases may be benign, the area of agricultural land that is considered "polluted," i.e., contaminated to an extent that crop production or food safety is endangered, is in the order of tens of millions of hectares (Evangelou and Conesa, 2012; Evangelou et al., 2015). Most contaminated soils

DOI: 10.1201/9781003093565-4

are co-contaminated with several TEs (Bourennane et al., 2010). The severity of contamination may be assessed using pollution indices, many of which weigh individual TEs based on their (eco)toxicity and the degree to which their concentration exceeds background levels (Bourennane et al., 2010; Kowalska et al., 2018). TE-contaminated agricultural soils can endanger human health through the consumption of contaminated agricultural produce or associated dust. Some contaminating TEs in agricultural soil may migrate into drinking water, aquatic, or marine environments affecting other food sources.

Earth's finite resources of agricultural land are relied upon to support the needs of a population that is expected to reach 11.2 billion by 2100 (United Nations, 2022). Whereas much research has focused on losses of agricultural land through desertification, erosion, and soil covering, there is as yet no overarching assessment of the sustainability of our agricultural practices with regard to TE. This review aims to determine the extent that TE deficiencies or contamination that may become limiting factors for agricultural production in the next few decades and identify strategies to mitigate these factors.

4.2 THE RELATIVE IMPORTANCE OF TOTAL AND BIOAVAILABLE TRACE ELEMENTS

The total TE concentration in soils is a function of the TE concentration of the parent material whence the soil was derived, human and natural additions, and losses through leaching, erosion, and removal via crops. The background concentrations of TEs vary widely (Mason and Moore, 1985; Percival et al., 1996), and depending on the environment, they may be insufficient (Jensen et al., 2019) or excessive (Vithanage et al., 2019) for safe agricultural production. However, for an individual soil, the total element concentration provides little information on the effect that the TE will have on agricultural production; only some fraction of the total TE concentration is available for plant uptake. This "bioavailable" fraction is dependent on the physicochemical and biological properties of the soil as well as the receptor organism (Violante et al., 2010).

Most TEs bind strongly to the solid phase in soil through ion exchange, specific adsorption, occlusion, precipitation, and sorption into soil colloids (Sposito, 2016). For most TEs in agricultural soils, K_D values (i.e., soil matrix/soil solution concentration quotients) are typically in the 100s–1,000s (Gartler et al., 2013; Reiser et al., 2014; Jensen et al., 2018). The exceptions are neutral (e.g., $B(OH_3)$) or some anionic (e.g., CrO_4^{2-} but not AsO_4^{3-}) TE species as well as Li^+, which can have K_D values <10 (Robinson et al., 2007, 2018; Sposito, 2016). The K_D values of TEs are affected by the physicochemical characteristics of the soil: sorption of TEs typically increases with increasing organic matter, phyllosilicate clays, and metal oxides (Robinson et al., 2005). Sorption of cationic TEs invariably increases with increasing pH; however, anionic TEs may become more mobile at high soil pH (Tschan et al., 2009).

The implications of the typically high K_D values for most TEs are that relatively high total concentrations of TEs do not necessarily indicate that there will be adequate bioavailable TEs for agricultural production. Some contaminating TEs may be benign if they are unavailable for plant uptake. Given the low mobility of most TEs, they tend to accumulate in soil, potentially to unsafe concentrations.

Bioavailability is also a function of species. In the same environment, different plant species may have TE concentrations that vary by over an order of magnitude (Baker, 1981). This is because plants and their associated microorganisms have contrasting mechanisms that solubilize and take up essential elements from soil, including essential TEs. Solubilized TEs may accumulate in the root tissue, or be transported via the xylem tissues to the water sinks (leaves), or redistributed via the phloem to other parts of the plant, including the fruits and seeds. Uptake mechanisms often have poor specificity and contaminating TEs may be taken up and transported via the same pathways as essential elements (Reid and Hayes, 2003).

4.3 TRACE ELEMENT DEFICIENCIES IN AGRICULTURE

TE deficiencies in crops result from insufficient bioavailable TE in soil. This may arise when there is a low total TE concentration or when bioavailability is limited by the soil's physicochemical or biological parameters. In animal agriculture, TE deficiencies are more common because some essential TEs for animals, notably Se and I, are not essential plant nutrients and may not be taken up into fodder in sufficient quantities (Grace and Knowles, 2012). Essential TEs in animal fodder may not be *bioaccessible*, i.e., not absorbed by the animal. This depends on the speciation of the TE as well as other components of the fodder. It is notable that Se and I are present as anions in soil and are more readily leached than other TEs. Seaspray is a major source of these elements in soil, and severe deficiencies may occur in inland areas where the soils are highly leached (Jensen et al., 2019).

Some 30% of Earth's soils are calcareous (Marschner, 1995), and their high pH limits the solubility of TE cations, particularly Fe (Valentinuzzi et al., 2015). Low bioavailability, rather than low total concentrations, results in an insufficient plant supply of Fe, Mn, and Zn on 20%–50% of soils surveyed in India (Shukla et al., 2021). In such systems, the application of TEs to soil through fertilization may either result in a temporary increase in plant TE uptake or be entirely ineffective.

Normal agricultural practices may further reduce the bioavailability of essential TEs (Table 4.1): While liming is essential to maintain nutrient cycling in 30%–40% of agricultural soils (Marschner, 1991), raising the pH can reduce the plant uptake of Cu, Fe, and Zn to levels that result in reduced yields. Similarly, the application of phosphate fertilizers can induce Zn deficiencies in crops (Csathó et al., 2019; Sacristán et al., 2019). Fertilization with Mo to improve clover N fixation can induce Cu deficiencies in livestock (Grace et al., 2005). The removal of crop residues, ploughing, and widespread use of urea and inorganic fertilizers are depleting soil organic matter (Sanderman et al., 2017). The normal cycling of soil organic matter provides an important source of plant-available TEs, so such practices are further reducing the plant uptake of essential micronutrients (Dhaliwal et al., 2019).

Low uptake of essential TEs may be reversed through changing agricultural practices, for example, foliar and soil applications of N can increase Zn uptake by durum wheat (Kutman et al., 2010).

TABLE 4.1
Agricultural Practices That May Exacerbate Trace Element Deficiencies in Agricultural Systems

Action	Effect	Reference
Removal of crop residues	Bioavailable TEs not returned to the soil	Blanco-Canqui and Lal (2009)
Application of high rates of urea, inorganic nitrogen fertilizers and other inorganic fertilizers	Loss of organic matter, which over the long term, reduces plant-available TE	Dhaliwal et al. (2019)
	Accelerated growth reduces TE concentration in crops	Park et al. (2011)
Application of phosphate fertilizer	Reduced plant availability of Zn	Csathó et al. (2019; Sacristán et al. (2019)
	Formation of phytates reducing TE availability to animals and humans	Bindraban et al. (2020)
Addition of lime or liming agents	Can induce deficiencies in Cu, Fe, Mn, and Zn, particularly if pH raised beyond the target value	Brennan et al. (2005)
Addition of Mo	Reduced availability of Cu to livestock	Grace et al. (2005)
Ploughing	Loss of organic matter, which over the long term reduces plant-available TE	Dhaliwal et al. (2019)
	Loss of TE-rich topsoil through wind and water erosion	Olife et al. (2007)

Plant breeding (Nestel et al., 2006), genetic modification, and foliar sprays can also increase plant uptake (Cakmak, 2008). Providing grazing cattle with a diverse range of feeds and pasture species can lessen TE deficiencies that are present in animals grazing a restricted diet, e.g., of ryegrass and clover (Grace and Knowles 2012). Zinc solubilizing rhizobacteria may increase plant uptake on soils where physicochemical factors render Zn otherwise unavailable for plant uptake (Chakraborty et al., 2022). Arbuscular mycorrhizal fungi may improve plant growth under both Zn deficient and Zn-contaminated soils (Watts-Williams et al., 2017). Novel soil amendments, such as ground tires combined with sulfur oxidizing bacteria, may supply Zn to deficient crops (Asadollahzadeh et al., 2019). While repeated applications of wastes such as sewage sludge to soil can result in contamination, a single application to a deficient soil has been demonstrated to effectively increase the Zn concentration in a variety of crops (Gartler et al., 2013).

4.4 THE UNSUSTAINABLE ACCUMULATION OF TES IN AGRICULTURAL SOILS

This strong bonding to the soil matrix results in the accumulation of TEs that are added to soil, either intentionally or unintentionally. For many TEs, the rate of application to soils exceeds the rate of loss. There are a plethora of non-agricultural anthropogenic activities that can contaminate agricultural soils (Table 4.2), with soil contamination reported in thousands or articles. For most contaminated soils, the most important contaminants in terms of concentration and toxicity (to plants or humans) are As, Cd, Cu, Cr, Hg, Ni, Pb, and Zn. Micó et al. (2006) reported the largest increases caused by anthropogenic activity were for Cd, Cu, and Pb across a range of soils. These TEs may reduce productivity (especially Cu and Zn) or endanger human health (As, Cd, Cr, Hg, Ni, and Pb). The TEs may enter agricultural produce via plant uptake (especially Cd and Ni), be present in attached (or incorporated) soil particles, or become attached to the surfaces of agricultural produce via atmospheric deposition.

New technologies, particularly relating to the semiconductor industry, are resulting in soils becoming contaminated, especially following disposal with novel TE including Li, Ga, and Ge (Robinson, 2009; Jensen et al., 2018; Bolan et al., 2021). The mass of electronic waste, currently at around 50,000,000 t/yr, is rapidly increasing (Andeobu et al., 2021); therefore, there is an imperative to identify novel contaminating TEs and determine their behavior in agricultural systems.

While non-agricultural activities have caused severe contamination of agricultural soil and, in some cases, resulted in human illness, the area of land that has been affected is small compared with the total area of agricultural land on Earth. Peri-urban agricultural land can be repurposed to other non-food producing land uses without greatly affecting humanity's ability to produce food. Better controls on anthropogenic emissions of TEs can reduce or prevent further contamination.

TABLE 4.2
Non-Agricultural Activities That Can Contaminate Agricultural Soils with Trace Elements (Excluding Nuclear)

Activity	Predominant Trace Elements	Example Reference
Mining	As, Cd, Cu, Cr, Hg, Ni, Pb, Tl, Zn	Moreno et al. (2005); Domínguez et al. (2008); Bhuiyan et al. (2010); Barjoee et al. (2021)
Smelting	As, Cd, Cu, Cr, Hg, Ni, Pb, Zn	Sterckeman et al. (2002); Islam et al. (2019)
Industry and energy	As, Cd, Cu, Cr, Hg, Ni, Pb, Zn	Chouaieb et al. (2018)
Transport	Cd, Pb, Zn	Wei and Yang (2010); Taylor and Kruger (2020)
Waste disposal	As, Cu, Cr, Hg, Ni, Pb, Zn	Petrović and Fiket (2022)
E-waste reprocessing	Cd, Cu, Cr, Ga, Ge, Hg, In, Li, Ni, Pb, Sb, Tl, Zn	Robinson (2009)
Military	Hg, Pb, U (depleted), W	Robinson et al. (2008); Tschan et al. (2009)

TABLE 4.3
Agricultural Practices That Add Trace Elements to Soil, Not Including Trace Element Fertilizers

Material	TE	Concentration (mg/kg) (Unless Otherwise Indicated)	Typical Application Rates (kg/ha/yr) Unless Otherwise Indicated	Reference (s)
		TE as Active Ingredient		
Roxarsone (growth promoter for poultry)	As	c.a. 40 mg/kg in litter	0.12 (at 3t/ha litter application)	Gupta et al. (2018)
Copper (fungicide)	Cu	Cu salts	Up to 25	Mills et al. (2005)
Mancozeb (fungicide)	Zn	$C_4H_6N_2S_4Zn$	2.4–7.5	
	Mn	$C_4H_6N_2S_4Mn$		
Crop growth promoters (used in China)	La	n.a.	0.23–2	Tommasi et al. (2021)
	Ce			
Facial eczema prophylaxis	Zn	ZnO	Up to 6.7	Mills et al. (2005); Anderson et al. (2012)
TE as Unwanted Contaminant in Soil Fertilizer, Soil Conditioner, or Irrigation Water				
Phosphate fertilizers at 20–40 kg P/ha/yr	Cd	50–270 mg Cd/kg P	1.8–14 (g/ha/yr)	Nziguheba and Smolders, (2008); Belon et al. (2012); Taylor et al. (2016)
	F	10–75 g F/kg P	0.5–3.8	
	U	150–590 mg U/kg P	7.5–30 (g/ha/yr)	
	Zn	Up to 2,300 mg Zn/kg P	0.5–1.3	
Composts, including animal wastes at 20 t/ha/yr	As	2–46	0.04–0.92	McBride and Spiers (2001); Hseu (2004); Møller et al. (2007); Ko et al. (2008); Smith (2009)
	Cd	0.2–8.4	4–168 (g/ha/yr)	
	Cu	60–640	1.2–13	
	Pb	Up to 190	Up to 3.8	
	Zn	81–770	1.6–15	
Sewage sludge at 20 t/ha/yr	As	6–38	0.12–0.76	Merrington et al. (2003); Speir et al. (2003); Bolan et al. (2004); Smith (2009); Pathak et al. (2009)
	Cd	2.3–10	46–200 (g/ha/yr)	
	Cu	180–2,300	3.6–46	
	Pb	26–500	0.52–10	
	Zn	350–2,400	7.0–48	
Ash at 40 t/ha/yr	As	16–19	0.64–0.76	Nable et al. (1997); Goodarzi and Huggins (2001); Sushil and Batra (2006); Singh et al. (2010)
	B	Up to 1,000	Up to 40	
	Cr	49 – 103	2.0–4.0	
	Cu	50–83	2.0–3.3	
	Ni	26–63	1.5–2.5	
	Pb	20–800	8.0–32	
Treated timber at 580 posts/ha (vineyard), with 4% annual replacement	As	3,000	2.8–4.2	Robinson et al. (2006); Clothier et al. (2006); Greven et al. (2007)
	Cr	5,000	1–2	
	Cu	2,500	3.5–5	
Irrigation water at 500 mm/yr	As	Up to 0.1	Up to 0.5	Robinson et al. (1995); Bhatti et al. (2013); Saber et al. (2021)
	Cd, Zn, Pb	n.a.	Reports of soil contamination	

In terms of sustainability, a greater concern may be TEs that are added to soil as part of normal agricultural practices (Table 4.3). Even in industrial areas, TEs added to soil through agriculture may exceed industrial inputs: In an industrial area in China, atmospheric deposition was responsible for 43–85% of As, Cr, Hg, Ni, and Pb inputs, while livestock manures accounted for 55%, 69%, and 51% of Cd, Cu, and Zn inputs, respectively (Luo et al., 2009). TEs applied in agriculture may be an active ingredient in a pesticide or animal remedy, or they may be present as a contaminant in a fertilizer or soil conditioner. While there is a wide variation in the rates of TE application for individual farming systems, Table 4.3 shows that TE inputs into agricultural soil may well exceed losses, given the high K_D values for many TEs.

Repeated measures of TEs in agricultural soil have shown increases (20%–150%) in Cd, Cu, F, Pb, U, and Zn concentrations in agricultural soil over a few decades (Taylor, 2007; Atafar et al., 2010; Kim et al., 2016; Zhou et al., 2021; Drewry et al., 2021; Gray and Cavanagh, 2022). In some cases, these increases have led to reduced soil fertility, especially due to Cu and Zn (Wightwick et al., 2013), and/or agricultural produce exceeding food safety standards, especially due to Cd (Al Mamun et al., 2016).

TE contamination of agricultural soils can increase antibiotic resistance of pathogenic bacteria through co-selection (Seiler and Berendonk, 2012; Heydari et al., 2022). Much of the soil microbiota is more sensitive to elevated concentrations of TEs than plants (Giller et al., 1998). This has implications for soil functioning, including the degradation of SOM and nitrogen transformations. Because of this, bacteria may be an effective indicator of the severity of soil contamination (Hermans et al., 2017).

Assuming minimal TE losses from agricultural soils, which is reasonable based on the aforementioned repeated measures, Mills et al. (2005) calculated times until soil concentrations reached guideline values for intensive farming practices. For all the land uses tested (market gardens, tree fruit, viticulture, sheep and beef, and dairy), concentrations of either Cd, Cu, or Zn would exceed guideline values in surface (7.5 cm) soils after 10–50 years. Extensive farming systems would take longer to exceed soil guideline values.

4.5 PROLONGING THE LIFE OF AGRICULTURAL SOILS

The accumulation of TEs in agricultural soil may be slowed by reducing inputs. These may include alternative pesticides and animal remedies that do not have TEs as an active ingredient and sources of plant nutrients that contain low concentrations of contaminants. Cadmium and some other contaminants can be removed from fertilizer (Benredjem and Delimi, 2009); however, this is not currently financially feasible for most agricultural systems. Moreover, sources of low-Cd phosphate are becoming more expensive as Earth's phosphate resources become depleted (Li et al., 2019). There is an imperative to find alternative phosphate sources, particularly through the reuse of biological wastes, which can be applied to soil as composts or sludges. However, Zhou et al. (2021) reported that, in China, the long-term application of organic fertilizers increased Cu, Zn, and Cd by 26%, 19%, and 67%, respectively, whereas the increase after long-term inorganic fertilizer use was just 6%. Clearly, this depends on the TE concentrations in the organic amendments. Some composts have TE concentrations (Table 4.3) that are within the ranges found in agricultural soils.

Ploughing soil reduces TE concentrations through dilution but does not arrest their rate of increase. In some cases, ploughing may release TEs through the degradation of soil organic matter to which the TEs are bound. Increasing soil pH can reduce the bioavailability of many TEs and improve soil health. At $pH > 7$, moderately contaminated soils (<100 mg/kg for a suite of TEs) did not have significantly reduced microbial biomass compared to a control (Marzaioli et al., 2022); however, liming may induce Fe deficiencies and, in some cases, increase plant Cd uptake (Al Mamun et al., 2017). The liming effect of fly ash can immobilize TEs in contaminated soils (Dermatas and Meng, 2003). The ash may be used to synthesize TE-absorbing zeolite materials (Querol et al., 2006).

Organic amendments that are low in TEs can immobilize TEs in soil and reduce plant uptake, including lignite coal (Simmler et al., 2013), composts (Bolan et al., 2003; Al Mamun et al., 2016), and animal manures. These studies highlight the importance of maintaining soil organic matter. In addition to reducing soil fertility and releasing greenhouse gases, the degradation of soil organic matter increases the solubility of contaminating TEs. Soil organic matter helps support plant growth-promoting bacteria, which can mitigate some of the toxic effects of TEs as well as reduce plant uptake (Kumar et al., 2008). Correcting a Zn deficiency in a contaminated soil can reduce the plant uptake of contaminating Cd (Chaney et al., 2009; Zhen et al., 2021).

Biochar may reduce plant TE uptake in contaminated soils (Beesley et al., 2011; Kim et al., 2015); however, biochar may introduce contaminating TEs if it is manufactured with timber treated with either Cu, Cr, and As or B (sap stain prevention). Charring concentrates TEs (Wang et al., 2021) and may lead to further soil contamination. Clay minerals, such as sepiolite, palygorskite, and bentonite, have been used for the immobilization of Cd in paddy soils and some farmland soils irrigated with wastewater (Xu et al., 2017). However, clay minerals may reduce soil fertility and their application on a large scale is impractical.

Crops may be selected or bred that exclude TEs or do not translocate TEs to edible portions (Neu et al., 2020). However, the ingestion of contaminated dust on vegetables may still provide an important exposure pathway for TEs to enter humans (Fernández-Caliani et al., 2019). This exposure pathway can be minimized by avoiding low-growing or root crops and instead cultivating tree crops and crops where the edible portion is contained within a pod or skin.

4.6 CONCLUSIONS

TE deficiencies in agriculture are widespread, reduce yields, and result in deficiencies in human diets. Drivers to increase crop yields, particularly the application of urea and inorganic fertilizers, are exacerbating TE deficiencies through loss of soil organic matter and dilution through growth. Nevertheless, there are a suite of techniques that can be used to reverse this situation and increase essential TEs in crop plants, especially the development of new plant varieties and maintaining soil organic carbon. A more intractable problem is the inexorable increase of TEs, particularly Cd, in agricultural soils through their presence in fertilizers and soil conditioners. Current rates of accumulation of Cu, Zn, and Cd in intensive farming systems will lead to their concentrations in soil exceeding guideline values in a few decades at most. As with TE deficiencies, some of the negative effects of TE contamination can be mitigated by maintaining soil organic matter. There is an urgent need to develop low-cost solutions to reduce the TE inputs into agricultural systems. Some TEs, such as F, are rapidly increasing through the application of phosphate fertilizers, yet their effect on agricultural soils is poorly understood. There is also a critical knowledge gap on the effect of TEs on the functioning of the soil microbiota.

REFERENCES

Al Mamun, S., G. Chanson, E. Benyas, M. Aktar, et al. 2016. Municipal composts reduce the transfer of Cd from soil to vegetables. *Environ. Pollut.* 213: 8–15.

Al Mamun, S., N.J. Lehto, J. Cavanagh, R. McDowell, M. Aktar, et al. 2017. Effects of lime and organic amendments derived from varied source materials on cadmium uptake by potato. *J. Environ. Qual.* 46(4): 836–844.

Andeobu, L., S. Wibowo, and S. Grandhi. 2021. A systematic review of E-waste generation and environmental management of Asia Pacific countries. *Int. J. Environ. Res. Public Health* 18(17). doi: 10.3390/ijerph18179051.

Anderson, C.W.N., B.H. Robinson, D.M. West, L. Clucas, and D. Portmann. 2012. Zinc-enriched and zinc-biofortified feed as a possible animal remedy in pastoral agriculture: Animal health and environmental benefits. *J. Geochem. Explor.* 121: 30–35.

Asadollahzadeh, M.J., A.H. Khoshgoftarmanesh, and R.L. Chaney. 2019. Ability of sulfur-oxidising bacteria to hasten degradation of ground rubber particles in soil for release of zinc as a fertiliser to correct deficiency in wheat. *Crop Pasture Sci.* 70(1): 26–35.

Atafar, Z., A. Mesdaghinia, J. Nouri, M. Homaee, M. Yunesian, et al. 2010. Effect of fertilizer application on soil heavy metal concentration. *Environ. Monit. Assess.* 160(1–4): 83–89.

Baker, A.J.M. 1981. Accumulators and excluders-strategies in the response of plants to heavy metals. *J. Plant Nutr*. https://www.tandfonline.com/doi/abs/10.1080/01904168109362867.
Barjoee, S.S., S.Z.M. Abadi, M.R. Elmi, V.T. Varaoon, and M. Nikbakht. 2021. Evaluation of trace elements pollution in deposited dust on residential areas and agricultural lands around Pb/Zn mineral areas using modified pollution indices. *J Environ Health Sci Eng*. 19(1): 753–769.
Beesley, L., E. Moreno-Jiménez, J.L. Gomez-Eyles, E. Harris, B. Robinson, et al. 2011. A review of biochars' potential role in the remediation, revegetation and restoration of contaminated soils. *Environ. Pollut*. 159(12): 3269–3282.
Belon, E., M. Boisson, I.Z. Deportes, T.K. Eglin, I. Feix, et al. 2012. An inventory of trace elements inputs to French agricultural soils. *Sci. Total Environ*. 439: 87–95.
Benredjem, Z., and R. Delimi. 2009. Use of extracting agent for decadmiation of phosphate rock. *Phys. Procedia* 2(3): 1455–1460.
Bhatti, S.M., C.W.N. Anderson, R.B. Stewart, and B.H. Robinson. 2013. Risk assessment of vegetables irrigated with arsenic-contaminated water. *Environ. Sci. Process. Impacts* 15(10): 1866–1875.
Bhuiyan, M.A.H., L. Parvez, M.A. Islam, S.B. Dampare, and S. Suzuki. 2010. Heavy metal pollution of coal mine-affected agricultural soils in the northern part of Bangladesh. *J. Hazard. Mater*. 173(1–3): 384–392.
Bindraban, P.S., C.O. Dimkpa, and R. Pandey. 2020. Exploring phosphorus fertilizers and fertilization strategies for improved human and environmental health. *Biol. Fertil. Soils* 56(3): 299–317.
Blanco-Canqui, H., and R. Lal. 2009. Crop residue removal impacts on soil productivity and environmental auality. *CRC Crit. Rev. Plant Sci*. 28(3): 139–163.
Bolan, N.S., D.C. Adriano, P. Duraisamy, and A. Mani. 2003. Immobilization and phytoavailability of cadmium in variable charge soils. III. Effect of biosolid compost addition. *Plant Soil* 256: 231–241.
Bolan, N.S., D.C. Adriano, and S. Mahimairaja. 2004. Distribution and bioavailability of trace elements in livestock and poultry manure by-products. *Crit. Rev. Environ. Sci. Technol*. 34(3): 291–338.
Bolan, N.S., S.A. Hoang, M. Tanveer, L. Wang, and S. Bolan. 2021. From mine to mind and mobiles–Lithium contamination and its risk management. *Environmentalist*. https://www.sciencedirect.com/science/article/pii/S0269749121016493.
Bourennane, H., F. Douay, T. Sterckeman, E. Villanneau, H. Ciesielski, et al. 2010. Mapping of anthropogenic trace elements inputs in agricultural topsoil from Northern France using enrichment factors. *Geoderma* 157(3–4): 165–174.
Brennan, R.F., M.D.A. Bolland, and R.W. Bell. 2005. Increased risk of zinc deficiency in wheat on soils limed to correct soil acidity. *Soil Res*. 43(5): 647–657.
Cakmak, I. 2008. Enrichment of cereal grains with zinc: Agronomic or genetic biofortification? *Plant Soil* 302(1–2): 1–17.
Chakraborty, S., M. Mandal, A.P. Chakraborty, and S. Majumdar. 2022. Zinc solubilizing rhizobacteria as soil health engineer managing zinc deficiency in plants. In: *Rhizosphere Engineering*. Amsterdam: Elsevier. pp. 215–238.
Chaney, R.L., C.E. Green, H.A. Ajwa, and R.F. Smith. 2009. Zinc Fertilization Plus Liming to Reduce Cadmium Uptake by Romaine Lettuce on Cd-Mineralized Lockwood Soil. https://escholarship.org/uc/item/5js5s736 (accessed 9 July 2022).
Chouaieb, L., A. Hatira, and N. Gabteni. 2018. Assessment of selected metal trace elements from industrial activities in the agricultural surface soil of Menzel Bourguiba (Tunisia). *Environ. Forensics* 19(4): 277–287.
Clothier, B.E., S.R. Green, I. Vogeler, M.M. Greven, R. Agnew, et al. 2006. CCA transport in soil from treated-timber posts: Pattern dynamics from the local to regional scale. *Hydrol. Earth Syst. Sci. Discuss*. 3(4): 2037–2061.
Csathó, P., T. Árendás, A. Szabó, R. Sándor, P. Ragályi, et al. 2019. Phosphorus-induced zinc deficiency in maize (*Zea mays* L.) on a calcareous chernozem soil. *Agrokem. Talajtan* 68: 40–52.
Dermatas, D., and X. Meng. 2003. Utilization of fly ash for stabilization/solidification of heavy metal contaminated soils. *Eng. Geol*. 70(3–4): 377–394.
Dhaliwal, S.S., R.K. Naresh, A. Mandal, R. Singh, and M.K. Dhaliwal. 2019. Dynamics and transformations of micronutrients in agricultural soils as influenced by organic matter build-up: A review. *Environ. Sustain. Indic*. 1–2: 100007.
Domínguez, M.T., T. Marañón, J.M. Murillo, R. Schulin, and B.H. Robinson. 2008. Trace element accumulation in woody plants of the Guadiamar Valley, SW Spain: A large-scale phytomanagement case study. *Environ. Pollut*. 152(1): 50–59.
Drewry, J.J., J.-A.E. Cavanagh, S.J. McNeill, B.A. Stevenson, D.A. Gordon, et al. 2021. Long-term monitoring of soil quality and trace elements to evaluate land use effects and temporal change in the Wellington region, New Zealand. *Geoderma Reg*. 25: e00383.

Evangelou, M.W.H., and H.M. Conesa. 2012. Biomass production on trace element–contaminated land: A review. *Environmentalist*. https://www.liebertpub.com/doi/abs/10.1089/ees.2011.0428.

Evangelou, M.W.H., E.G. Papazoglou, B.H. Robinson, and R. Schulin. 2015. Phytomanagement: phytoremediation and the production of biomass for economic revenue on contaminated land. In: Ansari, A.A., Gill, S.S., Gill, R., Lanza, G.R., and Newman, L., editors, *Phytoremediation: Management of Environmental Contaminants*, Volume 1. Springer International Publishing, Cham. pp. 115–132

Fernández-Caliani, J.C., M.I. Giráldez, and C. Barba-Brioso. 2019. Oral bioaccessibility and human health risk assessment of trace elements in agricultural soils impacted by acid mine drainage. *Chemosphere* 237: 124441.

Gartler, J., B. Robinson, K. Burton, and L. Clucas. 2013. Carbonaceous soil amendments to biofortify crop plants with zinc. *Sci. Total Environ.* 465: 308–313.

Giller, K.E., E. Witter, and S.P. Mcgrath. 1998. Toxicity of heavy metals to microorganisms and microbial processes in agricultural soils: A review. *Soil Biol. Biochem.* 30(10–11): 1389–1414.

Goodarzi, F., and F.E. Huggins. 2001. Monitoring the species of arsenic, chromium and nickel in milled coal, bottom ash and fly ash from a pulverized coal-fired power plant in western Canada. *J. Environ. Monit.* 3(1): 1–6.

Grace, N.D., and S.O. Knowles. 2012. Trace element supplementation of livestock in New Zealand: Meeting the challenges of free-range grazing systems. *Vet. Med. Int.* 2012: 639472.

Grace, N.D., P.R. Wilson, and A.K. Quinn. 2005. Impact of molybdenum on the copper status of red deer (*Cervus elaphus*). *N. Z. Vet. J.* 53(2): 137–141.

Gray, C.W., and J.-A.E. Cavanagh. 2022. The state of knowledge of cadmium in New Zealand agricultural systems: 2021. *N. Z. J. Agric. Res.* 1–51.

Greven, M., S. Green, B. Robinson, B. Clothier, I. Vogeler, et al. 2007. The impact of CCA-treated posts in vineyards on soil and ground water. *Water Sci. Technol.* 56(2): 161–168.

Gupta, S.K., X.C. Le, G. Kachanosky, M.J. Zuidhof, and T. Siddique. 2018. Transfer of arsenic from poultry feed to poultry litter: A mass balance study. *Sci. Total Environ.* 630: 302–307.

Hermans, S.M., H.L. Buckley, B.S. Case, F. Curran-Cournane, M. Taylor, et al. 2017. Bacteria as emerging indicators of soil condition. *Appl. Environ. Microbiol.* 83(1). doi: 10.1128/AEM.02826-16.

Heydari, A., N.D. Kim, J. Horswell, G. Gielen, A. Siggins, et al. 2022. Co-selection of heavy metal and antibiotic resistance in soil bacteria from agricultural soils in New Zealand. *Sustainability (Switzerland)* 14(3). doi: 10.3390/su14031790.

Hseu, Z.-Y. 2004. Evaluating heavy metal contents in nine composts using four digestion methods. *Bioresour. Technol.* 95(1): 53–59.

Islam, M.A., M.A. Akber, M.B. Rahman, M.A. Rahman, M.A. Haque, et al. 2019. Trace elements in rice grain and agricultural soils: Assessment of health risk of inhabitants near a former secondary lead smelter in Khulna, Bangladesh. *Environ. Geochem. Health* 41(6): 2521–2532.

Jaishankar, M., T. Tseten, N. Anbalagan, B.B. Mathew, and K.N. Beeregowda. 2014. Toxicity, mechanism and health effects of some heavy metals. *Interdiscip. Toxicol.* 7(2): 60–72.

Jensen, H., S. Gaw, N.J. Lehto, L. Hassall, and B.H. Robinson. 2018. The mobility and plant uptake of gallium and indium, two emerging contaminants associated with electronic waste and other sources. *Chemosphere* 209: 675–684.

Jensen, H., B. Orth, R. Reiser, D. Bürge, N.J. Lehto, et al. 2019. Environmental parameters affecting the concentration of iodine in New Zealand pasture. *J. Environ. Qual.* 48(5): 1517–1523.

Kim, H.-S., K.-R. Kim, H.-J. Kim, J.-H. Yoon, J.E. Yang, et al. 2015. Effect of biochar on heavy metal immobilization and uptake by lettuce (*Lactuca sativa* L.) in agricultural soil. *Environ. Earth Sci.* 74(2): 1249–1259.

Kim, N.D., Taylor, and J.J. Drewry. 2016. Anthropogenic fluorine accumulation in the Waikato and Bay of Plenty regions of New Zealand: Comparison of field data with projections. *Environ. Earth Sci.* 75(2): 1–15.

Ko, H.J., K.Y. Kim, H.T. Kim, C.N. Kim, and M. Umeda. 2008. Evaluation of maturity parameters and heavy metal contents in composts made from animal manure. *Waste Manage.* 28(5): 813–820.

Kowalska, J.B., R. Mazurek, M. Gąsiorek, and T. Zaleski. 2018. Pollution indices as useful tools for the comprehensive evaluation of the degree of soil contamination–A review. *Environ. Geochem. Health* 40(6): 2395–2420.

Kumar, K.V., N. Singh, H.M. Behl, and S. Srivastava. 2008. Influence of plant growth promoting bacteria and its mutant on heavy metal toxicity in *Brassica juncea* grown in fly ash amended soil. *Chemosphere* 72(4): 678–683.

Kutman, U.B., B. Yildiz, L. Ozturk, and I. Cakmak. 2010. Biofortification of durum wheat with zinc through soil and foliar applications of nitrogen. *Cereal Chem.* 87(1): 1–9.

Li, B., K.B. Bicknell, and A. Renwick. 2019. Peak phosphorus, demand trends and implications for the sustainable management of phosphorus in China. *Resour. Conserv. Recycl.* 146: 316–328.

Luo, L., Y. Ma, S. Zhang, D. Wei, and Y.-G. Zhu. 2009. An inventory of trace element inputs to agricultural soils in China. *J. Environ. Manage.* 90(8): 2524–2530.

Marschner, H. 1991. Mechanisms of adaptation of plants to acid soils. *Plant Soil* 134(1): 1–20.

Marschner, H. 1995. *Mineral Nutrition of Higher Plants.* Amsterdam: Elsevier.

Marzaioli, R., M. Trifuoggi, and F.A. Rutigliano. 2022. Soil microbial biomass, activities and diversity in Southern Italy areas chronically exposed to trace element input from industrial and agricultural activities. *Appl. Soil Ecol.* 174. doi: 10.1016/j.apsoil.2022.104392.

Mason, B., and C.B. Moore. 1985. Principles of Geochemistry. https://inis.iaea.org/search/search.aspx?orig_q=RN:21063393.

McBride, M.B., and G. Spiers. 2001. Trace element content of selected fertilizers and dairy manures as determined by ICP-MS. *Commun. Soil Sci. Plant Anal.* 32(1–2): 139–156.

Merrington, G., I. Oliver, R.J. Smernik, and M.J. McLaughlin. 2003. The influence of sewage sludge properties on sludge-borne metal availability. *Adv. Environ. Res.* 1(8): 21–36.

Micó, C., L. Recatalá, M. Peris, and J. Sánchez. 2006. Assessing heavy metal sources in agricultural soils of an European Mediterranean area by multivariate analysis. *Chemosphere* 65(5): 863–872.

Mills, T.M., B.H. Robinson, S. Sivakumaran, B. Arnold, B.E. Clothier, et al. 2005. Current practice and future land-use: The sustainability of productive sector environments. *Acta Hortic.* 694: 159–164.

Møller, H.B., H.S. Jensen, L. Tobiasen, and M.N. Hansen. 2007. Heavy metal and phosphorus content of fractions from manure treatment and incineration. *Environ. Technol.* 28(12): 1403–1418.

Moreno, F.N., C.W.N. Anderson, R.B. Stewart, and B.H. Robinson. 2005. Mercury volatilisation and phytoextraction from base-metal mine tailings. *Environ. Pollut.* 136(2): 341–352.

Nable, R.O., G.S. Bañuelos, and J.G. Paull. 1997. Boron toxicity. *Plant Soil* 193(1/2): 181–198.

Nestel, P., H.E. Bouis, J.V. Meenakshi, and W. Pfeiffer. 2006. Biofortification of staple food crops. *J. Nutr.* 136(4): 1064–1067.

Neu, S., I. Müller, and E.G. Dudel. 2020. Management of trace element-contaminated agricultural land by in situ stabilization combined with phytoexclusion over a three years crop rotation. *Int. J. Phytoremediation* 22(10): 1059–1067.

Nziguheba, G., and E. Smolders. 2008. Inputs of trace elements in agricultural soils via phosphate fertilizers in European countries. *Sci. Total Environ.* 390(1): 53–57.

Olife, I.C., A.N. Okaka, C.E. Dioka, S.C. Meludu, and O.E. Orisakwe. 2007. Iodine status and the effect of soil erosion on trace elements in Nanka and Oba towns of Anambra State, Nigeria. *Ann. Chim.* 97(9): 895–903.

Park, J.H., G.K. Choppala, N.S. Bolan, J.W. Chung, and T. Chuasavathi. 2011. Biochar reduces the bioavailability and phytotoxicity of heavy metals. *Plant Soil* 348(1–2): 439–451.

Pathak, A., M.G. Dastidar, and T.R. Sreekrishnan. 2009. Bioleaching of heavy metals from sewage sludge: A review. *J. Environ. Manage.* 90(8): 2343–2353.

Percival, H.J., T.H. Webb, and T.W. Speir. 1996. Assessment of background concentrations of selected determinands in Canterbury soils. Landcare Research Contract Report LC9596/133, Landcare Research Ltd, New Zealand.

Petrović, M., and Ž. Fiket. 2022. Environmental damage caused by coal combustion residue disposal: A critical review of risk assessment methodologies. *Chemosphere* 299: 134410.

Pfeiffer, W.H., and B. McClafferty. 2007. HarvestPlus: Breeding crops for better nutrition. *Crop Sci.* 47: S-88–S-105.

Querol, X., A. Alastuey, N. Moreno, E. Alvarez-Ayuso, A. García-Sánchez, et al. 2006. Immobilization of heavy metals in polluted soils by the addition of zeolitic material synthesized from coal fly ash. *Chemosphere* 62(2): 171–180.

Reid, R., and J. Hayes. 2003. Mechanisms and control of nutrient uptake in plants. *Int. Rev. Cytol.* 73–114. doi: 10.1016/s0074-7696(03)29003-3.

Reiser, R., M. Simmler, D. Portmann, L. Clucas, R. Schulin, et al. 2014. Cadmium concentrations in new zealand pastures: Relationships to soil and climate variables. *J. Environ. Qual.* 43(3): 917–925.

Robinson, B.H. 2009. E-waste: An assessment of global production and environmental impacts. *Sci. Total Environ.* 408(2): 183–191.

Robinson, B.H., G. Bañuelos, H.M. Conesa, M.W.H. Evangelou, and R. Schulin. 2009. The phytomanagement of trace elements in soil. *CRC Crit. Rev. Plant Sci.* 28(4): 240–266.

Robinson, B.H., S. Bischofberger, A. Stoll, D. Schroer, G. Furrer, et al. 2008. Plant uptake of trace elements on a Swiss military shooting range: Uptake pathways and land management implications. *Environ. Pollut.* 153(3): 668–676.

Robinson, B., N. Bolan, S. Mahimairaja, and B. Clothier. 2005. *Solubility, Mobility, and Bioaccumulation of Trace Elements: Abiotic Pro cesses in the Rhizosphere.* https://books.google.ca/books?hl=en&lr=&id=TRzNBQAAQBAJ&oi=fnd&pg=PA97&ots=vY6PR8qQMu&sig=0yb-bnbejGRSik1qOLLyLg2q90E.

Robinson, B.H., S.R. Green, B. Chancerel, T.M. Mills, and B.E. Clothier. 2007. Poplar for the phytomanagement of boron contaminated sites. *Environ. Pollut.* 150(2): 225–233.

Robinson, B., M. Greven, S. Green, S. Sivakumaran, P. Davidson, et al. 2006. Leaching of copper, chromium and arsenic from treated vineyard posts in Marlborough, New Zealand. *Sci. Total Environ.* 364(1–3): 113–123.

Robinson, B., H. Outred, R. Brooks, and J. Kirkman. 1995. The distribution and fate of arsenic in the Waikato River system, North Island, New Zealand. *Chem Speciation* 7(3): 89–96.

Robinson, B.H., R. Yalamanchali, R. Reiser, and N.M. Dickinson. 2018. Lithium as an emerging environmental contaminant: Mobility in the soil-plant system. *Chemosphere* 197: 1–6.

Saber, S., N. Benlkhoubi, A. Lebkiri, E.H. Rifi, and E. Elfahime. 2021. Bioaccumulation of trace elements in agricultural crops and soils irrigated by the surface waters of Sebou and Beht Rivers. *Arab Gulf J. Sci. Res.* 39(2): 59–68.

Sacristán, D., A. González–Guzmán, V. Barrón, J. Torrent, and M.C. Del Campillo. 2019. Phosphorus-induced zinc deficiency in wheat pot-grown on noncalcareous and calcareous soils of different properties. *Arch. Agron. Soil Sci.* 65(2): 208–223.

Sanderman, J., T. Hengl, and G.J. Fiske. 2017. Soil carbon debt of 12,000 years of human land use. *Proc. Natl. Acad. Sci. U. S. A.* 114(36): 9575–9580.

Seiler, C., and T.U. Berendonk. 2012. Heavy metal driven co-selection of antibiotic resistance in soil and water bodies impacted by agriculture and aquaculture. *Front. Microbiol.* 3: 399.

Shukla, A.K., S.K. Behera, C. Prakash, A. Tripathi, A.K. Patra, et al. 2021. Deficiency of phyto-available sulphur, zinc, boron, iron, copper and manganese in soils of India. *Sci. Rep.* 11(1). doi: 10.1038/s41598-021-99040-2.

Simmler, M., L. Ciadamidaro, R. Schulin, P. Madejón, R. Reiser, et al. 2013. Lignite reduces the solubility and plant uptake of cadmium in pasturelands. *Environ. Sci. Technol.* 47(9): 4497–4504.

Singh, R., D.P. Singh, N. Kumar, S.K. Bhargava, and S.C. Barman. 2010. Accumulation and translocation of heavy metals in soil and plants from fly ash contaminated area. *J. Environ. Biol.* 31(4): 421–430.

Smith, S.R. 2009. A critical review of the bioavailability and impacts of heavy metals in municipal solid waste composts compared to sewage sludge. *Environ. Int.* 35(1): 142–156.

Speir, T.W., A.P. Van Schaik, H.J. Percival, M.E. Close, and L. Pang. 2003. Heavy metals in soil, plants and groundwater following high-rate sewage sludge application to land. *Water Air Soil Pollut.* 150(1/4): 319–358.

Sposito, G. 2016. *The Chemistry of Soils.* Oxford University Press, New York.

Sterckeman, T., F. Douay, N. Proix, H. Fourrier, and E. Perdrix. 2002. Assessment of the contamination of cultivated soils by eighteen trace elements around smelters in the north of France. *Water Air Soil Pollut.* 135(1): 173–194.

Sushil, S., and V.S. Batra. 2006. Analysis of fly ash heavy metal content and disposal in three thermal power plants in India. *Fuel* 85(17): 2676–2679.

Taylor, M. 2007. Accumulation of uranium in soils from impurities in phosphate fertilisers. *Landbauforschung Volkenrode* 57(2): 133–139.

Taylor, M., N. Kim, G. Smidt, C. Busby, S. McNally, et al. 2016. Trace element contaminants and radioactivity from phosphate fertiliser. In: Schnug, E. and De Kok, L.J., editors, *Phosphorus in Agriculture: 100% Zero.* Springer Netherlands, Dordrecht. pp. 231–266

Taylor, M., and N. Kruger. 2020. Tyre weights an overlooked diffuse source of lead and antimony to road run-off. *Sustainability (Switzerland)* 12(17). doi: 10.3390/SU12176790.

Tommasi, F., P.J. Thomas, G. Pagano, G.A. Perono, R. Oral, et al. 2021. Review of rare earth elements as fertilizers and feed additives: A knowledge gap analysis. *Arch. Environ. Contam. Toxicol.* 81(4): 531–540.

Tschan, M., B.H. Robinson, and R. Schulin. 2009. Antimony in the soil–plant system–a review. *Huanjing Huaxue.* http://www.publish.csiro.au/EN/EN08111.

United Nations. 2022. Population | United Nations. https://www.un.org/en/global-issues/population (accessed 6 July 2022).

Valentinuzzi, F., T. Mimmo, S. Cesco, S. Al Mamun, J. Santner, et al. 2015. The effect of lime on the rhizosphere processes and elemental uptake of white lupin. *Environ. Exp. Bot.* 118: 85–94.

Violante, A., V. Cozzolino, L. Perelomov, A.G. Caporale, and M. Pigna. 2010. Mobility and bioavailability of heavy metals and metalloids in soil environments. *J. Soil Sci. Plant Nutr.* 10(3): 268–292.

Vithanage, M., P. Kumarathilaka, C. Oze, S. Karunatilake, M. Seneviratne, et al. 2019. Occurrence and cycling of trace elements in ultramafic soils and their impacts on human health: A critical review. *Environ. Int.* 131: 104974.
Wang, C., N. Zheng, S. Wan, and J. Wang. 2021. Assessment of the modes of occurrence of trace elements in agricultural crop residues and their enrichments and bioavailability in bio-chars. *Biomass Convers. Biorefin.* 11(5): 2065–2077.
Watts-Williams, S.J., S.D. Tyerman, and T.R. Cavagnaro. 2017. The dual benefit of arbuscular mycorrhizal fungi under soil zinc deficiency and toxicity: Linking plant physiology and gene expression. *Plant Soil* 420(1–2): 375–388.
Wei, B., and L. Yang. 2010. A review of heavy metal contaminations in urban soils, urban road dusts and agricultural soils from China. *Microchem. J.* 94(2): 99–107.
White, P.J., and M.R. Broadley. 2009. Biofortification of crops with seven mineral elements often lacking in human diets--iron, zinc, copper, calcium, magnesium, selenium and iodine. *New Phytol.* 182(1): 49–84.
Wightwick, A.M., S.A. Salzman, S.M. Reichman, G. Allinson, and N.W. Menzies. 2013. Effects of copper fungicide residues on the microbial function of vineyard soils. *Environ. Sci. Pollut. Res. Int.* 20(3): 1574–1585.
Xu, Y., X. Liang, Y. Xu, X. Qin, Q. Huang, et al. 2017. Remediation of heavy metal-polluted agricultural soils using clay minerals: A review. *Pedosphere* 27(2): 193–204.
Zhen, S., H. Shuai, C. Xu, G. Lv, X. Zhu, et al. 2021. Foliar application of Zn reduces Cd accumulation in grains of late rice by regulating the antioxidant system, enhancing Cd chelation onto cell wall of leaves, and inhibiting Cd translocation in rice. *Sci. Total Environ.* 770: 145302.
Zhou, S., S. Su, L. Meng, X. Liu, H. Zhang, et al. 2021. Potentially toxic trace element pollution in long-term fertilized agricultural soils in China: A meta-analysis. *Sci. Total Environ.* 789: 147967.

5 Chromium Contamination in Soil from Mining Activities

A Constraint for Crop Production and Potential Mitigation Strategies

Arati Ray[1,2], *Manish Kumar*[1,2], *Adnan Asad Karim*[2], *Kushalindu Biswas*[2], *and Binoy Sarkar*[3]
[1]Academy of Scientific and Innovative Research (AcSIR), CSIR-Human Resource Development Centre (CSIR-HRDC) Campus, Ghaziabad, Uttar Pradesh, India
[2]Environment & Sustainability Department, CSIR-Institute of Minerals and Materials Technology, Bhubaneswar, Odisha, India
[3]Future Industries Institute, University of South Australia, Mawson Lakes, Australia

CONTENTS

5.1 INTRODUCTION

The heavy metal chromium (Cr) was discovered by the French chemist Louis Vauquelin in 1797. The name was coined from the Greek word "Chroma" due to the appearance of different colors from Cr-bearing compounds. Properties of Cr are as follows: steely grey, lustrous, hard, and brittle; takes high polish; resists tarnishing; and has a high melting point. There are six valence states of Cr out of which the metallic forms can exist as Cr(0), Cr(III), or Cr(VI) (Coetzee et al. 2020). Cr(III) is less toxic and insoluble, whereas Cr(VI) is extremely toxic and highly soluble. According to Du et al. (2022), Cr(VI) is 100 times more toxic and 1,000 times more mutagenic than Cr(III).

DOI: 10.1201/9781003093565-5

Compounds of Cr commonly are formed with iron (Fe), magnesium (Mg), or aluminium (Al), and they are called as chromate (Fe/Mg/Al) Cr_2O_4, which is the single commercially available ore of Cr. The Cr_2O_3 content of chromite of economic importance varies between 15% and 65% (Shtiza et al. 2005). The world's highest reserves of chromite ore are present in South Africa (72%–80%), Russia, and Kazakhstan (Coetzee et al. 2020). Different industries utilize Cr as a component of various products. Estimated annual world consumption of Cr is about 12.5 million tonnes, which are primarily utilized as follows: metallurgical (85%); leather tanning, textiles, mechanical and mining engineering, chemical (8%); metal finishing, ferrochrome, pigments, electroplating, photography, and refractories (7%) (Gu et al. 2021). Although Cr is a highly required element in many industries, Cr^{6+} causes toxic metal contamination in nearby air and water as well as soil systems (Whalley et al. 1999).

Both natural and anthropogenic activities contribute to soil Cr contamination. In this chapter, "mining" as one of the important contributors to soil Cr contamination will be described. Along with many advantages, Cr mining has several disadvantages for the environment and ecosystems. Chromite mining is primarily done as opencast and underground mining. Drilling and blasting procedures are commonly utilized in opencast mining. Mining is done by removal of an upper layer of soil, which is the vital part of soil for physical, chemical, and biological activities. The excavation activities during mining cause stripping of the soil surface, which leads to soil erosion followed by loss of plant diversity in that area. Continuous deposition of mine wastes (with mineral Cr and sterile waste) causes soil pollution in nearby areas. Mining causes landscape degradation, ecosystem disturbances, and biodiversity loss. According to Susaj (2017), the Cr content of soil taken from 500 and 1,500 m away from chromite mining site were 615 and 448 mg/kg, respectively, which means food crops for human and feed materials grown for animals in that area will be poisonous when used. According to Iyer and Mastorakis (2009), chromite mining soil is generally contaminated with 6,000–18,000 mg/kg of total Cr affecting the whole surrounding environment. Maximum allowable levels of Cr in agricultural soils were suggested as 50–100 mg/kg (Sasmaz and Yaman 2006), but it can also vary up to 400 mg/kg for agricultural soil as the maximum admissible concentration (Fordyce et al. 2017). Considering different plant species (on a dry-weight basis), Cr > 5.0 mg/kg is considered as excessive and toxic (Kabata-Pendias and Pendias 2001). Other than chromite mining, nickel, coal, iron, and gold mining also cause agricultural soil Cr contamination, which is schematically presented in Figure 5.1.

Ladwani et al. (2012) collected soil from near lignite coal mine tailings in Surat in Gujarat State in India and tested it for the presence of Cr using the toxicity characteristic leaching procedure (TCLP) method. They got Cr concentrations around 3–8.0 mg/kg. Gunkel-Grillon et al. (2014) reported that the Cr concentration in water near soil by a nickel opencast ore mine in New Caledonia was 0.73 mg/L. Gold mining also contributes to the Cr contamination of agricultural soil. Soil collected from an area surrounding a gold mine in Ijana, southwestern Nigeria, had a Cr concentration 0.148 mg/kg (Abiya et al. 2018). Concentration Cr in soil near an iron mining area in Khouribga Province, Central Morocco, was 51–222 mg/kg (Madani et al. 2015).

High Cr content in agricultural soil adversely affects animals as well as human beings. Nasal inhalation of Cr(VI) is more dangerous in human beings than oral intake. It induces skin allergies and is carcinogenic to humans via inhalation (Shtiza et al. 2005). It also causes Cr poisoning, ulcers, lung cancer, liver necrosis, brain damage, premature death, and kidney problems. Because of all the negative impacts of Cr along with its high toxicity at each trophic level of the ecosystem, Cr is considered to be a priority pollutant for monitoring and control. According to China's "Outline of the 13th Five-Year Plan for Land and Resources (2016–2020)," more than 5,000 km^2 of historical mines need to be restored, and methods for reclamation of mining wastelands are in increasing demand (Wang et al. 2022). Since soil is a non-renewable resource, the United Nations has set eight sustainable developmental goals for restoring a healthy soil environment by 2030 (Hou et al. 2020). The goals stress the prevention and reclamation of soils contaminated through anthropogenic activities such as mining.

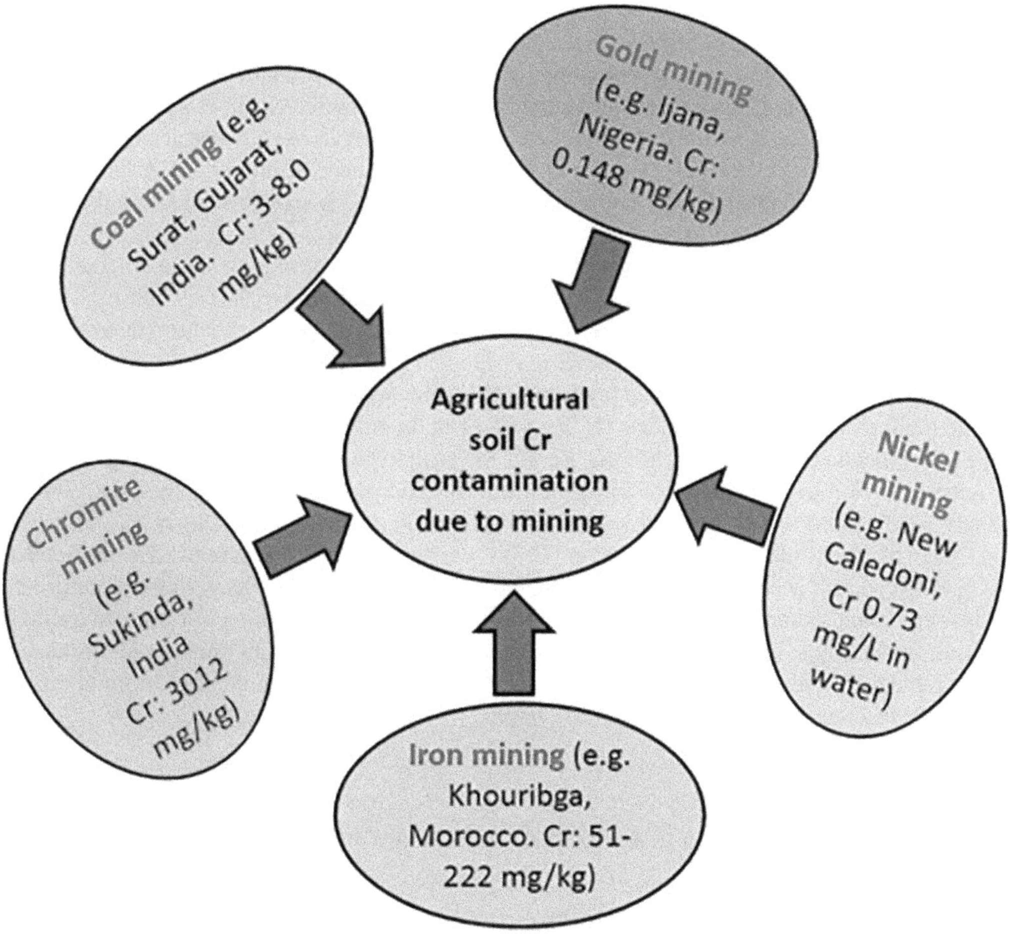

FIGURE 5.1 Cr contamination in agricultural soils due to different mining activities.

This chapter aims to give an overview of negative impacts of chromite mining on agricultural soils, illustrates some potential mitigation strategies to reduce exposure risks to humans and animals, and provides a perspective for future scientific development to tackle the Cr contamination issue arising from mining activities. The chapter also briefly describes a case study in India where soil Cr contamination, and its effects on soil physicochemical and biological parameters were evaluated recently.

5.2 NEGATIVE IMPACTS OF Cr(VI) CONTAMINATION IN AGRICULTURAL SOILS FROM MINING ACTIVITIES

Mining activities include metal ore mining, refinement, transport of ores, smelting, and metal finishing followed by recycling of metals. In Cr mining areas, the overburden produced from mining activities is sometimes dumped near rivers and streams. Leachates of Cr(VI) along with its runoff from the overburden are transported as either dissolved species in water or an integral part of suspended sediments, which show deleterious effects on surrounding ecosystems. The potential of Cr(VI) contamination increases when mining activities expose metal-bearing ores, and dumping of mined ores on the earth surface increases the chances of contamination to surrounding ecosystems. Due to all these activities, mining areas are highly prone to Cr(VI) pollution.

Metal toxicity has a high impact on plants growing in metal-polluted sites. They exhibit metal accumulation and disruption of cell membranes and chloroplasts, altered metabolism, growth reduction (particularly root), low seed germination, and low biomass production (Shahid et al. 2017). Various physiological (e.g., photosynthesis) and biochemical (impaired enzymes such as superoxide dismutase and catalase) activities of plants are affected. Reactive oxygen species are generated (Eleftheriou et al. 2015). Not only does Cr mining affect the surrounding air, water, and soil, but it also completely contaminates the entire ecosystem. The Balochistan Province of Pakistan has many chromite reserves (~2,000 km^2). It is well known for being a chromite hub and contains 325 open-cast and underground mines. It generates millions of tonnes of mining waste (Chandio et al. 2021). Chandio et al. (2021) reported that the Cr content of that area (6,900–19,500 mg/kg) was above the acceptable limit and caused carcinogenic and non-carcinogenic effects on human health. According to Das et al. (2013), a chromite mining site and nearby overburden soil samples showed altered physicochemical and microbial properties. Soil samples collected from nearby mining areas showed low concentrations of nutrients (N, P, and K) and organic carbon content as well as high Cr along with other heavy metals (Fe, Mn, Ni, Zn, Pb, Sr), which exceeded the eco-toxicological limits. Extreme reduction in the number of different useful microorganisms (45×10^5 bacteria, 20×10^5 fungus) were caused due to lowered vegetation in the contaminated areas (Das et al. 2013: Dhal et al. 2011). Mahimairaja et al. (2011) reported Cr(VI) contamination in soil caused 25%–40% reduction of crop yields over several years. Contamination of soil with Cr(VI) affects plants directly, and, because ruminant animals feed on the plants, it was present in blood plasma of ruminant animals, resulting in the whole ecosystem to be contaminated (Khan et al. 2020b: Otomo and Shikazono 2008). Table 5.1 highlights the variable presence of Cr in agricultural soils near chromite mining areas from different countries.

TABLE 5.1
Chromium (Cr) Concentration in Agricultural Soils Near Chromite Mining Areas of Different Countries

Mining Area	Cr-Contaminated Agricultural Soil	Cr Concentration	Reference
Baghjar chromite mine, Sabzevar ophiolite belt, Northeastern Iran	Soil samples taken from 0 to 20 cm depth	156.19 mg/kg	Solgi and Parmah (2015)
Co Dinh, chromite mine, Vietnam	Soil samples collected from 2.5 to 6 km away from mining site	100–5,750 mg/kg	Kien et al. (2011)
Pangkajene River, Kepulauan regency, Indonesia	Soil samples were collected 5 km away from the river, at a depth of 0 – 20 cm	92.9 mg/kg	Astuti et al. (2021)
Soil collected from sites close to mines or industrial sites, China	Soil samples were collected from 0 to 15/20 cm	1.48–820.24 mg/kg	Li et al. (2020)
Rustenburg mining town, North West Province, South Africa	Agricultural soil was collected near the mining site	Cr_2O_3 content 2224 mg/kg	Mosai et al. (2017)
Away from mining area, Sinop province, Turkey	Agricultural topsoil collected from 0 to 10 cm depth	23.64 and 761.36 mg/kg	Baltas et al. (2020)
Mining area, China	Soil collected from 0 to 15/20 cm of topsoil	Cr content 67.3 mg/kg	Huang et al. (2019)
Sukinda mine, Odisha, India	In situ and overburden mining soil collected	3,012 and 1,991 mg/kg	Pradhan et al. (2020)
Sukinda chromite mine, India	Soil collected from mine, roadside, agricultural, and control sites	52.35 and 244.8 mg/kg Cr(VI)	Naz et al. (2018)
Urban areas of India (away from mining areas)	Average Cr content of soil	412.28 mg/kg	Adimalla et al. (2022)

The situation of soil Cr contamination in China is alarming. The Ministry of Land Resources and the Ministry of Environmental Protection of China reported that ~2.9% of the arable land in China was moderately or highly contaminated with different pollutants (Khan et al. 2021). About 20,000,000 acres (25% of the total farmland area) of Chinese agricultural land have been polluted by Cr and other toxic elements (Pb, Cd, Cr, Sn, and Zn) (Khan et al. 2021).

5.3 POTENTIAL Cr(VI) MITIGATION STRATEGIES

Various strategies have been followed to get rid of high Cr content in mining wastes. Different physical, chemical, physicochemical, and biological methods have been adopted to lower the impact of mobile Cr(VI) contamination in agricultural soil near mining sites (Table 5.2).

5.3.1 Physical Methods

Physical methods of Cr extraction from contaminated agricultural soil near mines primarily depend on soil physical properties such as soil type, particle size, and bond strength between contaminant and soil particles. Physical methods such as excavation, soil washing/flushing, solidification, stabilization, vitrification, froth flotation, jigs, shaking tables, and spirals are used for Cr(VI) removal. Many of these methods are traditionally utilized techniques (Higgins et al. 1997).

5.3.1.1 Excavation and Off-Site Disposal

This is a process of removing Cr(VI) contaminated soil from a mining site and then transporting the waste to a dumping site. The primary disadvantages are the relatively high cost of excavation and transportation, and the fact that the material can become airborne during excavation and transportation.

5.3.1.2 Soil Washing/Flushing

Soil washing is a process of cleaning the excavated soil with water or other solvents. Different chemicals (chelating agents, acids, or bases) are added with a washing solution to increase the solubility of the contaminants like Cr(VI). Solubility of Cr is more in an alkaline medium, i.e., at $pH > 11$. Cleaning contaminated soil under in-situ conditions is called soil flushing. The main disadvantage of soil flushing is that it does not immobilize or destroy the contaminants but only transfers them from one phase to other. Soil flushing is not possible where Cr is deeply embedded in soil particles.

5.3.1.3 Solidification/Stabilization (Ex Situ and In Situ)

Solidification is binding of a waste into a solid mass to reduce the contaminant leaching potential, whereas stabilization means reduction in solubility and/or chemical reactivity of a waste. The solidifying or stabilizing materials generally used are cement, fly ash, lime, and cement kiln dust to form complex insoluble silicates or hydrates. Although the process is less expensive, it may cause Cr(VI) to become airborne and may generate a high volume of solidified mass that requires disposal.

5.3.1.4 Vitrification

This is a procedure in which Cr(VI) contaminated soil is heated to a molten state and allowed to cool, forming a material similar to volcanic glass. The process causes the reduction of mobile Cr(VI) to immobile Cr(III) or direct immobilization of Cr(VI). The product, which is characterized by reduced toxicity, mobility, volume, and yield, is generally less hazardous. The disadvantage of this process is that it is seldom applied for heavy metal immobilization due to the high-energy requirement and greater overall unit cost compared with other technologies.

TABLE 5.2
Mitigation Strategies for Chromium-Contaminated Soils

Mitigation Method	Details of Method	Experimental Details	Research Finding	Reference
Adsorption	Magnetic nanocomposite studied as adsorbent	*Study type:* Lab scale *Soil:* Synthetic soil *Experiment:* Immobilization of Cr(VI) by magnetic ceramsite coated with nanocarbon spheres (MCCS) and polyethylene glycol (PG) for possible adsorption	Adsorption capacity of MCCS-PG for Cr(VI) is 10.05 mg/g, which is six times efficient than basic magnetic separation *Mechanism*: Box–Behnken design	Maqbool et al. (2020)
Surfactant-based removal	Biodegradable plant (soapberry)-based surfactant i.e., saponin used for Cr adsorption	*Study type:* Lab-scale batch study *Soil:* Industrial site soil (Hai-Pu, Taiwan) *Experiment:* Remediation of Cr(VI) using biosurfactant	Removal of Cr was 73% in the presence of saponin (0.150 g/L) at pH 5 *Mechanism*: Bond between Cr and soil particle was destabilized and removal by saponin	Maity et al. (2013)
Adsorption	Application of biochar for reduction of Cr uptake through lettuce	*Study type:* Lab scale *Soil:* Natural mining soil (Pakistan) *Experiment:* Examine the effect of sugarcane bagasse biochar (SCBB) and wood-based biochar for reduction of Cr uptake by lettuce plant	Immobilization of Cr was 7% for the application of SCBB (69%) and WB (73.7%) *Mechanism*: Adsorption of Cr biochar surface	Khan et al. (2020a)
Adsorption	Nano zero-valent iron (nZVI) suspension is a remedial option for in situ treatment of Cr decontamination	*Study type:* Pilot scale *Soil:* Synthetic soil *Experiment:* Evaluate the efficiency of iron nanoparticles (GT-nZVI), synthesized by a green synthesis route, for Cr(VI) removal in a pilot scale study	GT-nZVI can be an efficient and environmentally friendly option for the treatment of Cr(VI) *Mechanism*: In-situ injection	Mystrioti et al. (2018)
Sorption	Rice straw biochars doped with iron and (3-Aminopropyl) triethoxysilane (APTES), subsequently mixed at 3% W/W and the prepared composite biochar utilized for Cr(VI) adsorption	*Study type:* Lab scale *Soil:* Natural mining soil (West Bengal) *Experiment:* Study the effect of iron and triethoxysilane composites towards enhanced Cr(VI) removal	Iron and biochar composite shows maximum removal of 100.59 mg/g Cr (VI). *Mechanism*: Langmuir-Freundlich isotherm was followed	Medha et al. (2021)

(*Continued*)

TABLE 5.2 (*Continued*)
Mitigation Strategies for Chromium-Contaminated Soils

Mitigation Method	Details of Method	Experimental Details	Research Finding	Reference
Physical remediation	Nano-magnetic $MnFe_2O_4$ utilized for Cr (VI) immobilization.	*Study type:* Series batch test and Fixed-bed column experiments *Soil:* Cr-contaminated spiked soil (200 mg/kg) was prepared *Experiment:* Dosage of $MnFe_2O_4$ was 2 g/L for 192 hours. Nanoparticle-mediated immobilization of Cr	TCLP, Cr(VI) leaching from treated soil reduced 70.95% to 4.22%. Plant bioavailable Cr was reduced from 83.72% to 5.53%	Eyvazi et al. (2019)
Physical remediation	Combined sorption and magnetic separation of Cr by zero-valent iron particles (ZVI)	*Study type:* laboratory-based study *Soil:* Contaminated soil *Experiment:* Contaminated soil mixed with water at 1: 1 ratio (m/v). ZVI particles added at 2.5% (m/m) and experiment carried out at neutral pH	92.25% of Cr were transferred from contaminated soils to zero-valent iron particles, followed by 99% recovery by magnetic separation	Zhou et al. (2020)
Magnetic separation	Zero-valent iron (ZVI) particles and magnetic separation techniques for Cr remediation	*Study type:* Lab scale *Soil:* Natural soil (Gainesville) *Experiment:* Removal of Cr from soil by initial binding of it to ZVI particles followed by its magnetic separation	ZVI as a sorbent for Cr with removal efficiencies of 92.25% in contaminated soils *Mechanism*: Immobilization of Cr caused by sorption mechanism	Zhou et al. (2020)
Co-precipitation	Novel composite of montmorillonite studied as precipitants	*Study type:* Lab scale *Soil:* Synthetic soil (Cr: 424.6 mg/kg) *Experiment:* Remediation of Cr(VI) by composite of montmorillonite-supported carboxymethyl cellulose-stabilized nanoscale iron sulphide (CMC@ MMT-FeS)	Cr(VI) content decreased with increasing CMC@ MMT-FeS loading, leads to decreasing Cr by 90.7% *Mechanism*: Co-precipitation of Cr(VI) with FeS-particle	Zhang et al. (2020a)
Ion exchange mechanism	Montmorillonite, attapulgite, and sepiolite clay minerals studied as adsorbents	*Study type:* Lab scale *Soil:* Synthetic soil *Experiment:* Investigated immobilization of Cr(VI) in soil by different clay minerals, adsorption isotherm models used to describe the possible mechanism	Immobilization of Cr(VI) was highest by 15% application of attapulgite, montmorillonite, and sepiolite *Mechanism*: Immobilisation caused by ion exchange mechanism	Zhang et al. (2020b)

(*Continued*)

TABLE 5.2 (*Continued*)
Mitigation Strategies for Chromium-Contaminated Soils

Mitigation Method	Details of Method	Experimental Details	Research Finding	Reference
Bioremediation	Soil-borne bacterium B2-DHA of genus *Enterobacteria cloacae*	*Study type*: Field study *Soil:* Contaminated soil (Cr 15.4 μg/g) samples collected from landfills of leather manufacturing tannery industries, Bangladesh *Experiment*: The analysis carried out at pH: 7 and temperature: 37 and mitigation was achieved through Cr reduction	120 hours of bacterial exposure to contaminant caused 320 mg/g of Cr accumulation in its dry biomass and hence caused 81% Cr reduction in liquid medium	Kozliak and Paca (2012)
Phytoremediation	In situ phytoremediation (phytoextraction and rhizofiltration mechanisms) using paragrass (*Brachiaria mutica*)	*Study type*: 125 days long plot experiment *Soil:* Soil total Cr: 11,170 mg/kg and mining wastewater (Cr^{6+}: 0.645 and total Cr: 0.74 mg/L) Cr content *Experiment:* Flow rate of wastewater: 950 mL/min. Mitigation of Cr by accumulation in plant parts	Paragrass plant roots accumulated 1,000 times more Cr than shoots and ultimately lowered the content > 60% after 100 days of plant transplantation	Mohanty and Patra (2012)
Phytoremediation	*L. hexandra* (grass) studied as phytoextraction of contaminated soil	*Study type:* Lab scale *Soil:* Synthetic soil *Experiment:* Examined the effect of oxalic acid (OA) on phytoremediation of Cr-contaminated soil	Soil application of OA for plant growth resulted to increased plant biomass to 78% compared to control (39%), concluding direct boost up of Cr uptake *Mechanism*: Translocation	Tan et al. (2013)
Phytoremediation	Cerium dioxide (CeO_2) nanoparticles (NPs) and *Staphylococcus aureus* studied for phytoextraction	*Study type:* Lab scale *Soil:* Synthetic soil *Experiment:* Efficacy study of CeO_2 NPs and Staphylococcus aureus for alleviating Cr toxicity through sunflower plant	Application of nanoparticles and bacteria enhanced enzymatic activities, chlorophyll a, b in sunflower plant and simultaneous Cr removal *Mechanism*: Biounavailability	Ma et al. (2022)
Photocatalysis	In situ photoreduction of Cr(VI) to Cr(III) in presence of Ti O_2/ZnO, UV light, and alkaline agent	*Study type:* Ptotocatalytic reduction of Cr in presence of catalyst and alkaline *Soil:* Soil Cr content was 112–651 mg/kg *Experiment:* Catalyst TiO_2/ZnO, and alkaline agent $Ca(OH)_2$ was used for the photoreduction of Cr	In presence of 1% and 2% TiO_2 caused 50 and 70% Cr(VI) reduction, respectively. Addition of alkaline agent $Ca(OH)_2$ showed no visible effect on the process	López–Vásquez et al. (2014)

(*Continued*)

TABLE 5.2 (*Continued*)
Mitigation Strategies for Chromium-Contaminated Soils

Mitigation Method	Details of Method	Experimental Details	Research Finding	Reference
Electro-remediation	Seven days (168 hours) electrokinetic (EK) treatment for removal of contaminants	*Study type:* Lab scale *Soil:* Artificially Cr(VI) contaminated in soil *Experiment:* Examines electroremediation's efficiency for artificially contaminated Cr(VI) in soil samples	Decrease in Cr(VI) level of soil sample by reduction mechanism and presence of organic matter enhanced the process *Mechanism*: Reaction time directly related to Cr reduction	Sakellariou and Papassiopi (2018)
Electrokinetic remediation	Polarity exchange and use of conventional electrokinetic techniques	*Study type:* Lab scale *Soil:* Industrial site soil of China (Cr content: 510 mg/kg) *Experiment:* Investigation of Cr immobilization among polarity exchange and conventional electrokinetic	Conventional electrokinetic remove only 57% of initial Cr, whereas polarity exchange technique removed 70% and 88% Cr in 96 and 48 hour interval time, respectively *Mechanism*: Exchanging polarity	Lu et al. (2012)
Enhanced electrokinetic (EK) method	EK method using a permeable reactive barrier (PRB) and chelating agents	*Study type:* Lab scale *Soil:* Synthetic soil *Experiment:* Investigates the effect of chelating agents (EDTA and citric acid) on EK coupled with PRB to remove Cr(VI) from the contaminated soil	Immobilization of Cr by chelating agent: EDTA was 78% whereas in presence of PRB-EDTA Cr removal reduced to 70% and 66% *Mechanism*: Immobilization of Cr by sequential extraction method	Nasiri et al. (2020)

5.3.1.5 Froth Flotation

In this method, the difference in hydrophobic properties of desired heavy metals and the contaminated soil matrix is taken into account. The separation principle is based on the degree of hydrophobicity of the heavy metals and, hence, their attachment to air bubbles injected into a heavy metal–contaminated soil slurry.

Three steps are followed for heavy metal separation by the froth flotation method (Dermont et al. 2008): (i) specific attachment of the desired metal-bearing particles to the air bubbles, (ii) metal-containing bubbles are gathered in a foam portion, and (iii) foam fractions which float up in the slurry are removed and heavy metals extracted.

5.3.1.6 Jigs, Shaking Tables, and Spirals

These are physical methods for Cr extraction from contaminated soil, and they fall under the broader category of "gravity concentration." The desired metals are separated from the soil matrix by preparing a slurry of both. This technique works best when there is a distinct gravitational difference (density, size, shape, and weight) among the particles to be separated. Jigs are commonly used to treat coarse sand fractions (size: 800–2,000 μm) or gravel fractions (2,000–6,000 μm). Shaking tables and spirals are more suitable to treat fine to medium/coarse sand fractions (63–2,000 μm) (Dermont et al. 2008).

5.3.2 Chemical Methods

In chemical extraction technology, chemical reagents are used to bring the heavy metals from the soil matrix into soil solution. Chemical reagents commonly used in this process are acids/bases, surfactants, chelating agents, salts, or redox agents. The choice of extracting agents to be used depends on the metal type, metal concentration, metal fractionation/speciation, and soil characteristics. The use of chloride salt solution as a chemical reagent causes formation of metal chloro-complexes to extract metals from the soil. Specific surfactants also are used for desorption of target metals from the soil matrix. Similarly, redox reagents can be used to enhance metal solubilization through a valence change reaction (Gupta and Mukherjee 2017).

Clay minerals: Sepiolite, montmorillonite, and attapulgite applied at 15% in contaminated soil decreased Cr(VI) concentrations in leachates by 16.8%, 18.9%, and 15.9%, respectively (Zhang et al. 2020b). Immobilization of Cr(VI) was caused by an ion exchange mechanism. According to Zhang et al. (2020a), montmorillonite-supported carboxymethyl cellulose-stabilized nanoscale iron sulphide (CMC@MMT-FeS) particles added at 5% to soil caused Cr(VI) co-precipitation and resulted in 90.7% reduction (from 424.6 to 39.4 mg/kg) of Cr(VI) concentration in the soil.

Organic and inorganic agents: Soapberry derived saponin is a natural plant-based surfactant, and it was utilized for Cr(VI) removal from contaminated soil in Hai-Pu, Taiwan. Saponin at 0.15 g/L removed Cr(VI) maximally (73%) at pH 5 (Maity et al. 2013). Similarly, $FeCl_2$ or $FeSO_4$ is commonly used as reducing agents for Cr(VI) remediation. Removal of Cr(VI) is achieved through the reduction of highly soluble Cr(VI) to insoluble Cr(III) using a reducing agent such as ferrous sulphate, sulphur dioxide, or sodium disulphite.

Carbon coated composites: This is an in situ Cr(VI) remediation strategy. Magnetic ceramsite coated with nanocarbon spheres (MCCS) and functionalized by polyethylene glycol (PG) was utilized for the removal of Cr(VI). Presence of –OH and –COOH functional groups on the surface of the material resulted in maximum adsorption of Cr(VI), which was ~ 10.05 mg/g. Factors such as pH, initial concentration of Cr(VI), and composite influenced the adsorption process. The efficiency of adsorption was 92.97%. Further the adsorbent was regenerated with an efficiency of ≥95% (Maqbool et al. 2020).

Magnetic separation technique: Magnetic separation is a separation and extraction process for the recovery of Cr from different mining wastes. Zero-valent iron (ZVI)-mediated Cr extraction from contaminated soil was carried out where contaminated soil and water were mixed in a 1:1 ratio (m/v) to produce a thick slurry. The ZVI particles were added at 2.5% (mass/mass) ratio and reacted for 3 hours. In the tested soil, Cr(VI) content was 85.76 mg/kg, which was lowered to 6.65 mg/kg with 92.25% removal efficiency (Zhou et al. 2020; Bai et al. 2015).

Adsorption: This is an adhesion process where adsorbates selectively attach on to adsorbents' surface in order to get separated. Adsorbents such as minerals, organic/biological agents, zeolites, industrial byproducts, agricultural wastes, biomass, biochar, and polymeric materials are utilized for the immobilization of target elements. For example, nano-magnetic adsorbent $MnFe_2O_4$ was utilized to immobilize Cr(VI) so that soil leachability was decreased. Hence, plant bioavailability and human bioaccessibility were decreased (Eyvazi et al. 2019). According to Ji et al. (2022), the removal rate of Cr(VI) increased with decreasing particle size of a pyrrhotite/magnetite composite and acidic pH enhanced the process. One gram of composite removed 45.8 mg Cr(VI) at pH 2.

Biochar is a pyrolyzed biomass, which is carbon rich and porous, and it can be used as an adsorbent for the immobilization of different heavy metals in soil including Cr(VI) in contaminated soil due to the high adsorption capacity of biochar materials. Biochar has the ability to improve physicochemical properties of soil, including cation exchange capacity (CEC), organic carbon content (OC), and bulk density. It also helps to overcome Cr stress to soil microbes (Khan et al. 2020a). Oxygen-containing functional groups on the surface of biochar are attributed to binding and reduction of Cr(VI) to Cr(III). Lyu et al. (2017) reported a synergistic effect of biochar and a nanoscale FeS composite on Cr(VI) immobilization in soil. The composite achieved Cr(VI) adsorption capacity of 130.5 mg/g at pH 5.5. Different contaminated agricultural soils near mines were treated with sugarcane bagasse biochar at 7% (wt/wt) application rate, which increased plant biomass and reduced Cr(VI) bioaccumulation in plants (Khan et al. 2020a). Similarly, triethoxysilane and iron rice straw biochar composites resulted in the adsorption of Cr(VI) with maximum removal of 100.59 mg/g in mine soil (Medha et al. 2021). $FeSO_4$-treated corn cob biochar was utilized for Cr removal from mine overburden. The modified biochar caused the reduction of leachable Cr from 4.88 to 0.87 mg/L and enhanced the phytoavailable Cr fraction from 5.73 to 7.41 mg/L at a 5% dose within the 90 days of the experimental period (Mohapatra et al. 2020). The treatment facilitated phytoremediation of the metal.

The primary disadvantages of various chemical treatment methods are that they are expensive, consume a lot of resources, leave a large carbon footprint, and may generate a large amount of secondary contaminants, leading to secondary environmental problems.

5.3.3 Biological Methods

The utilization of plants or microorganisms for the conversion of soluble and toxic Cr(VI) to insoluble and less toxic Cr(III) is called phytoremediation or bioremediation. Transformation of toxic Cr(VI) to less toxic Cr(III) can occur in aerobic or anaerobic conditions using in situ/ex situ technology. In the in situ technique, contaminated soil or water is treated on site. In contrast, in the ex situ technique, contaminants are treated in an alternative controlled environment away from the contaminated site.

Equation 5.1 represents a biotransformation mechanism for the conversion of toxic and soluble Cr(VI) to less toxic and insoluble Cr(III) in microorganisms.

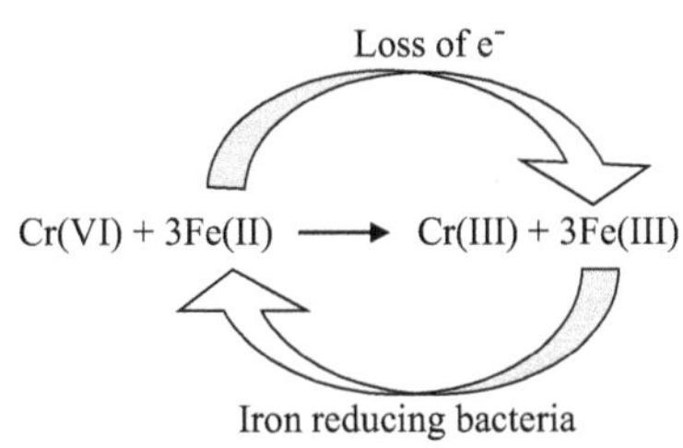

Iron-reducing bacteria reduce Fe(III) to Fe(II); then Fe(II) releases electrons and is converted to Fe(III). Simultaneously, the electrons are accepted by Cr(VI), reducing it to Cr(III). The reaction in Equation 1 indicates that iron is cycled within the system and acts as a biocatalyst to transfer electrons from Fe(II) to Cr(VI) (Bishop et al. 2014).

Primarily, there are three mechanisms behind reduction of Cr(VI) to Cr(III). These are (i) in aerobic conditions, the enzyme chromate reductase reduces in the presence of the cofactor NADH/NADPH; (ii) anaerobic metal reducing bacteria mediated reduction; and (iii) reduction in the presence of amino acids, nucleotides, sugars, vitamins, organic acids, or glutathione. In many bacteria, the reduction process also follows a plasmid resistance mechanism, which is a potentially useful detoxification process (Dhal et al. 2013).

Phytoremediation can be regarded as involving the following processes: phytoextraction, phytodegradation, phytovolatalization, rhizodegradation, or rhizofiltration (Nayak and Kale 2020). Mohanty and Patra (2012) studied in situ phytoextraction and rhizofiltration of Cr using paragrass (*Brachiaria mutica*) plants. Roots of this plant accumulated 1,000 times more Cr than shoots and ultimately lowered the content by > 60% after 100 days of plant transplantation (Mohanty and Patra 2012).

Bioremediation and phytoremediation processes have advantages over physical and chemical methods of soil Cr remediation, because both are simple processes, are low cost, and produce no or less secondary pollutants. All the above Cr remediation processes aim to bring the level of soil Cr content to a permissible limit. According to the World Health Organization, the permissible value of Cr in agricultural soil is 100 mg/kg (Astuti et al. 2021).

5.4 CHALLENGES AND FUTURE PERSPECTIVES

Detailed studies are required to understand the nature of Cr species, their half-life, and the mechanisms behind transformation from highly toxic Cr(VI) to less toxic Cr(III). Cr(VI) toxicity can impair physiology and biochemical pathways in plants and also can be carcinogenic to animals and human beings. Therefore, there is a need for extensive studies to determine the percentage of agricultural soils that are contaminated by Cr, primarily due to mining activities in India and other countries.

Among different methods for Cr mitigation, phytoremediation is an environmentally acceptable approach for Cr(VI) contamination reduction. The role of different types of genes and proteins of plants need to be explored in detail, which help in Cr uptake in different plant parts. Cr then has subsequent effects on plant physiology and metabolism, which need to be controlled. There are few plant species that are effective in Cr hyper-accumulation in root and shoot tissues. Therefore, different types of plant species must be explored for Cr accumulation.

Cr causes toxicity to plants by producing ROS (reactive oxygen species) that interfere in biological activities of plants (e.g., metabolism, physiology, and different biochemical pathways). Mechanisms behind the harmful action of Cr in plants are still not well known. More experimental studies are required to understand why the storage of Cr in plants is restricted to some specific cells in roots and shoots.

In the microbial bioremediation method, the transformation of Cr(VI) to less toxic and less soluble Cr(III) is considered safe, according to Indian standards. During remediation, however, abiotic factors (e.g., pH, temperature, and sources of nutrients) play an important role for Cr(VI) reduction. Efficient bacterial species need to be identified and examined in order to acclimatize specific bacterial strains to work under stressful conditions. Collection of different species from different contaminated sites and their utilization for soil Cr remediation, after series of adaptation studies in diverse media, could be a feasible approach.

Although bioremediation is the best option for Cr(VI) remediation, many external factors (e.g., pH, temperature, nutrients) of soil may limit the performance of bioremediation by controlling microbial growth and proliferation. Microbial immobilization technology should be implemented to reduce microbial loss and to improve overall bioremediation efficiency.

5.5 A CASE STUDY IN INDIA FOR MANAGING SOIL Cr CONTAMINATION

Soil samples contaminated with Cr from the Sukinda chromite mine in India were taken both from in situ sites and from overburden sites and investigated using the "Biolog eco-plate" approach to analyse for Cr eco-toxicity (Pradhan et al., 2020). Biolog eco-plates contained 31 carbon sources collected from Cr-contaminated sites and were grouped into six categories.

Soil Cr content in the in situ and in the overburden contaminated material were 3,012 and 1,991 mg/kg, respectively (Pradhan et al., 2020). The Cr content in an adjacent forest soil (taken as control) was 745 mg/kg. Higher soil Cr stress negatively affected the soil pH (caused alkalinity of soil) and carbon content. The activities of dehydrogenase and acid- and alkaline-phosphatase enzymes were studied. For in situ and overburden soil, the enzyme activities were lower than in the forest soil [dehydrogenase activity: 30.32 and 30.95<31.33 µg TPF/g/h (TPF stands for triphenylformazan); acid phosphatase: 7.39 and 14.00<23 µg *p*-nitrophenol/g/h; alkaline-phosphatase: 1.59 and 4.06<15.52 µg *p*-nitrophenol/g/h). Lowered enzymatic activity due to Cr contamination reduced the carbon utilization capacity of the inhabitant microorganisms. Increased Cr content also lowered soil respiration and nutrient mobilization and increased microbial richness but reduced evenness. All these led to a significant decrease in soil microbial metabolic activities and, hence, soil fertility due to Cr-mediated alteration of soil physicochemical properties. Decreased soil fertility due to Cr contamination can lead to reduction of crop productivity because microorganisms play a crucial role in soil nutrient cycling and mobilization (Pradhan et al. 2020).

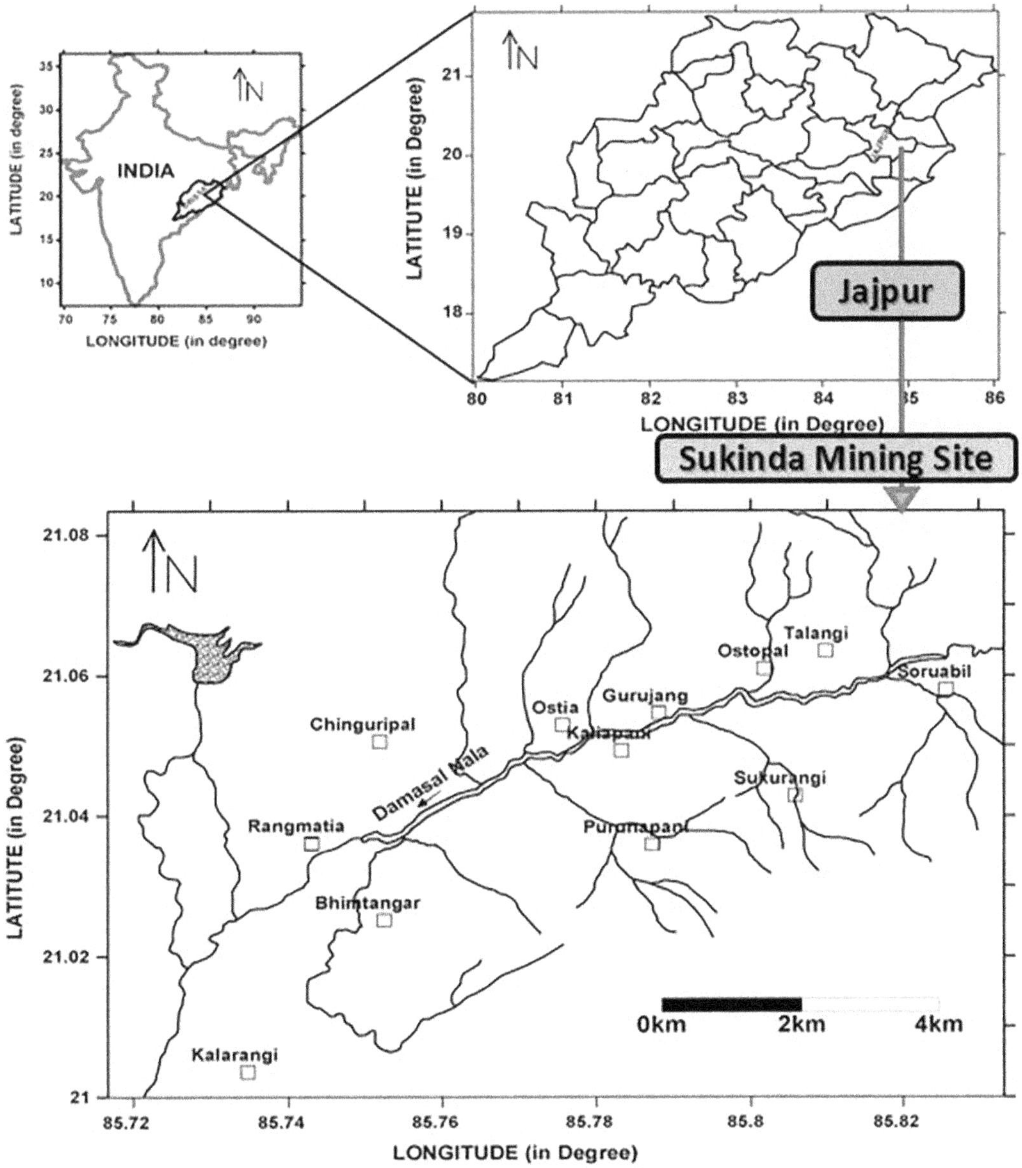

FIGURE 5.2 Location map of the study site. (Retrieved from Pradhan et al., 2020.)

5.6 CONCLUSIONS

Mining of chromium, nickel, gold, iron, and coal can cause chromium contamination of the surrounding environment. Mining activities such as excavation of ore, refinement, transport of ores, smelting, metal finishing, and recycling of metals all contribute to Cr contamination. At mining sites, chromium can leach from soil or be carried by air following wind erosion. The airborne particles are then precipitated onto nearby agricultural lands. According to the World Health Organization, the permissible value of Cr in agricultural soil is 100 mg/kg. High levels of Cr in soil are translocated to plants by root systems. The Cr then contaminates the food chain by affecting herbivore animals that eat the plants, and finally, the Cr reaches human beings who eat the animals. Various carcinogenic and non-carcinogenic effects of Cr have been reported in the literature. Although there are different physical and chemical treatment processes for Cr in soil, methods in use often result in a high amount of waste, which limits their wide application. Biological control of Cr contamination, such as phytoremediation and bioremediation, is easily accessible, low cost, and simple and is a better option for Cr treatment than physical and chemical treatments. Although Cr has many harmful effects, the metal is required in various industries (e.g., metallurgical, leather tanning, textiles, mechanical and mining engineering, chemical and metal finishing, ferrochrome, pigments, electroplating, photography, and refractories). Therefore, there is a need to explore efficient ways to manage Cr contamination during mining operations in order to have sustainable development of both the mining and agricultural sectors.

REFERENCES

Abiya, S.E., Odiyi, B.O., Ologundudu, F.A., Akinnifesi, O.J. and Akadiri, S., 2018. Assessment of heavy metal pollution in a gold mining site in Southwestern Nigeria. *Journal of Genetics and Cell Biology*, *1*(2), pp. 30–35.

Adimalla, N., Manne, R., Zhang, Y. and Qian, H., 2022. Appraisal of vulnerable zones of non carcinogenic and carcinogenic causing health risks associated with exposure of potentially toxic elements in soils of India: a meta-analysis. *Geocarto International*, *37*(25), pp. 10619–10635.

Astuti, R.D.P., Mallongi, A. and Rauf, A.U., 2021. Natural enrichment of chromium and nickel in the soil surrounds the karst watershed. *Global Journal of Environmental Science and Management*, *7*(3), pp. 383–400.

Bai, Z.T., Zhang, Z.A., Guo, M., Hou, X.M. and Zhang, M., 2015. Magnetic separation and extraction chrome from high carbon ferrochrome slag. *Materials Research Innovations*, *19*(2), pp. 113–118.

Baltas, H., Sirin, M., Gökbayrak, E. and Ozcelik, A.E., 2020. A case study on pollution and a human health risk assessment of heavy metals in agricultural soils around Sinop province, Turkey. *Chemosphere*, *241*, p. 125015.

Bishop, M.E., Glasser, P., Dong, H., Arey, B. and Kovarik, L., 2014. Reduction and immobilization of hexavalent chromium by microbially reduced Fe-bearing clay minerals. *Geochimica et Cosmochimica Acta*, *133*, pp. 186–203.

Chandio, T.A., Khan, M.N., Muhammad, M.T., Yalcinkaya, O., Turan, E. and Kayis, A.F., 2021. Health risk assessment of chromium contamination in the nearby population of mining plants, situated at Balochistan, Pakistan. *Environmental Science and Pollution Research*, *28*(13), pp. 16458–16469.

Coetzee, J.J., Bansal, N. and Chirwa, E., 2020. Chromium in environment, its toxic effect from chromite-mining and ferrochrome industries, and its possible bioremediation. *Exposure and Health*, *12*(1), pp. 51–62.

Das, S., Ram, S.S., Sahu, H.K., Rao, D.S., Chakraborty, A., Sudarshan, M. and Thatoi, H.N., 2013. A study on soil physico-chemical, microbial and metal content in Sukinda chromite mine of Odisha, India. *Environmental Earth Sciences*, *69*(8), pp. 2487–2497.

Dermont, G., Bergeron, M., Mercier, G. and Richer-Laflèche, M., 2008. Soil washing for metal removal: a review of physical/chemical technologies and field applications. *Journal of Hazardous Materials*, *152*(1), pp. 1–31.

Dhal, B., Das, N.N., Pandey, B.D. and Thatoi, H.N., 2011. Environmental quality of the Boula-Nuasahi chromite mine area in India. *Mine Water and the Environment*, *30*(3), pp. 191–196.

Dhal, B., Thatoi, H.N., Das, N.N. and Pandey, B.D., 2013. Chemical and microbial remediation of hexavalent chromium from contaminated soil and mining/metallurgical solid waste: a review. *Journal of Hazardous Materials*, *250*, pp. 272–291.

Du, X., Hong, M., Yang, Y., Li, J., Su, Y. and Liu, N., 2022. Removal mechanisms of aqueous Cr (VI) by anaerobic fermentation sludge. *Environmental Technology*, pp. 1–26.

Eleftheriou, E.P., Adamakis, I.D.S., Panteris, E. and Fatsiou, M., 2015. Chromium-induced ultrastructural changes and oxidative stress in roots of Arabidopsis thaliana. *International Journal of Molecular Sciences*, *16*(7), pp. 15852–15871.

Eyvazi, B., Jamshidi-Zanjani, A. and Darban, A.K., 2019. Immobilization of hexavalent chromium in contaminated soil using nano-magnetic $MnFe_2O_4$. *Journal of Hazardous Materials*, *365*, pp. 813–819.

Fordyce, F.M., Everett, P.A., Bearcock, J.M. and Lister, T.R., 2017. Soil metal/metalloid concentrations in the Clyde Basin, Scotland, UK: implications for land quality. *Earth and Environmental Science Transactions of the Royal Society of Edinburgh*, *108*(2–3), pp. 191–216.

Gu, F., Zhang, Y., Su, Z., Tu, Y., Liu, S. and Jiang, T., 2021. Recovery of chromium from chromium-bearing slags produced in the stainless-steel smelting: A review. *Journal of Cleaner Production*, *296*, p. 126467.

Gunkel-Grillon, P., Laporte-Magoni, C., Lemestre, M. and Bazire, N., 2014. Toxic chromium release from nickel mining sediments in surface waters, New Caledonia. *Environmental Chemistry Letters*, *12*(4), pp. 511–516.

Gupta, C.K. and Mukherjee, T.K., 2017. *Hydrometallurgy in Extraction Processes*. CRC Press, Boca Raton, FL.

Higgins, T.E., Halloran, A.R. and Petura, J.C., 1997. Traditional and innovative treatment methods for Cr (VI) in soil. *Soil and Sediment Contamination*, *6*(6), pp. 767–797.

Hou, D., O'Connor, D., Igalavithana, A.D., Alessi, D.S., Luo, J., Tsang, D.C., Sparks, D.L., Yamauchi, Y., Rinklebe, J. and Ok, Y.S., 2020. Metal contamination and bioremediation of agricultural soils for food safety and sustainability. *Nature Reviews Earth & Environment*, *1*(7), pp. 366–381.

Huang, Y., Wang, L., Wang, W., Li, T., He, Z. and Yang, X., 2019. Current status of agricultural soil pollution by heavy metals in China: A meta-analysis. *Science of the Total Environment*, *651*, pp. 3034–3042.

Iyer, V.G. and Mastorakis, N.E., 2009. September. Unsafe chromium and its environmental impact assessment. In *Proceedings of the WSEAS 2nd International Conference on Environmental and Geological Science and Engineering (EG'09)*, Transilvania University of Brasov, Romania, pp. 24–26.

Ji, X., Zhou, C., Chen, L., Li, Y., Hua, T., Li, Y., Wang, C., Jin, S., Ding, H. and Lu, A., 2022. Reduction, mineralization, and magnetic removal of chromium from soil by using a natural mineral composite. *Environmental Science and Ecotechnology*, *11*, p. 100181.

Kabata-Pendias, A. and Pendias, H., 2001. *Trace Elements in Soils and Plants*. CRC Press Inc., Boca Raton, FL.

Khan, A.Z., Khan, S., Ayaz, T., Brusseau, M.L., Khan, M.A., Nawab, J. and Muhammad, S., 2020a. Popular wood and sugarcane bagasse biochars reduced uptake of chromium and lead by lettuce from mine-contaminated soil. *Environmental Pollution*, *263*, p. 114446.

Khan, S., Naushad, M., Lima, E.C., Zhang, S., Shaheen, S.M. and Rinklebe, J., 2021. Global soil pollution by toxic elements: Current status and future perspectives on the risk assessment and remediation strategies–A review. *Journal of Hazardous Materials*, *417*, p. 126039.

Khan, Z.I., Ahmad, K., Siddique, S., Ahmad, T., Bashir, H., Munir, M., Mahpara, S., Malik, I.S., Wajid, K., Ugulu, I. and Nadeem, M., 2020b. A study on the transfer of chromium from meadows to grazing livestock: an assessment of health risk. *Environmental Science and Pollution Research*, *27*(21), pp. 26694–26701.

Kien, C.N., Tanaka, S., Van, N.N., Son, L.T., Phuong, N.M., Trang, B.T.Q., Noda, M., Ueno, D. and Iwasaki, K., 2011. Heavy metal concentrations in rice (Oryza sativa L.) plants grown in a chromite mining area in Vietnam. *Tropical Agriculture and Development*, *55*(4), pp. 135–141.

Kozliak, E.I. and Paca, J., 2012. Journal of Environmental Science and Health, Part A. Toxic/hazardous substances and environmental engineering. Foreword. *Journal of Environmental Science and health. Part A, Toxic/hazardous Substances & Environmental Engineering*, *47*(7), pp. 919–919.

Ladwani, K.D., Ladwani, K.D., Manik, V.S. and Ramteke, D.S., 2012. Assessment of heavy metal contaminated soil near coal mining area in Gujarat by toxicity characteristics leaching procedure. *International Journal of Life Sciences Biotechnology and Pharma Research*, *1*(4), pp. 73–80.

Li, X., Zhang, J., Ma, J., Liu, Q., Shi, T., Gong, Y., Yang, S. and Wu, Y., 2020. Status of chromium accumulation in agricultural soils across China (1989–2016). *Chemosphere*, *256*, p.127036.

López–Vásquez, A., Ramirez, N. and López–Vásquez, R., 2014. In situ photocatalytic reduction for contaminated soil with hexavalent chromium by titanium dioxide. In *Transactions on Engineering Technologies*, Haeng Kon Kim, Sio-long Ao, Mahyar A. Amouzegar, Eds, Springer, pp. 107–119.

Lu, P., Feng, Q., Meng, Q. and Yuan, T., 2012. Electrokinetic remediation of chromium-and cadmium-contaminated soil from abandoned industrial site. *Separation and Purification Technology*, *98*, pp. 216–220.

Lyu, H., Tang, J., Huang, Y., Gai, L., Zeng, E.Y., Liber, K. and Gong, Y., 2017. Removal of hexavalent chromium from aqueous solutions by a novel biochar supported nanoscale iron sulfide composite. *Chemical Engineering Journal*, *322*, pp. 516–524.

Ma, J., Alshaya, H., Okla, M.K., Alwasel, Y.A., Chen, F., Adrees, M., Hussain, A., Hameed, S. and Shahid, M.J., 2022. Application of Cerium Dioxide Nanoparticles and Chromium-Resistant Bacteria Reduced Chromium Toxicity in Sunflower Plants. *Frontiers in Plant Science*, *13*, pp. 876119.

Madani, S., Coors, A., Haddioui, A., Ksibi, M., Pereira, R., Sousa, J.P. and Römbke, J., 2015. Effects of contaminated soils from a former iron mine (Ait Amar, Morocco) on enchytraeids (Enchytraeus bigeminus) and predatory mites (Hypoaspis aculeifer) in standard laboratory tests. *Ecotoxicology and Environmental Safety*, *119*, pp. 90–97.

Mahimairaja, S., Shenbagavalli, S. and Naidu, R., 2011. Remediation of chromium-contaminated soil due to tannery waste disposal: potential for phyto-and bioremediation (symposium 3.5. 1 heavy metal contaminated soils,<special issue> international symposium: soil degradation control, remediation, and reclamation, Tokyo metropolitan university symposium series no. 2, 2010). *Pedologist*, *54*(3), pp. 175–181.

Maity, J.P., Huang, Y.M., Fan, C.W., Chen, C.C., Li, C.Y., Hsu, C.M., Chang, Y.F., Wu, C.I., Chen, C.Y. and Jean, J.S., 2013. Evaluation of remediation process with soapberry derived saponin for removal of heavy metals from contaminated soils in Hai-Pu, Taiwan. *Journal of Environmental Sciences*, *25*(6), pp. 1180–1185.

Maqbool, A., Hui, W., Xin, X., Khaliq, A. and Sarwar, M.T., 2020. Removal of heavy metals from urban soil using functionalized carbon-coated composite. *International Journal of Environmental Science and Technology*, *17*(12), pp. 4787–4802.

Medha, I., Chandra, S., Vanapalli, K.R., Samal, B., Bhattacharya, J. and Das, B.K., 2021. (3-Aminopropyl) triethoxysilane and iron rice straw biochar composites for the sorption of Cr (VI) and Zn (II) using the extract of heavy metals contaminated soil. *Science of the Total Environment*, *771*, p. 144764.

Mohanty, M. and Patra, H.K., 2012. Phytoremediation potential of paragrass—an in situ approach for chromium contaminated soil. *International Journal of Phytoremediation*, *14*(8), pp. 796–805.

Mohapatra, S., Kumar, M., Karim, A.A. and Dhal, N.K., 2020. Biochars evaluation for chromium pollution abatement in chromite mine wastewater and overburden of Sukinda, Odisha, India. *Arabian Journal of Geosciences*, *13*(13), pp. 1–14.

Mosai, A.K., Bakatula, E.N. and Tutu, H., 2017. Adsorption of cadmium, copper and chromium by an agricultural soil impacted by mining activities. *Water, Air, & Soil Pollution*, *228*(8), pp. 1–12.

Mystrioti, C., Toli, A., Papasiopi, N., Dermatas, D. and Thimi, S., 2018. Chromium removal with environmentally friendly iron nanoparticles in a pilot scale study. *Bulletin of Environmental Contamination and Toxicology*, *101*(6), pp. 705–710.

Nasiri, A., Jamshidi-Zanjani, A. and Darban, A.K., 2020. Application of enhanced electrokinetic approach to remediate Cr-contaminated soil: effect of chelating agents and permeable reactive barrier. *Environmental Pollution*, *266*, p. 115197.

Nayak, S. and Kale, P., 2020. A review of chromite mining in Sukinda Valley of India: impact and potential remediation measures. *International Journal of Phytoremediation*, *22*(8), pp. 804–818.

Naz, A., Chowdhury, A., Mishra, B.K. and Karthikeyan, K., 2018. Distribution of heavy metals and associated human health risk in mine, agricultural and roadside soils at the largest chromite mine of India. *Environmental Geochemistry and Health*, *40*(5), pp. 2155–2175.

Otomo, K. and Shikazono, N., 2008, February. Characteristic of soils and behavior of hexavalent chromium in soil-water interaction. In *AIP Conference Proceedings*, 987, pp. 62–65.

Pradhan, S.K., Kumar, U., Singh, N.R. and Thatoi, H., 2020. Functional diversity and metabolic profile of microbial community of mine soils with different levels of chromium contamination. *International Journal of Environmental Health Research*, *30*(4), pp. 461–473.

Sakellariou, L. and Papassiopi, N., 2018. An approach to electrokinetic removal of Cr (VI) from soil and Kaolin samples. *Bulletin of Environmental Contamination and Toxicology*, *101*(6), pp. 718–724.

Sasmaz, A. and Yaman, M., 2006. Distribution of chromium, nickel, and cobalt in different parts of plant species and soil in mining area of Keban, Turkey. *Communications in Soil Science and Plant Analysis*, *37*(13–14), pp. 1845–1857.

Shahid, M., Shamshad, S., Rafiq, M., Khalid, S., Bibi, I., Niazi, N.K., Dumat, C. and Rashid, M.I., 2017. Chromium speciation, bioavailability, uptake, toxicity and detoxification in soil-plant system: A review. *Chemosphere*, *178*, pp. 513–533.

Shtiza, A., Swennen, R. and Tashko, A., 2005. Chromium and nickel distribution in soils, active river, overbank sediments and dust around the Burrel chromium smelter (Albania). *Journal of Geochemical Exploration*, *87*(3), pp. 92–108.

Solgi, E. and Parmah, J., 2015. Analysis and assessment of nickel and chromium pollution in soils around Baghejar Chromite Mine of Sabzevar Ophiolite Belt, Northeastern Iran. *Transactions of Nonferrous Metals Society of China*, *25*(7), pp. 2380–2387.

Susaj, E., 2017. Environmental impacts assessment of chromium minings in Bulqiza Area, Albania. *European Journal of Interdisciplinary Studies*, *3*(4), pp. 35–41.

Tan, L.L., Zhang, Q.J., Xu, M.Z., Zhang, H. and You, S.H., 2013. Oxalic acid enhances remediation of Cr contaminated soil in Leersia Hexandra Swartz. In *Advanced Materials Research*, Zili Liu, Xinfa Dong, Zhaotie Liu, Qihai Liu, Eds, Scientific. Net, 807,pp. 1559–1563.

Wang, J., Hu, X., Shi, T., He, L., Hu, W. and Wu, G., 2022. Assessing toxic metal chromium in the soil in coal mining areas via proximal sensing: Prerequisites for land rehabilitation and sustainable development. *Geoderma*, *405*, p. 115399.

Whalley, C., Hursthouse, A., Rowlatt, S., Iqbal-Zahid, P., Vaughan, H. and Durant, R., 1999. Chromium speciation in natural waters draining contaminated land, Glasgow, UK. *Water, Air, and Soil Pollution*, *112*(3), pp. 389–405.

Zhang, D., Xu, Y., Li, X., Liu, Z., Wang, L., Lu, C., He, X., Ma, Y. and Zou, D., 2020a. Immobilization of Cr (VI) in soil using a montmorillonite-supported carboxymethyl cellulose-stabilized iron sulfide composite: effectiveness and biotoxicity assessment. *International Journal of Environmental Research and Public Health*, *17*(17), p. 6087.

Zhang, D., Xu, Y., Li, X., Wang, L., He, X., Ma, Y. and Zou, D., 2020b. The immobilization effect of natural mineral materials on Cr (VI) remediation in water and soil. *International Journal of Environmental Research and Public Health*, *17*(8), p. 2832.

Zhou, Z., Alhadidi, Q., Quiñones Deliz, K., Yamaguchi Greenslet, H. and Bonzongo, J.C., 2020. Removal of oxyanion forming elements from contaminated soils through combined sorption onto zero-valent iron (ZVI) and magnetic separation: Arsenic and chromium as case studies. *Soil and Sediment Contamination: An International Journal*, *29*(2), pp. 180–191.

6 Heavy Metals in Agricultural Soils

From Bioavailability to Productivity

Zhen Li[1,2], Lingzi Meng[1], Haoming Chen[3], and Fei Jin[4]
[1]College of Resources and Environmental Sciences, Nanjing Agricultural University, Nanjing, China
[2]Jiangsu Provincial Key Lab for Organic Solid Waste Utilization, Nanjing Agricultural University, Nanjing, China
[3]School of Environmental and Biological Engineering, Nanjing University of Science and Technology, Nanjing, China
[4]Queen's Buildings, School of Engineering, Cardiff University, Cardiff, Wales

CONTENTS

DOI: 10.1201/9781003093565-6

6.1 CURRENT SITUATION OF HEAVY METAL (HM) POLLUTION IN AGRICULTURAL SOILS

Soil is a complex open system with multiple functions (Nicolodi and Gianello, 2014). It is a medium for plant growth and engineering purposes, habitat for organisms, and where cycling of nutrients and organics happens. Soil is considered to be a non-renewable resource. Heavy metal (HM) pollution is one of the most serious problems in agricultural soils, where the most commonly encountered HMs include cadmium (Cd), copper (Cu), chromium (Cr), mercury (Hg), nickel (Ni), lead (Pb), and zinc (Zn) (FAO and UNEP, 2021; Hou et al., 2020). HM pollution has been observed in most countries on Earth (Su et al., 2014), and Cd is one of the most "dangerous" metals in soil. For instance, 21% of agricultural soils in Europe have Cd levels above the regulatory limit (FAO and UNEP, 2021). The United States, Europe, China, and India are all being threatened by Cd pollution in agricultural soils (Su et al., 2014). In both agricultural and urban soils, Cu, Pb, Cd, Cr, and As have been widely reported (Imperato et al., 2003; Morton-Bermea et al., 2009).

HM pollution not only reduces crop yield but also negatively affects human and animal health. There are many sources of HMs in agricultural soils, e.g., fossil fuel or waste combustion, atmospheric deposition, cement production, mining, and utilization of metal-bearing fertilizers (Hou et al., 2020) (Figure 6.1). In Europe, atmospheric deposition causes Pb pollution to be more widely distributed and more difficult to control (Nziguheba and Smolders, 2008). The same phenomenon has been observed in the UK. It is reported that atmospheric deposition is the major contributor of HMs in many agricultural areas. For example, more than 50% of Hg, Pb, Ni, and Cd are deposited annually in agricultural soils (Nicholson et al., 2003).

HMs are usually nonvolatile, nonbiodegradable, and toxic to microorganisms and plants, leading to degradation of soil health and reduced crop productivity and quality. HMs can strongly interact with soil components including soil minerals, organic matters, and microorganisms as well as plants, leading to different mobility, bioavailability, translocation, and ecotoxicity of HMs. These factors in turn affect soil and plant health differently in agricultural land. It is hence urgent to understand how HMs are transformed and transported within the soil environment and to the plants. Then, suitable and sustainable remediation techniques can be designed and implemented for different purposes under varying conditions.

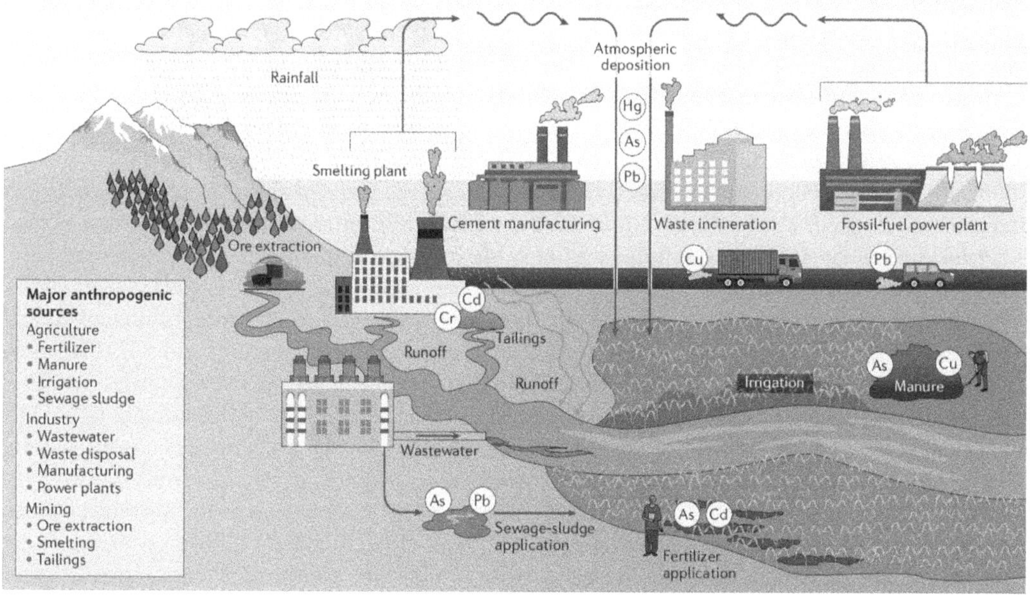

FIGURE 6.1 Sources of heavy metal(loid)s pollution in agricultural soils. Major anthropogenic sources can be classified into three categories: agricultural, industrial, and mining. Heavy metal(loid)s can enter agricultural soil through atmospheric deposition, following release into the atmosphere from fossil-fuel burning, waste incineration, or cement manufacture. Heavy-metal(loid)s-contaminated runoff from mining and industry can enter waterways and reach agricultural land. The use of manure or sewage contaminated with heavy metal(loid)s to fertilize crops can also contaminate agricultural land. (Hou et al. (2020), copyright 2020 Nature.)

6.2 HM TOXICITY TO SOIL MICROORGANISMS

The first observation on soil microbial processes influenced by HMs has been traced back to the early 19th century (Brown and Minges, 1916; Lipman and Burgess, 1914). In the 1960s–1970s, industrial emissions caused huge negative impacts on a wide range of ecosystems. Afterward, scientists were apprehensive that soil microorganisms were seriously disturbed by HM toxicity.

Soil HMs can be divided into two categories considering their biological influences. One is harmful to crops and living creatures, such as Pb, Cd, and Hg, whereas the other is beneficial and even essential to them at certain levels, but toxic when the amounts are excessive, such as Cu, Zn, and Mn. The chemical forms of HMs in different soil components vary significantly. Their total amount and dissolution in soil are not necessarily positively correlated. Moreover, HMs dissolved in soil solution cannot predict their toxicity directly (Wolt, 1994), as solubility is not equal to bioavailability. HMs exist in some chemical forms that may not induce direct uptake by plants (e.g., chelated and/or polymerized). There are also different bioaccumulation rates in plants and animals. Nonetheless, the toxicity of one specific HM is defined as the degree of its bioavailability, which is defined as the abundance that is eventually taken up by organisms via migration and sorption (Checkai et al., 1987; Sunda et al., 1978).

6.2.1 Microbial Community Structure and Diversity

Although many studies have proposed that HM pollution is able to reduce microbial diversity (Feng et al., 2018; Golebiewski et al., 2014; Wu et al., 2018), there is unlikely to be a simple linear relationship between them. In a stable environment with abundant resources, it is usually the highly competitive species that dominate (Austin, 1987; Austin and Smith, 1989; Grime, 1979). However, as the moderate interference hypothesis indicates, moderate pressure will reduce competitive exclusion and eventually result in increased diversity (Giller et al., 1998).

It has been well accepted that HM stresses change the soil microbial community structure (Fernandez-Calvino et al., 2011; Yang et al., 2014). For example, under long-term Cr pollution, the dominant family of bacterial community changed from *Proteobacteria* to *Firmicutes* (Desai et al., 2009). In Cr and As contaminated soil, dominant microorganisms changed from *Actinobacteria* and *Acidobacteria* to *Proteobacteria* (Sheik et al., 2012). Soil microbial populations should also be considered when studying the influences of HMs on the soil microbial community structure. For example, Cu accumulation induced changes in the composition of a bacterial community (Cavani et al., 2016). However, there was no significant influence on the composition of fungal communities, indicating the contrast between bacterial and fungal communities under HM stresses. Li et al. (2017) observed that bacteria were more sensitive to pollution than archaea, probably due to the low biomass of bacteria in the typical soil system. Moreover, the relative abundance of archaea always has a strong positive correlation with soil Cd, Pb, As, and Zn. In contrast, bacteria exhibit various responses to Cd. For instance, *α-Proteobacteria* are negatively correlated, whereas *β-Proteobacteria* are positively correlated to Cd levels. Some bacteria can be even irrelevant with the Cd stress, e.g., *γ-Proteobacteria* and *δ-Proteobacteria* (Li et al., 2017). This may be due to the enriched organic matters in the phyla *Proteobacteria*, which can be utilized by microorganisms for nutrients and energy (Bouskill et al., 2010). This enables them to be suitable for a wide range of environments (Li et al., 2017).

Microbial diversity and structure should also be addressed under HM stresses, as they are the "frame" of soil microorganisms, which is critical for sustainable agriculture. It is widely accepted that the reorganization of microbial communities and the improvement of diversity can prevent the decline of community function (Cavani et al., 2016). Previous studies have observed that, in environments polluted with HMs caused by mining minerals, the microbial diversity did not change, which is the opposite to the community structure (Pereira et al., 2014; Yin et al., 2015). Under the Cd stress, the relative abundance of soil microorganisms changed substantially, resulting in a reduction of richness and a change of community structure.

Species replacement determines the change of microbial communities under HM pollution. Community richness will be altered because some tolerant species will become dominant (Tipayno et al., 2018). For example, Zhang et al. (2015) showed that the concentrations of HMs such as Mn may significantly affect the richness of bacterial species. Species evenness also induces changes of community dynamics (Wittebolle et al., 2009). However, it does not necessarily change together with the species richness under the stress of HMs.

6.2.2 Microbial Physiological Processes

Microbial biomass contains oxygen (O), carbon (C), nitrogen (N), and phosphorus (P) (Xiao et al., 2017; Zhang et al., 2016). The difference in the cellular energy budget is thought to be related to the physiological processes (Gordon et al., 1993). This pathway is usually considered as a detoxification mechanism. However, this, for sure, will also cause reduction of the substrate for microbial growth. For example, a decrease in growth under HM stresses was demonstrated in a chemostat study by using bacterium *Vibrio alginolyticus*. Killham (1985) found that microorganisms divert energy from growth to cell maintenance, and this may be a plausible explanation for the decreased biomass under HM stresses. However, it increased respiration in HM-contaminated soils.

Microbial biomass in soils is commonly decreased under HM stresses. However, base respiration, as an indicator of microbial activities, could be either unaffected, slightly decreased, or increased in HM-contaminated soils (Bardgett and Saggar, 1994; Brookes and McGrath, 1984; Witter et al., 1993; Yeates et al., 1994). In most cases, low concentrations of HMs in soils are conducive to the release of CO_2. It is due to that living organisms need to consume more energy to survive in the environment with HMs (Leita et al., 1995). However, high HM concentrations significantly inhibit microbial activity (e.g., soil respiration) which may be due to the decreased abundance of substrates available for microbial respiration. Moreover, the influence of HM toxicity on soil respiration may depend on the nature of the mineralized substrates in the soil system. For example, Fliessbach et al. (1994) found that base respiration increased in most HM-contaminated soil, which can be explained by the above discussion regarding energy consumption. Meanwhile, the rate of degradation of cellulosic material was decreased, suggesting that microbial functions declined (Obbard and Jones, 1993). Therefore, HMs seriously threaten soil ecosystem functions.

6.2.3 Enzymes and Gene Regulation of Soil Microorganisms

6.2.3.1 Soil Enzymes

There is a close relationship between soil enzymes and microorganisms. Soil enzymes are related to nutrient cycling and HM contamination (Wen et al., 2021b). They are the main participants in soil biochemical processes, which have been well addressed regarding C, N, and P recycling (Adetunji et al., 2017). Enzyme activities can drive the biochemical reactions in soil. Many enzymes can also be applied as a potential biological indicator for evaluating soil health, as their abundance can be ascribed to microbial activities (Lebrun et al., 2012; Liang et al., 2014). Usually, the activities of enzymes have a positive correlation with soil quality. On the other hand, a negative correlation between enzyme activities and toxicity of pollutants has been proposed (Lebrun et al., 2012).

Many previous studies have shown that soil enzymes are sensitive to HMs (Hu et al., 2014b; Song et al., 2018). Their responses can be divided into three types, i.e., activation, inhibition, and no significant influence. Chen et al. (2014) indicated that dehydrogenase activity decreased sharply in HM-contaminated (Cd, Cu, Pb, Zn) soils. Xiao et al. (2017) showed that a low concentration (254 mg/kg) of vanadium (V) enhanced urease activity. Yang et al. (2006) indicated that different enzymes had various sensitivities to HMs based on studies of Cd, Zn, and Pb. For instance, urease activity was much more sensitive than calatase, invertase, and akalin phosphatase to Cd, Zn, and Pb. However, a later study by Yang et al. (2016a) showed that HMs (As, Cd, Cr, Cu, Ni, Pb, Zn) had insignificant effects on urease activity.

The type and concentration of HMs will affect their toxicity to soil enzymes and will induce synergistic or antagonistic effects. The situation is complicated with the co-existence of HMs. Pan and Yu (2011) showed an inhibitory effect under the co-existence of Cd and Pb. The results indicated that the activities of enzymes (urease, acid phosphatase, and dehydrogenase) were greater than those of Cd or Pb acting singly, which is in agreement with Yang et al. (2006). However, Wyszkowska et al. (2006) indicated that the toxicity of Cu to enzymes (dehydrogenases, urease, and phosphatase) was greater than that induced by the co-existence of Cu, Cd, Cr, Ni, Pb, and Zn. Moreover, previous studies have demonstrated that experimental results do not always agree with field studies (Yang et al., 2016a).

6.2.3.2 Gene Regulation

Ammonia oxidation is a key process in the biogeochemical N cycling, which is biologically driven by ammonia-oxidizing bacteria (AOB) and archaea (AOA) (Wang et al., 2018b). It is widely agreed that HMs notably affect biological processes related to the N cycle (Subrahmanyam et al., 2014). Soils with enriched As and Pb usually have a low abundance of the AOB and AOA *amoA* genes (Ollivier et al., 2012). The abundance of the *amoA* gene was negatively correlated with Cu levels in soils. Meanwhile, AOB showed a higher sensitivity than AOA (Li et al., 2009). Similar results were also proposed in soils undergoing Cd contamination (Zhang et al., 2017). The AOA *amoA* gene was more resistant than AOB because AOA have the ability to reduce the accumulation of metal ions, which is beneficial to microbial detoxification (Li et al., 2017). In addition, AOA have a higher resistance than AOB as they have more rigid and stable membranes (Valentine, 2007). Generally, due to the special chemical structure of lipids in the AOA membrane, its permeability to ions is not as functional as that of bacterial membranes (Kandler and Konig, 1998). However, in Zn-contaminated soil, there is no consensus about the sensitivity of AOA and AOB *amoA* genes. It was reported that the *amoA* gene of AOA was more sensitive to Zn than AOB in soil, and its abundance decreased faster (Mertens et al., 2009). Other studies found that the number of AOB *amoA* gene copies and gene transcripts increased during the development of long-term Zn stress. Additionally, the AOB community structure changed, while AOA did not respond to Zn (Ruyters et al., 2010; Ruyters et al., 2013).

The functional genes encoding denitrifying enzymes were also observed to be very sensitive to HMs. Many studies have explored the correlation between HMs and soil denitrification genes. It has been demonstrated that HMs not only reduced gene abundance and genetic diversity but also caused delays before gene expression (Gui et al., 2017; Moffett et al., 2003; Throback et al., 2007). It is reported that different denitrifying enzymes have different tolerance to Cu (Magalhaes et al., 2011). For example, the expression of *nirS*, *norB*, and *nosZ* was increased at a low level of Cu (0.15 mM). However, the expression of *nosZ* was inhibited during the time when Cu level increased (0.50 mM) (Black et al., 2016). Subsequently, the diversity of these genes decreased gradually (Magalhaes et al., 2011). Moreover, in Pb-polluted soil, the diversity of the *nirK* gene community was reduced (Sobolev and Begonia, 2008). However, sometimes the impact of HMs on the diversity of the denitrification gene is not completely negative. Although the increased content of Ag reduced the copy number of the *nirK* gene, the diversity of the *nirK* gene increased (Throback et al., 2007). *nirS* gene diversity was enhanced in Hg-contaminated soil (Zhou et al., 2012). This can be explained by the moderate interference hypothesis. Due to HMs reducing the inherent competitive exclusion of bacterial populations, species diversity was increased (Giller et al., 1998).

6.2.4 Microorganisms as HM Pollution Indicators

It is necessary to select appropriate indicators to diagnose HM pollution. Soil biological characteristics are considered to be reliable indicators of soil quality, because they are more dynamic and are more sensitive than soil physicochemical characteristics (Brookes, 1995; Hinojosa et al., 2004). Soil microorganisms are involved in almost all biochemical reactions in soil. In order to prevent soil degradation, they are usually applied as important indicators of soil environmental quality, because they have greater sensitivity than mollusks, megafauna, or plants (Zhang et al., 2010, 2016). By measuring soil microbial changes, the effect and extent of pollution can be evaluated. Advantages of microbiological diagnosis include the following aspects: (i) microorganisms are the most active organisms in most soil ecosystems (Feng et al., 2018); (ii) the process of microbial participation is an indispensable part of elemental cycles and is closely involved in the transformation processes of HMs in soil (Song et al., 2017); and (iii) microbial diagnosis requires only a small amount of sample and can be continuously monitored. In addition, it has the advantages of low cost and batch analysis (Song et al., 2016).

6.2.5 Soil Animals

HMs have direct toxicity and indirect toxicity (e.g., inducing changes in food resources and habitat quality) to soil animals (e.g., earthworms and nematodes). Earthworms are effective accumulators of metals in the soils. Haimi and SiiraPietikainen (1996) observed that HMs reduced the number of species of earthworms. Lukkari et al. (2004) found that HMs entered the earthworms. The sorption of HMs by earthworms leads to the separation of HM ions in most pathways. Meanwhile, earthworms have the ability to regulate HMs in their bodies, especially Cu, Zn, and other biologically essential metals (Dallinger, 1996; Morgan and Morgan, 1999). Subsequently, some earthworm species are able to live even in highly HM-contaminated areas due to their regulatory capacity and probably also due to their high biomass (compared with microorganisms) (Langdon et al., 2001;

Sivakumar, 2015). The sensitivity of the earthworms to HMs differs among species and also can be changed at different stages of their lives. For instance, the sensitivity of their growth to HMs is lower than that of reproduction (Owojori et al., 2009; Sivakumar et al., 2009; Žaltauskaitė and Sodienė, 2010).

Soil animals are usually less sensitive to HMs than soil microorganisms. However, they can also elicit behavioral responses at very low concentrations of HMs (Lukkari et al., 2004). Furthermore, because of their visibility, morphological abnormalities can readily serve as definitive evidence of HM toxicity (Sivakumar et al., 2006). Behavioral responses and morphological symptoms of soil animals as a result being exposed to various HMs have been recorded in various species of earthworms. These physiological changes upon exposure to HMs include failure to burrow, curling of the body, slow or swift movements, shortening or elongation of the body, oozing out of coelomic fluid, preclitellar swelling, body constriction, and segmental bulging (Sivakumar et al., 2006).

6.3 HM TOXICITY TO PLANTS

6.3.1 Availability of HMs to Plants

Despite the high total concentrations of HMs in soils, only a small fraction (i.e., bioavailable or mobile fraction) can be taken up by plants and other living organisms (Sammut et al., 2010; Sun et al., 2009; Vega et al., 2010) due to their binding with different soil components (Rascio and Navari-Izzo, 2011; Sammut et al., 2010; Uzu et al., 2009). Most of the HMs (such as Pb and Cu) are only marginally soluble or available for plant uptake (Pourrut et al., 2011; Punamiya et al., 2010; Sammut et al., 2010; Vega et al., 2010). But a few HMs, such as Cd, predominantly exist in the soil solution (Verbruggen et al., 2009), leading to higher bioavailability. HMs may cause damage to normal biological processes of plants because of their analogy to essential elements (Krämer et al., 2007).

6.3.2 Uptake of HMs

Many farmlands have been polluted by HMs, especially by Pb, Cd, As, and Hg (Bai et al., 2011; Foucault et al., 2013; Nabulo et al., 2010; Shahid et al., 2014a; Uzu et al., 2010). These toxic HMs can be accumulated in crops and then decrease crop productivity, threatening animal and human health (Foucault et al., 2013; Järup, 2003; Sabir et al., 2014b; Ullah et al., 2015).

HMs are mainly taken up by plants through their root systems (Pourrut et al., 2011; Uzu et al., 2009). Ullah et al. (2009) pointed out that several steps are involved in the uptake of HMs by plants, including desorption of HMs from soil particles, transport of them toward plant roots, uptake by roots, and translocation to shoots. Pourrut et al. (2011) observed that the pathway of HMs transported from soil to roots is complicated, i.e., HMs are first adsorbed by plant roots through binding to the mucilage polysaccharides on cell surfaces or binding to carboxyl groups of uronic acid around the roots. Bondada et al. (2004) found that the aerial organs of plants such as leaves, fruits, and flowers can all take up HMs. HMs can be accumulated in plant leaves through foliar transfer after deposition of atmospheric particles (Schreck et al., 2012; Xiong et al., 2014a). Many plant aerial organs can assist roots in taking up HMs. However, the resistance of these organs to HM toxicity shows significant differences. Soudek et al. (2014) found that low concentrations of HMs were less toxic to shoots than roots. Vegetables grown in HM-contaminated soil have severe HM accumulation in leaves, especially for leafy vegetables (e.g., lettuce and endive) (Al Jassir et al., 2005; Peijnenburg et al., 2000).

Identifying the relationships between soil properties and plant sorption of HMs can be referenced to set guideline values for soils. Detailed investigations will ensure the protection of human health from crops with toxic HM accumulation. This promotes food safety for sustainable agriculture (Toth et al., 2016; Yang et al., 2016b).

6.3.3 Plant Growth

6.3.3.1 Germination

HMs, such as Pb, typically have a negative effect on plant growth and germination (Kopittke et al., 2007). Germination is strongly inhibited even in an environment with low concentrations of Pb^{2+} (Islam et al., 2007; Tomulescu et al., 2004). The negative effect of Pb^{2+} on seed germination has been reported in many plant species, such as *Hordeum vulgare*, *Elsholtzia argyi*, *Spartina alterniflora*, *Pinus halepensis*, *Oryza sativa*, and *Z. mays* (Islam et al., 2007; Sengar et al., 2008; Tomulescu et al., 2004), which results from the interference of Pb on protease and amylase (Sengar et al., 2008). However, Islam et al. (2007) reported that high concentrations of Pb^{2+} may speed up germination and simultaneously induce a shortened length of radicals and hypocotyls in *E. argyi*. This stimulation effect of Pb was also reported in fungi (Tian et al., 2019).

6.3.3.2 Stem

Negative influences of HMs on plant height and shoot growth have been addressed by Rout et al. (1997). Anderson et al. (1973) observed that oat-plant height decreased when Cr was added to nutrient solutions. Reductions in the height of *Cucumis sativus*, *Lactuca sativa*, and *Panicum miliaceum* due to the Cr stress were reported by Gorsuch et al. (1995) and Sharma and Sharma (1993). This might be induced by degraded root growth, leading to decreased availability of nutrients and water.

6.3.3.3 Root

HMs are widely reported to inhibit root growth in trees and other plants. For example, Cr can inhibit root cell division, root elongation, and the cell cycle in roots (Shanker et al., 2005). Cd toxicity decreases root length and dry mass, although root diameter can be larger (Gratao et al., 2009). Inhibition of root elongation is one of the earliest and most distinct symptoms of Cd toxicity (Lux et al., 2011). As mentioned above, these shortened roots will cause a decrease in the supply of nutrients and water to plants. In addition, root browning has been observed in many plants under the Cd stress, probably due to the relatively high toxicity of Cd compared to other metals (Chang et al., 2013; Lux et al., 2011).

6.3.4 Physiology of Whole Plants

Toxic metal ions enter plant tissues along with other nutrients, usually from the soil solution (Dalcorso et al., 2013; Hossain and Komatsu, 2012). Inhibition in growth is the most common physiological outcome of HM exposure to plants (Hu et al., 2013). HMs can change leaf structure and physiology, reduce photosynthetic and respiration rate, and inhibit metabolism and energy production of plants. Meanwhile, HMs can also influence transpiration and transportation of materials among different plant organs (Ying et al., 2010). Furthermore, HMs inhibit nutrient and water uptake of plant roots (Poschenrieder and Barceló, 2004). Therefore, various physiological processes (e.g., flowering, embryogenesis, seed formation, etc.) can be influenced by HMs due to their great impact on plant roots and leaves.

6.3.4.1 Photosynthetic Activities of Plants

Photosynthesis provides energy to plants. Certain metals, such as Fe, Cu, and Zn, are related to the formation of many enzymes and are essential for plant photosynthesis (Kováčik et al., 2010; Shanmugam et al., 2011). However, it has been shown that excessive amounts of Cd, Cu, Zn, and Pb inhibit photosynthesis (Burzyński and Kłobus, 2004; Shahid et al., 2012, 2014b).

The abundances of light-harvesting pigments (i.e., chlorophyll and carotenoids) are reduced under HM stress. Then, disruption occurs in energy transfer in light-harvesting antennae

(Kuzminov et al., 2013). HMs can also affect the composition or structure of proteins, many of which are involved in photosynthesis, such as Rubisco (Franco et al., 1999), causing malfunction of both light and dark photosynthetic reactions. Thylakoid membranes are a well-organized system, which contain chlorophyll and protein complexes. The chlorophyll and protein complexes (PSI and PSII) are present hierarchically inside the thylakoid membranes (Croce and van Amerongen, 2011; Dekker and Boekema, 2005). They are usually surrounded by external light-harvesting complexes. Devi and Majeti (2004) observed that HMs may impede electron transport on the acceptor side of PSII. In addition, electron transport between PSI and PSII declines due to the damage to membrane lipids (Appenroth, 2010; Bertrand and Poirier, 2005; Kuzminov et al., 2013).

High amounts of HMs can decrease chlorophyll synthesis but increase chlorophyll degradation, thereby affecting photosynthetic electron transport and inhibiting plant photosynthesis. The ultra-structure of plant chloroplasts was distorted under the Pb stress (Kalaji and Loboda (2007). HMs can inhibit chlorophyll synthesis, electron transfer, and the activity of enzymes involved in Calvin cycle, eventually leading to reduction of the photosynthetic capacity. Chlorophyll fluorescence is sensitive to Cd. Marques and do Nascimento (2013) observed the changes in the absorption peaks of chlorophyll fluorescence spectra under the Cd stress, and the lowest fluorescence reabsorption corresponded to a high level of Cd.

6.3.4.2 The Production of Reactive Oxygen Species

Necrosis and programmed cell death have been detected in many plants under HM stresses. The production of reactive oxygen species (ROS) is known to be a major cause of HM toxicity in plants (Cambrolle et al., 2011; Deng et al., 2015; Shahid et al., 2014a). ROS production is considered as one of the earliest biochemical changes, when plants are subjected to HM stresses (Grover et al., 2010; Jasinski et al., 2008; Lushchak, 2011; Opdenakker et al., 2012; Yadav, 2010). The increased ROS can disrupt the redox status of cells, resulting in oxidative stress to live cells (Wang et al., 2020). Then, it induces membrane dismantling, biological macromolecule deterioration, ion leakage, lipid peroxidation, and DNA-strand cleavage (Carrasco-Gil et al., 2012; Chen et al., 2012; He et al., 2011). Previous studies have shown that HMs, such as Cu, Fe, Pb, Cd, Cr, As, Hg, Cr, and Zn, all are able to induce the production of ROS (Duquesnoy et al., 2010; Körpe and Aras, 2011; Márquez-García et al., 2011; Vanhoudt et al., 2010). ROS have been detected in *Pisum sativum* (Rodriguez-Serrano et al., 2006), *Triticum durum* (Hao et al., 2006), and *Vicia faba* (Pourrut et al., 2008), under the Cd, Ni, and Pb stress, respectively. The toxicity of ROS on plant macromolecules depends on the duration of exposure, types of plants, stages of development, and concentrations of HMs. In addition, HMs can deplete many antioxidants in cells (Lee et al., 2012; Lopareva-Pohu et al., 2011; Pourrut et al., 2013). This further intensifies the toxicity to plants, and a serious imbalance from the production and elimination of ROS occurs. Subsequently, dramatic physiological challenges to plants, called "oxidative stress," will be induced (Kováčik et al., 2010; Morina et al., 2010). Plants have evolved defense mechanisms to resist stress (Sabir et al., 2014a; Shahid et al., 2014c; Xiong et al., 2010; Xu et al., 2011). They fight for the stable cellular redox state via various pathways. This resistance is critical for maintaining productivity (particularly for crops) under the HM stress.

6.3.4.3 Hormonal Balance

HMs are able to change the hormonal balance in plants (Pető et al., 2011). For instance, Sudhakar et al. (2013) observed that As caused toxicity in *Brassica juncea* via changing auxin levels and the expression of many microRNAs. Plant hormones such as abscisic acid (ABA), auxins, brassinosteroids (BR), gibberellic (GA), and cytokinin (CK) play a significant role in adapting plants to HM toxicity through regulating a wide range of adaptive responses (Peleg and Blumwald, 2011). Srivastava et al. (2013) demonstrated that the growth of *Brassica juncea* plants treated with IAA significantly increased compared with untreated plants under As stress, suggesting that the As stress can be managed by adjusting the level of hormone.

6.3.4.4 Proteins

Some proteins are critical to the tolerance of a plant to HMs, such as proteins involved in the cell redox maintenance that defend a plant against the Pb stress (Gupta et al., 2010). Although a low concentration of Pb may increase total proteins (Mishra et al., 2006), high concentrations may decrease the protein pool (Chatterjee et al., 2004; Garcia et al., 2006; Mishra et al., 2006; Piotrowska et al., 2009). Pb is considered to be one of the most active HMs that binds or chelates proteins. It is able to reduce the quantity of total proteins, downregulate gene expression, increase ribonuclease activity, and lower amino acid content (Gupta et al., 2009; Kovalchuk et al., 2005; Piotrowska et al., 2009). Moreover, the protein profile of root cells can be altered after Pb exposure based on a study of bean seedlings (Beltagi, 2005).

6.4 INFLUENCES OF HMs ON CROP YIELD AND QUALITY

6.4.1 Crop Quality

6.4.1.1 Nutrient Uptake

HMs can influence plant sorption and utilization of mineral nutrients differently across different crop species. For example, Cd addition can reduce the uptake of other HMs such as Cu, Mn, and Zn, by the shoots of pea crops (Sandalio et al., 2001). It also causes a decline in Ca and K utilization by crops (Li et al., 2012, 2016; Shao et al., 2007). In castor beans, Li et al. (2016) showed that a low concentration of Cd promoted P uptake by roots, while a high concentration of Cd inhibited P transport to leaves. The possible mechanism is that most cations including HMs enter the same transport channels and compete with essential elements such as Fe, Mn, Ca, and Zn (Cheong et al., 2007; Tian et al., 2016). Therefore, HMs might inhibit the uptake and transport of these mineral nutrients, especially when they share the same electrovalence.

6.4.1.2 Shoot and Leaf

Leaf growth, area development, and total leaf number largely determine the yield of crops. HMs can influence plant leaves, thereby affecting plant growth. Navarro-Pedreño et al. (1997) found that Cr inhibited the growth of tomato plant leaves, especially the growth of young leaves. Sharma and Sharma (1993) reported that Cr reduced the leaf number per plant by 50% in wheat. Tripathi and Tripathi (1999) found that the leaf area and biomass of *Albizia lebbek* seedlings decreased greatly after adding 200 ppm Cr (VI). Singh (2001) reported that Cr reduced leaf size, causing burning of leaf tips. Long et al. (2003) observed the effects of Zn on crop growth based on three representative vegetables, i.e., Chinese cabbage, celery, and pakchoi. They showed that excessive Zn in the medium was toxic to all three vegetable crops, severely inhibiting their growth and causing young leaves to etiolate. Under the Cd stress, crop leaves will show toxic symptoms of stunting, chlorosis, necrosis, and dryness. Moreover, Li et al. (2014) showed that Cd caused a significant decrease of chlorophyll and carotenoids in rice, potato, and rape leaves. A normal concentration of Cd ranges from 0.05 to 0.2 mg/kg in leaf for most crop species (Solis-Dominguez et al., 2007). Studies suggest that leaves may serve as suitable bioindicators of HM pollution due to their sensitivity, easy access, and straightforward exposure. In addition, they are suitable as the indicator for the selection of resistant species.

6.4.1.3 Seed

Seed germination is also sensitive to different HMs in the environment, such as Cu, Cd, Cr, Mn, Zn, and Ni (Chigbo and Batty, 2013; Kalai et al., 2014; Ko et al., 2012; Léon et al., 2005; Márquez-García et al., 2013). For example, high levels of Ni induced the remobilization of macronutrients (Léon et al., 2005). High concentrations of Cd and Cu reduced the activities of acid phosphatase, α-amylase, and alkaline phosphatase in endosperm during germination (Kalai et al., 2014). Malan and Farrant (1998) reported that even low concentrations of Cd reduced the size of soybean

(*Glycine max*) seeds, although the average number of seeds per pod was not significantly influenced. Furthermore, decline of seed yields was also reported in chickpea (*Cicer arietinum*) and mung bean (*Vigna radiata*) in Cd-contaminated soils (Wani et al., 2007a, 2007b).

6.4.2 Productivity

HMs have different effects on crop productivity (Kavamura and Esposito, 2010). Some HMs (such as Zn, Cu, Mn, Ni, Se, Co, Cr, and Mo) have positive biological effects on crop productivity because they are essential for multiple biological processes (Kavamura and Esposito, 2010; Rascio and Navari-Izzo, 2011; Salla et al., 2011). However, these HMs and other non-essential HMs (e.g., Pb, Zr, Cd, As, Hg, and Sb) will reduce crop productivity if their concentrations rise beyond optimal levels (Foucault et al., 2013; Park et al., 2011; Pierart et al., 2015; Rascio and Navari-Izzo, 2011; Shahid et al., 2013; Xiong et al., 2014b).

The relationship between HM pollution and crop yield depends on the type and concentration of the HM. Wheat yield showed an upward trend under Cd, Pb, As, and Hg stresses (Nie et al., 2012). However, under Cr stress, it showed a downward trend. Wu et al. (2011) found the dry matter content of shoots decreased significantly under the Pb stress, resulting in the reduction of tea yield. Xia et al. (2010) observed that not all concentrations of Pb inhibit the growth of plants. Chlorophyll content was first stimulated and then inhibited with the increasing content of Pb in tea plants. Similarly, peanut yield also showed a trend of initial increasing and then decreasing when raising the Cd level (Fan et al., 2009). Zhou et al. (2017) concluded that, as the concentration of HMs increased (i.e., Pb, Cd, Zn, Cu), the total aboveground dry matter content of wheat showed a "slow-fast-slow" growth trend.

Excessive HMs in soils result in various problems in crops, inducing poor quality products and reduced yields (Austruy et al., 2014; Cambrolle et al., 2011; Guo et al., 2007). For example, Cd and Pb have negative effects on crop production due to the reduction in photosynthesis, root length, biomass, and diameter (Yourtchi and Bayat, 2013). At a finer scale, they damaged the growth apex, destroyed root hairs, and altered the structures of hypodermis and endodermis (Mohanpuria et al., 2007; Qin et al., 2020; Wojcik and Tukiendorf, 2004). Therefore, under severe HM stresses, crops showed significantly reduced uptake of water and essential nutrients. Subsequently, crop density, yield, and quality of harvested products are negatively influenced.

It has been widely recognized that HMs exhibit their toxicity not only as individual entities but also when they co-exist with others (Wang and Fowler, 2008). The co-existence of multiple HMs could exert higher toxicity to crops (Guo et al., 2007). For example, Huang et al. (2009) observed that single Cd or Pb stress caused less toxicity to the growth of *Zizania latifolia* than they are combined.

6.5 REMEDIATION TECHNIQUES FOR HEAVY METAL-CONTAMINATED AGRICULTURAL SOILS

HM pollution in agriculture is often dominated by low to medium metal concentrations. Its remediation is different from that for industrial sites and must comply with agricultural regulations and ecological requirements. Remediation techniques should be cost-effective and environmentally friendly, in addition to reduction of mobility and bioavailability of HMs in soil. Remediation can be executed in-situ or ex-situ via a single step or multiple stages, using physical, chemical, biological, and combined techniques. Nevertheless, the source and extent of agricultural soil pollution are complex, and remediation targets are stringent for agricultural soils. The search for an optimal remediation technique should consider cost, performance, and sustainability, which is an urgent and challenging task for researchers, practitioners, farmers, and authorities dealing with HM pollution in agriculture. In the following sections, we summarize the main remediation techniques for HM contamination in agricultural soils, aiming to provide reference and direction for designing the most suitable remediation strategy in the future.

6.5.1 Physical Remediation

Physical remediation refers to various physical methods to remove contaminants from soil, including soil replacement (El-Radaideh et al., 2014; Yao et al., 2012), thermal- and electric-assisted desorption or removal (Wei et al., 2001), and soil barriers (Bolan and Duraisamy, 2003; Dhaliwal et al., 2020). The advantages of these physical remediation methods are high efficiency and easy operation.

6.5.1.1 Soil Replacement

Agricultural soil replacement is usually applied to remediate a certain amount of heavily contaminated shallow (tilled) soil in a small area. It involves replacing or diluting contaminated soil with uncontaminated soil to reduce total HM levels in the soil (Song et al., 2022). The soil replacement method can be divided into three types: (i) total replacement and direct use of imported clean soil, which is suitable for a small volume of contaminated soil. It must be ensured that the contaminated soil that has been replaced will not cause secondary pollution; (ii) replacement of contaminated shallow soils with deep soils, which allows pollutants to be spread deeply into the soil and dilute the pollution; (iii) dilution with imported clean soils. Different types of clean soils are added to a contaminated area to dilute HM concentrations and enhance the soil purification capacity. The application of a soil replacement should consider the physiochemical properties of the clean soil and contaminated soil. The mixture of clean soil and contaminated soil will greatly change soil properties, such as clay content, mineral composition, soil organic matter, pH value, and density.

Soil replacement can help alleviate the contamination of HMs quickly and efficiently. In the 1980s, soil replacement was widely applied to remediate soil contamination (Gong et al., 2018), and it improved soil quality and fertility significantly (El-Radaideh et al., 2014). Recently, it has been found that some HMs in the contaminated soil might be activated during wetting and migrate to the upper layer of the clean soil through capillary action (Jiang et al., 2020a). Moreover, the high cost restricts its wide usage. Therefore, current research should focus on developing an economical soil replacement method while preventing the migration of HMs effectively (Jiang et al., 2022).

6.5.1.2 Thermal Treatment

Thermal desorption is a technique that separates contaminants with relatively low boiling points from the soil, and then removes them by sorption (Dazy et al., 2009; Teng et al., 2020). Soil heat treatment temperature is usually 150°C–500°C. This method usually has high efficiency, a short cycle, and no secondary pollution (Zhao et al., 2019). However, the equipment for thermal desorption is expensive and is time-consuming, so it is most suitable for the remediation of volatile pollutants, e.g., As and Hg (Liu et al., 2018; Wang et al., 2012). Thermal fixation is a method to fix nonvolatile metals in the solid phase by heating. For instance, the chemical forms of some HMs (e.g., Pb, Zn, Cu, Mn, Ni) might change at high temperatures, turning into insoluble carbonates or other relatively insoluble and immobile phases (Antilén et al., 2006).

Due to high-temperature pyrolysis, thermal treatments not only cause significant energy loss but could also kill functional microorganisms in soil. Then, the heated soils are not suitable for agricultural management (Akcil et al., 2015; Huang et al., 2011). Many studies have attempted to add reagents to improve the efficiency of HM removal at low temperatures. For example, the addition of $FeCl_3$ can effectively reduce the Hg content in the soil to <1 mg/kg at low temperatures. It can accelerate the evaporation of Hg by reducing the evaporation temperature of Hg (Ma et al., 2014). The addition of citric acid before heat treatment can reduce Hg concentrations and the energy cost in this process, which can restore the agricultural sustainability of soil (Ma et al., 2015).

Thermal desorption methods can be applied ex- or in-situ. Remote heat treatment largely depends on the efficiency of a rotary kiln-type reactor (Park et al., 2020). Figure 6.2 is a schematic diagram that shows combining chlorination and heating to improve the remediation efficiency of HMs (Pb, Cu, Cd, Cr, Ni, and Zn) in a full-scale industrial experiment (Li et al., 2019). However, remote heat

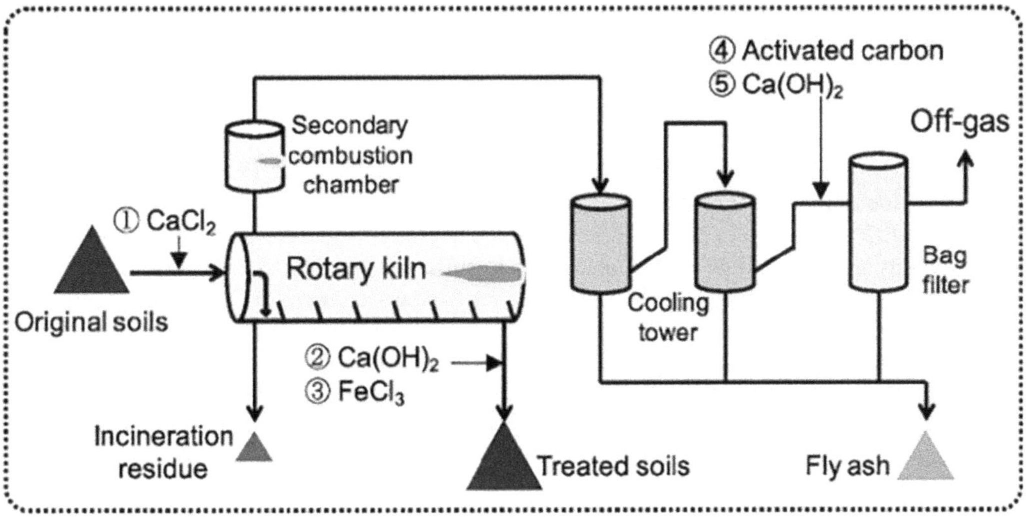

FIGURE 6.2 Schematic diagram of chlorination and thermal treatment. (Reproduced with permission from Ref. Li et al. (2019). Copyright 2019 Elsevier.)

treatment needs dredging and transportation, which greatly increases the cost of remediation (He et al., 2015). In-situ remediation can heat and vacuum subsurface soil at temperatures up to 600°C, but it is primarily used to remove volatile organic compounds.

6.5.2 Chemical Remediation

In chemical remediation, chemical reagents are injected into the soil. They reduce the mobility, bioavailability, and toxicity of HMs by sorption, precipitation, oxidation or reduction, polymerization, and complexation mechanisms. Many techniques exist, including chemical stabilization, soil washing, chemical oxidation or reduction, and electrokinetic-enhanced leaching.

6.5.2.1 Soil Stabilization

Soil stabilization can limit the bioavailability of HMs by fixing them in the soil. It is a simple operation, with low cost and environmentally friendliness, and it can remediate various HM pollutants simultaneously. It is different from the stabilization/solidification technique for industrial soils (Jin, 2020) as solidification would result in the degradation of ecological functions of agricultural soils (Shen et al., 2019). However, the addition of chemical reagents may increase soil volume, and contaminants could be reactivated when environmental conditions (such as soil moisture and pH) change (Chen et al., 2009). Therefore the efficiency, cost-effectiveness, and possible negative impact on the overall soil health need to be considered before practical applications (Xu et al., 2021). The efficiency of remediation is largely determined by the chemical stabilizers, which include organic, inorganic, and composite types (Ashraf et al., 2017; Austruy et al., 2014; Guo et al., 2019; Khalid et al., 2017; Shahid et al., 2014d).

Alkaline substances (liming agents), such as lime, steel slag, fly ash, and blast furnace slag, can effectively stabilize HMs in soil via cation exchange, precipitation, sorption, and electrostatic interaction (Xu et al., 2021). Among these substances, lime or limestone is widely used to stabilize HMs by increasing soil pH. Subsequently, lime or limestone promotes precipitation of metal carbonates or hydroxides with reduced solubility. However, excessive or continuous application of lime (stone) could damage the soil aggregate structure. Then, it may cause soil hardening, decrease the abundance of microorganisms, and ultimately affect the growth of crops.

Zeolites and clay minerals, such as montmorillonite, palygorskite, and sepiolite, have been widely used as stabilizers in numerous laboratory and field experiments because of their wide availability, high specific surface areas, and high cation exchange capacities (Fernandes and Baeyens, 2019; Otunola and Ololade, 2020). However, their natural forms usually have limited sorption capacities so the application rates need to be relatively high.

Industrial by-products and wastes have been applied to stabilize HMs in polluted soils, such as steel slag, mine tailings, and blast furnace slag, sewage sludge, and biomass ash (Ding et al., 2017; Yang et al., 2020; Yuan et al., 2018). These materials not only induce elevated soil pH, decreasing their solubility and bioavailability, but also provide nutrients (e.g., Mg, Si, and micronutrients) to promote crop growth (Ding et al., 2017; Lei et al., 2020). However, extreme caution should be taken due to the large compositional or mineralogical variability of these materials and the potential hazardous elements contained in them (Jin, 2022).

There is a growing interest in carbon-based materials (e.g., biochar and activated carbon) due to their wide variety, high availability, and cost-effectiveness. As a farmland soil conditioner, carbon-based materials can improve the water retention capacity and the nutrient cycling, and reduce HM sorption by plants (Chen et al., 2018; Fu et al., 2021a; Kavitha et al., 2018). Modification of carbon-based materials is also relatively easy. Modified carbon-based materials can increase the specific surface area and pore volume and greatly improve the sorption performance of materials (Chen et al., 2019, 2022). For example, magnetic separation has been successfully applied to recover HM complexes from sediment treated with magnetic biochars (Rizwan et al., 2020). The methods for resolving magnetic loading of carbon-based materials are simple and diverse with various suitable reagents such as Fe_2O_3, $FeCl_2$, and $FeCl_3$. Raw materials with enriched magnetic metals such as Fe and scrap metal minerals can be also adopted (Lin et al., 2022). In addition, according to different soil conditions, modification methods suitable for biomaterials, e.g., acid and alkali modifications, can be used before magnetic modification to achieve a better sorption effect (Rizwan et al., 2020). Moreover, magnetic materials can often be regenerated after use and hence reusability is important for assessing the performance of magnetic materials.

Phosphate is an essential element for plant growth. The solubility products of metal phosphates are usually the lowest. Therefore, the addition of synthetic hydroxylapatite, phosphate rock, bio-apatite, and other phosphates are used for treating HM-contaminated agricultural soil. Moreover, synthetic minerals and organic chemical reagents (Shen et al., 2018; Xu et al., 2021) used in green chemistry have been widely studied as soil stabilizers or amendments, taking advantage of their adjustable characteristics, fast, superior, and selective sorption capacities, and sometimes the slow nutrient-releasing properties. Given the complexity and multi-functions of agricultural soils, it is becoming more common that combinations of stabilizers are used to fulfill different purposes. These materials usually exhibit superior performance compared to the application of a sole stabilizer (Friesl-Hanl et al., 2009).

6.5.2.2 Soil Washing

Soil washing involves injecting chemical reagents (acids, alkalis, salts, chelating agents, and surfactants) into contaminated soil or sediment to dissolve or extract the HMs. These chemical reagents can destroy the complexes formed by HMs and functional groups on the soil surface via complexation, acidolysis, or cation/anion exchange. Soil washing has attracted extensive interest because of its simple operation and straightforward mechanism (Hu et al., 2021). However, it only can be used in sites close to water. Meanwhile, the wastewater must be treated to meet discharge standards, which might be tough for agricultural management (either for arid lands or paddy soils). It can be applied in- or ex-situ, and the key is to choose the appropriate leaching agent that can extract HMs without destroying the soil structure.

The efficiency of soil washing is determined by various factors, including soil pH, texture, organic matter content, forms of HMs, leaching conditions, and solid-to-liquid ratio (Zhu et al., 2021). HMs can be strongly adsorbed in loam and clay with high viscosity, but with limited sorption in sandy soil (Grzywaczyk et al., 2021). Based on fractionation using sequential extraction, HMs in soil are mainly grouped into water-soluble, carbonate-bound, iron-manganese oxide-bound, organic-bound, and residue fractions with decreased mobility and extractability in sequence. Only the first two fractions (i.e., water-soluble and carbonate-bound HMs) are easily extracted from soils (Dermont et al., 2008). Strong acidic conditions facilitate HM extraction, but they will significantly change soil physicochemical and biological properties and damage the structure of macro-aggregates (Hu et al., 2021).

6.5.2.3 Electrokinetic Remediation

Electrokinetic remediation is a low-cost technique for the treatment of HM-contaminated soils (Cameselle and Reddy, 2012). Electrodes are installed in the ground. By applying a low-level DC potential or an alternating current, an electric field is created between the electrodes, causing metal anions to move toward the anode and cations toward the cathode (Liu et al., 2017). The mechanisms of HM transportation include electromigration, electroosmosis, and electrophoresis (Bahemmat et al., 2016; Wen et al., 2021a). HMs accumulated on the electrodes are then removed or extracted via washing by water, sorption, precipitation, complexation with adsorbents, or ion exchange by resins.

Electrokinetic remediation equipment is simple and low-cost and can be deployed both in- and ex-situ. It can remove a variety of HMs at the same time, and it can be combined with bioremediation, phytoremediation, and other traditional remediation techniques such as permeable reactive barriers (Yuan and Chiang, 2007). Soil type, temperature, pH, HM concentration, and voltage and current levels are all important factors affecting the remediation performance (Cameselle and Reddy, 2012; Castillo et al., 2012). Moreover, the chemical forms of HMs influence the effectiveness of their removal, especially in some aged soils. HMs are precipitated as insoluble minerals or bound strongly to soil components, significantly inhibiting electrokinetic performance. Therefore, surfactants, chelating agents, and complexing agents might be added to facilitate the desorption or dissolution of HMs into soil solutions. In-situ electrokinetic remediation is usually not well controlled, and HM ions are unlikely to directly migrate to the electrodes via the shortest path, which results in excessive consumption of energy and time. In contrast, ex-situ application is simpler and can speed up HM removal with agitation, although with increased transportation costs. In addition, the potential negative impact on soil animals should be considered when using electrokinetic remediation in situ.

6.5.2.4 Chemical Oxidation and Reduction

Chemical oxidation or reduction uses the redox reactions between chemical reagents and HMs to change their oxidation states in order to reduce their mobility, environmental availability, and toxicity (Marques et al., 2011). The essence of redox reactions is the gain or loss of electrons, whereas species that give electrons are referred to as oxidants, and those that lose electrons as reductants. Commonly used oxidants include hydrogen peroxide (H_2O_2), potassium permanganate ($KMnO_4$), and chlorine dioxide (ClO_2) (Hamby, 1996; Marques et al., 2011). In contrast, calcium polysulfide, zero-valent iron, hydrogen sulfide, sodium bisulfite, ferrous iron, and ferrous sulfate are common reductants (Tang et al., 2019). The oxidation and reduction processes can be affected by various factors, including physical and chemical properties of soil, type and concentration of HMs, and dosage and type of the oxidants or reductants. Generally, sorption oxidation and catalytic oxidation are dominant in alkaline environments, while dissolution oxidation is dominant in acidic environments (Liang et al., 2022). Although chemical redox is fast and efficient to remediate contaminated soils, the high cost and the possibility of secondary pollution restrict the wide use of this method in agricultural soils.

6.5.3 Bioremediation

Bioremediation is an efficient, economical, and ecofriendly method to remediate the soil contaminated by HMs (Lai et al., 2022). It usually applies microorganisms or plants to treat

HM-contaminated soils, making them less hazardous to the environment and free from secondary pollution. This method can also facilitate the restoration of the ecosystem of contaminated sites. Bioremediation has a wide range of applications, and it is often used at sites with low to medium levels of HMs.

6.5.3.1 Microbial Remediation

Microorganisms are abundant on Earth, and they can live in a wide range of environmental conditions. In particular, many of them are functional during HM remediation, e.g., via biomineralization or sorption. Microbial remediation refers to the process of mobilizing, immobilizing, or eliminating HMs in the soil through biotic activities of soil microorganisms (bacteria, fungi, arbuscular mycorrhizal fungi) (Gadd et al., 2003; White et al., 1995). Since microorganisms can greatly affect the migration and transformation of HMs in soils, they are considered as important biological receptors (Gadd, 2010; Gadd et al., 2003). Although microbial remediation has attracted extensive attention, it is still mainly carried out in laboratories rather than in the field because the performance of microorganisms can be affected by multiple factors, e.g., soil pH, dissolved organic carbon, oxygen levels, and pollutant complexity (Qin et al., 2021; Wang et al., 2022b).

Bacteria can mineralize and immobilize HMs through extracellular polymers and then take them up into cells (Jiang et al., 2020b), further completing biodegradation by binding HMs to intracellular proteins (Ye et al., 2017). The extracellular macromolecules and functional groups of bacteria can also facilitate the uptake of HMs (Qu et al., 2022). However, HMs can affect bacterial enzyme secretion (e.g., oxidoreductases and oxygenase) and induce self-defense mechanisms, both of which can change cell morphology and affect the efficiency of bioremediation (Jacob et al., 2018; Jiang et al., 2020b).

Fungi are widely used as biosorbents for the removal of toxic metals (Igiri et al., 2018; Tian et al., 2019). The resistance of fungi to HMs is related to a synergistic action of multiple components, which can be divided into intracellular and extracellular actions (Wang et al., 2022c). Intracellular actions include the protection by antioxidant systems and the formation of chelates. The compartmentalization of vesicles can reduce or eliminate the HM toxicity. Meanwhile, fungi can distribute HM ions into different parts of a cell through compartmentalization, which confines or transforms them to less toxic forms (Yin et al., 2019). Extracellular ligands and cell wall binding are critical for HM sorption. Accordingly, due to the larger cell-to-surface ratio, fungi are more inclined to make physical and enzymatic contact with their surroundings. In particular, filamentous fungi, such as *Penicillium* and *Aspergillus* species, have been widely applied (Meng et al., 2022; Okolie et al., 2020; Tang et al., 2022).

Microbial remediation is usually low-cost and environmentally friendly for in-situ applications. However, it is sensitive to environmental conditions and may only work for specific HMs. Future research may include genetic modification of microorganisms to tolerate aggressive environments and develop synergistic techniques with microbial remediation to enhance the overall efficiency and performance.

6.5.3.2 Phytoremediation

Phytoremediation uses plants to treat polluted soils and includes phytoextraction, phytoaccumulation, phytostabilization, phytofiltration, phytovolatilization, and phytodegradation (Ali et al., 2013; Ashraf et al., 2019; Salas-Luevano et al., 2017; Tang et al., 2005). It is a green technology with the advantages of less disturbance to soil system and no damage to soil structure. In phytoaccumulation, a hyperaccumulator is a plant that absorbs and accumulates HMs and transfers them to the upper part of the plant. For example, Ma et al. (2001) found that the fern *Pteris vittata* (brake fern) is efficient in absorbing As from soils and translocating it into its aboveground part. However, it takes a relatively long time to completely remediate a HM-contaminated soil. Meanwhile, for deep soils (>1 m), phytoremediation is not useful under field conditions (Adams et al., 2014). Current research efforts focus on improving the enrichment capability of plants to HMs and safe disposal of the HM-enriched plants after remediation.

6.5.3.3 Soil Animal (Earthworm) Remediation

There are relatively fewer studies about soil animal remediation. As the most common soil animal, earthworms have been confirmed to be able to remove contaminants from soil with low cost (Zeb et al., 2020). Earthworms' removal of HMs is via two mechanisms: intrinsic and extrinsic. Intrinsically, earthworms accumulate HMs by ingestion or skin sorption of metal-containing substances in soil and then assimilation of them (Becquer et al., 2005; Mark et al., 2005). Extrinsically, earthworms can alter soil structure and physicochemical properties, which in turn change the distribution and bioavailability of HMs in soils (Curry and Schmidt, 2007). On the other hand, earthworms may live in groups, producing more slime, which will react with metals to form organometallic compounds. This enhances the transport of HMs and facilitates their efficient uptake by plants (Lemtiri et al., 2016). As earthworms have distinct response mechanisms upon HM exposure, they are also considered as ecological indicators for HM pollution in soils (Suthar et al., 2008; Usmani, 2015; Wang et al., 2018a).

6.5.4 Combined or Synergistic Remediation Techniques

In complex situations, the HM removal efficiency by any single remediation technique may not be satisfactory. Hence combining different and synergistic remediation techniques usually becomes feasible and more effective in soil remediation. Combined remediation usually refers to a strategy that incorporates more than two remediation techniques implemented at once or sequentially. The application of combined remediation to complex or mixed HM-contaminated soil has become a new direction, including the development of soil remediation technologies that consist of combined bioremediation, physicochemical-biological combined remediation, physical-chemical combined remediation, and many others.

6.5.4.1 Physical-Chemical Remediation

Physical-chemical combined remediation has already been widely reported and utilized. For instance, Hu et al. (2014a) used physical separation and EDTA leaching to perform ex-situ soil remediation. Physical separation divided the soil into different fractions according to the concentrations of HMs, and then soil washing with EDTA was applied to treat HM-enriched fractions. In this way, the consumption of the leaching agents is significantly reduced as well as the risk of secondary pollution. Moreover, chemical-biological remediation is one of the most applied methods for remediating HM-contaminated soils, which combines the advantages of (physico-) chemical and biological techniques, achieving high efficiency, low environmental or ecological impact, and long-term effectiveness. For instance, soil washing followed by phytoremediation utilizes organic chelating agents to raise the concentration of HMs in the soil solution, which improves their bioavailability, inducing and intensifying their uptake by plants (Fu et al., 2021b). Sulfur-oxidizing bacteria and electrokinetic remediation have also been combined to enhance the removal of Cu (Maini and Sharman, 2000).

6.5.4.2 Microorganism-Animal-Plant Remediation

Microorganism (bacteria, fungi)-plant and animal (earthworm)-plant remediation are emerging technologies in agricultural soil remediation. With the advantage of negligible disturbance to soil, microbe-plant remediation can enhance the removal efficiency of HMs from contaminated soils by screening for microorganisms with greater degradative capacity and plants that are hyperaccumulators. For example, microorganisms such as plant growth promoting bacteria (PGPB) belonging to *Bacillus*, *Pseudomonas*, *Flavobacteriaceae*, *Acinetobacter*, and *Enterobacteriaceae* have been used. They have the ability to alleviate HM toxicity, improve soil conditions, and secrete phytohormones and siderophores. These help to promote plant growth and improve the efficiency of HM removal by plants (Wang et al., 2022a). Meanwhile, plant roots secrete nutrients and provide favorable living environments and materials for reproduction and growth of microorganisms. It is envisaged that using the rhizosphere bacteria or fungi to promote the growth of plants, developing animal-microorganism combined remediation as well as rhizosphere-enhanced technology will be extensively studied in the near future.

6.6 CONCLUSIONS

This work has highlighted some key aspects of heavy-metal pollution in agricultural soils, including its current status, the toxicity to soil microorganisms and plants, effects on crop quality and yield, and various remediation technologies. HM contamination is one of the most severe challenges for agricultural soils, and its remediation is usually different from that used for industrial sites, due to the multiple and stringent ecological requirements. To achieve food safety and security and manage soils more sustainably in a more populated world, the designing, implementation, and assessment of soil remediation should be carried out by concerted efforts from soil scientists, botanists, microbiologists, material scientists, and engineers.

ACKNOWLEDGMENTS

This work was supported by National Key R&D Program of China (2020YFC1808000). This work was also partially supported by Fundamental Research Funds for the Central Universities (KYCYXT2022004) and Research Project of Higher Education Reform in Jiangsu Province (2021JSJG031). We would like to thank Program of Cooperation Among Industries, Universities, and Technologies from Jiangsu Provincial Department of Science and Technology (BY2021165) for the financial support.

REFERENCES

Adams, G., Tawari-Fufeyin, P., Igelenyah, E., Odukoya, E., 2014. Assessment of heavy metals bioremediation potential of microbial consortia from poultry litter and spent oil contaminated site. *International Journal of Environmental Bioremediation & Biodegradation* 2(2), 84–92.

Adetunji, A.T., Lewu, F.B., Mulidzi, R., Ncube, B., 2017. The biological activities of beta-glucosidase, phosphatase and urease as soil quality indicators: A review. *Journal of Soil Science and Plant Nutrition* 17(3), 794–807.

Akcil, A., Erust, C., Ozdemiroglu, S., Fonti, V., Beolchini, F., 2015. A review of approaches and techniques used in aquatic contaminated sediments: Metal removal and stabilization by chemical and biotechnological processes. *Journal of Cleaner Production* 86, 24–36.

Al Jassir, M.S., Shaker, A., Khaliq, M.A., 2005. Deposition of heavy metals on green leafy vegerables sold on roadsides of Riyadh City, Saudi Arabia. *Bulletin of Environmental Contamination and Toxicology* 75(5), 1020–1027.

Ali, H., Khan, E., Sajad, M.A., 2013. Phytoremediation of heavy metals-concepts and applications. *Chemosphere* 91(7), 869–881.

Anderson, A., Meyer, Mayer, F.K., 1973. Heavy metal toxicities: Levels of nickel, cobalt and chromium in the soil and plants associated with visual symptoms and variation in growth of an oat crop. *Crop & Pasture Science* 24, 557–571.

Antilén, M., Araya, N., Briceno, M., Escudey, M., 2006. Changes on chemical fractions of heavy metals in Chilean soils amended with sewage sludge affected by a thermal impact. *Soil Research* 44(6), 619–625.

Appenroth, K.-J., 2010. *Definition of "Heavy Metals" and Their Role in Biological Systems, Soil Heavy Metals.* Springer: Berlin Heidelberg, pp. 19–29.

Ashraf, A., Bibi, I., Niazi, N.K., Ok, Y.S., Murtaza, G., Shahid, M., Kunhikrishnan, A., Li, D.W., Mahmood, T., 2017. Chromium(VI) sorption efficiency of acid-activated banana peel over organo-montmorillonite in aqueous solutions. *International Journal of Phytoremediation* 19(7), 605–613.

Ashraf, S., Ali, Q., Zahir, Z.A., Ashraf, S., Asghar, H.N., 2019. Phytoremediation: Environmentally sustainable way for reclamation of heavy metal polluted soils. *Ecotoxicology and Environmental Safety* 174, 714–727.

Austin, M.P., 1987. Models for the analysis of species' response to environmental gradients. *Vegetatio* 69(1–3), 35–45.
Austin, M.P., Smith, T.M., 1989. A new model for the continuum concept. *Vegetatio* 83, 35–47.
Austruy, A., Shahid, M., Xiong, T.T., Castrec, M., Payre, V., Niazi, N.K., Sabir, M., Dumat, C., 2014. Mechanisms of metal-phosphates formation in the rhizosphere soils of pea and tomato: Environmental and sanitary consequences. *Journal of Soils and Sediments* 14(4), 666–678.
Bahemmat, M., Farahbakhsh, M., Kianirad, M., 2016. Humic substances-enhanced electroremediation of heavy metals contaminated soil. *Journal of Hazardous Materials* 312, 307–318.
Bai, J.H., Xiao, R., Cui, B.S., Zhang, K.J., Wang, Q.G., Liu, X.H., Gao, H.F., Huang, L.B., 2011. Assessment of heavy metal pollution in wetland soils from the young and old reclaimed regions in the Pearl River Estuary, South China. *Environmental Pollution* 159(3), 817–824.
Bardgett, R.D., Saggar, S., 1994. Effects of heavy-metal contamination on the short-term decomposition of labeled [C-14] glucose in a Pasture Soil. *Soil Biology and Biochemistry* 26(6), 727–733.
Becquer, T., Dai, J., Quantin, C., Lavelle, P., 2005. Sources of bioavailable trace metals for earthworms from a Zn-, Pb- and Cd-contaminated soil. *Soil Biology & Biochemistry* 37(8), 1564–1568.
Beltagi, M., 2005. Phytotoxicity of lead (Pb) to SDS-PAGE protein profile in root nodules of Faba bean (*Vicia faba* L.) plants. *Pakistan Journal of Biological Sciences* 8, 687–690.
Bertrand, M., Poirier, I., 2005. Photosynthetic organisms and excess of metals. *Photosynthetica* 43(3), 345–353.
Black, A., Hsu, P.C.L., Hamonts, K.E., Clough, T.J., Condron, L.M., 2016. Influence of copper on expression of nirS, norB and nosZ and the transcription and activity of NIR, NOR and N_2OR in the denitrifying soil bacteria Pseudomonas stutzeri. *Microbial Biotechnology* 9(3), 381–388.
Bolan, N.S., Duraisamy, V.P., 2003. Role of inorganic and organic soil amendments on immobilisation and phytoavailability of heavy metals: A review involving specific case studies. *Australian Journal of Soil Research* 41(3), 533–555.
Bondada, B.R., Tu, S., Ma, L.Q., 2004. Absorption of foliar-applied arsenic by the arsenic hyperaccumulating fern (*Pteris vittata* L.). *Science of the Total Environment* 332(1), 61–70.
Bouskill, N.J., Barker-Finkel, J., Galloway, T.S., Handy, R.D., Ford, T.E., 2010. Temporal bacterial diversity associated with metal-contaminated river sediments. *Ecotoxicology* 19(2), 317–328.
Brookes, P.C., 1995. The use of microbial parameters in monitoring soil pollution by heavy-metals. *Biology and Fertility of Soils* 19(4), 269–279.
Brookes, P.C., McGrath, S.P., 1984. Effects of metal toxicity on the size of the soil microbial biomass. *Journal of Soil Science* 35(2), 341–346.
Brown, P.E., Minges, G., 1916. The effect of some manganese salts on ammonification and nitrification. *Soil Science* 2, 67–86.
Burzyński, M., Kłobus, G., 2004. Changes of photosynthetic parameters in cucumber leaves under Cu, Cd, and Pb stress. *Photosynthetica* 42(4), 505–510.
Cambrolle, J., Mateos-Naranjo, E., Redondo-Gomez, S., Luque, T., Figueroa, M.E., 2011. Growth, reproductive and photosynthetic responses to copper in the yellow-horned poppy, Glaucium flavum Crantz. *Environmental and Experimental Botany* 71(1), 57–64.
Cameselle, C., Reddy, K.R., 2012. Development and enhancement of electro-osmotic flow for the removal of contaminants from soils. *Electrochimica Acta* 86, 10–22.
Carrasco-Gil, S., Estebaranz-Yubero, M., Medel-Cuesta, D., Millán, R., Hernández, L.E., 2012. Influence of nitrate fertilization on Hg uptake and oxidative stress parameters in alfalfa plants cultivated in a Hg-polluted soil. *Environmental and Experimental Botany* 75, 16–24.
Castillo, A.N., García-Delgado, R.A., Rivero, V.C., 2012. Electrokinetic treatment of soils contaminated by tannery waste. *Electrochimica Acta* 86, 110–114.
Cavani, L., Manici, L.M., Caputo, F., Peruzzi, E., Ciavatta, C., 2016. Ecological restoration of a copper polluted vineyard: Long-term impact of farmland abandonment on soil bio-chemical properties and microbial communities. *Journal of Environmental Management* 182, 37–47.
Chang, Y.S., Chang, Y.J., Lin, C.T., Lee, M.C., Wu, C.W., Lai, Y.H., 2013. Nitrogen fertilization promotes the phytoremediation of cadmium in Pentas lanceolata. *International Biodeterioration & Biodegradation* 85, 709–714.
Chatterjee, C., Dube, B.K., Sinha, P., Srivastava, P., 2004. Detrimental effects of lead phytotoxicity on growth, yield, and metabolism of rice. *Communications in Soil Science and Plant Analysis* 35(1–2), 255–265.
Checkai, R.T., Corey, R.B., Helmke, P.A., 1987. Effects of ionic and complexed metal concentrations on plant uptake of cadmium and micronutrient metals from solution. *Plant Soil* 99(2–3), 335–345.
Chen, F., Gao, J., Zhou, Q., 2012. Toxicity assessment of simulated urban runoff containing polycyclic musks and cadmium in *Carassius auratus* using oxidative stress biomarkers. *Environmental Pollution* 162, 91–97.

Chen, H.M., Ma, J.Y., Wei, J.X., Gong, X., Yu, X.C., Guo, H., Zhao, Y.W., 2018. Biochar increases plant growth and alters microbial communities via regulating the moisture and temperature of green roof substrates. *Science of the Total Environmen* 635, 333–342.

Chen, H.M., Tang, L.Y., Hu, Y.S., Geng, Y.Y., Meng, L.Z., Li, W.A., Wang, Z.J., Li, Z., Huo, Z.L., 2022. Investigating the pathways of enhanced Pb immobilization by chlorine-loaded biochar. *Journal of Cleaner Production* 344, 131097.

Chen, H.M., Zhang, J.W., Tang, L.Y., Su, M., Tian, D., Zhang, L., Li, Z., Hu, S.J., 2019. Enhanced Pb immobilization via the combination of biochar and phosphate solubilizing bacteria. *Environment International* 127, 395–401.

Chen, J.H., He, F., Zhang, X.H., Sun, X., Zheng, J.F., Zheng, J.W., 2014. Heavy metal pollution decreases microbial abundance, diversity and activity within particle-size fractions of a paddy soil. *FEMS Microbiology Ecology* 87(1), 164–181.

Chen, Q.Y., Tyrer, M., Hills, C.D., Yang, X.M., Carey, P., 2009. Immobilisation of heavy metal in cement-based solidification/stabilisation: A review. *Waste Manage* 29(1), 390–403.

Cheong, Y.H., Pandey, G.K., Grant, J.J., Batistic, O., Li, L., Kim, B.G., Lee, S.C., Kudla, J., Luan, S., 2007. Two calcineurin B-like calcium sensors, interacting with protein kinase CIPK23, regulate leaf transpiration and root potassium uptake in Arabidopsis. *Plant Journal* 52(2), 223–239.

Chigbo, C., Batty, L., 2013. Effect of combined pollution of chromium and benzo(a)pyrene on seed growth of *Lolium perenne*. *Chemosphere* 90(2), 164–169.

Croce, R., van Amerongen, H., 2011. Light-harvesting and structural organization of photosystem II: From individual complexes to thylakoid membrane. *Journal of Photochemistry and Photobiology B* 104(1–2), 142–153.

Curry, J.P., Schmidt, O., 2007. The feeding ecology of earthworms: A review. *Pedobiologia* 50(6), 463–477.

Dalcorso, G., Fasani, E., Furini, A., 2013. Recent advances in the analysis of metal hyperaccumulation and hypertolerance in plants using proteomics. *Frontiers in Plant Science* 4, 280.

Dallinger, R., 1996. Metallothionein research in terrestrial invertebrates: Synopsis and perspectives. *Comparative Biochemistry and Physiology Part C* 113(2), 125–133.

Dazy, M., Férard, J.-F., Masfaraud, J.F., 2009. Use of a plant multiple-species experiment for assessing the habitat function of a coke factory soil before and after thermal desorption treatment. *Ecological Engineering* 35(10), 1493–1500.

Dekker, J.P., Boekema, E.J., 2005. Supramolecular organization of thylakoid membrane proteins in green plants. *Biochimica et Biophysica Acta* 1706(1–2), 12–39.

Deng, L.J., Zeng, G.M., Fan, C.Z., Lu, L.H., Chen, X.F., Chen, M., Wu, H.P., He, X.X., He, Y., 2015. Response of rhizosphere microbial community structure and diversity to heavy metal co-pollution in arable soil. *Applied Microbiology and Biotechnology* 99(19), 8259–8269.

Dermont, G., Bergeron, M., Mercier, G., Richer-Laflèche, M., 2008. Soil washing for metal removal: A review of physical/chemical technologies and field applications. *Journal of Hazardous Materials* 152(1), 1–31.

Desai, C., Parikh, R.Y., Vaishnav, T., Shouche, Y.S., Madamwar, D., 2009. Tracking the influence of long-term chromium pollution on soil bacterial community structures by comparative analyses of 16S rRNA gene phylotypes. *Research in Microbiology* 160(1), 1–9.

Devi, S., Majeti, P., 2004. Membrane lipid alterations in heavy metal exposed plants: Heavy metal stress in plants. In: M.N.V. Prasad (Ed.), *From Molecules to Ecosystems*, 2nd edition. Springer-Verlag: Heidelberg, pp. 99–116.

Dhaliwal, S.S., Singh, J., Taneja, P.K., Mandal, A., 2020. Remediation techniques for removal of heavy metals from the soil contaminated through different sources: A review. *Environmental Science and Pollution Research* 27, 1319–1333.

Ding, Y.Z., Wang, Y.J., Zheng, X.Q., Cheng, W.M., Shi, R.G., Feng, R.W., 2017. Effects of foliar dressing of selenite and silicate alone or combined with different soil ameliorants on the accumulation of As and Cd and antioxidant system in Brassica campestris. *Ecotoxicology and Environmental Safety* 142, 207–215.

Duquesnoy, I., Champeau, G.M., Evray, G., Ledoigt, G., Piquet-Pissaloux, A., 2010. Enzymatic adaptations to arsenic-induced oxidative stress in Zea mays and genotoxic effect of arsenic in root tips of *Vicia faba* and Zea mays. *Comptes Rendus Biologies* 333(11), 814–824.

El-Radaideh, N., Al-Taani, A.A., Al-Momani, T., Tarawneh, K., Taani, A., 2014. Evaluating the potential of sediments in Ziqlab Reservoir (NW Jordan) for soil replacement and amendment. *Lake and Reservoir Management* 30(1), 32–45.

Fan, Z.X., Shan, S.H., Yang, Z.Y., Wan, S.B., Xiao, L.V., Liu, X., 2009. Effects of Cd stress on distribution characteristics at different parts of five peanut genotypes and their yields. *Journal of Agricultural Science and Technology* 5, 102–107.

FAO, UNEP, 2021. Global assessment of soil pollution: Summary for policy makers, Rome, FAO.
Feng, G., Xie, T., Wang, X., Bai, J.Y., Tang, L., Zhao, H., Wei, W., Wang, M.L., Zhao, Y., 2018. Metagenomic analysis of microbial community and function involved in cd-contaminated soil. *BMC Microbiology* 18, 11.
Fernandes, M.M., Baeyens, B., 2019. Cation exchange and surface complexation of lead on montmorillonite and illite including competitive adsorption effects. *Applied Geochemistry* 100, 190–202.
Fernandez-Calvino, D., Arias-Estevez, M., Diaz-Ravina, M., Baath, E., 2011. Bacterial pollution induced community tolerance (PICT) to Cu and interactions with pH in long-term polluted vineyard soils. *Soil Biology and Biochemistry* 43(11), 2324–2331.
Fliessbach, A., Martens, R., Reber, H.H., 1994. Soil microbial biomass and microbial activity in soils treated with heavy-metal contaminated sewage-sludge. *Soil Biology and Biochemistry* 26(9), 1201–1205.
Foucault, Y., Leveque, T., Xiong, T., Schreck, E., Austruy, A., Shahid, M., Dumat, C., 2013. Green manure plants for remediation of soils polluted by metals and metalloids: Ecotoxicity and human bioavailability assessment. *Chemosphere* 93(7), 1430–1435.
Franco, E., Alessandrelli, S., Masojídek, J., Margonelli, A., Giardi, M.T., 1999. Modulation of D1 protein turnover under cadmium and heat stresses monitored by [35S]methionine incorporation. *Plant Science* 144(2), 53–61.
Friesl-Hanl, W., Platzer, K., Horak, O., Gerzabek, M.H., 2009. Immobilising of Cd, Pb, and Zn contaminated arable soils close to a former Pb/Zn smelter: A field study in Austria over 5 years. *Environmental Geochemistry and Health* 31(5), 581–594.
Fu, H., Ma, S., Xu, S., Duan, R., Cheng, G., Zhao, P., 2021a. Hierarchically porous magnetic biochar as an efficient amendment for cadmium in water and soil: Performance and mechanism. *Chemosphere* 281, 130990.
Fu, L.J., Feng, A.X., Xiao, J.J., Wu, Q., Ye, Q.Y., Peng, S., 2021b. Remediation of soil contaminated with high levels of hexavalent chromium by combined chemical-microbial reduction and stabilization. *Journal of Hazardous Materials* 403, 123847.
Gadd, G.M., 2010. Metals, minerals and microbes: Geomicrobiology and bioremediation. *Microbiology* 156, 609–643.
Gadd, G.M., Burford, E.P., Fomina, M., 2003. Biogeochemical activities of microorganisms in mineral transformations: Consequences for metal and nutrient mobility. *Journal of Microbiology and Biotechnology* 13(3), 323–331.
Garcia, J.S., Gratao, P.L., Azevedo, R.A., Arruda, M.A.Z., 2006. Metal contamination effects on sunflower (*Helianthus annuus* L.) growth and protein expression in leaves during development. *Journal of Agricultural and Food Chemistry* 54(22), 8623–8630.
Giller, K.E., Witter, E., McGrath, S.P., 1998. Toxicity of heavy metals to microorganisms and microbial processes in agricultural soils: A review. *Soil Biology and Biochemistry* 30(10–11), 1389–1414.
Golebiewski, M., Deja-Sikora, E., Cichosz, M., Tretyn, A., Wrobel, B., 2014. 16S rDNA pyrosequencing analysis of bacterial community in heavy metals polluted soils. *Microbial Ecology* 67(3), 635–647.
Gong, Y., Zhao, D., Wang, Q., 2018. An overview of field-scale studies on remediation of soil contaminated with heavy metals and metalloids: Technical progress over the last decade. *Water Research* 147, 440–460.
Gordon, A.S., Harwood, V.J., Sayyar, S., 1993. Growth, copper-tolerant cells, and extracellular protein-production in copper-stressed chemostat cultures of vibrio-alginolyticus. *Applied and Environmental Microbiology* 59(1), 60–66.
Gorsuch, J.W., Ritter, M.H., Anderson, E., 1995. Comparative toxicities of six heavy metals using root elongation and SHOOT growth in three plant species.
Gratao, P.L., Monteiro, C.C., Rossi, M.L., Martinelli, A.P., Peres, L.E.P., Medici, L.O., Lea, P.J., Azevedo, R.A., 2009. Differential ultrastructural changes in tomato hormonal mutants exposed to cadmium. *Environmental and Experimental Botany* 67(2), 387–394.
Grime, J.P., 1979. *Plant Strategies and Vegetation Processes*. John Wiley & Sons, Ltd: New York.
Grover, P., Rekhadevi, P.V., Danadevi, K., Vuyyuri, S.B., Mahboob, M., Rahman, M.F., 2010. Genotoxicity evaluation in workers occupationally exposed to lead. *International Journal of Hygiene and Environmental Health* 213(2), 99–106.
Grzywaczyk, A., Smulek, W., Smulek, G., Slachciński, M., Kaczorek, E., 2021. Application of natural surfactants for improving the leaching of zinc and copper from different soils. *Environmental Technology & Innovation* 24(5), 101926.
Gui, M.Y., Chen, Q., Ma, T., Zheng, M.S., Ni, J.R., 2017. Effects of heavy metals on aerobic denitrification by strain Pseudomonas stutzeri PCN-1. *Applied Microbiology and Biotechnology* 101(4), 1717–1727.
Guo, C.M., Tian, W.T., Wang, Z.J., Han, F.Y., Su, M., Wu, Y.L., Li, Z., Hu, S.J., 2019. Reduction of Pb availability during surficial leaching in different types of soils with addition of apatite and oxalic acid. *Journal of Soils and Sediments* 19(2), 741–749.

Guo, T.R., Zhang, G.P., Zhang, Y.H., 2007. Physiological changes in barley plants under combined toxicity of aluminum, copper and cadmium. *Colloid Surface B* 57(2), 182–188.

Gupta, D.K., Huang, H.G., Yang, X.E., Razafindrabe, B.H.N., Inouhe, M., 2010. The detoxification of lead in *Sedum alfredii* H. is not related to phytochelatins but the glutathione. *Journal of Hazardous Materials* 177(1–3), 437–444.

Gupta, D.K., Nicoloso, F.T., Schetinger, M.R.C., Rossato, L.V., Pereira, L.B., Castro, G.Y., Srivastava, S., Tripathi, R.D., 2009. Antioxidant defense mechanism in hydroponically grown Zea mays seedlings under moderate lead stress. *Journal of Hazardous Materials* 172(1), 479–484.

Haimi, J., SiiraPietikainen, A., 1996. Decomposer animal communities in forest soil along heavy metal pollution gradient. *Fresenius' Journal of Analytical Chemistry* 354(5–6), 672–675.

Hamby, D.M., 1996. Site remediation techniques supporting environmental restoration activities: A review. *Science of the Total Environment* 191(3), 203–224.

Hao, F., Wang, X., Chen, J., 2006. Involvement of plasma-membrane NADPH oxidase in nickel-induced oxidative stress in roots of wheat seedlings. *Plant Science* 170(1), 151–158.

He, F., Gao, J., Pierce, E., Strong, P.J., Wang, H., Liang, L., 2015. In situ remediation technologies for mercury-contaminated soil. *Environmental Science and Pollution Research* 22(11), 8124–8147.

He, J., Qin, J., Long, L., Ma, Y., Li, H., Li, K., Jiang, X., Liu, T., Polle, A., Liang, Z., Luo, Z.B., 2011. Net cadmium flux and accumulation reveal tissue-specific oxidative stress and detoxification in Populus×canescens. *Physiol Plant* 143(1), 50–63.

Hinojosa, M.B., Carreira, J.A., Garcia-Ruiz, R., Dick, R.P., 2004. Soil moisture pre-treatment effects on enzyme activities as indicators of heavy metal-contaminated and reclaimed soils. *Soil Biology and Biochemistry* 36(10), 1559–1568.

Hossain, Z., Komatsu, S., 2012. Contribution of proteomic studies towards understanding plant heavy metal stress response. *Frontiers in Plant Science* 3, 310.

Hou, D.Y., O'Connor, D., Igalavithana, A.D., Alessi, D.S., Luo, J., Tsang, D.C.W., Sparks, D.L., Yamauchi, Y., Rinklebe, J., Ok, Y.S., 2020. Metal contamination and bioremediation of agricultural soils for food safety and sustainability. *Nature Reviews Earth & Environment* 1(7), 366–381.

Hu, P.J., Yang, B.F., Dong, C.X., Chen, L.K., Cao, X.Y., Zhao, J., Wu, L.H., Luo, Y.M., Christie, P., 2014a. Assessment of EDTA heap leaching of an agricultural soil highly contaminated with heavy metals. *Chemosphere* 117, 532–537.

Hu, W., Niu, Y., Zhu, H., Dong, K., Wang, D., Liu, F., 2021. Remediation of zinc-contaminated soils by using the two-step washing with citric acid and water-soluble chitosan. *Chemosphere* 282, 131092.

Hu, X.F., Jiang, Y., Shu, Y., Hu, X., Liu, L.M., Luo, F., 2014b. Effects of mining wastewater discharges on heavy metal pollution and soil enzyme activity of the paddy fields. *Journal of Geochemical Exploration* 147, 139–150.

Hu, Y.F., Zhou, G., Na, X.F., Yang, L., Nan, W.B., Liu, X., Zhang, Y.Q., Li, J.L., Bi, Y.R., 2013. Cadmium interferes with maintenance of auxin homeostasis in Arabidopsis seedlings. *Journal of Plant Physiology* 170(11), 965–975.

Huang, K., Yang, K., Jiang, J., 2009. Effect of cadmium and lead on the growth of different cultivars of *Zizania latifolia. Journal of Changjiang Vegetables* 12, 27–30.

Huang, Y.T., Hseu, Z.Y., Hsi, H.C., 2011. Influences of thermal decontamination on mercury removal, soil properties, and repartitioning of coexisting heavy metals. *Chemosphere* 84(9), 1244–1249.

Igiri, B.E., Okoduwa, S.I.R., Idoko, G.O., Akabuogu, E.P., Adeyi, A.O., Ejiogu, I.K., 2018. Toxicity and bioremediation of heavy metals contaminated ecosystem from tannery wastewater: A review. *Journal of Toxicology* 9, 2568038.

Imperato, M., Adamo, P., Naimo, D., Arienzo, M., Stanzione, D., Violante, P., 2003. Spatial distribution of heavy metals in urban soils of Naples city (Italy). *Environmental Pollution* 124(2), 247–256.

Islam, E., Yang, X., Li, T.Q., Liu, D., Jin, X.F., Meng, F.H., 2007. Effect of Pb toxicity on root morphology, physiology and ultrastructure in the two ecotypes of *Elsholtzia argyi. Journal of Hazardous Materials* 147(3), 806–816.

Jacob, J.M., Karthik, C., Saratale, R.G., Kumar, S.S., Prabakar, D., Kadirvelu, K., Pugazhendhi, A., 2018. Biological approaches to tackle heavy metal pollution: A survey of literature - ScienceDirect. *Journal of Environmental Management* 217, 56–70.

Järup, L., 2003. Hazards of heavy metal contamination. *British Medical Bulletin* 68, 167–182.

Jasinski, M., Sudre, D., Schansker, G., Schellenberg, M., Constant, S., Martinoia, E., Bovet, L., 2008. AtOSA1, a member of the Abc1-like family, as a new factor in Cadmium and oxidative stress response. *Plant Physiology* 147(2), 719–731.

Jiang, S., Dai, G., Zhou, J., Zhong, J., Liu, J., Shu, Y., 2022. An assessment of integrated amendments of biochar and soil replacement on the phytotoxicity of metal(loid)s in rotated radish-soya bean-amaranth in a mining acidy soil. *Chemosphere* 287, 132082.

Jiang, Y.L., Ruan, X.L., Ma, J.H., 2020a. Characteristics and classification management of heavy metal pollution in a polluted irrigated farmland near a battery factory in Xinxiang City. *Acta Scientiae Circumstantiae* 40, 645–654.

Jiang, Z.Q., Jiang, L., Zhang, L., Su, M., Tian, D., Wang, T., Sun, Y.L., Nong, Y., Hu, S.J., Wang, S.M., Li, Z., 2020b. Contrasting the Pb (II) and Cd (II) tolerance of Enterobacter sp. via its cellular stress responses. *Environmental Microbiology* 22(4), 1507–1516.

Jin, F., 2020. Long-term effectiveness of in situ solidification/stabilization. In: Hou, D. (Ed.) *Sustainable Remediation of Contaminated Soil and Groundwater*. Butterworth-Heinemann: Oxford, Cambridge, MA, pp. 247–278.

Jin, F., 2022. Sustainable utilization of slags. In: Tsang, D.C.W. and Wang, L. (Eds.), *Low Carbon Stabilization and Solidification of Hazardous Wastes*. Elsevier: Amsterdam, Netherlands. pp. 321–341.

Kalai, T., Khamassi, K., Teixeira da Silva, J.A., Gouia, H., Bettaieb Ben-Kaab, L., 2014. Cadmium and copper stress affect seedling growth and enzymatic activities in germinating barley seeds. *Archives of Agronomy and Soil Science* 60(6), 765–783.

Kalaji, H.M., Loboda, T., 2007. Photosystem II of barley seedlings under cadmium and lead stress. *Plant, Soil and Environment* 53(12), 511–516.

Kandler, O., Konig, H., 1998. Cell wall polymers in Archaea (Archaebacteria). *Cellular and Molecular Life Sciences* 54(4), 305–308.

Kavamura, V.N., Esposito, E., 2010. Biotechnological strategies applied to the decontamination of soils polluted with heavy metals. *Biotechnology Advances* 28(1), 61–69.

Kavitha, B., Reddy, P.V.L., Kim, B., Lee, S.S., Pandey, S.K., Kim, K.H., 2018. Benefits and limitations of biochar amendment in agricultural soils: A review. *Journal of Environmental Management* 227, 146–154.

Khalid, S., Shahid, M., Niazi, N.K., Murtaza, B., Bibi, I., Dumat, C., 2017. A comparison of technologies for remediation of heavy metal contaminated soils. *Journal of Geochemical Exploration* 182, 247–268.

Killham, K., 1985. A physiological determination of the impact of environmental-stress on the activity of microbial biomass. *Environmental Pollution A* 38(3), 283–294.

Ko, K.S., Lee, P.K., Kong, I.C., 2012. Evaluation of the toxic effects of arsenite, chromate, cadmium, and copper using a battery of four bioassays. *Applied Microbiology and Biotechnology* 95(5), 1343–1350.

Kopittke, P.M., Asher, C.J., Kopittke, R.A., Menzies, N.W., 2007. Toxic effects of Pb^{2+} on growth of cowpea (*Vigna unguiculata*). *Environmental Pollution* 150(2), 280–287.

Körpe, D.A., Aras, S., 2011. Evaluation of copper-induced stress on eggplant (*Solanum melongena* L.) seedlings at the molecular and population levels by use of various biomarkers. *Mutation Research/Genetic Toxicology and Environmental Mutagenesis* 719(1), 29–34.

Kováčik, J., Klejdus, B., Hedbavny, J., Bačkor, M., 2010. Effect of copper and salicylic acid on phenolic metabolites and free amino acids in Scenedesmus quadricauda (Chlorophyceae). *Plant Science* 178(3), 307–311.

Kovalchuk, I., Titov, V., Hohn, B., Kovalchuka, O., 2005. Transcriptome profiling reveals similarities and differences in plant responses to cadmium and lead. *Mutation Research - Fundamental and Molecular Mechanisms* 570(2), 149–161.

Krämer, U., Talke, I.N., Hanikenne, M., 2007. Transition metal transport. *FEBS Letters* 581(12), 2263–2272.

Kuzminov, F.I., Brown, C.M., Fadeev, V.V., Gorbunov, M.Y., 2013. Effects of metal toxicity on photosynthetic processes in coral symbionts, Symbiodinium spp. *Journal of Experimental Marine Biology and Ecology* 446, 216–227.

Lai, W.W., Wu, Y.Y., Zhang, C.N., Dilinuer, Y., Pasang, L., Lu, Y.Q., Wang, Y.H., Chen, H.M., Li, Z., 2022. Combination of biochar and phosphorus solubilizing bacteria to improve the stable form of toxic metal minerals and microbial abundance in lead/Cadmium-contaminated soil. *Agronomy-Basel* 12(5). doi: 10.3390/agronomy12051003.

Langdon, C.J., Piearce, T.G., Meharg, A.A., Semple, K.T., 2001. Resistance to copper toxicity in populations of the earthworms *Lumbricus rubellus* and *Dendrodrilus rubidus* from contaminated mine wastes. *Environmental Toxicology and Chemistry* 20(10), 2336–2341.

Lebrun, J.D., Trinsoutrot-Gattin, I., Vinceslas-Akpa, M., Bailleul, C., Brault, A., Mougin, C., Laval, K., 2012. Assessing impacts of copper on soil enzyme activities in regard to their natural spatiotemporal variation under long-term different land uses. *Soil Biology and Biochemistry* 49, 150–156.

Lee, J.C., Son, Y.O., Pratheeshkumar, P., Shi, X., 2012. Oxidative stress and metal carcinogenesis. *Free Radical Biology and Medicine* 53(4), 742–757.

Lei, C., Chen, T., Zhang, Q.Y., Long, L.S., Chen, Z., Fu, Z.P., 2020. Remediation of lead polluted soil by active silicate material prepared from coal fly ash. *Ecotoxicology and Environmental Safety* 206, 111409.

Leita, L., Denobili, M., Muhlbachova, G., Mondini, C., Marchiol, L., Zerbi, G., 1995. Bioavailability and effects of heavy-metals on soil microbial biomass survival during laboratory incubation. *Biology and Fertility of Soils* 19(2–3), 103–108.

Lemtiri, A., Liénard, A., Alabi, T., Brostaux, Y., Cluzeau, D., Francis, F., Colinet, G., 2016. Earthworms *Eisenia fetida* affect the uptake of heavy metals by plants *Vicia faba* and Zea mays in metal-contaminated soils. *Applied Soil Ecology* 104, 67–78.

Léon, V., Rabier, J., Notonier, R., Barthelémy, R., Moreau, X., Bouraïma-Madjèbi, S., Viano, J., Pineau, R., 2005. Effects of three nickel salts on germinating seeds of Grevillea exul var. rubiginosa, an endemic serpentine Proteaceae. *Annals of Botany* 95(4), 609–618.

Li, J., Ge, Y., Wang, M.X., Zhao, Y., Zhao, X.Q., 2016. Effect of Cd on tolerance physiology, nutrients uptake and translocation *in Ricinus communis L. Acta Scientiae Circumstantiae* 36(8), 3081–3087.

Li, L., Li, J.R., He, Q.L., Zhu, S.J., Chen, J.H., 2014. Nutrients, ultrastructures, and Cd subcellular localization in the cottonseeds of three upland cotton cultivars under Cd stress. *Journal of Soil Science and Plant Nutrition* 14(2), 276–289.

Li, S., Yu, J., Zhu, M., Zhao, F., Luan, S., 2012. Cadmium impairs ion homeostasis by altering K^+ and Ca^{2+} channel activities in rice root hair cells. *Plant, Cell & Environment* 35(11), 1998–2013.

Li, W.T., Ni, P.P., Yi, Y.L., 2019. Comparison of reactive magnesia, quick lime, and ordinary Portland cement for stabilization/solidification of heavy metal-contaminated soils. *Science of the Total Environment* 671, 741–753.

Li, X., Zhu, Y.-G., Cavagnaro, T.R., Chen, M., Sun, J., Chen, X., Qiao, M., 2009. Do ammonia-oxidizing archaea respond to soil Cu contamination similarly asammonia-oxidizing bacteria? *Plant Soil* 324(1–2), 209–217.

Li, X.Q., Meng, D.L., Li, J., Yin, H.Q., Liu, H.W., Liu, X.D., Cheng, C., Xiao, Y.H., Liu, Z.H., Yan, M.L., 2017. Response of soil microbial communities and microbial interactions to long-term heavy metal contamination. *Environmental Pollution* 231, 908–917.

Liang, Q., Gao, R.T., Xi, B.D., Zhang, Y., Zhang, H., 2014. Long-term effects of irrigation using water from the river receiving treated industrial wastewater on soil organic carbon fractions and enzyme activities. *Agricultural Water Management* 135, 100–108.

Liang, W., Wang, G., Peng, C., Tan, J., Wan, J., Sun, P., Li, Q., Ji, X., Zhang, Q., Wu, Y., Zhang, W., 2022. Recent advances of carbon-based nano zero valent iron for heavy metals remediation in soil and water: A critical review. *Journal of Hazardous Materials* 426, 127993.

Lin, H., Wang, Z., Liu, C., Dong, Y., 2022. Technologies for removing heavy metal from contaminated soils on farmland: A review. *Chemosphere* 305(2022), 135457.

Lipman, C.B., Burgess, P.S., 1914. Effect of copper, zinc, iron and lead salts on ammonification and nitrification in soils. University of California Publications in Agricultural Science 1, 127–139.

Liu, G., Li, X., Wang, J., Chen, M., 2017. Medium-scale ex suit electrokinetic remediation of Cd and Zn polluted soil enhanced by EDTA, tartaric acid, citric acid and acetic acid, *International Conference on Energy & Environmental Protection* 143, 1348–1356.

Liu, L.W., Li, W., Song, W.P., Guo, M.X., 2018. Remediation techniques for heavy metal-contaminated soils: Principles and applicability. *Science of the Total Environment* 633, 206–219.

Long, X.X., Yang, X.E., Ni, W.Z., Ye, Z.Q., He, Z.L., Calvert, D.V., Stoffella, J.P., 2003. Assessing zinc thresholds for phytotoxicity and potential dietary toxicity in selected vegetable crops. *Communications in Soil Science and Plant Analysis* 34(9–10), 1421–1434.

Lopareva-Pohu, A., Pourrut, B., Waterlot, C., Garçon, G., Bidar, G., Pruvot, C., Shirali, P., Douay, F., 2011. Assessment of fly ash-aided phytostabilisation of highly contaminated soils after an 8-year field trial: Part 1. Influence on soil parameters and metal extractability. *Science of the Total Environment* 409(3), 647–654.

Lukkari, T., Taavitsainen, M., Soimasuo, M., Oikari, A., Haimi, J., 2004. Biomarker responses of the earthworm *Aporrectodea tuberculata* to copper and zinc exposure: differences between populations with and without earlier metal exposure. *Environmental Pollution* 129(3), 377–386.

Lushchak, V.I., 2011. Environmentally induced oxidative stress in aquatic animals. *Aquatic Toxicology* 101(1), 13–30.

Lux, A., Martinka, M., Vaculik, M., White, P.J., 2011. Root responses to cadmium in the rhizosphere: A review. *Journal of Experimental Botany* 62(1), 21–37.

Ma, F., Peng, C., Hou, D., Wu, B., Zhang, Q., Li, F., Gu, Q., 2015. Citric acid facilitated thermal treatment: An innovative method for the remediation of mercury contaminated soil. *Journal of Hazardous Materials* 300, 546–552.

Ma, F., Zhang, Q., Xu, D., Hou, D., Li, F., Gu, Q., 2014. Mercury removal from contaminated soil by thermal treatment with FeCl3 at reduced temperature. *Chemosphere* 117, 388–393.
Ma, L.Q., Komar, K.M., Tu, C., Zhang, W.H., Cai, Y., Kennelley, E.D., 2001. A fern that hyperaccumulates arsenic: A hardy, versatile, fast-growing plant helps to remove arsenic from contaminated soils. *Nature* 409(6820), 579–579.
Magalhaes, C.M., Machado, A., Matos, P., Bordalo, A.A., 2011. Impact of copper on the diversity, abundance and transcription of nitrite and nitrous oxide reductase genes in an urban European estuary. *FEMS Microbiology Ecology* 77(2), 274–284.
Maini, G., Sharman, A.K., 2000. An integrated method incorporating sulfur-oxidizing bacteria and electrokinetics to enhance. *Environmental Science & Technology* 34(6), 1081–1087.
Malan, H.L., Farrant, J.M., 1998. Effects of the metal pollutants cadmium and nickel on soybean seed development. *Seed Science Research* 8(4), 445–453.
Mark, C., Hodson, M.E., Arnold, R.E., Langdon, C.J., 2005. Single versus multiple occupancy: Effects on toxicity parameters measured on Eisenia fetida in lead nitrate-treated soil. *Environmental Toxicology & Chemistry* 24(1), 110–116.
Marques, A.P.G.C., Rangel, A.O.S.S., Castro, P.M.L., 2011. Remediation of heavy metal contaminated soils: An overview of site remediation techniques. *Critical Reviews in Environmental Science & Technology* 41(10), 879–914.
Marques, M.C., do Nascimento, C.W.A., 2013. Analysis of chlorophyll fluorescence spectra for the monitoring of Cd toxicity in a bio-energy crop (*Jatropha curcas*). *Journal of Photochemistry and Photobiology B: Biology* 127, 88–93.
Márquez-García, B., Horemans, N., Cuypers, A., Guisez, Y., Córdoba, F., 2011. Antioxidants in Erica andevalensis: A comparative study between wild plants and cadmium-exposed plants under controlled conditions. *Plant Physiology and Biochemistry* 49(1), 110–115.
Márquez-García, B., Márquez, C., Sanjosé, I., Nieva, F.J.J., Rodríguez-Rubio, P., Muñoz-Rodríguez, A.F., 2013. The effects of heavy metals on germination and seedling characteristics in two halophyte species in Mediterranean marshes. *Marine Pollution Bulletin* 70(1), 119–124.
Meng, L.Z., Pan, S., Zhou, L.M., Santasup, C., Su, M., Tian, D., Li, Z., 2022. Evaluating the survival of Aspergillus niger in a highly polluted red soil with addition of Phosphogypsum and bioorganic fertilizer. *Environmental Science and Pollution Research* 29(50), 76446–76455.
Mertens, J., Broos, K., Wakelin, S.A., Kowalchuk, G.A., Springael, D., Smolders, E., 2009. Bacteria, not archaea, restore nitrification in a zinc-contaminated soil. *ISME Journal* 3(8), 916–923.
Mishra, S., Srivastava, S., Tripathi, R.D., Kumar, R., Seth, C.S., Gupta, D.K., 2006. Lead detoxification by coontail (*Ceratophyllum demersum* L.) involves induction of phytochelatins and antioxidant system in response to its accumulation. *Chemosphere* 65(6), 1027–1039.
Moffett, B.F., Nicholson, F.A., Uwakwe, N.C., Chambers, B.J., Harris, J.A., Hill, T.C.J., 2003. Zinc contamination decreases the bacterial diversity of agricultural soil. *FEMS Microbiology Ecology* 43(1), 13–19.
Mohanpuria, P., Rana, N.K., Yadav, S.K., 2007. Cadmium induced oxidative stress influence on glutathione metabolic genes of *Camellia sinensis* (L.) O. Kuntze. *Environmental Toxicology* 22(4), 368–374.
Morgan, J.E., Morgan, A.J., 1999. The accumulation of metals (Cd, Cu, Pb, Zn and Ca) by two ecologically contrasting earthworm species (*Lumbricus rubellus* and *Aporrectodea caliginosa*): Implications for ecotoxicological testing. *Applied Soil Ecology* 13(1), 9–20.
Morina, F., Jovanovic, L., Mojovic, M., Vidovic, M., Pankovic, D., Veljovic Jovanovic, S., 2010. Zinc-induced oxidative stress in *Verbascum thapsus* is caused by an accumulation of reactive oxygen species and quinhydrone in the cell wall. *Physiol Plant* 140(3), 209–224.
Morton-Bermea, O., Hernández-Álvarez, E., González-Hernández, G., Romero, F., Lozano, R., Beramendi-Orosco, L.E., 2009. Assessment of heavy metal pollution in urban topsoils from the metropolitan area of Mexico City. *Journal of Geochemical Exploration* 101(3), 218–224.
Nabulo, G., Young, S.D., Black, C.R., 2010. Assessing risk to human health from tropical leafy vegetables grown on contaminated urban soils. *Science of the Total Environment* 408(22), 5338–5351.
Navarro-Pedreño, J., Gómez Lucas, I., Moral, R., Mataix-Solera, J., 1997. Heavy metals and plant nutrition and development. *Recent Research Developments in Phytochemistry* 1, 173–179.
Nicholson, F.A., Smith, S.R., Alloway, B.J., Carlton-Smith, C., Chambers, B.J., 2003. An inventory of heavy metals inputs to agricultural soils in England and Wales. *Science of the Total Environment* 311(1), 205–219.
Nicolodi, M., Gianello, C., 2014. Understanding soil as an open system and fertility as an emergent property of the soil system. *Sustainable Agriculture Research* 4(1). doi: 10.22004/ag.econ.230408.

Nie, S.W., Huang, S.M., Zhang, S.Q., Guo, D.D., Zhang, Q.P., 2012. Effects of varieties heavy metals stress on wheat grain yields of two genotypes and the main ingredients. *Journal of Agro-Environment Science* 31(3), 455–463.

Nziguheba, G., Smolders, E., 2008. Inputs of trace elements in agricultural soils via phosphate fertilizers in European countries. *Science of the Total Environment* 390(1), 53–57.

Obbard, J.P., Jones, K.C., 1993. The use of the cotton-strip assay to assess cellulose decomposition in heavy metal-contaminated sewage sludge-amended soils. *Environmental Pollution* 81(2), 173–178.

Okolie, C.U., Chen, H.M., Zhao, Y.X., Tian, D., Zhang, L., Su, M., Jiang, Z.Q., Li, Z., Li, H.X., 2020. Cadmium immobilization in aqueous solution by Aspergillus niger and geological fluorapatite. *Environmental Science and Pollution Research* 27(7), 7647–7656.

Ollivier, J., Wanat, N., Austruy, A., Hitmi, A., Joussein, E., Welzl, G., Munch, J.C., Schloter, M., 2012. Abundance and diversity of ammonia-oxidizing prokaryotes in the root-rhizosphere complex of *Miscanthus x giganteus* grown in heavy metal-contaminated soils. *Microbial Ecology* 64(4), 1038–1046.

Opdenakker, K., Remans, T., Keunen, E., Vangronsveld, J., Cuypers, A., 2012. Exposure of arabidopsis thaliana to Cd or Cu excess leads to oxidative stress mediated alterations in MAPKinase transcript levels. *Environmental and Experimental Botany* 83, 53–61.

Otunola, B.O., Ololade, O.O., 2020. A review on the application of clay minerals as heavy metal adsorbents for remediation purposes. *Environmental Technology & Innovation* 18, 1–14.

Owojori, O.J., Reinecke, A.J., Rozanov, A.B., 2009. Role of clay content in partitioning, uptake and toxicity of zinc in the earthworm Eisenia fetida. *Ecotoxicology and Environmental Safety* 72(1), 99–107.

Pan, J., Yu, L., 2011. Effects of Cd or/and Pb on soil enzyme activities and microbial community structure. *Ecological Engineering* 37(11), 1889–1894.

Park, J.H., Lamb, D., Paneerselvam, P., Choppala, G., Bolan, N., Chung, J.-W., 2011. Role of organic amendments on enhanced bioremediation of heavy metal(loid) contaminated soils. *Journal of Hazardous Materials* 185(2), 549–574.

Park, M.O., Kim, M.H., Hong, Y., 2020. The kinetics of mercury vaporization in soil during low-temperature thermal treatment. *Geoderma* 363, 114150.

Peijnenburg, W., Baerselman, R., de Groot, A., Jager, T., Leenders, D., Posthuma, L., Van Veen, R., 2000. Quantification of metal bioavailability for lettuce (*Lactuca sativa* L.) in field soils. *Archives of Environmental Contamination and Toxicology* 39(4), 420–430.

Peleg, Z., Blumwald, E., 2011. Hormone balance and abiotic stress tolerance in crop plants. *Current Opinion in Plant Biology* 14(3), 290–295.

Pereira, L.B., Vicentini, R., Ottoboni, L.M.M., 2014. Changes in the bacterial community of soil from a neutral mine drainage channel. *PLoS One* 9(5), e96605.

Pető, A., Lehotai, N., Lozano-Juste, J., León, J., Tari, I., Erdei, L., Kolbert, Z., 2011. Involvement of nitric oxide and auxin in signal transduction of copper-induced morphological responses in Arabidopsis seedlings. *Annals of Botany* 108(3), 449–457.

Pierart, A., Shahid, M., Séjalon-Delmas, N., Dumat, C., 2015. Antimony bioavailability: Knowledge and research perspectives for sustainable agricultures. *Journal of Hazardous Materials* 289, 219–234.

Piotrowska, A., Bajguz, A., Godlewska-Zylkiewicz, B., Czerpak, R., Kaminska, M., 2009. Jasmonic acid as modulator of lead toxicity in aquatic plant *Wolffia arrhiza* (Lemnaceae). *Environmental and Experimental Botany* 66(3), 507–513.

Poschenrieder, C., Barceló, J., 2004. Water relations in heavy metal stressed plants. In: Prasad, M.N.V. (Ed.), *Heavy Metal Stress in Plants: From Biomolecules to Ecosystems*. Springer: Berlin Heidelberg, pp. 249–270.

Pourrut, B., Jean, S., Silvestre, J., Pinelli, E., 2011. Lead-induced DNA damage in *Vicia faba* root cells: Potential involvement of oxidative stress. *Mutation Research* 726(2), 123–128.

Pourrut, B., Perchet, G., Silvestre, J., Cecchi, M., Guiresse, M., Pinelli, E., 2008. Potential role of NADPH-oxidase in early steps of lead-induced oxidative burst in *Vicia faba* roots. *Journal of Plant Physiology* 165(6), 571–579.

Pourrut, B., Shahid, M., Douay, F., Dumat, C., Pinelli, E., 2013. Molecular mechanisms involved in lead uptake, toxicity and detoxification in higher plants. In: Gupta, D.K., Corpas, F.J., Palma, J.M. (Eds.), *Heavy Metal Stress in Plants*. Springer: Berlin Heidelberg, pp. 121–147.

Punamiya, P., Datta, R., Sarkar, D., Barber, S., Patel, M., Das, P., 2010. Symbiotic role of Glomus mosseae in phytoextraction of lead in vetiver grass [*Chrysopogon zizanioides* (L.)]. *Journal of Hazardous Materials* 177(1–3), 465–474.

Qin, J., Jiang, B., Nan, X., Xu, R., He, Y., Jiang, G., Qin, J., Yang, Y., Li, Q., 2021. Study on microbial remediation of heavy metal contaminated soil. *IOP Conference Series: Earth and Environmental Science* 687(1), 012041.

Qin, S.Y., Liu, H.G., Nie, Z.J., Rengel, Z., Gao, W., Li, C., Zhao, P., 2020. Toxicity of cadmium and its competition with mineral nutrients for uptake by plants: A review. *Pedosphere* 30(2), 168–180.

Qu, C., Yang, S., Mortimer, M., Zhang, M., Chen, J., Wu, Y., Chen, W., Cai, P., Huang, Q., 2022. Functional group diversity for the adsorption of lead(Pb) to bacterial cells and extracellular polymeric substances. *Environmental Pollution* 295, 118651.

Rascio, N., Navari-Izzo, F., 2011. Heavy metal hyperaccumulating plants: How and why do they do it? And what makes them so interesting? *Plant Science* 180(2), 169–181.

Rizwan, M., Lin, Q., Chen, X., Li, Y., Li, G., Zhao, X., Tian, Y., 2020. Synthesis, characterization and application of magnetic and acid modified biochars following alkaline pretreatment of rice and cotton straws: ScienceDirect. *Science of the Total Environment* 714, 136532.

Rodriguez-Serrano, M., Romero-Puertas, M.C., Zabalza, A., Corpas, F.J., Gomez, M., Del Rio, L.A., Sandalio, L.M., 2006. Cadmium effect on oxidative metabolism of pea (*Pisum sativum* L.) roots. Imaging of reactive oxygen species and nitric oxide accumulation in vivo. *Plant, Cell & Environment* 29(8), 1532–1544.

Rout, G.R., Samantaray, S., Das, P., 1997. Differential chromium tolerance among eight mungbean cultivars grown in nutrient culture. *Journal of Plant Nutrition* 20(4–5), 473–483.

Ruyters, S., Mertens, J., T'Seyen, I., Springael, D., Smolders, E., 2010. Dynamics of the nitrous oxide reducing community during adaptation to Zn stress in soil. *Soil Biology and Biochemistry* 42(9), 1581–1587.

Ruyters, S., Nicol, G.W., Prosser, J.I., Lievens, B., Smolders, E., 2013. Activity of the ammonia oxidising bacteria is responsible for zinc tolerance development of the ammonia oxidising community in soil: A stable isotope probing study. *Soil Biology and Biochemistry* 58, 244–247.

Sabir, M., Hanafi, M.M., Zia-Ur-Rehman, M., Ullah, S., Ahmad, H.R., Hakeem, K.R., Aziz, T., 2014a. Comparison of low-molecular-weight organic acids and ethylenediaminetetraacetic acid to enhance phytoextraction of heavy metals by Maize. *Communications in Soil Science and Plant Analysis* 45(1), 42–52.

Sabir, M., Waraich, E., Hakeem, K., Ozturk, M., Ahmad, H., Shahid, M., 2014b. Phytoremediation: Mechanisms and adaptations. In: Hakeem, K. R., Sabir, M., Ozturk, M., and Murmut, A. (Eds.), *Soil Remediation and Plants*. Academic Press: San Diego, pp. 85–105.

Salas-Luevano, M.A., Mauricio-Castillo, J.A., Gonzalez-Rivera, M.L., Vega-Carrillo, H.R., Salas-Munoz, S., 2017. Accumulation and phytostabilization of As, Pb and Cd in plants growing inside mine tailings reforested in Zacatecas, Mexico. *Environmental Earth Sciences* 76(23). doi: 10.1007/s12665-017-7139-y.

Salla, V., Hardaway, C.J., Sneddon, J., 2011. Preliminary investigation of *Spartina alterniflora* for phytoextraction of selected heavy metals in soils from Southwest Louisiana. *Microchemical Journal* 97(2), 207–212.

Sammut, M.L., Noack, Y., Rose, J., Hazemann, J.L., Proux, O., Depoux, M., Ziebel, A., Fiani, E., 2010. Speciation of Cd and Pb in dust emitted from sinter plant. *Chemosphere* 78(4), 445–450.

Sandalio, L.M., Dalurzo, H.C., Gómez, M., Romero-Puertas, M.C., del Río, L.A., 2001. Cadmium-induced changes in the growth and oxidative metabolism of pea plants. *Journal of Experimental Botany* 52(364), 2115–2126.

Schreck, E., Foucault, Y., Sarret, G., Sobanska, S., Cécillon, L., Castrec-Rouelle, M., Uzu, G., Dumat, C., 2012. Metal and metalloid foliar uptake by various plant species exposed to atmospheric industrial fallout: Mechanisms involved for lead. *Science of the Total Environment* 427–428, 253–262.

Sengar, R.S., Gautam, M., Sengar, R.S., Garg, S.K., Sengar, K., Chaudhary, R., 2008. Lead stress effects on physiobiochemical activities of higher plants. *Reviews of Environmental Contamination and Toxicology* 196, 73–93.

Shahid, M., Austruy, A., Echevarria, G., Arshad, M., Sanaullah, M., Aslam, M., Nadeem, M., Nasim, W., Dumat, C., 2014a. EDTA-enhanced phytoremediation of heavy metals: A review. *Soil and Sediment Contamination* 23(4), 389–416.

Shahid, M., Dumat, C., Aslam, M., Pinelli, E., 2012. Assessment of lead speciation by organic ligands using speciation models. *Chemical Speciation & Bioavailability* 24(4), 248–252.

Shahid, M., Dumat, C., Pourrut, B., Sabir, M., Pinelli, E., 2014b. Assessing the effect of metal speciation on lead toxicity to *Vicia faba* pigment contents. *Journal of Geochemical Exploration* 144, 290–297.

Shahid, M., Dumat, C., Pourrut, B., Silvestre, J., Laplanche, C., Pinelli, E., 2014c. Influence of EDTA and citric acid on lead-induced oxidative stress to *Vicia faba* roots. *Journal of Soils and Sediments* 14(4), 835–843.

Shahid, M., Ferrand, E., Schreck, E., Dumat, C., 2013. Behavior and impact of zirconium in the soil–plant system: Plant uptake and phytotoxicity. In: Whitacre, D. M. (Ed.), *Reviews of Environmental Contamination and Toxicology*, Vol. 221. Springer: New York, pp. 107–127.

Shahid, M., Xiong, T.T., Masood, N., Leveque, T., Quenea, K., Austruy, A., Foucault, Y., Dumat, C., 2014d. Influence of plant species and phosphorus amendments on metal speciation and bioavailability in a smelter impacted soil: A case study of food-chain contamination. *Journal of Soils and Sediments* 14(4), 655–665.

Shanker, A.K., Cervantes, C., Loza-Tavera, H., Avudainayagam, S., 2005. Chromium toxicity in plants. *Environment International* 31(5), 739–753.

Shanmugam, V., Lo, J.C., Wu, C.L., Wang, S.L., Lai, C.C., Connolly, E.L., Huang, J.L., Yeh, K.C., 2011. Differential expression and regulation of iron-regulated metal transporters in Arabidopsis halleri and Arabidopsis thaliana--the role in zinc tolerance. *New Phytologist* 190(1), 125–137.

Shao, G.S., Chen, M.X., Wang, W.X., Mon, R.X., Zhang, G.P., 2007. Iron nutrition affects cadmium accumulation and toxicity in rice plants. *Plant Growth Regulation* 53(1), 33–42.

Sharma, D.C., Sharma, C.P., 1993. Chromium uptake and its effects on growth and biological yield of wheat. *Cereal Research Communications* 21(4), 317–322.

Sheik, C.S., Mitchell, T.W., Rizvi, F.Z., Rehman, Y., Faisal, M., Hasnain, S., McInerney, M.J., Krumholz, L.R., 2012. Exposure of soil microbial communities to chromium and arsenic alters their diversity and structure. *PLoS One* 7(6), e40059.

Shen, Z.T., Jin, F., O'Connor, D., Hou, D.Y., 2019. Solidification/stabilization for soil remediation: An old technology with new vitality. *Environmental Science & Technology* 53(20), 11615–11617.

Shen, Z.T., Tian, D., Zhang, X.Y., Tang, L.Y., Su, M., Zhang, L., Li, Z., Hu, S.J., Hou, D.Y., 2018. Mechanisms of biochar assisted immobilization of Pb^{2+} by bioapatite in aqueous solution. *Chemosphere* 190, 260–266.

Singh, A., 2001. Effect of trivalent and hexavalent chromium on Spinach. *Environment and Ecology* 19(4), 807–810.

Sivakumar, S., 2015. Effects of metals on earthworm life cycles: A review. *Environmental Monitoring and Assessment* 187(8), 4742.

Sivakumar, S., Prabha, D., Subbhuraam, C.V., 2006. Effect of cadmium and mercury on the survival, morphology and burrowing behaviour of the earthworm *Lambito mauritii* (Kinberg). *The 8th International Symposium on "Earthworm Ecology"*, Poland.

Sivakumar, S., Young-Chae, S., Prabha, D.S., Subbhuraam, C.V., Karmegam, N., 2009. Life cycle parameters of the earthworm Eisenia fetida exposed to Cr(III) and Cr(VI) amended organic substrates.

Sobolev, D., Begonia, M., 2008. Effects of heavy metal contamination upon soil microbes: Lead-induced changes in general and denitrifying microbial communities as evidenced by molecular markers. *International Journal of Environmental Research and Public Health* 5(5), 450–456.

Solis-Dominguez, F.A., Gonzalez-Chavez, M.C., Carrillo-Gonzalez, R., Rodriguez-Vazquez, R., 2007. Accumulation and localization of cadmium in Echinochloa polystachya grown within a hydroponic system. *Journal of Hazardous Materials* 141(3), 630–636.

Song, B., Zeng, G.M., Gong, J.L., Liang, J., Xu, P., Liu, Z.F., Zhang, Y., Zhang, C., Cheng, M., Liu, Y., Ye, S.J., Yi, H., Ren, X.Y., 2017. Evaluation methods for assessing effectiveness of in situ remediation of soil and sediment contaminated with organic pollutants and heavy metals. *Environment International* 105, 43–55.

Song, B., Zhang, C., Zeng, G.M., Gong, J.L., Chang, Y.N., Jiang, Y., 2016. Antibacterial properties and mechanism of graphene oxide-silver nanocomposites as bactericidal agents for water disinfection. *Archives of Biochemistry and Biophysics* 604, 167–176.

Song, J.W., Shen, Q.L., Wang, L., Qiu, G.Y., Shi, J.C., Xu, J.M., Brookes, P.C., Liu, X.M., 2018. Effects of Cd, Cu, Zn and their combined action on microbial biomass and bacterial community structure. *Environmental Pollution* 243, 510–518.

Song, P.P., Xu, D., Yue, J.Y., Ma, Y.C., Dong, S.J., Feng, J., 2022. Recent advances in soil remediation technology for heavy metal contaminated sites: A critical review. *Science of the Total Environment* 838, 165–174.

Soudek, P., Petrova, S., Vankova, R., Song, J., Vanek, T., 2014. Accumulation of heavy metals using Sorghum sp. *Chemosphere* 104, 15–24.

Srivastava, S., Srivastava, A.K., Suprasanna, P., D'Souza, S.F., 2013. Identification and profiling of arsenic stress-induced microRNAs in Brassica juncea. *Journal of Experimental Botany* 64(1), 303–315.

Su, C., Jiang, L.Q., Zhang, W.J., 2014. A review on heavy metal contamination in the soil worldwide: Situation, impact and remediation techniques. *Environmental Skeptics & Critics* 3(2), 24–38.

Subrahmanyam, G., Hu, H.W., Zheng, Y.M., Archana, G., He, J.Z., Liu, Y.R., 2014. Response of ammonia oxidizing microbes to the stresses of arsenic and copper in two acidic alfisols. *Applied Soil Ecology* 77, 59–67.

Sudhakar, S., Kumar, S.A., Penna, S., S.F., D.S., 2013. Identification and profiling of arsenic stress-induced microRNAs in Brassica juncea. *Journal of Experimental Botany* 1, 303–315.

Sun, Y.-B., Zhou, Q.-X., An, J., Liu, W.-T., Liu, R., 2009. Chelator-enhanced phytoextraction of heavy metals from contaminated soil irrigated by industrial wastewater with the hyperaccumulator plant (Sedum alfredii Hance). *Geoderma* 150(1), 106–112.

Sunda, W.G., Engel, D.W., Thuotte, R.M., 1978. Effect of chemical speciation on toxicity of cadmium to grass shrimp, Palaemonetes pugio: Importance of free cadmium ion. *Environmental Science & Technology* 12(4), 409–413.

Suthar, S., Singh, S., Dhawan, S., 2008. Earthworms as bioindicator of metals (Zn, Fe, Mn, Cu, Pb and Cd) in soils: Is metal bioaccumulation affected by their ecological category? *Ecological Engineering* 32(2), 99–107.

Tang, D.Z., Christiansen, K.M., Innes, R.W., 2005. Regulation of plant disease resistance, stress responses, cell death, and ethylene signaling in Arabidopsis by the EDR1 protein kinase. *Plant Physiology* 138(2), 1018–1026.

Tang, F., Yue, J., Tian, J., Ge, F., Li, F., Liu, Y., Deng, S., Zhang, D., 2022. Microbial induced phosphate precipitation accelerate lead mineralization to alleviate nucleotide metabolism inhibition and alter Penicillium oxalicum's adaptive cellular machinery. *Journal of Hazardous Materials* 439, 129675.

Tang, Q., Shi, P., Yuan, Z., Shi, S., Xu, X., Katsumi, T., 2019. Potential of zero-valent iron in remediation of Cd(II) contaminated soil: From laboratory experiment, mechanism study to field application. *Soils and Foundations – Tokyo* 59(6), 2099–2109.

Teng, D., Mao, K., Ali, W., Xu, G., Huang, G., Niazi, N.K., Feng, X., Zhang, H., 2020. Describing the toxicity and sources and the remediation technologies for mercury-contaminated soil. *RSC Advances* 10(39), 23221–23232.

Throback, I.N., Johansson, M., Rosenquist, M., Pell, M., Hansson, M., Hallin, S., 2007. Silver (Ag^+) reduces denitrification and induces enrichment of novel nirK genotypes in soil. *FEMS Microbiology Letters* 270(2), 189–194.

Tian, D., Jiang, Z.Q., Jiang, L., Su, M., Feng, Z.Y., Zhang, L., Wang, S.M., Li, Z., Hu, S.J., 2019. A new insight into lead (II) tolerance of environmental fungi based on a study of Aspergillus niger and Penicillium oxalicum. *Environmental Microbiology* 21(1), 471–479.

Tian, S., Xie, R., Wang, H., Hu, Y., Ge, J., Liao, X., Gao, X., Brown, P., Lin, X., Lu, L., 2016. Calcium deficiency triggers phloem remobilization of cadmium in a hyperaccumulating species. *Plant Physiology* 172(4), 2300–2313.

Tipayno, S.C., Truu, J., Samaddar, S., Truu, M., Preem, J.K., Oopkaup, K., Espenberg, M., Chatterjee, P., Kang, Y., Kim, K., Sa, T., 2018. The bacterial community structure and functional profile in the heavy metal contaminated paddy soils, surrounding a nonferrous smelter in South Korea. *Ecology and Evolution* 8(12), 6157–6168.

Tomulescu, I.M., Edith, R., Merca, V., Tuduce, A., 2004. Effect of copper, zinc and lead and their combinations on the germination capacity of two cereals. *Acta Agraria Debreceniensis,* 39–42. doi: 10.34101/actaagrar/15/3355.

Toth, G., Hermann, T., Da Silva, M.R., Montanarella, L., 2016. Heavy metals in agricultural soils of the European Union with implications for food safety. *Environment International* 88, 299–309.

Tripathi, A.K., Tripathi, S., 1999. Changes in some physiological and biochemical characters in Albizia lebbek as bio-indicators of heavy metal toxicity. *Journal of Environmental Biology* 20, 93–98.

Ullah, S., Meers, E., Qadir, M., de Caritat, P., Tack, F.M., Du Laing, G., Zia, M.H., 2009. EDTA-assisted Pb phytoextraction. *Chemosphere* 74(10), 1279–1291.

Ullah, S., Shahid, M., Zia-ur-Rehman, M., Sabir, M., Ahmad, H., 2015. Phytoremediation of Pb-contaminated soils using synthetic chelates, In: Hakeem, K. R., Sabir, M., Ozturk, M., and Murmut, A. (Eds.), *Soil Remediation and Plants.* Academic Press: San Diego, pp. 397–414.

Usmani, Z., 2015. Role of earthworms against metal contamination: A review. *Journal of Biodiversity and Environmental Sciences* 6(1), 414–427.

Uzu, G., Sobanska, S., Aliouane, Y., Pradere, P., Dumat, C., 2009. Study of lead phytoavailability for atmospheric industrial micronic and sub-micronic particles in relation with lead speciation. *Environmental Pollution* 157(4), 1178–1185.

Uzu, G., Sobanska, S., Sarret, G., Munoz, M., Dumat, C., 2010. Foliar lead uptake by lettuce exposed to atmospheric fallouts. *Environmental Science & Technology* 44(3), 1036–1042.

Valentine, D.L., 2007. Adaptations to energy stress dictate the ecology and evolution of the Archaea. *Nature Reviews Microbiology* 5(4), 316–323.

Vanhoudt, N., Vandenhove, H., Horemans, N., Wannijn, J., Bujanic, A., Vangronsveld, J., Cuypers, A., 2010. Study of oxidative stress related responses induced in Arabidopsis thaliana following mixed exposure to uranium and cadmium. *Plant Physiology and Biochemistry* 48(10), 879–886.

Vega, F.A., Andrade, M.L., Covelo, E.F., 2010. Influence of soil properties on the sorption and retention of cadmium, copper and lead, separately and together, by 20 soil horizons: Comparison of linear regression and tree regression analyses. *Journal of Hazardous Materials* 174(1–3), 522–533.

Verbruggen, N., Hermans, C., Schat, H., 2009. Molecular mechanisms of metal hyperaccumulation in plants. *New Phytologist* 181(4), 759–776.

Wang, G.S., Fowler, B.A., 2008. Roles of biomarkers in evaluating interactions among mixtures of lead, cadmium and arsenic. *Toxicology and Applied Pharmacology* 233(1), 92–99.

Wang, J., Anderson, C.W.N., Xing, Y., Fan, Y., Xia, J., Shahen, S.M., Rinklebe, J., Feng, X., 2018a. Thiosulphate-induced phytoextraction of mercury in Brassica juncea: Spectroscopic investigations to define a mechanism for Hg uptake. *Environmental Pollution* 242, 986–993.

Wang, J.H., Wang, L.J., Zhu, L.S., Wang, J., 2018b. Individual and combined effects of enrofloxacin and cadmium on soil microbial biomass and the ammonia-oxidizing functional gene. *Science of the Total Environment* 624, 900–907.

Wang, M.X., Ma, J.X., Wang, X.W., Wang, Z.J., Tang, L.Y., Chen, H.M., Li, Z., 2020. Detoxification of Cu(II) by the red yeast Rhodotorula mucilaginosa: from extracellular to intracellular. *Applied Microbiology and Biotechnology* 104(23), 10181–10190.

Wang, Q., Zhou, Q., Huang, L., Xu, S., Fu, Y., Hou, D., Feng, Y., Yang, X., 2022a. Cadmium phytoextraction through *Brassica juncea* L. under different consortia of plant growth-promoting bacteria from different ecological niches. *Ecotoxicology and Environmental Safety* 237, 113541.

Wang, T., Zhang, L., Li, S.S., Meng, L.Z., Su, M., Wang, Z.J., Nong, Y., Sun, Y.L., Wang, S.M., Li, Z., 2022b. Weakened Cd toxicity to fungi under coexistence of Pb in solution. *Journal of Hazardous Materials* 426, 127984.

Wang, Y., Sikora, S., Kim, H., Dubey, B., Townsend, T., 2012. Mobilization of iron and arsenic from soil by construction and demolition debris landfill leachate. *Waste Manage* 32(5), 925–932.

Wang, Z.J., Zhang, Y., Jiang, L., Qiu, J.J., Gao, Y.N., Gu, T.T., Li, Z., 2022c. Responses of Rhodotorula mucilaginosa under Pb(II) stress: Carotenoid production and budding. *Environmental Microbiology* 24(2), 678–688.

Wani, P.A., Khan, M.S., Zaidi, A., 2007a. Cadmium, chromium and copper in greengram plants. *Agronomy for Sustainable Development* 27(2), 145–153.

Wani, P.A., Khan, M.S., Zaidi, A., 2007b. Impact of heavy metal toxicity on plant growth, symbiosis, seed yield and nitrogen and metal uptake in chickpea. *Australian Journal of Experimental Agriculture* 47(6), 712–720.

Wei, Y.L., Yang, Y.W., Cheng, N., 2001. Study of thermally immobilized Cu in analogue minerals of contaminated soils. *Environmental Science & Technology* 35(2), 416–421.

Wen, D.D., Fu, R.B., Li, Q., 2021a. Removal of inorganic contaminants in soil by electrokinetic remediation technologies: A review. *Journal of Hazardous Materials* 401, 123345.

Wen, E.G., Yang, X., Chen, H.B., Shaheen, S.M., Sarkar, B., Xu, S., Song, H., Liang, Y., Rinklebe, J., Hou, D.Y., Li, Y., Wu, F.C., Pohorely, M., Wong, J.W.C., Wang, H.L., 2021b. Iron-modified biochar and water management regime-induced changes in plant growth, enzyme activities, and phytoavailability of arsenic, cadmium and lead in a paddy soil. *Journal of Hazardous Materials* 407, 124344.

White, C., Wilkinson, S.C., Gadd, G.M., 1995. The role of microorganisms in biosorption of toxic metals and radionuclides. *International Biodeterioration & Biodegradation* 35(1–3), 17–40.

Wittebolle, L., Marzorati, M., Clement, L., Balloi, A., Daffonchio, D., Heylen, K., De Vos, P., Verstraete, W., Boon, N., 2009. Initial community evenness favours functionality under selective stress. *Nature* 458(7238), 623–626.

Witter, E., Martensson, A.M., Garcia, F.V., 1993. Size of the soil microbial biomass in a long-term field experiment as affected by different N-fertilizers and organic manures. *Soil Biology and Biochemistry* 25(6), 659–669.

Wojcik, M., Tukiendorf, A., 2004. Phytochelatin synthesis and cadmium localization in wild type of Arabidopsis thaliana. *Plant Growth Regulation* 44(1), 71–80.

Wolt, J., 1994. *Soil Solution Chemistry*. Wiley: New York.

Wu, B., Hou, S.Y., Peng, D.H., Wang, Y., Wang, C., Xu, F., Xu, H., 2018. Response of soil micro-ecology to different levels of cadmium in alkaline soil. *Ecotoxicology and Environmental Safety* 166, 116–122.

Wu, Y.S., Liang, Q.H., Tang, Q., 2011. Effect of Pb on growth, accumulation and quality component of tea plant. *Procedia Engineering* 18, 214–219.

Wyszkowska, J., Kucharski, J., Lajszner, W., 2006. The effects of copper on soil biochemical properties and its interaction with other heavy metals. *Polish Journal of Environmental Studies* 15(6), 927–934.

Xia, J.G., Lan, H.X., De-Yong, W.U., 2010. Lead stress on growth of tea trees and physiological index in leaves of tea. *Journal of Agro-Environment Science* 29(1), 43–48.

Xiao, X.Y., Wang, M.W., Zhu, H.W., Guo, Z.H., Han, X.Q., Zeng, P., 2017. Response of soil microbial activities and microbial community structure to vanadium stress. *Ecotoxicology and Environmental Safety* 142, 200–206.

Xiong, J., Fu, G., Tao, L., Zhu, C., 2010. Roles of nitric oxide in alleviating heavy metal toxicity in plants. *Archives of Biochemistry and Biophysics* 497(1–2), 13–20.

Xiong, T., Leveque, T., Shahid, M., Foucault, Y., Mombo, S., Dumat, C., 2014b. Lead and cadmium phytoavailability and human bioaccessibility for vegetables exposed to soil or atmospheric pollution by process ultrafine particles. *Journal of Environmental Quality* 43(5), 1593–1600.

Xiong, T.-T., Leveque, T., Austruy, A., Goix, S., Schreck, E., Dappe, V., Sobanska, S., Foucault, Y., Dumat, C., 2014a. Foliar uptake and metal(loid) bioaccessibility in vegetables exposed to particulate matter. *Environmental Geochemistry and Health* 36(5), 897–909.

Xu, D.M., Fu, R.B., Wang, J.X., Shi, Y.X., Guo, X.P., 2021. Chemical stabilization remediation for heavy metals in contaminated soils on the latest decade: Available stabilizing materials and associated evaluation methods: A critical review. *Journal of Cleaner Production* 321, 128730.

Xu, H., Song, P., Gu, W., Yang, Z., 2011. Effects of heavy metals on production of thiol compounds and antioxidant enzymes in Agaricus bisporus. *Ecotoxicology and Environmental Safety* 74(6), 1685–1692.

Yadav, S.K., 2010. Heavy metals toxicity in plants: An overview on the role of glutathione and phytochelatins in heavy metal stress tolerance of plants. *South African Journal of Botany* 76(2), 167–179.

Yang, H.F., Zhang, G., Fu, P., Li, Z., Ma, W.K., 2020. The evaluation of in-site remediation feasibility of Cd-contaminated soils with the addition of typical silicate wastes. *Environmental Pollution* 265, 114865.

Yang, J.S., Yang, F.L., Yang, Y., Xing, G.L., Deng, C.P., Shen, Y.T., Luo, L.Q., Li, B.Z., Yuan, H.L., 2016a. A proposal of "core enzyme" bioindicator in long-term Pb-Zn ore pollution areas based on topsoil property analysis. *Environmental Pollution* 213, 760–769.

Yang, J.Y., Huang, J.H., Lazzaro, A., Tang, Y., Zeyer, J., 2014. Response of soil enzyme activity and microbial community in vanadium-loaded soil. *Water, Air, & Soil Pollution* 225(7). doi: 10.1007/s11270-014-2012-z.

Yang, Y., Chen, W.P., Wang, M.E., Peng, C., 2016b. Regional accumulation characteristics of cadmium in vegetables: Influencing factors, transfer model and indication of soil threshold content. *Environmental Pollution* 219, 1036–1043.

Yang, Z.X., Liu, S.Q., Zheng, D.W., Feng, S.D., 2006. Effects of cadium, zinc and lead on soil enzyme activities. *Journal of Environmental Sciences* 18(6), 1135–1141.

Yao, Z., Li, J., Xie, H., Yu, C., 2012. Review on remediation technologies of soil contaminated by heavy metals. *Procedia Environmental Sciences* 16, 722–729.

Ye, S., Zeng, G., Wu, H., Zhang, C., Dai, J., Liang, J., Yu, J., Ren, X., Yi, H., Cheng, M., Zhang, C., 2017. Biological technologies for the remediation of co-contaminated soil. *Critical Reviews in Biotechnology* (1062–1076), 1–15.

Yeates, G.W., Orchard, V.A., Speir, T.W., Hunt, J.L., Hermans, M.C.C., 1994. Impact of pasture contamination by copper, chromium, arsenic timber preservative on soil biological-activity. *Biology and Fertility of Soils* 18(3), 200–208.

Yin, H.Q., Niu, J.J., Ren, Y.H., Cong, J., Zhang, X.X., Fan, F.L., Xiao, Y.H., Zhang, X., Deng, J., Xie, M., He, Z.L., Zhou, J.Z., Liang, Y.L., Liu, X.D., 2015. An integrated insight into the response of sedimentary microbial communities to heavy metal contamination. Scientific Reports UK, 5.

Yin, K., Wang, Q., Lv, M., Chen, L., 2019. Microorganism remediation strategies towards heavy metals: ScienceDirect. *Chemical Engineering Journal* 360, 1553–1563.

Ying, R.R., Qiu, R.L., Tang, Y.T., Hu, P.J., Qiu, H., Chen, H.R., Shi, T.H., Morel, J.L., 2010. Cadmium tolerance of carbon assimilation enzymes and chloroplast in Zn/Cd hyperaccumulator Picris divaricata. *Journal of Plant Physiology* 167(2), 81–87.

Yourtchi, M.S., Bayat, H.R., 2013. Effect of cadmium toxicity on growth, cadmium accumulation and macronutrient content of durum wheat (Dena CV.).

Yuan, C., Chiang, T.S., 2007. The mechanisms of arsenic removal from soil by electrokinetic process coupled with iron permeable reaction barrier. *Chemosphere* 67(8), 1533–1542.

Yuan, X.Z., Xiong, T., Wang, H., Wu, Z.B., Jiang, L.B., Zeng, G.M., Li, Y., 2018. Immobilization of heavy metals in two contaminated soils using a modified magnesium silicate stabilizer. *Environmental Science and Pollution Research* 25(32), 32562–32571.

Žaltauskaitė, J., Sodienė, I., 2010. Effects of total cadmium and lead concentrations in soil on the growth, reproduction and survival of earthworm *Eisenia fetida*. *Ekologija* 56, 10–16.

Zeb, A., Li, S., Wu, J., Lian, J., Liu, W., Sun, Y., 2020. Insights into the mechanisms underlying the remediation potential of earthworms in contaminated soil: A critical review of research progress and prospects. *Science of the Total Environment* 740, 140145.

Zhang, F.P., Li, C.F., Tong, L.G., Yue, L.X., Li, P., Ciren, Y.J., Cao, C.G., 2010. Response of microbial characteristics to heavy metal pollution of mining soils in central Tibet, China. *Applied Soil Ecology* 45(3), 144–151.

Zhang, J., Wang, L.H., Yang, J.C., Liu, H., Dai, J.L., 2015. Health risk to residents and stimulation to inherent bacteria of various heavy metals in soil. *Science of the Total Environment* 508, 29–36.

Zhang, M.Y., Bai, S.H., Tang, L., Zhang, Y.L., Teng, Y., Xu, Z.H., 2017. Linking potential nitrification rates, nitrogen cycling genes and soil properties after remediating the agricultural soil contaminated with heavy metal and fungicide. *Chemosphere* 184, 892–899.

Zhang, W., Chen, L., Zhang, R., Lin, K.F., 2016. High throughput sequencing analysis of the joint effects of BDE209-Pb on soil bacterial community structure. *Journal of Hazardous Materials* 301, 1–7.

Zhao, C., Dong, Y., Feng, Y., Li, Y., Dong, Y., 2019. Thermal desorption for remediation of contaminated soil: A review. *Chemosphere* 221, 841–855.

Zhou, Q., Mu, Y., Zhao, J., Zhang, G., 2017. Heavy metal stress on growth and development and physiological indexes of wheat. *Chinese Agricultural Science Bulletin* 33, 1–8.

Zhou, Z.F., Zheng, Y.M., Shen, J.P., Zhang, L.M., Liu, Y.R., He, J.Z., 2012. Responses of activities, abundances and community structures of soil denitrifiers to short-term mercury stress. *Journal of Environmental Sciences* 24(3), 369–375.

Zhu, Z., Wang, J., Liu, X., Yuan, L., Deng, H., 2021. Comparative study on washing effects of different washing agents and conditions on heavy metal contaminated soil. *Surfaces and Interfaces* 27, 101563.

7 Silent Alienation of Soils through Microplastic in the Anthropocene – A Constraint for Soil Productivity?

Ewald Schnug[1,2], *Xijuan Chen*[2], *Zhuang Jie*[3], *and Silvia Haneklaus*[4]

[1]Department of Plant Physiology and Biochemistry, Faculty of Life Sciences, Technical University, Braunschweig, Germany
[2]State Key Laboratory of Pollution Ecology and Environmental Engineering, Institute of Applied Ecology (IAE), Chinese Academy of Sciences (CAS), Shenyang, China
[3]Department of Biosystems Engineering and Soil Science, The University of Tennessee, Knoxville, Tennessee
[4]Institute for Crop and Soil Science, Federal Research Centre for Cultivated Plants (JKI), Braunschweig, Germany

CONTENTS

ABBREVIATIONS

C	carbon
MP	microplastic
MP-C	C in MP
NP	nanoplastic
PD	plastic debris
PFA	per- and polyfluoroalkyl substances
PFC	perfluorocarbons
SD	standard deviation

DOI: 10.1201/9781003093565-7

7.1 INTRODUCTION

The production of synthetic carbon polymers, often trivially called plastics, is a characteristic signature of the Anthropocene. More than 70 years into the new geological age plastics are ubiquitously abundant in every ecosystem. Plastics have been replacing traditional materials in nearly every daily life application. Primarily physical forces crush plastic litter, macroplastic, into plastic debris and finally reduce it to yet smaller particles, so-called microplastic, and ultimately nanoplastic. In comparison to physical fragmentation, plastics are stable against biological and chemical degradation (Büks and Kaupenjohann, 2021; de Souza et al., 2018b; Fraunhofer, 2021; Watteau et al., 2018).

Although in bright sight since the first synthesis of plastics, it took almost half a century to recognize that plastic debris transforms in steadily smaller particles, which is addressed as MP (Hartmann et al., 2019). Below a particle size of ≤5 mm plastic fragments are referred to as particulate plastic or microplastic (MP), a terminus coined in 2004 by Thompson (Bradney et al., 2019, WIKI, 2022). It becomes nanoplastic (NP) below a particle size of ≤1 µm (ISO TR 21960). MP and NP cover various geometric shapes, for example, spherical particles, films, chips, and fibers depending on the morphology.

Since millions of years, soils evolve from the same sources: minerals and organic compounds of biological origin. Since the beginning of the Anthropocene in 1950 (NGS, 2022), soils have been charged with numerous xenobiotics (Zalasiewicz et al., 2015), which affected their natural functionality, very rarely to the better. Over infinite time, almost all these substances or their metabolites are released from their carrier compounds and migrate into neighboring ecosystems (Teuten et al., 2009). Well-visible carbon polymers are commonly known as plastics, the invisible ones are so-called "foever chemicals," PFAs and PFCs (Cousins et al., 2016; Falk et al., 2019; Salvige, 2022), which are not in the focus of this chapter.

In terms of a sustainable soil management, input and output of such substances need to be balanced on a socially and ecologically acceptable level to avoid a potentially dangerous accumulation in soils. Synthetic carbon polymers are an exemption because of their durability and omnipresence.

It took more than a decade until environmental issues of plastics gained substantial recognition in the science community. Although publications about MP issues are exponentially increasing in numbers over the last decade (Akdogan and Guven, 2019; Qin et al., 2020), only less than 10% of them are dealing with soil-related issues (Kim et al., 2020a). Actually, numerous, comprehensive papers have been published, which review, summarize, and balance different types of plastics and their fate in the environment, some with special attention to MP in soils (e.g., Allouzi et al., 2021; Blasing and Amelung, 2018; Bolan et al., 2020; Büks and Kaupenjohann 2020; de Souza, 2018a, b; Fraunhofer, 2018, 2021; Jepsen et al., 2020, Kim et al., 2020a; Kumar et al., 2020, 2021; Nizzetto, 2016a, 2021; O'Kelly et al., 2021, Rillig, 2012; Vollertsen and Vianello, 2019; Wang et al., 2020b; Wu et al., 2021). It should be noted for further thematic segregation that primarily the typical characteristics of plastics, which are durability, and chemical and biological inertness, are in the scope of this treatise. For features and ecotoxicology of additives used in manufacturing plastics, one may refer to specialized sources (e.g., Barrick et al., 2021; Hahladakisa et al., 2017).

The objective of this chapter is to interpret the collective knowledge of studies on MP in soil science with respect to the impact on physical, chemical, and biological soil features, to feature the idea of the alienation of soils through the accumulation of plastics and to present mitigation strategies.

7.2 PLASTIC – THE NEMESIS OF THE ANTHROPOCENE

Parkesine was the first plastic, synthesized by the Briton Alexander Parkes in 1855 from cellulose nitrate. He invented celluloid as a replacement for ivory in billiard balls (Eschner, 2017). In 1907, the mass production of plastic began with the invention of Bakelite by the Belgian-American chemist Leo Baekeland (Plastics Europe, 2020a). Since then, plastics and plastic litters are more or less ubiquitously abundant (Jepsen et al., 2020).

Plastic comes in a wide range of products. It can be categorized from a physical/technical point of view in two major groups: thermoplastics (plastics that can be reversibly melted when heated and hardened when cooled) and thermosets (plastics that undergo a chemical change when heated, creating a three-dimensional network; after heating and formation, these plastics cannot be remelted and reformed (Plastics Europe, 2020b).[1] Besides the major elemental components C, H, O, and S, all these plastics may contain a broad variety of chemicals that was identified to be of potential concern for health and environment. Among others these are the organic forever chemicals phthalates and brominated flame retardants and their technical alternatives butyrate TXIB, citrate ATBC (Aurisano et al., 2021), and PBT (persistent, bioaccumulative, toxic) substances containing for instance PIP 3:1 (phenols, isopropylenated phosphates in a 3:1 relation (Echa, 2021; Rankin, 2021). The organic polymer additive triclosane has been used since 1960 as a common antimicrobial agent in personal hygiene products such as soaps, shampoos, deodorants, toothpastes, and mouthwashes, but also in consumer products, including kitchen utensils, toys, bedding, socks, and trash bags (EPA, 2010).

Formally, plastics include only thermoplastics and thermosets, which come in a wide and versatile range of carbon structures. Each of it has its special technical features addressing the usage of the product. A continuously increasing number of new types of plastics is to be expected matching the needs for new applications.

Long before plastic was considered of posing a threat to the environment, it was hailed to improve soil productivity, for instance by incorporating polystyrene foams (Styromull®) or polyacrylates (superabsorbers, e.g., Stockosorb®) into soils in order to increase pore volume and water-holding capacity (Anonymous, 2004; USTPO, 1989; Wirth, 1967). Superabsorbers are an example how resourceful industry can be to solve the disposal problem of their products at the end of their life cycle. Soon after superabsorbers appeared on the international market relevant industries tried to get them registered as carrier materials for fertilizer nutrients like potassium. This move seemed initially to be economically not viable due to the high prices of superabsorbers, but the insidious idea behind was simply that once a superabsorber containing product has been used and charged with "nutrients" from human effluents (e.g., baby nappies or sanitary towels) it could start a new life as a legal nutrient source in recycled fertilizers.

Plastics make everyday life easier but can become very quickly life threatening in fires, i.e., when used for insulation and cladding purposes (Burke, 1999; Symonds, 2021; Hahladakisa et al., 2017). There are many hazardous substances used in plastic manufacturing, which are prone to be released into the environment (Bradney et al., 2019). A historic example for such a hazardous inorganic element is the plastic additive cadmium (Cd) in the well-known LEGO® toy bricks, which were made since 1963 from PA with Cd compounds providing its typical yellow color (Silver, 2018). As mentioned before, this chapter focuses exclusively on the features and characteristics of the pure polymer without addressing eco-toxicological effects of additives.

A special type among plastic is rubber. Natural rubber is a polymer of the organic compound isoprene sourced from a natural organic source: latex (Bode et al., 2000; Sparovek et al., 2001; Wuttke, 1989). Latex is the exudate of *Hevea brasiliensis* trees. In 1839, Charles Goodyear invented a process called vulcanization in which sulfur cross-links the isoprene polymers in latex into rubber. Although listed in the group of plastics, rubber is an elastomer (Alemán et al., 2007). Besides natural rubber made from latex synthetic rubber like chloroprene (Neoprene®) or styrole-butadien-rubber exist. Approximately, 70% of the world rubber production is used for tires (Deutschle, 2021) and used tires are a major nuisance for the environment. van der Heuvel (2021) estimated that 70%

[1] Common plastic types and their abbreviations: polyacrylates (PA); polyethylene (PE, includes HDPE [high-density] and LDPE [low-density] PE), polypropylene (PP); polyvinyl-chloride (PVC); polyethylene-terephthalate (PET); polystyrene (PS); expanded-polystyrene (EPS); acrylnitril-butadiene-styrol (ABS); styrol-acrylnitril (SAN); polyamides (PA; common product names Nylon, Perlon, Kevlar); polycarbonates (PC); poly-methyl-methacrylate (PMMA); thermoplastic-elastomers (TPE); polyarylsulfone (PSU); polyether-ether-ketone (PEEK); polyoxymethylene (POM); polybutylene-terephthalate (PBT); ethylene-vinyl-alcohol (EVOH); polylactic acid (PLA); poly(butylene-adipate-co-terephthalate) (PBAT), only used in scientific experiments are fluorescent polymers (FP; NSP, 2023).

of the rubber particles deriving from wear and tear of tires end up in soils compared to only 20% in water bodies. Burning tires was a common, meanwhile prohibited method of disposing tires on farms. Tires were re-used successfully for repairing erosion damages in fields (Sparovek et al., 2001). Today tires are recognized as a secondary resource material, and recycling technologies for tires are widespread (Etra, 2021; Zafar, 2022).

It can be assumed that organisms exist which are able to degrade latex as rubber derives from this natural material (Brill, 2010; Linos et al., 2002; Rose and Steinbuechel, 2005). Espinosa et al. (2020) recently claimed the isolation of a *Pseudomonas* species that grows on polyurethane.

Biodegradable polymers (bioplastics) claim to solve the main problem of conventional plastic, which is its persistence. Among these substances are the aliphatic–aromatic co-polyesters polybutylene-terephthalat (PBT) and ethylene-vinyl-alcohol (EVOH), poly lactic acid (PLA), poly(butylene-adipate-co-terephthalate) (PBAT), "Ecoflex," a combination of adipic acid, butanediol and terephthalic acid (BASF, 2001) and "Mater-Bi," an aliphatic–aromatic copolyester (Novamont, 2022). Genetically engineered enzymes are one escape route for the difficulty durability of plastics. Tournier et al. (2020) manufactured a PET depolymerase, which they propose to be used for future plastic recycling.

The problem of the degradation of bioplastics is that this process requires the exposure to high temperatures to trigger hydrolysis, a condition not expected in natural soil environments, or during composting. Thus, bioplastics need to be addressed as contaminants in composting. The shortcomings of biodegradable plastics are the reason for a decline in the credibility of industry and standardization authorities according to WEL (2022).

Industry focuses on plastic recycling because of increasing prices for energy and raw materials and increasing demands toward environmental sustainability (RAM, 2022). However, the durability of plastics bars the way of a solution, rather it shifts time and location of the problem closer to soils, as the following examples will show: Plastic recycling has led to several new products of which for instance surface covering materials and here especially synthetic turf is becoming most popular (Figure 7.1a and b).

In the UK, there are actually more than 10,000 multi-use synthetic turf sports fields, for example in schools, parks, and professional sport arenas (Benson, 2017; Etra, 2021). Wear and tear from these materials are a source for environmental pollution with MP. A vivid example showing that recycling products from plastic bear the same environmental risk as the original products. Another environmental risky application of recycled plastics are weed barriers installed in soils underneath rolled turfs or gravel covers (Monteiro and Santos, 2022; Subrahmaniyan et al., 2011; Figure 7.2).

In 2020, the world plastic production was higher than 335 million tons; in the 1950s, the corresponding value was <1 million tons (Plastics Europe, 2020; Zalasiewicz et al., 2016). In parallel, the world rubber production increased during the same time to >14 million tons per year (Deutschle, 2021).

(a) (b)

FIGURE 7.1 (a) Plastics replace natural ground covers on playing grounds (Kristiansand, Norway, 2020) and (b) greens (Whitby, UK, 2021).

FIGURE 7.2 Plastic weed barriers, the ultimate shortcut for plastics entering soils (Whitby, UK, 2021).

FIGURE 7.3 Living with plastics in the suburbs of Cairo, Egypt, 2022.

When plastic commodities become litter and are disposed in the environment, it primarily sores personal esthetic senses (Figure 7.3). However, even this is depending on cultural circumstances. In many regions of the planet, especially heavy populated developing countries, humans live in obvious or apparent harmony with their plastic litter (Figure 7.3), sometimes being a welcome distraction for younger individuals (Figure 7.4). The reason behind is not only negligence but often also lack of an appropriate waste management, which makes it an indicator for poverty (Figure 7.3).

FIGURE 7.4 As early as in the 1960s PD advanced to a welcomed environmental enrichment (Morley-Fletcher et al., 2003) distracting kids from stress in everyday life routine (Altenkirchen, Germany, 1962).

FIGURE 7.5 Fly-tipping along a roadside near Bari, Italy, 2015.

Lumby (2018) observed polar bear cubs, which enjoyed plastic debris as an environmental enrichment in the remote and extreme ecosystem of Svalbard (Allen, 2022; Morley-Fletcher, 2003) making it, like their human siblings (Figure 7.4) a welcome distraction from everyday life routine.

For onlookers from outside a littered community or environment may be a humiliation of esthetics, but it is only an immediate danger for life or health of humans if it is leaking toxic compounds into the environment or releasing harmful volatiles when used as replacement for fuel. But not only in third-world countries plastic litter in the environment has become a frequent and increasing every day's nuisance: disposing plastic litter through fly-tipping (Figure 7.5) is meanwhile evolving from an occasional private offence to a high-rising organized crime (Crowe and Lynch, 2020).

Plastic litter in terrestrial and aquatic ecosystems (Figure 7.6) pose a severe danger for animal life and health when it is getting caught in plastic fragments (Daly, 2018; Hester, 2021; Lavers et al., 2019) or by confusing it with feed (Savoca et al., 2016; WWF, 2018).

Part of plastic litter moves via lakes and rivers and finally reaches the sea (Galloway et al., 2017). Globally plastic occurs virtually everywhere in the world's oceans (R4WO, 2015). Here, it concentrates in five regions, the so-called marine garbage belt with the North Pacific garbage patch being the most prominent one (Wiki, 2021). Beside these continuous accumulating processes, single accidents spill vast amounts of plastic into marine environments. Examples are the sinking of the

FIGURE 7.6 The place at the Nile in Cairo (2019) where Moses' basket once stranded.

vessel X-Press Pearl in the Indian Ocean in 2021 close to Sri Lanka. It has been named as the single largest plastic spill in history, where about 1,680 tons of so-called nurdles, an upstream product of most plastic goods were lost to the sea (Partow et al., 2021). Another large plastic loss on sea was the "Great Lego Spill," when in 1997 the cargo ship Tokio Express dumped 62 sea containers with 5 million plastic toys in front of the south coast of England (Weisberger, 2022). Pyroplastics are a new form of plastic debris recently detected in the sea. They look like stone pebbles and form when pieces of plastic are melted or burned before thrown into the sea. During their long ocean voyages, pyroplastics slowly weather and obtain the characteristic gray color and smooth surface (Turner et al., 2019).

7.3 OCCURRENCE OF MICROPLASTIC IN THE SOIL ENVIRONMENT

Plastic only degrades mechanically through physical forces, eventually supported by UV radiation to ever smaller particles (Büks and Kaupenjohann, 2021; Meides et al., 2021).

Biochemical pathways have been suggested to support the disintegration of plastic (Huerta et al., 2018; Sánchez, 2020; Wei and Zimmermann, 2017), but at present knowledge their contribution to the general breakdown process is far behind that of the physical one and neglectable. An exception might be rubber made from latex.

MP was first detected in aquatic environments, later also in terrestrial environments and atmospheric depositions (Bao et al., 2022; Gestoso et al., 2019; Klein and Fischer, 2019; Ziajahromi et al., 2016). MP stored in glaciers (Ambrosini et al., 2019) is considered a more important stratigraphic indicator of the Anthropocene in sediments than other indirect components such as sediment flux changes (Zalasiewicz et al., 2011, 2016).

The average figure for the annual input of MP to agricultural land in Europe and North America was estimated to about 156,000 tons a decade ago (Nizzetto et al., 2016b) compared to an average of 209,000 tons for oceanic and surface waters (Cara, 2019). Although the plastic masses in aquatic ecosystems look frightening impressive, the plastic inventory of soils is believed to be still at least 20 times higher than in oceans (Carstens, 2018; Petersen, 2020; UNEP, 2022a).

The figures reported for plastic inputs to soils in terms of masses are extremely variable and difficult to compare between different sources, because there are no harmonized or standardized investigation methods. In this respect, the various data from the different publications are difficult to compare (He et al., 2018). Research on plastic and MP in the environment started in the water sector with methods analyzing particle size and numbers. Büks and Kaupenjohann (2020) found in general that data on plastic masses in environmental media was rare, and often estimated by means of questionable shape-to-mass models instead of direct measurements. In recent years, thermoanalytical methods for the overall determination of plastics have been developed (BMBF, 2020; Dümichen et al., 2017, 2019; Eisentraut et al., 2018; Han et al., 2019; Leslie et al., 2022; Radford et al., 2021) and applied to analyze plastics in soils.

The data about the occurrence of plastic in soils reported in the literature reviewed for this book chapter were subjected to a metadata analysis. This analysis claims not to be complete; rather, it presents a freeze frame for a time window approximately for the time between 2010 and 2020. The analysis revealed that MP was abundant in all countries, in all soils, in any size, shape, and number from any plastic material and from any possible source. In addition, it shows that numerical ranges for the prevalence of plastic in the environment are already much larger than expected (Büks and Kaupenjohann, 2020). For a randomly selected number of countries and environments, the metadata analysis provided a range of 10–356,000 plastic particles/kg in topsoils with an average of 16,889 and a median of 612 plastic particles/kg. This corresponds with a figure averaged from several reports and countries around the world, which is about 13,000–18,500 particles per kg of soil (Büks and Kaupenjohann, 2020; Fuller and Gautam, 2016; Mintenig et al., 2017; Zhang and Liu, 2018; Zubris and Richards, 2005). More than 98% of all plastic reported in these investigations was MP. However, for a realistic assessment of the ecological problem caused by MP, figures on data about masses are more important than counting of particle numbers. Büks and Kaupenjohann (2020) estimated in their global review a median of 0.045% (w/w) for plastic concentrations in soils of agricultural and horticultural production systems with substantial plastic inputs. Fuller and Gautam (2016) reported extreme plastic concentrations of 6.7% (w/w) in a soil from the industrial zone of Sydney, which indicates that in some cases plastics have already become a substantial part of soils.

As this book focuses on the issue plastic and soil productivity, the following sections will assess the presently known effects of MP on soil features related to productivity, which are soil physical characteristics, soil fauna, and soil microorganisms. A quantitative assessment of the studies dealing with effects of plastics on soils reviewed here revealed that 8% address issues of soil physical properties, 15% effects on plant growth, 18% on soil microorganisms, and a majority of 59% on soil fauna.

7.4 EFFECTS OF MICROPLASTIC ON SOIL PHYSICAL PROPERTIES

MP enters the soil environment through long-term agricultural film residues, application of organic fertilizers and sludge, reclaimed water irrigation, and atmospheric deposition (Blasing and Amelung, 2018; UNEP, 2022a). The structure of MP diminishes in size during long-term weathering, UV irradiation, and physical forces from other components in soils (Horton et al., 2017). As the particle size becomes smaller, the affinity to heavy metals, polyaromatic hydrocarbons, polychlorinated biphenyls, pesticides, and antibiotics might be enhanced (Horton et al., 2017; Teuten et al., 2009). The number of studies determining the impact of MP on physical soil properties is limited. de Souza et al. (2018a) found that MP alters soil bulk density and hydraulic characteristics and agglomerates size distribution and stability at environmentally relevant concentrations. This effect seems to be different between plastic types: polyesters for instance significantly reduced, while polyethylenes significantly increased the amount of soil water-stable aggregates (de Souza et al., 2018a). A field study in China (Zhang and Liu, 2018) revealed that more than 70% of MP particles agglomerated with soil aggregates. However, there is still a lack of long-term and *in situ* studies in fields on the interaction of MP pollution with soil aggregates, so that it is not yet possible to assess its significance for water conservation and soil productivity (Nizzetto, 2018).

In a short-term experiment with different types of MP, de Souza et al. (2018) noticed a reduced bulk density. Wan et al. (2019) found that in particular smaller particles favored the rate of desiccation shrinkage and by this enhanced evaporation. Polyester decreased bulk density and enhanced the water-holding capacity, while polyacrylic fibers and polyethylene fragments unfolded no effect. Here, it can be speculated that the outcome of the experiment depends strongly on the scale of experimentation as results from small-scale pot experiments seldom reflect the situation in fields. The phenomenon is well-known from spiking soils with heavy metals compared to naturally contaminated soils (Haneklaus et al., 2007). Accordingly, Zhang et al. (2019a) stated an increase of water-stable aggregates in a pot, however, not in a field trial.

Last, but not least, effects on soil physical properties are very much the result of interactions between MP and soil life, mainly soil fauna. In this context, Rillig et al. (2017) highlighted the significance of anecic earthworms for the transfer of MP along different soil depths.

7.5 EFFECTS OF MICROPLASTIC ON SOIL FAUNA

The majority of the literature reviewed investigated the effects of MP on soil fauna with distinct emphasis on the phylum Annelida (segmented worms) to which the family Lumbricidae (earthworms) belongs. Plastics may like other hard materials obstruct movement of soil fauna thus limiting access to feed and mating partners, or when swallowed becoming a risk for obstructions and alterations of the microbiome of intestines and nutrient resorption.

MP affects soil animals of various trophic levels (Rillig, 2012) down to the size of so-called protists. According to Rillig and Bonkowsky (2018), soil protists are "a phylogenetically and functionally diverse group of eukaryotic, unicellular soil organisms. They are major consumers of bacteria in soils and are potentially important vehicles for the delivery of microplastics into the soil food chain and include filter feeders as well as amoebae which engulf their prey."

MP indirectly affects the activity of soil animals by altering their habitat. The effect of MP on soil animals seems to be relatively unspecific and related to physical factors, mainly particle size. Rillig (2012) found that MP could block soil pores and affect the activity of smaller soil animals (size smaller than, e.g., Annelida or Collembola). Depending on abrasive geometric features, any plastic particles have the potential to affect the skin surface of these animals (Baeza et al., 2020). However, there are obviously species, which seem to be able to biodegrade plastics like the caterpillars of the wax moth *Galleria mellonella* (Bombelli et al., 2017).

In general, animals take up, without much selection and discrimination, any soil substrate, which passes their mouth opening in size. On a macro and mesoscale of soil animals, the ratio of particle size to the size of the mouth opening is related to the feeding of MP by animals (Horton et al., 2017). MP particles with a particle size >1 mm could remain in the intestine or be excreted, while smaller particles are more likely to become encapsulated (Luttikhuizen et al., 2006), accumulated, and phagocytosed by the organism, which may cause physical damage of organs and tissues (Browne et al., 2013).

Rillig et al. (2017) summarized that soil protozoa such as ciliates, flagellates, and amoeba were able to ingest MP particles. In mites, Zhu et al. (2018b) observed that with increasing concentrations of NP in the diet the NP intake initially increases before it declines.

Studies on the model nematode *Caenorhabditis elegans* revealed that MP may cause size-dependent neurotoxicity, oxidative damage, and intestinal damage among other adverse effects (Lei et al., 2018a, b). In a study with soil collemboles, Zhu et al. (2018a, b) found that an exposure to MP over 28 days (1 g/kg MP, dry soil) not only significantly reduced their growth and reproduction, but also enhanced carbon and nitrogen isotope fractionation in their tissues. Effects, which were also reported about experiments with *Daphnia* species, which, however, live in waters (Kelpsiene et al., 2020). Recent studies on Oligochaeta have shown that acute dietary exposure of NP at low concentrations (0.5% NP in oatmeal) significantly inhibited their productive rate and biomass (Zhu et al., 2018a, b).

A reduced growth and reproduction rate of Lumbricidae fed with MP might have been related to a lower nutrient content of the substrate (Cao et al., 2020; Gaylor et al., 2013). Galloway et al. (2017) stated sub-lethal effects on earthworms and linked these to internal injuries such as lacerations and reduced feed intake, which resulted in reduced reproduction and individual growth. This weakens the adaptability of the animal (Wright et al., 2013). Alomar et al. (2017) add inflammatory responses and hepatic as well as oxidative stress to the list of effects.

Ingested MP affects the biodiversity of the intestines' microbiome (Ding et al., 2020a; Rodriguez-Seijo et al., 2017; Zhao et al., 2017; Zhu et al., 2018c). This is obviously a general effect that can also be observed in other soil animals like Achinata (Song et al., 2019; Chae and An, 2020), Collembola (Zhu et al., 2018a, b), and Nematoda (Judy et al., 2019; Kim et al., 2020 a–d; Lei et al., 2018a). Some studies have shown that MP not only remains in the intestine but can also pass through the intestinal wall so that it enters tissues and organs (Browne et al., 2013; Farrell and Nelson, 2013).

One major risk arising from MP for soil biota is that they can enter the food chain threatening health of humans and animals (Allouzi et al., 2021; Bradney et al., 2019; Mai et al., 2018; Leslie et al., 2022; Nizzetto et al., 2016a; Ziajahromi et al., 2016). This risk increases when MP is further broken down to NP, which may enter functional organs or even cellular structures (Jenner et al., 2022; Leslie et al., 2022; Lim, 2021).

Considering the long-term accumulation and persistence in organisms, MP can move along food chains and food webs, causing damage to predators (Steinmetz et al., 2016). A survey of traditional Mayan family gardens in southeastern Mexico (Lwanga et al., 2017a) revealed that MP concentrations increased from soil (0.9 particles/g soil), to earthworm casts (15 particles/g soil) to chicken feces (130 particles/g feces), indicating that MP indeed enters and accumulates in the terrestrial food chain, a fact confirmed by Rillig (2012). However, due to limitations of analytical methods, the understanding of the transfer of MP in the soil food chain is still very poor.

7.6 EFFECTS OF MICROPLASTIC ON SOIL MICROORGANISMS

Common understanding is that in general plastics are resistant against microbial decomposition. An exception might be rubber, which is a polymer of the organic compound isoprene (Bode et al., 2000). Tire wear contributes largely to the MP loads to soils (van der Heuvel, 2021). Although microorganisms capable in metabolizing rubber have been identified (Linos et al., 2002; Rose and Steinbuechel, 2005), chemicals added for the vulcanization of tires are highly antimicrobial (Wuttke 1989), and therefore, tire wear is unlikely to be less resistant against biochemical degradation.

There is only a finite number of studies on the effects of plastic pollution and its chemical composition on soil microbial communities (Awet et al., 2018; Chen et al., 2020a, b; de Souza et al., 2018b, 2019; Fei et al., 2020; Harrison et al., 2014; Huang et al., 2019; Liu et al., 2017; Salvador et al., 2017; Sun et al., 2018; Wang et al., 2020a; Yan et al., 2020; Yang et al., 2018; Yi et al., 2020). Sun et al. (2018) studied the antibiotics and phage-associated resistance of surfactants and MP in cultivated soils and found that the presence of MP significantly inhibited the dissipation of antibiotic resistant genes in the soil. Sophorolipid, which is a surface-active glycolipid that can be synthesized by non-pathogenic yeast species promoted this effect. Recent studies from Qian et al. (2018) found that the occurrence of PD and MP seemingly hindered the expression of microbial genes related to soil C and nitrogen cycles, which may affect the diversity of microbial communities and plant performance. This effect was coined the "neighbor avoidance effect" by Hong et al. (2021) but seems more related to chemicals released by plastics, rather than the plastic itself. Plasticizers, in particular phthalate esters, have been frequently detected contaminants in soils, which have mutagenic, teratogenic, and carcinogenic effects (Douglas et al., 1986; Kong et al., 2018; Wang et al., 2018).

With the tendency for long-term accumulation and resistance against degradation, MP may become a carrier for all sorts of microorganisms of which some may be pathogenic bacteria, affecting the health of soil ecosystems. MP may provide sites for colonies of microorganisms, turning into biofilms, affecting the ecological functions and diversity of soil microbial communities (Harrison et al., 2014; Hong et al., 2021; Zettler et al., 2013). In addition, with the migration of MP, microorganisms can like other xenobiotics be disseminated and eventually transported into the groundwater (Hu et al., 2021).

7.7 SILENT ALIENATION OF SOILS THROUGH MICROPLASTIC ACCUMULATION

In all likelihood, there is no alternative to an expanding use of plastics in daily life. It is inevitable that soils are continuously and *ad infinitum* charged with plastics if current human consumer behavior and demands do not change globally. The ban, also partial, of single-use plastic products in the EU might improve the appearance of our visible environment (Figures 7.3 and 7.6) but cannot prevent the accumulation of MP in soils, at best it can slow down the process. As there is virtually no biological degradation of plastics, their accumulation in soils is unstoppable.

The following thought experiment demonstrates what consequences the accumulation of MP may have for soils in the near and distant future. Without question, the range of variation with a view to plastic sources, physical conditions, and loads to soils is high. As a result, calculations underlie broad uncertainty ranges. Based on a simple linear model with arbitrary, but realistic figures from the EU about 125–850 tons of MP per million inhabitants are discharged annually on soils (Nizzetto et al., 2016a, b; Plastics Europe, 2020b; Ziajahromi et al., 2016). This implies that the 448 million people living actually in the EU discharge theoretically every year 0.04–0.43 million tons of MP on 173 million ha of farmland (Eurostat, 2022). By doing so, soils may gain hypothetically an average of 2.5 mg MP per kg of soil annually assuming that an amount of 0.43 million tons of MP is distributed evenly over the farmland within the EU and MP accumulates in the 0–20 cm layer of the topsoil with an average specific weight of 1.5 g/cm^3. For comparison, Fraunhofer (2021) estimated an enrichment value of 3.5 mg MP per kg of soil annually for Germany, a figure not much different from the calculation example. Assuming further an average C content of 86% in MP, 2.5 mg MP per kg of soil equals a dose of 2.1 mg MP-C per kg to soils annually. Setting this figure in relation to the accuracy of C analytics, it turns out that the average SD for C analysis amounts to 0.25% C (Chatterjee et al., 2009; Chen et al., 2015; Dijk et al., 2000; DIN 36845, 2008; Lesschen, 2020; Lorenz and Lal, 2016; Nätscher et al., 2013; Slepetiene et al., 2008; Walinga et al., 1992). This means that before any changes in the C concentration between two soils turn out to be statistically significant, the difference in the C content must exceed at least three times the SD, equaling 0.75% C (Wellmitz and Gluschke, 2005). According to UBA (2020), arable soils in Germany have a total C content of about 1%. Thus, the C content of a soil with 1% C needs to exceed 1.75% C by the application of MP before the increase in total C can be proven statistically significant. This simple math assumes that the natural soil C content is in a long-term equilibrium and neglects the considerable spatial variability of C in soils (Lesschen, 2020; Rühling, 1999). In terms of a time span the data suggest that it would take more than 3,572 years in a "business as usual scenario of plastic management" (tantamount with adding 2.1 mg MP-C per kg of soil annually) before a change in the total C content of soils becomes perceptible. With other words: the change in total C through MP accumulation stays statistically "silent" for more than 3,572 years. The time span may be much longer if a factor of ten instead of three is applied for the determination of the practical detection limit (Armbruster et al., 1994). Yet, another problem obscuring the verification of MP accumulation in soils to this day is the lack of a quantitative determination method for MP-C in soils (Möller et al., 2020; Watteau et al., 2018).

The continuous accumulation of MP will go along with a substantial change of the quality of the soil carbon inventory, which will entail the fraction inert carbon. This inert carbon fraction will inevitably dilute and weaken soil functionalities linked to organic matter as MP-C is more or less physically and biochemically inactive. For instance, approximately one third of the mean soil cation-exchange capacity is allocated to organic matter (Solly et al., 2020), a function MP cannot fulfill.

MP will become an ubiquitous, completely new soil constituent, a legacy of the Anthropocene and soils, which are a natural heritage and basis for the production of safe food will need to be categorized in future with respect to their MP status in a new order of alienated soils. Although anthropogenic influences on soil genesis have been addressed since the early 1960s for urban soils, the alienation through the accumulation of plastics is a process, which summons soil science in general, and soil taxonomy and classification in particular. Although 3,572 years exceed imagination of most humans (it is approximately the elapsed time since the Neolithic or the last ice age), MP accumulation in soils is a problem caused by the recent generation challenging all future generations.

7.8 EFFECT OF MICROPLASTIC ON CROP PERFORMANCE

Most significant indicator of soil productivity in terms of agriculture is the performance of crop plants. The effects of MP on plants reported in the recent literature are inconsistent and difficult to allocate to typical features of plastics. A similar dilemma was found for the impact of MP on soil fauna and soil microorganisms. Often the effects were indirectly induced: for instance, MP may affect plant growth indirectly via changing the physical characteristics of a soil and by this its ability to store and supply water and nutrients to plant roots, or simply affecting root growth by mechanical features of PD and MP such as bulk density and aggregate stability.

Nanoparticles are discussed as novel vectors for supplying plants with essential nutrients (El-Ramady et al., 2018), which implies that nanoparticles are able to cross the Casparian strip (Chen et al., 2011; Nawrath et al., 2012). Metalloid nanoparticles interfered with metabolic processes and mostly caused toxic effects (Ameh and Sayes, 2019; Pohanka, 2019; Seltenrich, 2013). Silica is a mineral in which chemical inertness and durability resemble similar features of plastics. Silica is applied in the form of NPs for plant nutrition, and toxic side effects have been observed (Datnoff, 2012; Olle and Schnug, 2016; Napierska et al., 2010). The studies of Li et al. (2020a, 2021) suggest that plants are able to take up MP particles of 0.2–2 µm as long as the Casparian band is not fully developed.

In general, positive effects on crop growth parameters outnumber that of negative impacts (Chae and An, 2020; Meng et al., 2021; Qi et al., 2018). Studies on the effects of MP and NP on germination report from negative (Boots et al., 2019; Bosker et al., 2019; Chae and An, 2020; Meng et al., 2021; Qi et al., 2018) over no effect at all (Giorgetti et al., 2020; Lian et al., 2019, 2020)) to positive influences (Boots et al., 2019; de Souza et al., 2019; Li et al., 2019, 2020b; Lozano and Rillig, 2020; Lozano et al., 2020).

A major handicap for the interpretation and extrapolation of data is, as mentioned above that research on the effects of MP and NP on crop performance is usually carried out in growth chambers or greenhouse experiments employing small pots. Consequently, artifacts in the outcome may occur and require a sensitive interpretation of data (Qi et al., 2018). Haneklaus et al. (2007) describe the problem of verifying causal relationships when going from laboratory to field level.

With a view to the applicability of experimental data on field level, two variables need to be considered. These are scale of experimentation and experimental design, as both of them tend to overestimate the validity of treatment effects. Quantum and type of MP/NP contamination will govern the physicochemical and biological behavior in soils and their carry-over into soil fauna and crop plants. In case of MP/NP, no data exist which reflect changes over time in the content and fractionation of plastics in relation to their origin in soils. Commonly experimentation assumes steady state rather than aging and transformation. Yet another problem is that it is analytically not possible to determine individual types of plastic quantitatively. When it comes to production fields, two other confounding variables need to be kept in mind, which may obscure treatment effects. Firstly, in agricultural research the biggest source of error poses sampling (Buczko et al., 2012), and secondly, increasing the scale of experimentation from lab to field will multiply responses (Haneklaus, 1989). In greenhouse and field experimentation, interactions of MP/NP with any other factor affecting plant growth might modify metabolic response and finally plant growth so that beneficial effects might be obscured and detrimental effects mitigated or compensated (Haneklaus et al., 2007). Keeping this in mind for instance changes in the concentration and activity of stress-related enzymes in plants (Giorgetti et al., 2020; Jiang et al., 2019a; Liao et al., 2019) cannot be attributed automatically to the factor MP/NP, as there are various other parameters, which may have caused the effect.

Based on the previous perceptions, there is a risk that MP and NP enter the food chain by plant uptake and cause serious environmental problems and health issues for humans as well as for animals (Allouzi et al., 2021; Bradney et al., 2019; Leslie et al., 2022; Mai et al., 2018; Heather et al., 2022; Nizzetto et al., 2016a; Ziajahromi et al., 2016). MP and NP in soils need to be regarded as a constraint for the ability of soils to produce healthy feedstuff and food.

7.9 FUTURE PERSPECTIVES FOR MITIGATING EFFECTS OF MICROPLASTIC ON SOILS

Plastic pollution is a global challenge as plastics are omnipresent. Völker et al. (2020) state that most public press articles (93%) reporting on the impacts of MP imply that risks of MP exist and harmful consequences are highly probable. At the very beginning of the Anthropocene, it was Arendt (1958), who anticipated this nemesis and predicted that regardless of any regulatory rules and politics that human nature is a strong opponent of a zero waste community. Based on the presented comprehensive literature, it is evident that the effect of plastics on soils and their productivity after 70 years in the Antropocene are small compared, for instance, with the issues of soil compaction and loss of organic matter (Sanderman et al., 2017; Sonderegger and Pfister, 2021). However, the key difference is that loss of organic matter and soil compaction are reversible processes, while plastic accumulation is not because of its persistence.

Though there is no immediate risk for soil productivity by plastic accumulation in soils action is needed, simply because it is an ongoing and irreversible process. The doctrine of sustainable development to which the United Nations committed themselves in 1987 is independent of the time scale so that any changes over time, which have the potential to become harmful, are counterproductive to the goals of sustainability (Brundtland, 1987).

For waterbodies there is a chance to turn the clock back, because the insolubility of plastic and differences in specific weight allows to harvest it from waters (Boffey, 2019) and re-use it as raw-material or even as fuel (Belden et al., 2021). However, separating MP from soils in true life is impossible.

Effects of an ever-increasing soil content of MP may stay for a long time statistically undetected but will silently diminish total and available water and nutrient reserves, cation-exchange capacity and compromise other vital ecological functionalities of soils. MP accumulation is a typical example where actions based on the "precautionary principle" apply (Bourguignon, 2015; Cousins et al., 2016; Pinto-Bazurco, 2020). Against this background and as a guideline to combat further plastic accumulation in soils, the Federal German Governments Environment Agency's Commission for Soil Protection released in 2021 a list of recommendations for actions in politics (Grathwohl et al., 2021):

- Establish methods for the identification of precautionary and guidance levels for plastics in soils, similar to the NOEL (No Observed Effect Level) in toxicology (Crane and Newman, 2000).
- Avoid use of PFAs and other forever chemicals in manufacturing and their uncontrolled release into environment.
- Establish methods and criteria for assessing the use of biodegradable materials with view to preventing plastic accumulation in soils.
- Control and regulate plastics and PFA containing materials in recycling, and expel them form manufacturing of recycling products (e.g., composts, digestate fertilizers, soil conditioners).
- Use municipal wastes and sludges, which still contain MPs and PFAs exclusively for energy production through incineration.

Silent alienation through MP accumulation bears no immediate threat to soil productivity but definitively has the potential in the future. The reason is that MP accumulation is an irreversible process that diminishes vital soil functions. These imply for example the storage and buffering capacity of genuine soil organic matter for plant nutrients and safeguarding of aggregate stability. With other silent changes of soil features, like the loss of infiltration capacity through silent sealing (Fink, 2005; Schnug and Haneklaus, 2002), the accumulation of MP has in common that it has a lapse much longer than the validity of political decisions, rendering commitments to sustainable development to hypocritical services. However, silent sealing has the prospective to be brought to a halt, if not reversed, while there exists no such option for the silent alienation of soils through MP accumulation. Although countermeasures are included basically in the European Green Deal (EU, 2022), they remain a lip service unless they are enshrined in law as passed for fertilizing products (EU, 2019). To this day, expert recommendations to politics do not address with one single word the problems caused by plastics and forever chemicals in organic fertilizers and organic wastes (Wieseler et al., 2015).

The result of the presented appraisal is that the accumulation of plastic in soils is not an immediate threat to soil productivity, but a challenge for man's commitment to sustainable development.

REFERENCES

Akdogan, Z. and B. Guven. 2019. Microplastics in the environment: A critical review of current understanding and identification of future research needs. *Environmental Pollution* 245:113011.

Alemán, J. V., A. V. Chadwick, J. He, M. Hess, K. Horie, R. G. Jones, P. Kratochvíl, I. Meisel, I. Mita, G. Moad, S. Penczek and R. F. T. Stepto. 2007. Definitions of terms relating to the structure and processing of sols, gels, networks, and inorganic–organic hybrid materials (IUPAC Recommendations 2007). *Pure and Applied Chemistry* 79:1801–1829. doi: 10.1351/pac200779101801.

Allen, G. 2022. Enrichment portfolio for bears. https://wildwelfare.org/wp-content/uploads/2016/05/The-Captive-Bear_Enrichment-Ideas.pdf (accessed April, 12, 2022).

Allouzi, M. M. A., D. Y. Y. Tang, K. W. Chew, J. Rinklebe, N. Bolan, S. M. A. Allouzi and P. L. Show. 2021. Micro (nano) plastic pollution: The ecological influence on soil-plant system and human health. *Science of the Total Environment* 788:147815. doi: 10.1016/j.scitotenv.2021.147815.

Alomar, C., A. Sureda, X. Capó, B. Guijarro, S. Tejada and S. Deudero. 2017. Microplastic ingestion by *Mullussurmuletus* L. fish and its potential for causing oxidative stress. *Environmental Research* 159:135142.

Ambrosini, R., R. S. Azzoni, F. Pittino, G. Diolaiuti, A. Franzetti and M. Parolini. 2019. First evidence of microplastic contamination in the supraglacial debris of an alpine glacier. *Environmental Pollution* 253:297–301. doi: 10.1016/j.envpol.2019.07.005.

Ameh, H. and C. M. Sayes. 2019. The potential exposure and hazards of copper nanoparticles: A review. *Environmental Toxicology and Pharmacology* 71:103220. doi: 10.1016/j.etap.2019.103220.

Anonymous. 2004. Grüne Wüste dank Windeltechnik. *Der Spiegel* 31/2004. https://www.spiegel.de/wissenschaft/gruene-wueste-dank-windeltechnik-a-a95bd4e4-0002-0001-0000-000031617139 (accessed April 5, 2022).

Arendt, A. 1958. *The Human Condition.* The University of Chicago Press: Chicago https://monoskop.org/images/e/e2/Arendt_Hannah_The_Human_Condition_2nd_1998.pdf (accessed April 8, 2022).

Armbruster, D. A., D. Tillman, and M. Hubb. 1994. Limit of detection (LQD)/limit of quantitation (LOQ): Comparison of the empirical and the statistical methods exemplified with GC-MS assays of abused drugs. *Clinical Chemistry* 40:1233–1238.

Aurisano, N., H. Lei, M. C. Llorenç, O. Jolliet and P. Franke. 2021. Chemicals of concern in plastic toys. *Environment International* 146. doi: 10.1016/j.envint.2020.106194.

Awet, T. T., Y. Kohl, F. Meier, S. Straskraba, A. L. Grün and T. Ruf. 2018. Effects of polystyrene nanoparticles on the microbiota and functional diversity of enzymes in soil. *Environmental Sciences Europe* 30:11. doi: 10.1186/s12302-018-0140-6.

Baeza, C., C. Cifuentes, P. González, A. Araneda and R. Barra. 2020. Experimental exposure of *Lumbricus terrestris* to microplastics. *Water, Air & Soil Pollution* 231:308.

Bao, M., Q. Huang, Z. Lu, F. Collard, M. Cai, P. Huang, Y. Yu, S. Cheng, L. An, A. Wold and G. W. Gabrielsen. 2022. Investigation of microplastic pollution in Arctic ford water: A case study of Rijpforden, Northern Svalbard. *Environmental Science and Pollution Research* 25:28725–28729 doi: 10.1007/s11356-022-19826-3.

Barrick, A., O. Champeau, A. Chatel, N. Manier, N. Northcott and L. A. Tremblay. 2021. Plastic additives: Challenges in ecotox hazard assessment. *PeerJ* 9:e11300. doi: 10.7717/peerj.11300.

BASF. 2001. BASF: Fertilizer from plastic waste. *Chemeurope* 26.02.2001, https://www.chemeurope.com/en/news/1491/basf-fertilizer-from-plastic-waste.html (accessed March 14, 2022).

Belden, E., N. K. Kazantzis, C. M. Reddy, H. Kite-Powell, M. T. Timko, E. Italliani and D. Herschbach. 2021. Thermodynamic feasibility of shipboard conversion of marine plastics to blue diesel for self-powered ocean cleanup. *Proceedings of the National Academy of Sciences* 118. doi: 10.1073/pnas.2107250118.

Benson, H. 2017. Synthetic turf crumbs: Nuisance or noxious? https://www.parentmap.com/article/artificial-turf-crumbs-cancer-risk (accessed March 22, 2022).

Blasing, M. and W. Amelung 2018. Plastics in soil: Analytical methods and possible sources. *Science of the Total Environment* 612:422–435. doi: 10.1016/j.scitotenv.2017.08.086.

BMBF. 2020. Analysis of microplastics: Sampling, preparation and detection methods. Status Report. https://bmbf-plastik.de/en/publication/analysis-microplastics-sampling-preparation-and-detection-methods (accessed 22 March, 2022).

Bode, H. B., A. Zeeck, K. Plückhahn and D. Jendrossek. 2000. Physiological and chemical investigations into microbial degradation of synthetic poly(cis-1,4-isoprene). *Applied and Environmental Microbiology* 66:368–3685. doi: 10.1128/AEM.66.9.3680-3685.2000.

Boffey, D. 2019. Ocean cleanup device successfully collects plastic for first time. https://amp.theguardian.com/environment/2019/oct/03/ocean-cleanup-device-successfully-collects-plastic-for-first-time (accessed April 6, 2022).

Bolan, N., M. B. Kirkham, C. Halsband, D. Nugegode and Y. S. Ok. 2020. *Particulate Plastics in Terrestrial and Aquatic Environments*. Taylor & Francis: London. doi: 10.1201/9781003053071.

Bolan, N. S., M. B. Kirkham, C. Halsband, D. Nugegoda and Y. S. Ok. 2020. *Particulate Plastics in Terrestrial and Aquatic Environments*. CRC Press: Boca Raton, FL. doi: 10.1201/9781003053071.

Bombelli, P., C. J. Howe and F. Bertocchini. 2017. Polyethylene bio-degradation by caterpillars of the wax moth *Galleria mellonella*. *Current Biology* 27. doi: 10.1016/j.cub.2017.02.060.

Boots, B., C. W. Russell and D. S. Green. 2019. Effects of microplastics in soil ecosystems: Above and below ground. *Environmental Science & Technology* 53:11496–11506.

Bosker, T., L. J. Bouwman, N. R. Brun, P. Behrens and M. G. Vijver. 2019. Microplastics accumulate on pores in seed capsule and delay germination and root growth of the terrestrial vascular plant *Lepidium sativum*. *Chemosphere* 226:774–781.

Bourguignon, D. 2015. The precautionary principle: Definitions, applications and governance. European Parliament Think Tank. https://www.europarl.europa.eu/thinktank/en/document/EPRS_IDA(2015)573876.

Bradney, L., H. Wijesekara, K. N. Palansuriya, N. Obadamudalige, N. S. Bolan, Y. S. Ok, J. Rinklebe, K.-H., Kim and M. B. Kirkham. 2019. Particulate plastics as a vector for toxic trace-element uptake by aquatic and terrestrial organisms and human health risk. *Environment International* 131:104937. doi: 10.1016/j.envint.2019.104937.

Brill, F. H. H. 2010. Mikrobielle Besiedlung und Materialzerstörung von Gummimaterialien in wasserführenden Systemen. Dissertation Fakultät für Chemie, Uni Duisburg, Essen. https://bibliographie.ub.uni-due.de/servlets/DozBibEntryServlet?id=ubo_mods_00029909 (accessed June 13, 2021).

Browne, M. A., S. J. Niven, T. S. Galloway, S. J. Rowland and R. C. Thompson. 2013. Microplastic moves pollutants and additives to worms, reducing functions linked to health and biodiversity. *Current Biology* 23:2388–2392.

Brundtland, G. H. 1987. Report of the World Commission on Environment and Development - Our Common Future. https://www.are.admin.ch/dam/are/en/dokumente/nachhaltige_entwicklung/dokumente/bericht/our_common_futurebrundtlandreport1987.pdf.download.pdf/our_common_futurebrundtlandreport1987.pdf (accessed April 5, 2022).

Buczko, U., R. O. Kuchenbuch, W. Ubelhor and L. Natscher. 2012. Assessment of sampling and analytical uncertainty of trace element contents in arable field soils. *Environmental Monitoring and Assessment* 184:4517–4538.

Büks, F. and M. Kaupenjohann. 2020. Global concentrations of microplastics in soils: A review. *Soil* 6:649–662. doi: 10.5194/soil-6-649-2020.

Büks, F. and M. Kaupenjohann 2021. The impact of microplastic weathering on interactions with the soil environment: A review. *Soil Discussions* 67. doi: 10.5194/soil-2021-67.

Burke, R. 1999. Plastics & polymerization: What firefighters need to know, Firehouse 28. February 1999. https://www.firehouse.com/rescue/article/10544313/plastics-polymerization-what-firefighters-need-to-know (accessed March 22, 2022).

Cao, D. D., X. Wang, X. X. Luo, G. Liu and H. Zheng. 2020. Effects of polystyrene microplastics on the fitness of earthworms in an agricultural soil. *IOP Conference Series: Earth and Environmental Science*, 61:012148.
Cara E. F. 2019. Plastic garbage in the ocean. https://institucional.us.es/blogimus/en/2019/02/plastic-waste-in-the-ocean/.
Carstens, P. 2018. Mikroplastik: Böden noch stärker belastet als Ozeane. https://www.geo.de/natur/nachhaltigkeit/18364-rtkl-plastikmuell-mikroplastik-boeden-noch-staerker-belastet-als-ozeane (accessed March 23, 2022).
Chae, Y. and Y. J. An. 2020. Nanoplastic ingestion induces behavioral disorders in terrestrial snails: Trophic transfer effects via vascular plants. *Environmental Science: Nano* 7:975–983.
Chatterjee, A., R. Lal, L. Wielopolski, M. Z. Martin and M. H. Ebinger. 2009. Evaluation of different soil carbon determination methods. *Critical Reviews in Plant Science* 28:164–178. doi: 10.1080/07352680902776556.
Chen, H. P., Y. Wang, X. Sun, Y. Peng and X. Lin. 2020a. Mixing effect of polylactic acid microplastic and straw residue on soil property and ecological function. *Chemosphere* 243:125271.
Chen, L., D. F. B. Flynn, X. Jing, P. Kühn, T. Scholten and J.-S. He. 2015. A comparison of two methods for quantifyings soil organic carbon of Alpine grasslands on the Tibetan Plateau. *PLoS One* 10:e0126372. doi: 10.1371/journal.pone.0126372.
Chen, T., X. Cai, X. Wu, I. Karahara, L. Schreiber and J. Lin. 2011. Casparian strip development and its potential function in salt tolerance. *Plant Signaling & Behavior* 6:1499–1502. doi: 10.4161/psb.6.10.17054.
Chen, Y. L., Y. Leng, X. Liu and J. Wang. 2020b. Microplastic pollution in vegetable Farmlands of suburb Wuhan, Central China. *Environmental Pollution* 257:113449.
Cousins, I. T., R. Vestergren, Z. Wang, M. Scheringer and M. S. McLachlan. 2016. The precautionary principle and chemicals management: The example of perfluoroalkyl acids in groundwater. *Environment International* 94:331–340. doi: 10.1016/j.envint.2016.04.044.
Crane, M. and M. C. Newman. 2000. What level of effect is a no observed effect? *Environmental Toxicology and Chemistry* 19:516–519. doi: 10.1002/etc.5620190234.
Crowe, P. and P. Lynch. 2020. Fly-tipping: Organised crime behind large rise. BBC Shared Data Unit. https://www.bbc.com/news/uk-england-50660138 (accessed March 22, 2022).
Daly, N. 2018. For animals, plastic is turning the ocean into a minefield. *National Geographic*. https://www.nationalgeographic.com/magazine/article/plastic-planet-animals-wildlife-impact-waste-pollution (accessed March 22, 2022).
Datnoff, L. E. 2012. *Silicon in Agriculture*. Amsterdam, Elsevier.
de Souza, M. A. A., C. W. Lau, J. Till, W. Kloas, A. Lehmann and R. Becker 2018a. Impacts of microplastics on the soil biophysical environment. *Environmental Science & Technology* 52:9656–9665.
de Souza, M. A. A., C. W. Lau, W. Kloas, J. Bergmann, J. B. Bachelier, E. Faltin, R. Becker, A. S. Görlich and M. C. Rillig. 2019. Microplastics can change soil properties and affect plant performance. *Environmental Science & Tech*nology 53:6044–6052.
de Souza, M. A. A., W. Kloas, C. Zarfl, S. Hempel and M. C. Rillig. 2018b. Microplastics as an emerging threat to terrestrial ecosystems. *Global Change Biology* 24:1405–1416. doi: 10.1111/gcb.14020.
Deutschle, T. 2021. Kautschuk - das weisse Gold. https://www.faszination-regenwald.de/info-center/zerstoerung/kautschuk/ (accessed March 22, 2022).
Dijk, D. V. and V. J. G. Houba. 2000. Homogeneity and stability of materials distributed within the Wageningen evaluating programmes for analytical laboratories. *Communications in Soil Science and Plant Analysis* 31:11–14. doi: 10.1080/00103620009370534.
DIN 32645. 2008. 11 Chemische Analytik; Nachweis-, Erfassungs- und Bestimmungsgrenze; Ermittlung unter Wiederholbedingungen; Begriffe, Verfahren, Auswertung. https://www.beuth.de/de/norm/din-32645/110729574 (accessed Febraury 03, 2023).
Ding, J., D. Zhu, H. T. Wang, S. B. Lassen, Q.-L. Chen, M. Lv and Y.-G. Thu. 2020a. Dysbiosis in the gut microbiota of soil fauna explains the toxicity of tire tread particles. *Environmental Science & Technology* 54:7450–7460.
Douglas, G. R., A. P. Hugenholtz and D. H. Blakey. 1986. Genetic toxicology of phthalate esters: Mutagenic and other genotoxic effects. *Environmental Health Perspectives* 65:255–262. doi: 10.1289/ehp.8665255.
Dümichen, E., P. Eisentraut, C. G. Bannick, A. K. Barthel, R. Senz and U. Braun. 2017. Fast identification of microplastics in complex environmental samples by a thermal degradation method. *Chemosphere* 174:572–584. doi: 10.1016/j.chemosphere.2017.02.010.
Dümichen, E., P. Eisentraut, M. Celina and U. Braun. 2019. Automated thermal extraction-desorption gas chromatography mass spectrometry: A multifunctional tool for comprehensive characterization of polymers and their degradation products. *Journal of Chromatography* 1592:133–142. doi: 10.1016/j.chroma.2019.01.033.

ECHA. 2021. Phenol, isopropylated, phosphate (3:1). https://echa.europa.eu/substance-information/-/substanceinfo/100.066.404 (accessed March 22, 2022).

Eisentraut, P., E. Dümichen, A. S. Ruhl, M. Jekel, M. Albrecht, M. Gehde and U. Braun. 2018. Two birds with one stone: Fast and simultaneous analysis of microplastics: Microparticles derived from thermoplastics and tire wear. *Environmental Science & Technology Letters* 5:608–613.

El-Ramady, H., N. Abdalla, T. Alshaal, A. El-Henawy, M. Elmahrouk, Y. Bayoumi, T. Shalaby, M. Amer, M. Shehata, M. F. Fari, E. Domokos-Szabolcsy, A. Sztrik, J. Prokisch, E. A. H. Pilon-Smits, M. Pilon, D. Selmar, S. Haneklaus and E. Schnug. 2018. Plant nano-nutrition: Perspectives and challenges. In: Gothandadam, K. M. et al. (Eds.), *Nanotechnology, Food Security and Water Treatment*. Springer: Heidelberg, pp. 129–161. doi: 10.1007/978-3-319-70166-0_4.

EPA. 2010. Triclosan facts. https://archive.epa.gov/pesticides/reregistration/web/html/triclosan_fs.html (accessed March 22, 2022).

Eschner, K. 2017. Once upon a time, exploding billiard balls were an everyday thing. *Smithsonian Magazine* 6. https://www.smithsonianmag.com/smart-news/once-upon-time-exploding-billiard-balls-were-everyday-thing-180962751/ (accessed April 5, 2022).

Espinosa, M. J. C., A. C. Blanco, T. Schmidgall, A. K. Atanasoff-Kardjalieff, U. Kappelmeyer, D. Tischler, D. Pieper, H. J. Heipieper, and C. Eberlein, 2020. Toward biorecycling: Isolation of a soil bacterium that grows on a polyurethane oligomer and monomer. *Frontiers in Microbiology* 11, doi: 10.3389/fmicb.2020.00404.

ETRA. 2021. The European Tyre Recycling Association. https://www.etra-eu.org/.

EU. 2019. Document 32019R1009: Regulation (EU) 2019/1009 of the European Parliament and of the Council of 5 June 2019 laying down rules on the making available on the market of EU fertilising products and amending Regulations (EC) No 1069/2009 and (EC) No 1107/2009 and repealing Regulation (EC) No 2003/2003 (Text with EEA relevance). https://eur-lex.europa.eu/eli/reg/2019/1009/oj (accessed April 5, 2022).

EU. 2022. Green Deal: New proposals to make sustainable products the norm and boost Europe's resource independence. Press Release March 30, 2022. https://ec.europa.eu/commission/presscorner/detail/en/IP_22_2013 (accessed April 5, 2022).

Eurostat. 2022. Eurostat: Your to key European statistics. https://ec.europa.eu/eurostat (accessed April 5, 2022).

Falk, S., T. Stahl, A. Fliedner, H. Rüdel, K. Tarricone, H. Brunn and J. Koschorreck. 2019. Levels, accumulation patterns and retrospective trends of perfluoroalkyl acids (PFAAs) in terrestrial ecosystems over the last three decades. *Environmental Pollution* 246:921–931. doi: 10.1016/j.envpol.2018.12.095.

Farrell, P. and K. Nelson. 2013. Trophic level transfer of microplastic: *Mytilus edulis* (L.) to *Carcinusmaenas*. *Environmental Pollution* 177:1–3. doi: 10.1016/j.envpol.2013.01.046.

Fei, Y. F., S. Y. Huang, H. B. Zhang, Y. Tong, D. Wen, X. Xia, H. Wang, Y. Luo and D. Barcelo. 2020. Response of soil enzyme activities and bacterial communities to the accumulation of microplastics in an acid cropped soil. *Science of the Total Environment* 707:135634.

Fink, M. 2005. Silent but deadly! Organic farming combats flood dangers. IDW-online 125134. https://idw-online.de/en/news125134 (accessed April 5, 2022).

Fraunhofer. 2018. Kunststoffe in der Umwelt: Mikro- und Makroplastik. doi: 10.24406/UMSICHT-N-497117.

Fraunhofer. 2021. Mikroplastik und Makroplastik in der Umwelt. https://www.umsicht.fraunhofer.de/content/dam/umsicht/de/dokumente/publikationen/2021/umsicht-studie-plastikemissionen-landwirtschaft.pdf (accessed March 22, 2022).

Fuller, S. and A. Gautam. 2016. A procedure for measuring microplastics using pressurized fluid extraction. *Environmental Science and Technology* 50:5774–5780. doi: 10.1021/acs.est.6b00816.

Galloway, T. S., M. Cole and C. Lewis. 2017. Interactions of microplastic debris throughout the marine ecosystem. *Nature Ecology Evolution* 1:0116. doi: 10.1038/s41559-017-0116.

Gaylor, M. O., E. Harvey and R. C. Hale. 2013. Polybrominated diphenyl ether (PBDE) accumulation by earthworms (Eisenia fetida) exposed to bosolids-, polyurethane foam microparticle-, and penta-BDE-amended soils. *Environmental Science & Technology* 47:13831–13839.

Gestoso, I., E., P. Cacabelos, J. Ramalhosa and J. Canning-Clode. 2019. Plasticrusts: A new potential threat in the Anthropocene's rocky shores. *Science of the Total Environment* 687:413–415. doi: 10.1016/j.scitotenv.2019.06.123.

Giorgetti, L., C. Spano and S. Muccifora. 2020. Exploring the interaction between polystyrene nanoplastics and *Allium cepa* during germination: Internalization in root cells, induction of toxicity and oxidative stress. *Plant Physiology and Biochemistry* 149:170–177.

Grathwohl, P., E. Schnug, U. Köpke, and J.-U. Fischer. 2021. Plastik und andere persistente "neue" Stoffe im Boden. Position der Kommission Bodenschutz beim Umweltbundesamt (KBU). https://www.umweltbundesamt.de/sites/default/files/medien/5750/publikationen/kbu_plastik_u_andere_persistente_neue_stoffe_2aufl_bf.pdf (accessed March 25, 2022).

Hahladakisa, J. N., C. A. Velisa, R. Weber, E. Iacovidou and P. Purnella. 2017. An overview of chemical additives present in plastics: Migration, release, fate and environmental impact during their use, disposal and recycling. *Journal of Hazardous Materials* 344:179–199. doi: 10.1016/j.jhazmat.2017.10.014.

Han, X., X. Lu and R. D. Vogt. 2019. An optimized density-based approach for extracting microplastics from soil and sediment samples. *Environmental Pollution* 254:113009. doi: 10.1016/j.envpol.2019.113009 (accessed April 5, 2022).

Haneklaus, S. 1989. Strontiumgehalte in Pflanzen und Böden Schleswig-Holsteins und Bewertung von Düngungsmaßnahmen zur Verminderung der Strontiumaufnahme von Kulturpflanzen. PhD thesis, University, Kiel, Germany, p. 169.

Haneklaus, S., E. Bloem and E. Schnug. 2007. Sulfur and plant disease. In: Datnoff, L., Elmer, W., and Huber, D. (Eds.), *Mineral Nutrition and Plant Diseases*. APS Press: St. Paul, MN, pp. 101–118.

Harrison, J. P., M. Schratzberger, A. Sapp and M. Osborn. 2014. Rapid bacterial colonization of low-density polyethylene microplastics in coastal sediment microcosms. *BMC Microbiology* 14(1):1–15. doi: 10.1186/s12866-014-0232-4.

Hartmann, N. B., T. Hüffer, T. C. Thompson, M. Hasselöy, A. Verschoor, A. Daugaard, S. Rist, T. Karlsson, N. Brennholt, M. Cole, M. P. Herrling, M. C. Hess, N. P. Ivleva, A. L. Lusher and M. Wagner. 2019. Are we speaking the same language? Recommendations for a definition and categorization framework for plastic debris. *Environmental Science Technology* 53:1039–1104. https://pubs.acs.org/doi/10.1021/acs.est.8b05297 (accessed April 8, 2022).

He, D., Y. Luo, S. Lu, M. Liu, W. Song and L. Lei. 2018. Microplastics in soils: Analytical methods, pollution characteristics and ecological risks. *Trends in Analytical Chemistry* 109:163–172.

Heather, A. L., M. J. M. van Velzen, H. S. H. Brandsma, D. Vethaak, J. J. Garcia-Vallejo and M. H. Lamoree. 2022. Discovery and quantification of plastic particle pollution in human blood. *Environment International* 107199. doi: 10.1016/j.envint.2022.107199.

Hester, J. L. 2021. How could a crocodile spend years cinched by a tire? *Atlas Obscura* https://www.atlasobscura.com/articles/crocodile-trapped-in-trash?utm_source=Atlas+Obscura+Daily+Newsletter&utm_campaign=704447ae9a-EMAIL_CAMPAIGN_2021_07_21&utm_medium=email&utm_term=0_f36db9c480-704447ae9a-69239377&mc_cid=704447ae9a&mc_eid=c57b327973 (accessed March 22, 2022).

Hong, Y., Y. Zhang and W. Tan. 2021. The "neighbor avoidance effect" of microplastics on bacterial and fungal diversity and communities in different soil horizons. *Environmental Science and Ecotechnology* 8:100121. doi: 10.1016/j.ese.2021.100121.

Horton, A. A., A. Walton, D. J. Spurgeon, E. Lahive and C. Svendsen. 2017. Microplastics in freshwater and terrestrial environments: Evaluating the current understanding to identify the knowledge gaps and future research priorities. *Science of the Total Environment* 586:127–141. doi: 10.1016/j.scitotenv.2017.01.190.

Hu, E., H. Yuan, Y. Du and X. Chen. 2021. LDPE and HDPE microplastics differently affect the transport of tetracycline in saturated porous media. *Materials* 14:1757. doi: 10.3390/ma14071757.

Huang, Y., Y. R. Zhao and J. Wang. 2019. LDPE microplastic films alter microbial community composition and enzymatic activities in soil. *Environmental Pollution* 254:112983.

Huerta, L. E., B. Thapa, X. Yang, H. Gertsen, T. Salanki and V. Geissen. 2018. Decay of low-density polyethylene by bacteria extracted from earthworm's guts: A potential for soil restoration. *Science of the Total Environment* 624:753–757.

Jenner, L. V., J. M. Rotchell, R. T. Bennett, M. Cowen, V. Tentzeris and L. R. Sadofsky. 2022. Detection of microplastics in human lung tissue using μFTIR spectroscopy. *Science of the Total Environment* 831:154907. doi: 10.1016/j.scitotenv.2022.154907.

Jepsen, D. T., L. Zimmermann, L. Spengler, R. Rödig, J. Bliklen, K. Wagner, L. Struck, H. Hiestermann and H. Schulz. 2020. Kunststoffe in der Umwelt – Erarbeitung einer Systematik für erste Schätzungen zum Verbleib von Abfällen und anderen Produktenaus Kunststoffen in verschiedenen Umweltmedien. UBA Texte 189. https://www.umweltbundesamt.de/sites/default/files/medien/5750/publikationen/2020_11_24_texte_198_2020_kunststoffe_in_der_umwelt.pdf (accessed March 22, 2022).

Jiang, X. F., H. Chen, Y. C. Liao, Z. Ye, M. Li and G. Klobucar. 2019a. Ecotoxicity and genotoxicity of polystyrene microplastics on higher plant *Vicia faba*. *Environmental Pollution* 250:831–838.

Judy, J. D., M. Williams, A. Gregg, D. Oliver, A. Kumar, R. Kookana and J. A. Kirby. 2019. Microplastics in municipal mixed-waste organic outputs induce minimal short to long-term toxicity in key terrestrial biota. *Environmental Pollution* 252:522–531.

Kelpsiene, E., O. Torstensson, M. T. Ekvall, L.-A. Hansson and T. Cedervall. 2020. Long-term exposure to nanoplastics reduces life-time in Daphnia magna. *Scientific Reports* 5979:2045–2322. doi: 10.1038/s41598-020-63028-1.

Kim, S. W., D. Kim, S. W. Jeong and Y.-J. An. 2020b. Size-dependent effects of polystyrene plastic particles on the nematode *Caenorhabditis elegans* as related to soil physicochemical properties. *Environmental Pollution* 258:113740.

Kim, S. W., W. R. Waldman and M. C. Rillig. 2020c. Effects of different microplastics on nematodes in the soil environment: Tracking the extractable additives using an ecotoxicological approach. *Environmental Science & Technology* 54: 3868–13878.

Kim, Y.-N., J.-H. Yoon and K.-H. Kim. 2020a. Microplastic contamination in soil environment: A review. *Soil Science Annual* 71:300–308. doi: 10.37501/soilsa/131646.

Klein, M. and E. K. Fischer. 2019. Microplastic abundance in atmospheric deposition within the Metropolitan area of Hamburg, Germany. *Science of the Total Environment* 685:96–103. doi: 10.1016/j.scitotenv.2019.05.405.

Kong, X., D. Jin, S. Jin, Z. Wang, H. Yin, M. Xu and Y. Deng. 2018. Responses of bacterial community to dibutyl phthalate (DBP) pollution in a soil-vegetable ecosystem. *Journal of Hazardous Materials* 353:142–150. doi: 10.1016/j.jhazmat.2018.04.015.

Kumar, M., H. Chen, S. Sarsaiya, S. Qin, H. Liu, M. K. Awasthi, S. Kumar, L. Singh, Z. Zhang, N. S. Bolan, A. Pandey, S. Varjani and M. J. Taherzadeh. 2021. Current research trends on micro- and nano-plastics as an emerging threat to global environment: A review. *Journal of Hazardous Materials* 409. doi: 10.1016/j.jhazmat.2020.124967.

Kumar, M., X. Xiong, M. He, D. C. W. Tsang, J. Gupta, E. K. Stuart, H. Deyi, H. Yong, S. Ok, S. and N. S. Bolan. 2020. Microplastics as pollutants in agricultural soils. *Environmental Pollution* IF8.071. doi: 10.1016/j.envpol.2020.114980.

Lavers, J. L., P. B. Sharp, S. Stuckenbrock and A. L. Bond. 2019. Entrapment in plastic debris endangers hermit crabs. *Journal of Hazardous Materials* 387:121703. doi: 10.1016/j.jhazmat.2019.121703.

Lei, L., S. Wu, S. Lu, M. Liu, Y. Song, Z. Fu and H. Shi. 2018b. Microplastic particles cause intestinal damage and other adverse effects in zebrafish Danio rerio and nematode Caenorhabditis elegans. *Science of the Total Environment* 619/620:1–8. doi: 10.1016/j.scitotenv.2017.11.103.

Lei, L. L., M. T. Liu, Y. Song, S. Lu, C. Cao, B. Xie, H. Shi and D. He. 2018a. Polystyrene (nano) microplastics cause size-dependent neurotoxicity, oxidative damage and other adverse effects in *Caenorhabditis elegans*. *Environmental Science: Nano* 5:2009–2020.

Leslie, H. A., M. J. M. van Velzen, S. H. Brandsma, D. Vethaak, J. J. Garcia-Vallejo and M. H. Lamoree. 2022. Discovery and quantification of plastic particle pollution in human blood. *Environment International* 107199. doi: 10.1016/j.envint.2022.107199.

Lesschen, P. 2020. One step closer to a new system for measuring carbon. https://www.wur.nl/en/newsarticle/One-step-closer-to-a-new-system-for-measuring-carbon-sequestration-in-dairy-production.htm (accessed April 8, 2022).

Li, C. J., Y. Gao, S. He, H. Y. Chi, Z. C. Li, X. X. Zhou and B. Yan. 2021. Quantification of nanoplastic uptake in cucumber plants by pyrolysis gas chromatography/mass spectrometry. *Environmental Science and Technology Letters* 8:633–638. doi: 10.1021/acs.estlett.1c00369.

Li, L. Z., Q. Zhou and N. Yi. 2019. Eating vegetables can absorb and accumulate microplastics. *Chinese Science Bulletin* 64:928–934.

Li, L. Z., Y. M. Luo, R. J. Li, Q. Zhou, W. J. G. M. Peihnenburg, N. Yin, J. Yang, C. Tu and Y. C. Zhang. 2020a. Effective uptake of submicrometre plastics by crop plants via a crack-entry mode. *Nature Sustainability* 3:929–937.

Li, Z. X., Q. F. Li, R. J. Li, Y. Zhao, J. Geng and G. Wang. 2020b. Physiological responses of lettuce (*Lactuca stiva L.*) to microplastic pollution. *Environmental Science and Pollution Research* 27:30306–30314.

Lian, J. P., J. N. Wu, H. X. Xiong, A. Zeb, T. Yang, X. Su, L. Su and W. Liu. 2020. Impact of polystyrene nanoplastics (PSNPs) on seed germination and seedling growth of wheat (*Triticum aestivum* L.). *Journal of Hazardous Materials* 385:121620.

Lian, J. P., M. M. Shen and W. T. Liu. 2019. Effects of microplastics on wheat seed germination and seedling growth. *Journal of Agro-Environment Science* 38:737–745.

Liao, Y. C., M. Li and X. L. Wang and L.-J. Jiang. 2019. Effects of microplastics on the growth, physiology, and biochemical characteristics of wheat (*Triticum aestivum*). *Environmental Science* 40:4661–4666.

Lim, X. 2021. Microplastics are everywhere: But are they harmful? *Nature News Feature.* https://www.nature.com/articles/d41586-021-01143-3 (accessed March 28, 2022).

Linos, A., M. M. Berekaa, A. Steinbüchel, K. K. Kim, C. Sproer and R. M. Kroppenstedt. 2002. *Gordonia westfalica* sp. nov., a novel rubber-degrading actinomycete. *International Journal of Systematic and Evolutionary Microbiology* 52:1133–1139.

Liu, H., X. Yang, G. Liu, C. Liang, S. Xue, H. Chen, C. J. Ritsema and V. Geissen. 2017. Response of soil dissolved organic matter to microplastic addition in Chinese loess soil. *Chemosphere* 185:907–917.

Lorenz, K. and R. Lal. 2016. Soil organic carbon: An appropriate indicator to monitor trends of land and soil degradation within the SDG Framework? UBA TEXTE 77/2016.

Lozano, Y. M. and M. C. Rillig. 2020. Effects of microplastic fibers and drought on plant communities. *Environmental Science & Technology* 54:6166–6173.

Lozano, Y. M., T. Lehnert, L. T. Linck, A. Lehmann and M. C. Rillig. 2020. Microplastic shape, concentration and polymer type affect soil properties and plant biomass. *Frontiers in Plant Science*. doi: 10.3389/fpls.2021.616645.

Lumby, T. 2018. Polar bears pictured playing with plastic on remote Arctic Island. Independent. https://www.independent.co.uk/climate-change/news/polar-bears-play-plastic-svalbard-arctic-microplastics-pollution-a8444361.html (accessed March 22, 2022).

Luttikhuizen, D. T., M. C. Harmsen and M. J. A. van Luyn. 2006. Cellular and molecular dynamics in the foreign body reaction. *Tissue Engineering* 12:1955–1970. doi: 10.1089/ten.2006.12.1955.

Lwanga, E. H., H. Gertsen, H. Gooren, P. Peters, T. Salanki, M. van der Ploeg, E. Besseling, A. A. Koelmans and V. Geissen. 2017a. Incorporation of microplastics from litter into burrows of *Lumbricus terrestris*. *Environmental Pollution* 220:523–531.

Mai, L., L. J. Bao, L. Shi, S. Charles, C. S. Wong and E. Y. Zeng. 2018. A review of methods for measuring microplastics in aquatic environments. *Environmental Science and Pollution Research* 25:11319–11332. doi: 10.1007/s11356-018-1692-0.

Meides, M., T. Menzel, B. Poetzschner, M. G. J. Löder, U. Mansfeld, P. Strohriegl, V. Altstaedt, and J. Senker. 2021. Reconstructing the environmental degradation of polystyrene by accelerated weathering. *Environmental Science & Technology* 55:7930–7938. doi: 10.1021/acs.est.0c07718.

Meng, F., X. Yang, M. Riksen, M. Xu and V. Geissen. 2021. Response of common bean (*Phaseolus vulgaris* L.) growth to soil contaminated with microplastics. *Science of the Total Environment* 755:142516.

Mintenig, S. M., I. Int-Veen, M. G. J. Löder, S. Primpke and G. Gerdts. 2017. Identification of microplastic in effluents of waste water treatment plants using focal plane array-based micro-Fourier-transform infrared imaging. *Water Research* 108:365–372. doi: 10.1016/j.watres.2016.11.015.

Möller, J. N., M. G. J. Löder and C. Laforsch. 2020. Finding microplastics in soils: A review of analytical methods. *Environmental Science & Technology* 54:2078–2090. doi: 10.1021/acs.est.9b04618.

Monteiro, A. and S. Santos. 2022. Sustainable approach to weed management: The role of precision weed management. *Agronomy* 118. doi: 10.3390/agronomy12010118.

Morley-Fletcher, S., M. Rea, S. Maccari and G. Laviola. 2003. Environmental enrichment during adolescence reverses the effects of prenatal stress on play behaviour and HPA axis reactivity in rats. *European Journal Neuroscience* 18:3367–3374. doi: 10.1111/j.1460-9568.2003.03070.x.

Napierska, D., L. Thomassen, D. Lison, J. A. Martens and P. H. Hoet. 2010. The nanosilica hazard: Another variable entity. *Particle and Fibre Toxicology* 39:1743–8977. doi: 10.1186/1743-8977-7-39.

Nätscher, L., W. Übelhör and S. Grimm. 2013. Ringversuch zur Bestimmung von Corg bei 550°C. Fachgruppe II Bodenuntersuchung Endbericht. https://www.vdlufa.de/Dokumente/Ringversuche/Endbericht_RV_Corg_2012_2013.pdf (accessed April 8, 2022).

Nawrath, C., L. Schreiber, R. B. Franke, N. Geldner, J. J. Reina-Pinto and L. Kunst. 2012. Apoplastic diffusion barriers in Arabidopsis. *Arabidopsis Book*. doi: 10.1199/tab.0167.

NGS. 2022. Encyclopedia national geographic society: Anthropocene. https://www.nationalgeographic.org/encyclopedia/anthropocene/1101 (accessed March 23, 2022).

Nizzetto, L. 2018. Microplastics in agricultural soils: A reason to worry? https://www.niva.no/en/news/microplastics-in-agricultural-soils-a-reason-to-worry (accessed March 22, 2022).

Nizzetto, L., A. Ricu, K. van Gestel, S. Selonen, J. Crossman, M. Futter and A. J. Kokalj. 2021. Impacts of microplastic in agrosystems and stream environments: Results from an international research project on sources, behaviour and ecological impacts of microplastics from sewage sludge application to agricultural soils. NIVA-rapport 7632–2021. https://kommunikasjon.ntb.no/pressemelding/grunn-til-bekymring-for-mikroplastinnholdet-i-europeisk-jordbruksjord?publisherId=17847668&releaseId=17910798 (accessed March 22, 2022).

Nizzetto, L., M. S. Futter and M. Langaas. 2016a. Are agricultural soils dumps for microplastics of urban origin? *Environmental Science and Technology* 50:10777–10779. doi: 10.1021/acs.est.6b04140.

Nizzetto, L., S. M. Langaas, and M. Futter. 2016b. Pollution: Do microplastics spill on to farm soils? *Nature* 537:488. doi: 10.1038/537488b.

Novamont. 2022. Mater-Bi. https://www.novamont.com/eng/mater-bi (accessed March 28, 2022).

NSP. 2023. Fluorescent polymers. https://www.nanosoftpolymers.com/product-category/fluorescent-dyes/ (accessed Febraury 03, 2023).
O'Kelly, B., A. El-Zein, X. Liu, A. Patel, X. Fei, S. Sharma, A. Mohammad, V. Goli, J. J. Wang, D. Li, Y. Shi, L. Xiao, G. Kuntikana, B. S. Shashank, T. S. Sarris, R. B. Hanumantha, A. M. O. Mohamed, E. K. Paleologos, M. M. Nezhad and D. N. Singh. 2021. Microplastics in soils: An environmental geotechnics perspective. *Journal of Environmental Geotechnics* 8:586–618. doi: 10.1680/jenge.20.00179.
Olle, M. and E. Schnug. 2015. The influence of foliar applied silicic acid on N, P, K, Ca and Mg concentrations in field peas. *Journal of Cultivated Plants* 68:7–10. doi: 10.5073/JFK.2016.01.02.
Partow, H., C. Lacroix, S. Le Floch and L. Alcaro. 2021. X-press pearl maritime disaster Sri Lanka: Report of the UN Environmental Advisory Mission. UN-Joint Environment Unit, July 2021. https://postconflict.unep.ch/Sri%20Lanka/X-Press_Sri%20Lanka_UNEP_27.07.2021_s.pdf (accessed March 22, 2022).
Petersen, K. S. 2020. Microplastics in farm soils: A growing concern ocean. *Environmental Health News*. https://www.ehn.org/plastic-in-farm-soil-and-food-2647384684.html (accessed March 28, 2022).
Pinto-Bazurco, J. F. 2020. The precautionary principle. IISD International Institute for Sustainable Development. *Deep Dive*. https://www.iisd.org/articles/deep-dive/precautionary-principle (accessed April 5, 2022).
Plastics Europe. 2020a. Plastics: The history. https://plasticseurope.org/plastics-explained/history-of-plastics/ (accessed March 22, 2022).
Plastics Europe. 2020b. Plastics: The facts. https://plasticseurope.org/knowledge-hub/plastics-the-facts-2020/ (accessed March 22, 2022).
Pohanka, M. 2019. Copper and copper nanoparticles toxicity and their impact on basic functions in the body. *Bratisl Lek Listy* 120, 397–409. doi: 10.4149/BLL_2019_065.
Qi, Y. L., X. M. Yang, A. M. Pelaez, E. H. Lwanga, N. Beriot, H. Gertsen, P. Garbeva and V. Geissen. 2018. Macro- and microplastics in soil-plant system: Effects of plastic mulch film residues on wheat (*Triticum aestivum*) growth. *Science of the Total Environment* 645:1048–1056.
Qian, H., M. G. Zhang, T. Liu, Q. Lu, B. Qu, B. Du and X. L. Pan. 2018. Effects of soil residual plastic film on soil microbial community structure and fertility. *Water Air & Soil Pollution* 229:261. doi: .org/10.1007/s11270-018-3916-9.
Qin, F., J. Du, J. Gao, G. Liu, Y. Song, A. Yang, H. Wang, Y. Ding and Q. Wang 2020. Bibliometric profile of global microplastics research from 2004 to 2019. *International Journal of Environmental Research and Public Health* 17:5639. doi: 10.3390/ijerph17165639.
R4WO. 2015. Race for water odyssey. An environmental and scientific expedition. https://www.raceforwater.org/en/race-for-water-odyssey-2015/ (accessed April 3, 2022).
Radford, F., L. M. Zapata-Restrepo, A. A. Horton, M. D. Hudson, P. J. Shaw and I. D. Williams. 2021. Developing a systematic method for extraction of microplastics in soils. *Analytical Methods* 13:1695–1705. doi: 10.1039/D0AY02086A.
RAM. 2022. Plastic recycling: Global market trajectory & analytics. https://www.researchandmarkets.com/reports/5028078/plastic-recycling-global-market-trajectory-and?utm_source=CI&utm_medium=PressRelease&utm_code=7fh2cv&utm_campaign=1626322+-+Global+Plastic+Recycling+Markets%2c+2021-2026+-+Government+Regulations+%26+Efforts+to+Promote+Use+of+Recycled+Plastic+Products+Drives+Market+Growth&utm_exec=chdo54prd (accessed April 2, 2022).
Rankin, D. 2021. PBT Stoffe und Verkaufsverbote in den USA: Vermeidbare Überraschungen? https://www.umco.de/de/blog/artikel/PBT-Stoffe-Verkaufsverbote-USA.html (accessed March 22, 2022).
Rillig, M. C. 2012. Microplastic in terrestrial ecosystems and the soil. *Environmental Science & Technology* 46:6453–6454. doi: 10.1021/es302011r.
Rillig, M. C. and M. Bonkowski 2018. Microplastic and soil protists: A call for research. *Environmental Pollution* 241:1128–1131.
Rillig, M. C., L. Ziersch and S. Hempel. 2017. Microplastic transport in soil by earthworms. *Scientific Reports* 7:1362. doi: 10.1038/s41598-017-01594-7.
Rodriguez-Seijo, A., J. Lourenco, T. A. P. Rocha-Santos, J. da Costa, A. C. Duarte, H. Vala and R. Pereira. 2017. Histopathological and molecular effects of microplastics in *Eisenia andrei* Bouché. *Environmental Pollution* 220:495–503.
Rose, K. and A. Steinbuechel. 2005. Biodegradation of natural rubber and related compounds: Recent insights into a hardly understood catabolic capability of microorganisms. *Applied and Environmental Microbiology* 71(6):2803–2812. doi: 10.1128/AEM.71.6.2803-2812.2005.
Rühling, I. 1999. Räumlich-zeitliche Variabilität von physikochemischen Bodeneigenschaften und Pflanzenmerkmalen in einer norddeutschen Landschaft. *Landbauforschung Völkenrode SH205, Bundesforschungsanstalt für Landwirtschaft (FAL)*, Braunschweig.

Salvador, C. F., A. Turra and J. Baruque-Ramos. 2017. Synthetic fibers as microplastics in the marine environment: A review from textile perspective with a focus on domestic washings. *Science of the Total Environment* 598:1116–1129.

Salvige, R. 2022. 'Forever chemicals': What are PFAS and what risk do they pose? *The Guardian* February 8, 2022. https://www.theguardian.com/environment/2022/feb/08/what-are-pfas-forever-chemicals-what-risk-toxicity (accessed March 23, 2022).

Sánchez, C. 2020. Fungal potential for the degradation of petroleum-based polymers: An overview of macro- and microplastics biodegradation. *Biotechnology Advances* 40:107501. doi: 10.1016/j.biotechadv.2019.107501.

Sanderman, J., T. Hengl and G. J. Fiske. 2017. Soil carbon debt of 12,000 years of human land use. *PNAS* 114:9575–9580.

Savoca, M. S., M. E. Wohlfeil, S. E. Ebeler and G. A. Nevitt. 2016. Marine plastic debris emits a keystone info chemical for olfactory foraging seabirds. *Science Advances* 2:1–8. doi: 10.1126/sciadv.1600395.

Schnug, E. and S. H. Haneklaus. 2002. Beitrag des ökologischen Landbaus zum vorbeugenden Hochwasserschutz. *Landbauforschung Völkenrode* 52:197–203.

Seltenrich, N. 2013. Nanosilver: Weighing the risks and benefits. *Environmental Health Perspectives* 121:220–225. doi: 10.1289/ehp.121-a220.

Silver, K. 2018. Second-hand toys could harm children, scientists discover. https://www.bbc.com/news/health-42832978 (accessed March 22, 2022).

Slepetiene, A., J. Slepetys and I. Liaudanskiene. 2008. Standard and modified methods for soil organic carbon determination in agricultural soils. *Agronomy Research* 6:543–554 https://agronomy.emu.ee/vol062/p6211.pdf (accessed April 8, 2022).

Solly, E. F., V. Weber, S. Zimmermann, W. Lorenz, F. Hagedorn and M. W. I. Schmidt. 2020. A critical evaluation of the relationship between the effective cation exchange capacity and soil organic carbon content in Swiss forest soils. *Frontiers in Forests and Global Change* 3 doi: 10.3389/ffgc.2020.00098.

Sonderegger, T. and S. Pfister. 2021. Global assessment of agricultural productivity losses from soil compaction and water erosion. *Environment Science & Technology* 55:12162–12171.

Song, Y., C. Ca, R. Qiu, J. Hu, M. Liu, H. Shi, K. M. Raley-Susman and D. He. 2019. Uptake and adverse effects of polyethylene terephthalate microplastic fibers on terrestrial snails (*Achatina fulica*) after soil exposure. *Environment Pollution* 250:447–455.

Sparovek, G., S. Hornink and E. Schnug. 2001. Solution for worn-out tires, gully erosions, forests and dengue fever in Brazil. *Landbauforschung Volkenrode* 51:95–99.

Steinmetz, Z., C. Wollmann, M. Schaefer, C. Buchmann, J. David, J. Troeger, K. Munoz, O. Fror and G. E. Schaumann. 2016. Plastic mulching in agriculture. Trading short-term agronomic benefits for long-term soil degradation? *Science of the Total Environment* 550:690–705. doi: 10.1016/j.scitotenv.2016.01.153.

Subrahmaniyan, K., W. Zhou and P. Veeramani 2011. Weed control through degradable, herbicidal plastics and organic mulches and their effect on crop growth and yield of winter rapeseed (*Brassica napus*). *Indian Journal of Agricultural Sciences* 81:348–352.

Sun, M., M. Ye, W. Jiao, Y. Feng, P. Yu, M. Liu, J. Jiao, X. He, K. Liu, Y. Zhao, J. Wu, X. Jiang and F. Hu. 2018. Changes in tetracycline partitioning and bacteria/phage-comediated ARGs in microplastic-contaminated greenhouse soil facilitated by sophorolipid. *Journal of Hazardous Materials* 345:131–139. doi: 10.1016/j.jhazmat.2017.11.036.

Symonds, T. 2021. Grenfell cladding: Did it fail safety tests 13 years before fire? https://www.bbc.com/news/uk-59434012 (accessed April 6, 2022).

Teuten, E. L., J. M. Saquing, D. R. U. Knappe, M. A. Barlaz, S. Jonsson, A. Björn, S. J. Rowland, R. C. Thompson, T. S. Galloway, R. Yamashita, D. Ochi, Y. Watanuki, C. Moore, P. H. Viet, T. S. Tana, M. Prudente, R. Boonyatumanond, M. P. Zakaria, K. Akkhavong, Y. Ogata, H. Hirai, S. Iwasa, K. Mizukawa, Y. Hagino, A. Imamura and M. Saha. 2009. Transport and release of chemicals from plastics to the environment and to wildlife. *Philosophical Transactions of the Royal Society B:Biological Sciences* 364(1526):2027–2045. doi: 10.1098/rstb.2008.0284.

Tournier, V., C. M. Topham, A. Gilles, B. David, C. Folgoas, E. Moya-Leclair, E. Kamionka, M.-E. Desrousseaux, H. Texier, S. Gavalda, M. Cot, E. Guémard, M. Dalibey, J. Nomme, G. Cioci, S. Barbe, M. Chateau, I. André, S. Duquesne and A. Marty. 2020. An engineered PET depolymerase to break down and recycle plastic bottles. *Nature* 580:216–219. doi: 10.1038/s41586-020-2149-4.

Turner, A., C. Wallerstein, R. Arnold and D. Webb. 2019. Marine pollution from pyroplastics. *Science of the Total Environment* 694:133610. doi: 10.1016/j.scitotenv.2019.133610.

UBA. 2020. Humusstatus der Böden. https://www.umweltbundesamt.de/daten/flaeche-boden-land-oekosysteme/boden/humusstatus-der-boeden#humusfunktionen-und-gehalte-von-boden.

UNEP. 2022a. How plastic is infiltrating the world's soils. https://www.unep.org/news-and-stories/story/how-plastic-infiltrating-worlds-soils (accessed April 3, 2022).

USTPO. 1989. Stockosorb. Trademark a serial number of 73797493. https://alter.com/trademarks/stocko-sorb-73797493 (accessed March 22, 2022).

van der Heuvel, M. 2021. Überraschende Spuren: Wo Reifenabrieb landet. https://www.ingenieur.de/fachmedien/umweltmagazin/abfall-und-kreislauf/ueberraschende-spuren-wo-reifenabrieb-landet-und-was-hersteller-unternehmen-sollten/ (accessed March 22, 2022).

Völker, C., J. Kramm and M. Wagner. 2020. On the creation of risk: Framing of microplastics risks in science and media. *Global Challenges* 4:1900010. doi: 10.1002/gch2.201900010.

Vollertsen, J. and A. Vianello. 2019. Knowledge gaps in microplastics research: Wastewater treatment plant sludge and application on farm lands. Aalborg University. https://plastikviden.dk/media/212209/knowledge-gaps-in-microplastics-research.pdf (accessed March 22, 2022).

Walinga, I., M. Kithome, I. Novozamsky, V. J. G. Houba and V. D. J. J. Lee. 1992. Spectrophotometric determination of organic carbon in soil. *Communications in Soil Science and Plant Analysis* 23:1935–1944. doi: 10.1080/00103629209368715.

Wan, Y., C. X. Wu, Q. Xue and X. Hui. 2019. Effects of plastic contamination on water evaporation and desiccation cracking in soil. *Science of the Total Environment* 654:576–582.

Wang, J., M. K. Huang, Q. Wang, Y. Sun, Y. Zhao and Y. Huang. 2020a. LDPE microplastics significantly alter the temporal turnover of soil microbial communities. *Science of the Total Environment* 726:138682.

Wang, L., W.-M. Wu, N. S. Bolan, D. C. W. Tsang, Y. Li, M. Qin and D. Hou. 2020b. Environmental fate, toxicity and risk management strategies of nanoplastics in the environment: Current status and future perspectives. *Journal of Hazardous Materials* 401:123415. doi: 10.1016/j.jhazmat.2020.123415.

Wang, Z. G., Y. M. You, W. H. Xu, W. J. Chen, J. Zeng, X. S. Zhao, Y. P. S and L. Y. Wu. 2018. Dimethyl phthalate altered the microbial metabolic pathways in a Mollisol. *European Journal of Soil Science* 69(3):439–449. doi: 10.1111/ejss.12545.

Watteau, F., M.-F. Dignac, A. Bouchard, A. Revallier, and S. Houot. 2018. Microplastic detection in soil amended with municipal solid waste composts as revealed by transmission electronic microscopy and pyrolysis/GC/MS. *Frontiers in Sustainable Food Systems* 2. doi: 10.3389/fsufs.2018.00081.

Wei, R. and W. Zimmermann. 2017. Microbial enzymes for the recycling of recalcitrant petroleum-based plastics: How far are we? *Microbial Biotechnology* 10:1308–1322.

Weisberger, M. 2022. 5 million shipwrecked Legos still washing up 25 years after falling overboard. *Life Sience*. https://www.livescience.com/great-lego-spill-25th-anniversary (accessed March 22, 2022).

WEL. 2022. Biodegradability & compostability. Woods End Laboratories. https://woodsend.com/biodegradibility/biodegradability/ (accessed March 28, 2022).

Wellmitz, J. and M. Gluschke. 2005. Leitlinie zur Methodenvalidierung. Texte 01/05 Fachgruppe II Bodenuntersuchung, Methodenvalidierung. https://www.umweltbundesamt.de/sites/default/files/medien/publikation/long/2832.pdf (accessed April 8, 2022).

Wieseler, F., K. Hund-Rinke, S. Gäth, E. George, J. Greef, L. Hoelzl, K. J. Huelsbergen, F. Holz, R. Pfeil, K. Severin, H.-G. Frede, B. Blum, H. Schenkel, W.-J. Horst, K. Dittert, T. Ebertseder, B. Osterburg, W. Philipp and M. Pietsch. 2015. Use of organic fertilizers and organic wastes in agriculture. Application of Organic Fertilisers and Organic Residual Materials in Agriculture. *Position Paper of the Scientific Advisory Board on Fertiliser Issues at the Federal Ministry of Food and Agriculture*. https://www.researchgate.net/publication/303568501_Use_of_organic_fertilizers_and_organic_wastes_in_agriculture (accessed April 12, 2022).

WIKI. 2021. Great Pacific garbage patch. https://en.wikipedia.org/wiki/Great_Pacific_garbage_patch (accessed March 22, 2022).

WIKI. 2022. Richard Thompson. https://en.wikipedia.org/wiki/Richard_Thompson_(marine_biologist) (accessed March 22, 2022).

Wirth, H. 1967. New applications for polystyrene foam. *Journal of Cellular Plastics* 3:463. doi: 10.1177/0021955X6700301005.

Wright, S. L., D. Rowe, R. C. Thompson and T. S. Galloway. 2013. Microplastic ingestion decreases energy reserves in marine worms. *Current Biology* 23:R1031–R1033. doi: 10.1016/j.cub.2013.10.068.

Wu, X., J. Lu, M. Du, X. Xu, J. Beiyuan, B. Sarkar, N. S. Bolan, W. Xu, S. Xu, X. Chen, F. Wu and H. Wang 2021. Particulate plastics-plant interaction in soil and its implications: A review. *Science of the Total Environment* 792:148337. doi: 10.1016/j.scitotenv.2021.148337.

Wuttke, S. 1989. Möglichkeiten zur Verbringung industrieller Produktionsabfälle auf landwirtschaftlicher Böden am Beispiel von Gummi. Diplomarbeit, Institut für Pflanzenernährung und Bodenkunde der Landwirtschaftlichen Fakultät der Christian Albrechts Universität zu Kiel. https://www.researchgate.net/publication/352465669_Diplomarbeit_Sabine_Wuttke_Moglichkeiten_zur_Verbringung_industrieller_Produktionsabfalle_auf_landwirtschaftlicher_Boden_am_Beispiel_von_Gummi_Diplomarbeit_bei_Ewald_Schnug_am_Institut_fur_Pflanzenern (accessed June17, 2021).

WWF. 2018. The mediterranean at risk of becoming 'a sea of plastic', WWF warns. https://wwf.panda.org/wwf_news/?329099/The-Mediterranean-at-risk-of-becoming-a-sea-of-plastic-WWF-warns (accessed March 22, 2022).

Yan, Y. Y., Z. H. Chen, F. X. Zhu, C. Zhu, C. Wang and C. Gu. 2020. Effect of polyvinyl chloride microplastics on bacterial community and nutrient status in two agricultural soils. *Bulletin of Environmental Contamination and Toxicology* 7:602–609. doi: 10.1007/s00128-020-02900-2.

Yang, X., E. H. Lwanga, A. Bemani, H. Gertsen, T. Salanki, X. Guo, H. Fu, S. Xue, C. Ritsema and V. Geissen. 2018. Biogenic transport of glyphosate in the presence of LDPE microplastics: A mesocosm experiment. *Environmental Pollution* 245:829–835.

Yi, M. L., S. H. Zhou, L. L. Zhan and S. Ding. 2020. The effects of three different microplastics on enzyme activities and microbial communities in soil. *Water Environment Research* 93:24–32.

Zafar 2022. Pyrolysis of scrap tires. https://www.bioenergyconsult.com/pyrolysis-of-scrap-tires/ (accessed April 5, 2022).

Zalasiewicz, J., C. N. Waters, J. A. Ivar do Sul, P. L. Corcoran, A. D. Barnosky, A. Cearreta, M. Edgeworth, A. Galuszka, C. Jeandel, R. Leinfelder, J. R. McNeill, W. Steffen, C. Summerhayes, M. Wagreich, M. Williams, A. P. Wolfe and Y. Yonan. 2016. The geological cycle of plastics and their use as a stratigraphic indicator of the Anthropocene. *Anthropocene* 13:4–17. doi: 10.1016/j.ancene.2016.01.002.

Zalasiewicz, J., C. N. Waters, M. Williams, A. D. Barnosky, A. Cearreta, P. Crutzen, E. Ellis, M. A. Ellisblan, J. Fairchild, J. Grinevald, P. K. Haff, H. Hajdas, R. Leinfelder, J. McNeill, E. O. Odada, C. Poirier, D. Richter, W. Steffen, C. Summerhayes, J. P. M. Syvitski, D. Vidas, M. Wagreich, S. L. Wing, A. P. Wolfe, A. Zisheng and N. Oreskes. 2015. When did the Anthropocene begin? A mid-twentieth century boundary level is stratigraphically optimal. *Quaternary International* 383:196–203. doi: 10.1016/j.quaint.2014.11.045.

Zalasiewicz, J., M. Williams, R. Fortey, A. Smith, T. L. Barry, A. L. Coe, P. R. Bown, P. F. Rawson, A. Gale, P. Gibbard, F. J. Gregory, M. W. Hounslow, A. C. Kerr, P. Pearson, R. Knox, J. Powell, C. Waters, J. Marshall, M. Oates and P. Stone. 2011. Stratigraphy of the Anthropocene. *Philosophical Transactions of the Royal Society* 369:1036–1055. doi: 10.1098/rsta.2010.0315.

Zettler, E. R., T. J. Mincer and L. A. Amaralzettler. 2013. Life in the "plastisphere" microbial communities on plastic marine debris. *Environmental Science and Technology* 47:7137–7146. doi: 10.1021/es401288x.

Zhang, G. S., F. X. Zhang and X. T. Li. 2019a. Effects of polyester microfibers on soil physical properties: Perception from a field and a pot experiment. *Science of the Total Environment* 670:1–7.

Zhang, G. S. and Y. F. Liu. 2018. The distribution of microplastics in soil aggregate fractions in Southwestern China. *Science of the Total Environment* 642:12–20. doi: 10.1016/j.scitotenv.2018.06.004.

Zhao, L., M. Qu, G. Wongand, and D. Wang. 2017. Transgenerational toxicity of nanopolystyrene particles in the range of μg L-1 in the nematode *Caenorhabditis elegans*. *Environmental Science: Nano* 4:2356–2366.

Zhu, B. K., Y. M. Fang, D. Zhu, P. Christie, X. Ke, and Y. G. Zhu. 2018c. Exposure to nanoplastics disturbs the gut microbiome in the soil oligochaete Enchytraeuscrypticus. *Environmental Pollution* 239:408–415. doi: 10.1016/j.envpol.2018.04.017.

Zhu, D., Q. F. Bi and Q. Xiang. 2018b. Trophic predator-prey relationships promote transport of microplastics compared with the single *Hypoaspis aculeifer* and *Folsomia candida*. *Environmental Pollution* 235:150–154.

Zhu, D., Q. L. Chen, X. L. An, X. R. Yang, P. Christie, X. Ke, L. H. Wu and Y. G. Zhu. 2018a. Exposure of soil collembolans to microplastics perturbs their gut microbiota and alters their isotopic composition. *Soil Biology and Biochemistry* 116:302–310. doi: 10.1016/j.soilbio.2017.10.027.

Ziajahromi, S., P. A. Neale and F. Leusch. 2016. Wastewater treatment plant effluent as a source of microplastics: Review of the fate, chemical interactions and potential risks to aquatic organisms. *Water Science & Technology* 74:2253–2269. doi: 10.2166/wst.2016.41.

Zubris, K. A. V. and B. K. Richards. 2005. Synthetic fibers as an indicator of land application of sludge. *Environmental Pollution* 138:201–211 doi: 10.1016/j.envpol.2005.04.013.

8 Elemental Loads with Phosphate Fertilizers

A Constraint for Soil Productivity?

Ewald Schnug[1,4], Yajie Sun[2], Liankai Zhang[3], Heike Windmann[1,4], Bernd G. Lottermoser[5], Andrea E. Ulrich[4], Roland Bol[2], Miyuki Maekawa[4], and Silvia H. Haneklaus[4]

[1]Faculty of Life Sciences, Department of Plant Physiology and Biochemistry, Technical University, Braunschweig, Germany
[2]Institute of Bio- and Geosciences, Agrosphere (IBG-3), Forschungszentrum Jülich, Jülich, Germany
[3]Institute of Karst Geology (CAGS), Guilin, P.R. China
[4]Institute for Crop and Soil Science, Julius-Kühn-Institut, Federal Research Centre for Cultivated Plants, Braunschweig, Germany
[5]MRE – Institute of Mineral Resources Engineering, RWTH Aachen University, Aachen, Germany

CONTENTS

ABBREVIATIONS

Cv Coefficient of variation
EU European Union
LLD Lower limit of detection
REE Rare earth elements
PCA Principal component analysis
PG Phosphogypsum

DOI: 10.1201/9781003093565-8

8.1 INTRODUCTION

Phosphorus (P) is essential for any kind of life, and P fertilization is pivotal for food quality and food security (Guedes et al., 2017). However, better plant growth and higher crop yields may come at the expense of adding unwanted chemicals to the soil with potential damage for health and the environment. For longer than a century, mineral fertilizers based on phosphate rock have been the most important P source in plant production (Cardoso, 2013), but not much attention has been paid to the fact that rock phosphates, and consequently, all fertilizers derived from their processing contain almost all known chemical elements, which makes them the most important source for soil pollution. Contamination by heavy metals originating from phosphate fertilizers has become a concern in several countries (Azzi et al., 2017; Kratz et al., 2016; Taylor, 2017) and, thus, needs to be considered as a significant chemical constraint for soil productivity.

Since fertilization became more widespread in agriculture during the 20th century, in 1964 Weber (1964) depicted the devil fertilizing while death is holding the fertilizer bag – an artist's imagination and vision of soil constraints caused by fertilization.

Agricultural production strategies combine the economic, social, and environmental expectations of various stakeholder groups (Andrew and Hubert, 2000). The so-called 4R Nutrient Stewardship is a conceptual approach to tackle chemical constraints for soil productivity, and it provides a framework to achieve increased production and profitability, enhanced environmental protection, and improved sustainability in order to keep collateral damages at bay (Four Rs, 2021). The concept addresses (i). the right fertilizer source; (ii). the right fertilizer rate; (iii). the right fertilization time; and (iv). the right fertilizer placement (Four Rs, 2021). The present chapter concentrates on issues of the right P fertilizer source and on the challenges of inorganic contaminants in fertilizers. The chemical composition of different P fertilizers will be discussed with respect to the geochemical genesis of their sources and their contribution to elemental balances in different production systems.

The large variety of elements in phosphate rocks is a challenge when trying to produce clean fertilizers. However, if certain elements with distinctive commercial value can be harvested during fertilizer processing, they will cover the increased production costs.

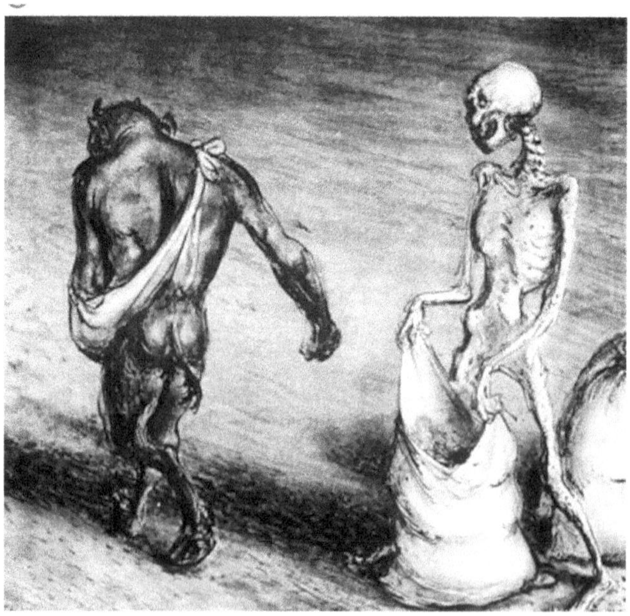

FIGURE 8.1 Death and Devil – devil fertilizing while death is holding the fertilizer bag (Weber, 1964).

Last, but not least, any human activity that interferes with the sustainable use of limited natural resources, like productive soils, requires a constant development and adaption of a legal framework, which will be the subject of the last section of this chapter.

8.2 HEAVY METALS IN ROCK PHOSPHATES AND P FERTILIZERS

Phosphates are the most versatile input to agricultural soils when it comes to the number of elements accompanying the main nutrient in a fertilizer. The heavy metal concentration in mineral P fertilizers is dependent on the phosphate rock used as raw material (Freitas et al., 2009). All phosphates originate primarily from igneous rock-forming processes, which take part in the biological cycle as dissolved orthophosphate after mineralization (Nash, 1984). Following biological turnover, phosphates re-enter the geo-biochemical cycle in the form of defected, sedimentary apatite deposits. The process of apatite formation has a high affinity for heavy metals (Kölle, 2017), so that, in the course of time, apatites contain virtually every heavy metal present in ambient waters. This makes phosphates not only a source of an essential plant nutrient but also a source of pollution and potential harm to living organisms.

Windmann (2020) compiled one of the largest databases worldwide for elemental compositions of rock P and P fertilizer sources, and it stems from the fertilizer collection maintained by the Institute for Crop and Soil Science (JKI-PB) in Braunschweig, Germany. The oldest mineral fertilizer specimen dates back to the 1970s. The collection comprises data from P fertilizers, as defined by the German Fertilizer Act (Anonymous, 2019). The Act is a mirror of the corresponding law of the European Union (EU, 2019; Kuhn, 2021). The database contains concentrations for 45 of the natural elements of the periodic system. One hundred and fifty specimens from the dataset are phosphate rocks, 485 are commercial mineral P fertilizers, and 135 are organic-mineral P fertilizers (Windmann, 2020).

Table 8.1 shows the elemental concentrations in 150 phosphate rock samples, of which 22 are of igneous origin and 128 are of sedimentary origin. Phosphate rocks of sedimentary origin contain more than 10 times higher concentrations of Bi, Cd, and Ca; 5–10 times higher concentrations of U and Cr; and 2–5 times more S, Tl, Zn, Sb, and B than igneous phosphates. Twofold higher concentrations occur for B, As, Se, and Ni. In comparison, the samples of igneous origin contain >400 times more Fe, Mg, Al, and Na, 5–10 times more Co, Mn, Nd, Sm, Ce, Pr, and >2 times higher concentrations of Eu, Gd, Tb, Ti, K, La, Dy, Sr, Ho, Cu and Er (Table 8.1).

The coefficient of variation (cv) in Table 8.1 reveals the variability of the elements in the phosphate rock samples. The highest cv values (>200) were found in sedimentary rock phosphates, and, sorted in decreasing order, were determined for Mn, Sn, Ca, K, Fe, P, Co, and Pb, and the lowest (< 100) were determined for Gd, La, Na, Lu, Yb, Tb, Tm, Er, Dy, Zn, Sr, U, Cr, Sb, B, Ni, and Be. In contrast, the highest cv values (>200) were found for Ni, Cd, Sn, S, Sb, Th, and Fe and the lowest (<100) were found for Pb, Al, Li, Dy, Mn, Cu, Ho, Er, Lu, Ti, Yb, Tm, Be, Zn, and Mg in igneous rock phosphates (Table 8.1).

Almost all mineral P fertilizers are manufactured from phosphate rocks (Freitas et al., 2009; Kratz and Schnug, 2006). With increasing public awareness and political pressure on the sustainable use of globally limited P resources, farmyard manures and mixtures of both, so-called organo-mineral fertilizers, are replacing more and more the industrially manufactured P fertilizers from phosphate rocks as sources for fertilizer P (Kratz et. al., 2014).

Table 8.2 compares the elemental concentrations in manufactured mineral and organo-mineral ($n=65$) fertilizers, according to definitions of the German Fertilizer Ordinance (Anonymous, 2019; Kuhn, 2021), and farmyard manures ($n=126$). The reason for including organic P sources and animal manures in Table 8.2 is that the rational use of organic wastes on farms is in the limelight of discussions, and heavy metal loads discharged by fertilizer practices are a centerpiece in the discussions concerning sustainable agricultural management.

TABLE 8.1
Geometric Mean[a] and Coefficients of Variation (cv)[b] of the Mineral Content in 150 Igneous and Sedimentary Rock Phosphates (Windmann, 2020 Revised)

	Igneous Phosphate Rocks		Sedimentary Phosphate Rocks		
Element	**mean (mg/kg)**	**cv (%)**	**mean (mg/kg)**	**cv (%)**	**Significance[c]**
Al	2,726	91	2,642	113	ns
As	3.33	102	5.59	106	ns
B	400	138	398	65	***
Be	0.488	65	1.21	41	***
Bi	<LLD[d]	109	0.090	159	ns
Ca	201,817	48	2,162,246	317	*
Cd	0.397	285	7.96	108	***
Co	12.0	145	1.20	248	***
Cr	14.7	157	87.3	68	***
Cu	34.5	83	13.7	179	**
Fe	7,160	206	2,466	269	*
K	11,393	131	3,189	304	**
Li	8.30	90	6.65	105	**
Mg	6,630	12	4,499	127	*
Mn	377	86	43.9	597	ns
Mo	5.04	163	3.80	91	***
Na	841	105	1,902	86	ns
Ni	14.6	305	19.8	62	ns
P	141,801	29	114,373	265	**
Pb	6.25	97	5.78	239	ns
S	3,251	256	14,146	125	*
Sb	0.446	232	1.16	67	**
Se	1.60	243	2.48	171	ns
Sn	0.387	282	0.219	480	ns
Sr	2,609	132	878	70	***
Ti	350	67	94.9	109	***
Tl	0.100	139	0.329	126	**
V	34.9	54	53.8	90	*
Zn	58.7	24	163.9	71	***
			REE		
Ce	198	141	40.3	145	**
Dy	19.7	87	6.41	76	***
Er	10.0	75	4.31	78	**
Eu	7.23	119	1.62	114	***
Gd	31.9	120	7.39	97	***
Ho	3.63	77	1.41	129	***
La	122	155	36.3	94	**
Lu	0.690	69	0.513	86	ns
Nd	151	132	30.1	127	***
Pr	34.1	134	7.03	126	***
Sm	30.4	128	6.12	120	***
Tb	4.16	108	1.11	82	***

(Continued)

TABLE 8.1 (*Continued*)
Geometric Mean[a] and Coefficients of Variation (cv)[b] of the Mineral Content in 150 Igneous and Sedimentary Rock Phosphates (Windmann, 2020 Revised)

	Igneous Phosphate Rocks		Sedimentary Phosphate Rocks		
Element	mean (mg/kg)	cv (%)	mean (mg/kg)	cv (%)	Significance[c]
Tm	0.96	65	0.546	81	ns
Yb	5.47	67	3.36	84	ns
			Actinides		
Th	5.08	219	2.75	92	**
U	7.56	119	59.6	69	***

[a] The geometric mean indicates the central tendency or typical value of a set of data as it is based on the product of the measured values in contrast to the arithmetic mean which uses their sums. The geometric mean is defined as the nth root of the product of n entries. The impact of outliers is greatly reduced when the geometric mean is calculated.

[b] The cv is defined as the ratio of the standard deviation to the arithmetic mean; it can be calculated, even if the geometric mean is zero because the standard deviation is divided by the arithmetic mean;

[c] Significance levels (F-test) between the groups igneous and sedimentary phosphate rocks: ns = $p > 0.05$, * = $p < 0.05$, ** = $p < 0.01$, *** = $p < 0.001$;

[d] LLD - Lower Limit of Detection; REE – rare earth element.

TABLE 8.2
Geometric Mean[a] and Coefficients of Variation (cv)[b] of Elemental Concentrations in 419 Mineral, 126 Farmyard Manures, and 65 Organo-Mineral Fertilizers (Windmann, 2020, Revised)

	Mineral P Fertilizers		Farmyard Manure		Organo-Mineral Fertilizers		
Element	mean (mg/kg)	cv (%)	mean (mg/kg)	cv (%)	mean (mg/kg)	cv (%)	Significance[c]
Al	2,159	133	466	97	1680	153	***
As	4.67	726	0.345	134	1.68	148	ns
B	297	129	31.6	49	59.4	214	***
Be	0.495	85	0.096	40	< LLD	140	***
Bi	< LLD[d]	62	0.116	67	0.180	160	***
Ca	56,116	82	13,663	71	60,065	73	***
Cd	2.87	136	0.195	50	0.379	113	***
Co	1.32	438	1.67	312	0.918	142	ns
Cr	46.3	211	8.22	205	8.45	93	***
Cu	18.9	333	48.7	71	26.0	166	ns
Fe	2,809	244	1,223	98	3,311	157	***
K	12,416	125	25,640	63	16,559	105	***
Li	2.06	141	0.880	31	2.12	91	***
Mg	5,434	136	4,806	60	8,372	11	***

(*Continued*)

TABLE 8.2 ***(Continued)***
Geometric Mean[a] and Coefficients of Variation (cv)[b] of Elemental Concentrations in 419 Mineral, 126 Farmyard Manures, and 65 Organo-Mineral Fertilizers (Windmann, 2020, Revised)

	Mineral P Fertilizers		Farmyard Manure		Organo-Mineral Fertilizers		
Element	mean (mg/kg)	cv (%)	mean (mg/kg)	cv (%)	mean (mg/kg)	cv (%)	Significance[c]
Mn	92.0	496	288	55	93.4	199	ns
Mo	3.65	600	2.91	57	1.64	455	ns
Na	1,720	88	1,306	80	2,109	100	ns
Ni	14.5	194	6.07	165	6.94	109	***
P	89,151	533	9,854	55	19,199	75	***
Pb	2.79	335	1.49	95	3.44	155	*
S	< LLD	98	3,530	58	< LLD	121	***
Sb	0.722	165	0.367	67	0.308	230	**
Se	< LLD	300	1.07	57	1.09	207	ns
Sn	0.273	358	0.770	45	0.802	168	**
Sr	398	196	31.9	66	129	86	***
Ti	118	110	18.2	67	79.8	771	***
Tl	0.171	173	0.075	67	0.064	59	***
V	56.2	102	1.62	65	5.21	110	***
Zn	121	234	231	60	130	117	ns
			REE				
Ce	29.3	233	-	-	3.44	91	*
Dy	3.72	142	-	-	< LLD	83	**
Er	2.65	110	-	-	< LLD	86	***
Eu	1.05	208	-	-	< LLD	67	*
Gd	4.57	202	-	-	< LLD	122	*
Ho	0.862	120	-	-	< LLD	57	***
La	24.6	222	-	-	2.10	93	*
Lu	< LLD	91	-	-	< LLD	44	***
Nd	18.3	231	-	-	1.63	82	*
Pr	4.39	228	-	-	0.556	80	*
Sm	3.56	236	-	-	< LLD	79	*
Tb	< LLD	177	-	-	< LLD	51	**
Tm	< LLD	91	-	-	< LLD	44	***
Yb	2.29	92	-	-	< LLD	91	***
			Actinides				
Th	2.46	270	-	-	0.158	120	ns
U	23.8	106	0.429	62	1.00	170	***

[a] The geometric mean indicates the central tendency or typical value of a set of data as it is based on the product of the measured values in contrast to the arithmetic mean which uses their sums. The geometric mean is defined as the nth root of the product of n entries. The impact of outliers is greatly reduced when the geometric mean is calculated.

[b] The cv is defined as the ratio of the standard deviation to the arithmetic mean; it can be calculated, even if the geometric mean is zero because the standard deviation is divided by the arithmetic mean;

[c] Significance levels (FS test) between the groups mineral P fertilizers, farmyard manures and organo-mineral P fertilizers: ns = $p > 0.05$, * = $p < 0.05$, ** = $p < 0.01$, *** = $p < 0.001$;

[d] LLD - Lower Limit of Detection; REE – rare earth element; - no data.

In terms of heavy metal loads, manufactured mineral fertilizers are at a disadvantage compared to farmyard manures, because mineral fertilizers contain >10 times more U, Cd, and As. Elements found in distinctly higher concentrations in farmyard manures than in manufactured fertilizers are S and Se. Mixtures of organic and mineral materials (organo-mineral fertilizers) reveal intermediate elemental concentrations (Table 8.2). Expectedly, the Cu and Zn contents in farmyard manure are roughly twice as high as in mineral and organo-mineral fertilizers. The reason is the use of these minerals as feed supplements in pig farming. With respect to mineral P fertilizers, the B content in mineral P fertilizers is about 10 times higher than in farmyard manures (Table 8.2).

Knowledge about relationships between elements in a P fertilizer can provide helpful information for identifying elements that may become in the future a chemical constraint for soil productivity. Out of more than 1,000 possible pairings of the 45 elements quantified by Windmann (2020), 36% were significant, although weak (correlation coefficient > 0.1). Only the combinations Be/Cr, Cu/Co, Cr/V, Se/Zn, and U/V and all combinations of the light rare earth elements (REE) (La, Ce, Pr, Nd, Sm, and Eu) show correlation coefficients > 0.8. In other words, the variability of any REE explains the variability of its pairing partner by > 64%.

Simple correlation coefficients derived from such complex datasets are highly prone to spuriousness (Gänsslen and Schubö, 1976; Schuchard-Ficher et al., 1982). Principal component analysis (PCA) provides a bias-free view to determine the true relationship between elements. PCA is a multivariate statistical procedure that reduces the variability observed in a large number of parameters. For example, different elemental concentrations in mineral P fertilizers are made into to a smaller, straightforward number of so-called nameless principal components (PC) or factors (Gänsslen and Schubö, 1976; Jackson, 1991; Joliffe, 1986; Schuchard-Ficher et al., 1982). The smaller number of PC carries the information about the variability of a much larger set of variables. This makes an interpretation easier, as a cluster of variables emerge clearer (Husson et al., 2009).

Figure 8.2 shows the interrelation of 45 elements analyzed in 150 rock phosphate samples in a plot of the first and second main components extracted from the entire data set.

In depositional environmental studies, the relative enrichment of heavy metals, such as Be, Cd, Cr, and U, is a geochemical indicator of sedimentary phosphate rocks, while igneous materials tend to be enriched in Sr and Ti (Sattouf, 2007). Accordingly, the first main component in Figure 8.2 reflects the composition of igneous materials, and the second main component reflects the composition of sedimentary rock phosphates. These findings suggest that trace-element patterns could be useful to distinguish between sedimentary or igneous phosphate rocks. However, it was impossible to identify the precise origin of sedimentary phosphate rocks, because there is a large variation in trace-metal concentrations, even in sedimentary phosphates from the same deposit (Table 8.1).

In Figure 8.2, the majority of REE is found in a cluster of high loading (>0.8) on the first main component, with the light REE (La, Ce, Pr, Nd, Sm, and Eu) separated in clusters from their heavy relatives (Gd, Dy, Ho, Er, Tm, Yb, and Lu) together with the actinide Th. The core finding of the PCA is that, in phosphogenesis, light and heavy REE, as well as Th, behave differently. In contrast, actinide U, together with Be, Cd, and Cr, is in a cluster with high loading (>0.8) in the second main component (Figure 8.2).

Knowledge about elemental relationships in P fertilizers is of particular interest when it comes to the evaluation of mitigation strategies for chemical constraints for soil productivity through P fertilization. Cadmium, for instance, has long been identified as such a constraint and legal limits for fertilizers have been proposed (Dharma-Wardana, 2018; Smolders and Six, 2013; Ulrich, 2018). Later, when U was recognized as a chemical constraint for soil productivity (Schnug, et al., 2018), people setting EU fertilizer policies were reluctant to furnish another hazardous element in P fertilizers with legal limits. This was based on the common opinion that heavy-element concentrations in phosphate rocks are all similarly related, and therefore, the already ongoing regulation for reducing Cd in P fertilizers would affect U in the same direction. Windmann (2020) analyzed the regression between both elements in different sample sets (Equations 8.1–8.3) with the result that, even if the

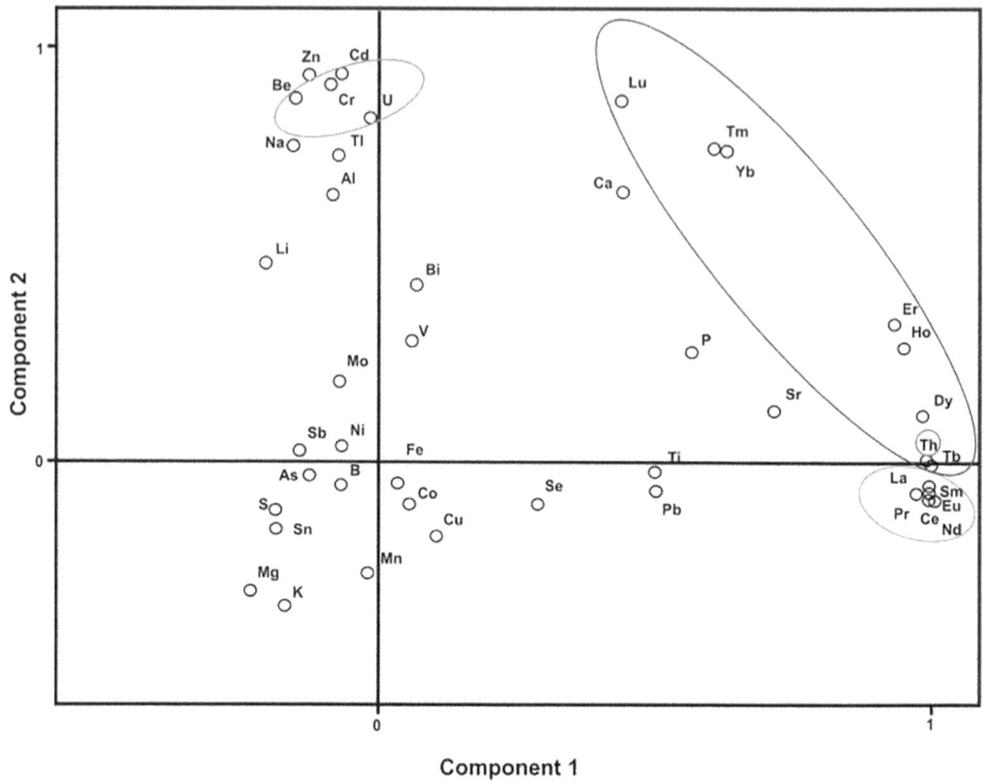

FIGURE 8.2 Varimax-rotated scattergram of the loads of 45 elements analyzed in 150 igneous and sedimentary phosphate rocks on the first two main components out of 11 with an eigenvalue > 1, representing 34% (component 1) and 20% (component 2) of the variability observed in the entire data set (Windmann, 2020).

Cd content in a mineral P fertilizer could be reduced close to zero, it still will contain significant U concentrations, because the constant of the regression is positive:

$$\text{Mineral P fertilizers}: \text{U}(\text{mg/kg}) = 4.62\ \text{Cd}(\text{mg/kg}) + 12.5(n = 372; r = 0.825^{***}) \quad (8.1)$$

$$\text{Igneous rock phosphates}: \text{U}(\text{mg/kg}) = 1.34\ \text{Cd}(\text{mg/kg}) + 12.7(n = 22; r = 0.530^{**}) \quad (8.2)$$

$$\text{Sedimentary rock phosphates}: \text{U}(\text{mg/kg}) = 0.654\ \text{Cd}(\text{mg/kg}) + 75.7(n = 128; r = 0.206^{*}) \quad (8.3)$$

With a view to mineral P fertilizers, the regression analysis conducted by Windmann (2020) reveals that, even if the Cd content in a mineral P fertilizer product is zero, the U content in the product is about 12.5 mg/kg and would still exceed the suggested legal limits for U (Schnug and Haneklaus, 2008; UBA, 2012).

8.3 ELEMENTAL SIGNATURES FOR IDENTIFYING ORIGINS OF PHOSPHATE ROCKS AND FERTILIZERS

The geographical origin of phosphate rock reveals to some extent its chemical constituents (Table 8.1; Dechema, 2017; Mortvedt and Beaton, 1995; Sun, 2021). Sedimentary phosphate rocks are by far the most common sources used for fertilizer manufacturing. Their deposits are located all around the world, mainly in Algeria, Brazil, Brazil, the People's Republic of China, Curacao, Egypt, Israel, Maroc, Tunisia, Togo, Senegal, and the USA. In contrast, igneous phosphate rocks are mined mainly in Russia and in much smaller quantities in Finland and South Africa. Even if the location of manufacturing of P fertilizers is known, this does not necessarily reveal from which region the phosphate rock originates, because P fertilizer factories may process rocks from different geographic sources or change sources in the course of time.

Sun (2021), Killiches (2013), Sattouf (2007), and Van Kauwenbergh (1997) give the chemical composition of phosphate rocks from different origins. However, this information does not necessarily enable the reconstruction of the origin of the rock from which a particular P fertilizer has been manufactured.

Sun (2021) attempted isotope analysis to trace the original source of P fertilizers. The three most commonly used elements for stable isotope analysis are O, C, and S, all being major structural components of francolite, which is the main mineral in phosphate rocks. Two other relevant radiogenic isotope tracers, Sr and U, are constituents in the lattice of francolite that replace major elements (Jarvis et al., 1995).

Stable oxygen isotopes are commonly used as geochemical indicators. The major advantage of using the oxygen-isotope ratios in phosphate ($^{18}O_p$) (the subscript p refers to phosphate) is that isotope exchange between oxygen in phosphate and water is extremely slow in inorganic systems (Kolodny et al., 1983). Therefore, it is fair to assume that the $\delta^{18}O_p$ of sedimentary phosphate rocks provides information of its original paleo-environment (Longinelli and Nuti, 1973). So, the application of $\delta^{18}O_p$ to sedimentary phosphate rocks has been used to estimate water paleo-temperatures and diagenetic conditions. The following phosphate-water temperature equation was formulated (Longinelli and Nuti, 1973):

$$T = 111.4 - 4.3\left(\delta^{18}O_p - \delta^{18}O_w\right),$$

where T is the average environmental temperature (°C), $\delta^{18}O_p$ is the natural oxygen-isotope composition of phosphate, and $\delta^{18}O_w$ is that of water. When comparing phosphate rocks of different ages, a temporal trend of decreasing $\delta^{18}O_p$ values (24–9 ‰) was found with increasing rock age (0.01–2,000 Ma) (Shemesh et al., 1983 and 1988). This $\delta^{18}O_p$ time trend might be useful to help identify the deposit of phosphate rocks.

The residence time of Sr in seawater is about 3 Myr, which is several orders of magnitude longer than oceanic mixing times; thus, it may be concluded that Sr in ocean water is more or less homogenously mixed (Turner and Whitfield, 1979). The Sr ($^{87}Sr/^{86}Sr$) isotope composition of phosphate is in general agreement with the Sr isotope ratio ($^{87}Sr/^{86}Sr$) in contemporary seawater (Kolodny and Luz, 1992). The Sr isotope content in seawater has changed during the geological ages so that the Sr isotope-age curve can be applied to date the age of marine sediments (Peterman, 1979), particularly for the Cenozonic (0–20 Ma) time period (Hodell et al., 1991). This procedure is most likely less precise when used to estimate the date of more ancient rocks. Attempts to date phosphogenesis by Sr isotopes have been made by McArthur et al. (1990), who found that age determinations for phosphorites from four different locations were in good agreement with their geological context and previous estimates. Subsequent research was conducted by Sattouf et al. (2008), who successfully used the $^{87}Sr/^{86}Sr$ isotope ratios to separate phosphate rock samples into four distinct groups: (i) igneous phosphate rock from Russia (Kola) and sedimentary rock from (ii) Algeria, Israel, Morocco, Tunisia, and Syria, (iii) the USA, and finally (iv) those from Senegal and Togo.

The U-series isotopes have the advantage in dating young phosphorites, because the modern ocean is isotopically uniform with a $^{234}U/^{238}U$ activity ratio of 1.144±0.004 (Jarvis et al., 1995). In U-series studies of phosphorite deposits in the Peru-Chile and Namibian margins (Baturin et al., 1972), Western South Africa (Birch et al., 1983), and East Australia (Kress and Veeh, 1980), the $^{234}U/^{238}U$ activity ratios were close to that of present seawater, which indicates a forming during the same period of time. The more recent study by Sattouf (2007) revealed that, based on their $^{234}U/^{238}U$ activity ratios, phosphate rock from the USA can be distinguished from that of the Middle East and North Africa. One reason is presumably the relatively younger age of the phosphate rocks originating from the USA.

It is important to note that, when isotopic signatures are applied to trace directly the origin of P fertilizers, changes in the overall elemental composition and isotopic ratio may occur during the fertilizer manufacturing process (Rutherford et al., 1994; El-Mrabet et al., 2003). The isotope ratio signatures in phosphate rocks are less susceptible to such alterations than the total elemental concentrations during the P fertilizer production process. As mentioned earlier, $^{18}O_p$ normally cannot be altered in abiotic systems. This means that $^{18}O_p$ in P fertilizers keeps its inherent values of phosphate rocks during the manufacturing process. Sattouf (2007) tested the reliability of Sr and U isotope signatures for detecting the origin of P fertilizers. The study showed that there was a linear, almost 1:1 regression ($r_2=0.989$, $p<0.05$) between the Sr isotope ratio in phosphate rocks and in P fertilizers originating from these materials. The slightly higher values of the $^{87}Sr/^{86}Sr$ isotope ratios in P fertilizers were most likely due to a kinetic isotope effect that happens during the digestion with sulfuric acid. The same author used the $^{234}U/^{238}U$ isotope ratio to verify the provenance of a certain fertilizer that was manufactured from phosphate rock mined in the USA; further differentiation down to a specific P deposit proved to be not feasible because of the spatial variation of the isotope ratio within the deposit (Sun, 2021). In conclusion, it is evident that origin and genesis do shape the large elemental and isotopic patterns of a particular phosphate rock. However, the inherent high variability in elemental concentrations makes source tracking difficult using specific elements or elemental ratios. An actual example is the EU survey on Cd (DGIP, 2017).

8.4 ENVIRONMENTAL LOADS OF HEAVY METALS ADDED TO AGRICULTURAL SOILS WITH P FERTILIZATION

Mineral fertilizers manufactured from phosphate rock are a significant source of elemental loads to soils, of which Cd and U by far are the most serious constraints for soil productivity. Based on data from Kratz et al. (2008), FAO (2015), Hohmann (2020), Sun (2021), and Windmann (2020), it can be estimated that annually 236–428 t Cd and 707–1,462 t U are applied with mineral P fertilizers to agricultural land in the 28 member countries of the European Union (EU, 2020; EU, 2021). For Germany, the share of heavy metal concentrations in soils caused by the application of mineral P fertilizers during the last 70 years is estimated to be 21% for Cd and 13% for U, but less than 3% for As, Cu, Ni, Pb, and Zn. Similarly, a study in Switzerland showed that the mean Cd and U concentrations were 58% and 9% higher in fertilized, arable top-soils compared to grassland top-soils (Bigalke et al., 2017). Sun (2021) found the highest fertilizer-derived U loads in an Icelandic grassland soil, caused by very high P fertilizer rates combined with a very low bulk density of investigated soils. Such low bulk densities are typical for most Andosols and may thus have a wider concern of U levels in these types of soils worldwide when P is fertilized.

Both elements, Cd and U, are rated as hazardous for humans unequivocally (Keith et al., 2013; WHO, 2021) but differ significantly in their entrance pathways to the food chain: Cd is highly immobile in soils and can be easily taken up by plants. Thus, the main constraint for productivity of soil coming from Cd in P fertilizers is its ability to produce edible plants with low Cd content for healthy food. In comparison, the plant uptake of U derived from P fertilizers is only small (Lamas, 2005; Rivas, 2005; Windmann, 2020), but its mobility in soils is fairly high (Zewainy, 2008), making losses through leaching the highest of all heavy metals in such fertilizers (Table 8.3). Thus the main constraints through Cd and U from P fertilizers for soil productivity are that Cd, which is taken up by plants, may compromise food quality and that U, which is translocated in soils, jeopardizes the filter function of a soil and, consequently, its ability to replenish clean groundwater (Hassoun and Schnug, 2011).

Different models have been developed to calculate heavy metal loads to agricultural soils by P fertilization over time. Table 8.3 shows the estimated mean annual loads of As, B, Cd, Cu, Li, Mo, Ni, U, and Zn with mineral P fertilization to German agricultural soils, environmental background concentrations, plant uptake capability, and susceptibility of elements to leaching. The calculations

TABLE 8.3
Estimated Mean Annual Loads of As, B, Cd, Cu, Li, Mo, Ni, U, and Zn with Mineral P Fertilization to German Agricultural Soils, Environmental Background Concentrations, Plant Uptake Capability, and Susceptibility of Elements to Leaching [Mean Values calculated from Data of Hassoun and Schnug (2011) and Kratz et al. (2011)]

Element	Mean annual Load (t) by Mineral P Fertilizers	Proportion of Soil Background Concentration from Mineral P Fertilization (%)	Element-Specific Uptake Capacity	Element Susceptibility to Leaching
B	1,477	4.12	High	High
Zn	518	1.21		Medium
Cu	133	1.12	Medium	Low
Mo	38.5	8.05		Medium
Pb	12.5	0.18	Low	Very low
Ni	63	0.55		Medium
Cd	47.5	20.9		Medium
As	39.0	2.70		High
Li	10	-		High
U	177	12.9	Very Low	High

are based on long-term data of German agricultural statistics and include data from years in the past when high P rates were applied because the common understanding in textbooks at that time was that the overall long-term utilization of fertilizer P by plants is limited to 75% (Schnug and Haneklaus, 2016). Meanwhile, it was shown that this figure relies on a systematic error based on inappropriate experimental designs and/or lack of analytical tools (Schnug and Haneklaus, 2016). State of the art is now that fertilizer P can be taken up (and utilized) by plants completely within an infinite period of time provided P has been applied previously in soluble forms. As a result of this paradigm change, P fertilizer rates have been leveling down during the last 30 years.

Historically, the common terminus for P in fertilizers is not phosphorus but phosphates, and the content is stated in concentration units of an arbitrary "P_2O_5" and not in "P". The reason goes back to the old, traditional way of P analysis by gravimetry, where the final ignition step yields P oxide, which is quantified by weighing and expressed as P_2O_5 (Schnug, 1985).

Schnug et al. (2018) used fertilizer consumption data provided by FAO (2015) and assumed the average U content was 259 mg/kg P_2O_5 to find that about 707 t U per year are added to agricultural soils in the 28 European countries with the application of mineral P fertilizers (average from 2002 to 2013). Only six countries account for three quarters of this load with France > Spain > Poland > Germany > Italy > UK. Tulsidas et al. (2019) calculated from European rock phosphate import data a similar magnitude of recoverable U.

8.5 MITIGATION OF CHEMICAL SOIL CONSTRAINTS CAUSED BY HEAVY METAL CONTAMINATION WITH P FERTILIZATION – FERTILIZER REGIMES

The comparison of heavy metal concentrations in different fertilizer types (Table 8.2) hints at a mitigation strategy that can be applied on many farms to limit chemical constraints for soil productivity: the wise selection of fertilizer sources for P. From the data provided for U as an example in Table 8.4, it is evident that the fertilizer regime on a farm significantly affects the load of potentially hazardous elements to agricultural soils. Kratz et al. (2008 and 2011) calculated that U loads to agricultural soils are highest when only mineral P fertilizers are employed and can be cut by 50% when a manure-based fertilization strategy is applied.

TABLE 8.4
U Loads from P Fertilization and U Plant Uptake; Comparison of Different Fertilizer Regimes (Kratz et al., 2008 and 2011 revised)

		U Loads (g/ha*yr)		U Uptake by Plants
		Mean	Range	(g/ha*yr)
Method 1: GAP[a] (22 kg/ha*yr P)	Mineral fertilization	15	7–23	<0.35
	Organic fertilization	2.8	2.3–3.2	
Method 2: German model regions	Mineral fertilization	10	7–15	
	Manure-based fertilization	7	2.8–16	
Method 3: based on official P balance for Germany	Combined mineral and organic fertilization	6	7.6–4.4	0.15

[a] Good Agricultural Practice (BMEL, 2020).

8.6 MITIGATION OF CHEMICAL SOIL CONSTRAINTS CAUSED BY HEAVY METAL CONTAMINATION WITH P FERTILIZATION – VALORIZATION OF FERTILIZER MANUFACTURING

Minimizing chemical constraints for soil productivity should imply the strategic implementation of the precautionary principle from the "Wingspread statement" (named after a conference held in 1998 at Wingspread, the headquarters of the Johnson Foundation in Racine, Wisconsin), which reads: "When an activity raises threats of harm to human health or the environment, precautionary measures should be taken even if some cause and effect relationships are not fully established scientifically" (Montague, 2001). The main obstacle for employing environmental and sound health measures in agricultural production techniques is that they almost always add costs to the production chain. The inevitable social weighing of goods almost always goes on the expenses of the vulnerable part, which is health and environment. A smart way out would be the valorization of the culprit: make the hazardous components pay for their removal from the fertilizer.

Valorization of mineral resources is the process of exploiting resource ingredients with a view to optimize the value of the geological ore deposits and to create added-value products for market applications. Thus, valorization of fertilizer production aims (i) to find alternative uses for wastes that accumulate during phosphate mining, mineral processing, and fertilizer manufacturing; and (ii) to extract resource ingredients other than phosphate from phosphate rock, for instance, by-products like U. In the next paragraphs, the recovery of additional resources during phosphate-fertilizer production is reviewed (Figure 8.3), insights into valorization paths of fertilizer production are provided, and the major challenge of achieving valorization of fertilizer production is critically described.

Since the early 1900s, global annual mine production of phosphate rock has steadily increased to a current peak of 263 Mt (USGS, 2018). The phosphate content of phosphate rock varies greatly from ~4% to ~20% P_2O_5, and some of the mined phosphate is lost during mining, mineral processing, and fertilizer manufacturing. Therefore, only a small proportion of the mined phosphate ore is being used for phosphate-fertilizer production, whereas the great majority (~80%–96%) of the mined phosphate rock is presently discarded as waste throughout the whole fertilizer industry value chain. In fact, the quantity of total solid waste produced can be high, but it varies significantly between mine sites and fertilizer plants. One study reports findings where, for 1 t of phosphoric acid produced, 9.5 t of phosphate ore are required, and 21.8 t of diverse wastes and 6.5 t of tailings are produced (Villalba et al., 2008). The growth in phosphate production has caused an ever-increasing volume of wastes to accumulate during mining and mineral processing. At mine

FIGURE 8.3 In the early days of phosphate mining, extracting U from phosphate rock for nuclear fuel and weaponry was a highly welcomed added value to the production of P fertilizers (Schnug, 2014); IMC Agrico Phosphate Processing Plant, Florida - photograph by Connett (2001).

sites, significant overburden and waste-rock volumes need to be removed to access the phosphate ore, and fine-grained tailings are generated during phosphate processing. These wastes are commonly placed into large, engineered repositories. Such growing waste volumes and associated environmental concerns have stimulated much research into additional resource recovery at phosphate mines. Research around the valorization of phosphatic mine wastes has demonstrated that wastes can serve as (i) dry covers in the capping and revegetation of mine waste repositories; (ii) additives in acid mine water treatment; (iii) feedstock for the ceramic industry; and (iv) resource materials in the building industry (Hakkou et al., 2016). Hence, phosphate mine wastes can be recycled and reused for a range of applications. Such activities would minimize potential environmental damage, reduce the footprint of phosphate mines, and lead to the development of other local industries.

Much larger waste volumes accumulate at fertilizer plants, with PG being by far the major waste product, which accrues during the hydrometallurgical treatment of phosphate ores (Lottermoser, 2010). Global production of PG has been estimated as high as 1,239 Mt per year (Cánovas et al., 2018). PG is largely comprised of calcium sulfate crystals, making it an attractive alternative to gypsum and suitable for building and agricultural uses and as a source of valuable products (e.g., sulfuric acid, ammonium sulfate). Over the last few decades, different PG valorization routes have been developed in the agriculture, building, environmental, and energy sectors (Cánovas et al., 2018). However, some PG may also contain residual phosphoric and hydrofluoric acid as well as metals (e.g., Cd, Pb) and radio nuclides (e.g. U-238, Th-232, Ra-226). In the USA, the EPA has focused on the presence of metals and radionuclides in PG and has classified PG as "Technologically Enhanced Naturally Occurring Radioactive Material." This has prevented the large-scale reuse of PG (Wing, 2016). Hence in some countries like the USA, PG is stored in large stacks due to regulation of radioactivity levels or because the alternatives (natural gypsum and flue gas gypsum) are more competitive. To be used for agricultural applications, PG needs to be highly pure, or PG needs to be reprocessed and purified from its contaminants. Otherwise, contaminants may be added to agricultural soils, become available to plants, and leach into ground and surface waters. Presently, only 15% of the worldwide PG produced is recycled, whereas the remaining 85% is usually stacked, landfilled, or discharged to surface waters (Zhang et al., 2017). Regardless, PG is increasingly utilized, with the People's Republic of China, Brazil, Indonesia, and India leading the way and promoting its reuse, partly facilitated through supportive governmental policies and constructive regulation (Wing, 2016). Phosphate deposits may contain elevated concentrations of valuable elements, including U, Th, and REE (rare earth elements). Trace-element concentrations of phosphate rock (e.g., U, REE, B, Cr, Cu, Ga, Ni, Sc, Th, V, and Zn) differ from region to region, with sedimentary phosphate ores containing, in general, higher trace-element concentrations than igneous phosphate ores with the exception of lanthanides, which show higher concentrations in igneous than in sedimentary rock phosphates (Table 8.1). Some U concentrations are high enough to be of commercial interest, and U has been recovered during fertilizer manufacturing at some locations in the past (Cánovas et al., 2017; Haneklaus et al., 2017; Khaledi et al., 2019; Ulrich et al., 2014). In wet processing, U and Cd largely migrate into phosphoric acid, whereas Th accumulates in PG. REE are present in both, phosphoric acid and PG. Metal recovery is possible through acid leaching or the application of organic solvents and subsequent recovery of valuable constituents by applying ion exchange technology (Valkov et al., 2014; Cánovas et al., 2018; Wu et al., 2018; Khaledi et al., 2019). Thus, the extraction of by-products during fertilizer manufacturing is possible using established technologies. Such resource recovery would minimize potential environmental damage, reduce the environmental footprint of fertilizer plants, lead to the development of additional local industries, and generate added-value products for market applications.

Next, a simplified example is given for the added value of valorization of U in phosphate fertilizers. It might be assumed (Table 8.3) that in Germany 177 t U has been applied annually with mineral P fertilizers during the past 10 years. This amount of U equals the U inventory (calculated as natural U) of the German repository for nuclear waste "Bergwerk ASSE II" (Assewasser, 2021) and would contain enough energy to supply 2,491,000 average-sized German households. Alternatively, firewood from 5,935,000 ha of forest would deliver the same amount of energy.

From 10 g U (corresponding to a P fertilization of 22 kg/ha P according to GAP (Code for Good Agricultural Practice, BMEL (2020)) 500 kW of energy can be produced. Compared to the same amount of energy derived from coal, this saves a total of 500 kg CO_2. At a CO_2–tax of 0.25 €/kg (Tagesschau, 2020), this equals a value of 125 €/ha to which a farmer could be entitled when using P fertilizers with low U concentration. In comparison, the monetary value of 10 g U (as yellow cake (Cameco, 2021)) is actually less than 1 €; the costs for 22 kg P per hectare are 49 € (Agrarheute, 2021). Hu et al. (2008) estimated that the U in world phosphate reserves could feed the nuclear energy cycle for another 350 years; without this secondary U resource, conservative U reserves may be exhausted in just 50 years.

To date, scientists have identified diverse valorization pathways for phosphate rock. Unfortunately, many of the proposed reprocessing, reuse, and recycling routes have remained only research outcomes that have not been taken up by industry. Consequently, much of the U, Th, and REE still end up in the waste stream or the fertilizer-value chain. Thus, agricultural soils and underlying drinking water aquifers are becoming increasingly contaminated by U and other elements (Liesch et al., 2015; Schnug et al., 2019). Many research outcomes and concepts of resource recovery remain ideas that cannot be taken up in the real world, because neither existing government regulations allow or stimulate such activities, nor does industry consider them economical to be implemented. In particular, legislation by governmental agencies in Europe and North America prevents the reprocessing and reuse of wastes. These authorities are obsessed with setting environmental risk limits for organic and inorganic substances. Changes in regulatory attitude are needed to ensure that regulations are based on scientific evidence and best practices. In particular, supportive governmental policies and constructive regulation are required (Hilton et al., 2010), which will allow the safe use of pure PG for suitable applications. Most of all, governmental authorities should not only set legal limits for contaminants, such as Cd and U, in fertilizers, but also thresholds for resource ingredients that, when present above a certain threshold concentration, have to be extracted from the mined ores. For phosphate rock in particular, thresholds for the compulsory recovery of by-products, such as U and REE, should be given to enforce the valorization of phosphate rock.

8.7 MITIGATION OF CHEMICAL SOIL CONSTRAINTS CAUSED BY HEAVY METAL CONTAMINATION WITH P FERTILIZATION – LEGAL MEASURES FOR FERTILIZERS

Financial incentives for industry, such as subsidies for cleaner fertilizers, may support the economic profitability of U and REE extraction from phosphate rock. However, such incentives would likely be ineffective and wasteful. Studies of incentives for other industries have established that the majority of them only achieve temporary compliance or give industry extra benefits to do things it would have done even without incentives. When it comes to producing lasting changes, incentives are strikingly ineffective. Phosphate rock has diverse compositions and geological properties that control its site-specific characteristics and opportunities for resource recovery and valorization. In the end, governmental authorities need to be more proactive, legislate valorization of fertilizer production, and enforce the manufacturing of added-value products for market applications.

It cannot be expected that P fertilizers with low concentrations of hazardous elements will appear on the market without significant financial advantages in their production through the valorization of by-products, because benefits for health and the environment simply do not pay off yet. The EU has assessed available legal measures and concluded that, as a classical steering tool in environmental policymaking, it needs to control Cd and provide a limit value, which would be the most practical and effective tool (Ulrich, 2018). The EU Cd rule, which is part of the revised EU Fertilizer Regulation, is among the most controversial regulations applied to the agricultural sector in recent years, because the rule is more a matter of trade and geopolitics than of environmental and public health protection. In 1997, the European Commission proposed limit values for Cd in mineral fertilizers, with the aim to reduce chemical constraints for soil productivity and to safeguard food, profitability for farmers, and human health. For many years, hope remained high for a progressive reduction of the Cd content in fertilizers from 60 to 40 to 20 mg Cd/kg P_2O_5 over a transition period of 12–16 years. But, retrospectively, this hope was unrealistic, because the political and industrial realms were divided between fear of overly high costs, insufficient supply, geopolitical concerns, and environmental and public health concerns (Ulrich, 2018). At this point, it shall be noted that in the EU fertilizer laws Cd concentrations are traditionally related to the phosphate (P_2O_5) content of a fertilizer (see above). A preliminary agreement on the limit was finally reached in November 2018, to be confirmed by the EU Council of Ministers in 2019. As the lowest common denominator, a single limit value of 60 mg Cd/kg P_2O_5 will be set 3 years after the start of the new EU Fertilizer Regulation, and a voluntary green label for fertilizers that contain $\leq$20 mg Cd/kg P_2O_5 will be put in place (European Parliament, 2018). The European Commission will then review this welcoming foundation for soil and public health protection from fertilizer-derived Cd 7 years after the rule will have come into force. The EU bodies will then decide if there is a need to lower the limit to 40 mg and eventually 20 mg as, for instance, Römkens et al. (2017) indicated. In addition, the issue of U will be monitored to learn if its content needs to be limited in the future.

The Cd limit is an historical achievement that opens the way for demonstrating sufficient political courage and consistency in setting higher standards in the future. Notably, market realities had already surpassed the EU norm of 60 mg/kg Cd even before the implementation of the rule. Trade-group data suggest that most fertilizers already comply with this limit (Ulrich, 2018). However, the agreement will only affect about 8% of products traded in the EU, which will have an impact on the exports of Senegal, Togo, and Tunisia and, to a lesser degree, Algeria, Israel, and Morocco (Financial Times, 2018). Swiss regulators demonstrated considerable political courage in 2018 by passing a Cd limit of 11 mg Cd/kg P_2O_5 in phosphate fertilizers from secondary sources starting in 2019 (Zimmermann, 2018); a limit of 25 mg Cd /kg P_2O_5 for mineral P fertilizers has been in place in Switzerland since 1986. Although in much higher concentration (Tables 8.1 and 8.2) and potentially much more harmful for human health and the environment, industry and lawmakers fear to regulate U in P fertilizers; they are afraid of regulation, following the proverbial saying that the devil fears holy water. Since the problem of U in fertilizers became public in 2009, German politics would not approve any further research on this topic in its departmental research institutes (Schnug and Haneklaus, 2008). A common disguise strategy to avoid another can of worms to be opened was the misbelief that, once Cd content in fertilizers would be limited, the U concentration would automatically decline. It has been outlined in a previous section of this chapter that, even if Cd can be eliminated from mineral fertilizers completely, it can be expected that the products will still contain significant amounts of U. The Commission for Soil Protection of the German Environmental Agency (UBA, 2012) proposed, based on a scientific model by Schnug and Haneklaus (2008), that P fertilizers with a phosphate content of $>5\%$ P_2O_5 should have a limit for declaration of 20 mg U/kg P_2O_5 and for trading of 50 mg U/kg P_2O_5 (UBA, 2012). In compliance with the anticipated regulation of Cd for fertilizers with $<5\%$ P_2O_5, a limit value of 1.5 mg/kg U in the fertilizer was proposed in order to protect agricultural soils from the chemical constraints on productivity through U accumulation.

The new EU Fertilizer Regulation considers nine pollutants in fertilizers (Cr(VI), Hg, Ni, Pb, As, Cu, Zn, biuret ($[H_2NC(O)]_2NH$), and perchlorate (ClO_4^-) and monitors the risk of toxicity and carcinogenicity of these contaminants, including fertilizer-derived U (Ulrich et al., 2014). At this point, it is important to recall that the Cd limit will not automatically coregulate U content in fertilier, although occasionally proposed. Depending on the chosen technological process, synergies may exist between removing two or more potentially harmful trace elements at the same time. However, a separate limit value for U needs to be established to ensure a safe, long-term application of mineral P fertilizers.

In the future, additional measures will be required to make the limit values for hazardous elements in fertilizers operational. Control and monitoring of such limits, if not globally, but at least in the customs union of the EU, are central for enforcement in the next years, although they are notoriously complicated issues. Commonly, the national authorities in which the products are being marketed perform the inspections of P fertilizer products. At this stage, the different member states' control regimes are somewhat opaque and appear very heterogeneous. The EU, and also its global trading partners, need to establish guidelines for enforcement assistance that will define (i) inspection frequencies, (ii) minimum inspection numbers, (iii) procedures for representative fertilizer sampling, (iv) analytical methods and permitted error ranges, and (v) a certification protocol for controlling laboratories. Taking these steps will help to ensure a minimum of comparability and consistency (and thus acceptability of the results). Labeling of hazardous elemental concentrations, beside nutritive ones, should become mandatory in order assist farm-control measures for all relevant mineral elements.

Other tools are in use in addition to these measures, notably via the EU's Common Agricultural Policy, which is currently in the reform process. An increasing number of consumers are asking for environmentally friendly agricultural products and production. The use of subsidies for fertilizers low in hazardous elements would encourage safe fertilization and would contribute to limiting chemical constraints for soil productivity.

At the moment, P regulation is often pursued in an indirect and poorly linked manner. One intervention that bears the potential to overcome the dispersed nature of legislation and reflects the complexity of real-world P governance, would be an EU "Phosphate Directive," similar to the Nitrates Directive (EU, 1991). Some EU member states already have passed P legislation, which is focused exclusively on P losses and the subsequent risk of eutrophication (Amery and Schoumans, 2014; Barreau et al., 2018). However, an EU directive is required that regulates proper P governance and stewardship.

Cleaner fertilizers result in cleaner soils, healthier food, cleaner waters, a cleaner atmosphere, and last, but not least, make fertilizer manufacturing more profitable.

REFERENCES

Agrarheute. 2021. Marktpreise Phosphordünger. [https://markt.agrarheute.com/duengemittel-4/phosphor-duenger-21] (last accessed 01.06.2021).

Amery, F. and O. F. Schoumans. 2014. *Agricultural Phosphorus Legislation in Europe*, 45 p. Merelbeke: ILVO.

Andrew, S. and T. Hubert. 2000. Phosphorus research strategies to meet agricultural and environmental challenges of the 21st century. *Journal of Environment Quality* 29: 176–181.

Anonymous. 2019. Düngemittelverordnung vom 5. Dezember 2012 (BGBl. I S. 2482), die zuletzt durch Artikel 3 der Verordnung vom 26. Mai 2017 (BGBl. I S. 1305) geändert worden ist. [https://www.gesetze-im-internet.de/d_mv_2012/DüMV.pdf] (last access 24.05.2021).

Assewasser. 2021. Was befindet sich eigentlich alles in der ASSE? [http://assewasser-nein-danke.de/washbefindet-sich-in-der-asse/] (last access 21.05.2021).

Azzi, V., A. El Samrani, B. Lartiges, A. Kobeissi, A. Kanso, and V. Kazpard. 2017. Trace metals in phosphate fertilizers used in eastern Mediterranean countries. *CLEAN - Soil, Air, Water* 45 (1). DOI: 10.1002/clen.201500988

Barreau, S., J. Magnier and C. Alcouffe. 2018. *Agricultural phosphorus regulation in Europe – Experience-sharing for 4 European countries*. International Office for Water (IOWater).

Baturin, G., K. Merkulova and P. Chalov. 1972. Radiometric evidence for recent formation of phosphatic nodules in marine shelf sediments. *Marine Geology* 13: 37–41.

Bigalke, M., A. E. Ulrich, A. Rehmus and A. Keller. 2017. Accumulation of Cadmium and Uranium in arable soils in Switzerland. *Environmental Pollution* 221, 85–93.

Birch, G., J. Thomson, J. McArthur and W. Burnett. 1983. Pleistocene phosphorites off the west coast of South Africa. *Nature* 302:601.

BMEL. 2020. Düngung. [https://www.bmel.de/DE/themen/landwirtschaft/pflanzenbau/ackerbau/duengung.html] (last accessed 01.06.2021).

Cameco. 2021. Uranium price. [https://www.cameco.com/invest/markets/uranium-price] (last accessed 01.06.2021).

Cánovas, C. R., R. Pérez-López, F. Macías, S. Chapron, J. M. Nieto and S. Pellet-Rostaing. 2017. Exploration of fertilizer industry wastes as potential source of critical raw materials. *Journal of Cleaner Production* 143:497–505.

Cánovas, C. R., F. Macías, R. Pérez-López, M. Basallote and R. Millán-Becerro. 2018. Valorization of wastes from the fertilizer industry: Current status and future trends. *Journal of Cleaner Production* 174:678–690.

Cardoso, E. J. B. N. 2013. Soil health: Looking for suitable indicators. What should be considered to assess the effects of use and management on soil health? *Scientia Agricola* 70:274–290.

Connett, M. 2001. Photographs of the Florida phosphate industry. [https://fluoridealert.org/content/phosphate-photos/] (last accessed on 20.03.2019).

Dechema. 2017. Phosphatrückgewinnung - Statuspapier der ProcessNet-Fachgruppe "Rohstoffe". [https://dechema.de/dechema_media/Downloads/Positionspapiere/Statuspap_Phosphat_2017_FINAL_NOV-p-20003290.pdf] (last accessed 31.05.2021).

DGIP. 2017. Directorate general for internal policies – policy department A: Economic and scientific policy. Scientific aspects underlying the regulatory framework in the area of fertilisers – state of play and future reforms. [http://www.europarl.europa.eu/RegData/etudes/IDAN/2017/602032/IPOL_IDA(2017)602032_EN.pdf] (last access 25.03.2019).

Dharma-Wardana, M. W. C. 2018. Fertilizer usage and cadmium in soils, crops and food. *Environmental Geochemistry and Health* 40: 2739–2759.

El-Mrabet, R., J. M. Abril, R. Perianez, G. Manjon, R. García-Tenorio, A. Delgado and L. Andreu. 2003. Phosphogypsum amendment effect on radionuclide content in drainage water and marsh soils from southwestern Spain. *Journal of Environmental Quality* 32: 1262–1268.

EU. 1991. The Nitrates Directive. [http://ec.europa.eu/environment/water/water-nitrates/] (last accessed 20.03.20199).

EU. 2019. Regulation (EU) 2019/1009 of the European Parliament and of the Council of 5 June 2019 laying down rules on the making available on the market of EU fertilising products and amending Regulations (EC) No 1069/2009 and (EC) No 1107/2009 and repealing Regulation (EC) No 2003/2003 (Text with EEA relevance). PE/76/2018/REV/1. [https://eur-lex.europa.eu/legal-content/EN/TXT/?uri=CELEX%3A32019R1009&qid=1622135537631] (last access 25.05.2021).

EU. 2020. Farms and farmland in the European Union - statistics. [https://ec.europa.eu/eurostat/statistics-explained/index.php?title=Farms_and_farmland_in_the_European_Union_-_statistics] (last accessed 03.02.2023).

EU. 2021. Agri-environmental indicator - risk of pollution by phosphorus. [https://ec.europa.eu/eurostat/statistics-explained/index.php?title=Archive:Agri-environmental_indicator_-_risk_of_pollution_by_phosphorus] (last accessed 31.05.2021).

European Parliament. 2018. Fertilisers/cadmium: Parliament and Council negotiators reach provisional deal. Press release, November 20, 2018. [http://www.europarl.europa.eu/news/en/press-room/20181119IPR19407/fertilisers-cadmiumhparliament-and-council-negotiators-reach-provisional-deal] (last access 13.12.2018).

FAO. 2015. World fertilizer trends and outlook to 2018. In: *Food and Agriculture Organization of the United Nations*, Rome, ISBN 978-92-5-108692-6.

Financial Times. 2018. EU agrees to provisional deal on cadmium fertilizer limit. [https://www.ft.com/content/b7bd1934-ecde-11e8-89c8-d36339d835c0] (last access 22.3.2019).

Four Rs. 2021. 4R nutrient stewardship. [https://nutrientstewardship.org/4rs/] (last access 02.06.2021).

Freitas, E. V. S., C. W. A. Nascimento, D. F. Goulart, and J. P. S. Silva. 2009. Disponibilidade de cádmio e chumbo para milho em solo adubado com fertilizantes fosfatados. *Revista Brasileira de Ciência do Solo, Viçosa*, 33:1899–1907.

Gänsslen, H. and W. Schubö. 1976. *Einfache und komplexe statistische Analyse*. Basel: Verlag Ernst Reinhardt.

Guedes, P., N. Couto, E. P. Mateus and A. P. Ribeiro. 2017. Phosphorus recovery in sewage sludge by electrokinetic based technologies: A Multivariate and Circular Economy View. *Waste & Biomass Valorization* 8: 1587–1596.

Hakkou, R., M. Benzaazoua and B. Bussiere. 2016. Valorization of phosphate waste rocks and sludge from the Moroccan phosphate mines: Challenges and perspectives. *Procedia Engineering* 138:110–118.

Haneklaus, N., Y. Sun, R. Bol, B. G. Lottermoser and E. Schnug. 2017. To extract, or not to extract uranium from phosphate rock, that is the question. *Environmental Science and Technology* 51:753–754.

Hassoun, R. and E. Schnug. 2011. Contribution of mineral and tap water to the dietary intake of As, B, Cu, Li, Mo, Ni, Pb, U and Zn by humans. In: *The New Uranium Mining Boom*, eds. B. Merkel and M. Schipek, pp. 795–802. Heidelberg: Springer Verlag.

Hilton, J., B. Birky and A. E. Johnston. 2010. The 'constructive regulation' of phosphates and phosphogypsum. A new, evidence-based approach to regulating a NORM industry vital to the global community. Proceedings of IRPA12: 12. Congress of the International Radiation Protection Association: Strengthening Radiation Protection Worldwide - Highlights, Global Perspective and Future Trends. International Atomic Energy Agency, Vienna.

Hodell, D. A., P. A. Mueller and J. R. Garrido. 1991. Variations in the strontium isotopic composition of seawater during the Neogene. *Geology* 19: 24–27.

Hohmann, M. 2020. Verbrauch von Düngemitteln in der Landwirtschaft in Deutschland nach Nährstoffarten in den Jahren 1990 bis 2019. [https://de.statista.com/statistik/daten/studie/161842/umfrage/verbrauchdaus-gewaehlter-duenger-in-der-landwirtschaft-in-deutschland/] (last accessed 31.05.2021).

Hu, Z., Zhang, H., Wang, Y., Haneklaus, S. and E. Schnug. 2008. Combining energy and fertilizer production – vision for China's future. In: *Loads and Fate of Fertilizer-Derived Uranium*, eds. L.J. De Kok and E. Schnug, pp. 127–134. Leiden: Backhys Publishers.

Husson, F., L. Sébastien and J. Pagès. 2009. *Exploratory Multivariate Analysis by Example Using*, 224 p. London: R. Chapman & Hall/CRC The R Series.

Jackson, J. E. 1991. *A User's Guide to Principal Components*. John Wiley & Sons. DOI: 10.1002/0471725331. [https://en.wikipedia.org/wiki/Principal_component_analysis] (last access 12.02.2018).

Jarvis, I., W. C. Burnett, Y. Nathan, F. S. M. Almbaydin, A. K. M. Attia, L. N. Castro, R. Flicoteaux, M. E. Hilmy, V. Husain, A. A. Qutawnah, A. Serjani and Y. N. Zanin. 1995. Phosphorites geochemistry: State-of-the-art and environment cancers. *Ecologae Geologicae Helvetie* 87: 643–700.

Jolliffe, I. T. 1986. *Principal Component Analysis*. Springer Series in Statistics. Heidelberg: Springer-Verlag.

Keith, S, O. Faroon and N. Roney. 2013. Toxicological Profile for Uranium. Agency for Toxic Substances and Disease Registry (US). [https://www.ncbi.nlm.nih.gov/books/NBK158798/] (last accessed 01.06.2021).

Khaledi, N. Al, M. Taha, A. Hussein, E. Hussein, N. Yahyaoui and N. Haneklaus. 2019. Direct leaching of rare earth elements and uranium from phosphate rocks. *IOP Conf. Series: Materials Science and Engineering* 479 (2019). [https://www.researchgate.net/publication/331623496_Direct_leaching_of_rare_earth_elements_and_uranium_from_phosphate_rocks] (last access 20.03.2019).

Killiches, F. 2013. Phosphat. Mineralischer Rohstoff und unverzichtbarer Nährstofffür die Ernährungssicherheit weltweit. [Report].– Hannover. Bundesanstalt für Geowissenschaften und Rohstoffe (BGR). [https://www.bgr.bund.de/DE/Themen/Zusammenarbeit/TechnZusammenarbeit/Politikberatung_SV_MER/Downloads/phosphat.pdf?__blob=publicationFile&v=4] (last access 30.05.2021).

Kölle, W. 2017. *Wasseranalysen - richtig beurteilt: Grundlagen, Parameter, Wassertypen, Inhaltsstoffe*. Weinheim: John Wiley & Sons.

Kolodny, Y., B. Luz and O. Navon. 1983. Oxygen isotope variations in phosphate of biogenic apatites, I. Fish bone apatite—rechecking the rules of the game. *Earth and Planetary Science Letters* 64: 398–404.

Kolodny, Y. and B. Luz. 1992. Isotope signatures in phosphate deposits: Formation and diagenetic history. In: *Isotopic Signatures and Sedimentary Records*, eds. N. Clauer, and S. Chaudhuri, pp. 69–121. Heidelberg: Springer.

Kratz, S. and E. Schnug. 2006. Rock phosphates and P fertilizers as sources of U contamination in agricultural soils. In: *Uranium in the Environment: Mining Impact and Consequences*, eds. B. Merkel, and A. Hasche-Berger, pp. 57–67. Heidelberg: Springer Verlag.

Kratz, S., F. Knappe, J. Rogasik and E. Schnug. 2008. Uranium balances in agroecosystems. In: *Loads and Fate of Fertilizer-Derived Uranium*, eds. L.J. de Kok, and E. Schnug, editors, pp. 179–190. Leiden: Backhuys Publisher.

Kratz, S., F. Godlinski and E. Schnug. 2011. Heavy metal loads to agricultural soils in Germany from the application of commercial phosphorus fertilizers and their contribution to background concentrations in soils. In: *The New Uranium Mining Boom – Challenges and Lessons Learned*, eds. B. Merkel and M. Schipek, pp. 755–76. Heidelberg: Springer Verlag.

Kratz, S., J. Schick, R. Shwiekh and E. Schnug. 2014. Schätzung des Potentials erneuerbarer P-haltiger Rohstoffe in Deutschland zur Substitution rohphosphathaltiger Düngemittel: 8 Estimating the potential of renewable P containing raw materials in Germany as a substitute for fertilizers made from rock phosphate. *Journal fuer Kulturpflanzen* 66: 261–275.

Kratz, S., J. Schick and E. Schnug 2016. Trace elements in rock phosphates and P containing mineral and organo-mineral fertilizers sold in Germany. *Science of the Total Environment* 542: 1013–1019.

Kress, A. and Veeh, H. 1980. Geochemistry and radiometric ages of phosphatic nodules from the continental margin of northern New South Wales, Australia. *Marine Geology* 36:143–157.

Kuhn, T. 2021. The revision of the German Fertiliser Ordinance in 2017. [http://www.ilr.uni-bonn.de/agpo/publ/dispap/download/dispap17_02.pdf] (last access 25.05.2021).

Lamas, M. C. 2005. Factors affecting the availability of uranium in soils. Landbauforschung Völkenrode, Special issue 278: FAL Braunschweig, Germany. [https://literatur.thuenen.de/digbib_extern/bitv/zi036810.pdf] (last accessed 32.05.2021).

Liesch, T., S. Hinrichsen and N. Goldscheider. 2015. Uranium in groundwater–fertilizers versus geogenic-sources. *Science of the Total Environment* 536: 981–995.

Longinelli, A. and S. Nuti. 1973. Revised phosphate-water isotopic temperature scale. *Earth and Planetary Science Letters* 19: 373–376.

Lottermoser, B. G. 2010. *Mine Wastes: Characterisation, Treatment and Environmental Impacts*. Heidelberg: Springer-Verlag.

McArthur, J., A. Sahami, M. Thirlwall, P. Hamilton and A. Osborn. 1990. Dating phosphogenesis with strontium isotopes. *Geochimica et Cosmochimica Acta* 54:1343–1351.

Montague, M. 2001. The precautionary principle in the real world. *Environmental Research Foundation*. [http://www.rachel.org/lib/pp_def.htm] (last access 17.03.2019).

Mortvedt, J. J. and J. D. Beaton. 1995. Heavy metal and radionuclide contaminants in phosphate fertilizers. *Scope-Scientific Committee on Problems of the Environment International Council of Scientific Unions* 54:93–106.

Nash, W. P. 1984. Phosphate minerals in terrestrial igneous and metamorphic rocks. In: *Phosphate Minerals*, eds. J.O. Nriagu, and P.B. Moore, pp. 215–241. Heidelberg: Springer Verlag.

Peterman, Z. E. 1979. Strontium isotope geochemistry of late Archean to late Cretaceous tonalites and trondhjemites. *Developments in Petrology* 6: 133–147.

Rivas, M. C. 2005. Interactions between soil uranium contamination and fertilization with N, P and S on the uranium content and uptake of corn, sunflower and beans, and soil microbiological parameters. Landbauforschung Völkenrode, Sonderheft 287. [https://literatur.thuenen.de/digbib_extern/bitv/zi038337.pdf] (last accessed 01.06.2021).

Römkens, P. F. A. M., J. C. Voogd, J. Kros, R. P. J. J. Rietra, J.-H. Yoo and G. Louagie. 2017. A spatially explicit model to calculate cadmium balances in agro-ecosystems in the EU-27: Model description and scenario analysis. [https://www.asktheeu.org/en/request/5352/response/17034/attach/6/Paper%20metal%20balances%20Europe%20final%20concept%2014%2011.pdf Accessed date 10.12.2018] (last access 01.02.2019).

Rutherford, P., M. Dudas and R. Samek. 1994. Environmental impacts of phosphogypsum. *Science of the Total Environment* 149:1–38.

Sattouf, M. 2007. Identifying the origin of rock phosphates and phosphorous fertilisers using isotope ratio techniques and heavy metal patterns. PhD thesis, TU Braunschweig, pp. 194. [https://nbnh resolving.org/urn:nbn:de:gbv:084-13987] (last access 25.03.2019).

Sattouf, M., S. Kratz, K. Diemer, J. Fleckenstein, D. Rienitz, D. Schiel and E. Schnug. 2008. Significance of uranium and strontium isotope ratios for retracing the fate of uranium during the processing of phosphate fertilizers from rock phosphates. In: *Loads and Fate of Fertilizer-Derived Uranium*, eds. L. J. de Kok, and E. Schnug, pp. 978–990. Leiden: Backhuys Publishers.

Schnug, E. 1985. Phosphor Bestimmung in P-Düngemitteln. In: *Praktikum der Pflanzenernährung*, pp. 43–46. Institut für Pflanzenernährung und Bodenkund der Agrarwissenschaftlichen Fakultät der ChristianI Albrechts-Universität zu Kiel. [https://www.researchgate.net/profile/Ewald-Schnug/publication/301788224_Praktikum_der_Pflanzenernahrung/links/5728565008aef9c00b8b5a4b/Praktikum-der-Pflanzenernaehrung.pdf] (last access 27.05.2021).

Schnug, E. 2014. Where has all the uranium gone? Or what feeds Dimona – circumstantial evidence for an illicit fate of uranium from rock phosphate processing. In: *Uranium - Past and Future Challenges*, eds. B. J. Merkel, and A. Arab, pp. 731–738. Heidelberg: Springer Verlag.

Schnug, E. and N. Haneklaus. 2008. Contribution of uranium in rock phosphates to environmentally-sound and cost-efficient energy production. In: *The New Uranium Mining Book*, eds. B. Merkel, and M. Schipek, pp. 789–793. Heidelberg: Springer Verlag.

Schnug, E. and S. H. Haneklaus. 2016. The enigma of fertilizer phosphorus utilization. In: *Loads and Fate of Fertilizer-Derived Uranium*, eds. L. J. de Kok, and E. Schnug, pp. 7–26. Leiden: Backhuys Publishers. [https://www.springer.com/de/book/9789401776110] (last access 14.03.2019).

Schnug, E., S. H. Haneklaus, M. Maekawa, Y. Sun and R. Bol. 2018. Uranium, a new villain among the dirty dozen in soil protection. *Global Symposium on soil pollution (GSOP18)*, May 2–4, Rome, Italy, [https://www.researchgate.net/publication/331824830_Uranium_a_new_villain_among_the_dirty_dozen_in_soil_protection] (last access 18.03.2019).

Schnug, E., S. H. Haneklaus, U. Hundhausen, F. Knolle, F. Jacobs and M. Birke. 2019. Significance of geographical, hydro-geological and hydro-geochemical origin for the elemental composition of bottled German mineral waters. In: *Bottled and Packaged Water*: Volume 4: The Science of Beverages, eds. A. Grumezescu, and A. M. Holban, pp. 277–309. Sawston: Woodhead Publishing.

Schuchard-Ficher, C., K. Backhaus, U. Humme, W. Lohrberg, W. Plinke and W. Schreiner. 1982. *Multivariate Analysenmethoden*. Berlin: Springer Verlag.

Shemesh, A., Y. Kolodny and B. Luz. 1983. Oxygen isotope variations in phosphate of biogenic apatites, II. Phosphorite rocks. *Earth and Planetary Science Letters* 64:405–416.

Shemesh, A., Y. Kolodny and B. Luz. 1988. Isotope geochemistry of oxygen and carbon in phosphate and carbonate of phosphorite francolite. *Geochimica et Cosmochimica Acta* 52:2565–2572.

Smolders, E. and L. Six. 2013. Revisiting and updating the effect of phosphate fertilizers to cadmium accumulation in European agricultural soils. [https://ec.europa.eu/health/scientific_committees/environmental_risks/docs/scher_o_168_rd_en.pdf] (last access 30.05.2021).

Sun, Y. 2021. Uranium accumulation in agricultural soils as derived from long-term phosphorus fertilizer applications. PhD thesis, University of Bonn. [https://bonndoc.ulb.uni-bonn.de/xmlui/bitstream/handle/20.500.11811/8951/6140.pdf?sequence=1&isAllowed=y] (last accessed 31.05.2021).

Tagesschau. 2020. Neue CO_2-Steuer. [https://www.tagesschau.de/wirtschaft/verbraucher/co2-abgabe-tanken-Nheizen-101.html] (last accessed 01.06.2021).

Taylor, M. D. 2017. The fate and impact of fertiliser derived contaminants in New Zealand soils – development of a risk assessment model. [http://publikationsserver.tu-braunschweig.de/get/64107].

Tulsidas, H., S. Gabriel, K. Kiegiele and N. Haneklaus. 2019. Uranium resources in EU phosphate rock imports. *Resource Policy* 61. [https://www.researchgate.net/publication/331223589_Uranium_resources_in_EU_phosphate_rock_imports] (last accessed 20.03.2019).

Turner, D. and M. Whitfield. 1979. Control of seawater composition. *Nature* 281:468.

UBA. 2012. Positionspapier der Kommission Bodenschutz beim Umweltbundesamt: Uran-Einträge in landwirtschaftliche Böden durch Düngemittel. 20. März 2012. [https://www.umweltbundesamt.de/sites/default/files/medien/421/publikationen/positionspapier_kbu_uraneintraege_in_landwirtschaftliche_boeden_durch_duengemittel_0.pdf] (last access 20.03.2019).

Ulrich, A. E. 2018. Cadmium governance in Europe's phosphate fertilizers: Not so fast? *Science of the Total Environment* 650: 541–545.

Ulrich, A. E., E. Schnug, H. M. Prasser, E. Frossard. 2014. Uranium endowments in phosphate rock. *Science of the Total Environment* 478: 226–234.

USGS. 2018. Phosphate rock. *U.S. Geological Survey, Mineral Commodity Summaries, January 2018*: 122–123.

Valkov, A. V., V. A. Andreev, A. V. Anufrieva, Y. N. Makaseev, S. A. Bezrukova and N. V. Demyanenko. 2014. Phosphogypsum technology with the extraction of valuable components. *Procedia Chemistry* 11:176–181.

Van Kauwenbergh, S. 1997. Cadmium and other minor elements in world resources of phosphate rock. Proceedings, The Fertilizer Society.

Villalba, G., Y. Liu, H. Schroder and R. Ayres. 2008. Global phosphorus flows in the industrial economy from a production perspective. *Journal of Industrial Ecology* 12: 557–569.

Weber, A. P. 1964. Tod und Teufel. [https://oamenicasobolani.files.wordpress.com/2014/04/1964_duenger.jpg and http://www.artnet.com/artists/andreas-paul-weber/tod-und-teufel-QEJJFaewk02KrKnidRjjUA2] (last access 21.05.2021).

WHO. 2021. International Programme on Chemical Safety-cadmium. [https://www.who.int/ipcs/assessment/public_health/cadmium/en/#:~:text=Cadmium%20exerts%20toxic%20effects%20on, media%20relevant%20to%20population%20exposure] (last accessed 01.06.2021).

Windmann, H. 2020. Ein Beitrag zur Gefährdungsabschätzung der Kontamination der Nahrungskette durch Uran aus phosphorhaltigen Düngemitteln, Lebensmittel- und Futtermittelzusatzstoffen. PhD thesis Technical University of Braunschweig. [https://publikationsserver.tu-braunschweig.de/receive/dbbs_mods_00068165] (last access 25.05.2021.)

Wing, J. 2016. The phosphogypsum era - utilizing the phosphate industry's most abundant co-product. *AIChE Clearwater Phosphate Conference*, Clearwater Beach, Florida, USA, 11 June 2016, 9pp.

Wu, S., L. Wang, L. Zhao, P. Zhang, H. El-Shall, B. Moudgil, X. Huang and L. Zhang. 2018. Recovery of rare earth elements from phosphate rock by hydrometallurgical processes – A critical review. *Chemical Engineering Journal* 335:774–800.

Zewainy, R. M. M. 2008. Evaluation of soil extraction methods for uranium. PhD thesis Technical University of Braunschweig. [https://publikationsserver.tu-braunschweig.de/servlets/MCRFileNodeServlet/dbbs_derivate_00004860/Raafat%20Zewainy%20thesis.pdf] (last accessed 31.05.2021).

Zhang, F., Q. Wang, J. Hong, W. Chen, C. Qi and L. Ye. 2017. Life cycle assessment of diammonium- and mono-ammonium-phosphate fertilizer production in China. *Journal of Cleaner Production* 141:1087–1094.

Zimmermann, M. 2018. Herleitung von Grenzwerten für die neue Düngerkategorie «Mineralische Recyclingdünger». Bundesamt für Landwirtschaft BLW Fachbereich Agrarumweltsysteme und Nährstoffe. [https://www.blw.admin.ch/dam/blw/de/dokumente/Nachhaltige%20Produktion/Produktionsmittel/Duenger/Grenzwerte%20MinRec_Vernehmlassungsbericht.pdf.download.pdf/Bericht%20zur%20Herleitung%20von%20MinRec-Grenzwerten_2.pdf] (last access 25.03.2019).

9 Soil Crust
Formation, Influence on Soil Productivity, and Management

Nanthi Bolan[1,2], Gerhardus Nortje[3], Shiv Bolan[1,2], Prashant Srivastava[4], Manish Kumar[5], Hailong Wang[6], DeAnn Presley[7], and M.B. Kirkham[7]

[1]UWA School of Agriculture and Environment, The University of Western Australia, Perth, Australia
[2]The UWA Institute of Agriculture, The University of Western Australia, Perth, Australia
[3]Department of Environmental Sciences, College of Agriculture & Environmental Sciences, UNISA Florida Science Campus, Johannesburg, South Africa
[4]The Commonwealth Scientific and Industrial Research Organisation (CSIRO), Environment Business Unit, Waite Campus, Urrbrae, Australia
[5]Engineering Department, Palermo University, Viale delle Scienze, Palermo, Italy
[6]Biochar Engineering Technology Research Center of Guangdong Province, School of Environmental and Chemical Engineering, Foshan University, Foshan, P.R. China
[7]Department of Agronomy, Kansas State University, Manhattan, Kansas, USA

CONTENTS

9.1 INTRODUCTION

Soil crusts are formed from specific modifications (structural breakdown) in the topsoil resulting from natural events such as raindrop impact followed by the drying process (Sumner and Miller, 1992). Soil crusts are formed as dense layers at the soil surface and are common in arid and semiarid regions. The thickness of soil crusts usually ranges from 1 mm to 5 cm; in specific cases, they are less than 1 mm, and under dry conditions, the crust layers become more compact, hard, and brittle

DOI: 10.1201/9781003093565-9

FIGURE 9.1 Thin soil crust in warm, dry area of South Africa.

than underlying soil components (Figure 9.1). Soil crust formation decreases both the size and the number of pores and also modifies the arrangement and distribution of the pore system.

Soil crusts are grouped into three morphological types that include physical crusts, chemical crusts, and biological crusts. While physical crusts are the result of modifications of topsoil caused by physical perturbation such as raindrop impacts or sedimentation, chemical crusts are formed by salt encrustation, and biological crusts are formed by soil biological communities such as lichens and algae-binding soil particles together (Coxson, 2002). Physical crusts are widespread and include structural crusts and depositional crusts.

While in temperate areas, soil crusts are formed in unstable loamy soils, especially when cultivated, in tropical areas, soil crusting develops on a wider range of soil types and becomes a serious soil constraint throughout the range of climatic regimes including drier regions. In relation to agricultural production, the major negative impacts of soil crusts are the restrictions on seedling emergence and water infiltration (Awadhwal and Thierstein, 1985). The decrease of water infiltration in crusted soils not only impacts crop production but also increases the surface runoff resulting in soil erosion (Figure 9.2) and environmental risks due to the contamination of surface waters from nutrient and pesticide losses by the enhanced runoff (Moore and Singer, 1990).

Soil crusting can be avoided by providing surface mulching to protect the soil from the impacts of raindrops leading to the formation of surface crust layers (Laker and Nortjé, 2019). The addition of organic matter sources such as composts and manures is likely to promote stable aggregate formation, thereby preventing crust formation. Soil constraints resulting from crusting can be managed by controlling cultivation practices including surface tilling. This chapter covers the major kinds of soil crusts, the formation and factors affecting their formation, soil constraints resulting from crust formation, and the agronomic practices to manage soil crust formation.

9.2 KINDS OF SOIL CRUSTS AND THEIR FORMATION

9.2.1 Kinds of Soil Crusts

Soil crusts are grouped into three morphological types that include physical crusts, chemical crusts and biological crusts. Two main groups of crusts are found in soils, namely mineral and biological crusts that include structural crusts, erosion and depositional crusts (Figure 9.3), which are widespread and formed by dense packing of soil particles (Laker & Nortjé 2019).

Structural crusts include slaking crusts, infiltrating crusts, coalescing crusts, and sieving crusts. Slaking crusts are formed due to the dispersion (or physical disruption) of weakly aggregated soils such as sodic soils by the impact of raindrops (Laker and Nortjé, 2019). Textural separation between coarse fractions (skeleton) and fine fractions (plasma) cannot be readily noticed within the crust layer. Infiltrating crusts are generally formed on medium-textured loamy soils due to the impact of raindrops, which promote the erosion of the surface layer aggregates, thereby resulting in the

FIGURE 9.2 Soil erosion due to soil surface crusting.

FIGURE 9.3 Soil crusts on bare soils due to raindrop impact.

separation of silt particles and sealing of void space at the surface. Textural separation is clearly observed within a crust layer with silt-sized grains forming net-like infillings on the surface of the crust layer and clay coatings on soil aggregates. Porosity is reduced due to the clogging of inter-aggregate as well as inter-grain void spaces by an infiltrating silt-size fraction. Coalescing crusts are formed in wet soils when the soil particles are in a viscous state caused by the raindrop energy-driving force. They are generally thick (thickness up to 20 mm), more densely packed than

FIGURE 9.4 Soil pedestals as result of soil crusting and erosion.

the underlying horizon, and exhibit a gradual transition to the underlying layer for which reason they are sometimes called transitional micro-horizons. Porosity is low at the surface, and it gradually increases with depth. Sieving crusts are formed by the impact of raindrops on sandy soils with unstable aggregates, leading to redistribution of soil particles with coarser fractions at the top and finer fractions in a deeper position. Thus sieving crusts are made up of two contrasting layers: the upper layer consisting of loosely packed coarser skeleton grains, and the underlying layer containing a high amount of fine particles resulting in very low porosity.

Erosional crusts are developed from the erosion of sieving crusts when the loose coarse fraction textured surface layer is stripped away by overland flow, exposing the underlying sheet of fine fractions at the surface layer. The formation of definite features, such as plants on pedestals (Figure 9.4), are clear characteristics of erosional soil crust formation. Depositional crusts are formed after sealing of the surface by a structural crust and accumulation of detached fractions from eroded aggregates. Depositional crusts are made up of a 5–10 mm or thicker sedimentary layer over a previously developed structural crust. Clear textural variation is noticed between the depositional crust layer and the underlying, previously developed structural crust.

Soil crusting is also linked to biological and chemical processes in soil. A biological crust is a soil surface layer consisting of a living community of various organisms including lichens, cyanobacteria, algae, and mosses that bind the soil particles together (Menon et al., 2011). Biological soil crusts are formed by an intimate association between soil particles and cyanobacteria, green algae, micro-fungi, bacteria, lichens, and bryophytes in the surface layers of soil (Figure 9.5). Surface crusts made of these microbial organisms that lie over any of the physical and chemical crusts are known as cryptogamic. They are also known as microphytic or biological crusts as they consist of a microbiological layer over one of the various types of physical and chemical crusts (Figure 9.3). Cryptogams are also formed due to the multiplication of various microbes on structural and depositional crusts. Chemical crusts are formed due to the precipitation of salts in soils with high salt content such as saline soils and calcareous soils.

9.2.2 Formation of Soil Crusts

Soil crusting occurs through physical, chemical, and biological processes. As discussed above, soil crusts formed by physical processes are more widespread. Generally, soil crusts are formed in soils where the physicochemical stability of surface aggregates is low. The major processes involved in the formation of soil crusts include: (i) mechanical disintegration of soil surface aggregates by raindrop and other physical impacts such as animal tramping and machinery traffic (Nortjé et al., 2012);

FIGURE 9.5 Biological soil crust under sprinkler irrigation in an avocado orchard (South Africa).

(ii) dispersion and transport of fine fractions and their subsequent accumulation in the underlying pores ("washing in"); (iii) compaction of the soil surface, thereby resulting in the formation of a thin compact layer which inhibits the entry of water and the movement of fine fractions in the soil pores (Nortjé, 2014; Laker and Nortjé, 2020); and (vi) cementation of the slaked soil particle fractions at the surface caused by drying and reorientation (Laker and Nortjé, 2019; Stoops et al., 2018). During subsequent drying, the orientation of fine fractions is likely to contribute to the rigidity and physical stability of the soil crusts.

A number of binding agents have been identified that are related to soil aggregation and soil structural stability. These play a major role in soil crust formation and include organic substances, silica, sesquioxides, liming materials, and exchangeable cations (Laker and Nortjé, 2019). Water and dissolved solutes determine the spatial arrangement of the soil particles in the soil. Therefore, a saturated or near-saturated soil is essential for crust formation. As soil dries out, soil water tension increases, bringing the particles closer together. Thus the combined effects of increased soil water tension, reduced pore size, and increased number of contact points enhance bond formation. The nature of exchangeable cations also affects the particle distribution and packing.

Horizontal stratification causes peeling of the crusts from the underlying soil layer, and during drying, the cracking of crust occurs. Soil crusting is also considered as a form of aggregation. However, soil crusts may lose their strength on the application of water to soil due to swelling and softening of binding or cementing agents.

9.3 FACTORS AFFECTING SURFACE CRUST FORMATION

The susceptibility of soils to crust formation not only depends on external environmental factors, such as raindrop impact and climatic conditions, but also on intrinsic soil properties that include soil particle size fractions (i.e., texture), pH, and organic matter and salt contents.

Soil texture and clay mineralogy: Soil particle size distribution, especially the amount and nature of clay content, affects soil crusting. Soils with high clay contents are likely to promote aggregation, thereby reducing the potential for crust formation, although clay mineral composition and the nature of exchangeable cations can impact the effect of clay content on crust formation (Laker and Nortjé, 2019). Soils dominated by kaolinite clay minerals are less susceptible to crust formation than soils dominated by smectite, illite, and micaceous minerals, which promote crust formation. Specific soil components, such as organic matter and Fe and Al hydrous oxides, can serve as stabilizing agents, thereby controlling aggregate formation and crust development. Soils with relatively low clay content (i.e., medium-textured soils with <20% clay content) are generally more susceptible to crusting than soils with high clay contents. However, soils with fine fractions (i.e., fine-textured soils or high fine sand+silt content) are likely to be subjected to crust formation, especially in the presence of the splashing effect of raindrops, thereby resulting in increased runoff and erosion of fine fractions (Laker and Nortjé, 2019). In extremely sandy soils, the low amount of dispersion clay may not be sufficient to block the conducting pores at the soil surface. Coarse fragments including sand and gravel are likely to protect the surface aggregates formed by fine fractions from raindrop impact, thereby promoting infiltration and reducing erosion. Thus, in coarse-textured sandy soils, soil crust layers are generally thin and unstable (Figure 9.1).

Organic matter content: Soil crusting is related to the stability of the soil structure in the surface soil. Soil organic matter and sesquioxides are the main aggregate stabilizing agents in soils (Laker and Nortjé, 2019). Generally the susceptibility to crusting increases especially when soils are intensively cultivated, which can partly be attributed to the progressive decrease of organic matter content with soil cultivation. Low organic matter contents, especially in coarse-textured sandy soils, can lead to poor soil structure, reduced pore space, and weak, unstable aggregates that disintegrate from the impact of raindrop splashing. Physical disintegration of soil aggregates resulting from rainfall and runoff can disperse the soil aggregates into individual particles that block soil pores and the seal soil surface with crust formation. Even in the presence of high organic matter contents, soils with high sodium content, such as sodic and saline soils, are more likely to promote surface crusts since these soils are more readily subjected to dispersion with rainfall and irrigation.

Exchangeable cations: Both the amount and nature of cations in the soil solution and exchange sites play an important role in crust formation, which is attributed to their effects on the dispersion and flocculation of soil colloids. It has been demonstrated that a high percentage of exchangeable sodium, as in the case of sodic and saline-sodic soils, favors clay dispersion, thereby resulting in crust formation, whereas soils with high exchangeable calcium content favor flocculation, as in the case of limed soil, which can overcome crust formation. However, calcareous soils can promote the development of surface crusts, especially under dry conditions, which is attributed to the precipitation of salts, thereby resulting in "chemical crust" formation.

Soil water content: Aggregates disintegrate more readily when they are initially dry and then undergo rapid and sudden wetting resulting from rainfall impact. Therefore, air-assisted slaking and dispersion of soil fractions occur more rapidly when raindrops hit dry soil aggregates compared to soil aggregates that are already wet (Laker and Nortjé, 2019) (Figure 9.3). It has often been observed that when the raindrops spread on a relatively dry aggregate, its periphery is wetted rapidly and water moves into the aggregate, resulting in the compression of the air trapped inside the aggregate. As the moistened aggregate zone is weakened by the swelling and the pressure of entrapped air, the aggregate disintegrates and the trapped air escapes as bubbles (i.e., air slaking), resulting in the collapse and complete disintegration of aggregates. Thus, the raindrop impact can lead to mechanical disintegration of soil aggregates, which is attributed to the kinetic energy of raindrops and the air slaking process.

9.4 INFLUENCE OF SOIL CRUSTS ON SOIL PRODUCTIVITY

Soil crusts can lead to reduced infiltration and air exchange between the soil and atmosphere, and poor seedling emergence. Surface sealing resulting from crust formation significantly reduces infiltration and increases runoff and erosion. Increased runoff can result in less water available in soil for plant growth and increase the loss of nutrients and pesticides, thereby leading to surface water pollution. Thus, surface crusts are considered one of the reasons for inducing runoff and favoring interrill soil erosion (Figure 9.2). Stable crusts with relatively high bulk densities can impede aeration under moist conditions by preventing effective diffusion of air into the soil, and thereby significantly reducing the subsequent oxygen diffusion into the soil profile if the soil crust is wet. Diffusion of oxygen is not restricted in case of dry crusts because of cracks.

The relatively smooth surface of a crusted soil initially tends to promote wind erosion, especially in the case of sandy soils. Loose sand particles are carried away by wind erosion leading to the roughening of the surface crust with eventual reduction in wind erosion. However, in the case of soils with a small amount of sand, hard crusts can protect the soil surface from wind erosion.

Since surface crusts are relatively hard and difficult to break, soil crust formation restricts seed germination and seedling emergence, especially in non-grass small-seeded crops such as canola and alfalfa. Thus, soil crusts can cause a serious physical barrier for seedling emergence. This is not just because of the strength of the crust, but also because the crust limits aeration below the soil crust (Laker and Nortjé, 2019). Furthermore, upon drying, a hard soil crust can crack and tear tender seedling roots. It has been demonstrated that the sunlight reflectance of a surface crust is higher than that of a tilled soil, thereby resulting in low soil temperature, which is likely to impact seed germination and the subsequent establishment of healthy seedlings, especially in cooler weather conditions (Laker and Nortjé, 2019; Nortjé, 2014) (Figure 9.3). Lack of aeration becomes a problem for germination of seeds, which also accounts for a decrease in seedling emergence. Thus, crust development soon after seeding can result in poor germination and crop emergence, thereby leading to the requirement for reseeding of the crop. The effect of soil crusting on seedling emergence depends on crust thickness and strength, as well as on the size of the seeds and vigor of the seedlings. For example, small-sized seeds, such as canola and lentil, are more sensitive to soil crusting.

Soil crusts can contribute to some limited benefits for soil productivity. Soil crust formation decreases evaporation loss of water, which is attributed to the exposure of less surface to air in the presence of a surface crust compared to a tilled soil. In addition, a crust formation can act as physical barrier to the evaporation of soil moisture. A reduction in evaporation of soil moisture can contribute to higher potentially plant available water content in soils. Biological crusts can contribute a number of benefits to soil that include stabilization of soil aggregates, thereby preventing erosion, promoting biological nitrogen fixation, providing a habitat and food for invertebrates, and assisting with water infiltration (Coxson, 2002; Elliott et al., 2014)).

9.5 MANAGEMENT OF SOIL CRUSTS

Some of the major practices that lead to soil crusting include the following: removal of plant residues through harvesting, burning (Mills and Fey, 2004), or burying, thereby leading to the exposure of bare soil surface for an extended period of time, and soil disturbing activities such as cultivation and animal tramping that can result in the decomposition of organic matter and damage soil structure and aggregation (Laker and Nortjé, 2019; Nortjé, 2014). Thus, agricultural management practices that can lead to the depletion of soil organic matter and exposure of bare soil to the direct impact of water droplets are likely to increase dispersion of soil fractions, surface sealing, and crusting. Excessive tillage and animal tramping can lead to the breaking of soil clods into smaller fractions, incorporate most plant residues with active soil fractions, and accelerate the loss of organic matter through microbial decomposition. In the absence of a crop's protective canopy, crop residue removal can lead to the exposure of soil to direct sunlight, which increases soil temperature, thereby accelerating the drying of the soil surface and the formation of a hard crust.

Soil crusting can be controlled by various approaches that include: (i) surface mulches (Figure 9.6), which provide protection to the soil from the impact of rain drops, thereby minimizing the formation of soil crusts; (ii) addition of organic matter sources such as composts and manures, which promote stable aggregate formation and resist dispersion, thereby preventing crust formation;

FIGURE 9.6 Soil mulching under macadamia tree.

(iii) application of some synthetic soil conditioners such as polyacrylamide (PAM), which reduce soil crust formation by producing stable aggregation; (iv) in sodic soils, lowering of the exchangeable sodium percentage of the soil by application of gypsum, which minimizes the formation of soil crust; and (v) a light tillage while the soil is still moist, which will break-up the crust before it hardens (Laker and Nortjé, 2019).

Farming practices that protect the soil structure and organic matter, and provide a protective vegetative canopy or residue cover on the soil surface, inhibit the development of soil crusts or minimize their negative impacts. These management practices that inhibit the formation of soil crusts include cover cropping, residue return, conservative tillage, and amelioration of saline and sodic soils.

No-till or reduced tillage of cropland helps to reduce or eliminate crust formation. Crop residues on the soil surface intercept the kinetic force of falling raindrops and also serve as a source of organic matter. Organic matter binds soil fractions and promotes and stabilizes soil aggregates making them more resistant to disintegration resulting from the physical impact of raindrops. Improved soil structure resulting from stable aggregate formation results in decreased bulk density and increased pore space, thereby improving water infiltration and water movement through soil.

In the case of soils high in sodium content (i.e., saline and sodic soils), the development of surface crusts can be overcome through improved irrigation water management practices that prevent sodium accumulation at the surface, and the application of gypsum to promote flocculation and inhibit dispersion of soil particles. Shallow, light tillage operations, such as with a rotary hoe or row cultivator preferably when the soil is still moist, can be practiced to break soil crusts. Light tillage also promotes seed germination, seedling emergence, and weed control.

Among soil management practices, soil tillage practices are likely to have the major influence on soil crusting. In long-term, intensively cultivated soils, the decrease in soil organic matter content is likely to weaken the stability of soil aggregates, thereby increasing the susceptibility of these soils for crust formation. However, soil crust formation occurs when the soil is bare, thereby subjecting surface aggregates to the direct action of raindrop impact. In cultivated soils, depositional crusts are more likely to occur because, after the mechanical destruction of soil aggregates through cultivation, the dispersed soil fractions can be transported by runoff and erosion. During the subsequent drying event, the deposition of the eroded fraction can occur on a previously formed crust, presenting a layer of different texture.

Thus, continuous conventional tillage is one of the major factors contributing to soil crust formation. It has often been demonstrated that conservation tillage practices, including zero tillage, minimum tillage, surface mulching (Figure 9.6), and contour ploughing, minimize soil loss through reduced runoff, thereby preventing and controlling surface crust formation. Surface crusts are formed more frequently in tilled soils than those under no-tillage, minimum tillage, and chisel ploughing, which indicates that soil tilling inhibits soil aggregate formation, thereby resulting in dispersion and promoting surface crust formation. Therefore, the adoption of reduced tillage practices and the introduction of farming rotations, including cover crops, instead of continuous monocultures, help prevent soil degradation including soil crusting, which, consequently, reduces surface runoff and soil erosion. Regular application of organic materials including manures, composts, and biosolids is an important management practice for the prevention of crust formation. These organic materials can contribute to the binding of soil fractions to form stable aggregates that are resistant to rainfall impacts, thereby reducing runoff, soil erosion, and surface crust formation. In the case of sodic soils, the addition of soil amendments such as gypsum and phospho-gypsum inhibits formation of crusts, especially when the exchangeable sodium content (i.e., sodicity) is high or the electrolyte concentration is low, which can lead to dispersion (Laker and Nortjé, 2019). Paterson (2014) obtained positive results with palm mat mulches to control erosion.

Irrigation management is also likely to influence the development of soil surface crusts. The electrolyte composition (i.e., salt composition) of irrigation water and the kinetic energy of water applied through overhead irrigation technologies are considered as the major factors in controlling soil crust formation. While irrigation water dominated by sodium salt promotes crust formation, calcium-dominated irrigation water inhibits and prevents crust formation. Furthermore, a number of studies have demonstrated that the application of synthetic soil conditioners, including dextran, polyvinyl alcohol, and polyacrylamide, help to increase the aggregate stability of soils, thereby reducing soil erosion and crust formation.

9.6 CONCLUSIONS

Soil crusts include physical crusts, chemical crusts, and biological crusts. While physical crusts are the result of modifications of topsoil caused by physical perturbation such as raindrop impacts or sedimentation, chemical crusts are formed by salt encrustation, and biological crusts are formed by soil biological communities such as lichens and algae binding to soil particles. Physical crusts are the most common, especially under managed farming conditions, and include structural crusts and depositional crusts. Soil crusts restrict seedling emergence and water infiltration. While the restriction of seedling emergence impacts crop growth, the restriction of water infiltration can lead to soil erosion. Soil crusting can be controlled by providing surface mulching to protect the soil from the impacts of raindrops leading to the formation of surface crusts. The addition of organic materials such as manures and composts, which promote stable aggregate formation, and minimizing cultivation that destroys stable aggregates, are methods to prevent crust formation.

REFERENCES

Awadhwal, N.K. and Thierstein, G.E. (1985). Soil crust and its impact on crop establishment: A review. *Soil Tillage Res.*, 5, 289–302.

Coxson, D. (2002). Biological soil crusts: Structure, function, and management. *Bryologist,* 105, 500–501.

Elliott, D.R., Thomas, A.D., Hoon, S.R., and Sen, R. (2014). Niche partitioning of bacterial communities in biological crusts and soils under grasses, shrubs and trees in the Kalahari. *Biodivers. Conserv.*, 23, 1709–1733.

Laker, M.C. and Nortjé, G.P. (2019). Review of existing knowledge on soil crusting in South Africa. *Adv. Agron.*, 155, 190–240.

Laker, M.C. and Nortjé, G.P. (2020). Review of existing knowledge on subsurface soil compaction in South Africa. *Adv. Agron.,* 162, 143–197.

Menon, M., Yuan, Q., Dougill, A.J., Hoon, S.R., Thomas, A.D., and Williams, R.A. (2011). Assessment of physical and hydrological profiles of biological soil crusts using microtomography and modelling. *J. Hydrol.*, 397, 47–54.

Mills, A.J. and Fey, M.V. (2004). Frequent fires intensify soil crusting: Physicochemical feedback in the pedoderm of long-term burn experiments in South Africa. *Geoderma*, 121, 45–64.

Moore, D.C. and Singer, M.J. (1990). Crust formation effects on soil erosion processes. *Soil Sci. Soc. Am. J.*, 54, 1117–1123.

Nortjé, G.P. (2014). Studies on the impacts of off-road driving and the influence of tourists' consciousness and attitudes on soil compactionand associated vegetation in the Makuleke Contractual Park, Kruger National Park. Doctoral dissertation, University of Pretoria.

Nortjé, G.P., Van Hoven, W., and Laker, M.C. (2012). Factors affecting the impact of off-road driving on soils in an area in the Kruger National Park, South Africa. *Environ. Manage.*, 50(6), 1164–1176.

Paterson, D.G. (2014). The use of palm leaf mats in soil erosion control. Doctoral dissertation, University of Pretoria.

Stoops, G., Marcelino, V., and Mees, F. (Eds.) (2018). *Interpretation of Micromorphological Features of Soils and Regoliths*. Elsevier: Amsterdam, Netherlands.

Sumner, M.E., and Miller, W.P. (1992). Soil crusting in relation to global soil degradation. *Am. J. Altern. Agric.*, 7(1/2), 56–62.

10 Hydrochar Production for Soil Environmental Improvement

Shiyu Xie and Tao Zhang
Beijing Key Laboratory of Farmland Soil Pollution Prevention and Remediation, Key Laboratory of Plant-Soil Interactions of Ministry of Education, College of Resources and Environmental Sciences, China Agricultural University, Beijing, P.R. China

CONTENTS

Biomass waste is waste generated from the growth and reproduction processes of plants and animals, and it includes manure, sludge, straw, and aquatic algae (Chew and Doshi, 2011; Deng et al., 2019; Li et al., 2022b; Wang et al., 2017; Zhang et al., 2021). The conversion of biomass waste into a sustainable resource is currently a reliable method of waste treatment. At present, biomass waste resource utilization methods include direct return to the field, anaerobic fermentation, aerobic composting (Li et al., 2020), and incineration for power generation. Livestock manure contains a high proportion of unstable carbon, which is easily and rapidly consumed by soil microbial respiration. Direct application has limited use to increase soil carbon sequestration; it leads to increased soil carbon emissions (Xia et al., 2017). Crop return to the field may increase methane emissions as well as intensify pest and disease occurrence. Livestock manure composting technology has problems, such as nutrient loss and greenhouse gas emissions, although it can increase a portion of the carbon pool. Anaerobic fermentation technology has drawbacks, such as a long cycle time, large land area requirements, and seasonal influences. The carbon contained in compost and anaerobic

DOI: 10.1201/9781003093565-10

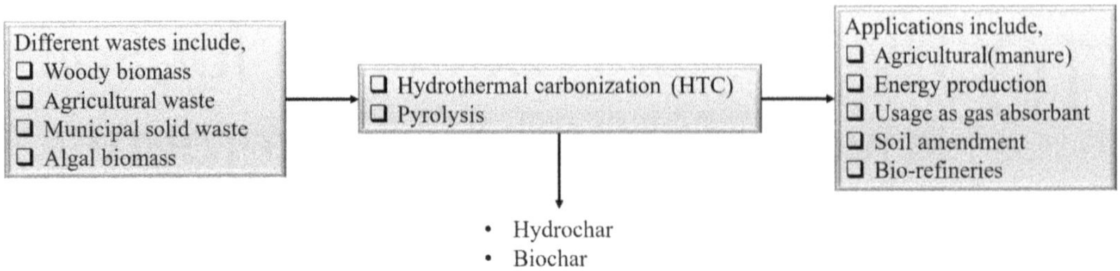

FIGURE 10.1 Hydrothermal carbonization process and application.

fermentation products is not sufficiently stable in the soil and has a short half-life (Gross et al., 2022; Lanza et al., 2018). Incineration for power generation can rapidly reduce biomass waste and generate energy, but the cost of power generation is high due to various factors, such as the nature of the raw materials, collection, transportation, and operating costs. Incineration wastes nutrients and organic matter. High-temperature pyrolysis technology greatly improves the stability of carbon in the material, but it consumes energy and causes carbon loss in the material, and it is not suitable for wet materials with high moisture content.

The hydrothermal carbonization process (HTC) has little loss of material carbon, and the half-life of its carbonization products in soil can be decades, increasing carbon sequestration rate (Deng et al., 2020; Malghani et al., 2015) and nutrient fixation. These products can be used as good soil amendments to supplement nutrients and achieve maximum recycling of organic carbon and nutrients in biomass. Based on the structure and properties of hydrothermal charring products, their return to fields is an effective way to improve soil fertility, reduce greenhouse gas emissions, and remediate polluted soil (Figure 10.1).

10.1 PRODUCTION OF HYDROCHAR

10.1.1 Hydrothermal Carbonization Process

Hydrothermal carbonization is a thermochemical reaction that converts wet materials into solid products in a liquid environment at a relatively low temperature and pressure. Reactions include hydrolysis, dehydration, decarboxylation, polycondensation, and aromatization (Funke and Ziegler, 2010; Kruse et al., 2013; Zhang et al., 2020a, b). Compared with pyrolysis technology (Table 10.1), hydrothermal carbonization technology has advantages, including mild hydrothermal carbonization reaction conditions and low-energy consumption. It is the preferred method for wet waste biomass treatment because it does not require pre-treatment of drying and dewatering (Marcus, 1999; Savage, 1999; Thunshirn et al., 2021).

The hydrothermal carbonation reaction medium is an aqueous solution, as shown in Figure 10.2. Reactions usually occur in the subcritical region, below 374°C and 22.1 MPa, where ionic reactions occur, so increasing the ionic strength is the main way to promote hydrothermal carbonation. Under subcritical region conditions, the dielectric constant decreases and the hydrogen bonding of water weakens, resulting in a high ionization constant, which causes the decomposition of water into acidic hydronium ions (H_3O^+) and basic hydroxide ions (OH^-) (Marcus, 1999; Savage, 1999). H_3O^+ has the effect of increasing the saturation of organic materials and promoting the breaking of chemical bonds, such as hydrogen bonds, which can catalyze and accelerate the hydrolysis of the raw material, intermediate dehydration, polymerization, and other reactions (Wu et al., 2021), with, ultimately, the formation of hydrochar. At the same time, subcritical water itself has a higher H^+ concentration than liquid water, which is an excellent medium for acid-catalyzed reactions of organic compounds without the addition of acid (Ruiz et al., 2013).

TABLE 10.1
Comparison of the Thermal Technologies for Production of Hydrochar and Biochar (Kambo and Dutta, 2015; Khosravi et al., 2022; Liu et al., 2021; Wang et al., 2018)

	Hydrothermal Carbonization	Pyrolysis
Temperature	180°C–250°C	300°C–700°C
Residence time	Minutes-hours	Hours-weeks
Pressure	Autogenous pressure	0.1 MPa
Main products	Hydrochar	Biochar
Advantages	It can be directly applied to biomass with high moisture	High efficiency and flexibility
Disadvantages	Difficult to collect the products and high requirements for equipment	Not suitable for high moisture biomass waste

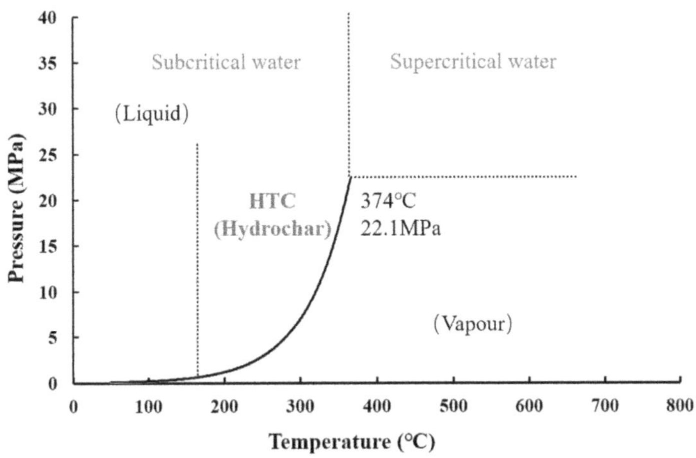

FIGURE 10.2 Classification of hydrothermal processing of biomass with reference to the pressure–temperature phase diagram of water (Kambo and Dutta, 2015).

10.1.2 Feedstock for Hydrothermal Carbonization

The fundamental prerequisite that determines the nature of hydrochar is the biomass raw material. The raw materials are mainly composed of carbon, hydrogen, and oxygen (Shi et al., 2018). Raw materials suitable for hydrothermal carbonization contain different contents of cellulose, hemicellulose, lignin, carbohydrates, and fats, resulting in different strengths and pathways of reactions, such as dehydration and decarboxylation in the hydrolysis process. The different reactions make the nature of hydrochar different. Biomass feedstock can be simply classified as (i) woody biomass, (ii) agricultural waste, (iii) municipal solid waste (MSW), and (iv) algal biomass. Woody biomass includes tree bark, leaves, and wood chips. Agricultural waste includes crop straw, rice husks, corn cobs, and livestock manure. Only the lignocellulosic fraction of MSW, such as coffee residues and food waste, can be hydrothermally carbonized (He et al., 2022; Jung et al., 2019). Algal biomass is a subcategory of lignocellulosic biomass, which comes from the artificial cultivation of algae and microalgae in bioreactors (Khan et al., 2021). Hydrothermal carbonization feedstock can be hydrothermally carbonized not only with a single biomass waste but also with a mixture of multiple biomasses to increase the solid-phase yield and calorific value (Bardhan et al., 2021). Studies have shown that, compared to non-lignocellulosic feedstock hydrochar from livestock manure and sludge, hydrochar from lignocellulosic feedstock, such as straw and sawdust, has a relatively lower yield, a poorer pore structure, and higher carbon content, calorific value, aromatization, and hydrophobicity (Nzediegwu et al., 2021).

10.1.3 Effect of Process Parameters

The main process conditions affecting the properties of hydrochar are temperature, residence time, pH, and substrate concentration. Table 10.2 lists the effects of hydrothermal carbonization process conditions on the different properties of the hydrochar produced (Wang et al., 2018).

10.1.3.1 Temperature

The H/C and O/C atomic ratios of hydrochar steadily decrease with increasing temperature. The increase in temperature deepens the degree of reactions, such as dehydration, decarboxylation, and condensation, releasing more volatile substances and thus increasing the pore structure of hydrochar. Studies have shown that the solid-phase yield increases in the temperature range of 200°C–250°C and decreases at a temperature of 280°C (Sun et al., 2010). Although high temperature leads to a decrease in solid yield, it increases the energy density of the hydrochar. An increase in temperature during hydrothermal carbonization increased the content of carbon, phosphorus, and other nutrients in the solid phase of pig manure hydrochar (Khosravi et al., 2022).

10.1.3.2 Residence Time

For pig manure, an increase in hydrothermal temperature and the time it was given increased the solid-phase C content and reduced oxygen-containing functional groups, which are unstable structures (Fu et al., 2022; Song et al., 2018). The results showed that there was a decrease in H/C and O/C and enhanced thermal stability of hydrochar. The hydrothermal carbonization temperature (T) and time (t) change the basic properties of hydrochar (Table 10.2) (Wang et al., 2018). Therefore, researchers describe changes in temperature and time in terms of hydrothermal reaction severity (R_0) with the equation $R_0 = \mathrm{t} \times \exp[(T - 100)/14.75]$ (Zhou et al., 2010). However, the effect of residence time on solid-phase yield is relatively weak compared to temperature. The decrease in solid-phase yield with increasing residence time is due to the fact that longer residence time leads to the polymerization of dissolved macromolecular fragments in the liquid phase, producing secondary products with aromatic structures (He et al., 2013).

10.1.3.3 pH

The hydrothermal carbonization process is considered to be autocatalytic (Zhang et al., 2020b). The process generates a large amount of organic acids, including formic acid and acetic acid, which

TABLE 10.2
Effect of Process Parameters When They Are Increased on Hydrochar Characterization (Wang et al., 2018)

Characterization \ Parameters	Temperature	Residence Time	pH	Substrate Concentration
Solid yield	–	–	ND	ND
BET surface area	+	+	–	ND
Porosity	+	+	–	ND
H/C	–	–	+	+
O/C	–	–	+	+
Energy densification	+	+	ND	ND

Note: "+": increase effect. "–": decrease effect. ND: not determined.

lead to a decrease in pH (Hoekman et al., 2013). The addition of acid or base to the process leads to an increase in H^+ and OH^- concentrations, which increases the ionic strength and acts as a catalyst. Acidic conditions accelerate the degradation of organic matter, thus reducing the total organic carbon (TOC) content in hydrochar. Both acidic and alkaline conditions increase the solubility of nutrients such as nitrogen and phosphorus, causing their content to decrease in solid-phase hydrochar (Song et al., 2020).

10.1.3.4 Substrate Concentration

When the concentration of biomass feedstock is high, the polymerization reaction will occur in a short reaction time, so that most of the reaction substances are not hydrolyzed. Meanwhile, due to incomplete hydrolysis, the hydrochar will have high O/C and H/C and the particle size of the hydrochar will become smaller (Sevilla and Fuertes, 2009). However, the biomass feedstock may affect the final result. When easily soluble substrates, such as glucose and sucrose, are used, the diameter of the hydrochar particles increases when the substrate concentration is high (Romero-Anaya et al., 2014).

10.2 CHARACTERIZATION OF HYDROCHAR

The characterization of hydrothermal carbonization products helps to understand the carbonization process and its mechanism. In characterization, the product properties are analyzed and their selective utilization can be determined. The analytical methods for charring raw materials and their product properties mainly include yield analysis, coalification degree analysis, elemental analysis, and chemical structure analysis.

10.2.1 Solid Yield

Khan et al. (2021) showed that the solid yield of hydrochar produced by hydrothermal carbonization of different biomass feedstocks ranged from 28.6% to 86.9%, and that hydrothermal temperature played a key role in hydrothermal carbonization. During the HTC reaction, the hydrothermal carbonization yield decreased with increasing temperature for all biomass feedstocks except algal biomass because the increase in temperature led to the generation of a large amount of organic acids (e.g., acetic acid, formic acid, and lactic acid). Dehydration and decarboxylation promoted the dissolution of mineral components in the feedstock and thus reduced the solid yield (Kambo and Dutta, 2015). In addition, Khan et al. (2021) found that woody biomass produced a hydrochar with the widest range of temperature regulation and the largest solid yield, while algal biomass produced a hydrochar with the smallest range of temperature regulation and the opposite trend of solid yield.

10.2.2 The Coalification Degree

The elemental composition (H/C, O/C) of hydrochar changes as a result of the hydrothermal reaction, indicating a change in the degree of carbonization of the hydrochar. Generally, hydrothermal carbonization performed at high temperatures produces low mass yields of solid phases; however, an increase in temperature increases the proportion of fixed carbon and decreases the H/C and O/C ratios in the hydrochar, indicating that a significant amount of carbon is retained in the hydrochar, which increases its stability. It can be further seen from the van-Krevelen plot (Figure 10.3) that the O/C and H/C atomic ratios of biochar are lower than those of hydrochar, indicating that hydrochar is less carbonized than biochar. This shows that hydrochar is less stable than biochar.

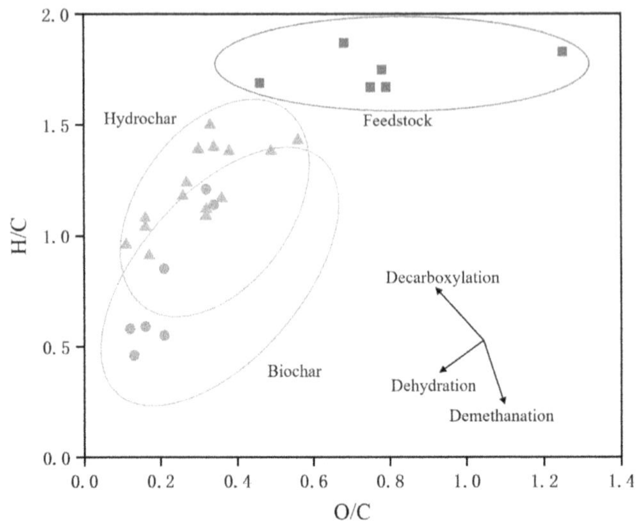

FIGURE 10.3 van-Krevelen diagram for biochar (circles) and hydrochar (triangles) produced from different feedstocks (squares) reported in the literature (Cheng et al., 2022; Cárdenas-Aguiar et al., 2019; Gascó et al., 2018; Li et al., 2022a).

10.2.3 Elemental Composition

The elements in hydrochar depend upon the biomass feedstock, and they consist mainly of C, H, O, and N and mineral elements (K, Ca, Mg, Fe, and Al). In hydrothermal carbonization, temperature has a great influence on the elemental composition of hydrochar. With an increase in hydrothermal temperature, the degree of carbonization gradually deepens, and the carbon content of hydrochar generally increases, while the content of hydrogen and oxygen decreases due to decarboxylation and dehydration reactions (Zhang et al., 2019). The carbon content of hydrochar also depends on the type of feedstock, with hydrochar from lignocellulosic biomass containing a higher carbon content than hydrochar from non-lignocellulosic biomass (Kumar et al., 2020; Wang et al., 2020). The content of elements such as N, K, and Na decreases during the hydrothermal process. The reduction of N-containing components is mainly through conversion to gaseous substances volatilized and dissolved into the liquid solution, while K and Na, as water-soluble ions, are mainly dissolved in the liquid produced by hydrothermal carbonization (Fu et al., 2022). Due to precipitation and adsorption of metal cations, pig manure hydrothermal char solids always sequestered most or even all of the phosphorus in the manure raw material, and the total phosphorus content increased with the concentration of the solid-phase volume (Fu et al., 2022; Huang et al., 2017; Heilmann et al., 2014). Most of the pig manure hydrochar phosphorus exists in the form of calcium-bound phosphorus (Xiong et al., 2021), which is mainly a highly stable hydroxyapatite (HAP), accounting for 48.4%–75.5% of the total phosphorus in hydrochar (Huang et al., 2018). Thus, effective phosphorus content is low, which is unfavorable for crop uptake and utilization when applied to soil. Further research is necessary to enhance the effective phosphorus percentage in hydrochar.

10.2.4 Surface Functional Groups

The structural characteristics of hydrochar produced by different feedstock types and process conditions vary greatly. Due to the complex composition of hydrochar and the complicated hydrothermal parameters and raw material changes during the hydrothermal process, most studies have used FTIR techniques to analyze the functional groups on the surface of hydrochar and how they change during hydrothermal processes (Kumar et al., 2020). O-containing and N-containing functional groups are the two most important groups in hydrochar, including hydroxyl (–OH), carboxyl (–COOH), and carbonyl (–C=O). Cheng et al. (2022) and Zhao et al. (2013) showed that the hydroxyl or carboxyl stretching vibrations were weaker in hydrochar made from sludge and straw due to the dehydration condensation reaction that occurred during hydrothermal carbonization. They also showed that the stretching vibration peaks of aliphatic (C–H, C=C) were stronger in hydrochar, which indicated the cleavage of the polymer during HTC, the decarboxylation reaction, and the appearance of aliphatic structures. Studies using ^{1}H NMR to analyze the functional groups in pig manure hydrochar indicate that the main components of hydrochar are aliphatic compounds and that the relative content of aliphatic groups decreases with increasing hydrothermal temperature (Song et al., 2020). The aliphatic structure of the branch chain and the structure of the polymethylene chain were gradually transformed into aromatic substances with the progress of hydrothermal carbonization, due to the polymerization and aromatization of monomers obtained from the initial hydrolysis (Song et al., 2020).

10.3 APPLICATION OF HYDROCHAR FOR SOIL IMPROVEMENT

In early research, the application of hydrochar mainly focused on soil pollution remediation. With the deepening of research and technological progress, the application scope of hydrochar has been continuously expanded, involving the use of hydrochar to sequester carbon and as a soil amendment.

10.3.1 Soil Pollution Remediation

Compared to biochar, hydrochar has a smaller porosity and specific surface area, but it also has a high adsorption capacity due to its surface rich in oxygen-containing functional groups (Zhou et al., 2017). The current excess of nitrogen and phosphorus in soil leads to the risk of loss, which pollutes groundwater and causes eutrophication of water bodies. Hydrochar has a good adsorption capacity due to its adsorption surface area. Zhang et al. (2014) prepared hydrochar using poplar wood shreds, and the amount of phosphate adsorbed reached 386.4 mg/g under Mg/Al-LDH (layered double hydroxide) modification. Using different biomass wastes to adsorb nitrogen and phosphorus, the results showed that hydrochar with high-specific surface area did not adsorb ammonium well compared to hydrochar with low-specific surface area, indicating that surface area is not the most important factor affecting the adsorption capacity of hydrochar. Calcium and magnesium content may affect phosphate ion adsorption (Takaya et al., 2016). Hydrochar can also adsorb heavy metals, such as lead and chromium. Sun et al. (2018) produced hydrochar by hydrothermal carbonization of microalgae raw materials and showed high adsorption for Cr(VI) ions. Similarly, hydrochar can also adsorb organic matter. Chen et al. (2017) used KOH to catalyze the activation of hydrochar, and the specific surface area of hydrochar increased from 1.7 to 1,710 m^2/g, which increased the adsorption capacity of tetracycline by organic matter to 423.7 mg/g.

10.3.2 Carbon Sequestration

Carbon sequestration refers to the process of storing biomass feedstock in soil after it has been converted to hydrochar. Hydrochar has a half-life in the soil of several decades and is effective in increasing soil carbon storage, but it is less stable compared to biochar (Lanza et al., 2018). Malghani et al. (2013) used corn silage to produce biochar and hydrochar, which were applied to soil and compared for carbon sequestration capacity and greenhouse gas emissions. They showed that hydrochar reduced N_2O emissions, but increased CO_2 emissions compared to biochar. Sun et al. (2020) showed that the addition of hydrochar increased the respiration rate of the soil, but, after one year, two-thirds of the hydrochar decomposed at a low rate with an estimated half-life of 19 years. The addition of hydrochar increased the carbon content of soil organic carbon by 1.0%–3.0%, with a decrease of 15.6%–33.6% in unstable soil organic carbon and an increase of 10.3%–27.0% in stable soil organic carbon. Washing of hydrochar before application to soil may reduce surface instability and reduce biological accessibility, thus increasing the carbon sequestration potential of hydrochar (Breulmann et al., 2018). Further research on the stability of hydrochar in soil is needed to exploit its advantages, which biochar does not possess.

10.3.3 Soil Amendment

Hydrochar acts as a slow-release fertilizer, providing essential N, P, and other nutrients to plants. Phosphorus in hydrochar is mainly in the form of calcium-bound phosphorus, which can be converted to plant-available phosphorus over time after application to the soil (Shi et al., 2019; Xiong et al., 2021). Fei et al. (2019) made phosphorus-rich hydrochar from sludge, in which most of the fractions were bound phosphorus, but the phosphorus fraction directly available after applying the hydrochar to the soil was as high as 417 mg/kg, which was still higher than the soil requirement. In contrast, hydrochar made from microalgae can convert soil soluble and exchangeable phosphorus to moderately effective phosphorus. Compared with chemical fertilizers, hydrochar fertilizers have a slower release of phosphorus and a longer lasting increase in phosphorus effectiveness, which helps to improve soil phosphorus utilization (Chu et al., 2021). Similarly, hydrochar can improve plant production by increasing the rate of nitrogen uptake due to the nitrogen-rich raw material of hydrothermal carbonization and the rich oxygen-containing functional groups (–OH, –COOH, –C=O) on the surface of hydrochar, which have a strong adsorption capacity for NH_4^+ (Chu et al., 2020). The nitrogen-rich hydrothermal char can be used as a slow-release fertilizer for crop uptake and utilization after application to the soil.

10.4 CONCLUSION AND AGRICULTURAL ENVIRONMENT IMPLICATIONS

HTC of biomass waste is beneficial in order to control environmental pollution and for advantageous utilization of its products. The distribution and transformation of elements in the original material are influenced by the process conditions, thus affecting the utilization of hydrochar. An increase in reaction temperature and time, although decreasing the solid-phase yield, improves the carbonation and increases the content of the nutrient elements carbon and phosphorus, the stability is also increased. Acidic conditions can catalyze the hydrothermal process and increase the instability of nitrogen and phosphorus in hydrochar. As a carbon-rich material, it can increase soil carbon content to offset carbon emissions and participate in soil carbon sequestration. The richness of oxygen-containing functional groups makes hydrochar have a strong adsorption capacity, which can help in situ remediation of soil contamination. The richness of hydrochar with nutrients (N/P/K) lets the HTC be used as a slow-release fertilizer to support crop production. Therefore, hydrochar has encouraging prospects for use in agriculture. But there are some issues that need to be further explored. (i) Compared with biochar, the carbon sequestration capacity of hydrochar needs to be further improved to reduce soil greenhouse gas emissions. (ii) Hydrochar has certain toxic effects on crop growth, and how to eliminate such toxic effects should be studied. (iii) The application of hydrochar with organic fertilizers and other soil conditioners has been not been studied to any extent, and it is necessary to determine better strategies for the application of hydrochar in agricultural environments.

ACKNOWLEDGMENTS

This research was supported by grants from the National Key Technology Research and Development Program of China (Grant number 2022YFE012907) and the National Natural Science Foundation of China (Grant Number 31401944).

REFERENCES

Bardhan, M., Novera, T.M., Tabassum, M., Islam, M.A., Islam, M.A., Hameed, B.H., 2021. Co-hydrothermal carbonization of different feedstocks to hydrochar as potential energy for the future world: A review. *J. Clean. Prod.* 298, 126734.

Breulmann, M., Kuka, K., van Afferden, M., Buscot, F., Fuhner, C., Muller, R., Schulz, E., 2018. Labile water soluble components govern the short-term microbial decay of hydrochar from sewage sludge. *Arch. Agron. Soil Sci.* 64(6), 873–880.

Cárdenas-Aguiar, E., Gascó, G., Paz-Ferreiro, J., Méndez, A., 2019. Thermogravimetric analysis and carbon stability of chars produced from slow pyrolysis and hydrothermal carbonization of manure waste. *J. Anal. Appl. Pyrol.* 140, 434–443.

Chen, S.Q., Chen, Y.L., Jiang, H., 2017. Slow pyrolysis magnetization of hydrochar for effective and highly stable removal of tetracycline from aqueous solution. *Ind. Eng. Chem. Res.* 56(11), 3059–3066.

Cheng, C., He, Q., Ismail, T.M., Mosqueda, A., Ding, L., Yu, J.Q., Yu, G.S., 2022. Hydrothermal carbonization of rape straw: Effect of reaction parameters on hydrochar and migration of AAEMs. *Chemosphere* 291, 132785.

Chew, J.J., Doshi, V., 2011. Recent advances in biomass pretreatment: Torrefaction fundamentals and technology. *Renew. Sust. Energ. Rev.* 15(8), 4212–4222.

Chu, Q.N., Xue, L.H., Singh, B.P., Yu, S., Muller, K., Wang, H.L., Feng, Y.F., Pan, G., Zheng, X.B., Yang, L.Z., 2020. Sewage sludge-derived hydrochar that inhibits ammonia volatilization, improves soil nitrogen retention and rice nitrogen utilization. *Chemosphere* 245.

Chu, Q., Lyu, T., Xue, L., Yang, L., Feng, Y., Sha, Z., Yue, B., Mortimer, R.J.G., Cooper, M., Pan, G., 2021. Hydrothermal carbonization of microalgae for phosphorus recycling from wastewater to crop-soil systems as slow-release fertilizers. *J. Clean. Prod.* 283, 124627.

Deng, Y., Zhang, T., Sharma, B.K., Nie, H., 2019. Optimization and mechanism studies on cell disruption and phosphorus recovery from microalgae with magnesium modified hydrochar in assisted hydrothermal system. *Sci. Total Environ.* 646, 1140–1154.

Deng, Y., Zhang, T., Clark, J., Aminabhavi, T., Kruse, A., Tsang, D.C.W., Sharma, B.K., Zhang, F., Ren, H., 2020. Mechanisms and modelling of phosphorus solid–liquid transformation during the hydrothermal processing of swine manure. *Green Chem.* 22(17), 5628–5638.

Fei, Y.H., Zhao, D., Cao, Y.D., Huot, H., Tang, Y.T., Zhang, H.G., Xiao, T.F., 2019. Phosphorous retention and release by sludge-derived hydrochar for potential use as a soil amendment. *J. Environ. Qual.* 48(2), 502–509.

Fu, H.B., Wang, B.Y., Wang, H.H., Liu, H., Xie, H.F., Han, L.F., Wang, N., Sun, X.L., Feng, Y.F., Xue, L.H., 2022. Assessment of livestock manure-derived hydrochar as cleaner products: Insights into basic properties, nutrient composition, and heavy metal content. *J. Clean. Prod.* 330, 129820.

Funke, A., Ziegler, F., 2010. Hydrothermal carbonization of biomass: A summary and discussion of chemical mechanisms for process engineering. *Biofuel. Bioprod. Bior.* 4(2), 160–177.

Gascó, G., Paz-Ferreiro, J., Álvarez, M.L., Saa, A., Méndez, A., 2018. Biochars and hydrochars prepared by pyrolysis and hydrothermal carbonisation of pig manure. *Waste Manage.* 79, 395–403.

Gross, C.D., Bork, E.W., Carlyle, C.N., Chang, S.X., 2022. Biochar and its manure-based feedstock have divergent effects on soil organic carbon and greenhouse gas emissions in croplands. *Sci. Total Environ.* 806(3), 151337.

He, C., Giannis, A., Wang, J.Y., 2013. Conversion of sewage sludge to clean solid fuel using hydrothermal carbonization: Hydrochar fuel characteristics and combustion behavior. *Appl. Energ.* 111, 257–266.

He, X., Zhang, T., Niu, Y., Xue, Q., Ali, E.F., Shaheen, S.M., Tsang, D.C.W., Rinklebe, J., 2022. Impact of catalytic hydrothermal treatment and Ca/Al-modified hydrochar on lability, sorption, and speciation of phosphorus in swine manure: Microscopic and spectroscopic investigations. *Environ. Pollut.* 299, 118877.

Heilmann, S.M., Molde, J.S., Timler, J.G., Wood, B.M., Mikula, A.L., Vozhdayev, G.V., Colosky, E.C., Spokas, K.A., Valentas, K.J., 2014. Phosphorus reclamation through hydrothermal carbonization of animal manures. *Environ. Sci. Technol.* 48(17), 10323–10329.

Hoekman, S.K., Broch, A., Robbins, C., Zielinska, B., Felix, L., 2013. Hydrothermal carbonization (HTC) of selected woody and herbaceous biomass feedstocks. *Biomass Convers. Biorefin.* 3(2), 113–126.

Huang, R.X., Fang, C., Lu, X.W., Jiang, R.F., Tang, Y.Z., 2017. Transformation of phosphorus during (hydro) thermal treatments of solid biowastes: Reaction mechanisms and implications for p reclamation and recycling. *Environ. Sci. Technol.* 51(18), 10284–10298.

Huang, R., Fang, C., Zhang, B., Tang, Y., 2018. Transformations of phosphorus speciation during (hydro) thermal treatments of animal manures. *Environ. Sci. Technol.* 52(5), 3016–3026.

Jung, H., Sewu, D.D., Ohemeng-Boahen, G., Lee, D.S., Woo, S.H., 2019. Characterization and adsorption performance evaluation of waste char by-product from industrial gasification of solid refuse fuel from municipal solid waste. *Waste Manage.* 91, 33–41.

Kambo, H.S., Dutta, A., 2015. A comparative review of biochar and hydrochar in terms of production, physico-chemical properties and applications. *Renew. Sust. Energ. Rev.* 45, 359–378.

Khan, N., Mohan, S., Dinesha, P., 2021. Regimes of hydrochar yield from hydrothermal degradation of various lignocellulosic biomass: A review. *J. Clean. Prod.* 288, 125629.

Khosravi, A., Zheng, H., Liu, Q., Hashemi, M., Tang, Y., Xing, B., 2022. Production and characterization of hydrochars and their application in soil improvement and environmental remediation. *Chem. Eng. J.* 430, 133142.

Kruse, A., Funke, A., Titirici, M.M., 2013. Hydrothermal conversion of biomass to fuels and energetic materials. *Curr. Opin. Chem. Biol.* 17(3), 515–521.

Kumar, A., Saini, K., Bhaskar, T., 2020. Hydochar and biochar: Production, physicochemical properties and techno- economic analysis. *Bioresour. Technol.* 310, 123442.

Lanza, G., Stang, A., Kern, J., Wirth, S., Gessler, A., 2018. Degradability of raw and post-processed chars in a two-year field experiment. *Sci. Total Environ.* 628–629, 1600–1608.

Li, H., Zhang, T., Tsang, D.C.W., Li, G., 2020. Effects of external additives: Biochar, bentonite, phosphate, on co-composting for swine manure and corn straw. *Chemosphere* 248, 125927.

Li, F., Jiang, Z., Ji, W., Chen, Y., Ma, J., Gui, X., Zhao, J., Chunhuo, Z., 2022a. Effects of hydrothermal carbonization temperature on carbon retention, stability, and properties of animal manure-derived hydrochar. *Int. J. Agric. Biol. Eng.* 15(1), 124–131.

Li, H., Zhang, T., Shaheen, S.M., Abdelrahman, H., Ali, E.F., Bolan, N.S., Li, G., Rinklebe, J., 2022b. Microbial inoculants and struvite improved organic matter humification and stabilized phosphorus during swine manure composting: Multivariate and multiscale investigations. *Bioresour. Technol.* 351, 126976.

Liu, H., Basar, I.A., Nzihou, A., Eskicioglu, C., 2021. Hydrochar derived from municipal sludge through hydrothermal processing: A critical review on its formation, characterization, and valorization. *Water Res.* 199, 117186.

Malghani, S., Gleixner, G., Trumbore, S.E., 2013. Chars produced by slow pyrolysis and hydrothermal carbonization vary in carbon sequestration potential and greenhouse gases emissions. *Soil Biol. Biochem.* 62, 137–146.

Malghani, S., Jüschke, E., Baumert, J., Thuille, A., Antonietti, M., Trumbore, S., Gleixner, G., 2015. Carbon sequestration potential of hydrothermal carbonization char (hydrochar) in two contrasting soils; results of a 1-year field study. *Biol. Fert. Soils* 51(1), 123–134.

Marcus, Y., 1999. On transport properties of hot liquid and supercritical water and their relationship to the hydrogen bonding. *Fluid Phase Equilibr.* 164(1), 131–142.

Nzediegwu, C., Naeth, M.A., Chang, S.X., 2021. Carbonization temperature and feedstock type interactively affect chemical, fuel, and surface properties of hydrochars. *Bioresour. Technol.* 330, 124976.

Romero-Anaya, A.J., Ouzzine, M., Lillo-Rodenas, M.A., Linares-Solano, A., 2014. Spherical carbons: Synthesis, characterization and activation processes. *Carbon* 68, 296–307.

Ruiz, H.A., Rodriguez-Jasso, R.M., Fernandes, B.D., Vicente, A.A., Teixeira, J.A., 2013. Hydrothermal processing, as an alternative for upgrading agriculture residues and marine biomass according to the biorefinery concept: A review. *Renew. Sust. Energ. Rev.* 21, 35–51.

Savage, P.E., 1999. Organic chemical reactions in supercritical water. *Chem. Rev.* 99(2), 603–621.

Sevilla, M., Fuertes, A.B., 2009. The production of carbon materials by hydrothermal carbonization of cellulose. *Carbon* 47(9), 2281–2289.

Shi, Y., Zhang, T., Ren, H., Kruse, A., Cui, R., 2018. Polyethylene imine modified hydrochar adsorption for chromium (VI) and nickel (II) removal from aqueous solution. *Bioresour. Technol.* 247, 370–379.

Shi, Y., Luo, G., Rao, Y., Chen, H.H., Zhang, S.C., 2019. Hydrothermal conversion of dewatered sewage sludge: Focusing on the transformation mechanism and recovery of phosphorus. *Chemosphere* 228, 619–628.

Song, C., Shan, S., Müller, K., Wu, S., Niazi, N.K., Xu, S., Shen, Y., Rinklebe, J., Liu, D., Wang, H., 2018. Characterization of pig manure-derived hydrochars for their potential application as fertilizer. *Environ. Sci. Pollut. R.* 25(26), 25772–25779.

Song, C., Shan, S., Yang, C., Zhang, C., Zhou, X., Ma, Q., Yrjälä, K., Zheng, H., Cao, Y., 2020. The comparison of dissolved organic matter in hydrochars and biochars from pig manure. *Sci. Total Environ.* 720, 137423.

Sun, P., Heng, M., Sun, S., Chen, J., 2010. Direct liquefaction of paulownia in hot compressed water: Influence of catalysts. *Energy* 35(12), 5421–5429.

Sun, Y.Y., Liu, C., Zan, Y.F., Miao, G., Wang, H., Kong, L.Z., 2018. Hydrothermal carbonization of microalgae (Chlorococcum sp.) for porous carbons with high Cr(VI) adsorption performance. *Appl. Biochem. Biotech.* 186(2), 414–424.

Sun, K., Han, L., Yang, Y., Xia, X., Yang, Z., Wu, F., Li, F., Feng, Y., Xing, B., 2020. Application of hydrochar altered soil microbial community composition and the molecular structure of native soil organic carbon in a paddy soil. *Environ. Sci. Technol.* 54(5), 2715–2725.

Takaya, C.A., Fletcher, L.A., Singh, S., Anyikude, K.U., Ross, A.B., 2016. Phosphate and ammonium sorption capacity of biochar and hydrochar from different wastes. *Chemosphere* 145, 518–527.

Thunshirn, P., Wenzel, W.W., Pfeifer, C., 2021. Pore characteristics of hydrochars and their role as a vector for soil bacteria: A critical review of engineering options. *Crit. Rev. Env. Sci. Technol.* 52(4), 1–25.

Wang, Y., Zhang, T., Westerhoff, P., Jiang, R., Ren, H., Yang, Y., Li, Z., 2017. Microwave-assisted digestion and NaOH treatment of waste-activated sludge to recover phosphorus by crystallizing struvite. *Environ. Technol.* 38(10), 1211–1222.

Wang, T., Zhai, Y., Zhu, Y., Li, C., Zeng, G., 2018. A review of the hydrothermal carbonization of biomass waste for hydrochar formation: Process conditions, fundamentals, and physicochemical properties. *Renewable Sustainable Energy Rev.* 90, 223–247.

Wang, G.W., Zhang, J.L., Lee, J.Y., Mao, X.M., Ye, L., Xu, W.R., Ning, X.J., Zhang, N., Teng, H.P., Wang, C., 2020. Hydrothermal carbonization of maize straw for hydrochar production and its injection for blast furnace. *Appl. Energ.* 266, 114818.

Wu, L., Wei, W., Wang, D.B., Ni, B.J., 2021. Improving nutrients removal and energy recovery from wastes using hydrochar. *Sci. Total Environ.* 783, 146980.

Xia, L., Lam, S.K., Yan, X., Chen, D., 2017. How does recycling of livestock manure in agroecosystems affect crop productivity, reactive nitrogen losses, and soil carbon balance? *Environ. Sci. Technol.* 51(13), 7450–7457.

Xiong, J., Chen, S., Wang, J., Wang, Y., Fang, X., Huang, H., 2021. Speciation of main nutrients (N/P/K) in hydrochars produced from the hydrothermal carbonization of swine manure under different reaction temperatures. *Materials* 14(15), 4114.

Zhang, M., Gao, B., Fang, J., Creamer, A.E., Ullman, J.L., 2014. Self-assembly of needle-like layered double hydroxide (LDH) nanocrystals on hydrochar: Characterization and phosphate removal ability. *RSC Adv.* 4(53), 28171–28175.

Zhang, T., He, X., Deng, Y., Tsang, D.C.W., Jiang, R., Becker, G.C., Kruse, A., 2020a. Phosphorus recovered from digestate by hydrothermal processes with struvite crystallization and its potential as a fertilizer. *Sci. Total Environ.* 698, 134240.

Zhang, T., He, X., Deng, Y., Tsang, D.C.W., Yuan, H., Shen, J., Zhang, S., 2020b. Swine manure valorization for phosphorus and nitrogen recovery by catalytic–thermal hydrolysis and struvite crystallization. *Sci. Total Environ.* 729, 138999.

Zhang, T., Wu, X., Shaheen, S.M., Rinklebe, J., Bolan, N.S., Ali, E.F., Li, G., Tsang, D.C.W., 2021. Effects of microorganism-mediated inoculants on humification processes and phosphorus dynamics during the aerobic composting of swine manure. *J. Hazard. Mater.* 416, 125738.

Zhang, Y., Jiang, Q., Xie, W.L., Wang, Y.F., Kang, J.M., 2019. Effects of temperature, time and acidity of hydrothermal carbonization on the hydrochar properties and nitrogen recovery from corn stover. *Biomass Bioenerg.* 122, 175–182.

Zhao, P.T., Chen, H.F., Ge, S.F., Yoshikawa, K., 2013. Effect of the hydrothermal pretreatment for the reduction of NO emission from sewage sludge combustion. *Appl. Energ.* 111, 199–205.

Zhou, N., Chen, H.G., Xi, J.T., Yao, D.H., Zhou, Z., Tian, Y., Lu, X.Y., 2017. Biochars with excellent Pb(II) adsorption property produced from fresh and dehydrated banana peels via hydrothermal carbonization. *Bioresource Technol.* 232, 204–210.

Zhou, Y., Li, Y., Wan, C., Li, D., Mao, Z., 2010. Effect of hot water pretreatment severity on the degradation and enzymatic hydrolysis of corn stover. *T. Asabe* 53(6), 1929–1934.

11 Using Biochar-Based Nitrogen Fertilizers to Enhance the Productivity of Low Fertility Soils

Yurong Gao[1–3], Zheng Fang[1], Hanbo Chen[1–3], Lukas Van Zwieten[4], Nanthi Bolan[5,6], Da Dong[7], Bert F. Quin[8], Jun Meng[2,3], Wenfu Chen[2,3], and Hailong Wang[1,2]

[1]School of Environmental and Chemical Engineering, Foshan University, Foshan, China
[2]Agronomy College, Shenyang Agricultural University, Shenyang, China
[3]Key Laboratory of Biochar and Soil Improvement, Ministry of Agriculture and Rural Affairs, China
[4]New South Wales Department of Primary Industries, Wollongbar, Australia
[5]School of Agriculture and Environment, The University of Western Australia, Perth, Australia
[6]The UWA Institute of Agriculture, The University of Western Australia, Perth, Australia
[7]School of Environmental and Resource Sciences, Zhejiang A&F University, Hangzhou, China
[8]Quin Environmentals (NZ) Ltd., Auckland, New Zealand

CONTENTS

DOI: 10.1201/9781003093565-11

11.1 INTRODUCTION

For agricultural systems, nitrogen (N) is an extremely important nutrient and is the most significant nutrient-limiting factor for productivity and constraining economic sustainability. To solve this limitation, about 135 million Ha synthetic N fertilizers are applied on agricultural land. Nevertheless, these synthetic N fertilizers cause significant N loss off-farm due to excessive or irrational application, resulting in groundwater contamination (Wang et al. 2019); eutrophication of lakes, rivers, and shallow marine water ecosystems (Ayele and Atlabachew 2021); impaired coral reefs (Lapointe et al. 2019); greenhouse gas emissions (Puga et al. 2020); and gas losses leading to acid rain (Penuelas et al. 2020). Approximately 50 million tons of reactive N are reported to be released into the environment each year (Bodirsky et al. 2014). The loss of N from soils to surface and groundwater through surface runoff and subsurface seepage from agricultural lands has become one of the research hotspots in many countries.

Nitrogen use efficiency (NUE) is widely defined as the amount of fertilizer N utilized by plants in the current and subsequent seasons (Hirose 2011). Providing higher NUE is critical to ensure food security with the reduced arable land area (Zhang et al. 2015) while minimizing off-site environmental impacts. Enhancing the synchronization between N supply and crop demand is one of the key strategies to improve NUE (Fageria and Baligar 2005). In recent years, slow-release nitrogen fertilizers have been developed rapidly both domestically and internationally and have been shown to achieve this result (Wen et al. 2017). Some new efficiency-enhancing fertilizers, such as lignin-based controlled-release fertilizers (Chen et al. 2020b), nanostructured slow-release fertilizers (Gurusamy et al. 2017), lignite-based slow-release N fertilizers (Saha et al. 2021), carbon-based slow-release fertilizer (Rashid et al. 2021), and slow-release fertilizer hydrogels (Ramli 2019), can better control the release rate of nutrients from fertilizers and can simplify fertilizer application techniques while providing benefits to crop production and less environmental impact.

While fully demonstrating the production and environmental benefits of enhanced efficiency N fertilizers (EENFs), their economic benefits should also be considered. There is little information and research on the economic costs or benefits of EENFs, and Khakbazan et al. (2013) have reported that the controlled-release urea as a beneficial management practice typically has a lower net revenue of production compared to conventional urea application in a range of agro-environments across western Canada. To address this consideration, research on the use of biochar as a cost-effective and environmentally friendly carrier for N fertilizer has received a lot of attention. Biochar, with a large surface area with a highly porous structure and rich surface functional groups (Chen et al. 2019a, Bolan et al. 2022), becomes an effective material for removing contaminants from water and the atmosphere (Zhang et al. 2021a), remediating soils contaminated with organic pollutants and potentially toxic elements (Yang et al. 2022), and improving the physical and chemical properties and micro-ecological environment of soil (Sun et al. 2021). Besides, biochar can be used for C sequestration and mitigation (Lu et al. 2019, Song et al. 2019) and is of great significance in adapting to the economic development model of the new era and promoting the recycling and sustainable development of the environment and resources.

The biochar properties, such as adsorption capacity and high porosity, have also been suggested as an effective N-carrier to realize agricultural sustainable development (Chen et al. 2019b). The use of biochar-based N fertilizers (BBNFs) has been considered among the most effective strategy for reducing N loss and improving NUE (Liu et al. 2019, Chen et al. 2019b, Dong et al. 2020). For example, Jia et al. (2021) reported that compared with urea BBNFs better improve NUE by about

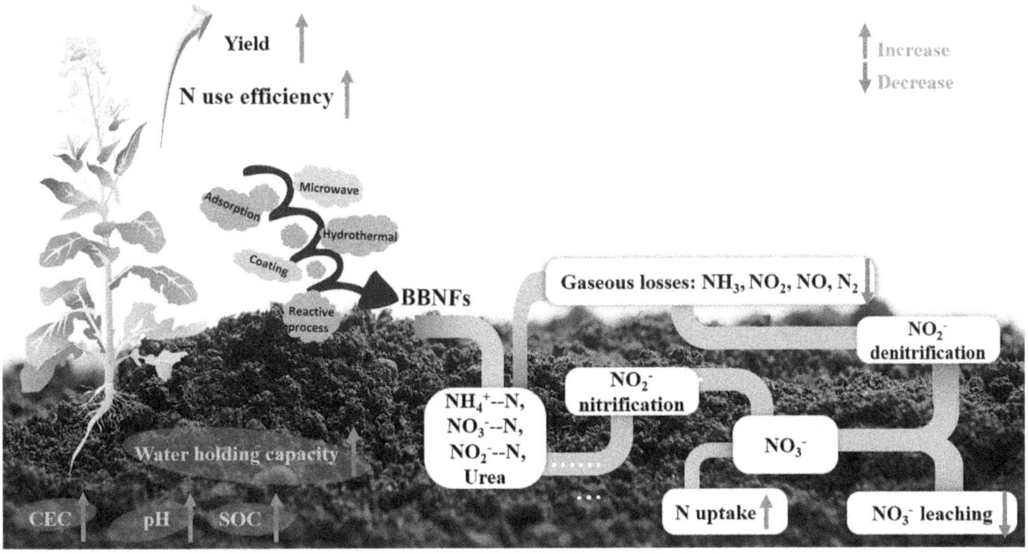

FIGURE 11.1 Simplified schematic showing methods used for preparations of BBNFs and the modification of N cycling processes in soil following BBNFs application.

20%. However, there is still room for improvement, and researchers probably should investigate the preparation of BBNFs and their role in soil improvement and remediation, as well as their impact on plant growth and crop yield. Thus, in this review, we aim to highlight the recent progress on the preparation of BBNFs and the mechanisms by which it improves NUE and crop production, and its role in crop production (Figure 11.1).

11.2 PREPARATION MATERIALS AND METHODS OF BBNFs

Biochar-based N fertilizers normally use a biochar/polymer coating of a synthetic fertilizer granule or through utilizing the structure of biochar (either adsorbed or reacted via mixing, solubilization of the fertilizer, or melting of urea into the structure) for preparation. The product also can then be physically altered to facilitate easier material handling via processes such as extrusion and pelletization (Table 11.1) (Gao et al. 2022). Up until now, physicochemical methods, which include physical mixing, coating, microwave, graft co-polymerization, and infiltration, have been tested during trials as mainly promising technologies for BBNFs.

11.2.1 The Blending of Biochar with Fertilizers

Perhaps blending is the simplest and most widely used method of formulating BBNFs. It is mainly the blending and mixing of basic granular fertilizers with close particle size and strength (essentially without chemical reactions between them) by a certain ratio. For better application, other physical preparation techniques such as pellet or extrusion formation are also included in the blended material. Also, to address specific crop/ pasture nutrient requirements, the formulation of BBNFs can be simply adjusted on this basis. Various organic and inorganic N fertilizers sources have been used to prepare BBNFs, including urea (Shi et al. 2020), $(NH_4)_2HPO_4$ (Dong et al. 2020), vinasse (Amaro et al. 2016), and swine manure compost (Zhou et al. 2021).

During the physical blending process, adding the clay minerals such as sepiolite (Shi et al. 2020), kaolin (Chen et al. 2018a) and bentonite (Yao et al. 2015) can contribute to the pelletization of BBNFs and improve the direct binding of N fertilizer and biochar. The granular structure was

TABLE 11.1
Production Methods for BBNFs

Method	Feedstock	Pyrolysis Temperature of Biochar (°C)	N Fertilizer	References
Physical mixing	Wood chips	700	Vinasse	Amaro et al. (2016)
	Wheat straw	450	Urea, $NH_4H_2PO_4$	Yao et al. (2015)
	Urban green waste (<2 cm)	450–540	Urea	Shi et al. (2020)
	Eucalyptus wood	400	Urea	Puga et al. (2020)
Coating	Maize straw, rice straw, forest litter	500	Urea	Chen et al. (2018b)
	Wheat straw	400	KNO_3	Ye et al. (2019)
	Vinasse	600	Urea	Jia et al. (2021)
	Corn straw	500	NH_4NO_3	Khajavi-Shojaei et al. (2020)
Adsorption	Oak wood	600	–	Khan et al. (2008)
	Agricultural waste biomass	450	NH_4NO	Zhang et al. (2018)
	Corn stover	550	$(NH_4)_2SO_4$	Cen et al. (2021)
	Eucalyptus wood	400	Urea	Puga et al. (2020)
Microwave	–	–	NH_4^+ aqueous solutions	Wen et al. (2017)
Hydrothermal	Oilseed rape straws	400	Urea	Liu et al. (2019)
Reactive process	Agricultural waste biomass	450	NH_3 H_2O, HNO_3	Zhang et al. (2018)
Graft co-polymerization	Cotton stalks	600	NH_4Cl	An et al. (2020)

protected from collapse due to the association between Al–OH of bentonite and N–H of fertilizer by adding bentonite to enhance the mechanical strength of BBNFs (Shi et al. 2020). It was reported that using bentonite promoted beneficial microorganisms' growth (Yao et al. 2015) and contribute to water retention and higher water-holding capacity (WHC%) (Wen et al. 2017). Furthermore, Wen et al. (2017) found lamellar bentonite could be stripped to form a complex physical network, in which the N release diffusion path from the BBNF was further extended. Besides, N leaching from BBNFs was effectively reduced by the interaction between N–H of N fertilizer and Mg–OH of sepiolite (Shi et al. 2020). Binding agents such as kaolin have also been used during the pelletization of BBNFs (Chen et al. 2018a). And Yang et al. (2018) showed the stability of biochar can be enhanced by kaolinite in the soil by reducing biological degradation and chemical oxidation, thus promoting long-term C sequestration. As an example, a BBNF product using crop residues, which was physically activated using carbon dioxide and then mixed with macro (N, P, K) and micro-nutrients, had a lower N loss (by up to 63%) compared to either NH_4NO_3 or urea (González-Cencerrado et al. 2020).

11.2.2 Coating of Fertilizers with Biochar

Biochar coating of mineral fertilizers is an entirely promising approach in the development of controlled-release fertilizers, providing a barrier to the dissolution and transport of nutrients thus limiting N loss pathways and mitigating environmental problems (Wang et al. 2015, Naz and Sulaiman 2016). The use of polymer films composited with biochar to coat commercial N fertilizers has been demonstrated to increase N release characteristics even further. To make biochar-based polymer coating for BBNFs, researchers used waterborne copolymer (made of PVA (polyvinyl alcohol) and polyvinylpyrrolidone (PVP) (Chen et al. 2018b), waterborne polymer (polyacrylate) (Zhou et al. 2015), biodegradable polymers (such as starch, and ethyl cellulose) (González et al. 2015) and water-retention polymers (acrylic acid and 2-acrylamide-2-methylpropanesulfonic acid) (Wen et al. 2017).

Waterborne polymers (i.e., water-dissolvable) have been developed as coatings to make BBNFs due to their low cost and low environmental effect. The incorporation of biochar in a polyacrylate coating, for example, allowed for a regulated release of nutrients over a 12-month period while having no effect on the dominant soil bacterial community (Zhou et al. 2015). Chen et al. (2018b) created BBNFs by coating urea with biochar-based waterborne copolymers, showing a slow-release formulation that released 65% of total N over 22 days. BBNFs with biochar-based waterborne copolymer films had lower water absorbency and were more successful in retaining N than copolymers due to the cross-linking networks between biochar and copolymers (Chen et al. 2018b). Biochar was combined with modified starch, bentonite, and humic acid in another coating technique (Dong et al. 2020). The SEM and FT-IR data indicated the formation of a thick layer that inhibited the release of N from the granule.

Considering water supply is so essential in agricultural production, polymers that improve soil water-retention and water-retaining capacity could help BBNFs perform better. BBNFs were synthesized from raw materials such as bentonite, acrylic acid (AA), NH_4^+-loaded biochar, and 2-acrylamide-2-methylpropanesulfonic acid (AMPS). Due to its highly porous structure, SEM research confirmed that biochar implanted in BBNFs could hold a significant amount of water. When BBNFs were applied at a rate of 2% (w/w), the water-retention capacity of the soil increased from 29% to 61%, compared to the untreated without BBNFs. Furthermore, after 12 days, the soil without the samples had lost all of its absorbed water, whereas BBNFs had kept 10.6% of the soil moisture on the 30th day (Wen et al. 2017). However, most biochar-based BBNFs only have a minor impact on soil water retention. As a result, the development of BBNFs that retain a lot of water represents a significant opportunity.

Another BBNF formulation was made by impregnating biochar with urea in a batch reactor utilizing aqueous media at 150°C±5°C and 1.0 bar, and then encapsulating the biochar with biodegradable polymers such as ethyl cellulose (EC), cellulose acetate (CA), and sodium alginate (SA) (González et al. 2015). Leaching of urea-N was highest with urea (4.1 mg/kg dry soil) and lowest with BBNF (2.4 mg/kg dry soil) in soil columns planted with wheat (dry soil). Another mechanism that could explain the higher NUE is that the BBNF products slow down urea hydrolysis.

11.3 PROPERTIES OF BBNFs

11.3.1 Development of Sorptive Properties of BBNFs

The sorptive properties of biochar are due to its porous and reactive surface, and these properties have been used in the construction of controlled-release fertilizer formulations. To develop these products, various papers discussed procedures such as solid-liquid adsorption and permeation. Khan et al. (2008) made BBNF by putting charcoal in a rotary vacuum evaporator with N fertilizer solution and rotating it at 100°C for 24 hours. As a result, a BBNF with significantly delayed N release into soil leachate was fabricated. After a 21-day desorption experiment, a comparable BBNF production procedure employing biochar (pyrolyzed at 200°C) and $(NH_4)_2SO_4$ demonstrated that more than 90% of the NH_4^+ was preserved (Cai et al. 2016). An adsorption approach was used to generate BBNFs from biochar derived from rice husks in a urea-hydrogen peroxide (UHP) solution. Due to the oxidation of carbonized surfaces by H_2O_2, more C-O and C=O groups could be obtained in the products compared to urea, resulting in better slow-release of N (Chen et al. 2018a). Through graft co-polymerization with superabsorbent hydrogels, biochar was incorporated into semi-interpenetrating polymer networks to make a BBNF (An et al., 2020). After 25 days, the product had a high water-retention capacity of 73%, which is significantly higher than fertilizer without charcoal. It was discovered that adding biochar to BBNFs improves the pore structure and increases the number of cross-linking points. In another method, molten urea (about 155°C) was utilized to synthesize BBNFs as both a binder and a source of N (Xiang et al. 2020). Fused urea penetrates the micropores of the biochar and reacts completely with the carbonyl and amino groups on the surface to form BBNF, preventing volatilization or leaching of nitrogen from the soil (Xiang et al. 2020).

Another method uses hydrothermal charring or leaching from soil to react urea, charcoal, bentonite, and polyvinyl alcohol in a sealed reactor near the melting point of urea (132°C–135°C) (Liu et al. 2019). When incubated in water, this BBNF released 61% of its total N in 28 days, and 54% in 98 days when incubated in silt loam soil. And adsorption to hydrogen bonding, oxygen-containing functional groups, and electrostatic interactions of the N–C=O and bentonite were shown to influence the rate of N release from BBNF (Liu et al. 2019).

In addition to the physical properties of biochar for the preparation of BBNFs, its surface functional groups also play a crucial role in the preparation of BBNFs through adsorption and chemical interactions. For instance, BBNFs were created using a reaction between rice husk-derived biochar, NH_3 and H_3PO_4, which exhibited a substantially slower release rate of N (NH_4^+ and NO_3^-) in sandy soil than $(NH_4)_3PO_4$ (El Sharkawi et al. 2018). FTIR has shown that aromatic C, C–O, and O–H functional groups are amenable to adsorption and, as a result, react with NH_3 and phosphate (El Sharkawi et al. 2018).

11.3.2 Influencing Factors for N Release

Factors such as biochar properties, component ratio, organic N content, and types of N precursor can make important contributions to the properties of BBNFs.

The surface functional groups and dosages of biochar utilized in BBNFs have a significant impact on the structure and characteristics of the resulting BBNFs. Urea particles contained by a rice biochar-based copolymer with fewer OH groups, for example, had a more compact and denser structure than maize biochar and forest litter biochar, indicating that they could be released slowly over time (66% of N released from the product over 22 days) (Chen et al. 2018a).

Biochar pyrolyzed at low temperatures had a low specific surface area and was found to be unsuitable for making BBNFs because it did not supply N in a slow-release form (Jia et al. 2021). The characteristics of BBNFs may also be influenced by pyrolysis conditions. For example, Ye et al. (2019) discovered that BBNF generated in an H_2O (g) and CO_2 environment efficiently maintained the N obtained from the feedstock compared to BBNF produced in an N_2 atmosphere.

The slow-release feature of BBNFs is also affected by the type of N fertilizer used. Organic N released by BBNFs was significantly lower (39%–56%) than that released by urea alone (Shi et al. 2020). When NH_4HCO_3 was utilized to make a BBNF, it was discovered that it had limited stability and decomposed at temperatures exceeding 36°C–60°C (Lee et al. 2013).

In conclusion, BBNFs made with simple physical mixing have the advantages of low cost and simple production techniques, as well as the ability to manipulate fertilizer ratios batch to batch. BBNFs with a biochar-based chemical fertilizer coating are more difficult to make, but they can give higher fertilizer rates per weight of the product. Both strategies have been well researched and have generally resulted in products with higher NUE, decreased nitrogen losses to the environment, and, in some cases, increased soil water-retention qualities. Other techniques of manufacturing, including solid-liquid adsorption, hydrothermal production, and chemical reaction, have received less attention but have shown promise in the development of BBNFs. Table 11.1 provided a full review of the BBNF preparation procedures.

11.3.3 Slow Release of N from BBNFs

Slow-release BBNF formulations are intended to better match N fertilizer availability in the soil to plant demand (Jia et al. 2021). Large mineral N pools that can arise after mineral fertilizer application are avoided by supplying this regulation, therefore limiting N loss pathways such as denitrification (Puga et al. 2020), leaching (Cen et al. 2021), and volatilization (Shi et al. 2020). The kinetics and processes of N release from BBNFs are described in next sections.

11.3.3.1 Kinetic Models of N Release and Slow-Release Mechanisms of BBNFs

Pseudo-first-order and pseudo-second-order kinetic models were used to investigate the cumulative N release from BBNFs. Both models produced good fits to the data, with a high level of confidence, according to the findings (Figure 11.2) (Gao et al. 2022). Surprisingly, the pseudo-second-order kinetics model can better fit the cumulative N release data. This could be because the pseudo-first-order kinetics process is thought to be controlled by diffusion, whereas the pseudo-second-order kinetics process is thought to be controlled by chemical adsorption (Fang et al. 2018). This shows that diffusion and chemical adsorption mechanisms are likely to play a role in the gradual release of N from BBNFs.

By comparing the constant "k_2" of the fitted pseudo-second-order kinetics model to the slow-release properties of BBNFs, the slow-release properties of BBNFs can be further investigated. When compared to urea or artificial N fertilizer, the k_2 of BBNFs can be reduced by 59% (Chen et al. 2018b) and about 92% (Jia et al. 2020), illustrating the delayed release of BBNFs. The k2 of BBNFs made by the coating method was higher than that of BBNFs made by the adsorption approach, indicating that the adsorption method may have a better slow-release capability (Wen et al. 2017). However, the variability of k_2 among studies, which is due to a wide range of soil and climatic conditions, severely restricts its practical application.

The most frequent source of nitrogen in the formation of BBNFs is urea, which must be hydrolyzed by the urease enzyme before being released into the soil. Shi et al. (2020) found that the DOC concentration utilizing urea fertilizer declined after day 3, whereas the DOC concentration using urea-based

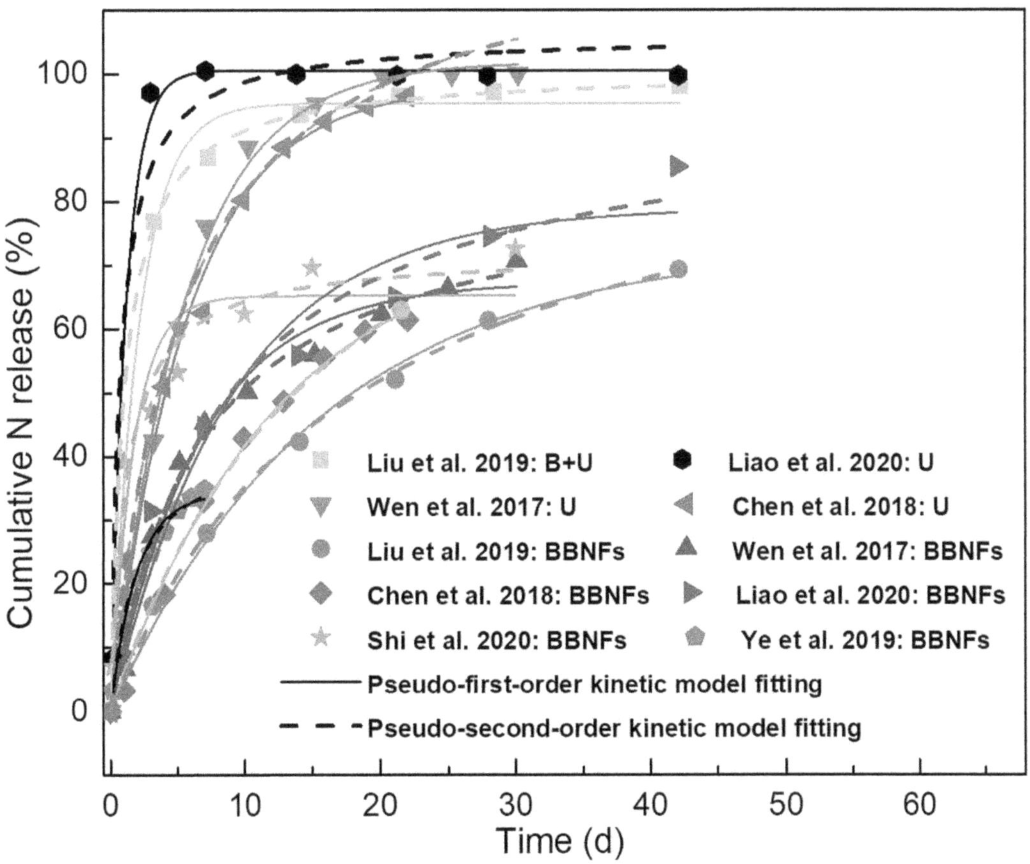

FIGURE 11.2 N release rate comparing urea to BBNFs across published studies. U represents urea alone, and B + U represents urea in combination with biochar.

BBNF remained steady for over 10 days. This shows that organic matter and mineral additions interact in BBNFs, preventing urea from rapidly hydrolyzing. González et al. (2015) also discovered that polymeric coating materials could delay urea hydrolysis, resulting in slower N release.

Hydrogen bonding and electrostatic interactions dictated the delayed release of NH_4^+ in BBNFs, according to Wen et al. (2017). Incomplete carbonization resulted in more oxygen functional groups when pyrolysis temperatures for biochar synthesis were in the range of 200–400°C. Cai et al. (2016) found that when this biochar was employed to make the BBNF, it may supply NH_4^+ sorption sites, resulting in a delayed release of N. Furthermore, diffusion mechanisms may potentially play a role in the gradual release of nitrogen by BBNFs. Sealing BBNFs with a bio-oil covering can also slow the diffusion of nitrogen, resulting in less N being released before they're applied to soil (Ye et al. 2019).

11.4 IMPACT OF BBNFs ON N CONVERSION PROCESS FOR AGRICULTURE SOIL

In recent years, there has been a growing focus on the impact of BBNFs on N cycling, with implications for both agronomic and environmental effects (Table 11.2) (Gao et al. 2022). Figure 11.1 depicted the variations in N cycling as a result of BBNF treatment.

Agriculture's long-term viability depends on boosting NUE and, as a result, minimizing N losses from various sources such as NO_3^- leaching, NH_3 volatilization, and N_2O emissions. When compared to conventional chemical fertilizers, BBNFs could reduce N loss, giving a variety of environmental benefits (Puga et al. 2020). Because adding biochar to soil can reduce N loss due to leaching (Yoo et al. 2014, Hossain et al. 2020), volatilization (Nguyen et al. 2017), or denitrification (Cayuela et al. 2015).

TABLE 11.2
Effects on N Loss from BBNFs Used in Field Studies

Experimental Condition	BBNFs Source	Method	The Rate of BBNFs Added	Soil Type	N Loss	Total N	References
Field trial	Wheat straw	Pelletization	225 kg N/ha	A calcareous soil	N_2O emission 13% ↓	6.98% ↑	Zheng et al. (2017)
	Municipal waste	Pelletization	450 kg N/ha	A paddy soil	N_2O emission 39.4% ↓	-	Qian et al. (2014)
	Eucalyptus wood	Pelletization	408 kg N/ha	A clay loam soil	N_2O emission 8% ↓, NH_3 volatilization 33% ↑	-	Puga et al. (2020)
Pot experiment	Cotton stalks	Microwave	-	-	N leaching loss 74% ↓	80%↑	Wen et al. (2017)
	Oilseed rape straws	Hydrothermal	1 g N, 200 g of dry soil	A silt loam texture soil	Leaching rate ↓	48%↑	Liu et al. (2019)
	Corn straw	Coating	250 kg N/ha	-	Cumulative nitrate concentration 61% ↓	65%↑	Khajavi-Shojaei et al. (2020)
	Vinasse	Coating	450 kg N/ha	-	NO_3^- leaching loss 24.7% ↓	30%↑	Jia et al. (2021)
	Urban green waste	Pelletization	92 kg N/ha	-	N leaching loss 44% ↓	-	Shi et al. (2020)

11.4.1 Leaching Loss

N is lost primarily through leaching (González-Cencerrado et al. 2020). Urea has the largest nitrogen content of all the solid nitrogenous fertilizers, at 46%, and is thus commonly utilized. However, because urea is very soluble, it can easily migrate with the flow of water. Jia et al. (2021) found that replacing urea with BBNFs generated by depositing biochar on urea particles reduced soil nitrogen loss by roughly 30%. To prepare BBNFs, Cen et al. (2021) coated the particles with polylactic acid (PLA). Soil study revealed that up to 66% N from BBNFs stayed in the soil available for crop absorption after a 25-day trial evaluating this BBNF, but only 41% N from $(NH_4)_2SO_4$ remained in the soil available for crop uptake.

In another investigation, the peak of NH_4^+ leaching loss with urea was delayed from day 17 to 25 with BBNFs (Jia et al. 2021), most likely due to the urea's delayed hydrolysis (Shi et al. 2020). The application of BBNFs greatly reduced the cumulative leaching loss of NH_4^+ (39%–50%) when fertilized at the same level (Jia et al. 2021). After replacing compound fertilizers with BBNF, Dong et al. (2020) discovered that maximum concentrations of NH_4^+-N in percolation water could be lowered by 55%. Biochar can adsorb NH_4^+ because it has cationic sorption sites, lowering NH_4^+ losses in soils with pH below 6.3 (Mandal et al. 2018).

Because NO_3^- is a soluble anion, it is easily leached out of agricultural soils. In general, BBNFs with negatively charged biochar has a low capacity for NO_3^- adsorption. Even though NO_3^- leaching contributed to 83%–84% of total N loss from BBNFs (Jia et al. 2021), NO_3^- leaching losses were still 20% lower than traditional fertilizers. Wang et al. (2015) found that NO_3^- leaching loss was the most common pathway of N loss, accounting for 52%–96% of total N lost from BBNFs. However, when compared to urea and NH_4NO_3, BBNFs reduced NO_3^- leaching loss by over 60% (González-Cencerrado et al. 2020). According to Jia et al. (2021), the cumulative leaching loss of NO_3^- with BBNFs was 25% lower than urea. The numerous oxygen functional groups on the surface of BBNFs, including carboxylic, hydroxyl, and lactone groups, play a major role in adsorbing NH_4^+ and NO_3^-, minimizing N loss (Zheng et al. 2017).

11.4.2 NH_3 Volatilization

Agricultural NH_3 emissions account for up to 90% of global NH_3 emissions (Brentrup et al. 2000), and it is a pollutant in aquatic systems (Park et al. 2019) as well as an indirect greenhouse gas (Ma et al. 2021). Using BBNFs, the volatilization of NH_3 was shown to be greatly reduced (by 50%) (González-Cencerrado et al. 2020). The NH_3 dissolving rate was lowest in BBNFs containing 51% biochar and 10% N (Puga et al. 2020). The cumulative NH_3 emitted by BBNFs may, however, exceed that of urea due to the extended NH_3 volatilization duration (from 23 to 26 days) (Jia et al. 2021). However, since BBNFs have a higher total NUE, they require less N, which means NH_3 volatilization can be decreased to below that of urea (Jia et al. 2021). Shi et al. (2020) discovered that NH_3 volatilization loss from BBNFs was comparable to that of urea, implying that the alkaline pH microenvironment around urea adsorbed into BBNF may be similar to that surrounding hydrolyzing urea granules in the soil (Puga et al. 2020).

Soil pH, temperature, and fertilizer level can all influence NH_3 volatilization (González-Cencerrado et al. 2020). The alkaline pH of BBNFs is usually due to the alkaline nature of biochar, which could result in increased NH_3 volatilization (Puga et al. 2020). Because of the cationic charge associated with BBNFs, NH_3 volatilization can be reduced (Puga et al. 2020). Furthermore, Puga et al. (2019) found that acidifying biochar before coating it with urea granules reduced NH_3 volatilization from BBNFs.

The possibility for NH_3 volatilization from BBNFs is determined by the BBNFs' manufacturing conditions. Cen et al. (2021) combined biochar with dissolved $(NH_4)_2SO_4$ to assess the volatilization losses associated with the manufacture of BBNFs. Due to the chemisorption of polar and nonpolar chemicals, such as hydrophobic bonding, electron donor-acceptor interactions originating from fused aromatic C structures, and weak unconventional H-bonds, NH_4^+ ions were attached on surfaces and confined within mesopores and micropores of biochar. The dissolved NH_4^+ ions are transported to the biochar-coating layer interface by diffusion, concentration or pressure gradient, or a combination of these as the driving mechanism. As a result, the rate of NH_3 emission slowed.

BBNFs have also been demonstrated to reduce soil N_2O leakage. When BBNFs compared to urea, Puga et al. (2020) observed that cumulative N_2O emissions from BBNFs were reduced by 8%, while Qian et al. (2014) found a 40% reduction. The lower N_2O emissions may be attributable to a lower population of denitrifies in soils treated with BBNFs compared to urea (Liao et al. 2020). Biochar-associated organic nitrogen would be less accessible to denitrifying bacteria than inorganic nitrogen. The adsorption of mineralized nitrogen on the surfaces of biochar particles may also slow down N consumption and transformation by bacteria that create N_2O (Wang et al. 2013). However, there is still a scarcity of evidence on the use of BBNFs to reduce N_2O losses from soil.

11.5 THE IMPACT OF BBNFs ON AGRICULTURAL SOILS

Water contamination, air pollution, soil quality degradation, and other undesirable consequences come from fertilizer loss. By changing soil characteristics with biochar products, the retention of nutrients in the soil can be extended (Table 11.2). The use of BBNFs, for example, can modify soil physical qualities like porosity, water storage, bulk density, and chemical properties like cation and anion exchange capacity, pH, and organic matter.

11.5.1 Physical Properties

Because of the low bulk density of biochar used in the production of BBNFs, their large specific surface area, porous structure, and the abundance of oxygen functional groups in biochar such as lactone groups, hydroxyl, and carboxylic, BBNFs can reduce the bulk density of soil, resulting in improved aeration and permeability (Chen et al. 2018b, Zhang et al. 2018, Puga et al. 2020). The water-holding capacity of soils may be affected if the bulk density is reduced. For example, Liu et al. (2019) discovered that BBNFs (1 g N, 200 g dry soil) may directly raise the water-holding ratio from 40% to 70% of the soil moisture in 30 days.

In this research, BBNF was sprayed at a rate of 2% w/w, and the presence of bentonite in the fertilizer formulation was credited with a significant increase in water-retention capacity. The water loss (%) of soil samples was utilized by Cen et al. (2021) to measure water-retention capacity, and they found that BBNFs had much lower water loss than all other samples until day 3. The use of polymers in the production of BBNF has been proven to reduce the rate of nitrogen release due to the physical barrier of hydrogel (Wen et al. 2017). In comparison to unamended soils, the water-holding capacity and water-retention period of soils with BBNFs generated by adsorption and polymer techniques increased and lasted 12–18 days and 30 days, respectively (Wen et al. 2017). However, it should be noted that the BBNFs were applied at a rate of 2% w/w, which is much higher than typical fertilizer application rates. When lower concentrations of BBNFs are used, these impacts on soil water-holding capacity are anticipated to be minimal.

11.5.2 Chemical Properties

Soil pH: One of the most crucial chemical qualities of soils is pH. BBNFs, which contain N as well as mineral components like K, Ca, Mg, and others, can exchange H^+ in soil, lowering H^+ concentration and thereby raising soil pH. Jia et al. (2021) discovered that soil pH was higher with BBNFs treatment (pH range 6.03–6.10) than soil treated with urea (5.89–5.95). Zheng et al. (2017)

observed that after BBNF treatment, soil pH changed slightly from 7.89 to 8.06 in alkaline soil. Puga et al. (2020) observed a significant increase in pH (to pH 9.0) after applying BBNF to another alkaline soil. Alkaline BBNFs may be unsuitable for application in alkaline soils due to increased NH_3 volatilization and the formation of toxicities in the soil. Because NH_4HCO_3 works as a pH buffer, Lee et al. (2013) discovered that combining 50/50 biochar with NH_4HCO_3 reduced the pH of biochar material from 9.85 to 7.89. The impact of BBNFs on soil pH may vary depending on the soil type.

The impact of BBNFs on soil pH may vary depending on the soil type. The influence of BBNFs on the pH of calcareous soils, for example, was shown to be insignificant in several investigations (Zheng et al. 2017, Shi et al. 2020). In contrast, due to an excessive increase in soil pH, the use of BBNFs derived from orange peel, residual wood, and water-treatment sludge impregnated with anaerobic digestion slurry reduced lettuce yield (Oh et al. 2014).

CEC: The cation-exchange capacity (CEC) of soil is a crucial measure of its fertility. The use of BBNFs may boost soil CEC, which is beneficial to the retention of cationic nutrients in the soil (Lee et al. 2013). The charge carried by biochar transformed some stable state elements into active states (Domene et al. 2014), for example, maize-cob-based BBNF raised the exchangeable cations from 16 to 24 cmol/kg (Zheng et al. 2017). In terms of soil CEC, however, Zhang et al. (2018) pointed out that no significant differences between any of the BBNFs and fertilizer treatments. As a result, the favorable impacts of BBNFs on soil CEC should be explored further under continuous BBNF application settings.

SOC: BBNF application has been proven in numerous studies to be one of the most promising techniques for increasing soil organic carbon (SOC). When compared to inorganic compound N fertilizers, the C content of BBNFs (225 kg N/ha) can result in a 16% rise in SOC, contributing to higher crop output. BBNF administration enhanced SOC by 20% compared to urea treatments during the harvest phases, according to Liao et al. (2020). Because of the recalcitrant C given by biochar, the long-term SOC content of the soil would likewise increase with the addition of BBNFs (Zheng et al. 2017, Chen et al. 2020a). Furthermore, since BBNFs have primarily stable C and are resistant to breakdown, Puga et al. (2020) showed that methane fluxes did not surpass 5.71 g C/m^2d when employing them (Grutzmacher et al., 2018). However, other circumstances, such as the overuse of fertilizers, might lower the concentration of SOC. When administered at high rates, even BBNFs with little organic C can reduce SOC (Qian et al. 2014). However, when employed at fertilizer rates, the impacts of adding BBNFs to soil on SOC levels are anticipated to be minimal (Shi et al. 2020).

11.5.3 Microbial Influence

Microorganisms play an important part in the soil N cycle, which can be influenced by biochar (Zhu et al. 2017, Siedt et al. 2021). The pores of BBNFs can absorb a lot of water and nutrients, which means that bacteria can get enough nourishment (Shi et al. 2019). BBNFs can affect the physical and chemical properties of soil, altering soil processes in the process (Amaro et al. 2016). In comparison to urea, BBNFs had increased urease activity (about 10%) during the harvest stages (Liao et al. 2020). Under BBNF treatment, the relative abundance of nitrifiers such as *Sphingomonadaceae* and *Acidothermaceae* increased (Liao et al. 2020). According to Dong et al. (2020), BBNF amendment can boost the relative abundance of nitrifying bacteria such as *Nitrospira*, *Nitrospora*, and *Nitrobacter*, as well as denitrifying bacteria like *Bradyrhizobium*, *Cupriavidus*, and *Burkholderia*. The decreased leaching or other forms of N loss from BBNFs could be linked to soil N cycling microorganisms. The study of the impact of BBNFs on soil bacterial communities and soil characteristics, on the other hand, is still in its early phases.

11.6 PROMOTING PLANT GROWTH AND CROP YIELD BY BBNFs

11.6.1 BBNFs FOR NUE IMPROVEMENT

Increasing fertilizer use efficiency, particularly nitrogen, which is a crucial factor limiting agricultural productivity in China (Cui et al., 2010), remains a major problem for agriculture around the world. Zhang et al. (2021b) examined 13 N budget datasets and showed that during the last six decades, NUE has decreased from 59% to 46%. This is partly related to the increased rate of N fertilizer application, which lowers NUE intrinsically. When BBNF was added to a 30-day pot experiment, the total NUE of cotton plants increased from 36% to 64%, and N leaching rates dropped by 74% compared to NH_4Cl. In a seven-month pot research, Liao et al. (2020) discovered that BBNFs had the highest N agronomic efficiency (NAE) and NUE, with a 28% higher NAE and 59% NUE than urea. Urea had an NUE of 28% in a field trial where maize was grown on clay loam soil, while BBNF increased its NUE to 43%. Lower N losses from leaching of NO_3^-, or NH_4^+, as well as gaseous N_2O loss, were linked to higher NUE (Puga et al. 2020).

The utilization efficiency of conventional nitrogen fertilizers in crop production is normally only 30%–40%. A mismatch in the rate of fertilizer N availability with N demand during crop growth is one of the key causes of low NUE. Cen et al. (2021) discovered that the amount of nitrogen released by the particles was significantly lower than that released by the particles in water. The soil retained at least 60% of the nitrogen content of BBNFs.

In oilseed rape, BBNFs outperformed urea and biochar mixed with urea in terms of N uptake at the same fertilization amount (Jia et al. 2021). The ^{15}N tracer approach revealed that when BBNFs were used during crop growth stages like bolting, blooming, and harvest, greater N uptake was attained than when urea was used (Liao et al. 2020). With the help of a one-year ^{15}N stable isotope study, Peng et al. (2021) discovered that utilizing BBNF may increase N retention (74%–80%) in brown soil, boosting maize N intake from 84 to 106 kg N/ha. Due to the delayed degradation of organic N limits as well as the adsorption of inorganic N by biochar, Shi et al. (2020) reported that BBNFs produced more inorganic N such as NH_4^+ and NO_3^-, which is conducive to improved N uptake of plants and, as a result, increased crop output. Liao et al. (2020) found that BBNFs improved NUE for rape growth by raising soil NO_3^- content via the slow-release characteristics of BBNFs, enhancing nitrification via AOB activity, reducing denitrification, and lowering nitrogen loss.

11.6.2 BBNFs CONCERNING CROP YIELD

The gradual release properties of BBNFs can help to increase the synchronization between N supply and crop demand. BBNFs application to soils has been reported in numerous studies to boost a wide range of agricultural yields (Table 11.3) (Gao et al. 2022). When BBNFs were used instead of matching quantities of chemical fertilizers, maize grain yield rose by 10% (Zheng et al. 2017) or 21% (Puga et al. 2020). After replacing $NH_4H_2PO_4$ with BBNFs, the fresh weight of Japanese mustard spinach plants increased by 33% (El Sharkawi et al. 2018). Green pepper yields increased from 9 t/ha to around 11 t/ha after BBNF treatment as compared to commercial chemical fertilizers (Yao et al. 2015). Despite a 20% reduction in fertilization application, the biomass of vegetables (oilseed rape) obtained with BBNF treatment was higher (~10%) than that of urea, demonstrating the technology's potential for usage in agriculture (Jia et al. 2021). When maize yield was normalized by mass of N fertilizers (N agronomic efficiency, AE_N), the improvement was improved to more than 40% with the treatment of BBNFs (Zheng et al. 2017). With the use of BBNFs, it was possible to attain a 10% increase in tea yield (He et al. 2019).

Other plant growth processes, such as germination rate and seedling growth characteristics, could be aided by BBNFs in addition to crop production. Cottonseed treated with BBNFs had a greater germination rate (about 94%) than cottonseed treated with NH_4Cl (approximately 84%) (Wen et al. 2017). On the other hand, there was a noticeable drop in turnip germination (Amaro et al. 2016).

TABLE 11.3
Effects of BBNFs on Crop Yield

Experimental Condition	Biochar Feedstock	Method	The Rate of BBNFs Added	Crop Type	N Productivity	Crop Yield	References
Pot experiment	Cotton stalks	Microwave	-	Cotton	NUE 79.78% ↑	Germination rate 94% ↑, the plant height 42.65% ↑, root length 25.78% ↑, fresh weight 68.94% ↑, dry weight 50.00% ↑	Wen et al. (2017)
	-	Pelletization	92 g N/kg	Maize	-	Biomass 25.1% ↑, the volume of root 37.7% ↑, shoot biomass 13.8% ↑	Shi et al. (2020)
	Wheat straw	Coating		Wheat	-	Dry matter mass 10% ↑	Ye et al. (2019)
	Corn straw	Coating	250 kg N/ha	Corn	N uptake 78% ↑, NUE 55.80% ↑	Leaf area 43.64% ↑, plant height 59.01% ↑, and root dry weight 95.28% ↑	Khajavi-Shojaei et al. (2020)
	Vinasse	Coating	450 kg N/ha	Rape (*B. campestris* L.)	NUE 67% ↑	Biomass 10% ↑	Jia et al. (2021)
	Oilseed rape straws	Hydrothermal	1 g N, 200 g of dry soil	Oilseed rape	NAE 27.7% ↑, NUE 58.79% ↑	16.6% ↑	Liao et al. (2020)
Field trial	Wheat straw	Pelletization	225 kg N/ha	Maize	NAE, 40%↑	10.7% ↑	Zheng et al. (2017)
	Wheat straw	Mixing	670 kg/ha	Green pepper	Total N in the fruit 11.8% ↑	18% ↑	Yao et al. (2015)
	Eucalyptus wood	Pelletization	80 kg N/ha	Maize	N grain uptake, 45% ↑, NUE 12% ↑	26% ↑	Puga et al. (2020)
	Municipal waste	Pelletization	450 kg N/ha	Rice	-	31.4% ↑	Qian et al. (2014)
	Agricultural waste biomass	Reactive process	220 kg/ha	Wheat	NAE 56.72% ↑, NAU 34.74% ↑	Productive tiller 5.50% ↑, 1000 grain weight 8.46% ↑, and biomass 23.85% ↑, grain yield 23.42% ↑	Zhang et al. (2018)

As a result, more research may be required to determine the important effects of BBNFs on plant germination.

According to a meta-analysis, the application of BBNFs to the soil boosted above ground productivity and crop output in general, but below ground productivity showed no significant response to biochar addition (Biederman and Harpole 2013). The root length of cotton plants treated with BBNFs increased by 26% compared to the NH_4Cl control (Wen et al. 2017). Like urea fertilizers, BBNFs boosted root biomass and volume by 26% and 38%, respectively, suggesting that improved plant tolerance to dry croplands (Shi et al. 2020) and drought may be a benefit (Bruun et al. 2014, Li and Tan, 2021). Furthermore, BBNFs treatment utilized much less N (58 kg N/ton grain) than traditional inorganic compound fertilizer treatment (85 kg N/ton grain), meaning higher plant NUE, lower N pollution, and higher economic output (Zheng et al. 2017).

11.7 CONCLUSIONS AND PERSPECTIVES

While there is good evidence that BBNFs can improve N use efficiency, more research is needed to determine their ideal application rates and how these affect their economic feasibility when compared to traditional fertilizers. BBNFs demonstrated a 15%–69% delay in cumulative N release and a 25%–65% percent improvement in fertilizer use efficiency when compared to typical chemical fertilizers in roughly 20 tests. The future task will be to improve the N efficiency of BBNFs, resulting in less pollution, while retaining their cost-effectiveness by lowering their production costs.

BBNFs can improve NUE and reduce fertilizer N loss to the environment through a variety of methods. Additional investigations using isotopic techniques, as well as modeling methodologies to understand the kinetics of fertilizer nutrient release, will allow a detailed knowledge of the benefits of BBNFs. It may be able to create BBNF formulations with optimum N release characteristics for certain crops and locations by combining this knowledge with an understanding of soil qualities, crop fertilizer demand, and seasonal climate (rainfall and temperature) projections. To maintain a balance between N release rate and N absorption, it is preferable to build BBNFs based on the ideal N absorption rate by crops or plants. More research is needed to determine the interaction between BBNF N release regulation and plant N absorption. The long-term environmental implications of the large-scale application of BBNFs, as well as the resulting ecological threats to biodiversity and ecosystem equilibrium, must also be considered. Finally, evidence suggests that BBNFs can improve fertilizer efficiency while also delivering additional soil benefits such as increased SOC. When compared to typical polymer coating, using biochar as a fertilizer carrier has various advantages, particularly in terms of preventing the introduction of plastics and microplastics into the environment.

This review demonstrates that BBNFs possess considerable potential as enhanced N fertilizers. Economic benefits, as well as reduced energy usage and greenhouse gas emissions, are all part of BBNFs' contribution to sustainable agriculture. Even though some biochar modification methods appear to be very expensive, due to the relatively low mass fraction of biochar in BBNFs (~ 29%) (Jia et al. 2020), increased economic output (Cen et al. 2021), and ecological and economic benefits resulting from reduced eutrophication, BBNFs may still be economically feasible. Biochar systems have the potential to reduce CO_2 emissions by 3.4–6.3 PgCO_2e (Lehmann et al., 2021), owing to the energy obtained during biochar formation as well as the stable C stored in the soil. When BBNFs are added, this could be further boosted by additional greenhouse gas (GHG) benefits. When chemical fertilizer is replaced with BBNF, GHG emissions are reduced significantly (e.g., 10 kg CO_2-eq emission per ton maize grain) (Zheng et al. 2017). By using a life cycle evaluation, Roberts et al. (2010) calculated the energy, climate change consequences, and economics of biochar systems (LCA). They discovered that returning biochar to the soil might cut CO_2 emissions by 800–900 kg per ton of pyrolyzed dry feedstock. González-Cencerrado et al. (2020) used an LCA technique to show that BBNFs have the least influence on acidification, terrestrial, and aquatic eutrophication when compared to traditional fertilizers. As a result, using BBNFs has the potential to help with carbon sequestration, reduced nitrogen consumption, and enhanced farmer revenue. The important benefits and challenges associated with the development of high-efficiency N fertilizers based on biochar technology have been addressed in our review. A framework for the rigorous examination of new goods will allow the technology to be quickly brought to market, facilitating a wide range of economic and environmental benefits, including the broader objective of carbon neutrality in primary industries.

REFERENCES

Amaro A., Bastos A. C., Santos M. J. G., Verheijen F. G. A., Soares A. M. V. M., Loureiro S. 2016. "Ecotoxicological assessment of a biochar-based organic N-fertilizer in small-scale terrestrial ecosystem models (STEMs)." *Applied Soil Ecology* 108:361–370.

An X. F., Yu J. L., Yu J. Z., Tahmasebi A., Wu Z. S., Liu X. C., Yu B. 2020. "Incorporation of biochar into semi-interpenetrating polymer networks through graft co-polymerization for the synthesis of new slow-release fertilizers." *Journal of Cleaner Production* 272:122731.

Ayele H. S., Atlabachew M. 2021. "Review of characterization, factors, impacts, and solutions of Lake eutrophication: Lesson for lake Tana, Ethiopia." *Environmental Science and Pollution Research* 28:14233–14252.

Biederman L. A., Harpole W. S. 2013. "Biochar and its effects on plant productivity and nutrient cycling: A meta-analysis." *Global Change Biology Bioenergy* 5:202–214.

Bodirsky B. L., Popp A., Lotze-Campen H., Dietrich J. P., Rolinski S., Weindl I., Schmitz C., Müller C., Bonsch M., Humpenöder F., Biewald A., Stevanovic M. 2014. "Reactive nitrogen requirements to feed the world in 2050 and potential to mitigate nitrogen pollution." *Nature Communications* 5:3858.

Bolan N., Hoang S. A., Beiyuan J., Gupta S., Hou D., Karakoti A., Joseph S., Jung S., Kim K.-H., Kirkham M. B., Kua H. W., Kumar M., Kwon E. E., Ok Y. S., Perera V., Rinklebe J., Shaheen S. M., Sarkar B., Sarmah A. K., Singh B. P., Singh G., Tsang D. C. W., Vikrant K., Vithanage M., Vinu A., Wang H., Wijesekara H., Yan Y., Younis S. A., Van Zwieten L. 2022. "Multifunctional applications of biochar beyond carbon storage." *International Materials Reviews* 67(2):150–200.

Brentrup F., Küsters J., Lammel J., Kuhlmann H. 2000. "Methods to estimate on-field nitrogen emissions from crop production as an input to LCA studies in the agricultural sector." *International Journal of Life Cycle Assessment* 5:349.

Bruun E. W., Petersen C. T., Hansen E., Holm J. K., Hauggaard-Nielsen H. 2014. "Biochar amendment to coarse sandy subsoil improves root growth and increases water retention." *Soil Use and Management* 30:109–118.

Cai Y., Qi H., Liu Y., He X. 2016. "Sorption/Desorption Behavior and Mechanism of NH4+ by Biochar as a Nitrogen Fertilizer Sustained-Release Material." *Journal of Agricultural and Food Chemistry* 64:4958–4964.

Cayuela M. L., Jeffery S., van Zwieten L. 2015. "The molar H:Corg ratio of biochar is a key factor in mitigating N_2O emissions from soil." *Agriculture, Ecosystems & Environment* 202:135–138.

Cen Z., Wei L., Muthukumarappan K., Sobhan A., Mcdaniel R. 2021. "Assessment of a biochar-based controlled release nitrogen fertilizer coated with polylactic acid." *Journal of Soil Science and Plant Nutrition* 21:2007–2019.

Chen L., Chen Q. C., Rao P. H., Yan L. L., Shakib A., Shen G. Q. 2018a. "Formulating and optimizing a novel biochar-based fertilizer for simultaneous slow-release of nitrogen and immobilization of cadmium." *Sustainability* 10:2740.

Chen S., Yang M., Ba C., Yu S., Jiang Y., Zou H., Zhang Y. 2018b. "Preparation and characterization of slow-release fertilizer encapsulated by biochar-based waterborne copolymers." *Science of the Total Environment* 615:431–437.

Chen W., Meng J., Han X., Lan Y., Zhang W. 2019a. Past, present, and future of biochar. *Biochar* 1:75–87.

Chen Y., Jiang Z., Wu D., Wang H., Li J., Bi M., Zhang Y. 2019b. "Development of a novel bio-organic fertilizer for the removal of atrazine in soil." *Journal of Environmental Management* 233:553–560.

Chen H., Yang X., Wang H., Sarkar B., Shaheen S. M., Gielen G., Bolan N., Guo J., Che L., Sun H., Rinklebe J. 2020a. "Animal carcass- and wood-derived biochars improved nutrient bioavailability, enzyme activity, and plant growth in metal-phthalic acid ester co-contaminated soils: A trial for reclamation and improvement of degraded soils." *Journal of Environmental Management* 261:110246.

Chen J., Fan X., Zhang L., Chen X., Sun S., Sun R. C. 2020b. "Research progress in lignin-based slow/controlled release fertilizer." *Chemsuschem* 13:4356–4366.

Cui Z., Chen X., Zhang F. 2010. "Current nitrogen management status and measures to improve the intensive wheat–maize system in China." *AMBIO* 39(5–6):376–384.

Domene X., Mattana S., Hanley K., Enders A., Lehmann J. 2014. "Medium-term effects of corn biochar addition on soil biota activities and functions in a temperate soil cropped to corn." *Soil Biology & Biochemistry* 72:152–162.

Dong D., Wang C., Van Zwieten L., Wang H., Jiang P., Zhou M., Wu W. 2020. "An effective biochar-based slow-release fertilizer for reducing nitrogen loss in paddy fields." *Journal of Soils and Sediments* 20(8):3027–3040.

El Sharkawi H. M., Tojo S., Chosa T., Malhat F. M., Youssef A. M. 2018. "Biochar-ammonium phosphate as an uncoated-slow release fertilizer in sandy soil." *Biomass & Bioenergy* 117:154–160.

Fageria N. K., Baligar V. C. 2005. "Enhancing nitrogen use efficiency in crop plants." *Advances in Agronomy* 88:97–185.

Fang Z., Hu Y., Wu X., Qin Y., Cheng J., Chen Y., Tan P., Li H. 2018. "A novel magnesium ascorbyl phosphate graphene-based monolith and its superior adsorption capability for bisphenol A." *Chemical Engineering Journal* 334:948–956.

Gao Y., Fang Z., Van Zwieten L., Bolan N., Dong D., Quin B. F., Meng J., Li F., Wu F., Wang H., Chen W. 2022. "A critical review of biochar-based nitrogen fertilizers and their effects on crop production and the environment." *Biochar* 4:36.

González M. E., Cea M., Medina J., González A., Diez M. C., Cartes P., Monreal C., Navia R. 2015. "Evaluation of biodegradable polymers as encapsulating agents for the development of a urea controlled-release fertilizer using biochar as support material." *Science of the Total Environment* 505:446–453.

González-Cencerrado A., Ranz J. P., Jiménez M. T. L., Gajardo B. R. 2020. "Assessing the environmental benefit of a new fertilizer based on activated biochar applied to cereal crops." *Science of the Total Environment* 711:134668.

Grutzmacher P., Puga A. P., Bibar M. P. S., Coscione A. R., Packer A. P., Andrade C. A. 2018. "Carbon stability and fertilizer induced N_2O emissions mitigation in soil treated with biochar." *Science of the Total Environment* 625:1459–1466.

Gurusamy U. M., Rajan R., Sundaram V. B., Selvaraj R. C. A., Mala R. 2017. "Evaluation of nano structured slow release fertilizer on the soil fertility, yield and nutritional profile of Vigna radiata." *Recent Patents on Nanotechnology* 11:50–62.

He T., Yuan J., Luo J., Liu D., Wang W. 2019. "Organic fertilizers have divergent effects on soil N2O emissions." *Biology Fertility of Soils* 55:685–699.

Hirose T. 2011. "Nitrogen use efficiency revisited." *Oecologia* 166:863–867.

Hossain M. Z., Bahar M. M., Sarkar B., Donne S. W., Ok Y. S., Palansooriya K. N., Kirkham M. B., Chowdhury S., Bolan N. 2020. "Biochar and its importance on nutrient dynamics in soil and plant." *Biochar* 2(4):379–420.

Jia Y., Hu Z., Mu J., Zhang W., Xie Z., Wang G. 2020. "Preparation of biochar as a coating material for biochar-coated urea." *Science of the Total Environment* 731:139063.

Jia Y. M., Hu Z. Y., Ba Y. X., Qi W. F. 2021. "Application of biochar-coated urea controlled loss of fertilizer nitrogen and increased nitrogen use efficiency." *Chemical and Biological Technologies in Agriculture* 8:3.

Khajavi-Shojaei S., Moezzi A., Norouzi M., Taghavi M. 2020. "Synthesis modified biochar-based slow-release nitrogen fertilizer increases nitrogen use efficiency and corn (Zea mays L.) growth." *Biomass Conversion and Biorefinery* 10:10.

Khakbazan M., Grant C. A., Finlay G., Wu R., Harker K. N. 2013. "An economic study of controlled release urea and split applications of nitrogen as compared with non-coated urea under conventional and reduced tillage management." *Canadian Journal of Plant Science* 93:523–534.

Khan M. A., Kim K. W., Wang M., Lim B. K., Lee W. H., Lee J. Y. 2008. "Nutrient-impregnated charcoal: An environmentally friendly slow-release fertilizer." *Environmentalist* 28:231–235.

Lapointe B. E., Brewton R. A., Herren L. W., Porter J. W., Hu C. 2019. "Nitrogen enrichment, altered stoichiometry, and coral reef decline at Looe Key, Florida Keys, USA: A 3-decade study." *Marine Biology* 166:108.

Lee J. W., Hawkins B., Li X., Day D. M. 2013. "Biochar fertilizer for soil amendment and carbon sequestration." In: Lee, J.W. (Ed.), *Advanced Biofuels and Bioproducts*. Springer New York, pp. 57–68.

Lehmann J., Cowie A., Masiello C. A., Kammann C., Woolf D., Amonette J. E., Cayuela M. L., Camps-Arbestain M., Whitman T. 2021. "Biochar in climate change mitigation." *Nature Geoscience* 14:883–892.

Li H., Tan Z. X. 2021. "Preparation of high water-retaining biochar and its mechanism of alleviating drought stress in the soil and plant system." *Biochar* 3(4):579–590.

Liao J. Y., Liu X. R., Hu A., Song H. X., Zhang Z. H. 2020. "Effects of biochar-based controlled release nitrogen fertilizer on nitrogen-use efficiency of oilseed rape (*Brassica napus* L.)." *Scientific Reports* 10:11063.

Liu X. R., Liao J. Y., Song H. X., Yang Y., Guan C. Y., Zhang Z. H. 2019. "A biochar-based route for environmentally friendly controlled release of nitrogen: Urea-loaded biochar and bentonite composite." *Scientific Reports* 9:9548.

Lu X., Li Y., Wang H., Singh B. P., Hu S., Luo Y., Li J., Xiao Y., Cai X., Li Y. 2019. "Responses of soil greenhouse gas emissions to different application rates of biochar in a subtropical Chinese chestnut plantation." *Journal of Agricultural Meteorology* 271:168–179.

Ma R., Zou J., Han Z., Yu K., Wu S., Li Z. F., Liu S. W., Niu S. L., Horwath W. R., Zhu-Barker X. 2021. "Global soil-derived ammonia emissions from agricultural nitrogen fertilizer application: A refinement based on regional and crop-specific emission factors." *Global Change Biology* 27:855–867.

Mandal S., Donner E., Vasileiadis S., Skinner W., Smith E., Lombi E. 2018. "The effect of biochar feedstock, pyrolysis temperature, and application rate on the reduction of ammonia volatilisation from biochar-amended soil." *Science of the Total Environment* 627:942–950.

Naz M. Y., Sulaiman S. A. 2016. "Slow release coating remedy for nitrogen loss from conventional urea: A review." *Journal of Controlled Release* 225:109–120.

Nguyen T., Xu C. Y., Tahmasbian I., Che R., Xu Z., Zhou X., Wallace H. M., Bai S. H. 2017. "Effects of biochar on soil available inorganic nitrogen: A review and meta-analysis." *Geoderma* 288:79–96.

Oh T. K., Shinogi Y., Lee S. J., Choi B. S. 2014. "Utilization of biochar impregnated with anaerobically digested slurry as slow-release fertilizer." *Journal of Plant Nutrition and Soil Science* 177:97–103.

Oladele S., Adeyemo A., Adegaiye A., Awodun M. 2019. "Effects of biochar amendment and nitrogen fertilization on soil microbial biomass pools in an Alfisol under rain-fed rice cultivation." *Biochar* 1(2):163-176.

Park M. H., Jeong S., Kim J. Y. 2019. "Adsorption of NH_3-N onto rice straw-derived biochar." *The Journal of Environmental Chemical Engineering* 7:103039.

Peng J. H., Zhang L., Meulien E. S., Bi X. T. T., Lim J. C., Chen W. H. 2021. "Waste plastics as an effective binder for biochar pelletization." *Energy & Fuels* 35(17):13840–13846.

Penuelas J., Gargallo-Garriga A., Jannssens I., Ciais P., Obersteiner M., Klem K., Urban O., Sardans J. 2020. "Could global intensification of nitrogen fertilisation increase immunogenic proteins and favour the spread of coeliac pathology?" *Foods* 9:1602.

Puga A. P., Queiroz M. C. D., Ligo M. A. V., Carvalho C. S., Pires A. M. M., Marcatto J. D. O., Andrade C. A. D. 2019. "Nitrogen availability and ammonia volatilization in biochar-based fertilizers." *Archives of Agronomy and Soil Science* 66:992–1004.

Puga A. P., Grutzmacher P., Cerri C., Ribeirinho V. S., Andrade C. 2020. "Biochar-based nitrogen fertilizers: Greenhouse gas emissions, use efficiency, and maize yield in tropical soils." *Science of the Total Environment* 704:135375.

Qian L., Chen L., Joseph S., Pan G. X., Li L. Q., Zheng J. W., Zhang X. H., Zheng J. F., Yu X. Y., Wang J. F. 2014. "Biochar compound fertilizer as an option to reach high productivity but low carbon intensity in rice agriculture of China." *Carbon Management* 5:145–154.

Ramli R. A. 2019. "Slow release fertilizer hydrogels: A review." *Polymer Chemistry* 10:6073–6090.

Rashid M., Hussain Q., Khan K. S., Alwabel M. I., Hayat R., Akmal M., Ijaz S. S., Alvi S. 2021. "Carbon-based slow-release fertilizers for efficient nutrient management: Synthesis, applications, and future research needs." *Journal of Soil Science and Plant Nutrition* 21:1144–1169.

Roberts K., Gloy B. A., Joseph S., Scott N. R., Lehmann J. 2010. "Life cycle assessment of biochar systems: estimating the energetic, economic, and climate change potential." *Environmental Science Technology* 44:827–833.

Saha B. K., Rose M. T., Van Zwieten L., Wong V. N. L., Patti A. F. 2021. "Slow release brown coal-urea fertilizer potentially influences greenhouse gas emissions, nitrogen use efficiency, and sweet corn yield in Oxisol." *ACS Agricultural Science Technology* 1(5):469–478.

Shi Y. L., Liu X. R., Zhang Q. W. 2019. "Effects of combined biochar and organic fertilizer on nitrous oxide fluxes and the related nitrifier and denitrifier communities in a saline-alkali soil." *Science of the Total Environment* 686:199–211.

Shi W., Ju Y. Y., Bian R. J., Li L., Pan G. 2020. "Biochar bound urea boosts plant growth and reduces nitrogen leaching." *Science of the Total Environment* 701:134424.

Siedt M., Schaeffer A., Smith K. E. C., Nabel M., Nickoll M. R., Dongen J. T. 2021. "Comparing straw, compost, and biochar regarding their suitability as agricultural soil amendments to affect soil structure, nutrient leaching, microbial communities, and the fate of pesticides." *Science of the Total Environment* 751:141607.

Song Y., Li Y., Cai Y., Fu S., Luo Y., Wang H., Liang C., Lin Z., Hu S., Li Y., Chang S. X. 2019. "Biochar decreases soil N_2O emissions in Moso bamboo plantations through decreasing labile N concentrations, N-cycling enzyme activities and nitrification/denitrification rates." *Geoderma* 348:135–145.

Sun H., Zhang Y., Yang Y., Chen Y., Jeyakumar P., Shao Q., Zhou Y., Ma M., Zhu R., Qian Q., Fan Y., Xiang S., Zhai N., Li Y., Zhao Q., Wang H. 2021. "Effect of biofertilizer and wheat straw biochar application on nitrous oxide emission and ammonia volatilization from paddy soil." *Environmental Pollution* 275:116640.

Wang Z., Zheng H., Luo Y., Deng X., Herbert S., Xing B. 2013. "Characterization and influence of biochars on nitrous oxide emission from agricultural soil." *Environmental Pollution* 174:289–296.

Wang H. H., Zheng-Yi H. U., Zhu X. Q. 2015. "Comparison of nitrogen loss after biochar coated urea and common urea fertilization in vegetable soil at Chaihe catchment of Dianchi Lake." *Journal of Anhui Agricultural Sciences* 43:104–107.

Wang Y., Ying H., Yin Y., Zheng H., Cui Z. 2019. "Estimating soil nitrate leaching of nitrogen fertilizer from global meta-analysis." *Science of the Total Environment* 657:96–102.

Wen P., Wu Z., Han Y., Cravotto G., Wang J., Ye B. C. 2017. "Microwave-assisted synthesis of a novel biochar-based slow-release nitrogen fertilizer with enhanced water-retention capacity." *ACS Sustainable Chemistry & Engineering* 5:7374–7382.

Xiang A. H., Qi R. Y., Wang M. F., Zhang K., Jiang E. C., Ren Y. Z., Hu Z. W. 2020. "Study on the infiltration mechanism of molten urea and biochar for a novel fertilizer preparation." *Industrial Crops and Products* 153:112558.

Yang F., Xu Z., Yu L., Gao B., Xu X. Y., Zhao L., Cao X. D. 2018. "Kaolinite enhances the stability of the dissolvable and undissolvable fractions of biochar via different mechanisms." *Environmental Science & Technology* 52:8321–8329.

Yang X., Shaheen S. M., Wang J., Hou D., Ok Y. S., Wang S. L., Wang H., Rinklebe J. 2022. "Elucidating the redox-driven dynamic interactions between arsenic and iron-impregnated biochar in a paddy soil using geochemical and spectroscopic techniques." *Journal of Hazardous Materials* 422:126808.

Yao C., Joseph S., Lianqing L. I., Pan G., Lin Y., Munroe P., Pace B. 2015. "Developing more effective enhanced biochar fertilisers for improvement of pepper yield and quality." *Pedosphere* 25:703–712.

Ye Z. X., Zhang L. M., Huang Q. Y., Tan Z. X. 2019. "Development of a carbon-based slow release fertilizer treated by bio-oil coating and study on its feedback effect on farmland application." *Journal of Cleaner Production* 239, 118085.

Yoo G., Kim H., Chen J., Kim Y. 2014. "Effects of biochar addition on nitrogen leaching and soil structure following fertilizer application to rice paddy soil." *Soil Science Society of America Journal* 78:852–860.

Zhang X., Davidson E. A., Mauzerall D. L., Searchinger T. D., Dumas P., Shen Y. 2015. "Managing nitrogen for sustainable development." *Nature* 528:51–59.

Zhang W., Jiang L., Xu C., He X., Geng Z. 2018. "Biochar and biochar-based nitrogenous fertilizers: Short-term effects on chemical properties of soils." *International Journal of Agriculture and Biology* 20:1555–1561.

Zhang X., Miao X., Xiang W., Zhang J., Cao C., Wang H., Hu X., Gao B. 2021a. "Ball milling biochar with ammonia hydroxide or hydrogen peroxide enhances its adsorption of phenyl volatile organic compounds (VOCs)." *Journal of Hazardous Materials* 403:123540.

Zhang X., Zou T., Lassaletta L., Mueller N. D., Davidson E. A. 2021b. "Quantification of global and national nitrogen budgets for crop production." *Nature Food* 2:529–540.

Zheng J., Han J., Liu Z., Xia W., Zhang X., Li L., Liu X., Bian R., Cheng K., Zheng J. 2017. "Biochar compound fertilizer increases nitrogen productivity and economic benefits but decreases carbon emission of maize production." *Agriculture, Ecosystems & Environment* 241:70–78.

Zhou Z. J., Du C. W., Li, T., Shen Y. Z., Zeng Y., Du J., Zhou J. M. 2015. "Biodegradation of a biochar-modified waterborne polyacrylate membrane coating for controlled-release fertilizer and its effects on soil bacterial community profiles." *Environmental Science and Pollution Research* 22:8672–8682.

Zhou J., Qu T., Li Y., Zwieten L., Wang H., Chen J., Song X., Lin Z., Zhang X., Luo Y., Cai Y., Zhong Z. 2021. "Biochar-based fertilizer decreased while chemical fertilizer increased soil N_2O emissions in a subtropical Moso bamboo plantation." *Catena*, 202:105257.

Zhu X., Chen B., Zhu L., Xing B. 2017. "Effects and mechanisms of biochar-microbe interactions in soil improvement and pollution remediation: A review." *Environmental Pollution* 227:98–115.

12 Biochar for Sustainable Soils, Agriculture, and Climate Change Mitigation

Amna Zulfqar[1], Nabeel Khan Niazi[1], Muhammad Mahroz Hussain[1], Rabia Amen[1,2], Irshad Bibi[1], Muhammad Shahid[3], Khalid Hussain[4], Fazila Younas[1], Hamna Bashir[1], and Muhammad Jahandad Khan[1]

[1]Institute of Soil and Environmental Sciences, University of Agriculture Faisalabad, Faisalabad, Pakistan
[2]Department of Sustainable Bioproducts, College of Forest Resources Mississippi State University, Starkville, Mississippi, USA
[3]Department of Environmental Sciences, COMSATS University Islamabad, Vehari, Pakistan
[4]Department of Agronomy, Faculty of Agriculture, University of Agriculture Faisalabad, Faisalabad, Pakistan

CONTENTS

DOI: 10.1201/9781003093565-12

12.1 INTRODUCTION

Soil health plays a vital role in sustainable agriculture, ecosystem restoration, food safety, and food security. All of these ultimately strengthen agriculture-based ecosystems and socio-economic situations at regional to global scales (Amundson et al., 2015; Hou et al., 2020). Soils are essential in providing various ecosystem functions, thereby supporting sustainable agriculture and environmental and human health (Keesstra et al., 2016; Robinson et al., 2017). Various soil-deteriorating factors are soil erosion (Hussain et al., 2021), contamination (Natasha et al., 2022; Natasha et al., 2021), poor drainage, and salinity (Pimentel and Burgess, 2013), which may be due to improper agricultural practices, deforestation, and unplanned urbanization (Borrelli et al., 2017). Soil degradation is caused by compaction, salinization, loss of water, and mineralization of soil organic matter (SOM). They decrease soil health and limit global food production (Gujre et al., 2021; Smith et al., 2016).

The effect of soil degradation is not restricted to fertility loss but also is associated with other impacts, such as sedimentation, eutrophication of streams, or increased flooding and siltation (Borrelli et al., 2017; Zahra et al., 2021). Erosion causes variations in soil carbon (C) cycling either by enhancing or diminishing CO_2 emission through mineralization and residue burial (Borrelli et al., 2017; Van Oost et al., 2007). Adding soil C can provide many benefits, such as sequestering C in soil, decreasing emission of greenhouse gases (GHGs), improving soil physicochemical properties, and mitigating climate change and soil degradation (Smith and Gregory, 2013; Zhang et al., 2016). Soil C sequestration is considered to be a key factor in the decline of CO_2 emissions, and it can increase the quality of soil in farmlands (Edwards et al., 2018). To increase soil C content, biochar is an emerging organic C enhancer that increases soil health and nutrient cycling (Hussain et al., 2022). Biochar application in different crops under various soil types and soil-plant systems has been explored (Diatta et al., 2020). The effect of biochar on crop yield is mostly governed by the type of crop, soil properties, and properties of biochar that rely on feedstock and pyrolytic temperature (slow or fast pyrolysis) (Purakayastha et al., 2019). Numerous recent studies have examined biochar's role in climate change mitigation (Minasny et al., 2017; Neogi et al., 2021), waste management (Mandal et al., 2018), agronomical advantages (Alvarez-Campos et al., 2018), water treatment (Shaheen et al., 2021; Shaheen et al., 2019), and contaminated soil remediation (O'Connor et al., 2018; Sun et al., 2021).

The current chapter provides a brief and concise overview of biochar production methods, biochar properties, and the role of biochar in soil C sequestration with nutritional properties affecting soil health and productivity of crops in the present perspective of climate change. We discuss the role of biochar in improving various physicochemical properties of soil and crop productivity; (ii) the potential benefits of biochar in terms of soil C sequestration and GHGs reduction for climate change mitigation; and (iii) biochar's role in ecosystem functioning and as a tool for soil C storage.

12.2 PRODUCTION AND CHARACTERISTICS OF BIOCHAR

Biochar is produced by pyrolysis of biomass (Figure 12.1). Pyrolysis is a relatively easy process where natural polymeric elements of biomass are converted to char (solid), bio-oil (condensable vapors), and non-condensable gases through thermal decay in a low-oxygen environment (Al-Wabel et al., 2018). Various biochar-producing techniques have been established, including slow and fast pyrolysis, torrefaction, and gasification. Biochar yield and physicochemical properties are dependent on pyrolytic temperature, feedstock type, and residence time (Niazi et al., 2016). According to the International Biochar Initiative (IBI), biochar is described by proximate analysis (fixed C, ash content, moisture content, and volatile matter), elemental analysis [C, nitrogen (N), hydrogen (H), and oxygen (O)], chemical analysis [cation exchange capacity (CEC), pH, and redox potential], and morphological and structural analysis (surface area, functional groups, and pore size) (Figure 12.2) (Al-Wabel et al., 2018).

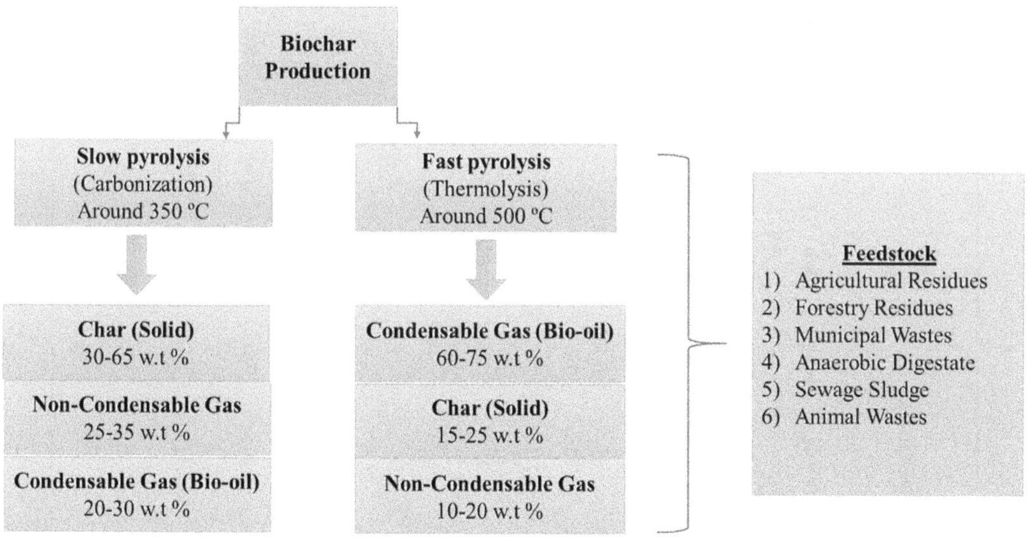

FIGURE 12.1 Schematic diagram of biochar production technologies.

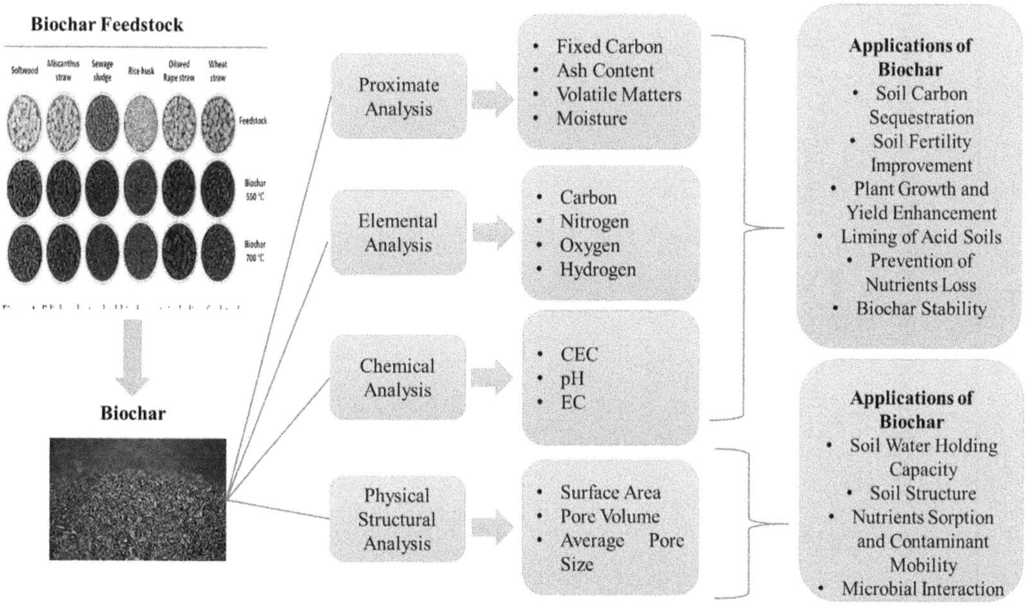

FIGURE 12.2 Biochar characterization and associated soil properties.

Biochar structure, rate of production, and thermal solidity are influenced by the type of feedstock (Table 12.1). Yuan et al. (2016) studied biochar derived from sewage sludge at an elevated temperature (700°C) and low temperature (300°C). They found that it contained low N content but high phosphorus (P) and potassium (K) contents when produced at the low temperature. Zornoza et al. (2016) found that at a pyrolytic temperature of >500°C biochar produced from crop residues, pig manure, and municipal solid waste had lower nutrient concentrations (K, P, and N), because of mineral crystallization, compared to that produced at a low temperature (<500°C). It has been reported that low-temperature biochar has low alkalinity and salinity but high CEC and hydrophobicity (Al-Wabel et al., 2018).

TABLE 12.1
Difference in Biochar Composition and Properties Induced by Feedstock Type and Pyrolytic Temperature

Feedstock	Temperature (°C)	pH	EC dS m^{-1}	SA $m^2 g^{-1}$	Pore Volume $mL^3 g^{-1}$	CEC cmol+kg^{-1}	C (%)	H (%)	O (%)	N (%)	Ash (%)	References
Sugarcane bagasse	300	4.97	—	224.1	—	—	—	—	—	—	4.2	Mubarik et al. (2016)
	400	5.93		361.8							4.2	
	500	6.12		291.4							4.1	
Rice husk	350	8.3	—	—	—	14	47.8	2.37	0.82			Obia et al. (2016)
Maize cob		8.8				22.2	53.8	2.36	0.65			
Corn stover	300	7.31	—	—	—	—	69.9	4.0.4	0.94	13.4	11.7	Chen et al. (2016)
	400	7.79					74.8	3.15	1.05	6.18	14.8	
	500	8.1					78.1	2.11	0.92	3.77	15.1	
Maize straw	350	7.9		6.71	—	—	58.9	3.36	1	21.3	15.4	Ren et al. (2016)
	700	10.6		265			43.8	1.47	1.27	31.6	21.9	
Maize	300	9.84	3.4	1	0.01	—	48.9	3.54	1.25	23.9	16.3	Wang et al. (2015)
	450	10.47	5.37	4	0.01		53.8	3.11	1.22	18.6	22.3	
	600	11.37	5.09	70	0.06		62.9	1.98	1.28	16.2	27.2	
Mulberry wood	350	—	—	—	—	—	67.9	4.53	25.2	2.16	7.52	Zama et al. (2017)
	450						70.8	3.32	23.8	1.92	7.72	
	550						77.0	2.41	18.8	1.68	9.82	
	650						80.1	1.63	16.6	1.58	9.87	
Jarrah wood	700	9.4	—	—	—	—	85	1.6	8.5	0.2	3.3	Jiang et al. (2016)
Pine Wood	700	7.8					79	3.0	15	0.2	2.2	
Lodgepole pine wood	1,000	9.7	—	—	—	—	—	—	—	—	—	Huggins et al. (2016)
Wood	200	4.6	—	—	—	—	48.8	—	—	0.2	—	Zhang et al. (2015a)
	400	6.9					42.7			0.3		
	600	9.5					45.5			0.4		

Depending on the feedstock type, Rehrah et al. (2016) found that biochar produced at high pyrolytic temperatures, i.e., paper, hardwood, soft material, and mixed materials, had higher pH, ash content, and total surface area compared to biochar produced at low temperatures. Hardwood yield recoveries at 750°C and 300°C were 23% and 94%, respectively. There was a lower surface area at 300°C. Carbon content of biochar was enhanced at a high pyrolytic temperature. At 750°C, for hardwood, there was 97% C, and for mixed materials, there was 62% C. For a critical understanding of pyrolytic temperature, which plays an important role in determining biochar's physical and chemical attributes, it is imperative to use advanced microscopic and spectroscopic techniques to investigate details about morphological and chemical characteristics (Al-Wabel et al., 2018).

12.3 IMPACT OF BIOCHAR ON PHYSICAL CHARACTERISTICS OF SOIL

12.3.1 Soil Porosity

The effect of biochar application on physical characteristics of soil has been studied, including soil porosity and soil aggregate development, which are crucial for organic C content determination. Biochar has shown high surface area, varying from 0.5 to 2,200 m^2/g, and it is a function of feedstock and heating treatment temperatures (HTTs) that may influence the soil physical properties (Downie et al., 2009; Han et al., 2020).

Soil porosity is crucial in determining the rate at which water can leach down the soil profile. It also affects the removal of nutrients from the soil required for plant growth and sustainable soil health. Capillaries in soil allow small pores to hold the soil solution, and they decrease leaching (Han et al., 2020; Major et al., 2009). When added to soil, biochar directly affects physical properties by changing surface area, soil mass and aggregation, and pore and particle size distribution (Yu et al., 2019). Because the physical structure of the soil matrix determines infiltration and water and air accessibility for roots, the effect of biochar on soil physical qualities has a direct impact on the plant (de Jesus Duarte et al., 2019).

Moreover, biochar application can enhance soil microbial health by providing C as a food source to increase the growth of microbes. The pores of biochar also serve as a shelter for colonizing microorganisms, shielding them from predators, or allowing for the production of less aggressive microbes in the soil environment (Han et al., 2020; Kookana et al., 2011).

12.3.2 Soil Aggregate Formation

Soil aggregate formation is responsible for soil structure and is considered to be essential for soil functioning. Aggregate quantity and structural stability can be affected by storage and composition of soil organic C (SOC) (Sarker et al., 2018). Biochar aggregates of diverse sizes and average weight diameter have an impact on aggregate stability. Han et al. (2020) found that biochar increased macro-sized aggregates (>200 μm) (mean: 24%), particularly small macro-sized aggregates (>200–2,000 μm). However, it reduced the quantity of micro-sized soil aggregates (mean: −7%).

Wang et al. (2016a) found that biochar derived from manure generated the greatest increase (mean: 42%) in little macro-aggregates, compared to other feedstocks (22.1% for biochar produced from crops and 16.3% for biochar derived from wood). This was probably because of the comparatively high calcium (Ca), iron (Fe), and magnesium (Mg) contents of biochar along with its numerous mineral components (e.g., calcite, whitlockite, and sylvite) (Han et al., 2018; Xu et al., 2017) and O-containing functional groups, particularly −COOH groups (Zhao and Zhou, 2019).

For the feedstock used, Wang et al. (2016b) found that the efficiency of biochar generated below 400°C (36%) was lower compared to biochar at high temperature (500–600°C; 9.6%). This may be due to biochar's high O-containing functional groups. Consequently, the higher O-rich functional groups of low-temperature biochar provide the possibility that minerals could be attached to the biochar, which will affect soil aggregation (Du et al., 2017; Zhang et al., 2015b).

12.3.3 Impact of Biochar on Chemical Properties of Soil

12.3.3.1 Soil pH

The application of biochar may influence the pH of soil by adding oxides, hydroxides, and carbonates of Ca, Mg, K, and Na directly into the soil, thus increasing the buffering capacity of acidic soils. Biochar has functional groups with various reactivity (alcoholic, –OH, phenolic-OH, and –COOH) on its surface, which can bind basic cations in soil. Liao and Thomas (2019) reported that small particles of biochar had the biggest liming effect, increasing substrate pH by ~0.3 pH unit. Tasneem and Zahir (2017) found that the addition of biochar to soil increased pH up to 50 days. Luo et al. (2017a) found that biochar did not affect microbial communities in soil at 10 and 30 ton/ha biochar application rates in a semi-arid farmland.

12.3.3.2 Cation Exchange Capacity of Soil

Soil CEC has the potential to attach micro- and macro-nutrients, specifically cations. High CEC allows greater nutrient holding capacity and a smaller loss of nutrients via leaching, which is related to microbial movement in soil (Bandara et al., 2017). Biochar behaves as a key material to hold cationic ions (e.g., Ca, Mg, K, and NH_4) in the soil because of its large surface area, negative charge surface, and high pore volume (Han et al., 2020; Houben et al., 2013). It increases soil CEC (Table 12.2). Biochar's potential to improve CEC of soil is also influenced by the type of feedstocks and pyrolytic temperature.

12.3.4 Impact of Biochar on Biological Attributes of Soil

The role of biochar on soil biological attributes is well documented. But the indirect impact on microbial organization and action by altering the physicochemical properties of soil can also be influenced by biochar. Table 12.3 shows how biochar affects bacterial and fungal abundance in various soil types. Fungal and bacterial populations are affected by the rate of application of biochar and physicochemical/biochemical pathways. The mechanisms for enhancing bacterial and fungal

TABLE 12.2
Summary on the Effect of Biochar Amendment on Soil pH and Cation Exchange Capacity (CEC)

Soil Type	Biochar	Addition rate (wt %)	Duration (days)	pH	CEC ($cmol_c$/kg)	References
Ferralsol	Commercial charcoal	0	45	5.14	5.40	Lehmann et al. (2003)
		20		5.89	28.50	
Red ferrosol	Paper mill at 550°C	0	56	4.20	4.03	Van Zwieten et al. (2010)
		2		5.93	10.50	
Calcarosol	Paper mill at 550°C	0	56	7.67	31.00	Van Zwieten et al. (2003)
		1.5		7.67	29.00	
Acrisol	Rice husk at unknown temperature	0	1460	5.50	0.94	Haefele et al. (2011)
		2		5.60	1.00	
Typic Hapludoll	Commercial charcoal	0	84	6.40	17.10	Laird et al. (2010)
		0.5		6.42	19.80	
		1		6.90	20.70	
		2		7.13	20.80	
Alfisol	Commercial charcoal	0	52	8.03	2.72	Luo et al. (2017b)
		0.5		7.70	3.03	
		1		7.72	3.19	
		2		7.84	3.15	

abundance mainly involve biochar-based carbon that acts as a substrate for microbial growth (Quan et al., 2020).

Biochar can help nutrients be released slowly into soils (Sun et al., 2018). Hence, it can support a slow release of nutrients from fertilizers which can increase crop productivity. Because of its high porosity, biochar may improve the accessibility of nutrients to the microbiota (Palansooriya et al., 2019). A decrease in fungal and bacterial populations has been ascribed to toxic compounds in biochar, which may reduce microbial activity in soil (Han et al., 2020; Lehmann et al., 2011). In biochar, certain compounds are identified as microbial inhibitors, including benzene, carboxylic

TABLE 12.3
Summary on the Effect of Biochar Amendment on Soil Microbial Community Composition

Soil	Biochar	Addition Rate (wt %)	Duration (Days)	Effects	Potential Mechanism	References
Phaozem, Luvisol, and Gleysol	Oak pellet at 550°C	0 and 1	369	In all soils, biochar increased bacterial and fungal abundance, but lowered the fungi-to-bacteria ratio	Biochar-derived OC served as a substrate for microbial activity	Gomez et al. (2014)
Psammaquent	Swine manure, fruit peels, *Phragmites australis* and *Brassica rapa* at 500°C	0, 1 and 3	90	Biochar boosted the number of bacteria and fungi. In the 1% biochar treatment, the ratio of fungal to bacteria was considerably reduced, but, in the 3% biochar treatment, it increased	Not given	Muhammad et al. (2014)
Fluvo-aquic soil	Maize straw at 450°C	0, 0.5, 1.0, 2.5 and 5.0	90	The amount of fungal growth decreased as the rate of biochar addition increased	An alkaline soil modified with a large amount of biochar may impede microbial activity in the soil	Wang et al., (2015)
Inceptisol	Sugar maple at 500°C	0, 0.24, 0.48 and 0.96	168	The bacteria and fungi concentration increased in the first 2 weeks, but decreased during 2–16 weeks	The increase was likely due to microbial acclimation to laboratory conditions. The decrease was because of the introduction of compounds in the biochar which were poisonous to microbes	Mitchell et al. (2015)
Fluvisol	Yeast at unknown temperature	0 and 30	182	Biochar promoted fungi growth	Biochar enhanced nutrient levels in soil	Elzobair et al. (2016)

acids, ketones, methoxyphenols, furans, polycyclic aromatic hydrocarbons (PAHS), and phenols (Ghidotti et al., 2017; Lyu et al., 2016). These compounds can be identified from water or organic solvent extractions of biochar. Qiu et al. (2015) found that biochar produced at moderate temperatures (300°C and 400°C) had higher concentrations of furans, PAHs, and polychlorinated dioxins than biochar produced above 400°C.

12.4 BIOCHAR AS A SOURCE OF NUTRIENTS

12.4.1 Biochar's Role as Fertilizer

High-molecular weight C compounds, such as humic and fulvic acids, and N, P, and K are important in the soil. Lin et al. (2012) demonstrated that biochar generated at 380°C from *Acacia saligna* and at 450°C from sawdust contained 17.7% humic substances and 16.2% fluvic substances. Biochar produced at 300°C using *Lantana camara* contained accessible P (0.64 mg/kg), K (711 mg/kg), Na (1,145 mg/kg), Ca (5,880 mg/kg), and Mg (1,010 mg/kg) (Masto et al., 2013). Newly produced biochar had the capability of delivering N (23–635 mg/kg) and P (46–1,664 mg/kg) (Ding et al., 2016; Mukherjee and Zimmerman, 2013; Zheng et al., 2013). This information demonstrates that biochar has the potential of providing nutrients.

Although total N, P, and K in biochar may not be completely accessible as nutrients to plants (Ding et al., 2016; Spokas et al., 2012), some of them might be available. For instance, total N in biochar was lost at high temperatures (Ding et al., 2016; Koutcheiko et al., 2007). Accessible K increases with total K in biochar (Ding et al., 2016; Zheng et al., 2013).

12.4.2 Availability of Nutrients and Uptake by Plants in Biochar-Applied Soils

The role of biochar as a soil ameliorant promotes the growth of crop plants by enhancing the accessibility of important nutrients (N, P, K) in soil-crop plants system (Al-Wabel et al., 2018; Lehmann, 2007) (Table 12.4). The availability of plant nutrients depends mainly on the biochar type, soil properties, plant attributes, and nutrient types. Biochar co-application with farmyard manure enhanced N content in soil and consequently plant nutrient uptake (Ali et al., 2015). Similarly, N uptake was improved by as much as 127% compared to control plots following the application of biochar. It was positively associated with the yield of rice, signifying that yield is dependent on nutrient uptake in rice (Partey et al., 2016).

Nitrogen uptake by roots of wheat plants improved after the application of biochar from deciduous wood (20 and 60 Mg/ha) because root growth increased (Prendergast-Miller et al., 2014). Abbasi and Anwar (2015) studied biochar generated from white clover residues and poultry manure. They added it to soil either alone or in a 50:50 combination. The biochar improved maize N uptake over untreated controls, although in wheat plants there was a negligible response.

12.4.3 Factors Affecting Nutrient Content and Availability in Biochar

The amount of nutrients in biochar depends upon the feedstock source and pyrolytic temperature (Table 12.1). For example, in three woody and four herbaceous biochars, it was found that loss of N starts at around 400°C and half of the N volatilizes at around 750°C (Ding et al., 2016; Lang et al., 2005). With the rise of pyrolytic temperature from 350°C to 600°C, available N content in biochar diminished from 39 to 8 mg/kg, which could be ascribed to total N loss and the heterocyclization of N through pyrolysis (Zheng et al., 2013).

Compared to total N content in biochar, the total P content increased from 0.12% to 0.17% with the rise of temperature from 300°C to 600°C (Ding et al., 2016; Zheng et al., 2013). The labile P in biochar, generated at a lower temperature, was higher than that in biochar produced at a higher temperature. This could be due to the fact that biochar contains low amounts of solidified P-related

TABLE 12.4
Influence of Biochar on Plant Yield and Nutrients Contents

Biochar	Biochar Rate	Biochar pH and Pyrolysis Temperature	Soil pH and Texture	Crop Yield	Nutrient Content in Plants	References
Tectona grandis (teak)-wood biochar	05 t/ha	pH: 7.5, 500°C	pH: 5.4, sandy-loam	For all cropping seasons, the average rice grain yield was 1.8 t/ha for biochar-treated plots and 1.3 t/ha for unamended plots	Biochar increased N uptake by 127% compared with unamended plots	Partey et al. (2016)
Wheat bran pellets biochar	14 t/ha	pH: 8.2	pH: 8.1, silty-clay	In biochar-treated plots, plant growth increased but tomato yield did not	Biochar plots had higher plant nutritional content (N, K, Mg, and P)	Vaccari et al. (2015)
Pine needle and *Lantana* biochar's	2 t biochar-C/ha, 5 t biochar-C/ha	Slow pyrolysis	pH: 6.5, loam	Biochars enhanced wheat grain production by 6.2–24.2% as compared to control. No important effect on rice yield	Biochars increased uptakes of N and P significantly in wheat and rice grains as compared to control	Bhattacharjya et al. (2016)
Castanea sativa wood biochar	20 t/ha, 50 t/ha	pH: 9.1, 350°C–400°C	pH: 6.3, sandy clay-loam	Biochar had no impact on the growth or harvest yield of field-grown spring barley, strawberry, and potato	Biochar had little influence on the tissue concentration of N, P, K, Ca, and Mg irrespective of crop. But biochar improved tissue molybdenum and a reduction in manganese, in strawberry	Jay et al. (2015)
Sawdust biochar	20 t/ha, NPK fertilizer (60–175-120 kg/ha)	pH: 5.21, 300°C–350°C	pH: 8.80	The use of biochar and NPK fertilizer considerably boosted the yield of all three soybean genotypes	On soil-accessible P, the interaction of biochar and fertilizer was significant	Mete et al. (2015)
Bamboo biochar	4.5 t biochar-C/ha	pH: 9.8, 600°C	pH: 6.16	Bamboo biochar did not considerably rise grain yield	Bamboo biochar and rice straw biochar supported K contents of rice grains in 2nd year	Liu et al. (2016)
Rice straw biochar		pH: 10.2, 550°C		Rice straw biochar improved grain yield by 8.5%–10.7% as compared to control	Rice straw biochar enhanced the total nutrient uptake by grain significantly (13.6%–16.4%) as compared to control	Liu et al. (2016)
Rice husk biochar	15 t/ha, 30 t/ha	pH: 9.1, 350°C–500°C	pH: 4.7, sandy-loam pH: 5.18, clay pH: 4.42, sandy-loam	In all soil types, rice husk biochar had no favorable influence on maize crop development	In two soils, rice husk biochar had no influence on maize N uptake. In one soil, rice husk biochar increased maize N uptake	Nguyen et al. (2016a)

minerals at low temperatures. Ding et al. (2016) and Zheng et al. (2013) showed that the content of total K was 3.7% at 300°C and 5.02% at 600°C, whereas the K (water-soluble) content increased with the rise in pyrolytic temperature (37%–47% at 300°C and 600°C).

Biochar fabricated from various feedstocks demonstrates different nutrient compositions. For example, at 400°C biochar generated from swine fertilizer had elevated N and P (3.2% and 6.1%, respectively) (Ding et al., 2016; Tsai et al., 2012). At 400°C, biochar generated from *Arundo donax* had low N and P (0.69% and 0.13%, respectively) (Zheng et al., 2013). At 350°C, biochar produced from poultry litter had an ash content of 30.7% (Cantrell et al., 2012), which was higher than that of biochar produced from pine wood chips at 350°C (1.5%) (Ding et al., 2016; Spokas et al., 2011). Soil pH is a significant factor influencing biochar nutrient accessibility (Ding et al., 2016; Silber et al., 2010).

12.5 EFFECT OF BIOCHAR ON CROP YIELD

Biochar application increases, decreases, or has no effect on crop yields depending upon the type of soil, feedstock, and pyrolytic conditions during the production of biochar (Table 12.4). Biochar application has improved the production of numerous crops by 4%–144%, whereas in some cases the crop production has been decreased by 4%–24% (Purakayastha et al., 2019; Rajkovich et al., 2012). Biochar generated at 500°C improves the growth of plants more than that generated at 300°C–400°C. Poultry litter biochar resulted in healthier plant growth, regardless of application rate and pyrolytic temperature (Macdonald et al., 2014; Purakayastha et al., 2019).

Biochar generated at 400°C from rice straw was applied at the 2.25 g/kg rate with 100% NPK (Bera et al., 2014). It increased the yield of rice in an Inceptisol by 24.3% and in an Alfisol by 31.3%. Yield of lowland rice was improved by the combined use of rice husk biochar and fly ash supplemented with chemical fertilizers (Munda et al., 2016). Biochar (produced at 750°C from 80% different hardwoods and 20% diverse coniferous wood chips) and biochar-compost treatments led to a slight impact on the productivity of vines in Switzerland's less fertile, alkaline, and temperate soils, and it was not economically important (Purakayastha et al., 2019; Schmidt et al., 2014). Even with the application of N fertilizer, yield was decreased at a high rate of charcoal (biochar) application (16 Mg/ha) due to N limitation (Asai et al., 2009; Purakayastha et al., 2019).

Shafiq (2016) applied biochar produced at 370°C–417°C from a weed (*Parthenium hysterophorus* L.) to a rice-wheat cropping system. It reduced the expense of manure by half while maintaining the same yield using the recommended quantity of compost. The bioavailability of N in a wheat-straw biochar generated at 400°C was observed to be extremely low with no increase in rice crop growth or nitrogen use efficacy from compost sources in the first year after application (Purakayastha et al., 2019; Xie et al., 2013).

12.6 BIOCHAR ROLE IN CLIMATE CHANGE MITIGATION

In 2015, the Paris Climate Agreement set an aim to limit temperature increase to 1.5°C above preindustrial levels (IPCC, 2014). Although traditional strategies to mitigate the emission of greenhouse gases (e.g., reducing fossil fuel consumption) are required to attain the Paris Agreement goal, immediate actions on CO_2 removal technologies and engineered improvement of natural carbon sinks are also required (Gasser et al., 2015). Current reports indicate that the aim of keeping global warming below 2°C is doubtful unless the emissions gap is not closed by 2030 (UNEP, 2017).

To attain declines in emission of GHGs, sequestering of C in the terrestrial sink is required. An increase in the input of OM into the soil or a decline in the rate of decay of SOM can increase the C stock in soil (Paustian et al., 2016). The introduced idea of 4 per mille for soils, which means increasing worldwide SOM stocks by 0.4% per year to compensate for emissions of global GHG because of human activities, has been recommended (Minasny et al., 2017).

The application of biochar to soil has been shown to increase C in soil and enhance plant biomass by increasing the supply of nutrients to plants and enhancing water use efficiency (Lehmann et al.,

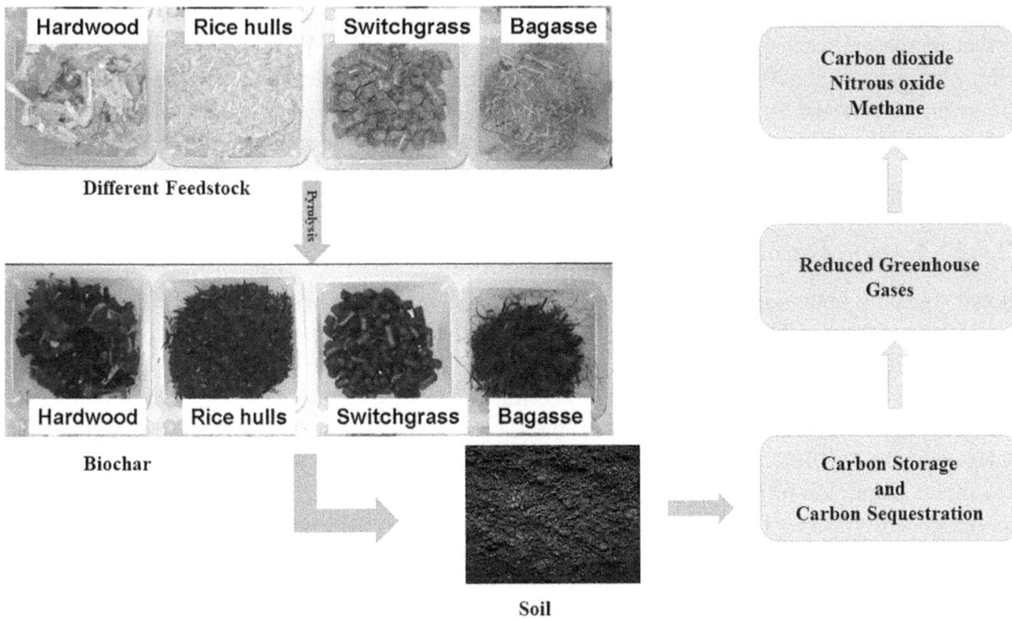

FIGURE 12.3 Impact of biochar on climate change.

2015; Minasny et al., 2017). Thus, biochar application to the soil has been suggested as an essential part of the pathway to climate-smart soil management (Paustian et al., 2016). Biochar helps to store C in soil, resulting in C sequestration and reducing GHGs (Figure 12.3).

12.7 GREENHOUSE GAS EMISSION FROM BIOCHAR-TREATED SOIL VS. CROP-RESIDUE TREATED SOIL

12.7.1 Methane (CH_4)

During decay, the volatile C fraction in biochar (vs. fixed) (Majumder et al., 2019; ANSI, 2013) becomes the main source of methanogenic substrates, thereby supporting CH_4 emission (Majumder et al., 2019; Zhang et al., 2012). Studies have been performed in paddy fields where waterlogged conditions play an essential role in CH_4 emission from soils. Methane is produced in waterlogged soils (Liu et al., 2011; Majumder et al., 2019). No variation in the emission of CH_4 has been noted with biochar application in some studies. This was ascribed to the deficiency of a priming impact which defends the SOM from mineralization by occlusion within biochar pores or sorption of labile organic matter released by biochar onto soil-mineral surfaces (Majumder et al., 2019; Mukherjee et al., 2014).

Both positive and negative priming impacts have been noted in biochar-amended soils depending on the biochar characteristics, soil type, and time of pyrolysis (Majumder et al., 2019; Zhang et al., 2011). Biochar application also reduces the release of CH_4 by increasing the abundance of methanotrophic proteobacteria (Feng et al., 2012; Majumder et al., 2019). Additionally, after biochar application, CH_4 oxidation can enhance the soil aeration status (Karhu et al., 2011; Majumder et al., 2019). The treated soil behaves as a sink for atmospheric CH_4, and it is dependent upon soil moisture, diurnal impacts of root respiration, or other microbial soil processes that alter the availability of oxygen in the soils (Majumder et al., 2019; Scheer et al., 2011).

The integration of crop residues increases CH_4 release, because of the decay of organic matter, which is the predominant source of methanogenic substrates. Additionally, labile C in some rice cultivars is released from roots harmed by extreme anaerobic conditions and is metabolized to its equivalent as CH_4. Methane release, resulting from C mineralization under anaerobic conditions, is further increased because of fertilization and accessibility of C resources for microbes (Liu et al., 2014; Majumder et al., 2019). In a study that focused on straw integration in a paddy field in China, CH_4 release increased by a factor of 1.6–10.5, although this was mostly incumbent upon the water regime, drainage system, and tillage practices. Poor drainage and conservation tillage practices significantly decreased the emission (Ma et al., 2009; Majumder et al., 2019).

12.7.2 Carbon Dioxide (CO_2)

Global warming potential (GWP) is the measure of the amount of heat caught in the environment by GHGs compared to CO_2. ^{14}C labeling is utilized to define the mineralization loss of C in biochar (Kuzyakov et al., 2009), and CO_2 can be derived from carbonate developed during pyrolysis. The modification of soil utilizing biochar may sometimes speed up the decay of SOC through increased microbial activity, increasing CO_2 release and total GWP (Deng et al., 2017).

Nguyen et al. (2016b) reported a decrease in CO_2 emission from two of three subtropical soils on a maize plantation following rice husk biochar application. Ethylene, which may be in biochar, can occasionally hinder this process (Mukherjee et al., 2014). Biochar application results in both positive (C mineralization) and negative (C stabilization) priming impacts. Positive priming impacts lead to CO_2 release and are typically from biochar fabricated at low temperatures (250°C–400°C), while negative priming impacts come from biochar generated at high temperatures (525°C–650°C) (Majumder et al., 2019; Zimmerman et al., 2011).

An increase in the release of CO_2 may also be ascribed to the high internal porosity and adsorbing ability of biochar, which give home for microorganisms, and, therefore, speed up the decay of adsorbed organic compounds. Reduced bulk density of biochar-amended soils also improves gas exchange and increases the partial pressure of oxygen in the soil pores, thus increasing the activity of aerobic microorganisms (Majumder et al., 2019; Rogovska et al., 2011). The release of CO_2 after crop residue incorporation happens because of a quick increase in microbial decay of the labile SOM pool. A low C/N ratio of the residues results in quicker mineralization, which, in turn, enhances microbial as well as plant root respiration leading to increased CO_2 release (Bhattacharyya et al., 2012; Majumder et al., 2019). Thus, release differs based on the type of crop residue, and it has a negative association with the C:N ratio of the applied plant residues. The elimination of crop residues from land, however, can enhance the risk of soil erosion, which eventually leads to a reduction of SOC, further augmenting CO_2 release (Lal, 2005; Majumder et al., 2019).

12.7.3 Nitrous Oxide (N_2O)

Biochar application can decrease N_2O release from agricultural fields (Mukherjee et al., 2014; Nguyen et al., 2016b). Improved N immobilization and diminished denitrification are two of the principal components for decreased N_2O (Majumder et al., 2019; Wang et al., 2013). A biochar amendment supports the action of N_2O reductase (which increases with an increase in soil pH) compared to other enzymes participating in the transformation of nitrate to N_2O, thus decreasing N_2O emission (Majumder et al., 2019; Mukherjee et al., 2014).

Another clarification is that biochar can work as an "electron shuttle," which encourages the transfer of electrons to soil-denitrifying microbes that decrease the conversion of N_2O to N_2 (Cayuela et al., 2013; Majumder et al., 2019). The inhibition of N_2O emission may possibly be driven by some abiotic factors, like increase in soil aeration and the sorption of ammonium and nitrate particles on biochar surfaces, as well as the existence of some inhibitory or harmful organic compounds (Majumder et al., 2019; Mukherjee et al., 2014). There are also some exceptions; for instance, the application of biochar to a calcisol resulted in a 54% increase in soil N_2O emission (Majumder et al., 2019; Sanchez-Garcia et al., 2014).

Nitrous oxide release resulting from the application of raw residues happens because of microbial nitrification during aerobic, fallow periods and denitrification when the soils are moistened in anaerobic growing seasons; this is extremely dependent on the type of crop residue and soil-mineral N content (Nguyen et al., 2016b; Wu et al., 2016a). However, the exclusion of residues has been demonstrated to enhance N_2O release under conservation tillage (Congreves et al., 2017). The accessibility of organic C, nutrients, and soil moisture are major components for soil denitrification (Dutta et al., 2015, 2016), which is an anaerobic microbial procedure including transfer of NO_3 to N gas. Denitrification is reliant on the amount of organic C readily accessible to denitrifiers because they are heterotrophs and need organic C for energy. When straw with large C:N ratios is used, the immobilization of fertilizer N happens, decreasing the accessibility of N (Wu et al., 2016b) and thus decreasing soil N_2O release. Additionally, crop residue return may encourage microbial growth and activity to a great extent and, thus, produce moderately anaerobic microsites, resulting in enhanced soil N_2O production.

12.8 BIOCHAR ROLE IN SOIL CARBON STORAGE

Biochar application to the soils resulting in C sequestration is presently being studied as one of the encouraging ways for mitigating the impacts of climate change (Jeffery et al., 2011; Palanivelu et al., 2021). Biochar is a path of getting C into soils as a C sink. Several researchers have suggested it as a long-term strategy for the storage of soil C (Lehmann et al., 2011; Palanivelu et al., 2021). Palanivelu et al. (2021) showed that modifications of biochar enhance the storage of C in soils. They utilized biochar generated from hydrothermal pyrolysis of glucose and yeast. They found that mean residence times of biochar in soils differed in 4 of 29 years. It was reliant on soil type and biochar amount. Microbial communities were stimulated by the biochar mixed with fertilizer and carbon storage function in the soil, resulting in nutrient discharge and fertilization, as well as adding to the decadal soil carbon pool. Biochar added to a refractory SOC pool can decrease atmospheric CO_2 emissions, thus reducing the impacts of global warming (Luo et al., 2011; Palanivelu et al., 2021; Steinbeiss et al., 2009). Wang et al. (2016a) explained that biochar can stay in the soil for a century, has a beneficial impact on the dynamics of SOM, and adds immediately to long-term C sequestration in soils. Minamino et al. (2019) showed that biochar addition to a leaf litter decay in a forest soil environment supported effective C sequestration.

12.9 CONCLUSIONS AND OUTLOOK

Biochar is a valuable amendment for enhancing soil health and physicochemical properties and reducing soil erosion and degradation. It acts as a nutrient source for plants that facilitates soil fertility, which eventually enhances crop yield and productivity. Biochar's effect on crop plants is associated with the feedstock type, pyrolytic temperature, soil type, environmental conditions, and crop species. Biochar produced at low temperatures may be used to treat alkaline soil, while biochar from high temperatures can be used for soil C sequestration. Biochar improves the physiochemical properties of soil and provides a habitat for microbes. Additionally, biochar increases water and nutrient availability for plants. Biochar also plays a paramount role in climate change mitigation, reduction of GHGs, and soil carbon storage.

In the future, there is a need to focus on the long-term effects of biochar on soils and the environment. Long-term studies should be done to observe the effect of specific biochars for sustainable crop yield and food security. Furthermore, the impact of biochar aging on microbial communities and SOM mineralization, soil carbon sequestration, and soil nutrients needs to be determined. Further investigation is necessary to observe the impact of modified biochars on soil properties, plant productivity, and climate change mitigation.

ACKNOWLEDGMENTS

Drs Nabeel Khan Niazi and Irshad Bibi are thankful to the Higher Education Commission, Pakistan. Dr. Irshad Bibi acknowledges the support from COMSTEQ-TWAS research grant 2018 (18–268 RG/EAS/AS_C).

REFERENCES

Abbasi, M.K., Anwar, A.A., 2015. Ameliorating effects of biochar derived from poultry manure and white clover residues on soil nutrient status and plant growth promotion-greenhouse experiments. *PLoS One* 10, e0131592.

Ali, K., Arif, M., Jan, M., Yaseen, T., Waqas, M., Munsif, F., 2015. Biochar: A novel tool to enhance wheat productivity and soil fertility on sustainable basis under wheatmaize-wheat cropping pattern. *Pakistan Journal of Botany* 47, 1023–1031.

Alvarez-Campos, O., Lang, T.A., Bhadha, J.H., McCray, J.M., Glaz, B., Daroub, S.H., 2018. Biochar and mill ash improve yields of sugarcane on a sand soil in Florida. *Agriculture, Ecosystems & Environment* 253, 122–130.

Al-Wabel, M.I., Hussain, Q., Usman, A.R., Ahmad, M., Abduljabbar, A., Sallam, A.S., Ok, Y.S., 2018. Impact of biochar properties on soil conditions and agricultural sustainability: A review. *Land Degradation & Development* 29, 2124–2161.

Amundson, R., Berhe, A.A., Hopmans, J.W., Olson, C., Sztein, A.E., Sparks, D.L., 2015. Soil and human security in the 21st century. *Science* 348, 1261071.

Asai, H., Samson, B.K., Stephan, H.M., Songyikhangsuthor, K., Homma, K., Kiyono, Y., Inoue, Y., Shiraiwa, T., Horie, T., 2009. Biochar amendment techniques for upland rice production in Northern Laos: 1. Soil physical properties, leaf SPAD and grain yield. *Field Crops Research* 111, 81–84.

Bandara, T., Herath, I., Kumarathilaka, P., Seneviratne, M., Seneviratne, G., Rajakaruna, N., Vithanage, M., Ok, Y.S., 2017. Role of woody biochar and fungal-bacterial co-inoculation on enzyme activity and metal immobilization in serpentine soil. *Journal of Soils and Sediments* 17, 665–673.

Bera, T., Purakayastha, T., Patra, A., 2014. Spectral, chemical and physical characterisation of mustard stalk biochar as affected by temperature. *Clay Research* 33, 36–45.

Bhattacharjya, S., Chandra, R., Pareek, N., Raverkar, K.P., 2016. Biochar and crop residue application to soil: Effect on soil biochemical properties, nutrient availability and yield of rice (*Oryza sativa* L.) and wheat (*Triticum aestivum* L.). *Archives of Agronomy and Soil Science* 62, 1095–1108.

Bhattacharyya, P., Roy, K.S., Neogi, S., Adhya, T.K., Rao, K.S., Manna, M.C., 2012. Effects of rice straw and nitrogen fertilization on greenhouse gas emissions and carbon storage in tropical flooded soil planted with rice. *Soil and Tillage Research* 124, 119–130.

Borrelli, P., Robinson, D.A., Fleischer, L.R., Lugato, E., Ballabio, C., Alewell, C., Meusburger, K., Modugno, S., Schütt, B., Ferro, V., 2017. An assessment of the global impact of 21st century land use change on soil erosion. *Nature Communications* 8, 1–13.

Cantrell, K.B., Hunt, P.G., Uchimiya, M., Novak, J.M., Ro, K.S., 2012. Impact of pyrolysis temperature and manure source on physicochemical characteristics of biochar. *Bioresource Technology* 107, 419–428.

Cayuela, M.L., Sánchez-Monedero, M.A., Roig, A., Hanley, K., Enders, A., Lehmann, J., 2013. Biochar and denitrification in soils: When, how much and why does biochar reduce N_2O emissions? *Scientific Reports* 3, 1–7.

Chen, T., Liu, R., Scott, N.R., 2016. Characterization of energy carriers obtained from the pyrolysis of white ash, switchgrass and corn stover: Biochar, syngas and bio-oil. *Fuel Processing Technology* 142, 124–134.

Congreves, K.A., Brown, S.E., Nemeth, D.D., Dunfield, K.E., Wagner-Riddle, C., 2017. Differences in field-scale N_2O flux linked to crop residue removal under two tillage systems in cold climates. *Gcb Bioenergy* 9, 666–680.

de Jesus Duarte, S., Glaser, B., Pellegrino Cerri, C.E., 2019. Effect of biochar particle size on physical, hydrological and chemical properties of loamy and sandy tropical soils. *Agronomy* 9, 165.

Deng, W., Van Zwieten, L., Lin, Z., Liu, X., Sarmah, A.K., Wang, H., 2017. Sugarcane bagasse biochars impact respiration and greenhouse gas emissions from a latosol. *Journal of Soils and Sediments* 17, 632–640.

Diatta, A.A., Fike, J.H., Battaglia, M.L., Galbraith, J.M., Baig, M.B., 2020. Effects of biochar on soil fertility and crop productivity in arid regions: A review. *Arabian Journal of Geosciences* 13, 1–17.

Ding, Y., Liu, Y., Liu, S., Li, Z., Tan, X., Huang, X., Zeng, G., Zhou, L., Zheng, B., 2016. Biochar to improve soil fertility: A review. *Agronomy for Sustainable Development* 36, 1–18.

Downie, A., Crosky, A., Munroe, P., 2009. Physical properties of biochar. In: Lehmann, J., Joseph, S. (Eds.), *Biochar for Environmental Management*. Earthscan: London, pp. 13–32.

Du, Z.-L., Zhao, J.-K., Wang, Y.-D., Zhang, Q.-Z., 2017. Biochar addition drives soil aggregation and carbon sequestration in aggregate fractions from an intensive agricultural system. *Journal of Soils and Sediments* 17, 581–589.

Dutta, T., Stehouwer, R.C., Dell, C.J., 2015. Linking organic carbon, water content, and nitrous oxide emission in a reclaimed coal mine soil. *Land Degradation & Development* 26, 620–628.

Dutta, T., Dell, C.J., Stehouwer, R.C., 2016. Nitrous oxide emissions from a coal mine land reclaimed with stabilized manure. *Land Degradation & Development* 27, 427–437.

Edwards, J.D., Pittelkow, C.M., Kent, A.D., Yang, W.H., 2018. Dynamic biochar effects on soil nitrous oxide emissions and underlying microbial processes during the maize growing season. *Soil Biology and Biochemistry* 122, 81–90.

Elzobair, K.A., Stromberger, M.E., Ippolito, J.A., Lentz, R.D., 2016. Contrasting effects of biochar versus manure on soil microbial communities and enzyme activities in an Aridisol. *Chemosphere* 142, 145–152.

Feng, Y., Xu, Y., Yu, Y., Xie, Z., Lin, X., 2012. Mechanisms of biochar decreasing methane emission from Chinese paddy soils. *Soil Biology and Biochemistry* 46, 80–88.

Gasser, T., Guivarch, C., Tachiiri, K., Jones, C.D., Ciais, P., 2015. Negative emissions physically needed to keep global warming below 2C. *Journal of Nature Communications* 6, 1–7.

Ghidotti, M., Fabbri, D., Hornung, A., 2017. Profiles of volatile organic compounds in biochar: Insights into process conditions and quality assessment. *ACS Sustainable Chemistry & Engineering* 5, 510–517.

Gomez, J., Denef, K., Stewart, C., Zheng, J., Cotrufo, M., 2014. Biochar addition rate influences soil microbial abundance and activity in temperate soils. *European Journal of Soil Science* 65, 28–39.

Gujre, N., Soni, A., Rangan, L., Tsang, D.C., Mitra, S., 2021. Sustainable improvement of soil health utilizing biochar and arbuscular mycorrhizal fungi: A review. *Environmental Pollution* 268, 115549.

Haefele, S., Konboon, Y., Wongboon, W., Amarante, S., Maarifat, A., Pfeiffer, E., Knoblauch, C., 2011. Effects and fate of biochar from rice residues in rice-based systems. *Field Crops Research* 121, 430–440.

Han, L., Ro, K.S., Wang, Y., Sun, K., Sun, H., Libra, J.A., Xing, B., 2018. Oxidation resistance of biochars as a function of feedstock and pyrolysis condition. *Science of the Total Environment* 616, 335–344.

Han, L., Sun, K., Yang, Y., Xia, X., Li, F., Yang, Z., Xing, B., 2020. Biochar's stability and effect on the content, composition and turnover of soil organic carbon. *Geoderma* 364, 114184.

Hou, D., Bolan, N.S., Tsang, D.C., Kirkham, M.B., O'Connor, D., 2020. Sustainable soil use and management: An interdisciplinary and systematic approach. *Science of the Total Environment* 729, 138961.

Houben, D., Evrard, L., Sonnet, P., 2013. Beneficial effects of biochar application to contaminated soils on the bioavailability of Cd, Pb and Zn and the biomass production of rapeseed (*Brassica napus* L.). *Biomass and Bioenergy* 57, 196–204.

Huggins, T.M., Haeger, A., Biffinger, J.C., Ren, Z.J., 2016. Granular biochar compared with activated carbon for wastewater treatment and resource recovery. *Water Research* 94, 225–232.

Hussain, K., Ilyas, A., Bibi, I., Hilger, T., 2021. Sustainable soil loss management in tropical uplands: Impact on maize-chili cropping systems. *Sustainability* 13. doi: 10.3390/su13116477.

Hussain, M.M., Bibi, I., Ali, F., Saqib, Z.A., Shahid, M., Niazi, N.K., Hussain, K., Shaheen, S.M., Wang, H., Shakil, Q., Rinklebe, J., 2022. The role of various ameliorants on geochemical arsenic distribution and CO_2-carbon efflux under paddy soil conditions. *Environmental Geochemistry and Health*. doi: 10.1007/s10653-021-01196-3.

IPCC, 2014. In: Edenhofer, O., et al. (Eds.), *Climate Change 2014: Mitigation of Climate Change*. Cambridge University Press: New York.

Jay, C.N., Fitzgerald, J.D., Hipps, N.A., Atkinson, C.J., 2015. Why short-term biochar application has no yield benefits: Evidence from three field-grown crops. *Soil Use and Management* 31, 241–250.

Jeffery, S., Verheijen, F.G., van der Velde, M., Bastos, A.C., 2011. A quantitative review of the effects of biochar application to soils on crop productivity using meta-analysis. *Agriculture, Ecosystems & Environment* 144, 175–187.

Jiang, S., Huang, L., Nguyen, T.A., Ok, Y.S., Rudolph, V., Yang, H., Zhang, D., 2016. Copper and zinc adsorption by softwood and hardwood biochars under elevated sulphate-induced salinity and acidic pH conditions. *Chemosphere* 142, 64–71.

Karhu, K., Mattila, T., Bergström, I., Regina, K., 2011. Biochar addition to agricultural soil increased CH_4 uptake and water holding capacity–Results from a short-term pilot field study. *Agriculture, Ecosystems & Environment* 140, 309–313.

Keesstra, S.D., Bouma, J., Wallinga, J., Tittonell, P., Smith, P., Cerdà, A., Montanarella, L., Quinton, J.N., Pachepsky, Y., Van Der Putten, W.H., 2016. The significance of soils and soil science towards realization of the United Nations sustainable development goals. *Soil* 2, 111–128.

Kookana, R.S., Sarmah, A.K., Van Zwieten, L., Krull, E., Singh, B., 2011. Biochar application to soil: Agronomic and environmental benefits and unintended consequences. *Advances in Agronomy* 112, 103–143.

Koutcheiko, S., Monreal, C., Kodama, H., McCracken, T., Kotlyar, L., 2007. Preparation and characterization of activated carbon derived from the thermo-chemical conversion of chicken manure. *Bioresource Technology* 98, 2459–2464.

Kuzyakov, Y., Subbotina, I., Chen, H., Bogomolova, I., Xu, X., 2009. Black carbon decomposition and incorporation into soil microbial biomass estimated by 14C labeling. *Soil Biology and Biochemistry* 41, 210–219.

Laird, D.A., Fleming, P., Davis, D.D., Horton, R., Wang, B., Karlen, D.L., 2010. Impact of biochar amendments on the quality of a typical Midwestern agricultural soil. *Geoderma* 158, 443–449.

Lal, R., 2005. World crop residues production and implications of its use as a biofuel. *Environment International* 31, 575–584.

Lang, T., Jensen, A.D., Jensen, P.A., 2005. Retention of organic elements during solid fuel pyrolysis with emphasis on the peculiar behavior of nitrogen. *Energy & Fuels* 19, 1631–1643.

Lehmann, J., 2007. A handbook of carbon. *Nature*, 447, 143–144.

Lehmann, J., da Silva, P.J., Steiner, C., Nehls, T., Zech, W., Glaser, B., 2003. Nutrient availability and leaching in an archaeological Anthrosol and a Ferralsol of the Central Amazon basin: Fertilizer, manure and charcoal amendments. *Plant and Soil* 249, 343–357.

Lehmann, J., Rillig, M.C., Thies, J., Masiello, C.A., Hockaday, W.C., Crowley, D., 2011. Biochar effects on soil biota: A review. *Soil Biology and Biochemistry* 43, 1812–1836.

Lehmann, J., Kuzyakov, Y., Pan, G., Ok, Y.S., 2015. Biochars and the plant-soil interface. *Plant and Soil,* 395(1), 1–5.

Liao, W., Thomas, S.C., 2019. Biochar particle size and post-pyrolysis mechanical processing affect soil pH, water retention capacity, and plant performance. *Soil Systems* 3, 14.

Lin, Y., Munroe, P., Joseph, S., Henderson, R., Ziolkowski, A., 2012. Water extractable organic carbon in untreated and chemical treated biochars. *Chemosphere* 87, 151–157.

Liu, Y., Yang, M., Wu, Y., Wang, H., Chen, Y., Wu, W., 2011. Reducing CH_4 and CO_2 emissions from waterlogged paddy soil with biochar. *Journal of Soils and Sediments* 11, 930–939.

Liu, C., Lu, M., Cui, J., Li, B., Fang, C., 2014. Effects of straw carbon input on carbon dynamics in agricultural soils: A meta-analysis. *Global Change Biology* 20, 1366–1381.

Liu, Y., Lu, H., Yang, S., Wang, Y., 2016. Impacts of biochar addition on rice yield and soil properties in a cold waterlogged paddy for two crop seasons. *Field Crops Research* 191, 161–167.

Luo, Y., Durenkamp, M., De Nobili, M., Lin, Q., Brookes, P., 2011. Short term soil priming effects and the mineralisation of biochar following its incorporation to soils of different pH. *Soil Biology and Biochemistry* 43, 2304–2314.

Luo, S., Wang, S., Tian, L., Li, S., Li, X., Shen, Y., Tian, C., 2017a. Long-term biochar application influences soil microbial community and its potential roles in semiarid farmland. *Applied Soil Ecology* 117, 10–15.

Luo, X., Liu, G., Xia, Y., Chen, L., Jiang, Z., Zheng, H., Wang, Z., 2017b. Use of biochar-compost to improve properties and productivity of the degraded coastal soil in the Yellow River Delta, China. *Journal of Soils and Sediments* 17, 780–789.

Lyu, H., He, Y., Tang, J., Hecker, M., Liu, Q., Jones, P.D., Codling, G., Giesy, J.P., 2016. Effect of pyrolysis temperature on potential toxicity of biochar if applied to the environment. *Environmental Pollution* 218, 1–7.

Ma, J., Ma, E., Xu, H., Yagi, K., Cai, Z., 2009. Wheat straw management affects CH_4 and N_2O emissions from rice fields. *Soil Biology and Biochemistry* 41, 1022–1028.

Macdonald, L.M., Farrell, M., Zwieten, L.V., Krull, E.S., 2014. Plant growth responses to biochar addition: An Australian soils perspective. *Biology and Fertility of Soils* 50, 1035–1045.

Major, J., Steiner, C., Downie, A., Lehmann, J., 2009. Biochar effects on nutrient leaching. In: Lehmann, J., Joseph, S. (Eds.) *Biochar for Environmental Management*. Earthscan: London, pp. 271–288.

Majumder, S., Neogi, S., Dutta, T., Powel, M.A., Banik, P., 2019. The impact of biochar on soil carbon sequestration: Meta-analytical approach to evaluating environmental and economic advantages. *Journal of Environmental Management* 250, 109466.

Mandal, S., Donner, E., Vasileiadis, S., Skinner, W., Smith, E., Lombi, E., 2018. The effect of biochar feedstock, pyrolysis temperature, and application rate on the reduction of ammonia volatilisation from biochar-amended soil. *Science of the Total Environment* 627, 942–950.

Masto, R.E., Ansari, M.A., George, J., Selvi, V., Ram, L., 2013. Co-application of biochar and lignite fly ash on soil nutrients and biological parameters at different crop growth stages of Zea mays. *Ecological Engineering* 58, 314–322.

Mete, F.Z., Mia, S., Dijkstra, F.A., Abuyusuf, M., & Hossain, A.S.M.I., 2015. Synergistic effects of biochar and NPK fertilizer on soybean yield in an alkaline soil. *Journal of Pedosphere* 25, 713–719.

Minamino, Y., Fujitake, N., Suzuki, T., Yoshitake, S., Koizumi, H., Tomotsune, M., 2019. Effect of biochar addition on leaf-litter decomposition at soil surface during three years in a warm-temperate secondary deciduous forest, Japan. *Scientific Reports* 9, 1–7.

Minasny, B., Malone, B.P., McBratney, A.B., Angers, D.A., Arrouays, D., Chambers, A., Chaplot, V., Chen, Z.-S., Cheng, K., Das, B.S., 2017. Soil carbon 4 per mille. *Geoderma* 292, 59–86.

Mitchell, P.J., Simpson, A.J., Soong, R., Simpson, M.J., 2015. Shifts in microbial community and water-extractable organic matter composition with biochar amendment in a temperate forest soil. *Soil Biology and Biochemistry* 81, 244–254.

Mubarik, S., Saeed, A., Athar, M., Iqbal, D.M., 2016. Characterization and mechanism of the adsorptive removal of 2, 4, 6-trichlorophenol by biochar prepared from sugarcane baggase. *Journal of Industrial and Engineering Chemistry* 33, 115–121.

Muhammad, N., Dai, Z., Xiao, K., Meng, J., Brookes, P.C., Liu, X., Wang, H., Wu, J., Xu, J., 2014. Changes in microbial community structure due to biochars generated from different feedstocks and their relationships with soil chemical properties. *Geoderma* 226, 270–278.

Mukherjee, A., Zimmerman, A.R., 2013. Organic carbon and nutrient release from a range of laboratory-produced biochars and biochar–soil mixtures. *Geoderma* 193, 122–130.

Mukherjee, A., Lal, R., Zimmerman, A., 2014. Effects of biochar and other amendments on the physical properties and greenhouse gas emissions of an artificially degraded soil. *Science of the Total Environment* 487, 26–36.

Munda, S., Nayak, A., Mishra, P., Bhattacharyya, P., Mohanty, S., Kumar, A., Kumar, U., Baig, M., Tripathi, R., Shahid, M., 2016. Combined application of rice husk biochar and fly ash improved the yield of lowland rice. *Soil Research* 54, 451–459.

Natasha, N., Shahid, M., Khalid, S., Bibi, I., Naeem, M.A., Niazi, N.K., Tack, F.M.G., Ippolito, J.A., Rinklebe, J., 2021. Influence of biochar on trace element uptake, toxicity and detoxification in plants and associated health risks: A critical review. *Critical Reviews in Environmental Science and Technology*, 1–41. 10.1080/10643389.2021.1894064.

Natasha, Bibi, I., Niazi, N.K., Shahid, M., Ali, F., Masoodul Hasan, I., Rahman, M.M., Younas, F., Hussain, M.M., Mehmood, T., Shaheen, S.M., Naidu, R., Rinklebe, J., 2022. Distribution and ecological risk assessment of trace elements in the paddy soil-rice ecosystem of Punjab, Pakistan. *Environmental Pollution*. doi: 10.1016/j.envpol.2022.119492.

Neogi, S., Sharma, V., Khan, N., Chaurasia, D., Ahmad, A., Chauhan, S., Singh, A., You, S., Pandey, A., Bhargava, P.C., 2021. Sustainable biochar: A facile strategy for soil and environmental restoration, energygeneration, mitigation of global climate change and circular bioeconomy. *Chemosphere*, 293, 133474.

Nguyen, D.H., Scheer, C., Rowlings, D.W., Grace, P.R., 2016a. Rice husk biochar and crop residue amendment in subtropical cropping soils: Effect on biomass production, nitrogen use efficiency and greenhouse gas emissions. *Journal of Biology Fertility of Soils* 52, 261–270.

Nguyen, D.H., Scheer, C., Rowlings, D.W., Grace, P.R., 2016b. Rice husk biochar and crop residue amendment in subtropical cropping soils: Effect on biomass production, nitrogen use efficiency and greenhouse gas emissions. *Biology and Fertility of Soils* 52, 261–270.

Niazi, N.K., Bashir, S., Bibi, I., Murtaza, B., Shahid, M., Javed, M.T., Shakoor, M.B., Saqib, Z.A., Nawaz, M.F., Aslam, Z., 2016. Phytoremediation of arsenic-contaminated soils using arsenic hyperaccumulating ferns. In: Ansari, A.A., Gill, S.S., Gill, R., Lanza, G.R., Newman, L. (Eds.) *Phytoremediation: Management of Environmental Contaminants*. Springer: Switzerland, pp. 521–545.

O'Connor, D., Peng, T., Zhang, J., Tsang, D.C., Alessi, D.S., Shen, Z., Bolan, N.S., Hou, D., 2018. Biochar application for the remediation of heavy metal polluted land: A review of in situ field trials. *Science of the Total Environment* 619, 815–826.

Obia, A., Mulder, J., Martinsen, V., Cornelissen, G., Børresen, T., 2016. In situ effects of biochar on aggregation, water retention and porosity in light-textured tropical soils. *Soil and Tillage Research* 155, 35–44.

Palanivelu, K., Ramachandran, A., Raghavan, V., 2021. Biochar from biomass waste as a renewable carbon material for climate change mitigation in reducing greenhouse gas emissions: A review. *Biomass Conversion and Biorefinery* 11, 2247–2267.

Palansooriya, K.N., Wong, J.T.F., Hashimoto, Y., Huang, L., Rinklebe, J., Chang, S.X., Bolan, N., Wang, H., Ok, Y.S., 2019. Response of microbial communities to biochar-amended soils: A critical review. *Biochar* 1, 3–22.

Partey, S.T., Saito, K., Preziosi, R.F., Robson, G.D., 2016. Biochar use in a legume–rice rotation system: Effects on soil fertility and crop performance. *Archives of Agronomy and Soil Science* 62, 199–215.

Paustian, K., Lehmann, J., Ogle, S., Reay, D., Robertson, G.P., Smith, P., 2016. Climate-smart soils. *Nature* 532, 49–57.

Pimentel, D., Burgess, M., 2013. Soil erosion threatens food production. *Agriculture* 3, 443–463.

Prendergast-Miller, M., Duvall, M., Sohi, S., 2014. Biochar–root interactions are mediated by biochar nutrient content and impacts on soil nutrient availability. *European Journal of Soil Science* 65, 173–185.

Purakayastha, T., Bera, T., Bhaduri, D., Sarkar, B., Mandal, S., Wade, P., Kumari, S., Biswas, S., Menon, M., Pathak, H., 2019. A review on biochar modulated soil condition improvements and nutrient dynamics concerning crop yields: Pathways to climate change mitigation and global food security. *Chemosphere* 227, 345–365.

Qiu, M., Sun, K., Jin, J., Han, L., Sun, H., Zhao, Y., Xia, X., Wu, F., Xing, B., 2015. Metal/metalloid elements and polycyclic aromatic hydrocarbon in various biochars: The effect of feedstock, temperature, minerals, and properties. *Environmental Pollution* 206, 298–305.

Quan, G., Fan, Q., Zimmerman, A.R., Sun, J., Cui, L., Wang, H., Gao, B., Yan, J., 2020. Effects of laboratory biotic aging on the characteristics of biochar and its water-soluble organic products. *Journal of Hazardous Materials* 382, 121071.

Rajkovich, S., Enders, A., Hanley, K., Hyland, C., Zimmerman, A.R., Lehmann, J., 2012. Corn growth and nitrogen nutrition after additions of biochars with varying properties to a temperate soil. *Biology and Fertility of Soils* 48, 271–284.

Rehrah, D., Bansode, R., Hassan, O., Ahmedna, M., 2016. Physico-chemical characterization of biochars from solid municipal waste for use in soil amendment. *Journal of Analytical and Applied Pyrolysis* 118, 42–53.

Ren, X., Zhang, P., Zhao, L., Sun, H., 2016. Sorption and degradation of carbaryl in soils amended with biochars: Influence of biochar type and content. *Environmental Science and Pollution Research* 23, 2724–2734.

Robinson, D.A., Panagos, P., Borrelli, P., Jones, A., Montanarella, L., Tye, A., Obst, C.G., 2017. Soil natural capital in Europe; a framework for state and change assessment. *Scientific Reports* 7, 1–14.

Rogovska, N., Laird, D., Cruse, R., Fleming, P., Parkin, T., Meek, D., 2011. Impact of biochar on manure carbon stabilization and greenhouse gas emissions. *Soil Science Society of America Journal* 75, 871–879.

Sanchez-Garcia, M., Roig, A., Sanchez-Monedero, M.A., Cayuela, M.L., 2014. Biochar increases soil N_2O emissions produced by nitrification-mediated pathways. *Frontiers in Environmental Science* 2, 25.

Sarker, T.C., Incerti, G., Spaccini, R., Piccolo, A., Mazzoleni, S., Bonanomi, G., 2018. Linking organic matter chemistry with soil aggregate stability: Insight from 13C NMR spectroscopy. *Soil Biology and Biochemistry* 117, 175–184.

Scheer, C., Grace, P.R., Rowlings, D.W., Kimber, S., Van Zwieten, L., 2011. Effect of biochar amendment on the soil-atmosphere exchange of greenhouse gases from an intensive subtropical pasture in northern New South Wales, Australia. *Plant and Soil* 345, 47–58.

Schmidt, H.-P., Kammann, C., Niggli, C., Evangelou, M.W., Mackie, K.A., Abiven, S., 2014. Biochar and biochar-compost as soil amendments to a vineyard soil: Influences on plant growth, nutrient uptake, plant health and grape quality. *Agriculture, Ecosystems & Environment* 191, 117–123.

Shafiq, M., 2016. Management of the Parthenium hysterophorus through biochar formation and its application to rice-wheat cultivation in Pakistan. *Agriculture, Ecosystems & Environment* 235, 265–276.

Shaheen, S.M., Niazi, N.K., Hassan, N.E., Bibi, I., Wang, H., Tsang, D.C., Ok, Y.S., Bolan, N., Rinklebe, J., 2019. Wood-based biochar for the removal of potentially toxic elements in water and wastewater: A critical review. *International Materials Reviews* 64, 216–247.

Shaheen, S.M., Mosa, A., El-Naggar, A., Hossain, M.F., Abdelrahman, H., Niazi, N.K., Shahid, M., Zhang, T., Tsang, Y.F., Trakal, L., 2021. Manganese oxide-modified biochar: Production, characterization and applications for the removal of pollutants from aqueous environments-a review. *Bioresource Technology*, 346, 126581.

Silber, A., Levkovitch, I., Graber, E.R., 2010. pH-dependent mineral release and surface properties of cornstraw biochar: Agronomic implications. *Environmental Science & Technology* 44, 9318–9323.

Smith, P., Gregory, P.J., 2013. Climate change and sustainable food production. *Proceedings of the Nutrition Society* 72, 21–28.

Smith, P., House, J.I., Bustamante, M., Sobocká, J., Harper, R., Pan, G., West, P.C., Clark, J.M., Adhya, T., Rumpel, C., 2016. Global change pressures on soils from land use and management. *Global Change Biology* 22, 1008–1028.

Spokas, K.A., Novak, J.M., Stewart, C.E., Cantrell, K.B., Uchimiya, M., DuSaire, M.G., Ro, K.S., 2011. Qualitative analysis of volatile organic compounds on biochar. *Chemosphere* 85, 869–882.

Spokas, K.A., Cantrell, K.B., Novak, J.M., Archer, D.W., Ippolito, J.A., Collins, H.P., Boateng, A.A., Lima, I.M., Lamb, M.C., McAloon, A.J., 2012. Biochar: A synthesis of its agronomic impact beyond carbon sequestration. *Journal of Environmental Quality* 41, 973–989.

ANSI. 2013. Standard test methods for proximate analysis of coal and coke by Macro thermogravimetric analysis, ASTM D7582-12.

Steinbeiss, S., Gleixner, G., Antonietti, M., 2009. Effect of biochar amendment on soil carbon balance and soil microbial activity. *Soil Biology and Biochemistry* 41, 1301–1310.

Sun, K., Qiu, M., Han, L., Jin, J., Wang, Z., Pan, Z., Xing, B., 2018. Speciation of phosphorus in plant- and manure-derived biochars and its dissolution under various aqueous conditions. *Science of the Total Environment* 634, 1300–1307.

Sun, Y., Xiong, X., He, M., Xu, Z., Hou, D., Zhang, W., Ok, Y.S., Rinklebe, J., Wang, L., Tsang, D.C., 2021. Roles of biochar-derived dissolved organic matter in soil amendment and environmental remediation: A critical review. *Chemical Engineering Journal* 424, 130387.

Tasneem, S., Zahir, S., 2017. Soil respiration, pH and EC as influenced by biochar. *Soil and Environment* 36, 77–83.

Tsai, W.-T., Liu, S.-C., Chen, H.-R., Chang, Y.-M., Tsai, Y.-L., 2012. Textural and chemical properties of swine-manure-derived biochar pertinent to its potential use as a soil amendment. *Chemosphere* 89, 198–203.

UNEP, 2017. The emissions gap report 2017. United Nations Environment Programme (UNEP), Nairobi.

Vaccari, F., Maienza, A., Miglietta, F., Baronti, S., Di Lonardo, S., Giagnoni, L., Lagomarsino, A., Pozzi, A., Pusceddu, E., Ranieri, R., 2015. Biochar stimulates plant growth but not fruit yield of processing tomato in a fertile soil. *Agriculture, Ecosystems & Environment* 207, 163–170.

Van Oost, K., Quine, T., Govers, G., De Gryze, S., Six, J., Harden, J., Ritchie, J., McCarty, G., Heckrath, G., Kosmas, C., 2007. The impact of agricultural soil erosion on the global carbon cycle. *Science* 318, 626–629.

Van Zwieten, L., Kimber, S., Morris, S., Chan, K., Downie, A., Rust, J., Joseph, S., Cowie, A., 2010. Effects of biochar from slow pyrolysis of papermill waste on agronomic performance and soil fertility. *Plant Soil* 327, 235–246.

Wang, Z., Zheng, H., Luo, Y., Deng, X., Herbert, S., Xing, B., 2013. Characterization and influence of biochars on nitrous oxide emission from agricultural soil. *Environmental Pollution* 174, 289–296.

Wang, X., Zhou, W., Liang, G., Song, D., Zhang, X., 2015. Characteristics of maize biochar with different pyrolysis temperatures and its effects on organic carbon, nitrogen and enzymatic activities after addition to fluvo-aquic soil. *Science of the Total Environment* 538, 137–144.

Wang, X., Song, D., Liang, G., Zhang, Q., Ai, C., Zhou, W., 2015. Maize biochar addition rate influences soil enzyme activity and microbial community composition in a fluvoaquic soil. *Applied Soil Ecology* 96, 265–272.

Wang, J., Xiong, Z., Kuzyakov, Y., 2016a. Biochar stability in soil: Meta-analysis of decomposition and priming effects. *Gcb Bioenergy* 8, 512–523.

Wang, Z., Han, L., Sun, K., Jin, J., Ro, K.S., Libra, J.A., Liu, X., Xing, B., 2016b. Sorption of four hydrophobic organic contaminants by biochars derived from maize straw, wood dust and swine manure at different pyrolytic temperatures. *Chemosphere* 144, 285–291.

Wu, H., Che, X., Ding, Z., Hu, X., Creamer, A.E., Chen, H., Gao, B., 2016a. Release of soluble elements from biochars derived from various biomass feedstocks. *Environmental Science and Pollution Research* 23, 1905–1915.

Wu, Y., Lin, S., Liu, T., Wan, T., Hu, R., 2016b. Effect of crop residue returns on N_2O emissions from red soil in China. *Soil Use and Management* 32, 80–88.

Xie, Z., Xu, Y., Liu, G., Liu, Q., Zhu, J., Tu, C., Amonette, J.E., Cadisch, G., Yong, J.W., Hu, S., 2013. Impact of biochar application on nitrogen nutrition of rice, greenhouse-gas emissions and soil organic carbon dynamics in two paddy soils of China. *Plant and Soil* 370, 527–540.

Xu, X., Zhao, Y., Sima, J., Zhao, L., Mašek, O., Cao, X., 2017. Indispensable role of biochar-inherent mineral constituents in its environmental applications: A review. *Bioresource Technology* 241, 887–899.

Yu, H., Zou, W., Chen, J., Chen, H., Yu, Z., Huang, J., Tang, H., Wei, X., Gao, B., 2019. Biochar amendment improves crop production in problem soils: A review. *Journal of Environmental Management* 232, 8–21.

Yuan, H., Lu, T., Wang, Y., Chen, Y., Lei, T., 2016. Sewage sludge biochar: Nutrient composition and its effect on the leaching of soil nutrients. *Geoderma* 267, 17–23.

Zahra, M.B., Fayyaz, B., Aftab, Z.-E., Haider, M.S., 2021. Mitigation of degraded soils by using biochar and compost: A systematic review. *Journal of Soil Science and Plant Nutrition* 21, 2718–2738.

Zama, E.F., Zhu, Y.-G., Reid, B.J., Sun, G.-X., 2017. The role of biochar properties in influencing the sorption and desorption of Pb (II), Cd (II) and As (III) in aqueous solution. *Journal of Cleaner Production* 148, 127–136.

Zhang, H., Xiao, R., Wang, D., He, G., Shao, S., Zhang, J., Zhong, Z., 2011. Biomass fast pyrolysis in a fluidized bed reactor under N_2, CO_2, CO, CH_4 and H_2 atmospheres. *Bioresource Technology* 102, 4258–4264.

Zhang, A., Bian, R., Pan, G., Cui, L., Hussain, Q., Li, L., Zheng, J., Zheng, J., Zhang, X., Han, X., 2012. Effects of biochar amendment on soil quality, crop yield and greenhouse gas emission in a Chinese rice paddy: A field study of 2 consecutive rice growing cycles. *Field Crops Research* 127, 153–160.

Zhang, H., Voroney, R., Price, G., 2015a. Effects of temperature and processing conditions on biochar chemical properties and their influence on soil C and N transformations. *Soil Biology and Biochemistry* 83, 19–28.

Zhang, Q., Du, Z., Lou, Y., He, X., 2015b. A one-year short-term biochar application improved carbon accumulation in large macroaggregate fractions. *Catena* 127, 26–31.

Zhang, D., Yan, M., Niu, Y., Liu, X., van Zwieten, L., Chen, D., Bian, R., Cheng, K., Li, L., Joseph, S., 2016. Is current biochar research addressing global soil constraints for sustainable agriculture? *Agriculture, Ecosystems & Environment* 226, 25–32.

Zhao, Z., Zhou, W., 2019. Insight into interaction between biochar and soil minerals in changing biochar properties and adsorption capacities for sulfamethoxazole. *Environmental Pollution* 245, 208–217.

Zheng, H., Wang, Z., Deng, X., Zhao, J., Luo, Y., Novak, J., Herbert, S., Xing, B., 2013. Characteristics and nutrient values of biochars produced from giant reed at different temperatures. *Bioresource Technology* 130, 463–471.

Zimmerman, A.R., Gao, B., Ahn, M.-Y., 2011. Positive and negative carbon mineralization priming effects among a variety of biochar-amended soils. *Soil Biology and Biochemistry* 43, 1169–1179.

Zornoza, R., Moreno-Barriga, F., Acosta, J., Muñoz, M., Faz, A., 2016. Stability, nutrient availability and hydrophobicity of biochars derived from manure, crop residues, and municipal solid waste for their use as soil amendments. *Chemosphere* 144, 122–130.

13 Nutrient Depletion in Salt-Affected Soils and Yield Reduction

Ghulam Murtaza[1], *Ayesha Abdul Qadir*[1], *Zia Ur Rahman Farooqi*[1], *and Behzad Murtaza*[2]
[1]Institute of Soil and Environmental Sciences, University of Agriculture, Faisalabad, Pakistan
[2]Department of Environmental Sciences, COMSATS University Islamabad-Vehari Campus, Vehari, Pakistan

CONTENTS

13.1 INTRODUCTION

The salt-affected area of the world is ≈932.2 Mha, with human-induced salinization and sodification affecting around 76 Mha (Shahid et al. 2018). Currently, one-fifth of irrigated land is impacted by salinity, and 1.5 Mha of land is rendered unfit for agricultural production each year due to excessive salinity levels (Shahid et al. 2018). If soil salinization continues at this rate, half of all cultivable land will be depleted by 2050. Australia, Bangladesh, China, Egypt, India, Iran, Iraq, Mexico, Pakistan, Russia, Syria, Turkey, and the United States are among the countries with substantial

DOI: 10.1201/9781003093565-13

salt-impacted soils. Most of the salt-affected soils on the planet are saline, accounting for 60% of total salt-affected soils. Sodic and saline-sodic soils make up 26% and 14% of the total, respectively (Wicke et al. 2011). Due to these constraints, nations face huge revenue losses every year. Salinity has resulted in a 25% decrease in agricultural yields in Pakistan (Zaman and Ahmad 2009). In 2017, Pakistan's agriculture sector faced the production loss of ≈ US$ 28.5 million due to waterlogging and salinity (Shahzad et al. 2017). According to an estimation, about 28,000–40,000 ha land is lost due to salinity, which results in an annual revenue loss of about US$ 230 million due to poor agricultural production (Aslam 2016). In the case of wheat, Pakistan has experienced yield losses from salt-affected soils ranging from 20% to 43% with an overall average loss of 32%, and, in rice, the crop yield losses range from 36% to 69% with an overall average loss of 48% compared with normal soil (Qadir et al. 2014; Murtaza et al. 2017a). In the Indo-Gangetic Basin (Indian part), the crop yield losses from salt-affected lands for wheat, rice, sugarcane, and cotton are 40%, 45%, 48%, and 63%, respectively (Qadir et al. 2014).

Nutrient depletion is a common issue that is exacerbated by excessive salts in the soil. Reduced biomass buildup, greater susceptibility to disease, and restricted plant development are all effects of nutrient shortages, which are directly related to the yield potential of most arable crops (Nawaz et al. 2020). Decreased salt tolerance is linked to inadequate nutrient absorption in agricultural plants. Around 135 Mha of soils worldwide have been reported to be vulnerable to nutrient depletion or exhaustion, with 97% of these occurring most in developing countries and least in developed countries (Tan et al. 2005). In the early part of the 20th century (the year 2000), the average yield of wheat, rice, and maize was only 1,372, 1,093, and 670 kg/ha/year, respectively, owing to a lack of potassium (K), phosphorus (P), and nitrogen (N) (Tan et al. 2005). According to van Beek et al. (2016), estimation of the average soil N, P, and K balances during the years 2012–2014 under diverse agro-ecological settings of Ethiopia were -23 ± 73, 9 ± 29, and -7 ± 64 kg/ha, respectively. During this period, the rate of N, P, and K depletion increased by 13, 3, and 10 kg/ha/year, respectively.

Micronutrient deficiency is not only common in humans, particularly in children, but it also exists in soil (De-Regil et al. 2013; Bailey et al. 2015; FAO 2015). In soil, micronutrient deficiency or availability is controlled by soil pH (Choudhary and Yaduvanshi 2016; Neina 2019). In alkaline salt-affected soils, especially sodic soils, high soil pH limits the solubility of most of the micronutrients, i.e., copper (Cu), iron (Fe), manganese (Mn), and zinc (Zn) (Aubert and Pinta 1980; Curtain and Naidu 1998). Drop in soil pH of these soils by any means, i.e., applied amendments, fertilizers, or flooding, increases the availability of these nutrients to the plant (Setia and Sharma 2004; Swarup and Yaduvanshi 2004; Qadir et al. 2022). Moreover, the complementary and antagonistic effect of micro- and macronutrients on one another also determines the availability of one element to plants (Dhaliwal et al. 2019).

This chapter focuses on yield reduction associated with the depletion of macro- and micronutrients under salt-stressed conditions and consequences of soil nutrient depletion.

13.2 BASIC CONCEPT OF NUTRIENT DEPLETION

Natural ecosystems often have balanced nutrient inputs and outputs (Figure 13.1). Rock weathering, atmospheric accession (precipitation), dry deposition of aeolian dust, gaseous absorption by plants and soil, and atmospheric fixation are all natural sources of nutrients. The use of manures and fertilizers provides further anthropogenic sources of nutrients in agroecosystems. Unproductive outputs such as erosion, runoff, leaching, gaseous losses, and crop waste burning decrease the pool of soil nutrients.

The agricultural system alters the natural nutrient balance by increasing annual nutrient removal through crops, which reduces the amount of nutrients stored in the soil (Williams 1991). Soil nutrient depletion has become a major issue because of extensive crop cultivation without the appropriate application of external nutrient inputs. Soil nutrient depletion directly impacts food security in emerging and least-developed countries (Jones et al. 2013). Continued depletion of nutrients from soils, along with wind- and water-erosion-based nutrient losses, aggravate soil deterioration and

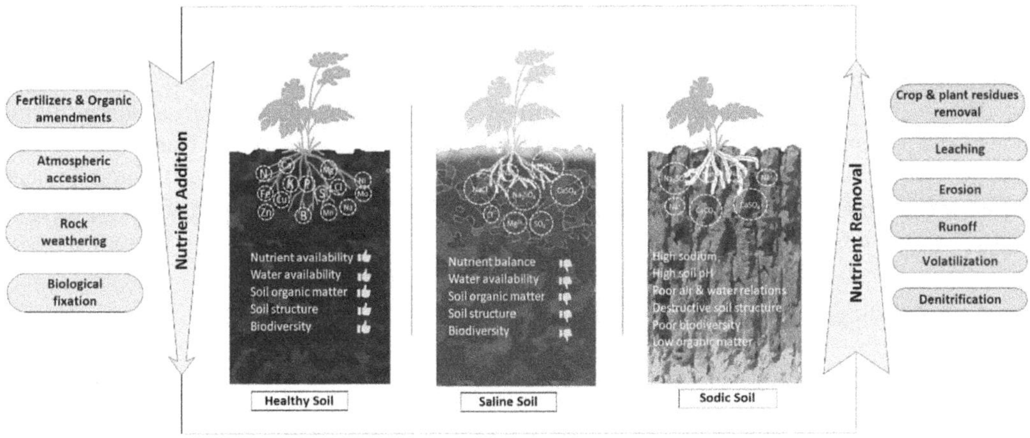

FIGURE 13.1 The nutrient balance system and salt-affected soils.

endanger agricultural viability in these areas (Jones et al. 2013). This may be seen in the long-term decrease in agricultural yields in many areas of Asia, Africa, and Latin America due to inadequate input and imbalanced fertilizer applications.

Average soil nutrient depletion rate in 1996 was estimated by Sheldrick et al. (2002). According to their estimation 12.1, 4.5, and 20.2 kg N, P, and K kg/ha/year was depleted (Sheldrick et al. 2002). Dobermann et al. (1995) calculated nutritional balances for rice-based agricultural systems in eastern Asian countries and found a K deficit ranging from 25 to 70 kg/ha/year. Lin et al. (1996) calculated nutrient balances on 71 rice-paddy farms in south China and found sufficient levels of N and P, while K was in short supply. Mutert (1996) concluded that the nutrient balance is negative for the main crops in many low-income countries with large and increasing populations, while there is a positive nutrient balance in high-income countries with small and stable populations. Henao and Baanante (1999) reviewed previous findings and concluded that soil nutrient depletion was occurring in the agricultural land of all African countries. From 2000 to 2009, Correndo et al. (2021) reported an average annual K removal of 62 kg/ha from an experimental area of the East-Central region of Argentina.

By contrast, a satisfactory level of N, P, and K is observed in Western European countries and Japan (Sibbesen and Runge-Metzger 1995). The findings revealed that Europe has an average yearly P surplus of 12.6 kg/ha/year, with the Netherlands having the greatest P surplus of 57.2 kg/ha/year. Some of the excess P applied to crops is retained in agricultural soils over multiple years of P application (Zhang et al. 2020). This increased soil-P stock can become a source of P in future years.

13.2.1 Salt-Affected Soils

Salts are present in all soils and are necessary for the normal development of plants. However, when the concentration of salts in the soil is too high, the productivity of most agricultural plants is reduced. Salt-impacted soils are those with excess salts that have a detrimental effect on plant growth and development (Figure 13.1). Salt-affected soils are classified into three groups based on the type and quantity of soluble salts present, which influence physicochemical properties, color, formation process, plant response, and management methods. These are saline, sodic, and saline-sodic soils.

13.2.2 Saline Soils

Saline soils are those that contain high soluble salts that inhibit the development of most agricultural plants. The electrical conductivity of a saturated paste extract (EC_e) in these soils is typically > 4 dS/m, the sodium adsorption ratio (SAR) is < 13, exchangeable sodium percentage (ESP) is < 15, and they have a pH < 8.5. These soils sustain heterogeneity in crop development and sometimes exhibit obvious indications of salt damage such as tip burn or chlorosis (pale yellow coloration). Salinity causes ionic toxicity, osmotic stress, oxidative stress in plants, deficiency of nutrients [e.g., N, calcium (Ca^{2+}), K, P, Fe, and Zn], and disturbed soil physicochemical properties, all resulting in reduced agricultural production and low economic returns (Shrivastava and Kumar 2015). The soluble salts found in these soils are mainly sodium (Na^{+}), Ca^{2+}, magnesium (Mg^{2+}), chlorides, and sulfates. The most common ions are Na^{+} and chloride (Cl), with a smaller amount of Ca^{2+} and Mg^{2+}, and sometimes nitrate (NO_3^-).

13.2.3 Saline-Sodic and Sodic Soils

Soil becomes sodic or saline-sodic when its ESP increases from 15 or SAR > 13 $(mmol/L)^{1/2}$. In addition to these conditions, if the EC_e of the soil is high (>4 dS/m), the soil stays flocculated. If salts are leached below a threshold level (EC_e < 4 dS/m), the effects of soil sodicity appear, i.e., soil disperses, soil structure is lost, and soil permeability is reduced (Rengasamy and Olsson 1991). Sodic soils contain sodium salts (mostly Na_2CO_3) that may result in alkaline soil water. This is often reflected in soils with high pH (>8.5) and SAR >13. When water evaporates via capillary action, dissolved organic matter in sodic soils is deposited on the surface, resulting in a dark black surface. That is why these soils are also known as black alkali soils (Richards 1954). Cropped fields on saline-sodic or sodic soils show signs of poor development, with crops exhibiting scorching and leaf burn due to salt toxicity, or leaf yellowing caused by poor soil physical structure (Osman 2018). Saline-sodic and sodic soils can become completely barren, with little or no flora, although some sodicity-tolerant grasses may be observed during the rainy seasons. Sodicity can result in restricted air and water entry in the soil causing poor root growth, restricted internal drainage resulting in temporary waterlogging, formation of a surface crust affecting seedling emergence, and compaction of the topsoil and subsoil causing problems with root penetration and soil tillage.

Chemical impacts on plants in saline-sodic and sodic soils vary from nutrient deficiency to excessive levels of Na^{+} and Cl^{-} (Qadir and Schubert 2002). Low ionic activity and high Na^{+}/K^{+}, Na^{+}/Ca^{2+}, Mg^{2+}/Ca^{2+}, and Cl^{-}/NO_3^{-} ratios of the soil solution result in specific ion toxicity, osmotic stress, and nutritional disorders for plants grown in sodic soil (Curtin and Naidu 1998; Grattan and Grieve 1999). Furthermore, plants face deficiency of metallic micronutrients (Cu, Fe, and Zn) in these soils (Curtin and Naidu 1998; Qadar 2002). In addition, the poor physical characteristics of sodic soils directly limit crop development by poor seedling emergence and root growth, while indirect impacts on plant nutrition include limiting water and nutrient absorption as well as gaseous exchange (Ibrahim 2016; Qadir et al. 2022).

13.2.4 Nutrient Depletion in Salt-Affected Soils and Its Impacts on Crop Yield

In salt-affected soils, nutrient imbalances or interactions may manifest themselves in a variety of ways. Complications happen due to the impacts of salinity on nutrient availability, competitive absorption, transport, or partitioning within the plant, or by physiological inactivation of a particular nutrient, which increases the plant's internal demand for that critical element (Hussain et al. 2021). These processes occur simultaneously; however, their impact on a plant depends on the crop species, salinity level, salt composition, nutrient, and several environmental variables (Isayenkov 2019).

13.3 MACRONUTRIENTS

13.3.1 Nitrogen

Nitrogen is a nutrient that often limits crop yield under most agroecological conditions, especially under salt stress (Fageria et al. 2011). Nitrogen makes up ≈ 80% of the total mineral nutrients received by plants (Marschner 1995). The lithosphere holds 98% N, while the atmosphere holds 50,000 times more N_2 than soils. Soil N is divided into two types: (i) organic (98%) and (ii) inorganic (2%). Hydrolyzable (mineralizable to mineral N) and non-hydrolyzable (resistant to mineralization) organic forms of N exist, while inorganic forms include NH_4^+ and NO_3^-. Rapid decomposition of soil organic matter (SOM) in salt-affected soils results in reduced organic carbon (OC) and increased N mineralization, resulting in soils that are deficient in total and accessible N (Swarup 1998). According to Fiers (2020), most areas of the world have a net N balance range of −10 to −100 kg/ha/year, which is substantial. Most of the net balances account for <10% of overall annual turnover. Agricultural regions under intensive cultivation in Europe, North America, and South America suffer the greatest net losses. In Australia [the most salt-affected continent, with 1/3 (260 Mha) of the total global salinized area and 1/2 of the alkaline area (Szabolcs 1989)], erosion losses of N from cultivated Vertisols were estimated to exceed 10 kg/ha/year, with losses from cereal cropping land commonly ranging from 10 to 50 kg/ha/year (Rose and Dalal 1988). Most of this N loss (typically >90%) occurs as organic N in eroded sediments and suspensions, because mineral forms of N account for just 1%–2% of soil N, and they contribute very little to N loss through erosion.

Organic matter and N are in short supply in sodic soil, and it is suggested to apply ~25% extra N during the first few years of soil reclamation compared to the recommended dosage for normal soils (Kaledhonkar et al. 2019; Murtaza et al. 2017b). In salt-affected soil N uptake, especially NO_3^- uptake, is reduced due to the antagonistic effects of Cl^- and SO_4^{2-} high leaching losses of NO_3^- (Singh and Singh 2022). In addition, these soils are characterized by poor symbiotic N-fixation, because a high salt level has an adverse effect on rhizobia leading to drastic reduction in nodulation and decreased rate of nitrification. Direct toxic effects of Cl^- on microbes also reduce the plant available N in saline soils. Furthermore, the inadequate transformation of N applied through fertilizers, ammonia volatilization and denitrification, and nutritional and cationic imbalances within plants all are possible factors that reduce the efficiency of N fertilizers under high soil sodicity or salinity.

Salt stress alters physiological processes and metabolic activities in plants including assimilation, uptake, and translocation of NO_3^- (Annunziata et al. 2017). Under high salinity or sodicity, a reduction in root permeability and the resulting decrease in water and nutrient uptake have been linked to reduce N absorption by plants (Ashraf et al. 2018). In sodic soils, around 32%–52% of applied N is lost by NH_3 volatilization (Bhardwaj and Abrol 1978), and the amount of NH_3 volatilization loss is dependent on pH, soil calcium carbonate ($CaCO_3$) content, and water alkalinity (Bajwa and Singh 1992). In saline and sodic soils, loss of N by volatilization follows first-order kinetics (Singh and Bajwa 1987; Bandhyopadhyay et al. 2000). Plants in waterlogged saline soils suffer water stress, which hinders efficient metabolization of absorbed N. Compared to volatilization losses, Bandhyopadhyay et al. (2000) found that denitrification losses of N from saline rice fields were comparatively low. In alkali soils, leaching may not be a significant source of N loss (Singh and Bajwa 1986).

Salinity in the soil may also impair symbiotic N_2-fixation systems in a variety of ways. Salts may reduce the number of rhizobia in the soil or inhibit their capacity to infect root hairs, limiting nodule development. Regardless of the impact of salts on the rhizobia and the nodulation process, direct impacts of salinity on the host plant may restrict N-fixation (Keck et al. 1984; Fageria 1992). The availability of photosynthate to the root nodules may be reduced if the host plant's development is stunted. This indirect impact may be significant since photosynthate availability is a key factor limiting N_2 fixation (Fageria 1992).

13.3.2 Phosphorus

Phosphorus is one of the main nutrients limiting crop growth in most arid and semiarid areas, and the availability and transformation of soil P differ significantly in salt-affected and normal soils. Almost all P loss occurs by erosion (>95%). Although a plant requires P for salinity tolerance in salt-stressed conditions, a high level causes toxicity in cells, which in turn decreases crop productivity

(L'taief et al. 2012). Inorganic calcium phosphate (Ca-P) is a major fraction of P in sodic soils, and the Ca/P ratio influences the conversion of soluble monocalcium phosphate to insoluble apatite (Chatterjee et al. 2019). Because of the high pH, the concentration of extractable P increases in sodic soils due to the conversion of insoluble Ca-P into soluble sodium forms (Na-P) (Singh and Singh 2022). At an initial stage of reclamation when the ESP of soil is greater than 20, sorption of P by sodic soils is small, and thus a surplus amount of P is present in soil solution for plant uptake (Chhabra 1985). However, in reclaimed sodic soils, the water soluble-P fraction decreases greatly while strongly adsorbed P fractions (NaOH) increase with the increasing rate of the reclaiming agent (gypsum). These changes occur due to increased sorption of P with the decrease in soil ESP and pH (Choudhary and Yaduvanshi 2016).

The availability of P is negatively affected by salinity due to precipitation of applied P, greater retention of soluble P, and the antagonistic effect of excess Cl^- and SO_4^{2-} in saline soils (Chhabra et al. 1976). Abd El-Rahim et al. (2021) reported that available P negatively correlates with EC_e, Ca^{2+}, Mg^{2+}, and Na^+ concentration of the soil solution. In saline soils, P is also less available to plants because salinity limits root development, which lowers the interaction between the root surface area of the crop and the soil P (Chhabra et al. 1986). As a result, the availability of P is largely determined by plant root properties and the antagonistic effects of excess Cl^- on root P absorption, so salinity reduces the bioavailability of P in most instances (Sharpley et al. 1992).

The soil environment and crop genotypes have a significant impact on P absorption (Kausar and Gull 2019; Chen et al. 2019). Phosphorus absorption in lowland rice decreases with increasing salt concentration, according to Fageria et al. (2011), with a 19% reduction at 10 dS/m increasing to a 93% reduction at 15 dS/m. Therefore, by using a combination of P fertilizers and organic amendments, such as crop residues, cattle manure, or press mud, P availability in saline and saline-sodic soils may be increased (Mahmood et al. 2013).

13.3.3 Potassium

A suitable quantity of potassium ion (K^+) in the growth medium is required for metabolic activities, ion transport, and osmotic adjustment in plants. High amounts of external Na^+ in saline-sodic or sodic soils not only interfere with K^+ absorption but also affect the integrity and permeability of root membranes and change their electrical membrane potential (Grattan and Grieve 1999). A sufficient supply of K^+ under salt-stressed conditions contributes to the low osmotic potential in the root stele, which is required to maintain the turgor pressure-driven solute transport in the xylem (Marschner 1995). Sodium-induced K^+ deficiency has been linked to decreased crop growth and yields (Shahzad et al. 2012; Munir et al. 2019). However, at suboptimal K^+ concentration, a relatively lower soil Na^+ level can be beneficial for plant growth (Wakeel et al. 2011; Ma and Bell 2016; Ferreira et al. 2020). The reduction in K^+ absorption by plants is due to competitive uptake mechanisms between K^+ and Na^+ that occur when the solution is dominated by Cl^- and/or SO_4^{2-} salts (Grattan and Grieve 1999). However, different plant species show different levels of K^+ limitation. The uptake of K^+ and Na^+ by plants is antagonistic because they have comparable physicochemical characteristics and ionic size (Gul et al. 2016). In plants, Na^+ promotes K^+ outflow, which results in an electrical potential gradient across the plasma membrane. The absorption of K^+ by the plant is reduced in sodic soils, because of high levels of Na^+ and low levels of Ca^{2+}. Instead of absolute K^+ concentration, the K^+/Na^+ ratio of the growth medium has a considerable effect on the K^+ content of plants (Chhabra 1985). Crop response to K^+ application is influenced by the ability of the cultivar to absorb K^+ in the presence of a significant amount of Na^+. Tao et al. (2021) found that tolerant wheat cultivars absorbed more K^+ and less Na^+ than sensitive wheat cultivars, indicating that the K^+/Na^+ ratio may be used as a predictor of varietal tolerance to high soil Na^+.

Medium to high levels of accessible K^+ are common in saline soils. However, during reclamation, K^+ may be lost due to leaching, resulting in limited availability. Soil salinity may alter, not only the absolute amount of K, but also its equilibrium with Na^+, Ca^{2+}, and Mg^{2+}. As a result, plants may act in response to the absorption of K^+. Under high salinity, changes in the Ca^{2+}: K^+ ratio may be a more significant factor in limiting yields, according to Francois and Bernstein (1964). The presence of sufficient Ca^{2+} in the substrate has a selective effect on the K^+/Na^+ ratio, changing it in favor of K^+ at the cost of Na^+ (Grattan and Grieve 1999). Improvements in Ca^{2+}-mediated membrane integrity result in less K^+ leakage from root cells and a better root K^+ status (Seifikalhor et al. 2019).

Potassium fertilizers may improve crop yields in moderately salty soils (Ma et al. 2015), either by directly supplying K^+ or by improving its ratio to Na^+, Ca^{2+}, and Mg^{2+}. However, when the salinity is very high, it may not be feasible to remove Na^+ from the plant by applying K^+ fertilizer. High amounts of Na^+ may displace Ca^{2+} from root membranes, causing them to lose their integrity and, consequently, altering K^+ selectivity (Cramer et al. 1987). Furthermore, a high concentration of Na^+ may hinder K^+ absorption (Zorb et al. 2019). The uptake of K^+ is highly selective at low external K^+ levels, while Na^+ may compete with K^+ inflow at greater K^+ concentrations. Moreover, it can also suppress K^+ influx at low K^+ concentrations (Munir et al. 2019).

13.3.4 Calcium and Magnesium

Leaching and crop removal can cause significant losses of Ca^{2+} and Mg^{2+} from soil. Maintaining a sufficient supply of Ca^{2+} in saline soil solution is critical to avoid specific ion toxicity, especially in crops sensitive to Cl^- and HCO_3^- (Maas 1993). Calcium plays an important role in ion transport, cell division, membrane integrity, selectivity control, and cell wall enzyme activities (Marschner 1995; Hadi and Karimi 2012). Calcium is required for the development of meristems and the proper operation of root tips. The concentration of Ca^{2+} in soil solution, soil pH, the Ca^{2+}: Na^+ ratio, and crop species affect Ca^{2+} absorption by crop plants. The Ca^{2+} requirement of plants increases when the salt content in the rhizosphere rises (Fageria et al. 1997; Seifikalhor et al. 2019). In addition, ion interactions, precipitation, and an increase in ionic strength reduce Ca^{2+} uptake in salt-affected soils (Hussain et al. 2016; Ashrafi et al. 2018).

Despite soil containing high levels of $CaCO_3$, the dominance of carbonate equilibria in sodic soils can lead to a shortage of accessible Ca^{2+} (both soluble and exchangeable forms). Because the solubility of $CaCO_3$ diminishes as pH rises, the amount of Ca^{2+} available from this source becomes inadequate to fulfill the requirements of the plant. Furthermore, Na^+ has an antagonistic impact on plant Ca^{2+} uptake. Plants growing in these soils are more likely to suffer from a shortage of Ca^{2+} than from the harmful consequences of excessive Na^+ (Chhabra and Abrol 1983). However, in extremely sodic soils and with sensitive crops, high amounts of Na^+ can accumulate in plant tissues, resulting in poor plant development and low yields. All of these factors decrease Ca^{2+} activity in solution and lower Ca^{2+} availability for plants (Cramer et al. 1986; Grattan and Grieve 1999; Farooq et al. 2015).

Plant development is inhibited and substantial morphological and anatomical alterations occur when the Na^+/Ca^{2+} ratio is high (Cramer 1992; Farooq et al. 2015; Kamran et al. 2019). NaCl-induced morphological and anatomical alterations in the main roots of plant have a negative impact on water transport characteristics. Supplementing the soil with Ca^{2+} tended to counteract these changes and avoid reduced root hydraulic conductivity (Du et al. 2022). The addition of a direct and indirect Ca^{2+} source to saline-sodic soils significantly improved rice yields (Qadir et al. 2022). Supplemental Ca^{2+} mitigated the harmful effects of salt stress in tomato plants, including oxidative stress, by reducing the uptake of Na^+ ions by 52% (Ahmad et al. 2018). Cachorro et al. (1994) found that applying Ca^{2+} to the saline medium improved bean root membrane integrity and reduced NO_3^- and $H_2PO_4^-$ leakage. The effectiveness of supplementary Ca^{2+} in reducing the negative effects of Na^+ is dependent on the plant species, Ca^{2+} content, and Na^+ supply (Grattan and Grieve 1999).

The influence of salinity or sodicity on Mg^{2+} availability or transformation is briefly described in the literature. Mg^{2+} ions are highly hydrated, and the binding sites on the root plasma membrane seem to have less affinity for Mg^{2+} ions than for Ca^{2+} ions (Marschner 1995). It is well known that Mg-dominated soil has poor physical properties that negatively affect plant growth (Qadir et al. 2018). For decades, it has been known that solutions with a Mg^{2+}/Ca^{2+} ratio >1, such as those produced by diluting seawater, inhibit maize development (Grattan and Grieve 1999). High substrate Ca^{2+} often leads to increased Ca^{2+} concentration in leaves with subsequent decrease in Mg^{2+} (Marschner 2011; Guo et al. 2016).

13.3.5 Sulfur

Sulfur (S) is a component of SOM, and most of the S in surface soils is in the organic form (Schoenau and Malhi 2008). However, in well-drained, well-aerated surface soils of neutral to alkaline pH, inorganic sulfate exists mainly in the soil solution. In a survey conducted in India, it was noted that the S availability (11–117 mg/kg) was adequate in salt-affected soils due to higher soluble salts in these areas (Kumar et al. 2013). The effect of salinity on S absorption and buildup in plants has received very little consideration. Although inorganic S comprised 5% of total S content in the soil, the contribution in plant absorption of inorganic S was >55% due to easy solubility and high bioavailability (Lu et al., 2020). Aziz et al. (2019) reported that a sufficient supply of S reduced Na^+ accumulation in plant tissues and electrolyte leakage, while an increase in K^+ and Ca^{2+} contents was observed with a subsequent increase in relative water content and photosynthetic rate of sunflower under salt stress. Mor and Manchanda (1992) observed that chloride salinity decreased the S content in the straw in an experiment that examined the effects of both Cl^- and SO_4^- salinity on peas. The application of 45 kg S/ha combined with an appropriate dose of NP fertilizer improved rice and wheat yields under sodic water conditions and reduced the negative effects of sodic water (Yaduvanshi et al. 2015).

13.4 MICRONUTRIENTS

The availability of micronutrients in salt-affected soils depends on the soil pH, SOM, crop species, salinity and sodicity level, and soil calcareousness (Singh and Singh 2022). In Australia, a shortage of Cu, Mn, and Zn has been observed in both sodic and non-sodic soils (Naidu and Rengasamy 1993). Therefore, it is unsatisfactory to link nutritional deficiency to sodicity solely. Because of their naturally low supply power and limited solubility under severe sodicity, alkaline soils suffer from Zn and Fe shortages (Meena et al. 2018).

13.4.1 Zinc

In the Indo-Gangetic Plain, total Zn concentration varies from 40 to 100 mg/kg soil in both normal and sodic soils (Singh and Singh 2022). High pH and soluble P, presence of $CaCO_3$, and little organic matter reduce plants' available Zn in salt-affected soils (Kumar et al. 2013). Sodic soils are often deficient in Zn (<0.6 mg DTPA extractable Zn/kg soil). Apart from the lack of accessible Zn, the efficiency and recovery of applied Zn are adversely affected in sodic soils because 85%–90% of the applied Zn becomes fixed (Chauhan et al. 1999). In salty soils, Zn deficiency is found, and a basal treatment of 20–35 kg $ZnSO_4$/ha was suggested by Kaledhonkar et al. (2019). For example, during the first few years of salt-affected soil reclamation, applying 25–35 kg $ZnSO_4$/ha to a pearl millet-mustard cropping sequence is recommended, and thereafter, it should be managed on a soil test basis (Meena et al. 2018). Zn solubility in sodic soils is regulated by $Zn(OH)_2$ and $ZnCO_3$. Takkar and Sidhu (1978) found that initially (up to 21 days after application) the Zn concentration in the soil solution was controlled by both $Zn(OH)_2$ and $ZnCO_3$ and later on by $ZnCO_3$ alone due to the buffering impact of soil carbonate equilibria. Sadana and Bajwa (1985) discovered that adding farm manure, pyrites, and gypsum to sodic soils reduced the time of predominant presence of the $Zn(OH)_2$ and increased Zn solubility and availability.

In sodic soil, the availability of native and added Zn slightly increased with flooding due to neutralization or drop in pH (Swarup and Yaduvanshi 2004). The applied Zn solubility and extractability declined with time but rose when the ESP of the soil increased (Singh and Singh 2022). The production of soluble sodium zincate accounts for the higher extractability of Zn at high ESP. Zinc, in addition to being an essential element, has an ameliorative impact in alkali soils by increasing Ca^{2+} and K^+ absorption, and therefore widening the Ca^{2+}/Na^+ and K^+/Na^+ ratios in plants (Minhas and Sharma 2003). In salt-affected soils, these physiological characteristics are combined and lead to higher crop yields. The content of Zn in plants has been shown to decrease with increasing sodicity but not always with increasing salinity (Fageria et al. 2011). Salinity decreases Zn content in shoot tissue in most studies, such as in rice (Huang et al. 2017) and wheat (Abbas et al. 2022). Although resistant to soil sodicity, the rice crop is susceptible to Zn deficiency (Shrestha et al. 2020), which may manifest itself 15–21 days after transplanting as dark patches on fully developed leaves and stunting. The optimal Zn dose is also determined by the degree of sodicity. In moderately sodic soil, Zn sprayed at 5.5–11.2 kg/ha was adequate to satisfy Zn requirements, while in very sodic soil, 22.4 kg Zn ha^{-1} was required (Nayyar et al. 1993).

13.4.2 Boron

Boron (B) is arguably the most challenging micro-element in sodic soils. Boron concentration in soil ranges from 2 to 200 mg/kg, but only around 5%–10% is accessible to plants (Diana 2006). Boron is found in soils as boric acid (H_3BO_3) when the pH is <7. The $B(OH)_4^-$ concentration is enhanced quickly as the soil pH rises to 9 and above. However, at these pH values, competition between increased OH^- and B ions occurs for adsorption sites that results in decreased B adsorption (Curtin and Naidu 1998). Low sorption and increased B concentrations in alkaline sodic soils may create B toxicity issues. Increased soil pH, EC, and ESP can cause B toxicity and limit plant growth on soils of arid and semiarid regions (Singh and Singh 2022). In most agricultural plants, B mobility after accumulating in a specific organ is limited. With increasing salinity, both increased and reduced tolerance of plants to B have been reported. Because of the negative relationship between B and SO_4^{2-} concentrations in plant tissues, SO_4^{2-} may be responsible for salinity-induced decrease in plant growth. In terms of wheat leaf B concentration, the interaction between salinity and B was negligible (Bingham et al. 1987). In contrast, Yadav et al. (1989) discovered that Na^+, Ca^{2+}, Cl^-, and SO_4^{2-} salts decreased leaf B concentration in chickpeas grown in loamy sand pots. High Ca^{2+} concentrations in the substrate, especially in calcareous circumstances, reduce B absorption and may cause B deficiency (Gupta et al. 1985). There is also evidence of a reduction in B absorption when Ca and SO_4 ions increase (Singh and Singh 2022). Due to the competition of NO_3^- and BO_3^{3-}, N application also is considered effective to reduce B toxicity (Hua et al. 2021).

13.4.3 Iron

Sodic soils in rice nurseries, especially those planted on raised beds, are prone to Fe deficiency (Singh and Singh 2022). The deficiency or unavailability was due to high soil pH and precipitation of Fe^{2+} with $CaCO_3$ in calcareous sodic soil (Kumar et al. 2013). Shortage of Fe is, however, minor in comparison to Zn deficiency in rice (Choudhary and Yaduvanshi 2016), since soils reclaimed by iron pyrites or organic additions contain sufficient Fe, and soil submergence releases significant quantities of Fe (Singh and Swarup 1990). Plants growing in salt-stressed soil may also mobilize Fe in the rhizosphere via a variety of mechanisms (Ferreira et al. 2019). The soil pH, $CaCO_3$ concentration, oxidation/reduction state, and organic matter content all influence Fe solubility (Choudhary and Yaduvanshi 2016). With submergence up to 60 days, Swarup (1980) found a significant rise in the extractable Fe and Mn status of an alkali soil. When organic ingredients, such as farmyard manure, were added to the soil, the rise was much greater. When rice nurseries exhibit signs of Fe deficiency, it is best to provide a lot of water or utilize green manures. Foliar treatment of Fe (1%–3% solution of $FeSO_4 \cdot 7H_2O$) is frequently recommended because Fe is immobile in plants. Plant absorption of Fe is variable when exposed to salt. More studies are required to fully comprehend the differences in Fe uptake in various crops grown in salt-affected soils.

13.4.4 Manganese

Manganese is one of those elements that are highly sensitive to pH and redox potential. Jatav and Sharma (2019) reported that soil Mn was negatively correlated to pH and $CaCO_3$ in most soils. Kumar et al. (2013) reported that the Mn content in salt-affected soils of India varied between 25.44 and 6.63 mg Mn kg^{-1}. The available Mn content is highly dependent on the OC content of the soil (Kumar et al. 2013). In wetland, rice deficiency of Mn is rare, but it may be a significant limiting factor for upland rice and other upland crops planted after rice. The concentrations of Mn decrease more rapidly than Fe when submerged, and it is leached to the deeper soil layers. Therefore, Mn deficiency is becoming more common in wheat crops that follow rice in a rice-wheat cropping system (Singh et al. 2012; Shukla et al. 2018). Following gypsum application, significant Mn leaching losses have also been observed (Watts and Torbert 2017). However, because of the poor permeability of un-reclaimed or partially reclaimed sodic soils, Mn deficiency is anticipated to be less severe than in normal soils owing to reduced leaching losses (Choudhary and Yaduvanshi 2016).

In maize, increased Mn absorption has been observed with both alkaline salts and neutral salts (Guo et al. 2017). Pandya et al. (2005) observed that supplemental Mn has a limited role in mitigating the adverse effect of salinity in barley. It is extremely difficult to reclaim Mn deficiency by applying $MnSO_4$ to the soil as Mn auto-oxidizes. Foliar Mn application is, therefore, more effective and cost-efficient than soil Mn application (Shukla et al. 2018). To rectify Mn shortage in upland crops, sprays of 1% $MnSO_4$ solution must be repeated three to four times.

13.4.5 Copper

The impact of salinity or sodicity on Cu absorption by agricultural plants has received very little attention. Phyto-availability of Cu is considerably reduced with increasing soil pH (Naidu and Rengasamy 1993; Page et al. 1990). Since a sodic soil is characterized with high soil pH (>8.5), the deficiency of Cu can be anticipated in these soils (Qadir and Schubert 2002; Sharma et al. 2009). Jassal et al. (2017) observed a significant positive correlation of EC with DTPA-Cu suggesting enhanced Cu availability with salt accumulation whereas a negative correlation of $CaCO_3$ with Cu fractions suggested formation of Cu complexes, which results in Cu deficiency in calcareous salt-affected soils. Matijevic et al. (2014) concluded that Cu solubility is adversely affected by NaCl toxicity. Sezer et al. (2021) reported that salinity significantly enhanced leaf-Cu content in maize.

13.4.6 Molybdenum

Like B, Mo becomes more soluble when the pH of the soil rises (Singh and Singh 2022). The Mo content of newly recovered alkali soils in the surface layer ranged from 0.01 to 0.44 mg/kg of soil, which is far beyond the critical level of deficiency (0.15 mg/kg) (Choudhary and Yaduvanshi 2016). As a result, forage crops cultivated on these soils are more prone to take up excessive levels of Mo, which may be harmful to the animals that eat them. Because of the antagonistic action of SO_4^{2-} on the absorption of Mo by the roots, the application of enough gypsum reduces the Mo concentration in the plants. Lopez-Zaplana et al. (2022) found that salinity stress decreased Mo concentration in roots and leaves of melon plants. The literature also indicates the importance of Mo application to increase salinity tolerance in plants by improving the photosynthetic rate and adjusting ionic homeostasis (Zhang et al. 2014).

13.5 CONCLUSION

In recent decades, salt-impacted soils are of global concern, and this concern is exacerbated with soil nutrient depletion. The deficiency of micro- and macronutrients reduces biomass growth, enhances disease susceptibility, and restricts plant development that result in decreased yield potential of most arable crops. Inadequate nutrient absorption by crop plants is linked with a decrease in salt tolerance. Although a recent estimate of soil nutrient depletion in salt-affected soils is not known, in salt-stressed soil, the availability of N and P is low. In salt-affected soil, antagonistic effects of Cl^- and SO_4^{2-}, high leaching losses, poor symbiotic N-fixation, inadequate nitrogenous fertilizer transformation, ammonia volatilization and denitrification, and nutritional and cationic imbalances within plants all are possible factors that reduce nitrogen uptake and directly influence crop yield. During initial reclamation processes of saline-sodic or sodic soils, higher levels of N (10%–30%) fertilizer than recommended increased the growth and yield of rice and wheat. Phosphorus availability to plants is reduced due to its precipitation in the form of calcium phosphate in saline-sodic soils. Sodium-induced potassium deficiency is also common under salt-stressed condition. Moreover, K^+ availability in these soils is limited due to its leaching losses during the reclamation of salt-affected soils. Supplementing saline-sodic and sodic soils with Ca^{2+} amendments not only counteracts the negative effects of excessive Na^+ on plants but also plays a significant role in soil reclamation. Similarly, a sufficient supply of S in saline-sodic soils helps to reduce Na^+ accumulation in plant tissues. Due to high soil pH, most of the micronutrients, i.e., Cu, Fe, Mn, and Zn, have poor solubility in sodic soils and thereby are poorly available for plant uptake.

ACKNOWLEDGMENT

The authors are grateful to Prof. Jo Smith from Institute of Biological and Environmental Sciences, School of Biological Sciences, University of Aberdeen for her helpful comments in composing the manuscript.

REFERENCES

Abbas, G., S. Rehman, M. Saqib, M. Amjad, B. Murtaza, M. H. Siddiqui, and Y. Chen. 2022. Resistance to NaCl salinity is positively correlated with iron and zinc uptake potential of wheat genotypes. *Crop and Pasture Science* 73:546–555.

Abd El-Rahim, W. M., H. Moawad, A. Z. A. Azeiz, and M. J. Sadowsky. 2021. Biodegradation of azo dyes by bacterial or fungal consortium and identification of the biodegradation products. *The Egyptian Journal of Aquatic Research* 47:269–276.

Ahmad, P., E. F. Abd-Allah, M. N. Alyemeni, L. Wijaya, P. Alam, R. Bhardwaj, and K. H. Siddique. 2018. Exogenous application of calcium to 24-epibrassinosteroid pre-treated tomato seedlings mitigates NaCl toxicity by modifying ascorbate–glutathione cycle and secondary metabolites. *Scientific Reports* 8:13515. doi: 10.1038/s41598-018-31917-1.

Annunziata, M. G., L. F. Ciarmiello, P. Woodrow, E. Maximova, A. Fuggi, and P. Carillo. 2017. Durum wheat roots adapt to salinity remodeling the cellular content of nitrogen metabolites and sucrose. *Frontiers in Plant Science* 7:2035.

Ashraf, M., S. M. Shahzad, M. Imtiaz, and M. S. Rizwan. 2018. Salinity effects on nitrogen metabolism in plants - focusing on the activities of nitrogen metabolizing enzymes: A review. *Journal of Plant Nutrition* 41:1065–1081. doi: 10.1080/01904167.2018.1431670.

Ashrafi, E., J. Razmjoo, and M. Zahedi. 2018. Effect of salt stress on growth and ion accumulation of alfalfa (*Medicago sativa* L.) cultivars. *Journal of Plant Nutrition* 41:818–831.

Aslam, M. 2016. Agricultural productivity current scenario, constraints and future prospects in Pakistan. *Sarhad Journal of Agriculture* 32:289–303.

Aubert, H. and M. Pinta, 1980. *Trace Elements in Soils*. Elsevier Scientific Publishing Company: Amsterdam, Netherlands.

Aziz, A., M. Ashraf, S. Sikandar, M. Asif, N. Akhtar, S. M. Shahzad, A. Wasaya, A. Raza, and B. H. Babar. 2019. Optimizing sulfur for improving salt tolerance of sunflower (*Helianthus annuus* L.). *Soil and Environment* 38:222–233.

Bailey, R. L., K. P. West Jr, and R. E. Black. 2015. The epidemiology of global micronutrient deficiencies. *Annals of Nutrition and Metabolism* 66:22–33.

Bajwa, M. S. and Singh, B. 1992. Fertilizer nitrogen management for rice and wheat crops grown in alkali soils. *Fertiliser News* 37:47–59.

Bandhyopadhyay, B. K., B. Maji, and H. S. Sen. 2000. Nutrient management in saline and alkali soils. *Fertiliser News* 45:33–40.

Bhardwaj, K. K. R. and I. P. Abrol. 1978. Nitrogen management in alkali soils. *Proceedings of National Symposium on Nitrogen Assimilation and Crop Productivity*, Haryana Agricultural University, Hissar, pp. 83–86.

Bingham, E. T., J. E. Strong, J. D. Rhoades, and R. Keren. 1987. Effects of salinity and varying boron concentration on boron uptake and growth of wheat. *Plant and Soil* 97:345–351.

Cachorro, P., A. Ortiz, and A. Cerda. 1994. Implications of calcium nutrition on the response of *Phaseolus vulgaris* L. to salinity. *Plant and Soil* 159:205–212.

Chatterjee, S., B. Sahu, and M. Barman. 2019. Response of major plant nutrients to salt-affected environment. *International Journal of Current Microbiology and Applied Sciences* 8:1244–1253.

Chauhan, R. P. S., B. B. Singh, R. K. Singh, and V. P. Singh. 1999. Strategic nutrient management for sustained rice production in sodic soil. *Fertilizer News* 44:13–26.

Chen, B., Q. Wang, H. Bücking, J. Sheng, J. Luo, Z. Chai, A. Kafle, Y. Hou, and G. Feng. 2019. Genotypic differences in phosphorus acquisition efficiency and root performance of cotton (*Gossypium hirsutum*) under low-phosphorus stress. *Crop and Pasture Science* 70:344–358.

Chhabra, R. 1985. Crop response to phosphorus and potassium fertilization of a sodic soil. *Agronomy Journal* 7:699–702.

Chhabra, R. and I. P. Abrol. 1983. Principles governing fertilization in salt-affected soils. *Fertilizer Indian Annual Review* 11:147–159.

Chhabra, R., A. Ringoet, and D. Lamberts. 1976. Kinetics and interaction of chloride and phosphate absorption by intact tomato plants (*Lycopercon esculentum*, Mill) from a dilute nutrient solution. *Zeitschrift fürPflanzenphysiologie* 78:253–261.

Chhabra, R., T. J. Flowers, and A. R. Yeo. 1986. Effect of pH and salinity on P sorption by rice. Report. Indo-British Tech Cooperation, UK and Central Soil Salinity Research Institute, Karnal.

Choudhary, O. P. and N. P. S. Yaduvanshi. 2016. Nutrient management in salt-affected soils. *Indian Journal of Fertilisers* 12:20–35.

Correndo, A. A., G. Rubio, F. O. García, and I. A. Ciampitti. 2021. Subsoil-potassium depletion accounts for the nutrient budget in high-potassium agricultural soils. *Scientific Reports* 11:1–10. doi: 10.1038/s41598-021-90297-1.

Cramer, G. R. 1992. Kinetics of maize leaf elongation. II. Response of a Na-excluding cultivar and a Na-including cultivar to varying Na/Ca salinities. *Journal of Experimental Botany* 43: 857–864.

Cramer, G. R., A. Lauchli, and E. Epstein. 1986. Effects of NaCl and $CaCl_2$ on ion activities in complex nutrient solutions and root growth of cotton. *Plant Physiology* 81:792–797.

Cramer, G. R., J. Lynch, L. Lauchli, and E. Epstein 1987. Influx of Na^+, K^+ and Ca^{2+} into roots of salt-stressed cotton seedlings. *Plant Physiology* 83:510–516.

Curtin, D. and R. Naidu. 1998. Chapter 7: Fertility constraints to plant production. In: Sumner, M. and R. Naidu (Eds.), *Sodic Soils: Distribution, Properties, Management and Environmental Consequences*. Oxford University Press: New York, pp. 107–123.

De-Regil, L. M., P. S. Suchdev, G. E. Vist, S. Walleser, and J. P. Peña-Rosas. 2013. Home fortification of foods with multiple micronutrient powders for health and nutrition in children under two years of age. *Evidence-Based Child Health: A Cochrane Review Journal* 8:112–201.

Dhaliwal, S. S., R. K. Naresh, A. Mandal, R. Singh, and M. K. Dhaliwal. 2019. Dynamics and transformations of micronutrients in agricultural soils as influenced by organic matter build-up: A review. *Environmental and Sustainability Indicators* 1:100007.

Diana, G. 2006. Boron in the soil, from deficit to toxicity. *Informatore Agrario* 62:54–58.

Dobermann, A., P. C. Santa Cruz, and K. G. Cassman. 1995. Potassium balances and soil potassium supplying power in intensive irrigated rice ecosystems. *In 24th Colloquium of IPI*, Chiang Mai, Thailand, pp. 199–229.

Du, X., S. Wang, H. Feng, Z. Xu, J. Liu, X. Ren, Z. Gao, and S. Hu. 2022. Morphological and physiological response of rice roots to the application of calcium polyamino acid under saline-alkali soil conditions. *Archives of Agronomy and Soil Science*. doi: 10.1080/03650340.2022.2045012.

Fageria, N. K. 1992. *Maximizing Crop Yields*. Marcel Dekker: New York.

Fageria, N. K., V. C. Baligar, and C. A. Jones. 1997. *Growth and Mineral Nutrition of Field Crops*, 2nd edition. Marcel Dekker: New York.

Fageria, N. K., H. R. Gheyi, and A. Moreira. 2011. Nutrient bioavailability in salt affected soils. *Journal of Plant Nutrition* 34:945–962.

FAO. 2015. The state of food insecurity in the world 2015. Meeting the 2015 International Hunger Targets: Taking Stock of Uneven Progress, FAO, Rome.

Farooq, M., M. Hussain, A. Wakeel, and K. H. Siddique. 2015. Salt stress in maize: Effects, resistance mechanisms, and management: A review. *Agronomy for Sustainable Development* 35:461–481.
Ferreira, M. J., H. Silva, and A. Cunha. 2019. Siderophore-producing rhizobacteria as a promising tool for empowering plants to cope with iron limitation in saline soils: A review. *Pedosphere* 29:409–420.
Ferreira, J. F. S., J. B. da Silva Filho, X. Liu, and D. Sandhu. 2020. Spinach plants favor the absorption of K^+ over Na^+ regardless of salinity, and may benefit from Na^+ when K^+ is deficient in the soil. *Plants* 9:507. doi: 10.3390/plants9040507.
Fiers, T. 2020. A New Global Assessment of Soil Nutrient Balances. MSc thesis, Soil Geography and Landscape Group. Wageningen University.
Francois, L. E. and L. Bernstein. 1964. Salt tolerance of safflower. *Agronomy Journal* 56:38–40.
Grattan, S. R., and C. M. Grieve. 1999. Salinity-mineral nutrient relations in horticultural crops. *Scientia Horticulturae* 78:127–157.
Gul, M., A. Wakeel, M. Saqib, and A. Wahid. 2016. Effect of NaCl-induced saline sodicity on the interpretation of soil potassium dynamics. *Archives of Agronomy and Soil Science* 62:523–532.
Guo, W., H. Nazim, Z. Liang, and D. Yang. 2016. Magnesium deficiency in plants: An urgent problem. *The Crop Journal* 4:83–91.
Guo, R., L. Shi, C. Yan, X. Zhong, F. Gu, Q. Liu, X. Xia, and H. Li. 2017. Ionomic and metabolic responses to neutral salt or alkaline salt stresses in maize (*Zea mays* L.) seedlings. *BMC Plant Biology* 17:1–13.
Gupta, U. C., Y. W. Jame, C. A. Campbell, A. J. Leyshon, and W. Nicholaichuk. 1985. Boron toxicity and deficiency: A review. *Canadian Journal of Soil Science* 65:381–409.
Hadi, M. R. and N. Karimi. 2012. The role of calcium in plants salt tolerance. *Journal of Plant Nutrition* 35:2037–2054.
Henao, J. and C. Baanante. 1999. *Estimating Rates of Nutrient Depletion in Soils of Agricultural Lands of Africa*. International Fertilizer Development Center (IFDC): Muscle Shoals, AL, 76 pp.
Hua, T., R. Zhang, H. Sun, and C. Liu. 2021. Alleviation of boron toxicity in plants: Mechanisms and approaches. *Critical Reviews in Environmental Science and Technology* 51:2975–3015.
Huang, L., X. Liu, Z. Wang, Z. Liang, M. Wang, M. Liu, and D. L. Suarez. 2017. Interactive effects of pH, EC and nitrogen on yields and nutrient absorption of rice (*Oryza sativa* L.). *Agricultural Water Management* 194:48–57.
Hussain, Z., R. A. Khattak, M. Irshad, Q. Mahmood, and P. An. 2016. Effect of saline irrigation water on the leachability of salts, growth and chemical composition of wheat (*Triticum aestivum* L.) in saline-sodic soil supplemented with phosphorus and potassium. *Journal of Soil Science and Plant Nutrition* 16:604–620.
Hussain, S., S. Hussain, B. Ali, X. Ren, X. Chen, Q. Li, M. Saqib, and N. Ahmad. 2021. Recent progress in understanding salinity tolerance in plants: Story of Na^+/K^+ balance and beyond. *Plant Physiology and Biochemistry* 160:239–256.
Ibrahim, E. A. 2016. Seed priming to alleviate salinity stress in germinating seeds. *Journal of Plant Physiology* 192:38–46. doi: 10.1016/j.jplph.2015.12.011.
Isayenkov, S. V. 2019. Genetic sources for the development of salt tolerance in crops. *Plant Growth Regulation* 89:1–17. doi: 10.1007/s10725-019-00519-w.
Jassal, H. S., O. P. Sandhu, and S. S. Dhaliwal. 2017. Distribution of copper fractions in the salt-affected soils of Muktsar District of Punjab. *Indian Journal of Fertilisers* 13:62–69.
Jatav, R. C. and Sharma Y. M. 2019. A review article: Status of micronutrients distribution in mix red and black soil of Rewa district of Madhya Pradesh, India. *International Journal of Chemical Studies* 7:3983–3990.
Jones, D. L., P. Cross, P. J. Withers, T. H. DeLuca, D. A. Robinson, R. S. Quilliam, I. M. Harris, D. R. Chadwick, and G. Edwards-Jones. 2013. Nutrient stripping: The global disparity between food security and soil nutrient stocks. *Journal of Applied Ecology* 50:851–862.
Kaledhonkar, M. J., B. L. Meena, and P. C. Sharma. 2019. Reclamation and nutrient management for salt-affected soils. *Indian Journal of Fertilisers* 15:566–575.
Kamran, M., A. Parveen, S. Ahmar, Z. Malik, S. Hussain, M. S. Chattha, M. H. Saleem, M. Adil, P. Heidari, and J. T. Chen. 2019. An overview of hazardous impacts of soil salinity in crops, tolerance mechanisms, and amelioration through selenium supplementation. *International Journal of Molecular Sciences* 21:148.
Kausar, A. and M. Gull. 2019. Influence of salinity stress on the uptake of magnesium, phosphorus, and yield of salt susceptible and tolerant sorghum cultivars (*Sorghum bicolor* L.). *Journal of Applied Biology and Biotechnology* 7:5–8.
Keck, T. J., R. J. Wagenet, W. F. Campbell, and R. E. Knighton. 1984. Effects of water and salt stress on growth and acetylene reduction in alfalfa. *Soil Science Society of America Journal* 48:1310–1316.

Kumar, M. V., G. V. Lakshmi, and P. Madhuvani, 2013. Appraisal of soil fertility status in salt-affected soils of Ongole division, Prakasam district, Andhra Pradesh. *Journal of the Indian Society of Soil Science* 61:333–340.

L'taief, B., S. Bouaziz, Z. Mainassara, H. Ralf, C. Molina, S. Beebe, P. Winter, G. Kahl, J. J. Drevon, and M. Lachaal. 2012. Genotypic variability for tolerance to salinity and P deficiency among N_2-dependent recombinant inbred lines of common bean (*Phaseolus vulgaris*). *African Journal of Microbiology Research* 6:4205–4213.

Lin, X., C. Yin, and D. Xu. 1996. Input and output of soil nutrients in high-yield paddy fields in South China. *In Proceedings of the International Symposium on Maximizing Rice Yields through Improved Soil and Environmental Management*, Khon Kaen, Thailand, pp. 93–97.

Lopez-Zaplana, A., N. Martinez-Garcia, M. Carvajal, and G. Bárzana. 2022. Relationships between aquaporins gene expression and nutrient concentrations in melon plants (*Cucumis melo* L.) during typical abiotic stresses. *Environmental and Experimental Botany* 195:104759.

Lu, Q., J. Bai, D. Yan, B. Cui, and J. Wu. 2020. Sulfur forms in wetland soils with different flooding periods before and after flow-sediment regulation in the Yellow River Delta, China. *Journal of Cleaner Production* 276:122969. doi: 10.1016/j.jclepro.2020.122969.

Ma, Q. and R. Bell. 2016. Canola, narrow-leafed lupin and wheat differ in growth response to low–moderate sodium on a potassium-deficient sandy soil. *Crop & Pasture Science* 67:1168–1178. doi: 10.1071/CP16220.

Ma, Q., R. Bell, C. Scanlan, G. Sarre, and R. Brennan. 2015. Growth and yield responses in wheat and barley to potassium supply under drought or moderately saline conditions in the south-west of Western Australia. *Crop and Pasture Science* 66:135–144. doi: 10.1071/CP14190.

Maas, E. V. 1993. Testing crops for salinity tolerance. In: Maranville, J. W. and V. C. Baligar (Eds.), *Proceedings of a Workshop on Adaptation of Plants to Soil Stresses*. University of Nebraska: Lincoln, NE, pp. 234–247.

Mahmood, I. A., A. Arshad, M. Aslam, A. Shahzad, T. Sultan, and F. Hussain. 2013. Phosphorus availability in different salt-affected soils as influenced by crop residue incorporation. *International Journal of Agriculture and Biology* 15:472–478.

Marschner, H. 1995. *Mineral Nutrition of Higher Plants*, 2nd edition. Academic Press: Amsterdam, The Netherlands.

Marschner, H. 2011. *Marschner's Mineral Nutrition of Higher Plants*. Elsevier/Academic Press: Amsterdam, The Netherlands, p. 684.

Matijevic, L., D. Romic, and M. Romic. 2014. Soil organic matter and salinity affect copper bioavailability in root zone and uptake by *Vicia faba* L. plants. *Environmental Geochemistry and Health* 36:883–896.

Meena, B. L., P. Kumar, A. Kumar, R. L. Meena, M. J. Kaledhonkar, and P. C. Sharma. 2018. Zinc and iron nutrition to increase the productivity of pearl millet-mustard cropping system in salt affected soils. *International Journal of Current Microbiology and Applied Sciences* 8:3201–3211.

Minhas, P. S. and O. P. Sharma. 2003. Management of soil salinity and alkalinity problems in India. *Journal of Crop Production* 7:181–230.

Mor, R. P. and Manchanda, H. R. 1992. Influence of phosphorus on the tolerance of table pea to chloride and sulfate salinity in a sandy soil. *Arid Soil Research and Rehabilitation* 6:41–52.

Munir, A., M. T. Shehzad, A. A. Qadir, G. Murtaza, and H. I. Khalid. 2019. Use of potassium fertilization to ameliorate the adverse effects of saline-sodic stress condition (ECw: SARw Levels) in rice (*Oryza Sativa* L.). *Communications in Soil Science and Plant Analysis* 50:1975–1985.

Murtaza, B., G. Murtaza, M. Imran, M. Amjad, A. Naeem, G. M. Shah, and A. Wakeel. 2017a. Yield and nitrogen use efficiency of rice-wheat cropping system in gypsum amended saline-sodic soil. *Journal of Soil Science and Plant Nutrition* 17(3):686–701.

Murtaza, B., G. Murtaza, M. Sabir, G. Owens, G. Abbas, M. Imran, and G. M. Shah. 2017b. Amelioration of saline–sodic soil with gypsum can increase yield and nitrogen use efficiency in rice–wheat cropping system. *Archives of Agronomy and Soil Science* 63:1267–1280.

Mutert, E. 1996. Plant nutrient balance in Asia and the Pacific region: Facts and consequences for agricultural production. *In APO-FFTC Seminar on Appropriate Use of Fertilizers,* Taiwan, pp. 73–112.

Naidu, R. and P. Rengasamy. 1993. Ion interactions and constraints to plant nutrition in Australian sodic soils. *Australian Journal of Soil Research* 31:801–819.

Nawaz, F., M. A. Shehzad, S. Majeed, K. S. Ahmad, M. Aqib, M. M. Usmani, and R. N. Shabbir. 2020. Role of mineral nutrition in improving drought and salinity tolerance in field crops. In: Hasanuzzaman, M. (Ed.), *Agronomic Crops*. Springer: Singapore. doi: 10.1007/978-981-15-0025-1_8.

Nayyar, V. K., I. M. Chhibba, and M. S. Bajwa. 1993. Management of Zn deficiency in sodic soils. *Indian Farming* 43:6–8.

Neina, D. 2019. The role of soil pH in plant nutrition and soil remediation. *Applied and Environmental Soil Science* 2019:5794869. doi: 10.1155/2019/5794869.

Osman, K. T. 2018. Saline and sodic soils. In: Osman, K. T. (Ed.), *Management of Soil Problems*. Springer: Cham. doi: 10.1007/978-3-319-75527-4_10.

Page, A. L., A. C. Chang, and D. C. Adriano. 1990. Deficiencies and toxicities of trace elements. In: Tanji, K. K. (Ed.), *Agricultural Salinity Assessment and Management*. American Society of Civil Engineers: New York, pp. 138–160.

Pandya, D. H., R. K. Mer, P. K. Prajith, and A. N. Pandey, 2005. Effect of salt stress and manganese supply on growth of barley seedlings. *Journal of Plant Nutrition* 27:1361–1379.

Qadar, A. 2002. Selecting rice genotypes tolerant to zinc deficiency and sodicity stresses. I. Differences in zinc, iron, manganese, copper, phosphorus concentrations, and phosphorus/zinc ratio in their leaves. *Journal of Plant Nutrition* 25:457–473.

Qadir, M. and S. Schubert. 2002. Degradation processes and nutrient constraints in sodic soils. *Land Degradation and Development* 13:275–294.

Qadir, M., E. Quillérou, V. Nangia, G. Murtaza, M. Singh, R. J. Thomas, P. Drechsel, and A. D. Noble. 2014. Economics of salt-induced land degradation and restoration. *Natural Resources Forum* 38:282–295.

Qadir, M., S. Schubert, J. D. Oster, G. Sposito, P. S. Minhas, S. A. M. Cheraghi, G. Murtaza, A. Mirzabaev, and M. Saqib. 2018. High-magnesium waters and soils: Emerging environmental and food security constraints. *Science of the Total Environment* 642:1108–1117.

Qadir, A. A., G. Murtaza, M. Z. Rehman, and E. A. Waraich. 2022. Application of gypsum or sulfuric acid improves physiological traits and nutritional status of rice in calcareous saline-sodic soils. *Journal of Soil Science and Plant Nutrition* 22:1846–1858.

Rengasamy, P. and K. A. Olsson. 1991. Sodicity and soil structure. *Soil Research* 29:935–952.

Richards, L. A. 1954. Diagnosis and improvement of saline and alkali soils. *Soil Science* 78:154.

Rose, C. W. and R. C. Dalal. 1988. Erosion and runoff of nitrogen. In: Wilson, J. R. (Ed.), *Advances in Nitrogen Cycling in Agricultural Systems*. CAB International: Wallingford, pp. 212–235.

Sadana, U. S. and M. S. Bajwa. 1985. Manganese equilibrium in submerged sodic soils as influenced by application of gypsum and green manuring. *Journal of Agricultural Science Cambridge* 104:257–261.

Schoenau, J. J. and S. S. Malhi. 2008. Sulfur forms and cycling processes in soil and their relationship to sulfur fertility. In: Jez, J. and American Society of Agronomy (Eds.), *Sulfur: A Missing Link between Soils, Crops, and Nutrition*. ASA-CSSA-SSSA: St. Louis, MI, vol. 50, pp. 1–10.

Seifikalhor, M., S. Aliniaeifard, A. Shomali, N. Azad, B. Hassani, O. Lastochkina, and T. Li. 2019. Calcium signaling and salt tolerance are diversely entwined in plants. *Plant Signaling and Behavior* 14:1665455.

Setia, R. K. and K. N. Sharma. 2004. Effect of continuous cropping and long-term differential fertilization on profile stratification of DTPA extractable micronutrients. *Journal of Food, Agriculture and Environment* 2:260–265.

Sezer, I., M. S. Kiremit, E. Öztürk, B. A. G. Subrata, H. M. Osman, H. Akay, and H. Arslan. 2021. Role of melatonin in improving leaf mineral content and growth of sweet corn seedlings under different soil salinity levels. *Scientia Horticulturae* 288:110376.

Shahid, S. A., M. Zaman, and L. Heng. 2018. Soil salinity: Historical perspectives and a world overview of the problem. In: Zaman, M., Shahid, S. A., and Heng, L. (Eds.), *Guideline for Salinity Assessment, Mitigation and Adaptation Using Nuclear and Related Techniques*. Springer: Cham, pp. 43–53. doi: 10.1007/978-3-319-96190-3_2.

Shahzad, M., K. Witzel, C. Zorb, and K. H. Muhling. 2012. Growth-related changes in subcellular ion patterns in maize leaves (*Zea mays* L.) under salt stress. *Journal of Agronomy and Crop Science* 198:46–56. doi: 10.1111/j.1439-037X.2011.00487.x.

Shahzad, R., S. Jamil, S. U. Rahman, and M. Z. Iqbal. 2017. Salt Stress and its impact on crop productivity in Pakistan. AGRIHUNT, May 10.

Sharma, B. D., R. Kumar, B. Singh, and M. Sethi. 2009. Micronutrients distribution in salt-affected soils of the Punjab in relation to soil properties. *Archives of Agronomy and Soil Science* 55:367–377.

Sharpley, A. N., J. J. Meisinger, J. F. Power, and D. L. Suarez. 1992. Root extraction of nutrients associated with long-term soil management. *Advances in Soil Science* 19:151–217.

Sheldrick, W. F., J. K. Syers, and J. Lingard. 2002. A conceptual model for conducting nutrient audits at national, regional, and global scales. *Nutrient Cycling in Agroecosystems* 62:61–72.

Shrestha, J., Kandel, M., Subedi, S., and Shah, K. K., 2020. Role of nutrients in rice (*Oryza sativa* L.): A review. *Agrica* 9:53–62.

Shrivastava, P. and R. Kumar. 2015. Soil salinity: A serious environmental issue and plant growth promoting bacteria as one of the tools for its alleviation. *Saudi Journal of Biological Sciences* 22:123–131.

Shukla, A. K., S. K. Behera, A. Pakhre, and S. K. Chaudhari. 2018. Micronutrients in soils, plants, animals and humans. *Indian Journal of Fertilisers* 14:30–54.

Sibbesen, E. and A. Runge-Metzger. 1995. Phosphorus balance in European agriculture–status and policy options. In: Tiessen, H. (Ed.), *Phosphorus in the Global Environment: Transfers, Cycles, and Management*. John Wiley & Sons: Chichester, pp. 43–57.

Singh, B. and M. S. Bajwa. 1986. Studies on leaching of urea in sodic soils. *Journal of Agricultural Science Cambridge* 106:323–330.

Singh, B. and M. S. Bajwa. 1987. Ammonia volatilization from salt affected soils fertilized with urea under submerged conditions. *Journal of the Indian Society of Soil Science* 35:358–364.

Singh, N. T. and A. Swarup. 1990. On reclaimed alkali soils urea before puddling benefits rice. *Indian Farming* 40:50.

Singh, S. and V. Singh. 2022. Nutrient management in salt affected soils for sustainable crop production. *Annals of Plant and Soil Research* 24:182–193.

Singh, V. K., V. Govil, S. K. Singh, B. S. Dwivedi, M. C. Meena, V. K. Gupta, K. Majumdar, and B. Gangwar. 2012. Precision nutrient management strategies using GIS-based mapping in western Uttar Pradesh. *Better Crops-South Asia* 6:15–8.

Swarup, A. 1980. Effect of submergence and farmyard manure application on the yield and nutrition of rice and sodic soil reclamation. *Journal of the Indian Society of Soil Science* 28:532–534.

Swarup, A. 1998. Soil fertility problems and their management. In: Tyagi, N. K., and P. S. Minhas (Eds.), *Agricultural Salinity Management in India*. Central Soil Salinity Research Institute: Karnal, pp. 145–158.

Swarup, A. and N. P. S. Yaduvanshi. 2004. Response of rice and wheat to organic and inorganic fertilizers and soil amendment under sodic water irrigated conditions. *IRRI News Letter* 29:49–51.

Szabolcs, I. 1989. *Salt Affected Soils*. CRC Press: Boca Raton, FL, p. 274.

Takkar, P. N. and B. S. Sidhu. 1978. Kinetics of zinc transformations in submerged alkaline soils in the rice growing tract of Punjab. *Journal of Agricultural Science Cambridge* 93:441–447.

Tan, Z. X., R. Lal, and K. D. Wiebe. 2005. Global soil nutrient depletion and yield reduction. *Journal of Sustainable Agriculture* 26:123–146.

Tao, R., J. Ding, C. Li, X. Zhu, W. Guo, and M. Zhu. 2021. Evaluating and screening of agro-physiological indices for salinity stress tolerance in wheat at the seedling stage. *Frontiers of Plant Science* 12:646175. doi: 10.3389/fpls.2021.646175.

van Beek, C. L., E. Elias, G. S. Yihenew, H. Heesmans, A. Tsegaye, H. Feyisa, M. Tolla, M. Melmuye, Y. Gebremeskel, and S. Mengist. 2016. Soil nutrient balances under diverse agro-ecological settings in Ethiopia. *Nutrient Cycling in Agroecosystems* 106:257–274. doi: 10.1007/s10705-016-9803-0.

Wakeel, A., M. Farooq, M. Qadir, and S. Schubert. 2011. Potassium substitution by sodium in plants. *Critical Reviews in Plant Sciences* 30:401–413.

Watts, D. B. and H. A. Torbert. 2017. Three annual flue gas desulfurization gypsum applications on macronutrient and micronutrient losses in runoff from bermudagrass fertilized with poultry litter. *Soil Science* 182:18–27.

Wicke, B., E. Smeets, V. Dornburg, B. Vashev, T. Gaiser, W. Turkenburg, and A. Faaij. 2011. The global technical and economic potential of bioenergy from salt-affected soils. *Energy and Environmental Science* 4:2669–2681.

Williams, J. 1991. Search for sustainability: Agriculture and its place in the natural ecosystem. 'Paper presented to the Australian Institute of Agricultural Science Symposium'Agriculture and the Ecosystem', Townsville, August 1990. *Agricultural Science (Melbourne)* 4:32–39.

Yadav, H. D., O. P. Yadav, O. P. Dhankar, and M. C. Oswal. 1989. Effect of chloride salinity and boron on germination, growth, and mineral composition of chickpea (*Cicer arietinum* L.). *Annals of Arid Zone, Jodhpur* 28:63–67.

Yaduvanshi, N. P. S., K. Lal, and A. Swarup. 2015. Effect of sodic water irrigation with sulphur through single super phosphate on yield, mineral composition and soil properties in rice-wheat system. *Journal of Soil Salinity and Water Quality* 7:35–39.

Zaman, S. B., and S. Ahmad. 2009. Salinity and waterlogging in the Indus Basin of Pakistan: Economic loss to agricultural economy. *Managing Natural Resources for Sustaining Future Agriculture* 1:4–8.

Zhang, M., C. Hu, X. Sun, X. Zhao, Q. Tan, Y. Zhang, and N. Li. 2014. Molybdenum affects photosynthesis and ionic homeostasis of Chinese cabbage under salinity stress. *Communications in Soil Science and Plant Analysis* 45:2660–2672.

Zhang, X., E. A. Davidson, T. Zou, L. Lassaletta, Z. Quan, T. Li, and W. Zhang. 2020. Quantifying nutrient budgets for sustainable nutrient management. *Global Biogeochemical Cycles* 34:e2018GB006060.

Zorb, C., C. M. Geilfus, and K. J. Dietz. 2019. Salinity and crop yield. *Plant Biology* 21:31–38.

14 Use of Biochar for the Amelioration of Problem Soils

Prakash C. Srivastava[1], *Rini Labanya*[1], *Satya P. Pachauri*[1], *and Prashant Srivastava*[2]

[1]Department of Soil Science, College of Agriculture, Govind Ballabh Pant University of Agriculture & Technology, Pantnagar, India

[2]Commonwealth Scientific and Industrial Research Organization (CSIRO) Environment Business Unit, Urrbrae, Australia

CONTENTS

14.1 INTRODUCTION

Maintenance of soil productivity and reclamation of problem soils are great challenges for humanity. The practice of charcoal application to build up soil fertility is a several thousand years old method. Ancient soils (Terra Preta de Indio) in the Amazon basin with high potential for soil fertility generated a new worldwide interest in the scientific community. A good amount of research on the role of biochar as a soil ameliorant has been carried out in the past few decades. The conversion of various wastes including crop residues that are not suitable as cattle feed and forest residues into biochar and their utilization as a soil ameliorant/amendment can turn them into wealth for enhancing soil health and crop productivity and improving the soils polluted by organic and inorganic pollutants to ensure a cleaner environment. This technology has societal benefits through the mitigation of global warming by carbon sequestration as well as the reduction in the emission of non-CO_2 greenhouse gases. In comparison to the burning of biomass in open, its controlled carbonization converts even larger quantities of biomass organic matter into stable C pools which are assumed to persist in the environment over centuries (Glaser et al., 1998; Schmidt and Noack, 2000; Glaser et al., 2001). These benefits provide the basis for upscaling the use of biochar in agriculture and environmental management. This chapter examines the biochars and their potential use as a soil ameliorant to manage soil-related constraints.

14.1.1 Production of Biochar and Its Chemical Structure

Biochar can be produced from a variety of lignocellulosic resources (Thomsen et al., 2011). A variety of thermal conversion processes can be used to prepare biochar (Deal et al., 2012). Pyrolysis systems employed to process unused and excess crop, and agroforestry residues for biochar production

DOI: 10.1201/9781003093565-14

can be categorized into four types: (i) slow pyrolysis, (ii) fast pyrolysis, (iii) flash pyrolysis, and (iv) gasification. Slow pyrolysis performed under lower temperatures (<400°C–500°C) and with long contact times often results in a high yield of biochar (35%) (Meyer et al., 2011). Faster pyrolysis or gasification operates at higher temperatures (<800°C) and gives a high yield of combustible gases in relation to the solid biochar (12%) (Laird et al., 2009). The most commonly employed method for biochar production is slow pyrolysis. This process involves direct thermo-chemical decomposition (exothermic reaction) to transform a low-density residue matrix into biochar at a temperature range of 450°C–500°C under oxygen-deprived conditions in a closed reactor (Jeffery et al., 2011; Gul et al., 2015). In commercial biochar pyrolysis systems, the process occurs in three steps: (i) loss of moisture and some volatiles; (ii) unreacted residues are converted to volatiles, gases, bio-oil, and biochar; and (iii) slow chemical rearrangement of the biochar (Demirbas, 2004).

Biochar is a fine, carbon-rich, porous product left after the thermo-chemical conversion process (pyrolysis) of feedstock biomass at a given temperature in a little or no oxygen environment (Amonette and Joseph, 2009). In biochars, C atoms are grouped into flat graphene-like aromatic sheets as basic structural units which are randomly cross-linked (Xiao and Chen, 2017). The mutual arrangement of these aromatic sheets is irregular which leaves free pores between the sheets making the biochar an excellent adsorbent. Their adsorption capacity is determined by the porous microtexture, but it is also strongly influenced by the chemical structure of the surface. Biochars invariably contain appreciable amounts of heteroatoms such as O and H, and also Cl, N, and S (Masek, 2009) derived from the feed material and become a part of the chemical structure of biochars due to imperfect carbonization or through chemical bonds on the surface. These heteroatoms either form surface groups at the fringe areas of the aromatic sheets, or they may be incorporated into the microtexture forming heterocyclic ring systems within the carbon layers or on the bridges connecting the basic structural units. Carbon–O surface functional groups (COOH, C–OH, C=O, O–C–O) are the main functional groups on biochar surfaces that influence the surface characteristics and behavior of biochars (Esfandbod et al., 2017a). In addition, C–N, C–S and C–X (halogen) surface groups also impart some specific adsorption properties to biochars. Heteroatoms, being more electronegative than C, have lower energies for their valence orbitals and result in a partial negative charge. The central quality of biochar that makes it important as a soil amendment is its high surface area and highly porous structure which is potentially responsible for improved water retention and adsorption capacity. The range of some general properties of biochars produced by low-temperature pyrolysis from cereal straw and forest residues which are of edaphological significance are presented in Table 14.1.

14.1.2 Biochar for Improvement of Soil Physical Properties

The major soil physical constraints affecting soil productivity in many geographical regions, including degraded soils, are high bulk density, poor soil water storage and aeration, and surface crusting. Owing to its internal structure and surface functionality, biochar can retain more soil water (Suliman et al., 2017). Different studies have shown that the application of biochar can reduce soil bulk density and increase total pore volume and water-holding capacity (Figure 14.1) (Abel et al., 2013; Chan et al., 2007). Fang et al. (2016) showed that biochar addition @26, 64 and 128 t/ha increased the average volumetric water content of soil by 10.4%, 24.6%, and 33.4%, respectively over the whole growing season. The same experiment revealed that the application of biochar at 64 and 128 t/ha significantly decreased soil bulk density by 14.6% and 32.5%, respectively, over the control. Cornelissen et al. (2013) also reported that maize cob biochar-treated soil resulted in higher plant available water as compared to control at five different farming conservation sites in Zambia. Khan et al. (2019) studied the effect of mesquite biochar prepared under slow-heating conditions at 575°C for 3 hours on soil physical conditions and soil fertility under onion crop and reported that the application of biochar at 10 t/ha lowered the bulk density from 1.23 to 1.10 g/cm^3 and increased the water-holding capacity from 0.46 to 0.55 mL water per g dry soil.

TABLE 14.1
The Range of Some General Properties of Biochars Produced[a] from Cereal Straw and Forest Residues (Labanya et al., 2022a)

Biochar Properties	Cereal Straw	Forest Residues
Mass yield (%)	17.3–36.3	18.5–47.8
Apparent density (g/cm^3)	0.20–0.30	0.27–0.51
Specific density (g/cm^3)	1.45–1.70	1.19–1.39
Ash content (%)	23.0–36.1	5.9–13.5
pH (1:10 water suspensions)	5.4–8.1	6.0–8.2
Specific electrical conductivity (dS/m at 25°C)	0.26–0.94	0.07–0.50
Cation-exchange capacity (cmol (p$^+$)/kg)	93–143	52–111
C (%)	21.3–27.3	59.9–62.2
N (%)	0.65–0.67	0.86–1.54
P (%)	0.23–0.44	0.27–0.42
K (%)	2.08–2.26	1.68–2.53
Ca (%)	0.45–0.59	0.57–1.17
Mg (%)	0.06–0.16	0.08–0.60
S (%)	0.11–0.12	0.10–0.12
Zn (mg/kg)	25.4–64.3	16.9–98.1
Cu (mg/kg)	9.4–12.1	5.8–21.9
Fe (mg/kg)	732.3–1129.7	63.5–493.4
Mn (mg/kg)	64.6–129.8	23.8–41.3
B (mg/kg)	6.0–12.5	8.6–47.7
Mo (mg kg^{-1})	0.6–1.0	1.3–2.6

[a] Pyrolysis temperature (300°C or 450°C).

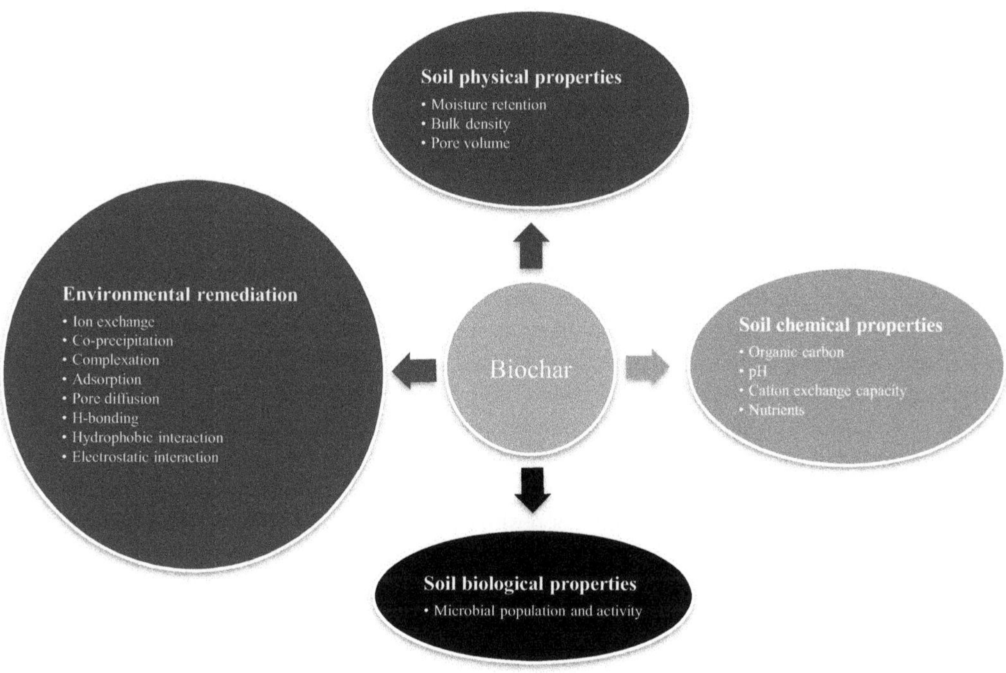

FIGURE 14.1 Effect of biochar on soil properties and its role in environmental remediation.

14.1.3 Biochar for Improvement of Soil Chemical Properties

The rapid decline in soil organic matter is a major concern for poor soil health and a decline in nutrient supplying capacity of soil in many agroecological regions. Faulty soil management with little use of organic manure results in the enhanced degradation of soil organic matter and the emission of increased quantities of CO_2. Many researchers have advocated regular incorporation of crops and animal wastes to increase the soil organic carbon content (Lou et al., 2011), but under tropical climate conditions, it is difficult owing to the increased decomposition rate of organic matter. By virtue of the recalcitrant nature of carbon present in biochar, soil application of biochar helps in isolating carbon in the soil to improve soil organic C content. Biochar has the potential to improve the soil chemical environment as well (Labanya et al., 2022a–c). Several researchers have reported that the application of different biochars brought significant changes in soil quality including an increase in the pH of acidic soils (Abewa et al., 2013; Huang et al., 2019), cation-exchange capacity, organic carbon, exchangeable cations, and extractable N and P (Figure 14.1) (Chan et al., 2008; Abewa et al., 2013; Bera et al., 2016; Glaser et al., 2002; Laird et al., 2010; Fang et al., 2016; Mohan et al., 2018; Agegnehu et al., 2019). The functional groups such as carboxylic, hydroxyl, and phenolic groups present in biochar interact with H^+ and increase soil pH. Carbonates, bicarbonates, and silicates present in biochar also neutralize soil acidity (Ahmad et al., 2012) and relieve Al toxicity prevalent in highly acidic soils.

The long-term effect of biochar application was examined by Tian et al. (2018) who reported that after 3 years of biochar applications, the total soil organic carbon content increased in the 0-40 cm soil layer. The experiment also revealed that biochar applications, years, and their interactions significantly influenced the soil NO_3^--N and NH_4^+-N content. The analysis of variance in the study showed that the addition of biochar, years, and biochar and years interactions had a significant ($p < 0.05$) effect on the soil available K and available P contents both in 0-20 cm and 20-40 cm soil layer. In another study, aged acidic biochar was found to increase nitrogen retention and decrease ammonia volatilization in alkaline bauxite residue sand (Esfandbod et al., 2017b).

14.1.4 Biochar for Improvement of Soil Biological Properties

Soil microorganisms play a key role in nutrient cycling and provide an important ecosystem service. Soil application of biochar induces different effects on soil biota compared to the addition of fresh organic matter, and this may affect the abundance, activity, and diversity of soil biotic communities (Figure 14.1). Biochar creates a more favorable habitat for microbes through improvement in the soil's physical and chemical environment rather than merely serving as a primary source of nutrients (Warnock et al., 2007; Lehmann et al., 2011; Mukherjee et al., 2022).

Smaller pores (<5 mm in diameter) present in biochar (Glaser, 2007) are likely to be accessible to bacteria and fungal hyphae but not to their larger predators such as mites and collembolans or most nematodes and protozoa (Swift et al., 1979; Wright et al., 1995). According to Rutigliano et al. (2014), microbial activity, often expressed as mean substrate-induced respiration, was significantly higher during 3 months following biochar incorporation compared with controls. Their results also revealed that 3 months after biochar treatment the differences of 16% (4/25 respiratory responses) and 20% (5/25 respiratory responses) of the catabolic fingerprints were observed in plots treated with 30 and 60 t biochar/ha, respectively, as compared to control. However, some organic pyrolytic products, such as phenolics and polyphenolics, present in biochar may negatively affect soil microorganisms (Warnock et al., 2010). For this reason, a negative, null, or positive effect of biochar on the soil microbial community has been reported depending on the type of biochar used and the type of soil (Galvez et al., 2012).

14.1.5 Biochar and Heavy Metal Contamination

Inorganic pollutants such as heavy metals pose a serious threat to all life forms, and the remediation of polluted soils is very difficult owing to their non-biodegradable nature. Some toxic heavy metals like cadmium, lead, nickel, mercury, copper, and zinc exist as cations, while others may exist as anions such as arsenic and chromium. These inorganic pollutants often find their way to agricultural soils through inadvertent disposal of industrial sludges/effluents or municipal wastewater (Zhang et al., 2013). The biochar produced by low-temperature pyrolysis, being rich in many functional groups, high organic carbon content, and porosity, are effective in retaining these toxic elements.

The mechanisms of contaminant/metal-biochar retention and release depend on the concentration of soil contaminants and physicochemical characteristics, as well as the stability of the metal-biochar complex, which may also be related to the conditions of the pyrolysis process. The main adsorption mechanism of biochar for the retention of heavy metal pollutants are (i) ion exchange, (ii) co-precipitation, (iii) complexation, and (iv) electrostatic adsorption (Figure 14.1) (Wang et al., 2018). According to Cui et al. (2011), Cd mobility in soil was significantly decreased under biochar application in 2 years of experimentation. The concentration of $CaCl_2$-extractable Cd in soil decreased significantly by 32.0%, 39.2%, and 52.5% in the first year and by 5.3%, 43.4%, and 39.8% in the subsequent year in plots treated with biochar at 10, 20, and 40 t/ha, respectively. This study also revealed that compared to no biochar amendment, total Cd uptake by rice was also significantly decreased. Komkiene and Baltrenaite (2016) noted that at the constant temperature, contact time, pH, and initial concentration of heavy metals, the adsorption of heavy metals increased with the increasing dosage of biochar. Biochar in combination with compost reduced Pb uptake and enhanced the growth of maize in the lead (Pb)-contaminated soil exposed to drought stress (Adejumo et al., 2020). Karami et al. (2011) added biochar to soil polluted by Pb and Cu due to mining activity. It was observed that biochar addition reduced Pb concentrations in the pore water by half as compared to the concentration prevailing in the mine soil. When biochar was combined along with the green waste compost, the levels of Pb concentrations in the pore water were 20 times less than the control.

14.1.6 Remediation of Soils Polluted by Organic Pollutants

Biochar has been shown to adsorb organic pollutants such as insecticides, herbicides, pesticides, and fungicides like imidacloprid (Jin et al., 2016), thiacloprid (Zhang et al., 2018), atrazine, simazine (Zheng et al., 2010), carbofuran (Yu et al., 2009), and different industrial chemicals such as polycyclic aromatic hydrocarbons (Gomez-Eyles et al., 2011), antibiotics (Chen et al., 2018), and drugs like tetracycline (Chen et al., 2018). Biochar material of finer size having a higher specific surface area is more effective for retaining these chemicals through adsorption. Both the adsorption and degradation of these exogenous chemicals were affected by the addition of biochars. Hence, the uptake of pesticide residues in the growing plants is also decreased in biochar-amended soils. The mechanism of removal is directly related to the possible interaction between biochar and pollutants. The principal mechanisms of adsorption of organic pollutants onto biochars are physisorption (electrostatic attraction/repulsion, pore diffusion, H-bonding, and hydrophobic) and chemisorption (electrophilic interaction) processes (Figure 14.1) due to the presence of various O-containing functional groups such as OH, COOH, etc. Other removal mechanisms include chemical transformation, partitioning, and biodegradation (Xiong et al., 2019).

14.2 CONCLUSIONS

Biochar is a good soil amendment with several desirable properties like higher porosity, surface area, and cation/anion exchange capacity which can improve soil health and also increase agricultural production. Biochar also contains essential nutrients which are often deficient or marginal in many agricultural soils and could supply these micronutrients in such a form that face lesser fixation in soil. Biochar, being a powerful adsorbent, can immobilize both inorganic and organic pollutants to reduce their bioavailability in the environment. Besides, biochar is an inexpensive material and has good potential for its use as a soil amendment to maintain good soil health and for the amelioration of soils polluted by both inorganic and organic pollutants.

REFERENCES

Abel, S., Peters, A., Trinks, S., Schonsky, H., Facklam, M., Wessolek, G. 2013. Impact of biochar and hydrochar addition on water retention and water repellency of sandy soil. *Geoderma* 202–203: 183–191.

Abewa, A., Yitaferu, B., Selassie, Y.G., Amare, T.T. 2013. The role of biochar on acid soil reclamation and yield of teff (*Eragrostistef [Zucc] trotter*) in Northwestern Ethiopia. *Journal of Agricultural Science* 6: 1–12.

Adejumo, S.A., Arowo, D.O., Ogundiran, M.B., Srivastava, P. 2020. Biochar in combination with compost reduced Pb uptake and enhanced the growth of maize in lead (Pb)-contaminated soil exposed to drought stress. *Journal of Crop Science and Biotechnology* 23: 273–288. doi: 10.1007/s12892-020-00035-8.

Agegnehu, G., Bass, A.M., Nelson, P.N., Muirhead, B., Wright, G., Bird, M.I. 2019. Biochar and biochar-compost as soil amendments: Effects on peanut yield, soil properties and greenhouse gas emissions in tropical North Queensland, Australia. *Agriculture, Ecosystems & Environment* 213: 72–85.

Ahmad, M., Lee, S.S., Dou, X., Mohan, D., Sung, J., Yang, J.E., Ok, Y.S. 2012. Effects of pyrolysis temperature on soybean stover- and peanut shell-derived biochar properties and TCE adsorption in water. *Bioresource Technology* 118: 536–544.

Amonette, J., Joseph, S. 2009. Characteristics of biochar Micro-chemical properties. In: Lehmann, J., Joseph, S. (Eds.), *Biochar for Environmental Management: Science and Technology Earth Scan*. Earthscan: London, pp. 33–52.

Bera, T., Collins, H.P., Alva, A.K., Purakayastha, T.J., Patra, A.K., 2016. Biochar and manure effluent effects on soil biochemical properties under corn production. *Applied Soil Ecology* 107: 360–367.

Chan, K., van Zwieten, L., Meszaros, I., Downie, A., Joseph, S. 2007. Agronomic values of greenwaste biochar as a soil amendment. *Australian Journal of Soil Research* 45: 629–634.

Chan, K.Y., van Zwieten, L., Meszaros, I., Downie, A., Joseph, S. 2008. Using poultry litter biochars as soil amendments. *Australian Journal of Soil Research* 46: 437–444.

Chen, W., Li, K., Xia, M., Yang, H., Chen, Y., Chen, X., Che, Q., Chen, H. 2018. Catalytic deoxygenation co-pyrolysis of bamboo wastes and microalgae with biochar catalyst. *Energy* 157: 472–482.

Cornelissen, G., Martinsen, V., Shitumbanuma, V., Alling, V., Breedveld, G.D., Rutherford, D.W., Sparrevik, M., Hale, S.E., Obia, S., Mulder, J. 2013. Biochar effect on maize yield and soil characteristics in five conservation farming sites in Zambia. *Agronomy* 3: 256–274.

Cui, L., Li, L., Zhang, A., Pan, G., Bao, D., Chang, A. 2011. Biochar amendment greatly reduces rice Cd uptake in a contaminated paddy soil: A two-year field experiment. *Bioresources* 6, 2605–2618.

Deal, C., Brewer, C.E., Brown, R.C., Okure, M.A.E., Amoding, A. 2012. Comparison of kiln-derived and gasifier-derived biochars as soil amendments in the humid tropics. *Biomass and Bioenergy* 37: 161–168.

Demirbas, A. 2004. Effects of temperature and particle size on biochar yield from pyrolysis of agricultural residues. *Journal of Analytical and Applied Pyrolysis* 72: 243–248.

Esfandbod, M., Merritt, C.R., Rashti, M.R., Singh, B., Boyd, S.E., Srivastava, P., Brown, C.L., Butler, O.M., Kookana, R.S., Chen, C. 2017a. Role of oxygen-containing functional groups in forest fire-generated and pyrolytic chars for immobilization of copper and nickel. *Environmental Pollution* 220: 946–954. doi: 10.1016/j.envpol.2016.10.080.

Esfandbod, M., Phillips, I.R., Miller, B., Rashti, M.R., Lan, Z.M., Srivastava, P., Singh, B., Chen, C.R. 2017b. Aged acidic biochar increases nitrogen retention and decreases ammonia volatilization in alkaline bauxite residue sand. *Ecological Engineering* 98: 157–165. doi: 10.1016/j.ecoleng.2016.10.077.

Fang, B., Lee, X., Zhang, J., Li, Y., Zhang, L., Cheng, J., Wang, B., Cheng, H. 2016. Impacts of straw biochar additions on agricultural soil quality and greenhouse gas fluxes in karst area, Southwest China. *Soil Science and Plant Nutrition* 62: 5–6, 526–533.

Fang, J., Zhan, L., Ok, Y.S., Gao, B. 2018. Mini review of potential applications of hydrochar derived from hydrothermal carbonization of biomass. *Journal of Industrial and Engineering Chemistry* 57: 15–21.

Galvez, A., Sinicco, T., Cayuela, M.L., Mingorance, M.D., Fornasier, F., Mondini, C. 2012. Short term effects of bioenergy by-products on soil C and N dynamics, nutrient availability and biochemical properties. *Agricultural Ecosystem and Environment* 160: 3–14.

Glaser, B. 2007. Prehistorically modified soils of central Amazonia: A model for sustainable agriculture in the twenty-first century. *Philosophical Transactions of the Royal Society B* 362: 187–196.

Glaser, B., Haumaier, L., Guggenberger, G., Zech, W. 1998. Black carbon in soils: The use of benzene carboxylic acids and specific markers. *Organic Geochemistry* 29: 811–819.

Glaser, B., Haumaier, L., Guggenberger, G., Zech, W. 2001. The 'Terra Preta' phenomenon: A model for sustainable agriculture in the humid tropics. *Naturwissenschaften* 88: 37–41.

Glaser, B., Lehmann, J., Zech, W. 2002. Ameliorating physical and chemical properties of highly weathered soils in the tropics with charcoal: A review. *Biology and Fertility of Soils* 35: 219–230.

Gomez-Eyles, J.L., Sizmur, T., Collins, C.D., Hodson, M.E. 2011. Effects of biochar and the earthworm *Eisenia fetida* on the bioavailability of polycyclic aromatic hydrocarbons and potentially toxic elements. *Environmental Pollution* 159: 616–622.

Gul, S., Whalen, J.K., Thomas, B.W., Sachdeva, V., Deng, H. 2015. Physico-chemical properties and microbial responses in biochar-amended soils: Mechanisms and future directions. *Agricultural Ecosystem and Environment* 206: 46–59.

Huang, M., Long, F., Li-geng, J., Shu-ying, Y., Ying-bin, Z., Uphoff, N. 2019. Continuous applications of biochar to rice: Effects on grain yield and yield attributes. *Journal of Integrative Agriculture* 18: 563–570.

Jeffery, S., Verheijen, F.G.A., van der Velde, M., Bastos, A.C. 2011. A quantitative review of the effects of biochar application to soils on crop productivity using meta-analysis. *Agricultural Ecosystem and Environment* 144: 175–187.

Jin, J., Kang, M., Sun, K., Pan, Z., Wu, F., Xing, B. 2016. Properties of biochar-amended soils and their sorption of imidacloprid, isoproturon, and atrazine. *Science of the Total Environment* 550: 504–513.

Karami, N., Clemente, R., Moreno-Jiménez, E., Lepp, N., Beesley, L. 2011. Efficiency of green waste compost and biochar soil amendments for reducing lead and copper mobility and uptake to ryegrass (*Lolium perenne*). *Journal of Hazardous Materials* 191: 41–48.

Khan, M.A., Khan, S., Ding, X., Khan, A., Alam, M. 2019. The effects of biochar and rice husk on adsorption and desorption of cadmium on to soils with different water conditions (upland and saturated). *Chemosphere* 193: 1120–1126.

Komkiene, J., Baltrenaite, E. 2016. Biochar as adsorbent for removal of heavy metal ions [Cadmium(II), Copper(II), Lead(II), Zinc(II)] from aqueous phase. *International Journal of Environmental Science and Technology* 13: 471–482.

Labanya, R., Srivastava, P.C., Pachauri, S.P., Shukla, A.K., Shrivastava, M., Srivastava, P. 2022a. Effect of three plant biomasses and two pyrolysis temperatures on structural characteristics of biochar. *Chemistry and Ecology* 38: 430–450. doi: 10.1080/02757540.2022.2075858.

Labanya, R., Srivastava, P.C., Pachauri, S.P., Shukla, A.K., Shrivastava, M., Srivastava, P. 2022b. Kinetics of micronutrients and S adsorption onto phyto-biochars: Influence of pyrolysis temperatures and properties of phyto-biochars. *Biomass Conversion and Biorefinery*. doi: 10.1007/s13399-022-02733-5.

Labanya, R., Srivastava, P.C., Pachauri, S.P., Shukla, A.K., Shrivastava, M., Srivastava, P. 2022c. Valorisation of phyto-biochars as slow release micronutrients and sulphur carrier for agriculture. *Environmental Technology*. doi: 10.1080/09593330.2022.2029953.

Laird, D.A., Brown, R.C., Amonette, J.E., Lehmann, J. 2009. Review of the pyrolysis platform for coproducing bio-oil and biochar. *Biofuels Bioproducts and Biorefining* 3(5): 547–562.

Laird, D.A., Fleming, P., Davis, D.D., Horton, R., Wang, B., Karlen, D.L. 2010. Impact of biochar amendments on the quality of a typical Midwestern agricultural soil. *Geoderma* 158: 443–449.

Lehmann, J., Rillig, M.C., Thies, J., Masiello, C.A., Hockaday, W.C., Crowley, D. 2011. Biochar effects on soil biota: A review. *Soil Biology and Biochemistry* 43: 1812–1836. doi: 10.1016/j.soilbio.2011.04.022.

Lou, L.P., Wu, B.B., Wang, L.N., Luo, L., Xu, X.H., Hou, J.A., Xun, B., Hu, B.L, Chen, Y.X. 2011. Sorption and ecotoxicity of pentachlorophenol polluted sediment amended with rice-straw derived biochar. *Bioresource Technology* 102: 4036–4041.

Masek, O. 2009. Biochar production technologies. http://www.geos.ed.ac.uk/sccs/biochar/documents/Biochar Launch-OMasek.pdf.

Meyer, S., Glaser, B., Quicker, P. 2011. Technical, economical, and climate-related aspects of biochar production technologies: A literature review. *Environmental Science and Technology* 45: 9473–9483.

Mohan, D., Abhishek, K., Saraswat, A., Patel, M., Singh, P., Pittman, C.U. 2018. Biochar production and applications in soil fertility and carbon sequestration sustainable solution to crop-residue burning in India. *Royal Society of Chemistry* 8: 508–520.

Mukherjee, S., Sarkar, B., Aralappanavar, V.K., Mukhopadhyay, R., Basak, B.B., Srivastava, P., Marchut-Mikołajczyk, O., Bhatnagar, A., Semple, K.T., Bolan, N. 2022. Biochar-microorganism interactions for organic pollutant remediation: Challenges and perspectives. *Environmental Pollution* 308: 119609. doi: 10.1016/j.envpol.2022.119609.

Rutigliano, F.A., Romano, M., Marzaioli, R., Baglivo, I., Baronti, S., Miglietta, S., Castaldi, S. 2014. Effect of biochar addition on soil microbial community in a wheat crop. *European Journal of Soil Biology* 60: 9–15.

Schmidt, M.W.I., Noack, A.G. 2000. Black carbon in soils and sediments: Analysis, distribution, implications, and current challenges. *Global Biogeochemical Cycles* 14: 777–793.

Suliman, W., Harsh, J.B., Abu-Lail, N.I., Fortuna, A.-M., Dallmeyer, I., Garcia-Pérez, M. 2017. The role of biochar porosity and surface functionality in augmenting hydrologic properties of a sandy soil. *Science of the Total Environment* 574: 139–147.

Swift, M.J., Heal, O.W., Anderson, J.M. 1979. *Decomposition in Terrestrial Ecosystems*. University of California Press: Berkeley, CA.

Thomsen, T., Nielsen, H.H., Bruun, E.W., Ahrenfeldt, J. 2011. The potential of pyrolysis technology in climate change mitigation: Influence of process design and parameters, simulated in SuperPro designer software. Riso-R-1764 (EN) National Laboratory for Sustainable Energy Technical University of Denmark. ISBN 978-87-550-3877-6. p 112.

Tian, X., Li, C., Zhang, M., Wan, W., Xies, Z., Chen, B., Li, W. 2018. Biochar derived from corn straw affected availability and distribution of soil nutrients and cotton yield. *PLos One* 13: 2–19.

Wang, S., Xu, Y., Norbu, N., Wang, Z. 2018. Remediation of biochar on heavy metal polluted soils. *IOP Conference Series: Earth and Environmental Science* 108: 042113.

Warnock, D.D., Lehman, J., Kuype, T.W., Rillig, M.C. 2007. Mycorrhizal responses to biochar in soil e concepts and mechanisms. *Plant and Soil* 300: 9–20.

Warnock, D.D., Mummey, D.L., McBride, B., Major, J., Lehmann, J., Rillig, M.C. 2010. Influences of non-herbaceous biochar on arbuscular mycorrhizal fungal abundances in roots and soils: Results from growth-chamber and field experiments. *Applied Soil Ecology* 46: 450–456.

Wright, D.A., Killham, K., Glover, L.A., Prosser, J.I. 1995. Role of pore size location in determining bacterial activity during predation by protozoa in soil. *Applied Environmental Microbiology* 61: 3537–3543.

Xiao, X., Chen, B. 2017. A direct observation of the fine aromatic clusters and molecular structures of biochars. *Environmental Science & Technology* 51: 5473–5482. doi: 10.1021/acs.est.6b06300.

Xiong, W., Zeng, Z., Li, X., Zeng, G., Xiao, R., Yang, Z., H. Xu, Chen, H., Cao, J., Zhou, C., Qin, L. 2019. Ni-doped MIL-53(Fe) nanoparticles for optimized doxycycline removal by using response surface methodology from aqueous solution. *Chemosphere* 232: 186–194.

Yu, X.Y., Ying, G.G., Kookana, R.S. 2009. Reduced plant uptake of pesticides with biochar additions to soil. *Chemosphere* 76: 665–671.

Zhang, M., Gao, B., Varnoosfaderani, S., Hebard, A., Yao, Y., Inyang, M. 2013. Preparation and characterization of a novel magnetic biochar for arsenic removal. *Bioresource Technology* 130: 457–462.

Zhang, P., Sun, H., Min, L., Ren, C. 2018. Biochars change the sorption and degradation of thiacloprid in soil: Insights into chemical and biological mechanisms. *Environmental Pollution* 236: 158–167.

Zheng, W., Guo, M.X., Chow, T., Bennett, D.N., Rajagopalan, N. 2010. Sorption properties of green waste biochar for two triazine pesticides. *Journal of Hazardous Materials* 181: 121–126.

15 Use of Arbuscular Mycorrhizal Fungi in Remediating Soils Constrained by Metal(loid) Contamination

Sebastián Meier[1,2], *Humberto Aponte*[3], *Naser Khan*[4], *Paula Cartes*[5], *Pablo Cornejo*[6], *Jorge Medina*[3], *and Alexander Neaman*[7]

[1]Instituto de Investigaciones Agropecuarias, INIA Carillanca, Temuco, Chile
[2]Escuela de Medicina Veterinaria, Facultad de Medicina y Ciencias de la Salud, Universidad Mayor, Temuco, Chile
[3]Laboratory of Soil Microbial Ecology and Biogeochemistry, Institute of Agri-Food, Animal and Environmental Sciences (ICA3), Universidad de O'Higgins, San Fernando, Chile
[4]Scarce Resources and Circular Economy (ScaRCE), UniSA STEM, University of South Australia, Mawson Lakes, Australia
[5]Scientific and Technological Bioresource Nucleus, BIOREN-UFRO, Universidad de La Frontera, Temuco, Chile
[6]Centro de Investigación en Micorrizas y Sustentabilidad Agroambiental, CIMYSA, Universidad de La Frontera, Temuco, Chile
[7]Instituto de Ingeniería Agraria y Suelos, Facultad de Ciencias Agrarias y Alimentarias, Universidad Austral de Chile, Valdivia, Chile

CONTENTS

DOI: 10.1201/9781003093565-15

15.1 INTRODUCTION

Metals are elements whose specific density is equal to or >5 g/mL or whose atomic number exceeds 20 but do not include elements in groups I and II of the periodic table (Adriano et al., 2004). However, classification based on density, atomic weight, atomic number, or other properties does not define the toxicity of these elements. The word metal(loid) means a chemical element that exhibits some properties of metals and some non-metals (Szefer et al., 1997) and will be the term used in this chapter.

Soils become polluted when toxic substances are present in amounts that exceed background levels and affect normal ecosystem functioning (Adriano et al., 2004). Metal(loid)s in soils at polluting levels may occur naturally or due to increasing industrialization and agricultural activities (Haroon et al., 2019), such as excessive application of agrochemicals, sewage sludge, industrial wastewater, biosolids, and manure (Hakeem et al., 2015). Metal(loid)s accumulation in soils has generated severe environmental pollution (Nagajyoti et al., 2010; Ramesh Kumar and Anbazhagan, 2018).

The introduction, dispersion, and migration of metal(loid)s into the environment usually cause severe health problems for plants, animals, and humans (Okereafor et al., 2020). Thus, recovery and remediation of polluted sites are required to prevent the negative impacts on ecosystems and human health. Historically, soil remediation techniques involved transferring contaminated soil to landfills, a practice now considered environmentally disruptive and economically unfeasible (Chibuike and Obiora, 2014). Modern remediation approaches increasingly focus on *in situ* environmentally friendly techniques, such as phytoremediation, often primed by the addition of soil amendments and the use of microorganisms, such as arbuscular mycorrhizal fungi (AMF), that could enhance and accelerate plant growth in metal(loid) polluted sites (Meier et al., 2012b). This chapter covers the contamination of soils with metal(loid)s and microbial-assisted phytoremediation, primarily through the symbiosis with AMF.

15.2 SOILS POLLUTED WITH METAL(LOID)S AND THEIR GLOBAL DISTRIBUTION

There are 5 million sites worldwide, covering a land area of ~500 million ha, contaminated with metal(loid)s at higher concentrations than the geo-baselines or regulatory levels (Liu et al., 2018). In soils, Metal(loid)s may occur naturally from lithogenic (geogenic) or anthropogenic sources. The lithogenic pools are derived from geological weathering of soil parent materials, including igneous, sedimentary rocks, and coals (Bolan et al., 2014). Several reports have indicated that, compared to anthropogenic activities, natural pools of metal(loid)s in the environment have less impact (Dixit et al., 2015). The main pool of metal(loid)s in the environment is derived from anthropogenic activities (Figure 15.1), primarily associated with industrial processes (Chibuike and Obiora, 2014). Over recent decades, the annual worldwide release of metal(loid)s derived from anthropogenic sources has reached 22,000 t for Cd; 939,000 t for Cu; 783,000 t for Pb; and 35,000 t for Zn (Thambavani and Prathipa, 2012).

In Europe, the most polluted regions with metal(loid)s include southwestern Germany, southern Poland, southern Spain, and northern Italy (Gil et al., 2018; Suschka, 1993; Yewale et al., 2017). In many African countries, such as Botswana, Burkina Faso, and Cameroon, drinking water concentration exceeds the maximum acceptable limit (0.01 mg/L, World Health Organization, 2014). In Mexico, As contamination in water is also predominant, with a high concentration beyond the maximum permissible limit of 25 μg/L (Hurtado-Jiménez and Gardea-Torresdey, 2006). In the central north of Chile, high contents of Cu, As, and Sb in soils have increased due to mining activities and erosion of As-rich epithermal ores (Aponte et al., 2021; De Gregori et al., 2003; González et al., 2014; Tapia-Gatica et al., 2020). Moreover, high soil metal(loid) contents have been found at several sites in Asia and Australia (Yewale et al., 2017), such as China and India.

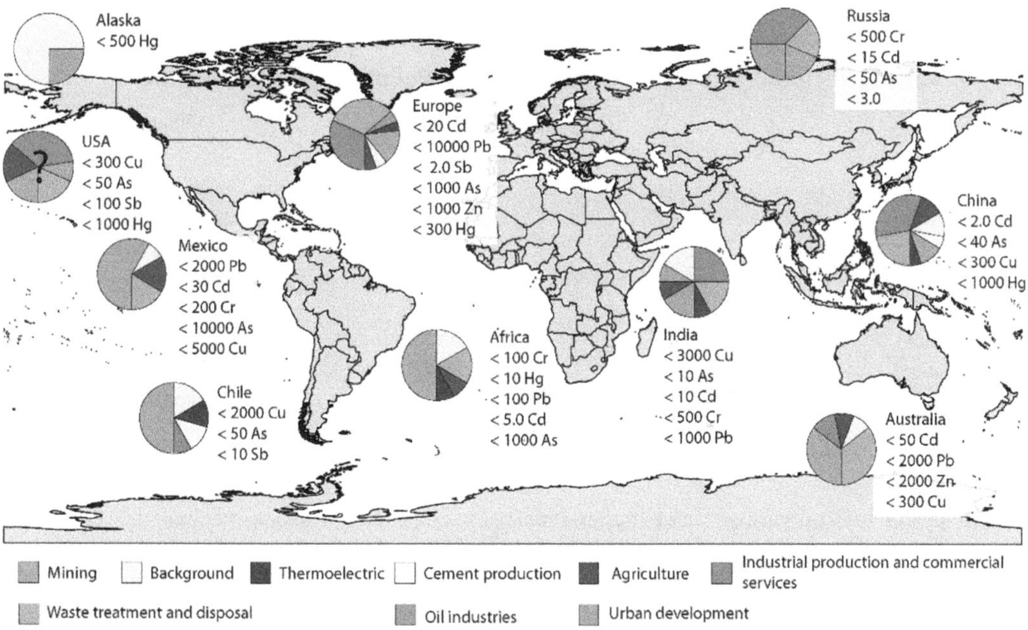

FIGURE 15.1 Charts presenting the total metal(loid) concentration (mg kg^{-1} soil) in contaminated areas of various continents and some countries. (The data are based on scientific literature and or official reports. Data from the USA were not available. Aponte et al. (2020), with permission from John Wiley & Sons and Copyright Clearance Center.)

In these locations, considerably high Pb, Cd, and Hg concentrations have also been found in the atmosphere (Hou et al., 2020). Thus, metal(loid)s are a global problem causing deleterious effects on ecosystems.

Metal(loid)s at a toxic level can cause irreversible soil degradation by modifying physical, chemical, and biological properties. This can eventually limit plant establishment for agriculture, while toxicity affects human health throughout the food chain (Jaishankar et al., 2014). Therefore, remediation of metal(loid) contaminated soils is needed to improve land resources for agricultural production, enhance food security, scale down land tenure problems arising from changes in land-use patterns, and reduce risks for human health (Wuana and Okieimen, 2011).

15.3 PHYTOREMEDIATION OF METAL(LOID) POLLUTED SOILS

Several conventional approaches to reclaiming metal(loid) contaminated sites are usually based on physicochemical techniques, including soil washing, electric field application (electrokinetics), excavation and reburial of contaminated matrices, and pumping and treating systems for polluted water (Dhaliwal et al., 2020). These approaches suffer from two main disadvantages: (i) high cost and (ii) inefficiency if metal(loid)s are present at low concentrations that occur commonly (Wan et al., 2016) (Table 15.1). Moreover, "brute force" approaches significantly affect soils' physicochemical and biological properties (Ashraf et al., 2019). Thus, effective technologies that are environmentally sensitive and also economical will be appreciated by land managers.

Phytoremediation is considered an economical and environmentally friendly alternative where plants detoxify a growing media by extracting contaminants, especially from soils (Mulligan et al., 2001, Table 15.1). Unlike organic contaminants, metal(loid)s cannot be degraded by biological or physical processes (Bolan et al., 2014); however, they can be immobilized or sequestered by biota,

TABLE 15.1
Cost of Various Technologies for Remediating Metal-Polluted Soils and Sediments.

Method	Cost (US $/ton)	Cost (US $/ton)	Other Factors to Be Considered
Containment	----	20–66	Transport/monitoring
Vitrification	75–425	90–870	Long-term monitoring
Land filling	100–500	----	Transport/excavation/ monitoring
Chemical treatment	100–500	60–290	Recycling of contaminants
Electrokinetics	20–200	70–170	Monitoring
Bioremediation	----	15–200	Long-term monitoring
Phytoremediation	5–4037.7/m^3	----	Long-term monitoring
References	Glass (1999) and Wan et al. (2016)	Mulligan et al. (2001)	

including soil microorganisms and higher plants, by employing various mechanisms (Jing and Kjellerup, 2018). Depending on the nature of the pollutant, soil conditions, plants, and rhizospheric microorganisms can be used for phytoremediation in different ways (Table 15.2).

a. *Phytostabilization or phytoimmobilization*: This technique reduces the bioavailability and dispersion of metal(loid)s in soils through different root/rhizosphere retention mechanisms (Alcalá and Clemente, 2020). Phytostabilization also refers to a soil vegetation cover to avoid direct interaction with humans or animals or metal(loid) dispersion by wind or water erosion. Therefore, phytostabilization reduces the possible risks of offsite contamination through leaching to groundwater or dissemination through rain and wind (Cristaldi et al., 2017). However, phytostabilization is a temporary management strategy because metal(loid) contaminants continue to persist in soils indefinitely (Ashraf et al., 2019).
b. *Phytoextraction*: It uses fast-growing plants to remove metal(loid)s from soils. Harvested plant biomass can be used for biogas production and combusted. In combustion, ashes can be processed for metal recovery, fixed in bricks, or disposed of in deserted lands. This technique is also known as phytomining or biomining (Ashraf et al., 2019).
c. *Phytovolatilization:* Phytovolatilization is the process that absorbs contaminants from soils, transports them via the xylem, converts them into less toxic and volatile forms, and releases them into the atmosphere. Phytovolatilization has been widely used to remove metals like Hg and Se, as these metals have high volatility (Cristaldi et al., 2017). However, it does not apply to most metal-polluted soils (Meier et al., 2012b).
d. *Phytofiltration*: It is the use of plant roots (rhizofiltration), seedlings (blastofiltration), or plant shoots (caulofiltration) to absorb/adsorb pollutants for reclamation of surface water, groundwater, and wastewater with a low level of contaminants (Prasad and De Oliveira Freitas, 2003). Plant roots or seedlings grown in aerated water absorb, precipitate, and concentrate toxic metal(loid)s from polluted effluents (Korjus, 2014). When roots become saturated, they are harvested (Korjus, 2014).
e. *Rhizodegradation or phytostimulation*: It is the process where plants facilitate the biodegradation of organic pollutants by microorganisms present in their rhizosphere via their enzymatic activities or by producing root exudates such as amino acids, carbohydrates, and flavonoids (Pilon-Smits, 2005).

Plants used for phytoremediation must be genetically capable of growing in soils with a high metal(loid) content and should have the ability to accumulate these elements in shoots and/or roots (Marchiol et al., 2004). Those plants are known as metallophytes (Awa and Hadibarata, 2020). Three types

TABLE 15.2
The Various Aspects of the Available Remediation Techniques for Metal(loid) Contaminated Soils

Type of Method	Method	Description	Applicability	Advantage	Limitations
Physical remediation	Soil replacement and/or isolation	Excavating contaminated soil and replacing it with non-contaminated soil. Isolating contaminated soil from uncontaminated soil using a barrier	Small-scale, short to long term	Effectively isolate and remove metal(loid) through the exclusion of contaminated layers of soil	Laborious, time-consuming, and not an economically viable method
	Vitrification	Reduction of metal(loid) availability by physically solidifying soil at high temperature	In situ and ex-situ, small-scale but long term	A highly efficient method, easy to apply to a variety of contaminants	High cost due to energy requirements. Limited to small area/volume
	Electrokinetics	Removal of metal(loid) from the soil via electrophoresis, electro-migration, or electro-osmotic flow by applying an electric field gradient	In situ, small-scale but long term	Minimal soil disturbance. Beneficial for low permeability soils with low cost and easy installation	Time-consuming, low efficiency, unable to control pH
Chemical remediation	Immobilization/ stabilization	Reduction of metal(loid) availability by immobilizing it using soil amendments that form stable complexes via adsorption	In situ large scale in less to medium term	Affordable and easy to implement	Temporary remediation depending on the type of soil, metal(loid), and soil amendment used
	Soil washing	Removal of metal(loid) by mechanical separation and chemical extraction	Ex situ moderate to high contamination	Cost-effective, fast, and complete removal of metal(loid)s	Extreme soil disturbance as washing extractants may cause environmental issues
Biological remediation	Phytoextraction	Use of metallophytes or hyperaccumulator plants to uptake, translocate and accumulate metal(loid) in harvestable plant parts	Large-scale and long term	Economical and less disruptive	The time required may be decades to hundreds of years (depending on the degree of contamination)
	Phytostabilization	Reduce if mobility and bioavailability of metal(loid)s in the environment using plants, thereby preventing their migration to groundwater or their entry into the food chain	Large-scale and long term	Economical and less disruptive	Temporary solution because metals are not removed, and effectiveness varies with soil, plant, and metal(loid)
	Phytofiltration	Absorb or adsorb metal(loid)s in plant roots (and also organic pollutants) from water and aqueous waste streams	Small-to-medium scale and long term		
	Phytovolatilization	Volatilize metal(loid)s or metabolites using plants	Small-to-medium scale and long term	Economical and less disruptive	Restricted to volatile metals; may cause other environmental issues; no control after metal released to the atmosphere
	Microbial-assisted phytoremediation	Enhance the metal(loid) phytoextraction capacity of plants by using microorganisms	Small-to-medium scale and long term	Economical and quick remediation that enhances plant growth and metal uptake and translocation	Difficulties for inoculation depending on microorganism, soil, plant, and metal(loid) type
	Soil amendment-assisted phytoremediation	Promote metal(loid) immobilization and plant establishment using organic and inorganic amendments	Small-to-medium scale and long term		

of metallophytes exist metal indicators, metal excluders, and metal hyperaccumulators (Pignattelli et al., 2012). Plants that take up a high concentration of metal(loid)s from the soil and accumulate them in the shoots and leaves are known as metal indicators (Awa and Hadibarata, 2020). Metal excluders are plants that interrupt the translocation of the absorbed metal(loid)s to their aboveground tissues but accumulate the metal(loid) in their roots (Fatnassi et al., 2015). Plants that accumulate high metal(loid) concentrations are hyperaccumulators (Memon and Schröder, 2009). The aerial biomass of hyperaccumulators may contain more than 0.01% (w/w) of Cd and Se; 0.1% of Ni, Co, Cu, and Cr; or 1% of Zn and Mn, irrespective of the metal concentration in soil (Whiting et al., 2004).

Nevertheless, phytoremediation is a prolonged process that requires several decades to reduce metal concentrations in the soil to harmless levels for humans and other organisms (Neaman et al., 2020). The above is related to the limited growth and biomass production of metallophyte and hyperaccumulator plants (Sugawara and Nikaido, 2014) plus to the lack of knowledge about the interactions among plants and microbial soil communities, which could limit metal availability and also accelerate plant growth (Meier et al., 2012b).

15.4 MICROBIAL-ASSISTED PHYTOREMEDIATION OF METAL(LOID) POLLUTED SOILS

Soil microorganisms involve diverse biochemical processes, such as soil formation, energy transfer, and nutrient cycling (Blagodatskaya and Kuzyakov, 2013). These processes promote or accelerate the establishment of a plant cover and the recovery of contaminated ecosystems (Moynahan et al., 2002); however, high metal(loid) concentrations are toxic to soil microorganisms, causing a reduction in biomass, activity, and diversity (Aponte et al., 2020; 2021; Giller et al., 1998; Niemeyer et al., 2012). Changes in the structure and functionality of microbial communities may eventually affect the whole soil-plant-microbe continuum (Ramsey et al., 2005) and limit plant establishment and development. Nevertheless, continuous exposure to high metal(loid) concentrations can induce microbial tolerance and promote the development of some specialized microbial populations (De La Iglesia et al., 2006), which can transform metal(loid)s by changing their chemical and physical properties (Dhaliwal et al., 2020).

Bacteria have developed several mechanisms for their survival under a metal(loid) stressed environment (Figure 15.2), by which they can immobilize or transform metal(loid)s to chemical species of lower/higher solubility for their removal (Choudhary et al., 2017; Hou et al., 2020; Iravani and Varma, 2020; Nies, 1999). Such mechanisms include (i) bioaccumulation of metal(loid) in the intracellular space; (ii) bioassimilation of metal(loid) via iron-assimilation pathways, e.g., formation of

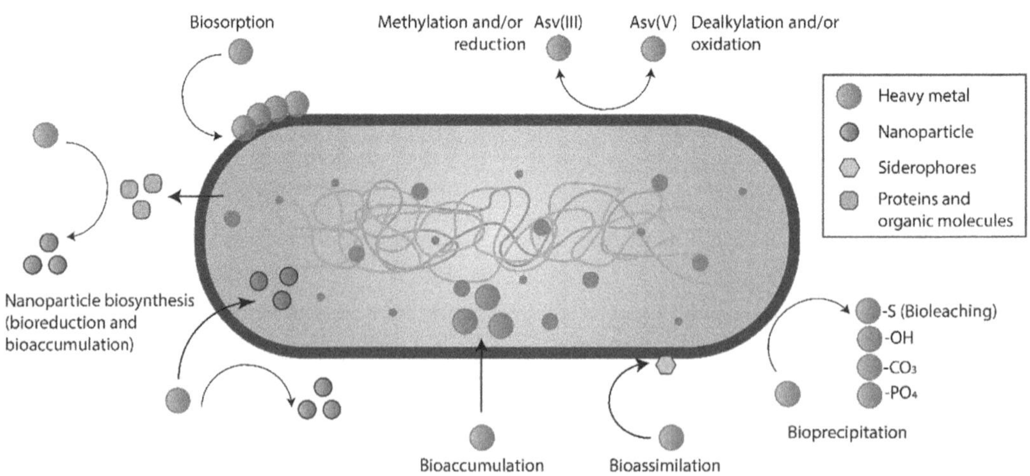

FIGURE 15.2 The mechanisms involved in bacterial remediation. (Modified from Hou et al. (2020).)

metal(loid) complexes with metal-binding proteins such as metallothioneins; (iii) bioprecipitation of metal(loid)s by the formation of low-solubility sulfides; (iv) methylation-demethylation; (v) biosorption and metal(loid) accumulation on the cell surface (extracellular cell-wall binding); and (vi) bioreduction of metal(loid)s via enzymatic reactions and interactions with peptides/proteins resulting in less toxic metal(loid) nanoparticles or lower toxic chemical species. In addition, bacteria are essential in bioremediation due to their size, ability to grow under controlled conditions, ubiquity, and resilience to environmental conditions (Srivastava et al., 2015).

A clear example of bacteria used in the bioremediation of metal(loid) contaminated soils is the use of plant growth-promoting rhizobacteria (PGPR), which are usually found in the rhizospheres of plants (Chibuike and Obiora, 2014). PGPR can stimulate plant growth through several mechanisms, such as (i) phytohormone production and nutrient supply (Glick et al., 1995); (ii) siderophore production and other chelating agents (Kamnev and Van Der Lelie, 2000); (iii) specific enzyme activity and N_2 fixation (Khan, 2005); (iv) reduction in ethylene production (Glick et al., 1998), among others. Moreover, PGPR have been shown to increase mycorrhizal efficiency, decrease metal accumulation, and increase plant growth on metal(loid) contaminated soils (Vivas et al., 2006). For instance, hyperaccumulator plants inoculated with *Bacillus* sp. have shown higher accumulation of metal(loid)s such as Cd and Ni (Sheng and Xia, 2006). Therefore, bacteria can enhance phytoremediation by directly affecting metal(loid)s and plants.

Regarding fungi, bioremediation by them of metal(loid) contaminated soils can be associated with extracellular and intracellular sequestration by metal(loid) chelation, cell-wall binding, and binding to proteins or other ligands to metal(loid)-sensitive cellular targets (Choudhary et al., 2017). The fungal cell wall is mainly composed of polysaccharides and chitin, which can act as a barrier against metal(loid) ions and other solutes and control their uptake into the cell (Ahalya et al., 2003). Moreover, free amino acids in the cell wall and hydroxylic, carboxylic, and other functional groups confer a negative charge to the structure, allowing it to bind to ionic elements, including most metals in soil (Meier et al., 2012b). For instance, some filamentous fungi like *Rhizopus arrhizus*, *Aspergillus niger*, *Mucor rouxxi*, and *Phanerochaete chryosporium* have a large capacity for adsorbing metal ions such as Cu^{2+}, Co^{2+}, Cd^{2+}, Zn^{2+}, Mg^{2+}, Ni^{2+}, and Pb^{2+}. Thus, they can serve as sorbents in bioremediation processes (Bhainsa and D'Souza, 2008). Dead fungal biomass of some fungi can convert metal(loid)s into a less toxic or non-toxic form (Park et al., 2005). A detailed list of bacteria and fungi used in the biological remediation of soil contaminated with metal(loid)s can be found in Dhaliwal et al. (2020).

Microbial bioremediation and microbial-assisted phytoremediation can use a vast number of microorganisms that can be inoculated to remediate metal(loid) contaminated soils; however, classical methods for microbial extraction are minimal since <1% of microorganisms can be cultured in the laboratory (Toju et al., 2018). Therefore, the recruitment and further inoculation of native microbiomes involved in plant fitness could improve plant establishment and development in metal(loid) contaminated soils. For this, a native microbiome with specialized traits can be recruited by several generations to promote plant alleviation under different stresses, which is known as the "*Cry for Help*" phenomenon (Bakker et al., 2018; Rolfe et al., 2019). In this context, AMF takes advantage of their cosmopolitan distribution and association with plants, high potential to remove metal(loid), and culture systems that allow the elaboration of inoculum with all the microbial community (including other microorganisms such as bacteria and non-mycorrhizal fungi). For the above, managing soil microorganisms in phytoremediation must include those forming symbiotic associations, like mycorrhizal fungi, as a prerequisite for any soil restoration program to succeed (Meier et al., 2015). In this respect, several studies have shown that ectomycorrhizal and ericoid mycorrhizal fungi play an essential role in the phytoremediation of metal(loid) contaminated sites (Agerer, 2001; Gadd, 1994). However, the most prominent symbiotic fungi for potential use in phytoremediation are the AMF due to their ubiquity in soil environments and because they have developed several strategies that allow the plant to tolerate high soil metal(loid) concentrations (Cornejo et al., 2013; Janoušková et al., 2006; Kothe and Turnau, 2018; Meier et al., 2012b; Ouziad et al., 2005).

15.5 ARBUSCULAR MYCORRHIZAL FUNGI IN METAL(LOID)-POLLUTED SOILS

The AMF is the primary rhizosphere organism that forms mutualistic associations with most higher plants (Smith and Read, 2008), forming the so-called arbuscular mycorrhizal symbiosis (Barea et al., 1987). The AMF symbiosis contributes to the establishment and growth of plants, especially under adverse conditions, such as in arid and low-fertility soils (Akbar Karimi, 2011), as well as in sites that are highly contaminated by metal(loid)s (Cornejo et al., 2013; Ma et al., 2019; Meier et al., 2012b; Yang et al., 2015). This is because AMF helps plants with nutrient acquisition, especially P (Bolan, 1991), as well as other nutrients such as nitrogen (Frey and Schüepp, 1993), calcium, sulfur (Bauch, 2000), potassium (Garcia and Zimmermann, 2014), and zinc (Audet and Charest, 2007). In addition, AMF could improve soil structure by increasing soil particle aggregation and preventing wind and water erosion through glomalin production (Rillig and Mummey, 2006). These functions contribute to the promotion of revegetation in ecosystems degraded by high metal(loid) concentrations (Hildebrandt et al., 2007; Leung et al., 2007; Meier et al., 2012c).

15.5.1 Adaptive Strategies Developed by Arbuscular Mycorrhizal Fungi and Plants to Alleviate Metal(loid) Stress

There are many similarities between the adaptive strategies developed by plants and AMF to cope with high concentrations of metal(loid)s because of the strict biotrophy of AMF (Barea et al., 2002). The main strategies used by AMF and plants are summarized in Figure 15.3: (i) metal(loid) dilution in plant tissues due to enhanced plant growth promoted mainly by AMF (Christie et al., 2004);

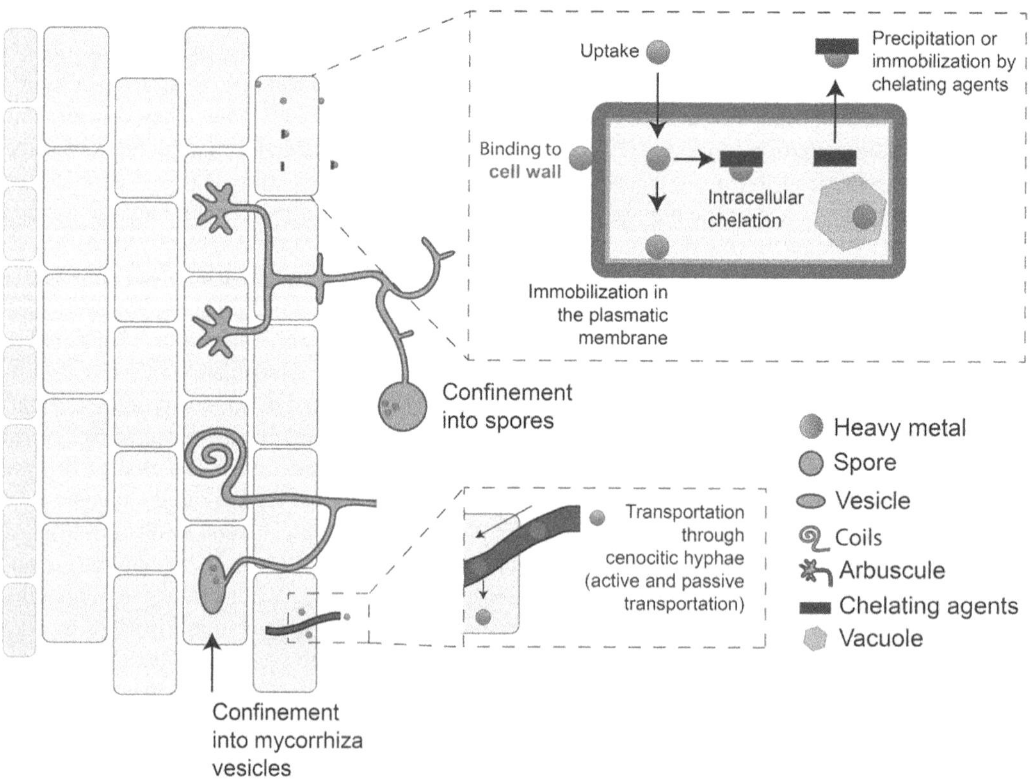

FIGURE 15.3 Adaptive strategies developed by arbuscular mycorrhizal fungi enhance the tolerance of plants to cope with metal(loid)s. (Modified from Meier et al. (2012b).)

(ii) exclusion from uptake through precipitation, binding, or immobilization by chelating substances secreted to soil (e.g., low molecular weight organic acid exudation by plants; Meier et al., 2012a); (iii) binding of metals to biopolymers in the cell wall, such as chitin and glomalin (González-Chávez et al., 2004); (iv) surface-immobilization in the plasmatic membrane once it crosses the cell wall (Ernst et al., 1992); (v) membrane transporters that mobilize metals from the soil to the cytosol including chelation through metallothionein and phytochelatins (MT, González-Guerrero et al., 2005); (vi) confinement of metal(loid)s in the vacuoles of plants and fungal cells (González-Guerrero et al., 2008); (vii) transportation through coenocytic hyphae (González-Guerrero et al., 2008); (viii) exportation by the fungus, and access into the plant cells, involving both active and passive transport by the fungus up to the root cell (Göhre and Paszkowski, 2006); and (ix) confinement of metal(oid)s into spores (Cornejo et al., 2013).

15.6 ROLE OF ARBUSCULAR MYCORRHIZAL FUNGI IN PROMOTING PHYTOREMEDIATION OF METAL(LOID) POLLUTED SOILS

15.6.1 Role Developed by AMF in Enhancing Metal(loid) Immobilization on Soils

Phytostabilization can reduce the spread of metal(loid)s in soils by decreasing their leaching and consequent contamination of aquifers and preventing dust dispersal through wind and water erosion (Göhre and Paszkowski, 2006). The AMF contributes to metal(loid) immobilization beyond the root area, creating an extensive influence zone known as the 'mycorrhizosphere' (Kothe and Turnau, 2018). In this zone, the fungus employs strategies to immobilize metal(loid)s similar to those developed by their host (Figure 15.3). Among them is the retention of metal(loid)s by fungal structures or by compounds secreted by the fungus and/or the chelation of metal(loid)s inside the fungus through intracellular homeostasis mechanisms (Figure 15.3 and Table 15.3) (Gaur and Adholeya, 2000).

15.6.1.1 Immobilization of Metal(loid)s by Fungal Structures

a. Metal(loid) binding due to glomalin production

Glomalin (*Bradford-related soil protein*, BRSP) is a glycoprotein produced apparently by AMF, which has diverse functions in soils (Gadkar and Rillig, 2006; Gao et al., 2019). Besides promoting particle aggregation (Cornejo et al., 2017), BRSP may have a physiological role by decreasing the palatability of AMF to soil microorganisms (Purin and Rillig, 2007). In addition, several studies confirmed that BRSP could immobilize metal(loid)s through functional groups in its structure (Cornejo et al., 2017; He et al., 2020; Vodnik et al., 2007). González-Chávez et al. (2004), using soils polluted with Cu, Cd, and Pb, showed that BRSP could immobilize up to 4300 mg of Cu, 1120 mg of Pb, and 80 mg of Cd per kg BRSP. Vodnik et al. (2007) reported that Pb-immobilization ranged from 690 to 23400 mg Pb/kg, corresponding to 0.8%–15.5% of the total Pb in the soil. Cornejo et al. (2008) observed a high correlation between the BRSP and the immobilization of Cu and Zn (r=0.89 and 0.76 for Cu and Zn, respectively, P <0.001); high Cu-binding to BRSP was later confirmed, and it represented up to 22% of the total Cu in that soil (Cornejo et al., 2017). He et al. (2020) showed that AMF increased the content of BRSP, which promoted soil aggregation and reduced Cd concentrations flowing in the soil, thus preventing Cd leaching.

Finally, a strong correlation between BRSP content and Al immobilization was found in agricultural soils (Seguel et al., 2017, 2013). Since there is a correlation between the amount of glomalin in the soil and the amount of metal(loid) bound, fungal strains with a significant secretion of glomalin should be more suitable for phytostabilization programs.

b. Immobilization of metal(loid)s by cell walls and hyphae

It is well-known that roots colonized by AMF can retain metal(loid)s in the cell walls of the hyphae and roots (Kaldorf et al., 1999). The immobilization of trace elements by fungal mycelium is likely the primary protection mechanism conferred to plants by AMF that

TABLE 15.3
List of Beneficial Effects of Arbuscular Mycorrhizal Fungi (AMF) Promoting Metal(loid) Immobilization (Studies Since 2011 to Date)

AMF	Metal	Host Plant	Beneficial Effect	References
Rhizophagus irregularis	Cr	*Daucus carota* (transformed roots)	Intraradical fungal structures and cell walls in mycorrhizal roots reduced Cr(VI) to Cr(III) mainly by carboxylic ligands and restrained Cr translocation to plant	Wu et al. (2016)
		Taraxacum platypecidum	AM fungi could immobilize Cr in mycorrhizal roots, besides enhancing plant P acquisition	Zhang et al. (2021)
	Pb	*Medicago truncculata*	AMF inoculation improved plant growth, increasing Pb immobilization in cell walls by increasing polysaccharide content	Wu et al. (2016)
	Cu	*Daucus carota* (transformed roots)	AMF accumulates Cu into spores, preventing the entrance of this metal to roots	Cornejo et al. (2013)
Funneliformis mosseae	Cd and Zn	*Helianthus annuus*	AMF-inoculated plants showed an ability to immobilize Cd and Zn concentrations in shoots and roots, decreasing the bioconcentration factor	Hassan et al. (2013)
Rhizophagus intraradices or Funneliformis mosseae	Cu	*Capsicum annuum*	Inoculation with AMF increased Cu tolerance, plant growth, stability of membranes, and photosynthetic activity	Ruscitti et al. (2017)
	Pb	*Robinia pseudoacacia* L.	AMF symbiosis decreased Pb concentrations in leaves and promoted the accumulation of biomass as well as photosynthetic pigment contents	Yang et al. (2015)
Claroideoglomus claroideum	Ni	*Helianthus annuus*	AMF improved plant growth, chlorophyll content, and physiological status (electrolyte leakage, proline, and malondialdehyde contents)	Ma et al. (2019)
	Cu	*Oenothera picensis*	AMF promoted plant establishment and enhanced the accumulation of Cu on roots tissues	Meier et al. (2015)
Funneliformis mosseae and *Diversispora spurcum*	S, Pb, Zn, Cd, As	*Zea mays*	AMF alleviated heavy metal toxicity in plants by increasing antioxidant defense in leaves and P and S uptake and heavy metal retention in roots of maize grown in heavy metal-polluted soils	Zhan et al. (2018)
Claroideoglomus claroideum and Funneliformis mosseae	Pb and Cd	*Calendula officinalis* L.	AM fungi modulated the metabolic activity of the whole plant, increasing the antioxidant capacity in front to high Pb and Cd concentrations	Hristozkova et al. (2016)
Glomus sp., *Acaulospora* and *Scutellospora*	Fe	*Sorghum bicolor*	AMF increased siderophores production and promoted Fe accumulation on plant roots	Mishra et al. (2016)
Glomus constrictum and *Glomus mosseae*	Cd	*Zea mays*	AMF decreased Cd translocation from roots to shoots in inoculated maize plants, indicating a potential stress tolerance mechanism in Cd-contaminated soils	Liu et al. (2014)

reduces the transference of metal(loid)s from soil to plants, promoting phytostabilization. The extraradical mycelium (ERM) has a larger capacity for sorbing metal(loid)s than root cells (Joner et al., 2007). In comparing the active sorption of Cd and Zn by several *Glomus* species, they could sorb 0.5 mg Cd/g, which was ten times higher than that observed with other fungal bio-sorbents such as *Rhizopus arrhizus* (Zhou, 1999). Similarly, González-Chávez et al. (2004) concluded that the ERM could sorb and accumulate high levels of Cu (3–14 mg Cu/g dry hyphae). In the tobacco roots (*Nicotiana tabacum*), Janoušková et al. (2006) found that the ERM can accumulate 10–20 times more Cd per unit of biomass than the non-mycorrhizal plants. The high metal sorption capacity might be related to free amino acids in the fungal cell wall and negatively charged hydroxyl, carboxyl, and other functional groups capable of binding metallic ions, such as Cu^{2+} and Zn^{+2} (He et al., 2020).

15.6.1.2 Intracellular Chelating Agents

The AMF have developed diverse strategies to immobilize metal(loid)s at the intracellular level (Ferrol et al., 2009). Under high metal(loid) concentrations, some elements may cross the cell wall and reach the fungal cytoplasm. In this case, chelating agents in the cytoplasm can deactivate metals and minimize cell damage due to metal intrusion (González-Guerrero et al., 2007).

Metallothioneins (MTs) are polypeptides consisting of 70–80 amino acids with a high content of cysteine, an amino acid capable of forming stable complexes with cations through its sulfhydryl groups (Kojima, 1991). MTs coordinate and sequester trace elements such as Cu, Zn, and Cd (da Silva Folli-Pereira et al., 2012). Different studies have shown that MTs have a complex role in trace element homeostasis and protection against oxidative stress caused by trace elements (Maret, 2003). Three MTs have been identified in mycorrhizal fungi: (i) *GrosMT1* in *Gigaspora rosea* (Stommel et al., 2001), (ii) *GmarMT1* in *Gigaspora margarita* (BEG 34) (Bergero et al., 2007; Lanfranco et al., 2005), and (iv) *GinMT1* (González-Guerrero et al., 2007).

15.6.2 Role Developed by AMF in Promoting Metal(loid) Translocation to Plants

In general, some biophysical determinations regarding metal(loid) concentrations in roots indicated that AMF-colonized plants contained considerably fewer metal(loid)s in their roots than uninoculated plants (Janeeshma and Puthur, 2020). Moreover, the uptake of significant amounts of metal(loid) fractions into cell compartments of both symbiotic partners might be prevented due to the action of the AMF, widely supporting the idea that most of the metal(loid)s are transferred to the cell walls and the vacuoles of the fungi (Cornejo et al., 2013). Meanwhile, the metal(loid)s that reaches the cytoplasm of the plant cells is deposited mainly into the inner root cortical cells where the fungal structures reside, with no translocation of these elements to shoots (Joner et al., 2007, 2000). However, some studies have highlighted the role of AMF in promoting metal(loid) translocation to the shoots (Khan et al., 2000; Davies et al., 2001; Janeeshma and Puthur, 2020). Some of them are detailed in Table 15.4. In this sense, despite the many studies carried out during the past two decades, the role of mycorrhiza in promoting metal(loid) mobilization in plants remains uncertain. Tonin et al. (2001) showed that AMF could enhance metal transfer from soil to plant (translocation factor), thus promoting phytoextraction. The above is supported by other studies (Hassan et al., 2013; He et al., 2020; Singh et al., 2019). However, Weissenhorn et al. (1993) and Meier et al. (2012b) indicated that high metal accumulation resulting from AMF colonization could inhibit plant growth and development.

There are several mechanisms proposed to explain the role of AMF on meta(loid) contaminated soils; most of them are summarized in Meier et al. (2012b). However, these mechanisms are not entirely understood yet and are still debated. On one side, some authors affirmed that AM fungi mainly promote the binding of metals on soils (Christie et al., 2004; Janoušková et al., 2006; Tiwari et al., 2020). In contrast, other studies have shown that AM increases the metal translocation factor, thus promoting phytoextraction (Singh et al., 2019; Trotta et al., 2006). There is also a model

TABLE 15.4
List of Beneficial Effects of Arbuscular Mycorrhizal Fungi (AMF) in Promoting Metal(loid) Translocation to Plant (Studies Since 2011 to Date)

AMF	Metal	Host Plant	Beneficial Effect	References
Rhizophagus irregularis	Cd and Zn	*Helianthus annuus*	AMF-inoculated plants increased Cd translocation from roots to shoots	Hassan et al., (2013)
Funneliformis mosseae	Cd	*Zea mays*	AMF increased glomalin production, preventing Cd leaching and promoting the uptake of Cd by plants	He et al. (2020)
Rhizophagus fasciculatus, *Rhizophagus intraradices*, *Funneliformis mosseae*, and *Glomus aggregatum*	Multi-metal contaminated soil	*Zea mays*	AMF inoculation increased plant growth, enhanced antioxidant compounds production (dehydrogenase, β-Glucosidase, acid, and alkaline phosphatase), and promoted bioaccumulation factor as well roots to shoots (translocation factor) of plant	Singh et al. (2019)
Funneliformis mosseae and *Rhizophagus intraradices*	Pb and Cd	*Calendula officinalis* L.	AMF increased roots and shoot accumulation of Pb and Cd	Tabrizi et al. (2015)
Glomus *versiforme*	Cd	Solanum nigrum	The presence of *G. versiforme* increased DTPA-extractable (phytoavailable) Cd concentrations in soils and promoted a higher Cd concentration in plant tissues	Liu et al. (2015)

proposed by Audet and Charest (2007) that states that the role of AM symbiosis in metal(loid) contaminated soils is characterized by (i) an increase in phytoextraction via mycorrhizospheric "Enhanced Uptake" at low soil-metal concentrations, and (ii) a reduced metal bioavailability via AM fungal "metal-binding" processes at high soil-metal(loid) concentrations, hence resulting in increased plant biomass and enhanced plant tolerance through metal(loid) stress-avoidance.

These contradictory reports suggest that further studies are needed to evaluate the role of AMF in the mobilization of metal(loid)s. The presence and diversity of metallophytes and hyperaccumulator plants, especially in geographical areas where the information is limited, deserve more comprehensive investigations.

15.7 CONCLUSIONS AND FUTURE PERSPECTIVES

A vast body of literature shows that AMF significantly promotes plant establishment and growth in metal(loid) polluted soils. However, the result of the utilization of AMF appears to be uncertain. This is partly due to opposing views about the path (process) AMF undertakes in contaminated soil: AMF can promote phytoextraction, promote phytostabilization, or embrace both of these processes. According to the third approach, AMF promotes phytoextraction when metal(loid) concentration in soil is low or phytostabilization when there is high metal(loid) concentration.

Recently, knowledge has enhanced our understanding of mycorrhizal biology and functioning regarding plants' metal(loid) tolerance and AMF. However, more studies are necessary to improve and implement AMF in phytoremediation projects.

New research on this topic should focus on process optimization, including determining physiological mechanisms involved in metal(loid) absorption, translocation, and metabolism by plants and understanding biochemical and physiological control mechanisms. Concerning the roles of AMF in phytoremediation, it is imperative to ascertain their presence and diversity in polluted soils and their functional compatibility with metallophytes and hyperaccumulators. Also, it is necessary to learn about AMF-related physiological mechanisms for metal(loid) tolerance, including intracellular metal chelation and the roles played by metallothioneins. In summary, future designs of phytoremediation programs should seriously consider AMF for improved efficiency and effectiveness.

ACKNOWLEDGMENTS

Sebastian Meier is acknowledged to the Fondecyt project 1220190, and Humberto Aponte is acknowledged to the Fondecyt postdoctoral project 3210752.

REFERENCES

Adriano, D.C., Wenzel, W.W., Vangronsveld, J., Bolan, N.S., 2004. Role of assisted natural remediation in environmental cleanup. *Geoderma* 122, 121–142. doi: 0.1016/j.geoderma.2004.01.003.

Agerer, R., 2001. Exploration types of ectomycorrhizae: A proposal to classify ectomycorrhizal mycelial systems according to their patterns of differentiation and putative ecological importance. *Mycorrhiza* 11, 107–114. doi: 10.1007/s005720100108.

Ahalya, N., Ramachandra, T.V., Kanamadi, R.D., 2003. Biosorption of heavy metals. *Res. J. Chem. Environ.* 7, 71–78. doi: 10.1021/bp00033a001.

Alcalá, I.M., Clemente, R., 2020. Phytoremediation of metalloid-contaminated soil. In: Deshmukh, R., D.K. Tripathi, G. Guerriero (Eds.), *Metalloids in Plants: Advances and Future Prospects*. John Wiley & Sons: Hoboken, NJ, pp. 27–46.

Aponte, H., Medina, J., Butler, B., Meier, S., Cornejo, P., Kuzyakov, Y., 2020. Soil quality indices for metal(loid) contamination: An enzymatic perspective. *Land Degrad. Dev.* 31, 2700–2719. doi: 10.1002/ldr.3630.

Aponte, H., Mondaca, P., Santander, C., Meier, S., Paolini, J., Buttler, B., Rojas, C., Diez, C., Cornejo, P., 2021. Enzyme activities and microbial functional diversity in metal(loid) contaminated soils near a copper smelter. *Sci. Total Environ.* 779, 146423. doi: 10.1016/j.scitotenv.2021.146423.

Ashraf, S., Ali, Q., Zahir, Z.A., Ashraf, S., Asghar, H.N., 2019. Phytoremediation: Environmentally sustainable way for reclamation of heavy metal polluted soils. *Ecotoxicol. Environ. Saf.* 174, 714–727. doi: 10.1016/j.ecoenv.2019.02.068.

Audet, P., Charest, C., 2007. Dynamics of arbuscular mycorrhizal symbiosis in heavy metal phytoremediation: Meta-analytical and conceptual perspectives. *Environ. Pollut.* 147, 609–614. doi: 10.1016/j.envpol.2006.10.006.

Awa, S.H., Hadibarata, T., 2020. Removal of heavy metals in contaminated soil by phytoremediation mechanism: A review. *Water Air Soil Pollut.* 231, 1–15. doi: 10.1007/s11270-020-4426-0.

Bakker, P.A.H.M., Pieterse, C.M.J., de Jonge, R., Berendsen, R.L., 2018. The soil-borne legacy. *Cell* 172, 1178–1180. doi: 10.1016/j.cell.2018.02.024.

Barea, J.M., Azcón-Aguilar, C., Azcón, R., 1987. Vesicular-arbuscular mycorrhiza improve both symbiotic N_2 fixation and N uptake from soil as assessed with a 15N technique under field conditions. *New Phytol.* 106, 717–725. doi: 10.1111/j.1469-8137.1987.tb00172.x.

Barea, J.M., Azcón, R., Azcón-Aguilar, C., 2002. Mycorrhizosphere interactions to improve plant fitness and soil quality. *Antonie van Leeuwenhoek* 81, 343–351. doi: 10.1023/A:1020588701325.

Bauch, J., 2000. The kinetics of calcium and magnesium entry into mycorrhizal spruce roots. *Planta* 210(3), 488–496.

Bergero, R., Lanfranco, L., Ghignone, S., Bonfante, P., 2007. Enhanced activity of the GmarMT1 promoter from the mycorrhizal fungus Gigaspora margarita at limited carbon supply. *Fungal Genet. Biol.* 44, 877–885. doi: 10.1016/j.fgb.2007.01.010.

Bhainsa, K.C., D'Souza, S.F., 2008. Removal of copper ions by the filamentous fungus, Rhizopus oryzae from aqueous solution. *Bioresour. Technol.* 99, 3829–3835. doi: 10.1016/j.biortech.2007.07.032.

Blagodatskaya, E., Kuzyakov, Y., 2013. Active microorganisms in soil: Critical review of estimation criteria and approaches. *Soil Biol. Biochem.* 67, 192–211. doi: 10.1016/j.soilbio.2013.08.024.

Bolan, N., Kunhikrishnan, A., Thangarajan, R., Kumpiene, J., Park, J., Makino, T., Kirkham, M.B., Scheckel, K., 2014. Remediation of heavy metal(loid)s contaminated soils: To mobilize or to immobilize? *J. Hazard. Mater.* 266, 141–166. doi: 10.1016/j.jhazmat.2013.12.018.

Bolan, N.S., 1991. A critical review on the role of mycorrhizal fungi in the uptake of phosphorus by plants. *Plant Soil* 134, 189–207. doi: 10.1007/BF00012037.

Chibuike, G.U., Obiora, S.C., 2014. Heavy metal polluted soils: Effect on plants and bioremediation methods. *Appl. Environ. Soil Sci.* 2014. doi: 10.1155/2014/752708.

Choudhary, M., Kumar, R., Datta, A., Nehra, V., Garg, N., 2017. Bioremediation of heavy metals by microbes. In: Arora, S., Singh, A.K., Singh, Y.P. (Eds.), *Bioremediation of Salt Affected Soils: An Indian Perspective*. Springer International Publishing: Cham, pp. 233–255. doi: 10.1007/978-3-319-48257-6_12.

Christie, P., Li, X., Chen, B., 2004. Arbuscular mycorrhiza can depress translocation of zinc to shoots of host plants in soils moderately polluted with zinc. *Plant Soil* 261, 209–217. doi: 10.1023/B:PLSO.0000035542.79345.1b.

Cornejo, P., Meier, S., Borie, G., Rillig, M.C., Borie, F., 2008. Glomalin-related soil protein in a Mediterranean ecosystem affected by a copper smelter and its contribution to Cu and Zn sequestration. *Sci. Total Environ.* 406, 154–160. doi: 10.1016/j.scitotenv.2008.07.045.

Cornejo, P., Pérez-Tienda, J., Meier, S., Valderas, A., Borie, F., Azcón-Aguilar, C., Ferrol, N., 2013. Copper compartmentalization in spores as a survival strategy of arbuscular mycorrhizal fungi in Cu-polluted environments. *Soil Biol. Biochem.* 57, 925–928. doi: 10.1016/j.soilbio.2012.10.031.

Cornejo, P., Meier, S., García, S., Ferrol, N., Durán, P., Borie, F., Seguel, A., 2017. Contribution of inoculation with arbuscular mycorrhizal fungi to the bioremediation of a copper contaminated soil using Oenothera picensis. *J. Soil Sci. Plant Nutr.* 17. doi: 10.4067/S0718-95162016005000070.

Cristaldi, A., Conti, G.O., Jho, E.H., Zuccarello, P., Grasso, A., Copat, C., Ferrante, M., 2017. Phytoremediation of contaminated soils by heavy metals and PAHs: A brief review. *Environ. Technol. Innov.* 8, 309–326. doi: 10.1016/j.eti.2017.08.002.

da Silva Folli-Pereira, M., Sant, L., Meira-Haddad, A., Mara Soares Bazzolli, D., Catarina Megumi Kasuya, M., 2012. Micorriza arbuscular e a tolerância das plantas ao estresse. *Revisão De Literatura* 36, 1663–1679.

Davies, F.T., Puryear, J.D., Newton, R.J., Egilla, J.N., Saraiva Grossi, J.A., 2001. Mycorrhizal fungi enhance accumulation and tolerance of chromium in sunflower (*Helianthus annuus*). *J. Plant Physiol.* 158, 777–786. doi: 10.1078/0176-1617-00311.

De Gregori, I., Fuentes, E., Rojas, M., Pinochet, H., Potin-Gautier, M., 2003. Monitoring of copper, arsenic and antimony levels in agricultural soils impacted and non-impacted by mining activities, from three regions in Chile. *J. Environ. Monit.* 5, 287–295. doi: 10.1039/b211469k.

De La Iglesia, R., Castro, D., Ginocchio, R., Van Der Lelie, D., González, B., 2006. Factors influencing the composition of bacterial communities found at abandoned copper-tailings dumps. *J. Appl. Microbiol.* 100, 537–544. doi: 10.1111/j.1365-2672.2005.02793.x.

Dhaliwal, S.S., Singh, J., Taneja, P.K., Mandal, A., 2020. Remediation techniques for removal of heavy metals from the soil contaminated through different sources: A review. *Environ. Sci. Pollut. Res.* 27, 1319–1333. doi: 10.1007/s11356-019-06967-1.

Dixit, R., Wasiullah, E., Malaviya, D., Pandiyan, K., Singh, U.B., Sahu, A., Shukla, R., Singh, B.P., Rai, J.P., Sharma, P.K., Lade, H., Paul, D., 2015. Bioremediation of heavy metals from soil and aquatic environment: An overview of principles and criteria of fundamental processes. *Sustainability* 7, 2189–2212. doi: 10.3390/su7022189.

Ernst, W.H.O., Verkleij, J.A.C., Schat, H., 1992. Metal tolerance in plants. *Acta Bot. Neerl.* 41, 229–248. doi: 10.1111/j.1438-8677.1992.tb01332.x.

Fatnassi, I.C., Chiboub, M., Saadani, O., Jebara, M., Jebara, S.H., 2015. Phytostabilization of moderate copper contaminated soils using co-inoculation of Vicia faba with plant growth promoting bacteria. *J. Basic Microbiol.* 55, 303–311. doi: 10.1002/jobm.201300323.

Ferrol, N., González-Guerrero, M., Valderas, A., Benabdellah, K., Azcón-Aguilar, C., 2009. Survival strategies of arbuscular mycorrhizal fungi in Cu-polluted environments. *Phytochem. Rev.* 8, 551–559. doi: 10.1007/s11101-009-9133-9.

Frey, B., Schüepp, H., 1993. Acquisition of nitrogen by external hyphae of arbuscular mycorrhizal fungi associated with *Zea mays* L. *New Phytol.* 124, 221–230. doi: 10.1111/j.1469-8137.1993.tb03811.x.

Gadd, G.M., 1994. Interactions of fungi with toxic metals. In: Powell, K.A., Renwick, A., Peberdy, J.F. (Eds.), *The Genus Aspergillus: From Taxonomy and Genetics to Industrial Application*. Springer US: Boston, MA, pp. 361–374. doi: 10.1007/978-1-4899-0981-7_28.

Gadkar, V., Rillig, M.C., 2006. The arbuscular mycorrhizal fungal protein glomalin is a putative homolog of heat shock protein 60. *FEMS Microbiol. Lett.* 263, 93–101. doi: 10.1111/j.1574-6968.2006.00412.x.

Gao, W.Q., Wang, P., Wu, Q.S., 2019. Functions and application of glomalin-related soil proteins: A review. *Sains Malaysiana* 48, 111–119. doi: 10.17576/jsm-2019-4801-13.

Garcia, K., Zimmermann, S.D., 2014. The role of mycorrhizal associations in plant potassium nutrition. *Front. Plant Sci.* 5, 1–9. doi: 10.3389/fpls.2014.00337.

Gaur, A., Adholeya, A., 2000. Effects of the particle size of soil-less substrates upon AM fungus inoculum production. *Mycorrhiza* 10, 43–48.

Gil, C., Boluda, R., Rodríguez Martín, J.A., Guzmán, M., del Moral, F., Ramos-Miras, J., 2018. Assessing soil contamination and temporal trends of heavy metal contents in greenhouses on semiarid land. *Land Degrad. Dev.* 29, 3344–3354. doi: 10.1002/ldr.3094.

Giller, K.E., Witter, E., Mcgrath, S.P., 1998. Toxicity of heavy metals to microoorganisms amd microbial processes in agricultural soils: A review. *Soil Biol. Biochem.* 30, 1389–1414.

Glass, D.J. (1999). *U.S. and International Markets for Phytoremediation*. Glass Associates: Needham, MA.

Glick, B.R., Karaturovíc, D.M., Newell, P.C., 1995. A novel procedure for rapid isolation of plant growth promoting pseudomonads. *Can. J. Microbiol.* 41, 533–536. doi: 10.1139/m95-070.

Glick, B.R., Penrose, D.M., Li, J., 1998. A model for the lowering of plant ethylene concentrations by plant growth-promoting bacteria. *J. Theor. Biol.* 190, 63–68. doi: 10.1006/jtbi.1997.0532.

Göhre, V., Paszkowski, U., 2006. Contribution of the arbuscular mycorrhizal symbiosis to heavy metal phytoremediation. *Planta* 223, 1115–1122. doi: 10.1007/s00425-006-0225-0.

González, I., Neaman, A., Rubio, P., Cortés, A., 2014. Spatial distribution of copper and pH in soils affected by intensive industrial activities in Puchuncaví and Quintero, central Chile. *J. Soil Sci. Plant Nutr.* 14. doi: 10.4067/S0718-95162014005000074.

González-Chávez, M.C., Carrillo-González, R., Wright, S.F., Nichols, K., 2004. The role of glomalin, a protein produced by arbuscular mycorrhizal fungi, in sequestering potentially toxic elements. *Environ. Pollut.* 130, 317–323. doi: 10.1016/j.envpol.2004.01.004.

González-Guerrero, M., Azcón-Aguilar, C., Mooney, M., Valderas, A., MacDiarmid, C.W., Eide, D.J., Ferrol, N., 2005. Characterization of a Glomus intraradices gene encoding a putative Zn transporter of the cation diffusion facilitator family. *Fungal Genet. Biol.* 42, 130–140. doi: 10.1016/j.fgb.2004.10.007.

González-Guerrero, M., Cano, C., Azcón-Aguilar, C., Ferrol, N., 2007. GintMT1 encodes a functional metallothionein in Glomus intraradices that responds to oxidative stress. *Mycorrhiza* 17, 327–335. doi: 10.1007/s00572-007-0108-7.

González-Guerrero, M., Melville, L.H., Ferrol, N., Lott, J.N.A., Azcón-Aguilar, C., Peterson, R.L., 2008. Ultrastructural localization of heavy metals in the extraradical mycelium and spores of the arbuscular mycorrhizal fungus Glomus intraradices. *Can. J. Microbiol.* 54, 103–110. doi: 10.1139/W07-119.

Hakeem, K.R., Sabir, M., Öztürk, M., Mermut, A.R., Surriya, O., Sarah Saleem, S., Waqar, K., Gul Kazi, A., 2015. *Soil Remediation and Plants, Soil Remediation and Plants*. Elsevier. doi: 10.1016/B978-0-12-799937-1.00001-2.

Haroon, B., Ping, A., Pervez, A., Faridullah, M., 2019. Characterization of heavy metal in soils as affected by long-term irrigation with industrial wastewater. *J. Water Reuse Desalin.* 9, 47–56. doi: 10.2166/wrd.2018.008.

Hassan, S.E., Hijri, M., St-Arnaud, M., 2013. Effect of arbuscular mycorrhizal fungi on trace metal uptake by sunflower plants grown on cadmium contaminated soil. *N. Biotechnol.* 30, 780–787. doi: 10.1016/j.nbt.2013.07.002.

He, Y., Yang, R., Lei, G., Li, B., Jiang, M., Yan, K., Zu, Y., Zhan, F., Li, Y., 2020. Arbuscular mycorrhizal fungi reduce cadmium leaching from polluted soils under simulated heavy rainfall. *Environ. Pollut.* 263. doi: 10.1016/j.envpol.2020.114406.

Hildebrandt, U., Regvar, M., Bothe, H., 2007. Arbuscular mycorrhiza and heavy metal tolerance. *Phytochemistry* 68, 139–146. doi: 10.1016/j.phytochem.2006.09.023.

Hou, D., O'Connor, D., Igalavithana, A.D., Alessi, D.S., Luo, J., Tsang, D.C.W., Sparks, D.L., Yamauchi, Y., Rinklebe, J., Ok, Y.S., 2020. Metal contamination and bioremediation of agricultural soils for food safety and sustainability. *Nat. Rev. Earth Environ.* 1, 366–381. doi: 10.1038/s43017-020-0061-y.

Hristozkova, M., Geneva, M., Stancheva, I., Boychinova, M., Djonova, E., 2016. Contribution of arbuscular mycorrhizal fungi in attenuation of heavy metal impact on Calendula officinalis development. *Appl. Soil Ecol.* 101, 57–63. doi: 10.1016/j.apsoil.2016.01.008.

Hurtado-Jiménez, R., Gardea-Torresdey, J.L., 2006. Arsenic in drinking water in the Los Altos de Jalisco region of Mexico. *Rev. Panam. Salud Publica* 20, 236–247. doi: 10.1590/S1020-49892006000900004.

Iravani, S., Varma, R.S., 2020. Bacteria in heavy metal remediation and nanoparticle biosynthesis. *ACS Sustain. Chem. Eng.* 8, 5395–5409. doi: 10.1021/acssuschemeng.0c00292.

Jaishankar, M., Tseten, T., Anbalagan, N., Mathew, B.B., Beeregowda, K.N., 2014. Toxicity, mechanism and health effects of some heavy metals. *Interdiscip. Toxicol.* 7, 60–72. doi: 10.2478/intox-2014-0009.

Janeeshma, E., Puthur, J.T., 2020. Direct and indirect influence of arbuscular mycorrhizae on enhancing metal tolerance of plants. *Arch. Microbiol.* 202, 1–16. doi: 10.1007/s00203-019-01730-z.

Janoušková, M., Pavlíková, D., Vosátka, M., 2006. Potential contribution of arbuscular mycorrhiza to cadmium immobilisation in soil. *Chemosphere* 65, 1959–1965. doi: 10.1016/j.chemosphere.2006.07.007.

Jing, R., Kjellerup, B.V., 2018. Biogeochemical cycling of metals impacting by microbial mobilization and immobilization. *J. Environ. Sci. (China)* 66, 146–154. doi: 10.1016/j.jes.2017.04.035.

Joner, E.J., Briones, R., Leyval, C., 2000. Metal-binding capacity of arbuscular mycorrhizal mycelium. *Plant Soil* 227, 227–234.

Joner, E.J., Briones, R., Leyval, C., 2007. Metal-binding capacity of arbuscular mycorrhizal mycelium. *Plant and Soil* 226, 227–234.

Kaldorf, M., Kuhn, A.J., Schröder, W.H., Hildebrandt, U., Bothe, H., 1999. Selective element deposits in maize colonized by a heavy metal tolerance conferring arbuscular mycorrhizal fungus. *J. Plant Physiol.* 154, 718–728. doi: 10.1016/S0176-1617(99)80250-8.

Kamnev, A.A., Van Der Lelie, D., 2000. Chemical and biological parameters as tools to evaluate and improve heavy metal phytoremediation. *Biosci. Rep.* 20, 239–258. doi: 10.1023/A:1026436806319.

Karimi, A., 2011. Arbuscular mycorrhizal fungi and heavy metal contaminated soils. *Afr. J. Microbiol. Res.* 5, 1571–1576. doi: 10.5897/ajmr11.465.

Khan, A.G., 2005. Role of soil microbes in the rhizospheres of plants growing on trace metal contaminated soils in phytoremediation. *J. Trace Elem. Med. Biol.* 18, 355–364. doi: 10.1016/j.jtemb.2005.02.006.

Khan, A., Kuek, C., Chaudhry, T.M., Khoo, C.S., Hayes, W.J., 2000. Role of plants, mycorrhizea, and phytochelators in heavy metal contaminated land remediation. *Chemosphere* 41, 197–207.

Kojima, Y., 1991. Definitions and nomenclature of metallothioneins. *Methods Enzym.* 205, 8–10.

Korjus, H., 2014. Polluted soils restoration. In: Arraiza Bermúdez-Cañete, M.P., Santamarta Cerezal, J.C., Ioras, F., García Rodríguez, J.L., Abrudan, I.V., Korjus, H., Borbála, G. (Eds.), *Climate Change and Restoration of Degraded Land.* Colegio de Ingenieros de Montes: Madrid, pp. 413–451

Kothe, E., Turnau, K., 2018. Editorial: Mycorrhizosphere communication: Mycorrhizal fungi and endophytic fungus-plant interactions. *Front. Microbiol.* 9, 1–4. doi: 10.3389/fmicb.2018.03015.

Lanfranco, L., Novero, M., Bonfante, P., 2005. The mycorrhizal fungus Gigaspora margarita possesses a CuZn superoxide dismutase that is up-regulated during symbiosis with legume hosts. *Plant Physiol.* 137, 1319–1330. doi: 10.1104/pp.104.050435.

Leung, H.M., Ye, Z.H., Wong, M.H., 2007. Survival strategies of plants associated with arbuscular mycorrhizal fungi on toxic mine tailings. *Chemosphere* 66, 905–915. doi: 10.1016/j.chemosphere.2006.06.037.

Liu, L., Gong, Z., Zhang, Y., Li, P., 2014. Growth, cadmium uptake and accumulation of maize (*Zea mays* L.) under the effects of arbuscular mycorrhizal fungi. *Ecotoxicology* 23, 1979–1986. doi: 10.1007/s10646-014-1331-6.

Liu, H., Yuan, M., Tan, S., Yang, X., Lan, Z., Jiang, Q., Ye, Z., Jing, Y., 2015. Enhancement of arbuscular mycorrhizal fungus (*Glomus versiforme*) on the growth and Cd uptake by Cd-hyperaccumulator *Solanum nigrum*. *Appl. Soil Ecol.* 89, 44–49. doi: 10.1016/j.apsoil.2015.01.006.

Liu, L., Li, W., Song, W., Guo, M., 2018. Remediation techniques for heavy metal-contaminated soils: Principles and applicability. *Sci. Total Environ.* 633, 206–219. doi: 10.1016/j.scitotenv.2018.03.161.

Ma, Y., Rajkumar, M., Oliveira, R.S., Zhang, C., Freitas, H., 2019. Potential of plant beneficial bacteria and arbuscular mycorrhizal fungi in phytoremediation of metal-contaminated saline soils. *J. Hazard. Mater.* 379, 120813. doi: 10.1016/j.jhazmat.2019.120813.

Marchiol, L., Assolari, S., Sacco, P., Zerbi, G., 2004. Phytoextraction of heavy metals by canola (*Brassica napus*) and radish (*Raphanus sativus*) grown on multicontaminated soil. *Environ. Pollut.* 132, 21–27. doi: 10.1016/j.envpol.2004.04.001.

Maret, W., 2003. Cellular zinc and redox states converge in the metallothionein/thionein pair. *J. Nutr.* 133, 1460–1462.

Meier, S., Alvear, M., Borie, F., Aguilera, P., Ginocchio, R., Cornejo, P., 2012a. Influence of copper on root exudate patterns in some metallophytes and agricultural plants. *Ecotoxicol. Environ. Saf.* 75, 8–15. doi: 10.1016/j.ecoenv.2011.08.029.

Meier, S., Borie, F., Bolan, N., Cornejo, P., 2012b. Phytoremediation of metal-polluted soils by arbuscular mycorrhizal fungi. *Crit. Rev. Environ. Sci. Technol.* 42, 741–775. doi: 10.1080/10643389.2010.528518.

Meier, S., Borie, F., Curaqueo, G., Bolan, N., Cornejo, P., 2012c. Effects of arbuscular mycorrhizal inoculation on metallophyte and agricultural plants growing at increasing copper levels. *Appl. Soil Ecol.* 61. doi: 10.1016/j.apsoil.2011.10.018.

Meier, S., Cornejo, P., Cartes, P., Borie, F., Medina, J., Azcón, R., 2015. Interactive effect between Cu-adapted arbuscular mycorrhizal fungi and biotreated agrowaste residue to improve the nutritional status of Oenothera picensis growing in Cu-polluted soils. *J. Plant Nutr. Soil Sci.* 178, 126–135. doi: 10.1002/jpln.201400092.

Memon, A.R., Schröder, P., 2009. Implications of metal accumulation mechanisms to phytoremediation. *Environ. Sci. Pollut. Res.* 16, 162–175. doi: 10.1007/s11356-008-0079-z.

Mishra, V., Gupta, A., Kaur, P., Singh, S., Singh, N., Gehlot, P., Singh, J., 2016. Synergistic effects of Arbuscular mycorrhizal fungi and plant growth promoting rhizobacteria in bioremediation of iron contaminated soils. *Int. J. Phytorem.* 18, 697–703. doi: 10.1080/15226514.2015.1131231.

Moynahan, O.S., Zabinski, C.A., Gannon, J.E., 2002. Microbial community structure and carbon-utilization diversity in a mine tailings revegetation study. *Restor. Ecol.* 10, 77–87. doi: 10.1046/j.1526-100X.2002.10108.x.

Mulligan, C.N., Yong, R.N., Gibbs, B.F., 2001. Remediation technologies for metal-contaminated soils and groundwater: An evaluation. *Eng. Geol.* 60, 193–207. doi: 10.1016/S0013-7952(00)00101-0.

Nagajyoti, P.C., Lee, K.D., Sreekanth, T.V.M., 2010. Heavy metals, occurrence and toxicity for plants: A review. *Environ. Chem. Lett.* 8, 199–216. doi: 10.1007/s10311-010-0297-8.

Neaman, A., Robinson, B., Minkina, T.M., Vidal, K., Mench, M., Krutyakov, Y.A., Shapoval, O.A., 2020. Feasibility of metal(loid) phytoextraction from polluted soils: The need for greater scrutiny. *Environ. Toxicol. Chem.* 39, 1469–1471. doi: 10.1002/etc.4787.

Niemeyer, J.C., Lolata, G.B., de Carvalho, G.M., Da Silva, E.M., Sousa, J.P., Nogueira, M.A., 2012. Microbial indicators of soil health as tools for ecological risk assessment of a metal contaminated site in Brazil. *Appl. Soil Ecol.* 59, 96–105. doi: 10.1016/j.apsoil.2012.03.019.

Nies, D.H., 1999. Microbial heavy-metal resistance. *Appl. Microbiol. Biotechnol.* 51, 730–750. doi: 10.1007/s002530051457.

Okereafor, U., Makhatha, M., Mekuto, L., Uche-Okereafor, N., Sebola, T., Mavumengwana, V., 2020. Toxic metal implications on agricultural soils, plants, animals, aquatic life and human health. *Int. J. Environ. Res. Public Health* 17, 1–24. doi: 10.3390/ijerph17072204.

Ouziad, F., Hildebrandt, U., Schmelzer, E., Bothe, H., 2005. Differential gene expressions in arbuscular mycorrhizal-colonized tomato grown under heavy metal stress. *J. Plant Physiol.* 162, 634–649. doi: 10.1016/j.jplph.2004.09.014.

Park, D., Yun, Y.-S., Hye, J., Park, J.M., 2005. Mechanism of hexavalent chromium removal by dead fungal biomass of Aspergillus niger. *Water Res.* 39, 533–540. doi: 10.1016/j.watres.2004.11.002.

Pignattelli, S., Colzi, I., Buccianti, A., Cecchi, L., Arnetoli, M., Monnanni, R., Gabbrielli, R., Gonnelli, C., 2012. Exploring element accumulation patterns of a metal excluder plant naturally colonizing a highly contaminated soil. *J. Hazard. Mater.* 227–228, 362–369. doi: 10.1016/j.jhazmat.2012.05.075.

Pilon-Smits, E., 2005. Phytoremediation. *Annu. Rev. Plant Biol.* 56, 15–39. doi: 10.1146/annurev.arplant.56.032604.144214.

Prasad, M.N.V., De Oliveira Freitas, H.M., 2003. Metal hyperaccumulation in plants: Biodiversity prospecting forphytoremediation technology. *Electron. J. Biotechnol.* 6, 110–146. doi: 10.2225/vol6-issue3-fulltext-6.

Purin, S., Rillig, M.C., 2007. The arbuscular mycorrhizal fungal protein glomalin: Limitations, progress, and a new hypothesis for its function. *Pedobiologia (Jena).* 51, 123–130. doi: 10.1016/j.pedobi.2007.03.002.

Ramesh Kumar, K., Anbazhagan, V., 2018. Analysis and assessment of heavy metals in soils around the industrial areas in Mettur, Tamilnadu, India. *Environ. Monit. Assess.* 190, 519. doi: 10.1007/s10661-018-6899-5.

Ramsey, P.W., Rillig, M.C., Feris, K.P., Gordon, N.S., Moore, J.N., Holben, W.E., Gannon, J.E., 2005. Relationship between communities and processes; new insights from a field study of a contaminated ecosystem. *Ecol. Lett.* 8, 1201–1210. doi: 10.1111/j.1461-0248.2005.00821.x.

Rillig, M.C., Mummey, D.L., 2006. Mycorrhizas and soil structure. *New Phytol.* 171, 41–53. doi: 10.1111/j.1469-8137.2006.01750.x.

Rolfe, S.A., Griffiths, J., Ton, J., 2019. Crying out for help with root exudates: Adaptive mechanisms by which stressed plants assemble health-promoting soil microbiomes. *Curr. Opin. Microbiol.* 49, 73–82. doi: 10.1016/j.mib.2019.10.003.

Ruscitti, M., Arango, M., Beltrano, J., 2017. Improvement of copper stress tolerance in pepper plants (*Capsicum annuum* L.) by inoculation with arbuscular mycorrhizal fungi. *Theor. Exp. Plant Physiol.* 29, 37–49. doi: 10.1007/s40626-016-0081-7.

Seguel, A., Cumming, J.R., Klugh-Stewart, K., Cornejo, P., Borie, F., 2013. The role of arbuscular mycorrhizas in decreasing aluminium phytotoxicity in acidic soils: A review. *Mycorrhiza* 23, 167–183. doi: 10.1007/s00572-013-0479-x.

Seguel, A., Cornejo, P., Ramos, A., Von Baer, E., Cumming, J., Borie, F., 2017. Phosphorus acquisition by three wheat cultivars contrasting in aluminium tolerance growing in an aluminium-rich volcanic soil. *Crop Pasture Sci.* 68, 305–316. doi: 10.1071/CP16224.

Sheng, X.-F., Xia, J.-J., 2006. Improvement of rape (Brassica napus) plant growth and cadmium uptake by cadmium-resistant bacteria. *Chemosphere* 64, 1036–1042. doi: 10.1016/j.chemosphere.2006.01.051.

Singh, G., Pankaj, U., Chand, S., Verma, R.K., 2019. Arbuscular mycorrhizal fungi-assisted phytoextraction of toxic metals by *Zea mays* L. from tannery sludge. *Soil Sediment Contam.* 28, 729–746. doi: 10.1080/15320383.2019.1657381.

Smith, S.E., Read, D.J., 2008. *Mycorrhizal Symbiosis*, Third edition, Academic Press, Amsterdam, The Netherlands.

Srivastava, S., Agrawal, S.B., Mondal, M.K., 2015. A review on progress of heavy metal removal using adsorbents of microbial and plant origin. *Environ. Sci. Pollut. Res.* 22, 15386–15415. doi: 10.1007/s11356-015-5278-9.

Stommel, M., Mann, P., Franken, P., 2001. EST-library construction using spore RNA of the arbuscular mycorrhizal fungus Gigaspora rosea. *Mycorrhiza* 10, 281–285. doi: 10.1007/s005720000090.

Sugawara, E., Nikaido, H., 2014. Properties of AdeABC and AdeIJK efflux systems of Acinetobacter baumannii compared with those of the AcrAB-TolC system of *Escherichia coli. Antimicrob. Agents Chemother.* 58, 7250–7257. doi: 10.1128/AAC.03728-14.

Suschka, J., 1993. Effects of heavy metals from mining and industry on some rivers in Poland: An already exploded chemical time bomb? *Land Degrad. Dev.* 4, 387–391. doi: 10.1002/ldr.3400040423.

Szefer, P., Ikuta, K., Kushiyama, S., Frelek, K., Geldon, J., 1997. Distribution of trace metals in the Pacific oyster, Crassostrea gigas, and crabs from the east coast of Kyushu Island, Japan. *Bull. Environ. Contam. Toxicol.* 58, 108–114. doi: 10.1007/s001289900307.

Tabrizi, L., Mohammadi, S., Delshad, M., Moteshare Zadeh, B., 2015. Effect of arbuscular mycorrhizal fungi on yield and phytoremediation performance of pot marigold (*Calendula officinalis L.*) under heavy metals stress. *Int. J. Phytorem.* 17, 1244–1252. doi: 10.1080/15226514.2015.1045131.

Tapia-Gatica, J., González-Miranda, I., Salgado, E., Bravo, M.A., Tessini, C., Dovletyarova, E.A., Paltseva, A.A., Neaman, A., 2020. Advanced determination of the spatial gradient of human health risk and ecological risk from exposure to As, Cu, Pb, and Zn in soils near the Ventanas Industrial Complex (Puchuncaví, Chile). *Environ. Pollut.* 258, 113488. doi: 10.1016/j.envpol.2019.113488.

Thambavani, D.S., Prathipa, V., 2012. Quantitative assessment of soil metal pollution with principal component analysis, geo accumulation index and enrichment index. *Asian J. Environ. Sci.* 7, 125–134.

Tiwari, J., Ma, Y., Bauddh, K., 2020. Arbuscular mycorrhizal fungi: An ecological accelerator of phytoremediation of metal contaminated soils. *Arch. Agron. Soil Sci.* doi: 10.1080/03650340.2020.1829599.

Toju, H., Peay, K.G., Yamamichi, M., Narisawa, K., Hiruma, K., Naito, K., Fukuda, S., Ushio, M., Nakaoka, S., Onoda, Y., Yoshida, K., Schlaeppi, K., Bai, Y., Sugiura, R., Ichihashi, Y., Minamisawa, K., Kiers, E.T., 2018. Core microbiomes for sustainable agroecosystems. *Nat. Plants.* doi: 10.1038/s41477-018-0139-4.

Tonin, C., Vandenkoornhuyse, P., Joner, E.J., Straczek, J., Leyval, C., 2001. Assessment of arbuscular mycorrhizal fungi diversity in the rhizosphere of Viola calaminaria and effect of these fungi on heavy metal uptake by clover. *Mycorrhiza* 10, 161–168. doi: 10.1007/s005720000072.

Trotta, A., Falaschi, P., Cornara, L., Minganti, V., Fusconi, A., Drava, G., Berta, G., 2006. Arbuscular mycorrhizae increase the arsenic translocation factor in the As hyperaccumulating fern *Pteris vittata* L. *Chemosphere* 65, 74–81. doi: 10.1016/j.chemosphere.2006.02.048.

Vivas, A., Biró, B., Ruíz-Lozano, J.M., Barea, J.M., Azcón, R., 2006. Two bacterial strains isolated from a Zn-polluted soil enhance plant growth and mycorrhizal efficiency under Zn-toxicity. *Chemosphere* 62, 1523–1533. doi: 10.1016/j.chemosphere.2005.06.053.

Vodnik, D., Gr, H., Ma, I., Elteren, J.T. Van, Kova, M., 2007. The contribution of glomalin-related soil protein to Pb and Zn sequestration in polluted soil 2. doi: 10.1016/j.scitotenv.2007.11.016.

Wan, X., Lei, M., Chen, T., 2016. Cost–benefit calculation of phytoremediation technology for heavy-metal-contaminated soil. *Sci. Total Environ.* 563–564, 796–802. doi: 10.1016/j.scitotenv.2015.12.080.

Weissenhorn, I., Leyval, C., Berthelin, J., 1993. Cd-tolerant arbuscular mycorrhizal (AM) fungi from heavy-metal polluted soils. *Plant Soil* 157, 247–256. doi: 10.1007/BF00011053.

Whiting, S.N., Reeves, R.D., Richards, D., Johnson, M.S., Cooke, J.A., Malaisse, F., Paton, A., Smith, J.A.C., Angle, J.S., Chaney, R.L., Ginocchio, R., Jaffré, T., Johns, R., McIntyre, T., Purvis, O.W., Salt, D.E., Schat, H., Zhao, F.J., Baker, A.J.M., 2004. Research priorities for conservation of metallophyte biodiversity and their potential for restoration and site remediation. *Restor. Ecol.* 12, 106–116. doi: 10.1111/j.1061-2971.2004.00367.x.

World Health Organization. (2014). Chemicals of public health concern and their management in the African Region. Regional assessment report Republic of Congo: WHO, Regional Office for Africa.

Wu, S., Zhang, X., Sun, Y., Wu, Z., Li, T., Hu, Y., Lv, J., Li, G., Zhang, Z., Zhang, J., Zheng, L., Zhen, X., Chen, B., 2016. Chromium immobilization by extra- and intraradical fungal structures of arbuscular mycorrhizal symbioses. *J. Hazard. Mater.* 316, 34–42. doi: 10.1016/j.jhazmat.2016.05.017.

Wuana, R.A., Okieimen, F.E., 2011. Heavy metals in contaminated soils: A review of sources, chemistry, risks and best available strategies for remediation. *ISRN Ecol.* 2011, 1–20. doi: 10.5402/2011/402647.

Yang, Y., Han, X., Liang, Y., Ghosh, A., Chen, J., Tang, M., 2015. The combined effects of arbuscular mycorrhizal fungi (AMF) and lead (Pb) stress on Pb accumulation, plant growth parameters, photosynthesis, and antioxidant enzymes in *Robinia pseudoacacia* L. *PLoS One* 10, 1–24. doi: 10.1371/journal.pone.0145726.

Yewale, P.P., Rahman, A., Nahar, N., Saha, A., 2017. Sources of metal pollution, global status, and conventional bioremediation practices. In: Das, S., Dash, H. (Eds.), *Handbook of Metal: Microbe Interactions and Bioremediation*. CRC Press, Taylor & Francis Group: Boca Raton, FL, pp. 25–40. doi: 10.1201/9781315153353.

Zhan, F., Li, B., Jiang, M., Yue, X., He, Y., Xia, Y., Wang, Y., 2018. Arbuscular mycorrhizal fungi enhance antioxidant defense in the leaves and the retention of heavy metals in the roots of maize. *Environ. Sci. Pollut. Res.* 25, 24338–24347. doi: 10.1007/s11356-018-2487-z.

Zhang, X., Hu, W., Xie, X., Wu, Y., Liang, F., Tang, M., 2021. Arbuscular mycorrhizal fungi promote lead immobilization by increasing the polysaccharide content within pectin and inducing cell wall peroxidase activity. *Chemosphere* 267, 128924. doi: 10.1016/j.chemosphere.2020.128924.

Zhou, J.L., 1999. Zn biosorption by Rhizopus arrhizus and other fungi. *Appl. Microbiol. Biotechnol.* 51, 686–693. doi: 10.1007/s002530051453.

16 Endophytic Microbes for Sustainable Crop Production and Soil Health Maintenance in Toxic Trace Element-Contaminated Soil

Asit Mandal[1], Abhijit Sarkar[1], Jyoti Thakur[1], A.B. Singh[1], Ashok K. Patra[1], and Binoy Sarkar[2]
[1]Indian Institute of Soil Science, Indian Council of Agricultural Research, Bhopal, India
[2]Future Industries Institute, University of South Australia, Mawson Lakes, Australia

CONTENTS

16.1 INTRODUCTION

Soil pollution with heavy loadings of toxic trace elements (TTE) is one of the major environmental concerns and a constraint to crop production. TTE are defined as non-essential elements for plants. Exposure to them causes toxicity effects in plants even in very minute quantities. Contamination of agricultural soils with TTE is ever-increasing across the globe due to various anthropogenic activities including enhanced use of agrochemicals, which pose adverse impacts on soil health and sustainability (Mandal et al., 2020). High concentrations of TTE, mainly Cd, Cr, Pb, Hg, and As, are frequently observed in terrestrial and aquatic ecosystems and are hazardous to organisms. In long-term, intensively cultivated soils with TTE contamination, TTE loads provide poor ecological functions, and they are a serious threat to soil health and crop production (Alengebawy et al., 2021).

Increased human population, food demand, industrialization, and improved residential convenience have resulted in an enormous amount of waste generation. These wastes pose a significant challenge to societies and increase the risk of environmental pollution with TTE, organic pollutants, and microbial contaminants because they can be disseminated from their source to the soil, waterways, crops, and finally human food chains (Minhas et al., 2022). Gastrointestinal problems

DOI: 10.1201/9781003093565-16

in humans due to consumption of TTE-contaminated crops grown on industrial wastes, or crops supplied with TTE-contaminated irrigation water or composts, are well documented (Vázquez et al., 2015). According to their toxicity in environments with biota, the following trace elements can be arranged in the order: Hg>Cu>Zn>Ni>Pb>Cd>Cr >Sn> Fe>Mn>Al (Filipiak-Szok et al., 2015). Some of these elements are essential for plants and human beings at low concentrations but cause contamination to the soil environment as (i) their rates of production via anthropogenic activities are more rapid compared to natural ones; (ii) they become randomly transported from ores and mines to the nearest lowland locations where the chances of direct exposure occur; (iii) the concentrations of elements in waste products are relatively high compared to those in the native receiving surroundings; and (iv) the chemical forms of elements in the receiving local system become increasingly bioavailable (D'Amore et al., 2005; Thakur et al., 2018). Based on the priority of contamination, major TTE follow the order As>Pb>Hg>Cd>Cr (ATSDR, 2015; Ishchenko, 2017). The literature suggests that Cd may act as a carcinogen at the cellular level, whereas Pb and Hg may act as neurotoxins. Arsenic acts as both a carcinogen and neurotoxin. Inhalation of Cr by humans may result in carcinogenic and mutagenic effects (Rahman and Singh, 2019).

To address the issues linked with soil and environmental pollution, food security, and ecological degradation, the role of endophytes has been explored recently as a potential technology that employs plant-associated microorganisms to restore the health of soil and aquatic systems. Biologically based approaches, especially microbial-assisted phytoremediation, is considered as an ecofriendly way to combat TTE contamination over conventional physicochemical methods (Deng and Cao, 2017; Srivastava et al., 2020). Endophytic microbes have unique roles in the remediation process by efficient biosorption and bioaccumulation of TTE from contaminated soils and with an improved element extraction capacity in plants. Because of a wide diversity of the host plants, endophytic microbes are resistant to multiple trace elements and are designated as important tools for the reclamation of TTE-contaminated soils. Limited studies have been conducted concerning the diversity and ecology of endophytes for overall soil health development. This chapter aims to highlight the source and impact of TTE on soil health and crop yield and synthesize information on the role of endophytes as a method to control soil TTE pollution.

16.2 SOURCE AND IMPACT OF TTE ON SOIL HEALTH

16.2.1 Sources of TTE

The Earth's crust is the mother of all TTE in terrestrial and aquatic ecosystems. In general, the abundance of As, Pb, Hg, Cd, and Cr in the Earth's crust is, respectively, 1.8–2.1, 10.0–14.0, 0.067–0.085, 0.15, and 102–140 mg/kg (Rahman and Singh, 2019). There are over 300 As-bearing minerals in the Earth's crust; among them are arsenopyrite (FeAsS), cobaltite (CoAsS), niccolite (NiAs), realgar (AsS), arsenolite (As_2O_3), and orpiment (As_2S_3); and they are rich sources of As (Drahota and Filippi, 2009). Cinnabar (HgS), corderoite ($Hg_3S_2C_{12}$), and livingstonite ($HgSb_4S_8$) are rich in Hg (USGS, 2016). Similar to As, the Earth's crust contains over 100 Pb-bearing minerals; among them galena (PbS), crocoite ($PbCrO_4$), anglesite ($PbSO_4$), pyromorphite ($Pb_5(PO_4)_3Cl$), coronadite ($PbMn_8O_{16}$), and cerussite ($PbCO_3$) are important (MRT, 2001). Greenockite and hawleyite (CdS), otavite ($CdCO_3$), niedermayrite ($Cu_4Cd(SO_4)_2(OH)_6.4H_2O$), and cadmoselite (CdSe) are Cd-rich minerals (LLC Books, 2010), whereas ferrochromite ($FeCr_2O_4$) and crocoite ($PbCrO_4$) are examples of Cr-rich minerals (ATSDR, 2012). Rocks like basaltic igneous rock are rich sources of Co, Cd, Cu, and Ni, whereas black shale contains an enormous quantity of Cd, Cu, Zn, Mn, and Pb (He et al., 2015). According to Göttingen University's database on Geochemistry of Rocks of the Oceans and Continents (https://georoc.eu/georoc/), mafic igneous rocks that stretch over Karoo, Parana, Deccan, North Atlantic Igneous Province, Siberian, Emeishan, and the Central Atlantic Magmatic Province contain Cr, Cu, Ni, and Zn in the range of 102.4–281.2, 92.7–191.0, 58.5–124.0, and 84.1–121.0 mg/kg, respectively. Ultramafic rocks such as Lherzolite, Dunite, and Harzburgite are also rich sources of Cr and Ni, which may reach beyond 2,649 and 2,176 mg/kg soil, respectively. In general, As diffuses from the Earth's crust to the surrounding ecosystem through rock weathering, microbial localization, and volcanic explosions (Mandal and Suzuki, 2002; Muller et al., 2007; López et al., 2012). Pb and Cd diffuse through natural deposits, sea salt spray, volcanic eruptions, dust storms, weathering of rocks, erosion, and wildfires (Richardson et al., 2001; Rahman and Singh, 2019). Mercury (Hg) diffuses through degassing, rock weathering, volcanic eruptions, wildfires, and re-emissions (Xu et al., 2015). Tectonic and hydrothermal events are responsible for the diffusion of Cr from the Earth's crust to the surrounding environment (Robles-Camacho and Armienta, 2000).

Apart from natural sources, anthropogenic sources of TTE pose a great risk to environmental sustainability. Anthropogenic sources may be classified as geogenic (e.g., mining, smelting, and combustion of fossil fuels), industrial (e.g., wastewater, landfilling, and atmospheric deposition), municipal (e.g., sewage-sludge, waste deposition), and agricultural-associated sources (e.g., pesticides, fertilizers, and irrigation water). Particularly, lowland deposition of mining ore or tailings can raise TTE concentration while flow of water can further influence the diffusion of TTE (Sun et al., 2019; Sharma et al., 2017). Even by-products of ore smelting contain large amounts of TTE, which pose a threat of TTE contamination during land deposition (Artiola et al., 2019). Csavina et al. (2014) reported that smelter effluent and TTE-bearing dust particles generated during smelting may travel a long distance and contaminate soil and waterways. Road traffic and different industrial sectors have also considerable effects on soil and water bodies. Industries such as tanning, mining, combusted coal, fly ash, manufacturing of textiles and dyes, preservation for wood treatment, metal anti-corrosive agents, electroplating, aircraft, and polishing of glassware pose a potential risk of Cr (Dhal et al., 2013; Rahman and Singh, 2019). Extractive metallurgy, such as from the mining industry, Pb-acid battery remaking, tetraethyl lead additives in gasoline, pipes, ammunitions, agrochemicals, dyes in paints, e-wastes, and ceramic glazes, acts as the supplier of Pb, whereas gases from fossil fuel combustion act as the aerial source of Pb (EU SCOOP, 2004). The pesticide industry acts as the supplier of Cu. Industries that make pigments, fertilizers, and paints pose a great risk for Cd. Mining, smelting, fossil-fuel combustion, glass manufacturing, semi-conductor materials, chemical fertilizers, and chemotherapeutic cancer drugs are the sources for As. The steel, automotive, and surgical industries pose a threat for Ni contamination (Zeng et al., 2017). Waste generated from municipalities is also a major source of TTE contamination in aquatic and terrestrial ecosystems. Household wastes, garbage, trash from commercial establishments, and health institutions contain high amounts of TTE along with toxins, micro-plastics, and organic pollutants (Okoh et al., 2007; Barrett et al., 2012; Du et al., 2020). Overall, there are numerous point and non-point sources of TTE for the soil system that potentially can impair soil health and crop productivity.

16.2.2 Impacts of TTE on Soil Health

TTE exist in soil as water-soluble-, exchangeable-, bound forms of oxides, carbonates, ligands, and sulfides, and residual fractions (Giacalone et al., 2005). The total concentration of TTE is not true representative of bioavailable TTE. For example, a proportion of ligand-bound TTE can be high but with low bioavailability (Violante et al., 2010). Soil temperature, moisture, texture, pH (soil reaction), electrical conductivity (EC), redox potential, exchangeable cations, clay type, carbon content, type and quantity of oxides, microbial communities, microbial metabolism, and abiotic transformation play key roles in TTE bioavailability in soil (Rieuwerts et al., 1998; Wenzel, 2013; Stanin and Pirnie, 2005; Rahman and Singh, 2019). Among different chemical species, the degree of toxicity of As follows the following trend: arsenine compounds > arsenites > arsenates > arsonium metals > native As species (Khanam et al., 2019). Chromium exists in Cr(VI) and Cr(III) forms, where Cr(VI) is more toxic than Cr(III) (Arita and Costa, 2011). As(V) and Cr(III) tend not to be toxic because Cr(III) is insoluble at higher pH (>4.0) and As(V) is relatively less soluble than As(III) (Sheik et al., 2012). In alkaline soil, Cd tends to be present as $CdCO_3$ in soil, whereas CdS predominates at low soil pH. Both forms of Cd may easily be converted to Cd^{2+} (bioavailable) (He et al.,

TABLE 16.1
Distinguishable Features between Soil Quality and Soil Health

Soil Health	Soil Quality
Soil health is the actual capacity of a particular soil to function, contributing to ecosystem services	Soil quality is the inherent capacity of a particular soil to function, contributing to ecosystem services
Soil health is the condition of the soil and its potential to sustain biological functions, maintain environmental quality, and promote plant and animal health	Soil quality is the capacity of each soil to function, within its natural or managed ecosystems, to sustain productivity, enhance water and air quality, support human and animal health, and habitation
Soil health has emphasis on dynamic quality linked with soil biology	Soil quality has greater emphasis on physical attributes
Soil health basically indicates the biological integrity of the soil community – the balance among soil organisms and their environment	Soil quality describes the physical attributes of soil. It also describes complex characteristics of soil such as soil carbon, mineral content, and soil aggregates

Source: Adapted from Bonfante et al. (2020).

2015). Pb^{2+} is the bioavailable form of Pb, and a small quantity of bioavailable Pb is phytotoxic. Fortunately, Pb is present as inorganic salts, and its organic ligand-bound forms are comparatively less soluble. Solubility of Pb and Cd is highly influenced by the pH of soil, but Pb has relatively less mobility than Cd and other TTE (ATSDR, 2012). Mercury (Hg) mostly occurs as HgS in soil; however, its low solubility in aqueous solution poses non-significant bioavailability of Hg. In contrast, chloride, hydroxide salts, or organic ligand-bound Hg species are present in low levels, but these forms are highly bioavailable. In terms of toxicity, availability, and mobility, Hg compounds generally show the following order: alkylmercury > Hg(II) > Hg(0) (Biester et al., 2002; Han et al., 2003).

Sustainability of soil health with TTE contamination is uncertain because TTE may have an adverse influence on soil processes and functions. The terms "soil health" and "soil quality" are used interchangeably, but there are distinguishable features between the two technical terminologies (Table 16.1). The quality and health of soil govern agricultural sustainability and soil environmental quality, which intimately are linked with the plant-, animal-, and human-health continuum (Haberern, 1992; Doran, 2002). Soil quality and soil health are the most vital components for achieving sustainable agroecosystem services.

Maintaining good soil quality is the backbone of environmental sustainability. Exposure to TTE adversely affects soil quality. In various ways, TTE cause risks to the soil environment through changes in soil color, porosity, pH, and its natural physicochemical and biological characteristics, which, in turn, moderate soil health and quality (Du et al., 2019). Soil chemical (e.g., soil pH, EC, pools and fractions of nutrients, and bioavailability), microbial (e.g., microbial biomass, cell viability, microbial diversity, and community structure), biochemical (e.g., enzymatic activities, basal respiration, particulate organic matter, and potentially mineralizable nitrogen) parameters, and soil fauna indicators, such as earthworm quantity and diversity, extensively influence nutrient cycling and crop production (Mahmood and Malik, 2014; Antisari et al., 2015). Exposure to TTE could result in a change of soil biogeochemical properties and plant parameters and, hence, human health.

16.3 NUTRIENT CYCLING AND CROP PRODUCTION IN TTE-CONTAMINATED SOIL

The prime role of microbial activity is to keep soil alive, and numerous microorganisms participate in the nutrient cycling processes (Paz-Ferreiro and Fu, 2016). Several authors reported that TTE contamination through natural and artificial sources negatively affects soil microorganisms

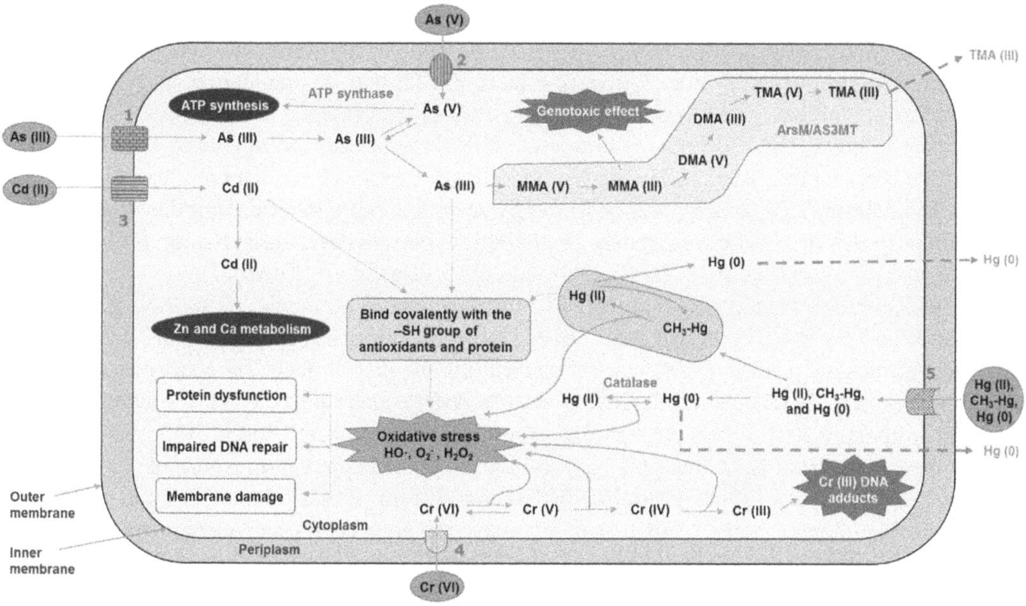

FIGURE 16.1 Schematic depiction of toxicity mechanisms of arsenic (As), cadmium (Cd), chromium (Cr), and mercury (Hg). Solid arrow indicates the direction of mechanism; dashed arrow indicates outer-cellular mechanisms; words indicate enzymes related to specific mechanism; MMA^{III} (monomethylarsonous acid), DMA^{III} (dimethylarsonous acid), MMA^{V} (monomethylarsonic acid), DMA^{V} (dimethylarsonic acid), TMA^{III} (trimethylarsine), TMA^{V} (trimethylarsine oxide); ArsM/AS3MT (arsenic methyltransferases). 1 (glycerol transporter); 2 (phosphate transporter); 3 (divalent cation transporter 1); 4 (sulfate/phosphate transport system); 5 (MerP and MerT in resistant bacteria; DMT1 and SLC11A2 in higher organisms). (Conceptualized from Ercal et al. (2001), Rosen (2002), O'Brien et al. (2003), Bridges and Zalups (2005), Mathema et al. (2011), Clemens and Ma (2016), Bridges and Zalups (2017), and Rahman and Singh (2019).)

(Sharma et al., 2017; Lin et al., 2021). A redox-inactive reaction is the main toxicity mechanism for Cd, Hg, and Pb, whereas redox-cycling is the mechanism of toxicity for Cr and As (Rahman and Singh, 2019). Plausible mechanisms of toxicity involved with TTE exposure are presented in Figure 16.1.

Microorganisms act as agents in the process of nutrient cycling and could ease the toxicity caused by TTE (Stuczynski et al., 2003). Endophytic microbes can adapt themselves in a stressed environment by regulating their own metabolism and growth, enzyme activities, and microbial community (Fang et al., 2017). The specific effects of TTE in soil on microbial metabolism are still unclear. A few studies have reported that TTE derived from municipal solid waste significantly influenced the performance of soil enzymes that are primarily involved in soil nutrient cycling (Xian et al., 2015). Reports have suggested that acid phosphomonoesterase and β-glycosidase enzymes showed comparatively low activity in soils having proximity to single- or multi-elemental mining. Soils in the proximity of a TTE mine had a lower cultural diversity of microorganisms than that of a control site (dos Santos et al., 2016). Open-cast coal mining also negatively impacted soil microbial richness (de Quadros et al., 2016). Yuan et al. (2015) recorded a negative correlation between prolonged Pb exposure and soil microbial sustainability. The negative impact of TTE on soil health is visible during residue decomposition in TTE-enriched soil or plant litter. In general, TTE exposure reduced litter decomposition rate in several soils; however, the rate of litter decomposition greatly depended on litter composition, duration, and intensity of TTE exposure. The greater the intensity and exposure of TTE are, the greater is the negative impact. Kozlov and Zvereva (2015) reported that micro fauna and soil respiration were adversely affected by the application of Pb and Ni-rich plant litter.

A reduced litter layer often resulted in an undecomposed litter layer in soil due to TTE contamination (Hogsden and Harding, 2012; Ferreira et al., 2016). Sudden environmental changes due to TTE exposure retard litter decomposition. Adverse effects of TTE may be non-significant in a low dose, but high doses significantly affect soil respiration, even in the presence of an adequate quantity of substrates (Dotaniya et al., 2017; Srivastava et al., 2017). Higher concentrations of TTE negatively affected the N-dynamics, particularly the N-mineralization process, *viz.* ammonification and nitrification. In addition, TTE toxicity was observed by several authors in terms of reduced abundance of certain enzymes such as dehydrogenase, acid/alkaline phosphatase, arylsulfatase, β-glycosidase, cellulase, and urease (Hu et al., 2014; Burges et al., 2015; Xian et al., 2015).

Like soil microorganisms, exposure of TTE to plants could adversely affect their growth and development (Sheng et al., 2008; Jan et al., 2019). Due to the mimicking properties of TTE, they can unwantedly enter the plant body via several nutrient uptake mechanisms. For example, As(V) has the properties of PO_4^{3-} (Kruger et al., 2013), because As(V) enters into the cell through phosphate transport, whereas As(III) enters through the glycerol transport system (Rosen, 2002). Similarly, Cr mimics the properties of SO_4^{2-} and PO_4^{3-}, and cellular uptake occurs via transporters of sulfate and phosphate (O'Brien et al., 2003). Cadmium (Cd^{2+}) and Pb^{2+} mimic Ca^{2+}, Fe^{2+}, Zn^{2+}, and other divalent cations (Handlogten et al., 2000); hence Pb^{2+} enters the plant cells via divalent metal transporter channels and carrier proteins like the divalent metal-ion transporter 1 (DMT1) (Bridges and Zalups, 2005; Clemens and Ma, 2016). Mercury (Hg^{2+}) follows the properties of divalent cations (Patra et al., 2004), but its uptake mechanism involves MerP- and MerT-specific transport proteins in resistant bacteria, and DMT1 and SLC11A2 in higher organisms (Mathema et al., 2011). Concentrations of As beyond 1.0 mg/kg (Anonymous, 1987), Pb beyond 2.5 mg/kg (Anonymous, 1987), Cr beyond 2.3 mg/kg (FAO/WHO, 2001), and Cd beyond 0.2 mg/kg (FAO/WHO, 2001) in vegetable tissues are considered toxic.

16.4 EFFECT OF ENDOPHYTES IN THE MAINTENANCE SOIL HEALTH UNDER TTE CONTAMINATION

Recently, microbe-assisted phytoremediation using endophytic microbes has been employed for the remediation of TTE-contaminated soils. Many endophytic microbes have been found to be resistant to multiple TTE and have a potential role in phytoremediation by virtue of TTE detoxification, metal-tolerance mechanisms, and plant-growth promotion under TTE stress (Yan et al., 2020; Li et al., 2012; Deng and Cao, 2017; Ma et al., 2016). Endophytic microbes directly enhance the biomass of host plants subjected to different TTE, indicating their effectiveness in protecting plants from several TTE. Endophytic microbes enhance plant health in TTE-polluted soils in two major ways: (i) directly by the release of plant growth-promoting substances with solubilizing ability of nutrient elements (P, N, and K) and secretion of phytohormones, siderophores, and specific enzymes; and (ii) indirectly through controlling phytopathogens or by inducing a systemic resistance of plants against phytopathogens (Table 16.2). Endophytes enhance the metal transformation and accumulation capacity in plants by excreting different metal-immobilizing extracellular polymeric substances and metal-mobilizing organic acids and biosurfactants (Azubuike et al., 2016; Yousaf et al., 2014).

Endophytic microbes that are present in association with roots and shoots of plants may enhance the plant metal accumulation capacity to a higher level through various strategies such as mobilization and mitigation of toxic elements, improving mineral nutrition, and growth promotion of the plants and enhanced stress tolerance (Table 16.3).

Many endophytic strains have been isolated from *Spartina maritima* tissue, which has been identified for its tolerance to salts and trace elements that help in plant growth and enzymatic activities (Mesa et al., 2015). Endophytic fungi associated with many species of grasses are found to have high TTE tolerance and ability to withstand other biotic and abiotic stresses (Wiewióra and Żurek, 2021). In the rhizosphere of TTE hyperaccumulating plants, greater abundance of resistant bacteria can be found (Mengoni et al., 2001). A large community of TTE-tolerating methylobacteria was

TABLE 16.2
Roles of Endophytic Microbes in Soil Health Maintenance

Endophytic Organism	Example	Function
Bacteria	*Bacillus* sp. *Pseudomonas* sp. *Enterobacter* sp. *Klebsiella* sp.	Biodegradation P solubilization Decrement of leaf transpiration, stomatal conductance Growth promotion Phyto-activation and/or phyto-inhibition Phytoaccumulation
Fungi	*Alternaria* sp. *Aspergillus* sp. *Fusarium* sp. *Penicillium* sp.	Enhanced seed germination Root growth and protection
Actinomycetes	*Streptomyces* sp.	Plant growth promotion and biocontrol Production of active secondary metabolites Phosphate solubilization

TABLE 16.3
Toxic Trace Element Alleviation Using Endophytes

Endophyte	Host Plant	Target Element	Role	References
Agrobacterium and *Bacillus* sp.	Chinese brake fern (*Pteris vittata*)	As	Tolerated elevated levels of As with high level of indoleacetic acid production and carried arsB/ACR3 genes Enhanced phytoremediation	Gu et al. (2018)
Paraphaeosphaeria sp., *Pyrenochaeta* sp., *Rhizopycnis vagum*, and *R. vagum*	*Salix variegate*	Cd	Augmented plant growth by reducing Cd toxicity Potential roles in Cd transformation in plant, soil, and aquatic system	An et al. (2015)
Bacillus, *Staphylococcus*, and *Aerococcus*	Mesquite (*Prosopis juliflora*)	Cr	Promoted plant biomass via TTE decontamination from tannery effluent Improved efficiency of TTE phytoremediation in degraded soils	Khan et al. (2015)
Acinetobacter, *Pseudomonas*, and *Rahnella*	Alfalfa (*Medicago sativa*) and chickpea (*Cicer arietinum*)	As	Presence of *arsC* gene siderophore, phosphate solubilization, N-fixation, protease, and lipase production Plant-growth promotion	Tashan et al. (2021)
Pseudomonas sp.				
Piriformospora indica	Rice (*Oryza sativa*)	Cd	Enhanced accumulation of Cd with more shoot and root growth but decreased accumulation of reactive oxygen species (ROS)	Dabral et al. (2019)
Piriformospora indica	Rice (*Oryza sativa*)	Cd	Higher plant biomass, chlorophyll synthesis, and antioxidant enzyme activity Decreased malondialdehyde and Cd translocation Upregulation of glutamine synthetase and downregulation of glycolysis cycle enzymes	Sagonda et al. (2021)

(*Continued*)

TABLE 16.3 (*Continued*)
Toxic Trace Element Alleviation Using Endophytes

Endophyte	Host Plant	Target Element	Role	References
Piriformospora indica	Rice (*Oryza sativa*)	As	Greater chlorophyll synthesis and fluorescence yield Increased electron transport rate and crop biomass More tolerance to As toxicity by downregulation of Lsi2 and upregulation of PCS1 and PCS2 genes Higher Fe accumulation by upregulating IRO2, YSL2, and FRDL1 genes	Ghorbani et al. (2021)
Piriformospora indica	Tobacco (*Nicotiana tabacum*)	Cd	Decreased production of malondialdehyde, and higher chlorophyll and proline contents, and other stress enzymatic activities Increased expression of Gsh2, TaPCS1, oas1, GPX, and Hsp70 genes	Hui et al. (2015)
Piriformospora indica	Alfalfa (*Medicago sativa*)	Cd	Enhanced dry weight Inhibition of Cd transport in shoot	Sepehri and Khatabi (2021)
Piriformospora indica	Sunfower (*Helianthus annus*)	Cd	Improved crop growth High chlorophyll a, chlorophyll b, and proline contents, and electron transport efficiency	Shahabivand et al. (2017)
Piriformospora indica	Tobacco (*Nicotiana tabaccum*)	Cd	Increased peroxidase and glutathione activity and genes involved in their expression	Su et al. (2021)
Piriformospora indica	Rice (*Oryza sativa*)	Cd	Higher root–shoot growth and biomass Less cell death and ROS accumulation	Dabral et al. (2019)
Piriformospora indica	Artemisia (*Artemisia annua*)	As	Augmented flavonoids and total biomass, antioxidant enzymes, secondary bioactive compounds, and signal molecules	Khalid et al. (2020)
Paneibacillus sp.	Tridax procumbens	As, Pb, Cu, Zn	Increased production of secondary bioactive compounds, indole acetic acid, siderophores, 1-aminocyclopropane-1-carboxylate deaminase, and biosurfactants Solubilized phosphate Survival under high TTE concentration	Govarthanan et al. (2016)
Trametes hirsuta	Chenopodium album	Pb	Improved survival of host plants in metal contaminated soils Improved Pb phytoextraction in wheat vegetative stage	Malik et al. (2020)
Pseudomonas fluorescens *Microbacterium sp.*	*Brassica napus*	Pb	Increased plant biomass production and Pb accumulation	Lally et al. (2017)

characterized from shoots of the Zn-hyperaccumulator *Thlaspi caerulescens*; however, they were not localized in the roots (Lodewyckx et al., 2002). The plant growth-promoting endophytic fungi *Penicillium ruqueforti* showed resistance toward various TTE such as Ni, Cd, Cu, Zn, and Pb, which was observed in inoculated wheat plants grown in TTE-contaminated soil, but not in non-inoculated plants (Ikram et al., 2018). Some endophytic bacteria also promote and support halophyte plants in the desalination of saline soils as well as phytoremediation of TTE-contaminated soil (Yasseen and Al-Thani, 2022). Plant-microbe interactions contributed to remediation of a TTE-contaminated site (Mandal et al., 2016). A few arbuscular mycorrhizal fungi (AMF) have important functions for enhancing resistance against biotic and abiotic stresses (Harrier, 2001). AMF have a great capacity to alleviate TTE stress of plants (Hildebrandt et al., 2007). AMF contribute substantially for the enhancement of mineral and water accumulation by host plants through their extraradical mycelial networks, which also protect host plants against TTE toxicity. Besides AMF, there are other beneficial microorganisms that may contribute to the tolerance of plants to TTE contamination. Khan et al. (2000) reported that mycorrhizal species changed the root rhizosphere system, and they can be used in decontamination of contaminated sites.

Plant growth-promoting rhizobacteria (PGPR) are bacteria that colonize plant rhizospheres and promote growth and yield of plants (Wu et al., 2006). Exact mechanisms of the effects of PGPR on plant growth are not completely understood. However, PGPR are known for improving root architecture that helps in the colonization of rhizobacteria, which facilitate nutrient absorption, mineralization, and chelation (Islam et al., 2014). For example, maize seeds inoculated with the rhizobacteria, *Pseudomonas cepacia*, *P. fluorescens*, and *Streptomyces aurantiacus*, in combination with nitrogen enhanced crop yield about 25% over non-rhizobacterial colonized maize (Islam et al., 2014). Soil and plants harbor and support millions of microorganisms that are involved in promoting plant growth, increasing nutrient use efficiency, and controlling disease-causing phytopathogens. Some specific groups of microorganisms play critical roles in adsorbing and desorbing mineral-nutrient elements to physical surfaces, as well as decomposition of organic residues followed by soil incorporation (Kumar and Dubey, 2020). Besides plant-growth promotion, endophytic microbes also aid in enhancing many aspects of plant resilience against a diverse array of biotic and abiotic stresses (Singh, 2016; Oleńska et al., 2020; Rai et al., 2020). Biotic stresses mainly include saprophyte groups of microorganisms and harmful microorganisms that cause plant diseases and harbor pest populations. Abiotic stresses include soil salinization, radiation, water logging, drought, extreme heat, and TTE.

Microbial metabolism of TTE and transformation met by several groups of microorganisms such as bacteria, fungi, and algae have a direct role for cleaning up TTE-contaminated environments. Groups of microbial strains singly, in consortium, or in immobilized forms are effective for TTE remediation (Kim et al., 2015). Rhizosphere-dwelling microorganisms have a great ability to control food-chain contamination with TTE. Metal-microbe interactions and microbial processes affect adsorption or desorption of metals on cell surfaces and redox phenomena that determine the fate and toxicity of TTE in the environment.

16.5 ENDOPHYTES TO ALLEVIATE ABIOTIC STRESS OF PLANTS

Besides the level of TTE, other soil factors influence the toxicity of TTE such as moisture, presence of salts in the rhizospheres, EC, temperature, pH, and organic matter content. Endophytic microbes could reduce the effect of TTE by moderating these soil factors (Table 16.4), which indirectly affects the toxicity of TTE. The presence of endophytes has positive effects on a plant's capacity to survive with various kinds of stresses such as drought, salinity, nitrogen deficiency, and pathogens (Giasson et al., 2008; Franco-Franklin et al., 2021). Sometimes endophytes confer entirely new traits to the host plant that were not present in same host without endophytes and add biotechnological value to the harboring plants. Among prokaryotes, Gammaproteobacteria constitute a major phylum of endophytes (26%) (Hardoim et al., 2015). A meta-analysis of 11 experiments that considered taxonomic studies of desert host endophytes showed that about 47.14% culturable endophytes belonged to the phylum Proteobacteria, 26.22% to Firmicutes, 22.55% to Actinobacteria, and 2.09% belonged to the phylum Bacteroides (Zhang and White, 2021). Endophytic bacteria from arable crops possess nutrient solubilization and fixation ability. Endophytic bacterial isolates obtained at different plant physiological growth stages of corn showed nitrogen fixation, phosphorus and potassium solubilizing ability, siderophore production, IAA production, and antifungal activity. Seed inoculation with the endophytic isolate *Burkholderia* sp. FDN2-1 enhanced root length and biomass of corn seedlings (Baghel et al., 2020). Endophytes have a high potential to mitigate different kinds of abiotic stresses via modulating local or systemic mechanisms and producing antioxidants to counteract reactive oxygen species (ROS) in plants (Verma et al., 2021). In desert soil, endophytes show a great potential

TABLE 16.4
Abiotic Stress (Heat, Drought, Salinity, Nutrient Stress) Alleviation in Plants Using Endophytes

Endophyte	Plant	Benefit Offered to Host	References
Aspergillus japonicus	Soybean and sunflower	Protection from heat stress	Hamayun et al. (2018)
Neocamarosporium sp. *Periconia macrospinosa*	Tomato	Drought stress amelioration	Moghaddam et al. (2022)
Streptomyces sp.	Sunflower	Drought and salt stress	Zahra et al. (2020)
Fungal endophytes	Antarctic pearlwort (*Colobanthus quitensis*)	UV radiation	Ramos et al. (2018)
Piriformospora indica	*Arabidopsis*	Salinity	Abdelaziz et al. (2017)
Piriformospora indica	*Brassica campestris ssp. chinensis*	Drought protection	Sun et al. (2010)
Gluconacetobacter diazotrophicus	*Saccharum officinarum* cv. SP70-1143 Shoots	Drought protection	Vargas et al. (2014)
Pantoea agglomerans	*Zea mays mexicana*	Salinity	Gond et al. (2015)
Burkholderia phytofirmans PsJN	*Arabidopsis thaliana*	*Low temperature*	*Su et al. (2015)*
Streptomyces coelicolor	Wheat (*Triticum aestivum*)	Drought protection	Yandigeri et al. (2012)
Burkholderia sp. strain FDN2-	Maize	P and K solubilization	Baghel et al. (2020)
Bacillus cereus PK6-15, B. subtilis PK5-26, and *B. circulans PK3-109*	*Arabidopsis thaliana*	Protection to salt stress	Bokhari et al. (2019)

to enhance nutrient uptake and plant resistance to drought and heat stress, salt stress, and pathogen attack (Zhang and White, 2021). Bokhari et al. (2019) reported plant growth promotion and enhanced fresh weight of host plants under salt stress after inoculation with the endophytic *Bacillus cereus* PK6–15, *B. subtilis* PK5–26, and *B. circulans* PK3–109, isolated from desert soil, compared to uninoculated plants. Induction of tissue-specific transcriptional changes of ion transporters and reduced Na^+/K^+ shoot ratios with inoculation of endophytic isolates were implicated as likely mechanisms for salt tolerance in *Arabidopsis* (Eida et al., 2019). Moghaddam et al. (2022) showed that halotolerant endophytic fungi, such as *Periconia macrospinosa*, *Neocamarosporium goegapense*, and *N. chichastianum*, isolated from the endo-rhizosphere of salt-lake plants, could alleviate harsh effects of extreme salinity and drought on barley under greenhouse conditions by increasing the proline content and antioxidant enzyme activity of catalase, peroxidase, and superoxide dismutase.

Endophytic association can mitigate temperature stress to plants. Khan et al. (2013) observed that *P. resedanum* inoculation led to improved plant growth in heat stress scenario. These plants having higher survival rate produced a greater above ground biomass compared to uninoculated plants. Inoculation with *P. resedanum*, the electrolytic leakage and lipid peroxidation decreased tremendously while reducing the adverse effect of temperature stress. Endophytes were reported to support plant survival in very harsh climatic conditions by several ways, such as the production of gibberellins by *P. funiculosum* LHL06 and *Exophiala* sp. LHL08, providing greater advantage to the host plants during salt and thermal stress.

Under abiotic stress conditions, plants overproduce ROS that can cause oxidative stress leading to degradation of protein molecules, lipid peroxidation, and interference with the physiology of plants. To counteract the oxidative stress due to ROS, glutathione and carotenoids are useful to sequester free radicals and protect the plant cell from oxidative stress (Verma et al., 2021). Endophytes influence these processes by improving nutrient acquisition, supporting photosynthesis, synthesis of pigments and phytohormones, ACC deaminase activity, and redox transformation of elements in plants

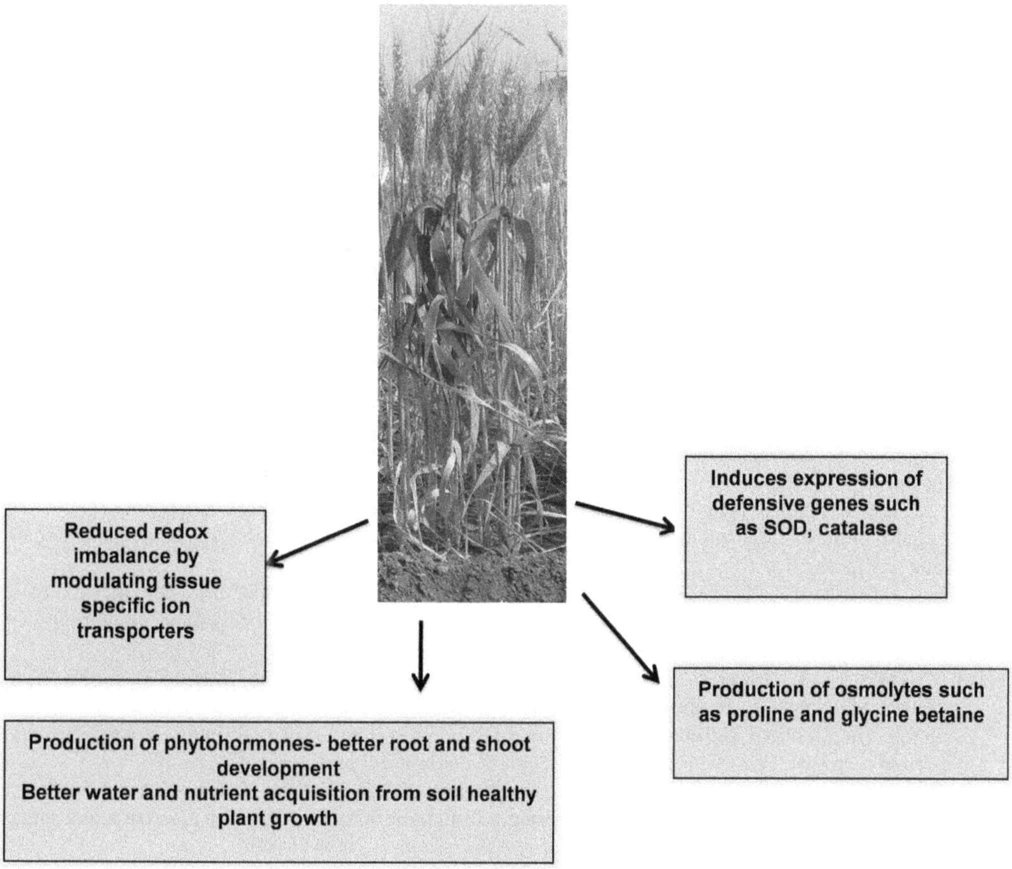

FIGURE 16.2 Mechanism of salinity and moisture stress alleviation by endophyte harboring plants.

in the vicinity of roots (Figure 16.2). Endophytes have also been observed to influence plant gene expression (Eida et al., 2019; Modi et al., 2020) to regulate physiological processes that overcome the effects of stress. Endophyte-associated plants grown in TTE-contaminated environments under abiotic stress can, thus, efficiently remediate TTE by combatting both TTE and abiotic stresses.

16.6 CONCLUSIONS

Due to various anthropogenic activities, soil pollution with TTE and consequent degradation of soil quality are increasing. The harmful effects of TTE eventually affect humans via consumption of edible plants grown on such contaminated sites. In the search for economically and environmentally viable techniques for TTE remediation, biological approaches that combine endophytic microbes and plants have shown promising outcomes in recent years. Endophytic microbes assist the plant partner in mitigating the effect of TTE and other biotic and abiotic stresses by transforming the redox state of elements; inducing defense molecules, enzymes, and hormones; and regulating the uptake of ionic elements by plants. Identification of suitable organisms and conditions for enhancing the TTE remediation potential will bring success to microbe-assisted phytoremediation. Hence, future studies should be carried out to find efficient organisms with high plant colonization efficiency and TTE detoxification potential for applying this promising remediation method under field conditions.

REFERENCES

Abdelaziz, M. E., Kim, D., Ali, S., Fedoroff, N. V., and Al-Babili, S. 2017. The endophytic fungus Piriformospora indica enhances Arabidopsis thaliana growth and modulates Na^+/K^+ homeostasis under salt stress conditions. *Plant Science* 263:107–115.

Alengebawy, A., Abdelkhalek, S. T., Qureshi, S. R., and Wang, M. Q. 2021. Heavy metals and pesticides toxicity in agricultural soil and plants: Ecological risks and human health implications. *Toxics* 9:42.

An, H., Liu, Y., Zhao, X., Huang, Q., Yuan, S., Yang, X., and Dong, J. 2015. Characterization of cadmium-resistant endophytic fungi from Salix variegata Franch in three Gorges Reservoir Region, China. *Microbiological Research* 176:29–37.

Anonymous. 1987. *Food Standards Code*. Australian Government Publishing Service: Canberra.

Antisari, L. V., Orsini, F., Marchetti, L., Vianello, G., and Gianquinto, G. 2015. Heavy metal accumulation in vegetables grown in urban gardens. *Agronomy for Sustainable Development* 35:1139–1147.

Arita, A., and Costa, M. 2011. Environmental agents and epigenetics. In: Tollefsbol, T. (Ed.), *Handbook of Epigenetics*. Academic Press: Cambridge, MA, pp. 459–476.

Artiola, J. F., Walworth, J. L., Musil, S. A., and Crimmins, M. A. 2019. Soil and land pollution. In: Brusseau, M. L., Pepper, I. L., and Gerba, C. (Eds.), *Environmental and Pollution Science*. Academic Press: Cambridge, MA, pp. 219–235.

ATSDR. 2012. Toxicological profile for chromium. http://www.atsdr.cdc.gov/toxprofiles/tp7.pdf.

ATSDR. 2015. Priority list of hazardous substances. http://www.atsdr.cdc.gov/spl/index.html#modalId String_myTable2015.

Azubuike, C. C., Chikere, C. B., and Okpokwasili, G. C. 2016. Bioremediation techniques–classification based on site of application: Principles, advantages, limitations and prospects. *World Journal of Microbiology & Biotechnology* 32:180.

Baghel, V., Thakur, J. K., Yadav, S. S., Manna, M. C., Mandal, A., Shirale, A. O., Sharma, P., Sinha, N. K., Mohanty, M., Singh, A. B. and Patra, A. K. 2020. Phosphorus and potassium solubilization from rock minerals by endophytic *Burkholderia* sp. strain FDN2-1 in soil and shift in diversity of bacterial endophytes of corn root tissue with crop growth stage. *Geomicrobiology Journal* 37:550–563.

Barrett, H. A., Ferraro, A., Burnette, C., Meyer, A., and Krekeler, M. P. 2012. An investigation of heavy metal content from disposable batteries of non-US origin from butler County, Ohio: An environmental assessment of a segment of a waste stream. *Journal of Power Sources* 206:414–420.

Biester, H., Müller, G., and Schöler, H. F. 2002. Binding andmobility of mercury in soils contaminated by emissions fromchlor-alkali plants. *Science of the Total Environment* 284:191–203.

Bokhari, A., Essack, M., Lafi, F. F., Andres-Barrao, C., Jalal, R., Alamoudi, S., Razali, R., Alzubaidy, H., Shah, K. H., Siddique, S., and Bajic, V. B. 2019. Bioprospecting desert plant Bacillus endophytic strains for their potential to enhance plant stress tolerance. *Scientific Reports* 9:1–13.

Bonfante, A., Basile, A., and Bouma, J. 2020. Targeting the soil quality and soil health concepts when aiming for the United Nations Sustainable Development Goals and the EU Green Deal. *Soil* 6(2):453–466.

Bridges, C. C., and Zalups, R. K. 2005. Molecular and ionicmimicry and the transport of toxic metals. *Toxicology and Applied Pharmacology* 204:274–308.

Bridges, C. C., and Zalups, R. K. 2017. Mechanisms involved in the transport of mercuric ions in target tissues. *Archives of Toxicology* 91:63–81.

Burges, A., Epelde, L., and Garbisu, C. 2015. Impact of repeated single-metal and multi-metal pollution events on soil quality. *Chemosphere* 120:8–15.

Clemens, S., and Ma, J. F. 2016. Toxic heavy metal and metalloid accumulation in crop plants and foods. *Annual Review of Plant Biology* 1–24:489–512.

Csavina, J., Taylor, M. P., Felix, O., Rine, K. P., Eduardo Saez, A., and Betterton, E. A. 2014. Size-resolved dust and aerosol contaminants associated with copper and lead smelting emissions: Implications for emissions management and human health. *Science of the Total Environment* 493:750–756.

D'Amore, J. J., Al-Abed, S. R., Scheckel, K. G., and Ryan, J. A. 2005. Methods for speciation of metals in soils: A review. *Journal of Environmental Quality* 34:1707–1745.

Dabral, S., Varma, A., Choudhary, D. K., Bahuguna, R. N., and Nath, M. 2019. Biopriming with Piriformospora indica ameliorates cadmium stress in rice by lowering oxidative stress and cell death in root cells. *Ecotoxicology Environmental Safety* 186:109741.

de Quadros, P. D., Zhalnina, K., Davis-Richardson, A. G., Drew, J. C., Menezes, F. B., Flávio, A. D. O., et al. 2016. Coal mining practices reduce the microbial biomass, richness and diversity of soil. *Appled Soil Ecology* 98:195–203.

Deng, Z., and Cao, L. 2017. Fungal endophytes and their interactions with plants in phytoremediation: A review. *Chemosphere* 168:1100–1106.

Dhal, B., Thatoi, H. N., Das, N. N., and Pandey, B. D. 2013. Chemical and microbial remediation of hexavalent chromium from contaminated soil and mining/metallurgical solid waste: A review. *Journal of Hazardous Materials* 250–251:272–291.

Doran, J. W. 2002. Soil health and global sustainability: Translating science into practice. *Agriculture, Ecosystem and Environment* 88:119–127.

dos Santos, J. V., Varón-López, M., Soares, C. R. F. S., Leal, P. L., Siqueira, J. O., and de Souza Moreira, F. M. 2016. Biological attributes of rehabilitated soils contaminated with heavy metals. *Environmental Science and Pollution Research* 23:6735–6748.

Dotaniya, M. L., Rajendiran, S., Meena, V. D., Saha, J. K., Coumar, M. V., Kundu, S., et al. 2017. Influence of chromium contamination on carbon mineralization and enzymatic activities in Vertisol. *Agricultural Research* 6:91–96.

Drahota, P., and Filippi, M. 2009. Secondary arsenic minerals in the environment: A review. *Environment International* 35:1243–1255.

Du, Y., Chen, L., Ding, P., Liu, L., He, Q., Chen, B., and Duan, Y. 2019. Different exposure profile of heavy metal and health risk between residents near a Pb-Zn mine and a Mn mine in Huayuan County, South China. *Chemosphere* 216:352–364.

Du, P., Zhang, L., Ma, Y., Li, X., Wang, Z., Mao, K., Wang, N., Li, Y., He, J., Zhang, X., et al. 2020. Occurrence and fate of heavy metals in municipal wastewater in Heilongjiang Province, China: A monthly reconnaissance from 2015 to 2017. *Water* 12:728.

Eida, A. A., Alzubaidy, H. S., de Zélicourt, A., Synek, L., Alsharif, W., Lafi, F. F., Hirt, H., and Saad, M. M. 2019. Phylogenetically diverse endophytic bacteria from desert plants induce transcriptional changes of tissue-specific ion transporters and salinity stress in Arabidopsis thaliana. *Plant Science* 280:228–240.

Ercal, N., Gurer-orhan, H., and Aykin-burns, N. 2001. Toxic metals and oxidative stress part I: Mechanisms involved in metal-induced oxidative damage. *Current Topics in Medicinal Chemistry* 1:529–539.

EU SCOOP. 2004. Assessment of the dietary exposure to arsenic, cadmium, lead and mercury of the population of the EU Member States. Reports on Tasks for Scientific Cooperation. Report of Experts Participating in Task 3.2.11. Directorate-General Health and Consumer Protection, European Commission, Brussels.

Fang, L., Liu, Y., Tian, H., Chen, H., Wang, Y., and Huang, M. 2017. Proper land use for heavy metal-polluted soil based on enzyme activity analysis around a Pb-Zn mine in Feng County, China. *Environmental Science and Pollution Research International* 24:28152–28164.

FAO/WHO. 2001. Food Additives and Contaminants. Joint Codex Alimentarius Commission. FAO/WHO Food Standards Programme, ALINORM 01/12A.

Ferreira, V., Koricheva, J., Duarte, S., Niyogi, D. K., and Guérold, F. 2016. Effects of anthropogenic heavy metal contamination on litter decomposition in streams–a meta-analysis. *Environmental Pollution* 210:261–270.

Filipiak-Szok, A., Kurzawa, M., and Szlyk, E. 2015. Determination of toxic metals by ICP-MS in Asiatic and European medicinal plants and dietary supplements. *Journal of Trace Elements in Medicine and Biology* 30:54–58.

Franco-Franklin, V., Moreno-Riascos, S., and Ghneim-Herrera, T. 2021. Are endophytic bacteria an option for increasing heavy metal tolerance of plants? A meta-analysis of the effect size. *Frontiers in Environmental Science* 8:603668.

Ghorbani, A., Tafteh, M., Roudbari, N., Pishkar, L., Zhang, W., and Wu, C. 2021. Piriformospora indica augments arsenic tolerance in rice (*Oryza sativa*) by immobilizing arsenic in roots and improving iron translocation to shoots. *Ecotoxicology Environmental Safety* 209:111793.

Giacalone, A., Gianguzza, A., Orecchio, S., Piazzese, D., Dongarra, G., Sciarrino, S., and Varrica, D. 2005. Metals distribution in the organic and inorganic fractions of soil: A case study on soils from Sicily. *Chemical Speciation and Bioavailability* 17:83–93.

Giasson, P., Karam, A., and Jaouich, A. 2008. Arbuscular mycorrhizae and alleviation of soil stresses on plant growth. In: Siddiqui, Z. A., Akhtar, M. S., and Futai, K. (Eds.), *Mycorrhizae: Sustainable Agriculture and Forestry*. Springer Science & Business Media: Berlin/Heidelberg, Germany pp. 99–134.

Gond, S. K., Torres, M. S., Bergen, M. S., Helsel, Z., and White, J. F. 2015. Induction of salt tolerance and up-regulation of aquaporin genes in tropical corn by rhizobacteriumPantoea agglomerans. *Letters in Applied Microbiology* 60(4):392–399.

Govarthanan, M., Mythili, R., Selvankumar, T., Kamala-Kannan, S., Rajasekar, A., and Chang, Y. C. 2016. Bioremediation of heavy metals using an endophytic bacterium Paenibacillus sp. RM isolated from the roots of Tridax procumbens. *3 Biotech* 6:1–7.

Gu, Y., Wang, Y., Sun, Y., Zhao, K., Xiang, Q., Yu, X., Zhang, X., and Chen, Q. 2018. Genetic diversity and characterization of arsenic-resistant endophytic bacteria isolated from Pteris vittata, an arsenic hyperaccumulator. *BMC Microbiology* 18:1–10.

Haberern, J. 1992. Viewpoint: A soil health index. *Journal of Soil and Water Conservation* 47:6.

Hamayun, M., Hussain, A., Iqbal, A., Khan, S. A., and Lee, I. J. 2018. Endophytic fungus Aspergillus japonicus mediates host plant growth under normal and heat stress conditions. *BioMed Research International* 2018:7696831.

Han, Y., Kingston, H. M., Boylan, H. M., Rahman, G. M. M., Shah, S., Richter, R. C., Link, D. D., and Bhandari, S. 2003. Speciation of mercury in soil and sediment by selective solvent and acid extraction. *Analytical and Bioanalytical Chemistry* 375:428–436.

Handlogten, M. E., Shiraishi, N., Awata, H., Huang, C., and Miller, R. T. 2000. Extracellular Ca($^{2+}$)-sensing receptor is a promiscuous divalent cation sensor that responds to lead. *American Journal of Physiology Renal Physiology* 279:F1083–F1091.

Hardoim, P. R., Van Overbeek, L. S., Berg, G., Pirttilä, A. M., Compant, S., Campisano, A., Döring, M., and Sessitsch, A. 2015. The hidden world within plants: Ecological and evolutionary considerations for defining functioning of microbial endophytes. *Microbiology and Molecular Biology Reviews* 79(3):293–320.

Harrier, L. A. 2001. The arbuscular mycorrhizal symbiosis: A molecular review of the fungal dimension. *Journal of Experimental Botany* 5:469–478.

He, Z., Yang, X., Baligar, V. C., Zhang, T., and Stoffella, P. J. 2015. Heavy metal contamination of soils: Sources, indicators, and assessment. *Journal of. Environmental Indicator* 9:17–18.

Hildebrandt, U., Regvar, M., and Bothe, H. 2007. Arbuscular mycorrhiza and heavy metal tolerance. *Phytochemistry* 68:139–146.

Hogsden, K. L., and Harding, J. S. 2012. Consequences of acid mine drainage for the structure and function of benthic stream communities: A review. *Freshwater Science* 31:108–120.

Hu, X. F., Jiang, Y., Shu, Y., Hu, X., Liu, L., and Luo, F. 2014. Effects of mining wastewater discharges on heavy metal pollution and soil enzyme activity of the paddy fields. *Jouranl of Geochemical Exploration* 147:139–150.

Hui, F., Liu, J., Gao, Q., and Lou, B. 2015. Piriformospora indica confers cadmium tolerance in Nicotiana tabacum. *Journal of Environmental Science* 37:184–191.

Ikram, M., Ali, N., Jan, G., Jan, F. G., Rahman, I. U., Iqbal, A., and Hamayun, M. 2018. IAA producing fungal endophyte Penicillium roqueforti Thom., enhances stress tolerance and nutrients uptake in wheat plants grown on heavy metal contaminated soils. *PLoS One* 13:e0208150.

Ishchenko, V. 2017. Soil contamination by heavy metal mobile forms near landfill. *International Journal of Environment and Waste Management* 20:66–74.

Islam, F., Yasmeen, T., Ali, Q., Ali, S., Arif, M. S., Hussain, S., and Rizvi, H. 2014. Influence of Pseudomonas aeruginosa as PGPR on oxidative stress tolerance in wheat under Zn stress. *Ecotoxicology and Environmental Safety* 104:285–293.

Jan, R., Khan, M. A., Asaf, S., Lee, I. J., and Kim, K. M. 2019. Metal resistant endophytic bacteria reduces cadmium, nickel toxicity, and enhances expression of metal stress related genes with improved growth of *Oryza sativa*, via regulating its antioxidant machinery and endogenous hormones. *Plants* 8(10):363.

Khalid, M., Kayani, S. I., and Tang, K. 2020. The ameliorative effects of exogenous inoculation of *Piriformospora indica* on molecular, biochemical and physiological parameters of *Artemisia annua* L. under arsenic stress condition. *Ecotoxicology and Environmental Safety* 206:111202.

Khan, A. G., Kuek, C., Chaudhry, T. M., Khoo, C. S., and Hayes, W. J. 2000. Role of plants, mycorrhizae and phytochelators in heavy metal contaminated land remediation. *Chemosphere* 41:197–207.

Khan, A. L., Kang, S., Dhakal, K. H., Hussain, J., Adnan, M., Kim, J., and Lee, I. 2013. Flavonoids and amino acid regulation in *Capsicum annuum* L. by endophytic fungi under different heat stress regimes. *Scientific Horticulture* 155:1–7.

Khan, M. U., Sessitsch, A., Harris, M., Fatima, K., Imran, A., Arslan, M., Shabir, G., Khan, Q. M., and Afzal, M. 2015. Cr-resistant rhizo-and endophytic bacteria associated with *Prosopis juliflora* and their potential as phytoremediation enhancing agents in metal-degraded soils. *Frontiers in Plant Science* 5:755.

Khanam, R., Kumar, A., Nayak, A. K., Shahid, M., Tripathi, R., Vijayakumar, S., Bhaduri, D., Kumar, U., Mohanty, S., Panneerselvam, P., Chatterjee, D., Satapathy, B. S. and Pathak, H. 2019. Metal(loid)s (As, Hg, Se, Pb and Cd) in paddy soil: Bioavailability and potential risk to human health. *Science of the Total. Environment* 699:134330.

Kim, I. H., Choi, J. H., Joo, J. O., Kim, Y. K., Choi, J. W., and Oh, B. K. 2015. Development of a microbe-zeolite carrier for the effective elimination of heavy metals from seawater. *Journal of Microbiology, Biotechnology* 25:1542–1546.

Kozlov, M. V., and Zvereva, E. L. 2015. Decomposition of birch leaves in heavily polluted industrial barrens: Relative importance of leaf quality and site of exposure. *Environmental Science and Pollution Research* 22:9943–9950.

Kruger, M. C., Bertin, P. N., Heipieper, H. J., and Arsène-Ploetze, F. 2013. Bacterial metabolism of environmental arsenic: Mechanisms and biotechnological applications. *Applied Microbiology and Biotechnology* 97:3827–3841.

Kumar, A., and Dubey, A. 2020. Rhizosphere Microbiome: Engineering bacterial competitiveness for enhancing crop production. *Journal of Advanced Research* 24:337–352.

Lally, R. D., Galbally, P., Moreira, A. S., Spink, J., Ryan, D., Germaine, K. J., and Dowling, D. N. 2017. Application of endophytic Pseudomonas fluorescens and a bacterial consortium to Brassica napus can increase plant height and biomass under greenhouse and field conditions. *Frontiers in Plant Science* 8:2193.

Li, H.-Y., Wei, D.-Q., Shen, M., and Zhou, Z.-P. 2012. Endophytes and their role in phytoremediation. *Fungal Diversity* 54:11–18.

Lin, H., Liu, C., Li, B., and Dong, Y. 2021. *Trifolium repens* L. regulated phytoremediation of heavy metal contaminated soil by promoting soil enzyme activities and beneficial rhizosphere associated microorganisms. *Journal of Hazardous Materials* 402:123829.

LLC Books. 2010. *Cadmium Minerals: Greenockite, Hawleyite, Otavite, Niedermayrite, Cadmoselite*, General Books, LLC: Memphis, TN, p. 20.

Lodewyckx, C., Mergeay, M., Vangronsveld, J., Clijsters, H., van der Lelie, D. 2002. Isolation, characterization, and identification of bacteria associated with the zinc hyperaccumulator Thlaspicaerulescens subsp. calaminaria. *International Journal of Phytoremediation* 4:101–115.

López, D. L., Bundschuh, J., Birkle, P., Armienta, M. A., Cumbal, L., Sracek, O., Cornejo, L., and Ormachea, M. 2012. Arsenicin volcanic geothermal fluids of Latin America. *Science of the Total Environment* 429:57–75.

Ma, Y., Rajkumar, M., Zhang, C., and Freitas, H. 2016. Beneficial role of bacterial endophytes in heavy metal phytoremediation. *Journal of Environmental Management* 174:14–25.

Mahmood, A., and Malik, R. N. 2014. Human health risk assessment of heavy metals via consumption of contaminated vegetables collected from different irrigation sources in Lahore, Pakistan. *Arabian Journal of Chemistry* 7:91–99.

Malik, A., Butt, T. A., Naqvi, S. T. A., Yousaf, S., Qureshi, M. K., Zafar, M. I., Farooq, G., Nawaz, I., and Iqbal, M. 2020. Lead tolerant endophyte Trametes hirsuta improved the growth and lead accumulation in the vegetative parts of *Triticum aestivum* L. *Heliyon* 6:04188.

Mandal, B. K., and Suzuki, K. T. 2002. Arsenic round the world: A review. *Talanta* 58, 201–235.

Mandal, A., Thakur, J. K., Sahu, A., Bhattacharjya, S., Manna, M. C., and Patra, A. K. 2016. Plant–microbe interaction for the removal of heavy metal from contaminated site. In: Choudhary, D. K., Varma, A., and Tuteja, N. (Eds.), *Plant-Microbe Interaction: An Approach to Sustainable Agriculture*. Springer: Singapore, pp. 227–247.

Mandal, A., Sarkar, B., Mandal, S., Vithanage, M., Patra, A. K. and Manna, M. C. 2020. Impact of agrochemicals on soil health. In: Vara Prasad, M. N. (Ed.), *Agrochemicals Detection, Treatment and Remediation*. Butterworth-Heinemann: Oxford, pp. 161–187.

Mathema, V. B., Thakuri, B. C., and Sillanpää, M. 2011. Bacterial mer operon-mediated detoxification of mercurial compounds: A short review. *Archieves in Microbiololgy* 193:837–844.

Mengoni, A., Barzanti, A., Gonnelli, C., Gabrielli, R., and Bazzicalupo, M. 2001. Characterization of nickel-resistant bacteria isolated from serpentine soil. *Environmental Microbiology* 3:691–698.

Mesa, J., Mateos-Naranjo, E., Caviedes, M. A., Redondo-Gómez, S., Pajuelo, E., and Rodríguez-Llorente, I. D. 2015. Endophytic cultivable bacteria of the metal bioaccumulator Spartina maritima improve plant growth but not metal uptake in polluted marshes soils. *Frontiers in Microbiology,* 6:1450.

Minhas, P. S., Saha, J. K., Dotaniya, M. L., Sarkar A., and Saha, M. 2022. Wastewater irrigation in India: Current status, impacts and response options. *Science of the Total Environment* 808:15.

Modi, A., Kanani, P., and Kumar, A. 2020. Fungal endophytes-induced gene expression studies in biotic and abiotic stress management. In: Kumar, A., and Radhakrishnan, E. K. (Eds.), *Microbial Endophytes*. Woodhead Publishing, Sawston, CA, pp. 251–271.

Moghaddam, M. S. H., Safaie, N., Rahimlou, S., and Hagh-Doust, N. 2022. Inducing tolerance to abiotic stress in *Hordeum vulgare* L. by halotolerant endophytic fungi associated with salt lake plants. *Frontiers in Microbiology* 13:906365.

MRT. 2001. Tasmanian geological survey record 2001/08, A mineralogical field guide for a Western Tasmania minerals and museums tour, Mineral Resources, Tasmania.

Muller, D., Médigue, C., Koechler, S., et al. 2007. A tale of two oxidation states: Bacterial colonization of arsenic-rich environments. *PLoS Genetics* 3518:530.

O'Brien, T. J., Ceryak, S., and Patierno, S. R. 2003. Complexities of chromium carcinogenesis: Role of cellular response, repair and recovery mechanisms. *Mutation Research* 533:3–36.

Okoh, A. T., Odjadjare, E. E., Igbinosa, E. O., and Osode, A. N. 2007. Wastewater treatment plants as a source of microbial pathogens in receiving water sheds. *African Journal of Biotechnology* 6:2932–2944.

Oleńska, E., Małek, W., Wójcik, M., Swiecicka, I., Thijs, S., and Vangronsveld, J. 2020. Beneficial features of plant growth-promoting rhizobacteria for improving plant growth and health in challenging conditions: A methodical review. *Science of Total Environment* 32:429–448.

Patra, M., Bhowmik, N., Bandopadhyay, B., and Sharma, A. 2004. Comparison of mercury, lead and arsenic with respect to genotoxic effects on plant systems and the development of genetic tolerance. *Environmental and Experimental Botany* 52:199–223.

Paz-Ferreiro, J., and Fu, S. 2016. Biological indices for soil quality evaluation: Perspectives and limitations. *Land Degradation and Development* 27:14–25.

Rahman, Z., and Singh, V. P. 2019. The relative impact of toxic heavy metals (THMs) (arsenic (As), cadmium (Cd), chromium (Cr)(VI), mercury (Hg), and lead (Pb)) on the total environment: An overview. *Environmental Monitoring Assessment* 191:419.

Rai, P.K., Singh, M., Anand, K., Saurabh, S., Kaur, T., Kour, D. et al. 2020. *New and Future Developments in Microbial Biotechnology and Bioengineering*. Elsevier: Netherland, pp. 49–60.

Ramos, P., Rivas, N., Pollmann, S., Casati, P., and Molina-Montenegro, M. A. 2018. Hormonal and physiological changes driven by fungal endophytes increase Antarctic plant performance under UV-B radiation. *Fungal Ecology* 34:76–82.

Richardson, G. M., Garrett, R., Mitchell, I., Mah-Poulson, M., and Hackbarth, T. 2001. Critical review on natural global and regional emissions of six trace metals to the atmosphere. Prepared for the International Lead Zinc Research Organisation, the International Copper Association, and the Nickel Producers Environmental Research Association.

Rieuwerts, J. S., Thornton, I., Farago, M. E., and Ashmore, M. R. 1998. Factors influencing metal bioavailability in soils: Preliminary investigations for the development of a critical loads approach for metals. *Chemical Speciation and Bioavailability* 10:61–75.

Robles-Camacho, J., and Armienta, M. A. 2000. Natural chromium contamination of groundwater at Leon Valley, Mexico. *Journal of Geochemical Exploration* 68:167–181.

Rosen, B. P. 2002. Biochemistry of arsenic detoxification. *Science* 529:86–92.

Sagonda, T., Adil, M. F., Sehar, S., Rasheed, A., Joan, H. I., Ouyang, Y., and Shamsi, I. H. 2021. Physio-ultrastructural footprints and iTRAQ-based proteomic approach unravel the role of Piriformospora indica-colonization in counteracting cadmium toxicity in rice *Ecotoxicology Environmental Safety* 220:12390.

Sepehri, M., and Khatabi, B. 2021. Combination of siderophore-producing bacteria and Piriformospora indica provides an efficient approach to improve cadmium tolerance in Alfalfa. *Microbial Ecology* 81:717–730.

Shahabivand, S., Parvaneh, A., and Aliloo, A. A. 2017. Root endophytic fungus Piriformospora indica afected growth, cadmium partitioning and chlorophyll fuorescence of sunfower under cadmium toxicity. *Ecotoxicology Environmental Safety* 145:496–502.

Sharma, B., Sarkar, A., Singh, P., and Singh, R. P. 2017. Agricultural utilization of biosolids: A review on potential effects on soil and plant grown. *Waste Management* 64:117–132.

Sheik, C. S., Mitchell, T. W., Rizvi, F. Z., Rehman, Y., Faisal, M., Hasnain, S., McInerney, M. J., and Krumholz, L. R. 2012. Exposure of soil microbial communities to chromium and arsenic alters their diversity and structure. *PLoS One* 7:e40059.

Sheng, X. F., Xia, J. J., Jiang, C. Y., He, L. Y., and Qian, M. 2008. Characterization of heavy metal-resistant endophytic bacteria from rape (Brassica napus) roots and their potential in promoting the growth and lead accumulation of rape. *Environmental Pollution* 156:1164–1170.

Singh, S. 2016. *Microbial Inoculants in Sustainable Agricultural Productivity*. Springer: Cham, pp. 69–83.

Srivastava, V., Sarkar, A., Singh, S., Singh, P., de Araujo, A. S. F., and Singh, R. P. 2017. Agroecological responses of heavy metal pollution with special emphasis on soil health and plant performances. *Frontiers in Environmental Science* 5:1–19.

Srivastava, S., Chaudhuri, M., and Pandey, V. C. 2020. Endophytes: The hidden world for agriculture, ecosystem, and environmental sustainability. In: Pandey, V. C., and Singh, V. (Eds.), *Bioremediation of Pollutants*. Elsevier: Amsterdam, Netherlands, pp. 145–159.

Stanin, F. T., and Pirnie, M. 2005. *The Transport and Fate of Cr (VI) in the Environment, Chromium (VI) Handbook.* CRC Press: Boca Ratón, FL, pp. 162–211.

Stuczynski, T. I., McCarty, G. W., and Siebielec, G. 2003. Response of soil microbiological activities to cadmium, lead, and zinc salt amendments. *Journal of Environmental Quality* 32:1346–135.

Su, F., Jacquard, C., Villaume, S., Michel, J., Rabenoelina, F., Clément, C., Barka, E. A., Dhondt-Cordelier, S., and Vaillant-Gaveau, N. 2015. Burkholderia phytofirmans PsJN reduces impact of freezing temperatures on photosynthesis in Arabidopsis thaliana. *Frontiers in Plant Science* 6:810.

Su, Z., Zeng, Y., Li, X., Perumal, A. B., Zhu, J., Lu, X., and Lin, F. 2021. The endophytic fungus *Piriformospora indica*-assisted alleviation of cadmium in tobacco. *Journal of Fungi* 7:675.

Sun, C., Johnson, J. M., Cai, D., Sherameti, I., Oelmüller, R., and Lou, B. 2010. Piriformospora indica confers drought tolerance in Chinese cabbage leaves by stimulating antioxidant enzymes, the expression of drought-related genes and the plastid-localized CAS protein. *Journal of Plant Physiology* 167(12):1009–1017.

Sun, L., Guo, D., Liu, K., Meng, H., Zheng, Y., Yuan, F., and Zhu, G. 2019. Levels, sources, and spatial distribution of heavy metals in soils from a typical coal industrial city of Tangshan, China. *Catena* 175:101–109.

Tashan, H., Harighi, B., Rostamzadeh, J., and Azizi, A. 2021. Characterization of arsenic-resistant endophytic bacteria from alfalfa and chickpea plants. *Frontiers in Plant Science* 12:696750.

Thakur, J. K., Mandal, A., Amat, D., and Manna, M. C. 2018. Bioprospecting of endophytes in agriculture. *HaritDhara* 1(1):13–14.

USGS. 2016. Mineral commodity summaries 2016, U.S. Geological Survey.

Vargas, L., Santa Brígida, A. B., Mota Filho, J. P., de Carvalho, T. G., Rojas, C. A., Vaneechoutte, D., Van Bel, M., Farrinelli, L. et al. 2014. Drought tolerance conferred to sugarcane by association with Gluconacetobacter diazotrophicus: A transcriptomic view of hormone pathways. *PLoS One* 9:e114744.

Vázquez, M., Calatayud, M., Jadán Piedra, C., Chiocchetti, G. M., Vélez, D., and Devesa, V. 2015. Toxic trace elements at gastrointestinal level. *Food and Chemical Toxicology* 86:163–175.

Verma, S. K., Sahu, P. K., Kumar, K., Pal, G., Gond, S. K., Kharwar, R. N., and White, J. F. 2021. Endophyte roles in nutrient acquisition, root system architecture development and oxidative stress tolerance. *Journal of Applied Microbiology* 131:2161–2177.

Violante, A., Cozzolino, V., Perelomov, L., Caporale, A., and Pigna, M. 2010. Mobility and bioavailability of heavy metals and metalloids in soil environments. *Journal of Soil Science and Plant Nutrition* 10:268–292.

Wenzel, W. W. 2013. Arsenic. In: Alloway, B. J. (Ed.), *Heavy Metals in Soils: Trace Metals and Metalloids in Soils and Their Bioavailability*, 3rd edition. Springer: New York, pp. 241–282.

Wiewióra, B., and Żurek, G. 2021. The response of the associations of grass and epichloë endophytes to the increased content of heavy metals in the soil. *Plants* 10:429.

Wu, S. C., Cheung, K. C., Luo, Y. M., and Wong, M. H. 2006. Effects of inoculation of plant growth-promoting rhizobacteria on metal uptake by Brassica juncea. *Environmental Pollution* 140:124–135.

Xian, Y., Wang, M., and Chen, W. 2015. Quantitative assessment on soil enzyme activities of heavy metal contaminated soils with various soil properties. *Chemosphere* 139:604–608.

Xu, J., Bravo, A. G., Lagerkvist, A., Bertilsson, S., Sjöblom, R., and Kumpiene, J. 2015. Sources and remediation techniques for mercury contaminated soil. *Environmental International* 74:42–53.

Yan, A., Wang, Y., Tan, S. N., Mohd Yusof, M. L., Ghosh, S., and Chen, Z. 2020. Phytoremediation: A promising approach for revegetation of heavy metal-polluted land. *Frontiers in Plant Science* 11:359.

Yandigeri, M. S., Meena, K. K., Singh, D., Malviya, N., Singh, D. P., Solanki, M. K., Yadav, A. K., and Arora, D. K. 2012. Drought-tolerant endophytic actinobacteria promote growth of wheat (*Triticum aestivum*) under water stress conditions. *Plant Growth Regulation* 68(3): 411–420.

Yasseen, B. T., and Al-Thani, R. F. 2022. Endophytes and halophytes to remediate industrial wastewater and saline soils: Perspectives from qatar. *Plants* 11:1497.

Yousaf, S., Afzal, M., Anees, M., Malik, R. N., and Campisano, A. 2014. Ecology and functional potential of endophytes in bioremediation: A molecular perspective. In: Verma, V. C., and Gange, A. C. (Eds.), *Advances in Endophytic Research.* Springer: New Delhi, pp. 301–320.

Yuan, L., Zhi, W., Liu, Y., Karyala, S., Vikesland, P. J., Chen, X., et al. 2015. Lead toxicity to the performance, viability, and community composition of activated sludge microorganisms. *Environmental Science Technology* 49:824–830.

Zahra, T., Hamedi, J., and Mahdigholi, K. 2020. Endophytic actinobacteria of a halophytic desert plant Pteropyrum olivieri: Promising growth enhancers of sunflower. *3 Biotech* 10:514.

Zeng, G., Jia, W., Huang, D., Liang, H., Chao, H., Min, C., Xue, W., Gong, X., Wang, R., and Jiang, D. 2017. Precipitation, adsorption and rhizosphere effect: The mechanisms for phosphate-induced Pb immobilization in soils-a review. *Journal of Hazardous Materials* 339:354–367.

Zhang, Q., and White, J. F. 2021. Bioprospecting desert plants for endophytic and biostimulant microbes: A strategy for enhancing agricultural production in a Hotter, drier future. *Biology* 10:961.

17 Allelopathy
A Natural Factor Constraining or Supporting the Productivity of Soils?

Beata Barabasz-Krasny[1], *Peiman Zandi*[2], *Joanna Puła*[3], *Alina Stachurska-Swakoń*[4], *Ewald Schnug*[5], *and Andrzej Danel*[6]

[1]Department of Botany, Institute of Biology and Earth Science, Pedagogical University of Kraków, Kraków, Poland
[2]International Faculty of Applied Technology, Yibin University, Yibin, China
[3]Department of Agroecology and Crop Production, Faculty of Agriculture and Economics, University of Agriculture, Kraków, Poland
[4]Institute of Botany, Jagiellonian University, Kraków, Poland
[5]Department of Life Sciences, Institute for Plant Biology, Technical University of Braunschweig, Braunschweig, Germany
[6]Faculty of Materials Engineering and Physics, Kraków University of Technology, Kraków, Poland

CONTENTS

17.1 ALLELOPATHY: A GENERAL INTRODUCTION

The term "allelopathy" is derived from the Greek language and comes from the two words allelon "mutual" and "pathos," which consequently means "mutual harm" or "mutual suffering" (Rizvi et al., 1992). It was first used in 1937 by the Austrian scientist Hans Molisch (1937) in a book called *Der Einfluss einer Pflanze auf die andere – Allelopathie* (the influence of one plant on the other – allelopathy) (Willis, 2010; Ferguson et al., 2013). Allelopathy as a science is a relatively young field (Weston, 2005). Although the phenomenon itself has been known since antiquity, it was not possible to explain its mechanisms at that time (Willis, 1985). Democritus (370 BC) described the destruction of trees by treating their roots with a mixture of *Lupinus* sp. flowers and sap of *Cicuta virosa* L. (Oleszek, 1996). The occurrence of interactions between plants was noticed by Theophrastus (285 BC and 300 BC), who stated that *Brassica* sp. negatively influenced the development of *Vitis vinifera* L., which may be related to the emanation of smells (odors) (Willis, 1985, 2010). The phenomenon of allelopathy

DOI: 10.1201/9781003093565-17

was also described by Pliny (ca. AD 1), who pointed to the influence of *Juglans regia* L. and *J. nigra* L. on other plant species, including *Quercus* sp. and *Pinus* sp. (Rice, 1984; Vyvyan, 2002).

A big problem, both in Greek and Roman times, was the so-called "soil disease," which caused significant reductions in crop yield. At this time, its probable causes were not even anticipated let alone understood (Khalid and Mohd, 2016; Mushtaq et al., 2020). Theophrastus noticed that *Pisum sativum* L. does not fertilize the soil but exhausts it (Willis, 2010). At the turn of the seventeenth and eighteenth centuries, Herman Boerhaave (1668–1738), a Dutch physician and professor of botany and chemistry, who investigated balanced fluid flow and pressure balance in the human body, suggested that root exudates may play a very important role in the interaction between plants. Influenced by the increasing amount of information about the effects of various compounds on plant growth, the Swiss botanist Augustin Pyrame de Candolle (1778–1841) developed a theory of plant interaction through root exudates (Rice, 1984), which became the basis for the later concept of allelopathy. In the late nineteenth century, many studies on interactions between plants were conducted, with results that could not be adequately explained by species-specific depletion of soil nutrients. Improved chemical techniques have identified organic toxins from non-productive soils (Willis, 1985, 2010; Ferguson et al., 2013) as a cause of these effects.

In 1996 in Cádiz (Spain), during the first World Allelopathy Congress entitled "Allelopathy – a science for the future," an official definition was formulated, terming allelopathy *as any process involving secondary metabolites produced by plants, microorganisms, and fungi, influencing the growth and development of biological and agricultural systems, including animals* (Oleszek, 1996). Definitions of this phenomenon have appeared before. Rice (1974) for instance defined allelopathy as any direct or indirect influence of one plant (including microorganisms) on another by producing chemicals that escape into the environment and then inhibit the growth and development of neighboring plants. The use of the term "allelopathy" in this sense, however, may be somewhat controversial. Chemicals that inhibit the growth of a species at a certain concentration can stimulate the growth of the same species or another at a lower concentration (Putnam and Tang, 1986; Narwal, 1994). Rice (1984) later modified the definition of allelopathy, calling it a "form of positive and negative interaction of organisms, caused by the action of chemical compounds called allelochemicals." Allelochemicals are mainly secondary metabolites of plants and microorganisms that can simultaneously influence many processes in natural ecosystems and agroecosystems (Rizvi et al., 1992; Seigler, 1996; Olofsdotter et al., 2002; Trezzi et al., 2016). Until recently, the allelopathic effect was considered primarily a negative phenomenon (Radosevich et al., 2007; De Albuquerque et al., 2011). It turned out, however, that there are also positive interactions, which depend on the type of allelochemical, its concentration, and the plant species (Eichenberg et al., 2014). Many significant ecological functions of allelopathy may have been overlooked due to the single-eyed focus on their harmful effects (Bezuidenhout, 2012; Puła et al., 2016; Lipniak and Kliszcz, 2020; Możdżeń et al., 2021a).

According to some researchers, allelochemicals are an important element in the defense of plants against herbivores (Fraenkel, 1959; Lovett and Ryuntyu, 1992; Stamp, 2003). Allelopathic effects cannot be considered as part of the competition, because they are not based on the elimination of environmental resources. Competition is the removal or reduction of a common resource, while allelopathy is the addition of a chemical compound to the environment in various processes (Rice, 1984; Putnam, 1985). To avoid confusion and recognize that competition and allelopathy are often involved at the same time, Müller (1969) defined the collective interaction of one plant with another as interference. Interference is a term that is now widely used in the literature, especially in the biology of weeds, which has generally focused on harmful effects on crops. Potential causes of interference include: (i) Allelospoly (competition) – depletion of one or more resources required for growth; (ii) Allelopathy – the addition of chemical toxins by one or more species together; (iii) Allelomediation – the selective inhibition of herbivores that may feed on one species, thus benefiting others species (Szczepański, 1977; Putnam and Tang, 1986).

Aldrich (1984) described two types of allelopathy: "true allelopathy" like the release of compounds into the environment that are toxic in the form in which they are produced, and "functional

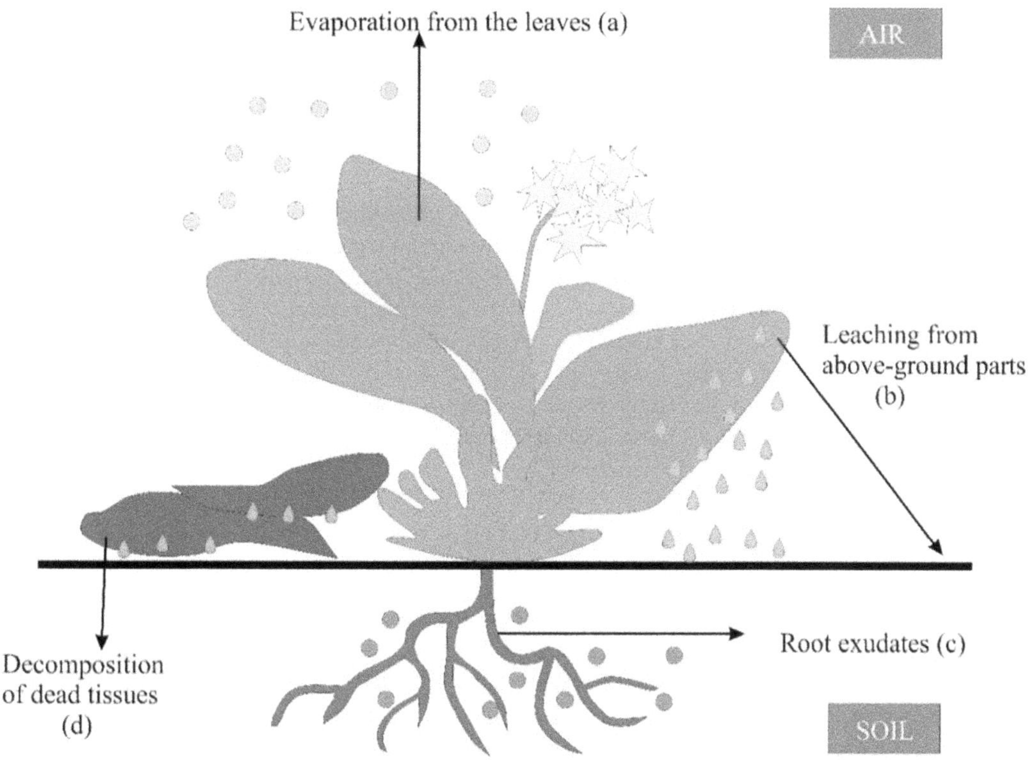

FIGURE 17.1 Routes of release of allelopathic compounds into the environment (Based on: Oleszek (1997), modified).

allelopathy," which is, for instance, the release of a toxic substance into the environment as a result of transformation by soil microorganisms. Compounds with allelopathic potential can be released into the environment, both through the aerial parts of plants and their roots (Gniazdowska et al., 2004; Gniazdowska and Bogatek, 2005). The richest source of allelopathic compounds, in terms of quality and quantity, are leaves (Bhadoria, 2011). The efficiency of the release process is largely determined by the prevailing external environmental conditions, including environmental stress (Rice, 1984), stage of development of the emitting organism, and its target (Qasem and Foy, 2001; Jilani et al., 2008).

Rice (1984) and Putnam (1985) have indicated that there are four main ways of releasing biosynthetic allelopathic substances (Figure 17.1):

a. *Evaporation, release to air*: volatile allelopathic compounds (most often essential oils) can be absorbed by surrounding plants in the form of steam, vapor condensate in dew, or, after condensation, become accessible to the roots, which are an important pathway in arid or semi-desert habitats (Bertin et al., 2003);
b. *Leaching from plant tissue by precipitation, dew, or irrigation*: allelochemicals washed from above-ground parts of plants may be deposited on other plants or accumulate in the soil (Zielińska, 1986). Leaching may also release allelopathic compounds from plant debris;
c. *Exudation (root exudate)*: allelopathic compounds released in this way probably play an important role in regulating the growth and development of neighboring plants (Bertin et al., 2003);

d. *Decomposition of plant debris*: it is difficult to unequivocally state whether the toxic substances contained in plant debris are released immediately after tissue decomposition, or whether microorganisms convert the resulting simple compounds into toxic products of higher complexity in the presence of specific enzymes (Aldrich, 1984; Mushtaq and Siddiqui, 2018, Mushtaq et al., 2020).

The source of allelopathic compounds may also be pollen from some crops (Cruz-Ortega et al., 1998).

Of all pathways through the atmosphere or soil, leaching by precipitation and decomposition of plant debris seems to be the most significant (Khalid et al., 2002; Mushtaq et al., 2020). In the case of exudation, due to the low degree of spreading in the soil – regardless of whether the secreted allelopathic compounds ooze, spill, or arise from the disintegration of plant debris, the point of action is mainly the root zone (rhizosphere) (Bertin et al., 2003; Khalid and Mohd, 2016). Allelopathy has, therefore, been considered a very complex phenomenon, involving a vast network of connections between ecological and physiological processes (Scognamiglio et al., 2013) that shape the immediate vicinity of plants, including soil conditions.

17.2 CHEMISTRY OF ALLELOPATHIC INTERACTIONS

Plants communicate with each other and with the environment using the oldest method of communication, which is called the "chemical language" (Macias et al., 2008, 2020). This "language" consists mainly of secondary metabolites as "words," which exceed the number of known words in human languages – it is estimated that there are over 400,000 (Swain, 1977; Ninkuu et al., 2021). Allelopathy deals with this chemical linguistics, a language that has the power to reduce or stimulate the growth of other plants or to repel predatory insects (Hussain and Abbas, 2021).

Among secondary metabolites with allelopathic action are alkaloids, cyanogenic glycosides, terpenes, terpenoids, waxes, tannins, steroids, alcohols, phenols, aldehydes, ketones, esters, or non-protein amino acids (Ostrumow, 1992; Table 17.1).

Quinones and phenolic compounds are the most common phytotoxins (Inderjit, 1996; Latif et al., 2017). However, the microbiological transformation of allelochemicals results in the formation of new products with other biological properties (Jilani et al., 2008; Aslam et al., 2017). Considering the richness of the plant world, we can certainly say that we only know a small fragment from the huge reservoir of substances produced by plants, and a meager fraction of which has been studied for allelopathic interactions (Einhellig and Leather, 1988).

There are several criteria for classifying allelopathic compounds. One of the most popular is the division of allelopathic compounds based on the donor-acceptor relationship (Stugren, 1978; Gniazdowska and Bogatek, 2005) – Figure (17.2).

In terms of the nature of the interaction, they can be divided into inhibitors and stimulators (Siegień et al., 2008). On the one hand, a classic example of an allelopathic inhibitor is juglone or more precisely 5-hydroxy-1,4-naphtoquinone – a chemical compound secreted by the roots and leaves of the walnut (Duran et al., 2018). Similarly, asparagine, secreted by *Medicago sativa* L. seedlings, inhibits the growth of *Triticum* sp. seedlings, and lucerne itself, which is a manifestation of autotoxicity in this plant (Oleszek et al., 1991). On the other hand, lepidimoid, a compound stimulating the growth of *Amaranthus* sp. hypocotyls, was detected in the root exudates of *Helianthus annuus* L. and *Lepidium sativum* L., which is an example of a positive effect of the allelopathic substances on plants (Yamada et al., 1995).

The criterion for the classification of allelopathic compounds may also be their ecological role. The following are distinguished in intraspecific interactions: autotoxins (they are toxic to the donor), adaptive autoinhibitors (inhibit the growth and development of the donor), and pheromones (they are responsible for communication between organisms). Among allelopathic compounds that interact between two different species, there are allomons (they have a positive effect on the donor), kairomones (use affects the acceptor), and depressants (they are neutral for the donor, but limit the growth and development of the acceptor) (Oleszek, 1996).

TABLE 17.1
Examples of Allelopathic Plant Substances

No.	Name of Substance	Chemical Structure	Mechanism of Action
		Carboxylic acids	
1.	Cinnamic acid	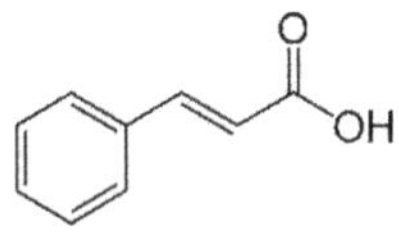	Effect on membrane permeability, hormonal activity, respiration, photosynthetic activity, and synthesis of organic compounds
2.	Benzoic acid	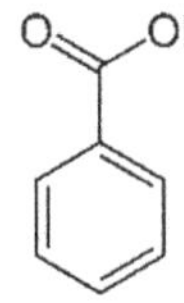	
		Terpenoids	
3.	1,8-Cineole	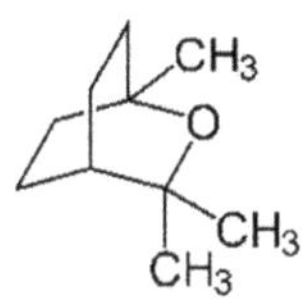	Effect on all mitosis stages, respiration, and membrane permeability
4.	Thymol	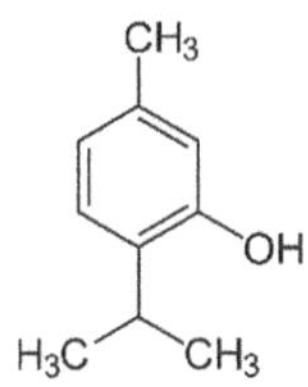	
		Alkaloids	
5.	Berberine	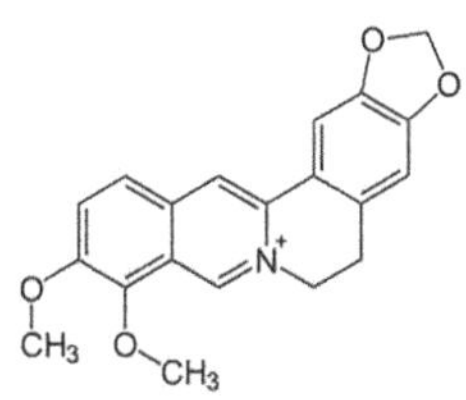	Changes in DNA structure, effect on enzyme activity, protein synthesis, and membrane permeability
6.	Papaverine	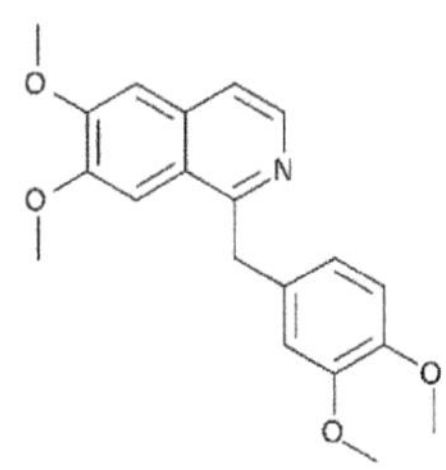	
		Cyclic Hydroxamic Acids	
7.	BOA – benzoxazoline-2(3*H*)-one	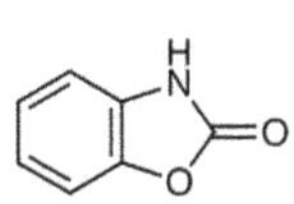	Effect on ATPase activity and covalent protein modification
8.	DIMBOA – 2,4-dihydroxy-7-methoxy-2*H*-1,4-benzoxazin-3-one	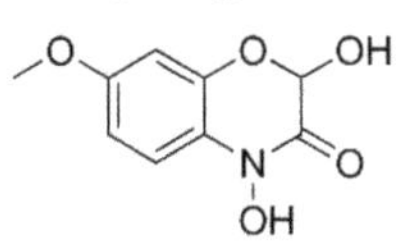	

Source: After Latif et al. (2017).

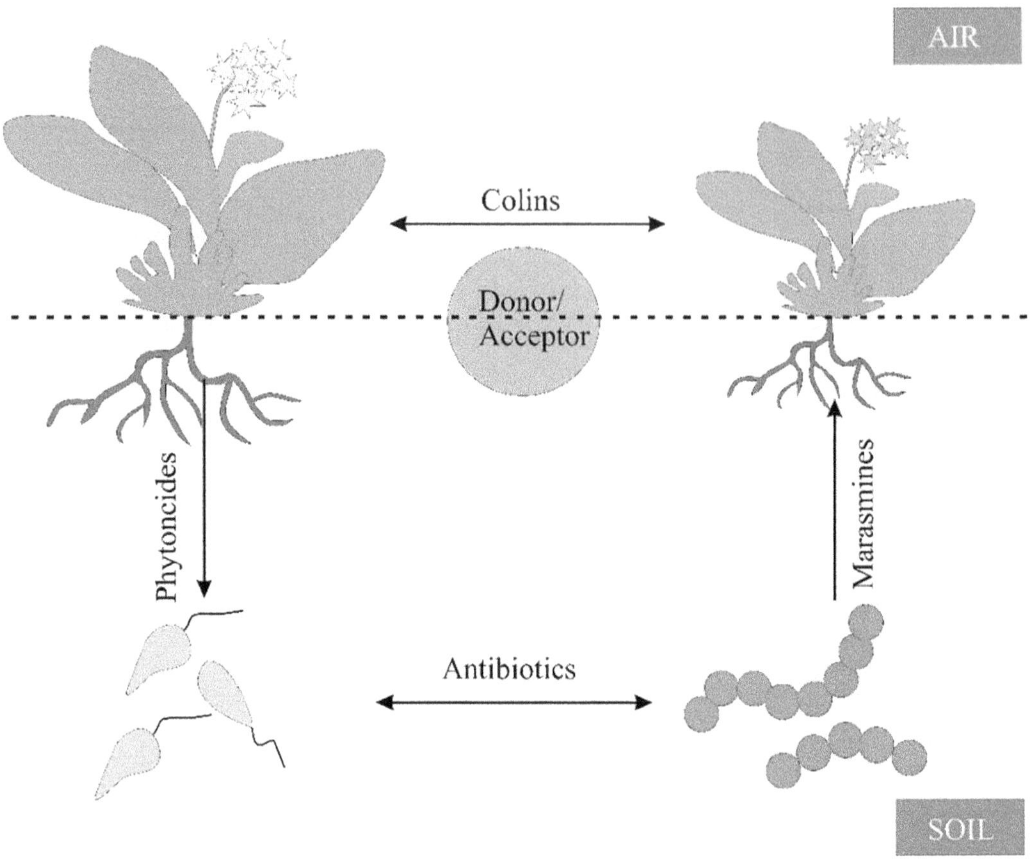

FIGURE 17.2 Division of allelopathic compounds according to the secreting (donor) and receiving (acceptor) organism (Based on: Gniazdowska et al., (2004); modified).

17.3 THE ROLE OF ALLELOPATHY IN NATURAL ECOSYSTEMS

Allelopathy research in wetland ecosystems is less common than in terrestrial ones, because of the assumption that allelochemicals are diluted and less effective in an aqueous environment (Ervin and Wetzel, 2003; Jarchow and Cook, 2009). Despite the importance and widespread occurrence of allelopathy in the opinion of many authors, there is difficulty in distinguishing these effects in natural communities from other factors, especially nutrient competition (Wardle et al., 1998; Pellissier and Souto, 1999; Kimmins, 2004; Inderjit et al., 2011; Hale and Kalisz, 2012). The phenomenon of allelopathy is widespread, especially in forest ecosystems (Blanco, 2007). It is especially known from eucalyptus communities in Australia, but it is also typical for many tree species in temperate and tropical forests. In terrestrial ecosystems, the relative importance of allelopathy has been found to increase under stressful conditions such as shading, water, and nutrient scarcity, excess herbivores, and resistance or susceptibility to microbiological effects (Weidenhamer et al., 1989; Seigler, 1996; Inderjit and del Moral, 1997; von Elert and Jüttner, 1997; Callaway et al., 1999; Callaway and Vivanco, 2007; Nazarov and Shirokov, 2014). Research on lianas has shown that early successional lianas from shaded environments had greater allelopathic potential than those growing in the sun (Ladwig et al., 2012). Additionally, lianas from mature forests (*Vitis* ssp., *Celastrus orbiculatus* Thunb.) were unresponsive to light conditions, a feature fostering the persistence of some liana species in an ecosystem. This leads to the assumption that the number of potential allelopathic compounds is not only species-specific but could also depend on environmental conditions.

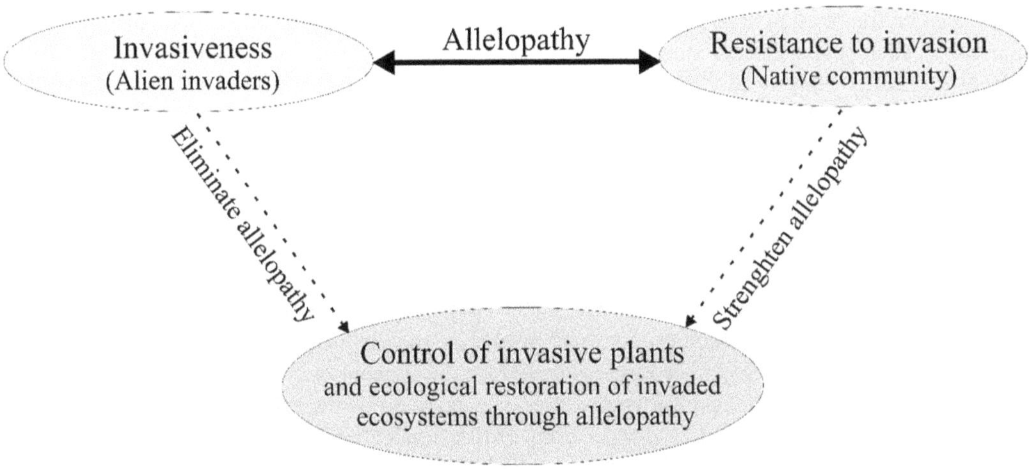

FIGURE 17.3 Role of allelopathy in the invasiveness of alien plant species and resistance of native communities; dashed lines indicate pathways to control invasive plants through allelopathy (Chen et al. (2017), modified).

The chemicals that organisms release into the environment can have substantial effects on the succession processes (Gómez-Aparicio and Canham, 2008; Thorpe et al., 2009; Kaligarič et al., 2011). For example, Jackson and Willemsen (1976) proved that the failure of *Ambrosia artemisiifolia* L. to recolonize bare soil in the second year of succession was attributed to chemical effects. In a young forest, *Alliaria petiolata* (M. Bieb.) Cavara et Grande inhibits the formation of mycorrhizae on associated plants (Roberts and Anderson, 2001), and the allelopathic substances released by *Solidago canadensis* L. could effectively reduce plant germination (Zandi et al., 2020). However, laboratory results do not always fully reflect the dominance patterns in the field (Pisula and Meiners, 2010a, b). In natural habitats, allelopathic effects often overlap with other growth factors (Pellissier and Souto, 1999; Inderjit et al., 2011).

Invasions of alien species are very strongly associated with the phenomenon of plant succession (Chen et al., 2012). Successful plant invasion depends on many factors, among other interactions between introduced species and the native community (Verhoeven et al., 2009; Qin et al., 2013; van Kleunen et al., 2015; Chen et al. 2017). In this respect, allelopathy is considered to be a very important mechanism that has a great influence on the spread of foreign taxa. Hence, there is an urgent need to understand the integrative role of allelopathy in plant invasion (invasiveness of alien plants, the resistance of a native community to invasive plants, and control/management of invasive plants using allelochemicals) – Figure 17.3.

Invasive species are frequently regarded as strong competitors. It has been suggested that with producing allelopathic compounds the plant could strongly outcompete native species (Hierro and Callaway, 2003; Jarchow and Cook, 2009; Gruntman et al., 2014; Możdżeń et al., 2021b; Tatoj et al., 2022). The "novel-weapons hypothesis" poses that the invasive introduced species produce novel allelopathic compounds to which the resident community is not adapted (Hierro and Callaway, 2003). The alien species can suppress the performance of neighboring species by directly releasing allelopathic substances into the soil (Inderjit and Callaway, 2003; Saul et al., 2013) or by inhibiting the growth of symbiotic mycorrhizal fungi (Stinson et al., 2006; Lankau, 2013). They can also release substances which, under the new conditions, become strong toxins after being processed by the soil flora (Aslam et al., 2017). Understanding of this particular dynamic in a soils biota is very important for maintaining soil cover, increasing soil fertility, and rehabilitating anthropogenically disturbed land (Polyak and Sukcharevich, 2019).

Studies on the resistance of native species against invasive species have provided also information about allelopathic effects among native species on forest invaders (Cummings et al., 2012; Hou et al., 2012). Ning et al. (2016) used ^{14}C labeled carbon sources added to grassland soil to demonstrate that allelopathy of a native plant community might increase its resistance against alien plants. The allelopathic potential of common meadow plants has also been proved, e.g., for *Dactylis glomerata* L. (Możdżeń et al., 2020), *Filipendula vulgaris* L. (Fecowicz et al., 2020), or *Cirsium oleraceum* L. (Barabasz-Krasny et al., 2017).

Soil microorganisms can effectively eliminate the effects of active allelopathic compounds excreted by fungi (Zhang et al., 2010; Xie and Dai, 2015; Liu et al., 2018) or bacteria (Müller et al., 2007; Zhang et al., 2010; Cheng and Cheng, 2015). They have the potential to assist in the management of allelopathy, related to soil productivity of natural and disturbed ecosystems (Svistova et al., 2004; Polyak et al., 2017). Soil microorganisms reduce the concentration of highly toxic substances in the soil, such as catechins and 8-hydroxyquinoline, secreted by the roots, e.g., *Centaurea stoebe* L. (syn. *C. maculosa* Lam.) and *C. diffusa* Lam. (Ehlers and Thompson, 2004; Perry et al., 2007; Inderjit et al., 2008). However, also microorganisms may enhance an already strong allelopathic potential (Cipollini et al., 2012). For example, *Acinetobacter calcoaceticus* Beijerinck can convert 2-benzoxazolinone into 2,2′-oxo-1,1′-azobenzene (AZOB) and its methoxy derivatives, which are more toxic to plants (Macias et al., 2008). The significance of soil microorganisms on allelopathy has also been discussed extensively by Blanco (2007), Zhang et al. (2010), Cheng and Cheng (2015), and Polyak and Sukcharevich (2019).

17.4 ALLELOPATHY IN AGROCENOSES

Knowledge of the positive or negative allelopathic effects of plants can be used in the selection of species for mixed cropping (Zeng, 2008, 2014; Możdżeń et al., 2019; Szafraniec et al. 2019). This mechanism may turn out to be of significant importance for management practices in organic farming. For example, substances secreted by *Artemisia absinthium* L. inhibit the growth of fennel, caraway, sage, and lemon balm, even when they are grown at a distance of 1 m apart (Kreuter, 2007). Some of the cultivated plants have significant allelopathic potential, such as *Brassica* sp., which produce, e.g., glucosinolates (Schnug, 1990). After being released by the plant, glucosinolates under the influence of the myrosinase enzyme break down into several compounds, including glucose and volatile isothiocyanates, which act as a defense against weeds and other plants (Haneklaus et al., 2006; Ishida et al., 2014). Proper selection of plants should, therefore, limit the negative allelopathic effects of crops and take into account plant interactions such as pest deterrence and disease resistance (Table 17.2).

In southern China, for instance, *Ageratum conyzoides* L. (white weed), which is native to tropical America, is grown as a floor layer in citrus orchards to make orchards a better habitat for *Amblyseius* spp. mites, which are natural enemies of the citrus pest *Panonychus citri* McGregor (Liang and Huang, 1994). Allelopathic compounds from *Lantana camara* L., a species native to Mexico and widely cultivated for ornamental purposes, combat a wide variety of harmful organisms, including microbes, insects, and other plants. In particular, in freshwater farming systems, substances derived from *L. camara* can be used, for example, against *Eichhornia crassipes* (Mart.) Solms (Kong, 2005; Zhang et al., 2005).

Producing sufficient quantities of food from soils is one objective of plant production, but taking care that this food is of high value for consumers' health and wellness is the second one of equal significance element (Schnug et al., 2006). Crop residues left behind in soils after harvest are also a potential reservoir of allelochemicals, in particular if derived from plants rich in secondary compounds, like *Brassica* species. Such type of residues may, due to allelopathic effects, contribute to the reduction of weeds (Jabran et al., 2015; Farooq et al., 2011). Residues from crop plants, like, for instance, barley, or rye may have allelopathic potential (Jones et al., 1999). Tabaglio et al. (2008) studied the allelopathic effect of 2,4-dihydroxy-1,4-(2*H*)-benzoxazin-3-one (DIBO)

TABLE 17.2
Some Examples of Positive and Negative between Plants (Kreuter, 2007)

Species	Positive Allelopathic Activity On	Negative Allelopathic Activity On
Beetroot	Broccoli, bean, pea, carrot, kohlrabi, cucumber, tomatoes, leek, spinach	Potatoes, corn, faba bean
Onion	Beetroot	Brassicaceae
Horse radish	Potatoes, fruit trees	–
Garlic	Strawberries, carrot, tomatoes, potatoes, cucumber, beetroot	Brassicaceae, bean, pea
Pumpkin	Cauliflower, bean, corn	Potato, strawberries
Bean	Pumpkin, cucumber, celeriac, spinach, tomatoes, carrots, Brassicaceae	Onion, garlic, leek, pea
Pea	Zucchini, kohlrabi, carrot, cucumber, celeriac, spinach, Brassicaceae, corn	Onion, garlic, leek, bean, potato
Brassicaceae	Beetroot, endive, pea, leaf cabbage, carrot, tomato, spinach, lettuce, celeriac, dill	Bean, radish, strawberries
Carrot	Onion, garlic, endive, pea, cucumber, leek, celeriac, dill, lettuce	Potato, Brassicaceae
Cucumber	Beetroot, onion, leek, bean, pea, dill, corn, carrot, leek, radish, spinach, celeriac	Potato, strawberries, tomato, Cucurbitaceae
Tomato	Beetroot, onion, zucchini, garlic, chicory, endive, napa cabbage, parsley, corn, leek, radish, celeriac, spinach	Potato, pea, corn, carrot
Lettuce	Beetroot, onion, carrot, leek, Brassicaceae, strawberries, bean, pea, dill	Radish, celeriac
Celeriac	Broccoli, beetroot, bean, pea, kohlrabi, dill, carrot, cucumber, tomato, leek, parsnip, turnip, spinach	Faba bean, potato, corn
Potato	Broccoli, horseradish, bean, kohlrabi, dill, corn, parsnip, spinach, faba bean, pea	Cucumber, sunflower, tomato, carrot, beetroot, onion, pumpkin

and benzoxazolin-2(3*H*)-one (BOA, Table 17.1) in rye mulch on *Amaranthus retroflexus* L. and *Portulaca oleracea* L., where they showed the ability to suppress weeds by 40%–52% and 40%–74%, respectively.

Although all plants produce specific allelochemicals, some of them are particularly suited for use in sustainable agriculture. These include, among others, *Secale cereale* L., *Sorghum vulgare* L., *Brassica* sp., *Fagopyrum esculentum* Moench., or *Helianthus annuus* L. (Jabran et al., 2015). Sunflower, for instance, has been successfully implemented into Indian wheat-rice crop rotations where it significantly reduced the level of weeds in the current and subsequent crops (Vyvyan, 2002; Rawat et al., 2017). Similarly, the use of cereal sorghum in a wheat-rice crop rotation, before sowing rice, resulted in a reduction of weed infestation of the crop, with a significant reduction in the number of herbicides used (Narwal, 2000; Hussain et al., 2021). In Nebraskan soybean and maize fields, sorghum reduced the growth of weeds (Einhellig and Rasmussen, 1989). Sorghum and sunflower both emit a whole range of allelochemicals, which include helianuols or annuolides in the case of sunflower or sorgoleone and dhurrin in the case of sorghum (Figure 17.4, Siquiera et al., 1991; Macias et al., 1993, 1999; Nimbal et al., 1996; Busk and Møller, 2002).

It should be mentioned, however, that crop rotation is not always an ideal solution to get rid of weeds, because allelochemicals secreted by one plant (e.g., sorghum) may also have a negative impact on the reduction of the yield of the succeeding plant (Roth et al., 2000), for instance, when wheat follows sorghum in a rotation (Ben-Hammouda et al., 1995). It indicates the need for further studies on the interactions of crop plants under various environmental conditions to use allelopathy in optimizing cropping systems and reducing pesticide inputs.

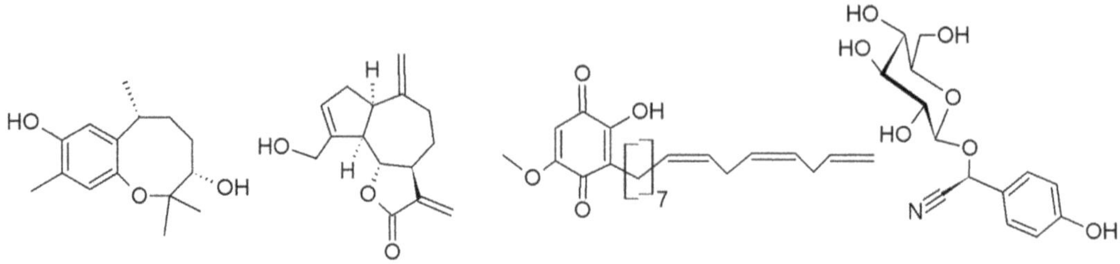

Heliannuol A Annuolide A Sorgoleone Dhurrin

FIGURE 17.4 Typical allelochemicals in sunflower and sorghum.

17.5 CAN ALLELOPATHY BE A MEANS TO ALLEVIATE CONSTRAINTS OF SOIL PRODUCTIVITY?

Powdered *Chrysanthemum* ssp. flowers have been used in the Caucasus and northwestern Iran for centuries as an insecticide. The preparation was called Persian powder. Information about this practice showed up in Europe in the 1820s (Abivardi, 2008). These compounds were later largely modified to a whole range of derivatives with a broad spectrum of applications that were successfully commercialized. Moreover, these derivatives show much lower toxicity than other chemical means, like, for instance, common organophosphorus insecticides (Goulson, 2013; Lybrand et al., 2020). In 1848, it was reported that the extracts of legume species had an insecticidal effect on leaf-eating caterpillars. At the end of the nineteenth century, the compound for the responsibility for these properties was isolated, and it is called rotenone (Figure 17.5), which is a pesticide used in organic farming, although banned in the USA and UK. Rotenone was used in the mid-twentieth century, and it has been used as a fish poison for a 100 years (Shepard, 1951).

For centuries, indigenous populations of New Zealand and Australia used an essential oil derived from *Leptospermum scoparium* J. R. Forst. and G. Forst. trees for wounds, cuts, sores, and fever treatment (Methew et al., 2020). Regardless of its healing properties, the essential oil has been used as a natural herbicide by itself or in combination with lemon grass oil – *Cymbopogon citratus* (DC.) Stapf (Subramanyam, 2013). Currently, there are preparations available on the American market containing black walnut extracts, which inhibit the growth of *Conyza canadensis* (L.) Cronquist and *C. bonariensis* (L.) Cronquist (Shrestha, 2009). Other examples of relatively old pesticides are two insecticides, namely nicotine from *Nicotiana tabacum* L. leaves and pyrethrines isolated from *Tanacetum cinerariifolium* (Trevir.) Sch. Bip. (syn. *Chrysanthemum cinerariifolium* (Trevir.)

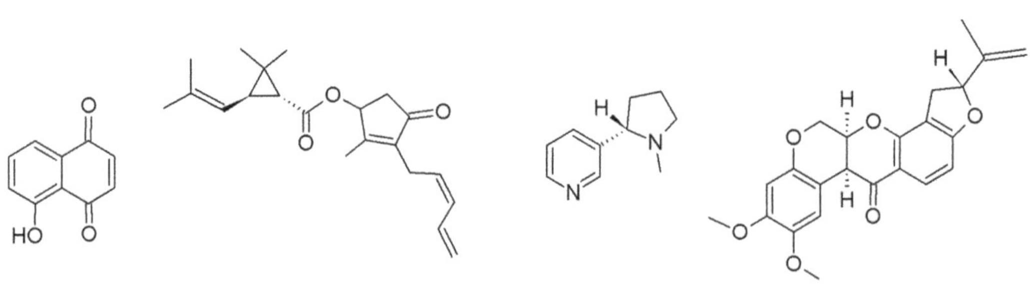

Juglone Pyrethrin I Nicotine Rotenone

FIGURE 17.5 Some natural secondary plant metabolites applied as natural pesticides.

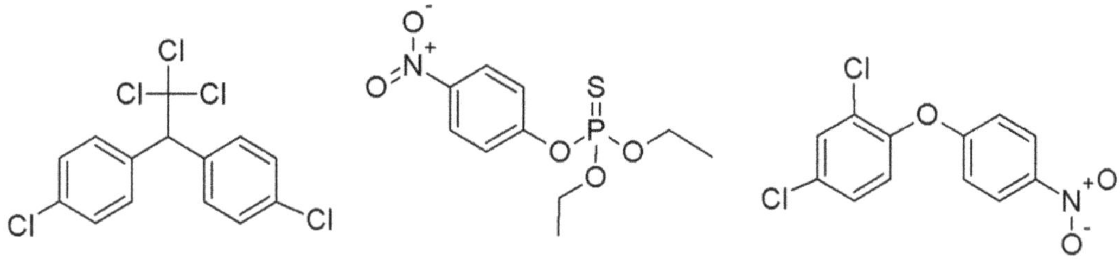

FIGURE 17.6 Chemical structures of some end-of-use pesticides.

Vis) flowers (Tomizawa and Casida, 2005; Housset and Dickmann, 2009) (Figure 17.5). Nicotine pesticides have not been commercially available in the US since 2014, and homemade pesticides are banned on organic crops and not recommended for small gardeners (Tharp, 2009). Nicotine pesticides have been banned in the EU since 2009. Foods are imported from countries in which nicotine pesticides are allowed, such as China, but foods may not exceed maximum nicotine levels (Michalski et al., 2017). An interesting case is the 1,8-cineole (eucalyptol) and its isomer 1,4-cineole which can be isolated from the essential oil extracted from *Eucalyptus* sp. or other plant species. Laboratory studies have shown that these compounds have the ability to inhibit the growth of some weeds; however, in field experiments, these properties turned out to be too weak due to significant volatility. Therefore, attempts were made to modify the parent system of eucalyptol isomer, 1,4-cineol, which resulted in the commercial herbicide cinmethylin invented in the Shell Chemicals laboratories. This herbicide exhibit phytotoxic effect in rice *Oryza sativa* L. seedlings at 75g a.i./ha is applied (Subramanyam, 2007). This proves that relatively simple chemical structure modifications have a powerful effect on biological activity. The examples cited above clearly show that nature can be an inspiration to look for preparations that will fit into sustainable agriculture.

On the one hand, with the increasing number of people on Earth, there is a need to intensify crop production to provide sufficient food for several billion people. On the other hand, there is a need to protect crops against weeds, fungi, insects, and rodents. It is estimated that without the use of pesticides, crop losses due to weeds in agriculture reach 33%–55% (Oerke and Dehne, 1997), which makes weeds a serious constraint for soil productivity. For quite a long time, the use of synthetic pesticides provided crop protection, but it was a double-edged sword. Suffice it to mention the infamous DDT, parathion, or herbicides such as nitrofen (Figure 17.6).

Excessive use of synthetic agrochemicals contributed to significant devastation of the natural environment and also the biological factors of soil productivity. Furthermore, it is worth mentioning the economics of the ecological costs caused by the excessive use of agrochemicals. For this reason, there is a growing trend to look for safer products to manage crops and their biological environment by taking advantage of the phenomenon of allelopathy. In light of the side effects of synthetic pesticides, it seems that active substances isolated from plants may be a beneficial alternative from an ecological point of view.

However, there are still problems to be solved. Allelochemicals are compounds that are quite complex in terms of structure, and chemical synthesis on an industrial scale is unprofitable from an economic point of view. Isolating the active substance from the plant is also unprofitable unless it is present in the tissue in high concentrations (e.g., isolates from essential oils). Islam et al. (2018) propose two solutions to the problem: (i) to isolate a single active ingredient from the plant material and prove its activity; to analyze its structure and develop large-scale synthesis and commercialize production; (ii) to use plant extracts as natural carriers for the active allelopathic compound.

Regardless of the approach, it is worth remembering Paracelsus maxim "*Sola dosis facit venenum – Solely the dose determines that a thing is not a poison*," and the fact that compounds of natural origin (when used as drugs or pesticides) also have side effects. Unfortunately, it often happens that, despite all precautions, pesticides hurt the natural environment, poisoning the soil or groundwater, and thus both wildlife and people. Mankind is exposed to the decomposition products of these compounds, which contributes to food chain poisoning with further consequences. In the era of the development of sustainable agriculture and the increasing concern for the preservation of a clean environment, it is worth considering the possibility of using allelopathy in finding safer alternatives to synthetic agents.

17.6 CONCLUSIONS

17.6.1 Is Allelopathy a Natural Factor Constraining or Supporting the Productivity of Soils?

The phenomenon of allelopathy is a common ecological phenomenon. It is a kind of language of communication between plants, herbivores, and soil microorganisms. It is a specific network of ecological connections existing in a given ecosystem. It often determines the success of the invasion of alien species and the victory in the competition for new habitats and their resources. The influence of plants and soil microorganisms has a direct impact on soil properties, their abundance, the content of various chemicals, and consequently the productivity of entire ecosystems. From the point of view of agriculture, knowledge of the interactions between plant species (both useful and weeds) allows for the proper selection of plants for crop rotation to obtain efficient and resistant crops to pests and diseases. The use of plant biomass remaining after cultivation in the form of mulch contributes to the improvement of chemical, biological, and physical properties of the soil and the reduction of weed infestation. At the present stage of agricultural development, various techniques using allelopathy will not solve the problem of weeds on a global scale but can be combined with other methods of reducing them. Moreover, their use can significantly contribute to the elimination of at least some synthetic pesticides and herbicides, and their replacement with biological ones. Active compounds isolated from plants can become model substances in the synthesis of bioherbicides. However, it should not be expected that soon there will be exact analogs of substances that occur in plants and exhibit allelopathic effects. This is due to, among other factors, their complex structure, and thus the need for multi-stage synthesis, which generates high industrial costs. In the future, biotechnological methods are likely to be used on a large scale, both for the synthesis of the target compound and for the synthesis of intermediates, which will reduce production costs.

REFERENCES

Abivardi, C. 2008. Pyrethrum and persian insect powder, pp. 3084–3089. In: J.L. Capinera (Ed.), *Encyclopedia of Entomonology*. Springer: Dordrecht, The Netherlands.

Aldrich, J.D. 1984. *Weed-Crop Ecology: Principles and Practices*. Breton Publishers: Sydney, NS, pp. 215–241.

Aslam, F., Khaliq, A., Matloob, A., Tanveer, A., S. Hussain, and Z.A. Zahir. 2017. Allelopathy in agro-ecosystems: A critical review of wheat allelopathy – concepts and implications. *Chemoecology* 27: 1–24.

Barabasz-Krasny, B., Możdżeń, K., A. Sołtys-Lelek, and A. Stachurska-Swakoń. 2017. The allelopathic potential of *Cirsium oleraceum* (L.) Scop. into the fodder meadow plants. *Not. Bot. Hortic. Agrobot.* 45(1): 255–261.

Ben-Hammouda, M., Kremer, J.R., H.C. Minor, and M. Sarwar. 1995. Chemical basis for differential allelopathic of sorghum hybrid on wheat. *J. Chem. Ecol.* 21(6): 75–786.

Bertin, C., X.H. Yang, and L.A. Weston. 2003. The role of root exudates and allelochemicals in the rhizosphere. *Plant Soil* 256(1): 67–83.

Bezuidenhout, S.R. 2012. Allelopathy: Allelopathy as a possible cause for crop yield reductions. Agricultural & Rural Development, pp. 1–10. Department of Agriculture and Environmental Affairs (DAEA).

Bhadoria, P.B.S. 2011. Allelopathy: A natural way towards weed management. *Am. J. Exp. Agric.* 1: 7–20.

Blanco, J.A. 2007. The representation of allelopathy in ecosystem-level forest models. *Ecol. Model.* 209: 65–77.

Busk, P.K., and B.L. Møller. 2002. Dhurrin synthesis in sorghum is regulated at the transcriptional level and induced by nitrogen fertilization in older plants. *Plant Physiol.* 129(3): 1222–1231.

Callaway, R.M., and J.M. Vivanco. 2007. Invasion of plants into native communities using the underground information super-highway. *Allelopathy J.* 19: 143–151.
Callaway, R.M., T. DeLuca, and W.M. Ridenour. 1999. Herbivores used for biological control may increase the competitive ability of the noxious weed *Centaurea maculosa*. *Ecology* 80: 1196–1201.
Chen, L., Peng, S., Li, J., Z. Lin, and Y. Zeng. 2012. Competitive control of an exotic mangrove species: Restoration of native mangrove forests by altering light availability. *Restor. Ecol.* 21: 215–223.
Chen, B.-M., Liao, H.-X., Chen, W.-B., H.-J. Wie, and S.-L. Peng. 2017. Role of allelopathy in plant invasion and control of invasive plants. *Allelopathy J.* 41(2): 155–166.
Cheng, F., and Z. Cheng. 2015. Research progress on the use of plant allelopathy in agriculture and the physiological and ecological mechanisms of allelopathy. *Front. Plant Sci.* 6: 1020.
Cipollini, D., C.M. Rigsby, and E.K. Barto. 2012. Microbes as targets and mediators of allelopathy in plants. *J. Chem. Ecol.* 38: 714–727.
Cruz-Ortega, R., Anaya, A.L., B.E. Hernández-Bautista, and G. Laguna-Hernández. 1998. Effects of allelochemical stress produced by *Sicyos deppei* on seedling root ultrastructure of *Phaseolus vulgaris* and *Cucurbita ficifolia*. *J. Chem. Ecol.* 24(12): 2039–2057.
Cummings, J., I. Parker, and G. Gilbert. 2012. Allelopathy: A tool for weed management in forest restoration. *Plant Ecol.* 213: 1975–1989.
De Albuquerque, M.B., Santos, R.C., Lima, L.M., Melo Filho, P.A., Nogueira, R.J.M.C., C.A.G. Da Câmara, and A.R. Ramos. 2011. Allelopathy, an alternative tool to improve cropping systems: A review. *Agron. Sustain. Dev.* 31: 379–395.
Duran, A.G., Chinchilla, N., J.M. Molinllo, and F.A. Macias. 2018. Influence of lipophilicity in *O*-acyl and *O*-alkyl derivatives of juglone and lawsone: A structure-activity relationship study in the search for natural herbicide models. *Pest Manag. Sci.* 74: 682–694.
Ehlers, B.K., and J. Thompson. 2004. Do co-occurring plant species adapt to one another? The response of *Bromus erectus* to the presence of different *Thymus vulgaris* chemotypes. *Oecologia* 141: 511–518.
Eichenberg, D., Ristok, C., W. Kroeber, and H. Bruelheide. 2014. Plant polyphenols: Implications of different sampling, storage and sample processing in biodiversity-ecosystem functioning experiments. *J. Chem. Ecol.* 30: 676–692.
Einhellig, F.A., and G.R. Leather. 1988. Potentials for exploiting allelopathy to enhance crop production. *J. Chem. Ecol.* 14: 1829–1844.
Einhellig, F.A., and T.A. Rasmussen. 1989. Prior cropping with grain sorghum inhibit weeds. *J. Chem. Ecol.* 15: 951–960.
Ervin, G.N., and R.G. Wetzel. 2003. An ecological perspective of allelochemical interference in land-water interface communities. *Plant Soil* 256: 13–28.
Farooq, M., Jabran, K., Cheema, Z.A., A. Wahid, and K.H.M. Siddique. 2011. The role of alleopathy in agricultural pest management. *Pest Manag. Sci.* 67: 493–506.
Fecowicz, M., Możdżeń, K., B. Barabasz-Krasny, and A. Stachurska-Swakoń. 2020. Allelopathic influence of medicinal plant *Filipendula vulgaris* Moench on germination process. *Not. Bot. Hortic. Agrobot.* 48(4): 2032–2049.
Ferguson, J.J., B. Rathinasabapathi, and C.A. Chase. 2013. Allelopathy: How plants suppress other plants. EDIS, The Horticultural Sciences Department, Institute of Food and Agricultural Sciences, University of Florida, Gainesville. HS944: 1–5.
Fraenkel, G.S. 1959. The raison d'ĕtre of secondary plant substances. *Science* 129: 1466–1470.
Gniazdowska, A., and R. Bogatek. 2005. Allelopathic interactions between plants. Multi-site action of allelochemicals. *Acta Physiol. Plant.* 27(3): 395–407.
Gniazdowska, A., K. Oracz, and R. Bogatek. 2004. Allelopatia: Nowe interpretacje oddziaływań pomiędzy roślinami. *Kosmos. Probl. Nauk Biol.* 53: 207–217 [In Polish].
Gómez-Aparicio, L., and C.D. Canham. 2008. Neighbourhood analyses of the allelopathic effects of the invasive tree *Ailanthus altissima* in temperate forests. *J. Ecol.* 96: 447–458.
Goulson, D. 2013. Review: An overview of environmental risks posed by neonicotinoid insecticides. *J. Appl. Ecol.* 50: 977–987.
Gruntman, M., Pehl, A., S. Joshi, and K. Tielbörger. 2014. Competitive dominance of the invasive plant *Impatiens glandulifera*: Using competitive effect and response with a vigorous neighbour. *Biol. Inv.* 16: 141–151.
Hale, A., and S. Kalisz. 2012. Perspectives on allelopathic disruption of plant mutualisms: A framework for individual- and population-level fitness consequences. *Plant Ecol.* 213: 1991–2006.
Haneklaus, S., E. Bloem, and E. Schnug. 2006. Sulfur interactions in crop ecosystems, pp. 17–58. In: M.J. Hawkesford, and L.J. DeKok (Eds.), *Sulfur in Plants: An Ecological Perspective*. Springer Verlag: Berlin/Heidelberg, Germany.

Hierro, J.L., and R.M. Callaway. 2003. Allelopathy and exotic plant invasion. *Plant Soil* 256: 29–39.
Hou, Y., Peng, S., G. Ni, and L. Chen. 2012. Inhibition of invasive species *Mikania micrantha* H.B.K. by native dominant trees in China. *Allelopathy J.* 29: 307–314.
Housset, P., and R.A. Dickmann. 2009. A promise fulfilled: Pyrethroid development and the benefits for agriculture and human health. *Bayer Crop Sci.* 62: 135–144.
Hussain, W.S., and M.M. Abbas. 2021. Application of allelopathy in crop production. In: M. Asaduzzaman, and M. Afroz (Eds.), *A Agricultural Development in Asia-Potential Use of Nano-Materials and Nano-Technology*. IntechOpen: London.
Hussain, M.I., Danish, S., Sanchez-Moreiras, A.M., Vicente, O., Jabran, K., Chaudhry, U.K., F. Branca, and M.J. Reigosa. 2021. Unravelling *Sorghum allelopathy* in agriculture: Concepts and implications. *Plants* 10(9): 1795.
Inderjit, 1996. Plant phenolics in allelopathy. *Bot. Rev.* 62(2): 186–202.
Inderjit, and D. del Moral. 1997. Is separating resource competition from allelopathy realistic? *Bot. Rev.* 63: 221–230.
Inderjit, and R.M. Callaway. 2003. Experimental designs for the study of allelopathy. *Plant Soil* 256: 1–11.
Inderjit, Pollock, J.L., R.M. Callaway, and W. Holben. 2008. Phytotoxic effects of (6)-catechin in vitro, in soil, and in the field. *PLoS One,* 3: e2536.
Inderjit, Wardle, D.A., R. Karban, and R.M. Callaway. 2011. The ecosystem and evolutionary contexts of allelopathy. *Trends Ecol. Evol.* 26: 655–662.
Ishida, M., Hara, M., Fukino, N., T. Kakizaki, and Y. Morimitsu. 2014. Glucosinolate metabolism, functinality and breeding for the improvement of Brassicaceae vegetables. *Breed Sci.* 64(1): 48–59.
Islam, A.K.M.M., Yeasmin, S., Quasem, J.R.S., A.S. Juraimi, and M.P. Anwar. 2018. Allelopathy of medicinal plants: Current status and future prospects in weed management. *Agric. Sci.* 9: 1569–1588.
Jabran, K., Mahajan, G., V. Sardana, and B.S. Chauhan. 2015. Allelopathy for weed control in agricultural systems. *J. Crop Prot.* 72: 57–65.
Jackson, J.R., and R.W. Willemsen. 1976. Allelopathy in the first stages of secondary succession on the Piedmont of New Jersey. *Am. J. Bot.* 63: 1015–1023.
Jarchow, M.E., and B. Cook. 2009. Allelopathy as a mechanism for the invasion of *Typha angustifolia*. *Plant Ecol.* 204: 113–124.
Jilani, G., Mahmood, S., Chaudhry, A.N., I. Hassan, and M. Akram. 2008. Allelochemicals: Sources, toxicity and microbial transformation in soil – A review. *Ann. Microbiol.* 58(3): 351–357.
Jones, E., Jessop, R.S., B.M. Sindel, and A. Hoult. 1999. Utilising crop residues to control weeds, pp. 373–376. In: A. Bishop, M. Boersma, and C.D. Barnes (Eds.), *Proceedings of the 12th Australian Weeds Conference,* Tasmanian Weeds Society, Devenlport. http://caws.org.nz/old-site/awc/1999/awc199913731.pdf.
Kaligarič, M., M.H. Meister, and S. Škornik. 2011. Grassland succession is mediated by umbelliferous colonizers showing allelopathic potential. *Plant Biosyst.* 145: 688–698.
Khalid, R.H., and S.A. Mohd. 2016. *Plant, Soil and Microbes: Mechanism and Molecular Interaction*, Vol. 2. Springer: Berlin/Heidelberg, Germany.
Khalid, S., T. Ahmad, and R.A. Shad. 2002. Use of allelopathy in agriculture. *Asian J. Plant Sci.* 1(3): 292–297.
Kimmins, J.P. 2004. *Forest Ecology: A Foundation for Sustainable Forest Management and Environmental Ethics in Forestry*. Prentice Hall: Hoboken, NJ.
Kong, C. 2005. Allelopathic research in China. The Regional Institute – online publishing. http://www.regional.org.au/au/allelopathy/2005/2/1/2420_kongch.htm.
Kreuter, M.L. 2007. Ogród w zgodzie z naturą. Warszawa, Wyd. Elipsa: 73–122 [In Polish].
Ladwig, L.M., Meiners, S.J., N.L. Pisula, and K.A. Lang. 2012. Conditional allelopathic potential of temperate lianas. *Plant Ecol.* 213: 1927–1935.
Lankau, R.A. 2013. Species invasion alters local adaptation to soil communities in a native plant. *Ecology* 94: 32–41.
Latif, S., G. Chiapusio, and L.A. Weston. 2017. Allelopathy and the role of allelochemicals in plant defense. *Adv. Bot. Res.* 82: 19–54.
Liang, W.G., and M.D. Huang. 1994. Influence of citrus orchard ground cover plants on arthropod communities in China: A review. *Agric. Ecosyst. Environ.* 50: 29–37.
Lipniak, K., and A. Kliszcz. 2020. Allelopathic effect of goosefoot on germination and early stage growth of triticale and radish. *AUPC Stud. Nat.* 5: 110–128.
Liu, S., F. Qin, and S. Yu. 2018. *Eucalyptus urophylla* root-associated fungi can counteract the negative influence of phenolic acid allelochemicals. *Appl. Soil Ecol.* 127: 1–7.

Lovett, J.V., and M. Ryuntyu. 1992. Allelopathy: Broadening the context, pp. 11–19. In: S.J.H. Rizvi, and V. Rizvi (Eds.), *Allelopathy: Basic and Applied Aspects*. Chapmann & Hall Publishers, Springer: Dordrecht, The Netherlands.
Lybrand, D.B., Xu, H., R.L. Last, and E. Pichersky. 2020. How plants synthesize pyrethrins: Safe and biodegradable insecticides. *Trends Plant Sci.* 25(12): 1240–1251.
Macias, F.A., Varela, R.M., A. Torres, and J.M. Molinillo. 1993. Potential allelopathic guaianolides from cultivar sunflower leaves, var. SH-222. *Phytochemistry* 34: 669–674.
Macias, F.A., Varela, R.M., A. Torres, and J.M.G. Molinillo. 1999. New bioactive plant heliannuols from cultivar sunflower leaves. *J. Nat. Prod.* 62: 1636–1639.
Macias, F.A., Oliveros-Bastidas, A., Marin, D., Carrera, C., N. Chinchilla, and J.M.G. Molinillo. 2008. Plant biocommunicators: Their phytotoxicity, degradation studies and potential use as herbicide models. *Phytochem. Rev.* 7: 179–194.
Macias, F.A., A.G. Durán, and J.M.G. Molinillo. 2020. Allelopathy: The chemical language of plants. *Prog. Chem. Org. Nat. Prod.* 112: 1–84.
Methew, C., Tesfaye, W., Rasmussen, P., Peterson, G.M., Bartholomaeus, A., M. Sharma, and J. Thomas. 2020. Mānuka oil: A review of antimicrobial and other medicinal properties. *Pharmaceuticals* 13(11): 343.
Michalski, B., M. Herrmann, and R. Solecki. 2017. How does a pesticide residue turn into a contaminant? *Bundesgesundheitsbl. Gesundheitsforsch. Gesundheitsschutz* 60(7): 768–773 [In German].
Molisch, H. 1937. *Der Einfluss einer Pflanze auf die andere: Allelopathie*. Gustav Fischer: Jena [In German].
Możdżeń, K., Barabasz-Krasny, B., Stachurska-Swakoń, A., I. Turisova, and P. Zandi. 2021a. Germination and growth of radish under influence of nipplewort aqueous extracts. *Not. Bot. Hortic. Agrobot.* 49(1): 1–16.
Możdżeń, K., Barabasz-Krasny, B., Stachurska-Swakoń, A., P. Zandi, and J. Puła. 2019. Effect of aqueous extracts of peppermint (*Mentha ×piperita* L.) on the germination and the growth of selected vegetable and cereal seeds. *Not. Bot. Hortic. Agrobot.* 47(2): 412–417.
Możdżeń, K., Barabasz-Krasny, B., Stachurska-Swakoń, A., Zandi, P., Puła, J., Y. Wang, and I. Turisova. 2020. Allelopathic interaction between two common meadow plants: *Dactylis glomerata* L. and *Trifolium pratense* L. *Biologia* 75(5): 653–663.
Możdżeń, K., Tatoj, A., Barabasz-Krasny, B., Sołtys-Lelek, A., W. Gruszka, and P. Zandi. 2021b. The allelopathic potential of *Rosa blanda* Aiton on selected wild-growing native and cultivated plants in Europe. *Plants* 10: 1806.
Müller, C.H. 1969. Allelopathy as a factor in ecological process. *Vegetatio* 18: 348–357.
Müller, N., Hempel, M., B. Philipp, and E.M. Gross. 2007. Degradation of gallic acid and hydrolysable polyphenols is constitutively activated in the freshwater plant associated bacterium *Matsuebacter* sp. FB25. *Aquat. Microb. Ecol.* 47(1): 83–90.
Mushtaq, W. and B.M. Siddiqui. 2018. Allelopathy in Solanaceae plants. *J. Plant Prot. Res.* 58(1): 1–7.
Mushtaq, W., B.M. Siddiqui, and K.R. Hakeem. 2020. *Allelopathy Potential for Green Agriculture*. Springer: Berlin, Germany.
Narwal, S.S. 1994. *Allelopathy in Crop Production*. Scientific Publishers: Jodhpur, Rajasthan, p. 288.
Narwal, S.S. 2000. Weed management in rice: Wheat rotation by *allelopathy. Crit. Rev. Plant Sci.* 19: 249–266.
Nazarov, V.V., and A.I. Shirokov. 2014. Preliminary research results of allelopathic interaction of some orchids *in vitro*. Scientific notes of the Taurida National University. V. I. Vernadsky Series Biology, Chemistry, Special issue No. 5, 27(66): 102–109 [In Russian].
Nimbal, C.J., Yerkes, C.N., L.A. Weston, and S.C. Weller. 1996. Herbicidal activity and site of action of the natural product sorgoleone. *Pestic. Biochem. Physiol.* 54: 73–83.
Ning L., F.-H. Yu, and M. von Kleunen. 2016. Allelopathy of a native grassland community as a potential mechanism of resistance against invasion by introduced plants. *Biol. Inv.* 18: 3481–3493.
Ninkuu, V., Zhang, L., Yan, J., Fu, Z., T. Yang, and H. Zeng. 2021. Biochemistry of terpenes and recent advances in plant protection. *Int. J. Mol. Sci.* 22: 5710.
Oerke, E.C., and H.W. Dehne. 1997. Global crop production and the efficacy of crop protection-current situation and future trends. *Eur. J. Plant Pathol.* 103: 203–215.
Oleszek, W. 1996. Allelopatia: rys historyczny, definicja, nazewnictwo, pp. 5–15. In: W. Oleszek (Ed.), *Teoretyczne i praktyczne aspekty allelopatii*. Materiały Konferencyjne IUNG: Puławy, Poland [In Polish].
Oleszek, W. 1997. Allelopatia, pp. 249–272. In: L. Jankiewicz (Ed.), *Regulatory wzrostu i rozwoju roślin*. Wydawnictwo Naukowe PWN: Warszawa [In Polish].
Oleszek, W., M. Jurzysta, and P. Górski. 1991. Alfalfa saponins: The allelopathic agents, pp. 151–167. In: S.J.H. Rizvi, and V. Rizvi (Eds.), *Allelopathy: Basic and Applied Aspects*. Chapman and Hall: London.
Olofsdotter, M., L.B. Jensen, and B. Courtois. 2002. Improving crop competitive ability using allelopathy: An example from rice. *Plant Breed.* 121: 1–9.

Ostrumow, S.A. 1992. *Wprowadzenie do ekologii biochemicznej*. Wydawnictwo Naukowe PWN: Warszawa [In Polish].

Pellissier, F., and X.C. Souto. 1999. Allelopathy in northem temperate and boreal semi-natural woodland. *Crit. Rev. Plant. Sci.* 18: 637–652.

Perry, L.G., Thelen, G.C., Ridenour, W.M., Callaway R.M., M.W. Paschke, and J.M. Vivanco. 2007. Concentrations of the allelochemical (+/–)-catechin in *Centaurea maculosa* soils. *J. Chem. Ecol.* 33: 2337–2344.

Pisula, N.L., and S.J. Meiners. 2010a. Relative allelopathic potential of invasive plant species in a young disturbed woodland. *J. Torrey Bot. Soc.* 137: 81–87.

Pisula, N.L., and S.J. Meiners. 2010b. Allelopathic effects of goldenrod species on turnover in successional communities. *Am. Midl. Nat.* 163: 161–172.

Polyak, Y.M., and V.I. Sukcharevich. 2019. Allelopathic interactions between plants and microorganisms in soil ecosystems. *Biol. Bull. Rev.* 9(6): 562–574.

Polyak, Y.M., Bakina, L.G., N.V. Mayachkina, and A.A. Galdiyants. 2017. Role of allolopathic interactions in the structural and functional changes of anthropogenically disturbed soils. *Gumus Pochvoobraz.* 21: 65–69.

Puła, J., Barabasz-Krasny, B., Możdżen, K., A. Sołtys-Lelek, and A. Lepiarczyk. 2016. Effect of aqueous extracts of sticky willy (*Galium aparine* L.) on the growth of seedlings of selected maize varieties (*Zea mays* L.). *Not. Bot. Hortic. Agrobot.* 44(2): 518–524.

Putnam, A.R., 1985. Weed allelopathy, pp. 131–155. In: S.O. Duke (Ed.), *Weed Physiology Volume 1: Reproduction and Ecophysiology*. CRC Press: Boca Raton, FL.

Putnam, A.R., and C.S. Tang. 1986. Allelopathy: State of the science, pp. 1–19. In: A.R. Putnam, and C.S. Tang (Eds.), *The Science of Allelopathy*. Wiley: New York.

Qasem, J.R., and C.L. Foy. 2001. Weed allelopathy, its ecological impacts and future prospects: A review. *J. Crop Prod.* 4(2): 43–119.

Qin, R.M., Zheng, Y.L., Valiente-Banuet, A., Callaway, R.M., Barclay, G.F., C.S. Pereyra, and Y.L. Feng. 2013. The evolution of increased competitive ability, innate competitive advantages, and novel biochemical weapons act in concert for a tropical invader. *New Phytol.* 197: 979–988.

Radosevich, S.R., J.S. Holt, and C. Ghersa. 2007. *Ecology of Weeds and Invasive Plants: Relationship to Agriculture and Natural Resource Management*. Wiley: New York, p. 454.

Rawat, L.S., Maikhuri, R.K., Bahuguna, Y.M., N.K. Jha, and P.C. Phondani. 2017. Sunflower allelopathy for weed control in agriculture systems. *J. Crop. Sci. Biotech.* 20(1): 45–60.

Rice, E.L. 1974. *Allelopathy*. Academic Press: New York.

Rice, E.L. 1984. *Allelopathy*, 2nd Edition. Academic Press INC: Orlando, FL.

Rizvi, S.J.H., H. Haque, V.K. Singh, and V. Rizvi. 1992. A discipline called allelopathy, pp. 1–8. In: S.J.H. Rizvi, and V. Rizvi (Eds.), *Allelopathy: Basic and Applied Aspects*. Chapmann & Hall Publishers: Springer, Dordrecht.

Roberts, K.J., and R.C. Anderson. 2001. Effect of garlic mustard [*Alliaria petiolata* (Bieb. Cavara & Grande)] extracts on plants and arbuscular mycorrhizal (AM) fungi. *Am. Midl. Nat.* 146: 146–152.

Roth, C.M., J.P. Shroyer, and G.M. Paulsen. 2000. Allelopathy of sorghum on wheat under several tillage systems. *Agron. J.* 92: 885–860.

Saul, W., J. Jeschke, and T. Heger. 2013. The role of eco-evolutionary experience in invasion success. *NeoBiota* 17: 57–74.

Schnug, E. 1990. Glucosinolates: Fundamental, environmental and agricultural aspects, pp. 97–106. In: H. Rennenberg (Ed.), *Sulfur Nutrition and Sulfur Assimilation in Higher Plants*. Academic Publishing: The Hague, The Netherlands.

Schnug E., Haneklaus S., G. Rahmann, and K. Walker. 2006. Organic farming: Stewardship for food security, food quality, environment and nature conservation. *Asp. Appl. Biol.* 79: 57–62.

Scognamiglio, M., D'Abrosca, B., Esposito, A., Pacifico, S., P. Monaco, and A. Fiorentino. 2013. Plant growth inhibitors: Allelopathic role or phytotoxic effects? Focus on mediterranean biomes. *Phytochem. Rev.* 12(4): 803–830.

Seigler, D.S. 1996. Chemistry and mechanism of allelopathic interactions. *Agron. J.* 88: 876–885.

Shepard, H. 1951. *The Chemistry and Action of Insecticides*. McGraw-Hill: New York.

Shrestha, A. 2009. Potential of a Black Walnut (*Juglans nigra*) extract product (NaturCur) as pre- and post-emergence bioherbicide. *J. Sustain. Agric.* 33(8): 810–822.

Siegień, I., Trocka, A., Bosa, K., R. Bogatek, and A. Gniazdowska. 2008. Potencjał allelopatyczny słonecznika. *Post. Nauk Roln.* 6: 55–71 [In Polish].

Siquiera, J.O., Nair, M.G., R. Hammerschmidt, and G.R. Safire. 1991. Significance of phenolic compounds implant-soil-microbial systems. *Crit. Rev. Plant Sci.* 10: 63–121.
Stamp, N. 2003. Out of the quagmire of plant defense hypotheses. *Q. Rev. Biol.* 78: 23–55.
Stinson, K.A., Campbell, S.A., Powell, J.R., Wolfe, B.E., Callaway, R.M., Thelen, G.C., Hallett, S.G., D. Prati, and J.N. Klironomos. 2006. Invasive plant suppresses the growth of native tree seedlings by disrupting belowground mutualisms. *PLoS Biol.* 4: e140.
Stugren, B. 1978. *Grundlagen der Allgemeinen Ökologie*. VEB Gustav Fischer Verlag: Jena [In German].
Subramanyam, D., R.C. Raghava, and R. Maheswara. 2007. Phytotoxicity of cinmetylin on transplanted rice (*Oryza sativa* L.) under varied puddling and water management practices. *Indian J. Weed Sci.* 39: 162–166.
Subramanyam, D., Srinivasulu, D., V. Sumath, and K. Sajitha. 2013. Plant origin herbicides: Chemistry, mode of action and weed management. *Int. J. Agric. Sci.* 9(1): 413–422.
Svistova, I.D., Shcherbakov, A.P., and Frolova, L.O. 2004. Toxins of chernozemic micromycetes: The range of antibiotic effect and their role in formation of the microbial community. *Eurasian Soil Sci.* 37(10): 1078–1085.
Swain, T. 1977. Secondary compounds as protective agents. *Ann. Rev. Plant Physiol.* 28: 479–501.
Szafraniec, R., Możdżeń, K., Barabasz-Krasny, B., P. Zandi, and Y. Wang. 2019. Influence of volatile peppermint (*Mentha × piperita* L.) compounds on germination and seedling of radish (*Raphanus sativus* L. var. radicula Pers.) Growth. *Not. Bot. Hortic. Agrobot.* 47(4): 1277–1284.
Szczepański, A.J. 1977. Allelopathic as a means of biological control of water weeds. *Aquat. Bot.* 3: 193–197.
Tabaglio, V., Gavazzi, C., M. Schulz, and A. Marocca. 2008. Alternative weed control using allelopathic effect of natural benzoxazinoids from rye mulch. *Agron. Sustain Dev.* 28: 397–401.
Tatoj, A., Możdżeń, K., Barabasz-Krasny, B., Sołtys-Lelek, A., W. Gruszka, and P. Zandi. 2022. Effect of *Rosa gorenkensis* Besser aqueous extracts on germination and early growth of native plant species. *Not. Bot. Hortic. Agrobot.* 50(1): 1–20.
Tharp, C. 2009. Safety for homemade remedies for pest control, pp. 1–2. In: *Montana Pesticide Bulletin.* Montana State University: Bozeman, Montana.
Thorpe, A.S., Thelen, G.C., A. Diaconu, and R.M. Callaway. 2009. Root exudate is allelopathic in invaded community but not in native community: Field evidence for the novel weapons hypothesis. *J. Ecol.* 97: 641–645.
Tomizawa, M., and J.E. Casida. 2005. Neonicotinoid insecticide toxicology: Mechanisms of selective action. *Annu. Rev. Pharmacol. Toxicol.* 45: 247–68.
Trezzi, M.M., Vidal, R.A., Balbinot A.A., H. von Hertwig Bittencourt, and A.P. da Silva Souza Filho. 2016. Allelopathy: Driving mechanisms governing its activity in agriculture. *J. Plant Interact.* 11(1): 53–60.
van Kleunen, M., W. Dawson, and N. Maurel. 2015. Characteristics of successful alien plants. *Mol. Ecol.* 24: 1954–1968.
Verhoeven, K.J.F., Biere, A., J.A. Harvey, and W.H. Van Der Putten. 2009. Plant invaders and their novel natural enemies: Who is naive? *Ecol. Lett.* 12: 107–117.
von Elert, E., and F. Jüttner. 1997. Phosphorus limitation and not light controls the extracellular release of allelopathic compounds by *Trichormus doliolum* (Cyanobacteria). *Limnol. Oceanogr.* 42(8): 1796–1802.
Vyvyan, J.R. 2002. Allelochemicals as leads for new herbicides and agrochemicals. *Tetrahedron* 58(9): 1631–1646.
Wardle, D.A., Nilsson, M.-C., C. Gallet, and O. Zackrisson. 1998. An ecosystem-level perspective of allelopathy. *Biol. Rev.* 73: 305–319.
Weidenhamer, J.D., D.C. Hartnett, and J.T. Romeo. 1989. Density-dependent phytotoxicity: Distinguishing resource competition and allelopathic interference in plants. *J. Appl. Ecol.* 26(2): 613–624.
Weston, L.A. 2005. History and current trends in the use of allelopathy for weed management. *Cornell University Turfgrass Times* 13: 529–534.
Willis, R.J. 1985. The historical bases of the concept of allelopathy. *J. Hist. Biol.* 18: 71–102.
Willis, R.J. 2010. *The History of Allelopathy*. Springer: Dordrecht, The Netherlands.
Xie, X.G., and C.C. Dai. 2015. Biodegradation of a model allelochemical cinnamic acid by a novel endophytic fungus *Phomopsis liquidambari*. *Int. Biodeter. Biodegrad.* 104: 498–507.
Yamada, K., T. Anai, and K. Hasegawa. 1995. Lepidimoide, an allelopatic substance in the exudates from germinated seeds. *Phytochemistry* 39: 1031–1032.
Zandi, P., Barabasz-Krasny B., Stachurska-Swakoń A., J. Puła, and K. Możdżeń 2020. Allelopathic effect of invasive Canadian goldenrod (*Solidago canadensis* L.) on early growth of red clover (*Trifolium pratense* L.). *Not. Bot. Hortic. Agrobot.* 48(4): 2060–2071.
Zeng, R. 2008. Allelopathy in Chinese ancient and modern agriculture, pp. 39–50. In: R. Zeng, A. Mallikand, and S. Luo (Eds.), *Allelopathy in Sustainable Agriculture and Forestry*. Springer New York Press: New York.

Zeng, R.S. 2014. Allelopathy: The solution is in direct. *J. Chem. Ecol.* 40: 515–516.
Zhang, M.X., Ling, B., Kong, C.H., G.W. Liang, and Y.Z. Dong. 2005. Allelopathic effect of Lantana (*Lantana camara*) extracts on growth of water hyacinth (*Eichhornia crassipes*). *Allelopathy J.* 15: 125–130.
Zhang, Z.Y., L.P. Pan, and H.H. Li. 2010. Isolation, identification and characterization of soil microbes which degrade phenolic allelochemicals. *J. Appl. Microbiol.* 108(5): 1839–1849.
Zielińska, E. 1986. Wybrane zagadnienia allelopatii. *Sylwan* 130(7): 37–42 [In Polish].

18 Managing Multiple Constraints to Crop Production in Dispersive Soils

Pichu Rengasamy[1], *Glenn McDonald*[1], *and Ehsan Tavakkoli*[1,2]
[1]School of Agriculture, Food and Wine,
The University of Adelaide, Adelaide, Australia
[2]NSW Department of Primary Industries (Agriculture),
Wagga Wagga Agricultural Institute, Wagga Wagga, Australia

CONTENTS

18.1 INTRODUCTION

Crop production in dispersive soils is limited because of the solid phase chemistry leading to poor soil physical conditions hindering air, water and nutrient movements and soil solution chemistry enhancing elemental toxicity and nutrient deficiency (McDonald et al., 2020). Dispersive soils adversely affect the geotechnical behaviour of clay soil under buildings and other infrastructure (Faulkner, 2006). Different types of soil erosion such as gully and tunnel erosions commonly occur

DOI: 10.1201/9781003093565-18

in these soils (Heinzen and Arulanandan, 1977). Physical breakdown of the particle aggregates and swelling of clays reduce the shear strength of the soil sediments resulting in structural failure and initiation of landslides (Baldermann et al., 2021). Estimates (Page et al., 2021) show that ~618 million ha of land globally are dispersive soils. Classification as "sodic soils" is the basis for these estimates. Soil surveys globally do not use clay dispersivity as a criterion in the soil classification systems.

Traditionally, the focus was on exchangeable sodium as the main factor causing swelling and dispersion of soil clays. Hence, commonly, the choice of parameters to identify swelling and dispersion in soils and to define soil sodicity is either exchangeable sodium percentage (ESP) of soils or sodium adsorption ratio (SAR) of soil solutions. These parameters remained as decisive factors in the reclamation of sodic soils. However, recent studies show that many factors other than ESP or SAR influence the behaviour of sodic soils (Rengasamy and Sumner, 1998). The first section of this chapter will deal with these soil factors and other issues associated with dispersive soils steering to the new concept proposed by Rengasamy et al. (2016).

Dispersive soils develop in saline environments with various types of salts. In addition to changing exchangeable cation composition and related dispersive reaction, a saline environment dictates the composition of soil solutions, which can induce nutrient deficiency and/or toxicity. The interaction of "salinity" effects and soil dispersivity leads to multiple constraints to crop production in dispersive soils and poses many challenges in the management of these soils. This chapter aims to discuss these issues and possible solutions.

18.2 FUNDAMENTAL FACTORS INVOLVED IN CLAY DISPERSION

To understand the role of several soil factors including the exchangeable cations in relation to clay dispersion in these soils, we detail, below, the fundamental chemistry factors involved in the processes leading to dispersion of clays from soil aggregates.

Water is a polar compound having positive and negative charges in its molecules. They react with soil aggregates possessing electrical charges. Non-polar solvents, such as benzene, do not react with charged soil particles. Similarly, soil aggregates whose net charge is zero do not react with water. The following table (Table 18.1) shows that soil aggregates do not slake or disperse in benzene and n-hexane, which have very low dielectric constant values. The dielectric constant is a measure of polarity (with electrical charge) of a solvent. In water, which has a high dielectric constant, the aggregates slake and disperse in proportion to ESP. In ethanol, whose dielectric constant is comparatively lower, both slaking and dispersion are low.

Chorom and Rengasamy (1996) reported that on heating at 300°C, cation-saturated smectitic clays did not disperse in water. They concluded that at high temperatures, cations form covalent bonds with clay surfaces, reducing the ionic charge and, hence, the water interaction. Covalently

TABLE 18.1
Slaking and Dispersion of Alfisol Aggregates (2–4 mm) at Two Levels of Sodicity in Various Solvents

Solvent	Dielectric Constant at 25°C	Slaking % <2 mm ESP 1	Slaking % <2 mm ESP 20	Dispersed Clay as % of Total Clay ESP 1	Dispersed Clay as % of Total Clay ESP 20
Water	78.5	67	80	0	26
Ethanol	24.3	24	12	0	6
Benzene	2.3	0	0	0	0
n-Hexane	1.9	0	0	0	0

Source: After Rengasamy and Sumner (1998).

TABLE 18.2
Soil Factors Controlling Swelling, Dispersion and Flocculation

Soil Factors	Mechanism
1. Clay mineralogy and clay content	Charge originates in clay structures because of isomorphous substitution and broken bonds. Location of charge in tetrahedral structure is not available for hydration reactions. Thus, the total charge depends on the mineralogy and the amount of clay in soils
2. Soil pH	Alters the charge on broken bonds by adsorption of H^+ or OH^- ions; with increasing concentration of carbonate anions, pH increases, and also negative charge on soil particles increases. When pH decreases, as observed in acidic soils, negative charge decreases
3. Organic matter	Organic molecules bonded to clays by covalent bonding reduce the hydration charge of clay particles. Unbound, charged organic molecules can increase the hydration charge. Soil particles covered by hydrophobic organic matter are not affected by water interaction
4. Inner sphere complexes	Cations such as Fe, Al, K fixed by clay minerals by inner sphere complexation (covalent bonding) reduces the hydration charge
5. Cementation	Cementation of soil particles by Fe and Al oxides or calcium carbonate can block the charge available for water interaction
6. Exchangeable cations	Exchangeable cations are attached to charged soil particles by a mixture of ionic and covalent bonding. The resultant ionicity of these bindings determines the net hydration charge. Dispersive charge depends on the dispersive power of the exchangeable cations
7. Electrolytes	Free (unbound) electrolytes in soil water contribute to the cationic flocculating charge, which is a function of the flocculating power and the concentration of individual cations

bonded molecules are hydrophobic (non-reactive with water) and electrostatically bound molecules facilitate hydration (reactive with water). Thus, ionic charge on soil particles and in the solvent molecules are the major factors in inducing clay dispersion from soils. Table 18.2 lists the several soil factors that control swelling, dispersion and flocculation.

18.3 IONIC CHARGE OF SOIL PARTICLES AND CLAY DISPERSIVITY

Soil aggregates are formed by the combination of clay (<2 µm), silt and sand particles; organic and biological compounds; and other inorganic compounds such as Fe and Al oxides and hydroxides. Sand particles are composed of silica (SiO_2), mostly in the form of quartz, and do not have charge. Clay and silt particles acquire charge contributed by clay minerals such as kaolinite, illite and smectite. Organic and biological products can have charge depending on the nature of the functional groups (carboxyl, amino, carbonyl, alcoholic OH, etc.) in their molecules. Other inorganic compounds also contribute to positive and negative charges depending on their molecular structure.

The charge in clay minerals arises because of their structure involving tetrahedral silicate and octahedral alumina layers. Si^{4+} balances the negative charge of co-ordinated O^{2-} ions in tetrahedral layers, and Al^{3+} balances the negative charge of co-ordinated OH^- or O^{2-} ions. However, during clay mineral formation, the isomorphous substitution of Si^{4+} by Al^{3+} (or Fe^{3+}) and the substitution Al^{3+} by Mg^{2+} or Ca^{2+} can occur in certain parts of the mineral structure. Such isomorphous substitutions lead to unbalanced negative charges in oxygen and hydroxyl ions in the mineral structure. This layer charge in minerals is known as 'permanent negative charge'. During the weathering process, the clay minerals can disintegrate leading to the breaking of bonds. The edges and broken bonds in clay minerals can expose charges that adsorb H^+ or OH^- ions from soil solutions. Under acidic pH conditions, H^+ is adsorbed inducing positive charge, and under alkaline conditions, OH^- is adsorbed adding negative charge to clay minerals. These charges are known as 'pH-dependent charge'.

Thus, total electrical charge on aggregates depends on the type and amount of clay minerals, organic matter and other inorganic constituents. Although the net charge can be calculated on the basis of the charge on the individual components, the charge available for hydration is altered because of the different types of bonding among these components (Rengasamy and Olsson, 1991). Covalent bonding among soil constituents and clays reduces the charge available for hydration. Soil clays are complex intergrowths of different clay structures intimately associated with inorganic and organic molecules and biopolymers; as a result, they do not have same charge characteristics as the pure clay mineral counterparts of the soil clay mineral identified by X-ray diffraction methods. The layer charge of clay minerals contributed by isomorphous substitution may be large, but the charge available for hydration may be quite different as observed in illite where K is bound to layer charge by covalent bonding and smectites with charge originating from tetrahedral substitution leading to covalent bonding of cations, including Na.

According to thermodynamic principles, the net charges have to be balanced by oppositely charged elements or molecules in order to be in a stable equilibrium. Generally, most of agricultural soils (except perhaps Ferrosols, equivalent of Oxisols) have net negative charge, when soil pH is above ZPC (zero point of charge, i.e. pH at which the net charge is zero). These charges are balanced by exchangeable cations, most commonly by Na, K, Mg and Ca. Exchangeable Al, Fe and Mn can be involved in acidic soils. While Na, K, Mg, Ca ions are hydrated (or solvated) by water molecules, Al, Fe and Mn hydrolyse water molecules and form different positively charged hydroxy cations, which are usually bound to clays by covalent bonding. In soils with net positive charge (as in Oxisols where soil pH is less than ZPC), the charge is balanced by exchangeable anions.

The bonds between exchangeable cations and clay particles were thought to be completely ionic. However, recent advances in inorganic chemistry have shown that any given heteronuclear bond found in natural systems has a mixture of covalent and ionic character. The degree of ionicity of clay-cation bonding determines the water reactivity. Increase in covalence decreases the water reaction. For example, the ionicity of Ca^{2+} in $CaCl_2$ is higher than in $CaCO_3$ where Ca^{2+} is bound to CO_3^{2-} by covalent bonding. This is also reflected in their water solubilities, $CaCl_2$ being highly soluble compared to the very low solubility of $CaCO_3$.

Marchuk and Rengasamy (2011) hypothesised that, because of both increasing charge and increasing size of a clay particle, the electron cloud in the bond formation will be less influenced by the clay anion, meaning the covalence or ionicity index of a cation alone will indicate the ionic character of a clay-cation bond. These authors derived the covalence index (CI) of a clay-cation bond defined by:

$$\mathrm{CI} = \left(I_z / I_{z+1}\right) Z^{0.5}$$

where Z is the valence of the cation whose ionisation potential is I_z, and I_{z+1} is the ionisation potential when the valence of the cation changes to $Z+1$.

The ionicity index (II) is then defined as II=1−CI. The ionicity index of a clay-cation bond indicates the reactivity of water molecules disrupting the bond. Marchuk and Rengasamy (2011) reported that the clay behaviour in aqueous suspensions such as turbidity (i.e. clay dispersion), zeta potential, or mean particle size of different soil clays made homoionic with Li^+, Na^+, K^+, Mg^{2+}, Ca^{2+}, Sr^{2+}, or Ba^{2+} was highly correlated to their respective ionicity indices (Figure 18.1). Ionicity of clay-cation bonds determines the dispersive power of the adsorbed cations. It can be concluded that the inherent ionic charge of a soil is a result of the unique combination of its components as well as pH and is balanced by exchangeable cations, which is then altered by the degree of ionicity of clay-cation bonds.

The degree of ionicity of clay-cation bonding indicates water interaction, but swelling and dispersion depend on the dispersive power of cations (Rengasamy et al., 2016). Rengasamy et al. (2016) defined the 'dispersive charge' of a soil as follows:

$$\text{Dispersive charge} = (\mathrm{Ca}) + 1.7(\mathrm{Mg}) + 25(\mathrm{K}) + 45(\mathrm{Na})$$

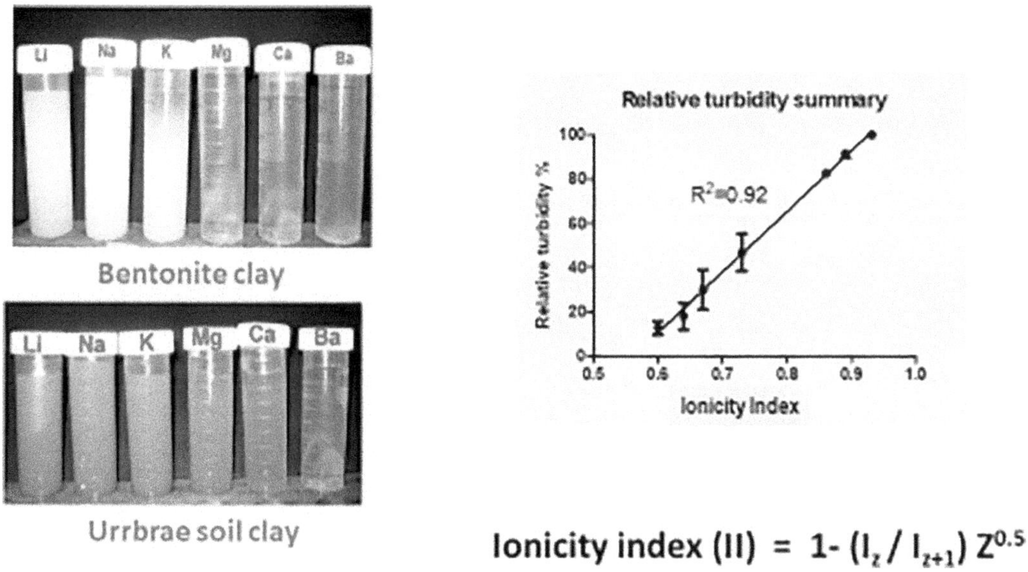

FIGURE 18.1 Clay dispersion (measured as turbidity) from homoionic bentonite clay and Urrbrae soil clay as related to ionicity of clay-cation bonds (Marchuk and Rengasamy, 2011).

TABLE 18.3
The Flocculating Power (as Related to Na) and Dispersive Power (as Related to Ca) of Cations Derived from Their Ionicity Indices

Cation	Flocculating Power	Dispersive Power	Ionicity Index
Na	1.0	45.0	0.89
K	1.8	25.0	0.86
Mg	27.0	1.7	0.73
Ca	45.0	1.0	0.67

Ionicity index $= 1-\left(I_z/I_{z+1}\right)Z^{0.5}$.
Flocculating power $= 100\left(I_z / I_{z+1}\right)^2 Z^3$.
Dispersive power is inverse of flocculating power.

where concentrations of exchangeable cations measured at the given soil pH are expressed as cmol_c/kg, and the coefficients of each cation are their respective dispersive powers relative to Ca and derived from the ionicity of clay-cation bonds.

The dispersive charge is reduced by the flocculating effects of cations present in electrolytes in soil solutions. Rengasamy et al. (2016) also defined the 'flocculating charge' as follows:

$$\text{Flocculating charge} = 45(\text{Ca})+27(\text{Mg})+1.8(\text{K})+(\text{Na})$$

where the concentration of soluble cations in the dispersed (or flocculated) soil-water suspension is expressed as cmol_c/kg (on soil basis). The weighting factors of the cations are based on the flocculating powers of cations as discussed in Rengasamy and Sumner (1998). Flocculating power and dispersive power of Na, K, Mg and Ca are summarised in Table 18.3.

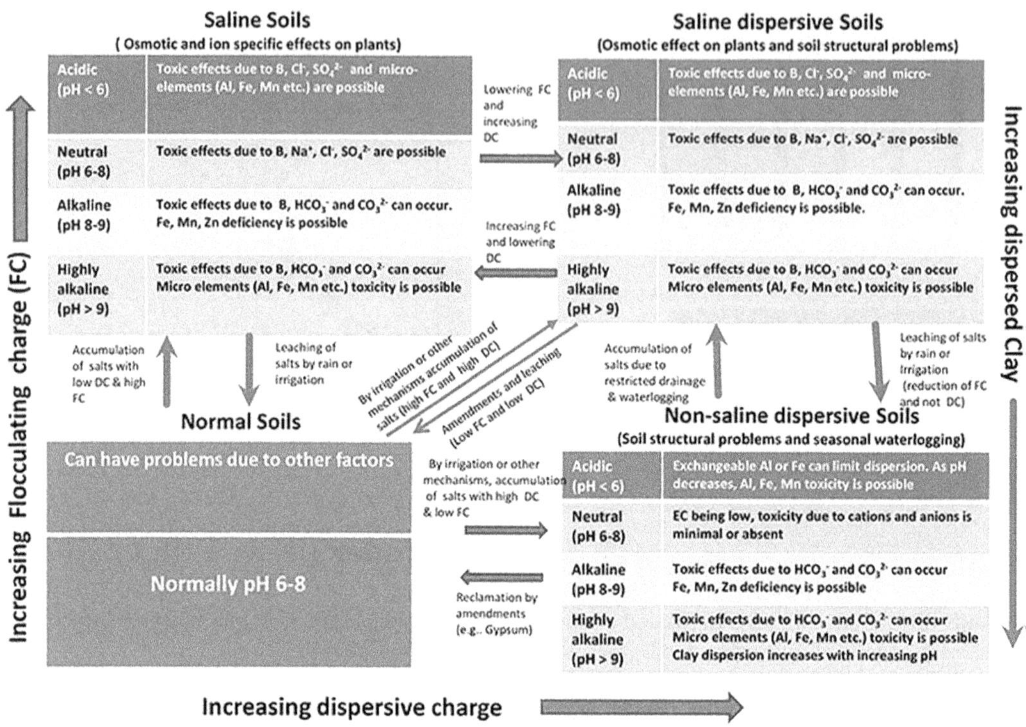

FIGURE 18.2 Categories and dynamics of dispersive soils as related to dispersive charge, flocculating charge and pH. (After Rengasamy (2016).)

In dispersed suspensions, the 'net dispersive charge' (Dispersive charge – Flocculating charge) determines the amount of clay dispersed (Rengasamy et al., 2016). The zeta potential of the dispersed clay is highly correlated with the net dispersive charge, indicating the importance of electrostatic forces involved in clay dispersion and flocculation processes. At zero point of dispersion (the point of complete flocculation), dispersive charge equals flocculating charge. Figure 18.2 describes the categories and dynamics of dispersive soils as related to dispersive charge, flocculating charge and pH.

18.3.1 Constraints to Crop Production in Dispersive Soils

The chemistry of solid and solution phases of dispersive soils determines the constraints to crop production. Solid phase chemistry is responsible for the problems due to the deterioration of soil physical conditions while solution phase chemistry controls soil pH, adequacy or deficiency of plant nutrients and excessive or toxic amounts of macro- and micro-nutrients.

18.3.2 Constraints Due to Solid Phase Chemistry

The chemistry of solid phase leading to clay dispersion plays a vital role in the destruction of all hierarchical orders of soil structure affecting water retention and movement, soil aeration, temperature and mechanical strength – all of which impair root growth, seedling emergence and water and nutrient uptake (Qadir et al., 2007; Rengasamy, 2018). Topsoil layers with dispersive character are affected by waterlogging, even during moderate rainfall events. When soil dries, it sets hard with high bulk density with crust formation impeding germination and seedling emergence. Dispersive subsoil layers experience compaction and reduction in porosity, which hinders water storage,

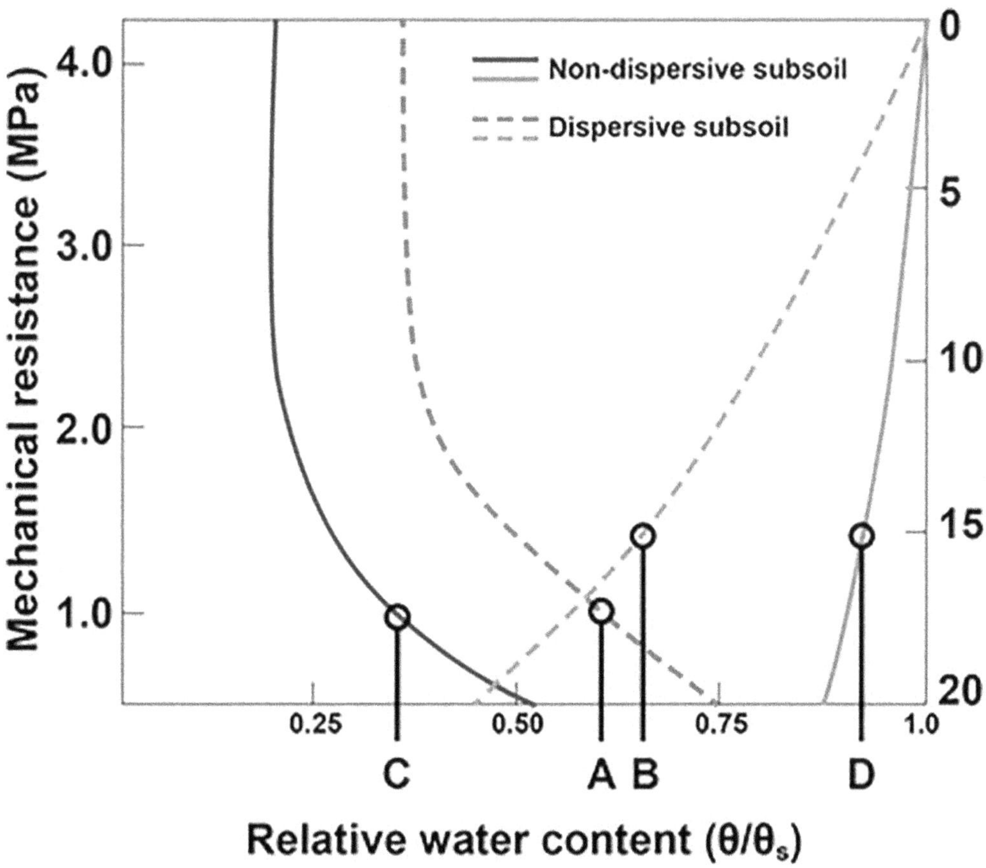

FIGURE 18.3 Non-limiting water range in a dispersive subsoil compared with a non-dispersive soil in relation to mechanical strength and aeration porosity. The points A and B denote the mechanical resistance and aeration porosity for dispersive subsoil, and C and D are for non-dispersive subsoil, respectively. (After Rengasamy (2016).)

movement of oxygen and water, the ability of roots to penetrate the soil and the uptake of water and nutrients. In dryland cropping regions, crops do not efficiently use the 'green water' (the water stored in subsoils) in these soils, and the yield is far less than the potential (Rengasamy et al., 2022).

The schematic diagram in Figure 18.3 shows that the 'non-limiting water range' (Letey, 1985) in dispersive soil is very low compared to non-dispersive normal soil. Aeration porosity of >15% and mechanical strength of <1 MPa are considered to be conducive to good crop performance, and both are related to soil water content. In this diagram, point B is the water content of a dispersive soil where aeration porosity starts to increase above 15%, and point B where the mechanical strength starts to decrease below 1 MPa. The water contents between B and A are the non-limiting water range of the dispersive soil, which is very narrow compared to the wide range between points D and C of the non-dispersive soil.

Marchuk et al. (2013) used non-destructive X-ray computed tomography scanning to characterise changes in pore architecture by the amount of clay dispersed, measured by the turbidity of the dispersed suspension of Na, K, Mg or Ca-saturated soil. They also measured the saturated hydraulic conductivity of the soils. As shown in Table 18.4, both active porosity and saturated hydraulic conductivity decreased with increasing turbidity (degree of clay dispersion). These results confirm that the degree of clay dispersion is the key factor determining the soil physical conditions in dispersive soils.

TABLE 18.4
Turbidity, Active Porosity and Saturated Hydraulic Conductivity of Soil with Dominant Exchangeable Cation Na, K, Mg or Ca

Treatments	Turbidity (NTU)	Active Porosity (%)	Sat. Hydraulic Conductivity (cm/hours)
Na soil	2,800	2.6	0.08
K soil	1,540	8.4	0.14
Mg soil	970	17.9	0.92
Ca soil	500	38.6	1.00

Source: Data from Marchuk et al. (2013).

18.3.3 Constraints Due to Soil Solution Chemistry

The aqueous components of soil at different water contents in the field determine the abiotic stress experienced by plants during their growth, consequently affecting the crop yield (Rengasamy, 2016). In salt-affected soils, the total concentration of dissolved salts in soil solutions, generally measured as the electrical conductivity (EC) of the soil solutions, is the primary criterion affecting the yield.

However, the composition of electrolytes, in addition to the total concentration, determines the multi-phase equilibria involving the solid phase comprising clay minerals and other inorganic and organic materials, the liquid phase comprising soil water, the gaseous phase comprising mainly oxygen and carbon dioxide, and the complex exchange phase. During crop growth in the field, soil water fluctuates from high to very low levels, and the chemical equilibria become more complex. The nature of cations and anions determines the exchange reactions and soil pH.

Hydrolysis of cations, anions and organic substances with different functional groups play a vital role in the equilibrium of soil pH. The acidic pH (<6) arises mainly because of the hydrolysis of the cations Fe and Al. By contrast, the soils of arid regions usually have alkaline pH (>7) and contain mainly Ca^{2+}, Mg^{2+}, K^+ and Na^+ as exchangeable cations. Tavakkoli et al. (2015) showed that a soil with >90% exchangeable Na^+ had near neutral pH when accompanying anions were Cl^- and SO_4^{2-} and alkaline pH (>8) when accompanying anions were HCO_3^- and CO_3^{2-}. The weak hydrolysis reactions of Cl^- and SO_4^{2-} result in a small change from neutral pH. However, the hydrolysis of HCO_3^- and CO_3^{2-} produce higher concentrations of OH^- resulting in higher pH. They also showed that by increasing the ionic strength of an alkaline soil (pH 9.82) with NaCl the pH was reduced to 8.32, whereas by increasing the ionic strength by adding $CaCl_2$, pH was reduced below eight; the precipitation of CO_3^{2-} as $CaCO_3$ was also involved in pH decrease. Thus, saline-alkaline soils have lower pH compared to alkaline soils with low ionic strength.

Bicarbonate and carbonate ions dominate in alkaline soils with pH > 8, which is related to 'carbonate alkalinity', as follows:

$$\text{Carbonate alkalinity} = \left[HCO_3^-\right] + 2\left[CO_3^{2-}\right]$$

Calcareous soils contain lime (calcite, $CaCO_3$), and irrespective of the concentration of lime, the soil pH is buffered between 8 and 8.5. Because of the low solubility of lime, soil pH levels are determined by the chemical buffering of calcite. Very alkaline environments (pH ≥ 9) are found in large areas of Western Asia, East Europe, Southern South America and Southern Australia (Rengasamy et al., 2022). High pH (≥ 9) environments are developed when concentrations of HCO_3^- and CO_3^{2-} species in solutions are higher than (Ca^{2+} + Mg^{2+}). Na^+ + K^+ also dominate in soil solutions, leading to

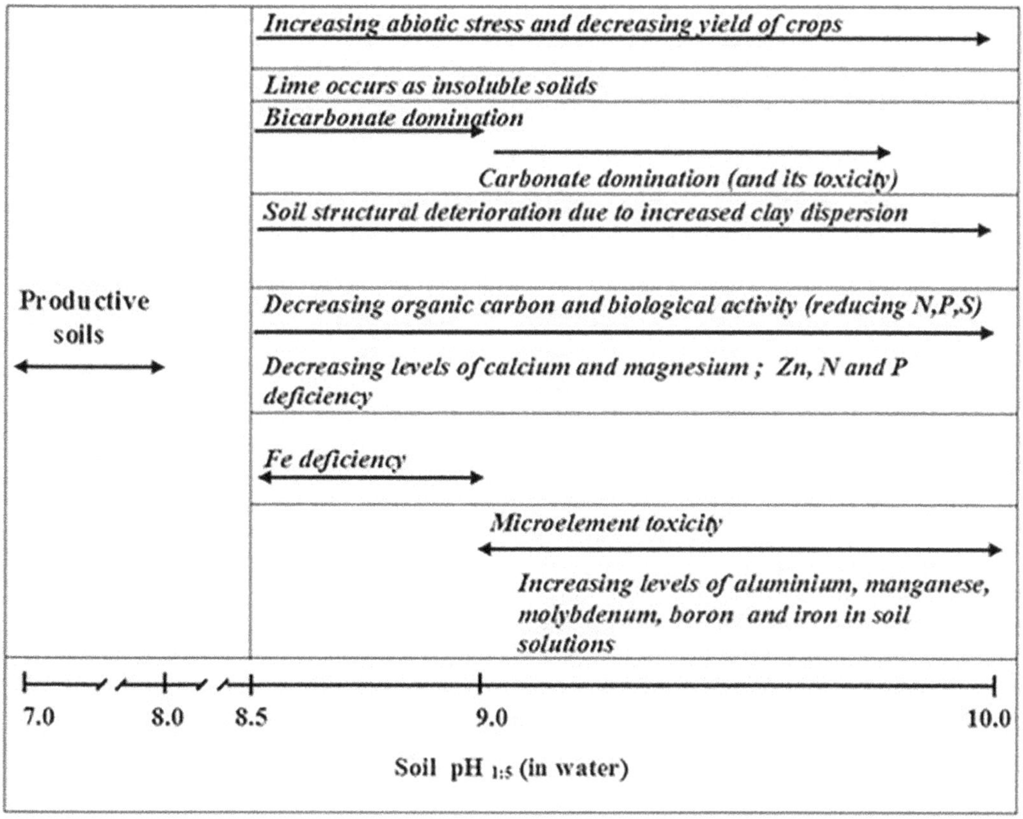

FIGURE 18.4 Abiotic stress and plant growth in alkaline soils in relation to soil pH.

the increase in exchangeable sodium and potassium. However, it is important to recognise that high alkalinity (or high pH) is caused by bicarbonate and carbonate concentrations and not by high ESP or excessive Na^+ in soil solutions (Tavakkoli et al., 2015). Jobbagy et al. (2017) cite the examples where seawater having high sodium (SAR>58) and low carbonate species causes the formation of saline soils and not alkaline-sodic soils. Determination of pH of alkaline soils using 0.01 M $CaCl_2$ always results in a pH value ≈ 8.2 reflecting that (Ca+Mg) > (HCO_3^- + CO_3^{2-}) in the solutions, because of the precipitation of carbonate species by $CaCl_2$.

Specific constraints in alkaline soils (pH in water between 8.5 and 10) depend on soil pH (Figure 18.4). For example, bicarbonate toxicity may prevail between pH values of 8 and 9, while carbonate toxicity dominates above pH 9. Iron deficiency may be a problem when pH is between 8 and 9, while Fe toxicity can occur above pH 9 (Setter et al., 2009). Similarly, boron deficiency and toxicity are dependent on soil pH. As the pH increases above 9, $B(OH)_4^-$ species increase and are adsorbed by soil particles (Rengasamy, 2016), perhaps leading to B deficiency. Recent studies have shown that Al toxicity can be a problem when soil pH increases above nine due to the formation of soluble anionic aluminate species (Brautigan et al., 2012). With increasing pH, monovalent ions, particularly Na^+, increase in soil solution, while concentrations of divalent Ca^{2+} and Mg^{2+} decrease, leading to high clay dispersivity and soil structural degradation. This also renders subsoil with transient waterlogging and anoxic conditions, where pH and pE interactions induce microelements (such as Mn, Fe, Al and B) toxicities due to their increasing concentrations in soil solutions and/or reduced energy of plant roots to exclude these ions (Setter et al., 2009). When alkaline soils are salinised by chloride and sulphate salts of sodium, high soil pH is considerably reduced, and toxicity due to carbonates and elements (Mn, Al and Fe) is diminished, and osmotic effects become dominant. Alkaline soils sequester less organic carbon, high pH being responsible for the increase in dissolved organic carbon (McDonald et al., 2017) and reduced microbial activity.

18.3.4 Principles Involved in the Management of Dispersive Soils

The management of dispersive soils should focus on the correction of the chemistry of the solid and solution phases to be suitable for crop production. Reduction of net dispersive charge both by decreasing the inherent dispersive charge and by increasing the flocculation charge should remove the constraints caused by the dispersivity of the solid phase. Ion toxicity and deficiency are generally due to the pH of the solution phase. Hence, amendment of acidic or alkaline pH of these soils is necessary to improve crop productivity. Reduction of salt concentrations below the plant tolerance limits may be necessary in saline-dispersive soils to avoid the constraints due to osmotic and ion imbalance effects of salinity. Table 18.5 summarises the problems encountered and the principles involved in the management of different categories of dispersive soils.

Globally, the choice of parameters to identify dispersive soils and to define soil sodicity is either ESP of soils or SAR of soil solutions. Many reports show high correlations between ESP and SAR with the amounts of clay dispersed in soils, but with varying trends. As discussed earlier, many factors other than ESP or SAR influence the behaviour of dispersive soils (Rengasamy et al., 2016), and, hence, they are not suitable factors in diagnosing the adverse soil physical conditions observed in sodic soils and in decisions related to reclamation or management of dispersive soils. All the research reports published, discussed in later sections, have used ESP or SAR as the criterion for the reclamation of sodic dispersive soils using inorganic and organic amendments. Similarly, the studies on plant-based solutions for sodic dispersive soils have used changes in ESP or SAR as reference points.

TABLE 18.5
Problems Encountered and the Principles Involved in the Management of Dispersive Soils

Soil Category	Problems Encountered	Principles of Management
	Dispersive Soils	
1. Acidic (pH<6)	Swelling and clay dispersion affecting soil physical conditions (including surface hard-setting and waterlogging)	Reduction of dispersive charge and increasing flocculating charge
	Acidic pH affecting crops via element toxicity and/or deficiency	Increasing soil pH (e.g. lime addition)
2. Neutral (pH 6–8)	Swelling and clay dispersion affecting soil physical conditions (including surface hard-setting and waterlogging)	Reduction of dispersive charge and increasing flocculating charge
3. Alkaline (pH 8–9) and highly alkaline (pH>9)	Swelling and clay dispersion affecting soil physical conditions (including surface hard-setting and waterlogging)	Reduction of dispersive charge and increasing flocculating charge
	Alkaline pH affecting crops via element toxicity and/or deficiency	Reduction of pH (<8.2) (e.g. phytoreduction, gypsum application)
	Saline-Dispersive Soils	
4. Acidic (pH<6) 5. Neutral (pH 6–8) 6. Alkaline (pH 8–9) and highly alkaline (pH>9)	The problems identified in corresponding dispersive soils will be encountered. In addition, osmotic effect of salt concentration can affect crop species	Leaching of salts to reduce osmotic effect (crop specific). Once salts are reduced, the soils become dispersive soils, and the principles involved in corresponding dispersive soils will apply

We emphasise that, in the future, the measurement of clay dispersion should be the first step in diagnosing soil physical constraints, because the amount of clay dispersed is highly correlated with net dispersive charge on soil particles (Rengasamy et al., 2016). Measurements of clay dispersion, electrolyte composition of soil solution and soil pH are essential to identify the categories of dispersive soils (Figure 18.2) and to devise the management strategies.

18.4 MECHANISMS THAT DRIVE THE AMELIORATION PROCESS IN THE SUBSOIL

Sodic soils are often poorly structured, have a high clay content and high bulk density and are dispersive. These factors result in poor subsoil structure that can impede drainage, promote waterlogging (low water infiltration) and increase de-nitrification (nutrient imbalance) and soil strength (Orton et al., 2018). These properties also impede the infiltration of water into and within the soil and reduce water and nutrient storage capacity, and ultimately the plant available water (PAW) content of the soil. Consequently, root growth and rooting depth are impeded, as is a crop's ability to access and extract deeper stored water and nutrients (Passioura and Angus, 2010). This is particularly problematic in environments where the reproductive phase can coincide with periods of water stress, and when the conversion of water to grain has the greatest effect both on yield and the likelihood and magnitude of a yield gap (Adcock et al., 2007).

In Australian winter cropping for example, crops commonly have sufficient water supply during their early growth stages either from stored soil water or from rainfall. However, the reproductive phase is often affected by water stress or terminal drought, and this is thought to be the major cause of variable grain yield (Chenu et al., 2013). The effect of water stress in the reproductive phase is further impacted by shallow root depth induced by subsoil sodicity. Under such conditions, a key to improving crop productivity is to improve root growth in and through sodic subsoils to enable use of deep subsoil water later in the growing season. Water use at this late stage has a 2–3 fold greater conversion efficiency into grain yield (Kirkegaard et al., 2007) than seasonal average based conversions efficiencies (e.g. 20–25 kg/mm verses 50–60 kg/mm).

Sodic subsoils (e.g. >20 cm) are widespread and are particularly difficult to manage in arid and semi-arid regions, due to the insufficient rainfall to leach excess sodium (Na^+) from the root zones (Rengasamy and Olsson, 1991). However, most studies have been focused on ameliorating the surface layers (i.e. <10 cm) of sodic soils by the application of amendments (Barzegar et al., 2002; Leogrande and Vitti, 2019; Uddin et al., 2022). For example, gypsum amendment ($CaSO_4 \cdot 2H_2O$) can decrease clay dispersion and increase flocculation by replacing Na^+ with calcium (Ca^{2+}) on soil clay surfaces (Rengasamy et al., 2016; Shainberg and Letey, 1984). Organic amendments (OA), such as crop residues, can improve soil chemical, physical and biological properties. Yet, compared to the soil surface, amelioration of sodic subsoils through a combination of organic and inorganic amendments has received much less attention, and there are few studies that have explored this area (Clark et al., 2007; Jayawardane and Chan, 1994; Sale et al., 2021).

Gypsum is used almost exclusively to manage soil dispersion worldwide (Loveday, 1974). Gypsum reduces dispersion in three ways. First, the Ca^{2+} ions can replace exchangeable Na^+ ions bonded to the clay surface, and the amount of gypsum required for this to occur is known as the 'gypsum requirement' (Rengasamy et al., 2016). Gypsum can also reduce dispersion by increasing the electrolyte concentration of the soil solution, which prevents dispersion by the so-called 'electrolyte effect'. Third, gypsum can reduce the extreme alkalinity of soils which in turn reduces the soil dispersion. Highly alkaline soils contain CO_3^{2-} and HCO_3^- salts with accompanying cations such as Na^+ and K^+. The reaction of gypsum with these salts (e.g. Na_2CO_3) may result in the precipitation of $CaCO_3$. This precipitation will reduce the concentration of soluble CO_3^{2-} (in the form of Na and/or K) which will reduce the soil pH (Tavakkoli et al., 2015).

It is now accepted that the recommended rate of gypsum that needs to be added to a dispersive soil should be based on its 'electrolyte effect' rather than its 'gypsum requirement'. Electrolyte effect refers to the concept of 'threshold electrolyte concentration' (Quirk and Schofield, 1955) and 'flocculating charge' (Rengasamy et al., 2016). The rate based on gypsum requirement is considerably larger and is considered to be uneconomic for many dryland cropping systems. It also requires considerable rainfall to dissolve the large amount of gypsum that may be required. One important consequence of the lower gypsum rate to deliver the electrolyte effect is that the effect will be relatively short term (Loveday, 1974) and will disappear when the gypsum in the soil solution is leached from the soil by infiltrating rainfall.

18.4.1 Use of Gypsum to Ameliorate Topsoil Dispersion

The responses of crops to gypsum application is variable. While several studies reported positive yield responses to gypsum application (Dang et al., 2010; Doyle et al., 1979), several other trials reported no yield responses to gypsum (Gill et al., 2008; Hamza and Anderson, 2003). Part of the reason is that the metric used to identify dispersive soils worldwide is based on an ESP of 6%–15%. Several studies (Marchuk and Rengasamy, 2011; Nuttall et al., 2003) showed that the use of ESP alone may not be an appropriate metric of soil dispersion. The extent to which clay particles in a soil will disperse depends on other soil factors such as exchangeable potassium and exchangeable magnesium (Rengasamy and Marchuk, 2011), the organic matter in the soil (Sale et al., 2021), clay mineralogy (Goldberg and Forster, 1990), soil pH (Chorom and Rengasamy, 1995; Tavakkoli et al., 2015) and the electrolyte concentration in the soil solution (Quirk and Schofield, 1955). Thus, the conditions that lead to dispersive behaviour are complex. The challenge for growers is to determine which soils will, or will not, respond to gypsum application. Loveday (1974) who developed a procedure for identifying 'gypsum responsive' soils, which involved a direct measure of dispersion and soil pH, first considered this issue. More recently, researchers have developed a simple test for use by land managers and producers, without the need for complex analytical procedures in the laboratory (Kelly and Rengasamy, 2006). The basis for the test is the actual dispersive behaviour of the clay itself. This is likely to be more accurate that any critical ESP value, which requires laboratory analyses. Recently, the concept of 'Net Dispersive Charge' for functional characterisation of dispersive soils has been introduced to replace traditional approaches to define clay dispersion of soils and resolved important controversies in the literature (Rengasamy et al., 2016). This work defined the net dispersive charge as the difference between the dispersive charge and the flocculating charge and focuses on the correction of the chemistry of the solid and solution phases to be suitable for crop production (for more details, see sections below).

18.4.2 Use of Gypsum to Ameliorate Subsoil Dispersion

In most field experiments that have assessed gypsum to ameliorate soil dispersion problems, there were no physical measurements made on subsoil layers. In many cases where the gypsum application increased crop yield, there was no indication whether this yield response resulted from improvements in subsoil structure. Nevertheless, a number of comprehensive investigations were undertaken, and these provide insights into the effect of gypsum on the structure of sodic subsoils.

i. The benefit of gypsum in reducing clay dispersion in the subsoils is a result of the 'electrolyte effect' which reduces dispersion, and not to the replacement of exchangeable sodium by calcium ions. The continuing presence of the dissolved gypsum was necessary for the reduced dispersion to continue. However, in high rainfall or irrigated conditions, the electrolyte effect will cease when the soluble gypsum is leached from the subsoil layer. In highly alkaline-dispersive subsoils, gypsum application led to significant reduction of pH which in turn reduced clay dispersion. The impact of gypsum on reducing soil pH and dispersion is now being considered as main driving factors for the management of dispersive subsoils (Rengasamy et al., 2016; Tavakkoli et al., 2022b; Uddin et al., 2022).

ii. Adding gypsum to a dispersive subsoil matrix consistently increases its intrinsic permeability. This will increase the hydraulic conductivity through the flocculated clay matrix (Loveday, 1974). There will be benefits in water infiltration and redistribution, and improved water storage and improved drainage. However, there will be no increase in macro-porosity or the air-filled porosity (Blackwell et al., 1991). The impact on root growth in the subsoil may therefore be limited due to a lack of large transmission pores.

iii. Most studies show the value of deep ripping with surface gypsum application to assist the movement of gypsum into the subsurface layers (Blackwell et al., 1991). In the absence of deep ripping, it may take considerable time for the gypsum to leach into the subsoil layers, with the length of this period depending on rainfall. However, this will depend on the permeability of the upper layers of the soil profile.

iv. It is almost impossible to distribute the gypsum evenly through the subsoil. This was apparent with the gypsum slotting technology developed by CSIRO in 1980s and when gypsum was incorporated at depth in rip-lines (Gill et al., 2008; Sale et al., 2019). There were significant benefits via improved structure in the gypsum-amended slots, but the benefit did not extend into the subsoil between the slots. This restricted the extent of the structural improvement in the subsoil that resulted from the gypsum that was mixed through the soil in the slot. If gypsum can be physically placed throughout the subsoil, then this should improve the hydraulic conductivity across the upper layers of the subsoil. Improvements in subsoil permeability are then possible, and this would improve profile drainage.

18.4.3 Use of Organic Amendments

It is now nearly 50 years since the first field trial was undertaken using organic amendments to ameliorate poorly structured subsoils. In 1973, chopped hay was incorporated into the topsoil of sodic cracking clay soils in NSW (Doyle et al., 1979). In 1986, Project Sexton demonstrated how high rates of lawn clippings, supplemented with nutrients, placed in the subsoil, resulted in large and long-lasting yield increases across a variety of sites in the wheat belt of South Australia (Graham et al. 1992). However, most studies have been focused on ameliorating the surface layers (i.e. <10 cm) of sodic soils by applying amendments (Barzegar et al., 2002; Leogrande and Vitti, 2019). For example, the application of OA, such as crop residues, showed significant improvements in soil chemical, physical and biological properties (Abiven et al., 2009). Yet, compared to the soil surface layer, there have been a few studies on using a combination of organic and inorganic amendments to ameliorate sodic subsoils (Clark et al., 2007). In recent years, the subsoil placement of OAs expanded in attempts to improve soil properties deep in the profile (Celestina et al., 2019). For example, deep banding into the subsoil of amendments has been tested in Solonetz soils which are characterised by high alkalinity, dispersion, slaking, high bulk density and overall poor physical structure (Jayawardane and Chan, 1994; Sale et al., 2019). Several studies reported significant grain yield increases for several years following the incorporation of high rates of nutrient-rich organic amendments such as poultry manure or Lucerne pellets (Gill et al., 2008; Leskiw et al., 2012). While the cause of these yield benefits remain unclear, it has been attributed to the amelioration of subsoil constraints and increased water use and nutrient supply (Gill et al., 2009). For example, in an experiment near Ballan, Victoria, substantial improvements in soil aggregation was observed in the 30–40 cm clay subsoil, which was treated with banded OA (Figure 18.5). The incubation study by Clark (2010) used similar sodic clay subsoil from Ballan and reported that the percentage of macro aggregates (diameters>2mm) in the clay increased from 1% to 42% when finely ground wheat straw and N fertiliser were added.

A recent study at Rand in southern NSW (clay subsoil with bulk density of 1.55 g/cm, pH_{H_2O} of 9 and ESP of 17%) reported significant yield responses of deep-placed amendments from 2017

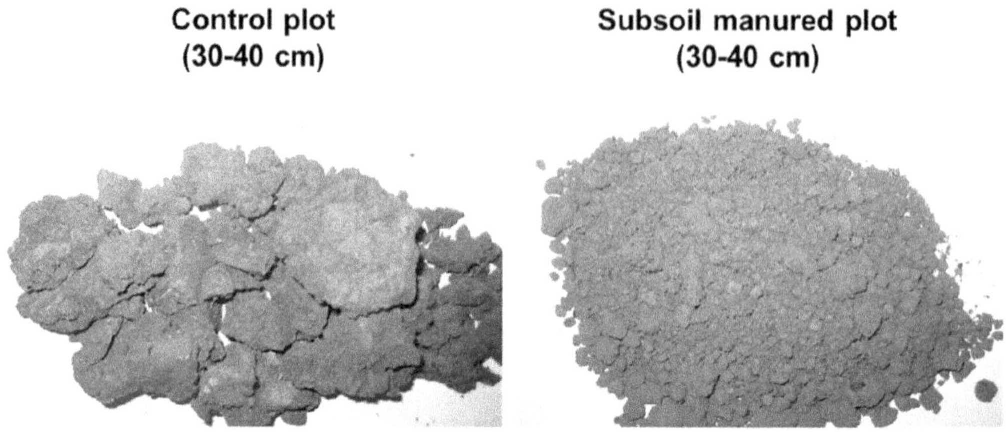

FIGURE 18.5 The effect of subsoil (30–40 cm) placement of organic amendment (lucerne stubble) on structural improvements of soil aggregates. The amendments were applied in 2005 at Ballan, Victoria, Australia (Gill et al., 2009; Sale et al., 2021).

to 2021. The amendment that consistently resulted in significant yield increases above the control was the deep-placed pea or wheat stubble pellets and the combination of pea straw pellets, gypsum and some liquid fertiliser nutrients (Tavakkoli et al., 2022a). Over a 4-year period, this treatment resulted in a 27% improvement in cumulative yield. Improvements in subsoil chemical and physical properties were measured in the winter of 2019. The deep crop residue amendments significantly increased macro aggregation, as measured on the rip-line at a depth of 20–40 cm. Similarly, deep gypsum and the deep gypsum/pea straw/nutrient combination markedly increased water infiltration into the soil profile, with higher saturated hydraulic conductivities measured on the rip-line (Sale et al., 2021). Compared to the control, the deep placement of gypsum reduced the soil pH by 0.86 units at 20–40 cm depth. However, pH was not affected by OAs. These results indicate that independent modes of action of various amendments (e.g. crop residue vs gypsum) are required in the amendment mix in order to ameliorate these subsoils. For example, adding gypsum reduced pH in the amended subsoil from 8.99 to below 8.13. This indicates that significant changes in soil pH can occur with realistic application rates of gypsum in the subsoil. Since high alkalinity also increases negative charges on the surfaces of clay particles (Rengasamy et al., 2016), which increases clay dispersion, a reduction in pH following gypsum application also resulted in significant reduction in soil dispersion (Tavakkoli et al., 2015; Tavakkoli et al., 2022b). In alkaline-sodic soils, high ESP and high pH are always linked and it is difficult to apportion their effects on the resulting poor soil physicochemical conditions and consequently on crop growth. The addition of pea straw and nutrients provides substrate for enhanced biological activity resulted in increased macro aggregation and improved subsoil structure. When combined, organic and inorganic amendments may result in additive effects to improve soil physical and chemical properties (Fang et al., 2021; Niaz et al., 2022). These limited data suggest that the direct application of OAs or the co-application of OAs and gypsum in sodic subsoils may improve soil structure (i.e. aggregate stability). However, mechanistic understanding between the development of soil structure (e.g. aggregate stability) and the decomposition of OAs in the presence of Ca^{2+} (from gypsum) remains elusive, particularly under poorly structured alkaline-sodic subsoils. Future research should identify the key properties of OAs that are most important for soil aggregate formation (e.g. functional groups, quality of C and ability to generate microbial biomass) as well as the interaction between OAs and inorganic amendments and its impact on the correction of the chemistry of the solid and solution phase (e.g. manipulation of zeta potential and net dispersive charge).

18.4.4 Plant-Based Solutions for Dispersive Soils

The two approaches to improve productivity on sodic soils are to ameliorate the soil to reduce the severity of a soil's physical and chemical constraints or to use plant-based approaches to either modify soil properties through phytoremediation or to develop varieties that tolerate or escape the major soil constraints. The largest improvements in productivity are likely to come from a multi-faceted approach which combines a number of approaches to address the limitations associated with the solid and solution phases rather than relying on a single mechanism of plant tolerance (McDonald et al., 2020). The physical limitations of dispersive soils and a narrowing in the range of PAW are important factors that restrict yield. The ability of plants to maintain root growth and water uptake in a constrained environment is fundamental to improving productivity, and plant-based solutions should be viewed in this context. In this section, the role of plant-based approaches to manage dispersive soils will be described.

18.4.4.1 The Use of Primer Plants to Ameliorate Poorly Structured Subsoils

Primer plants are used to modify the structure of deep soil layers by extending their roots into these layers, then dying and decaying, leaving pores or channels (Yunusa and Newton, 2003). Plant roots preferentially grow in pores and cracks in dense subsoils (White and Kirkegaard, 2010); therefore, on face value, the use of primer plants to perforate poorly structured subsoil layers with 'biopores' appears to be an attractive low-cost amelioration option. Primer crops can produce structural improvements to dense clay subsoils (McCallum et al., 2004), resulting in subsequent crops extracting more subsoil water (McCallum et al., 2004; Nuttall et al., 2008). However, the research conducted to date in Australia suggests that there was minimal improvement in the productivity of crops grown following the 'primer plant' phase that could be attributed to improved subsoil structure. Furthermore, it is possible that the including a primer plant phase in a cropping rotation might have some unanticipated negative effects.

One disadvantage of primer plants is that they may not provide a cost-effective phase in the cropping rotation. For example, the inclusion of a lucerne phase to improve subsoil structure is only likely to benefit a farm business if a financial benefit can be gained from the lucerne phase in its own right. This is more difficult for a farm where continuous cropping is practised, and there are no grazing animals. It is also likely that an extended period of time will be required for the primer phase to improve subsoil structure. McCallum et al. (2004) reported that 10 years of a phalaris pasture, and 3–4 years of lucerne were used in their study to generate improved macro-porosity in a dense B horizon in a Sodosol soil in central western NSW. Even in a non-constrained soil, it may take 6 years before a stable system of biopores is established (Lucas, 2022). Any reduction in income during the primer crop phase could have significant financial implications for many cropping businesses due to the opportunity costs of foregoing cropping.

Another negative effect of using a primer plant strategy in drier areas is that the primer plant might de-water the soil profile, resulting in less PAW for following crops. This was the case when perennial *Acacia* species were used to ameliorate the dense clay subsoil in a Chromosol in NE Victoria (Yunusa et al., 2002). The first oat crop grown after these woody perennials suffered a yield loss of 23% due to the dry profile (Mele and Yunusa, 2001). Modelling also found that crops that follow an *Acacia* primer phase would on average suffer a yield reduction of 74%, due to the dried soil profile resulting from the woody perennials. Similar results have been reported by Angus et al. (1996), Hirth et al. (2001) and McCallum et al (2004) where crop yields immediately following a lucerne phase were reduced in dry years due to the suspected drying of the soil profile by the lucerne plants. Researchers in the Victorian Mallee (Nuttall et al., 2008) also concluded that the water deficits generated by the primer plants would be unlikely to be replenished before the following crop was grown, and this would limit any benefits from the amelioration of poorly structured subsoils in the first year.

18.4.4.2 Phytoremediation of Dispersive Soils

Changing soil properties by phytoremediation has been used for over 100 years to improve productivity of sodic soils (Pessarakli and Szabolcs, 2011; Qadir et al., 2007). Phytoremediation uses gypsum, incorporation of organic matter and salt-tolerant species or legumes to alter soil properties (FAO, 1976; Gupta and Abrol, 1990; Ilyas et al., 1997; Qadir et al., 2007). The basis of phytoremediation is an increase in the dissolution of calcium carbonate with the higher Ca^{2+} concentration in the soil solution displacing Na^{+} from the cation exchange complex. As a result, there are improvements in soil structure, water infiltration, hydraulic conductivity and pH. Phytoremediation can reduce the net dispersive change because the increase in Ca^{2+} in the soil solution will increase the flocculating charge while the reduction in exchangeable Na^{+} will reduce the dispersive charge. However, for phytoremediation to be effective, Na^{+} needs to be leached from the root zone by irrigation or high rainfall.

Plants have a number of chemical and physical mechanisms which increase the availability of Ca^{2+} and improve the movement of water through the soil (Qadir et al., 2007). These include (i) increasing the partial pressure of CO_2 in the root zone by root respiration and microbial respiration, (ii) altering root zone pH by the release of H^{+} by the roots, (iii) improving soil aggregate stability and soil porosity and increasing the hydraulic properties of the soil and (iv) removing Na^{+} in harvested shoots. A strategy to improve soils by phytoremediation involves growing a crop or pasture for one or more seasons and leaching the Na^{+} from the root zone by rainfall and irrigation. Gypsum can also be applied during the cropping phase. Phytoremediation can be as effective as chemical amelioration in improving soil properties, especially on coarse- tomedium-textured soils (Kumar, 1988; Kumar et al., 1994; Qadir et al., 2005; Qadir et al., 2007; Qadir et al., 2002). However, in arid and semi-arid regions, phytoremediation may be less effective than chemical amelioration because leaching of Na^{+} from the root zone is low (Qadir et al., 2005) and other approaches are needed.

18.5 BREEDING APPROACHES

Developing varieties that are tolerant to the major constraints of sodic soil may be important when soil amelioration is not possible or when management practices are costly or ineffective. The success of this approach will depend on having sufficient genetic variation for the traits of interest and having the tools and techniques to identify superior lines. The approaches that can be used are to select for improved tolerance to sodicity or to breed for tolerance to the individual soil constraints of dispersive soils, such as tolerance to various nutrient toxicities, waterlogging or high soil strength.

18.5.1 Improved Tolerance to Sodicity

There are relatively a few reports of breeding specifically for tolerance to sodicity. Much of the published work on screening for sodicity tolerance concerns rice where field selection at sites with sodic soils has been a common approach (Ali et al., 2013; Singh et al., 2002, 2016). Genetic variability in tolerance in the field has also been reported in wheat and barley (Dang et al. (2016), Dowla et al. (2021)). Khokhar et al. (2017) reported large Genotype × Environment interactions for yield in wheat over a number of sodic and saline sites in India with some varieties yielding well at sites with sodic soils. These results indicate that selection for improved tolerance to sodicity is feasible and in rice has resulted in the development of more tolerant varieties.

18.5.2 Surface Crusting and Waterlogging

The poor structure of soils with dispersive topsoils leads to waterlogging after rainfall or irrigation and surface crusting when the soil dries. Both of these reduce seedling emergence and seedling vigour. Dispersive topsoils are most commonly treated by applications of gypsum (Gupta and Abrol,

1990; Qadir et al., 2002; Wheaton et al., 2002); however, there is also evidence of genetic variation in wheat for tolerance to anoxia at germination (Setter and Waters, 2003; Tavakkoli et al., 2017) and in the ability to penetrate surface crusts (Anzooman et al., 2018). This opens up opportunities to select for varieties with improved establishment in dispersive soil.

18.5.3 High Soil Strength

Soils with dispersive clay have low porosity and high densities resulting in strong soils with high resistance to root growth. While there has been a considerable amount of work on examining genetic differences in root growth under high soil strength, little progress has been made in developing varieties with improved tolerance to high bulk density that would be suitable for dispersive soils. Recent advances in understanding how plant roots respond to soil compaction through their sensitivity to ethylene (Pandey et al., 2021b), in imaging roots in soil (Pandey et al., 2021a) and in analysing field data on root distributions (Wasson et al., 2017) may improve the ability to screen for tolerance to high soil strength and to assess genotypes in the field.

18.5.4 Drought Tolerance

The smaller range in water availability in dispersive soils compared to non-dispersive soils means that crops will be more prone to drought stress in sodic soils. Drought tolerance and tolerance to sodicity are linked traits. High salinity will exacerbate the effect by decreasing the osmotic potential of the soil solution. Other constraints of sodic soils, such as high boron or high alkalinity, which restrict root growth may increase the sensitivity of crops to drought. This was highlighted by the observation that selection for drought tolerance in barley resulted in lines with improved levels of boron tolerance (Yau and Ryan, 2008). Osmotic adjustment, which helps maintain growth as soil water availability declines (Turner, 2018), is also associated with the maintenance of growth under high salinity (Genc et al., 2019; Munns et al., 2016).

Drought-escape traits, such as quick development and early flowering, may help to improve productivity on dispersive soils where the small non-limiting range in soil moisture can cause drought stress to occur early in the growing season. For example, early flowering was found to be an important trait in an analysis of wheat breeding trials on alkaline soils in the winter rainfall regions of Australia where dispersive soils predominate (McDonald et al., 2012), and days to anthesis was also found to be associated with high yields among varieties of wheat grown at sites with saline and sodic soil in India (Khokhar et al., 2017).

18.5.5 Tolerance to Nutrient Imbalances

The chemical properties of dispersive soils influence the pH of the solution phase which in turn affects the occurrence of ion toxicities and deficiencies (Table 18.6 and Figure 18.4). Soil amendments that

TABLE 18.6
Reported Nutritional Problems in Dispersive Soils

Soil Category	Nutritional Constraints
Acidic-dispersive	Toxicities: Al, Mn, Fe
	Deficiencies: Ca, Mg, Mo
Alkaline-dispersive	Toxicities: B
	Deficiencies: P, Cu, Zn, Mn, Fe
Saline-dispersive	Toxicities: Na, Cl
	Deficiencies: Ca, K

change pH will help overcome some of these nutrient imbalances; however, there will be situations where soil amelioration is costly or ineffective. For example, dispersive subsoils may be difficult or uneconomic to change and increased productivity may rely on improving the genetic tolerance to the nutritional problems.

18.5.6 Salinity Tolerance

Breeding for salinity tolerance may improve yields on dispersive soils when salinity is a major limitation to growth. Selection for improved salt tolerance has been reported to increase yields on sodic soils in rice (Ali et al., 2013; Singh et al., 2016; Vasuki and Geetha, 2016), while salt tolerance is being targeted to improve tolerance to sodic soils in cotton (Liu et al., 2020). However, focussing on improved salinity tolerance may not provide consistent improvements in productivity where other chemical and physical constraints are important. For example, high pH occurs in many saline-dispersive soils due to the presence of sodium carbonates, and high pH may reduce plant growth both directly and by exacerbating the effects of salinity (Bui, 2013).

Success in improving salinity tolerance has been variable. Varieties of rice with improved tolerance to salinity have been developed (Qin et al., 2020), but progress has been slow in wheat and barley because of the debate about defining the most appropriate criteria for salt tolerance (sodium exclusion vs tissue tolerance), selection of the most appropriate screening method (soil vs hydroponic) and possible interactions with other soil constraints (Asif et al., 2019; Barrett-Lennard et al., 2016; Bromham et al., 2013; Bui, 2013; Genc et al., 2019; Tavakkoli et al., 2012; Wani et al., 2020).

18.5.7 Tolerance to Ion Toxicities

Depending on pH, plants growing in dispersive soils can be exposed to a number of ion toxicities (Table 18.6 and Figure 18.4). In acidic soils, tolerance to high Al has been a major focus of genetic improvement (Kochian et al., 2015; Rahman and Upadhyaya, 2021), and Al-tolerant varieties can form an important component in managing acidic-dispersive soils. However, the use of Al-tolerant varieties alone does not address the fundamental cause of low pH, and their use needs to be supplement with amelioration practices that control acidification.

In Australia and west Asia, alkaline soils can contain significant amounts of boron, and B toxicity can be an important limiting factor on alkaline-dispersive soils. Breeding for improved boron tolerance has been successful in a number of crops (Yau and Erskine, 2002; Yau and Ryan, 2008) and has led to improved production (Schnurbusch et al., 2010).

18.5.8 Tolerance to High pH

High pH can also substantially reduce shoot and root growth in alkaline-dispersive soils (Millar et al., 2007; Patil et al., 2020). However, the analysis of the importance of subsoil constraints by McDonald et al. (2012) found that tolerance to high pH explained little of the observed variation in wheat yields. A reason for this was that there was little genetic variation in tolerance to high pH among the varieties used. Recent work has shown that more tolerant germplasm than the best current cultivars can be identified (Figure 18.6), and this opens up new opportunities to improve tolerance to high pH in wheat. There is little information on tolerance to high pH in many of the important pulse and oilseed crops.

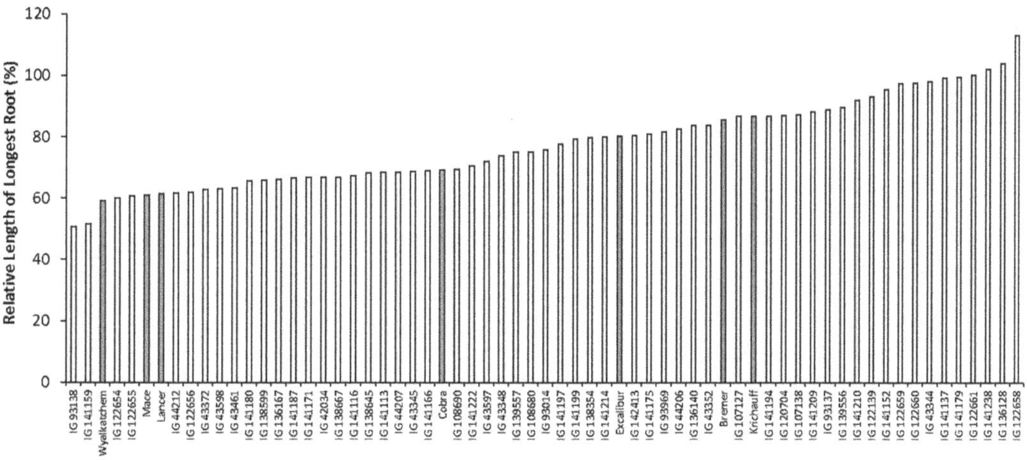

FIGURE 18.6 Variation in tolerance to high pH among a selections of landraces of wheat from the salinity Focussed International Germplasm (FIG) set and Australian cultivars. Tolerance was assessed as the relative root length of seedlings when grown at pH 7.2 and 9.5. Current cultivars are indicated as shaded bars. The variety Krichauff is classified as a tolerant Australian cultivar (Schilling and McDonald, unpublished data).

REFERENCES

Abiven, S., Menasseri, S., and Chenu, C. (2009). The effect of organic inputs over time on soil aggregate stability: A literature analysis. *Soil Biology & Biochemistry* **41**, 1–12.

Adcock, D., McNeill, A.M., McDonald, G.K., and Armstrong, R.D. (2007). Subsoil constraints to crop production on neutral and alkaline soils in south-eastern Australia: A review of current knowledge and management strategies. *Australian Journal of Experimental Agriculture* **47**, 1245–1261.

Ali, S., Gautam, R.K., Mahajan, R., Krishnamurthy, S.L., Sharma, S.K., Singh, R.K., and Ismail A.M. (2013). Stress indices and selectable traits in SALTOL QTL introgressed rice genotypes for reproductive stage tolerance to sodicity and salinity stresses. *Field Crops Research* **154**, 65–73.

Angus, J.F., Peoples, M.B., Gault, R.R., Gardner, P.A., and Hpwe, G.N. (1996). Improved pastures and their effects on subsequent crops. *Proceedings of the 8th Australian Agronomy Conference*, Toowoomba. http://www.regional.org.au/au/asa/.

Anzooman, M., Christopher, J., Mumford, M., Dang, Y.P., Menzies, N.W., and Kopittke, P.M. (2018). Selection for rapid germination and emergence may improve wheat seedling establishment in the presence of soil surface crusts. *Plant and Soil* **426**, 227–239.

Asif, M.A., Pearson, A.S., Schilling, R.K., and Roy, S.J. (2019). Opportunities for developing salt-tolerant wheat and barley varieties. *Annual Plant Reviews* **2**, 157–218.

Baldermann, A., Reinprecht, V., and Dietzel, M. (2021). Chemical weathering and progressing alteration as possible controlling factors for creeping landslides. *Science of the Total Environment* **778**, 146300.

Barrett-Lennard, E.G., Anderson, G.C., Holmes, K.W., and Sinnott, A. (2016). High soil sodicity and alkalinity cause transient salinity in south-western Australia. *Soil Research* **54**, 407–417.

Barzegar, A.R., Yousefi, A., and Daryashenas, A. (2002). The effect of addition of different amounts and types of organic materials on soil physical properties and yield of wheat. *Plant and Soil* **247**, 295–301.

Blackwell, P., Jayawardane, N., Green, T., Wood, J., Blackwell, J., and Beatty, H. (1991). Subsoil macropore space of a transitional red-brown earth after either deep tillage, gypsum or both.II. Chemical effects and long term changes. *Soil Research* **29**, 141–154.

Brautigan, D.J., Rengasamy, P., and Chittleborough, D.J. (2012). Aluminium speciation and phytotoxicity in alkaline soils. *Plant and Soil* **360**, 187–196.

Bromham, L., Saslis-Lagoudakis, C.H., Bennett, T.H., and Flowers, T.J. (2013). Soil alkalinity and salt tolerance: Adapting to multiple stresses. *Biology Letters* **9**, 20130642.

Bui, E. (2013). Possible role of soil alkalinity in plant breeding for salt-tolerance. *Biology Letters* **9**, 20130566.

Celestina, C., Hunt, J.R., Sale, P.W.G., and Franks, A.E. (2019). Attribution of crop yield responses to application of organic amendments: A critical review. *Soil and Tillage Research* **186**, 135–145.

Chenu, K., Deihimfard, R., and Chapman, S.C. (2013). Large-scale characterization of drought pattern: A continent-wide modelling approach applied to the Australian wheatbelt – spatial and temporal trends. *New Phytologist* **198**, 801–820.
Chorom, M., and Rengasamy, P. (1995). Dispersion and zeta potential of pure clays as related to net particle charge under varying pH, electrolyte concentration and cation type. *European Journal of Soil Science* **46**, 657–665.
Chorom, M. and Rengasamy, P. (1996). Effect of heating on swelling and dispersion of different cationic forms of a smectite. *Clays and Clay Minerals* **44**, 783–790.
Clark, G.J. (2010). Managing subsoil constraints for increased productivity and water use efficiency, La Trobe University.
Clark, G.J., Dodgshun, N., Sale, P.W.G., and Tang, C. (2007). Changes in chemical and biological properties of a sodic clay subsoil with addition of organic amendments. *Soil Biology and Biochemistry* **39**, 2806–2817.
Dang, Y.P., Christopher, J.T., and Dalal, R. (2016). Genetic diversity in barley and wheat for tolerance to soil constraints. *Agronomy* **6**, 55.
Dang, Y.P., Dalal, R.C., Buck, S.R., Harms, B., Kelly, R., Hochman, Z., Schwenke, G.D., Biggs, W., Ferguson, N.J., Norrish, S., Routley, R., McDonald, M., Hall, C., Singh, D.K., Daniells, I.G., Farquharson, R., Manning, W., Speirs, S., Grewal, H.S., Cornish, P., and Orange, D. (2010). Diagnosis, extent, impacts, and management of subsoil constraints in the northern grains cropping region of Australia. *Australian Journal of Soil Research* **48**, 105–119.
Dowla, M.A.N.N.U., Sharma, D.L., Reeves, K., and Smith, R. (2021). Tolerance of wheat to soil sodicity can be better detected through an incremental crop tolerance approach and ascertained through multiple sowing times. *Agronomy* **11**, 1571.
Doyle, A., Tayler, D., Yates, W., So, H., and McGarity, J. (1979). Amelioration of structurally unstable grey soils in the north-western wheat belt of New South Wales. *Australian Journal of Experimental Agriculture* **19**, 590–598.
Fang, Y., Singh, B.P., Van Zwieten, L., Collins, D., Pitt, W., Armstrong, R., and Tavakkoli, E. (2021). Additive effects of organic and inorganic amendments can significantly improve structural stability of a sodic dispersive subsoil. *Geoderma* **404**, 115281.
FAO. (1976). Sodic soils and their management. In Soils Bulletin No. 31, Prognosis of salinity and alkalinity.
Faulkner, H. (2006). Piping hazard on collapsible and dispersive soils in Europe. In *Soil Erosion in Europe* (eds. J. Boardman and J. Poesen). Chichester: John Wiley & Sons, Ltd, pp. 537–562.
Genc, Y., Taylor, J., Lyons, G., Li, Y., Cheong, J., Appelbee, M., Oldach, K., and Sutton, T. (2019). Bread wheat with high salinity and sodicity tolerance. *Frontiers in Plant Science* **10** (1280).
Gill, J.S., Sale, P.W.G., and Tang, C. (2008). Amelioration of dense sodic subsoil using organic amendments increases wheat yield more than using gypsum in a high rainfall zone of southern Australia. *Field Crops Research* **107**, 265–275.
Gill, J.S., Sale, P.W.G., Peries, R.R., and Tang, C. (2009). Changes in soil physical properties and crop root growth in dense sodic subsoil following incorporation of organic amendments. *Field Crops Research* **114**, 137–146.
Goldberg, S., and Forster, H.S. (1990). Flocculation of reference clays and arid-zone soil clays. *Soil Science Society of America Journal* **54**, 714–718.
Graham, R.D., Turner, N.C., and Ascher, J.S. (1992). Evidence for subsoil constraints and potential benefits from amelioration. *Proceedings of the National Workshop on Subsoil Constraints to Root Growth and High Soil Water Use by Plants*, August-September 1992, Tanunda, South Australia.
Gupta, R.K., and Abrol, I.P. (1990). Salt-affected soils: Their reclamation and management for cop production. *Advances in Soil Science* **11**, 223–288.
Hamza, M., and Anderson, W. (2003). Response of soil properties and grain yield to deep ripping and gypsum application in a compacted loamy sand soil contrasted with a sandy clay loam in Western Australia. *Crop and Pasture Science* **54**, 273–282.
Heinzen, R.T. and Arulanandan, K. (1977). Factors influencing dispersive clays and methods of identification. *ASTM Special Technical Publication,* **623**, 202–217.
Hirth, J.R., Haines, P.J., Ridley, A.M., and Wilson, K.F. (2001). Lucerne in crop rotations on the Riverine Plains. 2. Biomass and grain yields, water use efficiency, soil nitrogen, and profitability. *Australian Journal of Agricultural Research* **52**, 279–293.
Ilyas, M., Qureshi, R.H., and Qadir, M.A. (1997). Chemical changes in a saline-sodic soil after gypsum application and cropping. *Soil Technology* **10**, 247–260.
Jayawardane, N.S., and Chan, K.Y. (1994). The management of soil physical properties limiting crop production in Australian sodic soils: A review. *Australian Journal of Soil Research* **32**, 13–44.

Jobbagy, E.G., Toth, T., Nosetto, M.D., et al. (2017). On the fundamental causes of high environmental alkalinity (pH > 9): An assessment of its drivers and global distribution. *Land Degradation & Development.* doi: 10.1002/ldr.2718.

Kelly, J., and Rengasamy, P. (2006). Diagnosis and management of soil constraints: Transient salinity, sodicity and alkalinity. In *The University of Adelaide and Grain Research and Development Corporation*, Australia. Available at http://www.arris.com.au/index.php?id=39 [Verified 24 May 2010].

Khokhar, J.S., Sareen, S., et al. (2017). Characterising variation in wheat traits under hostile soil conditions in India. *PLoS One* **12**(6), e0179208.

Kirkegaard, J.A., Lilley, J.M., Howe, G.N., and Graham, J.M. (2007). Impact of subsoil water use on wheat yield. *Australian Journal of Agricultural Research* **58**, 303–315.

Kochian, L.V., Piñeros, M.A., Liu, J., and Magalhaes, J.V. (2015). Plant adaptation to acid soils: The molecular basis for crop aluminum resistance. *Annual Review of Plant Biology* **66**, 571–598.

Kumar, A. (1988). Long-term forage yields of five tropical grasses on an extremely sodic soil and the resultant soil amelioration. *Experimental Agriculture* **24**, 89–96.

Kumar, A., Batra, L., and Chhabra, R. (1994). Forage yield of sorghum and winter clovers as affected by biological and chemical reclamation of a highly alkaline soil. *Experimental Agriculture* **30**, 343–348.

Leogrande, R., and Vitti, C. (2019). Use of organic amendments to reclaim saline and sodic soils: A review. *Arid Land Research and Management* **33**, 1–21.

Leskiw, A., Welsh, M., and Zeleke, B. (2012). Effect of subsoiling and injection of pelletized organic matter on soil quality and productivity. *Canadian Journal of Soil Science* **92**, 269–276.

Letey, J. (1985). Relationship between soil physical properties and crop production. *Advances in Soil Science* **1**, 277–294.

Liu, S., Constable, G., and Stiller, W. (2020). Using leaf sodium concentration for screening sodicity tolerance in cotton (*Gossypium hirsutum* L.). *Field Crops Research* **246**, 107678.

Loveday, J. (1974). Recognition of gypsum-responsive soils. *Soil Research* **12**, 87–96.

Lucas, M. (2022). Perspectives from the Fritz-Scheffer Awardee 2020: The mutual interactions between roots and soil structure and how these affect rhizosphere processes. *Journal of Plant Nutrition and Soil Science* **185**, 8–18.

Marchuk, A., and Rengasamy, P. (2011). Clay behaviour in suspension is related to the ionicity of clay–cation bonds. *Applied Clay Science* **53**, 754–759.

Marchuk, A., Rengasamy, P., McNeill, A., and, Kumar, A. (2013). Nature of the clay–cation bond affects soil structure as verified by X-ray computed tomography. *Soil Research* **50**, 638–644.

McCallum, M., Kirkegaard, J., Green, T., Cresswell, H., Davies, S., Angus, J., and Peoples, M. (2004). Improved subsoil macroporosity following perennial pastures. *Australian Journal of Experimental Agriculture* **44**, 299–307.

McDonald, G.K., Taylor, J.D., Verbyla, A.P., and Kuchel, H. (2012). Assessing the importance of subsoil constraints to yield of wheat and its implications for yield improvement. *Crop and Pasture Science* **63**, 1043–1065.

McDonald, G.K., Tavakkoli, E., Cozzolino, D., and Rengasamy, P. (2017). A survey of total and dissolved organic carbon in alkaline soils of southern Australia. *Soil Research* **55**, 617–629.

McDonald, G.K., Tavakkoli, E., and Rengasamy, P. (2020). Commentary: Bread wheat with high salinity and sodicity tolerance. *Frontiers in Plant Science* **11**, 1194–1194.

Mele, P.M., and Yunusa, I.A.M. (2001). Cereal yields associated with changes in soil characteristics following six years of acacia. *Proceedings of the 10th Australian Agronomy Conference*, Hobart. http://www.regional.org.au/au/asa/.

Millar, A.L., Rathjen, A.J., and Cooper, D.S. (2007). Genetic variation for subsoil toxicities in high pH soils. In *Wheat Production in Stressed Environments* (eds. H.T. Buck, J.E. Nisi, and N. Salomon). Dordrecht: Springer, Vol. 12, pp. 395–401.

Munns, R., James, R.A., Gilliham, M., Flowers, T.J., and Colmer, T.D. (2016). Tissue tolerance: An essential but elusive trait for salt-tolerant crops. *Functional Plant Biology* **43**, 1103–1113.

Niaz, S., Wehr, J.B., Dalal, R.C., Kopittke, P.M., and Menzies, N.W. (2022). Organic amendments and gypsum reduce dispersion and increase aggregation of two sodic Vertisols. *Geoderma* **425**, 116047.

Nuttall, J.G., Armstrong, R.D., and Connor, D.J. (2003). Evaluating physicochemical constraints of Calcarosols on wheat yield in the Victorian southern Mallee. *Australian Journal of Agricultural Research* **54**, 487–497.

Nuttall, J., Davies, S., Armstrong, R., and Peoples, M. (2008). Testing the primer-plant concept: Wheat yields can be increased on alkaline sodic soils when an effective primer phase is used. *Australian Journal of Agricultural Research* **4**, 331–338.

Orton, T.G., Mallawaarachchi, T., Pringle, M.J., Menzies, N.W., Dalal, R.C., Kopittke, P.M., Searle, R., Hochman, Z., and Dang, Y.P. (2018). Quantifying the economic impact of soil constraints on Australian agriculture: A case-study of wheat. *Land Degradation & Development* **29**, 3866–3875.

Page, K.L., Dang, Y.P., Dalal, R.C., Kopittke, P.M., and Menzies, N.W. (2021). The impact, identification and management of dispersive soils in rainfed cropping systems. *European Journal of Soil Science* **72**, 1655–1674.

Pandey, B.K., Atkinson, J.A., and Sturrock, C.J. (2021a). Non-invasive imaging of rice roots in non-compacted and compacted soil. *Bio Protocol* **11**(24), e4252.

Pandey, B.K., Huang, G., et al. (2021b). Plant roots sense soil compaction through restricted ethylene diffusion. *Science* **371**(6526), 276–280.

Passioura, J.B., and Angus, J.F. (2010). Chapter 2: Improving productivity of crops in water-limited environments. In *Advances in Agronomy* (ed. D.L. Sparks). Cambridge, MA: Academic Press, Vol. 106, pp. 37–75.

Patil, S.B., Rajendran, K., Kumar, J., Sen Gupta, D., Singh, S., Hamwieh, A., Mansur, C.P., and Kumar, S. (2020). Adaptation of food legumes to problem soils using integrated approaches. *Euphytica* **216**, 190.

Pessarakli, M., and Szabolcs, I. (2011). Soil salinity and sodicity as particular plant/crop stress factors. In *Handbook of Plant and Crop Stress*, 3rd edition, (ed. M. Pessarakli). Boca Raton, FL: CRC Press, pp. 23–40.

Qadir, M., Noble, A.D., Oster, J.D., Schubert, S., and Ghafoor, A. (2005). Driving forces for sodium removal during phytoremediation of calcareous sodic and saline–sodic soils: A review. *Soil Use and Managment* **21**, 173–180.

Qadir, M., Qureshi, R.H., and Ahmad, N. (2002). Amelioration of calcareous saline sodic soils through phytoremediation and chemical strategies. *Soil Use and Management* **18**, 381–385.

Qadir, M., Oster, J.D., Schubert, S., Noble, A.D., and Sahrawat, K.L. (2007). Phytoremediation of sodic and saline-sodic soils. *Advances in Agronomy* **96**, 197–247.

Qin, H., Li, Y., and Huang, R. (2020). Advances and challenges in the breeding of salt-tolerant rice. *International Journal of Molecular Sciences* **21**(21), 8385.

Quirk, J.P., and Schofield, R.K. (1955). The effect of electrolyte concentration on soil permeability. *Journal of Soil Science* **6**, 163–178.

Rahman, R., and Upadhyaya, H. (2021). Aluminium toxicity and its tolerance in plant: A review. *Journal of Plant Biology* **64**(2), 101–121.

Rengasamy, P. (2016). Soil chemistry factors confounding crop salinity tolerance: A review. *Agronomy* **6**, 53.

Rengasamy, P. (2018). Irrigation water quality and soil structural stability: A perspective with some insights. *Agronomy* **8**, 72.

Rengasamy, P., and Marchuk, A. (2011). Cation ratio of soil structural stability (CROSS). *Australian Journal of Soil Research* **49**, 280–285.

Rengasamy, P. and Olsson, K.A. (1991). Sodicity and soil structure. *Soil Research* **29**, 935–952.

Rengasamy, P. and Sumner, M.E. (1998). Processes involved in sodic behaviour. In *Sodic Soils: Distribution, Properties, Management, and Environmental Consequences* (eds. M.E. Sumner and R. Naidu). New York: Oxford University Press, pp. 35–50.

Rengasamy, P., Tavakkoli, E., and McDonald, G.K. (2016). Exchangeable cations and clay dispersion: Net dispersive charge, a new concept for dispersive soil. *European Journal of Soil Science* **67**, 659–665.

Rengasamy, P., Feitosa de Lacerda, C., and Gheyi, H.R. (2022). Chapter 4: Salinity, sodicity and alkalinity. In *Subsoil Constraints for Crop Production* (eds. T.S.D. Oliveira and R.W. Bell). Cham: Springer, pp. 83–107. doi: 10.1007/978-3-031-00317-2_4.

Sale, P.W., Gill, J.S., Peries, R.R., and Tang, C. (2019). Crop responses to subsoil manuring. I. Results in south-western Victoria from 2009 to 2012. *Crop and Pasture Science* **70**, 44–54.

Sale, P., Tavakkoli, E., Armstrong, R., Wilhelm, N., Tang, C., Desbiolles, J., Malcolm, B., O'Leary, G., Dean, G., Davenport, D., Henty, S., and Hart, M. (2021). Chapter 6: Ameliorating dense clay subsoils to increase the yield of rain-fed crops. In *Advances in Agronomy* (ed. D.L. Sparks). Cambridge, MA: Academic Press, Vol. 165, pp. 249–300.

Schnurbusch, T., Hayes, J., and Sutton, T. (2010). Boron toxicity tolerance in wheat and barley: Australian perspectives. *Breeding Science* **60**, 297–304.

Setter, T.L., and Waters, I. (2003). Review of prospects for germplasm improvement for waterlogging tolerance in wheat, barley and oats. *Plant and Soil* **253**, 1–34.

Setter, T.L., Waters, I., Sharma, S.K. et al. (2009). Review of wheat improvement for waterlogging tolerance in Australia and India: The importance of anaerobiosis and element toxicities associated with different soils. *Annals of Botany* **103**, 221–235.

Shainberg, L., and Letey, J. (1984). Response of soils to sodic and saline conditions. *Hilgardia* **52**, 1–57.

Singh, R.K., Mishra, B., Chauhan, M., Yeo, A.R., Flowers, S.A., and Flowers, T.J. (2002). Solution culture for screening rice varieties for sodicity tolerance. *The Journal of Agricultural Science* **139**, 327–333.

Singh, Y.P., Mishra, V.K., Singh, S., Sharma, D.K., Singh, D., Singh, U.S., Singh, R.K., Haefele, S.M., and Ismail, A.M. (2016). Productivity of sodic soils can be enhanced through the use of salt tolerant rice varieties and proper agronomic practices. *Field Crops Research* **190**, 82–90.

Tavakkoli, E., Fatehi, F., Rengasamy, P., and McDonald, G.K. (2012). A comparison of hydroponic and soil-based screening methods to identify salt tolerance in the field in barley. *Journal of Experimental Botany* **63**, 3853–3867.

Tavakkoli, E., Rengasamy, P., Smith, E., and McDonald, G.K. (2015). The effect of cation–anion interactions on soil pH and solubility of organic carbon. *European Journal of Soil Science* **66**, 1054–1062.

Tavakkoli, E., Schilling, R.K., and McDonald, G.K. (2017). Improving wheat yields on hostile soils. In *GRDC Updates*. Wagga Wagga: GRDC.

Tavakkoli, E., Pitt, W., Armstrong, D., Hilderbrand, S., Fang, Y., Uddin, S., and Armstrong, R.D. (2022a). Amelioration of hostile subsoils via incorporation of organic and inorganic amendments and subsequent changes in soil properties, crop water use and improved yield, in a medium rainfall zone of south-eastern Australia. In *GRDC Update Paper*. Wagga Wagga: GRDC.

Tavakkoli, E., Uddin, S., Rengasamy, P., and McDonald, G.K. (2022b). Field applications of gypsum reduce pH and improve soil C in highly alkaline soils in southern Australia's dryland cropping region. *Soil Use and Management* **38**, 466–477.

Turner, N.C. (2018). Turgor maintenance by osmotic adjustment: 40 years of progress. *Journal of Experimental Botany* **69**, 3223–3233.

Uddin, S., Williams, S.W., Aslam, N., Fang, Y., Parvin, S., Rust, J., Van Zwieten, L., Armstrong, R., and Tavakkoli, E. (2022). Ameliorating alkaline dispersive subsoils with organic amendments: Are productivity responses due to nutrition or improved soil structure? *Plant and Soil* 480, 227–244.

Vasuki, A., and Geetha, S. (2016). Validation of 'Saltol' QTL under sodicity. *Electronic Journal of Plant Breeding* **7**, 838–841.

Wani, S.H., Kumar, V., Khare, T., Guddimalli, R., Parveda, M., Solymosi, K., Suprasanna, P., and Kavi Kishor, P.B. (2020). Engineering salinity tolerance in plants: Progress and prospects. *Planta* **251**, 76.

Wasson, A.P., Chiu, G.S., Zwart, A.B., and Binns, T.R. (2017). Differentiating wheat genotypes by bayesian hierarchical nonlinear mixed modeling of wheat root density. *Frontiers in Plant Science* **8**, 282.

Wheaton, A.D., McKenzie, B.M., and Tisdall J.M. (2002). Management of a sodic soil for wine grape production. *Australian Journal of Experimental Agriculture* **42**, 333–339.

White, R.G., and Kirkegaard, J.A. (2010). The distribution and abundance of wheat roots in a dense, structured subsoil: Implications for water uptake. *Plant, Cell & Environment* **33**, 133–148.

Yau, S.K., and Erskine, W. (2002). Diversity of boron-toxicity tolerance in lentil growth and yield. *Genetic Resounrces and Crop Evolution* **47**, 55–62.

Yau, S.K., and Ryan, J. (2008). Boron toxicity tolerance in crops: A viable alternative to soil amelioration. *Crop Science* **48**, 854–865.

Yunusa, I.A.M., and Newton, P.J. (2003). Plants for amelioration of subsoil constraints and hydrological control: The primer-plant concept. *Plant and Soil* **257**, 261–281.

Yunusa, I., Mele, P., Rab, A., Schefe, C., and Beverly, C. (2002). Priming of soil structural and hydrological properties by native woody species, annual crops, and a permanent pasture. *Soil Research* **40**, 207–219.

19 Sandy Soil Constraints

Organic and Clay Amendments to Improve the Productivity of Sandy Soils

Amanda Schapel[1], *Richard Bell*[2], *Simon Yeap*[2,3], *and David Hall*[4]

[1]Soil and Land Management, South Australian Research and Development Institute, Adelaide, South Australia
[2]SoilsWest, Centre for Sustainable Farming Systems, Food Futures Institute, Murdoch University, Perth, West Australia
[3]Horticulture & Irrigated Agriculture, Department of Primary Industries and Regional Development, South Perth, West Australia
[4]Grains Directorate, Department of Primary Industries and Regional Development, Esperance, West Australia

CONTENTS

DOI: 10.1201/9781003093565-19

19.1 INTRODUCTION

Sandy soils are used for agriculture in many regions of the world but are generally not considered high-performance soils in terms of agricultural productivity (Usowicz and Lipiec 2017; Scanlan et al. 2022). Deep sands (Arenosols) cover ~900 million hectares (M ha) of the world's surface (FAO/UNESCO 1995) with the highest proportion in Africa (51%) followed by Australasia (21%) (Yost and Hartemink 2019). A recent revision suggests that there are 16.04 M ha of sandy soils (soils with < 10% clay in the 0–30 cm depth) in Southern Australia (Robinson et al. 2021). Profiles of other soil groups with deep sand over clay (e.g. Regosols, Leptosols, Fluvisols) often share similar characteristics to deep sandy soils in the upper layers of the profile.

Agricultural productivity on sandy soils is low, often only achieving 30%–40% of water-limited yield potential in rainfed agriculture (Hall et al. 2010; Macdonald et al. 2019). Constraints to production on sandy soils are most evident in the broadacre grains industry, mixed farming, and livestock grazing systems in water-limited environments. Multiple limitations may constrain productivity on sands, and it is common for more than one of these to exist on a particular site (e.g. Hall et al. 2020). With multiple limitations, it is critical to identify and address all key soil constraints in order to achieve significant improvements in productivity potential (Bell and Oliveira 2022). Many inherent limitations of sandy soils can be addressed or alleviated in agricultural systems. For example, diverse crop rotation can recycle nutrients, increase symbiotic nitrogen fixation, interrupt disease cycles, and suppress weeds. Some properties are only partly alleviated (e.g., nutrient supply and low organic carbon (OC)), while others deteriorate further due to management systems and practices used (e.g. water repellence and acidification).

In this chapter, we initially outline the range of characteristics of types of sand profile and then examine key production limitations on sands for crop and pasture production. The final section of the chapter focuses on organic and clay amendments that offer most promise for long-term improvement in the productivity of sandy soils.

19.2 CHARACTERIZING SANDY SOILS

Sandy soils have inherently low water, nutrient, and organic carbon-holding capacity, and are prone to compaction, water repellence, wind erosion, acidification, and leaching of nutrients beyond the rootzone. Therefore, sandy soils have restricted capacity to supply water and nutrient resources to the roots of crops. Sands often have multiple limitations (chemical, physical, and biological); hence diagnosis and treatment of the main constraints to production need to be a key focus for raising their productivity (Bell et al. 2015; Hall et al. 2020; Scanlan et al. 2022). With multiple limitations, there will generally be a weak response to inputs unless the key constraints are overcome at the same time. For example, a deep sand (Quartzipsamment) in Mali, West Africa, was very low in OC, pH (with high exchangeable Al), cation exchange capacity (CEC), and exchangeable Ca and K (Table 19.1). In central Vietnam, deep sands were deficient in K, S, B, and Cu (Tam et al. 2015) while in Northeast Thailand deficiencies of P, K, S, B, Cu, and Mo were reported on deep sands

TABLE 19.1
Multiple Limiting Factors in Sands (Quartzipsamment from Mali, West Africa, West et al. 1983)

Depth (cm)	Organic C (g/kg)	pH (H_2O)	CEC (cmol/kg)	Exch Ca (cmol/kg)	Exch K (cmol/kg)	Exch Al (cmol/kg)
0–15	2.5	5.1	1.5	0.2	0.1	0.6
15–27	1.5	4.9	1.5	0.1	0.1	0.7
28–44	1.1	4.8	1.3	0.1	-	0.6

(Bell et al. 1990). Deep sands on the south coast region of West Australia have limitations for crop and pasture production due to water repellence, acidity, low nutrient availability and retention (especially low K and CEC), and subsoil compaction (Hall et al. 2020). A subset of deep sands form from alkaline parent materials that result in multiple nutrient limitations including micronutrients.

For the purposes of this chapter, sandy soils are defined as soils with <10% clay content in the top 30 cm of soil and include sands (<5% clay), loamy sands (5%–8% clay), and clayey sands (8%–10% clay). By contrast, the Arenosol order of the World Reference Base includes soils with clay up to 15%, which includes sandy loam texture (IUSS Working Group WRB 2015).

Deep sandy soils have sand, loamy sand, or sandy loam textures to depths >100 cm (IUSS Working Group WRB 2015). Subsoils are differentiated from topsoils by lower organic carbon and stronger coloration caused by increases in clay, iron oxide, or fine carbonate contents in the B horizon. The depth of sands varies with some profiles having clayey subsoils (duplex soils) within the potential plant root zone. Duplex soils are defined by a sandy A horizon within 70 cm of the surface and a distinct color and texture boundary to the clayey or clay loamy subsoil. Depending on depth, these soils can have higher nutrient and water storage than deep sands, but clays are often poorly structured, restricting root growth and causing temporary waterlogging. Some sandy duplex soils feature a bleached subsurface layer (A2 horizon) which is generally 5–60 cm thick and especially infertile. Shallow variants are defined where the A horizon is <30 cm (Schoknecht and Pathan 2013).

Gravelly sands can occur in both deep sand and sandy duplex soils (Scanlan et al. 2022). They are considered to have >15% gravel (>2 mm in size). Gravel often sits above or is embedded within the clay layer in duplex soils. The gravel can be within the whole of the A horizon or in defined layers. The gravel can exist as discreet units or as a conglomerate layer. Gravel can affect hydraulic conductivity and water retention in sandy soils. Ironstone gravel has been reported to decrease plant-available water-holding capacity with the greatest effect expected in water-limited cropping regions where yield is constrained by restricted root growth (Scanlan et al. 2022). The presence of gravel also affects the degree of water repellence, nutrient adsorption and retention, and the efficacy of applied amendments and machinery used to incorporate them.

While the properties of sandy soils vary across a continuum, they have a number of distinctive properties that can be used for classification. Parent material is the most significant single determinant of properties on the soils formed (Hall et al. 2009). For example, Newsome (2000) examined the origins of the deep sand profiles on the Victoria Plateau in Western Australia (WA) and concluded that they were derived from in situ deep weathering. There was little evidence of aeolian reworking of these sand materials. Elsewhere, however, aeolian processes best explain the origin of the deep sands in southwest Australia (Commander et al. 2002). In Cambodia, deep sands appear to be mostly derived from in situ weathering of sandstone parent rock, plus some influence of colluvial processes (Hin 2018).

Geological origins and geomorphic processes affect the shape and size of sand particles. Grain shape (rounded or angular) can influence root growth, with lower wheat root length and higher soil strength observed in sands with angular particles (Lipiec et al. 2016; Scanlan et al. 2022). Sandy soils can be derived from siliceous or carbonaceous parent material. Siliceous sands are generally neutral to acidic in pH, and carbonaceous sands are often strongly alkaline.

Sandy soils have <10% clay, but the proportion and type of clay, silt, and sand (coarse, medium, and fine) particles determine many soil properties, including nutrient and water-holding capacity, soil structure, and storage and protection of organic carbon. The combined effect of clay and silt (fine fraction) is important for properties, such as water and organic carbon storage. The fine fraction has a substantially larger surface area than the coarse fraction (sand), and hence a dominating influence on water and nutrient retention. A sandy soil with 5% clay will have approximately five-fold greater reactive surface area for nutrient and water reactions than those with 1% clay.

The fine fraction is a key driver of soil organic carbon storage through stabilization of organic matter via interaction with the mineral surface (Wiesmeier et al. 2019). In sandy soils, silt particles may be particularly important as they have been reported to double water storage with a doubling of the clay plus silt fraction (Bell et al. 2015). The relative concentration of these fractions is also important with silt and clay potentially increasing cementation, while a mix of fine to coarse sand particles can lead to high packing density (Mullins and Panayiotopoulos 1985) both leading to high resistance to plant root penetration. Therefore, in sandy soils, small differences in composition of the fine fraction can deliver profound differences to soil properties and hence have a strong effect on management.

19.3 PRODUCTION INFLUENCES

19.3.1 Reactive Surface Area

Reactive surface area is a combination of surface area per unit mass and chemical reactivity per unit soil mass. CEC is the most common measure of chemical reactivity and captures the "activity" of chemical reactions at the surface of clay-sized minerals and organic matter. Other measures include anion exchange capacity, P buffering index, and soil organic matter content. A kaolinite clay particle has about 1,000 times the external surface area as the same weight of sand particles (Table 19.1). This greater reactive surface means that small proportions of clay and organic matter can have a dominating influence on the adsorption and release of nutrients and water.

19.3.2 Water-Holding Capacity and Plant-Available Water

Water availability is a major limitation in most rainfed agricultural systems, and hence soil water-holding capacity strongly affects plant productivity. Water-holding capacity is primarily controlled by reactive surface area which in turn depends on soil texture (particle size distribution) and organic matter. Sands have lower plant available water-holding capacity than clays (60 compared to 190 mm of water/mm of soil depth) because of the lower proportion of capillary and micropores (Table 19.2).

Heavier textured soils generally outperform sandy soils due to higher water storage and their ability to maintain crop water supply, despite unreliable rainfall, during critical growth periods. However, sandy soils often outperform heavier textured soils during low rainfall seasons with regular small rainfall events. This is because sandy soils are able to replenish plant-available water supply quickly after light showers due to their low wilting point (Maschmedt 2000). Therefore, the majority (60%–75%) of soil water is plant available, and significant volumes of water can penetrate deep enough to slow evaporative losses on the sands.

TABLE 19.2
Estimates of Surface Area (m^2/g) and Chemical Reactivity (CEC in cmol/kg) of Various Soil Minerals

Mineral/Amendment	Specific Surface Area (m^2/g)	Chemical Reactivity (cmol/kg)
Sand	0.01–0.1	<2
Silt	1.0	0–2
Kaolinite	5–100	1–20
Biochar	50–500	10–120
Illite	80–200	10–40
Iron and aluminum oxides	100–400	0–4
Smectite	280–750	60–150
Vermiculite	300–500	80–150
Zeolite	500–800	150–450
Organic matter	550–800	40–200
Humus	800–900	250–400
Allophane	100	25–150

Source: Adapted from Hall et al. (2009), Hoyle (2013), Lonappan et al. (2016), and Yang et al. (2019).

Low plant-available water storage in sandy soils can also be induced by restrictions to root growth resulting from factors, such as low pH, low nutrient availability, and compaction. The importance of overcoming these restrictions for increased root growth was demonstrated by a 0.62 t/ha increase in grain yield (56 kg/ha/mm efficiency) from an additional 10.5 mm of water extracted post-anthesis from the 1.35 to 1.85 m layer in a Kandosol subsoil (Kirkegaard et al. 2007).

As sandy soils have low clay and often low clay plus silt concentrations, organic matter can strongly influence their water-holding capacity. The effects of organic matter can be direct or indirect through increased infiltration resulting from improved structure.

19.3.3 Soil Structure and Soil Strength

Structure describes the arrangement and organization of the solid soil particles and the pore spaces surrounding them. In all but loose sands, individual clay, silt, and sand particles arrange themselves to form peds or aggregates. This creates intra- and inter-aggregate pore spaces that change the flow of air and water and storage of organic matter.

The amount and type of clay, the silt content, and the amount, shape, and size of sand grains affect the way the soil particles pack together (Mullins and Panayiotopoulos 1985). Sand in South Australia can be hard and massive due to high proportions of silt or very fine sand. These particles, when mixed with small amounts of clay, iron, manganese, and aluminum oxides, can cement soil particles forming a compacted layer (Mullins et al. 1990). The presence of materials such as calcium can also bind inorganic and organic compounds. Generally, stable aggregation is strongly influenced by iron and aluminum oxides in humid soils and by organic matter in semi-arid soils (Mullins et al. 1990).

High soil strength and bulk density reduce root growth, affecting crop yield. Values exceeding 2.5 MPa for soil strength and 1.6 g/cm^3 for bulk density (McKenzie et al. 2002) are commonly used as upper values indicating restrictions to root growth. High subsurface soil strength induced by repeated trafficking by heavy machinery is a major limitation in sandy soil (Kirkegaard 2009; Unkovich 2014). Tillage, which accelerates the breakdown of organic matter, high surface tension created when soils dry, and compressive force from the weight of machinery, all contribute to high soil strength.

A sandy soil with a wide distribution of sand to clay particles (broad range of particle sizes) will compress more readily than a sand with a narrow particle distribution (e.g. a coarse sand with little clay; Scanlan et al. 2022). Smoother sand grains (rounded) are more likely to compact than rough grains (Scanlan et al. 2022).

Organic matter plays a key structural role in soils, providing the "glue" between particles to maintain pore spaces between silt and sand particles. It supports the soil biological activity that generates exudates and fungal hyphae which are important for soil aggregation. In a pot experiment using a sandy soil, increases in aggregate stability were largely attributed to the growth of hyphae that cross-linked sand grains in the aggregates with little involvement of microbial polysaccharides (Degens et al. 1996). Fungal hyphae may be integral to accessing moisture stored in micropores and preventing a rapid haying off seen in plants which could enhance drought tolerance (Allen 2007). The mechanisms of how reverse flows (hydraulic redistribution from plant to fungus) work are not well understood, but the ultimate importance of mycorrhizae in plant–water relations depends on drying patterns, soil pore structure, and number of hyphal connections extending from the root into the soil (Allen 2007).

The low organic matter, clay, and water-holding capacity of sands generally limit aggregation, making them susceptible to wind erosion when the surface soil is dry if there is insufficient surface cover. Wind erosion can strip fertile topsoil, further limiting the available resources for plant growth in sandy soils.

19.3.4 Soil pH

The pH of sands in calcium chloride solution (pHca) can vary from less than 4 to greater than 9 depending on parent material, climate, and management. Sands derived from siliceous parent materials with low CEC (e.g. granites) in climates where there is sufficient rainfall to leach ions are more likely to be acidic (Gray et al. 2009). The process of acidification is affected by carbon and nitrogen cycling (Helyar and Porter 1989) and is most evident in agricultural subsoil sands which, in the absence of organic matter, have low buffering capacity (Gazey et al. 2014; Azam and Gazey 2020). Conversely, calcareous sands derived from carbonates associated with marine environments, and aeolian accessions are mainly alkaline.

Soil pH extreme values can affect nutrient availability or root growth directly and hence depress plant productivity and microbial activity. On calcareous sandy soils, phosphorus, manganese, zinc, and iron deficiencies occur with the degree largely determined by the fine carbonate content. On acidic sandy soils, soluble aluminum reaches toxic levels when pH in calcium chloride is <4.8 and can severely inhibit root growth (Butterly et al. 2022).

Soil pH also influences the soil's CEC through variable surface charge on the colloids. As pH increases, the density of negative charges on the colloids increases, effectively increasing CEC. The permanent component of CEC has been regarded as derived predominantly from the aluminosilicate clay fraction, while the variable charge component is mainly in the organic matter fraction, the broken edges of kaolinite clays, and the oxyhydroxides of iron and aluminum (Loveland and Webb 2003). Net positive charge can develop on acid sands dominated by variable charge (Wong and Wittwer 2009). Sorption of nitrate and sulfate increases when acid sands develop positive net charge, and conversely leaching of these anions is inhibited and delayed.

19.3.5 Nutrient-Holding Capacity

Surfaces of clay minerals and organic matter (colloids) are generally negatively charged and attract positively charged cations, such as calcium, magnesium, potassium, and sodium. The density of exchange sites in the soil, expressed as CEC, is an indicator of the inherent fertility of a soil. Low clay and organic matter concentrations in sands results in low CEC, poor pH buffering, and low nutrient retention. As a result, management practices can induce rapid and substantial changes in properties of sands and can have compounding effects, such as an inherently low CEC being further reduced by soil acidification (Noble et al. 2000).

Ions, solutes (including pesticides), and water can readily leach past the root zone and be lost from the root zone or cause groundwater contamination, due to the high permeability of sandy soils (e.g. Sitthaphanit et al. 2009). Phosphorus can leach on very pale sands even at low concentrations in the soil (Sharma et al. 2015, 2017, 2021). This is unique to some sandy soils as phosphorus leaching does not occur on any other soil texture class unless extractable P levels are well above critical levels for crops (Sharma et al. 2017).

19.3.6 Water Repellence

Sandy soils can be susceptible to water repellence which restricts the uniformity and rate of water infiltration into the soil (Roper et al. 2015). After rainfall, water repellence causes uneven and variable wetting of the soil, leading to increased run-off and soil erosion, increased leaching along preferential flow pathways, patchy plant germination and emergence, and poor utilization of nutrients as plants cannot access nutrients in the resultant dry patches of soil. As a result, plant growth, nutrition, and productivity can often become limited on these soils, particularly in dryland systems where soil water is limited to seasonal rainfall and stored soil moisture (Bond 1972; DeBano 1981; Müller et al. 2014; Roper et al. 2015; Hewelke et al. 2018).

Water repellence, which affects more than 10 M ha of arable sandy soils in Southern and Western Australia (Roper et al. 2015), is most pronounced on soils with <3% clay content (Harper and Gilkes 1994; Unkovich 2014). Average losses in crop and pasture production in Australia due to water repellence are estimated to be as high as 40% (Blackwell et al. 1994; Abadi Ghadim 2000),

In agricultural soils, water repellence is primarily caused by waxes produced during the breakdown of crop and pasture residues, causing it to be most severely expressed within the top 2–10 cm of the soil profile. Due to the low surface area of sands, these waxes coat a larger fraction of the soil surface than on a more clayey soil. Only 3% of sand grains need to be coated for repellence to be slightly expressed, and up to 5% for severe water repellence (Bauters et al. 1998). By contrast, a far greater density of hydrophobic molecules is needed to saturate the hydrophilic sites on clay particles (Daniel et al. 2019).

Although water repellence may not necessarily occur every year, it is more common in years of low, early winter rainfall (Maschmedt 2000). Seasonal variation in water repellence can also be observed, with water repellence increasing during dry summer periods and decreasing during wet winter periods (Crockford et al. 1991; Müller et al. 2014; Rye and Smettem 2015). However, given the low hydraulic conductivity of dry and repellent soils, preferential flow paths may persist throughout the winter, resulting in dry soil patches even after heavy rainfall (DeBano 1981; Ritsema et al. 1993; Ritsema and Dekker 1994). Although less commonly reported, water repellence has been reported in deep sands in tropical Savannah regions such as Northeast Thailand (Coker et al. 2004).

19.3.7 Organic Carbon

In sandy soil, a value of 1% OC appears to be an effective critical value, below which the soil's capacity to perform key functions is impaired (Loveland and Webb 2003; Kay and Angers 1999). One of the most important effects of low OC levels is the reduction in the soil's capacity to mineralize nutrients. This means an increased reliance on other forms of nutrient supply from inorganic or organic sources.

Studies on eroded soil have shown that a reduction in OC is linked to reduced crop yields; however, increasing OC does not necessarily lead to increased yield (Celestina et al. 2019). Increased crop yields over the last 50 years have occurred despite reported decreases in soil OC (Loveland and Webb 2003). Yield is often assumed to drive OC levels, but Oldfield et al. (2019) demonstrated that, while there was an initial linear relation between yield and OC, this was stabilized by 2% soil OC. Loveland and Webb (2003) proposed critical soil OC values of 3.5% for clay-textured soil and 1% for sandy soil. This corresponds to Kay and Angers (1999) critical value of 1% for soil OC, below which the soil's capacity to perform key functions is constrained, such as supply of nitrogen from plant residue, effectively preventing crops from reaching the water-limited yield potential.

Oldfield et al. (2019) found that increasing soil OC is not directly substitutable for using inorganic nitrogen fertilizers in high input systems; however, at lower rates of nitrogen input (<50 kg/ha), increasing soil OC from 0.5% to 1% could potentially maintain yield and reduce nitrogen fertilizer inputs by about 50% for wheat crops. Therefore, increasing the OC content could help reduce nitrogen inputs and potentially close the yield gap on sands in rainfed cereal cropping.

For micro-organisms to efficiently break down organic matter to humus, mineralize nutrients for plant growth, and increase soil OC concentration, the addition of supplementary nutrients may be required (Kirkby et al. 2013). For every 1,000 kg of humus, an extra 80 kg of N, 20 kg of P, and 14 kg of S will be sequestered in the structure of the humus formed (Kirkby et al. 2014).

An increase in soil OC is also important for the nutrient-holding capacity of the soil. Sandy soils can have very low (<2 cmol(+)/kg) CEC and an increase in OC by 1% can increase CEC by ~2.2 cmol(+)/kg, which can contribute to as much as 85% of the CEC (Hoyle 2013) and improve the buffering capacity of the sandy soil. Furthermore, a 1% increase in OC can increase the available water-holding capacity by 1.4–1.9 mm per 10 cm of soil depth, with a greater response in a sandy than clayey soil (Minasny and McBratney 2018).

19.3.8 Wind Erosion

Wind erosion is a production risk for soils with sand and loamy sand surface horizons, especially in semi-arid or arid regions with strong and persistent wind and when the sand is predominantly in the fine sand range (Moore et al. 2004). Maintaining adequate soil cover throughout the year, especially when soils are dry and erosive winds are likely, is generally and effective strategy to prevent wind erosion. The practice of Conservation Agriculture, involving minimal soil surface disturbance, maintaining permanent soil cover, and diversified crop rotations, is effective in mitigating the wind erosion risk. On the other hand, over-grazing of pastures or soil tillage during dry periods with erosive wind are high-risk activities on soils with fine sandy texture.

19.4 AMENDMENTS FOR SANDY SOILS

Improvements in the performance of sands have been achieved using balanced fertilizer inputs, ripping to alleviate compaction, liming to alleviate acidity, and zero tillage to slow down organic matter decomposition; however, these best management practices may still only achieve 30%–40% of water-limited yield potential (Hall et al. 2010; Macdonald et al. 2019). Improved yields by 5%–15% on sandy soil can occur using low-cost options mitigation options, including winged and paired-row seeding points, on-row seeding, and soil-wetting agents as observed in a Western Australian study (Davies 2016). However, mitigation practices do not permanently improve the reactive surface area of sandy soils. In the following section, we focus on one-off, long-lasting amelioration options, such as clay amendments, which despite higher cost have higher yield responses (30%–40%).

19.4.1 Addition of Organic Amendments

The application of organic amendments to sands can improve the reactive surface area along with nutrient and water use efficiencies. Fertilizers derived from organic sources vary in chemical and stoichiometric properties (Chen et al. 2018a), with nutrient content highly dependent on the source and quality of the amendment (Quilty and Cattle 2011). Farrell et al. (2017a) identified that generally only the nutrient-rich composts and manures enhanced nutrient availability for wheat compared to unamended control soils. They found a general lack of evidence for the efficacy of other biological inputs, such as microbial, humates, and biostimulants, for crop yield increase in rainfed, broadacre systems.

Biochars produced from manure, greenhouse waste, and grasses were more effective in providing nutrients than wood-based products but the latter were better for soil OC sequestration (Singh et al. 2014). The authors also found that increased yield was more likely in acidic rather than alkaline soil and that the high reactive surface area of biochars can decrease herbicide and pesticide efficacy. Pre-loading biochar with nutrients may restrict their adsorption of herbicides and pesticides but provide a source of nutrients while increasing the nutrient use efficiency of future fertilizer applications (Singh et al. 2014).

An understanding of the major soil constraint and the chemical properties of the organic amendment is critical to making an accurate prediction about their impact on soil properties (Farrell et al. 2017a) and plant productivity. For example, the addition of biochar and compost to sandplain soils of WA increased crop yield to a maximum of 8% in the first 3 years with no significant difference attributed to organic amendments thereafter (Hall and Bell 2015). The increases in yield were largely attributed to direct nutrient addition with the organic amendment and this dissipated over time (by 3 years). The attribution of benefits from organic materials added to soil depends on adequate controls that isolate the effects of nutrient supply and soil disturbance from other roles such as improved water retention of soil structure (Celestina et al. 2019).

Chen et al. (2018a) found that among the waste-derived organic fertilizers:

- Biosolids had a very strong effect on crop yield but were associated with high contaminant loads, making them more suitable for nutrient recovery than direct application.
- Poultry litter had a high nutrient load and pathogenicity but did not significantly increase crop yield when compared to addition of inorganic fertilizer. However, composting poultry litter decreased the pathogenicity and enabled direct application to crops, producing yields comparable to inorganic fertilizer addition.
- Cattle manure resulted in crop yield comparable to inorganic fertilizer addition and was suited to direct crop application without composting.

In these cases, waste-derived fertilizers yielded as well as conventional inorganic fertilizers (Chen et al. 2018a). In contrast to other soils, there is evidence that the addition of organic amendments to sands increases crop response above that of inorganic fertilizers alone (Chen et al. 2018b; Hijbeek et al. 2017; Loveland and Webb 2003). In the sandy soils of northeast Thailand, the response to inorganic fertilizer was poor unless organic materials were also added (Ragland and Boonpukdee 1987). The addition of organic matter has been suggested to increase the buffering function of soil, decrease the demand for mineral fertilizers (if organic inputs and inorganic fertilizers are combined), and reduce the variability in yield (Hijbeek et al. 2017).

Careful consideration needs to be given to the selection of organic amendments to ensure that nutrient use efficiencies are maximized while minimizing risk to the environment by leaching of nutrients or risk to human health through contamination of the food chain.

The low nutrient and water-holding capacity of sandy soil make it difficult to increase OC content (Hall et al. 2010) as the low reactive surface area of sandy soils enables only minor amounts of OC to be stabilized and protected. The remaining OC inputs are readily broken down and lost as CO_2 via microbial decomposition.

To ensure a consistent supply of benefits from organic amendments, frequent applications of organic amendments may be required (Farrell et al. 2017a; Oldfield et al. 2019). However, even with the addition of organic amendments, sandy soils have limited reactive surface area which limits the retention of nutrients and moisture and the ability to stabilize OC. Permanently increasing the reactive surface area of these soils will ensure sustained benefits. There might also be merit in the deep incorporation of organic amendments through mouldboard ploughing or spading to slow their rate of decomposition.

19.4.2 Addition of Clay Amendments

Clay addition to sandy soil can offer a long-term solution to increasing the reactive surface area, thereby improving the supply of resources to plant roots. In Australia, subsoil clay addition to sandy soil has been used as a farm management practice since the 1970s (Cann 2000). It is a practice commonly used in South Australia, Western Australia, and Victoria and was first used to overcome water repellence (Harper et al. 2000; Ma'shum et al. 1989; Ward and Oades 1993), but a number of other co-benefits have been realized. In Australia, it is estimated that 0.2 million hectares of sandy soils have already been amended with clay (Bell 2016). The most common methods of clay amendment include the addition of clay-rich subsoil to the surface of the sand (spreading) and then incorporating by spading to 20–40 cm depth, elevation of in-situ clay-rich subsoil by delving and then mixing in the surface layers, or inversion (mouldboard plough) of the topsoil and clay-rich subsoil layers (Davenport et al. 2011). The selection of the appropriate method is determined by the depth of clay-rich subsoil, the available machinery, and whether soil amendments such as lime are also to be incorporated by the operation. The amount of clay-rich subsoil added, its clay percentage, and the depth of incorporation into the sandy profile determine the final clay concentration, distribution of clay clods, and OC stabilization capacity of the re-engineered soil.

Added subsoil clay affects nutrient availability by changing soil pH, direct nutrient addition, and water and nutrient adsorption. In South Australia, subsoils are often alkaline due to the presence of calcium carbonate, and the application of subsoil clay often raises the pH of the clay-amended sands. In acidic sands, this can increase productivity due to the liming effect (B. Hughes personal commutation), but application of too much calcium carbonate can reduce yield through induced nutrient deficiencies of zinc, iron, copper, and manganese. Elsewhere, subsoil kaolinitic clay (pHca 6.1–7.2, 300 mg K/kg, 2 mg P/kg) applied at a rate of 300 t/ha to a deep sand soil increased; soil pHca from 5 to 5.6, soil K by 47 mg/kg, and soil P by 12 mg/kg to a depth of 0.1m over 10 years compared to no clay amendment (adapted from Hall et al. 2010). In this study, K was directly added by the clay whereas P increases are attributed to adsorption over time. Clay-amended sands can also decrease N leaching with the mechanisms being higher water adsorption resulting in reduced leaching rates (Dempster et al. 2012) and increased N adsorption where smectite is added (Beusch et al. 2019).

In clay-amended sands, water-limited yield increased to about 70% of the potential based on rainfall (Hall et al. 2010). Hall et al. (2020) argued that even after the surface clay amendment of sands at Esperance, subsoil compaction may remain as a yield limitation. The tillage processes required for clay amendment can overcome subsoil physical limitations, such as high bulk density and soil strength caused by compaction (Usowicz and Lipiec 2017; Hall et al. 2020). Increased root growth following clay amendment has been reported (Bailey et al. 2010; Hall et al. 1994), and this should lead to improvements in productivity.

Clay addition to sandy soil has been reported to increase water retention (Betti et al. 2015; Ogunniyi 2017; Schapel et al. 2018a) and decrease saturated hydraulic conductivity (Betti et al. 2016). Added subsoil clay in the field is not homogenously mixed but distributed as clods ranging in size from a few mm up to >200mm (Schapel et al. 2019) in a sandy matrix. Betti et al. (2016) found that clods <6mm in size added to sand retained more water than clods >6mm, as an intimate mixture of smaller pores was created compared to a bimodal mixture where clods existed as discrete entities in a sand matrix. Boldt-Burisch et al. (2013) demonstrated that clay clods embedded in sandy soil act as water and nutrient "hot spots," improving pea plant and root development. They found sandy soil had many large pore spaces and very few small pores due to insufficient OM to bind sand grains into larger aggregates. Conversely, the compacted clay clod had many small pores and a few large pores.

Modeling of water movement in sandy soil with the addition of various-sized clay clods demonstrated that water uptake by plant roots occurred from the sandy soil (Schapel et al. 2018b). The size of the clay clod affected the speed of water flow, with higher velocities for small clods (5cm) than for the large clods (10cm) because of the number of connected flowpaths. Clod size did not affect transpiration or drainage rates. The apparent discontinuous, stop/start flow of water and nutrients to plant roots in sandy soils is different from the continuous flow of water in loam or clay soils and could affect a number of properties including plant productivity.

Internationally, improved yields have been reported in Thailand with the addition of bentonite to sandy soil for forage sorghum (Noble et al. 2001) and Turkey with the addition of pumice to increase water-holding capacity (Ozhan et al. 2008). The addition of 60–120t bentonite (montmorillonite) per hectare to a sandy soil in Poland was found even after 37years to increase OC, total N, water, microbial activity, and structure (Czaban et al. 2014; Czaban and Siebielec 2013). Field trials comparing the effectiveness of bentonite and kaolin addition to sandy soil in the United Kingdom determined that both amendments increased crop growth and yield, increased CEC, OC, C:N ratio, microbial activity, nutrient efficiency, water retention, and soil pore characteristics (Ogunniyi 2017).

In Australia, yield increases between 20% and 130% (Davenport et al. 2011; Hall et al. 2010) have been reported following the addition of clay-rich subsoil to sandy soil. Consistent and significant yield increases around 50% were obtained from clay amendment of deep sands after 15years following the addition of 200–300t of subsoil (containing 42% clay) per hectare at Esperance, Western Australia (Hall et al. 2018). In South Australia, yield increases (68–97%) were obtained 8years following 200–250t clay concentration per hectare (Schapel et al. 2018a). Similar yield increases across three trial sites in South Australia (60%–95%) were recorded 5years following clay additions of 450–600t subsoil clay per hectare. A doubling of yield in the fifth year demonstrated the longevity of the treatments (M. Fraser personal communication 2019). Nevertheless, similar yield responses were obtained at a number of sites where soil mixing occurred to 30cm without added clay.

A number of studies have reported decreases in crop yield, due to the addition of clay or clay and OM to sandy soil (Bell 2016; Ogunniyi 2017; Schapel 2018). High levels of early biomass growth on clay-amended soils lead to "haying off" of the crop during critical spring periods with limited rainfall. This generally occurs on sites where clay addition and incorporation depth are not suited to the rainfall or farming system. Decreased yield can be attributed to unsuitable clay addition rates, particularly in the top 10cm, that can lead to changes to soil properties that negatively affect plant growth. These include decreased plant-available water, increased bulk density, surface sealing, high pH (Davenport et al. 2011; Schapel et al. 2018a), increased surface evaporation (G. Betti and G. Azam personal communication 2020), and nutrient deficiencies.

The addition of clay to sandy soil (clay amendment) can increase OC storage through binding to clay surfaces (Baldock 2007; Skjemstad et al. 1993) and by occlusion in aggregates formed by clay (Tisdall and Oades 1982). Subsoil clay should have high potential for OC stabilization as saturation of the mineral particles surfaces by bound C is unlikely to have occurred (Lützow et al. 2006). In field studies, OC was compared after subsoil clay addition to an untreated sandy soil. Hall et al. (2010) reported a 0.2% OC increase in the top 10 cm, 8 years after clay addition. Bailey and Hughes (2012) found a 0.4% increase in OC in the bleached A2 horizon up to 7 years after addition of subsoil clay. Schapel et al. (2017) reported OC increases of 0.1 to 0.5% in the surface 30 cm, equating to an increase between 1 and 22 t/ha (average 10 t/ha) in OC stock 3–9 years after addition of subsoil clay. A meta-analysis of 49 clay-amended sites in South Australia demonstrated an average increase of 4.9 t C/ha, above that of comparable unamended sandy soils (Schapel et al. 2018b). Annual rainfall influenced the size of the OC stock change with the largest increase of 8.2 t C/ha at 350–400 mm rainfall, and a decrease of 1 t C/ha where rainfall was >600 mm.

The incorporation of additional organic matter during the clay amendment process led to higher OC stock compared to treatments without OM addition (Schapel et al. 2018a) and may offset the loss of OC resulting from soil disturbance. Olchin et al. (2008) observed that the negative disruption to OC through tillage appeared to be offset by slower decomposition of residues when they were incorporated to 30 cm compared to 15 cm. Modeling has shown that sandy topsoils are often close to OC saturation and practices that enable OC to be stored deeper in the soil profile are needed to increase OC storage in the profile (Hoyle et al. 2013). There is potential for additional OC storage at 20–30 cm (Schapel et al. 2018a) in clay-amended, particularly delved, soil. Plant roots have an important role increasing OC at depth through increased root biomass, root exudates, and sloughing of cells. The OC in the area surrounding the root (rhizosphere) is much higher compared to the bulk of the soil. The root exudates encourage microbes deeper into the soil and in doing so, transform OC through the fractions from particulate to the more recalcitrant humic form. This is important for increasing reactive surface area and hence nutrient and water storage.

Physical inaccessibility of OM is an important factor to consider in sandy soils. The rate of sorption to minerals and occlusion in aggregates is determined by the mineral surface area (Woolf and Lehmann 2019). Microbial-produced OM is preferentially associated with the clay mineral (Sokol et al. 2019), so adding clay to sand will benefit by providing more sorption sites and creating an environment that is more conducive to microbial activity. It could be important to provide a continual source of OM (e.g. through repeated OM application or increased root growth) because, even in clay-amended sandy soils, the clay concentration is generally <10%, offering limited sorption sites and hence continued susceptibility to decomposition.

Stabilization of OC will depend on many factors, including sesquioxide, calcium, and clay concentrations, clay mineralogy, and the formation of complexes and stable aggregates (Rasmussen et al. 2018). In acidic soils, chemical stabilization can occur through the binding of OC to hydroxyl groups of iron and aluminum (Saidy et al. 2012). Iron oxides also improve aggregation that can limit access by microbes to OC within aggregates (Singh et al. 2018). In alkaline soils, chemical stabilization is largely through the formation of stable aggregates and complexation with calcium carbonate (Fernandez-Ugalde et al. 2011; Rowley et al. 2018; Rasmussen et al. 2018).

Field trials comparing the effectiveness of clay with different mineralogy found that kaolin is as, if not more, suitable than bentonite to improve sandy soil characteristics (Ogunniyi 2017). Ogunniyi (2017) found that kaolin amendment increased nutrient uptake and anion retention compared to bentonite, and when kaolin was combined with OM it produced higher CEC, available water, and spring wheat yield than bentonite and OM combined at the same application rate. The low pH in kaolin was found to reduce soil microbial activity and resulted in less carbon mineralization than bentonite with a higher pH.

Although kaolinitic clay may have lower capacity for OC stabilization than other clay minerals due to lower CEC and surface area, addition to siliceous sand could result in higher OC stabilization capacity and crop growth than other clay types by addressing water repellence and plant-available water limitations. Kaolinite and illite clays are more effective than smectite or vermiculite clays in alleviating water repellence (Ma'shum et al. 1989) as the kaolin spreads more readily over sand grains and remains evenly distributed after drying (Ward and Oades 1993). Consideration of the pH of the applied subsoil will also be important to optimize soil microbial activity and OM mineralization.

This demonstrates that sandy soil amendments cannot be chosen on a single characteristic but need to encompass an array of features. Identification of the sandy soil's key constraints to production (e.g., water repellence, soil pH, and annual rainfall) is critical to ensure that a suitable product and rate of addition is chosen to optimize yield and the mineralization and stabilization of OM.

19.4.3 Mechanisms to Improve Yield following Clay Addition

Clay addition to sandy soil can have positive productivity and OC impacts through mechanisms including (Figure 19.1):

1. *Overcoming water repellence*: increasing the clay content to over 3% in the topsoil overcomes water repellence, leading to more plant numbers with greater early vigor.
2. *Reducing frost damage*: due to the low heat and water retention of sands, crops on sands are prone to frost. A reduction in frost damage through higher heat storage has been reported to increase yields on clay-amended sands (Betti et al. 2019; Rebbeck et al. 2007) and may improve resilience in areas that experience more frequent and severe frosts due to climate change.
3. *Provision of plant essential nutrients*: applied subsoil clay can act as a medium to long-term pool to supply nutrient to plants when concentrations are high (Hall 2010; Bell 2016). This can correct a deficiency and help achieve the yield potential of sandy soils with low

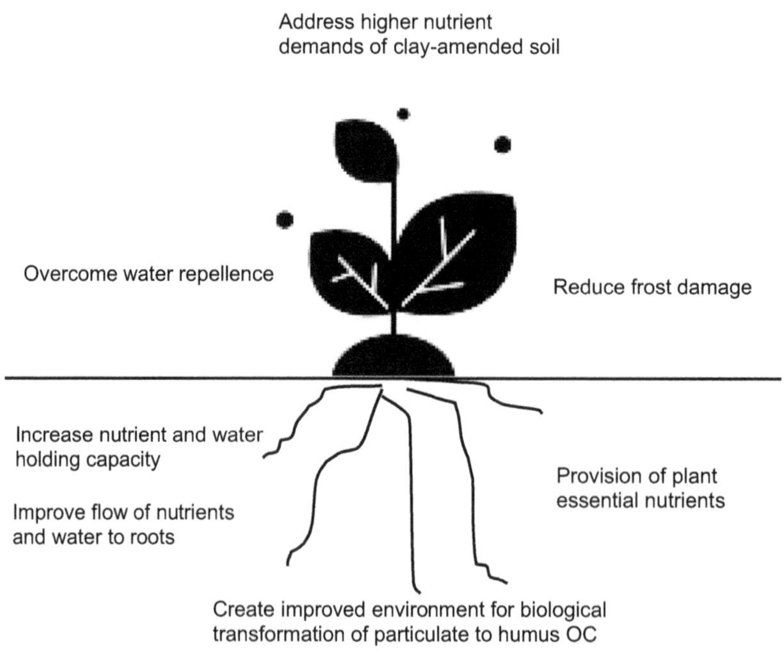

FIGURE 19.1 Mechanisms to improve plant productivity and soil organic carbon (OC) following clay addition.

initial fertility. Conversely, where subsoil clays are low in an important nutrient, such as potassium, the nutrient deficiency can limit yield potential and financial return from clay additions (Bell 2016). The chemical composition of the subsoil clay amendment must be determined to understand its effect on nutrient supply and plant growth.

4. *Increase nutrient-holding capacity*: clay amendments have been reported to increase soil CEC, OC, extractable S, Colwell P, and Colwell K (Bailey and Hughes 2012; Hall et al. 2010; Schapel et al. 2018a). This can be through increased retention of nutrients via increased CEC, P sorption and OC.
5. *Increase water-holding capacity*: clay addition to sandy soil has been reported to increase water retention (Betti et al. 2015; Ogunniyi 2017; Schapel et al. 2018a) and decrease saturated hydraulic conductivity (Betti et al. 2016). Increasing water storage capacity in sandy soils can improve plant productivity where low and seasonal rainfall limits growth (Usowicz and Lipiec 2017). Grain yield has been found to be more strongly correlated with water-holding capacity than OC stock (Schapel et al. 2018a) and can be specifically linked to soil moisture at grain filling (Ogunniyi 2017). Betti et al. (2016) found that the addition of as little as 10% clay-rich subsoil to sand significantly decreased saturated hydraulic conductivity but plant-available water only increased when more than 20% clay-rich subsoil was added.
6. *Improve flow of nutrients and water to roots*: sandy soils have a discontinuous, stop/start supply of water and nutrients to plant roots due to poor connectivity of pore water channels. The addition of clay in large clods can exacerbate the problem, creating water and nutrient "hot spots" in the clay clods and low concentrations in the sand. The addition of smaller clay clods (<5 cm) has the potential to increase the number of connected flowpaths (Schapel et al. 2018b; Boldt-Burisch et al. 2013) improving the supply of water and nutrients to plant roots.
7. *Address higher nutrient demands of amended soil*: it is critical to address the increased nutrient demands of a higher yielding crop on a clay-amended soil. An increase in nitrogen supply is predicted in clay-amended soil, but further research is needed to clarify how to modify nitrogen fertilizer practices to achieve yield potential while minimizing the risks of haying off (Bell 2016).
8. *Create an improved environment for biological transformation of OC:* enhance soil conditions to improve the transformation rate of particulate OC to the more stable humus fraction. Ensure that soils have sufficient nutrition, if necessary, through supplementation, to perform the biological process of transforming particulate OM to humus, thereby stabilizing soil OC.

Even with substantial improvements of sands after clay addition, they are often achieving a maximum of 70% of the water-limited yield (Hall et al. 2010, 2020). With current practices, the addition of subsoil clay to sandy soil will largely be restricted to systems with suitable in-situ subsoil clay and availability of appropriate machinery. As suitable subsoil clay can be difficult or expensive to find at the high rates required to overcome limitations, novel products are required.

19.5 NOVEL PRODUCTS

Currently, amendments for sandy soils are limited to using in-situ clays, on-farm OM, or materials from waste sources. Products that are as effective or more effective at lower rates than current methods have the potential to greatly increase the use of amendments. A review of national and international literature identified the following novel products for consideration.

19.5.1 Industrial Waste Products

The application of fly ash to a sandy soil effectively increased soil water-holding capacity, hydraulic conductivity, and turf plant nutrition and growth (Pathan et al. 2001; Pathan et al. 2003). Additions of fly ash (~10% clay and 50% silt) were at 0%, 5%, 10%, 20%, and 100% (by weight) to sand (~4% clay and 2% silt) and incorporated to 15 cm. While nutrient and water retention increased with fly ash addition, water infiltration was progressively reduced and limited the maximum application rate to between 10% and 20% (Pathan et al. 2001).

Churchman et al. (2014) reported increases in OC in the surface 30 cm ranging between 0.10% and 0.65%, resulting in increased stock of about 12 t/ha, 29 years after addition of fine-textured bauxite processing waste (85% silt, 11% clay, and 4% sand) to sand.

19.5.2 Minerals

Most of the research on clay amendment of sandy soil in southern Australia has used locally sourced, low activity (kaolinitic) clay. While these have improved productivity of sands in many cases, greater benefits may be realized from the use of high-activity clay minerals (bentonite), minerals (zeolite and hydrotalcite), and clay-organic mixes (clay-fortified compost). Hydrotalcite (HT) is a synthetic clay with high structural positive charge (~400 cmolc/kg) (Gillman 2006; Gillman et al. 2008; Jobbágy and Regazzoni 2005). The HT clays could be used in slow-release formulations of fertilizers alone or in mixtures with local clay to adsorb leached nutrient anions. Pre-loading phosphorus on HT could be explored to increase root proliferation in and around the HT granules.

A review by Nakhli et al. (2017) found that natural and surface modified zeolites have a large surface area and can hold nitrogen, phosphate, potassium, and sulfate within their structure and can act as slow-release fertilizers. Zeolites have a larger effect in sands compared to clay soils and decreased hydraulic conductivity while increasing nutrient retention.

19.5.3 Organic Products

Based on the review by Chen et al. (2018b), compost appears to be a promising organic material that could be used to enhance the water and nutrient supply to plant roots on sands. Some compost mixes contain bentonite and zeolite clays. These products require assessment for their suitability as sand amendments, including investigations on a greater range of rates and clay types (HT, bentonite, kaolin, and Fe/Al oxides). Their effects on cation and anion exchange capacity, water-holding capacity, and decomposition rates in sands need to be determined. These organic materials also need to be tested in granulated formulations, along with unamended clays and composts, initially under leaching conditions for nutrient retention, and then in plant growth experiments.

Biochar is variably charged, with a high surface area and high porosity. When incorporated into soil, biochar has been shown to increase CEC and the availability of macronutrients, such as N and P (Kookana et al. 2011). Biochars produced from manure, greenhouse waste, and grasses were more effective in providing nutrients than wood-based products that were better for carbon sequestration (Singh et al. 2014). Loading biochar with nutrients or products, such as goethite (L. Van Zwieten, personal communication), can further improve its use as a novel sandy soil amendment. Hydrolyzed wool added to a loamy sand increased zinc and iron uptake and doubled yield and grain protein by wheat plants grown in a pot experiment (Gogos et al. 2013). Kallenbach et al. (2019) evaluated "lactobionate" a dairy industry by-product added at a rate of 1.7% w/w to fine sandy loam in a laboratory experiment. They reported increased soil OC, water, and microbial content.

19.5.4 Super Absorbent Polymers

Hydrogels (super absorbent polymers) in agriculture are used to increase the amount of available moisture and nutrients in the rootzone in a controlled release formulation (Rajakumar and Sankar 2016). The effect is greater in sandy than clay-textured soils. Reported benefits include increased CEC, reactive surface area, water storage, fertility, microbial enzymatic activity, and decreased pH, bulk density, and macro-porosity (Rajakumar and Sankar 2016).

Zhou et al. (2016) reported that application of an amendment with a formulation of Na-bentonite (91%), humic acid (6%), Na_2CO_3 (2%), and plant cellulose (1%) at 30 t/ha to the soil prior to incorporation to 25 cm increased maize yield by 23%–41% over the 5-year monitoring period.

Soil enzymatic activity also improved over time with increases in soil water storage that peaked in year 3 but were still persistent in year 5. Youssef et al. (2018) reported that the application of 1% w/w biochar with 0.7% w/w superabsorbent polymer was the best treatment to alleviate yield decreases in response to water deficit in a green pea crop.

19.6 IDEAL AMENDED SANDY SOIL

There are a number of key factors in creating a high-performance sand for improved plant productivity and soil organic carbon (Figure 19.2). These include soil limitations that can be practically and economically addressed or overcome, nutrient and water efficiencies above and below ground, and best ways to support biological functioning of the soil.

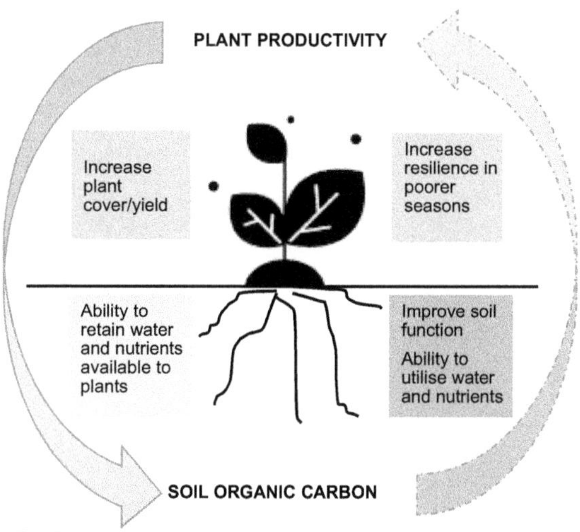

Increased plant biomass production	**Increased 'resilience' in poor seasons**
Address identified soil limitations Grow more on the same area Increase efficiencies for water and nutrients Decrease land degradation issues, leaching	Keep ground covered to reduce evaporation and erosion Reduce variability in yield between good and poor seasons

Retention of plant-available water	**Retention of nutrients in the profile**
Hold water in crop rooting zone with the majority in the plant-available range Increase rooting depth to increase water storage Increase AMF fungi to increase accessibility of water and nutrients	Increase reactive surface area to hold nutrients in rooting zone Improve N, P mineralisation efficiency Ensure nutrient supply for biology to transform particulate OC to humus
Utilisation of water and nutrients stored in the rootzone Increase water, N, and P efficiency	**Improved soil function** Increase OC to improve nutrient mineralisation and associated chemical, physical and biological benefits

FIGURE 19.2 Criteria to address to create a high-performing sandy soil.

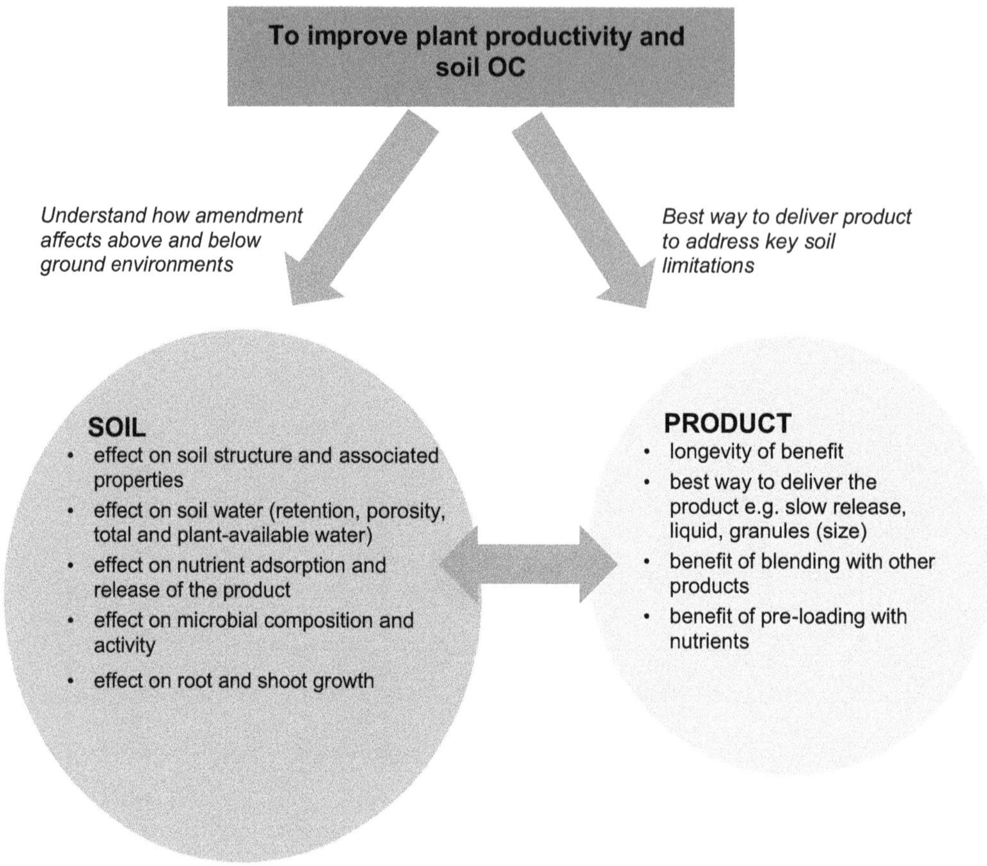

FIGURE 19.3 Factors to consider when selecting the most suitable amendment for improving plant productivity and soil OC in sandy soil.

Selection of the most suitable amendment to improve plant productivity and soil OC requires an understanding of the product's properties, mode of operation and duration, and how these will affect the soil. Consideration needs to be given to chemical, physical, and biological properties that are changed with a focus on nutrient and water availability, microbial activity, and function (Figure 19.3).

ACKOWLEDGMENTS

We are grateful to the Soil CRC for support (Project 2019/3_3.003) to prepare a report on which this article is based. The authors also thank GRDC for investing in various sand-related amelioration projects (UMU00041, DAW00093, and DAW00244).

REFERENCES

Abadi Ghadim, AK (2000) Water repellency: A whole-farm bio-economic perspective. *Journal of Hydrology* **231–232**, 396–405.

Allen M (2007) Mycorrhizal fungi: Highways for water and nutrients in arid soils. *Vadose Zone Journal* **6**, 291–297.

Azam G, and Gazey C (2020) Slow movement of alkali from surface-applied lime warrants the introduction of strategic tillage for rapid amelioration of subsurface acidity in south-western Australia. *Soil Research* **59**(1), 97–106.

Bailey G, and Hughes B (2012) An observational study of clay delving and its impact on the A2 horizon in sand over clay soils. *In 5th Joint SSA and NZSSS Soil Science Conference: Soil Solutions for Diverse Landscapes*, Hobart, Tasmania.

Bailey G, Hughes B, Tonkin R, Dowie R, and Watkins N (2010) Gross soil modification of duplex soils through delving and spading. *In 19th World Congress of Soil Science, Soil Solutions for a Changing World*, Brisbane, Australia.

Baldock J (2007) Composition and cycling of organic carbon in soil. In *Nutrient Cycling in Terrestrial Ecosystems* (Eds., P Marschner and Z Rengel). Springer-Verlag: Berlin, pp. 1–35.

Bauters TWJ, DiCarlo DA, Steenhuis TS, and Parlange JY (1998) Preferential flow in water-repellent sands. *Soil Science Society of America Journal* **62**, 1185–1190.

Bell RW (2016) Assessing the nutritional benefits of clay amendment and cultivation of sands. Final Report of Project UMU00041, GRDC, Canberra.

Bell RW, and Oliveira TS (2022) Chapter 16: Subsoil constraints for crop production: Recent advances, new technologies and priorities for further research. In *Subsoil Constraints for Crop Production* (Eds., TS Oliveira and RW Bell). Springer Nature: Switzerland, pp. 397–422. doi: 10.1007/978-3-031-00317-2_16.

Bell RW, Chon N, and Cong P (2015) Soil types, properties and limiting factors in south-central coastal Vietnam. In *Sustainable and Profitable Crop and Livestock Systems in South-Central Coastal Vietnam* (Eds., S Mann, M Webb, and R Bell). *Proceedings of the Final Workshop*, Quy Nhon, Vietnam, pp. 42–60. ACIAR Proceedings No. 143.

Bell RW, Rerkasem B, Keerati-Kasikorn P, Phetchawee S, Hiranburana N, Ratanarat S, Pongsakul P, and Loneragan, JF (1990). Mineral nutrition of food legumes in Thailand with particular reference to micronutrients. ACIAR Technical Report 19, p. 52.

Betti G, Edwards T, Biddulph B, Davies S, Van Burgel A, Hall D, and Turner C (2019) Mounting evidence that soil amelioration can contribute to reduced frost severity on water repellent soils. GRDC Update Papers, GRDC, Canberra.

Betti G, Grant C, Churchman G, and Murray R (2015) Increased profile wettability in texture-contrast soils from clay delving: Case studies in South Australia. *Soil Research* **53**(2), 125–136.

Betti G, Grant CD, Murray RS, and Churchman GJ (2016) Size of subsoil clods affects soil-water availability in sand–clay mixtures. *Soil Research* **54**(3), 276–290.

Beusch C, Cierjacks A, Böhm J, Mertens J, Bischoff WA, de Araújo Filho JC, and Kaupenjohann M (2019) Biochar vs. clay: Comparison of their effects on nutrient retention of a tropical Arenosol. *Geoderma* **337**, 524–535.

Blackwell P, Morrow G, Webster T, and Nicholson DF (1994) Improvement to crop production from wide furrow sowing in water repellent sand: A comparison to level sowing methods. In *Proceedings of the Second National Water Repellency Workshop*, Perth, 1–5 August 1994, Eds., DJ Carter and KMW Howes, pp. 106–113. Department of Agriculture and Food/Grains Research and Development Corporation: Perth, Western Australia.

Boldt-Burisch KM, Gerke HH, Nii-Annang S, Schneider BU, and Hüttl RF (2013) Root system development of *Lotus corniculatus* L. in calcareous sands with embedded finer-textured fragments in an initial soil. *Plant and Soil* **368**(1), 281–296.

Bond RD (1972) Germination and yield of barley when grown in a water-repellent sand. *Agronomy Journal* **64**, 402–403.

Butterly CR, Amado TJC, and Tang C (2022) Soil acidity and acidification. In *Subsoil Constraints for Crop Production* (Eds., TS Oliveira and RW Bell). Springer International Publishing: Cham, pp. 53–81.

Cann MA (2000) Clay spreading on water repellent sands in the south east of South Australia: Promoting sustainable agriculture. *Journal of Hydrology* **231–232**, 333–341.

Celestina C, Hunt JR, Sale PWG, and Franks AE (2019) Attribution of crop yield responses to application of organic amendments: A critical review. *Soil and Tillage Research* **186**, 135–145.

Chen C, Esfandbod M, Liu Y, and Mallavarapu M (2018a) Waste-derived fertiliser products, nano-porous materials for pesticide delivery, and moisture retention and microbial carrier technologies: Review and meta-analysis [Project 3.1.0.1]. Soil CRC: Australia.

Chen Y, Camps-Arbestain M, Shen Q, Singh BK, and Cayuela ML (2018b) The long-term role of organic amendments in building soil nutrient fertility: A meta-analysis and review. *Nutrient Cycling in Agroecosystems* **111**, 103–125.

Churchman GJ, Noble A, Bailey G, Chittleborough D, and Harper R (2014) Clay addition and redistribution to enhance carbon sequestration in soils. In *Soil Carbon* (Eds., AE Hartemink and K McSweeney). Springer International Publishing: Cham, pp. 327–335.

Coker J, Poss R, Hartmann C, and Buthorndharaj S. (2004) Effects of recycled bentonite addition on soil properties, plant growth and nutrient uptake in a tropical sandy soil. *Plant and Soil* **267**, 155–163.

Commander P, Schoknecht N, Verboom B, and Cacceta P (2002) The geology, physiology and soils of Wheatbelt valleys. In *Dealing with Salinity in Wheatbelt Valleys. Proceedings of the Conference, Processes, Prospects and Practical Options*, 31 July to 1 August 2001, Merredin, WA, pp. 1–27.

Crockford H, Topalidis S, and Richardson, DP (1991) Water repellency in a dry sclerophyll eucalypt forest: Measurements and processes. *Hydrological Processes* **5**, 405–420.

Czaban J, Czyż E, Siebielec G, and Niedźwiecki J (2014) Long-lasting effects of bentonite on properties of a sandy soil deprived of the humus layer. *International Agrophysics,* 28(3), 279–289.

Czaban J, and Siebielec G (2013) Effects of bentonite on sandy soil chemistry in a long-term plot experiment (II); effect on pH, CEC, and macro- and micronutrients. *Polish Journal of Environmental Studies* **22**, 1669–1676.

Daniel, NRR, Uddin, SMM, Harper, RJ, and Henry, DJ (2019) Soil water repellency: A molecular-level perspective of a global environmental phenomenon. *Geoderma* **338**, 56–66.

Davenport D, Hughes B, Davies S, and Hall D (2011) Spread, delve, spade, invert: A best practice guide to the addition of clay to sandy soils. Rural Solutions SA, Agricultural Bureau of South Australia, Caring for our Country, Grains Research and Development Corporation. Grains Research and Development Corporation (GRDC), Kingston ACT.

Davies S (2016) Delivering agronomic strategies for water repellent soils. DAW000204 Final Report, GRDC, Canberra.

DeBano LF (1981) Water repellent soils: A state-of-the-art. General Technical Report PSW-46. United States Department of Agriculture Forest Service, Pacific Southwest Forest and Range Experiment Station, Berkeley, CA.

Degens BP, Sparling GP, and Abbott LK (1996) Increasing the length of hyphae in a sandy soil increases the amount of water-stable aggregates. *Applied Soil Ecology* **3**(2), 149–159.

FAO/UNESCO (1995) *Digital Soil Map of the World and Derived Properties.* FAO: Rome, Italy.

Dempster DN, Jones DL, and Murphy DV (2012) Clay and biochar amendments decreased inorganic but not dissolved organic nitrogen leaching in soil. *Soil Research* **50**(3), 216–221.

Farrell M, Macdonald LM, Webb M, Wong M, Jenkins S, and Abbott LK (2017a) Understanding biological farming inputs. In GRDC Update Papers, GRDC, Canberra.

Fernandez-Ugalde O, Virto I, Barre P, Gartzia-Bengoetxea N, Enrique A, Imaz MJ, and Bescansa P (2011) Effect of carbonates on the hierarchical model of aggregation in calcareous semi-arid Mediterranean soils. *Geoderma* **164**(3/4), 203–214.

Gazey C, Davies S, and Master R (2014) *Soil Acidity: A Guide for Farmers and Consultants*, Second edition. Bulletin 4858, Department of Agriculture and Food: Perth, Western Australia.

Gillman GP (2006) A simple technology for arsenic removal from drinking water using hydrotalcite. *Science of the Total Environment* **366**(2), 926–931.

Gillman GP, Noble MA, and Raven MD (2008) Anion substitution of nitrate-saturated layered double hydroxide of Mg and Al. *Applied Clay Science* **38**(3), 179–186.

Gogos A, Evangelou MWH, Schäffer A, and Schulin R (2013) Hydrolysed wool: A novel soil amendment for zinc and iron biofortification of wheat. *New Zealand Journal of Agricultural Research* **56**(2), 130–141.

Gray JM, Humphreys GS, and Deckers JA (2009) Relationships in soil distribution as revealed by a global soil database. *Geoderma* **150**(3–4), 309–323.

Hall D, Edwards T, Davies S, Bell R, Farre I, Petersen L, and Dodge D (2018) Longer term effects of spading, mouldboard ploughing and claying on the south coast of WA. *In GRDC Update Papers*, GRDC, Canberra.

Hall DJ, McKenzie DC, Macleod DA, and Barrett A (1994) Amelioration of a hardsetting alfisol through deep moldboard plowing, gypsum application and double cropping. 1. Soil physical and chemical properties. *Soil & Tillage Research* **28**(3–4), 253–270.

Hall DJM, and Bell RW (2015) Biochar and compost increase crop yields but the effect is short term on sandplain soils of Western Australia. *Pedosphere* **25**(5), 720–728.

Hall DJM, Davies SL, Bell RW, and Edwards TJ (2020) Soil management systems to overcome multiple constraints for dryland crops on deep sands in a water limited environment on the South Coast of Western Australia. *Agronomy* **10**(12), 1881.

Hall DJM, Jones HR, Crabtree WL, and Daniels TL (2010) Claying and deep ripping can increase crop yields and profits on water repellent sands with marginal fertility in southern Western Australia. *Australian Journal of Soil Research* **48**(2), 178–187.

Hall J, Maschmedt D, and Billing B (2009) The soils of Southern South Australia. Department of Water, Land and Biodiversity Conservation, Adelaide.

Harper RJ, and Gilkes RJ (1994) Soil attributes related to water repellency and the utility of soil survey for predicting its occurrence. *Australian Journal of Soil Research* **32(5)**, 1109–1124.

Harper RJ, McKissock I, Gilkes RJ, Carter DJ, and Blackwell PS (2000) A multivariate framework for interpreting the effects of soil properties, soil management and landuse on water repellency. *Journal of Hydrology* **231–232**, 371–383.

Helyar KR, and Porter WM (1989) Soil acidification, its measurement and the processes involved. In *Soil Acidity and Plant Growth* (Ed. AD Robson). Academic Press: Sydney, pp. 61–101. doi: 10.1016/b978-0-12-590655-5.50007-4.

Hewelke E, Oktaba L, Gozdowski D, Kondras M, Olejniczak I, and Górska, E (2018) Intensity and persistence of soil water repellency in pine forest soil in a temperate continental climate under drought conditions. *Water* **10**, 1121–1137.

Hijbeek R, van Ittersum MK, ten Berge HFM, Gort G, Spiegel H, and Whitmore AP (2017) Do organic inputs matter: A meta-analysis of additional yield effects for arable crops in Europe. *Plant and Soil* **411**(1), 293–303.

Tam HM, Nhan DT, Thuong NT, Vinh H, Hoa HTT, Wen C, Thinh T, Le Dinh Q, Mann S, and Bell RW (2015) Diagnosing multiple nutrient deficiencies that limit crop growth and yields on sands in south central coast of Vietnam. In *Sustainable and Profitable Crop and Livestock Systems for South Central Coastal Vietnam - Proceedings*, (Eds., Mann S, Webb MC, and Bell RW), pp. 62–79. ACIAR Proceedings No. 143.

Hoyle FC (2013) *Managing Soil Organic Matter: A Practical Guide*. Grains Research and Development Corporation: Canberra. Available at https://grdc.com.au/resources-and-publications/all-publications/publications/2013/07/grdc-guide-managingsoilorganicmatter.

Hoyle FC, D'Antuono M, Overheu T, and Murphy DV (2013) Capacity for increasing soil organic carbon stocks in dryland agricultural systems. *Soil Research* **51**(8), 657–667.

Hin S (2018) Acid sands of South-Eastern Cambodia: Their origin, properties and management for upland crops. PhD thesis, Murdoch University, West Australia.

IUSS Working Group WRB (2015) World reference base for soil resources 2014, Update 2015. International Soil Classification System for Naming Soils and Creating Legends for Soil Maps. World Soil Resources Reports No. 106, Rome: FAO.

Jobbágy M, and Regazzoni AE (2005) Anion-exchange equilibrium and phase segregation in hydrotalcite systems: Intercalation of hexacyanoferrate(III) ions. *The Journal of Physical Chemistry B* **109**(1), 389–393.

Kallenbach CM, Conant RT, Calderón F, and Wallenstein MD (2019) A novel soil amendment for enhancing soil moisture retention and soil carbon in drought-prone soils. *Geoderma* **337**, 256–265.

Kay B, and Angers DA (1999) Soil structure. In *Handbook of Soil Science* (Ed. M Sumner). CRC Press: New York, pp. A-229-A-269.

Kirkby CA, Richardson AE, Wade LJ, Batten GD, Blanchard C, and Kirkegaard JA (2013) Carbon-nutrient stoichiometry to increase soil carbon sequestration. *Soil Biology and Biochemistry* **60**, 77–86.

Kirkby CA, Richardson AE, Wade LJ, Passioura JB, Batten GD, Blanchard C, and Kirkegaard JA (2014) Nutrient availability limits carbon sequestration in arable soils. *Soil Biology and Biochemistry* **68**, 402–409.

Kirkegaard J (2009) GRDC factsheet 2009 deep ripping https://grdc.com.au/resources-andpublications/all-publications/factsheets/2009/06/grdc-fs-deepripping.

Kirkegaard JA, Lilley JM, Howe GN, and Graham JM (2007) Impact of subsoil water use on wheat yield. *Australian Journal of Agricultural Research* **58**(4), 303–315.

Kookana RS, Sarmah AK, Van Zwieten L, Krull E, and Singh B (2011) Chapter 3: Biochar application to soil: Agronomic and environmental benefits and unintended consequences. In *Advances in Agronomy* (Ed. DL Sparks). Academic Press: Cambridge, MA, Vol. 112, pp. 103–143.

Lipiec J, Siczek A, Sochan A, and Bieganowski A (2016) Effect of sand grain shape on root and shoot growth of wheat seedlings. *Geoderma* **265**, 1–5.

Lonappan L, Rouissi T, Das DRK, Brar S, Avalos Ramirez A, Verma M, Surampalli R, and Valero J (2016) Adsorption of methylene blue on biochar microparticles derived from different waste materials. *Waste Management* **49**, 537–544.

Loveland P, and Webb J (2003) Is there a critical level of organic matter in the agricultural soils of temperate regions: A review. *Soil and Tillage Research* **70**(1), 1–18.

Lützow MV, Kögel-Knabner I, Ekschmitt K, Matzner E, Guggenberger G, Marschner B, and Flessa H (2006) Stabilization of organic matter in temperate soils: Mechanisms and their relevance under different soil conditions: A review. *European Journal of Soil Science* **57**(4), 426–445.

Ma'shum M, Oades J, and Tate M (1989) The use of dispersible clays to reduce water repellency of sandy soils. *Soil Research* **27**(4), 797–806.

Macdonald LM, McBeath TM, Llewellyn R, Moodie, M, Desbiolles J, Saunders C, Ugcul M, Fraser M, Wilhelm N, Davenport D, Trengrove S, Haskins B, and Whitworth R (2019) Underperforming sandy soils: Targeting constraints for cost effective amelioration. In *GRDC Updates*. GRDC, Canberra.

Maschmedt D (2000) Assessing Agricultural Land. Agricultural land classification standards used in South Australia's land resource mapping program. PIRSA Land Information, Adelaide, South Australia.

McKenzie N, Coughlan K, and Cresswell H (2002) *Soil Physical Measurement and Interpretation for Land Evaluation*. CSIRO: Melbourne.

Minasny B, and McBratney AB (2018) Limited effect of organic matter on soil available water capacity. *European Journal of Soil Science* **69**(1), 39–47.

Moore G, Findlater P, and Carter D (2004) Section 7.1 Wind erosion. In *Soil Guide: A Handbook for Understanding and Managing Agricultural Soils* (Ed., G Moore). Department of Agriculture, Western Australia, Bulletin 4343, pp. 211–222.

Müller K, Deurer M, Jeyakumar P, Mason K, van den Dijssel C, Green S, and Clothier B (2014) Temporal dynamics of soil water repellency and its impact on pasture productivity. *Agricultural Water Management* **143**, 82–92.

Mullins CE, MacLeod DA, Northcote KH, Tisdall JM, and Young IM (1990) Hardsetting soils: Behaviour, occurrence, and management. *Advances in Soil Science* **11**, 37–108.

Mullins CE and Panayiotopoulos KP (1985) Packing of sands. *European Journal of Soil Science* **36**(1), 129–139.

Nakhli SA, Delkash M, Ebrazi Bakhshayesh B, and Kazemian H (2017) Application of zeolites for sustainable agriculture: A review on water and nutrient retention. *Water, Air, & Soil Pollution* **228**, 464.

Newsome D (2000) Origin of sandplains in Western Australia: A review of the debate and some recent findings. *Australian Journal of Earth Sciences* **47**(4), 695–706.

Noble AD, Gillman GP, Nath S, and Srivastava RJ (2001) Changes in the surface charge characteristics of degraded soils in the wet tropics through the addition of beneficiated bentonite. *Soil Research* **39**(5), 991–1001.

Noble AD, Gillman GP, and Ruaysoongnern S (2000) A cation exchange index for assessing degradation of acid soil by further acidification under permanent agriculture in the tropics. *European Journal of Soil Science* **51**, 233–43. https://doi.org/10.1046/j.1365-2389.2000.00313.x.

Ogunniyi J (2017) Effects of clay and organic matter amendments on water and nutrient retention of sandy soils. PhD Thesis, University of Warwick, Coventry, UK.

Olchin GP, Ogle S, Frey S D, Filley, TR, Paustian K, and Six J (2008) Residue carbon stabilization in soil aggregates of no-till and tillage management of dryland cropping systems. *Soil Science Society of America Journal* **72**(2), 507–513.

Oldfield EE, Bradford MA, and Wood SA (2019) Global meta-analysis of the relationship between soil organic matter and crop yields. *Soil* **5**(1), 15–32.

Ozhan S, Ozcan M, and Gokbulak F (2008) Effect of pumice addition on available water capacity of different soil textural classes. BALWOIS 2008, Ohrid, Republic of Macedonia.

Pathan S, Aylmore G, and Colmer T (2001) Fly ash amendment of sandy soil to improve water and nutrient use efficiency in turf culture. *International Turfgrass Society Research Journal* **9**, 33–39.

Pathan SM, Aylmore LAG, and Colmer TD (2003) Properties of several fly ash materials in relation to use as soil amendments. *Journal of Environmental Quality* **32**(2), 687–693.

Quilty JR, and Cattle SR (2011) Use and understanding of organic amendments in Australian agriculture: A review. *Soil Research* **49**(1), 1–26.

Ragland J, and Boonpukdee L (1987) Fertilizer responses in northeast Thailand. I. Literature review and rationale. *Thai Journal of Soils and Fertilizers* **9**, 65–79.

Rajakumar R, and Sankar J (2016) Hydrogel: Novel soil conditioner and safer delivery vehicle for fertilizers and agrochemicals: A review. *International Journal of Applied and Pure Science and Agriculture* **2**, 163–172.

Rasmussen C, Heckman K, Wieder WR, Keiluweit M, Lawrence CR, and Berhe AA, et al. (2018) Beyond clay: Towards an improved set of variables for predicting soil organic matter content. *Biogeochemistry* **137**(3), 297–306.

Rebbeck M, Lynch C, Hayman PT, and Sadras VO (2007) Delving of sandy surfaced soils reduces frost damage in wheat crops. *Australian Journal of Agricultural Research* **58**(2), 105–112.

Ritsema CJ, and Dekker LW (1994) How water moves in a water repellent sandy soil: 2. Dynamics of fingered flow. *Water Resources Research* **30**, 2519–2531.

Ritsema CJ, Dekker LW, Hendrickx JMH, and Hamminga W (1993) Preferential flow mechanism in a water repellent sandy soil. *Water Resources Research* **29**, 2183–2193.

Robinson N, Pope R, Liddicoat C, Holmes K, Griffin E, Kidd D, Jenkins B, Rees D, and Searle R. (2021) Sandy soils: Organic and clay amendments to improve the productivity of sandy soils. Detailed plan for mapping and grouping of sands. Soil CRC Project 3.3.003. Cooperative Research Centre for High Performance Soils, March 28 2021.

Roper MM, Davies SL, Blackwell PS, Hall DJM, Bakker DM, Jongepier R, and Ward PR (2015) Management options for water-repellent soils in Australian dryland agriculture. *Soil Research* **53**(7), 786–806.

Rowley MC, Grand S, and Verrecchia ÉP (2018) Calcium-mediated stabilisation of soil organic carbon. *Biogeochemistry* **137**(1), 27–49.

Rye, CF, and Smettem, KRJ (2015) Seasonal and interannual variability of the effective flow cross-sectional area in a water-repellent soil. *Vadose Zone Journal* **14**, 1–12.

Saidy AR, Smernik RJ, Baldock JA, Kaiser K, Sanderman J, and Macdonald LM (2012) Effects of clay mineralogy and hydrous iron oxides on labile organic carbon stabilisation. *Geoderma* **173**, 104–110.

Scanlan CA, Holmes KW, and Bell RW (2022) Sand and gravel subsoils. In *Subsoil Constraints for Crop Production* (Eds., TS Oliveira and RW Bell). Springer Nature: Switzerland, pp. 179–198. doi: 10.1007/978-3-031-00317-2_8.

Schapel A (2018) Carbon storage in sandy soil amended with clay. Examining the relationship of organic carbon concentration to clay concentration, clod size and distribution. PhD thesis, The University of Adelaide, Adelaide, South Australia.

Schapel A, Davenport D, and Marschner P (2017) Increases in organic carbon concentration and stock after clay addition to sands: Validation of sampling methodology and effects of modification method. *Soil Research* **55**(2), 124–133.

Schapel A, Marschner P, and Churchman J (2018a) Clay amount and distribution influence organic carbon content in sand with subsoil clay addition. *Soil and Tillage Research* **184**, 253–260.

Schapel A, Marschner P, and Churchman J (2019) Influence of clay clod size and number for organic carbon distribution in sandy soil with clay addition. *Geoderma* **335**, 123–132.

Schapel A, Reseigh J, Wurst M, Mallants D, and Herrmann T (2018b) Offsetting greenhouse gas emissions through increasing soil organic carbon in SA clay-modified soils: Knowledge gap analysis. Goyder Institute for Water Research Technical Report Series No. 18/04, Adelaide, South Australia.

Schoknecht N and Pathan S (2013) *Soil Groups of Western Australia: A Simple Guide to the Main Soils of Western Australia*, Fourth Edition. Resource Management Technical Report 380. Department of Agriculture and Food: Perth, Western Australia.

Sharma R, Bell RW, and Wong MTF (2015) Phosphorus forms in soil solution and leachate of contrasting soil profiles and their implications for P mobility. *Journal of Soils and Sediments* **15**(4), 854–862.

Sharma R, Bell RW, and Wong MTF (2017) Dissolved reactive phosphorus played a limited role in phosphorus transport via runoff, throughflow and leaching on contrasting cropping soils from southwest Australia. *Science of the Total Environment* **577**, 33–44.

Sharma R, Wong MTF, Weaver DM, Bell RW, Ding X, and Wang K (2021) Runoff and leaching of dissolved phosphorus in streams from a rainfed mixed cropping and grazing catchment under a Mediterranean climate in Australia. *Science of the Total Environment* **771**, 145371.

Singh B, Macdonald LM, Kookana RS, van Zwieten L, Butler G, Joseph S, Weatherley A, Kaiudal BB, Regan A, Cattle J, Dijkstra F, Boersma M, Kimber S, Keith A, and Esfandbod M. (2014) Opportunities and constraints for biochar technology in Australian agriculture: Looking beyond carbon sequestration. *Soil Research* **52**(8), 739–750.

Singh M, Sarkar B, Sarkar S, Churchman J, Bolan N, Mandal S, Menon M, Purakayastha TJ, and Beerling DJ (2018) Chapter 2: Stabilization of soil organic carbon as influenced by clay mineralogy. In *Advances in Agronomy*. Elsevier, Vol. 148, pp. 33–84. doi: 10.1016/bs.agron.2017.11.001.

Sitthaphanit S, Limpinuntana V, Toomsan B, Panchaban S, and Bell RW (2009) Fertiliser strategies for sandy soils in a high rainfall regime. *Nutrient Cycling in Agroecosystems* **85**, 123–139.

Skjemstad JO, Janik LJ, Head MJ, and McClure SG (1993) High energy ultraviolet photo-oxidation: A novel technique for studying physically protected organic matter in clay- and silt-sized aggregates. *Journal of Soil Science* **44**(3), 485–499.

Sokol NW, Sanderman J, and Bradford MA (2019) Pathways of mineral-associated soil organic matter formation: Integrating the role of plant carbon source, chemistry, and point of entry. *Global Change Biology* **25**(1), 12–24.

Tisdall JM, and Oades JM (1982) Organic matter and water-stable aggregates in soils. *Journal of Soil Science* **33**(2), 141–163.

Unkovich M (2014) A review of the potential constraints to crop production on sandy soils in low rainfall south-eastern Australia and priorities for research. Technical report for the Grains Research and Development Corporation, Canberra.

Usowicz B, and Lipiec J (2017) Spatial variability of soil properties and cereal yield in a cultivated field on sandy soil. *Soil and Tillage Research* **174**, 241–250.

Ward P, and Oades J (1993) Effect of clay mineralogy and exchangeable cations on water repellency in clay-amended sandy soils. *Soil Research* **31**(3), 351–364.

Wiesmeier M, Urbanski L, et al. (2019) Soil organic carbon storage as a key function of soils: A review of drivers and indicators at various scales. *Geoderma* **333**, 149–162.

Woolf D, and Lehmann J (2019) Microbial models with minimal mineral protection can explain long-term soil organic carbon persistence. *Scientific Reports* **9**, 1–20.

Wong MTF and Wittwer K (2009) Positive charge discovered across Western Australian wheatbelt soils challenges key soil and nitrogen management assumptions. *Australian Journal of Soil Research* **47**, 127–135.

Yang X, Zhang S, Ju M, and Liu L (2019) Preparation and modification of biochar materials and their application in soil remediation. *Applied Sciences* **9**, 1365.

Yost J, and Hartemink A (2019) Soil organic carbon in sandy soils: A review. *Advances in Agronomy* **158**, 217–310.

Youssef S, Riad G, Abuelazm N, and Ahmed E (2018) Amending sandy soil with biochar or/and superabsorbent polymer mitigates the adverse effects of drought stress on green pea. *Egyptian Journal of Horticulture* **45**, 169–183.

Zhou L, Liu JH, Zhao BP, Xue A, an Hao GC (2016) Effects of soil amendment on soil characteristics and maize yield in Horqin Sandy Land. *IOP Conference Series: Earth and Environmental Science* **41**, 012005.

20 Subsoil Acidity and Aluminum Toxicity

Measurement, Formation, and Management Strategies in Conservation Farming Systems

Geoffrey C. Anderson[1], *Nanthi Bolan*[2], *James Easton*[3], *David J. M. Hall*[4], *Rajesh Sharma*[5], *Shahab Pathan*[6], *and Richard W. Bell*[7]

[1]Department of Primary Industries and Regional Development, SoilsWest, Northam, Australia

[2]School of Agriculture and Environment, The University of Western Australia, Perth, Australia

[3]CSBP Limited, Kwinana, Australia

[4]Department of Primary Industries and Regional Development, SoilsWest, Esperance, Australia

[5]ChemCentre, Resources and Chemistry Precinct Level 2, Bentley, Australia

[6]Department of Primary Industries and Regional Development, SoilsWest, South Perth, Australia

[7]SoilsWest, Centre for Sustainable Farming Systems, Food Futures Institute, Murdoch University, Murdoch, Australia

CONTENTS

DOI: 10.1201/9781003093565-20

20.1 INTRODUCTION

Soil acidity with toxic levels of soil aluminum (Al^{3+}) is a common problem limiting agricultural production worldwide which is managed by applying lime (Bolan et al., 2003a, Rheinheimer et al., 2018; Holland et al., 2021; Xu et al., 2022; Butterly et al., 2022). In south Western Australia, soil acidity or Al^{3+} toxicity is divided into topsoil (0–10 cm) and subsoil (soil layers below 10 cm) because the 0–10 cm depth has been used historically for making fertilizer recommendations (Hayes et al., 2022). Topsoils with relatively high soil organic carbon (SOC) contents result in the binding of soil solution Al^{3+} (Al_{Soln}) to form Org-Al_{Soln}, which is not toxic to plant root growth (Anderson et al., 2021). Hence, topsoil limitation for agricultural production is measured using soil pH_{CaCl_2}. In contrast, subsoils have low SOC content, and the production limitation is measured using soil Al_{CaCl_2}. Al_{CaCl_2}, when measured in the 0–30 cm layer using sampling increments of 10 cm, predicted wheat response to lime application (Anderson and Bell, 2019). Subsoil Al^{3+} toxicity occurs because some soil profiles are inherently acidic, while other soils have an Al^{3+} toxicity layer in the 10–30 cm due to the plant and soil processes of acidification (Dolling and Porter, 1994). More recently, commercial soil sampling techniques have been developed to enable sampling to 50 cm. Harries et al. (2015) observed in 12% of cases, pH_{CaCl_2} was less than 4.8, indicating the occurrence of Al^{3+} toxic layer in the 30–50 cm layer. However, the impact of Al^{3+} toxicity in the 30–50 cm soil layer on crop production is unknown.

The south Western Australia agricultural region has a high frequency of surface acidity and subsoil Al^{3+} toxicity soils due to high acidification rates (Gazey and Davies, 2009; Gazey et al., 2013). Soil acidification in the south Western Australia agricultural region is due to its Mediterranean climate (high winter rainfall), highly weathered soils with a low water-holding capacity (light texture or sandy soils), and productive farming systems. These factors combine to result in high rates of basic cation depletion resulting from nitrate leaching and product removal, which result in high rates of soil acidification (Bolan et al., 1991). Also, the region's topsoil has relatively low SOC content or a low pH buffer capacity (pH_{BC}) (Wong et al., 2013). This means that soil acidification and lime application will significantly change soil pH. The surface application of lime effectively treats topsoil acidity but may take years to correct subsoil Al^{3+} toxicity (Anderson and Bell, 2019; Azam and Gazey, 2021).

Subsoil Al^{3+} toxicity is managed by applying sufficient lime rates at frequencies to maintain the topsoil pH_{CaCl_2} within the range of 5.5–7.5 (Li et al., 2019). Gypsum is more soluble than lime. Hence, the applied calcium and sulfate are more readily leached into the subsoil to provide short-term treatment of subsoil Al^{3+} toxicity (Anderson et al., 2020a, b, 2021). The application of lime with gypsum also can increase lime alkalinity (bicarbonate) movement from the topsoil into the subsoil (Lauricella et al., 2021). Hence, applying lime and gypsum can be the most effective and profitable strategy for managing subsoil Al^{3+} toxicity in low rainfall environments (Anderson et al., 2020a).

Plant response to lime application results from increased root growth into the subsoil, improving water and nutrient uptake (Bolan et al., 2003a; Azam and Gazey, 2021). Lime application impacts manganese, cadmium, chromium, aluminum, iron, copper, cobalt, zinc, and nickel uptake with concentrations in barley shoots related to soil pH (Holland et al., 2021). In addition, barley response to the lime application is related to exchangeable soil manganese, calcium, and aluminum soil content. The application of lime increases molybdenum uptake due to decreased adsorption and increased availability (Bolan et al., 2003a, Anderson et al., 2020a). In contrast, increased soil sulfate content due to gypsum application results in decreased molybdenum uptake. Because molybdate uptake occurs through sulfate transporters with high soil sulfate content reduces molybdate uptake (Fitzpatrick et al., 2008, Dudev and Lim, 2004). Using gypsum increases plant calcium and sulfate concentrations. However, the impact of these changes on plant growth is small due to the application of the basal nutrients or the presence of sufficient nutrients based on soil and plant test measurements (Anderson et al., 2020a).

Conservation farming systems or no-tillage and crop residue retention are used worldwide (Aikins et al., 2020). However, the reduced cultivation results in the stratification of surface-applied lime in soil layers at or near the surface, reducing its effectiveness in treating subsoil Al^{3+} toxicity

(Anderson and Bell, 2019). Strategic tillage practices (surface and deep) are used in no-tillage farming and pasture systems to reduce lime stratification and increase lime's effectiveness in treating subsoil Al^{3+} toxicity (Rheinheimer et al., 2018). At the same time, strategic surface tillage has no detrimental effect on crop productivity and soil health (Dang et al., 2018). In contrast, annual tillage is more efficient than no-tillage systems in decreasing Al^{3+} toxicity and increasing soil pH in the subsoil but can negatively impact SOC, soil structure, and grain yields (Silva et al., 2022).

20.2 SOIL ACIDIFICATION

Soil acidification or soil pH decline is caused by the release of protons (H^+) due to reactions associated with the C, N, and S cycles that are associated with the soil and plant base process (Helyar and Porter, 1989; Bolan et al., 2003a; Bolan and Hedley, 2003b; Bolan et al., 2005; Butterly et al., 2022). Proton release and accumulation depend on many soil and plant reactions. Net soil acidification occurs when proton release is greater than proton accumulation resulting from nitrate leaching, use of ammonium-based fertilizers, and removal of a product with excess base cations (Ca^{2+}, Mg^{2+}, K^+, and Na^+) compared to anions. The Mediterranean environment of the south Western Australia agricultural region has some unique features which make the area prone to the development and occurrence of subsoil acidity (Gazey and Davies, 2009). These include highly weathered light texture soils or sands, low SOC, and winter-dominated rainfall, which result in nitrate leaching and acidification of the 10–30cm soil layer (Anderson et al., 1998b; Dolling and Porter, 1994).

Acidification due to grain removal based on the grain's nutrient content is in the order of 5.5–14.9kg $CaCO_3$/t/ha (Moore et al., 2001). The amount of soil acidification associated with 1kg N /ha is leached, requiring 3.6kg $CaCO_3$/ha to neutralize it. The calculation of acidification rates is based on the amount of pure $CaCO_3$. The amount of nitrate leached depends on the distribution of inorganic N within the soil profile at the start of the growing season, the soil's water-holding capacity, and rainfall amount and intensity (Anderson et al., 1998b). Soil profile inorganic N content at the beginning of the growing season is determined by the SOC content, rotation history, and frequency of summer rainfall events (Harries et al., 2015). High values of up to 140kg N/ha occur in the soil profile following growing legumes in the previous year, summer fallow periods with summer rainfall events, and soil with high SOC (Anderson et al., 1998a, b; Harries et al., 2015). Anderson et al. (1998b) found that the highest rate of nitrate leaching, up to 69kg N/ha, then occurred under wheat with intense April to July rainfall after summer fallow with rainfall events, which resulted in mineralization of soil and the previous legume phase organic nitrogen.

The relative importance of product removal and nitrate leaching on soil acidification rates varies according to annual rainfall, soil types, and land use. In a wet year (703mm), acidification rates were greatest under wheat grown following legumes (0.17–0.25t $CaCO_3$ /ha/yr), with nitrate leaching accounting for 95% of the soil acidification (Anderson et al., 1998a, b). Rates were lower under lupins following wheat (0.10–0.15t $CaCO_3$/ha/yr), with nitrate leaching accounting for 80%–90% of the soil acidification. Under pastures following wheat, soil acidification was between 0.05 and 0.15t $CaCO_3$/ha/yr, with 0%–60% due to nitrate leaching. The low acidification rates observed under pastures were due to the early growth of capeweed, which effectively took up nitrate before the leaching rainfall events (Anderson et al., 1998a).

The rate of soil acidification rates in the Anderson et al. (1998a, b) experiment is relatively high due to wet seasonal conditions and the low water-holding capacity of the soil. Lower nitrate leaching rates will occur in lower rainfall zones and for soils with a higher plant-available water-holding capacity (Asseng et al., 2001). These experiments used a two-year legume/cereal rotation, the standard cropping practice of the 1980s and early 1990s. However, the farming system's frequency of legumes (grain and pastures) has declined (Harries et al., 2015). Acidification rates are likely to be high due to the greater use of N fertilizers and higher grain yields due to improvements in the farming system (Harries et al., 2022). Growers tend to use split application of nitrogen fertilizer to increase nitrogen use efficiency and reduce the risk of nitrate leaching, which is likely to reduce the rate of soil acidification.

20.3 MEASUREMENT

Measuring the limitation imposed by soil acidity and Al^{3+} toxicity is undertaken using both pH_{CaCl_2} and Al_{CaCl_2} (Anderson and Bell, 2019). High SOC content of the surface soil layer means surface Al^{3+} toxicity can be low due to the formation of organically bound Al (Org-Al_{Soln}), which is not toxic to plant roots (Anderson et al., 2021). The SOC of subsoil layers is lower than the topsoil, but Org-Al_{Soln} is present due to the higher Al^{3+}. Despite the complexity of Al species' impact on root growth, measuring Al_{CaCl_2} in the 0–30 cm soil layer predicted field wheat response to the lime application (Anderson and Bell, 2019). In contrast, under glasshouse conditions which had received an application of phosphogypsum, the Al species' Al^{3+}, F-Al, and Org-Al_{Soln} provided a better prediction of lucerne (*Medicago Sativa*) response than Al_{CaCl_2} (Bouray et al., 2022).

Grouping acidity and Al^{3+} toxicity into topsoil and subsoil are related to management practices required to address these two soil constraints. In the case of topsoil acidity, the surface application of lime is an effective treatment strategy. In contrast, lime application to the soil surface is ineffective in treating subsoil acidity due to its low solubility, low capacity of the soil to dissolve lime, and low rate of alkalinity leaching (Anderson and Bell, 2019; Azam and Gazey, 2021). Furthermore, an increase in surface soil pH resulting from the lime application is likely to increase the cation exchange capacity of variable charged soil, thereby reducing the movement of basic cations to the subsoil to neutralize subsoil acidity (Bolan et al., 2003a). For these reasons, additional management strategies are required to address subsoil acidity. These practices include applying regular and relatively high lime rates to maintain the topsoil pH_{CaCl_2} between 5.5 and 7.5 and using strategic tillage to increase lime dissolution and alkalinity movement (Scanlan and Davies, 2019; Davies et al., 2019; Azam and Gazey, 2021). In conducting strategic tillage, it must be done to minimize the risk of wind erosion because of the high risk of a wind erosion event following tillage (Anderson, 2009). In addition, gypsum, which is more soluble than lime, can be used in combination with surface lime application to treat subsoil Al^{3+} toxicity (Anderson et al., 2020a)

Soil acidity (pH) and soil Al^{3+} toxicity measures are used to predict soil acidity's impact on agricultural production (Anderson and Bell, 2019). There are several methods for measuring soil pH and Al. In Australia, relationships with agricultural productivity responses are defined using 0.01 M $CaCl_2$ (pH_{CaCl_2} and Al_{CaCl_2}) with a soil:solution ratio of 1:5. Exchangeable Al (Al_{Ex}) is also used to measure soil Al, with calibration relationships used to convert these measurements to Al_{CaCl_2} (Anderson et al., 2020b). A KCl solution and different concentrations of $CaCl_2$ are also utilized for measuring pH and Al^{3+} content in New Zealand (Bouray et al., 2022). The pH_{CaCl_2}, Al_{CaCl_2}, and pH_{BC} measurements are combined with estimated acidification rates in models for making lime recommendations to growers (Kalkhoran et al., 2018).

As measured by pH_{CaCl_2}, soil acidity describes the general limitation imposed because it includes an implied lower availability of some nutrients (phosphorus and molybdenum) and interaction with soil-borne diseases (Tang et al., 2003; Bolan et al., 2003a; Anderson et al., 2020a). Al^{3+} toxicity, as measured by Al_{CaCl_2}, is the soil factor restricting root growth of sensitive plants in soils with pH_{CaCl_2} less than 5.5 (Tang et al., 2003; Anderson and Bell, 2019). Low pH directly reduces root growth in solution studies with barley and canola, which is reduced further when Al^{3+} is added (Du et al., 2022, Szurman-Zubrzycka et al., 2021). As soil pH_{CaCl_2} declines below 5.5, there is an exponential increase in Al_{CaCl_2} (Anderson and Bell, 2019). Across a range of soil types, pH_{CaCl_2} is poorly correlated to Al_{CaCl_2}, resulting in Al_{CaCl_2} providing a more accurate measurement of the wheat response to a lime application than pH_{CaCl_2} (Anderson and Bell, 2019). In evaluating the performance of soil test measurements to predict lime response, it is essential to use sites where the lime treatment had removed the subsoil Al^{3+} toxicity limitation and under relatively consistent seasonal rainfall. Wheat response to the lime application is greater under seasonal conditions characterized by high summer and low growing season rainfall than near-average rainfall due to the crop being more dependent on subsoil water supply (Anderson and Bell, 2019; Oliver et al., 2021). In addition, wheat response to lime is lower when insufficient lime has moved into the subsoil to ameliorate subsoil Al^{3+} toxicity (Anderson and Bell, 2019).

Soil Al_{CaCl_2} critical value (90% of maximum yield) for wheat ranges between 2.5 and 4.5 mg Al/kg depending on the method used to measure the Al_{CaCl_2} content of the 0–30 cm soil layer, measured in increments of 10 cm (Anderson and Bell, 2019). Lower critical values were defined using a method which places greater emphasis on measuring topsoil Al_{CaCl_2} (weighted root distribution) than methods which emphasize subsoil Al_{CaCl_2} (layer maximin). When Al_{CaCl_2} is above this critical value, there is reduced wheat biomass and grain yield due to reduced root growth, which restricts the ability of plants to take up water and nutrients. In addition, when subsoil pH_{CaCl_2} is less than 4.5, lime application to the soil surface results in a limited change in subsoil pH_{CaCl_2} compared to large changes in Al_{CaCl_2} (Anderson et al., 2020a; Lauricella et al., 2021). Hence, Al_{CaCl_2} provides better measurements of grain yield response to the application of lime than pH_{CaCl_2} and is the preferred measurement when assessing the impact of lime strategies on soil properties (Anderson and Bell, 2019; Anderson et al., 2020b; Lauricella et al., 2021).

The Al_{CaCl_2} indicates the soluble Al^{3+} content of the soil (Anderson et al., 2021). However, Al occurs in the soil in a range of chemical species, some of which are toxic to root growth, and others are not toxic (Anderson et al., 2021, Bouray et al., 2022). The most plant toxic Al species in soils is $Al(H_2O)_6^{3+}$ (Al^{3+}) followed by $AlOH^{2+}$. As the pH declines below 5.0, Al^{3+} becomes the more dominant Al species. Across a wide range of pH, $Al(H_2O)_6^{3+}$ undergoes hydrolysis, which breaks the H–O bond leading to the release of hydrogen ions and the formation of the amorphous Al hydroxide species $AlOH^{2+}$, $Al(OH)_2^{+}$, $Al(OH)_3$, and $Al(OH)_4^{-}$. In addition, Al forms complexes with soil solution sulfate, fluoride, phosphate, and dissolved organic carbon (DOC_{Soln}), which are not toxic to plants. Chemical equilibrium models (MINTEQ and PHREEQC) are used to understand the treatment effects of soil amendments on Al species using soil solution measurements (Anderson et al., 2021; Bouray et al., 2022).

SOC is a critical soil property which determines the Al species distribution between soil solution Al^{3+} (Al_{Soln}) and Org-Al_{Soln}. SOC content is highly variable across south Western Australia, driven mainly by rainfall, which determines net primary productivity and temperature (Harries et al., 2021). Most of the SOC is contained in the topsoil, with content declining with increasing soil depth (Harries et al., 2015). The southern cooler and higher rainfall areas generally have the highest SOC content, which decreases as the distance from the coast increases due to lower rain, and the northern regions have lower rainfall and higher temperature. For example, SOC is greater than 1.6% in the topsoil in rainfall zones, which receive more than 278 mm of annual rainfall, while areas with a daily mean temperature greater than 16.7°C are 13 times more likely to have SOC less than 1.6% (Harries et al., 2021).

Soil solution measures of DOC_{Soln}, Al_{Soln}, and pH_{Soln} over the narrow pH_{Soln} range of 4.0–5.5 are used to calculate soil Org-Al_{Soln} concentration (Anderson et al., 2021). For example, the topsoil, with SOC=0.8%, pH_{CaCl_2}=4.6, and DOC_{Soln}=26 mg /L, has 45% of the Al_{Soln} associated with Org-Al_{Soln}. Even though a significant amount of Al_{Soln} was bound by the DOC_{Soln}, the remaining Al_{Soln} was sufficient and resulted in a 14% wheat grain yield increase in the year when lime was applied (Anderson et al., 2020a). In the higher rainfall zone, which has greater SOC, the role of DOC_{Soln} in bonding with Al_{Soln} is expected to be higher, and the grain yield increase associated with Al toxicity in the 0–10 cm will be less.

20.4 LIME AND GYPSUM SOURCES

The south Western Australian agricultural region has coastal deposits of limesand and limestone (Gazey and Davies, 2009) and inland deposits of gypsum. Most of these limesand deposits are in the northern section of the region. The limesand deposits are mainly calcium carbonate ($CaCO_3$) with low amounts of magnesium carbonate ($MgCO_3$) formed by wind-blown shells and coral fragments from marine organisms to form coastal dunes over a very long period. The limesand has a fine particle size and only requires sieving large rocks and contaminants before it is ready for use by growers.

The region has inland deposits of dolomite consisting of calcium magnesium carbonates. The limestone deposits are located in the south coastal areas and are 1–2 million years old. The limestone deposits are formed by the cementation of limesand deposits due to the action of rainfall or derived from submerged times during periods of high sea level. Both the dolomite and limestone require crushing to reduce the particle size distribution before being used by growers.

The eastern wheatbelt has significant, heavy loam soil types, which are pink to brown and contain carbonate nodules with soil pH_{CaCl_2} between 7.0 and 9.0. Due to the high cost of lime in the eastern wheatbelt, growers have been looking for cost-effective alternative sources of lime. Recent research has focused on identifying on-farm lime sources and their use by growers (Cribb, 2019). The cost of the on-farm lime is higher due to additional mining and sieving costs. Also, on-farm lime sources have lower lime quality than limesand. However, low transport costs of on-farm lime enable growers to use high rates to counter the lower lime quality compared to limesand. On-farm lime sources can also contain significant amounts of potassium, which is beneficial when applied to acidic sands deficient in potassium.

Lime quality varies according to $CaCO_3$ percentage and particle size distribution (Gazey and Davies, 2009; Conyers et al., 2020; Conyers et al., 2022). Lime source purity is expressed as the $CaCO_3$ content compared to pure $CaCO_3$ or calcium carbonate equivalent (CCE). The CCE is measured on five sieved fractions (0–0.125 mm, 0.125–0.25 mm, 0.25–0.50, 0.50–1.00, and > 1.00 mm) multiplied by the average diameter of the sieved fractions to obtain the number of particles within the fraction (Gazey and Davies 2009). The number of particles in each fraction is then multiplied by the surface area and neutralizing values per cent to calculate the fraction's total surface area of $CaCO_3$. The lime quality is the total (sum of the individual fractions) surface area of $CaCO_3$. In Western Australia, coastal limesand has a better quality due to higher $CaCO_3$ content and finer particle size distribution than limestone and dolomite. The cost of purchasing lime is a relatively small component of the overall cost of using lime (Kalkhoran et al., 2018), but limesand has a cost advantage because limestone, dolomite, and on-farm lime have additional processing costs (sieving and grinding). Transport costs are the main cost of growers using lime. Hence lime use is higher in areas within 200 km of the lime deposits.

Gypsum deposits are associated with, and are a direct product of, evaporation from the salt lakes in the eastern zone of the south Western Agricultural region. Growers use gypsum as a source of sulfur, especially when growing canola, improving the soil structure of clay soil, and applying it for treating subsoil acidity in the low rainfall zones. Phosphogypsum, a by-product of the phosphoric acid industry, is used in New Zealand (Bouray et al., 2022), but it is no longer produced in south Western Australia (Easton, personal communication). In general, gypsum is more expensive than lime. However, because the gypsum deposits are located in the eastern region, the transport costs are less, making it cheaper to apply gypsum than lime.

20.5 MANAGEMENT STRATEGIES

Lime application is used worldwide to treat top and subsoil acidity (Gazey and Davies 2009, Li et al., 2019; Rheinheimer et al., 2018; Holland et al., 2021; Butterly et al., 2022). Lime application effectively manages topsoil acidity (Oliver et al., 2021). However, the surface lime application is ineffective at treating subsoil acidity due to the low solubility of lime and the low capacity of the soil to dissolve lime under conservation cropping and pasture systems (Anderson and Bell, 2019; Oliver et al., 2021). The acidification process results in a bulge of acidity in the 10–30 cm soil layer (Dolling and Porter, 1994). Hence, liming strategies need sufficient alkalinity leaching to prevent subsoil acidification. An effective surface lime strategy is applying lime before subsoil toxicity has developed at a rate which provides alkalinity movement to prevent subsoil acidity from developing (Conyers et al., 2003; Tang et al., 2003). For soils with subsoil Al^{3+} toxicity in the 10–30 cm layer or inherently acidic soil profiles, the surface lime application can become ineffective in treating subsoil Al^{3+} toxicity requiring a multi-pronged management approach. This may include growing

Al^{3+} tolerant plant species, applying sufficient lime to maintain topsoil pH_{CaCl_2} between 5.5 and 7.5, and using strategic tillage to increase lime dissolution and redistribution into the subsoil (Tang et al., 2003; Azam and Gazey, 2021).

20.5.1 Plant Al^{3+} Tolerance

Agricultural plant species have genetic variation in Al^{3+} tolerance, and growing Al^{3+} tolerant plants on Al^{3+} toxic soils is an important management strategy (Ryan, 2018). There are two main groups of tolerance mechanisms in plants. The first mechanism involves the exclusion of Al^{3+} from the root tips (apices), while the second mechanism involves the tolerance to Al^{3+} taken up by the root tips. Al^{3+} exclusion occurs due to the release of carboxylic anions (malate, citrate, and, to a lesser extent, oxalate) in the presence of Al^{3+} by root tips which bind Al^{3+} in the apoplast to reduce its toxicity effect on the root tip. Al^{3+} exclusion tolerance occurs due to the binding of Al^{3+} to the cell wall matrix, because of the activity of methylation of pectin residues or the O-acetylation of xyloglucan chains when plants take up Al^{3+}. The Al-activated malate transporter (ALMT) gene family controls the release of malate anions from wheat roots. At the same time, the multi-drug and toxic compound exudation (MATE) gene family encodes a proton co-transporter to release citrate anions from sorghum and barley roots (Ryan, 2018).

The current research focus is on enhancing the tolerance of Al-sensitive species, for example, durum wheat (*Triticum turgidum*) (Kawasaki et al., 2022), barley (*Hordeum vulgare* L.) (Szurman-Zubrzycka et al., 2021), chickpeas (*Cicer arietinum* L.) (Vance et al., 2021), and canola (*Brassica napus*) (Du et al., 2022). These studies have illustrated genetic variation for Al^{3+} resistance, which can be used for future breeding purposes to improve Al^{3+} tolerance. In addition, for field peas (*Pisum sativum*) and vetch (*Vicia* spp.), a new rhizobia strain (WSM4643) has been identified, which will enhance the robustness of the legume/rhizobia symbiosis in low pH soils compared to the SU303 strain (Yates et al., 2021). Further details of these mechanisms, the genes responsible for producing the response, and the use of genetic engineering to improve Al^{3+} tolerance in the future are discussed by Ryan (2018).

20.5.2 Farming System

Conservation farming systems or no-tillage and crop residue retention are used to minimize soil disturbance and maximize surface crop residue retention to prevent wind erosion and SOC decline (Anderson, 2009). The system also reduces machinery costs, improves the timeliness of operations, lowers other production costs, and is used worldwide (Llewellyn et al., 2012; Aikins et al., 2020). A no-tillage farming seeding system consists of components that handle residue, prepare a seeding row, open a furrow, place seed and fertilizer at control depths, and firm or cover the seed (Aikins et al., 2020).

A limitation of the low soil disturbance in no-tillage farming systems is a reduction in the effectiveness of surface-applied lime (Conyers et al., 2003; Anderson, 2009). The same issue would apply to pasture systems. In no-tillage farming systems, it is possible to adjust the amount of soil disturbance based on furrow openers' geometry, their setting, and interactions with soil properties and crop residue management (Aikins et al., 2020). As a result, the level of soil disturbance will range from very little associated with disc openers and likely to low lime effectiveness to wide tines with wings with deep seed placement, resulting in greater soil disturbance and potential higher lime effectiveness. Soil water content at seeding will also impact the degree of mixing lime because it is challenging to incorporate two insoluble components (lime and soil) when the main element (soil) is wet.

Seeding of winter crops occurs in south Western Australia between late April to early June following a dry summer. Dry seeding is a standard practice because the winter rains often arrive in late May or early June, growers manage large areas, and significant loss of yield potential is associated with delayed crop establishment (Fletcher et al., 2020). No-tillage seeding systems are used to enable timely sowing, mainly with knife-point furrow openers, which result in sufficient soil disturbance to allow the incorporation of pre-emergent soil-applied herbicides (Roper et al., 2015). Growers also use winged tines to form larger furrows compared to when using knife points to enable water harvesting to improve crop establishment on water-repellent sands under dry seasonal conditions (Roper et al., 2015). The amount of soil disturbance associated with winged tines varies greatly based on the wings' position, width, lift height, depth, and operating speed (Aikins et al., 2020). In addition, using long coleoptile wheat improves crop establishment by seeding deep onto subsoil water (Rebetzke et al., 2007; Zhao et al., 2022). It is hypothesized the combination of dry seeding using winged tines and long coleoptile wheat can potentially increase topsoil disturbance and mixing of lime. However, "no-tillage" seeding systems which result in adequate soil disturbance for lime incorporation are likely only suitable for specific climatic conditions, soil types, and management objectives. For most no-till systems, growers are beginning to use strategic surface tillage to increase lime dissolution and deep strategic tillage to incorporate lime into the subsoil and overcome other soil constraints of water repellence and subsoil compaction (Davies et al., 2019).

20.5.3 Strategic Tillage

Strategic surface tillage is recommended in conservation farming systems to increase the soil's capacity to dissolve lime, resulting in increased lime dissolution and alkalinity movement rates (Conyers et al., 2003; Conyers et al., 2019; Azam and Gazey, 2021). The approach effectively addresses topsoil acidity in the year of lime application (Anderson, 2009) while having no adverse impact on crop productivity and soil health in conservation farming systems (Dang et al., 2018). Similarly, a strategic surface tillage event incorporating 3.1 t lime/ha can provide a long-term (24 years) treatment of subsoil Al^{3+} toxicity in a no-tillage farming system in the humid subtropical climate of southern Brazil with an annual rainfall of 1,769 mm (Rheinheimer et al., 2018). However, strategic surface tillage does not always improve the effectiveness of lime application as measured by wheat grain yield response in the Mediterranean climate of south Western Australia, with an annual rainfall of 288 mm (Reynolds et al., 2021). This suggests that a no-tillage farming system where dry seeding is combined with the use of winged tines for lime incorporation and furrow formation for water harvesting in low rainfall regions is a topic for further research.

Strategic deep tillage benefits are grouped into overcoming physical, chemical, and biological constraints (Ning et al., 2022). A single strategic deep tillage event can overcome the multiple limitations of soil compaction, water repellence, and acidity while improving nutrient availability by transferring lime-treated topsoil into the 10–40 cm soil layer (Scanlan and Davies, 2019; Davies et al., 2019). The amount of soil transfer in a deep strategic tillage event depends on the equipment used. For example, only 5%–10% of topsoil is incorporated into the subsoil to a maximum depth of 15 cm when narrow straight-tyned deep rippers are used to remove soil compaction (Davies et al., 2019). Wings fitted to a straight tyne deep ripper increase subsoil mixing (Ning et al., 2022). Providing a shank attachment to the ripping tyne can deliver lime at selected depths to make a lime slot (Blumenschein et al., 2019; de Campos et al., 2022). The positioning of this lime slot below the seeding row increases crop response compared to the interrow placement of the lime slot (Damon et al., 2022). The use of shank attachments to deep ripping tynes for lime placement into the subsoil is a current research topic.

The use of parabolic, wider, or angled chisel-point tynes can delve subsoil to the soil surface and mix it into the topsoil layer (Spoor, 2006). Also, additional harrows and discs are used to provide strategic surface tillage when conducting the deep ripping operation. Rotary spaders can mix 50%–60% of the topsoil into the subsoil to a maximum depth of 30 cm, with slower speeds resulting in more uniform mixing than higher speeds (Ucgul et al., 2018). Soil inversion using a mouldboard plough transfers 80–96% into the subsoil to a maximum depth of 40 cm (Ucgul et al., 2017). Other equipment includes large offset discs and modified one-way ploughs that transfer 60%–75% of the topsoil to the subsoil (Davies et al., 2019). However, in terms of the efficiency of lime dissolution, these incorporation techniques do not result in uniform mixing with spading resulting in the formation of pockets of topsoil, while the mouldboard plough results in bands of topsoil within the 0–40 cm soil layer (Scanlan and Davies, 2019). The cost associated with increased mixing range from $45/ha for shallow ripping to $150 for spading and mouldboard ploughing (Davies et al., 2019).

20.5.4 Lime

Subsoil Al^{3+} toxicity is treated by surface application of lime at a sufficient rate and frequency to maintain topsoil pH_{CaCl_2} between 5.5 and 7.5, which provides a pool of excess alkalinity which is leached into the subsoil (Li et al., 2019). The soil's capacity to dissolve lime limits the effectiveness of this approach because lime stops dissolving when the soil pH_{CaCl_2} surrounding the dissolving lime particle equals 7.5 (Nye and Ameloko, 1987). Hence, the target pH_{CaCl_2} range for the topsoil layer is 5.5–7.5. The approach's effectiveness requires maximizing the pool of excess alkalinity in the surface soil layers by mixing the lime into a greater soil volume by surface cultivation (Azam and Gazey, 2021).

Soil testing practices aim to develop a site-specific lime management strategy that growers can use to achieve to apply a recommended lime rate and frequency to maintain pH_{CaCl_2} within this range based on the soil's pH_{BC} and acidification rates (Xu et al., 2022). To develop targeted zones for lime application, growers can undertake paddock scale surveys of site-specific measurements (Hayes et al., 2022). Measurement of soil properties, soil core diameter, site selection, soil increment sampling depth, and the number of samples collected all impact achieving a 10% level of precision. For soil pH measurements, there is an increased focus on measuring pH stratification by dividing the 0–10 cm layer into 0–5 cm and 5–10 cm increments (Condon et al., 2021). Also, acceptable precision is obtained using a 19 mm coring tube and collecting 30–40 composite cores or a 25 mm coring tube and collecting 20–30 cores (Hayes et al., 2022). Inexpensive colourimetric plater reader measurements of pH, Al, and Colwell phosphorus are used to reduce the laboratory cost of measuring these soil properties (Weligama et al., 2022).

New technologies can produce 3D-like digital soil pH maps to accurately define subsoil acidity zones within a paddock (Malone et al., 2022). The cost of developing 3D-like digital maps is relatively high, and cheaper alternatives are available for mapping topsoil, pH, SOC (OpticMapper, red LED, wavelength: 660 nm; NIR LED, wavelength: 940 nm), and texture (ECmapper) using on-the-go equipment (Vogel et al., 2022). Frequent soil pH measurements using on-the-go equipment can effectively define a variable lime rate map. The OpticMapper readings improved the lime rate predictions because they are used to estimate base neutralizing capacity or pH_{BC}. These paddock maps can then be combined with land-use practices (produce removal, fertilizer input rates, and crop rotations) to upscale lime requirement paddock maps to a farm and regional scale (Xu et al., 2022).

Equation 20.1 describes the soil's lime dissolution capacity (LDC, t pure $CaCO_3$/ha), as determined by soil and management factors (Anderson, 2009; Xu et al., 2022). The soil factors include initial pH_{CaCl_2}, pH_{BC}, bulk density (BD in g soil/cm^3), and a conversion factor (CF=0.05) to report the values calculated in the units of t pure $CaCO_3$/ha. The management factors include cultivation depth (CD, in cm), mixing efficiency (MF), and soil redistribution efficiency (RF). Strategic surface and deep tillage for lime incorporation in conservation farming systems have become more widespread (Conyers et al., 2019; Davies et al., 2019). Hence, mixing and redistribution efficiencies are recent additions to the equation. The equation uses the law of the minimum approach to combine index values (MF and RF) to determine the overall impact of management practice on lime dissolution (van der Ploeg et al., 1999). At the same time, environmental (rainfall frequency and amount) and lime quality factors are additional factors which will impact lime dissolution and movement.

$$LDC = \left(7.5 - pH_{CaCl_2}\right) \times pH_{BC} \times BD \times CD \times \min\left(MF \times RF\right) \times CF \qquad (20.1)$$

In Equation 20.1, the SOC determines the topsoil's pH_{BC} ($pH_{BC} = 0.73 + (0.66 \times SOC\%), r^2 = 0.89$; Wong et al., 2013) of the topsoil, while the subsoil's pH_{BC} is determined by the SOC and exchangeable Al content ($pH_{BC} = 0.42 + (0.73 \times SOC\%) + (0.40 \times Al_{Ex}), r^2 = 0.82$, where Al_{Ex} is in the units of cmol/kg; Moore et al., 2001).

Equation 20.1 illustrates that when lime is applied to the soil surface, the soil will have a low capacity to dissolve lime due to the concentration of lime in a thin soil layer at the soil surface (Azam and Gazey, 2021). Mixing lime into the soil to a greater depth using a technique with a high mixing efficiency will increase the soil's capacity to dissolve lime. For example, rotary hoeing of historical lime applications of up to 8.5 t lime/ha to a depth of 25 cm increased soil pH_{CaCl_2} within that soil layer by 0.63 (Azam and Gazey, 2021). The mixing efficiency will differ according to the cultivation technique used and the water content at the time of the mixing event. In the case of Azam and Gazey (2021), the mixing efficiency of the April cultivation event would be relatively high due to the soil being dry, with only a significant rainfall event of 44 mm in January 2018. Surface tillage (ploughing and harrowing) after 3.1 t lime/ha application to raise topsoil pH_{H_2O} to 6.0, followed by long-term (24 years) use of no-tillage with the additional lime application, was effective in managing subsoil acidity and Al^{3+} toxicity (Rheinheimer et al., 2018).

20.5.5 Gypsum

Gypsum is more soluble than lime, with the dissolved calcium and sulfate leaching readily into the subsoil (Anderson et al., 2020b; 2021). The gypsum rate required to overcome subsoil Al^{3+} toxicity depends on the soil's sulfate adsorption properties (Sumner, 1990). The rate to apply is calculated as the amount of sulfate to maintain a level of 10 mg S/mL in the soil solution (Zoca and Penn, 2017) using a defined sulfate adsorption curve (Anderson, 2020). Gypsum's impact on subsoil Al^{3+} toxicity is due to its effect on soil solution ionic strength (IS_{Soln}), soil pH_{Soln}, and Al species distribution. Gypsum application can have an impact on pH_{Soln} due to two possible reactions. First, the increase in soil solution sulfate (S_{Soln}) results in greater sulfate sorption, the displacement of hydroxide ions, and the associated increase in pH_{Soln} (Curtin and Syers, 1990; Wang et al., 1999a, b). Second, pH_{Soln} can decrease due to the increase in soil solution ionic strength (IS_{Soln}) and calcium (Ca_{Soln}), resulting in the release of hydrogen ions and the adsorption of hydroxide ions (Uehara and Gillman 1981). The balance between these two reactions determines the net effect on pH_{Soln}. For soils with a low capacity to adsorb sulfate, gypsum application did not affect pH_{CaCl_2} or pH_{Soln} (Anderson et al., 2020b; 2021).

Impacts of gypsum and gypsum plus lime applications on Al species distribution can be identified using soil solution measurements and chemical equilibrium models such as MINTEQ because Al–inorganic reactions are well defined (Gustafsson et al., 2018). Soil pH_{Soln}, chemical equilibrium reactions in the soil solution, and surface dissolution reactions of the soil minerals determine the Al speciation distribution (Gustafsson et al., 2018). Gypsum application increases (Ca_{Soln}), resulting in the displacement of Al^{3+} from the exchange sites (Wang et al., 1999a; Anderson et al., 2021). However, there is no change in Al_{Soln} soil because the displaced exchangeable Al is removed from the soil solution due to the formation of aluminum sulfate (Anderson et al., 2021). Hence, gypsum application increases IS_{Soln} due to the higher S_{Soln} and Ca_{Soln} concentrations, diluting the Al^{3+} toxic effect on root growth and increasing wheat grain yield (Anderson et al., 2020a; 2021; Wang et al., 1999b). However, over time S_{Soln} is leached from the soil profile, reducing IS_{Soln} and the effectiveness of gypsum in treating subsoil Al^{3+} toxicity (Anderson et al., 2021). When phosphogypsum is used, the formation of Al fluoride is an important mechanism for reducing Al^{3+} toxicity (Bouray et al., 2022).

20.5.6 Lime and Gypsum

Applying gypsum or phosphogypsum with lime can increase alkalinity movement from the topsoil to the subsoil (Anderson et al., 2020a, b; 2021, Lauricella et al., 2021; Bouray et al., 2022). Also, manures can be applied with lime and gypsum to increase alkalinity movement into the subsoil (Lauricella et al., 2021).

Lime application with gypsum reduces the complexity of the chemical reaction because alkalinity leaches from the topsoil to the subsoil to increase pH_{CaCl_2} and reduce Al_{Soln} and Al_{CaCl_2} (Anderson et al., 2020b; 2021). In addition, there was sufficient leaching alkalinity to remove the effect of Ca_{Soln} displacing Al^{3+} from the exchange site to increase Al_{Soln} (Anderson et al., 2021). Interestingly, combining gypsum application with 5.0 t lime/ha had a greater effect on Al_{CaCl_2} in the soil layers below 30 cm than when only lime was applied in the medium term (2 years 8 months). McLay et al. (1994) also observed that the combined application of gypsum and lime had a greater effect on subsoil Al_{CaCl_2} than only lime. However, the soil responses observed are inconsistent between experimental sites. More recently, the application of lime with gypsum and organic manures has shown the potential to increase alkalinity movement when Al_{CaCl_2} is used to assess the impact of the treatments (Lauricella et al., 2021). In contrast, combining lime with other soil amendments had minimal effect on subsoil pH_{CaCl_2}. This study highlights the importance of measuring Al_{CaCl_2} to assess the impact of lime and amendment treatments because a slight difference in pH_{CaCl_2} results in a considerable shift in Al_{CaCl_2}. Small increases in pH_{CaCl_2} below 4.5 are associated with significant changes in Al_{CaCl_2} due to the exponential relationship between pH_{CaCl_2} and Al_{CaCl_2} (Anderson et al., 2020b).

20.6 ALKALINITY MOVEMENT

Experiments report a range of alkalinity movement from no leaching to surface applied lime impacting soil pH to a depth of 55 cm (Nunes et al., 2019). Many factors affect alkalinity movement (Rheinheimer et al., 2018; Silva et al., 2022). These include lime properties; lime application rate, frequency, and quality; cultivation system; time; the amount and frequency of rainfall events; and the anion adsorption soil properties as influenced by soil texture and soil mineralogy. Rheinheimer et al. (2018) suggest two mechanisms for alkalinity movement in no-tillage farming systems. First, maintaining a straw cover of no-tillage systems improves the soil's physical and biological characteristics. Lime movement then occurs in the biopores formed by dead roots and macro soil fauna. Second, low molecular weight organic acids form when plant residue decomposition binds with Ca and Mg, which are leached to treat subsoil Al^{3+} toxicity. However, the small amount of excess alkalinity in the topsoils, bicarbonate adsorption, and the conversion to carbonic acid also restrict alkalinity movement.

Inorganic carbon occurs in soil as the dissolved species (carbon dioxide, carbonic acid, bicarbonate, and carbonate) and minerals (calcium carbonate and dolomite). Carbon dioxide reacts with water to form carbonic acid (H_2CO_3). The carbon dioxide concentration depends on the partial pressure of the atmosphere (412 ppm) and soil and plant root respiration (Bolan et al., 1991; Bolan et al., 2005; Bargrizan et al., 2020). Carbonic acid dissociates into bicarbonate and carbonate depending on soil pH. As the pH increases from 4.0 to 8.0, bicarbonate becomes more prevalent, and the carbonic acid fraction declines to zero. In alkaline soils, carbonate becomes the dominant species. Carbon dioxide reaction and the formation of carbonic acid and bicarbonate and carbonate ions in soils are described by Henry's law constant for CO_2 (K_H), the first and second dissociation constants of carbonic acid (K_1 and K_2), and water's self dissociation equilibrium constant (K_w) (Bargrizan et al., 2020). In acidic soils, carbonic acid is the dominant species. Based on these reactions, applying lime to increase pH_{CaCl_2} between 5.5 and 7.5 results in the formation of bicarbonate which leaches into the soil profile.

The concentration and distribution of inorganic carbon species require measurement of total alkalinity, pH, pCO_2, and dissolved inorganic carbon (DIC) (Bargrizan et al., 2020). Using at least two of these components enables a chemical equilibrium model (MINTEQ) to calculate the other components (Gustafsson, 2019). For example, using pH and alkalinity measurements will calculate the carbon dioxide partial pressure and carbon species distribution. Including DOC in the calculation has a relatively small effect on carbon species because organic matter does not strongly bind calcium.

Published lime experiments rarely discuss soil properties that impact the bicarbonate leaching process. In contrast, nitrate and sulfate leaching depends on the concentration in the soil solution, rainfall amount and intensity and the water-holding and anion adsorption properties of the soil profile (Anderson et al., 1998b; Anderson et al., 2006). Phosphate leaching is less common due to the high adsorption by soils. The exception is for soils with a low phosphorus buffer index (Summers and Weaver, 2022). In addition, anions are leached with an associated cation and the chemical the cation is associated with, impacting the rate of leaching of the anion form (Ritchey et al., 1980). For example, the highest rate of Ca^{2+} leaching occurs when applied as $CaCl_2$, followed by $CaSO_4$ and $CaCO_3$. The lower leaching rate of Ca^{2+} applied as $CaSO_4$ compared to $CaCl_2$ is due to sulfate adsorption by the soil. In comparison, there is limited Ca^{2+} leaching when using $CaCO_3$ due to the low solubility and restricted dissolution.

The soil's adsorption properties significantly impact the leaching of anions because they control the concentration in the soil solution. Anion adsorption capacity is related to the soil sulfate status of the soil (Wong and Wittwer, 2009). Soil sulfate adsorption is measured by the procedure of Anderson (2020), while the Burkitt et al. (2002) procedure calculates phosphate adsorption. Anions compete for adsorption sites. For one soil, phosphate adsorption was less due to bicarbonate and silicate, but it was the reverse for another soil (De and Datta, 2007). Bicarbonate solutions are used to measure plant-available phosphate, indicating soils more strongly adsorb bicarbonate than phosphate (Nagarajah et al., 1968). Also, phosphate solutions are widely used as soil testing solutions because soils adsorb phosphate more strongly than sulfate (Blair et al., 1991). Phosphate adsorption increases as pH decreases (De and Datta, 2007; Barrow et al., 2022). Maximum sorption of weak acid anions occurs when the soil pH equals the pK value of that weak acid (Hingston et al., 1968). Silicate affected phosphate adsorption most in soils with pH_{H_2O} between 7.2 and 7.8, and bicarbonate reduced phosphate adsorption in soil with a pH_{H_2O} of 5.4 (De and Datta, 2008). However, a silicate system was more effective in lowering P sorption, particularly at lower pH values, than the bicarbonate system (De and Datta, 2007). Understanding the soil properties which determine the amount of anion adsorption and the associated interactions will identify rates of alkalinity movement when lime is applied with gypsum and organic materials and the need for strategic tillage.

20.7 CONCLUSIONS

Subsoil acidity with toxic levels of soil Al^{3+} is a common problem limiting agricultural production worldwide. The presence of Al^{3+} in the soil reduces root growth and restricts plants' ability to take up water and nutrients resulting in lower biomass and grain production. Measuring soil Al_{CaCl_2} more accurately predicts wheat grain response and changes in soil properties to applied lime than pH_{CaCl_2} because Al_{CaCl_2} increases exponentially as pH_{CaCl_2} declines below 5.5, and Al_{CaCl_2} is poorly related to pH_{CaCl_2}. The surface application of lime manages topsoil acidity. However, subsoil Al^{3+} toxicity is difficult to address due to lime's low solubility and the slow rates of alkalinity movement from the topsoil to the subsoil. Gypsum is more soluble than lime, resulting in rapid leaching of sulfate and calcium into the subsoil to reduce the impact of subsoil Al^{3+} toxicity. Using the combined application of lime and gypsum can effectively manage subsoil Al^{3+} toxicity in the short term, but its effectiveness declines over time due to sulfate leaching. Strategic tillage (surface and deep) increases lime effectiveness in no-tillage and pasture systems. Surface strategic tillage can increase alkalinity movement by increasing the capacity of the surface soil layer to dissolve lime. Deep strategic tillage transfers lime-treated surface soil into the subsoil while overcoming the constraints of compaction and topsoil water repellence. A no-tillage seeding system remains the primary seeding system used by growers, but some "no-till" seeders can significantly disturb the topsoil. Hence, research opportunities exist to examine seeding systems which result in sufficient lime incorporation and do not require strategic surface tillage. At the same time, it may be possible to increase alkalinity movement by applying lime with gypsum and organic materials based on an improved understanding of the chemistry and mechanisms involved.

REFERENCES

Aikins, K.A., Barr, J.B., Mustafa Ucgul, M., Jensen, T.A., Antille, D.L., Desbiolles, J.M.A. 2020. No-tillage furrow opener performance: a review of tool geometry, settings and interactions with soil and crop residue. *Soil Res.* 58: 603–621.

Anderson, G.C. 2009. The impact of tillage practices and crop residue (stubble) retention in the cropping system of Western Australia. Department of Agriculture and Food, Bulletin Number, 4786. ISSN: 1833-7236.

Anderson, G.C. 2020. Sulfate sorption as measured by a buffering index over a range of soil properties from south Western Australia. *Soil Res.* 58: 651–661.

Anderson, G.C., Bell, R.W. 2019. Wheat grain-yield response to lime application: relationships with soil pH and aluminium in Western Australia. *Crop Pasture Sci.* 70: 295–305.

Anderson, G.C., Fillery, I.R.P., Dolling, P.J., Asseng, S. 1998a. Nitrogen and water flows under pasture-wheat and lupin-wheat rotations in deep sands in Western Australia. 1. Nitrogen fixation in legumes, net mineralisation, and utilisation of soil-derived nitrogen. *Aust. J. Agric. Res.* 49: 329–343.

Anderson, G.C., Fillery, I.R.P., Dolling, P.J., Asseng, S. 1998b. Nitrogen and water flows under pasture-wheat and lupin-wheat rotations in deep sands in Western Australia. 2. Drainage and nitrate leaching. *Aust. J. Agric. Res.* 49: 345–361.

Anderson, G.C., Fillery, I.R.P, Ripper, F.H., Leach, B.J. 2006. Sulfur mineralisation in a coarse-textured soil after different sulfate fertilisation histories and yield responses of wheat and lupin. *Aust. J. Soil Res.* 44: 165–174.

Anderson, G.C., Pathan, S, Easton, J, Hall, D.J.M., Sharma, R. 2020a. Short- and long-term effects of lime and gypsum application on acid soils in a water-limited environment: 1. Grain yield response and nutrient concentration. *Agronomy* 10: 1213.

Anderson, G.C., Pathan, S, Easton, J., Hall, D.J.M., Sharma, R. 2020b. Short- and long-term effects of lime and gypsum application on acid soils in a water-limited environment: 2. Soil chemical properties. *Agronomy* 10: 1987.

Anderson, G.C., Pathan, S., Easton, J., Hall, D.J.M., Sharma, R. 2021. Short- and long-term effects of lime and gypsum application on acid soils in a water-limited environment: 3. Soil solution chemistry. *Agronomy* 11: 826.

Asseng, S., Fillery, I.R.P., Dunin, F.X., Keating, B.A., Meinke, H. 2001. Potential deep drainage under wheat crops in a Mediterranean climate. 1. Temporal and spatial variability. *Aust. J. Agric. Res.* 52: 45–56.

Azam, G., Gazey, C. 2021. Slow movement of alkali from surface-applied lime warrants the introduction of strategic tillage for rapid amelioration of subsurface acidity in south-western Australia. *Soil Res.* 5v9: 97–106.

Bargrizan, S., Smernik, R.J., Mosley, L.M. 2020. Constraining the carbonate system in soils via testing the internal consistency of pH, pCO_2 and alkalinity measurements. *Geochem. Trans.* 21, Article number: 4

Barrow, N.J., Debnath, A., Sen, A. 2022. Effect of phosphate sorption on soil pH. *Eur. J. Soil Sci.* 73: e13172.

Blair, G.J., Chinoim, N., Lefroy, R.D.B., Anderson, G.C. 1991. A soil sulfur test for pastures and crops. *Aust. J. Soil Res.* 29: 619–626.

Blumenschein, T.G., Nelson, K.A., Motavalli, P.P. 2019. Changes in subsoil acidity with a new deep vertical lime placement practice for conservation tillage systems. *J. Soil Water Conserv.* 74: 426–438.

Bolan, N.S., Adriano, D.C., Curtin, D., 2003a. Soil acidification and liming interactions with nutrient and heavy metal transformation and bioavailability. *Adv. Agron.* 78: 215–272.

Bolan, N.S., Curtin, D., Adriano, D.C. 2005. Acidity. *Ency. Soils Environ.* 11–17.

Bolan, N.S., Hedley, M.J. 2003b. Role of carbon, nitrogen and sulfur cycles in soil acidification. pp. 29–52. In: Z. Rengel (Editor). *Handbook of Soil Acidity.* Marcel Dekker, New York.

Bolan, N.S., Hedley, M.J., White, R.E. 1991. Processes of soil acidification during nitrogen cycling with emphasis on legume based pastures. *Plant Soil* 134: 53–63.

Bouray, M., Moir, J.L., Condron, L.M., Lehto, N.J., Bayad, M., El Gharous, M., El Khalil, M. 2022. Effect of phosphogypsum application on aluminum speciation in acid pasture soils. *J. Soils Sediments* 22: 1959–1975.

Burkitt, L.L., Moody, P.W., Gourley, C.J.P., Hannah, M.C. 2002. A simple phosphorus buffering index of Australian soils. *Aust. J. Soil Res.* 40: 497–513.

Butterly, C.R., Amado, T.J.C., Tang, C. 2022. Soil acidity and acidification. pp. 53–81. In: T.S.d. Oliveira, and R.W. Bell (Editors). *Subsoil Constraints for Crop Production.* Springer Nature, Switzerland.

Condon, J., Burns, H., Li, G. 2021. The extent, significance and amelioration of subsurface acidity in southern New South Wales, Australia. *Soil Res.* 59: 1–11.

Conyers, M., Dang, Y.P., Kirkegaard, J.A. 2019. Strategic tillage within conservation agriculture. pp. 107–115. In: J. Pratley, and J. Kirkegaard (Editors). *Australian Agriculture in 2020: From Conservation to Automation.* Agronomy Australia and Charles Sturt University, Wagga Wagga.

Conyers, M.K., Heenan, D.P., McGhie, W.J., Poile, G.J. 2003. Amelioration of acidity with time by limestone under contrasting tillage. *Soil Tillage Res.* 72: 85–94.

Conyers, M.K., Scott, B.J., Whitten, M.G. 2020. The reaction rate and residual value of particle size fractions of limestone in southern New South Wales. *Crop Pasture Sci.* 71: 368–378.

Conyers, M.K., Scott, B.J., Whitten, M.G., Poile, G.J. 2022. The reaction rate and residual value in southern New South Wales of 12 commercial liming materials from around Australia. *Crop Pasture Sci.* 73: 1056–1069.

Cribb, M. 2019. Locating and assessing on-farm lime sources. Grains Research and Development Corporation, Canberra, Australia. https://grdc.com.au/resources-and-publications/all-publications/publications/2019/locating-and-assessing-on-farm-lime-sources (retrieved 5 Sept. 2022).

Curtin, D., Syers, J.K. 1990. Mechanism of sulphate adsorption by two tropical soils. *J. Soil Sci.* 41: 295–304.

Damon, P.M., Azam, G., Gazey, C., Scanlan, C.A., Rengel, Z. 2022. Response of wheat to the geometry and proximity of lime slotting in an acidic soil profile. *Soil Tillage Res.* 217: 105269.

Dang, Y.P, Balzer, A., Crawford, M., Rincon-Florez, V., Liu, H., Melland, A.R., Antille, D., Kodur, S., Bell, M.J., Whish, J.P.M., Lai, Y., Seymour, N., Carvalhais, L.C., Schenk, P. 2018. Strategic tillage in conservation agricultural systems of north-eastern Australia: why, where, when and how? *Environ. Sci. Pollut. Res.* 25: 1000–1015.

Davies, S.L., Armstrong, R., Macdonald, L., Condon, J., Petersen, E. 2019. Soil constraints: a role for strategic deep tillage. pp. 117–135 In: J. Pratley, and J. Kirkegaard (Editors). *Australian Agriculture in 2020: From Conservation to Automation.* Agronomy Australia and Charles Sturt University, Wagga Wagga.

de Campos, M., Penn, C.J., Gonzalez, J.M., Crusciol, C.A.C. 2022. Effectiveness of deep lime placement and tillage systems on aluminum fractions and soil chemical attributes in sugarcane cultivation. *Geoderma* 407: 115545.

De, N., Datta, S.C. 2007. Relationship between phosphorus sorption and soil acidity as affected by bicarbonate and silicate ions. *Commun. Soil Sci. Plant Anal.* 38: 679–694.

De, N., Datta, S.C. 2008. Modeling time-dependent phosphate buffering capacity in different soils as affected by bicarbonate and silicate ions. *Aust. J. Soil Res.* 46: 315–322.

Dolling, P.J., Porter, W.M. 1994. Acidification rates in the central wheatbelt of Western Australia. 1. On a deep yellow sand. *Aust. J. Exp. Agric.* 34: 1155–1164.

Dudev, T., Lim, C. 2004, Oxyanion selectivity in sulfate and molybdate transport proteins: an ab Initio/CDM Study. *J. Am. Chem. Soc.* 126: 10296–10305.

Du, H., Raman, H., Kawasaki, A., Perera, G., Diffey, S., Snowdon, R., Raman, R., Ryan, P.R. 2022. A genome-wide association study (GWAS) identifies multiple loci linked with the natural variation for Al^{3+} resistance in Brassica napus. *Funct. Plant Biol.* doi:10.1071/FP22073.

Fitzpatrick, K.L., Tyerman, S.D., Kaiser, B.N., 2008. Molybdate transport through the plant sulfate transporter SHST1. *FEBS Lett.* 582: 1508–1513.

Fletcher, A.L., Chen, C., Ota, N., Lawes, R.L., Oliver, Y.M., 2020, Has historic climate change affected the spatial distribution of water-limited wheat yield across Western Australia? *Clim. Change* 159: 347–364.

Gazey, C., Andrew, J., Griffin, E. 2013. Soil acidity. pp. 26–43. In: N. Schoknecht, D. Bicknell, J. Ruprecht, F. Smith and A. Massenbauer (Editors). *Report Card on Sustainable Natural Resource Use in Agriculture.* Department of Agriculture and Food, Perth. Available online: https://www.agric.wa.gov.au/report-card-conditions-and-trends/report-card-sustainable-natural-resource-use-agriculture-western (retrieved 10 Aug. 2022).

Gazey, C., Davies, S. 2009. Soil acidity: A guide for WA farmers and consultants. Department of Agriculture and Food, Perth, Western Australia, Bulletin 4784, Available online: https://researchlibrary.agric.wa.gov.au/cgi/viewcontent.cgi?article=1031&context=bulletins (retrieved 10 Aug. 2022).

Gustafsson, J.P. 2019. Visual MINTEQ 3.1 User Guide. 2019. Available online: https://vminteq.com/ (retrieved 22 Sept. 2022).

Gustafsson, J.P., Belyazid, S., McGivney, E., Löfgren, S. 2018. Aluminium and base cation chemistry in dynamic acidification models–Need for a reappraisal? *Soil* 4: 237–250.

Harries, M., Anderson, G.C., Hüberli, D. 2015. Crop sequences in Western Australia: what are they and are they sustainable? Findings of a four-year survey. *Crop Pasture Sci.* 66: 634–647.

Harries, M., Flower, K.C., Renton, M., Anderson, G.C. 2022. Water use efficiency in Western Australian cropping systems. *Crop Pasture Sci.* doi:10.1071/CP21745.

Harries, M., Flower, K.C., Scanlan, C.A. 2021. Sustainability of nutrient management in grain production systems of southwest Australia. *Crop Pasture Sci.* 72: 197–212.

Hayes, R.C., Rohan, M., Li, G.D., Orgill, S.E., Poile, G.J., Oates, A.A., Conyers, M.K. 2022. The nature of spatial variability of four soil chemical properties and the implications for soil sampling. *J. Soils Sediments.* doi:10.1007/s11368-022-03285-x.

Helyar, K.R., Porter, W.M. 1989. Soil Acidification: Its Measurements and the processes involved. pp. 61–101. In: A.D. Robson (Editor). *Soil Acidity and Plant Growth*. Academic Press, Sydney.

Hingston, F.J., Atkinson, R.J., Posner, A.M., Quirk, J.P. 1968. Specific adsorption of anions on goethite. *Trans. 9th Int. Congress Soil Sci.* 1: 669–678.

Holland, J.E., White, P.J., Thauvin, J.-N., Jordan-Meille, L., Haefele, S.M., Thomas, C.L., Goulding, K.W.T., McGrath, S.P. 2021. Liming impacts barley yield over a wide concentration range of soil exchangeable cations. *Nutr. Cycl. Agroecosyst.* 120: 131–144.

Kalkhoran, S.S., Pannell, D.J., White, B., Polyakov, M., Chalak, M., Mugera, A., Farre, I. 2018. A dynamic model of optimal lime applications for wheat production in Australia. *Agric. Applied Econ. Assoc.* 274340. https://econpapers.repec.org/paper/agsaaea18/274340.htm (retrieved 29 Aug. 2022).

Kawasaki, A., Wasson, A., Kielnhofer, E., Ryan, P., Delhaize, E. 2022. Assessing the effectiveness of the TaMATE1B and TaALMT1 genes to enhance the Al^{3+} tolerance of durum wheat (Triticum turgidum) grown under controlled conditions and in the field. *Res. Square*. doi:10.21203/rs.3.rs-1549253/v1.

Lauricella, D., Butterly, C.R., Weng, Z., Clark, G.J., Sale, PWG, Li, G., Tang, C. 2021. Impact of novel materials on alkalinity movement down acid soil profiles when combined with lime. *J. Soils Sediments* 21: 52–62.

Li, G., Conyers, M.K., Helyar, K.R., Lisle, C.J., Poile, G.J., Cullis, B.R., Heylar, K.R. 2019. Long-term surface application of lime ameliorates subsurface soil acidity in the mixed farming zone of south-eastern Australia. *Geoderma* 338: 236–246.

Llewellyn, R.S., D'Emden, F.H., Kuehne, G. 2012. Extensive use of no-tillage in grain growing regions of Australia. *Field Crops Res.* 132: 204–212.

Malone, B., Stockmann, U., Glover, M., McLachlan, G., Engelhardt, S., Tuomi, S. 2022. Digital soil survey and mapping underpinning inherent and dynamic soil attribute condition assessments. *Soil Security* 6: 100048.

McLay, C.D.A., Ritchie, G.S.P., Porter, W.M., Cruse, A. 1994. Amelioration of subsurface acidity in sandy soils in low rainfall regions. 2. Changes to soil solution composition following the surface application of gypsum and lime. *Aust. J. Soil Res.* 32: 847–865.

Moore, G., Dolling, P.J., Porter, W.M., Leonard, L. 2001. Soil acidity. pp. 127–141. In: G. Moore (Editor). *Soilguide. A Handbook for Understanding and Managing Agricultural Soils*. Agriculture Western Australia, Bulletin No. 4343.

Nagarajah, S., Posner, A.M., Quirk, J.P. 1968. Desorption of phosphate from kaolinite by citrate and bicarbonate. *Soil Sci. Soc. Am. Proc.* 32: 507–510.

Ning, T., Liu, Z., Hu, H., Li, G., Kuzyakov, Y. 2022. Physical, chemical and biological subsoiling for sustainable agriculture. *Soil Tillage Res.* 223: 105490.

Nunes, M.R., Denardin, J.E., Vaz, C.M.P., Karlen, D.L., Cambardella, C.A. 2019. Lime movement through highly weathered soil profiles. *Environ. Res. Commun.* 1: 115002.

Nye, P.H., Ameloko, A.Y. 1987. Predicting the rate of dissolution of lime in soils. *J. Soil Sci.* 38: 641–644.

Oliver, Y.M., Gazey, C., Fisher, J., Robertson, M. 2021. Dissection of the contributing factors to the variable response of crop yield to surface applied lime in Australia. *Agronomy* 11: 829.

Rebetzke, G.J., Richards R.A., Fettell, N.A., Long, M, Condon, A.G., Forrester, R.I., Botwrightad, T.L. 2007. Genotypic increases in coleoptile length improves stand establishment, vigour and grain yield of deep-sown wheat. *Field Crops Res.* 100: 10–23.

Reynolds, C., Azam. G., Davies, S., Gazey, C., Parker, W., Walker, J. 2021. Liming – it' a no brainer. https://grdc.com.au/resources-and-publications/grdc-update-papers/tab-content/grdc-update-papers/2021/02/liming-its-a-no-brainer (retrieved 31 Sept. 2022).

Rheinheimer, D.S., Tiecher, T., Gonzatto, R., Zafar, M., Brunetto, G. 2018. Residual effect of surface-applied lime on soil acidity properties in a long-term experiment under no-till in a Southern Brazilian sandy Ultisol. *Geoderma* 313: 7–16.

Ritchey, K.D., Souza, D.M.G., Lobato, E., Correa, O. 1980. Calcium leaching to increase rooting depth in a Brazilian savannah Oxisol. *Agron. J.* 72: 40–44.

Roper, M.M., Davies, S.L., Blackwell, P.S., Hall, D.J.M., Bakker, D.M., Jongepier, R., Ward, P.R. 2015. Management options for water-repellent soils in Australian dryland agriculture. *Soil Res.* 53: 786–806.

Ryan, P.R. 2018. Assessing the role of genetics for improving the yield of Australia's major grain crops on acid soils. *Crop Pasture Sci.* 69: 242–264.

Scanlan, C.A. Davies, S.L. 2019. Soil mixing and redistribution by strategic deep tillage in a sandy soil. *Soil Tillage Res.* 185: 139–145.

Silva, S.R., dos Santos, H.P., Lollato, R.P., Santi, A., Fontaneli, R.S. 2022. Long-term effects of tillage systems on liming efficiency, soil chemical properties and wheat yield in Southern Brazil. *Soil Res.* 60: 497–510.

Spoor, G. 2006. Alleviation of soil compaction: requirements, equipment and techniques. *Soil Use Manage.* 22: 113–122.

Summers, R., Weaver, D. 2022. Lowering the water solubility of phosphorus fertilisers impacts leaching, plant growth and residual soil phosphorus. *Soil Res.* doi:10.1071/SR22037.

Sumner, M.E. 1990. *Gypsum as an Ameliorant for the Subsoil Acidity Syndrome.* Florida Institute of Phosphate Research, Bartow, FL.

Szurman-Zubrzycka, M., Chwiałkowska, K., Niemira, M., Kwasniewski, M., Nawrot, M., Gajecka, M., Larsen, P.B., Szarejko, I. 2021. Aluminum or low pH – Which is the bigger enemy of barley? Transcriptome analysis of barley root meristem under Al and low pH stress. *Front. Genet.* 12: 675260.

Tang, C., Rengel, Z., Diatloff, E., Gazey, C. 2003. Response of wheat and barley to liming sandy soil with subsurface soil acidity. *Field Crops. Res.* 80: 235–244.

Ucgul, M., Saunders, C., Fielke, J.M. 2017. Discrete element modelling of top soil burial using a full scale mouldboard plough under field conditions. *Biosyst. Eng.* 160: 140–153.

Ucgul, M., Saunders, C., Li, P., Lee, S-H., Desbiolles, J.M.A. 2018. Analyzing the mixing performance of a rotary spader using digital image processing and discrete element modelling (DEM). *Comput. Electronics Agric.* 151: 1–10.

Uehara, G., Gillman, G.P. 1981. *The Mineralogy, Chemistry, and Physics of Tropical Soils with Variable Charge Clays.* Westview Press, Boulder, CO.

van der Ploeg, R.R., W. Böhm, W, Kirkham, M.B. 1999. On the origin of the theory of mineral nutrition of plants and the Law of the Minimum. *Soil Sci. Soc. Am. J.* 63: 1055–1062.

Vance, W., Pradeep, K., Strachan, S.R., Diffey, S., Bell, R.W. 2021. Novel sources of tolerance to aluminium toxicity in wild cicer (Cicer reticulatum and Cicer echinospermum) collections. *Front. Plant Sci.* 12: 678211.

Vogel, S., Bönecke, E., Kling, C., Kramer, E., Lück, K., Philipp, G., Rühlmann, J., Schröter, I., Gebbers, R. 2022. Direct prediction of site-specific lime requirement of arable fields using the base neutralising capacity and a multi-sensor platform for on-the-go soil mapping. *Precis. Agric.* 23: 127–149.

Wang, H.L., Hedley, M.J., Bolan, N.S., Horne, D.J. 1999a. The influence of surface incorporated lime and gypsiferous by-products on surface and subsurface soil acidity. I. Soil solution chemistry. *Aust. J. Soil Res.* 37: 165–180.

Wang, H.L., Hedley, M.J., Bolan, N.S., Horne, D.J. 1999b. The influence of surface incorporated lime and gypsiferous by-products on surface and subsurface soil acidity. II. Root growth and agronomic implications. *Aust. J. Soil Res.* 37: 181–190.

Weligama, C., Anton Wasson, A., Permalloo, G., Delhaize, E. 2022. Rapid colorimetric methods for analysis of pH, extractable aluminium and Colwell phosphorus in soils. *Soil Res.* doi:10.1071/SR22012.

Wong, M.T.F., Webb, M.J., Wittwer, K. 2013. Development of buffer methods and evaluation of pedotransfer functions to estimate pH buffer capacity of highly weathered soils. *Soil Use Manage.* 29: 30–38.

Wong, M.T.F., Wittwer, K. 2009. Positive charge discovered across Western Australian wheatbelt soils challenges key soil and nitrogen management assumptions. *Aust. J. Soil Res.* 47: 127–135.

Xu, D., Zhu, Q., Ros, G., Cai, Z., Wen, S., Xu, M., Zhang, F., de Vries, W. 2022. Calculation of spatially explicit amounts and intervals of agricultural lime applications at county-level in China. *Sci. Total Environ.* 806: 150955.

Yates, R.J., Harrison, R.J., Loi, A., Steel, E.J., Edwards, T.J., Nutt, B.J. 2021. Sourcing Rhizobium leguminosarum biovar viciae strains from Mediterranean centres of origin to optimise nitrogen fixation in forage legumes grown on acid soils. *Grass and Forage Sci.* 76: 33–43.

Zhao, Z., Wang, E., Kirkegaard, J.A. 2022. Novel wheat varieties facilitate deep sowing to beat the heat of changing climates. *Nat. Clim. Change* 12: 291–296.

Zoca, S.M., Penn, C. 2017. An important tool with no instruction manual: a review of gypsum use in agriculture. *Adv. Agron.* 144: 1–44.

21 Management of Acid and Acid Sulfate Soils

Sonia Mayakaduwage[1], *Chamindra L. Vithana*[2], *Luke M. Mosley*[3], *and Meththika Vithanage*[4,5]

[1]School of Biological Sciences, The University of Adelaide, Adelaide, Australia

[2]Faculty of Science and Engineering, Southern Cross University, Lismore, Australia

[3]Acid Sulfate Soils Centre, School of Biological Sciences, The University of Adelaide, Adelaide, Australia

[4]Faculty of Applied Sciences, Ecosphere Resilience Research Centre, University of Sri Jayewardenepura, Sri Lanka

[5]The Institute of Agriculture, The University of Western Australia, Perth, Australia

CONTENTS

DOI: 10.1201/9781003093565-21

LIST OF ABBREVIATIONS

Abbreviation	Definition
AS	Acid sulfate
DSR	Dissimilatory sulfate reduction
PASS	Potential acid sulfate soils
RIS	Reduced inorganic sulfur
MRD	Microbial reductive dissolution
IASS	Inland acid sulfate soils
CLASS	Coastal lowland acid sulfate soils
AASS	Actual acid sulfate soils
EUS	Epizootic ulcerative syndrome
AVS	Acid volatile sulfide

21.1 INTRODUCTION

Soil acidification is recognized as a key factor underlying poor agricultural production throughout the world. Acid soils cover 30%–40% of arable land and over 70% of potential arable land (Von Uexküll and Mutert, 1995). Furthermore, it has been estimated that acid soils occur on 38% of farmland in Southeast Asia, 31% in Latin America, 20% in East Asia, 56% in Africa, and parts of North America (Wood et al., 2000). In Australia and New Zealand, 239 million ha of agricultural land has been found to be acidic (Von Uexküll and Mutert, 1995). Soil acidification occurs as a result of natural (e.g., weathering of acidic parent materials, acid rain, and decay of organic matter) or anthropogenic activities (e.g., continuous application of acid-forming fertilizers and intensive farming) (Agegnehu et al., 2021). Toxicity of acid soils is often caused by their low pH, which increases the solubility of metals (including Al, Fe, Cu, and Mn), and their excessive concentrations in the soil solution (Foy et al., 1978). In addition, high amounts of H^+ pose an adverse effect on the availability of soil nutrients. For example, available phosphorus (P), which is a macro nutrient for plant growth, decreases with increasing pH (Binkley and Vitousek, 1989). Also, deficiencies of other plant nutrients (Ca, Mg, and K) and negative impacts on soil physical properties (e.g., lowering of water holding capacity and compaction) are potential results of soil acidification (Sumner and Noble, 2003).

Acid sulfate (AS) soils are considered as the "nastiest" soils in the world (Dent and Pons, 1995). They contain reduced inorganic sulfur minerals such as pyrite. Oxidation of pyrite can produce sulfuric acid in amounts that have a lasting effect on the main soil characteristics (Dent, 1986). Globally, AS soils occupy more than 17 million hectares including coastal and inland areas in Australia, Europe, Asia, North and South America, and Africa (Fanning et al., 2010). Once oxidized, sulfuric acid and mobilized metals from AS soils can cause detrimental effects on soil and water quality, resulting in acidification of surface and groundwater, destruction of aquatic habitats, and inhibition of vegetation growth (Fitzpatrick et al., 2009). Also, AS soils can have impacts on agricultural ecosystems through strong acidity, metal toxicity, and nutrient deficiency [e.g., phosphorus (P)], which may lead to loss of yield. In addition, release of sulfuric acid from AS soils can cause acidic corrosion of infrastructure (e.g., pipes, bridges, and roads), which are very expensive to be replaced or repaired (Fitzpatrick et al., 2009).

Due to the abovementioned negative impacts on agriculture and the environment, it is crucial to implement proper management strategies for acid soils and AS soils. To achieve this, several approaches can be used to treat acidity and prevent its formation. This includes the addition of alkaline materials (e.g., lime, organic matter, biochar), minimizing acid production by reducing acidifying inputs and product removal, and the use of species or varieties that have the potential of surviving under acidic conditions (Helyar, 1991). Among them, the addition of ameliorants is considered as the most common and effective strategy to overcome acidity. To implement this, however, other factors such as cost, soil properties, possible toxicities, and soil nutrient status should always be taken into account (Helyar, 1991). Despite the significant improvement of management of acid and AS soils, they are still a major challenge for many countries because the rate of acidification is faster than that of amelioration.

The major aim of this chapter is to provide a synopsis of acid and AS soils, emphasizing the available management options. Also, potential limitations of those management strategies will be discussed, highlighting future research needs.

21.2 FORMATION OF ACID AND ACID SULFATE (AS) SOILS

21.2.1 Development of Acid Soils

While various definitions exist, here we consider acid soils as those having pH values <5.5 for most of the year. Acid soils naturally occur in areas with poor pH buffer capacities or where rainfall exceeds evapotranspiration (Bolan et al., 2003; Artiola et al., 2019). The formation of an acid soil can be a complex process, resulting from interrelated actions of physical, chemical, and biological processes (Essington, 2015). Both natural and anthropogenic activities can lead to an increase in the H^+ concentration by directly adding H^+ to soil or, indirectly, by removing exchangeable cations such as Ca^{2+}, Mg^{2+}, K^+, and Na^+. Exchangeable cations buffer soil acidity and can be removed via leaching or runoff during heavy rainfall. During continuous leaching of such cations, they can be replaced by H^+ at soil adsorption sites (Essington, 2015). The uptake of cations and nitrogen by crops and legumes can result in charge imbalance, which is overcome by releasing H^+ ions into the soil by roots (Bolan et al., 2003). The charge balance in agricultural soils is also affected by the continuous removal of alkaline plant or animal products. Microbial decomposition of organic matter and conifer leaf litter can generate various acidic components such as H_2CO_3 (formed from CO_2) and weak organic acids (Bolan et al., 2003; Sparks, 2003). The excessive application of fertilizers, such as urea and ammonia, can release H^+ during nitrification (Bolan et al., 2003; Sparks, 2003). Acid rain containing H_2SO_4 and HNO_3 can acidify soils locally and in areas further away from the source. These acidic vapors are formed by reactions of water in the atmosphere with SO_2 and NOx derived from anthropogenic activities (e.g., burning of fossil fuel, vehicle exhausts, and smelting of ores) and natural activities (e.g., volcanoes, air-borne sea salt sprays, and lightning) (Bolan et al., 2003; Sparks, 2003). Weathering of acid rocks (e.g., granite and quartz), dissolution and hydrolysis of aluminium containing minerals (e.g., kaolinite and aluminosilicates), and formation and subsequent hydrolysis of AS soil materials (e.g., pyrite, iron monosulfides, and secondary iron minerals) can also release H^+ and H_2SO_4 to the soil (Essington, 2015; Sparks, 2003; Artiola et al., 2019). Figure 21.1 highlights several key factors that contribute to the formation of acid soils. The development and formation of AS soil materials are discussed in the following sections.

21.2.2 Development of Coastal and Inland Acid Sulfate Soils

A major development of coastal AS soils occurred during the Holocene period. About 18,000 years BP, sea level rose from 80–120 m below to approx. 1–2 m above the present level (Doyle et al., 2015). The submergence of soil on the coastal margins, coupled with a plentiful supply of sulfate from seawater, provided conducive conditions (further described below) for sulfate reduction to occur, leading to formation of iron sulfide minerals (Figure 21.2). The sea level receded and stabilized at the present level about 6,000 years BP, forming shallow, coastal water bodies (e.g., lakes, lagoons,

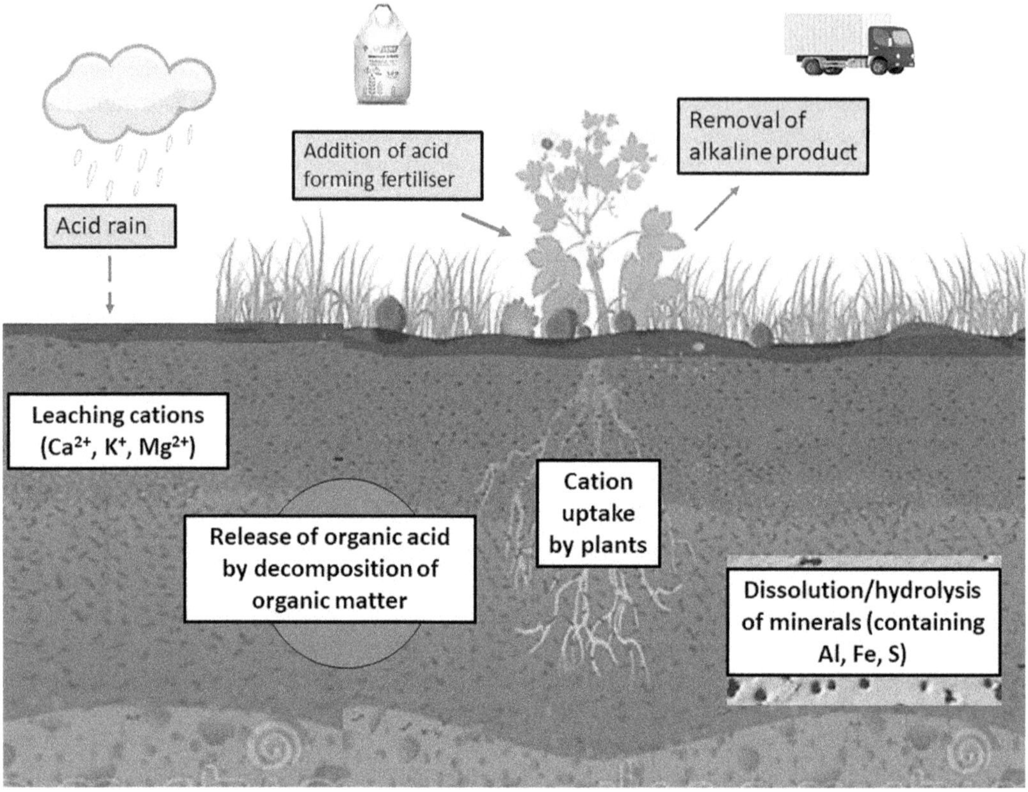

FIGURE 21.1 Key contributors to acid soil formation.

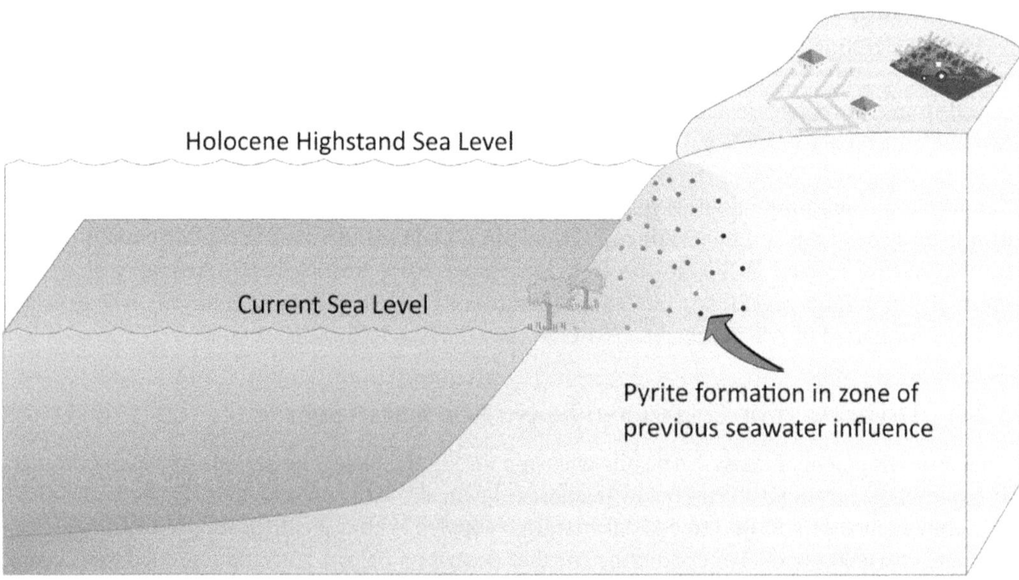

FIGURE 21.2 Development of coastal acid sulfate soils.

and estuaries) and low-lying coastal plains extending to a few kilometers from the coast. Over time, the sulfidic materials were often buried and covered by peat derived from the decomposed coastal vegetation and alluvial deposits. Consequently, iron sulfide containing soil and sediments can be found in many low-lying coastal ecosystems such as salt marshes, swamps, estuaries, and mud or tidal flats (van Breemen, 1973: Pons et al., 1982; Dent and Pons, 1995).

Iron sulfide formation requires the following five conditions to be met and occurs through dissimilatory sulfate reduction (DSR) (Berner, 1984; Dent and Pons, 1995; Fanning et al., 2002):

i. Source of dissolved sulfate (e.g., seawater) to form sulfide
ii. Source of reactive Fe
iii. Sulfate-reducing bacteria (SRB) (e.g., *Desulfovibrio desulfuricans*)
iv. An energy source for microorganisms (e.g., labile organic matter)
v. Anoxic, reducing conditions in the environment

Inland AS soils (IASS) are found in freshwater-brackish wetlands, rivers, lakes, streams, and groundwater systems containing sufficient SO_4^{2-} for DSR and the formation of iron sulfides (Hall et al., 2006; Fitzpatrick and Shand, 2008; Baldwin and Fraser, 2009). The SO_4^{2-} is typically provided by saline groundwater, and, in some cases, various anthropogenic inputs such as municipal waste and agricultural and irrigation discharges (Wong et al., 2015). Iron sulfide formation continues to occur in low-lying coastal plains where the necessary conditions are met.

21.2.3 Formation of Potential Acid Sulfate Soils (PASS)

Potential AS soils (PASS – also termed hypersulfidic materials) are characterized by the presence of reduced inorganic sulfur (RIS) species (van Breemen, 1973; Dent and Pons, 1995). The process of RIS formation is referred to as "sulfidization" (Fanning et al., 2002, 2017). The commonly found RIS phases in AS soil environments are iron monosulfides (makinawite-FeS), iron polysulfides (greigite-Fe_3S_4), elemental sulfur (S), and iron disulfides (pyrite-FeS_2 and marcasite-FeS_2). Under favorable conditions as mentioned above, SO_4^{2-} is reduced to H_2S by SRB that consume acidity and generate alkalinity (Equation 21.1). However, tidal flushing prevents the build-up of alkalinity in the pore water, thereby maintaining the continuation of DSR and promoting the formation of potentially acidic pyritic sediments.

$$CH_2O + SO_4^{2-} + 5H^+ \xrightarrow{SRB} H_2S + HCO_3^- + 2H_2O \tag{21.1}$$

Under moderately reducing conditions (Eh ~ +50 mV), iron (hydr)oxides (e.g., goethite-FeOOH) in coastal AS soils and sediments are subjected to microbial reductive dissolution (MRD) and release Fe^{2+} and more alkalinity (Equation 21.2) (Dent and Pons, 1995; Burton et al., 2007; Sullivan et al., 2012). Metastable iron monosulfides (FeS) are rapidly formed by H_2S reacting with dissolved Fe^{2+} (Equation 21.3) (Fanning et al., 2002; Sullivan et al., 2012).

$$FeOOH + 0.5CH_2O + 0.5H^+ \xrightarrow{MRD} Fe^{2+} + 0.5HCO_3^- + H_2O \tag{21.2}$$

$$Fe^{2+} + H_2S \rightarrow FeS + 2H^+ \tag{21.3}$$

Initial FeS phases precipitated in the abiotic pathway (Equation 21.3) have been characterized as nanoparticulate, disordered makinawite, and found in surface and subsurface drain sediments of coastal lowland AS waterways (Burton et al., 2006b; Boman et al., 2008; Wong et al., 2013).

The nanoparticulate FeS phases in bottom sediments have a characteristic strong-black color and gel like consistency and are also referred to as monosulfidic black ooze (MBO) and readily react with dissolved oxygen leading to an hypoxic situation (Bush et al., 2004a; Burton et al., 2009; Ward et al., 2010). Mackinawite, being unstable and highly reactive, partially is oxidized to an iron polysulfide phase, greigite (Equation 21.4) (Lennie et al., 1997). The occurrence of greigite has been reported in surface sediments in coastal AS soils that are subjected to groundwater fluctuations and oscillating redox conditions (Burton et al., 2008; Keene et al., 2011; Moon et al., 2017).

$$FeS + 0.125O_2 + 0.5H^+ \rightarrow 0.25Fe_3S_4 + 0.25Fe^{2+} + 0.25H_2O \quad (21.4)$$

Both makinawite and greigite are metastable and can rapidly be transformed into more thermodynamically stable iron disulfides via a sequence of reactions (Rickard, 1997). In many soil profiles, both iron mono- and disulfide phases coexist, where a thin layer of iron monosulfide commonly occurs on the surface and iron disulfide (e.g., pyrite) becomes the dominant RIS phase toward the bottom (Schoonen et al., 2004; Kraal et al., 2013). Pyrite is the main iron disulfide phase present in AS landscapes under neutral to alkaline pHs (Burton et al., 2010; Claff et al., 2011), and a few studies have also reported the occurrence of marcasite in organic-rich peaty soils having slightly acidic conditions ($pH<4$) (Ward et al., 2002; Bush et al., 2004b). Pyrite formation mainly occurs via polysulfide and H_2S pathways, involving dissolved clusters of FeS and dissolved Fe and S complexes (e.g., $FeS\text{-}SH_2$, $FeS_2.H_2$) which could have originated from iron mono- and polysulfide phases (Rickard, 1997; Schoonen et al., 2004; Rickard and Morse, 2005).

21.2.3.1 Polysulfide Pathway

In this pathway, pyrite is formed by the reaction of nanoparticulate disordered FeS with metastable elemental S involving several intermediates such as sulfur oxyanions and polysulfide ions (Equation 21.5). Sulfur isotope studies confirmed that polysulfide is the main source of sulfur in pyrite and not the FeS (Schoonen et al., 2004; Rickard and Morse, 2005).

$$FeS + S_n^{2-} \rightarrow FeS_2 + S_{n-1}^{2-} \quad (21.5)$$

21.2.3.2 H_2S Pathway

In this pathway, FeS in the aqueous phase is oxidized by H_2S, and pyrite is formed via the inner-sphere surface complex, which carries out the electron transfer (Equation 21.6). Unlike the polysulfide pathway, sulfur in pyrite is derived from aqueous FeS (Schoonen et al., 2004; Rickard and Morse, 2005).

$$FeS + H_2S \rightarrow FeS_2 + H_2 \quad (21.6)$$

Pyrite formation can be slowed down in the presence of high porewater Fe^{2+} (>2 mM) and low S^{2-} (< 0.1 μM) concentrations resulting in the favorable formation of disordered mackinawite rather than pyrite. In drain sediments of coastal lowland AS soils (CLASS), the high porewater Fe^{2+} concentration and carbonate alkalinity can lead to the precipitation of siderite ($FeCO_3$, Equation 21.7). Therefore, the pore water Fe^{2+} concentration in these sediments is regulated by the precipitation-dissolution of iron monosulfides (FeS) and siderite ($FeCO_3$) (Equations 21.4 and 21.7) (Burton et al., 2006b, 2007).

$$Fe^{2+} + HCO_3^- \leftrightarrow FeCO_3 + H^+ \quad (21.7)$$

In contrast, pyrite is often abundant in benthic sediments of AS waterways containing high dissolved S^{2-} concentration (up to 610 mg/L) and low dissolved Fe^{2+} (< 0.2 mg/L) (Ward et al., 2010).

21.2.4 Formation of Actual Acid Sulfate Soils (AASS)

Actual AS soils (AASS) are the soils containing sulfuric soil materials (pH<4) that are derived from the oxidation and partial hydrolysis of iron sulfidic soil materials when they are exposed to air, which is referred to as "sulfurization" (Nordstrom, 1982; Ritsema et al., 1992; Fanning et al., 2002). The sulfurization generates acidity and secondary iron minerals such as schwertmannite ($Fe_8O_8(OH)_6SO_4$) and jarosite ($KFe_3(SO_4)(OH)_6$) through a series of reactions producing several intermediate products (Nordstrom, 1982; Bigham and Nordstrom, 2000; Fanning et al., 2002).

Pyrite can be oxidized abiotically by O_2 and biotically by Fe^{3+} via a series of steps involving iron and sulfur oxidizing microorganisms (Nordstrom, 1982; Bigham and Nordstrom, 2000; Fanning et al., 2002). Under low Eh conditions (~ −200mV), the oxidation of S^{2-} is more favorable than that of Fe^{2+}. Therefore, in the presence of O_2, pyrite is initially oxidized to Fe^{2+} and SO_4^{2-} by releasing acidity (Equation 21.8).

$$FeS_2 + O_2 + 6H_2O \rightarrow Fe^{2+} + 2SO_4^{2-} + 12H^+ \tag{21.8}$$

In AS soils, the dissolved Fe^{2+} can be partially oxidized in-situ or moved away together with SO_4^{2-} and the acidity through the soil solution and runoff to nearby landscapes, waterways, or groundwater. Consequently, precipitates of iron (hydr)oxides can be seen further away from the pyrite source (Equation 21.9) (Cook et al., 2004; Mosley et al., 2014b).

$$Fe^{2+} + 0.5O_2 + H_2O \rightarrow FeOOH + H^+ \tag{21.9}$$

Furthermore, under dry, humid conditions, the Fe^{2+} derived from pyrite oxidation can also be precipitated as Fe (II) sulfate minerals such as szomolnokite ($FeSO_4.H_2O$), rozenite ($FeSO_4.4H_2O$), ferrohexahydrite ($FeSO_4.6H_2O$), and melanterite ($FeSO_4.7H_2O$) (Fanning et al., 2002; Sullivan et al., 2012). However, owing to greater solubility, their identification is very difficult in AS soils (Sullivan et al., 2012).

Under low pH (~4) conditions, Fe^{2+} derived from pyrite oxidation (Equation 21.8) is oxidized to Fe^{3+} by acidophilic chemolithotrophic microorganisms such as *Acidothiobacillus ferroxidans* because the direct chemical oxidation by O_2 is kinetically unfavorable and has a half-life of ~1,000 days (Equation 21.10) (Nordstrom, 1982; Cook et al., 2004).

$$Fe^{2+} + 0.5O_2 + 2H^+ \rightarrow Fe^{3+} + H_2O \tag{21.10}$$

The reactivity of Fe^{3+} is controlled by pH in the surrounding environment (Ahern et al., 2004; Sullivan et al., 2012). At pH<4, Fe^{3+} exists in the aqueous phase as dissolved Fe^{3+} (Appelo and Postma, 2005), and pyrite can be oxidized by generating acidity and Fe^{2+} (Equation 21.11) (Ahern et al., 2004; Cook et al., 2004). The pyrite oxidation by Fe^{3+} is fast and continued by the supply of Fe^{3+}, which is provided by the microbial re-oxidation of the Fe^{2+} (Equation 21.9) (Nordstrom, 1982; Cook et al., 2004).

At pH<4

$$FeS_2 + Fe^{3+} + 8H_2O \rightarrow 2Fe^{2+} + 2SO_4^{2-} + 16H^+ \tag{21.11}$$

At pH>4, Fe^{3+} is hydrolyzed and precipitated as ferric hydroxide ($Fe(OH)_3$) releasing acidity (Equation 21.12) (Fanning et al., 2002).

At pH>4

$$Fe^{3+} + 3H_2O \rightarrow Fe(OH)_3 + 3H^+ \quad (21.12)$$

The complete sulfurization of one mole of pyrite generates one mole of secondary iron (III) minerals and four moles of acidity (Dent and Pons, 1995; Ahern et al., 2004; Sullivan et al., 2012). Soil can be acidified from this acidity if it lacks sufficient inherent acid neutralization capacity (Equation 21.13).

$$FeS_2 + 3.75O_2 + 3.5H_2O \rightarrow Fe(OH)_3 + 2SO_4^{2-} + 4H^+ \quad (21.13)$$

Furthermore, iron monosulfide oxidation is rapid (half-life 40–60 minutes) and generates elemental S (S_8^0) as an intermediate apart from iron (hydr)oxide (FeOOH) (Equation 21.14) (Burton et al., 2009). A substantially higher elemental S_8^0 was observed in drain sediments of coastal AS soils waterways (Burton et al., 2006a).

$$FeS + 0.75O_2 + 0.5H_2O \rightarrow 0.125S_8^0 + FeOOH \quad (21.14)$$

The formed elemental S (S_8^0) subsequently is oxidized by O_2 or Fe^{3+} (Equations 21.15 and 21.16) (Burton et al., 2006c; Burton et al., 2009).

$$0.125S_8^0 + O_2 + 2H_2O \rightarrow SO_4^{2-} + 4H^+ \quad (21.15)$$

$$0.125S_8^0 + Fe^{3+} + 4H_2O \rightarrow SO_4^{2-} + Fe^{2+} + 8H^+ \quad (21.16)$$

The hydrolysis of Fe^{3+} following the oxidation of pyrite (Equations 21.8 and 21.9) can lead to formation of a range of secondary iron (III) minerals under low pH (< 4) and sulfate-rich (> 3,000 mg/L) conditions (Bigham and Nordstrom, 2000; Fitzpatrick and Shand, 2008, Sullivan et al., 2012). Schwermannite ($Fe_8O_8(OH)_6SO_4$) and jarosite ($KFe_3(SO_4)(OH)_6$) are two main secondary iron minerals commonly found in AASS soils, and their formation generates a substantial amount of acidity (Equations 21.17 and 21.18) (Sullivan and Bush, 2002; Fitzpatrick et al., 2010; Vithana et al., 2022).

$$8Fe^{3+} + SO_4^{2-} + 14H_2O \rightarrow Fe_8O_8(OH)_6SO_4 + 22H^+ \quad (21.17)$$

$$3Fe^{3+} + SO_4^{2-} + K^+ + 6H_2O \rightarrow KFe_3(SO_4)(OH)_6 + 6H^+ \quad (21.18)$$

Both schwertmannite and jarosite are metastable and, hence, are transformed into more stable mineral phases (e.g., goethite) over time releasing acidity (Equations 21.19 and 21.20) (Vithana et al., 2015; Trueman et al., 2020). This acidity can be transported to nearby waterways via runoff (Fitzpatrick et al., 2010; Sullivan et al., 2012).

$$Fe_8O_8(OH)_6SO_4 + 2H_2O \rightarrow 8FeOOH + SO_4^{2-} + 2H^+ \quad (21.19)$$

$$KFe_3(SO_4)(OH)_6(s) + 3H_2O(l) \rightarrow 3Fe(OH)_3(s) + SO_4^{2-}(aq) + K^+(aq) + 3H^+ \quad (21.20)$$

21.3 AGRICULTURAL AND ENVIRONMENTAL IMPACTS OF ACID AND ACID SULFATE SOILS

The main hazard associated with acid and AS soils is "acidification," which can lead to severe acidification of soils in the absence of inherent acid neutralization capacity (Dent, 1986; Sullivan et al., 2012). Acidification can significantly alter soil and water chemistry and can trigger another set of environmental issues such as metal (e.g. Al^{3+}, Fe^{2+}, Zn^{2+}) and metalloid (e.g., As) release, mobility, toxicity, and nutrient deficiency (Bolan et al., 2003; Sullivan et al., 2012). Soil scalding, deoxygenation, and generation of noxious gases are the other hazards associated with AS soils (Sullivan et al., 2012). Acid and AS soil hazards have greatly impacted agriculture, aquatic and terrestrial ecosystems, infrastructure, and human health (Figure 21.3) (Sammut et al., 1996; Powell and Martens, 2005; Ljung et al., 2009). These impacts and subsequent remediation activities can cost millions of dollars imposing a heavy economic burden on countries (Powell and Martens, 2005; Hemming et al., 2018).

21.3.1 Impact on the Environment

21.3.1.1 Soil and Sediment Acidification and Accumulation of Metals and Trace Elements

Acidification and accumulation of metals or trace elements to toxic levels can place soil and benthic ecosystems at risk, because a majority of microorganisms, fauna, and flora in these ecosystems survive and optimally perform at near-neutral conditions (pH 6–7). As a consequence, the long-term influence of acidification and associated stressors can decrease the biodiversity and species richness and inhibit the growth of crops and other vegetation, eventually forming infertile, unproductive soils (Dent, 1986). In AS soils, accumulation of acidity, metals (Al, Fe, and trace elements), soluble ions (Cl^-, SO_4^{2-}), secondary iron minerals (schwertmannite, jarosite), and salts such as gypsum and barite close to the surface form scalds indicating a severe land degradation. Drainage and evaporation are the main contributing factors of scald formation that lead to completely drying out of the soil. These scalded sites lack organic matter and are deficient in nutrients (PO_4^{3-}). Hence, vegetation

(a) (b)

FIGURE 21.3 Impact of acidification on aquatic and terrestrial ecosystems: (a) highly acidic drain with acidification impact on vegetation on the bank (b) acidic subsurface soils containing secondary iron minerals (schwertmannite-orange and jarosite-yellow color) upon exposure of sulfidic soil materials.

in these areas gradually disappears, eventually transforming the land into bare land (Lin et al., 2001; Rosicky et al., 2004).

Prolonged acidic drainage from AS soils resulted in higher concentrations of Al and trace elements (Co, Cd, Cu, Mn, Ni, and Zn) in the surface and subsurface of the sampled bottom sediments of the Vörå River estuary, Finland (Nordmyr et al., 2008) and the eastern Gulf of Bothnia in the northern Baltic Sea (Virtasalo et al., 2020). According to the studies by Wallin et al. (2015) and Vehanen et al. (2022), the trace elements deposited in benthic sediments of the estuaries on the western coast of Finland were bioavailable and toxic to benthic invertebrates and led to reducing their community structure and taxa composition, as opposed to those in rivers, which were unaffected by AS soil drainage.

21.3.1.2 Surface Water Acidification and Mobilization of Metals and Trace Elements

The acidic, metal-rich drainage of acid and AS soils can be exported to nearby waterways via runoff or by man-made drains or canals and to groundwater by leaching (Toivonen and Österholm, 2011; Nystrand and Österholm, 2013; Shahabi-Ghahfarokhi et al., 2022). Export of acidic drainage to waterways and its impact on aquatic ecosystems have been observed in western Finland (Åström and Åström, 1997; Åström, 2001; Roos and Åström, 2006), Sweden (Lindgren et al., 2022), Vietnam (Minh et al., 1997), Lower Lakes, South Australia, (Mosley et al., 2014a), northern Australia (Green et al., 2006), and Malaysia (Karananidi et al., 2022). The severity of the impact, chemical composition of the AS soil drainage, and the water chemistry are influenced by natural phenomena such as rainfall, drought, and snow melt (Åström, 2001).

Several studies have observed the destruction of aquatic habitats, including the spawning and breeding grounds and nurseries, in estuaries and tidal and subtidal areas due to acidification (Cook et al., 2000; Toivonen et al., 2020). Acidic discharges rich in trace elements have detrimentally impacted the highly sensitive marine ecosystems in and around the Great Barrier Reef, Australia (Powell and Martens, 2005). A majority of aquatic fauna survive at a near-neutral to slightly alkaline pH range (5.0–9.0) (ANZECC/ARMCANZ, 2000). However, prolonged exposure to acidified water containing toxic concentrations of Al and trace elements may result in a decrease in species diversity, poor growth at the larval stage, reduced reproduction, diseases, and death (Sammut et al., 1995). According to Joukainen and Yli-Halla (2003), the acidity concentration (> 0.3 m/mol and pH <5.5.) of the discharge from agricultural AS soils was detrimental and may affect the physiological functioning of aquatic species. Furthermore, acidic barriers can reduce fish migration, which in turn affects their reproduction (Sammut et al., 1996; Cook et al., 2000; Toivonen et al., 2020).

Many lowland streams on the south-western coast of Finland carrying acidic drainage of surrounding AS soils are discharged into the northern Baltic Sea, which is a spawning and nursery area for fish (Toivonen et al., 2020). Acidification and increased metal content reduced the number of fish species, community structure, and reproduction in those lowland streams (Vehanen et al., 2022) and also affected the larval abundance of burbot (*Lota lota L*) in the northern Baltic Sea (Toivonen et al., 2020). In the Manning River estuary, eastern Australia, acidification caused shell degradation of Sydney Rock oysters (*Saccostrea commercialis*), which led to an increase in their mortality (Dove and Sammut, 2007). A study by Amaral et al. (2011) observed a lower population of oysters (*Saccostrea glomerata*) and gastropods (*Bembicium auratume*) in three estuaries affected by AS soil drainage in NSW, Australia, compared to the reference sites. A massive number of fish kills occurred in Trinity Inlet estuary in East Trinity, northern Australia, as a result of prolonged exposure to acidic (pH<4) and trace-element rich export from disturbed AS soils in surrounding wetlands (Powell and Martens, 2005). Furthermore, chronic exposure to acidic, Al-rich waters can cause epizootic ulcerative syndrome (EUS) or "red-spot disease" in fish, and Fe and Al flocs formed during the oxidation of dissolved Fe and Al can block gills in fish and smother the benthic communities (Cook et al., 2000). The development of toxic algal blooms (*Lyngbya majuscula*) was observed in coastal waters in northern and southern Australia that receive Fe and organic matter rich AS soil drainage (Cook et al., 2000; Powell and Martens, 2005; Ahern et al., 2006).

21.3.1.3 Groundwater Acidification and Mobilization of Metals and Trace Elements

Groundwater contamination by AS soil drainage is also a significant environmental issue, because the contaminated groundwater can be transported away from the source through aquifers or pipes and pollute wetlands, and also be used for irrigation and drinking purposes in different areas (Appleyard et al., 2004, 2006). Submarine groundwater is a source of water in many wetlands adjacent to AS soil landscapes. Hence continuous discharge of acidic and metal-enriched groundwater can pose a threat to the sustainability of these wetlands (Santos and Eyre, 2011; Santos et al., 2011; McCance et al., 2022).

21.3.1.4 Deoxygenation

Heavy runoff during flooding can mobilize MBO containing acid volatile sulfides (AVS) to receiving waters. When MBO mixes with oxygenated water, it rapidly oxidizes releasing soluble iron, and the oxidation of MBO makes the water deoxygenated within a short period of time (Sullivan and Bush, 2002). Deoxygenation can cause detrimental impacts on aquatic fauna and flora, because a majority of these species require dissolved oxygen (DO) concentrations > 5 mg O_2/L (ANZECC/ARMCANZ, 2000). The formation of hypoxic water is the main contributor to massive fish mortality observed in 2001 in Richmond River in northern NSW, Australia (Bush et al., 2004c; Bush and Sullivan, 2002). The development of AS soils in back swamp areas in northern NSW has gradually changed the vegetation type from flood tolerant (reed and rushes) to flood intolerant (pasture grass), which can be rapidly decomposed in the event of flooding forming hypoxic blackwater with redox-sensitive ions such as Fe^{2+}, Mn^{2+}, and NH_4^+ (Wong et al., 2010). According to Wong et al. (2010), hypoxic blackwater was the major cause that led to the sudden fish kill observed in 2008 in the Richmond River, Australia.

21.3.1.5 Release of Greenhouse Gases (GHGs)

Apart from the major hazards such as acidification and trace-element mobilization, AS soils are considered a source of greenhouse gases (CH_4, SO_2, N_2O) in the atmosphere (van Breemen, 1973; Macdonald et al., 2004; Denmead et al., 2010; Šimek et al., 2014; Yli-Halla et al., 2020). Because AS soils are abundant in carbon and nitrogen stocks and sulfides, the mineralization of those buried stocks and partial oxidation of sulfides contribute to GHGs such as CO_2, CH_4, SO_2, and N_2O (Macdonald et al., 2004; Šimek et al., 2014; Yli-Halla et al., 2020). According to Macdonald et al. (2004), high concentrations of SO_2 in coastal areas in northern NSW, Australia, may be attributed to the AS soils in these areas. The global emission of SO_2 by AS soils was estimated as 3.0 Tg S/year, which is comparable to the total emission of S from terrestrial biogenic origin (0.2–4.0 Tg S/year) (Macdonald et al., 2004).

21.3.2 Impact on Agriculture

Agriculture is one of the main areas that has been significantly impacted by acidification and associated hazards (Dent, 1986; Kochian et al., 2004; Shamshuddin et al., 2014). In many countries, acid and AS soils have developed and been utilized for agriculture to meet the growing food demand owing to a rapid increase in global population (Österholm and Åström, 2002; Joukainen and Yli-Halla, 2003; Shamshuddin et al., 2004; Kawahigashi et al., 2008; Russell et al., 2011; Shamshuddin et al., 2014; Sukitprapanon et al., 2020; Minh et al., 2022). The optimum pH for many crops and native vegetation ranges from 5.5 to 6.5 (Islam et al., 1980). Acidification drops the pH < 5 and solubilizes trace elements, Al, Mn, and Fe bound to clay and other minerals (poorly crystalline oxides of Al, Fe, and Mn) (Dent, 1986; Bolan et al., 2003). Aluminium and Fe toxicity are the key constrains in acid and AS soils that limit crop production (Bolan et al., 2003; Shamshuddin et al., 2014).

Solubilized aluminium (Al^{3+}) can accumulate in tissues of roots to toxic levels and prevent their growth by inhibiting cell division and enzyme activity, resulting in a deformed root system and eventually leading to a substantial reduction in crop productivity (Dent, 1986; Kochian et al., 2004; Panhwar et al., 2016). The acidity, Al and Fe toxicity, and loss of cations such as Ca^{2+} have led to reducing the growth of rice (*Oryza sativa*) grown in AS soils in Thailand (Moore et al., 1990). Similarly, the growth of oil palm seedlings was retarded at Al^{3+} activity > 100 μM in AS soils in Malaysia (Auxtero and Shamshuddin, 1991).

Iron toxicity is commonly observed in crops grown in waterlogged AS soils and nutrient poor (Ca, K, P, Zn) acid soils with low organic matter content, where both soils have acidity and high levels of dissolved reducible iron (Fe^{2+} 10 to 2,000 mg/L) (Dent, 1986; Sahrawat, 2005). In waterlogged AS soils, ferrous ions (Fe^{2+}) are released during the reductive dissolution of secondary iron minerals. Iron toxicity in rice is manifested as a gradual change in leaf color to orange and stunted black colored roots that lead to poor growth (Sahrawat, 2005; Panhwar et al., 2016; Varghese et al., 2022). Iron toxicity in rice has led to reducing the yield by 15%–30% and in some cases up to 100%, which means a complete failure due to extreme toxicity (Dent, 1986; Panhwar et al., 2016; Varghese et al., 2022). Similar to Fe toxicity, hydrogen sulfide (H_2S) toxicity is also observed in waterlogged soils. Plants affected by H_2S toxicity are subjected to impaired root growth and functioning and are likely to be infected with diseases (Dent, 1986).

Also, soil solution P reacts with Fe and Al free cations or amorphous Fe and Al mineral oxides in AS soils (Fitzpatrick et al., 2017), resulting in formation of poorly soluble P salts and thereby reducing the availability of P to the plants (Pierzynski and McDowell, 2005). Particularly, minerals such as jarosite in AS soils play an important role in phosphate binding through anion or ligand exchange mechanisms (Wang et al., 2013). Due to the reason that even added P can be bound to those Fe minerals, efficiency of P fertilizer is low in AS soils (Mayakaduwage et al., 2020). This may lead to reduction of yield and economic loss in agricultural systems where AS soils are used for paddy rice and cereal cultivation (Nguyen et al., 2017).

Increased mobility of trace elements by acidification has resulted in bioaccumulation in crops grown in acid and AS soils. The indirect impact of bioaccumulation of trace elements in crops is a possible cause for the entering of these elements into grazing animals and humans through the food chain (Ljung et al., 2009). Elevated Mn, Co, and Ni concentrations were observed in oats (*Avena sativa* L.) grown in drained AS soils in Finland (Fältmarsch et al., 2009). According to Palko (1986), timothy grass (*Phleum pretense* L.) grown in AS soils in northern Finland contained five times greater Ni, Co, and Cr concentrations compared to that grown in non-AS soils in Finland. Furthermore, high concentrations of Zn, Fe, and Al were reported in cow milk obtained from dairy cattle fed on pasture in AS soils compared to those obtained from other dairy farms in Finland (Fältmarsch et al., 2008). High concentrations of Al, As, Cd, Fe, Pb, Hg, and Se compared to drinking or irrigation water quality standards were reported in groundwater in the Swan Coastal Plain, Western Australia. This groundwater was polluted from the oxidation of sulfidic soil materials in AS peaty soils as a result of urban development activities in the suburbs of Perth, and it has been used for irrigation of homegrown produce via domestic gardens by the residents (Appleyard et al., 2004). It is suggested that the continuous application of contaminated groundwater to homegrown crops can lead to bioaccumulation (Hinwood et al., 2006).

Moreover, agricultural infrastructures or other infrastructures such as pipes, roads bridges, canals, and boats, which are made from concrete, steel, and metals (Fe or Al), can be corroded by sulfuric acid drainage from AS soils (Dent, 1986). This corrosive effect reduces strength and life time of the infrastructures. To avoid this, surface structures may need specific precautions, such as extensive spreading of membranes or construction of piles, which increase construction costs. For example, the annual cost for treatment technologies in Queensland to avoid or prevent this infrastructure damage was estimated to around $100 million (COUNCIL, 2000).

21.4 MANAGEMENT PRACTICES

21.4.1 Chemical Amelioration

Chemical amelioration is defined as the use of chemical amendments to enhance chemical, physical, and other properties of soils on the basis of regulation of acid–base and ion-exchange soil processes (Maslov, 2009). Liming is the most common chemical amelioration practice to raise soil pH and minimize acidity-related constraints for crop growth (e.g., exchangeable Al) (Moreira and Fageria, 2010). There are several lime sources including agricultural limestone (mainly crushed and sieved calcium carbonate), lime sand (mined from natural deposits), dolomite (a mixture of magnesium and calcium carbonates), and a by-product of soda ash production (calcium carbonate with some calcium oxide and hydroxide) (Anderson et al., 2013). When lime (e.g., $CaCO_3$) is added to moist soil, firstly, it is dissolved by the moisture, as follows;

$$CaCO_3 + H_2O_{(in\ soil)} \rightarrow Ca^{2+} + 2OH^- + CO_{2(gas)}$$

The released Ca^{2+} exchanges with Al^{3+} and H^+ on the surface of the acid soil.

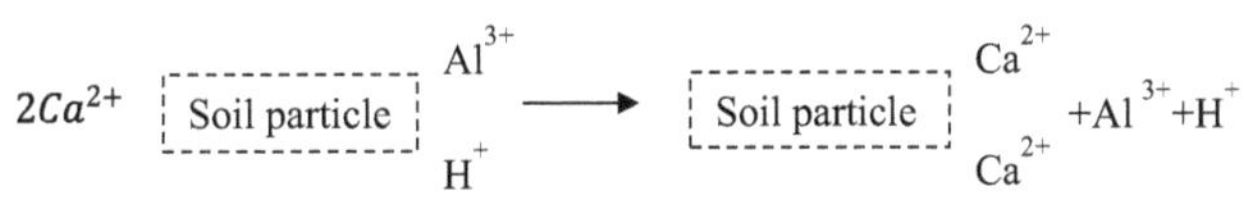

Also, the OH^- produced from lime dissolution will react with Al^{3+} or Fe^{3+} to precipitate Al or Fe hydroxides by eliminating toxic Al or Fe concentrations (Harter, 2007).

$$3OH^- + Al^{3+} \rightarrow Al(OH)_{3(solid)}$$

As a result of increasing pH and reduced Fe or Al solubility, nutrient availability of acid soils (e.g., P) is enhanced (Cronk et al., 2012), facilitating more favorable conditions for plant growth. Furthermore, it has been reported that liming increases microbial activity in soil by creating better environmental conditions for microbes (Neale et al., 1997). According to Shah et al. (1990), bacterial numbers in an acidic podzolic soil increased by 20-fold after the application of powdered limestone (at a rate of 7.5 Mg/ha). When lime increases pH in acid soils, it results in deprotonation of organic substances, and, consequently, organic matter becomes more available to microorganisms (Curtin et al., 1998). Under such favorable conditions (optimum environment and access to organic matter), microbes can enhance N mineralization in acid soils (Curtin et al., 1998). Moreover, increased Ca^{2+} concentration by lime addition improves soil structure and hydraulic conductivity and thereby repairs degraded soils (Kirkham et al., 2007).

However, large amounts of lime are required to correct soil acidity (e.g., 2–10 t/ha). Also, dissolution of lime is a slow process and is determined by several factors including particle size of lime, initial soil pH, and moisture content (Nye and Ameloko, 1987). Due to their low solubility, lime materials have very limited movement into the soil (Nye and Ameloko, 1987). Ameliorating subsoil acidity via deeper incorporation of lime in the soil is difficult, requires special machinery (e.g., via different equipment for cultivation, incorporation, and spading), and is very expensive. For AS soils, particularly those with reserves of pyrite or with a high clay content, liming was reported to be not economical (Dent, 1986). According to Lani (1997), a lime rate of 50 t/ha was required to raise the pH to near 6.0 for AS soils in Brazil. Nevertheless, it has been suggested that liming is effective for old AS soils where the oxidation of pyrite is complete, and it can achieve a significant crop response (Dent, 1986). Accurate calculations for the lime requirement are crucial, because over-liming of soil induces nutrient deficiencies (e.g., P and micronutrients) (Haynes, 1982).

Gypsum ($CaSO_4$) is another chemical ameliorant that has been used to reduce the toxic effects of Al in acid soils. It has been found that the surface incorporation of gypsum is more effective to overcome Al toxicity in acid soils compared to lime, due to its higher solubility. Gypsum does not provide alkalinity (i.e., acid-neutralizing capacity) in the form of carbonate like lime (Farina and Channon, 1988). However, it may increase pH by dislocating OH^- through ligand exchange on surface sites making them variable charge sites (Milne et al., 1995). Conversely, in some instances, gypsum can lead to a decreased pH in acid soil, because Ca^{2+} displaces H^+ (Butterly et al., 2021). In addition, materials such as phosphate rocks and some industrial by products (e.g., slags, refuse sludges, and flue dusts) are also used for the amelioration of acid soils (He et al., 2005; Alves et al., 2006). Garrido et al. (2003) found that gypsum by products can be used as effective alternatives to lime and mined gypsum for alleviating Al toxicity and soil acidity. Phosphate rocks have been found not only to increase P availability but also to decrease acidity through the dissolution of phosphate and subsequent association with protons (He et al., 2005; Loganathan et al., 2005). However, more than a single application could be required to achieve desired benefits for acid soils (Loganathan et al., 2005). Also, industrial by products require proper testing prior to use, because some of them are corrosive and harmful to the environment.

21.4.2 Use of Organic Matter

The application of organic materials to manage acid soils can be of significant benefit because they are often more readily available and economical compared to lime (Michael et al., 2015). It has been reported that organic materials such as plant residues, compost, and manure can increase soil pH, alleviate Al toxicity, and enhance favorable conditions for plant growth in acid soils (Hue and Amien, 1989; Alter and Mitchell, 1992; Hue, 1992). Organic materials increase soil pH by a mechanism that includes specific adsorption of organic anions and subsequent release of hydroxyl ions (Hue, 1992). Also, humic substances in organic matter can bind protons through their functional groups (e.g., carboxyl groups) (Stevenson and Vance, 1989). When organic matter is added to soil, Al toxicity is also eliminated by the formation of Al-organic complexes (Hue et al., 1986), adsorbing soluble Al onto the surface of organic matter (Hoyt and Turner, 1975), and through the production of hydroxyl ions released from ligand exchange reactions (Hue and Amien, 1989). Moreover, organic materials can be used to overcome P deficiency in acid soil, because they can influence the availability of P in several ways (Sharpley and Moyer, 2000). Firstly, they release inorganic P directly and after mineralization of organic P. Secondly, organic matter indirectly enhances P availability by the exchange of organic anions with bound P from anion exchange sites (Hue, 1991) and by the prevention of re-adsorption of phosphate by blocking P adsorption sites (Reddy et al., 1990). Also, organic materials have been shown to produce alkalinity below the layer of application, due to the leaching of organic compounds that generate alkalinity (Butterly et al., 2013; Tang et al., 2013). Several studies have reported an increase in pH below the top layer after amending soil with manure, compost, and plant residues (Lauricella et al., 2021; Butterly et al., 2021). Thus, the addition of organic materials can be used in ameliorating subsurface acidity with or without lime (Butterly et al., 2021).

Organic matter is considered as a promising remedial measure for AS soils because it consumes oxygen during decomposition and stimulates SRB, retards pyrite oxidation, and increases soil pH (Ward et al., 2004; Rigby et al., 2006; Kölbl et al., 2017). Sulfate reduction not only leads to the formation of pyrite, but also generates alkalinity through the release of bicarbonates that can neutralize acidity (Yuan et al., 2016). Additionally, organic matter addition can reduce acidity by proton consumption during decarboxylation of organic anions or proton binding to organic anions (Yan and Schubert, 2000). For example, Dang et al. (2016) demonstrated the high capacity of organic materials to retain protons in AS soil drainage water. Furthermore, organic matter can prevent the acidification of sulfidic materials if it is used as a mulch on the soil surface by maintaining anoxia and buffering acid (Cook et al., 2004; Baldwin and Fraser, 2009). Michael et al. (2015) observed larger increases in pH in sulfuric AS soils by incorporation of dry plant leaves under both anaerobic and aerobic conditions. According to Jayalath et al. (2016), the addition of plant residue stimulated a pH increase under submergence and maintained a higher pH during oxidation, compared to unamended soil. Also, Yuan et al. (2015b) showed that the addition of wheat straw can prevent acidification in sulfuric AS soils. According to Mayakaduwage et al. (2021), organic matter addition increased available P in AS soils, because it was a source of P and released bound P after Fe reduction.

However, the abovementioned effects of organic matter in the remediation of AS soils depend on the chemical composition of organic materials. For example, organic matter with a low to moderate C/N ratio may cause a higher pH increase during anaerobic conditions and reduce acidification under oxidizing conditions than those with a high C/N ratio (Jayalath et al., 2016). This is because the former can meet the N demand of reducing microbes, and therefore, decomposition is quicker compared to the latter (Vanlauwe et al., 1996). Also, organic materials with relatively low lignin contents (e.g., green manure, rice straw) have been found to increase pH during submergence, whereas those with a high lignin content (e.g., peat, sawdust) showed no significant effect on pH, which was attributed to the chemical recalcitrance of lignin (Muhrizal et al., 2006). Acid sulfate subsoils with lower concentration and bioavailability of organic carbon have been shown to have much longer recovery times from acidification due to limitations on microbial reduction reactions (Kölbl et al., 2017). Amendment with sufficient available organic matter can overcome these limitations and enable the recovery of AS soils (Kölbl et al., 2022).

However, remediation by organic matter can also be dependent on the properties of AS soils (Yuan et al., 2015a). During decomposition, soil microbes reduce NO_3^-, Mn^{4+}, Fe^{3+}, and SO_4^{2-} and, then lastly, CO_2 (in a sequence) (Ponnamperuma, 1972). If AS soils contain NO_3^- or Fe^{3+}, they are reduced at a less negative redox potential than sulfate by heterotrophs using organic matter. This may lead to a decrease in the amount of available organic matter for sulfate reducers, lowering the remediation success (Yuan et al., 2015a). Furthermore, the reduction of sulfate can be restricted by the clay content in AS soils because the binding of organic matter to clay particles and oxide surface limits its availability to reducing microorganisms (Yuan et al., 2015a). Additionally, the way in which organic matter is added (e.g., mulching or mixing) may also influence its effectiveness. As found by Yuan (2015a), this was due to the prevention of pyrite oxidation by complexation of Fe^{3+} with organic matter, which occurred more efficiently with mixing compared to a layer.

21.4.3 Biochar Amendment

The application of biochar, which is carbon-rich material obtained via the pyrolysis of biomass, has been recently found to be a successful remediation strategy for acid soils (Fidel et al., 2017). A number of studies have reported a significant increase in soil pH and reduction in soil acidity after incorporation of biochar with soil (Peng et al., 2011; Streubel et al., 2011; Yuan and Xu, 2011; Zong et al., 2018). During the pyrolysis process, cations in the feedstock material (Ca, K, Mg, Na, and Si) lead to the formation of carbonates and oxides that can react with H^+ and metals in acid soils (Brewer et al., 2012) and drainage waters (Mosley et al., 2015). Also, surface functional groups (e.g., –COO– and –O–) in biochar accept protons after incorporation with soil, enhancing soil pH and neutralizing acidity (Dai et al., 2017). These carbonates, oxides, and functional groups contribute to biochar alkalinity (Yuan et al., 2011).

Acid soil amelioration capacity of biochar is mainly determined by its alkalinity, which is significantly dependent on feedstock and production conditions (Brewer et al., 2011; Hass et al., 2012). For example, the alkalinity of biochar produced from legumes was higher than that of non-legumes due to the unbalanced uptake of cations and anions and subsequent higher accumulation of alkali in legume plant biomass in comparison with non-legumes (Yan and Schubert, 2000). However, it has been reported that the oxygen-containing functional groups on the surface of biochar decreased with increasing pyrolysis temperature, because they were transformed into aromatic groups (Chen et al., 2017).

In addition to neutralizing acidity, biochar is capable of inhibiting re-acidification of ameliorated acid soils (Xu et al., 2012; Shi et al., 2016). This has been explained by a protonation and deprotonation process of surface functional groups that change the soil pH buffer capacity (Xu et al., 2012). Dai et al. (2014) found that manure-derived biochar had a greater effect on soil pH buffer capacity than plant-based biochar. Also, the carbonate or oxide content, large surface area, and porous structure of biochar facilitate the adsorption and immobilization of toxic Al^{3+} and Fe^{3+}, reducing exchangeable acidity in acid soil (Qian and Chen, 2013). However, the formation of stable complexes with Al^{3+} or Fe^{3+} prevents their reaction with P and thereby enhances available P in the soil solution (Johan et al., 2021). Ch'ng et al. (2016) showed a higher concentration of available P with biochar amendment, compared to compost, due to a higher affinity of the former for Al and Fe, with a more negative charge. Another benefit of biochar for acid soil is the enhancement of water soluble and exchangeable base cations (e.g., K, Na, Ca, and Mg), which increase the capacity of the acid soil to retain nutrient base cations for plant growth (Yuan et al., 2022). Moreover, it has been suggested that, as a polyporous media, biochar affects soil biogeochemistry by changing soil physicochemical properties and also by mediating electron transfer process. Therefore, biochar is a potential amendment for AS soils, because it can serve as an electron shuttle for Fe reduction and, thereby, retard pyrite oxidation (Kappler et al., 2014).

Furthermore, fulvic and humic acid materials in biochar have a high capacity to bind protons and metals, which prevent toxic leaching from AS soils (Milne et al., 1995). According to Dang et al. (2016), biochar derived from wheat and eucalyptus significantly reduced leaching of protons and Al^{3+} from sulfuric ($pH<4$) AS soils. They further observed a positive correlation between Al^{3+} retention and oxygenated functional groups in biochar. However, further systematic investigations are required to evaluate the role of biochar in remediation of AS soils.

21.4.4 Other Strategies to Minimize Acidity

Among other techniques for managing acid soil and AS soils are effective irrigation and leaching systems that can help remove toxic acid drainage (Suswanto et al., 2007), although offsite risks need to be carefully considered. The oxidation of AS soils can be prevented by water management practices, to a certain extent, because the risk of pyrite oxidation can be controlled by maintaining a high water-table level (Mosley et al., 2014c; 2019). However, these methods may not much contribute to neutralizing existing acidity once it has formed in the soil profile (Susuwanto et al., 2007; Mosley et al., 2017). Also, it can be costly or not practically possible, particularly during drought, to manage water levels, due to the requirement for large amounts of water (Fitzpatrick et al., 2009).

Acidity development in agricultural soil can be reduced through the appropriate use of fertilizers. It is beneficial to reduce the use of more acidifying fertilizers (e.g., ammonium-based fertilizers). The use of N fertilizers increases acidity not only through the generation of H^+ during nitrification, but also when leaching of N is higher than plant uptake (Rahman et al., 2018). Therefore, fertilizer application rates should be adjusted according to the nutrient removal by the crops, or, instead, less acidifying fertilizers (e.g., nitrate or organic based) can be considered. Furthermore, crop rotations in farming systems can reduce effects of soil acidity. For example, because legumes are more acidifying, they can be replaced with a less acidifying crop or pasture. Also, crop rotations limit acidifying effects from N fertilizer applications (Fernåndez and Hoeft, 2009). In addition, diverse rotations with high-residue crops may lead to the improvement of soil pH buffer capacity. Another important strategy for acid soils is the use of acid and Al^{3+} tolerant crop species that have their origin in acid soil regions and have adaptations to acidity (maybe as a part of the evolutionary process) (Agegnehu et al., 2021). Acid-tolerant crop varieties of barley, maize, soybean, and potato have been developed and used in the world during last decade (Ngoune Tandzi et al., 2018). However, implementation of many of the abovementioned management practices may require the incorporation of lime to achieve desired benefits (Coventry et al., 2003).

21.5 CHALLENGES AND LIMITATIONS

One of the major limitations of management of acid soils and AS soils is lack of scientific knowledge and locally available technologies. For example, existence of acid soils is unknown for many farmlands, because soil testing is not easily available for farmers. Cost-effective, simple, and user-friendly soil testing devices are scarce. Also, some of the existing soil test methods are not reliable and confusing. Thus, a review of soil testing methods should be considered. In terms of the distribution of soil acidity, conventional field mapping is not appropriate to identify potential risk areas. To address this, advanced remote sensing technologies are in need for an accurate understanding of the distribution of acid soils and AS soils and developing suitable management strategies. Proximal soil sensing using infrared spectroscopy has also shown to have value for measuring pH and tracing lime movement at high resolution (Rossel et al., 2006; Sleep et al., 2022, Hume et al., 2022). Mid-infrared spectroscopy techniques have also been successfully applied to AS soils (Soriano-Disla et al., 2019).

Another challenge relating to AS soils is climate change, which is reducing rainfall and increasing temperature and evaporation in many regions of the world. Droughts have been shown to severely impact AS soils because these materials are exposed due to falling groundwater levels and an increase in unsaturated soil conditions (Stirling et al., 2020). The recovery of AS soils after drought has also been shown to be very prolonged (Mosley et al., 2017), creating the likelihood of soils developing and persisting in alternative stable states. Furthermore, proper acid soil or AS soil management requires a profound initial investigation that enables decision makers to assess the level of risk and decide suitable management strategies. However, site assessment, including excavation, dredging, sampling, and laboratory analysis, is associated with significant costs. Hence, these costs should be taken into the account at the earliest stage of the management plan. In addition to financial constraints, managing acid soils and AS soils in environmentally sensitive areas (e.g., wetlands) is another complicated task. Amelioration strategies, particularly chemical treatments, are not allowed under most environmental regulations. The challenge is, therefore, to establish appropriate buffer zones to separate disturbed areas from the sensitive ecosystems.

Also, lack of proper education and programs that raise awareness about acid soils and AS soils is another concern in addressing soil acidity. Many farmers, land owners, communities, and even some state and local governments have not understood the significance of soil acidity (e.g., socio-economic impact) or the importance of appropriate and timely management. This is mainly because of the communication gap between scientific communities and farmers, which has resulted in a limited transfer of knowledge and technologies. Thus, training and knowledge-sharing programs are required on a regular and sustainable basis.

21.6 FUTURE RESEARCH

In order to meet growing challenges of management of acid soils and AS soils, future research should focus on the development of appropriate scientific methods to control acidity and recovery of already affected soils. Particularly, many of the scientific studies about acid soils and AS soils have been carried out under laboratory conditions. However, there is a great scope for future research under field conditions to provide landholders with the information required to manage their acid soils and AS soils. Under such conditions, the accelerating effects on soil acidity by agricultural practices and land-use patterns should be identified. In addition, there is a lack of technology to optimize the lime requirement for different crops. Thus, the development of devices to determine the quantity of lime for acid neutralization based on soil properties and environmental factors would be beneficial. Also, the time lag between the application of lime and its effect on soil pH should be identified using appropriate modeling tools.

Although it has been widely reported about the potential of organic amendments to remediate acid soils and AS soils, relative effects of various types of organic amendments and biochar should be further investigated along with the mechanisms involved. However, combined use of different organic materials with inorganic fertilizer can be a potential alternative to liming. Thus, future research can focus on evaluating such integrated management approaches, which may allow farmers to determine the most effective combination.

Furthermore, the challenge of treating subsurface soil acidity can also be curbed through incorporating lime with organic amendments, as mentioned in previous sections. However, research can assist to find other cost-effective and efficient strategies to enhance the movement of lime throughout the soil profile. Also, the downward movement of lime products with or without other amendments should be investigated more thoroughly by using techniques such as advanced infrared spectroscopy. Studies for continued development and improvement of such techniques are required.

Finally, plant-based remediation or phytoremediation might be a promising solution for the remediation of acid soils and AS soils. Nevertheless, studies about the effect of plant root systems on soil pH and changes in redox potential are scarce. Future research may focus on studying those effects and the applicability of this ecofriendly technique to restore affected areas and prevent further acidity generation.

21.7 REMARKS

Globally, more than 30% of arable land, covering a large proportion of agricultural land, is affected by soil acidity. Although soil acidification occurs slowly in natural ecosystems, nowadays it has been accelerated by current agricultural practices, such as excessive use of ammonium-based fertilizers and removal of alkaline crop biomass. Soil acidification is also caused by oxidation of sulfidic materials in AS environments, which are commonly found in coastal and inland wetlands where conditions are favorable for sulfate reduction (waterlogged and sulfate and organic matter rich). Release of sulfuric acid and subsequent toxic metal release after oxidation of AS soils often pose detrimental effects on the environment. The major impacts associated with acid soils and AS soils include metal toxicity, plant nutrient deficiency, and surface and groundwater contamination, leading to subsequent loss of yield, infrastructure damage, and land degradation. These impacts and their remediation methods can ultimately pose an overwhelming financial burden on countries. Thus, it is necessary to use proper and effective strategies for the management of acid soils and AS soils. Current management practices include liming, addition of organic matter or biochar, crop rotation, and use of acid-tolerant crops. Nevertheless, these existing methods have a number of limitations related to their scope. Also, the rate of acidification is faster than that of amelioration, and timely and efficient implementation of management plans remains a challenge. One of the biggest challenges is lack of scientific knowledge and cost-effective technologies, which can be used to get details of existing soil acidity and to determine remediation requirements. Hence, further research is recommended to improve novel technologies to effectively address current acidity and to slow further acidification. In addition, the integrated use organic materials and phytoremediation for the management of acid soils and AS soils should be evaluated.

REFERENCES

Agegnehu, G., Amede, T., Erkossa, T., Yirga, C., Henry, C., Tyler, R., Nosworthy, M.G., Beyene, S., Sileshi, G.W., 2021. Extent and management of acid soils for sustainable crop production system in the tropical agroecosystems: A review. *Acta Agriculturae Scandinavica, Section B—Soil & Plant Science*, 71(9): 852–869.

Ahern, C.R., McElnea, A.E., Sullivan, L.A., 2004. *Acid Sulfate Soils Laboratory Methods Guidelines*. Queensland Department of Natural Resources, Mine and Energy, Indooroopilly, Queensland.

Ahern, K.S., Udy, J.W., Pointon, S.M., 2006. Investigating the potential for groundwater from different vegetation, soil and land uses to stimulate blooms of the cyanobacterium, *Lyngbya majuscula*, in coastal waters. *Marine and Freshwater Research*, 57: 177–186.

Alter, D., Mitchell, A., 1992. Use of vermicompost extract as an aluminium inhibitor in aqueous solutions. *Communications in Soil Science and Plant Analysis*, 23(3–4): 231–240.

Alves, M.C., Gonzalez, A.P., Colodro, G., Junior, H.P., Vázquez, E.V., 2006. Influence of biosolids rate on chemical properties of an oxisol in Sao Paulo, Brazil. *Communications in Soil Science and Plant Analysis*, 37(15–20): 2481–2493.

Amaral, V., Cabral, H., Bishop, M., 2011. Effect of runoff from acid-sulfate soils on pneumatophores of the grey mangrove, *Avicennia marina. Marine and Freshwater Research*, 62: 974–979.

Anderson, N.P., Hart, J.M., Sullivan, D.M., Horneck, D.A., Pirelli, G.J., Christensen, N.W., 2013. Applying lime to raise soil pH for crop production (western Oregon).

ANZECC/ARMCANZ, 2000. Australian and New Zealand Guidelines for Fresh and Marine Water Quality Australian and New Zealand Environment and Conservation Council, Agricultural and Resource Management Council of Australia and New Zealand.

Appleyard, S.J., Wong, S., Willis-Jones, B., Angeloni, J., Watkins, R., 2004. Groundwater acidification caused by urban development in Perth, Western Australia: Source, distribution, and implications for management. *Soil Research*, 42: 579–585.

Appleyard, S.J., Angeloni, J., Watkins, R., 2006. Arsenic-rich groundwater in an urban area experiencing drought and increasing population density, Perth, Australia. *Applied Geochemistry*, 21: 83–97.

Appelo, T., Postma, D., 2005. *Geochemistry, Ground Water and Pollution. Geochemistry, Groundwater and Pollution*, Second Edition, CRC Press, London, pp. 683.

Artiola, J.F., Walworth, J.L., Musil, S.A., Crimmins, M.A., 2019. Chapter 14: Soil and land pollution. In: Brusseau, M.L., Pepper, I.L., Gerba, C.P. (Eds.), *Environmental and Pollution Science*, Third Edition, Academic Press, San Diego, CA, pp. 219–235.

Åström, M., 2001. The effect of acid soil leaching on trace element abundance in a medium-sized stream, W. Finland. *Applied Geochemistry*, 16(3): 387–396.

Åström, M., Åström, J., 1997. Geochemistry of stream water in a catchment in Finland affected by sulfidic fine sediments. *Applied Geochemistry*, 12: 593–605.

Auxtero, E.A., Shamshuddin, J., 1991. Growth of oil palm (*Elaeis guineensis*) seedlings on acid sulfate soils as affected by water regime and aluminium. *Plant and Soil*, 137(2): 243–257.

Baldwin, D.S., Fraser, M., 2009. Rehabilitation options for inland waterways impacted by sulfidic sediments-A synthesis. *Journal of Environmental Management*, 91: 311–319.

Berner, R.A., 1984. Sedimentary pyrite formation: An update. *Geochimica et Cosmochimica Acta*, 48(4): 605–615.

Bigham, J.M., Nordstrom, D.K., 2000. Iron and aluminum hydroxysulfates from acid sulfate waters. *Reviews in Mineralogy and Geochemistry*, 40: 351–403.

Binkley, D., Vitousek, P., 1989. Soil nutrient availability. In: Pearcy, R.W., Ehleringer, J.R., Mooney, H.A., Rundel, P.W. (Eds.), *Plant Physiological Ecology*. Springer, Dordrecht, the Netherlands, pp. 75–96.

Bolan, N., Adriano, D., Curtin, D., 2003. Soil acidification and liming interactions with nutrient and heavy metal transformation and bioavailability. *Advances in Agronomy*, 78: 215–272.

Boman, A., Åström, M., Fröjdö, S., 2008. Sulfur dynamics in boreal acid sulfate soils rich in metastable iron sulfide-The role of artificial drainage. *Chemical Geology*, 255: 68–77.

Brewer, C.E., Unger, R., Schmidt-Rohr, K., Brown, R.C., 2011. Criteria to select biochars for field studies based on biochar chemical properties. *Bioenergy Research*, 4(4): 312–323.

Brewer, C.E., Hu, Y.Y., Schmidt-Rohr, K., Loynachan, T.E., Laird, D.A., Brown, R.C., 2012. Extent of pyrolysis impacts on fast pyrolysis biochar properties. *Journal of Environmental Quality*, 41(4): 1115–1122.

Burton, E.D., Bush, R.T., Sullivan, L.A., 2006a. Elemental sulfur in drain sediments associated with acid sulfate soils. *Applied Geochemistry*, 21: 1240–1247.

Burton, E.D., Bush, R.T., Sullivan, L.A., 2006b. Sedimentary iron geochemistry in acidic waterways associated with coastal lowland acid sulfate soils. *Geochimica et Cosmochimica Acta*, 70: 5455–5468.

Burton, E.D., Bush, R.T., Sullivan, L.A., 2006c. Acid-volatile sulfide oxidation in coastal flood plain drains: Iron sulfur cycling and effects on water quality. *Environmental Science and Technology*, 40: 1217–1222.

Burton, E.D., Bush, R.T., Sullivan, L.A., Mitchell, D.R.G., 2007. Reductive transformation of iron and sulfur in schwertmannite-rich accumulations associated with acidified coastal lowlands. *Geochimica et Cosmochimica Acta*, 71: 4456–4473.

Burton, E.D., Bush, R.T., Sullivan, L.A., Johnston, S.G., Hocking, R.K., 2008. Mobility of arsenic and selected metals during re-flooding of iron- and organic-rich acid-sulfate soil. *Chemical Geology*, 253: 64–73.

Burton, E.D., Bush, R.T., Sullivan, L.A., Hocking, R.K., Mitchell, D.R., Johnston, S.G., Fitzpatrick, R.W., Raven, M., McClure, S., Jang, L.Y., 2009. Iron-monosulfide oxidation in natural sediments: Resolving microbially mediated S transformations using XANES, electron microscopy, and selective extractions. *Environmental Science & Technology*, 43(9): 3128–3134.

Burton, E.D. Johnston, S.G., Watling, K., Bush, R.T., Keene, A.F., Sullivan, L.A., 2010. Arsenic effects and behaviour in association with the Fe(II)-catalysed transformation of schwertmannite. *Environmental Science and Technology*, 44: 2016–2021.

Bush, R.T., Sullivan, L.A., 2002. Iron monosulfides at the oxidation front of a coastal acid sulfate soil. In: Lin, C., Melville, M.D., Sullivan, L.A. (Eds.), *Acid Sulfates Soils in Australia and China*. Science Press, Beijing, pp. 131–142.

Bush, R.T., Sullivan, L.A., Fyfe, D.M., Johnston, S.G., 2004a. Redistribution of monosulfidic black oozes by floodwaters in a coastal acid sulfate soil floodplain. *Soil Research*, 42: 603–607.

Bush, R.T., McGrath, R., Sullivan, L.A., 2004b. Occurrence of marcasite in an organic-rich Holocene estuarine mud. *Soil Research*, 42: 617–621.

Bush, R.T., Fyfe, D.M., Sullivan, L.A., 2004c. Occurrence and abundance of monosulfidic black ooze in coastal acid sulfate soil landscapes. *Soil Research*, 42: 609–616.

Butterly, C., Baldock, J.A., Tang, C., 2013. The contribution of crop residues to changes in soil pH under field conditions. *Plant and Soil*, 366(1): 185–198.

Butterly, C.R., Costello, B., Lauricella, D., Sale, P., Li, G., Tang, C., 2021. Alkalinity movement down acid soil columns was faster when lime and plant residues were combined than when either was applied separately. *European Journal of Soil Science*, 72(1): 313–325.

Ch'ng, H.Y., Ahmed, O.H., Majid, N.M.A., 2016. Improving phosphorus availability, nutrient uptake and dry matter production of zea mays l. On a tropical acid soil using poultry manure biochar and pineapple leaves compost. *Experimental Agriculture*, 52(3): 447–465.

Chen, Y., Zhang, X., Chen, W., Yang, H., Chen, H., 2017. The structure evolution of biochar from biomass pyrolysis and its correlation with gas pollutant adsorption performance. *Bioresource Technology*, 246: 101–109.

Claff, S.R., Burton, E.D., Sullivan, L.A., Bush, R.T., 2011. Metal partitioning dynamics during the oxidation and acidification of sulfidic soil. *Chemical Geology*, 286: 146–157.

Cook, F.J., Hicks, W., Gardner, E.A., Carlin, G.D., Froggatt, D.W., 2000. Export of acidity in drainage water from acid sulfate soils. *Marine Pollution Bulletin*, 41: 319–326.

Cook, F.J., Dobos, S.K., Carlin, G.D., Millar, G.E., 2004. Oxidation rate of pyrite in acid sulfate soils: In-situ measurements and modelling. *Soil Research*, 42: 499–507.

COUNCIL, C., 2000. National strategy for the management of coastal acid sulfate soils. National Working Party on Acid Sulfate Soils, NSW Agriculture Wollongbar Agricultural Institute, Australia.

Coventry, D.R., Farhoodi, A., Ren-kou, K.X., 2003. Managing soil acidification through crop rotations in southern Australia. In: *Handbook of Soil Acidity*. CRC Press, Boca Raton, FL, pp. 421–444.

Cronk, J., Curry, J., Dent, D., Frankenberger, W., Good, J., Goswami, N., Hons, F., Hossner, L., Lal, R., Logan, T., 2012. *Advances in Soil Science: Soil Restoration*, Volume 17. Springer Science & Business Media.

Curtin, D., Campbell, C., Jalil, A. 1998. Effects of acidity on mineralization: pH-dependence of organic matter mineralization in weakly acidic soils. *Soil Biology and Biochemistry*, 30(1): 57–64.

Dai, Z., Wang, Y., Muhammad, N., Yu, X., Xiao, K., Meng, J., Liu, X., Xu, J., Brookes, P.C., 2014. The effects and mechanisms of soil acidity changes, following incorporation of biochars in three soils differing in initial pH. *Soil Science Society of America Journal*, 78(5): 1606–1614.

Dai, Z., Zhang, X., Tang, C., Muhammad, N., Wu, J., Brookes, P.C., Xu, J., 2017. Potential role of biochars in decreasing soil acidification-a critical review. *Science of the Total Environment*, 581: 601–611.

Dang, T., Mosley, L.M., Fitzpatrick, R., Marschner, P., 2016a. Addition of organic material to sulfuric soil can reduce leaching of protons, iron and aluminium. *Geoderma*, 271: 63–70.

Dang, T., Mosley, L.M., Fitzpatrick, R., Marschner, P., 2016b. Organic materials retain high proportion of protons, iron and aluminium from acid sulphate soil drainage water with little subsequent release. *Environmental Science and Pollution Research*, 23(23): 23582–23592.

Denmead, O.T., Macdonald, B.C.T., Bryant, G., Naylor, T., Wilson, S., Griffith, D.W.T., Wang, W.J., Salter, B., White, I., Moody, P.W., 2010. Emissions of methane and nitrous oxide from Australian sugarcane soils. *Agricultural and Forest Meteorology*, 150(6): 748–756.

Dent, D., 1986. *Acid Sulfate Soils: A Baseline for Research and Development*. International Institute for Land Reclamation and Improvement, Wageningen.

Dent, D.L., Pons, L.J., 1995. A world perspective on acid sulfate soils. *Geoderma*, 67: 225–225.

Dove, M.C., Sammut, J., 2007. Impacts of estuarine acidification on survival and growth of Sydeny rock oysters *Saccostrea glomerata* (Gould 1850). *Journal of Shellfish Research*, 26: 519–527.

Doyle, T.W., Chivoiu, B., Enwright, N.M., 2015. *Sea-Level Rise Modeling Handbook: Resource Guide for Coastal Land Managers, Engineers, and Scientists*. US Department of the Interior, US Geological Survey, Lafayette, LA.

Essington, M.E., 2015. *Soil and Water Chemistry: An Integrative Approach*. CRC Press, Boca Raton, FL.
Fältmarsch, R.M., Åström, M., Vuori, K.-M., 2008. Environmental risks of metal mobilised from acid sulfate soils in Finland: A literature review. *Boreal Environment Research*, 13: 444–456.
Fältmarsch, R., Österholm, P., Greger, M., Åström, M., 2009. Metal concentrations in oats (*Avena sativa* L.) grown on acid sulfate soils. *Agricultural and Food Science*, 18(1): 45–56.
Fanning, D.S., Rabenhorst, M.C., Burch, S.N., Islam, K.R., Tangren, S.A., 2002. Sulfides and sulfates. In: Dixon, J.B., Schulze, D.G. (Eds.), *Soil Mineralogy with Environmental Applications*. SSSA Book Series. Soil Science Society of America, pp. 229–260.
Fanning, D., Rabenhorst, M.C., Balduff, D., Wagner, D., Orr, R., Zurheide, P., 2010. An acid sulfate perspective on landscape/seascape soil mineralogy in the us mid-Atlantic region. *Geoderma*, 154(3–4): 457–464.
Fanning, D., Rabenhorst, M., Fitzpatrick, R., 2017. Historical developments in the understanding of acid sulfate soils. *Geoderma*, 308: 191–206.
Farina, M., Channon, P. 1988. Acid-subsoil amelioration: II. Gypsum effects on growth and subsoil chemical properties. *Soil Science Society of America Journal*, 52(1): 175–180.
Fernández, F.G., Hoeft, R.G., 2009. Managing soil pH and crop nutrients. *Illinois Agronomy Handbook*, 24: 91–112.
Fidel, R.B., Laird, D.A., Thompson, M.L., Lawrinenko, M., 2017. Characterization and quantification of biochar alkalinity. *Chemosphere*, 167: 367–373.
Fitzpatrick, R.W., Shand, P., 2008. Inland acid sulfate soil systems across Australia: Overview and conceptual models. In: Fitzpatrick, R.W., Shand, P. (Eds.), *Inland Acid Sulfate Soil Systems Across Australia.* CRCLEME (Corporate Research Centre for Landscape Environments and Mineral Exploration), Perth, pp. 6–74.
Fitzpatrick, R.W., Shand, P., Merry, R.H., 2009. Acid sulfate soils. In: Jennings, J.T. (Ed.), *Natural History of the Riverland and Murrayland*. Royal Society of South Australia (Inc.), Adelaide, pp. 65–111.
Fitzpatrick, R.W., Marvanek, S., Powell, B., Grealish, G., 2010. Atlas of Australian acid sulfate soils: Recent developments and future priorities. In: Gilkes, R.J., Prakongkep, N. (Eds.), *19th World Congress of Soil Science: Soil Solutions for a Changing World. Symposium WG 3.1 Processes in Acid Sulfate soil Materials: IUSS*, Brisbane, Australia, pp. 24–27.
Fitzpatrick, R., Mosley, L., Raven, M., Shand, P., 2017. Schwertmannite formation and properties in acidic drain environments following exposure and oxidation of acid sulfate soils in irrigation areas during extreme drought. *Geoderma*, 308: 235–251.
Foy, C., Chaney, R.T., White, M., 1978. The physiology of metal toxicity in plants. *Annual Review of Plant Physiology*, 29(1): 511–566.
Garrido, F., Illera, V., Vizcayno, C., García-González, M., 2003. Evaluation of industrial by products as soil acidity amendments: Chemical and mineralogical implications. *European Journal of Soil Science*, 54(2): 411–422.
Green, R., Waite, T.D., Melville, M.D., Macdonald, B.C.T., 2006. Characteristics of the acidity in acid sulfate soil drainage waters, McLeods Creek, Northeastern NSW, Australia. *Environmental Chemistry*, 3: 225–232.
Hall, K.C., Baldwin, D.S., Rees, G.N., Richardson, A.J., 2006. Distribution of inland wetlands with sulfidic sediments in the Murray-Darling Basin, Australia. *Science of the Total Environment*, 370(1): 235–244.
Harter, R.D., 2007. Acid soils of the tropics. ECHO Technical Note, ECHO, 11.
Hass, A., Gonzalez, J.M., Lima, I.M., Godwin, H.W., Halvorson, J.J., Boyer, D.G., 2012. Chicken manure biochar as liming and nutrient source for acid appalachian soil. *Journal of Environmental Quality*, 41(4): 1096–1106.
Haynes, R., 1982. Effects of liming on phosphate availability in acid soils. *Plant and Soil*, 68(3): 289–308.
He, Z., Yao, H., Calvert, D., Stoffella, P., Yang, X., Chen, G., Lloyd, G., 2005. Dissolution characteristics of central Florida phosphate rock in an acidic sandy soil. *Plant and Soil*, 273(1): 157–166.
Helyar, K., 1991. The management of acid soils. In: Wright, R.J., Baligar, V.C., Murrmann, R.P. (Eds.), *Plant-Soil Interactions at Low pH*, Kluwer Academic Publ., Dordrecht, the Netherlands, pp. 365–382.
Hemming, S., Rigney, D., Rigney, G., Trevorrow, L., Muller, S.L., Della-Sale, A., 2018. Ngarrindjeri Vision for the Ecological Character of the Coorong and Lower Lakes. Natural History Series: Natural History of the Coorong, Lower Lakes, and Murray Mouth Region (Yarluwar-Ruwe).
Hinwood, A.L., Horwitz, P., Appleyard, S.J., Barton, C., Wajrak, M., 2006. Acid sulfate soil disturbance and metals in groundwater: Implications for human exposure through home grown produce. *Environmental Pollution*, 143: 100–105.
Hoyt, P., Turner, R., 1975. Effects of organic materials added to very acid soils on pH, aluminum, exchangeable NH_4^+, and crop yields. *Soil Science*, 119(3): 227–237.

Hue, N., 1991. Effects of organic acids/anions on P sorption and phytoavailability in soils with different mineralogies. *Soil Science*, 152(6): 463–471.

Hue, N., 1992. Correcting soil acidity of a highly weathered ultisol with chicken manure and sewage sludge. *Communications in Soil Science and Plant Analysis*, 23(3–4): 241–264.

Hue, N., Amien, I., 1989. Aluminum detoxification with green manures. *Communications in Soil Science and Plant Analysis*, 20(15–16): 1499–1511.

Hue, N., Craddock, G., Adams, F., 1986. Effect of organic acids on aluminum toxicity in subsoils. *Soil Science Society of America Journal*, 50(1): 28–34.

Hume, R., Marschner, P., Schilling, R.K., Mason, S., Mosley, L.M., 2022. Detection of agriculturally relevant lime concentrations in soil using mid-infrared spectroscopy. *Geoderma*, 409: 115639.

Islam, A.K.M., Edwards, D. G., Asher, C.J., 1980. pH optima for plant growth: Results of a flowing solution culture experiment with six species. *Plant and Soil*, 54(3): 339–357.

Jayalath, N., Fitzpatrick, R.W., Mosley, L., Marschner, P. 2016. Type of organic carbon amendment influences pH changes in acid sulfate soils in flooded and dry conditions. *Journal of Soils and Sediments*, 16(2): 518–526.

Johan, P.D., Ahmed, O.H., Omar, L., Hasbullah, N.A. 2021. Phosphorus transformation in soils following co-application of charcoal and wood ash. *Agronomy*, 11(10): 2010.

Joukainen, S., Yli-Halla, M., 2003. Environmental impacts and acid loads from deep sulfidic layers of two well-drained acid sulfate soils in western Finland. *Agriculture, Ecosystems & Environment*, 95(1): 297–309.

Kappler, A., Wuestner, M.L., Ruecker, A., Harter, J., Halama, M., Behrens, S. 2014. Biochar as an electron shuttle between bacteria and Fe(III) minerals. *Environmental Science & Technology Letters*, 1(8): 339–344.

Karananidi, P., Valente, T., Braga, M.A.S., Reepei, M., Pechy, M.I.N.F., Wang, Z., Bachmann, R.T., Shamshuddin, J., Md Som, A., 2022. Acid sulfate soils decrease surface water quality in coastal area of West Malaysia: Quo Vadis? *Geoderma Regional*, 28: e00467.

Kawahigashi, M., Minh Do, N., Nguyen, V.B., Sumida, H., 2008. Effect of land developmental process on soil solution chemistry in acid sulfate soils distributed in the Mekong Delta, Vietnam. *Soil Science and Plant Nutrition*, 54: 342–352.

Keene, A.F., Johnston, S.G., Bush, R.T., Sullivan, L.A., Burton, E.D., McElnea, A.E., Ahern, C.R., Powell, B., 2011. Effects of hyper-enriched reactive Fe on sulfidization in a tidally inundated acid sulfate soil wetland. *Biogeochemistry*, 103: 263–279.

Kirkham, J., Rowe, B., Doyle, R., 2007. Persistent improvements in the structure and hydraulic conductivity of a ferrosol due to liming. *Soil Research*, 45(3): 218–223.

Kochian, L.V., Hoekenga, O.A., Pineros, M.A., 2004. How do crop plants tolerate acid soils? Mechanisms of aluminum tolerance and phosphorous efficiency. *Annual Review of Plant Biology*, 55: 459–493.

Kölbl, A., Marschner, P., Fitzpatrick, R., Mosley, L., Kögel-Knabner, I., 2017. Linking organic matter composition in acid sulfate soils to pH recovery after re-submerging. *Geoderma*, 308: 350–362.

Kölbl, A., Kaiser, K., Thompson, A., Mosley, L., Fitzpatrick, R., Marschner, P., Sauheitl, L., Mikutta, R., 2022. Rapid remediation of sandy sulfuric subsoils using straw-derived dissolved organic matter. *Geoderma*, 420: 115875.

Kraal, P., Burton, E.D., Bush, R.T., 2013. Iron monosulfide accumulation and pyrite formation in eutrophic estuarine sediments. *Geochimica et Cosmochimica Acta*, 122(0): 75–88.

Lani, J.L., 1997. Deltas dos rios doce e ttapemirim: Solos, com ênfase nos tiomórficos, água e impacto ambiental do uso.

Lauricella, D., Butterly, C.R., Weng, Z.H., Clark, G.J., Sale, P.W., Li, G., Tang, C. 2021. Impact of novel materials on alkalinity movement down acid soil profiles when combined with lime. *Journal of Soils and Sediments*, 21(1): 52–62.

Lennie, A.R., Redfern, S.A., Champness, P.E., Stoddart, C.P., Schofield, P.F., Vaughan, D.J., 1997. Transformation of mackinawite to greigite: An in-situ X-ray powder diffraction and transmission electron microscope study. *American Mineralogist*, 82: 302–309.

Lin, C., Rosicky, M.A., McConchie, D.M., Sullivan, L.A., Lancaster, G., 2001. Coastal land scalding in NSW, Australia: Soil chemical characteristics and their implications for remediation of the scalded lands. *Land Degradation and Development*, 12: 293–303.

Lindgren, A., Jonasson, I.K., Öhrling, C., Giese, M., 2022. Acid sulfate soils and their impact on surface water quality on the Swedish west coast. *Journal of Hydrology: Regional Studies*, 40: 101019.

Ljung, K., Maley, F., Cook, A., Weinstein, P., 2009. Acid sulfate soils and human health-A millennium ecosystem assessment. *Environmental International*, 35: 1234–1242.

Loganathan, P., Hedley, M., Bolan, N., Currie, L. 2005. Field evaluation of the liming value of two phosphate rocks and their partially acidulated products after 16 years of annual application to grazed pasture. *Nutrient Cycling in Agroecosystems*, 72(3): 287–297.

Macdonald, B.C.T., Denmead, O.T., White, I., Melville, M.D., 2004. Natural sulfur dioxide emissions from sulfuric soils. *Atmospheric Environment*, 38: 1473–1480.

Maslov, B.S., 2009. *Agricultural Land Improvement: Amelioration and Reclamation*, Volume ii. EOLSS publications.

Mayakaduwage, S., Mosley, L.M., Marschner, P., 2020. Threshold for labile phosphate in a sandy acid sulfate soil. *Geoderma*, 371: 114359.

Mayakaduwage, S., Mosley, L.M., Marschner, P. 2021. Phosphorus pools in acid sulfate soil are influenced by pH, water content, and addition of organic matter. *Journal of Soil Science and Plant Nutrition*, 21(2): 1066–1075.

McCance, W., Surapaneni, A., Hampton, J., Wang, S., Currell, M., 2022. Decoupling anthropogenic vs. natural impacts at a wastewater treatment plant situated on acid sulfate soils. *Science of the Total Environment*, 821: 153357.

Michael, P.S., Fitzpatrick, R., Reid, R., 2015. The role of organic matter in ameliorating acid sulfate soils with sulfuric horizons. *Geoderma*, 255: 42–49.

Milne, C., Kinniburgh, D., De Wit, J., Van Riemsdijk, W., Koopal, L. 1995. Analysis of proton binding by a peat humic acid using a simple electrostatic model. *Geochimica et cosmochimica acta*, 59(6): 1101–1112.

Minh, L.Q., Tuong, T.P., van Mensvoort, M.E.F., Bouma, J., 1997. Contamination of surface water as affected by land use in acid sulfate soils in the Mekong River Delta, Vietnam. *Agriculture, Ecosystems and Environment*, 61: 19–27.

Minh, V.Q., Khoa, L.V., Du, T.T., Mi, N.T.H., Dai, N.T.P., 2022. Impact of steel slag fertilizer on acid sulfate soils used for rice cultivation: A case study in An Giang Province, Vietnam. *Soil Science Annual*, 73(1): 1–7.

Moon, E.M., Bush, R.T., Gibbs, D.H.M., Mata, J.P., 2017. Divergent Fe and S mineralization pathways during the oxidative transformation of greigite, Fe_3S_4. *Chemical Geology*, 468: 42–48.

Moore Jr., P.A., Attanandana, T., Patrick Jr., W.H., 1990. Factors affecting rice growth on acid sulfate soils. *Soil Science Society of America Journal*, 54(6): 1651–1656.

Moreira, A., Fageria, N.K., 2010. Liming influence on soil chemical properties, nutritional status and yield of alfalfa grown in acid soil. *Revista Brasileira de Ciencia do solo*, 34: 1231–1239.

Mosley, L.M., Zammit, B., Jolley, A.-M., Barnett, L., Fitzpatrick, R., 2014a. Monitoring and assessment of surface water acidification following rewetting of oxidised acid sulfate soils. *Environmental Monitoring and Assessment*, 186(1): 1–18.

Mosley, L.M., Fitzpatrick, R.W., Palmer, D., Leyden, E., Shand, P., 2014b. Changes in acidity and metal geochemistry in soils, groundwater, drain and river water in the Lower Murray River after a severe drought. *Science of the Total Environment*, 485: 281–291.

Mosley, L.M., Palmer, D., Leyden, E., Cook, F., Zammit, B., Shand, P., Baker, A. and Fitzpatrick, R.W., 2014c. Acidification of floodplains due to river level decline during drought. *Journal of Contaminant Hydrology*, 161: 10–23.

Mosley, L.M., Willson, P., Hamilton, B., Butler, G., Seaman, R., 2015. The capacity of biochar made from common reeds to neutralise pH and remove dissolved metals in acid drainage. *Environmental Science and Pollution Research*, 22(19): 15113–15122.

Mosley, L.M., Biswas, T.K., Cook, F.J., Marschner, P., Palmer, D., Shand, P., Yuan, C. and Fitzpatrick, R.W., 2017. Prolonged recovery of acid sulfate soils with sulfuric materials following severe drought: Causes and implications. *Geoderma*, 308: 312–320.

Mosley, L.M., Thomas, B.P., Fitzpatrick, R.W., 2019. A Guide to Managing Acid Sulfate Soil Risks in South Australian River Murray Wetlands. Acid Sulfate Soils Centre, University of Adelaide. Acid Sulfate Soils Centre Report: ASSC_155.

Muhrizal, S., Shamshuddin, J., Fauziah, I., Husni, M. 2006. Changes in iron-poor acid sulfate soil upon submergence. *Geoderma*, 131(1–2): 110–122.

Neale, S., Shah, Z., Adams, W. 1997. Changes in microbial biomass and nitrogen turnover in acidic organic soils following liming. *Soil Biology and Biochemistry*, 29(9–10): 1463–1474.

Ngoune Tandzi, L., Mutengwa, C.S., Ngonkeu, E.L.M., Gracen, V. 2018. Breeding maize for tolerance to acidic soils: A review. *Agronomy*, 8(6): 84.

Nguyen, D.N., Nguyen, T.T., Tran, Q.N., Macdonald, B., To, T.P., Tran, D.V., Nguyen, Q.V. 2017. Soil and rice responses to phosphate fertilizer in two contrasting seasons on acid sulfate soil. *Communications in Soil Science and Plant Analysis*, 48(6): 615–623.

Nordmyr, L., Åström, M., Peltola, P., 2008. Metal pollution of estuarine sediments caused by leaching of acid sulfate soils. Estuarine. *Coastal and Shelf Science*, 76(1): 141–152.

Nordstrom, D.K., 1982. Aqueous pyrite oxidation and the consequent formation of secondary iron minerals. In: Kittrick, J.A., Fanning, D.S., Hossner, L.R. (Eds.), *Acid Sulfate Weathering*-SSSA Special Publication Number 10. Soil Science Society of America, Madison, WI, pp. 37–56.

Nye, P., Ameloko, A., 1987. Predicting the rate of dissolution of lime in soil. *Journal of Soil Science*, 38(4): 641–649.

Nystrand, M.I., Österholm, P., 2013. Metal species in a Boreal river system affected by acid sulfate soils. *Applied Geochemistry*, 31: 133–141.

Österholm, P., Åström, M., 2002. Spatial trends and losses of major and trace elements in agricultural acid sulfate soils distributed in the artificially drained Rintala area, W. Finland. *Applied Geochemistry*, 17(9): 1209–1218.

Palko, J., 1986. Mineral element content of Timothy (*Phleum pratense* L.) in an acid sulfate soil area of Tupos Village, Northern Finland. *Acta Agriculturae Scandinavica*, 36: 399–409.

Panhwar, Q., Naher, U., Shamshuddin, J., Othman, R., Hakeem, K., 2016. Management of Acid Sulfate Soils for Sustainable Rice Cultivation in Malaysia.

Peng, X., Ye, L., Wang, C., Zhou, H., Sun, B. 2011. Temperature-and duration-dependent rice straw-derived biochar: Characteristics and its effects on soil properties of an ultisol in southern China. *Soil and Tillage Research*, 112(2): 159–166.

Pierzynski, G.M., McDowell, R.W., 2005. Chemistry, cycling, and potential movement of inorganic phosphorus in soils. *Phosphorus: Agriculture and the Environment(Phosphorusagric)*, 46: 53–86.

Ponnamperuma, F.N., 1972. The chemistry of submerged soils. *Advances in Agronomy*, 24: 29–96.

Pons, L.J., van Breemen, N., Driessen, P.M., 1982. Physiography of coastal sediments and development of potential acidity. In: Kittrick, J.A., Fanning, D.S., Hossner, L.R. (Eds.), *Acid Sulfate Weathering*-SSSA Special Publication Number 10. Soil Science Society of America, Madison, WI, pp. 1–18.

Powell, B., Martens, M.A., 2005. A review of acid sulfate soil impacts, actions and policies that impact on water quality in Great Barrier Reef catchments, including a case study on remediation at East Trinity. *Marine Pollution Bulletin*, 51: 149–164.

Qian, L., Chen, B., 2013. Dual role of biochars as adsorbents for aluminum: The effects of oxygen-containing organic components and the scattering of silicate particles. *Environmental Science & Technology*, 47(15): 8759–8768.

Rahman, M.A., Lee, S.H., Ji, H.C., Kabir, A.H., Jones, C.S., Lee, K.W. 2018. Importance of mineral nutrition for mitigating aluminum toxicity in plants on acidic soils: Current status and opportunities. *International Journal of Molecular Sciences*, 19(10): 3073.

Reddy, K., D'angelo, E., DeBusk, T., 1990. Oxygen transport through aquatic macrophytes: The role in wastewater treatment. *Journal of Environmental Quality*, 19(2): 261–267.

Rickard, D., 1997. Kinetics of pyrite formation by the H_2S oxidation of iron (II) monosulfide in aqueous solutions between 25 and 125°C: The rate equation. *Geochimica et Cosmochimica Acta*, 61(1): 115–134.

Rickard, D., Morse, J.W., 2005. Acid volatile sulfide (AVS). *Marine Chemistry*, 97: 141–197.

Rigby, P.A., Dobos, S., Cook, F.J., Goonetilleke, A. 2006. Role of organic matter in framboidal pyrite oxidation. *Science of the Total Environment*, 367(2–3): 847–854.

Ritsema, C.J., Groenenberg, J.E., Bisdom, E.B.A., 1992. The transformation of potential into actual acid sulfate soils studied in column experiments. *Geoderma*, 55: 259–271.

Roos, M., Aström, M., 2006. Gulf of Bothnia receives high concentrations of potentially toxic metals from acid sulfate soils. *Boreal Environment Research*, 11: 383.

Rosicky, M.A., Sullivan, L.A., Slavich, P.G., Hughes, M., 2004. Factors contributing to the acid sulfate soil scalding process in the coastal floodplains of New South Wales, Australia. *Soil Research*, 42: 587–594.

Rossel, R.V., Walvoort, D.J.J., McBratney, A.B., Janik, L.J., Skjemstad, J.O., 2006. Visible, near infrared, mid infrared or combined diffuse reflectance spectroscopy for simultaneous assessment of various soil properties. *Geoderma*, 131(1–2), 59–75.

Russell, D.J., Preston, K.M., Mayer, R.J., 2011. Recovery of fish and crustacean communities during remediation of tidal wetlands affected by leachate from acid sulfate soils in north-eastern Australia. *Wetlands Ecology and Management*, 19(1): 89–108.

Sahrawat, K.L., 2005. Iron toxicity in wetland rice and the role of other nutrients. *Journal of Plant Nutrition*, 27(8): 1471–1504.

Sammut, J., Melville, M.D., Callinan, R.B., Fraser, G.C., 1995. Estuarine acidification: Impacts on aquatic biota of draining acid sulfate soils. *Australian Geographical Studies*, 33(1): 89–100.

Sammut, J., White, I., Melville, M.D., 1996. Acidification of an estuarine tributary in eastern Australia due to drainage of acid sulfate soils. *Marine and Freshwater Research*, 47: 669–684.

Santos, I.R., Eyre, B.D., 2011. Radon tracing of groundwater discharge into an Australian estuary surrounded by coastal acid sulfate soils. *Journal of Hydrology*, 396: 246–257.

Santos, I.R., de Weys, J., Eyre, B.D., 2011. Groundwater or floodwater? Assessing the pathways of metal exports from a coastal acid sulfate soil catchment. *Environmental Science and Technology*, 45: 9641–9648.

Schoonen, M.A.A., Amend, J.P., Edwards, K.J., Lyons, T.W., 2004. Mechanisms of sedimentary pyrite formation, sulfur biogeochemistry-past and present. *Geological Society of America*, 379: 117–134.
Shah, Z., Adamst, W., Haven, C., 1990. Composition and activity of the microbial population in an acidic upland soil and effects of liming. *Soil Biology and Biochemistry*, 22(2): 257–263.
Shahabi-Ghahfarokhi, S., Åström, M., Yu, C., Lindquisit, T., Djerf, H., Kalbitz, K., Ketzer, M., 2022. Extensive dispersion of metals from hemiboreal acid sulfate soil into adjacent drain and wetland. *Applied Geochemistry*, 136: 105170.
Shamshuddin, J., Muhrizal, S., Fauziah, I., Husni, M.H.A., 2004. Effects of adding organic materials to an acid sulfate soil on the growth of cocoa (*Theobroma cacao* L.) seedlings. *Science of the Total Environment*, 323: 33–45.
Shamshuddin, J. et al., 2014. Properties and management of acid sulfate soils in Southeast Asia for sustainable cultivation of rice, oil palm, and cocoa. *Advances in Agronomy*, 124: 91–142.
Sharpley, A., Moyer, B., 2000. Phosphorus forms in manure and compost and their release during simulated rainfall. *Journal of Environmental Quality*, 29(5): 1462–1469.
Shi, R.-Y., Li, J.-Y., Xu, R.-K., Qian, W., 2016. Ameliorating effects of individual and combined application of biomass ash, bone meal and alkaline slag on acid soils. *Soil and Tillage Research*, 162: 41–45.
Šimek, M., Virtanen, S., Simojoki, A., Chroňáková, A., Elhottová, D., Krištůfek, V., Yli-Halla, M., 2014. The microbial communities and potential greenhouse gas production in boreal acid sulfate, non-acid sulfate, and reedy sulfidic soils. *Science of the Total Environment*, 466–467: 663–672.
Sleep, B., Mason, S., Janik, L., Mosley, L., 2022. Application of visible near-infrared absorbance spectroscopy for the determination of Soil pH and liming requirements for broad-acre agriculture. *Precision Agriculture*, 23(1): 194–218.
Soriano-Disla, J.M., Janik, L.J., Forrester, S.T., Grocke, S.F., Fitzpatrick, R.W., McLaughlin, M.J., 2019. The use of mid-infrared diffuse reflectance spectroscopy for acid sulfate soil analysis. *Science of the Total Environment*, 646: 1489–1502.
Sparks, D.L., 2003. *Environmental Soil Chemistry*. Academic Press, San Diego, CA.
Stevenson, F., Vance, G., 1989. Naturally occurring aluminum-organic complexes. In: Spocito, G. (Ed.), *Environmental Chemistry of Aluminum*, CEC Press Inc., Florida, pp. 117–145.
Stirling, E., Fitzpatrick, R.W., Mosley, L.M., 2020. Drought effects on wet soils in inland wetlands and peatlands. *Earth-Science Reviews*, 210: 103387.
Streubel, J., Collins, H., Garcia-Perez, M., Tarara, J., Granatstein, D., Kruger, C., 2011. Influence of contrasting biochar types on five soils at increasing rates of application. *Soil Science Society of America Journal*, 75(4): 1402–1413.
Sukitprapanon, T., Suddhiprakarn, A., Kheoruenromne, I., Anusontpornperm, S., Gilkes, R.J., 2020. Nature of redox concentrations in a sequence of agriculturally developed acid sulfate soils in Thailand. *Pedosphere*, 30(3): 390–404.
Sullivan, L.A., Bush, R.T., 2002. Acid sulfate soil drain ooze: Distribution, behaviour and implications for acidification and deoxygenation of waterways. In: Lin, C., Melville, M.D., Sullivan, L.A. (Eds.), *Acid Sulfates Soils in Australia and China*. Science Press, Beijing, pp. 91–99.
Sullivan, L.A., Bush, R.T., Burton, E.D., Ritsema, C.J., van Mensvoort, M.E.F., 2012. Acid Sulfate Soils. In: Hung, P.M., Li, Y., Sumner, M.E. (Eds.), *Handbook of Soil Sciences: Resource Management and Environmental Impacts*. CRC Press, Taylor and Francis Group, Boca, Raton, FL.
Sumner, M.E., Noble, A.D., 2003. Soil acidification: The world story. In: Rengal, Z. (Ed.), *Handbook of Soil Acidity*. CRC Press, Boca, Raton, FL, pp. 15–42.
Suswanto, T., Shamshuddin, J., Syed Omar, S., Mat, P., Teh, C., 2007. Effects of lime and fertiliser application in combination with water management on rice (*Oryza sativa*) cultivated on an acid sulfate soil. *Malaysian Journal of Soil Science*, 11: 1–16.
Tang, C., Weligama, C., Sale, P., 2013. Subsurface soil acidification in farming systems: Its possible causes and management options. In: Xu, J., Sparks, D.L. (Eds.), *Molecular Environmental Soil Science*. Springer, Dordrecht, the Netherlands, pp. 389–412.
Toivonen, J., Österholm, P., 2011. Characterization of acid sulfate soils and assessing their impact on a humic boreal lake. *Journal of Geochemical Exploration*, 110: 107–117.
Toivonen, J., Hudd, R., Nystrand, M., Österholm, P., 2020. Climatic effects on water quality in areas with acid sulfate soils with commensurable consequences on the reproduction of burbot (*Lota lota L.*). *Environmental Geochemistry and Health*, 42(10): 3141–3156.
Trueman, A.M., McLaughlin, M.J., Mosley, L.M., Fitzpatrick, R.W., 2020. Composition and dissolution kinetics of jarosite-rich segregations extracted from an acid sulfate soil with sulfuric material. *Chemical Geology*, 543: 119606.

van Breemen, N., 1973. Soil forming processes in acid sulfate soils. In: Dost, H. (Editor), *International Symposium on Acid Sulfate Soils*. International Institute for Land Reclamation and Improvement, Wageningen, The Netherlands, pp. 66–129.

Vanlauwe, B., Nwoke, O., Sanginga, N., Merckx, R., 1996. Impact of residue quality on the C and N mineralization of leaf and root residues of three agroforestry species. *Plant and Soil*, 183(2): 221–231.

Varghese, E.M., Kour, B., Ramya, S., Kumar, N.S., Jisha, M.S., Ramakrishnan, B., 2022. Chapter 25: Rhizosphere microbe-mediated alleviation of aluminum and iron toxicity in acidic soils. In: Dubey, R.C., Kumar, P. (Eds.), *Rhizosphere Engineering*. Academic Press, pp. 499–526.

Vehanen, T., Sutela, T., Aroviita, J., Karjalainen, S-M., Riihimäki, J., Larsson, A., Vuori, K-M., 2022. Land use in acid sulfate soils degrades river water quality – Do the biological quality metrics respond? *Ecological Indicators*, 141: 109085.

Vithana, C.L., Sullivan, L.A., Burton, E.D., Bush, R.T., 2015. Stability of schwertmannite and jarosite in an acidic landscape: Prolonged field incubation. *Geoderma*, 239–240(0): 47–57.

Vithana, C.L. et al., 2022. Assessment of the acidification risk of the acid sulfate soil materials in a tropical coastal peat bog: Muthurajawela Marsh, Sri Lanka. *CATENA*, 216: 106396.

Virtasalo, J.J., Österholm, P., Kotilainen, A.T., Åström, M.E., 2020. Enrichment of trace metals from acid sulfate soils in sediments of the Kvarken Archipelago, eastern Gulf of Bothnia, Baltic Sea. *Biogeosciences Discuss*, 2020: 1–26.

Von Uexküll, H., Mutert, E., 1995. Global extent, development and economic impact of acid soils. *Plant and Soil*, 171(1): 1–15.

Wallin, J., Karjalainen, A.K., Schultz, E., Järvistö, J., Leppänen, M., Vuori, K-M., 2015. Weight-of-evidence approach in assessment of ecotoxicological risks of acid sulfate soils in the Baltic Sea river estuaries. *Science of the Total Environment*, 508: 452–461.

Wang, X., Liu, F., Tan, W., Li, W., Feng, X., Sparks, D.L., 2013. Characteristics of phosphate adsorption-desorption onto ferrihydrite: Comparison with well-crystalline fe (hydr) oxides. *Soil Science*, 178(1): 1–11.

Ward, N.J., Sullivan, L.A., Bush, R.T., 2002. Sulfide oxidation and acidification of acid sulfate soil materials treated with $CaCO_3$ and seawater-neutralised bauxite refinery residue. *Soil Research*, 40(6): 1057–1067.

Ward, N.J., Sullivan, L.A., Bush, R.T., 2004. Soil pH, oxygen availability, and the rate of sulfide oxidation in acid sulfate soil materials: Implications for environmental hazard assessment. *Soil Research*, 42(6): 509–514.

Ward, N.J., Bush, R.T., Burton, E.D., Appleyard, S., Wong, S., Sullivan, L.A., Cheeseman, P.J., 2010. Monosulfidic black ooze accumulations in sediments of the Geographe Bay area, Western Australia. *Marine Pollution Bulletin*, 60: 2130–2136.

Wong, V.N.L., Bush, R.T., Burton, E.D., Appleyard, S., Wong, S., Sullivan, L.A., Cheeseman, P.J., 2010. Spatial and temporal changes in estuarine water quality during a post-flood hypoxic event. *Estuarine, Coastal and Shelf Science*, 87: 73–82.

Wong, V.N.L., Johnston, S.J., Burton, E.D., Bush, R.T., Sullivan, L.A., Slavich, P.G., 2013. Seawater-induced mobilization of trace metals from mackinawite-rich estuarine sediments. *Water Research*, 47(2): 821–832.

Wong, V., Cheetham, M., Bush, R., Sullivan, L., Ward, N., 2015. Accumulation of sulfidic sediments in a channelised inland river system, southern Australia. *Marine and Freshwater Research*, 67: 1655–1666.

Wood, S., Sebastian, K., Scherr, S.J., 2000. *Pilot Analysis of Global Ecosystems: Agroecosystems*. World Resources Institute.

Xu, R.K., Zhao, A.-Z., Yuan, J.H., Jiang, J., 2012. pH buffering capacity of acid soils from tropical and subtropical regions of China as influenced by incorporation of crop straw biochars. *Journal of Soils and Sediments*, 12(4): 494–502.

Yan, F., Schubert, S., 2000. Soil pH changes after application of plant shoot materials of faba bean and wheat. *Plant and Soil*, 220(1): 279–287.

Yli-Halla, M., Virtanen, S., Regina, K., Österholm, P., Ehnvall, B., Uusi-Kämppä, J., 2020. Nitrogen stocks and flows in an acid sulfate soil. *Environmental Monitoring and Assessment*, 192(12): 751.

Yuan, J.H., Xu, R.K., 2011. The amelioration effects of low temperature biochar generated from nine crop residues on an acidic ultisol. *Soil Use and Management*, 27(1): 110–115.

Yuan, J.H., Xu, R.K., Qian, W., Wang, R.-H., 2011. Comparison of the ameliorating effects on an acidic ultisol between four crop straws and their biochars. *Journal of Soils and Sediments*, 11(5): 741–750.

Yuan, C., Fitzpatrick, R., Mosley, L.M., Marschner, P., 2015a. Sulfate reduction in sulfuric material after re-flooding: Effectiveness of organic carbon addition and ph increase depends on soil properties. *Journal of Hazardous Materials*, 298: 138–145.

Yuan, C., Mosley, L.M., Fitzpatrick, R., Marschner, P., 2015b. Amount of organic matter required to induce sulfate reduction in sulfuric material after re-flooding is affected by soil nitrate concentration. *Journal of Environmental Management*, 151: 437–442.

Yuan, C., Mosley, L.M., Fitzpatrick, R., Marschner, P., 2016. Organic matter addition can prevent acidification during oxidation of sandy hypersulfidic and hyposulfidic material: Effect of application form, rate and c/n ratio. *Geoderma*, 276: 26–32.

Yuan, J., S. E., Che, Z., 2022. Base cation-enhancing role of corn straw biochar in an acidic soil. *Soil Use and Management*, 38(2): 1322–1336.

Zong, Y., Wang, Y., Sheng, Y., Wu, C., Lu, S., 2018. Ameliorating soil acidity and physical properties of two contrasting texture ultisols with wastewater sludge biochar. *Environmental Science and Pollution Research*, 25(26): 25726–25733.

22 Mapping and Modelling Soil Constraints

Yuxin Ma[1] *and Budiman Minasny*[2]
[1]Landcare Research, Manawatu Mail Centre, Palmerston, New Zealand
[2]Sydney Institute of Agriculture, School of Life and Environmental Sciences, The University of Sydney, Sydney, Australia

CONTENTS

22.1 INTRODUCTION

Soil constraints significantly impact crop production, farm sustainability, and consequently environmental quality by limiting plants' ability to utilize soil water and nutrients or inhibiting growth via toxicities (Dang et al., 2011a). Therefore, recognizing areas affected by soil constraints and assessing their potential effects on productivity and profitability are the first steps in developing and optimizing management strategies for ameliorating the soil constraints.

Conventional soil maps usually do not depict soil constraints. They require manual interpretation. More detailed and accurate information on soil properties and their spatial distribution are required to delineate soil-constrained areas (Gobin et al., 2001). Furthermore, conventional soil maps cannot efficiently provide the quantitative information on soil properties that are needed (Mosleh et al., 2016).

Alternatively, digital soil mapping (DSM) has increasingly become practical to provide spatial information of soil properties at required resolutions by numerical models from soil observations coupled with environmental variables (Kidd et al., 2020; Minasny and McBratney, 2016). Such an operational approach is easy to reproduce and update, cost-effective, and quantifies prediction uncertainties over conventional soil mapping (Ma et al., 2019a).

High-resolution and low uncertainties spatial soil data and information produced through DSM facilitate assessing constraints to soil functions (Carré et al., 2007; van Zijl et al., 2014). This chapter will review how DSM can generate quantitative soil information, help understand soil conditions, and identify the spatial pattern of soil constraints, vertically down the soil profile and spatially across the landscape, and thus assess soil resources. In addition to DSM, digital soil assessment (DSA) can help in the practical application of mapping and managing soil constraints. Figure 22.1 shows the framework for soil constraint mapping and assessment using DSM and DSA approaches.

DOI: 10.1201/9781003093565-22

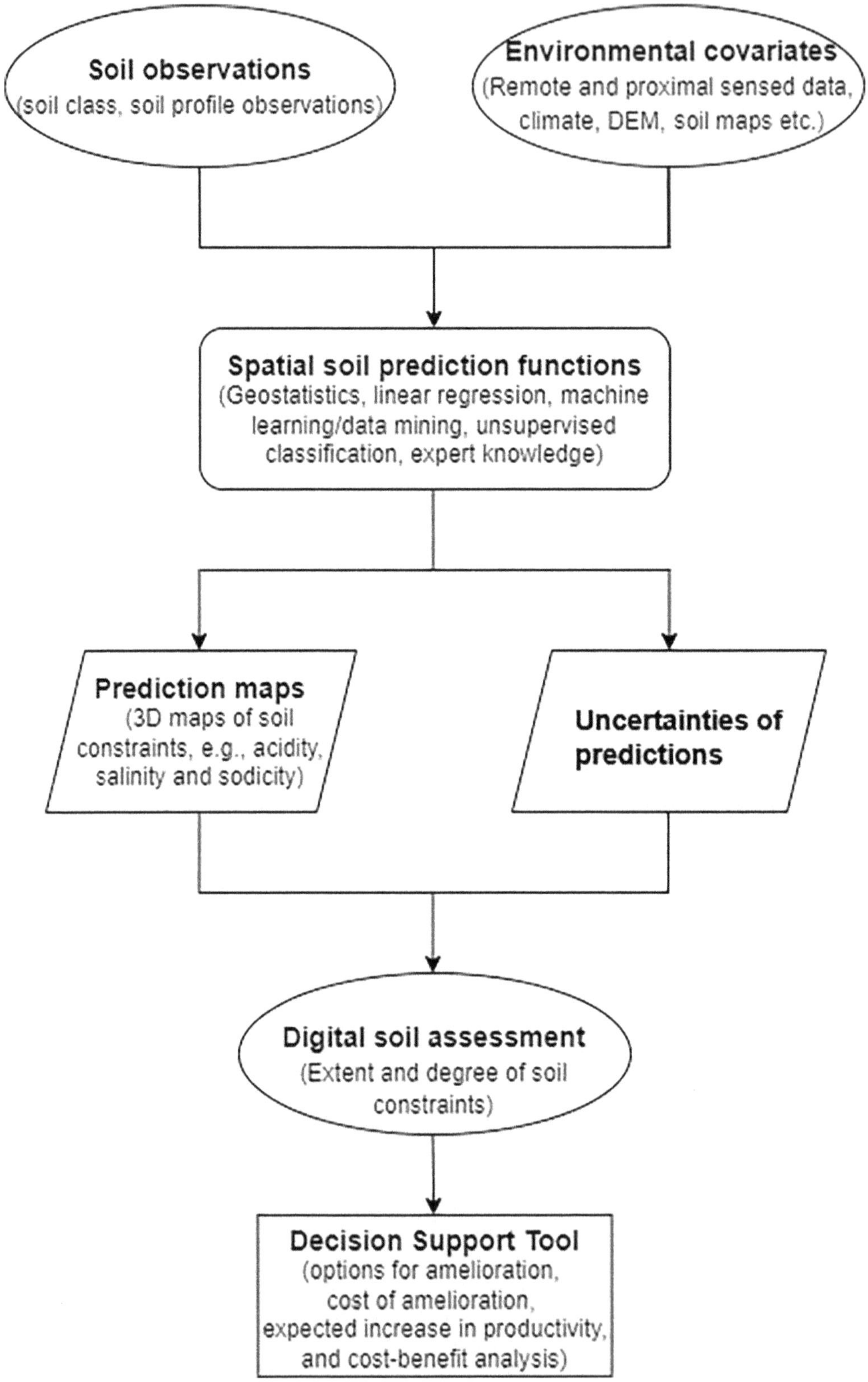

FIGURE 22.1 The framework for soil constraint mapping and assessment.

22.2 DIGITAL SOIL MAPPING AND DIGITAL SOIL ASSESSMENT

22.2.1 DSM and *SCORPAN* Model

Jenny's *clorpt* model (Jenny, 1941; Jenny, 1961) has been the scientific rationale for DSM, where soil (*S*) is a dynamic system and defined by five variables (state factors): climate (*cl*), organisms (*o*), relief (*r*), parent material (*p*), and time (*t*). Jenny's model was formulated to explain pedogenesis, and, in contrast, McBratney et al. (2003) reformulated and introduced an empirical model, *scorpan* for predicting soil attributes or soil classes ($S_{a,c}$) by seven covariates: soil factors (*s*), climate (*c*), organism (*o*), relief (*r*), parent material (*p*), age (*a*), and spatial position (*n*), with spatially correlated errors (*e*):

$$S_{a,c} = f\left(s, c, o, r, p, a, n\right) + e$$

Here, S_a can be soil constraint parameters such as pH or salinity level, or S_c can be a soil class which presents a constraint for crop production, e.g., sandy soils or sodic soils. The *scorpan* model for use in DSM facilitates the quantification of functional relationships between soil attributes or classes and environmental covariates and the estimation of prediction uncertainties in a spatial context.

Both continuous and categorical variables representing *scorpan* factors have been widely used in quantitative models, such as observational soil data, legacy soil maps, geology/geomorphology maps, digital elevation model (DEM) and derivatives, climate, remotely sensed imagery, and proximal soil sensor data (Heung et al., 2016; Ma et al., 2021a). In addition, pedological knowledge can also be incorporated into the predictive models to increase prediction efficiency (Walter et al., 2006).

A variety of mathematical or statistical approaches, from simple linear models to complex machine learning techniques, have been applied to link soil observations and auxiliary information for predicting the spatial distribution of soil properties (Ma et al., 2017; Ma et al., 2021b) including:

1. Geostatistical methods, such as ordinary kriging (Dai et al., 2014; Robinson and Metternicht, 2006) and regression kriging (Pouladi et al., 2019; Somarathna et al., 2016);
2. Linear models, such as generalized linear models (GLM) (Calzolari et al., 2021; Mosleh et al., 2016) and generalized additive models (GAM) (Poggio et al., 2013; Poggio and Gimona, 2014).
3. Machine learning models, such as random forest (RF) (Dharumarajan et al., 2017; Tziachris et al., 2019), cubist (Ma et al., 2019b; Mulder et al., 2016), artificial neural networks (Behrens et al., 2005; Malone et al., 2009), support vector machine (Ahmad et al., 2010; Ballabio, 2009), and deep learning (Padarian et al., 2019; Wadoux, 2019).
4. Unsupervised classification, such as the k-means algorithm to define conceptual soil classes with similar environmental conditions (Román Dobarco et al., 2021).
5. Expert knowledge models, such as the Bayesian inference model (Hawinkel et al., 2016; Poggio et al., 2016).

22.2.2 DSA

DSA extends the application of DSM (Carré et al., 2007). DSA is the translation of DSM outputs into decision-making, in this case, for identifying and managing soil constraints. DSA was initially used to assess soil threats and functions using simulation models and evaluate the risk to aid policy development. DSA has been applied for land suitability evaluation for crop production (Kidd et al., 2015). In the case of soil constraints, DSA can produce an inference where multiple data input sources are required, such as soil pH, sodicity, and structural stability. DSA can infer the types of constraint and the severity of each constraint.

22.2.3 What Soil Constraints Can Be Mapped?

DSM has been successfully applied for mapping soil physical constraints such as permeability (Thomas et al., 2015), rockiness (Harms et al., 2015), soil compaction (Veronesi et al., 2012), soil bulk density (low value of the topsoil for fluffy paddy soil), available water capacity (Hong et al., 2013; Malone et al., 2009), soil slaking index (Jones et al., 2021), and soil thickness (Malone and Searle, 2020). Chemical and biological constraints have also been mapped using DSM such as soil acidity (Chabala et al., 2014; Filippi et al., 2019), salinity (Taghizadeh-Mehrjardi et al., 2016), sodicity (Filippi et al., 2020), alkalinity (Okonkwo et al., 2018), soil nutrients (N, P and K) (Dong et al., 2019), toxic metals (Agyeman et al., 2021), and others.

22.3 MAPPING SOIL CONSTRAINTS

In DSM, both remote and proximal soil sensing approaches are useful to provide large amounts of quantitative information on soil measurements and environmental covariates rapidly and cheaply. However, remote sensing for DSM mainly accounts for the soil surface, while proximal soil sensing can measure both surface and subsurface soil constraints.

22.3.1 Mapping Soil Constraints at Field Scales

One of the most widely used proximal sensors is electromagnetic induction (EMI) (de Jong et al., 1979), a non-invasive sensor that can measure the apparent electrical conductivity (EC_a) at different depths of a soil profile at field and landscape scales (Doolittle and Brevik, 2014). EC_a is usually an indicator for assessing soil salinity initially but when the salt content is not excessive, it can characterize the spatial variability of other soil properties that influence soil EC_a, such as soil water content (Tromp-van Meerveld and McDonnell, 2009), clay content (Sudduth et al., 2005), soil compaction (Al-Gaadi, 2012), and cation exchange capacity (CEC) (Triantafilis et al., 2009).

EMI sensors commonly used in soil investigations include the DUALEM-2 and the Geonics EM38-MK2 for plant rooting depth measurement or the Geonics EM31-MK2 for deeper into the soil profile. These instruments can be towed on a sled behind a vehicle and integrated with GPS (global positioning systems). The fusion of technologies allows the rapid collection of spatially dense data sets using on-the-go or mobile platforms (Doolittle and Brevik, 2014).

EC_a measurements can be a linear function of soil water content given constant soil temperature, texture, mineralogy, and soil salinity, as shown by Sheets and Hendrickx (1995) and Brevik et al. (2006). Thus, EMI sensor data can be a proxy for soil moisture monitoring to improve on-farm irrigation and agricultural productivity (Badewa et al., 2018; Corwin and Lesch, 2003). Dang et al. (2011b) measured EC_b both at the drained upper limit of moisture (UL) and the lower limit of moisture (LL) extracted by the crop for three cropping fields in Australia's northern grain region. They showed strong negative relations between grain yields and EC_a at both UL and LL (stronger) and identified potential management classes formed from EC_a at LL. Al-Gaadi (2012) measured EC_a under four soil moisture contents (4.96%, 5.33%, 6.94%, and 8.0%) and a range of soil compaction levels (on average, from 220 to 2,070 kPa) on a sandy soil field. The results revealed that measured EC_a was proportional to soil compaction at all considered soil conditions except at the moisture content of 8.0%. Thus, EC_a measurement can be an indicator of soil compaction, given that the soil moisture content is below 7% in sandy soil.

High-resolution (resolution < 10 m) EMI sensor data for DSM can be used to characterize the short-range spatial variation of soil properties but also for the larger areas by spatially aggregating the sensor data (Viscarra Rossel et al., 2010). However, EMI results and interpretation of constraints are site-specific and can be influenced by the complex interactions among multiple and variable soil properties (Doolittle and Brevik, 2014).

Another proximal soil sensor commonly used in precision agriculture is gamma-ray spectrometry, a passive sensing technique. Gamma-ray radiometer measures the abundance of potassium (K), thorium (Th), and uranium (U) in soils from these elements' natural isotopic radioactive decay; thus they relate to the mineralogy and geochemistry of the bedrock and weathered materials. Most of the gamma rays emanated from the top 40 cm of soil (Wilford et al., 1997). The concentration of K is generally high in felsic rocks and low in mafic rocks, in contrast with U and Th, which are relatively rare. The concentration of U and Th in igneous rocks generally increases with rock acidity (i.e., rocks with higher silica content) (Wilford and Minty, 2007).

This radiometric technology can help to overcome the inability of current EMI methods to work in sufficiently weathered soil or sandy soil with gravel. Both airborne and ground-based gamma-ray techniques have shown promise for predicting soil properties, e.g., plant-available K, soil texture, and organic carbon (Pracilio et al., 2006; Wong and Harper, 1999).

22.3.2 Mapping Soil Constraints at Landscape and Regional Scales

Remote sensing imagery can be used to map spatial variation of soil constraints indirectly, i.e., using remote sensing data to detect the impact of soil constraints. For example, the temporal pattern of vegetation indices such as NDVI can depict crop productivity performance, indicating areas of lower production, implying constraints. Orton et al. (2018) presented the effects of soil constraints (sodicity, acidity, and salinity) on wheat yield across the wheat-cropping areas of Australia using Landsat and MODIS satellite data. They found an average yield gap of 0.13 t/(ha yr) due to sodicity, 0.04 t/(ha yr) (acidity), and 0.02 t/(ha yr) (salinity).

DSM can produce a 3D mapping of soils (Malone et al., 2009), which is important for mapping soil constraints at various depths. The equal-area spline function (Malone et al., 2009) for mapping soil properties with depth has been used to produce soil constraints within a soil profile. For example, Filippi et al. (2019) used such smoothing splines to resample vertical soil profile data, combined with airborne gamma radiometric and Landsat 7 data, to produce the depth-to-soil pH (H_2O) constraint (>9) map at a 1-cm vertical resolution to a 1-m depth on a dryland cropping farm. These predictions can be used to infer at what depth intervals or identify where peaks occur or where a soil property is above or below a threshold value of constraints.

Okonkwo et al. (2018) assessed the soil salinity or alkalinity or both constraints to rice production based on the spatial variation and associated uncertainties in soil electrical conductivity (EC_e) and pH to the 20 cm depth in soils across Punjab, India, by combining DSM and Monte Carlo simulation methods. They estimated far smaller areas, 1.2 Mha likely saline and 3.0 Mha likely alkaline, while the values were over 2 and 3.2 Mha, respectively, if not allowing for prediction uncertainties, which also showed the flexibility of DSM for setting constraint criteria thresholds.

22.3.3 Case Study: Identifying Yield Constraints in a Field

A field in the Nowley farm in Northern New South Wales in Australia was reported to have a low yield. The 7 ha seemingly uniform field's wheat yield map (Figure 22.2) shows an abrupt change in yield from the bottom half to the top half. And thus, an investigation was conducted to identify soil constraints. Proximal sensing was conducted in the field using an on-the-go measurement with EMI and gamma radiometric sensors.

The EMI technique is a non-invasive measurement of EC_a. Based on the EC_a measurement of a maximum of 100–200 mS/m (Figure 22.3a,b), salinity was not a limiting factor for wheat production. A general trend can be observed that on the east side of the field, high EC_a values were observed, which can be interpreted as greater moisture and/or clay content. However, no consistent pattern was observed between the wheat yield map and EC_a maps. The north side of the field was under chickpea stubbles during the EMI recording, whereas the bottom section was under wheat stubbles, which may influence the results in terms of moisture content.

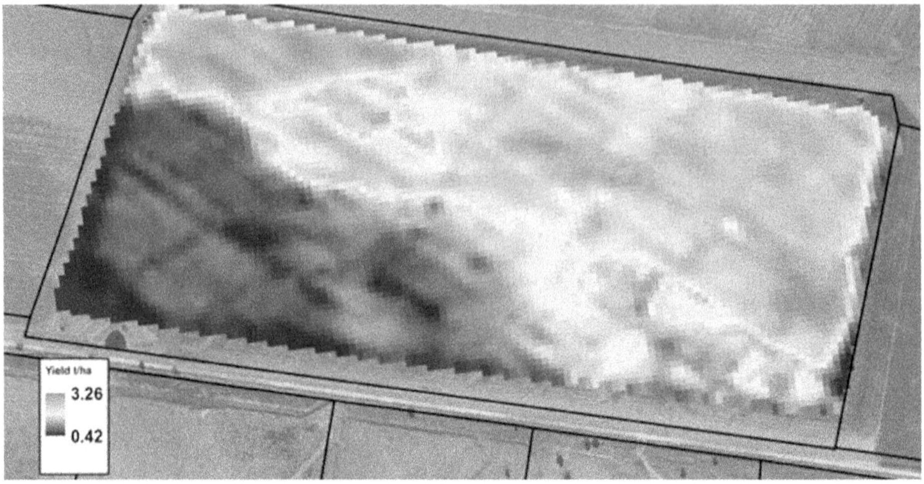

FIGURE 22.2 Wheat yield in tonnes per ha. of a field in the Nowley farm in New South Wales, Australia.

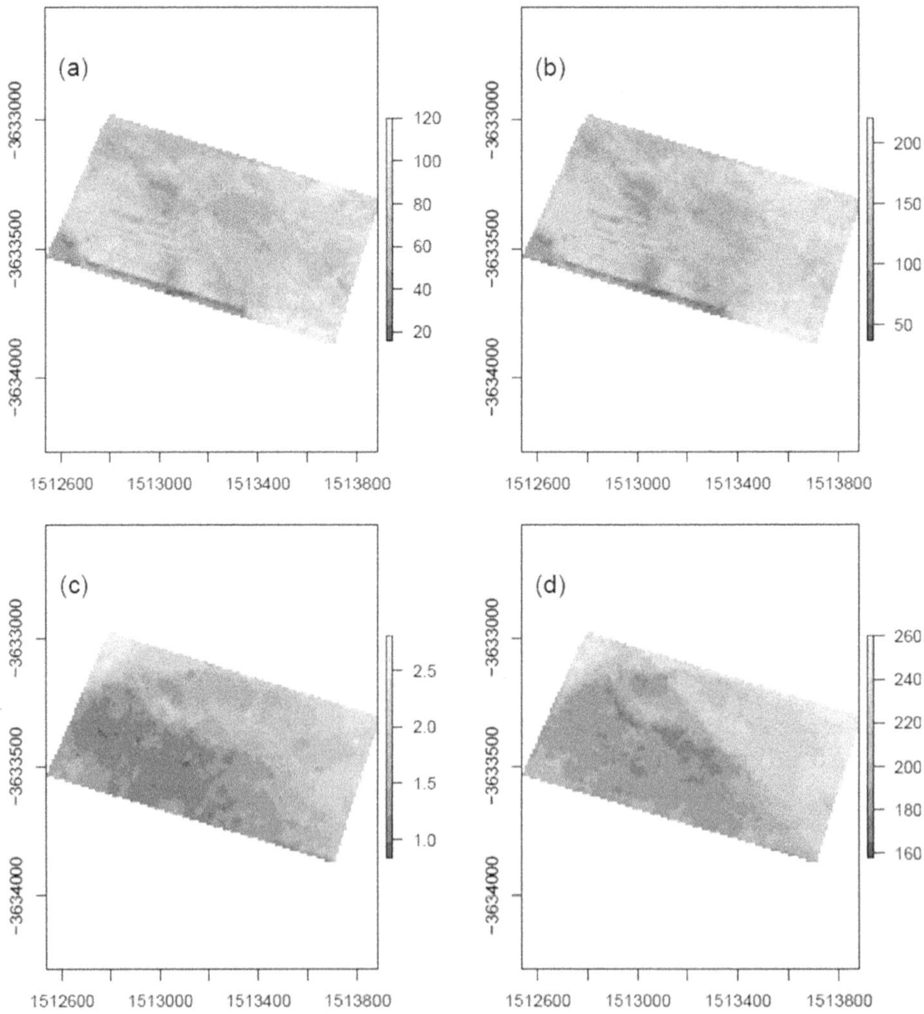

FIGURE 22.3 Proximal sensing to identify the cause of soil constraints on yield. (a) soil electrical conductivity (EC_a in mS/m) at 0–0.5 m; (b) EC_a at 0–1 m; (c) gamma radiometric K count; (d) total gamma radiometric count.

The gamma radiometric data allow the inference of soil's radioisotopic signature to be linked to soil physical and chemical properties. The total gamma count (Figure 22.3d) was similar to the wheat yield map, indicating soils with different parent materials within the field. The bottom section had lower radioelement concentration than the top section, suggesting that the bottom part of the field was influenced by soils derived from sandstone with lower soil quality. In contrast, the top section had soils derived from basalt colluvium with higher exchangeable bases (Stockmann et al., 2016). Further soil analysis revealed that the soils at the bottom section of the field were derived from sandstone that had lower soil pH, lower infiltration capacity, and higher penetration resistance. And thus, the field had to be rezoned and managed differently. This study demonstrates that there is a need for a combination of EMI, gamma radiometric, and soil observations to identify the source of yield constraints. Relying solely on satellite images and machine learning would not be able to infer the actual soil constraints.

22.4 DISCUSSION

DSM has become an essential tool for land resource management across the world and has already moved to an operational stage (Kidd et al., 2020). For example, in Australia, national DSM products are used for land suitability classification and erosion assessments; in France for site contamination assessment, fertilizer applications, and irrigation management; and in India for land degradation and food security assessments (Dash et al., 2022).

DSM operates at various spatial scales. For a regional assessment, we recommend a region-wide DSM of various soil parameters followed by a DSA to identify soil constraints for crop production. However, this regional-scale information is too coarse for farmers' intervention. In local areas, at the farm scale, proximal soil sensing, yield maps, and soil data can be used to pinpoint areas of constraints related to the type and severity of constraints.

For better information concerning site-specific soil constraint management practices, it is necessary to focus on the accuracy and precision of fully-fledged DSM products. For example,

- incorporating more legacy soil data, soil maps, geology maps
- optimizing soil sampling strategies
- fusing proximal and remote sensing sensor data to provide high-resolution information
- adequate independent validation
- mapping at fine spatial resolution (<10 m)
- applying robust predictive models and supplying associated uncertainty maps

While there are many studies on DSM of specific constraints, there is a lack of DSA that infer the types of constraint and the severity for each constraint. Ultimately DSM and DSA products will require a decision support tool to estimate expected costs of intervention to alleviate or address soil constraints. There are constraints that are natural and cannot be ameliorated or are too expensive to be ameliorated. A decision support tool will follow a sequence of the extent and severity of constraints, the options for amelioration, the cost of amelioration, the expected increase in productivity, and a cost-benefit analysis.

22.5 CONCLUSIONS

DSM overcomes the difficulty and expense of conventional soil mapping and facilitates the quantitative relationship between soil measurements and environmental variables. Therefore, future DSM attempts should focus on DSA of soil constraints that different stakeholders can effectively use, e.g., farmers, policymakers, crop planners, and researchers for sustainable management of soil-constrained areas.

REFERENCES

Agyeman, P.C., Ahado, S.K., Borůvka, L., Biney, J.K.M., Sarkodie, V.Y.O., Kebonye, N.M., Kingsley, J., 2021. Trend analysis of global usage of digital soil mapping models in the prediction of potentially toxic elements in soil/sediments: A bibliometric review. *Environmental Geochemistry and Health* 43(5), 1715–1739.

Ahmad, S., Kalra, A., Stephen, H., 2010. Estimating soil moisture using remote sensing data: A machine learning approach. *Advances in Water Resources* 33, 69–80.

Al-Gaadi, K.A., 2012. Employing electromagnetic induction technique for the assessment of soil compaction. *American Journal of Agricultural and Biological Sciences* 7(4), 425–434.

Badewa, E., Unc, A., Cheema, M., Kavanagh, V., Galagedara, L., 2018. Soil moisture mapping using multi-frequency and multi-coil electromagnetic induction sensors on managed podzols. *Agronomy* 8, 224.

Ballabio, C., 2009. Spatial prediction of soil properties in temperate mountain regions using support vector regression. *Geoderma* 151(3), 338–350.

Behrens, T., Förster, H., Scholten, T., Steinrücken, U., Spies, E.-D., Goldschmitt, M., 2005. Digital soil mapping using artificial neural networks. *Journal of Plant Nutrition and Soil Science* 168, 21–33.

Brevik, E.C., Fenton, T.E., Lazari, A., 2006. Soil electrical conductivity as a function of soil water content and implications for soil mapping. *Precision Agriculture* 7, 393–404.

Calzolari, C., Ungaro, F., Vacca, A., 2021. Effectiveness of a soil mapping geomatic approach to predict the spatial distribution of soil types and their properties. *Catena* 196, 104818.

Carré, F., McBratney, A.B., Mayr, T., Montanarella, L., 2007. Digital soil assessments: Beyond DSM. *Geoderma* 142(1), 69–79.

Chabala, L.M., Mulolwa, A., Lungu, O., 2014. Mapping the spatial variability of soil acidity in Zambia. *Agronomy* 4(4), 452–461.

Corwin, D.L., Lesch, S.M., 2003. Application of soil electrical conductivity to precision agriculture. *Agronomy Journal* 95, 455–471.

Dai, F., Zhou, Q., Lv, Z., Wang, X., Liu, G., 2014. Spatial prediction of soil organic matter content integrating artificial neural network and ordinary kriging in Tibetan Plateau. *Ecological Indicators* 45, 184–194.

Dang, Y.P., Pringle, M.J., Schmidt, M., Dalal, R.C., Apan, A., 2011a. Identifying the spatial variability of soil constraints using multi-year remote sensing. *Field Crops Research* 123(3), 248–258.

Dang, Y.P., Dalal, R.C., Pringle, M.J., Biggs, A.J.W., Darr, S., Sauer, B., Moss, J., Payne, J., Orange, D., 2011b. Electromagnetic induction sensing of soil identifies constraints to the crop yields of north-eastern Australia. *Soil Research* 49(7), 559–571.

Dash, P.K., Panigrahi, N., Mishra, A., 2022. Identifying opportunities to improve digital soil mapping in India: A systematic review. *Geoderma Regional* 28, e00478.

de Jong, E., Ballantyne, A.K., Cameron, D.R., Read, D.W.L., 1979. Measurement of apparent electrical conductivity of soils by an electromagnetic induction probe to aid salinity surveys. *Soil Science Society of America Journal* 43(4), 810–812.

Dharumarajan, S., Hegde, R., Singh, S.K., 2017. Spatial prediction of major soil properties using Random Forest techniques - A case study in semi-arid tropics of South India. *Geoderma Regional* 10, 154–162.

Dong, W., Wu, T., Luo, J., Sun, Y., Xia, L., 2019. Land parcel-based digital soil mapping of soil nutrient properties in an alluvial-diluvia plain agricultural area in China. *Geoderma* 340, 234–248.

Doolittle, J.A., Brevik, E.C., 2014. The use of electromagnetic induction techniques in soils studies. *Geoderma* 223–225, 33–45.

Filippi, P., Jones, E.J., Ginns, B.J., Whelan, B.M., Roth, G.W., Bishop, T.F.A., 2019. Mapping the depth-to-soil pH constraint, and the relationship with cotton and grain yield at the within-field scale. *Agronomy* 9(5), 251.

Filippi, P., Jones, E.J., Bishop, T.F.A., 2020. Catchment-scale 3D mapping of depth to soil sodicity constraints through combining public and on-farm soil databases – A potential tool for on-farm management. *Geoderma* 374, 114396.

Gobin, A., Campling, P., Feyen, J., 2001. Soil-landscape modelling to quantify spatial variability of soil texture. *Physics and Chemistry of the Earth, Part B: Hydrology, Oceans and Atmosphere* 26(1), 41–45.

Harms, B., Brough, D., Philip, S., Bartley, R., Clifford, D., Thomas, M., Willis, R., Gregory, L., 2015. Digital soil assessment for regional agricultural land evaluation. *Global Food Security* 5, 25–36.

Hawinkel, P., De Pauw, E., Deckers, J., 2016. Probabilistic soil mapping by Bayesian inference to assess suitability for derocking in northwest Syria. *Soil Use and Management* 32(1), 137–149.

Heung, B., Ho, H.C., Zhang, J., Knudby, A., Bulmer, C.E., Schmidt, M.G., 2016. An overview and comparison of machine-learning techniques for classification purposes in digital soil mapping. *Geoderma* 265, 62–77.

Hong, S.Y., Minasny, B., Han, K.H., Kim, Y., Lee, K., 2013. Predicting and mapping soil available water capacity in Korea. *PeerJ* 1, e71–e71.

Jenny, H., 1941. *Factors of Soil Formation: A System of Quantitative Pedology*. McGraw-Hill, New York.

Jenny, H., 1961. Derivation of state factor equations of soils and ecosystems. *Soil Science Society of America Proceedings* 25, 385–388.

Jones, E.J., Filippi, P., Wittig, R., Fajardo, M., Pino, V., McBratney, A.B., 2021. Mapping soil slaking index and assessing the impact of management in a mixed agricultural landscape. *Soil* 7(1), 33–46.

Kidd, D., Webb, M., Malone, B., Minasny, B., McBratney, A., 2015. Digital soil assessment of agricultural suitability, versatility and capital in Tasmania, Australia. *Geoderma Regional* 6, 7–21.

Kidd, D., Searle, R., Grundy, M., McBratney, A., Robinson, N., O'Brien, L., Zund, P., Arrouays, D., Thomas, M., Padarian, J., Jones, E., Bennett, J.M., Minasny, B., Holmes, K., Malone, B.P., Liddicoat, C., Meier, E.A., Stockmann, U., Wilson, P., Wilford, J., Payne, J., Ringrose-Voase, A., Slater, B., Odgers, N., Gray, J., van Gool, D., Andrews, K., Harms, B., Stower, L., Triantafilis, J., 2020. Operationalising digital soil mapping – Lessons from Australia. *Geoderma Regional* 23, e00335.

Ma, Y.X., Minasny, B., Wu, C.F., 2017. Mapping key soil properties to support agricultural production in Eastern China. *Geoderma Regional* 10, 144–153.

Ma, Y.X., Minasny, B., Malone, B., Mcbratney, A.B., 2019a. Pedology and digital soil mapping (DSM). *European Journal of Soil Science* 70(2), 216–235.

Ma, Y.X., Minasny, B., Welivitiya, W.D.D.P., Malone, B.P., Willgoose, G.R., McBratney, A.B., 2019b. The feasibility of predicting the spatial pattern of soil particle-size distribution using a pedogenesis model. *Geoderma* 341, 195–205.

Ma, Y.X., Minasny, B., McBratney, A.B., 2021a. Identifying soil provenance based on portable X-ray fluorescence measurements using similarity and inverse-mapping approaches – A case in the Lower Hunter Valley, Australia. *Geoderma Regional* 25, e00368.

Ma, Y.X., Minasny, B., McBratney, A.B., Poggio, L., Fajardo, M., 2021b. Predicting soil properties in 3D: Should depth be a covariate? *Geoderma* 383, 114794.

Malone, B., Searle, R., 2020. Improvements to the Australian national soil thickness map using an integrated data mining approach. *Geoderma* 377, 114579.

Malone, B.P., McBratney, A.B., Minasny, B., Laslett, G.M., 2009. Mapping continuous depth functions of soil carbon storage and available water capacity. *Geoderma* 154, 138–152.

McBratney, A.B., Mendonca Santos, M.L., Minasny, B., 2003. On digital soil mapping. *Geoderma* 117, 3–52.

Minasny, B., McBratney, A.B., 2016. Digital soil mapping: A brief history and some lessons. *Geoderma* 264, 301–311.

Mosleh, Z., Salehi, M.H., Jafari, A., Borujeni, I.E., Mehnatkesh, A., 2016. The effectiveness of digital soil mapping to predict soil properties over low-relief areas. *Environmental Monitoring and Assessment* 188(3), 195.

Mulder, V.L., Lacoste, M., Richer-de-Forges, A.C., Martin, M.P., Arrouays, D., 2016. National versus global modelling the 3D distribution of soil organic carbon in mainland France. *Geoderma* 263, 16–34.

Okonkwo, E.I., Corstanje, R., Kirk, G.J.D., 2018. Digital soil assessment for quantifying soil constraints to crop production: A case study for rice in Punjab, India. *Soil Use and Management* 34(4), 533–541.

Orton, T.G., Mallawaarachchi, T., Pringle, M.J., Menzies, N.W., Dalal, R.C., Kopittke, P.M., Searle, R., Hochman, Z., Dang, Y.P., 2018. Quantifying the economic impact of soil constraints on Australian agriculture: A case-study of wheat. *Land Degradation & Development* 29(11), 3866–3875.

Padarian, J., Minasny, B., McBratney, A.B., 2019. Using deep learning for digital soil mapping. *Soil*, 5, 79–89.

Poggio, L., Gimona, A., Brewer, M.J., 2013. Regional scale mapping of soil properties and their uncertainty with a large number of satellite-derived covariates. *Geoderma* 209–210, 1–14.

Poggio, L., Gimona, A., 2014. National scale 3D modelling of soil organic carbon stocks with uncertainty propagation — An example from Scotland. *Geoderma* 232–234, 284–299.

Poggio, L., Gimona, A., Spezia, L., Brewer, M.J., 2016. Bayesian spatial modelling of soil properties and their uncertainty: The example of soil organic matter in Scotland using R-INLA. *Geoderma* 277, 69–82.

Pouladi, N., Møller, A.B., Tabatabai, S., Greve, M.H., 2019. Mapping soil organic matter contents at field level with Cubist, Random Forest and kriging. *Geoderma* 342, 85–92.

Pracilio, G., Adams, M.L., Smettem, K.R.J., Harper, R.J., 2006. Determination of spatial distribution patterns of clay and plant available potassium contents in surface soils at the farm scale using high resolution gamma ray spectrometry. *Plant and Soil* 282(1–2), 67–82

Robinson, T.P., Metternicht, G., 2006. Testing the performance of spatial interpolation techniques for mapping soil properties. *Computers and Electronics in Agriculture* 50(2), 97–108.

Román Dobarco, M., McBratney, A., Minasny, B., Malone, B., 2021. A modelling framework for pedogenon mapping. *Geoderma* 393, 115012.
Sheets, K.R., Hendrickx, J.M.H., 1995. Noninvasive soil water content measurement using electromagnetic induction. *Water Resources Research* 31(10), 2401–2409.
Somarathna, P.D.S.N., Malone, B.P., Minasny, B., 2016. Mapping soil organic carbon content over New South Wales, Australia using local regression kriging. *Geoderma Regional* 7(1), 38–48.
Stockmann, U., Cattle, S.R., Minasny, B., McBratney, A.B., 2016. Utilizing portable X-ray fluorescence spectrometry for in-field investigation of pedogenesis. *Catena* 139, 220–231.
Sudduth, K.A., Kitchen, N.R., Wiebold, W.J., Batchelor, W.D., Bollero, G.A., Bullock, D.G., Clay, D.E., Palm, H.L., Pierce, F.J., Schuler, R.T., Thelen, K.D., 2005. Relating apparent electrical conductivity to soil properties across the north-central USA. *Computers and Electronics in Agriculture* 46(1), 263–283.
Taghizadeh-Mehrjardi, R., Nabiollahi, K., Kerry, R., 2016. Digital mapping of soil organic carbon at multiple depths using different data mining techniques in Baneh region, Iran. *Geoderma* 266, 98–110.
Thomas, M., Clifford, D., Bartley, R., Philip, S., Brough, D., Gregory, L., Willis, R., Glover, M., 2015. Putting regional digital soil mapping into practice in Tropical Northern Australia. *Geoderma* 241–242, 145–157.
Triantafilis, J., Lesch, S.M., La Lau, K., Buchanan, S.M., 2009. Field level digital mapping of cation exchange capacity using electromagnetic induction and a hierarchical spatial regression model. *Australian Journal of Soil Research* 47, 651–663.
Tromp-van Meerveld, H.J., McDonnell, J.J., 2009. Assessment of multi-frequency electromagnetic induction for determining soil moisture patterns at the hillslope scale. *Journal of Hydrology* 368(1), 56–67.
Tziachris, P., Aschonitis, V., Chatzistathis, T., Papadopoulou, M., 2019. Assessment of spatial hybrid methods for predicting soil organic matter using DEM derivatives and soil parameters. *Catena* 174, 206–216.
van Zijl, G.M., Bouwer, D., van Tol, J.J., le Roux, P.A.L., 2014. Functional digital soil mapping: A case study from Namarroi, Mozambique. *Geoderma* 219–220, 155–161.
Veronesi, F., Corstanje, R., Mayr, T., 2012. Mapping soil compaction in 3D with depth functions. *Soil and Tillage Research* 124, 111–118.
Viscarra Rossel, R.A.V., McKenzie, N.J., Grundy, M.J., 2010. Using proximal soil sensors for digital soil mapping. In: J.L. Boettinger, D.W. Howell, A.C. Moore, A.E. Hartemink, S. Kienast-Brown (Eds.), *Digital Soil Mapping: Bridging Research, Environmental Application, and Operation*. Springer Netherlands, Dordrecht, pp. 79–92.
Wadoux, A.M.J.C., 2019. Using deep learning for multivariate mapping of soil with quantified uncertainty. *Geoderma* 351, 59–70.
Walter, C., Lagacherie, P., Follain, S., 2006. Integrating pedological knowledge into digital soil mapping. In: P. Lagacherie, A.B. McBratney, M. Voltz (Eds.), *Developments in Soil Science*, Elsevier, pp. 281–300.
Wilford, J., Minty, B., 2007. The use of airborne gamma-ray imagery for mapping soils and understanding landscape processes. In: P. Lagacherie, A.B. McBratney, M. Voltz (Eds.), *Digital Soil Mapping, an Introductory Perspective. Developments in Soil Science*. Elsevier, Amsterdam, pp. 207–218.
Wilford, J.R., Bierwirth, P.N., Craig, M.A., 1997. Application of airborne gamma-ray spectrometry in soil/regolith mapping and applied geomorphology. *AGSO Journal of Australian Geology and Geophysics* 17, 201–216.
Wong, M.T.F., Harper, R.J., 1999. Use of on-ground gamma-ray spectrometry to measure plant-available potassium and other topsoil attributes. *Australian Journal of Soil Research* 37(2), 267–277.

23 A Critique of the Language of 'Constraints' and Its Consequences for Soil Management

Catherine Allan
School of Agricultural, Environmental and Veterinary Sciences, Charles Sturt University, Albury, Australia
Charles Sturt University, Co-operative Research Centre for High Performance Soils, Callaghan, NSW Australia

CONTENTS

23.1 INTRODUCTION

We can improve our practice (i.e., the things we do) through reflecting (Schön 1983). Reflecting is a complex activity that requires, among numerous other facilities and inclinations, the ability to identify assumptions and prior thinking (Fazey and Schultz 2009). One way to assist reflecting in and on practice is to consider how particular situations are 'framed' in our professions and beyond (Ison, Allan, and Collins 2015). We all use frames to help us understand and relate to the world around us; they mediate what is observed and what it means (see, for example, Bednarek 2005; Goffman 1975). Framing is rarely a conscious act, it is just that when we observe things happening or encounter new ideas we make sense of the information by referring to what we already know, or understand, or expect. We build up these mental frames over time and reinforce them through our own telling and retelling of situations. The frames we use reflect physical contexts, but the main influences on them are social; they are developed, maintained and revised via discourse, and in particular, spoken and written language (Penman, Fill, and Muhlhausler 2001). Lakoff (2010, 72) notes that all thinking and talking involves frames and reminds us that these frames '...come in systems, a single word typically activates not only its defining frame, but also much of the system its defining frame is in'. So, the words we use bring along a host of related ideas (entailments). Consciously examining our frames by considering the words we use, and their associated systems or entailments, provides opportunity to consider consequences of actions and potential alternatives.

DOI: 10.1201/9781003093565-23

This book is about soil constraints and addressing them; that is, it is focused on a particular way of framing soil and soil management. By reflecting on when and how we use the term 'soil constraints', soil scientists and managers can develop a deeper understanding of practice. In this chapter I do not suggest that soil constraints are imaginary, or that they are unimportant, but I do want to highlight that using the discourse of constraints to frame soil and its management has real consequences. I start by considering the words constraint and constraints and the general ideas behind those words. I then briefly examine the history of using constraints as a frame for soils research and management, and three of the potential consequences of this choice. I conclude with some alternate framings of soil and soil management that could be considered to broaden practice in response to the complexity of soil managing in the Anthropocene.

23.2 CONSTRAINTS

'Constraint' is a commonly used word that has specific nuances in particular fields of endeavour, but all relate to the idea of holding something back – preventing or slowing it down. In much traditional scientific research, 'constraints' are something imposed by humans, being used to provide experimental conditions by reducing complexity in the situation being studied. Similarly, setting constraints, or boundaries, is a key part of science-based modelling. The world of business, on the other hand, has a designated Theory of Constraints Management, which considers constraints as being imposed **on** human endeavour or desires, rather than **by** humans in experiments and models. Because the focus in business is on production and throughput, constraints become somewhat synonymous with constrictions of flow. In this theory, optimisation becomes important, and multiple constraints can lead to 'trade-offs'.

Using the term 'constraints' in relation to soils mirrors the concept used in the business world, that is, it is about working with constraints imposed on human desires such as maximum crop yields, or the provision of ecosystem services, often with an aim of optimisation, or trading off the impacts of multiple constraints. This is reflected in the introduction to this book which notes that 'It is important to recognize soil constraints that limit crop productivity, farm sustainability, and environmental quality… Soil constraints include any physical, chemical, or biological characteristics that limit or inhibit plant access to resources such as moisture and nutrients, thereby impacting farm productivity'.

So, soil constraints impose limits on our human goals. Focusing on constraints imposed by features of the soil is an entirely reasonable way of considering the influence of a soil on productivity; but it is historically quite a recent framing choice. It is possible to visualise the use of the term 'soil constraints' in published books over time, using the Google Ngram (Michel et al. 2011), as shown in Figure 23.1. The Ngram, and my subsequent review of the usage within the listed volumes, suggests there was little use of the term 'soil constraints' before the 1960s.

Similarly, Figure 23.2 shows that searching using the term 'soil constraint*' in the journal database Web of Science (WOS) yielded 262 papers, book chapters and conference proceedings that discussed soil constraints. There was some use of the term in the 1970s (too little to show on this scale), a steady growth from the late 1980s, and by the 2000s, the term is appearing in the title of some of these papers.

Figures 23.1 and 23.2 are searches made without context, but because 'constraints' relates to holding something back we need to consider what that something is, by understanding the use of the term in various contexts in these publications. The thing being constrained may be root growth or crop yields, or it may be a more overarching concept such as productivity or profitability. A closer look at the use of soil constraint in WOS shows the term is predominantly used in papers discussing soil science, agronomy, and plant sciences, as well as general agriculture and environmental science (Figure 23.3).

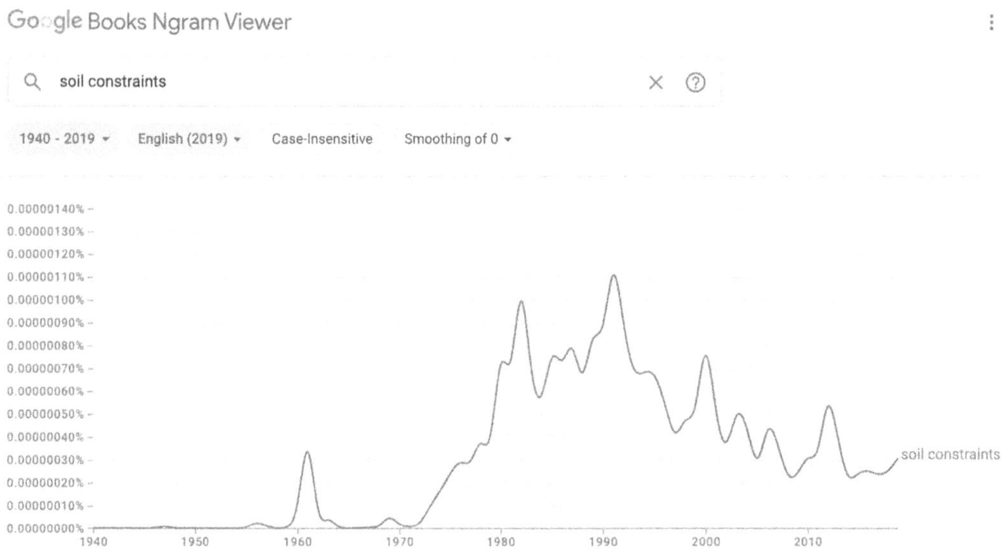

FIGURE 23.1 Google Books Ngram showing use of the term 'soil constraints' in scientific books between 1940 and 2019 as a percentage of all terms.

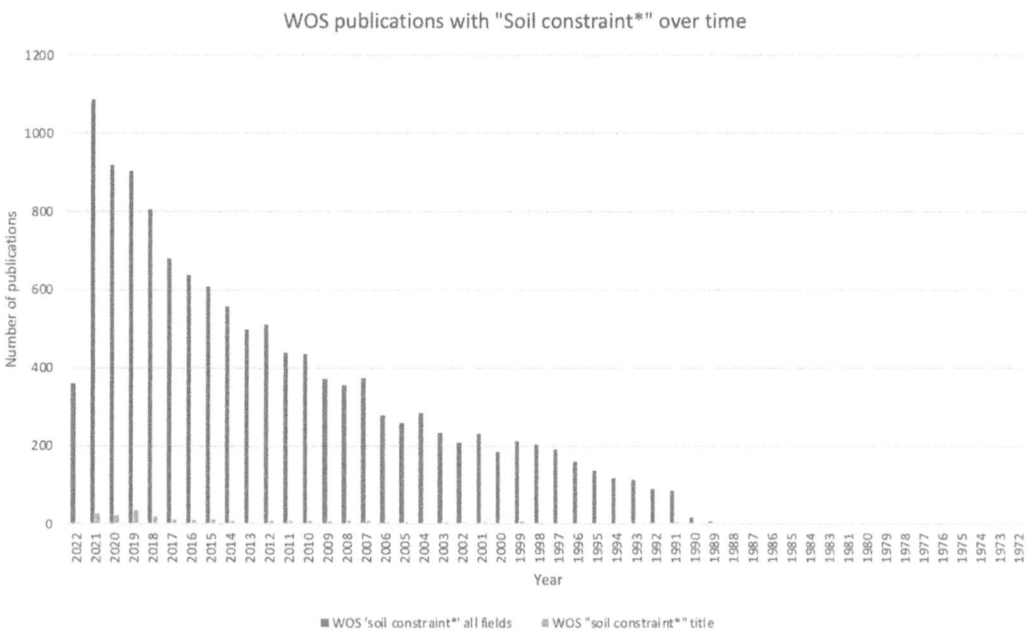

FIGURE 23.2 Summary of search for 'soil constraint' in all fields, and in titles, of papers in all WOS journals from the first mention in 1972 to part way through 2022.

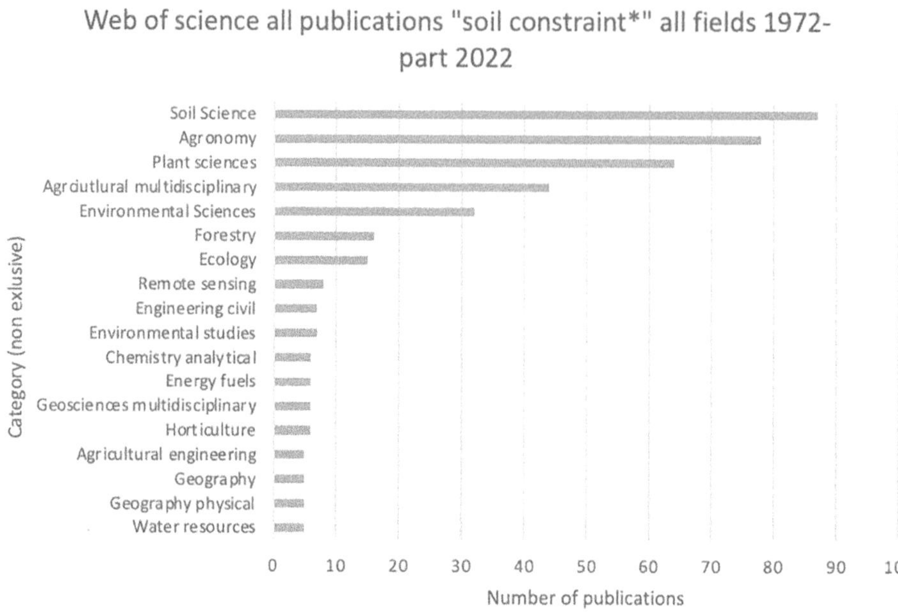

FIGURE 23.3 The category of WOS papers in which 'soil constraint' is used, from the first mention in 1972 to part way through 2022.

23.3 CONSEQUENCES

The figures suggest that 'soil constraints' is currently found to be useful for guiding action, that is, many soil scientists, agronomists and plant scientists are choosing to frame some of their work within the constraints frame. But framing a topic highlights one view of a situation over others (Jorgenson and Steier 2013). With the privileging of the constraints frame for soil, what might soil scientists and others be giving less time to? In other words, how else could soil and its management be framed? An opposite concept to constraint could be 'letting something go', captured by the ideas of opportunity, or potential. The term 'soil opportunity' does not appear as a regular way of framing soils, but 'soil potential' does and it has a long history. Figure 23.4 indicates that, in published

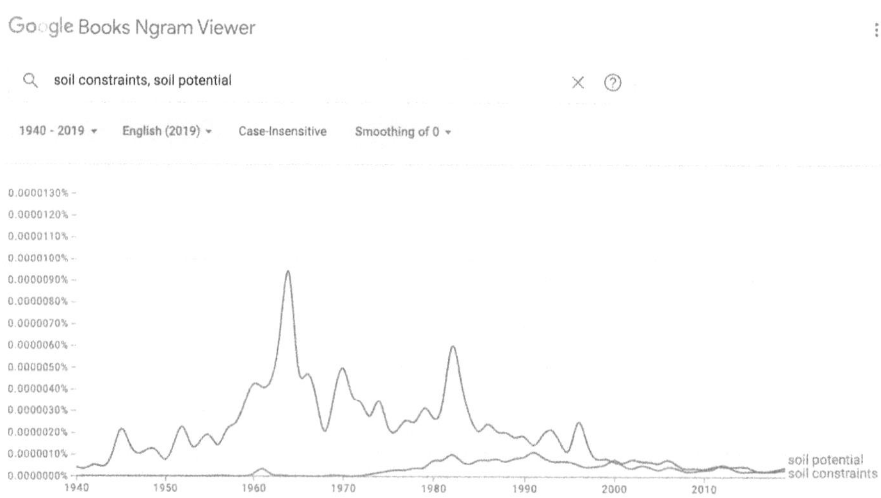

FIGURE 23.4 Google Books Ngram showing the use of the terms 'soil constraints' and 'soil potential' in scientific books between 1930 and 2019 as a percentage of all terms.

books, potential was a term that used to be used more when framing discussions about soils, and that the use declined around the time the idea of constraints began to rise.

Figure 23.4 indicates that the increased use of the term soil constraint coincided with the decline in use of the term soil potential. It is impossible from this to speculate on the reasons behind this pattern; it merely indicates that these two frames have been employed differently over time. Whatever the underlying cause, there are a number of potential consequences of this shift in discourse. Three of these are discussed below.

23.3.1 Potential Consequence 1: Reduced Options for On-Farm Adaptation

Both longer-term global warming and local weather shocks prompt farmers to adapt their practices to maintain total farm productivity (Sheng, Zhao, and Yang 2021). Adaptation at any scale relies on having a range of real management options (Gunderson 2015), but reducing the frames in use around a situation reduces options for action (Lakoff 2010). A constraints frame focuses on what a soil cannot do, while a focus on the potential of a soil leads to considering a soil for what it can do. Keeping the additional frame increases the number of management options. For example, perennial grasses are an important component of long-term profitability and soil health in grazing systems (e.g. Damsma, Rose, and Cavagnaro 2015). For many decades, talking of perennial grasses in southern Australia has meant talking about Phalaris (*Phalaris aquatica*) and other introduced and 'improved'[1] pasture species, most of which grow poorly in acidic soils (see, for example, Lattimore and McCormick 2012). One understanding of the acidic soils in the higher rainfall grazing zone of south east NSW is that growing Phalaris is constrained by the acid soils, and amelioration through incorporating lime is desirable (Hayes et al. 2016). Looking at the situation through the 'potential' frame might suggest that a soil could be really good for growing the perennial native grass *Microaleana stipoides*. Remember, the ultimate aim in this example is to successfully grow perennial grass for profitable grazing. *Microaleana stipoides* thrives in acidic soils and produces high quality, if relatively low mass, feed (Allan and Whalley 2004). Managing this native grass requires lower inputs, as lime and its specialist equipment are not needed, but additional stock management may be required to, for example, reduce downgrading of sheep fleece from seed contamination. Neither of these framings of the soil situation is 'right', or complete, but by just considering two, rather than one, frame the range of management options has increased. Each of these two framings of the situation suggest a different set of soil, pasture and stock management practices for profiting from grazing, that may or may not suit particular farmers at particular times. This is a very simplistic example, but my point is that considering the soil potential, as well as the soil constraints, broadens the range of management options available for achieving the same endpoint of profitable grazing from perennial pastures. I stress that I am not suggesting that considering soil constraints and their amelioration is wrong, or inappropriate, but that it is often not the **only** useful frame, to consider.

23.3.2 Potential Consequence 2: Creating More Issues through Simplification

As noted earlier, frames come in systems, and bring a set of related ideas and entailments with them when a word is used. One entailment of the constraints frame is the expectation of full knowledge and the possibility of optimisation. With soil management and agronomy, we are working with complexity and uncertainty, but a focus on constraints suggests an inclination to treat soil like an ideal machine with an optimum throughput that can be achieved through tinkering with the parts. This focus on 'fixing' a single constraint almost inevitably leads to new issues emerging. Choosing optimisation of soil function as the goal is likely to set up a chain of new constraints and advice about addressing them.

---

[1] Another frame with consequences…but for another book.

23.3.3 Potential Consequence 3: Reduced Social Goodwill towards Soils

To be constantly focusing on soil constraints is to be talking down the soils with those constraints. For some practitioners, this may lead to a combative relationship with 'hostile' or even 'villainess' soil (see, for example Grains Research and Development Corporation 2022), with soil becoming an enemy or a challenge, and conquering it becoming the goal. Even more problematic, the negative discourse towards some soils may be shared within districts and practitioner cohorts. Treating soils as an enemy to be conquered may reduce the standing of soils within a society. Rather than being something to cherish, and be thankful for, or to exploit or even something to protect, soils may become despised and derided. This, in turn, may have impacts on public interest in, and funding for, soils.

23.4 SOME OTHER CURRENT FRAMINGS FOR CONSIDERATION

I have shown how framings such as constraint and potential have waxed and waned over time, mostly in response to new knowledge and academic fashions rather than any conscious decision by soil scientists. There has, however, been a recent, and very deliberate, attempt by some soil scientists to frame soils as an actor in global environmental issues (Bouma and McBratney 2013), through the actively chosen term 'soil security'. Soil security is a metaphor with a range of entailments (Allan 2019), but of most interest in this discussion is the focus it brings to the frames of soil 'capability' and soil 'condition'. These are not new ideas but have been assigned specific meaning in the soil security literature. Capability addresses the functions that a soil could be expected to perform, and what it could produce, with any 'answer' always related to the context of the soil's reference state (McBratney, Field, and Koch 2014). Capability can be considered as part of the soil potential frame, as it is centred on what a soil can do.

Another concept within the wider soil security frame is that of 'soil condition', which is assessed at contemporary and short-term management scales (McBratney et al. 2016). The concept of soil condition is similar to that of soil 'health'. The metaphor of health is frequently evoked to aid understanding of complex, dynamic entities such as rivers or wetlands (Allan 2007), and is often applied to soils (see, for example, Lobry de Bruyn and Andrews 2016; Bennett and Cattle 2013). A soil health frame brings entailments from situations of human health. In this general frame, a soil constraint is similar to sickness or injury and requires diagnosis and treatment. However, the end goal of health and vitality relates to soil potential.

23.5 IN CONCLUSION

I have presented this brief exploration of the use of the concept of soil constraints and its potential consequences to demonstrate the usefulness of the frame as an object to aid reflecting on and in practice. Framing is a choice, whether conscious or otherwise; and there is great reflective power in considering the 'what can be done if we frame this situation as X' (Ison and Straw 2020, 150). Next time you think or say or write the words 'soil constraint'...I encourage you to pause to reflect. Constraint on what? What is the context? What are the goals? Is considering soil as a machine appropriate in this instance? What might be the consequences of using a constraints frame when sharing insights with your peers? Or with next and end users? Is the soil an enemy with hostile intent? Or a friend who can support your dreams? And, importantly, are other frames of equal or more use in this particular situation? How might you approach the situation with a different frame? Through questions such as these we can all deepen our soil research and management practice.

REFERENCES

Allan, C. 2007. Exploring natural resource management with metaphor analysis. *Society & Natural Resources* 20 (4):351–362.

———. 2019. The opportunities and risks of the soil security metaphor: A review. *Sustainability* 11 (16):4464.

Allan, C., and R. D. B. Whalley. 2004. Some factors influencing landholder opinion of the native grass Microlaena stipoides. *Rangeland Journal* 26 (2):178–189.

Bednarek, M. A. 2005. Frames revisited—the coherence-inducing function of frames. *Journal of Pragmatics* 37 (5):685–705.

Bennett, J. M., and S. R. Cattle. 2013. Adoption of soil health improvement strategies by Australian farmers: I. Attitudes, management and extension implications. *The Journal of Agricultural Education and Extension* 19 (4):407–426.

Bouma, J., and A. McBratney. 2013. Framing soils as an actor when dealing with wicked environmental problems. *Geoderma* 200–201:130–139.

Damsma, K. M., M. T. Rose, and T. R. Cavagnaro. 2015. Landscape scale survey of indicators of soil health in grazing systems. *Soil Research* 53 (2):154–167.

Fazey, I., and L. Schultz. 2009. Adaptive people for adaptive management. In *Adaptive Environmental Management: A Practitioners Guide*, edited by C. Allan, and G. Stankey. Dordrecht: Springer.

Goffman, E. 1975. *Frame Analysis: An Essay on the Organization of Experience*. Harmondsworth: Penguin.

Grains Research and Development Corporation. 2022. *Physical soil constraints fact sheet; The villains of hard subsurface layers in sandy soils* 2022 [cited July 12 2022]. Available from GRDC_FS2022_PhysicalSoilConstraints_v02-002.pdf.

Gunderson, L. 2015. Lesson from adaptive management: Obstacles and outcomes. In *Adaptive Management of Social-Ecological Systems*, edited by C. R. Allen, and A. S. Garmestani. Netherlands: Springer.

Hayes, R. C., G. D. Li, M. K. Conyers, J. M. Virgona, and B. S. Dear. 2016. Lime increases productivity and the capacity of lucerne (Medicago sativa L.) and phalaris (Phalaris aquatica L.) to utilise stored soil water on an acidic soil in south-eastern Australia. *Plant and Soil* 400 (1–2):29–43.

Ison, R.L., C. Allan, and K. Collins. 2015. Reframing water governance praxis: Does reflection on metaphors have a role? *Environment and Planning C: Government and Policy* 33 (6):1697–1713.

Ison, R.L., and E. Straw. 2020. *Crafting Governance Fit for the Anthropocene. The Hidden Power of Systems Thinking in Practice*. Abingdon: Routledge.

Jorgenson, J., and F. Steier. 2013. Frames, framing, and designed conversational processes. *The Journal of Applied Behavioral Science* 49 (3):388–405.

Lakoff, G. 2010. Why it matters how we frame the environment. *Environmental Communication: A Journal of Nature and Culture* 4 (1):70–81.

Lattimore, M.A., and L. McCormick. 2012. *Pasture Varieties used in New South Wales, 2012–2013*. NSW: NSW DPI and The Grassland Society of NSW.

Lobry de Bruyn, L., and S. Andrews. 2016. Are Australian and United States farmers using soil information for soil health management? *Sustainability* 8 (4):304.

McBratney, A., D.J. Field, and A. Koch. 2014. The dimensions of soil security. *Geoderma* 213:203–213.

McBratney, A., D.R. Field, C.L.S. Morgan, and L.E. Jarrett. 2016. Soil Security: A rationale. In *Global Soil Security*, edited by D.R. Field, C.L.S. Morgan, and A. McBratney. Switzerland: Springer.

Michel, J.B., Y.K. Shen, A.P. Aiden, et al. 2011. Quantitative analysis of culture using millions of digitized books. *Science* 331 (6014):176–182.

Penman, R., A. Fill, and P. Muhlhausler. 2001. *The Ecolinguistics Reader: Language, Ecology and Environment, Environmental Matters and Communication Challenges*. London: Continuum.

Schön, Donald A. 1983. *The Reflective Practitioner: How Professionals Think in Action*. New York: Basic Books.

Sheng, Yu, Shiji Zhao, and Sansi Yang. 2021. Weather shocks, adaptation and agricultural TFP: A cross-region comparison of Australian Broadacre farms. *Energy Economics* 101:105417.

24 Soil Constraints on Ecosystem Productivity and Their Management in Oil Sands Reclamation

Min Duan[1,2] *and Scott X. Chang*[2]
[1]Key Laboratory of Ecology of Rare and Endangered Species and Environmental Protection (Guangxi Normal University), Ministry of Education, Guilin, P.R. China
[2]Department of Renewable Resources, University of Alberta, Edmonton, Alberta, Canada

CONTENTS

24.1 OIL SANDS RECLAMATION IN CANADA

Oil sands are a mixture of sand, clay, other minerals, water, and bitumen (Alberta, 2022a). Oil sands deposits are distributed in many countries such as Venezuela, the United States, Russia, and Canada (Alberta, 2022a). In Alberta, Canada, it is estimated that there are 165.4 billion barrels of bitumen that can potentially be economically extracted (Alberta, 2022b). The oil sands resource accounts for 96% of Canada's oil reserve [Canadian Association of Petroleum Producers (CAPP), 2022]. These Albertan oil sands deposits are primarily distributed in the watersheds of the Athabasca River, Peace River, and Cold Lake, which combined cover an area of more than 142,000 km^2 (CAPP, 2022). Oil sands deposits within 70 m of the surface can be recovered by surface mining, but 80% of the deposits are not shallow enough to have them economically extracted by surface mining, so they are extracted using energy-intensive in-situ methods (CAPP, 2022). The Athabasca oil sands region along the Athabasca River is the largest single bitumen deposit in the world and is also the only one suitable for surface mining in Alberta (Mossop et al., 1982). In 2017, crude bitumen production in

DOI: 10.1201/9781003093565-24

Alberta reached over 2.8 million barrels per day, of which 45% came from surface mining, and 54% came from in-situ extraction (Alberta Energy Regulator, 2022). It is expected that by 2027 bitumen production will increase to 3.9 million barrels per day (Alberta, 2022a).

Prior to surface mining, the land is drained if the water table is high, and forests are clear-cut harvested. The muskeg or peat material and surface mineral soils are stripped and either placed directly at reclamation sites or stockpiled for future reclamation uses (Fung and Macyk, 2000). Below the surface soil and above the oil sands is the overburden material, which is physically removed and used for later reclamation efforts (Fung and Macyk, 2000). Oil sands are mined and transported to the plant for hot water extraction. The bitumen is recovered for subsequent upgrading and refining (CAPP, 2022). The tailings sand material, one main byproduct of the extraction process, settles in tailings ponds; it is used as a reconstruction substrate in reclamation (Fung and Macyk, 2000).

For in-situ extractions, production facilities, including power lines, pipelines, roads, and well pads, are built. In the extraction, wells are drilled into deep oil sands deposits, and high-pressure steam, combustion, or other sources of heat are injected into the reservoir to heat up the bitumen, so bitumen can flow to recovery wells and is pumped to the surface (CAPP, 2022). The disturbances before and during the extraction fragment the landscape and disturb wildlife habitats, causing significant harm to boreal forest ecosystems (Dyer and Huot, 2010).

Oil sands mining can significantly disrupt the landscape and alter ecosystem structure and function. It is estimated that surface mining will disturb 4,800 km^2 of land, 65% of which are wetlands (Raine et al., 2002), while *in situ* operations will result in a potential loss of 6,500 km^2 of boreal forest by direct disturbance (Dyer and Huot, 2010). To date, oil sands extraction activities have disturbed about 1,055 km^2 of land since the first commercial mineable oil sands operations began in 1967, and more than 90% of disturbances are from surface mining activities (CAPP, 2022). According to the Alberta Environmental Protection and Enhancement Act, oil sands mining operators are required to create reclaimed soils and landforms that are "capable of supporting and establishing a self-sustaining, locally common boreal forest, whose productivity is equivalent to or better than the pre-disturbance level" (Alberta Environment, 2010; Cumulative Environmental Management Association, 2009). However, the diverse boreal forest ecosystems have developed over millennia under the influences of parent material, topography, climate, and natural disturbances (Beckingham and Archibald, 1996). After mining, the ecosystem is totally disturbed. It is not a simple task to restore disturbed boreal ecosystems in the oil sands to their previous state within a short time. It is estimated that it takes more than 200 years to develop an old-growth forest following oil sands mining (Thompson, 2009). Restoring boreal forest ecosystems after disturbance from oil sands mining is a complex and multistep process, which involves the creation of landform, the reconstruction of functioning soil, the establishment of plant communities, the formation of mutually beneficial soil–vegetation relationships, the establishment of animal, insect, and microorganism communities, and the development of new hydrogeological and nutrient cycles and vegetation succession patterns (Macdonald et al., 2012). Up to 2020, only roughly 1 km^2 of surface disturbed mineable oil sands and 0.4 km^2 of in-situ extraction sites have been successfully reclaimed and certified (CAPP, 2022).

Wetlands are an integral component of the boreal ecosystem in the oil sands region. Wetlands have special ecological functions and values, such as accumulating carbon, slowing water runoff, recharging groundwater, attenuating floods, purifying water, and maintaining biodiversity and aesthetics or cultural values (Foote, 2012). The common wetland types in the Athabasca boreal forest region include bogs, fens, swamps, marshes, and ponds (Grant et al., 2008). It is estimated that existing, approved, and proposed surface mining and in-situ recovery projects can result in a loss of as much as 4,600 km^2 of peatlands (bogs and fens) (Pembina Institute, 2022). However, the reclamation of wetlands in the oil sands region has not yet been demonstrated because of unsophisticated techniques, gaps in reclamation knowledge, and the tremendous cost involved (Grant et al., 2008). For example, peatlands, as the primary class of wetlands in the oil sands region, cannot feasibly be restored due to the long time requirement for peat formation. Natural peat accumulation is a complex nonlinear process (Clymo, 1992). Restoring peatlands to a socially acceptable condition will require 1–3 centuries under a favorable climate (Foote, 2012). Neither oil sands companies nor their insurers have wanted obligations for centuries. In addition, there is little clear and consistent guidance and criteria to assess the success of wetland reclamation, and there is no "no net loss of wetland" policy in Alberta; for quicker results, the oil sands companies prefer to replace the disturbed peatlands to upland forests, open-water wetlands, and end pit lakes, which are less challenging for oil sands operators and have more chances to succeed within a meaningful timeframe (Foote, 2012). Currently, there is no closure plan calling for the restoration of lost peatlands (Rooney et al., 2012); only a few oil sands companies are trying to restore wetlands, but no demonstrated success has been achieved in reclaiming wetlands (Foote, 2012). Since there is no legal requirement for type-for-type replacement of disturbed lands, operators tend to construct upland forests and shallow open-water wetlands instead of returning peatlands under the concept of equivalent capability (Foote, 2012). The approaches currently used will result in a massive loss of wetlands in Alberta. However, this chapter will focus on the reclamation of oil sands disturbed lands to upland forest ecosystems.

The general sequence of reclaiming upland forests after oil sands mining is (i) surveying natural soils to delineate suitable reclamation materials; (ii) salvaging, placing, and preparing reclamation materials on-site; (iii) seeding of a grass/legume mixture before tree planting to control soil erosion; (iv) planting trees; (v) maintaining and monitoring the established plant communities; and (vi) certification of the reclaimed land (Fung and Macyk, 2000).

The reconstruction of functioning soil is an important step of reclamation (Figure 24.1) (Macdonald et al., 2012) because the soil provides the base for many other ecosystem services, including anchorage support to vegetation, water storage and supply, habitat for microbial and faunal diversity, and nutrient supply (Dominati et al., 2010). In the oil sands region, overburden materials, as well as tailings sand, are used as substrates to rebuild landforms. Surface soils and near-surface geological materials (to a depth of 3 m) salvaged before mining activities are used as the cover soil for reconstructing the reclaimed landscape (Turchenek and Lindsay, 1982). Peat (due to its availability across large portions of the mined area) salvaged from pre-mining wetland areas or the forest floor from pre-mining upland areas is usually mixed with the cover soil to increase the nutrient availability and water-holding capacity (Mackenzie and Naeth, 2010; Dhar et al., 2018).

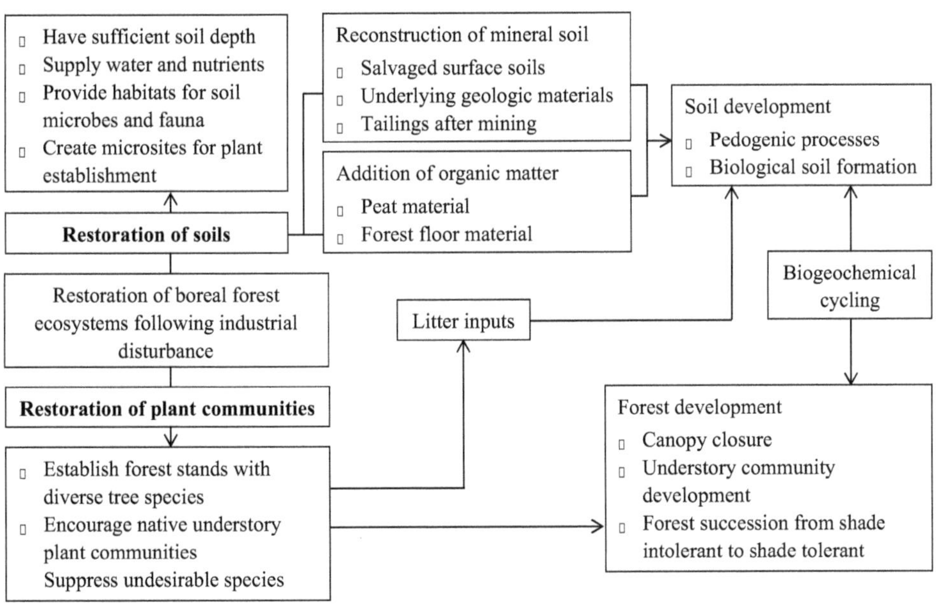

FIGURE 24.1 Multistep processes to restore boreal forest ecosystems after disturbance from oil sands mining. (Modified from Macdonald et al., 2012.)

Another important step of reclamation is the re-establishment of plant communities (Figure 24.1) (Macdonald et al., 2012; Dhar et al., 2018). After the reconstruction of the landscape, native tree species, primarily trembling aspen (*Populus tremuloides* Michx.), balsam poplar (*Populus balsamifera* L.), white spruce (*Picea glauca* (Moench) Voss), black spruce (*Picea mariana* (P. Mill.) B.S.P.), and jack pine (*Pinus banksiana* Lamb.), are planted to the harsher boreal reclamation sites according to the guidelines for reclamation to forest vegetation in the Athabasca oil sands region (Alberta Environment, 2010). The rapid establishment of a closed tree canopy is expected to suppress the establishment of weedy forb and grass species (Maundrell and Hawkins, 2004) and encourage the production of litter and forest floor layers, which will facilitate soil redevelopment (Klinka et al., 1990). Apart from the establishment of overstory tree species, understory plant communities are restored as well. The most common shrubs planted in reclaimed sites include nitrogen-fixing shrubs such as green alder (*Alnus crispa* (Ait.) Pursh) and Canadian buffaloberry (*Shepherdia canadensis* (L.) Nutt.), and other shrubs such as red osier dogwood (*Cornus stolonifera* Michx.), wild roses (*Rosa acicularis* Lindl. and *Rosa woodsii* Lindl.), pin cherry (*Prunus pensylvanica* L.f.), choke cherry (*Prunus virginiana* L.), and saskatoon berry (*Amelanchier alnifolla* (Nutt.) Nutt.) (Fung and Macyk, 2000).

The soil development is then typically assessed every five years to monitor changes in the morphological and physicochemical properties (Fung and Macyk, 2000). The aboveground productivity is typically assessed using measures such as tree and shrub survival rate, and height and diameter growth one year after planting and at 5-year intervals thereafter (Fung and Macyk, 2000). When a site meets or exceeds the regulatory re-vegetation guidelines for oil sands reclamation and is monitored over 15 years, a certificate is issued by the government of Alberta, and the company's responsibility for land reclamation is released (Alberta Environment, 2010).

24.2 SOIL CONSTRAINTS ON FOREST GROWTH IN THE RECLAIMED LANDSCAPE IN THE OIL SANDS

Forest growth depends on environmental conditions such as light availability, temperature, topography, the supply of water and nutrients from soils, and natural disturbances (Peace and Grubb, 1982; Beckingham and Archibald, 1996). Reconstructed soils placed for reclamation following surface mining in the oil sands region may exhibit unfavorable physical, chemical, and biological properties, which can negatively affect the early growth of trees planted during the land reclamation process (Yan et al., 2012; Farnden et al., 2013; Duan et al., 2015; Pinno and Errington, 2015).

24.2.1 Physical Factors Affecting Forest Growth

In oil sands reclamation, soil construction using peat and/or forest floor, mineral soil, overburden, and/or tailings sand leads to the formation an artificial soil profile. With abrupt changes between those materials, soil properties may be greatly changed (Li et al., 2014; Jung et al., 2014). Some soil physical properties of reclaimed soils, such as water availability and soil compaction, may adversely affect plant growth.

24.2.1.1 Water Availability

Soil water storage and transmission characteristics and water availability are influenced by many factors, such as soil structure, soil texture, and soil layering (Barnhisel and Hower, 1997; Li et al., 2014). In oil sands reclamation, the reconstruction of multi-layered soils using different materials (such as peat and/or forest floor as the soil amendment, surface soils, and near-surface geological materials as the cover soil, overburden, and/or tailings sand as the substrate) resulted in different distributions of soil water (Li et al., 2014; Jung et al., 2014), thus markedly changing soil water availability and leading to varied root production and tree growth (Jamro et al., 2014a; Jung et al., 2014).

Duan et al. (2015) studied the limiting factors for the growth of lodgepole pine (*Pinus contorta*) on a peat mineral soil mix over tailings sand in the Athabasca oil sands region, Alberta, and found that the mean annual growth of height, diameter at breast height, and aboveground biomass of pine trees were significantly correlated with soil water content in the peat mineral soil mix and foliar $\delta^{13}C$ (which indicates the degree of water stress and water use efficiency) in current-year needles; and sites with high productivities had the lowest foliar $\delta^{13}C$ (indicates a lower level of water stress) relative to sites with low and medium productivities.

Tailings sand (with more than 90% sand) as a substrate has a low water-holding capacity, thus precipitation can be easily drained into deeper soil layers and decrease soil water availability in upper soil layers, which is likely to limit fine root growth, thus restricting the ability of roots to absorb water and nutrients from the soil (Jamro et al., 2014a; Jung et al., 2014). Meanwhile, overburden material usually has a sandy clay loam or clay soil texture, which is often highly compacted due to the use of large equipment during reclamation (Fung and Macyk, 2000). Overburden material as a substrate has less macro-porosity and low permeability, thus precipitation can easily run away along a slope, and only a small part of precipitation is retained in upper soil layers, which potentially affects root penetration into deeper soil layers and limit tree growth (Jung et al., 2014; Duan et al., 2015). Moreover, the interface between the reclamation capping soil and the substrate and abrupt changes of soil texture induced from the reconstruction of multi-layered soils restrict water movement between layers because of the low capillary action of tailings sand or low permeability of overburden (Naeth et al., 2011; Jung et al., 2014). Although previous studies showed that placement of a thick capping soil layer with finer materials over coarser-textured tailings sand layers could enhance water-holding capacity due to the nature of the organic material and the capillary barrier effect (Moskal, 1999; Barbour et al., 2007; Li et al., 2014), there is no standard prescription for reclamation, and oil companies adopt different capping depths based on the substrate used (Rowland et al., 2009) and the availability of cover soils. For example, the peat mineral soil mix layer is usually less than 50 cm deep at the soil reconstruction stage (Rowland et al., 2009), and soil erosion may occur because of low vegetation coverage, which could reduce the thickness of the peat mineral soil mix layer. Duan et al. (2015) surveyed the thickness of the peat mineral soil mix layer in several sites that had been reclaimed for more than 20 years in the Athabasca oil sands region and reported that the thickness of the peat mineral soil mix layer ranged from 12 to 30 cm, suggesting that the capacity of the capping layer to increase water availability in the soil might have been very limited.

24.2.1.2 Soil Compaction

Soil compaction of reclaimed mined lands is very common due to material handling operations for restoration (e.g., land forming and spreading of topsoil material) using large equipment during reclamation (Fung and Macyk, 2000). Soil compaction causes the following problems: (i) reduction in soil porosity and aeration, (ii) loss of soil structure, (iii) reduction in hydraulic and gas conductivity, (iv) changes in thermal properties, (v) increases in strength of dry soil, (vi) increases in the depth of soil freezing, and (vii) reduction in the growth of soil organisms (Omi, 1986; Chong and Cowsert, 1997). Previous studies also showed that soil compaction reduced soil water-holding capacity, restricted water infiltration, increased runoff and erosion, resisted root penetration to deep soil layers, and changed nutrient cycling (Hatchell et al., 1970; Greacen and Sands, 1980; Corns, 1988; Cullen et al., 1991; Kozlowski, 1999; Startsev and Mcnabb, 2001; Tan et al., 2005). Tyner et al. (1948) reported that compacting surface layers down to 45 cm created survival problems for tree seedlings. Planting trees into compacted soils generally results in high tree mortality and reduced hardwood forest production (Ashby, 1991; Chaney et al., 1995; Torbert, 1995; Rodrigue and Burger, 2004), as the changes induced by soil compaction can reduce root growth, which limits the tree's ability to explore the soil volume for nutrients and moisture, thus contributing to short-term reductions in tree growth or long-term reductions in site productivity. However, less intensive grading left looser, albeit rougher, surfaces which enhanced moisture retention and root penetration, thereby promoting tree growth (Phillips, 2006).

In oil sands reclamation, soil compaction resulting in high bulk density in overburden by heavy equipment or from highly compacted reconstructed materials had less macro-porosity, which would influence the growth of roots into the overburden layer (Jamro et al., 2014a; Jung et al., 2014). Duan et al. (2015) compared the soil bulk density among white spruce sites with different productivities and found that high productivity sites had the lowest soil bulk density, followed by medium and low productivity sites, and the bulk density of overburden in low and medium productivity sites reached 1.46 and 1.55 Mg/m^3, respectively. However, Jung et al. (2014) reported that the compaction of sub-layers could reduce water infiltration, increase water storage in soil layers above, and increase the amount of available water for plant uptake. This result was further illustrated by Duan et al. (2015) who showed that soil water availability was not the limiting factor for white spruce growth on a peat mineral soil mix over overburden in the Athabasca oil sands region with high bulk density in the overburden layer and increased water availability in the peat mineral soil mix layer.

24.2.2 Chemical Factors Affecting Forest Growth

Reconstructed soils for reclamation following surface mining in the oil sands region may have unfavorable chemical properties such as low availability of nutrients (Fung and Macyk, 2000; Yan et al., 2012), unfavorable soil pH (Howat, 2000), high concentration of salts (Barbour et al., 2007), heavy metals (Bicalho et al., 2017; Abdolahnezhad and Lindsay, 2022), and residual bitumen (Scott et al., 2020). Some of those are inherent properties of boreal forest soils, and those factors can negatively affect tree performance in reclaimed sites.

24.2.2.1 Nutrient Availabilities

One of the most important issues for restoring functional ecosystems in post-mining landscapes is the reconstruction of functional soils, in which the accumulation of organic carbon in surface soil layers is critical (Jochimsen, 1996; Bradshaw, 1997). Nitrogen availability is often reported to be one of the major limiting factors for tree growth and site productivity in the boreal forest region (Yan et al., 2012; Hu et al., 2014), as boreal forests are generally nitrogen-limited because of higher nitrogen immobilization than nitrogen mineralization (Reich et al., 1997; Carmosini et al., 2003; Jerabkova et al., 2006). Immobilization makes nutrients unavailable for plants and suppresses plant growth (Zimmerman et al., 1995). The situation is even worse in the oil sands region (Fung and Macyk, 2000; Duan et al., 2015), because oil sands mining and related activities can lead to substantial loss of soil organic matter due to the loss of fertile topsoil and mechanical mixing of surface soils with infertile subsurface soils (Lal et al., 1998; Ussiri and Lal, 2005; Shrestha and Lal, 2011). Although organic amendments, including a peat mineral soil mix or forest floor-mineral soil mix, are incorporated into the reconstructed cover soil above overburden or tailings sand substrates to improve nutrient availability of the reclaimed soil, in some cases, the slow release of organic nitrogen from organic materials may affect nitrogen availability (McMillan et al., 2007; Hemstock et al., 2010; Jamro et al. 2014b; Kwak et al. 2015) and limit plant growth (Reeder and Sabey, 1987; Yan et al., 2012; Farnden et al., 2013). McMillan et al. (2007) and Hemstock et al. (2010) compared nitrogen transformation rates in reclaimed and natural boreal forest soils and found that net nitrogen mineralization, ammonification, and nitrification rates in peat-amended soils were significantly lower than that in natural soils. The differences could be attributed to inhibited nitrification by lower soil pH (Ste-Marie and Paré, 1999; Myrold, 2005), higher soil salinity (Duan et al., 2018), slow release of organic nitrogen by reduced microbial activity (Davidson et al., 1992; Horwath and Paul, 1994), and lower nitrogen reserve and labile nitrogen content in reclaimed soils (Yan et al., 2012).

Low plant-available nitrogen often limits the re-establishment and maintenance of vegetation on disturbed lands (Reeder and Sabey, 1987; Yan et al., 2012). To overcome low nitrogen availability, planted seedlings are usually fertilized in the initial five years after planting (Fung and Macyk, 2000). As the planted stands develop, the demand for nutrients may increase (Hu et al., 2014). Farnden et al. (2013) showed that nitrogen availability was limiting tree growth in a 21-year-old jack pine stand with a peat mineral soil mix over tailings sand in a reclaimed oil sands area and found a positive relationship between tree height growth and soil organic matter content. They attributed the relationship to organic matter's contribution to increasing soil nitrogen availability. Duan et al. (2015) also found that nitrogen availability was limiting white spruce growth on some reconstructed sites using peat mineral soil mix as the cover soil over overburden substrate in a 2-year field study conducted on representative reclaimed sites in the Athabasca oil sands region. They found positive relationships between tree growth and soil inorganic nitrogen concentrations. Foliar nitrogen concentrations in current-year needles in all sites were less than the critical value of 13 mg/g, which indicates nitrogen deficiency for spruce (Ballard and Carter, 1986) except for the high productivity sites (Duan and Chang, 2015). They also reported that the application of nitrogen or complete fertilizers (nitrogen, phosphorus, potassium, and micronutrients) promoted height, diameter at breast height, and aboveground biomass growth of white spruce two growing seasons after fertilization in sites with low productivity, further demonstrating the limitation of soil nitrogen availability to tree growth in reclaimed lands in the oil sands.

24.2.2.2 Acidity and Salinity

Soil pH values near neutral are favorable for root development and biomass production in different land uses (Haynes and Naidu, 1998). However, reclaimed landscapes often have low or high soil pH (Howat, 2000), which can reduce plant growth. The pH of mine soils is largely determined by their parent material and weathering conditions. Hearing et al. (2004) found that soil materials consisting of weathered brown sandstone tended to be more acidic (4.0–5.5) while unweathered gray sandstone materials were generally much more alkaline (7.5–8.0) in a surface mine in West Virginia. Low-base-content overburden causes growth-limiting acidity in the surface-mined soils, and acidic conditions limit root growth and the establishment of plants (Mays et al., 2000). The alkaline pH of reclaimed material has also been reported in many reclamation studies (Mummey et al., 2002; McMillan et al., 2007). Salvaged mineral materials of the Athabasca oil sands region were typically associated with an alkaline pH (Fung and Macyk, 2000). Mixing this alkaline material with surface organic material during over-stripping produced a more alkaline reclaimed material compared to that found in natural boreal soils.

Soil salinity is often a major abiotic stress influencing plant survival, growth and development, and site productivity (Boyer, 1982; Andrews et al., 1998; Renault et al., 1999; Rodrigue and Burger, 2004; Dluzniewska et al., 2007; Munns and Tester, 2008). According to the guidelines for reclamation to forest vegetation in the Athabasca oil sands region, if soil electrical conductivity is greater than 4 dS/m, the soils are unfavorable for the survival and growth of plants and the establishment of a forest ecosystem (Alberta Environment, 2010). High soil salinity (EC >4 dS/m) can decrease osmotic potential, induce nutrient imbalance, and cause ion toxicity (Munns, 1993; Jacoby, 1994; Cornillon and Palloix, 1997; Munns and Tester, 2008). In addition, high salinity influences plant water uptake and nitrogen metabolism, including nitrogen uptake and assimilation, and amino acid and protein synthesis (Aslam et al., 1984; Hawkins and Lewis, 1993; Parida and Das, 2004; Rubinigg et al., 2005; Ehlting et al., 2007; Botella et al., 1997; Dluzniewska et al., 2007; Duan and Chang, 2017). High salinity can also suppress soil microbial growth and activity (Yuan et al., 2007; Elmajdoub et al., 2014), thus it may markedly influence nitrogen availability in soils. High salinity was reported to influence nitrogen cycling processes (Inubushi et al., 1999; Irshad et al., 2005; Cheng et al., 2013). Studies showed that nitrification and nitrogen mineralization rates were inhibited in soils with high salinity, leading to reduced nitrogen availability (McClung and Frankenberger, 1987; Inubushi et al., 1999; Cheng et al., 2013). Boreal forests are typically nitrogen-limited, and the negative effect of salinity on nitrogen mineralization and nitrification could further lower nitrogen availability in the mined area in the boreal forest region (Duan et al., 2018).

The reconstruction of the landscape after oil sands surface mining may increase salinity in reclaimed soils in the oil sands region (Fung and Macyk, 2000; Barbour et al., 2007). Salts, mainly sodium chloride (NaCl) and sodium sulfate (Na_2SO_4), may migrate from the underlying saline tailings sand or marine shale overburden to peat mineral soil mix layers (Barbour et al., 2007). Sometimes salts are introduced into the cover soil when the saline material is improperly mixed with the non-saline cover soil. The salinity of the reclaimed soil varied depending on the substrate used and the site condition, but the electrical conductivity in tailings sand or saline overburden materials was often several orders of magnitude higher than those in natural boreal forest soils (Barbour et al., 2007; Jung et al., 2014). The upward movement of salts from the saline overburden layers into the cover soil can raise the electrical conductivity level up to 6.0 dS/m in the lower part of the peat mineral soil mix used as a cover soil (Kessler et al., 2010). Boreal tree species have considerable variability in salt tolerance between and within species, and most are salt-sensitive (Allen et al., 1994; Lilles et al., 2012; Khasa et al., 2002). Duan et al. (2015) showed that although the average electrical conductivity in all reclaimed white spruce sites was 2.12 and 3.05 dS/m in the peat mineral soil mix and overburden layer, respectively, soil electrical conductivity in the peat mineral soil mix or overburden layer was more than two times higher in low than in medium and high productivity sites, and high salinity limited the growth of white spruce trees.

24.2.2.3 Heavy Metals

After reclamation, the reclaimed soil may be affected by heavy metals, which may accumulate in the tissue of growing plants and affect plant growth (Shotyk, 2020). The main toxicity effects of heavy metals are disruptions of photosynthesis, functions of the stomata, respiration, and biomass production, which may substantially inhibit growth. Brams and Fiskell (1971) reported that copper tended to be very strongly bound to soil organic matter; although it was nontoxic to plants, it hindered the uptake of essential nutrients and thus retarded plant growth when it accumulated in plant roots.

Following oil sands mining, an overburden which contains low concentrations of hydrocarbons (usually accounting for <8% of bulk soil volume) is buried under suitable soils in some reclamation practices (Visser, 2008; La Flèche et al., 2021). The organic fraction of bitumen is enriched in vanadium (V), nickel (Ni), and molybdenum (Mo) (Bicalho et al., 2017; La Flèche et al., 2021). Moreover, large stockpiles of petroleum coke, one of the bitumen upgrading byproducts, can be incorporated into mine closure landscapes, which is also a potential source of V, Ni, and Mo in the Athabasca oil sands region (Kessler and Hendry, 2006; Nesbitt and Lindsay, 2017; Abdolahnezhad and Lindsay, 2022). Approximately 1.13×10^7 t of coke were generated during bitumen upgrading in 2019, and coke stockpiles in the Athabasca oil sands region have steadily increased over time, reaching 1.32×10^8 t by the end of 2019 (AER, 2019). Studies showed that plants and invertebrates take up and accumulate V, Ni, and Mo in the Athabasca oil sands region (Puttaswamy et al., 2010; Nakata et al., 2011; Puttaswamy and Liber, 2011, 2012). Previous studies have reported dissolved V, Ni, and Mo concentrations up to 3 mg/L, 120 μg/L and 2.0 mg/L, respectively, within fluid coke deposits (Nesbitt and Lindsay, 2017; Nesbitt et al., 2018; Robertson et al., 2019). Plants can take up those metals in different forms based on soil pH and aeration conditions (Pendias and Kabata-Pendias, 2000; Nieminen et al., 2007; Huang et al., 2015; Chen et al., 2021). Influent water chemistry also strongly influenced heavy metal releases. Properties such as higher surface area, circumneutral to alkaline pH, and increased ionic strength are generally conducive to V and Mo release, and mildly and extremely acidic conditions may favor Ni release (Abdolahnezhad and Lindsay, 2022). If roots passively take up those metals with the uptake of water, those metals enriched in reclaimed soils would be present in higher concentrations in the woody tissues of trees (La Flèche et al., 2021).

24.2.3 Biological Factors Affecting Forest Growth

In addition to the physical and chemical soil constraints on forest growth in oil sands reclamation discussed above, some biological factors, such as vegetation competition and low microbial activities, can negatively affect tree performance in reclaimed sites. In some reclaimed sites, the damage from small mammals or rodents was a problem for tree establishment. Skousen et al. (2007)

reported that a heavy ground cover of grasses and legumes during reclamation provided cover for small mammals, which may girdle trees and feed on tree bark. Tyner et al. (1948) found that compacted surface layers down to 45 cm created survival problems for tree seedlings; hardwoods such as oaks (*Quercus* spp.) and cherry (*Prunus* spp.) did not grow well because of rodent damage.

24.2.3.1 Vegetation Competition

The development of plant communities (including trees, shrubs, and forbs) with diverse species compositions, a suitable stand structure, and spatial heterogeneity is important in forest reclamation and restoration (Grant and Koch, 2007; Macdonald et al., 2012; Melnik, 2017; Duan et al., 2019). The development of plant communities on reclaimed sites is influenced by many biotic factors, such as the size of seed and bud banks, seed dispersal and wildlife impacts, and abiotic factors, including cover soil type, site preparation, climatic conditions, topography, soil properties, and human disturbances (Burger and Fannon, 2009; Tropek et al., 2013; Dhar et al., 2018). The Surface Mining Control and Reclamation Act (SMCRA) in the US requires coal operators to establish "a diverse, effective, and permanent vegetative cover of the same seasonal variety native to the area…that is capable of self-regeneration and plant succession…" (Section 816.111 of SMCRA). Coal operators generally establish an aggressive, dense ground cover to comply with these standards. In oil sands reclamation in Alberta, grass species and legumes are often seeded after cover soil placement to stabilize exposed soil and prevent soil erosion (Naeth and Wilkinson, 2004). Commonly used grass species include *Hordeum vulgare* (L.), *Avena sativa* (L.), *Elymus trachycaulus* Gould ex Shinners, *Elymus lanceolatus* (Scribn. & J.G.Sm.) Gould., *Trisetum spicatum* (L.) K. Richt., *Deschampsia cespitosa* (L.) P. Beauv., *Agrostis scabra* Willd., *Festuca saximontana* Rydb., *Festuca rubra* L., *Bromus ciliatus* L., *Poa palustris* L., *Poa compressa* L., *Elymus canadensis* L., and *Koeleria macrantha* (Ledeb.) Schult. (Canadian Natural Resources Limited, 2012; Shell Canada Energy, 2012a, 2012b; Suncor Energy Inc., 2012; Syncrude Canada Limited, 2012). A commonly used legume is *Medicago sativa* L. (Dhar et al., 2018). However, the use of grass species and legumes as cover crops can have detrimental effects on tree, shrub, and forb establishment after reclamation (Parrotta et al., 1997; Grant and Koch, 2007). This is largely due to those cover crops being highly competitive and could impede the establishment of seedlings of trees and shrubs, and decrease tree growth in the reclamation of surface-mined forestland (Burger and Torbert, 1990; Torbert, 1995; Ashby, 1997; Holl, 2002), thus influencing site productivity. These grasses and legumes compete with tree seedlings for light, water, and nutrients. Some annual or perennial grasses can grow much taller than the young seedlings in the early stage of development and smother seedlings causing the aerial part of the seedlings to die back in winter. Consequently, the survival and growth of trees planted on reclaimed mine sites are often poor. Duan et al. (2015) reported that the total understory plant community cover was greater in low than in high productivity sites, suggesting that the understory vegetation could be a strong competitor with trees for the water and nutrients stored in the soil, which could significantly decrease water and nutrient availabilities and negatively affect tree growth (Nambiar and Sands, 1993).

24.2.3.2 Effects of Reclamation on Microbial Activities

Oil sands reclamation often influences soil microbial structure, composition, and function (Ma et al., 2017), and soil microbial biomass and activities typically decrease following disturbance (Hemstock et al., 2010). Ma et al. (2017) showed that capping material type (overburden and tailings sand) affected bacteria community structure in the cover soil in oil sands reclamation, and soil pH, electric conductivity, and water-soluble organic carbon were responsible for the difference in bacterial community structure. McMillan et al. (2007) and Hemstock et al. (2010) reported that dramatic changes in soil physical and chemical properties after oil sands reclamation would influence microbial activity, organic matter decomposition, and lower net nitrogen transformation rates in reconstructed soils as compared to those in natural boreal forest soils. They were attributed to decreased osmotic water potential, changed soil pH, and toxicity of the chloride ion under saline conditions

(Heilman, 1975; Martikainen, 1985; Low et al., 1997; Inubushi et al., 1999). Decreasing soil osmotic potential with increasing electrical conductivity has decreased water availability to microbes and reduced microbial activity and biomass (Chowdhury et al., 2011). Salt addition could increase or decrease soil pH depending on the salt type, thus changing the microbial community composition. Sodium chloride and sodium sulfate usually slightly increase soil pH. Fungi are favored over bacteria under low soil pH (Högberg et al., 2003; Firer and Jackson, 2006; Rousk et al., 2009); an increase in soil pH after salt addition may reduce fungi metabolism. Mummey et al. (2002) also found that microbial carbon was 56% lower in reclaimed surface-mined soils than in adjacent undisturbed soils 20 years after land reclamation; explained that soil stockpiling was extremely detrimental to fungi, in particular mycorrhizal fungi; and stated that reclaimed soils are not able to provide any inoculant for regenerating plants. This was potentially a large problem in the boreal forest, where most tree species rely on mycorrhizal associations to secure nutrients (Jonsson et al., 1999). McMillan et al. (2007) reported that lower microbial carbon and nitrogen in reclaimed sites compared to a natural forest was the result of lower moisture content and/or differences in organic matter composition. The return of microbial populations led to improvements in the establishment of the native forest vegetation on reclaimed surface mine sites by mineralization of nutrients, nitrogen fixation, and symbiotic relationships with tree roots (Showalter et al., 2010). The establishment of vesicular-arbuscular fungi within roots of plants growing on mining disturbed lands was beneficial for enhancing phosphorus uptake by vegetation. However, the establishment of vesicular-arbuscular fungi has not been readily achieved during reclamation in arid regions compared to regions with mesic moisture regimes (Medve, 1984). In addition to phosphorus, vesicular-arbuscular fungal colonization in plants may provide nutrients such as N, Ca, S, Mg, Mn, Fe, Zn, and K (Stribley, 1987). Reclamation with relatively infertile mine soils required rapid rates of vesicular-arbuscular fungal colonization to satisfy the nutritional requirements of dependent plant species as well as more obligate species (White et al., 1992).

24.3 POTENTIAL MANAGEMENT MEASURES TO ALLEVIATE SOIL CONSTRAINTS

Reclamation of disturbed oil sands sites should consider multiple factors that may limit the success of land reclamation to ensure the sustainable development of reconstructed oil sands sites. Restoration of soil quality and ecosystem function of reclaimed soils depended on the reclamation methods and re-establishment of vegetation, which could enhance and sustain plant growth (Shrestha and Lal, 2006). Understanding what factor or factors affect the re-establishment of plant communities and designing strategies to deal with factors limiting plant growth will help to ensure the success of reclamation to upland forests in the oil sands region (Macdonald et al., 2012). The interactions among these factors determine the success or failure of reforestation efforts. In land reclamation, information on soil-site-tree growth relationships can be useful in constructing mine soils best suited to the tree species of interest. Activities such as spoil selection, grading practices, fertilization, and soil amendments can profoundly affect long-term soil productivity and tree growth (Torbert et al., 1988; Burger and Torbert, 1990).

Since limiting factors vary in different reclaimed sites, site-specific treatments need to be developed to manage the limitations. Take compacted sites as an example. The reconstructed landscape can be severely compacted due to the use of heavy equipment for stockpiling soil (Jung et al., 2014; Duan et al., 2015). Tree survival and growth on compacted mine soils were invariably poor. Tillage practices could be used to ameliorate the detrimental effects of compaction and improve plant growth. Also, the soil could be loosened by other methods, including the use of rippers, winged rippers, winged subsoilers, excavators with a bucket, or excavators with a fork. James and Evans (2010) studied the effects of ripping on tree survival, height, and diameter of sycamore (*Platanus occidentalis*), sweetgum (*Liquidambar styraciflua*), yellow poplar (*Liriodendron tulipifera*), loblolly pine

(*Pinus taeda*), and white pine (*Pinus strobus*) on mine land after it was graded and reclaimed, and reported that ripping significantly increased average tree survival and improved the growth of all species except white pine.

For salinity problems in reclaimed sites, gypsum has been used extensively to improve infiltration rates and enhance the desalinization of the soil. The application of gypsum to sodic soils significantly improves physical soil conditions by promoting flocculation, enhancing aggregate stability, and increasing the infiltration rate (Lebron et al., 2002). The use of amendments could also decrease the detrimental effects of salinity. In central Ohio, Kost et al. (1998) reported that survival rates of *F. pennsylvanica*, *P. nigra*, and *Acer saccharinum* increased by 10%, 62%, and 171%, respectively, when 30 cm of topsoil was applied over alkaline cast overburden spoil compared with the overburden alone, and tree height of these species increased 33%, 69%, and 26%, respectively, with the topsoil addition.

Disturbance during mining generally results in low soil nutrient availabilities. Fertilization is an important management practice to supply plant nutrients and enhance tree growth and stand productivity (Yang, 1985a, 1985b, 1998). Fertilization could offset the nutrient loss from the soil, stimulate biological activity, speed up the process of soil structure restoration, and accelerate ground cover development. Long-term fertilization could affect the size of the microbial community that mediates ammonium oxidation, nitrate reduction, and denitrification in soils, thus affecting nitrogen cycling (Hallin et al., 2009). Many studies have proved the positive effects of fertilization on seedling establishment and growth in oil sands reclamation (Khasa et al., 2005; Pinno et al., 2012; Sloan and Jacobs, 2013). Studies showed that two growing seasons after fertilization, both nitrogen alone and complete fertilizer (nitrogen, phosphorus, potassium plus micronutrients), increased the height, diameter at breast height, and aboveground biomass growth of white spruce planted in soils with overburden material as a substrate in the oil sands region (Figure 24.2), and fertilization effects on tree growth could be attributed to increased growth of needle length and surface area and enhanced photosynthetic rate (Chandler and Dale, 1995; Murthy et al., 1996; Albaugh et al., 2004; Amponsah et al., 2005). Duan and Chang (2017) also reported that nitrogen fertilization improved the growth of lodgepole pine and white spruce seedlings under low salt stress through enhancing photosynthesis and plant nutrition, which provides valuable insights into strategies to manage nitrogen nutrition in saline boreal forest soils in oil sands reclamation. Shukla et al. (2004) showed that fertilizer application also improved soil aggregation, structure, and water transmission properties in the reclaimed soils. Reclamation with topsoil and fertilizer application doubled the soil organic carbon concentration in the 0–10 cm layer.

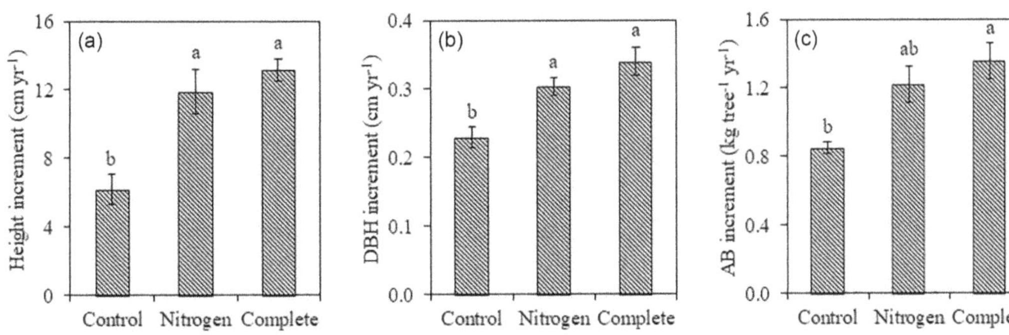

FIGURE 24.2 Increments of (a) height, (b) diameter at breast height (DBH), and (c) aboveground biomass (AB) of white spruce trees planted in soils with overburden material as a substrate in the oil sands region two growing seasons after fertilization with nitrogen alone or complete fertilizer (nitrogen, phosphorus, potassium plus micronutrients). Different letters represent significant differences among different fertilization treatments at $p<0.05$. Vertical bars are standard errors of means ($n=3$). (Reproduced from Duan and Chang, 2015, with permission from the publisher.)

REFERENCES

Abdolahnezhad, M., Lindsay, M.B.J. 2022. Geochemical conditions influence vanadium, nickel, and molybdenum release from oil sands fluid petroleum coke. *Journal of Contaminant Hydrology*, 245: 103955.

AER. 2019. ST39: Alberta Mineable Oil Sands Plant Statistics. Alberta Energy Regulator (AER), Government of Alberta, Calgary, Canada, p. 53.

Albaugh, T.J., Allen, H.L., Dougherty, P.M., Johnsen, K.H. 2004. Long term growth responses of loblolly pine to optimal nutrient and water resource availability. *Forest Ecology and Management*, 192: 3–19.

Alberta Energy Regulator. 2022. Crude bitumen-Mining. https://www.aer.ca/providing-information/data-and-reports/statistical-reports/st98/crude-bitumen/production/mining (Accessed 25 February 2022).

Alberta Environment. 2010. *Guidelines for Reclamation to Forest Vegetation in the Athabasca Oil Sands Region*, 2nd edition. Alberta Environment, Edmonton, Alberta.

Alberta. 2022a. Oil sands-Overview. https://www.alberta.ca/oil-sands-overview.aspx (Accessed 25 February 2022).

Alberta. 2022b. Oil sands facts and statistics. https://www.alberta.ca/oil-sands-facts-and-statistics.aspx (Accessed 25 February 2022).

Allen, J.A., Chambers, J.L., Stine, M. 1994. Prospects for increasing the salt tolerance of forest trees: A review. *Tree Physiology*, 14: 843–853.

Amponsah, I.G., Comeau, P.G., Brockley, R.P., Lieffers, V.J. 2005. Effects of repeated fertilization on needle longevity, foliar nutrition, effective leaf area index, and growth characteristics of lodgepole pine in interior British Columbia, Canada. *Canadian Journal of Forest Research*, 35: 440–451.

Andrews, J.A., Johnson, J.E., Torbert, J.L., Burger, J.A., Kelting D.L. 1998. Minesoil and site properties associated with early height growth of eastern white pine. *Journal of Environmental Quality*, 27: 192–199.

Ashby, W.C. 1991. Surface mine tree planting in the midwest pre- and post-public law 95-87. *Proceedings of the American Society for Surface Mining and Reclamation Meeting*. American Society for Surface Mining and Reclamation. Durango, CO.

Ashby, W.C. 1997. Soil ripping and herbicides enhance tree and shrub restoration on stripmines. *Restoration Ecology*, 5: 169–177.

Aslam, M., Huffaker, R.C., Rains, D.W. 1984. Early effects of salinity on nitrate assimilation in barley seedlings. *Plant Physiology*, 76: 321–325.

Ballard, T.M., Carter, R.E. 1986. *Evaluating Forest Stand Nutrient Status*. Ministry of Forests, Victoria, British Columbia. p. 33.

Barbour, S.L., Chanasyk, D., Hendry, J., Leskiw, L., Macyk, T., Mendoza, C., Naeth, A., Nichol, C., O'KaneO'Kane, M., Purdy, B., Qualizza, C., Quideau, S., Welham, C. 2007. *Soil Capping Research in the Athabasca Oil Sands Region*. Volume 1: Technology Synthesis. Syncrude Canada Ltd., FortMcMurray, Alberta. pp. 20–88.

Barnhisel, R.I., Hower, J.M. 1997. Coal surface mine reclamation in the eastern United States: The revegetation of disturbed lands to hayland/pasture or cropland. *Advances in Agronomy*, 61: 233–275.

Beckingham, J.D., Archibald. J.H. 1996. Field Guide to Ecosites of Northern Alberta. Special report 5. Canadian Forest Service, Northern Forestry Centre, Edmonton, Alberta.

Bicalho, B., Grant-Weaver, I., Sinn, C., Donner, M. W., Woodland, S., Pearson, G., Larter, S., Duke, J., Shotyk, W. 2017. Determination of ultratrace (<0.1mg/kg) elements in Athabasca Bituminous Sands mineral and bitumen fractions using inductively coupled plasma sector field mass spectrometry (ICPSFMS). *Fuel*, 206: 248–257.

Botella, M.A., Martínez, V., Nieves, M., Cerdá, A. 1997. Effect of salinity on the growth and nitrogen uptake by wheat seedlings. *Journal of Plant Nutrition*, 20: 793–804.

Boyer, J.S. 1982. Plant productivity and environment. *Science*, 218: 443–448.

Bradshaw, A. 1997. Restoration of mined lands-using natural processes. *Ecological Engineering*, 8: 255–269.

Brams, E.A., Fiskell, J.G.A. 1971. Copper accumulation in citrus roots and desorption with acid. *Soil Science Society of America Journal*, 35: 772–775.

Burger, J.A., Fannon, A.G. 2009. Forest land capability of reclaimed mined land for seven Appalachian hardwood species. *Proceedings of the American Society for Mining and Reclamation Meeting*. American Society for Surface Mining and Reclamation. Billings, MT.

Burger, J.A., Torbert, J.L. 1990. Mined land reclamation for wood production in the Appalachian region. *Proceedings of the 1990 Mining and Reclamation Conference and Exhibition*. American Society of Mining and Reclamation. Lexington, KY.

Canadian Association of Petroleum Producers (CAPP). 2022. What are the oil sands? https://www.capp.ca/oil/what-are-the-oil-sands/ (Accessed 25 February 2022).

Canadian Natural Resources Limited. 2012. 2011 conservation and reclamation annual report for the Horizon Oil Sands. Alberta Environment and Water, Edmonton, Alberta. p. 111.

Carmosini, N., Devito, K.J., Prepas, E.E. 2003. Net nitrogen mineralization and nitrification in trembling aspen forest soils on the Boreal Plain. *Canadian Journal of Forest Research*, 33: 2262–2268.

Chandler, J.W., Dale, J.E. 1995. Nitrogen deficiency and fertilization effects on needle growth and photosynthesis in Sitka spruce (*Picea sitchensis*). *Tree Physiology*, 15: 813–817.

Chaney, W.R., Pope, P.E., Byrnes, W.R. 1995. Tree survival and growth on land reclaimed in accord with public law 95-87. *Journal of Environmental Quality*, 24: 630–634.

Chen, L., Liu, J., Gao, J., Hu, W., Yang, J. 2021. Vanadium in soil-plant system: Source, fate, toxicity, and bioremediation. *Journal of Hazardous Materials*, 405: 124200.

Cheng, Y., Wang, J., Mary, B., Zhang, J.B., Cai, Z.C., Chang, S.X. 2013. Soil pH has contrasting effects on gross and net nitrogen mineralizations in adjacent forest and grassland soils in central Alberta, Canada. *Soil Biology & Biochemistry*, 57: 848–857.

Chong, S.K., Cowsert, P.T. 1997. Infiltration in reclaimed mine land ameliorated with deep tillage treatments. *Soil & Tillage Research*, 44: 255–264.

Chowdhury, N., Marschner, P., Burns, R.G. 2011. Response of microbial activity and community structure to decreasing soil osmotic and matric potential. *Plant & Soil*, 344: 241–254.

Clymo, R.S. 1992. Models of peat growth. *Suo*, 43: 127–136.

Cornillon, P., Palloix, A. 1997. Influence of sodium chloride on the growth and mineral nutrition of pepper cultivars. *Journal of Plant Nutrition*, 20: 1085–1094.

Corns, I.G.W. 1988. Compaction by forestry equipment and effects on coniferous seedling growth on four soils in the Alberta foothills. *Canadian Journal of Forest Research*, 18: 75–84.

Cullen, S.J., Montagne, C., Ferguson, H. 1991. Timber harvest trafficking and soil compaction in western Montana. *Soil Science Society of America Journal*, 55: 1416–1421.

Cumulative Environmental Management Association. 2009. *A Framework for Reclamation Certification Criteria and Indicators for Mineable Oil Sands*. Cumulative Environmental Management Association, Fort McMurray, Alberta.

Davidson, E.A., Hart, S.C., Firestone, M.K. 1992. Internal cycling of nitrate in soils of a mature coniferous forest. *Ecology*, 73: 1148–1156.

Dhar, A., Comeau, P.G., Karst, J., Pinno, B.D., Chang, S.X., Naeth, A., Vassov, R., Bampfylde, C. 2018. Plant community development following reclamation of oil sands mine sites in the boreal forest: A review. *Environmental Reviews*, 26: 286–298.

Dluzniewska, P., Gessler, A., Dietrich, H., Schnitzler, J.P., Teuber, M., Rennenberg, H. 2007. Nitrogen uptake and metabolism in *Populus* x *canescens* as affected by salinity. *New Phytologist*, 173: 279–293.

Dominati, E., Patterson, M., Mackay, A. 2010. A framework for classifying and quantifying the natural capital and ecosystem services of soils. *Ecological Economics*, 69: 1858–1968.

Duan, M., Chang, S.X. 2015. Responses of lodgepole pine (*Pinus contorta*) and white spruce (*Picea glauca*) to fertilization in some reconstructed boreal forest soils in the oil sands region. *Ecological Engineering*, 84: 354–361.

Duan, M., Chang, S.X. 2017. Nitrogen fertilization improves the growth of lodgepole pine and white spruce seedlings under low salt stress through enhancing photosynthesis and plant nutrition. *Forest Ecology and Management*, 404: 197–204.

Duan, M., House, J., Chang, S.X. 2015. Limiting factors for lodgepole pine (*Pinus contorta*) and white spruce (*Picea glauca*) growth differ in some reconstructed sites in the Athabasca oil sands region. *Ecological Engineering*, 75: 323–331.

Duan, M., House, J., Chang, S.X. 2019. Understory plant communities vary with tree productivity in two reclaimed boreal upland forest types in Canada. *Forest Ecology and Management*, 453: 117577.

Duan, M., House, J., Liu, Y., Chang, S.X. 2018. Contrasting responses of gross and net nitrogen transformations to salinity in a reclaimed boreal forest soil. *Biology and Fertility of Soils*, 54: 385–395.

Dyer, S., Huot, M. 2010. Mining vs. *in situ*: What is the highest environmental impact oil? http://www.pembina.org/pub/2017 (Accessed 25 February 2022).

Ehlting, B., Dluzniewska, P., Dietrich, H., Selle, A., Teuber, M., Hänsch, R., Nehls, U., Polle, A., Schnitzler, J.P., Rennenberg, H., Gessler, A. 2007. Interaction of nitrogen nutrition and salinity in grey poplar (*Populus tremula* x *alba*). *Plant, Cell & Environment*, 30: 796–811.

Elmajdoub, B., Barnett, S., Marschner, P. 2014. Response of microbial activity and biomass in rhizosphere and bulk soils to increasing salinity. *Plant & Soil*, 381: 297–306.

Farnden, C., Vassov, R.J., Yarmuch, M., Larson, B.C. 2013. Soil reclamation amendments affect long term growth of jack pine following oil sands mining. *New Forests*, 44: 799–810.

Firer, N., Jackson, R.B. 2006. The diversity and biogeography of soil bacterial communities. *Proceedings of the National Academy of Sciences*, 103: 626–631.

Foote, L. 2012. Threshold considerations and wetland reclamation in Alberta's Alberta's mineable oil sands. *Ecology and Society*, 17: 35–46.

Fung, M.Y.P., Macyk, T.M. 2000. Reclamation of oil sands mining areas. In: Barnhisel, R.I., Darmody, R.G., Daniels, W.L. (Eds.), *Reclamation of Drastically Disturbed Lands*. ASA, CSSA, and SSSA, Madison, WI. pp. 755–774.

Grant, C., Koch, J. 2007. Decommissioning Western Australia's first bauxite mine: Co-evolving vegetation restoration techniques and targets. *Ecological Management and Restoration*, 8: 92–105.

Grant, J., Dyer, S., Woynillowicz, D. 2008. Fact or fiction: oil sands reclamation. The Pembina Institute. https://www.pembina.org/reports/fact-or-fiction-report-rev-dec08.pdf. Accessed 12 September, 2022.

Greacen, E.L., Sands, R. 1980. Compaction of forest soils: A review. *Australian Journal of Soil Research*, 18: 163–189.

Hallin, S., Jones, C.M., Schloter, M., Philippot, L. 2009. Relationship between N-cycling communities and ecosystem functioning in a 50-year-old fertilization experiment. *ISME Journal*, 3: 597–605.

Hatchell, G.E., Ralston, C.W., Foil, R.R. 1970. Soil disturbances in logging: Effects on soil characteristics and growth of loblolly pine in the Atlantic Coastal Plain. *Jourmal of Forestry*, 68: 772–775.

Hawkins, H.J., Lewis, O.A.M. 1993. Effect of NaCl salinity, nitrogen form, calcium and potassium concentration on nitrogen uptake and kinetics in *Triticum aestivum* L. cv. Gamtoos. *New Phytologist*, 124: 171–177.

Haynes, R.J., Naidu, R. 1998. Influence of lime, fertilizer, and manure applications on soil organic matter content and soil physical conditions: A review. *Nutrient Cycling in Agroecosystems*. 51: 123–137.

Hearing, K.C., Daniels, W.L., Galbraith, J.M. 2004. Appalachian mine soil morphology and properties: Effects of weathering and mining method. *Soil Science Society of America Journal*, 68: 1315–1325.

Heilman, P. 1975. Effect of added salts on nitrogen release and nitrate levels in forest soils of the Washington coastal area. *Soil Science Society of America Journal*, 39: 778–782.

Hemstock, S.S., Quideau, S.A., Chanasyk, D.S. 2010. Nitrogen availability from peat amendments used in boreal oil sands reclamation. *Canadian Journal of Soil Science*, 90: 165–175.

Högberg, M.N., Bååth, E., Nordgren, A., Arnebrant, K., Höberg, P. 2003. Contrasting effects of nitrogen availability on plant carbon supply to mycorrhizal fungi and saprotrophs: A hypothesis based on field observations in boreal forest. *New Phytologist* 160: 225–238.

Holl, K.D. 2002. Long-term vegetation recovery on reclaimed coal surface mines in the eastern USA. *Journal of Applied Ecology*, 39: 960–970.

Horwath, W.R., Paul, E.A. 1994. Microbial biomass. In: Weaver, R.W., Angle, S., Bottomley, P., Bezdicek, D., Smith, S., Tabatabai, A., Wollum, A. (Eds.), *Methods of Soil Analysis: Part 2-Microbiological and Biochemical Properties*. SSSA, Madison, WI. p. 1121.

Howat, D.R. 2000. *Acceptable Salinity, Sodicity and pH Values for Boreal Forest Reclamation*. Alberta Environment, Edmonton, Alberta.

Hu, Y.L., Yan, E.R., Choi, W.J., Salifu, F., Tan, X., Chen, Z.C., Zeng, D.H., Chang, S.X. 2014. Soil nitrification and foliar $\delta^{15}N$ declined with stand age in trembling aspen and jack pine forests in northern Alberta, Canada. *Plant & Soil*, 376: 399–409.

Huang, J.H., Huang, F., Evans, L., Glasauer, S. 2015. Vanadium: Global (bio)geochemistry. *Chemical Geology*, 417: 68–89.

Inubushi, K., Barahona, M.A., Yamakawa, K. 1999. Effects of salts and moisture content on N_2O emission and nitrogen dynamics in Yellow soil and Andosol in model experiments. *Biology & Fertility of Soils*, 29: 401–407.

Irshad, M., Honna, T., Yamamoto, S., Eneji, A.E., Yamasaki, N. 2005. Nitrogen mineralization under saline conditions. *Communication in Soil Science and Plant Analysis*, 36: 1681–1689.

Jacoby, B. 1994. Mechanisms involved in salt tolerance by plants. In: Pessarakli, M. (Ed.), *Handbook of Plant and Crop Stress*. Marcel Dekker Inc., New York. pp. 97–145.

James, A.B., Evans, D.M. 2010. Ripping compacted mine soils improved tree growth 18 years after planting. *Proceedings of the Meeting of the American Society of Mining and Reclamation*. American Society of Mining and Reclamation. Pittsburgh, PA.

Jamro, G.M., Chang, S.X., Naeth, M.A. 2014b. Organic capping type affected nitrogen availability and associated enzyme activities in reconstructed oil sands soils in Alberta, Canada. *Ecological Engineering*, 73: 92–101.

Jamro, G.M., Chang, S.X, Naeth, M.A, Duan, M., House, J. 2014a. Fine root dynamics in lodgepole pine and white spruce stands along productivity gradients in reclaimed oil sands sites. *Ecology and Evolution*, 5: 4655–4670.

Jerabkova, L., Prescott, C.E., Kishchuk, B.E. 2006. Nitrogen availability in soil and forest floor of contrasting types of boreal mixedwood forests. *Canadian Journal of Forest Research*, 36: 112–122.

Jochimsen, M.E.A. 1996. Reclamation of colliery mine spoil founded on natural succession. *Water, Air and Soil Pollution*, 91: 99–108.

Jonsson, L., Dahlberg, A., Nilsson, M.C., Zackrisson, O., Karen, O. 1999. Ectomycorrhizal fungal communities in late successional Swedish boreal forests, and their composition following wildfire. *Molecular Ecology*, 8: 205–215.

Jung, K., Duan, M., House, J., Chang, S.X. 2014. Textural interfaces affected the distribution of roots, water, and nutrients in some reconstructed forest soils in the Athabasca oil sands region. *Ecological Engineering*, 64: 240–249.

Kessler, S., Barbour, S.L., van Rees, K.C.J., Dobchuk, B.S. 2010. Salinization of soil over saline-sodic overburden from the oil sands in Alberta. *Canadian Journal of Soil Science*, 90: 637–647.

Kessler, S., Hendry, M.J. 2006. *Geochemistry and Leaching of Coke from Syncrude and Suncor Sites*. University of Saskatchewan, Saskatoon.

Khasa, D.P., Fung, M., Logan, B. 2005. Early growth response of container-grown selected woody boreal seedlings in amended composite tailings and tailings sand. *Bioresource Technology*, 96: 857–864.

Khasa, P.D., Hambling, B., Kernaghan, G., Fung, M., Ngimbi, E. 2002. Genetic variability in salt tolerance of selected boreal woody seedlings. *Forest Ecology and Management*, 165 (1–3): 257–269.

Klinka, K., Wang, Q., Carter, R.E. 1990. Relationships among hunus forms, forest floor nutrient properties, and understory vegetation. *Forest Science*, 36: 564–581.

Kost, D.A., Vimmerstedt, J.P., Brown, J.H. 1998. Topsoiling, ripping and fertilizing effects on tree growth and nutrition on calcareous mine soils. *Forest Ecology and Management*, 103: 307–319.

Kozlowski, T.T. 1999. Soil compaction and growth of woody plants. *Scandinavian Journal of Forest Research*, 14: 596–619.

Kwak, J.H., Chang, S.X., Naeth, M.A., Schaaf, W. 2015. Coarse woody debris increases microbial community functional diversity but not enzyme activities in reclaimed oil sands soils. *PLoS One*, 10: e0143857.

La Flèche M., Cuss C.W., Noernberg T., Shotyk W., Karst J. 2021. Trace metals as indicators of tree rooting in bituminous soils. *Land Degradation and Development*, 32: 1970–1980.

Lal, R., Kimble, J.M., Follett, R.F. 1998. Land use and soil C pools in terrestrial ecosystem. In: Lal, R., Kimble, J.M., Follett, R.F., Stewart, B.A. (Eds.), *Management of Carbon Sequestration in Soil*. CRC Press Inc., Boca Raton, FL. pp. 1–8.

Lebron, I., Suarez, D.L., Yoshida, T. 2002. Gypsum effect on the aggregate size and geometry of three sodic soils under reclamation. *Soil Science Society of America Journal*, 66: 92–98.

Li, X.P., Chang, S.X., Salifu, K.F. 2014. Soil texture and layering effects on water and salt dynamics in the presence of a water table: A review. *Environmental Reviews*, 22: 41–50.

Lilles, E.B., Purdy, B.G., Macdonald, S.E., Chang, S.X. 2012. Growth of aspen and white spruce on naturally saline sites in northern Alberta: Implications for development of boreal forest vegetation on reclaimed saline soils. *Canadian Journal of Soil Science*, 92: 213–227.

Low, A.P., Stark, J.M., Dudley, L.M. 1997. Effects of soil osmotic potential on nitrification, ammonification, N-assimilation, and nitrous oxide production. *Journal of Soil Science*, 162: 16–27.

Ma, B., Li, X.P., Chang, S.X. 2017. Capping material type affects rhizosphere bacteria community structure in the cover soil in oil sands reclamation. *Journal of Soils & Sediments*, 17: 2516–2523.

Macdonald, E., Quideau, S., Landhäusser, S. 2012. Rebuilding boreal forest ecosystems after industrial disturbance. In: Vitt, D.H., Nhatti, J.S. (Eds.), *Restoration and Reclamation of Boreal Ecosystems: Attaining Sustainable Development*. Cambridge University Press, New York. pp. 123–160.

Mackenzie, D.D., Naeth, M.A. 2010. The role of the forest soil propagule bank in assisted natural recovery after oil sands mining. *Restoration Ecology*, 18: 418–427.

Martikainen, P.J. 1985. Nitrification in forest soil of different pH as affected by urea, ammonium sulfate and potassium sulfate. *Soil Biology & Biochemistry*, 17: 363–367.

Maundrell, C., Hawkins, C. 2004. Use of an aspen overstory to control understory herbaceous species, bluejoint grass (*Calamagrotis canadensis*), and Fireweed (*Epilobium angustifolium*). *Northern Journal of Applied Forestry*, 21: 74–79.

Mays, D.A., Sistani, K.R., Soileau, J.M. 2000. Lime and fertilizer needs for land reclamation. In: Barnhisel, R.I., Darmody, R.G., Daniels, W.L. (Eds.), *Reclamation of Drastically Disturbed Lands*. ASA, CSSA, and SSSA, Madison, WI. pp. 217–271.

McClung, G., Frankenberger, W.T. 1987. Nitrogen mineralization rates in saline vs salt-amended soils. *Plant & Soil*, 104: 13–21.

McMillan, R., Quideau, S.A., MacKenzie, M.D., Biryukova, O. 2007. Nitrogen mineralization and microbial activity in oil sands reclaimed boreal forest soils. *Journal of Environmental Quality*, 36: 1470–1478.

Medve, R.M. 1984. The mycorrhizae of pioneer species in disturbed ecosystems in western Pennsylvania. *American Journal of Botany*, 71: 787–794.

Melnik, K. 2017. The role of microtopography in the expression of soil propagule banks on reclamation sites. M.Sc. thesis, Department of Renewable Resources, University of Alberta. Edmonton, Alberta.

Moskal, T. 1999. Moisture characteristics of coarse textured soils and peat: Mineral mixtures. M.Sc. Thesis. Department of Renewable Resources, University of Alberta, Edmonton, Alberta.

Mossop, G.D., Flach, P.D., Pemberton, S.G., Hopkins, J.C. 1982. Field excursion 22: Athabasca oil sands, sedimentology and development technology. *11th International Congress on Sedimentology*. Hamilton, ON.

Mummey, D.L., Stahl, P.D., Buyer, J.S. 2002. Soil microbiological properties 20 years after surface mine reclamation: Spatial analysis of reclaimed and undisturbed sites. *Soil Biology & Biochemistry*, 34: 1717–1725.

Munns, R. 1993. Physiological processes limiting plant growth in saline soils: Some dogmas and hypotheses. *Plant, Cell & Environment*, 16: 15–24.

Munns, R., Tester. M. 2008. Mechanisms of salinity tolerance. *Annual Review of Plant Biology*, 59: 651–681.

Murthy, R., Dougherty, P.M., Zarnoch, S.J., Allen, H.L. 1996. Effects of carbon dioxide, fertilization, and irrigation on photosynthetic capacity of loblolly pine trees. *Tree Physiology*, 16: 537–546.

Myrold, D.D. 2005. Transformation of nitrogen. In: David, M.S., Jeffry, J.F., Peter, G.H., David, A.Z. (Eds.), *Principles and Applications of Soil Microbiology*. 2nd edition. Prentice Hall, Upper Saddle River, NJ. p. 640.

Naeth, M.A., Chanasyk, D.S., Burgers, T.D. 2011. Vegetation and soil water interactions on a tailings sand storage facility in the Athabasca oil sands region of Alberta Canada. *Physics & Chemistry of the Earth*, 36: 19–30.

Naeth, M.A., Wilkinson, S.R. 2004. *Revegetation Research in the Athabasca Oil Sands: a Literature Review*. Prepared for Suncor Energy Inc. p. 75. http://www.barbau.ca/content/revegetation-research-athabasca-oil-sands-literature-review

Nakata, C., Qualizza, C., MacKinnon, M., Renault, S. 2011. Growth and physiological responses of *Triticum aestivum* and *Deschampsia caespitosa* exposed to petroleum coke. *Water, Air and Soil Pollution*, 216: 59–72.

Nambiar, E.K.S., Sands, R. 1993. Competition for water and nutrients in forests. *Canadian Journal of Forest Research*, 23: 1955–1968.

Nesbitt, J.A., Lindsay, M.B.J. 2017. Vanadium geochemistry of oil sands fluid petroleum coke. *Environmental Science & Technology*, 51: 3102–3109.

Nesbitt, J.A., Robertson, J.M., Swerhone, L.A., Lindsay, M.B.J. 2018. Nickel geochemistry of oil sands fluid petroleum coke deposits, Alberta, Canada. *Facets (Ottawa)*, 3: 469–486.

Nieminen, T., Ukonmaanaho, L., Rausch, N., Shotyk, W. 2007. Biogeochemistry of nickel and its release into the environment. In Sigel, A., Sigel, H., Sigel, R.K.O. (Eds.). *Nickel and Its Surprising Impact in Nature*, Volume 2.. John Wiley & Sons Inc., Hoboken, NJ, pp. 1–30. https://doi.org/10.1002/9780470028131.ch1

Omi, S.K. 1986. Soil compaction: Effects on seedling growth. Report M-125. USDA Forest Service.

Parida, A.K., Das, A.B. 2004. Effects of NaCl stress on nitrogen and phosphorous metabolism in a true mangrove *Bruguiera parviflora* grown under hydroponic culture. *Journal of Plant Physiology*, 161: 921–928.

Parrotta, J.A., Knowles, O.H., Wunderle, J.M., Jr. 1997. Development of floristic diversity in 10-year-old restoration forests on a bauxite mined site in Amazonia. *Forest Ecology and Management*, 99: 21–42.

Peace, W.J.H., Grubb, P. 1982. Interaction of light and mineral nutrient supply in the growth of *Impatiens parviflora*. *New Phytologist*, 90: 127–150.

Pembina Institute. 2022. Alberta provincial wetland policy. http://www.pembina.org/pub/2448 (Accessed 25 February 2022).

Pendias, H., Kabata-Pendias, A. 2000. *Trace Elements in Soils and Plants*. 3rd edition. CRC Press, Boca Raton, FL.

Phillips, M. 2006. Procedures to insure reclamation success. *Proceedings of the American Society for Surface Mining and Reclamation Meeting*. American Society for Surface Mining and Reclamation. Billings, MT.

Pinno, B.D., Errington, R.C. 2015. Maximizing natural trembling aspen seedling establishment on a reclaimed boreal oil sands site. *Ecological Restoration*, 33: 43–50.

Pinno, B.D., Landhäusser, S.M., MacKenzie, M.D., Quideau, S.A., Chow, P.S. 2012. Trembling aspen seedling establishment, growth and response to fertilization on contrasting soils used in soil sands reclamation. *Canadian Journal of Soil Science*, 92: 143–151.

Puttaswamy, N., Liber, K. 2011. Identifying the causes of oil sands coke leachate toxicity to aquatic invertebrates. *Environmental Toxicology & Chemistry*, 30: 2576–2585.

Puttaswamy, N., Liber, K., 2012. Influence of inorganic anions on metals release from oil sands coke and on toxicity of nickel and vanadium to *Ceriodaphnia dubia*. *Chemosphere*, 86: 521–529.

Puttaswamy, N., Turcotte, D., Liber, K. 2010. Variation in toxicity response of *Ceriodaphnia dubia* to Athabasca oil sands coke leachates. *Chemosphere*, 80: 489–497.

Raine, M., Mackenzie, I., Gilchrist, I. 2002. Terrestrial vegetation, wetlands and forest resources baseline. CNRL Horizon Project Environmental Impact Assessment Volume 6, Appendix B. Report #012-2220. Golder Associates, Alberta.

Reeder, J.D., Sabey, B. 1987. Nitrogen. In: Williams, R.D., Schuman, G.E. (Eds.), *Reclaiming Mine Soils and Overburden in the Western United States*. Soil Conservation Society of America, Ankeny, IA. pp. 155–184.

Reich, P.B., Grigal, D.F., Aber, J.D., Gower, S.T. 1997. Nitrogen mineralization and productivity in 50 hardwood and conifer stands on diverse soils. *Ecology*, 78: 335–347.

Renault, S., Paton, E., Nilsson, G., Zwiazek, J.J., MacKinnon, M.D. 1999. Responses of boreal plants to high salinity oil sands tailings water. *Journal of Environmental Quality*, 28: 1957–1962.

Robertson, J.M., Nesbitt, J.A., Lindsay, M.B.J. 2019. Aqueous- and solid-phase molybdenum geochemistry of oil sands fluid petroleum coke deposits, Alberta, Canada. *Chemosphere*, 217: 715–723.

Rodrigue, J.A., Burger, J.A. 2004. Forest soil productivity of mined land in the mid-western and eastern coalfield regions. *Soil Science Society of America Journal*, 68: 833–844.

Rooney, R.C., Bayley, S.E., Schindler, D.W. 2012. Oil sands mining and reclamation cause massive loss of peatland and stored carbon. *Proceedings of the National Academy of Sciences*, 109: 4933–4937.

Rousk, J., Brookes, P.C., Bååth, E. 2009. Contrasting soil pH effects on fungal and bacterial growth suggest functional redundancy in carbon mineralization. *Applied & Environmental Microbiology*, 75: 1589–1596.

Rowland, S.M., Prescott, C.E., Grayston, S.J., Quideau, S.A., Bradfield, G.E. 2009. Recreating a functioning forest soil in reclaimed oil sands in northern Alberta: An approach for measuring success in ecological restoration. *Journal of Environmental Quality*, 38: 1580–1590.

Rubinigg, M., Posthumus, F.S., Elzenga, J.T.M. 2005. Effect of NaCl salinity on nitrate uptake in *Plantago maritima* L. *Phyton*, 45: 295–302.

Scott, A.C., Zubot, W., Davis, C.W., Brogly, J. 2020. Bioaccumulation potential of naphthenic acids and other ionizable dissolved organics in oil sands process water (OSPW)-A review. *Science of the Total Environment*, 712: 134558.

Shell Canada Energy. 2012a. 2011 environment report-Jackpine Mine. Alberta Environment and Water. Edmonton, Alberta. p. 426.

Shell Canada Energy. 2012b. 2011 environment report-Muskeg River Mine. Alberta Environment and Water. Edmonton, Alberta. p. 628.

Shotyk, W. 2020. Trace elements in wild berries from reclaimed lands: Biomonitors of contamination by atmospheric dust. *Ecological Indicators*, 110: 105960.

Showalter, J.M., Burger, J.A., Zipper, C.E. 2010. Hardwood seedling growth on different mine spoil types with and without topsoil amendment. *Journal of Environmental Quality*, 39: 483–491.

Shrestha, R.K., Lal, R. 2006. Ecosystem carbon budgeting and soil carbon sequestration in reclaimed mine soil. *Environment International*, 32: 781–796.

Shrestha, R.K., Lal, R. 2011. Changes in physical and chemical properties of soil after surface mining and reclamation. *Geoderma*, 161: 168–176.

Shukla, M.K., Lal, R., Ebinger, M. 2004. Soil quality indicators for reclaimed minesoils in southeastern Ohio. *Soil Science*, 169: 133–142.

Skousen, J.G., Gorman, J., Emerson, P. 2007. Survival of commercial hardwoods on reclaimed minesoils in West Virginia: 6th year results. *Proceedings of the American Society for Surface Mining and Reclamation Meeting*. American Society for Surface Mining and Reclamation. Gillette, WY. pp. 229–237.

Sloan, J.L., Jacobs, D.F. 2013. Fertilization at planting influences seedling growth and vegetative competition on a post-mining boreal reclamation site. *New Forests*, 44: 687–701.

Startsev, A.D., Mcnabb, D.H. 2001. Skidder traffic effects on water retention pore size distribution, and van Genuchten parameters of boreal forest soils. *Soil Science Society of America Journal*, 65: 224–231.

Ste-Marie, C., Paré, D. 1999. Soil, pH, and N availability effects on net nitrification in the forest floor of a range of boreal forest stands. *Soil Biology & Biochemistry*, 31: 1579–1589.

Stribley, D.P. 1987. Mineral nutrition. In: Safir, G.R. (Ed.), *Ecophysiology of v/a Mycorrhizal Plants*. CRC Press Inc., Boca Raton, FL. pp. 59–70.

Suncor Energy Inc. 2012. Reclamation progress tracking report for the year 2011. Alberta Environment and Water. Edmonton, Alberta. p. 156.

Syncrude Canada Limited. 2012. 2011 annual reclamation progress tracking report-Mildred Lake and Aurora North Oil Sands mines. Alberta Environment and Water. Edmonton, Alberta. p. 126.

Tan, X., Chang, S.X., Kabzems, R. 2005. Effects of soil compaction and forest floor removal on soil microbial properties and N transformations in a boreal forest long-term soil productivity study. *Forest Ecology and Management*, 217: 158–170.

Turchenek, L.W., Lindsay, J.D. 1982. Soils inventory of the Alberta Oil Sands environmental research program study area. AOSERP Report 122. Alberta Environment, Edmonton, AB. 221 p.+ 8 maps. https://open.alberta.ca/dataset/51da1449-ce10-4e22-8778-e4a33c8cce5a/resource/825a2149-b546-45de-887d-5129be0a2f95/download/ab42_report.pdf. Accessed 12 December. 2022.

Thompson, D. 2009. *The Facts on Oil Sands Reclamation*. The Oil Sands Developers Group, Fort McMurray, Alberta.

Torbert, J.L. 1995. Reclamation of surface-mined forestland in the southern Appalachians. Ph.D. Dissertation. Virginia Polytechnical Institute and State University, Blacksburg, VA.

Torbert, J.L., Tuladhar, A.R., Burger, J.A., Bell, J.C. 1988. Mine soil property effects on the height of ten-year-old white pine. *Journal of Environmental Quality*, 17: 189–192.

Tropek, R., Hejda, M., Kadlec, T., Spitzer, L. 2013. Local and landscape factors affecting communities of plants and diurnal Lepidoptera in black coal spoil heaps: Implications for restoration management. *Ecological Engineering*, 57: 252–260.

Tyner, E.H., Smith, R.M., Galpin, S.L. 1948. Reclamation of strip-mined areas in West Virginia. *Agronomy Journal*, 40: 313–323.

Ussiri, D.A.N., Lal, R. 2005. Carbon sequestration in reclaimed mine soils. *Critical Reviews in Plant Sciences*, 24: 151–165.

Visser, S. 2008. Petroleum Hydrocarbons (PHCs) in Lean Oil Sand (LOS): Degradation potential and toxicity to ecological receptors. Report prepared for Cumulative Environmental Management Association. 121.

White, J.A., DePuit, E.J., Smith, J.L., Williams, S.E. 1992. Vesicular-arbuscular mycorrhizal fungi and irrigated mined land reclamation in southwestern Wyoming. *Soil Science Society of America Journal*, 56: 1464–1469.

Yan, E.R., Hu, Y.L., Salifu, F., Tan, X., Chen, Z.C., Chang, S.X. 2012. Effectiveness of soil N availability indices in predicting site productivity in the oil sands region of Alberta. *Plant & Soil*, 359: 215–231.

Yang, R.C. 1985a. Ten-year growth response of 70-year-old lodgepole pine to fertilization in Alberta. Canadian Forestry Service, Northern Forest Research Centre, Edmonton, Alberta. Information Report NOR-X-266.

Yang, R.C. 1985b. Effects of fertilization on growth of 30-year-old lodgepole pine in West-central Alberta. Canadian Forestry Service, Northern Forest Research Centre, Edmonton, Alberta. Information Report NOR-X-268.

Yang, R.C. 1998. Foliage and stand growth responses of semi-mature lodgepole pine to thinning and fertilization. *Canadian Journal of Forest Research*, 28: 1794–1804.

Yuan, B.C., Li, Z.Z., Liu, H., Gao, M., Zhang, Y.Y. 2007. Microbial biomass and activity in salt affected soils under arid conditions. *Applied Soil Ecology*, 35: 319–328.

Zimmerman, J.K., Pulliam, W.M., Lodge, D.J., Quinones-Orfila, V., Fetcher, N., Guzmán-Grajales, S., Parrotta, J.A., Asbury, C.E., Walker, L.R., Waide, R.B. 1995. Nitrogen immobilization by decomposing woody debris and the recovery of tropical wet forest from hurricane damage. *Oikos*, 72: 314–322.

25 Organic Carbon Turnover as Affected by Agricultural Activities in a Semitropical Zone of China

Bo Pan and Fangfang Li
Faculty of Environmental Science and Engineering, Yunnan Provincial Key Lab of Soil Carbon Sequestration and Pollution Control, Kunming University of Science and Technology, Kunming, Yunnan, P.R. China

CONTENTS

DOI: 10.1201/9781003093565-25

LIST OF ABBREVIATIONS

Some Abbreviations	Definition
	Sample names
NS	undisturbed natural soil
TP	terrace paddy soil
TD	dry farmland soil
	Lignin parameters
S/V	ratio of syringl unit to vanillyl unit
C/V	ratio of cinnamyl phenols to vanillyl phenols
$(Ad/Al)_v$	ratio of vanillic acid to vanillin
VSC	the sum of vanillyl unit, syringyl unit, and cinnamyl unit
	Lipid parameters
CPI_{AF}	carbon preference index of alkanoic acids (≥20)
CPI_{ALK}	carbon preference index of alkanes (≥20)
RLS_{AF}	the ratio of long-chain vs. short-chain alkanoic acids
ACL_{AF}	average chain length of alkanoic acids
	Parameters of black carbon
BPCAs	benzene-polycarboxylic acids
B3CA	trimellitic acid
$B4CA_{py}$	pyromellitic acid
$B4CA_{me}$	mellophanic acid
$B4CA_{pr}$	prehnitic acid
B5CA	benzenepenta-carboxylic acid
B6CA	mellitic acid

25.1 INTRODUCTION

Soil organic matter (SOM) is a heterogeneous and complex mixture of organic molecules predominantly originating from plant and microbial residues (Feng and Simpson, 2011). SOM contains more than twice as much carbon as the atmosphere and plays a key role in soil fertility and carbon cycling in the terrestrial ecosystem (Schmidt et al., 2011). In the last two decades, increased human activities such as deforestation (e.g., conversion of natural and stable ecosystems to fragile agricultural systems) and cultivation have substantially accelerated carbon flux in terrestrial ecosystems, thereby partly contributing to global climate change and soil degradation (Guo and Gifford, 2002). Recently, an increasing number of studies have reported a significant loss of SOM in agrosystems compared to natural ecosystems because of intensive tillage, suggesting a great demand for carbon sequestration (Chen et al., 2019). Therefore, increasing SOM contents is imperative for maintaining soil fertility and food production as well as mitigating climate change.

Carbon skeletons retained from parent precursor molecules during SOM decomposition serve as molecular biomarkers (Otto and Simpson, 2007). Therefore, they are recognized as a powerful tracer for SOM turnover and stabilization. For example, extractable lipids are used to distinguish the source and degradation of plant and microorganism residues, while lignin is used to differentiate the source and decomposition of gymnosperm and angiosperm vegetation (Otto and Simpson, 2007). Several studies have applied the molecular biomarker method to investigate the vertical distributions and decomposition stage of SOM in natural soils (e.g., forest and grassland soils) affected by climate change and vegetation alteration (Zhu et al., 2019). For agricultural soils, SOM turnover and stabilization remain poorly understood. Elucidating SOM turnover and stabilization in agricultural soils is a prerequisite step for increasing SOM stocks and consequently promoting sustainable agriculture and mitigating global climate change.

Different SOM components have distinct turnover time, which is mainly attributed to their chemical recalcitrance and mineral protection (Hall et al., 2016; Hemingway et al., 2019). Studies have reported that plant-derived macromolecules, such as lignin, are stable because of their intrinsic stability against biodegradation (Pisani et al., 2014); however, recent studies observed that turnover of these components in soils is faster than that of the bulk SOM (Thevenot et al., 2010). An increasing number of studies have realized the role of organo-mineral associations in long-term SOM stabilization (Hemingway et al., 2019). However, minerals selectively protect different SOM components. Some studies reported that polymethylene chains, such as cutin and suberin, can be preferentially adsorbed to kaolinite and montmorillonite surfaces more than other SOM components, whereas aromatic compounds, including lignin, are preferentially preserved by amorphous Fe oxides (Hall et al., 2016; Huang et al., 2018). Some also reported that plant-derived aliphatic compounds are enriched mainly in mineral fractions compared to microorganism-derived compounds (Wiesenberg et al., 2010). Consequently, the presence of minerals could alter the extractability of related molecular biomarkers, such as lipids, thus obfuscating our understanding of SOM turnover based on molecular biomarker analysis.

The turnover of SOM in subtropical or tropical regions is fast and complex due to the warm and humid climate. However, there is little information on the effects of human activity on SOM turnover and stabilization in this region. In previous studies (Li et al., 2015, Li et al., 2018a, 2018b), we investigated the extractability and protection of free lipids affected by reactive minerals as well as SOM turnover and stabilization in agricultural soils subjected to different human activities. In this chapter, we summarize the results from those previous studies.

25.2 MATERIALS AND METHODS

25.2.1 Sample Collection and Pretreatment

We sampled soils in a rubber plantation area in Xishuangbanna (100°46′–100°47′E, 21°53′–21°54′N) and paddy soils in Yuanyang (102°27′–103°13′E, 22°49′–23°19′N) of Yunnan province. The pH values ranged from 4.4 to 6.2, indicating acidic soils. In the rubber plantation area, bamboo was the native plant, and soils cultivated with rubber trees for 0, 10, and 60 years were sampled. We obtained soil samples from the surface and from different depths (0–20 cm, S; 20–40 cm, Z; and 40–60 cm, X). In addition, soils from two areas of different cultivation patterns, a terrace paddy soil (TP) and a dry farmland soil (TD), were also collected from Yuanyang. All soil samples were dried, ground, and passed through a 60-mesh sieve. Aliquots of subsamples were washed with a mild acid mixture (10% hydrofluoric acid and 1 M hydrochloric acid) to remove reactive minerals, as described by Lin and Simpson (2016).

25.2.2 Biomarker Analysis

Molecular biomarkers were extracted and analyzed, including solvent-extractable lipids (mainly alkanoic acids, alkanols, and alkanes), lignin-derived phenols (namely syringyl, vanillyl, and cinnamyl units [S, V, and C units, respectively]), and benzene-polycarboxylic acids (BPCAs). All these biomarkers were extracted as described in previous studies (Otto et al., 2005). In brief, the solvent-extractable lipids were directly extracted by continuous organic extraction (dichloromethane, dichloromethane: methanol [1:1; v/v], and methanol). The lignin-derived phenols and BPCAs were small byproducts mainly obtained by alkaline cupric oxide (CuO) oxidation (the mixture of CuO, ammonium iron sulfate hexahydrate [$Fe(NH_4)_2(SO_4)_2 \cdot 6H_2O$], and 2 M NaOH) and HNO_3 oxidation, respectively. All extracts were derivatized by trimethysilyl and analyzed by gas chromatography–mass spectrometry (GC–MS) (Agilent, 7890A GC equipped with a 5975C quadrupole mass selective detector).

25.2.3 Elemental Analysis

All soil samples were characterized to determine their bulk elements (MicroCube, Elementar, Langenselbold, Germany). Mineral compositions were analyzed by X-ray diffraction (D/Max 2200, Rigaku, Beijing, China).

25.3 RESULTS AND DISCUSSION

25.3.1 Extraction Enhancement of Molecular Biomarkers after Acid Treatment

25.3.1.1 Increased Lipid Extraction after the Removal of Reactive Minerals

The removal of reactive minerals led to more than 50% mass loss (Table 25.1), which was mainly attributed to the dissolution of minerals. Minerals in the investigated soil samples mainly included quartz, kaolinite, and mica. After acid treatment, kaolinite was completely removed, the contents of other minerals sharply decreased, and the total free lipid content increased. Because of the different properties of various lipid types (n-alkanoic acids, n-alkanes, and n-alkanols), their interactions with the minerals differed. Thus, a general correction of the extraction efficiency before mineral removal could not be applied.

To compare the extractability of individual lipids after acid treatment, all sample concentrations were normalized to those in the original soil mass. N-alkanoic acids were the most abundant type of free lipid. After acid treatment, the amount of alkanoic acids increased by 6-times. In other words, more than 80% of the total alkanoic acids were protected by minerals. The extractabilities of alkanols and alkanes increased only slightly compared to alkanoic acids, which was attributed to

TABLE 25.1
The Elemental Compositions and the Mass of Solvent Extracts before and after Acid Treatment

	Before Acid Treatment					After Acid Treatment							
Sample	C%	N%	C/N[a]	TSE/C[b]	TSE/S[c]	C%	N%	C/N[a]	TSE/C[b]	TSE/O[d]	TSE/S[c]	ML[e]	MC[e]
0-S	2.31	0.20	13.8	10.6	0.25	3.51	0.31	13.3	53.6	1.88	0.87	53.5	29.4
0-Z	1.68	0.19	10.2	6.79	0.11	1.93	0.19	11.6	48.9	0.94	0.39	58.5	52.2
0-X	1.12	0.20	6.7	4.93	0.06	1.45	0.16	10.3	40.4	0.59	0.26	55.3	42.2
10-S	1.93	0.17	13.2	13.7	0.27	2.64	0.24	12.8	22.4	0.59	0.29	50.5	32.2
10-Z	1.70	0.12	16.8	13.7	0.23	2.22	0.20	13.2	20.8	0.46	0.21	54.7	39.6
10-X	1.25	0.14	10.4	10.6	0.13	1.47	0.15	11.4	19.6	0.29	0.13	55.1	47.1
60-S	2.58	0.21	14.3	20.7	0.54	3.15	0.25	14.7	46.1	1.45	0.78	46.6	34.8
60-Z	1.48	0.25	6.8	12.0	0.18	1.60	0.19	10.0	36.8	0.59	0.32	45.0	40.6
60-X	1.05	0.15	8.2	13.0	0.14	1.21	0.18	7.8	33.1	0.40	0.22	44.8	36.3

Source: Reprinted from Li et al. (2018b), with permission from John Wiley and Sons.
-*S:* 0–20 cm depth; -Z: 20–40 cm depth; -X: 40–60 cm depth. 0, 10, and 60: different years planted rubber
[a] Atomic ratios;
[b] Total solvent extracts content (mg/g C);
[c] Total solvent extracts (mg/g original soil);
[d] Total solvent extracts (mg/g particle mass of the corresponding soil);
[e] ML indicates soil particle mass loss after acid treatment, and CL represents soil carbon loss after acid treatment.

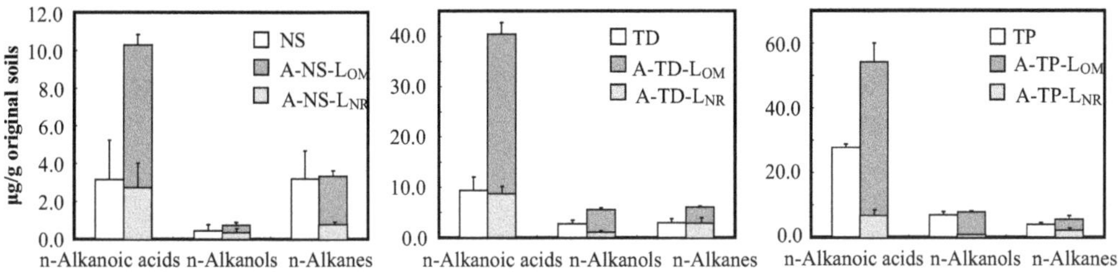

FIGURE 25.1 The contents of total n-alkanoic acids, n-alkanols, and n-alkanes before and after acid treatment based on the original soil mass. NS: for undisturbed natural soil; TD: for dry farming soil; TP: for terrace paddy soil; A: acid treatment; L_{OM} and L_{NR}: OM-enriched layer and non-reactive mineral-enriched layer in acid-treatment soils, respectively. Reprinted from Li et al. (2018a), with permission from Elsevier.

the strong interactions between alkanoic acids and reactive minerals via ligand exchange in acidic soils (pH 4.4–4.6 in this study) (von Luetzow et al., 2006). Early studies suggested that alkanes could wedge into the interlamellar space of minerals like montmorillonite (Eltantawy and Arnold, 1972). However, the minerals in this study did not contain montmorillonite, which may be the cause of the weak interaction between n-alkanes and reactive minerals. It should be noted that the molecular biomarkers detected after mineral removal suggested their true contents, and the comparison between molecular biomarkers before and after mineral removal indicated the extent of mineral protection.

In general, intensive human activities partially break up soil aggregates and may reduce the protection of SOM by minerals (Wiesenberg et al., 2010). However, lipid extractability after the removal of reactive minerals in TD did not decrease compared to that in undisturbed natural soil (NS) (Figure 25.1). This phenomenon may be attributed to the following reasons: (i) the input of farmyard manure or plant residues introduced a large amount of OM in cultivated soils, and the organic manure can form complexes with the free iron to increase the content of amorphous Fe oxide, which in turn may adsorb more OM (Huang et al., 2016); and (ii) montmorillonite that was detected only in the cultivated soils was beneficial for the protection of SOM.

25.3.1.2 Active Mineral Removal Ensures Accurate Biomarker Information

The removal of reactive minerals enhanced the extractability of n-alkanoic acids in free lipids. It may also alter the related parameters and source analysis of n-alkanoic acids. For example, because most long-chain alkanoic acids could not be detected before acid treatment, the carbon preference index (CPI) of alkanoic acids (CPI_{AF}) in some soils could not be calculated (Table 25.2). The calculable CPI_{AF} values in the cultivated soils decreased with soil depth before acid treatment, whereas the opposite trend was observed after acid treatment (Table 25.2). In addition, acid treatment changed the ratio of long-chain carbon to short-chain carbon (RLS). RLS ranged from 0 to 0.4 before acid treatment and increased to 0.1–0.5 after treatment, suggesting more release of long-chain alkanoic acids. The above results showed that the poor detectability of long-chain alkanoic acids before acid treatment was due to the decreased extractability in the presence of reactive minerals. Therefore, if the influence of reactive minerals on free lipid extraction was not considered, then inaccurate results could be obtained regarding the source and degradation state of free lipids.

Acid treatment may also result in the dissolution of lignin-derived phenols (about 60%, data unpublished), especially those that have compositions with a high oxidation state. This dissolution process may result in underestimating the ratio of acids to aldehydes in vanillyl ($[Ad/Al]_V$) after acid treatment. However, acid treatment did not alter the plots of S/V and C/V that signify the source of lignin, indicating no selective protection of S, V, and C units by minerals. Therefore, the detection and comparison of lipid biomarkers should be based on the concentration after the removal of reactive minerals.

TABLE 25.2
The Comparison of Biomarker Parameters Regarding SOM Source and Degradation before and after Acid Treatment

	Before Acid Treatment					After Acid Treatment						
Samples	CPI_{AF}[a]	CPI_{ALK}[a]	RLS_{AF}[b]	ACL_{AF}[b]	R[c]	CPI_{AF}[a]	CPI_{ALK}[a]	RLS_{AF}[b]	ACL_{AF}[b]	R[c]	AR[d]	CR[d]
0-S	7.9	1.7	0.1	17.8	18.0	4.5	2.0	0.5	19.8	15.0	4.5	4.4
0-Z	NC	1.7	0.0	16.9	15.6	6.0	2.7	0.3	18.7	23.8	8.9	3.4
0-X	NC	1.9	0.0	16.8	34.8	9.7	2.5	0.2	18.1	72.3	7.6	3.0
10-S	5.3	1.7	0.4	19.6	2.4	5.7	1.6	0.4	19.3	10.6	4.5	1.2
10-Z	1.7	1.3	0.1	17.6	2.3	5.6	1.3	0.3	19.0	9.1	4.4	1.3
10-X	0.0	1.8	0.0	17.1	4.5	16.7	1.3	0.1	17.6	8.3	4.1	3.0
60-S	11.1	1.8	0.4	20.2	2.8	5.6	1.0	0.2	18.5	13.6	5.3	1.3
60-Z	5.1	1.4	0.2	19.3	2.5	6.5	1.6	0.1	17.8	13.2	7.2	1.8
60-X	2.3	1.9	0.1	17.4	3.3	13.1	1.6	0.2	18.0	8.5	3.0	1.3

Source: Reprinted from Li et al. (2018b), with permission from John Wiley and Sons.
NC: Not calculated. CPI for the n-alkanols could not be determined because no odd-numbered alkanols were detected.
[a] $CPI_{AF}(\geq 20)$ indicated the characteristic even-over-odd chain-length predominance of n-alkanoic acid with over C20 carbon, $CPI_{AF}(\geq 20) = ((C20–C30) + (C22–C32))_{even}/2 \times (C21–C31)_{odd}$; $CPI_{ALK}(\geq 20) = ((C21–C31) + (C23–C33))_{odd}/2 \times (C22–C32)_{even}$: the characteristic odd-over-even chain-length predominance of n-alkanes with over C20 carbon (Jansen et al., 2008);
[b] RLS_{AF} indicated ratio of long chain vs. short chain acids (≥C20/<C20), ACL_{AF} indicated that average chain length;
[c] The ratio of aliphatic lipids/cyclic lipids;
[d] AR indicated ratio of aliphatic lipids before and after acid treatment; CR indicated ratio of cyclic lipids before and after acid treatment.

25.3.2 SOM Turnover after the Conversion of Natural Bamboos to Rubber Trees

25.3.2.1 Changes in Soil Elemental Composition

Elemental analysis revealed no changes in SOC content after long-term cultivation (Table 25.1). This was mainly attributed to the fact that an input of new organic matter restored SOC after deforestation. SOC contents and C/N ratios decreased with soil depth. In addition, the composition of minerals included quartz, kaolinite, and muscovite. The presence of reactive minerals such as kaolinite can protect SOM through organo-mineral complexes or soil aggregates.

25.3.2.2 SOM Turnover Indicated by Lipid Biomarkers

Among the lipid biomarkers, n-alkanoic acids were predominant compared to alkanes and alkanols. Previous studies reported that alkanoic acids were mainly accumulated in acidic soils, while alkanes were more abundant in alkaline soils, and the difference was related to microorganism activity (Bull et al., 2000). The pH of Yunnan soil was acidic (4.4–4.6 in this study). The n-alkanoic acids in the range of C_{14}–C_{32} showed significant even-over-odd dominance, indicating that plant waxes were the main sources of lipids (Figure 25.2).

The distribution patterns of alkanoic acids in soil with rubber trees were similar to those in the original soil, which may be caused by similar patterns between rubber trees and native vegetation (bamboo). However, the amount of alkanoic acids from the soil surface of the 60-year rubber tree plantation was increased compared to that of the original soils, suggesting that new input from rubber trees greatly contributed to soil alkanoic acid contents. In addition, we collected the plant leaves to extract free lipids. As shown in Figure 25.3, the free lipids in rubber tree and bamboo leaves included alkanoic acids, alkanes, and alkanols, among which alkanoic acids were predominant;

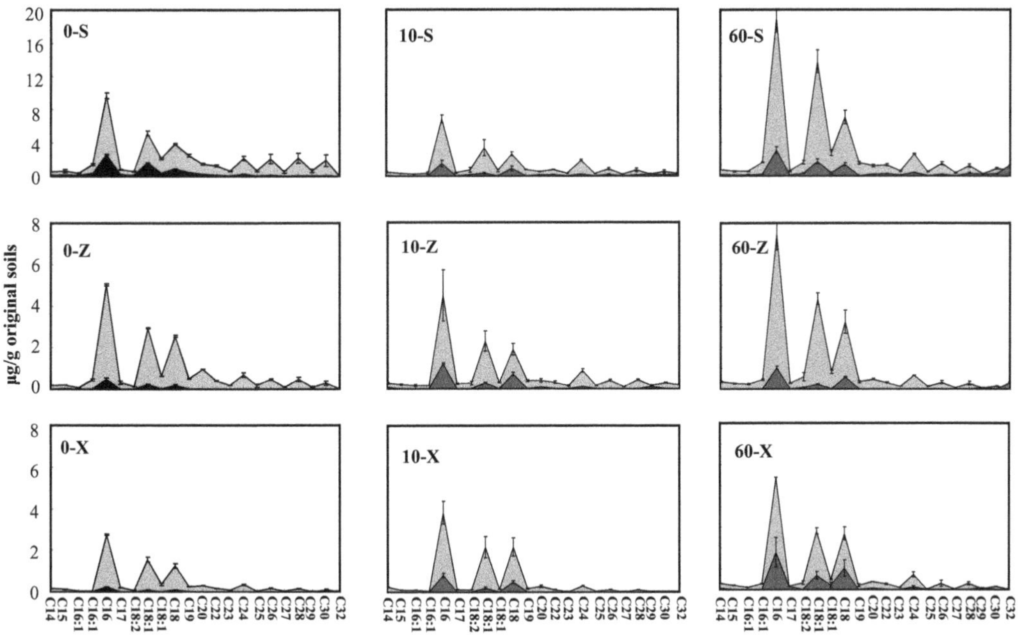

FIGURE 25.2 The distribution of n-alkanoic acids in soils planted with rubber trees for 0, 10, and 60 years. -S: 0–20 cm soil depth; -Z: 20–40 cm soil depth; -X: 40–60 cm soil depth. Gray background: after acid treatment; Black background: before acid treatment. (Reprinted from Li et al., 2018b, with permission from John Wiley and Sons.)

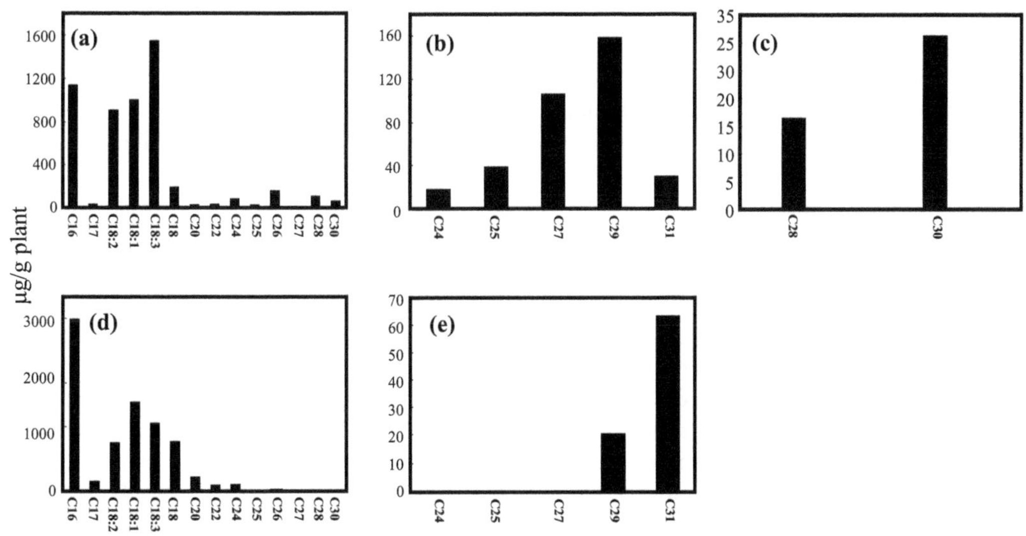

FIGURE 25.3 The distribution of n-alkanoic acids, n-alkanes, and n-alkanols in local vegetation. Panels (a), (b), and (c) are for n-alkanoic acids, n-alkanes, and n-alkanols in bamboo leaves, respectively. Panels (d) and (e) are for n-alkanoic acids and n-alkanes in rubber tree leaves, respectively. No n-alkanols were detected in rubber tree leaves. (Reprinted from Li et al., 2018b, with permission from John Wiley and Sons.)

furthermore, C_{16} and C_{18} alkanoic acids were most common with even-number carbons having predominance. The concentrations of a series of n-alkanoic acids in rubber tree leaves were higher than those in bamboos. This phenomenon further explained why the concentrations of soil alkanoic acids were much higher after 60 years than those in original soils.

Moreover, n-alkanols and n-alkanes were hardly detected in the rubber leaves, and their distribution did not match that observed in soils. A series of alkanes and some alkanols (C_{28} and C_{30}) were detected in bamboo leaves. In addition, both alkanols and alkanes showed similar distribution patterns in soil before and after rubber tree replacement. Therefore, we concluded that alkanols and alkanes were mainly derived from the original bamboo vegetation.

Lipid degradation can be reflected by lipid-related degradation parameters (e.g., CPI and RLS). The CPI of alkanes (CPI_{ALK}) decreased after replacement by rubber trees, indicating extended lipid degradation. However, the CPI_{AF} and RLS_{AF} values in surface soil increased and suggested lower degradation. This seemed to conflict with the earlier result (Table 25.2), which indicated that the concentration of alkanoic acids in rubber tree leaves was higher than that in bamboo leaves, and their distributions showed a similar pattern. Thus, the larger CPI_{AF} and smaller RLS_{AF} values before acid treatment in the surface soil of rubber tree areas were mainly caused by greater input of alkanoic acids from rubber leaves compared to the original soil. Therefore, since the application of related parameters of alkanoic acids, such as CPI_{AF}, was unable to determine the degradation of the original SOM in surface soils after cultivation, other complementary technologies (e.g., compound-specific isotope analysis) should be taken into consideration. However, we still observed that the CPI_{AF} values increased and the RLS_{AF} values decreased with soil depth before and after replacement with rubber trees, because the main source of plant-derived alkanoic acids were leaf waxes in surface soils rather than roots in deep soils (Reiffarth et al., 2016). This phenomenon thus indicated that plant-derived alkanoic acids in deep soils might be mainly from the OM of surface leaves and were relatively better-preserved compared to surface soil because of limited microorganism activity or oxygen.

25.3.2.3 SOM Turnover as Suggested by Lignin-Derived Phenol Biomarkers

The content of the total lignin-derived phenols (VSC, the sum of V, S, and C units) in soil in the 10-year rubber tree plantation was reduced by three times when compared to the original soil (data not shown), which might be due to the decreased input of lignin-derived phenols from rubber trees during short-term replacement. The VSC content also decreased with soil depth without significant changes in S/V and C/V values. This result indicated that the sources of lignin in deep and surface soils were the same, suggesting that the lignin in deep soils might be from the vertical migration of lignin in surface soil.

The S/V values in soils ranged from 0.25 to 0.5, while those in the leaves of local trees were greater than 0.5, especially for fresh bamboo (S/V up to 2). The C/V values in soils and leaves were also different. Thus, the sources of lignin in soils did not well match those in local vegetation. Considering the relatively stable input of local vegetation types, we can exclude the input of other plants. Lignin degradation in soils was also suggested by higher $(Ad/Al)_V$ values compared to those in leaves (data not shown). Moreover, the S and C units were preferentially degraded compared to the V unit, which was relatively recalcitrant, resulting in decreases in S/V and C/V after lignin degradation. This could explain the difference in S/V or C/V values between soil and plants. Notably, $(Ad/Al)_V$ values of lignin in fresh vegetation ranged from 0.32 to 0.75, which was higher than the previously reported range of 0.1 to 0.5. Dittmar and Lara (2001) also reported that the leaves of *Rhizophora mangle* in mangrove forests had a high $(Ad/Al)_V$ value (about 0.6). We speculated that the relatively high degradation state of SOM was closely related to the plant origin. Because of local climate and vegetation characteristics, we observed that the $(Ad/Al)_V$ values in soils were higher than 0.8. After replacement with rubber trees, the $(Ad/Al)_V$ values increased in soils growing rubber trees for 10 years. However, after 60 years, the $(Ad/Al)_V$ values in soils decreased. Considering the decreased lignin content compared to original soils, this phenomenon was attributed to a large loss of the previously preserved lignin with high degradation and considerable input of fresh OM. According to our local survey, after replacement with rubber trees, chemical fertilizers were widely applied; along with the high temperature in Xishuangbanna, this led to highly accelerated SOM degradation. Therefore, the degradation status of lignin in local soil depends on its source and surrounding environment.

25.3.3 SOM Turnover in Yuanyang Terrace

25.3.3.1 Changes in Soil Elemental Compositions

We studied the influence of different cultivation patterns on soil elemental compositions in a terraced area. TD and TP soils under human activity showed 5-fold higher organic carbon contents than those of NS (Table 25.3). Human activities introduced fresh OM into the soil during ploughing or compost application, which resulted in an increase in organic carbon content. In addition, the content of organic carbon in tilled soils (TD and TP) was comparable (data not shown). The mineral compositions in all soils contained quartz, muscovite, and kaolinite. In addition, montmorillonite was detected in TD and TP. Amorphous Fe oxides were detected in all investigated soils, and the contents of Fe_o in TD and TP were two times higher than that in NS. These reactive minerals with a large specific surface area played an important role in SOM stabilization.

25.3.3.2 SOM Turnover as Suggested by Lipid Biomarkers

The lipid content under human activity also increased along with the increase in organic carbon (Figure 25.1). The n-alkanoic acids were predominant, accounting for 70% of the total lipid content. Similar to the distribution of alkanoic acids in soil with rubber trees, alkanoic acids in the range of C_{12}–C_{32} showed significant even-over-odd carbon dominance. The results showed that lipids were mainly derived from plant waxes in the Yuanyang soil, as also suggested by even-carbon dominance of long-chain n-alkanols and odd-over-even carbon dominance of long-chain n-alkanes.

The degradation parameters based on lipid biomarkers were applied to investigate SOM degradation under different cultivation patterns. The CPI_{AF} values in TD (~0.97) and TP (~1.16) were

TABLE 25.3
Carbon Contents and Mass Loss before and after Acid Treatment

Samples	OC%	OC[a] % L_{OM}	OC[a] % L_{NR}	Fe_o (g/kg Soil)	Fe_d (g/kg Soil)	ΔMass [b] (%)	ΔOC[c] (%)	Mass [d] (%) L_{OM}	Mass [d] (%) L_{NR}
NS	0.45	–	–	4.32 ± 0.03	40.6 ± 3.1	65.7	68.5	35.4	64.6
TD	1.94	–	–	8.67 ± 0.3	18.0 ± 0.8	62.2	46.8	25.4	74.6
TP	1.93	–	–	8.15 ± 0.3	19.5 ± 5.9	63.4	29.8	27.7	72.3
A-NS	0.41 ± 0.00	0.83 ± 0.06	0.18 ± 0.03	0.07 ± 0.01	143.1 ± 3.1	–	–	–	–
A-TD	2.72 ± 0.44	9.49 ± 0.09	0.44 ± 0.01	0.19 ± 0.02	220.1 ± 7.1	–	–	–	–
A-TP	3.70 ± 1.04	12.28 ± 0.12	0.38 ± 0.01	0.09 ± 0.03	165.1 ± 2.7	–	–	–	–

Source: Reprinted from Li et al. (2018a), with permission from Elsevier.

NS: undisturbed natural soil; TD: dry-framing soil; TP: terrace paddy soil; A: acid treatment; L_{OM} and L_{NR}: OM-enriched layer and non-reactive mineral-enriched layer in acid-treatment soils, respectively. Fe_d: the content of dithionite-extractable iron, representing both crystalline and amorphous crystalline Fe oxides; Fe_o: the content of oxalate-extractable iron, representing amorphous Fe oxides

[a] the contents of organic carbon for different layer in acid-treatment soils;

[b] the total mass loss before and after acid treatment;

[c] the loss of total organic carbon before and after acid treatment;

[d] the mass percentage of different layer in acid treatment soils.

lower than those in NS (~1.2), suggesting higher lipid degradation in cultivated soils. In addition, the CPI_{AF} value of TP soil was much higher than that of TD, probably because the lipids in TP were not easily oxidized in the absence of oxygen. This result suggested that lipids were better preserved in TP than in TD.

25.3.3.3 SOM Turnover as Suggested by Lignin-Derived Phenols

The contents of the total lignin-derived phenols (VSC) after human activity increased by more than 20 times compared to those in NS (more than 300 μg/g soils for TD and TP; about 12 μg/g soils for NS; data unpublished). The S/V ratios in TD and TP (>0.8) were higher compared to that in NS (~0.23), suggesting that human activity introduced much angiosperm-derived OM (S/V>0.6) into soils. Based on the local vegetation (pine trees) in the sampling area, the low S/V value was mainly attributed to the abundant source of gymnosperm tissue rather than higher degradation of the S unit than the V unit. Similarly, the C/V values in TD and TP increased by up to 0.1, which was higher than that of NS (<0.02). This phenomenon was attributed to the input of fresh OM rich in C units (e.g., rice or corn stalks). Moreover, compared to TD and NS, TP showed a lower degradation state of lignin, as suggested by lower $(Ad/Al)_V$ values (~0.8) (data unpublished). Considering both lipid and lignin biomarkers, we thus concluded that the water chemistry conditions in paddy soil facilitate the preservation of SOM compared to upland soils. The phenomenon was attributed to two reasons: (i) the reducing environment of paddy soils inhibited the oxidation of SOM, and (ii) the SOM in upland soils may be subjected more to photochemical degradation in regions with intensive UV irradiation.

25.3.4 Role of Organo-Mineral Association in SOM Stabilization

Understanding the interaction mechanism between SOM and minerals is crucial for SOM stabilization. As discussed above, the removal of reactive minerals could enhance the extractability of lipids. In other words, the presence of reactive minerals can promote OM stabilization. Although previous studies found different degrees of protection by minerals for various types of SOM, such as aromatic and aliphatic organic compounds, knowledge about the protection mechanism for different SOM molecules was limited. Here, we explored the protection of different SOM components by minerals at a molecular level. We observed that during acid treatment, soil particles were separated into two layers: a dark-gray top layer with enriched organic matter (L_{OM}) and a light-gray bottom layer with enriched non-reactive minerals (L_{NR}). The SOM in L_{OM} and L_{NR} were then analyzed based on lipid and lignin biomarkers. Most SOM was enriched in L_{OM} (Figure 25.4). However, although the SOM content in the L_{NR} layer was limited, significant amounts of lipids such as alkanes were detected, accounting for up to 50% of the detected lipid biomarkers. Lignin-derived phenols in L_{NR} were hardly detected (<20%) (Figure 25.4). This difference indicated that reactive minerals protected lignin more than lipid compositions. Moreover, oxidized lignin could form stronger organo-mineral complexes, which was reflected by oxidation degradation parameters of lignin. For example, the $(Ad/Al)_V$ values of L_{OM} (>1.5) in agricultural soils were generally higher than those of L_{NR} (~1.0) (data unpublished). It should be noted that lipids may be also protected by soil aggregates to some extent. Some researchers reported that the turnover time of lignin in agricultural soils was slower than that of the bulk SOM (Thevenot et al., 2010); recently, studies found that lignin in soil clay fractions had higher oxidation states than that in sand fractions (Heim and Schmidt, 2007). Therefore, the turnover of lignin, especially for highly oxidized lignin in soils, was mainly controlled by reactive minerals. In addition, red soils in the Yunnan region were enriched with iron and aluminum minerals, which could complex these organic molecules to form organo-mineral complexes.

Crop straw burning is very common during agricultural management (Lehndorff et al., 2014). During this process, black carbon (BC), also called pyrogenic carbon or biochar, was introduced into soils as the product of incomplete combustion of vegetation, which is rich in carbon with a

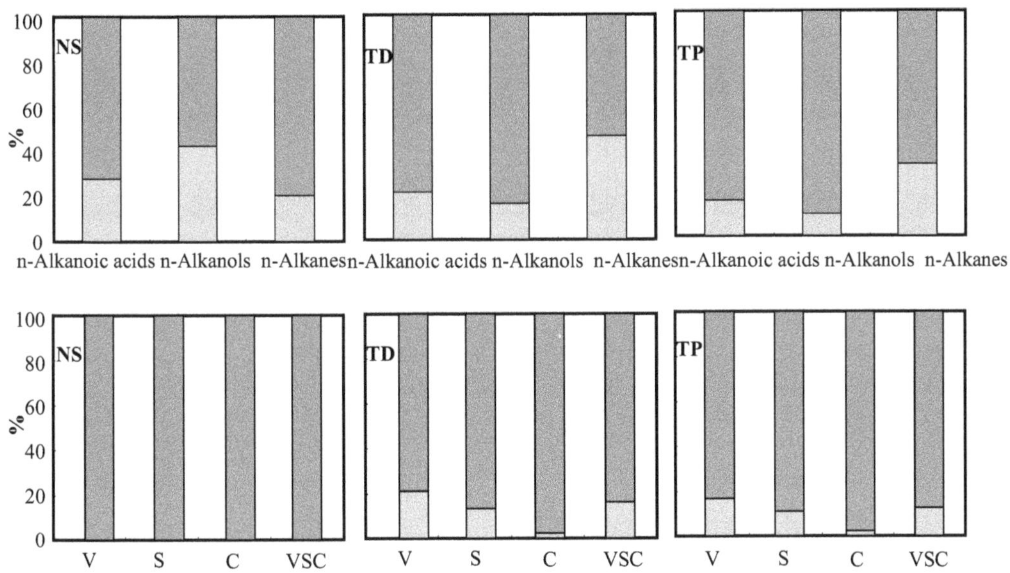

FIGURE 25.4 The percentage of biomarkers in L_{OM} and L_{NR} of acid-treatment soils. L_{OM} and L_{NR}: OM-enriched layer (dark-gray) and non-reactive mineral-enriched layer (light-gray) in acid-treatment soils, respectively. NS: for un-disturbed natural soil; TD: for dry farming soil; TP: for terrace paddy soil. The lignin-derived phenols of L_{NR} in NS were not detected. (Reprinted from Li et al., 2018a, with permission from Elsevier.)

porous structure (Sohi et al., 2010). The turnover time of BC is generally longer than that of other non-pyrogenic OM due to its condensed aromatic structure that is highly recalcitrant to biotic and abiotic decomposition. However, this traditional viewpoint has come into question (Singh et al., 2014). Some studies reported that BC can be quickly degraded (Brodowski et al., 2006). Therefore, BC decomposition may be not only related to its chemical structure but also related to environmental conditions, such as minerals and climate.

Accumulating evidence has suggested that organo-mineral complexes also regulated BC decomposition (Hemingway et al., 2019). BPCAs have been used as molecular biomarkers to characterize BC in soils (Brodowski et al., 2005). BPCAs mainly include a series of individual BPCA molecules (B3CA: trimellitic acid; $B4CA_{py}$: pyromellitic acid; $B4CA_{me}$: mellophanic acid; $B4CA_{pr}$: prehnitic acid; B5CA: benzenepenta-carboxylic acid; B6CA: mellitic acid). The sum of BPCAs could be associated with the relative amount of BC. For example, we observed that the sum of BPCAs in tilled soil (TP and TD) increased after mineral removal (more than 80% and 30% protected by minerals, respectively), indicating that BC strongly interacted with soil minerals. In addition, the extent of mineral complexation of individual BPCAs was analyzed. The extent of BPCA complexation was higher for B3CA and B4CA (representing small aromatic cluster size) than for B5CA and B6CA (representing aromatic condensed BC) (Figure 25.5). This result indicated that smaller aromatic clusters more easily formed organo-mineral complexes than the aromatic condensed ones. Compared to more condensed BC, the less condensed BC might contain more partial oxidation fractions, which facilitate their interactions with mineral surfaces. However, increased BPCA levels after mineral removal were not observed in NS, probably because the highly condensed soot was hardly incorporated in soil mineral compositions. This phenomenon further demonstrated that the reactive minerals mainly regulated SOM stabilization and OM that had a highly oxidized aromatic structure.

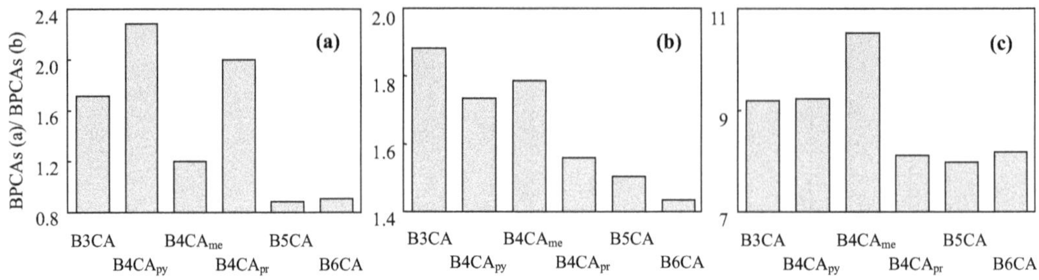

FIGURE 25.5 The ratios BPCA concentrations after (BPCAs (a)) and before (BPCAs (b)) mineral removal. (Reprinted from Li et al., 2015, with permission from Springer Nature.) B3CA: trimellitic acid; B4CApy: pyromellitic acid; B4CAme: mellophanic acid; B4CApr: prehnitic acid; B5CA: benzenepenta-carboxylic acid; B6CA: mellitic acid).

25.4 CONCLUSION AND PROSPECTS

This chapter summarizes previous studies that investigated two typical agricultural soil types, commercial rubber tree plantation soil and paddy soil, after the conversion of natural ecosystems to commercial plantations. Our results found that the removal of reactive minerals could enhance lipid extractability. Compared to alkanes and alkanols, the contents of alkanoic acids increased after acid treatment. Moreover, the lipid biomarker indexes (such as CPI, the carbon preference index, and RLS_{AF}, the ratio of long-chain carbon to short-chain carbon of alkanoic acids) that indicate the source and degradation of SOM were also altered substantially after acid treatment. Therefore, before determining the calibration factor of mineral protection in lipids, lipid biomarkers based on acid treatment would provide more reliable information to investigate SOM turnover as affected by human activities.

Based on the analysis of bulk chemicals (elemental compositions) and molecular biomarkers (lipids and lignin), we did not observe clear OC loss after tillage. However, we still observed high degradation of SOM, as suggested by lipid- and lignin-related degradation parameters [e.g., lower CPI_{ALK} (carbon preference index of alkanes) and higher $[Ad/Al]_V$ values (ratio of acids to aldehydes in vanillyl)]. The alkanoic acid biomarkers indicated new input from the plants into soils, while alkanol and alkane biomarkers indicated that SOM was preserved from the original soil. The related degradation parameters, such as low CPI_{ALK} values, indicated extended degradation status of lipids in soil after replacement with rubber trees. Moreover, the protection of lipids increased with soil depth. In the Yuanyang terraced area, organic carbon content in human-affected soils markedly increased compared to that in the original soil (pine forests). One important reason was that long-term input of OM (e.g., organic manure) probably resulted in an increase in SOM.

We further discussed the role of organo-mineral complexes in different SOM compositions. Compared to lipids, lignin-derived phenols were more dependent on reactive minerals, especially for highly oxidized lignin. In addition, we investigated the content and source of BC (black carbon) as indicated by BPCAs. BPCA analysis suggested that reactive minerals offered more protection for BC with minimally condensed aromatic structures than for BC with highly condensed aromatic structures. Overall, organo-mineral complexes played an important role in controlling the stabilization of SOM, especially for highly oxidized OM, in agricultural soils.

ACKNOWLEDGMENTS

This research was supported by the National Natural Scientific Foundation of China (41907300), Yunnan Fundamental Research Projects (202001AU070088).

REFERENCES

Brodowski, S., John, B., Flessa, H., Amelung, W., 2006. Aggregate-occluded black carbon in soil. *European Journal of Soil Science* 57, 539–546.

Brodowski, S., Rodionov, A., Haumaier, L., Glaser, B., Amelung, W., 2005. Revised black carbon assessment using benzene polycarboxylic acids. *Organic Geochemistry* 36, 1299–1310.

Bull, I.D., van Bergen, P.F., Nott, C.J., Poulton, P.R., Evershed, R.P., 2000. Organic geochemical studies of soils from the Rothamsted classical experiments - V. The fate of lipids in different long-term experiments. *Organic Geochemistry* 31, 389–408.

Chen, S.C., Arrouays, D., Angers, D.A., Martin, M.P., Walter, C., 2019. Soil carbon stocks under different land uses and the applicability of the soil carbon saturation concept. *Soil and Tillage Research* 188, 53–58.

Dittmar, T., Lara, R.J., 2001. Molecular evidence for lignin degradation in sulfate-reducing mangrove sediments (Amazonia, Brazil). *Geochimica Et Cosmochimica Acta* 65, 1417–1428.

Eltantawy, I.M., Arnold, P.W., 1972. Adsorption of n-alkanes by wyoming montmorillonite. *Nature* 237, 123–125.

Feng, X.J., Simpson, M.J., 2011. Molecular-level methods for monitoring soil organic matter responses to global climate change. *Journal of Environmental Monitoring* 13, 1246–1254.

Guo, L.B., Gifford, R.M., 2002. Soil carbon stocks and land use change: A meta analysis. *Global Change Biology* 8, 345–360.

Hall, S.J., Silver, W.L., Timokhin, V.I., Hammel, K.E., 2016. Iron addition to soil specifically stabilized lignin. *Soil Biology and Biochemistry* 98, 95–98.

Heim, A., Schmidt, M.W.I., 2007. Lignin is preserved in the fine silt fraction of an arable Luvisol. *Organic Geochemistry* 38, 2001–2011.

Hemingway, J.D., Rothman, D.H., Grant, K.E., Rosengard, S.Z., Eglinton, T.I., Derry, L.A., Galy, V.V., 2019. Mineral protection regulates long-term global preservation of natural organic carbon. *Nature* 570, 228–231.

Huang, C., Liu, S., Li, R., Sun, F., Zhou, Y., Yu, G., 2016. Spectroscopic Evidence of the Improvement of Reactive Iron Mineral Content in Red Soil by Long-Term Application of Swine Manure. *PLoS One* 11, e0146364.

Huang, X.L., Tang, H.Y., Kang, W.J., Yu, G.H., Ran, W., Hong, J.P., Shen, Q.R., 2018. Redox interface-associated organo-mineral interactions: A mechanism for C sequestration under a rice-wheat cropping system. *Soil Biology and Biochemistry* 120, 12–23.

Lehndorff, E., Roth, P.J., Cao, Z.H., Amelung, W., 2014. Black carbon accrual during 2000 years of paddy-rice and non-paddy cropping in the Yangtze River Delta, China. *Global Change Biology* 20, 1968–1978.

Li, F.F., Liang, N., Zhang, P.C., Xu, Y., Chang, Z.F., Wu, M., Duan, W.Y., Steinberg, C.E.W., Pan, B., 2018a. Protection of extractable lipid and lignin: Differences in undisturbed and cultivated soils detected by molecular markers. *Chemosphere* 213, 314–322.

Li, F.F., Pan, B., Zhang, D., Yang, X.L., Li, H., Liao, S.H., Ghaffar, A., Peng, H.B., Xing, B.S., 2015. Organic matter source and degradation as revealed by molecular biomarkers in agricultural soils of Yuanyang terrace. *Scientific Reports* 5:11074.

Li, F.F., Zhang, P.C., Wu, D.P., Xu, Y., Chen, F.Y., Chang, Z.F., Chu, G., Wang, L., Pan, B., Xing, B.S., 2018b. Acid pretreatment increased lipid biomarker extractability: A case study to reveal soil organic matter input from rubber trees after long-term cultivation. *European Journal of Soil Science* 69, 315–324.

Lin, L.H., Simpson, M.J., 2016. Enhanced extractability of cutin- and suberin-derived organic matter with demineralization implies physical protection over chemical recalcitrance in soil. *Organic Geochemistry* 97, 111–121.

Otto, A., Shunthirasingham, C., Simpson, M.J., 2005. A comparison of plant and microbial biomarkers in grassland soils from the Prairie Ecozone of Canada. *Organic Geochemistry* 36, 425–448.

Otto, A., Simpson, M.J., 2007. Analysis of soil organic matter biomarkers by sequential chemical degradation and gas chromatography - mass spectrometry. *Journal of Separation Science* 30, 272–282.

Pisani, O., Hills, K.M., Courtier-Murias, D., Haddix, M.L., Paul, E.A., Conant, R.T., Simpson, A.J., Arhonditsis, G.B., Simpson, M.J., 2014. Accumulation of aliphatic compounds in soil with increasing mean annual temperature. *Organic Geochemistry* 76, 118–127.

Reiffarth, D.G., Petticrew, E.L., Owens, P.N., Lobb, D.A., 2016. Sources of variability in fatty acid (FA) biomarkers in the application of compound-specific stable isotopes (CSSIs) to soil and sediment fingerprinting and tracing: A review. *Science of the Total Environment* 565, 8–27.

Schmidt, M.W.I., Torn, M.S., Abiven, S., Dittmar, T., Guggenberger, G., Janssens, I.A., Kleber, M., Kogel-Knabner, I., Lehmann, J., Manning, D.A.C., Nannipieri, P., Rasse, D.P., Weiner, S., Trumbore, S.E., 2011. Persistence of soil organic matter as an ecosystem property. *Nature* 478, 49–56.

Singh, N., Abiven, S., Maestrini, B., Bird, J.A., Torn, M.S., Schmidt, M.W.I., 2014. Transformation and stabilization of pyrogenic organic matter in a temperate forest field experiment. *Global Change Biology* 20, 1629–1642.

Sohi, S.P., Krull, E., Lopez-Capel, E., Bol, R., 2010. A review of biochar and its use and function in soil, in *Advances in Agronomy*, Ed. D.L. Sparks, San Diego: Academic Press Inc Elsevier Science, 47–82.

Thevenot, M., Dignac, M.F., Rumpel, C., 2010. Fate of lignins in soils: A review. *Soil Biology and Biochemistry* 42, 1200–1211.

von Luetzow, M., Koegel-Knabner, I., Ekschmitt, K., Matzner, E., Guggenberger, G., Marschner, B., Flessa, H., 2006. Stabilization of organic matter in temperate soils: Mechanisms and their relevance under different soil conditions - a review. *European Journal of Soil Science* 57, 426–445.

Wiesenberg, G.L.B., Dorodnikov, M., Kuzyakov, Y., 2010. Source determination of lipids in bulk soil and soil density fractions after four years of wheat cropping. *Geoderma* 156, 267–277.

Zhu, S.S., Dai, G.H., Ma, T., Chen, L.T., Chen, D.M., Lu, X.T., Wang, X.B., Zhu, J.T., Zhang, Y.J., Bai, Y.F., Han, X.G., He, J.S., Feng, X.J., 2019. Distribution of lignin phenols in comparison with plant-derived lipids in the alpine versus temperate grassland soils. *Plant and Soil* 439, 325–338.

26 Management of Soil Constraints in China with a Biochar Amendment

A Case Study in Paddy Fields in Subtropical China

Shen Jianlin, Wang Cong, Jiang Bingshen, Liu Ji, Li Yanyan, Li Yong, and Wu Jinshui
Institute of Subtropical Agriculture, Chinese Academy of Sciences, Changsha, P.R. China

CONTENTS

26.1 INTRODUCTION

Biochar is the product of the pyrolysis of organic materials under no oxygen or limited oxygen conditions (Lehmann and Joseph, 2009). It is rich in carbon, is highly porous, and is recalcitrant to degradation. Biochar application has been widely proposed as an option to improve soil fertility,

DOI: 10.1201/9781003093565-26

mitigate soil degradation, reduce soil greenhouse gas (GHG) emissions, and remediate soil contamination (Gul and Joann, 2016; Kuzyakov et al., 2014; Van Zwieten et al., 2010; Wang et al., 2018). Globally, soil biochar amendments have shown the potential to improve soil quality through changing the physical, chemical, and biological properties of soil and thus increasing agricultural productivity. Biochar research is now an important topic in the world for achieving sustainable agriculture and protecting the environment. In regions with soil constraints, e.g., soil degradation, low nutrient use efficiencies, or high greenhouse gas emissions, the amendment of soils with biochar has been used as a cost-beneficial technology to solve these problems.

The subtropical hilly region of China has high agricultural production, especially paddy rice production. In recent decades, the large use of chemical fertilizers and the reduction in the use of organic fertilizers have caused soil acidification, low soil organic matter, and a decrease in nutrient use efficiencies (Jiang et al., 2021; Liu et al., 2021a; Shen et al., 2022). The chemical, physical, and biological properties of paddy soils are different from those of upland soils due to long-term waterlogging during the rice growing season, and they contribute to large quantities of CH_4 emissions (Shen et al., 2014; Liu et al., 2014; Wang et al., 2018). The alkaline, carbon-rich, and highly porous properties of biochar make it a potential soil amendment to manage soil constraints in paddy fields (Wang et al., 2021a). Here we conducted a field experiment to study the effects of a biochar amendment in paddy fields to manage soil constraints in a subtropical region of China. The effects of biochar on soil aggregates; soil pH; soil contents such as carbon, nitrogen, and phosphorus; and soil microbes were examined. In addition, the uptake of nitrogen and phosphorus by rice and greenhouse gas emissions were measured.

26.2 MATERIALS AND METHODS

26.2.1 Study Site and Experimental Design

The experiment was conducted in Changsha County, Hunan Province. A typical paddy field with double-rice cropping was chosen as the experimental field. The experiment was started in 2012 and continued till 2017. Five treatments were set up in the experiment: only chemical fertilizers (CK), chemical fertilizers plus straw incorporation at a low rate (3 t/ha per season, LS), chemical fertilizers plus straw incorporation at a high rate (6 t/ha per season, HS), chemical fertilizers plus straw-derived biochar at a low rate (24 t/ha applied only once, LB), and chemical fertilizers plus straw-derived biochar at a high rate (48 t/ha applied once, HB). Treatments were laid out in a randomized complete block design with triplicates. The area of each plot was $35\,m^2$ ($7\,m \times 5\,m$). The rice cultivars used in this study were "T-you 207" for the early rice and "Xiang 45" for the late rice. These were common rice cultivars that grew in the study region. Each year, rice seedlings (approximately 30 days old) were transplanted in late April and harvested in the middle of July for the early rice, while for the late rice, they were transplanted in July and harvested in late October. Nitrogen fertilizers (urea, applied at a rate of 150 kg N/ha for the late rice and 120 kg N/ha for the early rice) were applied three times a season: a basal application before transplanting, topdressing at the early tillering stage, and at the heading stage. Phosphorus (calcium superphosphate) and potassium (potassium sulfate) fertilizers were applied at 40 kg P_2O_5 and 100 kg K_2O/ha, respectively, as basal fertilizers.

26.2.2 Soil Properties Measurements

26.2.2.1 Soil Aggregates Measurement

Aggregate stability was analyzed using a wet sieving apparatus (TPT-100, Zhejiang Top Cloud-Agricultural Technology Co., Ltd., China) containing four sieves (2.0-, 1.0-, 0.25-, and 0.053-mm openings); 100 g of air-dried soil (<5 mm) was submerged in water for 5 min (amplitude 3 cm, vibration frequency 50 times/min); and soil samples were sieved through sieves of 2, 1, 0.5, and 0.053 mm in order, which were collected in aluminum boxes and then dried at 55°C. Soil aggregate content is expressed as a percentage (Cambardella and Elliott, 1993).

The weight proportion of the > 0.25 mm aggregate size fraction was calculated as follows (Hou et al., 2012):

$$R_{0.25} = \frac{M_{r>0.25}}{M_T} \tag{26.1}$$

where $R_{0.25}$ is the weight proportion of the > 0.25 mm aggregate size fraction (%), M (r>0.25) is the weight of the > 0.25 mm aggregate size fraction, and M_T is the total weight of aggregates.

The stability of soil water-stable aggregates was evaluated by the mean weight diameter (MWD) and the geometric mean diameter (GMD), which were calculated as follows (Kemper and Rosenau, 1986):

$$\text{MWD} = \sum_{i=1}^{n} w_i x_i \tag{26.2}$$

where x_i is the mean diameter (mm) and w_i is the weight proportion of the i size fraction, and

$$\text{GMD} = \exp\left\{\frac{\sum_{i=1}^{n} w_i \log x_i}{\sum_{i=1}^{n} w_i}\right\} \tag{26.3}$$

where w_i is the weight of aggregates in the i size fraction (g) and $log\ x_i$ is the natural logarithm of the mean diameter of the i size fraction.

The calculation formula of fractal dimensions (D) was as follows (Coskun, 2006)

$$\frac{M(r < x_i)}{M_T} = \left(\frac{x_i}{x_{\max}}\right)^{3-D} \tag{26.4}$$

The logarithm of both sides can be obtained as follows:

$$\lg\left[\frac{M(r < x_i)}{M_T}\right] = (3 - D)\lg\left(\frac{x_i}{x_{\max}}\right) \tag{26.5}$$

where x_i is the mean diameter (mm) and x_{max} is the maximum diameter (mm), M_T is the total weight of aggregates, and M (r<x_i) is the weight of aggregates of <x_i.

26.2.2.2 Soil Chemical Properties Measurement

The soil pH of each sample was measured in 1:2.5 mixtures of soil and deionized water with a pH meter (PHS-3C, Lei-ci). The soil available phosphorus (SAP) was determined by extracting the soil with 0.5 mol/L $NaHCO_3$ at pH 8.5 (Olsen and Sommers, 1982). Soil total organic carbon (STOC), soil total nitrogen (STN), soil total phosphorus (STP), and N and P concentrations in rice plants were determined using the methods described in Page et al. (1982), Kalra (1998), and Jiang et al. (2021).

26.2.2.3 Soil Biological Properties Measurement

Soil samples were stored at 4°C for measuring soil MBC and MBN. They were extracted using 0.5 mol/L K_2SO_4. The extracts were measured with a total organic carbon analyzer (TOC-VWP, Shimadzu Corporation, Japan) for MBC and with a flow-injection auto-analyzer (Tecator FIAstar

5000 Analyzer, Foss Tecator, Sweden) for MBN (Brookes et al., 1982, 1985; Vance et al., 1987; Wu et al., 1990). A conversion factor of 0.45 was used to calculate MBC and MBN contents (Wu et al., 2006).

Soil samples were stored at −80°C for DNA extraction. The DNA was isolated from each soil sample collected at tillering, flowering, filling, ripening, and the fallow season as described by Chen et al. (2010). The quality and concentration of the DNA were analyzed by a Nanodrop spectrophotometer (ND-1000, Thermo Scientific, USA). The abundance of the bacterial 16S ribosomal RNA (rRNA), archaeal 16S rRNA, and 18S rRNA genes was measured in triplicate by a quantitative real-time PCR (qPCR) detection system (ABI prism 7900, Applied Biosystem, USA), with the primer sets 1369F/1492R (Suzuki et al., 2000), 109F/934R (Großkopf et al., 1998), and NS1/Fung (Hoshino and Morimoto, 2008), respectively.

Each reaction solution (10 μL) contained 5 μL of 2× SYBR Green PCR Master Mix (TaKaRa Bio, Japan), 0.2 μL of Rox Reference DyeII, 0.4 μL of 10 μM each primer, and 5 ng DNA. The cycling conditions of bacterial 16S rRNA were used: 94°C for 31 s; 5 cycles of 92°C for 5 s, 57°C for 28 s, and 70°C for 25 s; 31 cycles of 96°C for 5 s, 58°C for 25 s, and 68°C for 25 s. The amplification conditions for archaeal 16S rRNA were as follows: 96°C for 31 s; 39 cycles of 96°C for 5 s, 65°C for 25 s, and 70°C for 20 s. The amplification conditions for 18S rRNA were used: 94°C for 31 s; 42 cycles of 94°C for 6 s and 62°C for 25 s. Serial tenfold stepwise dilutions of linearized plasmids were made to generate standard curves. The qPCR efficiencies were in the range of 97%–105%.

26.2.3 Rice Yield and Nutrient Uptake Measurements

The four rice strains were randomly sampled from each plot at harvest to measure plant N content. The samples were cleaned; divided into root, straw, and grain; sorted into sample bags; dried at 70°C to a constant weight; milled; and stored at room temperature in ziplock bags. Biomass was calculated using the root, straw, and grain DM content. Plant N was determined by digesting the samples on a heating block in H_2SO_4-H_2O_2 and measuring the N content in the digest in an automated injection system (AA3; Seal Analytical GmbH, Norderstedt, Germany). Rice grain and straw yields were determined by harvesting all rice plants in each plot at the end of each cropping season.

Rice nitrogen uptake rate (NU; kg N/ha) was calculated as follows:

$$NU = SB \times SNC + GB \times GNC \tag{26.6}$$

where SB is the straw biomass excluding the stubble (kg/ha), SNC is the straw N content (%), GB is the grain biomass (kg/ha), and GNC is the grain N content (%).

N fertilizer use efficiency (NUE, kg N/ha) was calculated as follows:

$$NUE = \frac{NU_T - NU_{PK}}{F} \times 100\% \tag{26.7}$$

where NU_T and NU_{PK} are the crop nitrogen uptake levels for the treatment and the PK treatment without N fertilizer, respectively (kg N/ha), and F is the amount of N fertilizer applied (kg N/ha).

26.2.4 Greenhouse Gas Emission Measurement

GHG fluxes were determined using the static opaque chamber-gas chromatograph method (Zheng et al., 2008). In each plot, two base frames made of stainless steel (covering an area of 0.64 m × 0.64 m) were inserted 20 cm into the soil. One frame allowed crop growth at normal density, for measuring CH_4 and N_2O fluxes, while the other frame was installed in the center of a bare area (1.5 × 1.5 m^2) for measuring Rh (Iqbal et al., 2009). Chambers (0.64 m long × 0.64 m wide × 1.00 m high) made of stainless steel sheet, and wrapped with insulating foam, were temporarily mounted onto the frames with water seal for gas flux measurements. Inside the top of each chamber was installed a circulating fan (12 V) for mixing air, a portable digital thermocouple (JM624, Tianjin Jinming Instrument Co. Ltd., China) to measure air temperatures, and a three-way stopcock. The air pressure was recorded at an automatic weather station near the study field. Gas sampling was performed between 9:00 am and 11:00 am every week (twice a week during fertilization and the aeration period). Five gas samples were withdrawn from each chamber, using a 60 mL plastic syringe, at intervals of 10 min (0, 10, 20, 30, and 40 min after enclosure). The air samples were then transferred into pre-evacuated 12 mL vacuum vials with butyl rubber stoppers for laboratory analysis. Concentrations of CH_4, N_2O, and CO_2 in the gas samples were analyzed with a gas chromatograph (Agilent 7890D, Agilent Technologies, USA) equipped with a hydrogen flame ionization detector (FID) for CH_4 and CO_2 detection at 250°C, and an electron capture detector (ECD) for N_2O detection at 350°C. For GHG analysis, N_2 was used as a carrier gas, and a mixture of CO_2 and N_2 (10% CO_2 in N_2) was used as a make-up gas (Zheng et al., 2008).

The hourly chamber fluxes were determined by the nonlinear (exponential) or linear method using the GHG concentration data during each chamber enclosure. First, several criteria were applied to detect significant nonlinear cases. The criteria were as follows: (i) all the five GHG concentration data were valid; (ii) the concentration-time relationship could be significantly ($p<0.05$) fitted with not only the linear function ($C=a_0+a_1t$, where C is the measured concentration, a_0 is the intercept, a_1 is the slope of the fitting line, and t is the time) but also the nonlinear (exponential) function following Kroon et al. (2008) and Valente et al. (1995) ($C=k_1/k_2+(C_0-k_1/k_2)\times\exp(-k_2t)$, where C_0 is the concentration at the beginning of the enclosure, and k_1 and k_2 are the fitting parameters); (iii) the correlation coefficient of the nonlinear regression was at least 0.001 greater than that of the linear regression; and (iv) the initial slope of the nonlinear fitting curve ($d_c/d_t|_{t=0}=k_1-k_2C_0$) was larger than the slope of the linear fitting line ($d_c/d_t=a_1$). If these criteria were satisfied, the hourly chamber flux was determined based on the initial slope of the nonlinear regression; otherwise, the slope of the linear regression was used to calculate the flux ($p<0.05$ and concentration number ≥3). The measurements with neither significant nonlinear nor linear relationships between the concentrations and time were discarded.

26.2.5 Statistical Analyses

The significant differences among the treatment were based on the least significant difference (LSD) method (Steel and Jh, 1981) using R software (R Core Team, 2022). A one-way ANOVA with 0.05 ($p<0.05$) level was performed to determine the significant differences between treatments.

26.3 RESULTS AND DISCUSSION

26.3.1 Effects of Biochar Amendment on Soil Aggregates

Straw and straw-derived biochar had significant effects on soil aggregates (Table 26.1). In the >2 mm size soil aggregates, straw incorporation (LS and HS) and high straw-derived biochar applications (HB) increased soil aggregates by 24.9%, 57%, and 37.9%, respectively, compared to the control (CK). The proportion of the 1–2 mm size soil aggregates of straw and straw-derived biochar applications (LS, HS, LB, and HB) was lower than that for the CK. The high straw and straw-derived biochar applications (HS and HB) did not increase the proportion of the 0.25–1 mm size soil aggregates when compared with the CK, while the low straw and straw-derived biochar applications (LS and LB) increased by 26.4% and 26.0%, respectively.

$R_{0.25}$ represents the > 0.25 mm soil aggregates mass percentage. The higher its value is, the more stable the soil aggregates are. Straw and straw-derived biochar had significant effects on the stable index ($R_{0.25}$) (Table 26.2). The low straw-derived biochar applications (LB) did not have a significant impact on $R_{0.25}$ as compared with the CK, while $R_{0.25}$ significantly increased with LS, HS, and HB,

TABLE 26.1
Size Distribution of Soil Water-Stable Aggregates in the Different Treatments

	Aggregate Size/%				
Treatments	>2 mm	1–2 mm	0.25–1 mm	0.053–0.25 mm	<0.053 mm
CK	22.6±1.0d	24.1±0.3a	29.1±0.9b	14.0±0.9a	10.2±1.6a
LS	28.2±0.7bc	20.8±1.0b	36.8±0.9a	8.9±1.8b	5.3±0.4bc
HS	35.5±3.5a	18.6±0.8bc	32.9±2.1ab	9.0±1.1b	4.1±1.5c
LB	25.6±1.4cd	17.4±2.3c	36.7±3.2a	11.7±4.2ab	8.6±2.1ab
HB	31.1±2.9b	16.3±1.2c	33.5±4.0ab	12.1±1.9ab	7.0±2.3abc

TABLE 26.2
The Stability Index ($R_{0.25}$), the Mean Mass Diameter (MWD), the Mean Geometric Diameter (GMD), and the Fractal Dimension (*D*) of Soil Aggregates in the Different Treatments

Treatments	$R_{0.25}$ (mm)	MWD (mm)	GMD (mm)	D
CK	75.77±1.3c	1.02±0.0c	0.64±0.0d	2.40±0.0a
LS	85.79±2.1a	1.12±0.0b	0.81±0.0ab	2.38±0.0a
HS	86.89±0.6a	1.21±0.0a	0.88±0.1a	2.24±0.1b
LB	79.74±2.3bc	1.03±0.0c	0.67±0.0cd	2.37±0.1a
HB	80.90±4.0b	1.10±.01b	0.74±0.1c	2.27±0.0b

indicating that straw incorporation and high straw-derived biochar applications can enhance the stability of soil aggregates. $R_{0.25}$ was not significantly different between low and high straw incorporation or between low and high straw-derived biochar applications.

The mean mass diameter (MWD) and mean geometric diameter (GMD) of soil aggregates reflect the distribution of the size of the soil aggregates. The higher the proportion of soil macroaggregates is, the greater the MWD value is, indicating that the soil aggregate stability will increase with the increase of the degree of soil particle agglomerates. The larger the GMD value is, the stronger is the stability and corrosion resistance of soil aggregates. Straw and straw-derived biochar applications had a significant effect on the MWD and GMD (Table 26.2). Compared with the CK, except for the low straw-derived biochar applications (LB), the MWD and GMD increased in the other treatments, indicating that the stability of the soil aggregates was increased by the application of straw and high straw-derived biochar.

The fractal dimension (*D*) can describe the quantitative composition of soil aggregates, and the smaller the value of *D* is, the better is the stability of the soil structure (Table 26.2). Compared with the CK, *D* was not significantly different from the LS and LB treatments, while there was a significant decrease in the HS and HB treatments, indicating that soil aggregate stability increased in the HS and HB treatments.

26.3.2 Effects of Biochar Amendment on Soil pH

Soil pH is an important parameter to characterize soil fertility, and it affects nutrient availability. Under the long-term application of N fertilizer, severe acidification has occurred in subtropical, double-cropped rice fields. Soil pH in some fields is below 5.0. Soil pH is an important regulator of soil nutrient availability and influences transformation processes of soil N and P. Soil pH was elevated under all straw treatments, and they averaged 0.13–0.16 pH units higher than the CK treatment

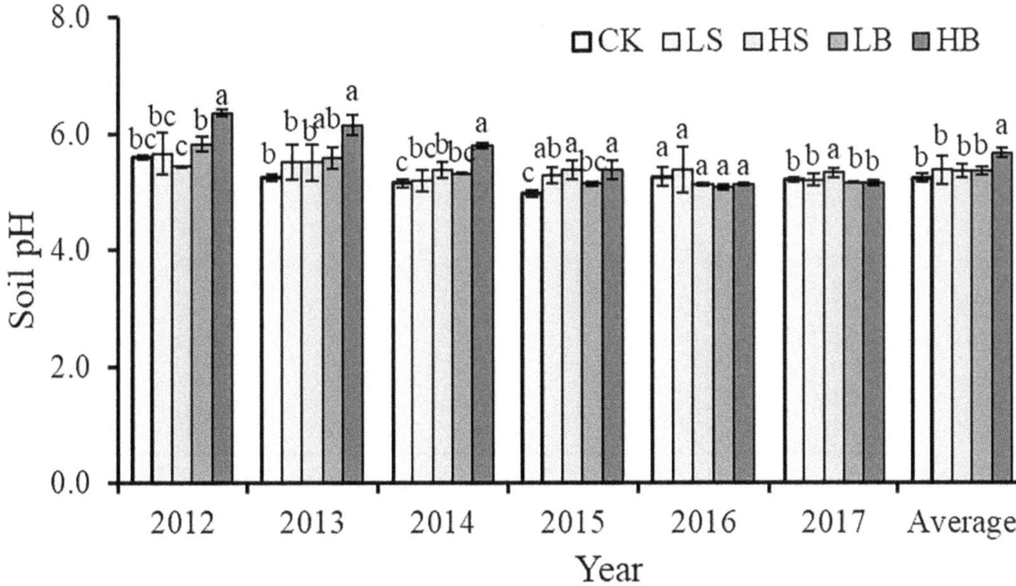

FIGURE 26.1 Variation in soil pH from 2012 to 2017.

over the 6-year period (Figure 26.1). Biochar significantly increased soil pH in the year of application, with the LB and HB treatments increasing by 0.52 and 1.03 pH units, respectively, compared to the CK treatment. As the biochar aged, the effect of biochar in raising soil pH decreased, and by the fifth year after application, the soil pH of the biochar treatment was no longer significantly different from the CK treatment. This is probably because biochar application brought in abundant cations, such as Ca^{2+}, K^{+}, and Mg^{2+}, which helped to alleviate soil acidity and raise pH. Due to crop harvesting resulting in cation loss, runoff loss, and leaching, the amount of cations in the biochar-treated soil was no longer significantly different from the CK, thus resulting in no significant difference in pH from the CK.

26.3.3 Effects of Biochar Amendment on Soil Carbon Sequestration

Straw incorporation and biochar (straw charring) application increased the soil organic carbon content (Figure 26.2). Biochar application increased soil organic carbon content compared to the CK treatment, with the biochar treatments increasing by 5.72 g C/kg (LB) and 9.42 g C/kg (HB) in the first year of application. Soil organic carbon increased slowly after straw incorporation and was significantly higher than the CK treatment in the third year after application. Although the amount of straw input was the same for the HS treatment and the LB treatment over the 6 years, the LB treatment still had significantly higher organic carbon than the HS treatment, indicating that biochar application significantly increased soil organic carbon content compared to straw incorporation. The trend of soil organic carbon change across the 6-year period showed that soil organic carbon increased slowly in all treatments, and the annual increase in soil organic carbon content ranged from 0.23 to 0.71 g C/kg, indicating that double-season rice fields are an important carbon sink area (Figure 26.3; Liu et al., 2021a).

26.3.4 Effects of Biochar Amendment on Soil Nitrogen Availability and Rice Nitrogen Uptake

Straw incorporation and biochar (straw charring) application had differential effects on soil N pools. As shown in Figure 26.4, the trend of soil total N content from 2012 to 2017 was generally similar

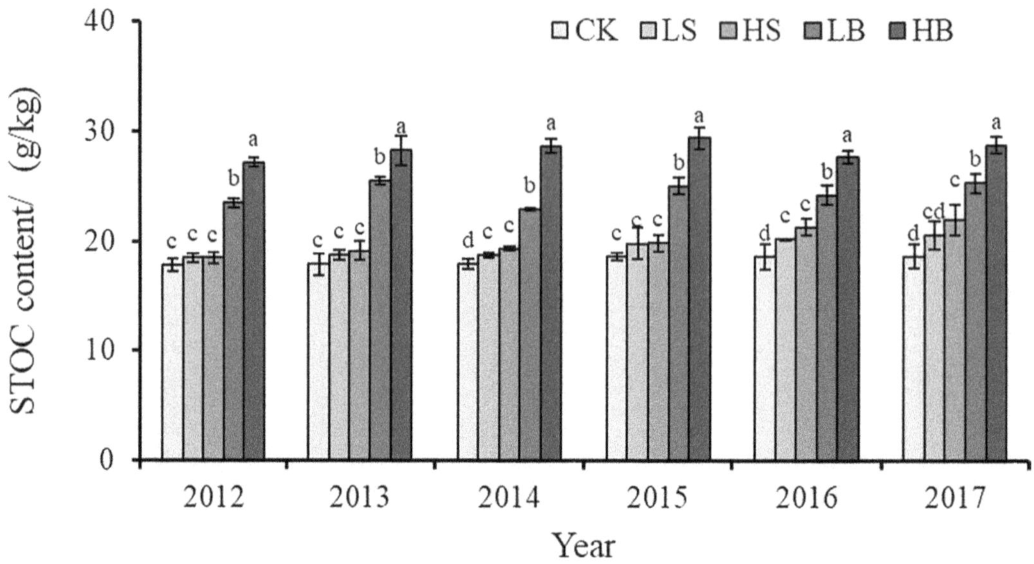

FIGURE 26.2 Variation in total soil organic carbon (STOC) from 2012 to 2017.

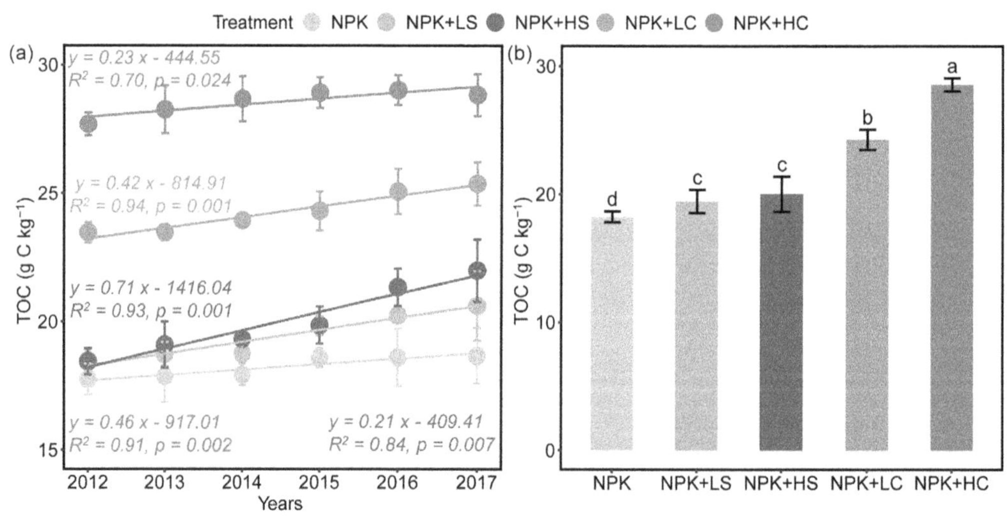

FIGURE 26.3 Interannual changes in the total organic carbon (TOC) in double-rice cropping systems with straw and straw-derived biochar treatments from 2012 to 2017 (a) and significance test of the difference in TOC (b).

among the treatments, with an average annual variation range of 1.9–2.1 g/kg. Soil total N content in paddy fields was stable. Compared with the CK treatment, the LS, HS, LB, and HB treatments increased STN content by an average of 4.3%, 4.5%, 8.1%, and 9.7%, respectively, over the 6 years.

During the 6-year observation period, the average annual N uptake of the aboveground part of rice (seed + straw) ranged from 63.6 to 77.6 kg N/hm^2, and the annual NUE ranged from 10.5% to 51.2% in the early seasons across the treatments (Figure 26.5; Liu et al., 2021b). From 2012 to 2014, the LB and HB treatments improved NUE by an average of 50.3% and 43.8%, respectively, compared to the CK, due to the additional available N from biochar application (Figure 26.5g). In contrast, the LS and HS treatments reduced NUE by up to 21.4% and 19.6%, respectively, compared

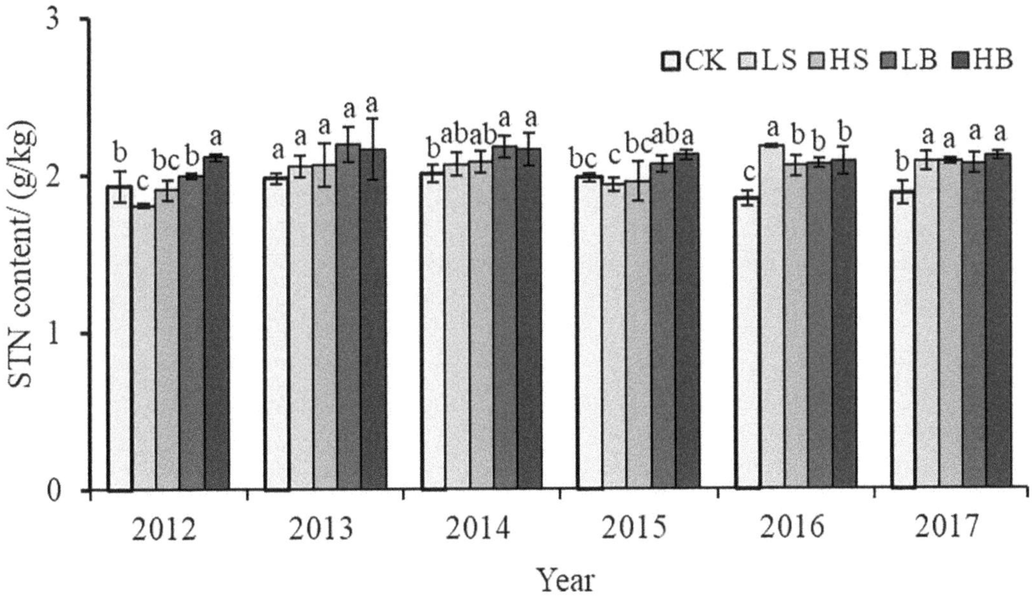

FIGURE 26.4 Variation in total soil nitrogen (STN) from 2012 to 2017.

to the CK in the first 3 years with straw incorporation, although the differences were not significant (Figure 26.5g), due to slow decomposition and low nutrient release from straw returned to the field in the early rice season, leading to competition between microorganisms and the crop for fertilizer nutrients, which resulted in reduced rice yields. In 2015–2017, there were no significant differences in NUE between CK and straw treatments or biochar treatments due to that the long-term straw return increased soil organic carbon and organic N contents, as well as soil N supply, while biochar, due to aging, may not have a significant difference in its own supply of effective nutrients (e.g., available N, Ca^{2+}, and Mg^{2+}) compared to the CK (Figure 26.5h). This resulted in nonsignificant differences in N fertilizer utilization between the straw return, biochar application, and CK treatments (Figure 26.5h).

The average annual N uptake of the aboveground part of rice ranged from 99.3 to 114 kg N/hm^2, and the annual NUE ranged from 24.2% to 60.2% in the late rice season across the treatments (Figure 26.6; Liu et al., 2021a). In 2012–2014, there was no significant difference in NUE among the treatments. In contrast, from 2015 to 2017, the NUE rate increased by 22.3% and 39.8% in the LS and HS treatments, respectively, compared to the CK treatment, due to the increase in soil organic carbon and N contents, and the increase in soil organic matter mineralization as a result of the long-term straw return.

Overall, long-term straw return improved NUE by enhancing soil organic carbon and N contents and improving soil microbial activity. Moreover, straw return to the field allowed straw N to replace 13%–34% of inorganic N fertilizer. Therefore, after straw return, the inorganic NUE increased compared to the control, and straw return reduced the amount of inorganic N fertilizer applied. Thus, rice straw return is an effective way to improve NUE in paddy fields. However, further research is needed on how to reduce the negative impact of straw return on early season yields.

26.3.5 Effects of Biochar Amendment on Soil Phosphorus Availability and Rice Phosphorus Uptake

For the late rice-harvesting stage from 2012 to 2017, the variation trend of STP was basically the same among the treatments and ranged from 0.56 to 0.63 kg P/ha (Figure 26.7). Moreover, the STP content of each treatment changed little over time, indicating that the STP content was stable in

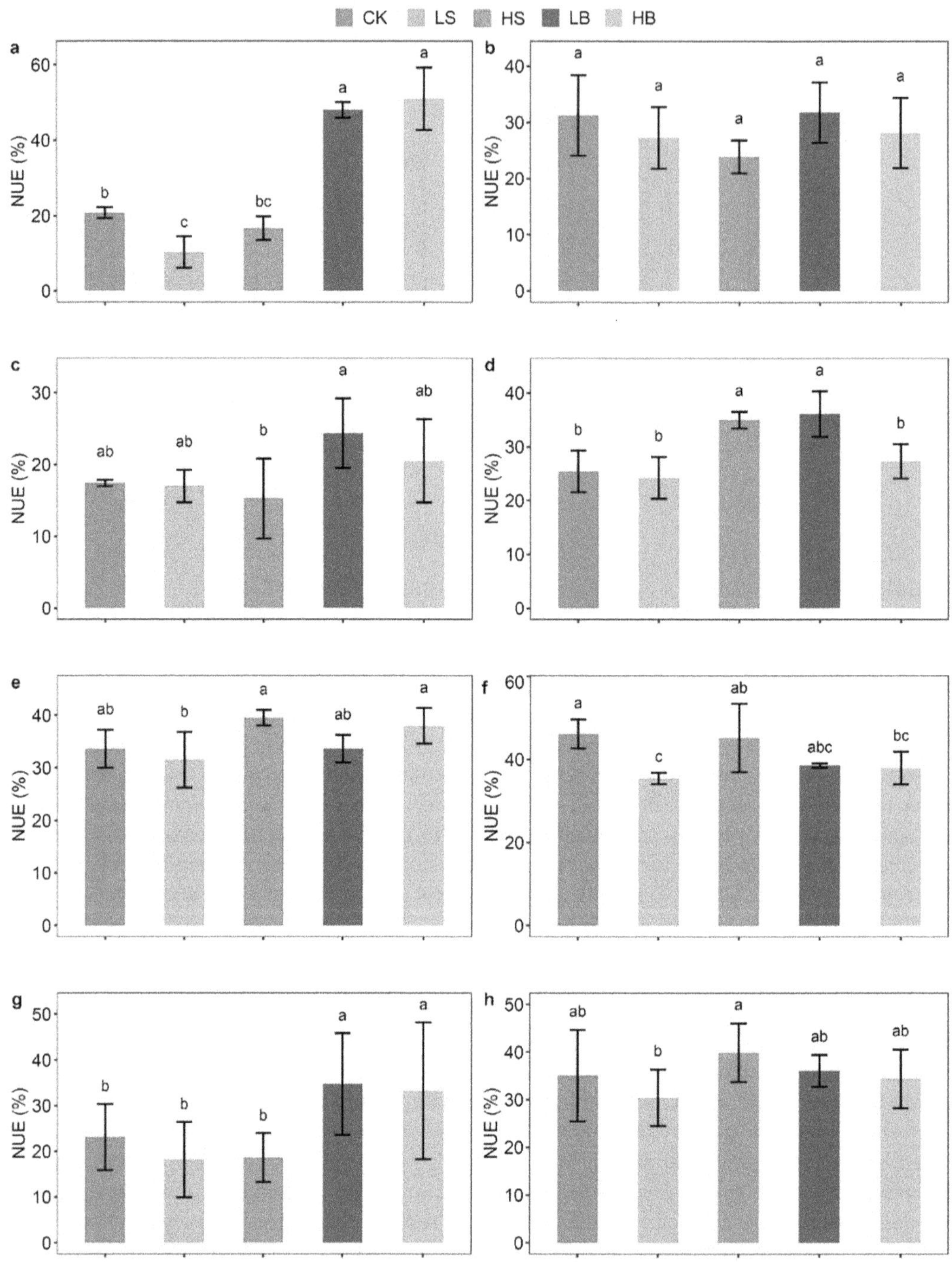

FIGURE 26.5 The N fertilizer use efficiency (NUE) in early rice season across the treatments. Notes: a, b, c, d, e, f, g and h refer to 2012, 2013, 2014, 2015, 2016, 2017, 2012~2014 average and 2015~2017 average, respectively.

the paddy soil. Compared with the CK, the average annual STP content had no significant increase in the LS and HS treatments, while it significantly increased by 8.3% and 9.2% in the LB and HB treatments, respectively. The annual variation trend of STP in each treatment from 2012 to 2017 showed that the STP content of the LB and HB treatments was significantly higher than that of the other treatments, and the higher the straw-derived biochar added, the greater the increase in the STP

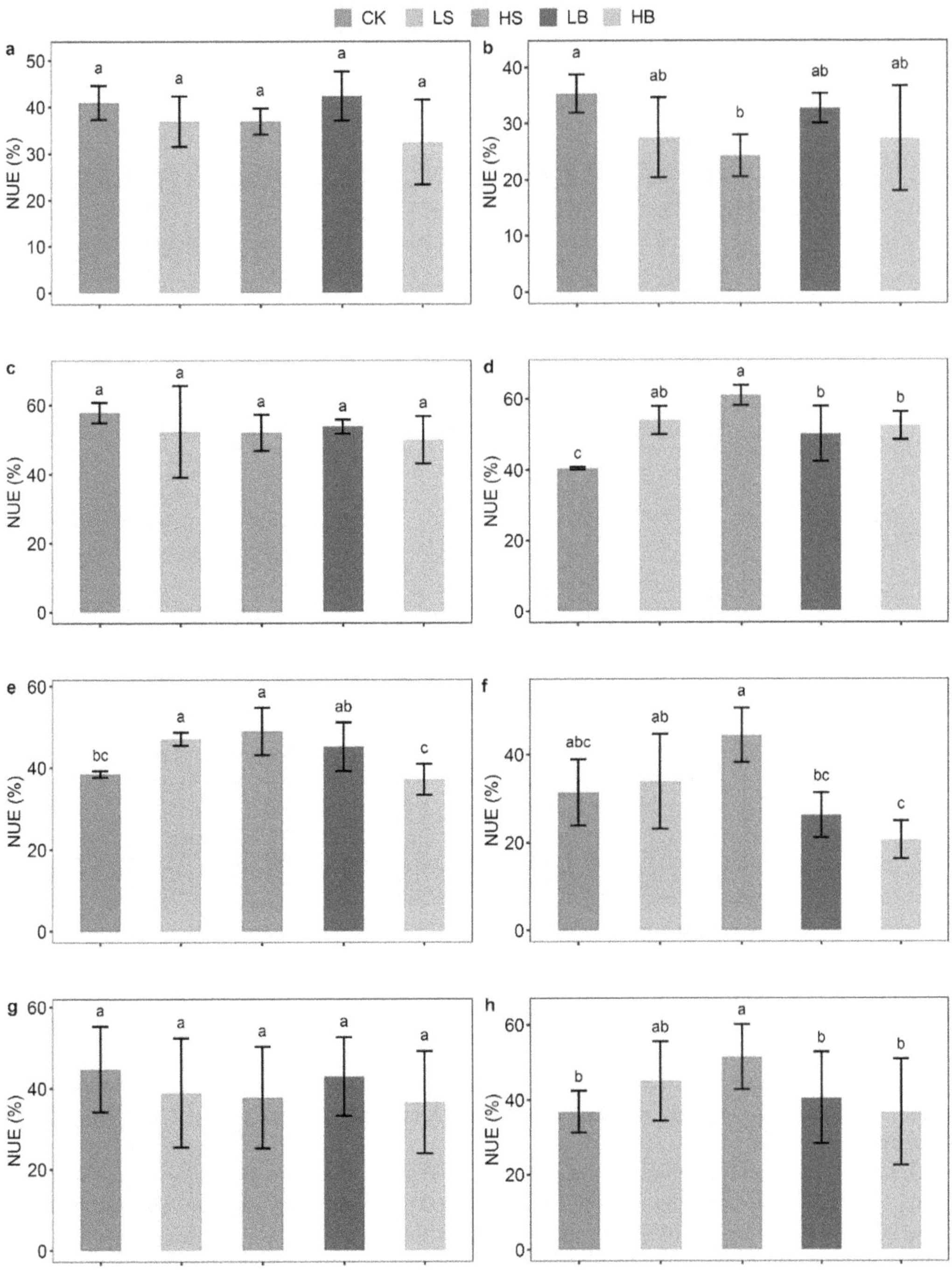

FIGURE 26.6 The N fertilizer use efficiency (NUE) in late rice season across the treatments. Notes: a, b, c, d, e, f, g, and h refer to 2012, 2013, 2014, 2015, 2016, 2017, 2012–2014, and 2015~2017 average, respectively.

content of paddy soil. The STP content of each treatment gradually increased from the late rice-harvesting stage of 2016. The main reason may be that phosphorus fertilizer application increased from 17.5 to 32.7 P kg/hm^2 per growing season from the beginning of the late rice season in 2016.

For the late rice-harvesting stage from 2012 to 2017, the variation trend of soil available phosphorus (SAP) was basically the same among the treatments and ranged from 7.5 to 12.8 mg/kg

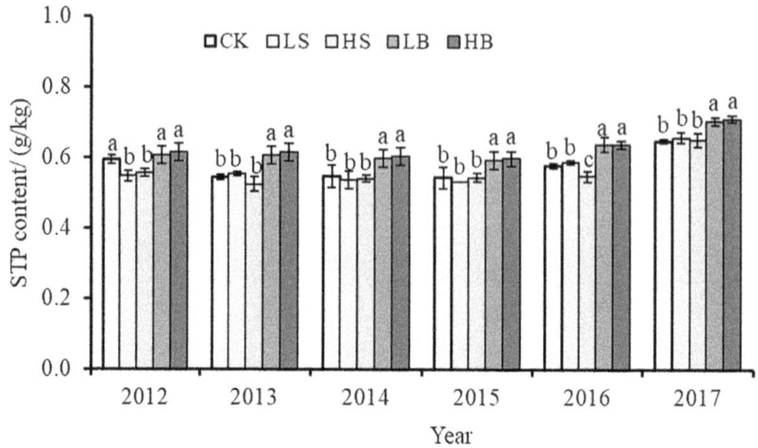

FIGURE 26.7 The change of soil total phosphorus (STP) from 2012 to 2017.

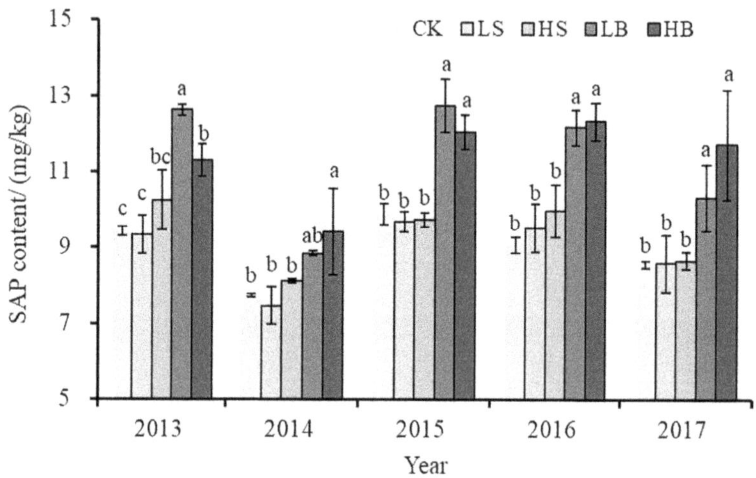

FIGURE 26.8 The change of soil available phosphorus (SAP) from 2013 to 2017.

(Figure 26.8). Compared with the CK, the average annual SAP content was significantly higher in the HS treatment. The LB and HB treatments significantly increased by 24.5% and 25.1%, respectively.

The effect of straw incorporation on rice grain and straw P uptake showed different trends in the early and late rice maturity season (Figure 26.9). The P uptake of rice grain and straw under straw incorporation was slightly lower than that under the CK in the early rice season, whereas the P uptake of rice grain and straw under straw incorporation was slightly higher than that under the CK in the late rice season. The P uptake of rice grain and straw under biochar application was not increased as compared with that under the CK in the early rice season, whereas the P uptake of rice straw under biochar application was lower than that under the other treatments in the late rice season. The high biochar application (the addition of 2% biochar) was the lowest.

Rice grain yield with straw incorporation had a decreasing trend during the early rice season and a slight increase during the late rice season as compared with the CK. Rice grain yield with biochar application had a small but not significant increase (Figure 26.10). The annual P output was 41.3–47.7 kg/ha under double-cropping in a paddy field (Table 26.3). Phosphorus use efficiency

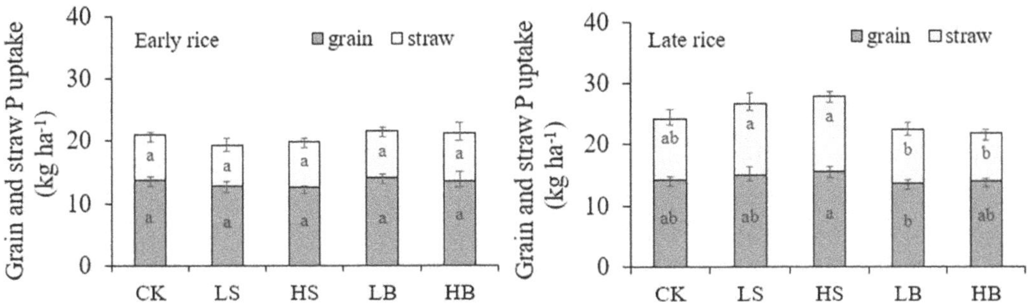

FIGURE 26.9 Effect of straw incorporation and biochar applications on straw and grain P uptake in the different treatments. Different lowercase letters on the column indicate significant differences at the 0.05 level.

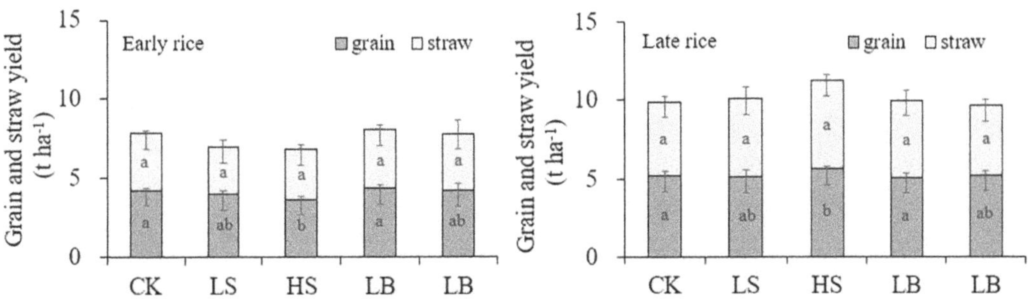

FIGURE 26.10 Effect of straw incorporation and biochar applications on straw and grain yield in the different treatments. Different lowercase letters on the column indicate significant differences at the 0.05 level.

TABLE 26.3
Phosphorus Input and Output Balance and P Fertilizer Use Efficiency across the Treatments (2016–2017)

	ATPI	APO	APAB	PUE
Treatments	**kg P/(ha yr)**	**kg P/(ha yr)**	**kg P/(ha yr)**	%
CK	42.6	43.8±1.1ab	−1.2±1.2c	103.26±1.93ab
LS	42.6	45.9±1.2a	−3.35±1.2d	107.86±2.28a
HS	42.6	45.6±0.9ba	−3.0±0.9d	107.58±2.88a
LB	42.6	42.5±0.8ab	−0.1±0.8b	99.71±1.84b
HB	42.6	41.3±0.6b	1.3±0.6a	97.05±1.51b

Note: ATPI: Annual total phosphorus input; APO: Annual phosphorus output; APAB: Annual phosphorus apparent balance; PUE: Phosphorus use efficiency (phosphorus output/phosphorus input).

with straw incorporation (P fertilizer output/P fertilizer input) was increased compared with the control without straw incorporation, and the output of P fertilizer exceeded the P input. This indicated that straw incorporation can activate the original soil available P in the paddy soil. Straw incorporation also replaced part of the chemical P fertilizer, so the partial productivity of chemical P fertilizer (grain yield/chemical P fertilizer input) in the paddy fields was higher than that of the other treatments. Thus, straw incorporation can be used as an effective measure for improving the

TABLE 26.4
Cumulative CH_4 (a), N_2O (b), and Soil Heterotrophic (Rh) (c) Emissions from 2012 to 2016

Rice Season	Treatment	2012–2013	2013–2014	2014–2015	2015–2016	Average 2012–2016
			CH_4 (kg C/ha)			
Early	CK	25.8±6.7a[a]	55.6±4.8a	50.0±1.0a	88.2±3.7a	54.9±2.2a
	LB	24.7±2.4a	25.4±2.6b	30.4±0.9c	68.2±1.6ab	37.2±0.5b
	HB	13.6±1.9a	30.9±3.0b	34.4±0.7b	65.0±9.4b	36.0±2.5b
Late	CK	71.2±1.7a	117.4±4.8a	135.2±6.7a	163.5±10.3a	121.8±4.8a
	LB	41.5±3.4b	78.0±3.2b	60.4±1.8c	132.8±7.9b	78.2±3.8b
	HB	45.6±2.0b	84.1±1.4b	95.7±4.1b	131.8±1.6b	89.3±1.3b
Fallow	CK	5.8±0.4a	0.5±0.0a	0.7±0.1a	2.3±0.2a	2.3±0.1a
	LB	1.1±0.0c	0.2±0.1b	0.2±0.0b	2.0±0.6a	0.9±0.2b
	HB	1.9±0.3b	0.2±0.0b	0.7±0.1a	2.0±0.3a	1.2±0.1b
Total	CK	102.7±7.8a	173.4±3.7a	185.9±5.7a	254.0±13.7a	179.0±6.3a
	LB	67.2±4.8b	103.5±5.7b	91.0±2.7c	203.0±5.9b	116.2±3.9b
	HB	61.15±3.2b	115.2±2.0b	130.7±3.4b	199.0±8.1b	126.5±1.9b
			N_2O (g N/ha)			
Early	CK	110.8±4.2b	135.8±6.2b	258.2±19.2a	108.3±6.4a	153.3±8.0b
	LB	124.8±1.7b	280.3±33.8a	254.6±27.0a	121.5±10.6a	195.3±8.3a
	HB	199.6±5.4a	266.3±8.3a	208.7±15.5a	91.3±10.0a	191.5±5.9a
Late	CK	−32.6±5.3b	72.0±5.4b	101.8±11.8a	168.5±11.0a	77.4±1.7c
	LB	168.9±14.0a	62.2±1.6b	128.2±16.7a	156.4±5.4a	128.9±0.7b
	HB	160.9±9.6a	95.7±4.3a	92.0±9.8a	196.6±25.6a	136.3±2.6a
Fallow	CK	78.0±11.0a	267.4±35.8a	51.1±9.8a	192.9±21.4a	147.4±5.4a
	LB	97.0±13.2a	82.2±6.0b	47.9±7.4a	158.5±11.0a	96.4±4.5b
	HB	93.3±9.5a	86.0±9.6b	26.9±6.7a	146.7±16.3a	88.2±8.4b
Total	CK	156.3±17.2c	475.2±31.2a	411.1±24.4ab	469.7±26.0a	378.1±7.2b
	LB	390.7±4.1b	424.7±29.2a	430.8±31.0a	436.5±17.2a	420.6±9.1a
	HB	453.7±24.0a	448.0±14.0a	327.6±27.1b	434.6±5.5a	416.0±10.4a
			Rh (kg C/ha)			
Early	CK	275.4±6.1b	528.9±14.0b	491.1±19.2b	489.3±10.7c	446.2±6.2c
	LB	390.0±14.3a	604.3±27.7a	555.9±38.0ab	540.3±8.9b	522.6±6.0b
	HB	388.7±1.2a	623.4±9.5a	584.2±12.7a	583.3±9.6a	544.9±3.3a
Late	CK	868.3±24.0b	940.7±7.3a	1,079.9±27.1a	873.8±21.8a	940.7±4.2b
	LB	772.1±25.6b	845.0±48.9a	1,052.5±32.5a	901.2±35.9a	892.7±17.0b
	HB	1,086.5±42.1a	932.2±48.9a	1,110.0±33.3a	931.0±9.1a	1,015.0±29.8a
Fallow	CK	656.2±21.2a	1,427.3±21.6b	2,039.2±35.8a	1,744.41±15.7b	1,466.8±5.1b
	LB	590.1±2.4b	1,497.2±39.0b	2,143.0±34.0a	1,740.4±48.2b	1,492.7±29.4b
	HB	668.8±12.3a	1,853.5±50.9a	2,149.7±61.9a	1,878.6±25.0a	1,637.6±18.0a
Total	CK	1,799.9±48.6b	2,896.9±26.5b	3,610.2±15.0a	3,107.5±40.7b	2,853.6±5.5b
	LB	1,752.2±27.6b	2,946.5±55.5b	3,751.4±99.8a	3,181.9±81.0b	2,908.0±49.9b
	HB	2,144.0±40.2a	3,409.1±81.7a	3,843.8±103.3a	3,392.9±19.4a	3,197.5±51.0a

Different lowercase (uppercase) letters indicate significance in the same season (cycle) between the treatments at the 5% level, according to Duncan's multiple range test. Early, late, and fallow: early rice, late rice, and fallow seasons, respectively.

[a] Mean ± SE in same column.

P use efficiency in the double-cropping rice system. The input and output of soil P were in balance under biochar application, and there was no significant difference in the utilization efficiency of P fertilizer compared with the control. Biochar application improved the availability of soil P, and rice grain yield increased under low biochar application.

26.3.6 Effects of Biochar Amendment on Greenhouse Gas Emissions

In a 4-year study, biochar amendment largely decreased CH_4 emissions from the paddy field with double rice cropping (Table 26.4). The CH_4 emissions were 13.6–88.2, 41.5–163.5, and 0.2–5.7 kg C/ha in the early rice season, late rice season, and fallow season, respectively, across the treatments. The mean reduction rates of CH_4 emissions from the treatments with biochar amendment were 32.3% and 34.5%, 35.8% and 26.7%, 62.0% and 47.4%, and 35.1% and 29.4% for the early rice season, late rice season, fallow season, and whole year in the 4-year experimental period as compared with the control treatment. The ratios of abundance of methanogens to methanotrophs decreased by 11%–31% in each of the 4 years and were positively correlated to CH_4 emissions. The reduction of CH_4 emissions owing to biochar was mainly due to the lesser suppression of abundance and activity of methanotrophs compared with methanogens (Wang et al., 2019), and this was due to the improvement in soil aeration (Chen et al., 2018). The reduction of CH_4 emissions from paddy soils with biochar amendments was found in other studies (Liu et al., 2021b; Qin et al., 2016).

The N_2O emissions were 91.3–280.3, –32.6–196.5, and 26.9–267.4 g N/ha in the early rice season, late rice season, and fallow season, respectively, across the treatments. Increase in N_2O emissions for the biochar treatments was only found in the first year when the biochar was applied in the field. This is because the amendment of biochar also caused an increase in nitrogen input due to the relatively high N content of the biochar used. There was no increase in N_2O emissions in the paddy field in the second and third years after biochar amendment, which means biochar amendment would not increase N_2O emissions in paddy fields if the additional N input from biochar was exhausted.

The soil heterotrophic respiration rates (Rh) were 275.4–623.4, 772.1–1110.0, and 590.1–2149.7 kg C/ha in the early rice season, late rice season, and fallow season, respectively, across the treatments. Increase in the Rh rate with biochar amendment was mainly found in the HB treatment with a biochar application rate of 48 t/ha, which may be caused by the degradation of the labile components of biochar. The average increase in the Rh rate for the HB treatment was 12.0% as compared to the CK in the 4-year experimental period, which means the biochar can be stable when applied in the paddy soil.

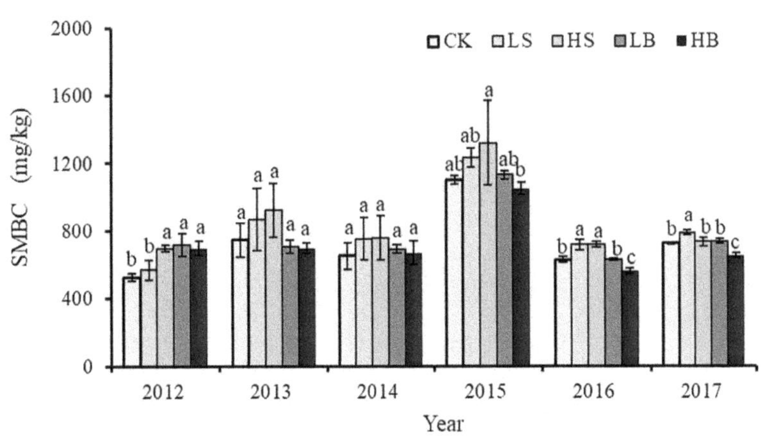

FIGURE 26.11 Soil microbial biomass carbon (SMBC) under different treatments from 2012 to 2017.

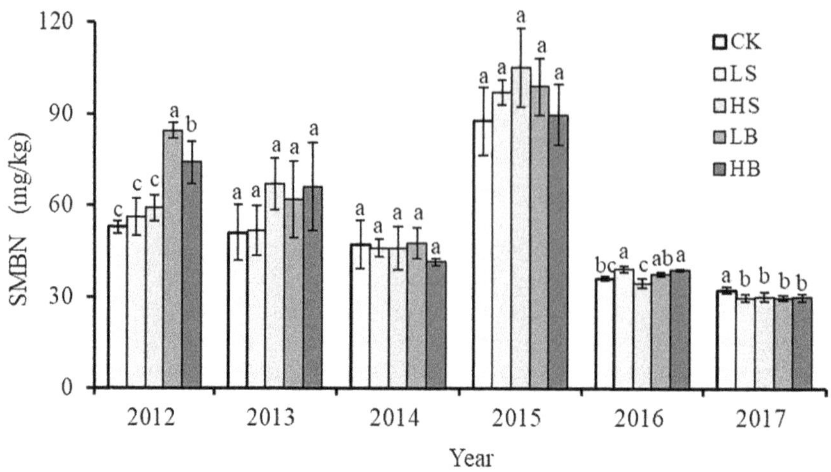

FIGURE 26.12 Soil microbial biomass nitrogen (SMBN) under different treatments from 2012 to 2017.

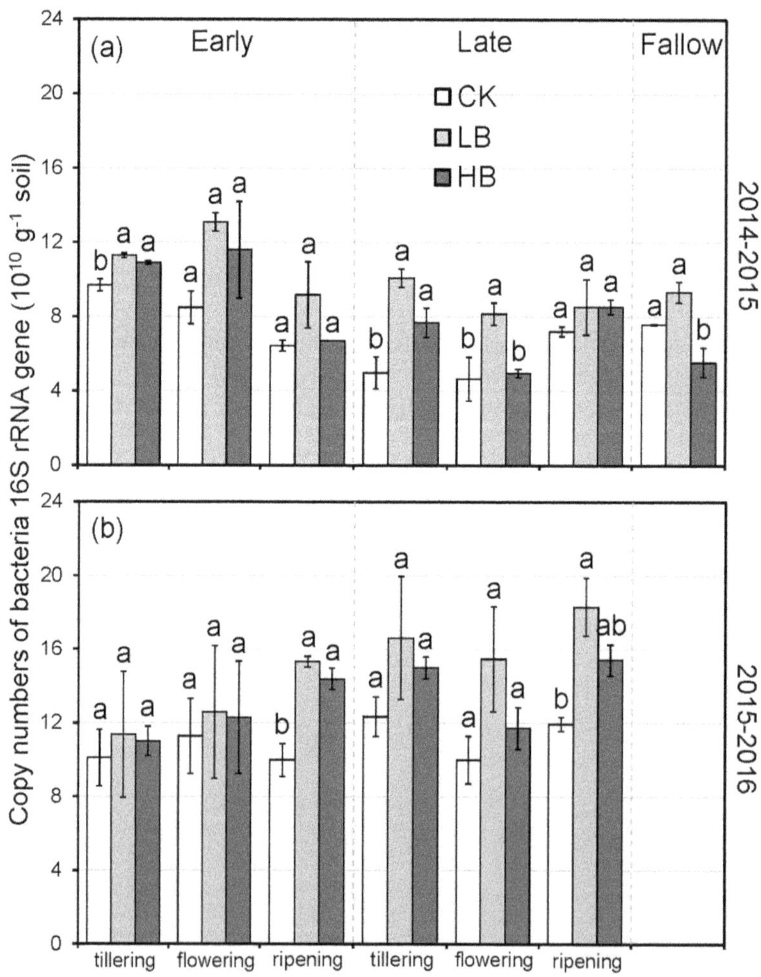

FIGURE 26.13 Abundance of the bacterial 16S rRNA gene during the 2014–2015 (a) and 2015–2016 (b) annual cycles. Early, late, and fallow: early rice season, late rice season, and fallow season, respectively.

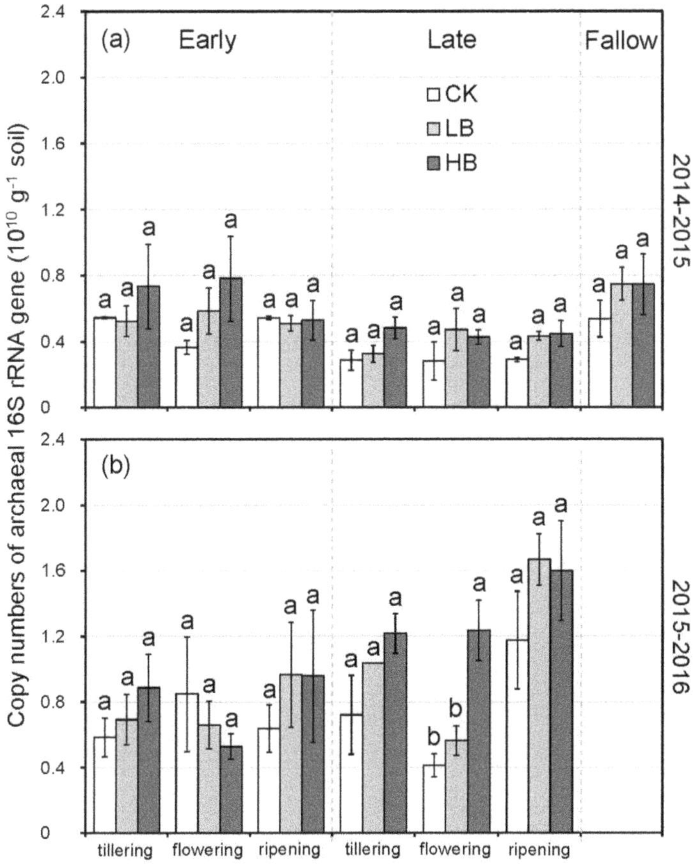

FIGURE 26.14 Abundance of the archaeal 16S rRNA gene during the 2014–2015 (a) and 2015–2016 (b) annual cycles. Early, late, and fallow: early rice season, late rice season, and fallow season, respectively.

26.3.7 Effects of Biochar Amendment on Soil Microbial Diversity

During the 2012–2017 annual cycles, the average annual change of soil microbial biomass carbon content across the treatments ranged from 721 to 858 mg/kg (Figure 26.11). Compared with the control treatment, the average annual soil microbial biomass carbon content during 2012–2017 was increased by 12.7%, 17.8%, and 5.5% in the treatments of low straw amendment rate (LS), high straw amendment rate (HS), and low straw biochar addition rate (LB), respectively. The addition of a high amount of straw biochar (HB) decreased the soil microbial biomass carbon content, which indicated that the addition of a high amount of straw biochar inhibited the effect of microbial carbon fixation. In 2015, soil microbial biomass carbon was higher than that in other years in the paddy fields. In 2013–2016, soil microbial biomass carbon in paddy fields with straw was higher than that under other treatments, but there was no significant difference in soil microbial biomass carbon content between the low straw and high straw treatments.

The average soil microbial biomass nitrogen content during 2012–2017 across all treatments ranged from 51.2 to 60.1 mg/kg (Figure 26.12). Compared with the control treatment, LS, HS, LB, and HB increased the average content of soil microbial biomass nitrogen by 4.2%, 11.4%, 17.3%, and 10.7%, respectively. In 2015, the soil microbial biomass nitrogen content of paddy fields under the different treatments was higher than that of other years, while in 2016 and 2017, the soil microbial biomass

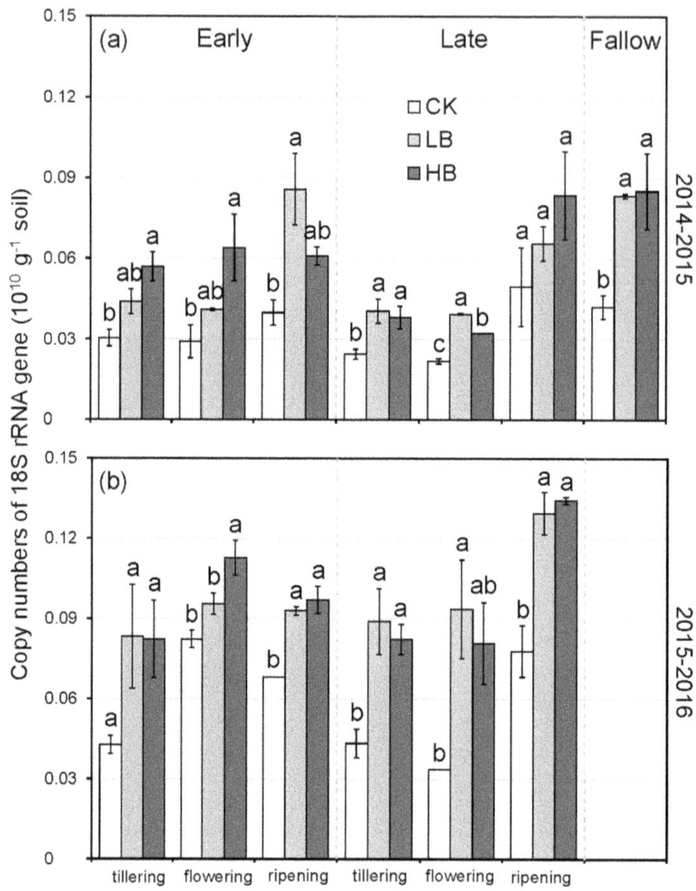

FIGURE 26.15 Abundance of the 18S rRNA gene during the 2014–2015 (a) and 2015–2016 (b) annual cycles. Early, late, and fallow: early rice season, late rice season, and fallow season, respectively.

nitrogen content of paddy fields under the different treatments was lower than that of other years. The soil microbial biomass nitrogen content of the paddy fields across the treatments changed in different years.

Biochar had a significant effect on microbial abundances in the rice field (Figure 26.13; Wang et al., 2021a). In the third and fourth years after biochar application, the copy numbers of 16S rRNA genes of bacteria and archaea and 18S rRNA genes of fungi showed that biochar application increased bacterial abundances. LB and HB treatments increased by 11.6%–101.8% and 4.2%–53.9%, respectively, compared with the control. However, the difference between the LB and HB was not significant (Figure 26.13). There was no significant difference in the copy number of archaea between biochar application and control treatments (Figure 26.14). The application of biochar increased fungal abundances, and those in the LB and HB treatments increased by 16.1%–177.8% and 36.9%–139.7%, respectively, compared with the control.

26.4 CONCLUSIONS

We conducted a 6-year field experiment with biochar amendment at the rates of 24 and 48 t/ha in a double-rice cropping system. Biochar amendment at a high rate increased the stability of soil aggregates by increasing the mean mass diameter and mean geometric diameter of soil aggregates. Biochar amendment also largely increased STOC, and the carbon sequestration rate in the biochar treatment was 2.6 times of that in the straw incorporation treatment in the 6 years when the straw and biochar treatments had the same amount of straw input considering that the biochar used in this study was also derived from straw. Biochar amendment enhanced soil pH by 0.5–1.0 unit. However, the increase in soil pH only occurred in the first 4 years with biochar amendment. Biochar amendment did not affect rice grain yield and nitrogen and phosphorus use efficiencies, although biochar amendment increased STN, total phosphorus, and available phosphorus. Biochar amendment reduced annual CH_4 emissions from the paddy field by 29%–35%. There was a slight increase in soil N_2O emissions and soil heterotrophic respiration in the first year and second year after biochar amendment in the paddy field but not in later years, indicating that, after 2 years, biochar amendment can be an effective measure to reduce net greenhouse gas emissions from paddy fields. Biochar amendment increased bacterial and fungal abundances. Overall, the biochar amendment improved soil fertility, reduced CH_4 emissions, and increased soil carbon sequestration. Thus, it is a useful measure to manage paddy soil constraints with decreased soil fertility and high CH_4 emissions in subtropical China.

REFERENCES

Brookes P.C., Landman A., Pruden G., Jenkinson D.S. 1985. Chloroform fumigation and the release of soil nitrogen: A rapid direct extraction method to measure microbial biomass nitrogen in soil. *Soil Biology and Biochemistry* 17: 837–842.

Brookes P.C., Powlson D., Jenkinson D.S. 1982. Measurement of microbial biomass phosphorus in soil. *Soil Biology and Biochemistry* 14: 319–329.

Cambardella C.A., Elliott E.T. 1993. Methods for physical separation and characterization of soil organic matter fractions. *Geoderma* 56(1–4): 449–457.

Chen D., Wang C., Shen J.L., Li Y., Wu J., 2018. Response of CH_4 emissions to straw and biochar applications in double-rice cropping systems: Insights from observations and modeling. *Environmental Pollution* 235: 95–103.

Chen Z., Luo X.Q., Hu R.G., Wu M.N., Wu J.S., Wei W.X. 2010. Impact of long term fertilization on the composition of denitrifier communities based on nitrite reductase analyses in a paddy soil. *Microbial Ecology* 60: 850–851.

Coskun G. 2006. Effect of forage cropping treatments on soil structure and relationships with fractal dimensions. *Geoderma* 131(1–2): 33–44.

Großkopf R., Janssen P.H., Liesack W. 1998. Diversity and structure of the methanogenic community in anoxic rice paddy soil microcosms as examined by cultivation and direct 16S rRNA gene sequence retrieval. *Application Environmental Microbiology* 64: 960–969.

Gul S., Joann K.W. 2016. Biochemical cycling of nitrogen and phosphorus in biochar-amended soils. *Soil Biology and Biochemistry* 103: 1–15.

Hoshino Y.T., Morimoto S. 2008. Comparison of 18S rDNA primers for estimating fungal diversity in agricultural soils using polymerase chain reaction-denaturing gradient gel electrophoresis. *Soil Science and plant Nutrition* 54(5): 701–710.

Hou X., Li R., Jia Z., Han Q., Wang W., Yang B. 2012. Effects of rotational tillage practices on soil properties, winter wheat yields and water-use efficiency in semi-arid areas of north-west China. *Field Crop Research* 129: 7–13.

Iqbal J., Hu R.G., Lin S., Hatano R., Feng M.L., Lu L., Ahamadou B., Du L.J. 2009. CO_2 emission in a subtropical red paddy soil (Ultisol) as affected by straw and N-fertilizer applications: A case study in Southern China. *Agriculture Ecosystems Environment* 131: 292–302.

Jiang B.H., Shen J.L., Sun M.H., Hu Y.J., Jiang W.Q., Wang J., Li Y., Wu J.S. 2021. Soil phosphorus availability and rice phosphorus uptake in paddy fields under various agronomic practices. *Pedosphere* 31(1): 103–115.

Kalra Y.P. 1998. *Handbook of Reference Methods for Plant Analysis*. CRC. Press, Boca Raton, FL.

Kemper W.D, Rosenau R.C., 1986. Aggregate Stability and Size Distribution. In *Methods of Soil Analysis*, Part 1. Physical and Mineralogical Methods (2nd Edition). Agronomy Monograph No. 9. American Society of Agronomy, Madison, WI, USA. pp. 425–442.

Kroon P.S., Hensen A., van den Bulk W.C.M., Jongejan P.A.C., Vermeulen A.T. 2008. The importance of reducing the systematic error due to non-linearity in N_2O flux measurements by static chambers. *Nutrient Cycling in Agroecosystems* 82: 175–186.

Kuzyakov Y, Bogomolova I, Glaser B, 2014. Biochar stability in soil: Decomposition during eight years and transformation as assessed by compound-specific 14C analysis. *Soil Biology and Biochemistry* 70: 229–236.

Lehmann J., Joseph S. 2009. Biochar for Environmental Management: An introduction. In: Lehmann, J., Joseph, S. (Eds.), *Biochar for Environmental Management*. Earthscan, London, UK, pp. 1–12.

Liu J., Jiang B.S., Shen J.L., Zhu X., Yi W.Y., Li Y., Wu J.S. 2021a. Contrasting effects of straw and straw-derived biochar applications on soil carbon accumulation and nitrogen use efficiency in double-rice cropping systems. *Agriculture, Ecosystems and Environment* 311: 107286.

Liu J.Y., Qiu H.S., Wang C., Shen J.L., Zhang W.Z., Cai J.M., Tang H., Wu J.S. 2021b. Effects of biochar amendment on greenhouse gas emission in two paddy soils with different textures. *Paddy and Water Environment* 19: 87–98.

Liu J.Y., Shen J.L., Li Y., Su Y.R., Ge T.D., Jones D. L., Wu J.S. 2014. Effects of biochar amendment on the net greenhouse gas emission and greenhouse gas intensity in a Chinese double rice cropping system. *European Journal of Soil Biology* 65: 30–39.

Olsen S.R., Sommers L.E. 1982. Phosphorus. In Page A.L, Miller R.H, Keeney D.R (eds.) *Methods of soil Analysis. Part 2*. 2nd edn. American Society of Agronomy, Madison, WI, pp. 403–430.

Page A.L, Miller R.H., Keeney D.R. 1982. *Methods of Soil Analysis. Part 2: Chemical and Microbial Properties*. American Society of Agronomy, Inc., Madison, WI, pp. 539–594.

Qin X.B., Li Y.E., Wang H., Liu C., Li J.L., Wan Y.F., Gao Q.Z., Fan F.L., Liao Y.L. 2016. Long-term effect of biochar application on yield-scaled greenhouse gas emissions in a rice paddy cropping system: A four-year case study in south China. *Science of the Total Environment* 569: 1390–1401.

R Core Team. 2022. R: A Language and Environment for Statistical Computing. R Foundation for Statistical Computing, Vienna, Austria. http://www.R-project.org/ (Accessed 25 May 2022).

Shen J.L., Li Y., Wang Y., Li Y.Y., Zhu X., Jiang W.Q., Liu Y.Y., Wu J.S. 2022. Soil nitrogen cycling and environmental impacts in the subtropical hilly region of China: Evidence from measurements and modeling. *Frontiers of Agricultural Science and Engineering* 9(3): 407–424.

Shen J.L., Tang H., Liu J., Wang C., Li Y., Ge T.D., Jones D., Wu J.S. 2014. Contrasting effects of straw and straw-derived biochar amendments on greenhouse gas emissions within double rice cropping systems. *Agriculture, Ecosystems and Environment* 188: 264–274.

Suzuki M.T., Taylor L.T., De Long E.F. 2000. Quantitative analysis of small-subunit rRNA genes in mixed microbial populations via 5'-nuclease assays. *Applied and Environmental Microbiology* 66(11): 4605–4614.

Valente R.J., Thornton F.C., Williams E.J. 1995. Field comparison of static and flowthrough chamber techniques for measurement of soil NO emission. *Journal of Geophysical Research* 100: 21147–21152.

Van Zwieten L., Kimber S., Morris S., Chan, K.Y., Rust, D.J., Joseph, S., Cowie, A. 2010. Effects of biochar from slow pyrolysis of papermill waste on agronomic performance and soil fertility. *Plant and Soil* 327: 235–246.

Vance E., Brookes P., Jenkinson D. 1987. An extraction method for measuring soil microbial biomass C. *Soil Biology and Biochemistry* 19: 703–707.

Wang C., Chen D., Shen J.L., Yuan Q., Fan F.L., Wei W.X., Li Y., Wu J.S. 2021a. Biochar alters soil microbial communities and potential functions 3–4 years after amendment in a double rice cropping system. *Agriculture, Ecosystems and Environment* 311: 107291.

Wang C., Liu J.Y., Shen J.L., Chen D., Li Y., Jiang B.S., Wu J.S. 2018. Effects of biochar amendment on net greenhouse gas emissions and soil fertility in a double rice cropping system: A 4-year field experiment. *Agriculture Ecosystems and Environment* 262: 83–96.

Wang C., Shen J.L., Liu J.Y., Qin H.L., Yuan Q., Fan F.L., Hu Y.J., Wang J., Wei W.X., Li Y., Wu J.S. 2019. Microbial mechanisms in the reduction of CH_4 emission from double rice cropping system amended by biochar: A four-year study. *Soil Biology and Biochemistry* 135: 251–263

Wang F., Harindintwali J.D., Yuan Z.Z., Wang M., Wang F.M., Li S., Yin Z.G., Huang L., Fu Y.H., Li L., Chang S.X., Zhang L.J., Rinklebe J., Yuan Z.Q., Zhu Q.G., Xiang L.L., Tsang, Daniel C.W., Xu L., Jiang X., Liu J.H., Wei N., Kästner M., Zou Y., Ok Y.S., Shen J.L., Peng D.L., Zhang W., Barceló D., Zhou Y.J., Bai Z.H., Li B.Q., Zhang B., Wei K., Cao H.J., Tan Z.L., Zhao L.B., He X., Zheng J.X., Bolan N., Liu X.H., Huang C.P., Dietmann S., Luo M., Sun N.N., Gong J.R., Gong Y.L., Brahushi F., Zhang T.T., Xiao C.D., Li X.F., Chen W.F., Jiao N.Z., Lehmann J., Zhu Y.G., Jin H.G., Schäffer A., Tiedje J.M., Chen J.M. 2021b. Technologies and perspectives for achieving carbon neutrality. *The Innovation* 2: 100180.

Wu J.S., Jorgensen R.G., Pommerening B., Chaussod R., Brookes P.C. 1990. Measurement of soil microbial biomass-C by fumigation-extraction-an automated procedure. *Soil Biology and Biochemistry* 22: 1167–1169.

Wu J.S., Lin Q.M., Huang Q.Y., Xiao H.A. 2006. Soil microbial biomass-methods and application, in: Wu, J.S., Lin, Q.M., Huang, Q.Y., Xiao, H.A. (eds.), *Soil Microbial Biomass Nitrogen*. China Meteorological Press, Beijing, pp. 65–74, in Chinese.

Zheng X.H., Mei B.L., Wang Y.H., Xie B.H., Wang Y.S., Dong H.B., Xu H., Chen G.X., Cai Z.C., Yue J., Gu J.X., Su F., Zou J.W., Zhu J.G. 2008. Quantification of N_2O fluxes from soil-plant systems may be biased by the applied gas chromatograph methodology. *Plant Soil* 311: 211–234.

27 Sugarcane Cultivation in Saline Soils of India

P. K. Karthikeyan and D. Gokul
Faculty of Agriculture, Department of Soil Science and Agricultural Chemistry, Annamalai University, Chidambaram, Tamil Nadu, India

CONTENTS

27.1 INTRODUCTION

Sugarcane (*Saccharum* spp.) is an important cash crop used in the production of sugar and biofuel in India. Sugarcane belongs to the grass family Poaceae. The juice from sugarcane is used for making white sugar, brown sugar, and jaggery. During sugar making, by-products of sugarcane, such as bagasse, molasses, and press mud, are produced. Press mud can be used as an organic material, which acts as a soil amendment for saline and sodic soils. Sugarcane has been largely cultivated in tropical and subtropical regions of the world. Cultivation of sugarcane needs high levels of water and nutrients in the soil to reach maximum yield, and cultivation under salinized soils can result in a significant reduction of growth that finally leads to low productivity (Picture 27.1).

PICTURE 27.1 Sugarcane variety "Co 11015" at Arachalur, Erode district, Tamil Nadu state, India.

DOI: 10.1201/9781003093565-27

27.2 SUGARCANE SITUATION

Globally, Brazil is the largest sugarcane producer, which in 2019–2020 produced 29.61 million metric tons of sugar followed by India which produced 27.4 million metric tons of sugar (Ogunsola et al., 2021) (Figure 27.1).

In India, sugarcane is cultivated in area of 48.57×10^5ha with production of 399.25 million tons and yield of 82.20 t/ha. As far as the average productivity is concerned, Tamil Nadu has the maximum (99.70 t/ha) followed by West Bengal (86.44 t/ha) and Karnataka (83.22 t/ha). Area, production, and productivity of major states of India are given in Figures 27.2–27.4 (Directorate of Economics and Statistics, New Delhi, 2020–2021). The states of Uttar Pradesh, Maharashtra, and Karnataka have the maximum production and area. According to Singh et al. (2020), these three states contribute about 73% of the area and 74% of the production for the country.

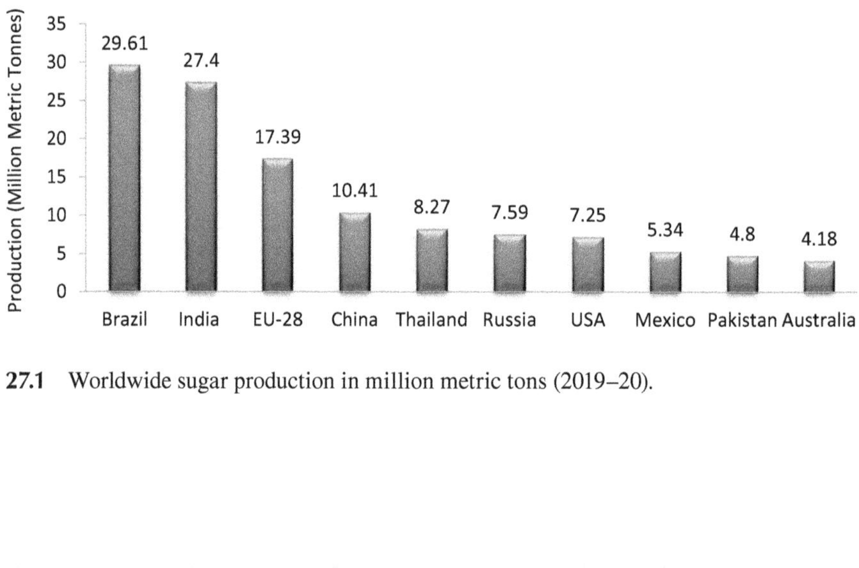

FIGURE 27.1 Worldwide sugar production in million metric tons (2019–20).

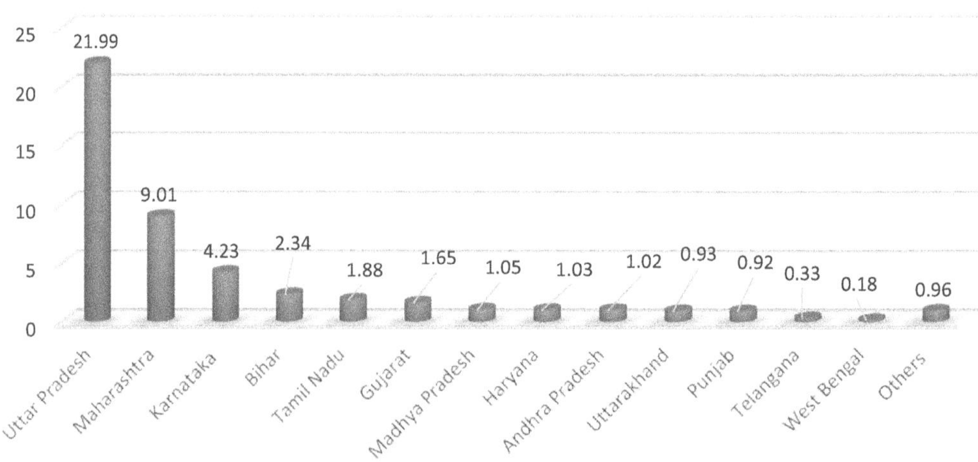

FIGURE 27.2 State-wise area of sugarcane in India (av. 2015–16 to 2019–20) in 105 ha.

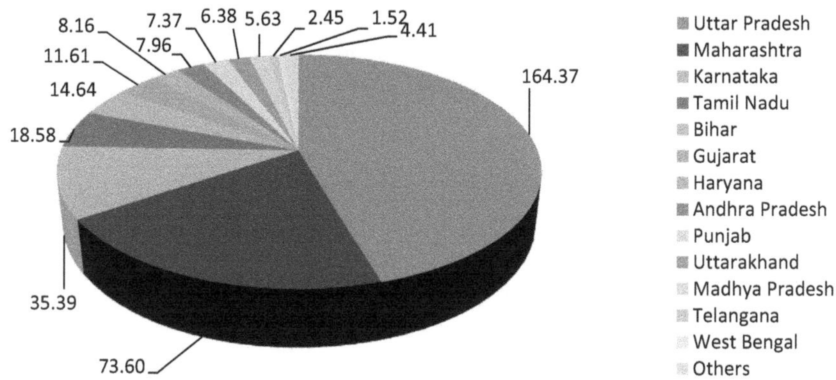

FIGURE 27.3 State-wise production of sugarcane in India (av. 2015–16 to 2019–20) in million tons.

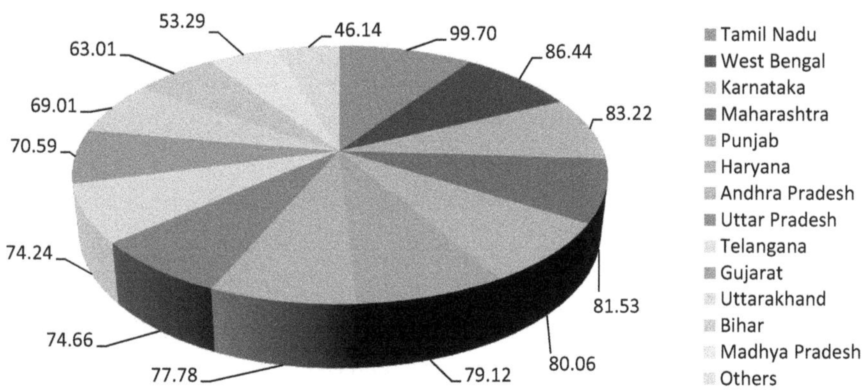

FIGURE 27.4 State-wise productivity of sugarcane in India (av. 2015–16 to 2019–20) in tons/ha.

27.3 SALT-AFFECTED SOILS

Soils which contain excess soluble salts that inhibit plant growth and development are called as salt-affected soils. These soils are problem soils in which productivity of the soil is lowered due to the inherent unfavorable soil conditions of the salt content. In arid and semiarid regions, this salinity creates major problems and restricts crop production from seed germination to crop harvest. Salts in saline soils often originate from the earth's crust. Water in coastal areas, a low rainfall, and a high rate of evaporation in arid regions cause increasing salt concentration in soil at various layers. Irrigation water that contains high concentrations of soluble salts also leads to soil salinity.

27.3.1 Classification of Salt-Affected Soils

Saline soils contain dissolved salts, such as NaCl, Na_2SO_4, $CaSO_4$, $MgSO_4$, $MgCl_2$, KCl, and Na_2CO_3, each of which contributes to saline conditions. But in most saline soil areas NaCl, followed by Na_2SO_4, is the dominant soluble salt. Saline soils are characterized by white or light brown crusts on the surface (Sabareshwari and Ramya, 2018). These soils are sometimes called white alkali soils to describe the white incrustation that develops at the soil surface. The white color on the surface is due to NaCl and Na_2CO_3, and the brown color is due to $NaNO_3$. These saline soils are generally deficient in P, Zn, Fe, and N. Saline soils are aggregated and highly permeable. Sodic soils are

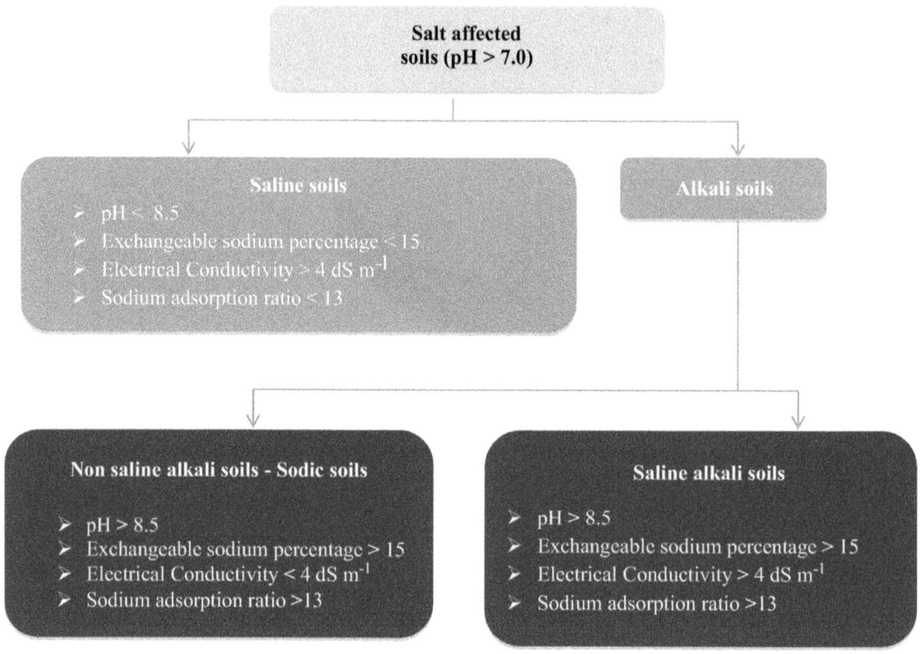

FIGURE 27.5 Classification of salt-affected soils.

relatively high in exchangeable sodium (have a high exchangeable sodium percentage, ESP) but are low in soluble salts. They mostly occur in arid and semiarid regions in small areas, and they are often termed as "slick spots". Sodium carbonates and bicarbonates contribute highly to the alkalinity of soil. Removal of excess salts in such soils tends to increase the rate of hydrolysis of exchangeable sodium which causes a rise in pH. Saline-sodic soils are formed mainly due to salinization and alkalinization. Saline-sodic soils are like saline soils but exhibit significantly higher concentrations of sodium salts relative to calcium and magnesium salts. Soil salinity creates toxicity of some ions like sodium, chlorine, and boron and creates deficiency of some ions like calcium and potassium (Figure 27.5).

27.3.2 Distribution of Salt-Affected Soils in India

The distributed areas under saline and sodic soils in India are shown in Figures 27.6 and 27.7. According to Kumar and Sharma (2020), the largest area with saline soils (1,680,570 ha) occurs in the state of Gujarat followed by West Bengal (441,272 ha), and the largest area with sodic soils (1,346,971 ha) occurs in the state of Utter Pradesh followed by Gujarat (541,430 ha).

27.3.3 Impact of Soluble Salts on Plants

As soils become more saline, plants find it difficult to draw water from the soil. Plant roots, which contain varying concentrations of salt ions, create a natural flow of water from soil into plant roots. When soil salinity is increased in the root zone, water becomes less available and cannot enter plant roots. When levels of soil salinity are high enough, water in the roots is pulled back into the soil

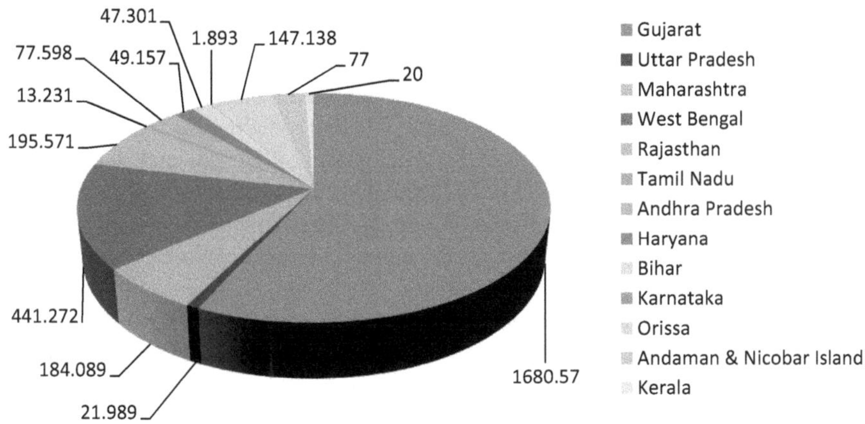

FIGURE 27.6 Extent of saline soils in India ('000 ha).

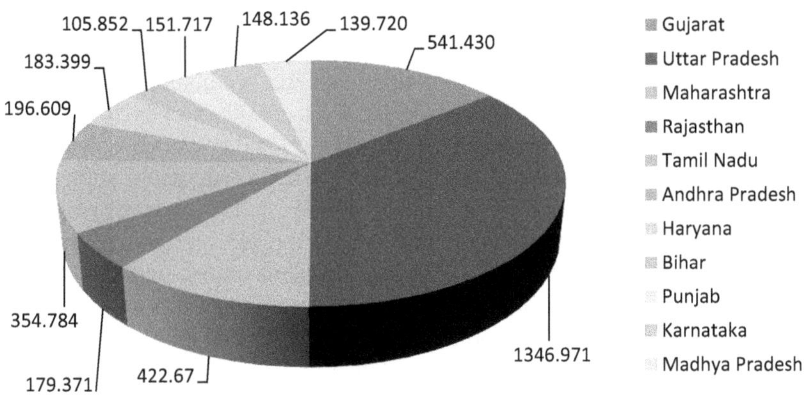

FIGURE 27.7 Extent of sodic soils in India ('000 ha).

TABLE 27.1
Soil Salinity Classes

Salinity Class	EC (dS/m)	Effect on Crop Plants
Non-saline	0–2	Salinity effects negligible
Slightly saline	2–4	Yields of sensitive crops may be restricted
Moderately saline	4–8	Yields of many crops are restricted
Strongly saline	8–16	Only tolerant crops yield satisfactorily
Very strongly saline	>16	Only a few very tolerant crops yield satisfactorily

(Table 27.1). The plant parts become unable to take in enough water for their growth. So, the plant exhibits wilting symptoms and finally growth is reduced. High pH levels in soil affect soil structure and change the ionic form of nutrients. Nutrients become unavailable to plants, and plants exhibit deficiency symptoms under saline conditions.

27.4 SUGARCANE RESPONSE AND SALINITY

Sugarcane response to high saline conditions leads to delayed germination, leaf yellowing and burning, reduced tillering, stunted growth, a poor field stand, and reduced cane yields. The salt interferes with sugar production by affecting cane growth and yield and reducing the sucrose content of stalk (Sundara and Vasantha, 2004). Salt increases the fiber content of the cane and electrical conductivity of juice, which have an antagonistic effect on the preparation of sugar and jaggery and their quality. Growth and yield of sugarcane are affected adversely, depending on the number of salts present in soil. During the dry season, excess soluble salts move to the soil surface through capillary rise, which makes the soil saline and unproductive for crop production. High concentrations of toxic ions such as Na^+ and Cl^- reduce the osmotic potential in the soil solution and limit the ability of plant roots to extract water from soil even under wet conditions causing drought like symptoms in the plant. Excess amounts of ions, especially Na^+ and Cl^-, in the soil lead to conformational changes in protein structure, membrane depolarization, and ion toxicity of plants.

In a greenhouse study, sugarcane cultivation on soils with different textures, *viz.*, sandy, clayey, and medium texture, and different salinity levels in the irrigation water ranging from 0.1 to 8 dS/m, affected the aerial and root dry mass (Santana et al., 2007). The application of nitrogen as calcium ammonium nitrate and potassium as sulfate of potash was recommended for salt-affected soils to enhance the yield of sugarcane. The maximum growth and yield attributes with combined application di-ammonium phosphate+sulpfate of potash+calcium ammonium nitrate on salt-tolerant varieties of sugarcane were recorded. If the salinity level of soil exceeds 10 dS/m, considerable yield reduction of sugarcane occurs (Ashraf et al., 2008). Quality parameters like polarization, Brix, purity, Mg^{2+}, and K^+ decreased with increasing electrical conductivity of the irrigation water (Golabi et al., 2009). Decreased chlorophyll content of leaves, reduction in turgor pressure, limited elongation and cell division, and accumulation of reactive oxygen species were related directly to decreased plant growth due to salt stress in sugarcane. Supplementation with GA_3 (gibberellic acid) at 150 ppm as a treatment when the plants were at the sett stage and foliar spray at the tillering phase resulted in better growth under saline conditions and thus enhanced the cane yields (Gomathi and Thandapani, 2014).

27.5 SALINITY TOLERANCE OF SUGARCANE

Soil salinity and drought individually or in combination adversely affect growth and yield of sugarcane. Vasantha et al. (2017) showed that physiological traits, *viz.*, chlorophyll fluorescence, chlorophyll index, biomass production, and leaf area index, were reduced greatly with a combined stress (salinity and drought). The stress treatments resulted in reduction of the essential elements of phosphorus, iron, and zinc, and responses differed with growth stage. Zhao et al. (2020) found that sugarcane plants treated with 75, 150, and 300 mM salt concentrations had lower leaf net photosynthetic rates of 12.7%, 18.7%, and 35.3%, respectively. But plants treated with a 38 mM salt concentration did not differ from the controls. Applications of high levels of salt concentration decreased sugar translocation from leaves to other tissues.

Vasantha et al. (2010a) assessed the relative performance of different sugarcane genotypes in terms of cane and sugar yield by plotting the cane yield under salinity treatments against cane yield under normal irrigated conditions. About 1,200 genotypes of *Saccharum* spp., which included advanced breeders' material, germplasm material, different species, and soma clones, were screened for salinity tolerance (Table 27.2).

For comparison purposes, an equal number of plots exposed to salinity and control plots was maintained. Soil was salinized prior to planting, and, from planting until harvest, the electrical conductivity of the saturated soil extract was maintained at 8 dS/m. The salinity treatment was obtained by irrigating with a 1% salt solution of $NaCl:CaCl_2:Na_2SO_4$ in the ratio of 2:2:1. Control plots were irrigated with normal water. The electrical conductivity of the soil was monitored at

TABLE 27.2
***Saccharum* Species Screened for Salinity Tolerance**

Species	Total Number Screened	Number of Tolerant Types
Saccharum officinarum	346	113
Saccharum robustum	59	15
Saccharum barberi	39	12
IND clones	155	21

Source: Vasantha et al. (2010a).

TABLE 27.3
List of Sugarcane Genotypes Identified as Tolerant to Salinity

Co 87002	Co 89010	Co 89027	Co 90010	Co 91002
Co 91005	Co 91011	Co 92102	Co 93005	Co 93009
Co 93011	Co 93015	Co 93016	Co 93018	Co 93019
Co 94004	Co 93021	Co G 93079	Co 94005	Co 94008
Co 94010	Co 94011	Co 94012	Co 94015	CoJn 86141
CoJn 94141	Co G 93076	Co 95003	Co 95006	MS 92121
Co 99004	Co 2001-13	Co 2001-15	Co 2000-10	Co 97008

Source: Vasantha et al. (2010a).

regular intervals. The experiment revealed that, among the different sugarcane varieties tested, Co 86011, Co 7717, Co 7219, Co 8208, Co 85004, CoC 671, Co 6806, Co 94008, Co 85019, Co 94012, Co 97008, and Co 99004 were found to be suitable varieties for salt-affected soils. However, the variety Co 97010 was found to be sensitive (Vasantha et al., 2010a). Several promising genotypes were identified as tolerant types to salinity (Table 27.3).

Commercial sugarcane clones, including the salinity tolerant types (Co 85019, Co 94012, Co 94008, Co 86032) and susceptible types (Co 97010, Co 95007, Co 97009, Co 9506), were used to study the effect of salinity over the life cycle of sugarcane (Vasantha et al., 2010b). Photosynthetic rate, transpiration rate, and leaf water potential showed little difference in early stages of crop growth. The failure of osmotic adjustment coupled with reduced photosynthetic rate resulted in poor dry matter production in sensitive clones (Co 97010, Co 95007, and Co 97009). The salt-tolerant clones exhibited a minimal reduction of photosynthesis, better osmotic adjustment, and a smaller reduction in biomass (Vasantha et al., 2010b). The genotype Co 85019, followed by Co 99004, was found to be the best-surviving genotype under EC 8 dS/m (Brindha et al., 2019). Karpe et al. (2012) used *in vitro* plantlets and treated them with increasing concentrations of NaCl (0, 50, 100, 150, 200, and 250 mM) and evaluated different responses of two sugarcane varieties (CoC 671 and Co 86032) to salinity stress. CoC 671 showed the highest tolerance to NaCl (up to 150 mM), while Co 86032 survived under NaCl salinity only up to 100 mM.

Patade et al. (2011) treated sugarcane (*Saccharum officinarum* L. cv. Co 86032) plants with NaCl (150 mM) or polyethylene glycol-PEG 8000 solutions (20% w/v) for 15 days. Both treatments reduced leaf width, chlorophyll stability index, and number of green leaves. Osmotic adjustment indicated that salt stress and polyethylene glycol stress led to the accumulation of osmolytes and sugars. Sugarcane plants that were salt stressed had an increase in the activities of superoxide dismutase and ascorbate peroxidase, while PEG stress led to an increase in superoxide dismutase but not ascorbate peroxidase activity (Patade et al., 2011). Singh and Sengar (2020) maintained salinity by irrigation with demineralized water mixed with a specific ratio of NaCl, Na_2SO_4, $CaCl_2.2H_2O$ at 3:1:2. Two salinity levels of electrical conductivities of the irrigation water were maintained: 10 and 20 dS/m. They found that commercial cane genotypes of Co Pant 97222, CoS 7250, Co 98014, Co 5009, and Co 5011 were salinity tolerant. CoLk 99270 and Co 8279 were moderately tolerant, and Co 0118, Co 0238 and CoSe 8457 were salinity susceptible.

Simoes et al. (2016) evaluated the effect of salinity on growth of sugarcane varieties. They used six levels of salinity (0, 1, 2, 4, 6, and 8 dS/m) and ten varieties of sugarcane. Various levels of salinity in the irrigation water were obtained by adding NaCl, $CaCl_2.2H_2O$, and $MgSO_4.7H_2O$ to achieve an equivalent ratio among Na:Ca:Mg of 7:2:1. Six days after irrigation, plant height, stem diameter, number of leaves, leaf area, and fresh and dry mass of aerial parts and roots of the different varieties showed similar responses. Growth was reduced as soil salinity increased, and the plants were considered to be moderately sensitive to salinity. Gandonou et al. (2012) grew three-week old sugarcane cultivars [CP66-346, CP65-357, CP70-321, CP59-73 (American cultivars) and NCo310 (Indian cultivar selected in South Africa)] with NaCl concentrations of 0, 17, 34, and 68 mM for two weeks in a hydroponic medium. These NaCl concentrations corresponded to electrical conductivities of 1.00, 2.83, 4.26, and 6.63 mS/cm, respectively. Different salinity concentrations impacted both plant growth and survival of sugarcane. Plant mortality rate increased with increasing salinity. Cultivars CP70-321 and CP65-357 exhibited high plant mortality. The effect of salinity on shoot growth was highest with cultivar CP70-321. CP66-346 was the most salt-tolerant cultivar followed by cultivar NCo310 and CP59-73. Cultivars CP70-321 and CP65-357 were the most sensitive to salinity in the hydroponic medium.

27.6 ORGANIC AMENDMENTS FOR SUGARCANE UNDER SALINE CONDITIONS

An integrated approach to manage saline soils has been suggested. It includes trench planting; using a higher seeding rate (25% is recommended to compensate for germination loss); use of organic manures, *viz.,* press mud (10–15 t/ha) and farm yard manure (FYM) (25 t/ha); irrigating the field with good quality water; trash mulching; green manures; crop rotation with salt-tolerant crops; growing salt-tolerant varieties or genotypes; gypsum application (3–6 t/ha); and application of nitrogen at a 25% additional dose. These procedures are recommended to manage moderately saline soils in agricultural lands (Sundara and Vasantha, 2004). Reclamation of saline soils improves the availability of nutrients, which enhances yield of sugarcane. These are management practices recommended for higher cane production. Choudhary et al. (2004) found that, under long-term saline-sodic irrigation, the application of farm yard manure was effective, and cane yield, sugar yield, and percent commercial cane sugar per unit area were improved. They said that, when long-term sodic irrigation occurs, gypsum should be applied to bring down the soil ESP to achieve sustainable cane and sugar yield. However, the application of gypsum alone was unable to improve cane and sugar yield.

Thus, profitable yields of good quality sugarcane are possible by treating sodic soils with gypsum and organic manure. Izhar-ul-Haq et al. (2007) found that gypsum, applied at 100% of the soil requirement, in combination with farm yard manure at 10 t/ha improved the growth, yield, and quality of sugarcane. Sugarcane growth was increased under sodic soils when amended with gypsum and manure by reducing soil pH and the sodium adsorption ratio. In a sandy loam soil, Udayasoorian et al. (2009) showed that the application of press mud at 15 t/ha along with 50% gypsum requirement on a sugarcane field reduced soil pH. A 1.33% reduction was observed under treatment with paper mill effluent, and a 1.25% reduction was recorded under saline ground water irrigation. The application of amendments reduced soil exchangeable sodium and increased the influence of the soil exchangeable cations Ca, Mg, and K.

Interaction between vermicompost and nitrogenous fertilizer may have synergistic effects on growth and uptake of nitrogen and potassium by sugarcane in saline soils. Djajadi et al. (2020) showed that the addition of vermicompost at 20 t/ha along with 50 kg N/ha reduced soil electrical conductivity and the Na/K ratio, which produced higher biomass of sugarcane after four months, when it was grown with a soil with electrical conductivity of 4.12 dS/m.

27.7 CONCLUSION

Salt accumulation limits sugarcane production. Increasing salinity levels reduce sugarcane yield. Salinity has a negative effect on both socioeconomic and environmental aspects of sugarcane production in India. Prevention and reclamation of saline soils are required for higher cane yield.

Integrated nutrient management practices, such as application of vermicompost, gypsum, and farm yard manure along with adequate nutrients, especially nitrogen and potassium, will improve the physical, chemical, and biological properties of soil and minimize the effects of salinity and maximize cane yield and quality.

REFERENCES

Ashraf, M.Y., Hussain, F., Akhter. J., Gul, A., Ross, M. and Ebert, G. (2008). Effect of different sources and rates of nitrogen and supra optimal level of potassium fertilization on growth, yield and nutrient uptake by sugarcane grown under saline conditions. *Pak. J. Bot.*, 40(4): 1521–1531.

Brindha, C., Vasantha, S. and Arunkumar, R. (2019). The response of sugarcane genotypes subjected to salinity stress at different growth phases. *J. Plant Stress Physiol.*, 5: 28–33.

Choudhary, O.P., Josana, A.S., Bajwaa, M.S. and Kapurb, M.L. (2004). Effect of sustained sodic and saline-sodic irrigation and application of gypsum and farmyard manure on yield and quality of sugarcane under semi-arid conditions. *Field Crops Res.*, 87: 103–116.

Directorate of Economics and Statistics, New Delhi (2020–2021). Department of Agriculture and Cooperation, Fourth Advance Estimates. http://sugarcane.dac.gov.in/pdf/StatisticsAPY.pdf.

Djajadi, D., Syaputra, R. and Hidayati, S.N. (2020). Effect of vermicompost and nitrogen on N, K, Na uptakes and growth of sugarcane in saline soil. *Agrivita J. Agric. Sci.*, 42(1): 110–119.

Gandonou, B., Gnancadja, L.S., Abrini, J. and Skali-Senhaji, N. (2012). Salinity tolerance of some sugarcane (*Saccharum* sp.) cultivars in hydroponic medium. *Int. Sugar J.*, 114(1359): 190–196.

Golabi, M., Naseri, A.A. and Kashkuli, H.A. (2009). Mathematical modeling of the relationship between irrigation water salinity and sugarcane juice quality. *J. Food Agric. Environ.*, 7(3&4): 600–602.

Gomathi, R. and Thandapani, P. (2014). Influence of salinity stress on growth parameters and yield of sugarcane. *J. Pharm. Biol. Sci.*, 9(3): 8–32.

Izhar-ul-Haq, Khan, M. and Iqbal, F. (2007). Management of salt-affected soils for sugarcane production. *Sarhad J. Agric.*, 23(2): 345–350.

Karpe, A., Nikam, A.A., Chimote, K.P., Kalwade, S.B., Kawar, P.G., Babu, H., Devarumath, R.M. and Suprasanna, P. (2012). Differential responses to salinity stress to two varieties (CoC 671 and Co 86032) of sugarcane (*Saccharum officinarum L.*). *Afr. J. Biotechnol.*, 11(37): 9028–9035.

Kumar, P. and Sharma, P.K. (2020). Soil salinity and food security in India. *Front. Sustain.*, 4: 1–15.

Ogunsola, A., Zancanner, A. and Seed, B. (2021). The sugar series: The top 10 sugar producing countries in the world. https://www.czarnikow.com/blog/the-sugar-series-the-top-10-sugar-producing-countries-in-the-world.

Patade, V.Y., Bhargava, S. and Suprasanna, P. (2011). Salt and drought tolerance of sugarcane under iso-osmotic salt and water stress: Growth, osmolytes accumulation, and antioxidant defense. *J. Plant Interact.*, 6(4): 275–282.

Sabareshwari, V. and Ramya, A. (2018). Coastal saline soils of India: A review. *Agric. Rev.*, 39(1): 86–88.

Santana, M.J., Carvalho, J.A., Souza, A.M.G., Sousa, K.J., Vasconcelos, C.L. and Andrade. L.A.B. (2007). Effects of irrigation water salinity upon the sprouting and initial development of sugarcane (*Saccharum* spp.) and on soils with different textural levels. *CienciaAgrotec., Lavras*, 31: 1470–1476.

Simoes, W.L., Calgaro, M., Coelho, D.S., Santos, D.B., and Souza, M.A. (2016). Growth of sugar cane varieties under salinity. *Rev. Ceres, Vicosa.*, 63(2): 265–271.

Singh, M., Waghmare, A.L. and Kumar, M. (2020). *Glimpses of Sugarcane Cultivation*. Directorate of Sugarcane Development, Government of India, Lucknow, pp. 25–36.

Singh, R. and Sengar, R.S. (2020). Effect of salinity stress on morphological and yield attributes of sugarcane (*Saccharum officinarum L.*) genotypes. *Int. J. Chem. Stud.*, 8(5): 2312–2316.
Sundara, B. and Vasantha, S. (2004). *Sugarcane Management in Saline Soils*. Sugarcane Breeding Institute, Coimbatore.
Udayasoorian, C., Paul Sebastian, S. and Jayabalakrishnan, R.M. (2009). Effect of amendments on problem soils with poor quality irrigation water under sugarcane crop. *American J. Agric. & Environ. Sci.*, 5(5): 618–626.
Vasantha, S., Gomathi, R. and Brindha, C. (2017). Growth and nutrient composition of sugarcane genotypes subjected to salinity and drought stresses. *Commun. Soil Sci. Plant Anal.*, 48(9): 989–998.
Vasantha, S., Gomathi, R., Rakkiyappan, P., RajulaShanthy, T., Vijayan Nair, N. (2010a). *Screening for Salinity Tolerance in Sugarcane*. Sugarcane Breeding Institute, Coimbatore.
Vasantha, S., Venkataramana, S., Gururaja Rao, P.N. and Gomathi, R. (2010b). Long term salinity effect on growth, photosynthesis, and osmotic characteristics in sugar cane. *Sugar Tech.*, 12(1): 5–8.
Zhao, D., Zhu, K., Momotaz, A. and Gao. X. (2020). Sugarcane plant growth and physiological responses to soil salinity during tillering and stalk elongation. *Agriculture*, 10(608): 1–13.

28 Soil Constraints and Management Options for Sustainable Rainfed Production Systems of India

Ch. Srinivasarao[1], C. Subha Lakshmi[1], S. Rakesh[1], M. Mohanty[2], G. Narayanaswamy[3], S. Kundu[4], G. Pratibha[4], J. V. N. S. Prasad[4], Rajbir Singh[5], and A. Arunachalam[6]

[1]ICAR-National Academy of Agricultural Research Management, Rajendranagar, India
[2]ICAR-Indian Institute of Soil Science, Bhopal, India
[3]Agricultural Research Station, Acharya NG. Ranga Agricultural University, Andhra Pradesh, India
[4]ICAR-Central Research Institute for Dryland Agriculture, Hyderabad, India
[5]ICAR-Agricultural Technology Application Research Institute Zone-1, Ludhiana, Punjab, India
[6]ICAR-Central Agroforestry Research Institute, Jhansi, India

CONTENTS

DOI: 10.1201/9781003093565-28

28.1 INTRODUCTION

Rainfed agriculture, which constitutes about 80% of global agriculture, holds pivotal importance to attain food and nutritional security. The prominence of rainfed agriculture varies from region to region across the world, but it caters toward meeting the major proportion of the dietary requirements of communities inhabiting developing countries. India is positioned first in the rainfed agriculture globally both in terms of area (86 Mha) and the value of produce. In India, rainfed areas contribute a substantial share to the food grain production including 44% of rice, 87% of coarse cereals, 85% of food legumes, 72% of oilseeds, 65% of cotton, and 90% of minor millets. They support two-thirds of the livestock population and assume significance as a source of livelihood security. For most of the major crops, viz., rice, wheat, and maize, the yield curve in irrigated areas has plateaued owing to stagnated responses to supplemented inputs. Rainfed areas, commonly referred to as gray areas, hold vast potential to elevate production levels and alleviate the concerns pertaining to malnutrition and food and income insecurity in developing nations. The gap between possible potential yield of rainfed crops and their actual yield still continues on a broad scale. Because of the limitations of moisture availability, rainfed ecosystems are at disadvantage owing to numerous constraints. The current chapter focuses on the production constraints in rainfed ecosystems, especially in India, with special emphasis on soil-related problems and their management.

28.2 PRODUCTION CONSTRAINTS IN RAINFED PRODUCTION SYSTEMS OF INDIA

The agriculture sector is highly fragile and risk-prone to climate change (Rakesh et al., 2019). Rainfed ecosystems have various constraints that limit enhancing crop yields. These ecosystems have high sensitivity to climate-induced vagaries. Climatic anomalies lead to altered rainfall patterns in the form of long dry spells and delay in onset and early withdrawal of monsoons, which interfere with agricultural operations that eventually affect the quantity and quality of economic produce. Rainfall also influences the amount of soil organic carbon (SOC) stocks, and considerable variation in carbon stocks can be observed with the variations in rainfall (Figure 28.1).

Low levels of SOC and nutrients, coupled with poor soil management and inappropriate land use practices, affect soil quality in these areas. The amount of soil organic matter (SOM) alters various soil properties that are important for soil health on a sustainable basis. SOM influences water-holding capacity, aggregate stability, compaction, and strength characteristics of soils. Escalating populations demand enhanced production levels, and, ultimately in the process of attaining food

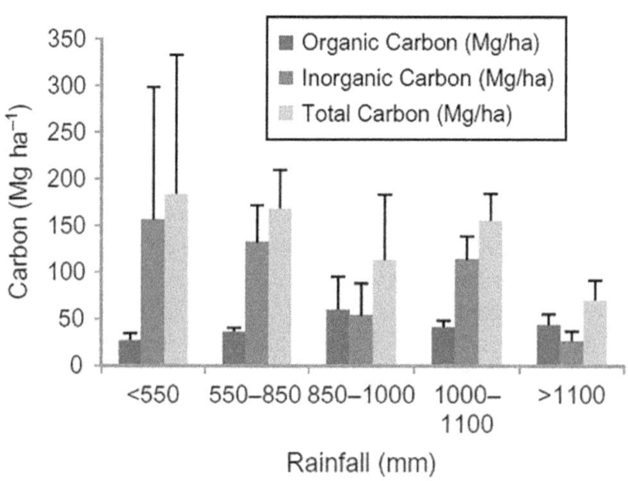

FIGURE 28.1 Influence of rainfall on soil carbon stocks. (Srinivasarao et al., 2009.)

needs, they lead to over exploitation of available resources, consequently leaving rainfed arable lands in arid, semi-arid, and sub-humid areas degraded. By 2030, it is estimated that India will need around 304 million tons (Mt) of food grain; 96 Mt of fruits; 175 Mt of vegetables; 21 Mt of meat, eggs, and fish; and 170 Mt of milk (Srinivasarao et al., 2015).

Among various constraints contributing to reduced productivity in rainfed regions, soil-related factors hold prominence. Soils in rainfed areas are highly affected by erosion and runoff resulting in loss of finer fractions of top-soil rich in organic matter. Nutrient additions to soil are meager, and the major reason for this fact is limited or no recycling of crop residues into the soil, as the crop residues are used as fodder for animals. Temperature-mediated rapid oxidation of organic matter is also a common phenomenon in these ecosystems. Moisture stress is a prominent factor on which cultivation of crops is reliant in these areas. Because moisture is limited, cropping intensity is low, which curtails the addition of organic matter to the soil through root biomass additions. Cultivation of crops is largely practiced on marginal and sub-marginal lands, and the adoption of crop management practices, viz., nutrient and pest management, is given little attention, which adds to the decline in soil fertility. Imbalanced and low fertilizer use lead to negative nutrient budgets (nutrient mining), which also are a cause for yield reduction in rainfed crops. The major types of soils in rainfed agro-ecological regions of India are Alfisols, Vertisols and Vertic subgroups, Oxisols, Inceptisols, Entisols, and Aridisols. They exhibit variations in moisture-holding capacity (Srinivasarao et al., 2017a). Except for Vertisols, most of the soils of rainfed regions are coarse-textured and exhibit poor water retention capacity (Table 28.1). This aggravates drought stress in plants and results in reduction of yields.

Increased yield with application of fertilizers or organic sources can be attained only with sufficient moisture, because there exists a synergistic relation between nutrients applied and soil moisture levels. Hence, nutrient (NUE) is low in the periods which experience drought stress. The estimated probability of drought occurrence in various meteorological subdivisions of India is presented in Table 28.2.

A wide variation in nutrient application occurs in rainfed farming systems in comparison to irrigated areas, the prime reason being uncertainty in yields. This leads to multi-nutrient deficiency. Among micronutrient deficiencies, zinc and boron limit sustainable production in rainfed ecosystems. The accrual of all these limiting factors contributes to a poor functional capacity of soils (Srinivasarao et al., 2019a; Srinivasarao and Vittal, 2007).

Most inhabitants of rainfed regions of India are resource-poor, small holder farmers with low socioeconomic status. Imperfect markets for inputs and products mean small profits and poor access to credit markets, which result in sub-optimal investment and cultivation practices. Lack of technological know-how is a general problem that confronts small holder farmers (Mahendra

TABLE 28.1
Water Storage Capacity of Major Soil Groups of India under Rainfed Situation

Soil Group	Depth (cm)	Subgroup	Moisture Storage Capacity (mm)
Inceptisols	>90	Deep	110–140/m
			140–180/m
			110–140/m
Aridisols	Up to 90	Medium to deep	80–90/90 cm
Alfisols and related soils	Up to 45	Shallow to medium	40–70/45 cm
Vertisols and related soils	Up to 45	Shallow to medium	135–145/45 cm
	45–90	Medium to deep	145–270/90 cm
	>90	Deep	300/m

Source: Adapted from Srinivasarao et al. (2013a).

TABLE 28.2
Probability of Drought Occurrence in Various Meteorological Subdivisions of India

Frequency of Deficient Rainfall (75% of Normal or Less)	Meteorological Subdivision
Very rare, once in 15 years	Assam
Once in 5 years	Bihar, Odisha, Madhya Pradesh, Konkan and West Bengal,
Once in 4 years	South interior Karnataka, Eastern Uttar Pradesh, and Vidarbha
Once in 3 years	East Rajasthan, Gujarat and Western Uttar Pradesh
Once in 2.5 years	West Rajasthan, Tamil Nadu, Jammu and Kashmir, Telangana

Source: Adapted from NRAA (2013).

Dev, 2011). Existence of a wide gap between producer and consumer prices, due to interference of middlemen who take a major share of the profit, is a problem. On account of poor investment capacity and infrastructure availability, the farming systems that are adopted are of the subsistence type. Management practices to enhance soil health are rarely practiced, and this contributes to declining soil health and fertility. The adoption of strategies to overcome soil-related constraints is necessary to enhance productivity in rainfed ecosystem.

28.3 SOIL PRODUCTIVITY CONSTRAINTS AND MANAGEMENT STRATEGIES

28.3.1 Land Degradation and Soil Erosion

A decline in ecosystem functions and productivity over a long period of time results in land and soil degradation. A comprehensive term to define the process is called desertification. In rainfed regions, land degradation is a major issue of great concern, and it has implications for environmental quality and food security. Climatic anomalies, shifts in composition of vegetation, accelerated soil erosion processes, and anthropogenic disturbances have rendered these landscapes highly vulnerable to rapid degradation that has resulted in desertification (Sujith et al., 2010). Water and wind erosion lead to removal and redistribution of top soils, where nutrient rich finer particles occur. This, in-turn, affects soil properties and eventually alters soil moisture and affects the establishment and growth of vegetation in rainfed areas (Puigdefabregas, 2005). Limited use of organic manures, uncontrolled and excessive grazing, deep tillage, mono-cropping, and poor recycling of crop residues play significant roles as contributors aggravating soil loss in regions practicing rainfed farming. The magnitude of soil loss depends on several factors, i.e., soil type, vegetation, and slope gradient. Soil erosion loss ranges from 5 to 150 Mg/ha/yr. The principal factors that result in soil degradation are wind and water erosion, acidity, salinity, and alkalinity. Erosion also assumes prominence as a process contributing to enhanced carbon emissions, because it reduces the potential of soil to store carbon. Enhancement in the SOC pool contributes to improvement in soil structure and tilth, reduction of soil erosion, and increased water-holding capacity and plant nutrient reserves. It acts as an energy source for soil fauna. It is also considered as an indicator of soil quality (Rakesh et al., 2020a, b). Increased SOC also reduces the level of pollutants in soils, boosts soil biodiversity, purifies water, augments the crop/crop residue ratio, and alleviates the impacts of climate change. Management strategies that contribute to minimizing land degradation need to be given utmost consideration.

28.3.1.1 Management Strategies to Control Land Degradation and Soil Erosion

28.3.1.1.1 Conservation Agriculture

Conservation agriculture (CA) comprises practices of agriculture that aid in conserving land while realizing sustainable yield levels (Sharma, 2019). It is gaining momentum as an efficient strategy to

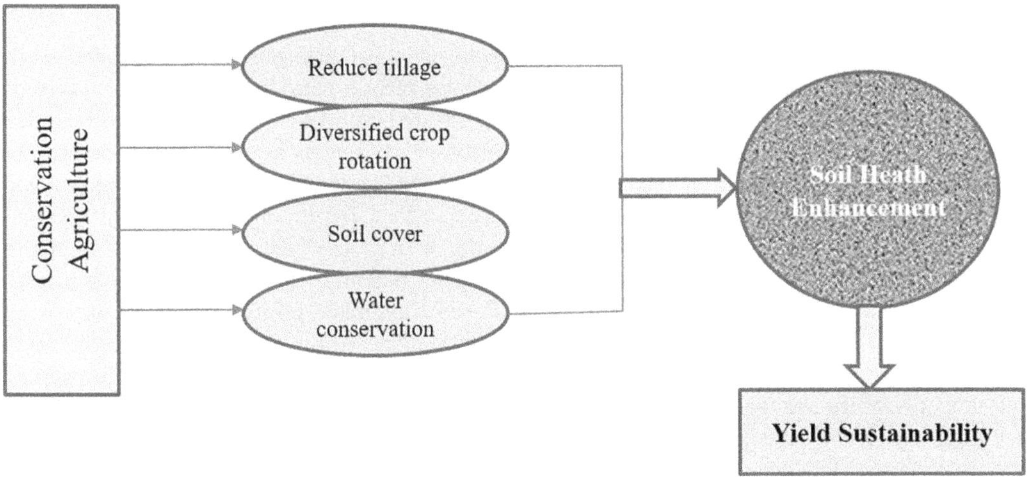

FIGURE 28.2 Principles of conservation agriculture.

attain sustainable yields through zero/no tillage, diversified cropping systems, and residue retention, CA practices also address the global warming issues (Figure 28.2).

Excessive long-term tillage practiced in conventional system results in rapid mineralization of organic matter and reduction in humus content, which lead to destruction of soil structure and aggregate stability. Eroded soils have decreased soil health. Hence, the situation demands adoption of suitable climate and location specific tillage practices that can aid in improving soil physical health and ultimately crop productivity. Conservation tillage has proved to be beneficial for soil moisture conservation, erosion control, and increased infiltration of water that aid in profile water recharge and improve soil organic matter and soil biological properties. Adopting no tillage (NT) or reduced tillage systems aids in enhancing the stabilization of SOC (Six et al., 2000). Studies conducted utilizing the CENTURY and Roth C models suggest that conversion to NT could elevate SOC stocks from 0.1 to 0.2 Mg/ha/yr under dryland conditions in sandy soils of West Africa (Farage et al., 2007). In the maize-horsegram sequence, CA under residue management in conjunction with balanced fertilization led to improvements in SOC (3.1–4.5 g/kg) in comparison to a conventional system (2.9–4.2 g/kg) (Kundu et al., 2013). The addition of crop residues to soil enhances the SOC content (Rakesh et al., 2020c). The major constraints limiting the adoption of CA in South Asia are small land holdings, non-availability of suitable farm implements, and inability of farmers to get technological information (Somasundaram et al., 2020). Region-specific CA options need to be identified and developed for implementation by resource-poor farmers (Fowler and Rockstrom, 2000). Even though the advantage of CA is realized in both irrigated and rainfed regions, less attention has been paid in the rainfed regions of the country. Key constraints in adoption of CA in rainfed areas are non-availability of crop residues, lack of proper machinery for sowing, termites in Alfisols, rodent problems in Vertisols, limited knowledge and capacity of farmers to implement CA systems, lack of profitable crop rotation systems, and lack of access to the inputs. Interventions to overcome these constraints would pave way toward adoption of CA on a larger scale in developing nations inhabited by small and marginal farm holders.

The unimodal, short rainy season has implications for the ability of smallholder farmers to grow crops and produce sufficient biomass for groundcover and feed (Mupangwa and Thierfelder, 2014). CA has a wide applicability when dealing with different rainfall amounts, soil types, and crops (Wall et al., 2014; Wall, 2007). Mupangwa et al. (2017) found that organic carbon build-up was observed as the tillage levels were reduced, but the degree of increase in organic carbon was depended on the length of study, the type of crop rotation used, and the quantity and quality of residue that could be retained in the soil.

28.3.1.1.2 Cover Cropping

Monocropping has been adopted in the rainfed regions owing to uncertainty in rainfall and availability of irrigation water. Thus, the land is left fallow for certain periods of time. Erosion sweeps away a large amount of soil from fallow lands leaving the land unfertile. Cover cropping is regarded as an effective practice that could be adopted in erosion-prone regions to minimize the severity of risks that these sensitive areas encounter. The jeopardy associated with wind erosion is minimized through cover cropping by physically defending the soil surface, improving soil structural aggregation, elevating SOC levels, and providing soil anchorage with roots when principal crops are not in position. Hence, cover crops lower rates of soil erodibility potential. The incorporation of a leguminous cover crop, horse gram (*Macrotyloma uniflorum*), exhibited positive effects in improving SOC, microbial biomass, and available nutrient status (Venkateswarlu et al., 2007). Another study involving conservation agriculture that included tillage (conventional and minimum) with cropping systems (sole maize and maize+cowpea intercrop) and followed by cover crops [fallow, horsegram, and toria (rapeseed)] revealed enhancement in SOC by inclusion of cover crops in comparison to no – cover crop (Mohanty et al., 2015).

28.3.2 Soil Physical Constraints and Management Strategies

Knowledge of physical properties of soils is important to define soil health and to attain high productivity levels. Physical characteristics of soil in a field vary spatially and temporally. The physical condition of soil determines crop yield. Soil physical constraints include high or slow permeability, compaction and hard pans, soil-water-related constraints, and surface crusting and hardening. Maintenance of soil physical conditions that permit adequate crop growth is essential. In order to realize a crop's genetic yield potential, it is important to sustain soil physical conditions at optimum levels, even when all other crop needs are fulfilled. Methods to manage soil physical conditions are site-specific, and advantages from their adoption are variable and depend on the intensity of rainfall, soil texture, slope of the land, and cropping systems (Indoria et al., 2016). Management technologies to alleviate soil physical constraints are discussed, as follows.

28.3.2.1 Soil Crust Formation

Formation of crust on the soil surface caused by the beating action of raindrops results in poor seed germination, reduction in infiltration rate, and increased runoff, especially in rainfed regions with Alfisols, Entisols, and Aridisols. Mechanical breaking of a crust using a hoe, or using crop residues or stubble incorporation, is beneficial in reducing the formation of crusts in these soils. The application of organic amendments and adopting the practice of integrated nutrient management is a viable option to overcome the impacts of crust formation. Farmyard manure (FYM) application in the form of mulch increased seedling emergence in comparison to crusted soil by ten-fold and three-fold in cotton and bajra, respectively (Phogat and Dahiya, 2002). Mulching enhanced the yield of cotton from 0.35 to 1.49 t/ha while that of bajra was increased from 2.63 to 3.42 t/ha, showing that the impact of rain drops was reduced with FYM mulch due to less crust formation, which led to the improvement in seedling emergence and crop yields (Phogat and Dahiya, 2002). Red *chalka* soils are red sandy loam soils that cover a large area in Andhra Pradesh and become very hard on drying. They usually encounter rapid and irreversible hardening upon drying, which is a chief limitation in rainfed crop production. The application of slow decomposing residues, viz., coir pith and paddy husk, has been found beneficial in these soils. Trials conducted in farmers' fields revealed that paddy husk application (@5 t/ha) enhanced the yield of castor and sorghum by 23% and 18%, respectively, in comparison to controls (Painuli and Yadav, 1988). Amendments directly or indirectly modify soil temperature and moisture by minimizing evaporation losses, thereby reducing the impact of raindrops and preventing crust formation. Oswal (2001) evaluated different amendments at varying rates for major crops. The results indicated that FYM at 10 t/ha was the best amendment followed by coir pith at 20 t/ha, powdered groundnut shell at 5 t/ha, gypsum at 4 t/ha, and paddy husk at 5 t/ha. Gypsum application at 2–5 t/ha enhanced the infiltration rate by 4–7 times when compared with the control in black soil (Oswal, 2001).

28.3.2.2 Subsurface Hardpans and Compactness

In areas experiencing dryness and in soils of very fine sand and coarse silt (Aridisols, Alfisols, and Entisols), the problem of subsurface hardpans and compactness is severe especially in surface horizons. In red soils, illuviation of clay to the subsurface horizons and cementing action of oxides of aluminum and iron and calcium carbonate elevate the soil bulk density (BD) > 1.8 Mg/m^3. Increased BD results in the formation of subsoil hard pans. High BD impedes root development in red Alfisols and also reduces the hydraulic conductivity and water diffusivity in black silty clay soils. Rainfed alluvial loam soils (Entisols, Aridisols, and Inceptisols) experience the problem of low infiltration rate due to compaction and reduced porosity that curb the entry of water into the soil. In Vertisols, the bulk density ranges from 1.44 to 1.8 Mg/m^3 with a slight increasing trend with depth (Murthy et al., 1982). Reduction in percentage of pores (> 0.5 mm) diameter without affecting capillary pore space results with increased bulk density. This reduction is drastic in heavy-textured soils (Vertisols) in comparison to light-textured soils (Alfisols, Aridisols). Root elongation is restricted. Shallow roots reduce a plant's capacity to draw moisture and nutrients from deeper layers, and the plant experiences drought stress. Under heavy rainfall, the plant is prone to lodging. Hard pans can be mechanically shattered by mold-board ploughing or chiseling, which improves infiltration and moisture-holding capacity of the soil, augmenting root length density, and yields of various crops in rainfed regions (Mohanty et al., 2007). The addition of gypsum could be regarded as viable strategy to enhance infiltration rate in black soils. Gypsum application at 2–5 t/ha enhanced the rate of infiltration 4–7 times in comparison to a control (0.25 cm/hr). A trial conducted at Coimbatore revealed that chiseling was beneficial in enhancing the yield of rainfed crops, viz., black gram, sorghum, groundnut, maize, and tomato. Yield was increased by 18.6%–64.1% in comparison to farmers' practice. In sandy loam soils in Hissar and Haryana, the yield increase was around 14% in wheat, 17% in cotton, and 41% in raya (mustard plant) by chiseling (up to 40 cm depth at 50 cm intervals). Subsoiling with a chisel plough (deep tillage) can facilitate infiltration and improve water-holding capacity, thus curtailing water stress in these soils. In comparison to conventional tillage, the rate of infiltration was higher after subsoiling. Reduction in BD and surface and subsurface penetration resistance were also observed with subsoiling (Mohanty et al., 2007).

28.3.2.3 Highly Permeable Rainfed Soils

Sandy and loamy sand-textured rainfed soils (Entisols, Aridisols, and Inceptisols) encounter problems like high permeability and poor nutrient retention capacity. A major proportion of rainwater is lost in deeper soil layers, and water is available only for a very short period in the upper soil profile. Around 25%–40% of rainfall is lost through percolation below the plant root zone in light-textured soils resulting in *in-situ* conservation of only negligible amounts of rainfall. Rainfed crops are less resistant to short (<10 days) dry spells under such conditions. Giving a few passes of heavy-duty rollers to achieve soil compaction has been tested by researchers (Majumdar, 1994). For rainfed cowpea, around 23% more seed yield and 9% more stover yield was achieved by passing of a 500 kg iron roller twice over uncompacted soil. The soil moisture content and saturated hydraulic conductivity also varied with compaction. Surface compaction to 30–40 cm depth lessened the infiltration rate and hydraulic conductivity and increased the soil moisture retention capacity (Agrawal et al., 1987). A study conducted on highly permeable loamy sand soils revealed reduction in hydraulic conductivity of compacted soil in comparison to an uncompacted one (Majumdar, 1994). Depth of the compacted soil layer impacts the hydraulic head gradient and alters water flux (Singh and Mishra, 1995). Increase in the BD exponentially reduces the water flux due to compaction and benefits rainfed crops. The application of clay and tank silt also serves as an efficient practice that could be adopted to minimize high permeability of soils. Available water content improved by 0.02, 0.07,

0.012, and 0.032 g/g soil by addition of tank silt at 50, 100, 150, and 375 tractor loads ha^{-1}, respectively, which might be due to the formation of aggregates (Osman, 2008). Integrating clay addition with compaction also has served as a viable technology. Increased compaction after addition of clay (2%) exhibited positive impacts on water and nutrient retention in arid regions of Rajasthan (Indoria et al., 2016). Mixing of 2% clay followed by four passes of a 500 kg iron roller augmented the moisture retention capacity and lessened the hydraulic conductivity and infiltration rate of loamy sandy soil in comparison to a control (Yadav and Majumdar, 1990).

28.3.3 Uncertainty of Rainfall

Indian agriculture is extremely vulnerable to climate related issues, specifically to drought conditions because around two-thirds of the agricultural land is under rainfed conditions in India. The suitability of rainfed lands for arable farming is governed by interactions among temperature, precipitation, and soil texture. Uncertainty in the distribution and amount of rainfall results in droughts. Droughts have multiple impacts on agricultural production. The after-effects of drought will also be experienced in subsequent years, because of (i) deficit in groundwater recharge, (ii) land degradation, (iii) low draught power for agricultural operations owing to distress sale of cattle, and (iv) drop in savings and investment capacity of farmers; hike in prices; and reduction in grain trade and power supply. Intermittent, short, or extended dry spells cause mid-season droughts. This situation leads to a decrease in soil moisture resulting in moisture stress, which causes frequent crop failures, accelerated soil degradation, and loss of arable land. Long gaps between two consecutive rain events results in soil moisture deficit causing mid-season droughts during critical crop growth stages that bring sharp declines in yields. Strategies to cope with mid-season droughts need to be adopted to minimize their deleterious impacts.

28.3.3.1 Management Strategies for Water Conservation

28.3.3.1.1 Farm Ponds

In rainfed areas receiving high rainfall, excess runoff could be harvested and stored in farm ponds and could be efficiently utilized during periods of mid-season drought for meeting the water requirements of *kharif* crops. (*Kharif* crops, also known as monsoon crops or autumn crops, are plants that are cultivated in India during the subcontinent's monsoon season, which lasts from June to October, depending on the area.) The water stored could also be used for irrigating the post-rainy season crop, which would contribute to enhancing cropping intensity (Srinivasarao, 2019b). In the Warangal district, Telangana State, India, thirteen off-stream and on-stream farm ponds were constructed with the objective of mitigating intermittent drought conditions and curtailing moisture stress in crops. Total volume of water harvested was 12,325 m^3. The water harvested was used to provide critical irrigation (5 cm depth) to a cotton crop cultivated on 32.87 acres resulted in an increase in yield from 4 to 6 quintal ha^{-1} (a quintal is equal to 100 kg) (Srinivasarao et al., 2017b). Farm ponds have been identified as efficient sources contributing to enhanced cropping intensity and additional net returns to farmers (Table 28.3).

28.3.3.1.2 In-situ Moisture Conservation

Conservation of water at the place where it falls is referred to as *in-situ* moisture conservation. In rainfed regions, implementing *in-situ* rainwater harvesting technologies are considered as essential components of the farming system for enhanced moisture storage capacity of the soil, crop productivity, and food security. The major objective of *in-situ* conservation practices is to increase the infiltration rate by reducing runoff. Water is temporarily impounded on the soil surface by altering the land configuration for inter-plot water harvesting (Muthamilselvan et al., 2006). Moisture conserved *in situ* is utilized by crops and, hence, is regarded as green water. Various *in-situ* soil moisture conservation techniques, including contour cultivation, contour bunds (a bund is an embankment), broad bed furrows, tied ridges, blind furrows, micro catchments, trenching, and stubble mulches,

TABLE 28.3
Impact of Low and High-Performing Farm Ponds in Chittoor, Andhra Pradesh, India

Major Crops	Plot Size (acres)	Pond Size (m³)	Digging Cost (Rs)	Cropping Pattern Changed (Y/N)	Cropping Intensity (%)	Bore Well (Y/N)	No. of Filings	Additional Returns (Rs.)
			Five Least Profitable Farm Ponds					
Groundnut	1	288	25,400	N	100	N	3	633
Tomato	1	200	12,800	Y	100	N	3	1,279
Paddy	0.5	72	4,000[a]	Y	100	N	3	2,585
Tomato	1	288	25,400	N	100	N	4	3,105
Groundnut +ridge gourd	2	128	8,300	N	100	N	3	3,396
Mean	1.1	195	15,180		100		3.2	2,200
			Five Most Profitable Farm Ponds					
Tomato+tomato	3	200	12,800	Y	200	Y	6	96,993
Mango	4	288	12,800	Y	100	N	3	65,735
Groundnut +tomato	3.5	200	12,800	Y	157	Y	2	59,853
Tomato	3.5	128	8,300	N	100	Y	3	59,523
Tomato	3.5	200	12,800	Y	100	Y	6	57,917
Mean	3.5	203.2	11,900		131.4		4	68,004

Source: Modified and adapted from Srinivasarao et al. (2017b).
[a] Farmer's own pond characterized by high seepage into it.

aid in achieving water conservation. Apart from supporting crop growth and groundwater recharge, conservation methods also aid in reducing runoff and soil loss (Mishra et al., 2018).

Deep digging and subsoiling reduce the soil penetration resistance and increase profile water use in comparison to conventional tillage. In the northern dry zone of Karnataka, workability and tillage-related constraints limit *kharif* cropping in medium to deep black soils. Also infiltration rate is low leading to a high rate of runoff. In June and July after the receipt of early rains, harrowing is done to remove germinating weeds. Compartmental bunds of 0.15 m height (size of bund varies from 3 m × 3 m to 4.5 m × 4.5 m, depending on slope) are formed using a bullock-drawn bund former. Bunds are retained until sowing of *rabi* crops (second fortnight of September to first fortnight of October). (*Rabi* crops, also known as winter crops, are agricultural crops that are sown in winter and harvested in the spring in India.) Compartmental bunding provides more time for water to infiltrate into the soil and aid in soil moisture conservation. In sodic and deep black soils of northern Karnataka that have low infiltration rates and high runoff, gravel and sand mulching is practiced for soil moisture conservation (Sreenivas Reddy, 2019).

28.3.3.1.3 Tank Silt Addition

Tank silt addition is a management practice that enhances soil organic carbon stocks and also augments soil water-holding capacity, which aid in mitigating intermittent droughts. The application of tank silt provides multiple benefits, viz., enhancing soil water-holding capacity, which helps to extend moisture availability to crops, and providing nutrients that contribute to increased crop growth and yield (Figure 28.3). Water storage of reservoirs and tanks is increased by desilting of tanks (Osman et al., 2015), and the silt is used. In the event of heavy rains where tank silt has been applied to soil, water could be harvested and utilized for irrigation.

(a) (b)

FIGURE 28.3 Performance of maize with (b) and without (a) tank silt application in low fertile red soils of Rangareddy district, Telangana, India. (Srinivasarao et al., 2013a.)

28.3.3.1.4 Mulch with Manuring

Tree species, viz. *Pongamia glabra*, *Peltophorum ferrugenum*, *Delonix regia*, and *Azadirachta indica*, are grown in various regions of India, and the foliage of these tree species could be utilized as efficient mulching material and as source of organic matter in rainfed agro-ecology (Srinivasarao et al., 2020a). Mulches reduce soil moisture evaporation. Soil moisture has significant impacts on nutrient availability, transport, and nutrient cycling, and therefore, curtailing its loss will enhance crop growth and productivity (Haokip et al., 2020). Also, the foliage upon decomposition provides adequate amounts of plant nutrients, viz., N, P, and K, to the growing crop. Mulches also alter the structure and diversity of bacterial and fungal communities by decreasing soil temperatures, retaining soil moisture, suppressing weed growth, and increasing SOM (Kader et al., 2017). Mulching is an ecofriendly and economic strategy that could be adopted in rain-dependent regions to minimize soil moisture loss, reduce expenses for nutrients, and maintain soil health.

28.3.4 Low Organic Carbon and Multi-Nutrient Deficiencies

Rainfed production systems are characterized by a set of features that affect their carbon sequestration rate. Accumulation of carbon and crop productivity are severely affected due to lack of water, which is the main characteristic of rainfed drylands. Rainfall is not only scanty but also erratic in these regions, which further aggravates the problem. Hence, efficient management of available water is essential. Also, the SOC pool tends to decline exponentially with temperature (Lal, 2002a). Subsequently, rainfed drylands contain lower quantities of C (between 1% and $<$0.5%) (Lal, 2002b), which result in low productivity of croplands (Srinivasarao et al., 2020b). Erosion by water and wind prevalent in rainfed regions, coupled with limited supplementation of nutrients to soil through external inputs, has resulted in a decline in soil nutrient levels. Moreover, nutrients are removed by crops, and unless their pools are replenished, the depletion of nutrient reserves occurs leading to multi-nutrient deficiencies. Opportunities for improving land management, along with enhancing carbon sequestration and alleviating multi-nutrient deficiencies, should be given emphasis in these areas.

28.3.4.1 Management Strategies for Enhancing Soil Organic Carbon and Other Nutrients

28.3.4.1.1 Integrated Nutrient Management

Integrated nutrient management (INM) is the concept of aiming to maintain soil fertility and the supply of plant nutrients at an optimum level to sustain the desired crop productivity. Optimization of the benefits from all possible plant nutrient sources in an integrated manner is regarded as a viable

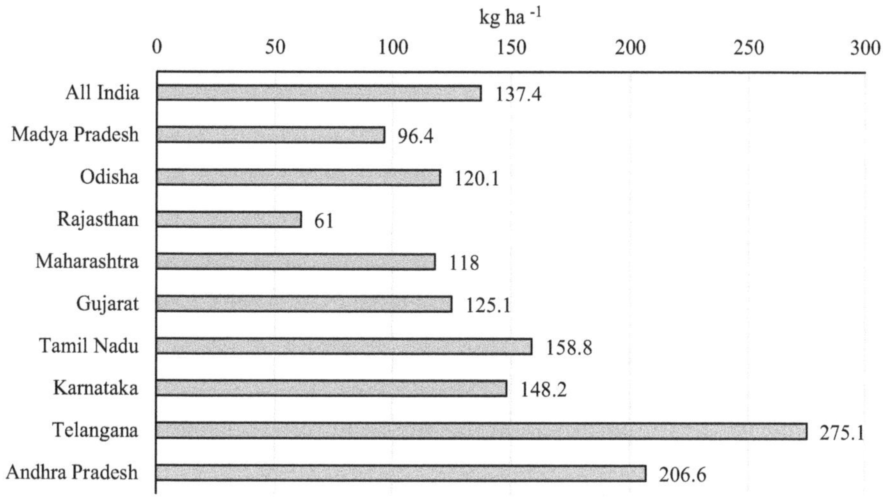

FIGURE 28.4 Fertilizer consumption in major rainfed states of India (2018–2019).

option for rainfed agro-ecologies (Srinivasarao et al., 2020a, d). INM curtails the excessive addition of synthetic fertilizers, minimizes the cost of purchasing fertilizers, and minimizes emissions of greenhouse gases. Fertilizer consumption in different rainfed states of India is shown in Figure 28.4.

In terms of total nutrient consumption, the country ranks second in the world following China. With regard to nutrient-wise consumption, it ranks second in N and P_2O_5 and fourth in K_2O. Fertilizers are essential for enhanced biomass production and crop yield, but, simultaneously, they exert negative impacts that hamper soil health and quality. The adoption of INM as a beneficial practice to enhance soil organic carbon and nutrient status has been reported by many researchers. A long-term (18 years) field trial was conducted on an Entisol in Western India involving a pearlmillet – cluster bean – castor sequence to evaluate the impacts of chemical fertilizers and manures on carbon pools in relation to crop productivity and carbon sequestration. The addition of 33.5 Mg C/ha through crop residues and farmyard manure (FYM) could not compensate for the depletion of SOC by oxidation, which led to net loss of 4.4 Mg C/ha. The integrated use [50% RDN (recommended dose of nitrogen) through chemical fertilizers+50% through FYM] proved beneficial in attaining higher yields and curtailed the rate of SOC depletion (Srinivasarao et al., 2011a). For several rainfed crops, each ton of SOC added to the soil resulted in a yield benefit of about 0.02– 0.16 t/ha (Srinivasarao et al., 2021). The combined application of NPK+FYM exerted a positive impact on increasing SOC (45.2%), mean weight diameter of water stable aggregates (13.8%), and hydraulic conductivity (95.8%), and decreasing bulk density (9.3%) and soil penetration resistance (42.6%) in comparison to the control (Bandopadhyay et al., 2010). A 43-year dataset was used in a process-based crop model (APSIM) to simulate long-term SOC dynamics under nitrogen and FYM practices for a soybean – wheat cropping system in the Vertisols of Central India. The results showed that improved N and FYM management augmented SOC sequestration. SOC concentration under different nitrogen management treatments improved with increasing N application rate. Combined application of nitrogen with FYM showed maximum increase in SOC (Mohanty et al., 2020).

28.3.4.1.2 Crop Residue Management

The availability of conventional organic manures is limited while the demand for fertilizers is increasing to meet the needs for food production. Excessive fertilizer application negatively impacts soil physical, chemical, and microbial properties, altering soil productivity. This situation calls for utilization of available, ecofriendly sources that can enhance soil health and crop productivity. Agricultural crops produce considerable amounts of leftover residues. Annual generation of crop residues accounts to around 500–550 Mt (MNRE, 2009). These residues, if left in the field, interfere with land preparation of the succeeding crop. To overcome this difficulty, crop residues are usually burnt. Burning of crop residues poses environmental concerns, viz., greenhouse gas emissions, which contribute to global warming, increased levels of particulate matter and smog, loss of biodiversity of agricultural lands, and soil fertility deterioration (Lohan et al., 2018). Incorporation of residues is an option to mitigate these effects, but it is both energy and time consuming. Temporary immobilization of nutrients, especially nitrogen, results from incorporation. The C:N ratio needs to be corrected by the application of nitrogen at the time of incorporation, which incurs additional cost and labor.

Conversion of crop residues to biochar (fine grained and carbon-rich product; it is produced by subjecting plant biomass to pyrolysis at low temperatures under little or no oxygen) could be regarded as a viable strategy to aid transformation of residues to valuable sources that have multiple benefits, viz., soil health improvement, crop productivity improvement, and climate change mitigation. The application of biochar at 20, 30, 40 t/ha resulted in a higher pH value, SOC, available P, total N, and C/N in comparison to a control (no biochar). Biochar application also proved beneficial in restraining cumulative SOC mineralization (Zhang et al., 2018). Composting of crop residues or vegetable wastes and their application to soil aids in improving physicochemical, as well as biological, properties of soil (Srinivasarao et al., 2020a). Vermicomposting is a non-thermophilic process by which organic materials are transformed by earthworms and microorganisms into rich soil amendments with high microbial activity and nutrient availability. Crop residues, farm wastes, and kitchen wastes could be subjected to vermicomposting and would be a rich source of many nutrients. Their application to crop lands could be advocated as a practice to cope with the low organic carbon of soils and multi-nutrient deficiencies (Srinivasarao et al., 2013b).

28.3.4.1.3 On-Farm Generation of Organic Matter

Strategies that contribute to on-farm generation of organic matter, viz., green leaf manuring, crop residue recycling, FYM, use of goat and sheep manure, vermicomposting, and tank silt application, at the village and farmer level should be given prominence. The organic matter contributes to enhanced soil health and crop yields, and it overcomes multi-nutrient deficiencies, mitigates climate change, and minimizes expenses incurred by buying synthetic fertilizers. *Glyricidia* is a tree species that could be easily grown in moist, dry, and acidic soils. It also grows even on marginal, degraded lands. Its cultivation on farm bunds could be regarded as a way to reduce soil erosion. The application of green leaf manure of *Glyricidia* enhanced soil quality and also proved beneficial in improving water-holding capacity. *Glyricidia* (1 t/ha) provided 21 kg N, 25 kg P, 18 kg K, 85 g Zn, 164 g Mn, 365 g Cu, 728 g Fe, as well as considerable quantities of S, Ca, Mg, B, and Mo. Decomposition of the green leaf material simultaneously produced carbon dioxide and organic acids, which enhanced the mobilization of native soil nutrients (Srinivasarao et al., 2011b; Srinivasarao, 2018).

The inclusion of pulses in cropping systems serves multiple benefits, viz., biological nitrogen fixation, organic matter addition through leaf fall and root biomass, soil erosion control, and improved microbial biomass, and all factors improve soil productive capacity. Low input requirement, coupled with capacity to withstand drought, makes cultivation of pulses a viable option in highly fragile arid and semi-arid regions. Legume cover crop incorporation supplements carbon inputs into the soil. Avoiding crop residue burning helps to attain high levels of soil organic carbon and water retention.

Livestock and poultry are major components contributing to the livelihood needs of farmers in rainfed regions. They serve as sources of organic manures (FYM, goat and sheep manure, poultry manure). They are rich in nutrients and can be used in crop nutrient management, either solely or in conjunction with fertilizers. They augment soil health, protect the soil, and enhance crop growth. Integrated farming systems encompass a judicious mix of organic and inorganic components.

28.3.5 Low Factor Productivity

Growth or decline in productivity is governed by investments in infrastructure. In rainfed regions, most farmers operate small or marginally sized farms. For the rural resource poor, low factor productivity is a major constraint that is attributed to low use efficiency of farm inputs, mainly water and nutrients. Adopting management practices to enhance the efficiency of these inputs would prove beneficial in augmenting productivity in these risk-prone areas.

28.3.5.1 Management Strategies for Increasing Factor Productivity

28.3.5.1.1 Micro-Irrigation

Drought has both direct effects (increased evapotranspiration) and indirect effects (decreased volume of groundwater extracted from wells due to declining groundwater levels from wells over time until wells get recharged), which ultimately alter crop productivity due to reduced irrigation. There is a possibility of wells going completely dry, if there is no recharge. Implementing micro-irrigation systems, viz., sprinkler and drip irrigation, might contribute to saving water, electricity, and labor, along with helping farmers to cope with the water scarcity in drought-prone regions. Fertigation, the practice of supplying fertilizers through irrigation systems, might enhance both nutrient and water-use efficiency by way of delivering water and nutrients directly into the root zone where they are most needed. This would save labor and enhance the profits of farmers. Micro-irrigation systems also aid in soil health management by avoiding crust formation, water logging, and soil erosion. According to the DFI Report (2018), a water saving of 40% could be achieved with micro-irrigation in comparison to conventional flood irrigation. Micro-irrigation also would augment crop yield and income to the extent of around 47% and 48%, respectively. In drought-prone soils, judicious application of irrigation water can increase the production of biomass (above ground and below ground), which enhances SOC concentration in soil (Sauerbeck, 2001).

28.3.5.1.2 Alternate Furrow Irrigation

Production of rainfed crops is governed by the degree to which soil moisture is conserved *in situ* or excess runoff is harvested, stored, and recycled for supplementary irrigation. Because of climate change, conservation of soil moisture is essential for sustainability of rainfed agriculture (Krishna Rao et al., 2019). This situation calls for evaluating and modifying existing irrigation practices. Better utilization of harvested rainwater needs to be carried out to augment productivity of rainfed areas. A modified form of deficit irrigation is the alternate furrow irrigation (AFI) method, wherein water is applied to every other furrow rather than to every furrow (Yao et al., 2012). The benefits through adoption of this system of irrigation include abatement of losses due to deep percolation, runoff, and evapotranspiration. There is enhancement in water use efficiency, uniform distribution of water in the entire field, and increased crop yield (Liu et al., 2013).

28.3.5.1.3 Improving Nutrient-Use Efficiency

Improving nutrient-use efficiency (NUE), along with improving water use efficiency of crops, is a major objective of agricultural scientists in rain-dependent regions, because these areas are major contributors in providing food to the nation. Adopting efficient management strategies, which would aid in augmenting nutrient-use efficiency, would maximize yield in water scarce environments while curtailing nutrient losses. Despite increasing consumption of fertilizers, the nutrient-use efficiency of major nutrients, secondary nutrients, and micronutrients is low. Nitrogen is the most important nutrient. Hegde et al. (2007) showed that enhancing its use efficiency through physical manipulation of the size of the granule, coating of granules with films, amendments to avoid rapid dissolution, or amendments to prevent bacterial mediated denitrification was not economically viable.

Inclusion of legumes in cropping systems contributes to biological N_2 fixation, which increases nitrogen use efficiency. Legumes aid in maintenance of soil organic matter, which, in turn, helps to enhance soil fertility and create a better soil physical and microbial environment.

Management options like utilization of on-farm generated organic resources such as crop residues, green leaf manures, farmyard manure, vegetable wastes, and composts if coupled with inorganic fertilizers, would help in enhancing nutrient-use efficiency. On-farm generated organic resources would minimize the use of inorganic fertilizers, enhance soil health, and mitigate the impacts of climate change. The inclusion of biofertilizers is a viable option to reduce the use of chemical fertilizers along with enhancing nutrient-use efficiency (Srinivasarao et al., 2020c).

Trials conducted for four consecutive *kharif* seasons on Alfisols in Hyderabad revealed that the application of FYM aided in enhancing nutrient-use efficiency in a senna (*Cassia angustifolia*) crop (Pratibha et al., 2010). Taking into consideration the widespread, low levels of SOC in rainfed regions of the country, supplementation of these fertility depleted soils with organic sources of nutrients is important to enhance nutrient-use efficiency. Converting crop residues into biochar through pyrolysis has proved advantageous in enhancing yield, soil organic carbon, and nutrient availability (Srinivasarao et al., 2013c). Nutrient-use efficiency could be augmented by split application of fertilizers, use of slow release and coated fertilizers, and site-specific nutrient management (FAI, 2017). Dwivedi et al. (2017) showed that, on Typic Haplustepts of the Western Indo-Gangetic Plain (IGP), the economic optimum rates of N and P for wheat in a pigeon pea-wheat system were lower in permanent raised beds in comparison to flat beds. This result could be attributed to increased N and P supply and better nutrient-use efficiency.

28.3.6 Small Holder Farmers

Farmers dwelling in rainfed regions are small holders and are less capable, in terms of financial, physical, human, and social capital, to acclimatize to climate change. Rainfed agriculture is practiced on degraded and fragile lands, which are both moisture and nutrient deficient and highly susceptible to erosion. The topographic conditions do not support good plant growth, because of shallow profile depths with low organic matter and poor water-holding capacity. These constraints impede technology implementation. Resource poor conditions of peasants in these regions restrict adoption of land leveling practices that incur high expenditure. In order to provide financial support and livelihood opportunities, various organization must be involved.

28.3.6.1 Management Strategies to Support Small Holder Farmers

28.3.6.1.1 Farmer Producing Organizations

A farmer producing organization (FPO) is a registered body formed by a group of farm producers, who are shareholders in the organization. Presently, there are 5000 FPOs functional in the country with 910 under the affiliation of Small Farmer's Agri-Business Consortium (SFAC), while 3000 are with the National Bank for Agriculture and Rural Development (NABARD). These organizations ensure that farmers get remunerative prices for their produce. FPOs act as aggregators of farmers, and they have higher bargaining power compared to individual farmers. They can trade directly with corporations, large processors, and millers, eliminating intermediaries, and they can pave the way for small and marginal farmers to get maximum value for their produce.

28.3.6.1.2 Custom Hiring Centers

Mechanization of agricultural field operations will help in attaining high farm productivity and enhancing farmers' income. Numerous constraints impede small farm mechanization. The major factors are non-adoption of improved technologies, a poor resource base, and lack of support services (Srinivas et al., 2017; Modi et al., 2020). An important strategy to meet mechanization requirements of farmers, especially in the small and marginal sectors, is the establishment of Custom Hiring Centers (CHCs) that have expensive farm implements. The lead role has been taken by the Central Research Institute for Dryland Agriculture (CRIDA) to provide farm implements needed by Indian farmers, and 151 CHCs have been established to provide farm implements to farmers in all ecological regions of the country. They have been established under the ICAR flagship platform, National Initiative on Climate Resilient Agriculture (NICRA). Also, various state governmental agencies and NGOs are operating CHCs in different regions of the country. CHCs in villages are instrumental in overcoming the perils associated with climate change. Under NICRA, CHC management through a Village Climate Risk Management Committee (VCRMC), which involves rural communities as decision makers in field operations, has proved to be a promising model.

28.3.6.1.3 Processing, Value Addition and Marketing

The perishable nature of produce has created a substantial gap between its production and availability. Post-harvest losses and a low level of processing of agricultural commodities account as major factors that slash a farmer's profits. Increasing shelf life of produce through processing and value addition into various products, viz., candy, jam, halwa (an Indian dessert that has a pudding-like texture), chips, papads (papadum in American English; it is a thin, crisp bread or wafer of India), and ready-to-make mixes, would aid in overcoming losses and provide an opportunity for increasing the income of farmers. Promotion and commercialization of processing, as well as value-addition technologies, are needed to enhance year-round availability of food products for consumers. They would serve as an income-enhancement source to rural communities engaged in farming. The absence of organized marketing strategies among growers leads to wastage of a major quantity of produce, which limits profits. Efficient marketing infrastructure and market information provide positive benefits to farmers. A knowledge of up-to-date information on prices and market-related factors empowers farmers in decision making and negotiating with buyers, and it enables distribution of products between markets (rural areas to urban areas).

28.4 CONCLUSION

Rainfed drylands are important as they contribute a major share to the food bowl of India. The potentiality of these areas remains untapped due to various reasons. The major limitations are high vulnerability to climatic aberrations, resource-poor inhabitants with a lack of technical knowledge, lack of focus on soil health management, poor infrastructure, and lack of market linkages. Sustaining crop productivity to achieve the food requirements of a burgeoning population continues to be an inevitable challenge. Hence, there is a direct need to protect and enhance the yield levels in these areas to attain food and nutritional security. Soil, which is the major component governing crop growth and productivity, should be protected and nourished in order to elevate food, fodder, fuel, and fiber productivity. The adoption of efficient strategies, viz., integrated nutrient management, crop residue recycling, addition of organic sources (compost, green leaf manures, bulky organic manures, vermicompost), mulching with manuring, and tank silt addition would enhance the nutrient status of impoverished soils. In rainfed areas, moisture is limiting, and it is the key factor governing the crop yield and soil health. Efficient utilization of water holds the key to sustain yields and soil health. Adoption of suitable moisture conservation practices would aid in the efficient water harvesting and water utilization during the periods of water scarcity. Enhancing nutrient and water-use efficiency, using micro-irrigation systems, and judicious management of available resources are interventions that would be beneficial to enhance the quantity of crops harvested and income security of inhabitants of arid and semi-arid regions. Post-harvest management of produce, efficient marketing channels, and collaboration with various organizations, viz., VCRMCs (Village Climate Risk Management Committees), FPOs (Farmer Producing Organizations), and CHCs (Custom Hiring Centers), would benefit the peasants so that they can overcome the perils associated with climate change. These organizations would enable them to attend to farm operations at the right time, which, in-turn, would enhance net profits and their livelihood status.

28.5 WAY FORWARD

- Creating awareness among the farmers regarding the significance of soil health and its impact on crop productivity and how a healthy soil can mitigate climate change in rainfed areas.

- Strengthening innovative farmers to experiment with novel technologies developed in their own farms, so that the benefits could be visualized and evaluated by them and later extended to their neighbors.
- Designing better-structured interventions that minimize soil degradation, protect the top soil, and use available resources efficiently. They eventually would benefit resource-poor farmers in terms of enhanced productivity and monetary returns.
- Developing infrastructure to educate farmers regarding soil testing, balanced fertilization, site-specific nutrient management, efficient residue recycling, post-harvest management, and marketing strategies.
- Developing strong linkages between farmers, soil scientists, and extension workers, which would narrow the gap between the scientific community and farmers. They would eventually enable farmers to gain knowledge about novel technologies, policies, and programs planned and initiated with a view to enhance productivity.
- Mainstreaming promising climate-resilient technologies into developmental programs to improve the adaptive capacities of communities located in regions highly vulnerable to climate aberrations.

REFERENCES

Agrawal, R.P., B.S. Jhorar., J.S. Dhankar, and M. Raj. 1987. Compaction of sandy soils for irrigation management. *Irrigation Science*. 8: 227–232.

Bandyopadhyay, K.K., A.K. Misra, P.K. Ghosh, and K.M. Hati. 2010. Effect of integrated use of farmyard manure and chemical fertilizers on soil physical properties and productivity of soybean. *Soil Tillage Research*. 110: 115–125.

DFI Report. 2018. Doubling Farmer's Income Report. Ministry of Agriculture and Farmers' Welfare. Government of India.

Dwivedi, B.S., V.K. Singh., K. Shekhawat, M.C. Meena, and A. De. 2017. Enhancing use efficiency of phosphorus and potassium under different cropping systems of India. *Indian Journal of Fertilisers*. 13(8): 20–41.

FAI. 2017. *Fertilizer Use and Environmental Quality*. Fertilizer Association of India, New Delhi.

Farage, P.K., J. Ardo, L. Olsson., E.A. Rienzi, A.S. Ball, and J.N. Pretty. 2007. The potential for soil carbon sequestration in three tropical dryland farming systems of Africa, Latin America: A modelling approach. *Soil Tillage Research*. 94: 457–472.

Fowler, R. and J. Rockstrom. 2000. Conservation Tillage for Sustainable Agriculture: An Agrarian Revolution Gathers Momentum in Africa. Keynote address, International Soil Tillage Research Organization (ISTRO), Fort Worth, USA.

Haokip, I.C., R.P. Premalatha, M. Homeshwari Devi, C. Jangir, B.H. Sunil, S. Kundu, K. Surekha, and Ch. Srinivasarao. 2020. Climate change adaptation and mitigation through soil management. In: Ch. Srinivasarao *et al.,* (Eds). *Climate Change and Indian Agriculture - Challenges and Adaptation Strategies*. ICAR-National Academy of Agricultural Research Management, Hyderabad, Telangana, India, pp. 105–130.

Hegde, D.M., S.N.S. Babu, A.A. Qureshi, and I.Y.L.N. Murthy. 2007. Enhancing nutrient use efficiency in crop production – A review. *Indian Journal of Agronomy*. 52(4): 261–274.

Indoria, A.K., K.L. Sharma, K. Sammi Reddy, and Ch. Srinivasarao. 2016. Role of soil physical properties in soil health management and crop productivity in rainfed systems – II. Management technologies and crop productivity. *Current Science*. 110(3): 320–328.

Kader, M.A., M. Senge, M.A. Mojid, and K. Ito. 2017. Recent advances in mulching materials and methods for modifying soil environment. *Soil Tillage Research*. 168: 155–166 doi: 10.1016/j.still.2017.01.001.

Krishna Rao, B., P.R. Bhatnagar, T. Kamble, R.S. Kurothe, P.K. Mishra, R. Sharma, A. Kumar, and V.C. Pande. 2019. Surge flow alternate furrow irrigation for enhancing water productivity in semi-arid regions. *Indian Journal of Agricultural Sciences*. 89(12): 1999–2002.

Kundu, S., Ch. Srinivasarao, R.B. Mallick, T. Satyanarayana, R.P. Naik, A. Johnston, and B. Venkateswarlu. 2013. Conservation agriculture in maize (*Zea mays* L.) horsegram (*Macrotyloma uniflorum* L.) system in rainfed Alfisols for C sequestration, climate change mitigation. *Journal of Agrometeorology*. 15: 144–149.

Lal, R., 2002a. Soil carbon dynamics in crop land and rangeland. *Environmental Pollution*. 116: 353–362.

Lal, R., 2002b. Carbon sequestration in dryland ecosystems of West Asia and North Africa. *Land Degradation and Development*. 13: 45–59.
Liu, L.J., T.T. Chen, Z.Q. Wang, H. Zhang, J.C. Yang, and J.H. Zhang. 2013. Combination of site-specific nitrogen management and alternate wetting and drying irrigation increases grain yield and nitrogen and water use efficiency in super rice. *Field Crops Research*. 154: 226–235.
Lohan, S.K., H.S. Jat, A.K. Yadav, H.S. Sidhu, M.L. Jat, M. Choudhary, P. Jyotsna Kiran, and P.C. Sharma. 2018. Burning issues of paddy residue management in north-west states of India. *Renewable and Sustainable Energy Review*. 81: 693–706.
Mahendra Dev, S. 2011. Small farmers in India: challenges and opportunities. In: *Paper Presented at "Emerging Economies Research Dialogue"*, Beijing, China, November 14e15, 2011 Organized by ICRIER. Accessed online at: http://www.igidr.ac.in/pdf/publication/WP-2012-014.pdf.
Majumdar, S.P. 1994. Soil physical constraints and their management in Rajasthan. In *Consolidated Report 1967–1992 AICRP on Improvement of Soil Physical Condition to Increase Agriculture Production of Problematic Area*. Rajasthan Agricultural University, Bikaner.
Mishra, P.K, S. Man, and G. Kumar. 2018. Water management and conservation innovations for doubling farmers' income. *Bulletin of the Indian Society of Soil Science*. 32: 23–47.
MNRE, Ministry of New and Renewable Energy Resources. 2009. Government of India, New Delhi.
Modi, U.R., K. Manjunatha, P.V. Gautam, T. Nageshkumar, R. Sanodiya, V. Chaudhary, G.R.K. Murthy, I. Srinivas, and Ch. Srinivasarao. 2020. Climate-smart technology-based farm mechanization for enhanced input use efficiency. In: Ch. Srinivasarao *et al.,* (Eds). *Climate Change and Indian Agriculture - Challenges and Adaptation Strategies*. ICAR-National Academy of Agricultural Research Management, Hyderabad, Telangana, India. pp. 325–358.
Mohanty, M., K.K. Bandyopadhyay, D.K. Painuli, P.K. Ghosh, A.K. Misra, and K.M. Hati. 2007. Water transmission characteristics of a Vertisol and water use efficiency of rainfed soybean (*Glycine max* (L.) Merr.) under subsoiling and manuring. *Soil Tillage and Research*. 93: 420–428.
Mohanty, A., K.N. Mishra., P.K. Roul, S.N. Dash, and K.K. Panigrahi. 2015. Effects of conservation agriculture production system (CAPS) on soil organic carbon, base exchange characteristics and nutrient distribution in a tropical rainfed agro-ecosystem. *International Journal of Plant, Animal and Environmental Sciences*. 5: 310–314.
Mohanty, M., N.K. Sinha, J. Somasundaram, S.S. McDermi, A.K. Patra, M. Singh, A.K. Dwivedi, K. Sammi Reddy, Ch. Srinivasrao, M. Prabhakar, K.M. Hatia, P. Jha, R.K. Singh, R.S. Chaudhary, S.N. Kumar, P. Tripathi, R.C. Dalal, D.S. Gaydong, and S.K. Chaudhari. 2020. Soil carbon sequestration potential in a vertisol in central India – results from a 43 year long term experiment and APSIM modelling. *Agricultural Systems*. 184: 1–11.
Mupangwa, W., and C. Thierfelder. 2014. Intensification of conservation agriculture systems for increased livestock feed and maize production in Zimbabwe. *International Journal of Agricultural Sustainability*. 12(4): 425–439.
Mupangwa, W., M. Mutenje, C. Thierfelder, and I. Nyagumbo. 2017. Are conservation agriculture (CA) systems productive and profitable options for smallholder farmers in different agro-ecoregions of Zimbabwe? *Renewable Agricultural Food Systems* 32(1): 87–103.
Murthy, R.S., J.C. Bhattacharjee, R.J. Landey, and R.M. Pofali. 1982. Distribution characteristics and classification of vertisols. In *Vertisols and Rice Soils of the Tropics. Twelfth International Congress of Soil Science*, New Delhi, India. pp. 3–22.
Muthamilselvan, M., R. Manian, and K. Kathirvel. 2006. In-situ Moisture Conservation Techniques in Dry farming-A Review, *Agricultural Review*. 24(1): 67–72.
National Rainfed Area Authority (NRAA). 2013. Contingency and compensatory agriculture plans for droughts and floods in India-2012. Position paper No.6. National Rainfed Area Authority, New Delhi, India, 87.
Osman, M. 2008. Recycling of tank silt for improving soil and water productivity in rainfed areas. In *Lecture Notes for the Winter School on Technological Advances in Conservation of Natural Resources in Rainfed Agriculture*. Central Research Institute for Dryland Agriculture (CRIDA), Hyderabad.
Osman, M., G.R. Chary, P.K. Mishra, M.A. Shankar, Y. Padmalata, R. Murali, K. Nimmaiah, K. Khasimpeera, A. Balamathi, and S. Haffis. 2015. Enhancing rainwater productivity and economic viability of rainfed crops through tank silt application. *Indian Journal of Dryland Agricultural Research and Development*. 30(2): 17–23.
Oswal, M.C. 2001. Characterizing physical environment of rainfed lands and their management for sustainable production. *Journal of Agricultural Physics*. 1: 71–75.
Painuli, D.K., and R.P. Yadav. 1988. Tillage requirements of Indian soils. In: Singh G.B. and Sharma B.R. (eds.) *50 Years of Natural Resource Management Research*. Indian Council of Agricultural Research, New Delhi. pp. 245–262.

Phogat, V.K., and S.S. Dahiya. 2002. Alleviating soil physical constraints for sustainable crop production through tillage, soil amendment and crop residue management. In: *12th International Soil Conservation Organization Conference*, Beijing.

Pratibha, G., G.R. Korwar, and S.K. Yadav. 2010. Productivity, quality, nutrient use efficiency and economics of Senna (*Cassia augustifolia*) as influenced by FYM. *Indian Journal of Agronomy*. 55(1): 79–83.

Puigdefabregas, J. 2005. The role of vegetation patterns in structuring runoff and sediment fluxes in drylands. *Earth Surface Processes and Landforms*. 30(2): 133–147.

Rakesh, S., D. Sarkar, A.K. Sinha, P.C. Abhilash, and A. Rakshit. 2019. Climate change and agricultural policy options: Indian story. *Climate Change and Environmental Sustainability*. 7(2): 208–11.

Rakesh S., A.K. Sinha, and P. Mukhopadhyay. 2020a. Vertical distribution of TOC, TN and other important soil attributes and their relationship in Alfisol and Entisol of West Bengal. *International Journal of Environment and Climate Change*. 10(1): 62–73.

Rakesh S., D. Sarkar, A. Sankar, A.K. Sinha, P. Mukhopadhyay, and A. Rakshit. 2020b. Chapter-6: Protocols for determination and evaluation of organic carbon pools in soils developed under contrasting pedogenic processes and subjected to varying management situations. In: *Soil Analysis: Recent Trends and Applications*. Springer Nature Singapore Pvt Ltd. Singapore. ISBN: 978-981-15-2039-6(eBook).

Rakesh S., A.K. Sinha, D. Sarkar, S. Sahoo, and D. Roy. 2020c. Key soil attributes as influenced by cropping systems in an Entisol of West Bengal, India. *Climate Change and Environmental Sustainability*. 8(2): 226–232.

Sauerbeck, D.R. 2001. CO_2 emissions and C sequestration by agriculture perspectives and limitations. *Nutrient Cycling in Agroecosystems*. 60: 253–266.

Sharma, K.L. 2019. Role of conservation agriculture in improving soil health and adaptation and mitigation of climate change. In: Vanaja M., Bal S.K., Nagasree K., Narsimlu B., Raju B.M.K., Reddy K.S., Prasad J.V.N.S., Ravindra Chary G., Goldi T., Mahantesh S., and Krishna Rao B. (eds.) *Climate Smart Agriculture, Feed The Future* - India Triangular Training (FTF-ITT), 20th August-3rd September, Indian Council of Agricultural Research (ICAR) - Central Research Institute for Dryland Agriculture, Hyderabad, India.

Singh, V.N., and B.B. Mishra. 1995. Land levelling of some Alfisols in toposequence occurring in Indo-Gangtic plains of Bihar. *Journal of Indian Society of Soil Science* 42: 626–632.

Six, J., E.T. Elliott., and K. Paustian. 2000. Soil macroaggregate turnover and microaggregate formation: A mechanism for C sequestration under no-tillage agriculture. *Soil Biology and Biochemistry*. 32: 2099–2103.

Somasundaram, J., N.K. Sinha, R.C. Dalal, R. Lal, M. Mohanty, A.K. Naorem, K.M. Hati et al. 2020. No-till farming and conservation agriculture in South Asia–issues, challenges, prospects and benefits. *Critical Reviews in Plant Sciences* 39(3): 236–279.

Sreenivas Reddy, K. 2019. Rain water management technologies for climate resilience in SAT regions of peninsular India. In: Vanaja M., Bal S.K., Nagasree K., Narsimlu Boini, Raju B.M.K., Reddy K.S., Prasad J.V.N.S., Ravindra Chary G., Goldi T., Mahantesh S., and Krishna Rao B. (eds.) *Climate Smart Agriculture, Feed The Future* - India Triangular Training (FTF-ITT), 20th August-3rd September, Indian Council of Agricultural Research (ICAR) - Central Research Institute for Dryland Agriculture, Hyderabad, India.

Srinivas, I., Ch. Srinivasarao, K. Sammi Reddy, B. Sanjeeva Reddy, R.V. Adake, A.S. Dhimate, M. Osman, J.V.N.S. Prasad, and C.A. Rama Rao. 2017. Upscaling farm machinery Custom hiring centres in India: A Policy Paper, Central Research Institute for Dryland Agriculture (CRIDA) Policy Paper No. 1/2017.

Srinivasarao, Ch., and K.P.R. Vittal. 2007. Emerging nutrient deficiencies in different soil types under rainfed production systems of India. *Indian Journal of Fertilizers* 3: 37–46.

Srinivasarao, Ch., K.P.R. Vittal, B. Venkateswarlu, S.P. Wani, K.L. Sahrawat, S. Marimuthu, and S. Kundu. 2009. Carbon stocks in different soil types under diverse rainfed production systems in tropical India. *Communications in Soil Science and Plant Analysis*. 40(15–16): 2338–2356.

Srinivasarao, Ch., B. Venkateswarlu, R. Lal, A.K. Singh, S. Kundu, K.P.R. Vittal, J.J. Patel, and M.M. Patel. 2011a. Long-term manuring and fertilizer effects on depletion of soil organic carbon stocks under pearl millet-cluster bean-castor rotation in western India. *Land Degradation and Development*. 25: 173–183. doi:10.1002/ldr.1158.

Srinivasarao, Ch., B. Venkateswarlu, M. Dinesh Babu, S.P. Wani, S. Dixit, K.L. Sahrawat, and S. Kundu. 2011b. *Soil Health Improvement with Gliricidia Green Leaf Manuring in Rainfed Agriculture, On Farm Experiences*. Central Research Institute for Dryland Agriculture, Hyderabad, Andhra Pradesh. pp. 1–16.

Srinivasarao, Ch., B. Venkateswarlu, R. Lal, A.K. Singh., and S. Kundu. 2013a. Sustainable management of soils of dryland ecosystems of India for enhancing agronomic productivity and sequestering carbon. *Advances in Agronomy*. 121: 253–329. doi:10.1016/b978-0-12-407685-3.00005-0.

Srinivasarao, Ch., B. Venkateswarlu, R. Veeraiah, S. Rammohan, V.S. Jakkula, S. Dixit, R.V. Rammohan, and B. Shivarudrappa. 2013b. *Vermicomposting for Efficient Crop Residue Recycling, Soil Health Improvement and Imparting Climate Resilience: Experiences from Rainfed Tribal Regions*. Central Research Institute for Dryland Agriculture, Hyderabad, Andhra Pradesh. 40 p.

Srinivasarao, Ch., K.A. Gopinath, G. Venkatesh, A.K. Dubey, H. Wakudkar, T.J. Purakayastha, H. Pathak, P. Jha, B.L. Lakaria, D.J. Rajkhowa, S. Mandal, S. Jeyaraman, B. Venkateswarlu, and A.K. Sikka. 2013c. *Use of Biochar for Soil Health Management and Greenhouse Gas Mitigation in India: Potential and Constraints.* Central Research Institute for Dryland Agriculture, Hyderabad, Andhra Pradesh. 51p.

Srinivasarao, Ch., R. Lal, J.V.N.S. Prasad, K.A. Gopinath, R. Singh., V.S. Jakkula, and S.M. Virmani. 2015. Potential and challenges of rainfed farming in India. *Advances in Agronomy.* 133: 113–181. doi: 10.1016/bs.agron.2015.05.004.

Srinivasarao, Ch., A.K. Indoria, and K.L. Sharma. 2017a. Effective management practices for improving soil organic matter for increasing crop productivity in rainfed agro ecology of India. *Current Science.* 112(7): 1497–1504.

Srinivasarao, Ch., R. Rejani, C.A. Rama Rao, K.V. Rao, M. Osman, K. Srinivas Reddy, M. Kumar, and P. Kumar. 2017b. Farm ponds for climate resilient rainfed agriculture. *Current Science.* 112(3): 471–477.

Srinivasarao, Ch. 2018. Soil carbon sequestration: Research, Technology and Policy Implementation Needs in India. *Journal of the Indian Society of Soil Science.* 66: 526–543.

Srinivasarao, Ch., C. Subha Lakshmi, and S. Kundu. 2019a. Critical role of nutrient management in climate adaptive dryland farming systems in India. *Indian Journal of Fertilizers.* 15(11): 1220–1233.

Srinivasarao, Ch. 2019b. Research, technology and policy implementation synergy for resilient dryland production systems in India. In: *Proceedings of 13th International Conference on Development of Drylands. Converting Dryland Areas from Grey to Green,* February 11–14, 2019, Indian Council of Agricultural Research (ICAR) – Central Arid Zone Research Institute, Jodhpur, India.

Srinivasarao, Ch., C. Subha Lakshmi, S. Kundu, G. Ranjith Kumar, R. Manasa, and S. Rakesh. 2020a. Integrated nutrient management strategies for rainfed agro-ecosystems of India. *Indian Journal of Fertilizers.* 16(4): 344–361.

Srinivasarao Ch., S. Kundu, B.S. Yashavanth, S. Rakesh, K.N. Akbari, G.S. Sutaria, V.D. Vora, D.S. Hirpara, K.A. Gopinath, G.R. Chary, J.V.N.S. Prasad, N.S. Bolan, and B. Venkateswarlu. 2020b. Influence of 16 years of fertilization and manuring on carbon sequestration and agronomic productivity of groundnut in vertisol of semi-arid tropics of Western India, *Carbon Management.* 12(1): 13–24.

Srinivasarao, Ch., S. Kundu, C. Subha Lakshmi, M. Vijay Sankar Babu, V.V. Gabhane, P.K. Sarma, S. Ayyappa, K.C. Nataraja, and H. Arunakumari. 2020c. Manures versus Fertilizers in rainfed dryland production systems of India. In: Rattan Lal (ed.) *Soil and Fertilizers – Managing the Environmental Footprint. Advances in Soil Science.* CRC Press, Taylor and Francis Group, Boca Raton, FL. pp. 131–168. doi:10.1201/9780429471049-6.

Srinivasarao, Ch., M. Ramesh Naik, C. Subha Lakshmi, G. Ranjith Kumar, R. Manasa, S. Rakesh, S. Kundu, and J.V.N.S. Prasad. 2020d. Economic and environmental benefits of integrated nutrient management in Indian agriculture. *Indian Journal of Fertilizers.* 16(11): 1124–1137.

Srinivasarao, Ch., S. Rakesh, G. Ranjith Kumar, R. Manasa, G. Somashekar, C. Subha Lakshmi, and S. Kundu. 2021. Soil degradation challenges for sustainable agriculture in tropical India. *Current Science.* 120(3): 492–500.

Sujith, R., D.D. Breshears, T.E. Huxman, and P.D. Odorico. 2010. Land degradation in drylands: Interactions among hydrologic Aeolian erosion and vegetation dynamics. *Geomorphology.* 116: 236–245.

Venkateswarlu, B., Ch. Srinivasarao, G. Ramesh, S. Venkateswarlu, and J.C. Katyal. 2007. Effects of long-term legume cover crop incorporation on soil organic carbon, microbial biomass, nutrient build-up and grain yields of sorghum/sunflower under rain-fed conditions. *Soil Use and Management.* 231: 100–107.

Wall, P.C. 2007. Tailoring conservation agriculture to the needs of small farmers in developing countries: An analysis of issues. *Journal of Crop Improvement.* 19(1/2): 137–155.

Wall, P.C., C. Thierfelder, A.R. Ngwira, B. Govaerts, I. Nyagumbo, and F. Baudron. 2014. Conservation agriculture in Eastern and Southern Africa. In: Jat RA, Sahrawat KL, Kassam AH (eds.) *Conservation Agriculture: Global Prospects and Challenges.* CABI, Wallingford, UK. pp. 263–292.

Yadav, B.L., and S.P. Majumdar. 1990. Studies on effect of different tillage systems on soil physical properties of highly permeable sandy soil, transport of nutrients and yield of Bajra. In *55th Annual Convention of Indian Society of Soil Science*, Udaipur, 22–25 December. T-5-111.

Yao, F.X., J.L. Huang., K.H. Cui, L.X. Nie, J. Xiang, X.J. Liu, W. Wu, M.X. Chen, and S.B. Peng. 2012. Agronomic performance of high-yielding rice variety grown under alternate wetting and drying irrigation. *Field Crops Research.* 126: 16–22.

Zhang, X., C. Chen, X. Chen, P. Tao, Z. Jin, and Z. Han. 2018. Persistent effects of biochar on soil organic carbon mineralizationand resistant carbon pool in upland red soil, China. *Environmental Earth Sciences.* 77: 177. doi:10.1007/s12665-018-7359-9.

29 Soil Salinity and Sodicity in India – Adaptation and Mitigation Options

Jawahar Durairaj[1] *and T. Sherene Jenita Rajammal*[2]
[1]Department of Soil Science and Agricultural Chemistry, Tamil Nadu Agricultural University, Coimbatore, India
[2]Department of Soil Science and Agricultural Chemistry, Horticultural College and Research Institute for Women, Tamil Nadu Agricultural University, Tiruchirappalli-620009, Tamil Nadu, India

CONTENTS

DOI: 10.1201/9781003093565-29

29.1 INTRODUCTION

Soil salinity is a major global issue owing to its adverse impact on agricultural productivity and sustainability. South Asia, including India, has about 52 million ha that are salt-affected. Most of the total global area (~85%) is only slightly to moderately affected by high salt concentrations, while the remaining 15% suffers from severe to extreme limitations for crop cultivation (Wicke et al., 2011).

In India, the area under salt-affected soils is about 6.73 million ha with the States of Gujarat (2.23 million ha), Uttar Pradesh (1.37 m ha), Maharashtra (0.61 m ha), West Bengal (0.44 m ha), and Rajasthan (0.38 million ha) together accounting for almost 75% of saline and sodic soils in the country (Kaledhonkar et al., 2019). In most of the salt-affected environments, the prevalence of poor quality (saline and sodic) waters is also noted. The States of Rajasthan, Haryana, and Punjab, lying in the North-Western arid part of the country, greatly suffer from the problem of marginal quality waters. According to estimates, the present area under salt-affected soils (6.73 million ha) in these states will almost treble to 20 million ha by 2050.

Salinity problems occur under all climatic conditions and can result from both natural and human-induced actions. Generally speaking, saline soils occur in arid and semi-arid regions, where rainfall is insufficient to meet the water requirements of the crops and leach mineral salts out of the root-zone. Soil salinity undermines the resource base by decreasing soil quality and can occur due to natural causes or from misuse and mismanagement, jeopardizing the integrity of soil's self-regulatory capacity. A good example is from Mesopotamia, where an early civilization first flourished and then failed due to human-induced salinization. A publication *Salt and Silt in Ancient Mesopotamian Agriculture* (Jacobsen and Adams, 1958) highlights the history of salinization in Mesopotamia. There are reports revealing that many societies based on irrigated agriculture have failed, for example, as noted, Mesopotamia, and the Viru Valley of Peru. Flooding, over-irrigation, seepage, silting, and a rising water-table have been reported to be the main causes of soil salinization (Sharma, 2015).

Soils affected by salinity and sodicity are not confined to arid and semi-arid regions, where rainfall is insufficient to leach salts from the soil. Saline and sodic soils have been recorded in a wide range of environments under many different hydrological and physiographic conditions. Such a wide distribution leads to infer that no single adaptation or mitigation technique will be applicable to all land areas. In the Tiruchirappalli District of Tamil Nadu State in India, out of the total geographical area of 4,40,412 ha, 7,362 ha are slightly salt affected and 10,729 ha are moderately salt affected. Diagnostics-based, site-specific recommendations are viable options to reclaim salt-affected soils. Integrated system of soil reclamation and management techniques are generally grouped into four mitigation and adaptation approaches to deal with salt-affected soils. The four approaches are (i) hydrological, (ii) physical, (iii) chemical, and (iv) biological.

29.2 MITIGATION AND ADAPTATION OPTIONS

Mitigation and adaptation are terms commonly used in the context of climate change, and many scientists believe that mitigation is primarily concerned with emission of greenhouse gases, while adaptation deals with water and agriculture. However, mitigation and adaptation also apply to how mankind must deal with salt-affected soils. In this chapter, we have defined both the terms in the context of salt-affected soils and their reclamation and management. Adaptation allows the continued use of salt-affected soils by making adjustments in response to the degree by which salinity and sodicity have affected the soil. In contrast, mitigation refers to the technologies that are adopted to stop further occurrence of salinization.

29.3 DIAGNOSTICS OF SOIL SALINITY

Soil salinity in agricultural fields is increasing worldwide, mainly due to poor farm management practices and the increasing demand for intensification of agriculture for the short-term benefits

(a) (b) (c)

FIGURE 29.1 Soil salinity development in agricultural and coastal fields. (a) Salinity in a furrow irrigated field, (b) salinity in a sprinkler irrigated field, and (c) salinity due to sea water intrusion in a coastal district. (Adopted from Shahid and Rahman (2011) and Shahid et al. (2011).)

of increased food production. Saline soils are defined as soils that have pH's usually <8.5, ECe ≥ 4 dS/m at 25°C, and an exchangeable sodium percentage (ESP) < 15 (USSL Staff, 1954). It is important to understand the salinity hazard, both spatially and temporally, at the regional and national farm levels.

29.4 VISUAL INDICATORS OF SOIL SALINITY

Once soil salinity develops in irrigated agricultural fields, its effects on soil properties and plant growth are obviously visible (Figure 29.1). Visual indicators of soil salinization (Shahid and Rahman, 2011) include:

- A white salt crust
- Fluffiness of the soil surface
- Salt stains on the dry soil surface
- Reduced or no seed germination
- Patchy crop establishment
- Reduced plant vigor
- Foliage damage – leaf burn
- Marked changes in leaf color and shape
- Increased occurrence of naturally growing halophytes, indicator plants
- Dead or dying trees
- Affected area increased after a rainfall
- Waterlogging

29.5 SOIL SODICITY AND ITS DIAGNOSTICS

Sodicity is a measure of sodium ions in soil water, in relation to calcium and magnesium ions. Sodic soils exhibit an ESP >15 and show an ECe < 4 dS/m. The pH generally ranges between 8.5 and 10 and may be even as high as 11. The low ECe and high ESP tend to de-flocculate soil aggregates and hence, lower their permeability to water.

29.6 VISUAL INDICATORS OF SOIL SODICITY

Soil sodicity can be predicted visually in the field in the following ways:

- Poorer vegetative growth than normal, with only a few plants surviving, or with many stunted plants or trees

FIGURE 29.2 Typical sodic soil view under a rice crop.

- Variable heights of the plants
- Poor penetration of rain water – surface ponding
- Raindrop splash action – surface sealing and crusting (hard setting)
- Cloudy or turbid water in puddles
- Shallow rooting depth of plants
- Black color of the soil due to the formation of a Na-humic substance complex
- High force required for tillage (especially in fine-textured soils)
- Difficulty in getting soil saturation extracts in laboratory due to a filter blockage with dispersed clay (Figure 29.2)

29.7 CLASSIFICATION OF SALT AFFECTED SOILS

The term salt-affected soil is used commonly to include saline, saline-sodic, and sodic soils (USSL Staff, 1954), as summarized in Table 29.1.

TABLE 29.1
Classification of Salt Affected Soils (USSL Staff, 1954)

Soil Class	ECe (dS/m)	ESP	pH
Saline	>4	<15	<8.5
Saline-sodic	>4	>15	>8.5
Sodic	<4	>15	>8.5

29.8 INTEGRATED SOIL RECLAMATION PROGRAM (ISRP)

Salt-affected soils are distributed across a wide range of hydrological and physiographical conditions, soil types, rainfall amounts, and irrigation regimes, as well as different socioeconomic settings. This diversity makes one realize that there is no single technique of soil reclamation applicable to all areas. The exploitation of these soils for agriculture will require an integrated reclamation and management plan. Fortunately, several approaches can be combined into an integrated system of soil reclamation and management (Shahid and Alshankiti, 2013). Thus, crop production on salt-affected lands can only be successful, if the soil is dealt with in a holistic manner, *i.e.*, in an 'integrated approach' including aspects of soil, water, plants, and climatic conditions. There is, unfortunately, a misconception about "biosaline agriculture," which states that this is a complete solution for using salt-affected lands and saline and saline-sodic waters. Rather, biosaline agriculture is just one of the components of an ISRP, which also includes physical, chemical, hydrological, and biological methods. No single approach can deliver a complete solution to fix the soil salinity problem. This means that a holistic approach is needed by using a combination of mitigation approaches that are site-specific and should only be used in other areas with similar soils and environmental conditions.

Strategies for soil reclamation can be grouped into the following approaches that cover almost all aspects of soil reclamation and management (adopted from Shahid and Rahman, 2011; Shahid et al., 2011):

- Physical
- Chemical
- Hydrological
- Biological
- Other options

29.8.1 Physical Methods of Soil Reclamation

There are several physical methods of soil reclamation, although not all of these methods are required for a given situation. Site-specific diagnostics allow one to select the most suitable method(s). The most commonly used physical methods of soil reclamation are listed below:

- Leveling
- Sub-soiling
- Mixing sand with heavy textured soil
- Scraping to reduce salinity effects
- Seed bed preparation

29.8.1.1 Leveling and Draining

Leveling of salt-affected lands prior to the implementation of a reclamation program is essential. This allows for a uniform water distribution, leading to effective leaching of salts. Unfortunately, the farmer usually accomplishes this task by plowing the field, followed by the use of a conventional planking tool. This practice usually leaves the land uneven, which means that when water is applied to the field to leach the salts, the water puddles in the depressions and heterogeneous conditions are formed. 'Laser land leveling' can be of use to level the land effectively. It is thus suggested that each farmer should contact extension services to access this modern tool, which will facilitate an effective initiation of the farmer's soil reclamation program.

The people of India have reclaimed land by the use of drainage. The Dutch have been continually enclosing and reclaiming smaller bodies (polders). After dikes are built around a polder, the area is drained by pumping out the water. Drainage channels and, in many places, subsurface drains are installed to drain the root zone of crops. After this, cropping is started as the last step in the reclamation process (Shahid, 2013).

The development of land-clearing machinery and surface-drainage techniques has made it possible to clear and drain tropical lands for agricultural production. A water-table management system, called the Controlled Drainage and Sub-irrigation system or Controlled and Reversible Drainage System, was designed and executed at the Agricultural Engineering College and Research Institute, Kumulur, Tiruchirapalli, Tamil Nadu. Selvaperumal et al. (2017) reported important parameters for the design of the water-table management system for both drainage and sub-irrigation modes. Further, they reported that, for both subsurface drainage and sub-irrigation, the spacing of 15 and 10 m, respectively, was recommended for water-table management systems in sandy clay loam sodic soils. A 15-m drain spacing at the 75 cm drain depth with 75 mm drain diameter was the economically viable water-table management system. They also added that the recommended level of press-mud at 2 t/ha combined with the water-table management system was found to reduce the ESP level below 15 in sodic soils (initial soil ESP 36.56) along with highest crop returns. (Press mud is the compressed sugarcane industry waste produced during filtration of cane juice.)

29.8.1.1.1 Subsurface Drainage Technology for Saline Soils

At the Central Soil Salinity Research Institute (CSSRI) of the Indian Council of Agricultural Research (ICART) in Karnal, India, the cumulative area reclaimed in 2011 was 70,084 ha by using the sub-surface drainage technology. During cultivation in subsequent years, an improvement of 25%–100% cropping intensity was observed. Yield increased 45% in rice, 111% in wheat, and 215% in cotton. An additional food grain production of 6 lakh tons was realized as a result of the reclamation (ICAR-CSSRI, 2013). (A lakh is a unit in the Indian numbering system equal to one hundred thousand, or 100,000.)

29.8.1.1.2 Laser Land Leveling

The Tungabhadra Command Area is located in Karnataka, which is a state located on the western coast of India. This command area has soil salinity, soil erosion, and shortage of irrigation water particularly during *Kharif* season. (*Kharif* season means the autumn season. This is when India has its fall harvest.) Soil moisture cannot be distributed uniformly due to improper leveling. Laser land leveling is used to enhance water productivity. After the land is levelled, direct seeded rice (DSR) is planted on the laser-leveled fields. If DSR on laser-leveled land is adopted by farmers in the head reach of command, proper distribution of irrigation water and fertilizer helps to reduce water logging and secondary salinization, especially at the tail end of the command. Twenty to twenty-five percent irrigation water is saved with DSR on laser leveled compared to puddled transplanted rice (PTR) on traditionally leveled land. Higher paddy yield (8.75t/ha) was recorded with PTR on laser-leveled land followed by DSR on laser-leveled land (7.88t/ha). Soil erosion was reduced, and soil moisture distribution was uniform over the entire area of the DSR laser-leveled land (ICAR-CSSRI, 2013).

29.8.1.1.3 Land Shaping for Controlling Water Logging and Salinity

The coastal areas of Andaman Islands, or the Andamans, located in the northeastern Indian Ocean, suffer from the twin problems of salinity and waterlogging. Sea water inundation during high tide makes life and livelihood difficult, as in the case of Padmaspahad village of the South Andaman Island (ICAR-CSSRI, 2017). An alternative land management system for waterlogged and saline soils was adopted. Water from ponds was drained to leach salts, and from the second year onward, rain water was harvested and stored. Within one year, salinity was reduced to <1.5 dS/m from the initial value of 4.6 dS/m and the pH improved to 6.4. Polyculture of the Indian major carps was practiced in the ponds with production of 520 kg/year. Banana cultivation gave a net return of Rs. 18,000. (The Indian unit of money is the Indian rupee, which is equal to about 0.013 United States dollar.)

After meeting home consumption of farmers, vegetable production returned Rs. 10,000 in one year. The benefit-cost ratio (B:C ratio) of 1:8 was observed in the first year and 2.0–2.5 from second year onward.

29.8.1.2 Sub-soiling

The soils affected by sodicity are usually underlain by dense clay-sodic layer(s). These dense layers are created by the dispersion of clay particles in the highly sodic water. The dispersed clay particles move to the subsurface of the soil, lodged on the ped faces next to conducting soil pores, thus, blocking the pores and preventing further water movement. It is particularly important to disrupt the dense layers deep in the soil in order to enhance permeability. This is especially important for reclaiming sodic soils, after the addition of amendments such as gypsum, which is then followed by watering the field. The addition of gypsum will enhance the removal of exchangeable sodium (which has already been exchanged by Ca^{2+}). It is moved into lower layers of the soil prior to finally moving into the main drainage system. In addition to dense sodic layers, the soils may be underlain by a plow layer or by other hard pans occurring during soil formation. These hard pans must be disrupted and broken in order to enhance drainage capacity and to facilitate the soil reclamation process (Gupta, 2002).

Deep tillage to break up sub-horizons high in adsorbed sodium is a common technique used for the reclamation of sodic soils. A combination of drainage and irrigation is used to reclaim large areas of land that have been abandoned because of salt accumulation. In this case sub-surface drainage systems must be installed, so that high water-tables are lowered and pure water flushed through the soil, dissolving the salts and carrying them away in the drainage water. Large areas in India, the United States, and the Middle East are potentially available for reclamation by this technique (Shakya and Singh, 2010).

29.8.1.2.1 Controlled Drainage

With controlled drainage, sub-surface drainage is provided with a device to maintain the water table, to reduce the drainage volume, and to prevent loss of nitrogen fertilizer. In this sub-surface drainage system, drainage water volume is controlled. Controlled drainage saves irrigation water and nitrogen fertilizer.

At the Agricultural College and Research Institute in Tiruchirappalli, a sub-surface drainage system using corrugated, perforated PVC pipe was installed at a cost of Rs. 12,900. The plot area was 33 × 60 m^2 with 100 m lateral drain and 100 m collector drain. The area was surrounded by a 5 cm thick envelope of 2–6 mm sized mineral components. The lateral drains were installed at 18 m spacing and 0.6 m depth, and the collector was at a depth of 0.7 m. The study revealed that provision of sub-surface drainage in sodic sandy loam soils improved yields of til (the Indian term for sesame), wheat, pearl millet, and chick pea, compared to the undrained condition, by 15%, 12%, 19%, and 18%, respectively. Yields were 500, 1,400, 530, and 1,180 kg/ha. The study further showed that pH, EC, and ESP in the drained condition were controlled, as compared to the undrained condition. The average values of pH, EC, and ESP were 7.7, 1.51 dS/m, and 39.4% in drained fields as compared to 9.28, 2.15 dS/m, and 51.9% in undrained fields. The water table depth was 65 cm from the ground surface in undrained fields, while, at the same time, no water table was observed in drained fields to 1m depth, when piezometers were used to measure the water table depth. The study also showed an improvement in pH, EC, and ESP of soils in the drained condition as compared to the undrained condition (ICAR-CSSRI, 2005) (Figure 29.3).

At Bapatla, a town of the Indian state of Andhra Pradesh, sub-surface drainage systems (SSD), viz., closed SSD, open SSD, tile drainage, and corrugated perforated pipelines, for reclamation of salt affected and waterlogged soils have been successfully tested in coastal agricultural systems. The sub-surface drainage system for reclamation of waterlogged saline soils also was tested at Uppu gunduru and Appikatla in the Krishna Western Delta, which is also in Andhra Pradesh state. The technology potentially can be used in 1.3 lakh ha of waterlogged saline soils of coastal regions

FIGURE 29.3 Construction of controlled drainage system. (Adopted from Selvaperumal et al. (2017).)

and major irrigation command areas of Andhra Pradesh. Where it has been used, soil salinity in the drainage area was reduced from 8.4 to 1.7 dS/m, and consequently, by the end of 7 years, paddy yields increased from an initial level of 1.8–6.75 t/ha (ICAR-CSSRI, 2017).

Technology for reclamation of abandoned aqua pond lands (aqua ponds are ponds used for fish farming) was developed and demonstrated in coastal Andhra Pradesh. Based on surveys, it is estimated that nearly 3.4 lakh ha area is affected by salinity on the eastern coast. These fields belonged to small and marginal farmers and through this technology, it was possible to get rice yields of 4–5 t/ha during initial first or second year of reclamation. The yields were increased by 15%–30% in paddy soils with the availability of sufficient irrigation water. To control salinity and water logging in undulating soils, a technology of multilayered interceptor drains was developed at Sindhanur near Gangavathi (ICAR-CSSRI, 2017). Interceptor drains, which are 10 cm in diameter and placed at the 1.7 m depth and 500 m away from distributary, intercept the incoming canal seepage and prevent secondary soil salinization in the low-lying area. Plants and grasses were identified for intercepting seepage. Bio-drainage studies done at the Agricultural Research Station (ARS) at Gangavathi revealed that planting of tree species, such as *Acacia nilotica*, *Delbergia sissoo*, and *Casuarina equisetifolia*, in 2–4 row spacings 5 m away and parallel to the canal were efficient in intercepting the canal seepage. Such plantations could intercept more than 80% of the seepage and relieve water logging problems along the canals (ICAR-CSSRI, 1998a, b).

After the implementation of sub-surface drainage technology, it has been established that reuse of drainage water is possible, and hence, it has become a source of irrigation water to many farmers located at the tail ends of the irrigation commands. Sub-surface drainage lowered the water table, reduced soil ECe, and increased the yield of crops. It has been shown that payback period of this technology is 2–3 years depending upon the initial status of the soil and crops grown following land reclamation. A sub-surface drainage system coupled with *nala*-cleaning is efficient in reclaiming salt affected soils, which in turn results in higher crop yields. (A *nala*-cleaning machine removes silt and solid materials from drains. *Nala* means "hollow reed" in Sanskrit.) Controlled drainage saves about 17.5% irrigation water (104 vs 126 cm) and 52.5% nitrogen (5.32 vs 11.20 t/ha) as compared to conventional drainage. Using controlled drainage in salt-affected soils, paddy rice can be successfully grown by adopting a continuous ponding method of irrigation. Intermittent ponding badly affects paddy yields. Continuous ponding assures better leaching of salts compared to intermittent ponding (ICAR-CSSRI, 1998a, b).

29.8.1.3 Mixing Sand

If the soil to be reclaimed has heavy texture (*i.e.*, a clay soil), the mixing of sand in an appropriate quantity can change the soil texture permanently. The soil becomes more permeable and is easier to reclaim. This practice also provides a favorable environment for plant growth compared to the original soil prior to sanding. Altering the soil texture is a difficult and costly task. When sand is readily available from a nearby source, this practice can be accomplished more easily (Mayalagu and Jawahar, 2000). A clay soil is considered to exist in an area that has the percentages of primary soil particles as follows: 10% sand, 20% silt, and 70% clay (textural class: clay). The clay soil is mixed with a calculated quantity of sand to develop the following percentages of soil particles: 60% sand, 15% silt, and 25% clay (textural class: sandy clay loam). In this way, the original soil texture (clay) is changed to another soil texture (sandy clay loam). Both soil textural classes can be located on the USDA-soil-textural triangle. The newly established texture, a sandy clay loam, has improved soil physical properties, for example, an increased drainage capacity and infiltration rate. This leads to an enhanced soil reclamation process and results in a much better leaching of salts (Bhatt and Khera, 2006).

29.8.1.4 Scraping

Salt accumulation in an irrigated field is common. Salts accumulate happens at certain zones based on the irrigation system in use and soil bed shape. When adopting the scraping practice for the purpose of removal salts, furrow irrigation system is selected. If this is done, the leveling process should be followed by sub-soiling or chiseling.

Both furrows on each side of the bed are irrigated, and the zone of salt accumulation appears in the center of the ridge. The salt crusts accumulating at the surface, due to capillary rise and subsequent evaporation, can be removed manually or through mechanical means. Mechanical removal is the simplest and the most economical way of reclaiming saline soils, if the area is small. This practice minimizes the salts temporarily. They will accumulate again if there is a continuous feed of saline water to the soil surface. Therefore, the scraping method is considered a temporary solution.

29.8.1.5 Seed Bed Preparation

Tillage can improve the soil's physical condition and can bring a saline layer to the surface. However, it can also create a plow layer through the continuous use of the plow. Thus, care must be exercised in using tillage. By forming the soil surface into different shapes and the selection of a specific irrigation system, a zone of low salinity can be achieved. Salts tend to accumulate on the ridges away from the wet zone, when furrow irrigation is adopted. Placing the seeds on the off-center slope (*i.e.*, shoulder) of a single row will position the seed in a location with a minimum salinity and an optimum moisture condition. Under high salinity conditions, the alternate row should be left un-irrigated. This will ensure maximum accumulation of salts in the un-irrigated area, thereby leaving the irrigated furrows free of salts and fit for planting seeds.

29.8.1.5.1 Raised Bed System

Raised beds with 405 cm x 30 cm size, along with farmyard manure (FYM) and pressmud at 12.5t ha^{-1}, was found suitable for reclamation of sodic-Vertisols at Tiruchirappalli, where sunflowers were grown under rainfed conditions.

29.8.2 Chemical Methods of Soil Reclamation

Chemical methods of reclamation are commonly used to reclaim sodic or saline-sodic soils. Saline soils are unable to be reclaimed by chemical methods. A soil with ESP > 15 is declared as a sodic soil (USSL Staff, 1954). At this threshold ESP value, the soil will show effects on both soil physical properties (structural damage) and also on plant growth. In such soils, the objective is to bring the soil ESP below the threshold value. This can be achieved by adding suitable amendments to increase the concentration of calcium ions (Ca^{2+}) in the soil. Chemical reclamation includes the following methods, and the methods used are based on the diagnostics of the problem.

- use of gypsum,
- elemental sulfur and acids (hydrochloric and sulfuric acids)

29.8.2.1 Use of Gypsum to Reclaim Sodic Soils

The most suitable method is to replace exchangeable sodium with calcium and the subsequent use of organic matter to bind the soil and improve its structure. Amendments are those materials that supply Ca^{2+} for the replacement of exchangeable sodium. They can furnish calcium indirectly by dissolving calcite ($CaCO_3$) that occurs naturally in many arid zone soils. Gypsum ($CaSO_4 \cdot 2H_2O$) and lime (CaO) can both add calcium and thus can overcome dispersion, because the calcium causes the inter-particle forces to more readily hold the particles together. The calcium causes particles to form clusters (flocculates), forming a clear puddle of water. Gypsum usually gives an immediate response as it dissolves (although it has low solubility) in water, although it leaches sooner than lime. Gypsum is reported to reduce the levels of exchangeable sodium in soil. It also improves both soil tilth and drainage and achieves better crop production. The addition of gypsum to a soil changes the soil chemistry in two ways:

i. through increase of salt concentration in solution. This is a short-term effect that occurs as the gypsum dissolves, and
ii. the calcium from the gypsum replaces the exchangeable sodium from specific sites. This process changes a sodic clay to a calcium clay. The displaced sodium is then leached into lower soil zones, below the plant root-zone. Mined gypsum (<2 mm particle size) of commercial grade (~70% purity) is commonly used for the reclamation of sodic soils.

The following reaction occurs when gypsum is added to a Na-clay soil.

$$2Na^+ - \text{Clay} + CaSO_4 \cdot 2H_2O \rightarrow Ca^{2+} - \text{Clay} + Na_2SO_4 + 2H_2O$$

Sodic soil Gypsum Normal soil Salts leach out

Out of the 21,404 ha of the total geographical area of the Manikandam Block of Trichirapalli, located in the state of Tamil Nadu, about 3,667 ha are salt-affected soils. The area under two types of salt-affected soils, 1,167 and 2,500 ha, is strongly or moderately salt affected, respectively. When gypsum at 50% of the gypsum requirement (GR) was added along with package of practices, grain yield of rice increased between 1,674 and 2,300 t/ha, when compared to the control. Due to the application of gypsum, the pH of soil was reduced to levels of 8.3–8.7 from initial levels of 8.7–10.0, and the ESP was reduced to levels of 13%–20% from the initial level of 23%–42%. Ezeaku et al. (2015) showed that rice and millet biomass and grain yield were higher when gypsum at 100% treatment was applied alone, followed by gypsum+farmyard manure. The combined use of gypsum, farmyard manure, and chiseling was best in improving the soil properties. Application of gypsum at 50% GR along with green manure at 6.25 t/ha increased grain yield of rice to between 1,674 and 2,300 t/ha, when compared to control. Due to the application of gypsum, the pH of soil was reduced to levels of 8.3–8.7 from initial levels of 8.7–10.0, and ESP was reduced to levels of 13%–20% from initial levels of 23%–42% (ICAR-CSSRI, 2017).

29.8.2.2 Use of Acids to Reclaim Calcareous Sodic Soils

The use of acids is recommended for calcareous-sodic soils. The acids react with the calcium carbonates present in soil to mobilize calcium, which ultimately replaces exchangeable sodium in the soil, thereby reducing the ESP.

The objectives of application of acids are as follows:

- Mobilize calcium from calcium carbonates
- Replace exchangeable sodium with calcium
- Bring about a reduction in soil pH, thereby enhancing nutrient uptake, and
- Improve soil health in order to obtain a better crop production

Both sulfuric and hydrochloric acids react rapidly with soil lime, because they do not have to go through an oxidation process. However, they are highly corrosive and dangerous to handle. Specialized equipment has recently become available to safely apply acids onto field soils, usually with irrigation water. The reaction of applied acids with naturally occurring soil $CaCO_3$ and exchangeable Na^+ is shown as follows:

$$CaCO_3 + H_2SO_4 \rightarrow CaSO_4 + CO_2 + H_2O$$

$$2Na^+ - Clay + CaSO_4 \rightarrow Ca^{2+} - Clay + Na_2SO_4 \downarrow$$

Sodic soil   Normal soil   Salts leach out

$$CaCO_3 + 2HCI \rightarrow CaCI_2 + CO_2 + H_2O$$

$$2Na^+ - Clay + CaCI_2 \rightarrow Ca^{2+} - Clay + 2NaCI \downarrow$$

Sodic soil   Normal soil   Salts leach out

The application of gypsum in saline-sodic soils containing lime does not improve the availability of phosphorus. The application of sulfuric acid with the irrigation water on a calcareous soil could increase the phosphorous concentration. It is important that, for improving phosphorous availability in calcareous-saline sodic soils, sulfuric acid is added to irrigation water instead of applying gypsum to the soil (Qadir and Oster, 2004).

29.8.2.3 Use of Elemental Sulfur to Reclaim Calcareous Sodic Soils

Elemental sulfur can also be used to reclaim calcareous-sodic soils under one condition – the sulfur must be completely oxidized. This occurs through the oxidation of sulfur by *Thiobacillus thiooxidans*, although in sodic soils sulfur oxidation is a very slow process. The complete oxidation of sulfur results in formation of sulfuric acid.

$$2S + 3O_2 + 2H_2O \rightarrow 2H_2SO_4$$

TABLE 29.2
Equivalent Amount of Various Amendments for Supplying Calcium in Terms of Pure Gypsum

Amendment	Ton (s) Equivalent to 1 ton of 100% Gypsum
Gypsum ($CaSO_4 \cdot 2H_2O$)	1.00
Calcium chloride ($CaCl_2 \cdot 2H_2O$)	0.86
Calcium nitrate ($Ca(NO_3)_2 \cdot 2H_2O$)	1.06
Press-mud (lime–sulfur 9%Ca +24%S)	0.78
Sulfuric acid (H_2SO_4)	0.61
Iron (ferrous) sulfate ($FeSO_4 \cdot 7H_2O$)	1.62
Ferrous sulfate ($Fe_2(SO_4)_3 \cdot 18H_2O$	1.09
Aluminium sulfate ($Al_2(SO_4)_3 \cdot 18H_2O$)	1.29
Sulfur (S)	0.19
Pyrites ($FeS_2 \cdot$ 30%S)	0.63
Limestone ($CaCO_3$)	0.58

Source: Adopted from USSL Staff (1954).

The H_2SO_4 formed through biological oxidation of sulfur reacts rapidly with lime (as shown above) and proceeds to reclaim the sodic soils. There is a significant financial outlay required from farmers when using chemical amendments in soil reclamation. It is, therefore, recommended that the benefit of any amendment should be tested first in field trials with regard to its cost, safety in use, and effectiveness in improving (reducing) soil sodicity and increasing crop production. The theoretical amounts of various amendments needed to supply an amount of Ca-equivalent to 1 ton of gypsum are shown in Table 29.2. Sharma and Gupta (1986) observed a similar change in ESP due to the application of either gypsum or H_2SO_4. However, they reported that there was a lower hydraulic conductivity and higher amounts of water-dispersible clay when H_2SO_4 was applied. This was caused by the Ca^{2+} being a stronger flocculent than the H^+ ion of sulfuric acid.

29.8.2.4 Acid Saline Pokkali Lands in Coastal Areas of Kerala

In the coastal area of Kerala, acid saline soils are dominant, and rice has poor productivity and profitability on them. Farmers have been advised to integrate the *Pokkali* variety of rice with prawn cultivation. It is a salt-tolerant rice. Farmers apply lime at the time of field preparation. Seeds of the *Pokkali* variety are sown on ridges. After one month when the conditions become favorable, the ridges are dismantled and the seedlings are distributed evenly in the field using a spade. Data have shown that there is an increase in all chemical parameters, except sulfur content, after the cultivation of prawn. Without applying chemical fertilizers or organic manure, farmers get an average yield of 2–3t for rice and 400 t/ha for prawn. The rice crop is grown without any fertilizers and pesticides. Average B:C ratios for rice, prawn, and the rice-prawn integrated system are 1.3, 3.5, and 2.02, respectively. The profit realized in a year on a farm is Rs.1,39,600 per hectare.

Soil moisture is an important factor affecting crop yields in rainfed agriculture. In addition, soil-sodicity affects crop performance, including germination and the availability of essential plant nutrients. In a project done under rainfed conditions on sodic-Vertisols in the Tungabadra project (TBP) command area, yield of sunflower and retention of rain water in soil were enhanced through constructing tied ridges in fields prior to the onset of the south-west monsoon. Soil ESP was reduced through the application of gypsum @ 50% of the gypsum requirement (ICAR-CSSRI, 2013, 2017).

29.8.3 Hydrological Methods of Soil Reclamation

Hydrological methods generally include irrigation, leaching and flushing, and the drainage of the leached water. Blending of water to reduce its salinity and sodicity and recycling of the water can also be included under this topic. The objectives of soil reclamation through hydrological methods are as follows: Efficient use of irrigation water; Leaching of salts into lower soil zones, below the root-zone; Improvement of water quality; Improvement of a waterlogged condition through drainage.

In irrigated agriculture, the salts in soil can be removed in two ways:

- Leaching of salts into a soil zone below the root-zone and subsequent drainage of the leached water to a safe place
- Surface flushing of dissolved salts.

The salt content within the root-zone is likely to be increased if the net downward movement of salts is less than the salt input from irrigation water. Therefore, the soil salt balance must be kept under control. Control of the salt balance is, thus, a function of irrigation water quality, the quantity of dissolved salts in the water, and the success of the soil drainage system.

29.8.3.1 Leaching

29.8.3.1.1 Leaching Requirement and Leaching Fraction

The quantity of water that must pass through the root-zone to maintain the EC level at or below a specified level defines the '*leaching requirement*'. The '*leaching fraction*' is the fraction of irrigation water that passes through the root-zone into lower soil zones.

Soils rich in soluble salts can be reclaimed through dissolving of these salts and their successful leaching. This can be accomplished through flooding or ponding of water at the surface for saline soils. In general, the depth of soil leached is roughly equal to the depth of water infiltrated during leaching. In order to leach salts from a soil, an understanding about the leaching requirement (LR) concept is important. The LR is the calculated fraction (depth) or quantity of water that must pass through the root-zone in order to maintain the EC of the drainage water at or below a specified level. Some soil scientists are of the opinion that the LR should be minimized to prevent raising the level of the groundwater table and also to reduce the load placed on the drainage system. Recently established guidelines for successful and economic leaching methods are described hereunder.

$$\text{Leaching Requirement (LR)} = \frac{\text{EC}_{iw}}{\text{EC}_{dw}} \times 100 \text{ or } \frac{D_{dw}}{D_{iw}} \times 100$$

Where EC = Electrical Conductivity in dS/m
iw = EC of irrigation water in dS/m
dw = EC of drainage water in dS/m
D_{dw} = Depth of drainage water in inches
D_{iw} = Depth of irrigation water in inches

29.8.3.1.2 Timing of Leaching Irrigation

Timing of leaching does not appear to be critical, provided that the crop salinity tolerance limit is not exceeded for extended periods of time or occurs during a critical stage of plant growth. A few important points to be considered are listed below.

- Leaching should be done when soil moisture is low and the water-table level is deep; leaching should precede the critical growing stages of the crop plant
- An optimal time for leaching would be during a period with a low evapotranspiration demand, at night, during high humidity, or in cooler weather

- Leaching can also be done at the end of the cropping season
- Soil and tissue analysis can help determine both the need and timing of leaching.

29.8.3.1.3 Leaching Requirement for Surface Irrigation

To determine the leaching requirement, it is essential to have information about two parameters: (i) salinity of the irrigation water to be used (dS/m) and (ii) crop tolerance to salinity (ECe in dS/m).

29.8.3.2 Flushing

Flushing is suitable for saline soils having surface salt crusts, which is a common situation in arid and semi-arid areas and where rainfall is insufficient to leach the salts. This practice flushes the salts from soil surface, and the flushed saline water then enters the drainage system, increasing the salt concentration. Flushing of surface salts is possible where soils are of a heavy texture and ponding can be accomplished easily. Once the water is ponded for a time that is sufficient to dissolve the salt crust, the ponded water can be flushed from the field, thus removing the surface salts. The following procedure will allow for successful flushing of surface salts.

A sufficient volume of good quality water is used to dissolve salts from the soil surface. The soil must possess the ability to allow for surface ponding, That is, its subsurface must have a heavy texture. The field must be capable of flushing the dissolved salts. This can be accomplished either by forming breaks in the sides of the field, so that the saline water will drain into adjacent channels. Alternatively, the ponded saline water can be removed by siphoning it, using long pipes, into adjacent channels. There must be ways to safely reuse the drained water or environmentally safe methods available for its disposal.

Sandy soils in a desert environment, such as Gulf Cooperation Council countries and other countries with similar environments are well drained. Therefore, leaching of salts by using irrigation water amounts in excess of evapotranspiration can maintain salts in the root-zone at a safe limit. However, one major problem for irrigated agriculture under hot desert conditions is the higher volume of drainage water that must be managed safely and sustainably, without compromising the environment.

29.8.4 Biological Methods of Soil Reclamation

Biological methods of soil reclamation include the use of organic material(s) to improve soil structure and to mobilize calcium from calcium carbonates through the decomposition process. Biosaline Agriculture (growing salt tolerant crops) is also part of biological reclamation.

29.8.4.1 Use of Organic Amendments

Soils of arid and semi-arid regions are generally deficient in organic matter where saline and sodic soils are commonly found. Organic matter can be added in the following ways: mixing previous crop stubble into the soil, addition of farmyard manure, addition of crop residues, use of mulch material(s), and growing of green manure crops, such as legumes. Development of soil structure prevents soil erosion and hastens soil reclamation, primarily due to increased infiltration. The decomposition of organic matter produces a high level of CO_2 and also increases organic acids (humic, fulvic), lowering the soil pH. Awan et al. (2015) reported that application of farmyard manure alone or in combination with inorganic N fertilizer had a significant effect on wheat yield on a saline-sodic soil both under monoculture and in an agro-forestry systems. The selection of the organic matter to be applied is important to avoid causing N deficiency (adding an amendment which has too high C:N ratio) and to avoid increasing the salinity (for example : animal dung slurry). Use of green manure crops may have a better chance than farmyard manure of being successfully integrated into a soil reclamation management package (Gharaibeh, 2014).

Organic matter added to saline sodic soils plays an important role in microbial activity. Garcia and Hernandez (1996) observed that an increase in soil salinity inhibited several soil enzymatic activities and microbial respiration. Many studies have been conducted dealing with the reclamation of saline-sodic soils, but only some of them have focused on calcareous soils (Qadir and Oster, 2004). Generally, calcareous saline sodic soils have problems related to the availability of macronutrients. Therefore, reclamation of such soils must involve improving soil fertility. In addition, there are different methods for ameliorating these soils that affect macronutrient re-distribution. In such soils, because of high osmotic pressure and ionic toxicity, microbial activity is reduced (Rietz and Haynes, 2003).After the reclamation process, the macronutrients nitrogen and potassium increase with soil depth, while the phosphorous concentration is higher in the soil surface. In other words, during reclamation of saline-sodic soils, as a consequence of leaching, nitrogen and potassium tend to be removed from the soil profile, whereas phosphorus is retained. Organic-matter-amended soils show the highest nitrogen concentration depending on the nitrogen content and decomposition rate of organic matter. Compared to cattle manure, pistachio residue is effective in providing nitrogen. It also has the additional benefit of increasing soil carbon sequestration.

At a study conducted at Tiruchirappalli, Tamil Nadu, among the organic amendments, application of FYM @ 12.5 t/ha produced the highest sunflower seed yield and was comparable with application of pressmud (12.5 t/ha) in sodic soils. Addition of urea @ 112.5 kgN/ha with green manures like *daincha* or *subabul* or FYM or rice husk ash @ 5, 15 and 5 t/ha respectively will result in higher grain yield and also improves soil fertility and apparent N recovery in sodic soils. The soil application of gypsum @ 50% GR significantly increased the yield of both okra (12.6% in 2009 – 10% and 12.8% in 2010–2011) and cluster bean (22.1% in 2009 – 10% and 20.9% in 2010–2011). Among the irrigation treatments, drip irrigation of spent wash treated water recorded the highest yield in okra and drip irrigation of gypsum bed treated water recorded highest yield in cluster bean crop (ICAR-CSSRI, 2013).

Among different crop residues and green manures that can be used for green manuring are daincha (*Sesbania bispinosa*, a legume) and sunhemp (*Crotalaria juncea*, a legume). They reduce pH and ESP and remove HCO_3^-, SO_4^{2-} and Cl^- which leads to the creation of a favorable soil environment for better crop growth. Use of crop residues and green manures allows gradual reclamation as compared to amendments like gypsum. In general, pistachio residue is a good amendment to enhance the availability of phosphorus and potassium. Amendment with cattle manure has a higher rate of carbon mineralization compared to pistachio residue despite of its higher C:N ratio. Following the addition of organic matter to the soil, the rate of microbial respiration rises. Subsequently, it decreases and reaches a steady-state condition. The rate of organic matter decomposition and mineralization is high for the first 20 days, and it is reduced afterward. Akthar et al. (2015) stated that, with biochar amendment, there was decreased Na^+, decreased Na^+/K^+ ratio, and increased K^+ content in the xylem of potato (*Solanum tuberosum*), which indicated the ameliorative effect of biochar in response to salinity stress. The results suggested that the incorporation of biochar might be a promising approach for enhancing crop productivity in salt-affected soils.

29.8.4.1.1 Sodic Soil Reclamation Using Distillery Spent Wash

Data have shown increased grain yield of rice of 1,560–2,428 t/ha due to the application of distillery spent wash (DSW) at 5 lakh liters/ha when compared to controls. With the application of DSW, pH of a soil was reduced to 8.3–8.8 from an initial level of 8.6–10.0, and the ESP was reduced to 14%–19% from an initial level of 24%–32%. The gypsum requirement was 9.5 t/ha (50% GR) costing around Rs.20,000. To reduce this reclamation cost, DSW can be used with gypsum. DSW for reclamation is available free of cost. The application of DSW on sodic soils for their reclamation can be done with 1.25 lakh liters/ha. The soil that was reclaimed was highly sodic and sandy clay loam in texture (Tripathi, 2008). The initial soil pH, EC, and ESP were 11.2, 1.7 dS/m, and 42%, respectively. After reclamation with DSW, the post-harvest soil pH was reduced to 8.46, and ESP was reduced to 17.2%. The grain yield obtained in control plots was 1.5 t/ha, whereas the yield in the DSW reclamation field was 4.5 t/ha. Five percent N and 100% of K fertilizer were substituted by the DSW. Farmers observed good crop growth in the DSW-treated fields (Tripathi, 2008).

29.8.4.2 Biosaline Agriculture

Biosaline agriculture is the economic utilization of salt-affected soils for agricultural purposes, for example, the growing of salt-tolerant crops of agricultural significance (Dagar, 2008). Biosaline agriculture includes the use of salty water for sustained agriculture. In the past, the term saline agriculture was used. However, a broader definition is now needed, one which includes the manipulation of desert and sea resources for both food and fuel (energy) production. Biosaline agriculture has a wide scope with diversified dimensions. These include the following: breeding for salt tolerance with appropriate plant species; selection of salt-tolerant cultivars; domestication of salt-tolerant plants for economically sound, but sustainable, exploitation of salt-affected lands; and climate-smart agricultural practices (land preparation, planting, irrigation, and fertilization) (Tripathi and Singh, 2010). Methods for screening and development of varieties and germplasm of various crops for salt-affected soils have been devised (ICAR-CSSRI, 2005).

a. Crops and varieties

   As reported by the status report of ICAR-CSSRI (2004), crops and varieties can contribute to biosaline cultivation. Sharma (2015) recorded the development of important food crops for the saline soils. They are as follows:

   - The coarse-grained rice variety *TRY 1* and fine-grained variety *ADT 45* tolerate up to an ESP of 26
   - The green gram variety *Pusa bold* tolerates an ESP of 16.
   - At a higher ESP level of 36, the pearl-millet cultivar *UCC 17* performed better with minimum yield loss.
   - Upto an ESP of 16, crops such as maize, sunflower, sesame, okra, cluster bean (*Pusa naubuhar*) performed better with minimum yield loss.
   - The threshold ESP for the sunflower variety CO4 and hybrid *TCSH1* were 16.5 and 13.0, respectively.
   - The ESP at 50% yield reduction for sunflower variety CO4 and hybrid *TCSH1* were 37 and 34, respectively.
   - The threshold ESP for okra varieties were 13.5, 15.5, and 16 for Hybrid No. 10, Arka Anamika, and Parbhani Kranti, respectively.
   - The ESP at 50% yield reduction for okra varieties were 31, 34, and 30 for Hybrid No. 10, Arka Anamika, and Parbhani Kranti, respectively.
   - Another study done at Bapatla showed that, when plants were screened upto 6 dSm^{-1}, salinity tolerance was in the order of cluster bean>tomato>radish>carrot>brinjal (eggplant)>okra>bitter gourd>chillies>palak (spinach)> capsicum (pepper).

b. Tree species

   - At the fifth year of planting, survival percentage was highest for *Tamarindus indica* and *Bambusa bamboo* followed by *Lueceana leucocephala*, *Acacia leucophloea*, and *Azardiracta indica*.
   - In general, the pit with auger hole (120 cm) was the best method of planting for establishment and growth of tree species (neem and tamarind) under rainfed, sodic-soil conditions. However, the amendments of DSW at 75 mL/kg soil with gypsum at 25% GR for neem and DSW at 150 mL/kg of soil for tamarind were found to be best for establishment and growth.

29.8.4.3 Bio-drainage

Potential technologies to mitigate dryland salinity include the pumping of saline groundwater and its safe disposal or use, as well as the development of alternate crop plant production systems to maximize saline groundwater use, such as deep-rooted trees. Deep-rooted trees utilize groundwater and lower the water-table. The process is called *bio-drainage* (or biological drainage). In Australia, dryland salinity is a major problem, costing Australia over $250 million a year from impacts on agriculture, water quality, and the natural environment (Tripathi and Singh, 2010). In Australia, to address dryland salinity issues, an approach, entitled *Options for the productive use of salinity – OPUS*, has been successfully used in a National Dryland Salinity Program. One of the options is the industrial use of harvested salts. There are a number of commercial uses for them in the pulp, paper, organic and inorganic chemical, glass, petroleum, plastic, and textile industries (Olsson Industries, 2001). Salts are used globally in the food industry, *i.e.*, in the preserving and preparing of canned and bottled foods, in cheese production, and in bakeries and cooking foods. In European countries or other countries, where heavy snow falls are frequent, salt is used to de-ice the roads to facilitate transport. Such a use accounts for 30% of the total salt use in Europe (European Salt Producers' Association, 2000).

29.8.5 Other Options with Silvi-Pasture

In the absence of a safe outlet of drainage water, the conventional methods of reclamation through leaching and drainage are not possible. The alternative is a silvi-pastural system utilizing trees in place of traditional crops to reclaim salt-affected soils. By removing water from lower layers of soil, the trees minimize the capillary rise and, therefore, shift the zone of salt accumulation from the surface to lower layers. Because trees use large quantities of water, they intercept the seepage from canals, which helps to control soil salinization by preventing a rise in the water-table. In addition, a good return can be earned through the sale of tree products (Siyal et al., 2002).

Tree species, such as *Casuarina equisetifolia*, *Dalbergia sissoo*, *Glyricidia*, and *Acacia auriculiformis*, have been found promising under saline, waterlogged conditions of the Tunga Bhadra Project (TBP) command area (ICAR-CSSRI, 2017). Fruit species, such as wood apple, jamun (Java plum), pomegranate, and sapota (a calorie-rich fruit like mango), have been found promising under saline, waterlogged conditions in the TBP area. The threshold water-table depth for cotton is found to be 95 cm in saline-Vertisols of TBP. Yields are drastically reduced when the water-table is shallower than 95 cm. On saline soils having a water-table deeper than 95 cm, cotton should be irrigated once in 12–15 days to obtain high yields. On saline soils, cotton and wheat can be successfully grown up with up to 5–7 dS/m salinity and safflower with 4–5 dS/m salinity. Brinjal (eggplant) and ridge gourd can be grown in saline-Vertisols with ECe of 4–6 dS/m. Beetroot can be successfully grown on saline soils with ECe up to 6–8 dS/m without any yield reduction. Tree species such as *Acacia nilotica*, *D. sissoo*, and *C. equisetifolia* were found efficient in intercepting canal seepage when grown parallel to distributaries and canals. Such tree plantations control waterlogging and salinity in low-lying areas. Salt-tolerant crops, such as cotton, wheat, and sunflower, can be grown successfully even in highly saline soils (12–15 dS/m) by providing pre-sowing irrigation (of 6 cm depth). Salts in the surface soil (0–30 cm) are leached out approximately at 60% leaching efficiency during pre-sowing irrigation. Due to continuous paddy cultivation, the water-table is rising at the rate of 10 cm per year in the TBP command area. In saline soils of the TBP canal command, growing daincha (the green manure plant) in a paddy-paddy cropping sequence resulted in higher paddy yields apart from improvement in soil fertility. A salinity tolerant, high yielding mid-early paddy genotype *CSR-22* was released for cultivation in saline soils of the TBP command area. Threshold soil salinity levels (ECe) were worked out for ashwagandha (4.87, 4.41 dS/m for seed and root, respectively), vetiver (a bunchgrass) (6.0 dS/m), palmarosa (a grass cultivated for its essential oil) (5.02 dS/m), khus grass (a bunchgrass) (6.0 dS/m), kamakasturi (basil) (4.8 dS/m), tulsi (a flowering plant in the mint family with aromatic leaves) (5.1 dS/m), citronella (a plant that produces an essential oil from different species of *Cymopogon*) (7.2 dS/m), and shatavari (a species of asparagus) (3.96 dS/m). Citronella could be grown in saline-Vertisols of ECe up to 7.2 dS/m. Rhodes grass, para rubber tree, and grazing guinea forage grasses could be grown in saline-Vertisols of ECe up to 4–8 dS/m.

REFERENCES

Akhtar, S. S., Andersen, M.N., and Liu, F. 2015. Biochar mitigates salinity stress in potato. *J. Agron. Crop Sci.* 201:368–378.

Awan, A. R., Siddiqui, M. T., Mahmood, K., Khan, R. A., and Maqsood, M. 2015. Interactive effect of integrated nitrogen management on wheat production in *Acacia nilotica-* and *Eucalyptus camaldulensis*-based ally cropping systems. *Int. J. Agric. Biol.* 17:1270–1127.

Bhatt, R. and Khera, K. L. 2006. Effect of tillage and mode of straw mulch application on soil erosion in the submontaneous tract of Punjab, India. *Soil Tillage Res.* 88:107–115.

Dagar, J. C. 2008. Drylandbiosaline agriculture: Hisar experience. Technical Bulletin No. 6/2008, Central Soil Salinity Research Institute, Karnal, 28 p.

European Salt Producers' Association. 2000. Salt in the EU. Website: www.eusalt.com.

Ezeaku, P., Mitchell, J. P., Shennan, C., Singer, M. J., and Peters, D. W. 2015. Impacts of gypsum and winter cover crops on soil physical properties and crop productivity when irrigated with saline water. *Agric. Water Manage.* 45(1):55–71. doi: 10.1016/S0378-3774(99)00070-0.

Garcıa, C. and Hernandez, T. 1996. Influence of salinity on the biological and biochemical activity of a calciorthird soil. *Plant Soil* 178(2):255–263. ISSN: 0032-079X.

Gharaibeh, M. A. 2014. Reclamation of highly calcareous saline sodic soil using low quality water and phosphor-gypsum. *Appl. Water Sci.* 4:223–230.

Gupta, S. K. 2002. A century of subsurface drainage research in India. *Irrig. Drain. Syst.* 16:69–84.

ICAR-CSSRI. 1998a. 25Years of research on management of salt-affected soil and use lime water in agriculture. Central Soil Salinity Research Institute. Archived from the original on 13 September 2014. Retrieved 1 July 2014. https://aicrp.icar.gov.in/sasusw/.

ICAR-CSSRI. 1998b. 50Years of natural resource management research. Central Soil Salinity Research Institute. Archived from the Original on 13 September 2014. Retrieved 1 July 2014. https://en.wikipedia.org/wiki/Central_Soil_Salinity_Research_Institute.

ICAR-CSSRI. 2004. Status report of AICRP on SAA &USW (1995–2004), Retrieved 1 July 2014, https://en.wikipedia.org/w/index.php?title=Central_Soil_Salinity_Research_Institute&oldid=1059956910.

ICAR-CSSRI. 2005. Crop improvement for management of salt affected soils, manual of lectures for the training program on 'Methods of Screening and Development of Varieties and Germplasm of various crops for Salt affected soils', CSSRI, Karnal, January 17–22. Central Soil Salinity Research Institute. Archived from the original on 13 September 2014.

ICAR-CSSRI. 2013. Annual report, 2012–2013 Central Soil Salinity Research Institute, Karnal, India, Retrieved 1 July 2014, http://krishi.icar.gov.in/jspui/handle/123456789/22967.

ICAR-CSSRI. 2017. Annual report of AICRP on SAA &USW 2016–2017. Central Soil Salinity Research Institute, Karnal, India. Retrieved 1 July 2014 http://krishi.icar.gov.in/jspui/handle/123456789/22967.

Jacobsen, T., and Adams, R. M. 1958. Salt and silt in ancient Mesopotamian agriculture. *Science* 128: 1251–1258.

Kaledhonkar, M. J., Meena, B. L., and Sharma, P. C. 2019. Reclamation and nutrient management for salt-affected soils. *Indian J. Fertil.* 15(5):566–575.

Mayalagu, K., and Jawahar, D. 2000. Effect of mixing black clay soil and organic amendment on properties of coarse-textured soil and rice yield. *Int. Rice Res. Notes* 25(3):30–31.

Olsson Industries. 2001. Olsson's salt information page. Retrieved 22 September 2022 www.olssons.com.au.

Qadir, M., and Oster, J. D. 2004. Crop and irrigation management strategies for saline-sodic soils and waters aimed at environmentally sustainable agriculture. *Sci. Total Environ.* 323:1–19.

Rietz, D.N., and Haynes, R.J. 2003. Effects of irrigation-induced salinity and sodicity on soil microbial activity. *Soil Biol. Biochem.* 35:845–854. doi: org/10.1016/S0038-0717(03)00125-1.

Selvaperumal, A., Muthuchamy, I., Vallalkannan, S., and Hemakumar, H. V. 2017. Drainage investigations for the design of water-table management systems in AEC&RI, Kumulur, Trichy District, Tamil Nadu. *Curr. J. Appl. Sci. Technol.* 24(1):1–9.

Shahid, S. A. 2013. Irrigation-induced soil salinity under different irrigation systems: Assessment and management, short technical note. *Clim. Chang. Outlook Adapt. Int. J.* 1(1):19–24.

Shahid, S. A., and Alshankiti, A. 2013. Sustainable food production in marginal lands: Case study GDLA member countries. *Int. Soil Water Conserv. Res.* 1(1):24–38.

Shahid, S. A., and Rahman, K. R. 2011. Soil salinity development, classification, assessment and management in irrigated agriculture. In: Passarakli, M. (ed.), *Handbook of Plant and Crop Stress*. CRC Press, Taylor & Francis Group: Boca Raton, FL, pp. 23–29.

Shahid, S. A., Taha, F. K., Ismail, S., Dakheel, A., and Abdelfattah, M. A. 2011. Turning adversity into advantage for food security through improving soil quality and providing production systems for saline lands: ICBA perspectives and approach. In: Behnassi, M., Shahid, S. A., and D'Silva, J. (eds.), *Sustainable Agricultural Development: Recent Approaches in Resources Management and Environmentally-Balanced Production Enhancement*. Springer: Dordrecht, pp. 43–67.

Shakya, S. K., and Singh, J. P. 2010. New drainage technologies for salt-affected waterlogged areas of southwest Punjab, India. *Curr. Sci.* 99(2):204–212.

Sharma, D. K. 2015. CSSRI vision. 2030. Central Soil Salinity Research Institute, Karnal, Water and Energy International, pp. 35–45.

Sharma, O. P., and Gupta, R. K. 1986. Comparative performance of gypsum and pyrites in sodic vertisols. *Ind. J. Agric. Sci.* 56:423–429.

Sharma, S. K. 2015. Success stories on development, spread and impact of salt: Tolerant varieties of rice, Wheat and Mustard in India. (PDF). Plant stress.com. Archived from the original (PDF) on 24 September 2015. Retrieved 9 May 2015.

Siyal, A., Siyal, G., and Abro, A. 2002. Salt affected soils. Their identification and reclamation. *J. Appl. Sci.* 2(5). doi: 10.3923/jas.2002.537.540.

Tripathi, K. P., and Singh, B. 2010. Impact of different forest plantations and cropping systems on reclamation of sodic soils. *Indian J. For.* 33:135–142.

Tripathi, V. K. 2008. Performance evaluation of subsurface drainage system with the strategy to reuse and disposal of its effluent for arid region of India. *J. Agric. Phys.* 8:43–50.

USSL Staff. 1954. Diagnosis and improvement of saline and alkali soils. USDA Handbook no 60, Washington, DC, 160 pp.

Wicke, B., Smeets, E., Dornburg, V., Vashev, B., Gaiser, T., Turkenburg, W., and Faaij, A. 2011. The global technical and economic potential of bioenergy from salt-affected soils. *Energy Environ. Sci.* 4:2669–2681.

30 Potassium in Soils of India
Strategies for Sustainable Crop Productivity, Food, and Nutritional Security

R. Santhi, S. Maragatham, R. Abishek, and R. Selvam
Department of Soil Science and Agricultural Chemistry
Directorate of Natural Resource Management, Tamil
Nadu Agricultural University, Coimbatore, India

CONTENTS

ABBREVIATIONS

AICRP-IFS	All India Coordinated Research Projects on Integrated Farming Systems
AICRP-STCR	All India Coordinated Research Project on Soil Test Crop Response
CEC	Cation-exchange capacity
DAC	Department of Agriculture and Cooperation
FFP	Farmers' fertilization practices

DOI: 10.1201/9781003093565-30

FTL	Fixation threshold level
GCA	Gross cropped area
GIS	Geographic information system
GPS	Global positioning system
IGP	Indo-Gangetic Plain
INR	Indian rupee
IPNI	International Plant Nutrition Institute
K	Potassium
KRB	Potassium-releasing bacteria
KUE	Potassium use efficiency
MIR	Mid-infrared
MOP	Muriate of potash
NH_4OAc-K	Ammonium acetate K
NaBPh4	Sodium tetraphenyl boron
NIC	National Informatics Centre
NSA	Net sown area
Q/I	Quantity-Intensity
RTL	Release threshold level
R-W	Rice–wheat
SOP	Sulfate of Potash
SR+M	Application of deficient secondary and micronutrients along with state recommended NPK
SR	State Recommendations
STCR-IPNS	Soil test crop response – integrated plant nutrition system
TGA	Thermo gravimetric analysis

30.1 INTRODUCTION

Potassium (K) is one of the three major essential nutrients required in crop nutrition, the other two being nitrogen and phosphorus (N&P). In the earth's crust in terms of relative abundance, K ranks in the 17th position. In India, with shrinking cultivable area year after year, by 2025, the expected population of around 1.4 billion would have a food grain demand of 300 million tons (Mt). To meet the additional food grain requirements, a huge demand for fertilizers including that for K is forecasted. Net negative nutrient balance of 8–10 Mt/year computed in 2004 (Tandon, 2004) is expected to reach around 15 Mt/year by 2025 (Srinivasarao et al., 2015). Therefore, it is imperative to assess the K fertility status of soils across India.

Potassium is generally absorbed by plants in amounts larger than any other nutrient except N. It plays a major role in plant metabolism and activates many enzymes involved in photosynthesis and metabolism. It is involved in the translocation of carbohydrates and proteins, membrane permeability, stomatal regulation, and water utilization. Other benefits ascribed to K include enhanced resistance of plants to pests and diseases and improved crop quality characteristics.

Besides plant nutrition, being a structural component of many important soil minerals, K also plays a crucial role in maintaining soil health. Because loss of K from interlayers of K-bearing 2:1 minerals might yield irreversible changes, which would have far-reaching adverse consequences in soil fertility, *e.g.*, nutrient and water-holding capacity and release and fixation of various nutrient ions. Under conditions of continuous plant K removal largely exceeding K input, K^+ present between subsequent layers of mica or illite might be replaced by other cations, *e.g.*, H_3O^+, hydrated Ca^{2+}, and NH_4^+ (from N fertilizers), causing transformation of mica or illite into vermiculite or some intergrade minerals (Das et al., 2019). Such processes have immense significance in soils, as they can alter soil CEC, retention of important elements, and also the K mobility, including its release and fixation (Bortoluzzi et al., 2013).

TABLE 30.1
Soil K Fertility Status of India over Years

No. of Soil Samples	No. of Districts Studied	Percent of the District			References
		Low	Medium	High	
1.3 million	184	20	53	27	Ramamoorthy and Bajaj (1969)
4.5 million	310	20	42	38	Ghosh and Hasan (1976)
–	361	13	53	34	Anonymous (1998)
3.65 million	–	13	37	50	Motsara (2002)
–	371	21	51	28	Hasan (2002)
–	500	9	42	49	Muralidharudu et al. (2011)

Source: Dey et al. (2017a).

30.2 POTASSIUM STATUS IN INDIAN SOILS

The sufficiency of K in Indian soils has been reported by the Royal Commission on Agriculture (1929), and later Stewart (1947) opined that most of the Indian soils (except laterite soils) were relatively well supplied with K. The majority of the Indian soils had 1%–10% of the total K in available form, which might be considered fairly high for crops (Raychaudhuri and Datta, 1964). Making use of the data of 34,015 soil samples from Soil Testing Laboratories generated up to 1959, Tamhane and Subbiah (1960) elucidated that 26.6%, 37.1%, and 36.35 of the samples were, respectively, low, medium, and high in available K with an overall nutrient index of 2.10.

Based on the compilation of data of 10 million samples from over 500 districts, the data revealed that the Indian soil K status was mainly under the medium-high rating. In a few regions of India, soils were low in potassium (Table 30.1).

30.3 RELATIVE DISTRIBUTION OF K IN MAJOR SOIL ORDERS OF INDIA

The relative abundance of K in *Inceptisols, Alfisols* and *Vertisols* representing alluvial, black, and red soils of India were listed by Dey et al. (2017a). The non-exchangeable K is generally more prevalent in illite dominant alluvial soils than smectitic black soils and kaolinitic red soils. Kaolinite dominant red and lateritic soils (broadly fall under the order *Alfisol*) have low contents of exchangeable and non-exchangeable K pools (Srinivasarao and Srinivas, 2017; Das et al., 2019). Light-textured alluvial soils with extensive mica contents (frequently categorized as *Inceptisols* if not as *Entisols*) show a medium level of exchangeable K and a high non-exchangeable K content. Smectitic black soils (mainly in the order *Vertisols* or vertic groups of any other order) with higher clay content and cation-exchange capacity (CEC) show higher levels of exchangeable K and medium to high non-exchangeable K content (Srinivasarao et al., 2003).

Srinivasarao et al. (2007) reported low exchangeable K (<50 mg/kg) in *Alfisols* of Bangalore, Anantapur, and Ranchi, and *Inceptisols* of Rakh Dinasar, Ballowal Saunkhri, and Agra. Exchangeable K was in the medium range (50–120 mg/kg) in an *Alfisol* of Phulbani, *Inceptisol* of Faizabad, and *Vertisols* of Ajira and Akola. They observed that an *Inceptisol* of Jhansi and *Vertisols* of Bijapur, Bellary, Solapur, Rajkot, Rewa, Indore, and Kovilapatti were high in exchangeable K(>120 mg/kg). Srinivasarao et al. (2007) also categorized soils across India based on non-exchangeable K extracted by HNO_3. They observed low contents of non-exchangeable K (<300 mg/kg) in *Alfisols* of Bangalore and Anantapur and *Vertisols* of Akola and Bijapur; medium non-exchangeable K (300–600 mg/kg) in *Inceptisols* of Rakh Dinasar and Ballowal Saunkhri, *Alfisol* of Phulbani, and *Vertisols* of Bellary, Solapur, Rajkot, Indore, and Kovilpatti; and high non-exchangeable K (>600 mg/kg) in some *Inceptisols* of Agra, Faizabad, and Jhansi and *Vertisols* of Ajira and Rewa (Table 30.2).

TABLE 30.2
Relative Distribution of Different Fraction of K

Location	Depth (cm)	Illite or Mica Content in Clay (%)	Water-Soluble Potassium (mg/kg)	Exchangeable Potassium	Non-Exchangeable Potassium	Total K
New Delhi, Delhi	0–15 (SCL)	54	17.8	122	1,407	18,135
Gosaba, West Bengal	0–15 (CL)	40	85.0	375	3,939	27,495
Cuttack, Orissa	0–15 (SCL)	36	20.9	72.5	635	20,635
Varanasi, Uttar Pradesh	0–15 (C)	–	8.0	66.0	1,947	–
Golaghat, Assam	0–15 (L)	42	20.0	68.8	1,641	19,900
LesserHimalaya[a], Jammu and Kashmir	0–15 (CL)	62	5.0[a]	56.0[a]	612[a]	14,221
Ranibundh, westbengal	0–15 (SCL)	38	19.9	79.9	2,636.4	12,909
Ranchi, Jharkhand	0–15 (SCL)	24	8.84	72.1	460.1	24,648
Md. Bazar, West Bengal	0–15 (SCL)	24	3.9	48.0	211	10,296
Phulbani, Orissa	0–15 (SCL)	–	4.0	81.0	725	–
Haveri, Karnataka	0–20 (SCL/CL)	–	2.9	52.3	582	–
Rahuri, Maharashtra	0–13 (C)	16(Coarse clay), 13 (fine clay)	–	365	1,430	9,900[b]
Coimbatore, Tamil Nadu	0–16 (C)	10 (Coarse clay) 9 (fine clay)	–	382	1,750	13,200e
Jabalpur, Madhya Pradesh	0–15 (C)	37	17.0	137	845	–
Rewa, Madya Pradesh	0–15 (C)	–	9.0	171	1051	–
Indore, Madhya Pradesh	0–15 (SC to C)	–	14.8	256	305	12,855

Source: Das et al. (2022).

SCL, Sandy Clay Loam; CL, Clay loam; C, Clay; L, Loam; SC, Sandy Clay.

[a] Data are averages of four locations (Kreeri, Kunzer, Dialgam, Ganasthan), [b] Structural K.

Total K content in soils may show variations over depth. Surface soils had lower K than sub-surface soils (Sharma et al., 2010). In contrast, relatively higher amounts of available K were reported on the surface than in sub-surface soils (Sahoo et al., 2021). However, no regular trend was observed in depth-wise variation of exchangeable K up to 60 cm from the surface in some soils of Assam, India, under varying land-use systems (Das et al., 2019).

30.4 DISTRIBUTION OF POTASSIUM FRACTIONS IN INDIAN SOILS

Traditionally, four pools of soil K include (i) structural K (K_{latt}), (ii) non-exchangeable K (K non-ex, also referred to as fixed K), (iii) exchangeable K (K_{ex}), and (iv) soil solution K (K_{sol}). Together, K_{ex} and K_{sol} constitute ~2% of total K and are often considered readily available to plants (Sparks, 1987), whereas K_{latt} and $K_{non\text{-}ex}$ constitute 98% of total K and are described as slowly (Pal et al., 1999) or potentially available to plants. Based on the assessment of the K status of major soil groups of Tamil Nadu, *viz*., alluvial (*Entisols*), black (*Vertisols*), red (*Ultisols*), and laterite (*Oxisols*) soils, Ramanathan (1977) opined that it is difficult to draw a sharp distinction among various K forms, and it is an equally difficult task to point out as to which is of more importance to plant growth.

Bolan (1976) studied the available K status of 123 soil samples representing 27 profiles of the Nilgiris District of Tamil Nadu and found that the total, water-soluble, exchangeable, and non-exchangeable K varied from 505–6975, 1.0–210, 14.0 to 750, and 13–1,120 mg/kg, respectively. Dhillon et al. (1985) reported that the average water-soluble, exchangeable, available, and boiling 1N HNO_3^- extractable K constituted 0.06%, 0.34%, 0.40%, and 7.75% of total K, respectively, in some benchmark soils of Punjab and Himachal Pradesh. Durairaj Muthiah (1986) observed that total K and available K ranged from 0.280 to 1.440 mg/kg and 80–538 mg/kg, respectively, in ten major soil series of the Thanjavur District of Tamil Nadu.

Studies by Brar et al. (1986) on some alluvial soils and on soils developed on different parent materials of *Vertisols* and red and lateritic soil zones showed that, in the short run, soils with coarse texture and high percentage K saturation would supply K more readily than soils with fine texture and low percentage K saturation. However, in long-term release, fine-textured, high K-saturated, and high buffer power soils can supplement K supply for a longer period of time than other soils. Soils of seven benchmark series from the states of Rajasthan (Masitawali), Gujarat (Pithvajal and Mazodar), Maharashtra (Shendvada and Kumbhave-5), and Madhya Pradesh (Sarol and Kamliakheri series) showed that the soils of Masitawali (alluvial) and Shendvada (swell-shrink) were high in non-exchangeable or reserve K (HNO_3^- soluble) while all others were low. The proportion of non-exchangeable to exchangeable form of K was relatively greater in soils dominated by illite and smectite in comparison to those rich in smectite alone. The influence of texture on available and reserve K status was quite prominent in the former group of soils compared to that observed in the smectite dominant ones (Subba Rao et al., 1988).

Distribution of soil K forms along with mineral composition in the major soil groups of India (Tandon and Sekhon, 1988) is furnished in Table 30.3. A summary on the status of K fractions,

TABLE 30.3
Distribution of Soil K Forms, Its K Content, and Mineral Composition in Major Soil Groups of India

		Percent to Total K as			Dominant Minerals	
Soil Group	**Total K (%)**	**Non-exch. K**	**Exch. K**	**Water-Soluble K**	**Clay**	**Silt**
Alluvial	2.77	5.67	0.32	0.05	Illite	Mica, Quartz
Black	2.45	2.85	0.87	0.08	Smectite	Feldspars
Red	1.96	1.92	0.41	0.10	Kaolinite	Quartz
Laterite	0.27	5.18	3.10	0.47	Kaolinite	Quartz

derived from the average data of 19 representative soil series of India, revealed that mineral K constituted up to 92.6% of total K; non-exchangeable K, exchangeable K, and soil solution K were 6.17%, 1.1% and 0.14%, respectively, of total K.

30.5 "MINING" OF SOIL K UNDER INTENSIVE CULTIVATION IN INDIA

The K requirement of crops, by and large, is as high as that of N, causing its substantial removal from the soil through harvested plant parts, especially under intensive cultivation without adequate K input (Das et al., 2019, 2020). The $N:P_2O_5:K_2O$ consumption ratio in 2019–2020 was 7.14:2.83:1 in India, and it has remained tilted toward N and P (more toward N between N and P) over the last many years despite the average crop uptake pattern being around 1.0:0.3:1.3 (Katyal, 2020).

Apart from the inadequate application rates, the common practice of removing crop residues from fields further aggravates the problem of K mining in Indian soils, as crop removal is the most significant output in K balance equations (Sanyal et al., 2014). Removing crop residues from fields accelerates nutrient depletion, especially K depletion (Goulding et al., 2021). Straw of small grain crops contains less N and P but higher amounts of K than grain. For instance, average straw: grain mass ratio for N and P, respectively, are 0.47 and 0.26 for wheat (*Triticum aestivum*) and 0.49 and 0.35 for barley (*Hordeum vulgare*); but for K they are 4.12 (wheat) and 5.04 (barley) (Tarkalson et al., 2009). Hence, removing straw and grain of these crops would extensively speed up soil K depletion compared to only removing the grain. Rice (*Oryza sativa*) straw contains over 85% of K present in the above ground parts of the plant, and 14–20 kg K_2O ha^{-1} is removed with the removal of 1 Mg of rice straw (Dobermann and Fairhurst, 2002). Potassium removal by rice straw is around six times more compared to removal by the same amount of rice grain. In general, K removal by crop residues in India has been estimated to be approximately five times that supplied through fertilizers.

Outcomes obtained from farmers' fields under the All India Coordinated Research Projects on Integrated Farming Systems (AICRP-IFS) showed the prevalence of soil K mining under several dominant cropping systems of India, including the most important one, *i.e.*, rice–wheat (Majumdar et al., 2017). For these reasons, K accounted for the major portion of reported negative N-P-K balance in Indian soils over the past few decades (Katyal, 2020).

Mining of K caused by imbalanced fertilization, including inadequate or no K input through fertilizer for prolonged periods, implies over-dependence of the prevailing cropping systems for K nutrition on a soil's native K reserves. The K demand of crops over the "depletion in the exchangeable pool" and "external K input" is met by the non-exchangeable K pool (Bilias and Barbayiannis, 2019). Consequently, continuous K mining alters the dynamic equilibrium: soil solution K $\rightleftharpoons$ exchangeable K $\rightleftharpoons$ non-exchangeable K and increases the net flux of K^+ in the backward direction. As a result, large portions of non-exchangeable K are released to compensate for the losses in solution and exchangeable pools. Due to the sluggish rate of weathering of primary minerals, *e.g.*, micas, feldspars, and associated slow K release, the structural pool of K has very little involvement in supporting plant K nutrition (Sanyal, 2014).

Hence, soil K mining is most likely to manifest itself as depletion in exchangeable and non-exchangeable K pools. The problem might be more severe in soils inherently poor in available and non-exchangeable K, *e.g.*, kaolinitic *Alfisols* of India (Srinivasarao and Vittal, 2007; Das et al., 2019). *Inceptisols* of the Indo-Gangetic Plains (IGP) contain large amounts of mica/illite in clay and silt fractions; consequently, they have large reserves of non-exchangeable and structural K (Majumdar et al., 2017). Most of the smectitic *Vertisols* of central and southern parts of India are high in exchangeable K and medium to high in non-exchangeable K content (Srinivasarao et al., 2007). Irrespective of soil orders, total K content is generally high. The K-rich soils may not show any noticeable effect of K mining on K pools and crop production for a brief timeframe, though their K fertility status may gradually decline.

Over a long time, plant-available K reserve (*i.e.*, exchangeable and non-exchangeable pool) could get exhausted due to K mining to the extent that even the inherently K-rich soils become unable to support desired levels of crop yields without adequate K input from outside (Das et al., 2018, 2021). Mining of soil K for prolonged periods has been found to adversely affect the K supplying capacity of a kaolinitic *Alfisol* (Ranchi), illitic *Inceptisols* (Cuttack and New Delhi), and a smectitic *Vertisol* (Jabalpur) of India (Das et al., 2021). Occasionally, mineral distortion was also observed due to long-term soil K mining in India (Das et al., 2019). Besides causing soil K mining, inadequate or no application of K has also been suspected as one of the primary reasons behind the persistent low N use efficiency (floating around 30%–40%, rarely 50%) despite the adoption of better agronomic practices and use of improved crop varieties over time (Katyal, 2020). Negative K balance in agricultural fields is a threat to food security in India and most of the developing and several developed nations (Lu et al., 2017). Soil K depletion has been marked as one of the main reasons for stagnation of crop yields and poor nutrient use efficiency in present-day intensive cultivation (Regmi et al., 2002). If allowed to go unchecked, soil K mining will eventually deteriorate soil health in general, soil K fertility in particular, and ultimately jeopardize agricultural sustainability as a whole.

30.6 REASONS BEHIND SOIL K "MINING" IN INDIA

Mining of soil K in agricultural lands occurs when K balance is negative, *i.e.*, removal by the harvested portion of crops exceeds the external K input, including the K recycled back to the ground with crop residues (Sanyal et al., 2014; Majumdar et al., 2017). In India, the removal of K is more or less equivalent to uptake for field crops, as their residues are mostly removed from the fields after harvesting. Crop residues and animal excreta (*e.g.*, cow dung) generally find use as a source of energy for cooking and heating in rural areas, rather than being recycled back to the fields. Crop residues have plenty of other uses in rural households in India, prompting their removal from the crop fields. On top of that, K fertilization is nil or severely inadequate in many parts of the country, including the intensively cultivated areas. Though local recommendations advocate proper balance among N, P, and K fertilizers, most farmers opt for N only or N and P with little or no K fertilizer. Surveys conducted in the upper and trans-Gangetic Plains of India revealed that most of the farmers apply N at more than recommended rates, along with P at nearly recommended rates, but apply minor amounts of K that are less than recommended rates, if applied at all (Singh et al., 2015).

One of the primary reasons behind the neglect of K fertilization in India is the general belief that Indian soils are inherently abundant in K reserves owing to K-containing minerals and can support crops without K fertilization. Secondly, as India is almost 100% dependent on imports for K fertilizers, there has always been an attempt to curtail K fertilization to reduce the burden on the national exchequer. Next comes the rise in K fertilizer prices after decontrol in August 1992 [from INR (Indian Rupee) 2.83 before decontrol to INR 7.50 kg^{-1} K_2O after decontrol]) and the introduction of a nutrient-based subsidy in April 2010 (from INR 8.43 in 2010–2011 to INR 28.33 kg^{-1} K_2O in 2012–2013), which further worsened the situation (Sanyal, 2014). Moreover, farmers are unaware that yield increment due to K fertilization can translate into economic gains, as discussed later. Also, farmers are most likely oblivious of the ongoing soil K mining caused by imbalanced fertilization and its harmful effects on soil fertility and crop productivity. The first two reasons are mainly at the level of policy makers; there stare at the farmers' level. For these reasons, fertilizer K consumption remained only ~10% of total fertilizer (NPK) consumption in India for the last five decades (Majumdar et al., 2017).

In reality, not all soils in India are rich in K-containing minerals, *e.g.*, red and lateritic soils (often classified as kaolinitic *Alfisols*). Compilation of soil test data of 500 districts across the country revealed that in ~51% of the cases, soils were low (<120 kg/ha) to medium (120–280 kg/ha) in available K (NH_4OAc-K) (Muralidharudu et al., 2011). Also, an abundance of K-containing minerals does not ensure adequate K supply to plants at required rates and at times of highest crop demand, especially when the micaceous minerals are mostly dioctahedral.

30.7 MAJOR CROPS AND CROPPING SYSTEMS OF INDIA AND K UPTAKE

As per the estimates for 2015–2016, net sown area (NSA) and gross cropped area (GCA) in India were 139.51 and 195.05 million ha (Mha), respectively, with a cropping intensity of ~141%. Cereals have the largest share (~62%) in GCA, followed by oilseeds (~14.4%), pulses (~11.6%), fiber crops (6.56%), vegetables (3.46%), and fruits (2.36%). Individually, rice has the highest share (~22%) in GCA, followed by wheat (~16%). Potassium uptake per unit mass of economic yield, expressed as kg K Mg^{-1}, varies from one crop to another, *e.g.*, ~16 for rice; ~19 for wheat (Chuan et al., 2013); ~7 for groundnut (*Arachis hypogaea*), ~1.2 for sugarcane (*Saccharum officinarum*), and ~12 for cotton (*Gossypium* spp.) (Dutta et al., 2013). However, due to the largest area coverage, annual N, P, and K uptakes (for 2015–2016) at the country scale were the highest for cereal crops. Table 30.4 shows annual N, P, and K uptake for 2015–2016 (at country scale) by different crop groups of India calculated based on crop production data (2015–2016) and nutrient content per unit mass of primary produce. Potassium uptake and balance are strongly influenced by cropping systems. Unless climate and other factors become limiting, two or more crops are usually grown in sequence within a year in much of India (Majumdar et al., 2021).

Over 60% of Indian soils fall under three soil orders, *viz.*, *Inceptisols*, *Alfisols*, and *Vertisols*. *Inceptisols* have the highest area coverage (39.4%) (Singh and Chandran, 2015). These soils are considered to be agriculturally productive and offer outstanding natural grazing grounds. The soils are cultivated chiefly for grain crops, *e.g.*, rice, wheat, maize, and barley; oilseeds, *e.g.*, mustard; improved pasture, and various vegetables and fruits. The range of crops grown varies with agro-climatic zones. A large part of the *Inceptisol* area belongs to IGP, which in turn occupies nearly 13% of the total geographical area (TGA) of the country.

Alfisols cover 42.2 Mha, which is around 13% of the TGA of the country (Bhattacharyya et al., 2013). These soils are generally low in available K content, making it difficult to grow crops without external K input (Srinivasarao et al., 2007). However, these soils are well suited for forestry, grazing, and, once cleared, a range of grain crops, fruits, and vegetables can be grown. *Alfisols* are mainly seen in Southern and Eastern India, supporting upland rice, wheat, groundnut, and fingermillet (*Eleusine coracana*) based systems (Srinivasarao et al., 2007). Among the major crops grown in *Alfisols* of Southern India, fingermillet is one of the high K demanding crops which can remove K as high as 50 kg/ha (in low rainfall areas) to 150 kg/ha (under optimum rainfall conditions) during its life cycle (Srinivasarao et al., 2014).

TABLE 30.4
Nutrient Uptake by Various Crops

	Nutrient Uptake (Mt)		
Crop Group	N	P	K
Cereals	6.56	1.344	7.71
Pulses	1.21	0.121	0.61
Oilseeds	1.38	0.244	0.93
Fiber	1.02	0.183	0.70
Fruits	0.51	0.059	0.88
Vegetables	0.56	0.089	0.73
Plantation	0.49	0.045	0.41
Forage	1.00	0.103	1.03
Spices	0.18	0.038	0.21
Sugarcane	0.82	0.205	1.11
Tuber	0.04	0.007	0.04
Others	0.02	0.004	0.03
Total	**13.78**	**2.44**	**14.37**

The *Vertisols* are found primarily in Peninsular India, covering 26.6 Mha constituting around 8.1% of the TGA (Singh and Chandran, 2015). Groundnut, soybean, cotton, maize, and pigeon pea are the most common crops grown during the rainy season, while crops like safflower (*Carthamus tinctorius*), chickpea (*Cicer arietinum*), sunflower (*Helianthus annuus*), and sorghum (*Sorghum bicolor*) are preferred during the post-monsoon period (Bhattacharyya et al., 2013). *Vertisols* are generally high in available K and medium to low in non-exchangeable K, which may get depleted under long-term cropping, especially without adequate K input (Srinivasarao et al., 2007). In addition, neglecting K fertilization in *Vertisols* (black soils) sometimes caused acute K deficiency in plants grown therein (Gurav et al., 2019).

30.8 SOIL K "MINING" WITH MAJOR CROPS AND CROPPING SYSTEMS

Tewatia et al. (2017) averaged all crop groups of India and calculated a net balance (unit area basis) of −42.2 kg K_2O ha^{-1} for 2015–2016. Negative K balance was the highest for sugarcane (−121.2 kg K_2O ha^{-1}), followed by fruits (−70.9 kg K_2O ha^{-1}), and lowest for pulses (−14.7 kg K_2O ha^{-1}). The negative balance for cereals (−45.9 kg K_2O ha^{-1}) was near the average of all crop groups, since cereals account for the most significant portion of GCA in India (Tewatia et al., 2017; Anonymous, 2019). Consequently, cereal crops have the most significant impact on the K balance in Indian agriculture. On-farm studies (trials conducted at farmers' fields) on major cropping systems conducted under the aegis of AICRP-IFS across India indicated that K among the primary nutrients is the most neglected in the fertilization schemes followed by farmers (Table 30.5). Those studies recorded large negative K balances under farmers' fertilizer practices (FFP), irrespective of cropping systems. Interestingly, existing state recommendations (SR) were also notable to prevent soil K mining. However, for most of the cropping systems studied, K mining under FFP was slight to extensively higher compared with SR. The application of deficient secondary and micronutrients along with state recommended NPK (SR+M) caused greater mining of K due to an increase in yield and associated K uptake.

However, due to continuous ignorance toward K fertilization, the soils in these areas face accelerated degradation (Majumdar et al., 2017). A rice–wheat (R-W) system, a dominant system in IGP, could remove K from soil to the amount of 325 kg K ha^{-1} (Singh et al., 2003). Moreover, deleterious effects of alternate flooding and drying adopted in the R-W system in these areas have further aggravated K loss from exchange complexes. Singh et al. (2013) observed a decrease in both exchangeable and non-exchangeable K in the IGP under the R-W system, where farmers did not add K. Shukla et al. (2005) assessed the NPK balances for the different sub-regions of the IGP, *viz.*, trans, upper, middle, and lower IGP. They found that the net balance was negative for each of the three nutrients in all sub-regions, except for N in the middle IGP. For K, the situation was much worse than N and P, as evidenced by its large share in the deficit (82%) compared with 12.5% and 5.5% in the case of P and N, respectively.

30.9 STRATEGIES FOR SUSTAINABLE MANAGEMENT OF POTASSIUM IN AGRICULTURE

Developing effective K management strategies that could help to reduce K losses and somehow return the depleted K to the soil to meet crop demand is vital for agricultural sustainability. Various aspects of K management are discussed here under:

30.9.1 Correct Assessment of Soil K Availability to Crops

An effective K fertilizer suggestion would provide optimal quantity and proportions of K supply to satisfy crop requirements, assist in developing the ideal K balance in soil, and ensure improved K and other nutrient usage efficiency (e.g., N). Currently, K fertilizer rate recommendations are

TABLE 30.5
Potassium Balance at Locations of Long-Term Fertilizer Experiments (LTFE) under Different Cropping Systems

Soil Type/LTFE Centre	Number of Years of Cultivation	Cropping System	Nutrients Applied	Total K (kg/ha)			
				Applied	Uptake	Apparent Balance	Apparent Balance/Year
			Inceptisols				
New Delhi	34	Maize- wheat	Control	0	3,672	–3,672	–108
			100% N	0	5,848	–5,848	–172
			100% NP	0	8,432	–8,432	–248
			100% NPK	2,244	9,996	–7,752	–228
			100% NPK+FYM	4,794	13,804	–9,010	–265
Coimbatore	44	Finger millet–maize	Control	0	5,601	–5,601	–127
			100% N	0	7,293	–7,293	–166
			100% NP	0	9,198	–9,198	–209
			100% NPK	1,892	9,893	–8,001	–182
			100% NPK+FYM	4,642	11,876	–7,234	–164
Jagtial	15	Rice–rice	Control	0	1,226	–1,226	–82
			100% N	0	1,738	–1,738	–116
			100% NP	0	2,246	–2,246	–150
			100% NPK	990	2,293	–1,303	–87
			100% NPK+FYM	1,740	2,507	–767	–51
			Vertisols				
Jabalpur	44	Soyabean–wheat	Control	0	3,667.84	–3,668	–83
			100% N	0	4,773.56	–4,774	–108
			100% NP	0	8,595.84	–8,596	–195
			100% NPK	2,200	11,402.6	–9,203	–209
			100% NPK+FYM	5,500	15,906.44	–10,406	–237
Raipur	20	Rice–wheat	Control	0	2,620	–2,620	–131
			100% N	0	4,300	–4,300	–215
			100% NP	0	4,900	–4,900	–245
			100% NPK	1,320	5,460	–4,140	–207
			100% NPK+FYM	1,820	5,540	–3,720	–186

(Continued)

TABLE 30.5 *(Continued)*
Potassium Balance at Locations of Long-Term Fertilizer Experiments (LTFE) under Different Cropping Systems

Soil Type/LTFE Centre	Number of Years of Cultivation	Cropping System	Nutrients Applied	Total K (kg/ha)			
				Applied	Uptake	Apparent Balance	Apparent Balance/Year
Parbhani	20	Safflower–wheat	Control	0	1,611	−1,611	−81
			100% N	0	1,831	−1,831	−92
			100% NP	0	2,730	−2,730	−137
			100% NPK	500	3,000	−2,500	−125
			100% NPK + FYM	1,500	3,426	−1,926	−96
Akola	20	Sorghum–wheat	Control	0	395	−395	−20
			100% N	0	3,552	−3,552	−178
			100% NP	0	1,443	−1,443	−72
			100% NPK	1,800	3,958	−2,158	−108
			100% NPK + FYM	2,800	5,446	−2,646	−132
			Alfisols				
Ranchi	44	Soyabean–wheat	Control	0	969	−969	−22
			100% N	0	1,222	−1,222	−28
			100% NP	0	2,769	−2,769	−63
			100% NPK	2,904	4,095	−1,191	−27
			100% NPK + FYM	5,104	5,528	−424	−10
Bangalore	22	Fingermillet–maize	Control	0	381	−381	−17
			100% N	0	456	−456	−21
			100% NP	0	568	−568	−26
			100% NPK	924	2,244	−1,320	−60
			100% NPK + FYM	2,574	2,845	−271	−12
Pattambi	20	Rice–rice	Control	0	1,042	−1,042	−52
			100% N	0	925.4	−925	−46
			100% NP	0	984.8	−985	−49
			100% NPK	1,520	1,196.2	324	16
			100% NPK + FYM	2,020	1,404.4	616	31

FYM, Farmyard manure.

based on the crop to be grown and soil K fertility as determined by neutral normal NH_4OAc, which extracts only the water-soluble and exchangeable components. Such suggestions have frequently been shown to be insufficient under intense agriculture and ineffective in emphasizing cropping-induced changes in soil K (Das et al., 2018, 2019). The dynamic equilibrium operating among various forms of K and the non-exchangeable pool has been critical in facilitating the luxurious K uptake by crops (Dey et al., 2017b). This is due to the non-exchangeable pool of K in soil contributing to plant uptake, which is particularly important in soils deficient in NH_4OAc-K (Ghorban, 2007). This is why some previous researchers used non-exchangeable K to properly analyze the available K. To include the non-exchangeable pool of K in defining soil K status, Srinivasarao et al. (2007) divided numerous Indian soils into low, medium, and high categories based on exchangeable and non-exchangeable K pool sizes, with the non-exchangeable K recovered using the boiling nitric acid technique. Li et al. (2015) discovered that the NaBPh4 (sodium tetraphenylborate) approach is far more selective to non-exchangeable K than the standard boiling nitric acid method, with the former releasing >99% of extracted K by cation exchange (the remainder through mineral structure dissolution). In the future, researchers should use the NaBPh4 approach rather than the traditional nitric acid method to classify soils for their non-exchangeable K status and provide adequate extraction time so that no interlayer K is lost. Furthermore, future studies should concentrate on separation of the entire non-exchangeable pool based on plant availability.

The NaBPh4 extraction technique, while more specific than boiling nitric acid toward non-exchangeable K (Li et al., 2015) and which has consistently shown superior predictions of plant K absorption than NH_4OAc (Bilias et al., 2021), is not without flaws. This approach cannot anticipate the real amount of diffusion-derived non-exchangeable K that plants may absorb, a limitation shared by all chemical extraction methods (Menon et al., 1989).

Cation-exchange resins may be beneficial in measuring the availability of soil K to plants in this situation. High CEC cation-exchange resins can operate as a sink for soil solution K and non-exchangeable K released for plant uptake (Helmke and Sparks, 1996). The quantity of K extracted using H-saturated resin may be equivalent to that extracted by acid treatment, although it is significantly less harmful in terms of mineral structure disintegration (Arvieu and Chaussidon, 1964). Cation-exchange resins have previously demonstrated their capacity to imitate plant K absorption by reflecting solution K activity and diffusion rate (Schaff and Skogley, 1982). Bilias et al. (2021) discovered that K recovered from H-resin could accurately predict plant K absorption from soils containing K-fixing 2:1 clay minerals for extraction times of 1 or 5 minutes (r^2=0.63 or 0.69, p = 0.001). With NaBPh4 and extraction times of 1 or 5 minutes, similar prediction accuracy (r^2=0.65 or 0.67, p=0.001) was found.

Thus, to target the plant-available portion of non-exchangeable K, one can use NaBPh4 with a short extraction duration (*i.e.*, 1–5 minutes) and H-saturated resin; conversely, to estimate the complete non-exchangeable K reserve, NaBPh4 with a substantial extraction time (10–20 days) should be used. Extraction of soil K with low-molecular weight organic acids, such as citric, oxalic, malic, and tartaric acid, can similarly simulate plant root activity.

Hydronium (H_3O^+) ions produced by organic acid dissociation can replace K^+ from planar sites and clay interlayers; the anionic component of the acid complexes the K^+ in solution and hinders sorption on the solid phase. Hosseinpur et al. (2014) used 0.01 M citric acid to study K release for up to 168 hours. The associated release rate constants were shown to be highly connected with bean (*Phaseolus vulgaris*) yield and K absorption in 15 distinct Iranian soils. K fixation and release play a crucial role in K supply to plants in soils with significant concentrations of 2:1 minerals in the clay fraction, which is not taken into account in current K fertilizer prescription systems (Das et al., 2018; Bilias and Barbayiannis, 2019). K fixation and release thresholds in the soil can assist to determine the rate of K fertilization for crop yield. Datta (2005) established the methodology for determining soil fixation and release threshold levels (FTL and RTL). The FTL is the soil solution or exchangeable K level over which K fixation into the non-exchangeable pool begins. In contrast, RTL is the level of soil solution or exchangeable K below which non-exchangeable pool release begins (Datta, 2005).

According to this technique, the K fertilization rate that might keep soil solution or exchangeable K at an ideal level between the FTL and RTL appears realistic. It would promote plant K nutrition without triggering mining from or fixing in the non-exchangeable pool (Das et al., 2018; Datta, 2005). This method might be enhanced further by taking into account the real K requirements of the crops to be produced. Based on plant growth indices and soil K release rate, Li et al. (2015) established a soil bioavailability grading system for K. To characterize the release of K in soil, they constructed a link between the theoretical soil capacity for sustainable plant development and the availability of K without additional K fertilizer.

Recently, exchangeable and non-exchangeable K extraction techniques were combined with quantity-intensity (Q/I) characteristics, CEC, and illite-K to develop a method for estimating K availability based on thermodynamic parameters acquired from Q/I correlations (Bilias and Barbayiannis, 2019). Despite the fact that the Q/I methodology provides greater accuracy than the existing method for forecasting soil K availability, creating multipoint-Q/I relationships is difficult and time-consuming and is not suitable for regular soil testing. Fortunately, it has been possible to predict different Q/I parameters for a wide variety of soils (concerning K status) using a single-point (using 10 mg K/L in 0.01 M $CaCl_2$ for equilibration) method, where CEC and the Gapon selectivity coefficient could further increase prediction accuracy (Bilias and Barbayiannis, 2020).

30.9.2 Sources of K in Agriculture

30.9.2.1 Fertilizers

Conventional fertilizer sources, such as muriate of potash (MOP) and sulfate of potash (SOP), are frequently used to address plants' immediate K needs. However, for K fertilizers, India is fully reliant on imports, putting a significant strain on the national budget. Santhi et al. (2013) reported that out of 11 high fertilizer consumption districts of Tamil Nadu, six belonged to "high K" and five districts fell under "medium-K" category Furthermore, while K from water-soluble fertilizers (e.g., MOP, SOP) is easily accessible to plants, it is also susceptible to leaching losses, particularly in sandy soils with low clay and organic matter content (Alfaro et al., 2004). Furthermore, the K mineral sources utilized in K fertilizer manufacturing are not limitless and may be depleted as a result of the increased demand caused by intensive agriculture. As a result, sustainable alternate sources of K should be sought in order for Indian agriculture to attain self-sufficiency in K nutrition. To restore soil K, this requires for a progressive transition from K fertilizers to indigenous K sources such as crop residues, manures, wood ash, distillery wastes, coir waste, and indigenous low-grade K-bearing minerals (Basak and Sarkar, 2017).

30.9.2.2 Crop Residues

Crop residues containing 1% or more K, such as rice straw (1.75% K), wheat straw (1% K), sugarcane bagasse (1.25% K), oat straw (1.55% K), and tobacco (*Nicotiana* spp.) stems (7% K), can serve as a sustainable alternative supply of K in agriculture (Basak and Sarkar, 2017). After harvesting grain crops, around 50%–70% of total biomass is left as residues, which would contain a significant quantity of K if returned to the soil. However, because decomposition and subsequent nutrient release from straw are time-consuming processes in the field, the positive impacts of straw-return on soil K status and crop yields may not be noticeable in the immediate term (Partey et al., 2011). Nonetheless, straw-return might reduce soil K depletion during heavy cropping, improve soil K status, and raise crop yields in the long run (Zhao et al., 2014). Crop left-overs might be integrated directly into the soil, processed to compost or biochar, before being applied to the soil, or left on the surface as part of conservation agriculture to be used as a K source (Katyal, 2020).

30.9.2.3 Manures and Composts

Organic manures and composts, while primarily utilized to increase organic matter content and overall soil fertility, can also serve as alternate sources of K depending on their K concentration, which varies mostly depending on their components and preparation process (Basak and Sarkar, 2017). For example, farmyard manure (FYM) (0.5% K), poultry manure (1.5%–3% K), pig manure (0.5%–1.2% K), urban/rural composts (0.5%–2% K), night soil (0.4% K), oil cakes (0.8%–1.4% K), and fish meal (1.5% K) could be used to boost soil fertility and supplement K fertilizers in agriculture (Basak and Sarkar, 2017). Furthermore, field tests have indicated that NPK+FYM have a greater favorable impact on soil K supply and accessible K fractions than unbalanced fertilization (without K) (Das et al., 2018, 2019).

30.9.2.4 Minerals

Suggestions for using silicate rocks with K-containing minerals as K sources are not new. Tectosilicates, such as orthoclase (16% K) and microcline (16% K), and phyllosilicates, such as muscovite (12% K), biotite (6%–10% K), phlogopite (7%–10% K), and glauconite (5%–8% K), are examples of K-containing minerals (Figure 30.1) that can be used as an alternative K source (Shirale et al., 2019). Instead of producing K fertilizers, the direct usage of these rocks is often regarded more acceptable and cost-effective. However, due to the relatively poor solubility of silicate rocks, the rate of release of K from such sources may be substantially less than what is necessary for optimum plant feeding if given directly. Bio-activation using K-solubilizing or silica-dissolving microorganisms might aid in achieving a desired K release rate from these sources.

Though K-containing minerals are significantly less effective than traditional fertilizers (e.g., MOP, SOP) in terms of supplying instantly available K, the former might give enough K over a lengthy period of time to provide agronomic advantages, particularly in heavily worn soils prone to leaching (Basak and Sarkar, 2017).

Another indigenous K source might be waste mica (6.5%–10% K), which is produced in large amounts as a by-product of the electrical sector. Waste mica may be efficiently used in agriculture as a K source with adequate treatment *via* biological or chemical techniques (Nishanth and Biswas, 2008).

FIGURE 30.1 Silicate rocks containing potassium minerals as K sources.

Waste mica applied straight after grinding might boost biomass production of Sudan grass (*Sorghum vulgare var.* Sudanese) compared to no fertilization; however, waste mica could further increase biomass output with appropriate bio-inoculation (Basak and Biswas, 2009). Aside from silicates, polyhalite, which is found in sedimentary marine evaporation, might be a viable K source for agricultural application. It has 13% K, 13% Ca, 6% Mg, and 19% S (Sirius Minerals, 2016). Polyhalite deposits in India are estimated to be over 16,000 Mt (Anonymous, 2018). This demonstrates the wide range of applications for this mineral as a K source in agriculture.

30.9.2.5 Potassium-Releasing Bacteria

Through complicated creation and exchange events from fixed K or structural pools, potassium-releasing bacteria (KRB) can enable the release of insoluble K. A diverse range of *Bacillus*, *Burkholderia*, *Acidithio bacillus*, *Paeni bacillus*, and *Pseudomonas* species have demonstrated encouraging results in releasing in accessible and slowly available K for crop K needs. These bacteria can be used with or without a K source, such as waste mica, to get good results. Several pot culture experiments (e.g., Han and Lee, 2005; Sugumaran and Janarthanam, 2007; Basak and Biswas, 2009) and a few field research studies (Supanjani et al., 2006) have indicated that KRB improve crop yields and K uptake. However, we believe that using KRB without an external K supply, such as waste mica, might have a negative impact on soil health since KRB will dissolve or modify the K-bearing minerals in the soil, resulting in soil K mining (Figure 30.2).

Furthermore, even when used in conjunction with an external K source, it is uncertain if the solubilization is confined to the extra K source. Future research should seek to quantify both the good and negative impacts of employing KRB alone or in combination with an external K source (*e.g.*, waste mica or glauconite) on soil health and crop performance.

30.9.3 Time and Method of Application

Application of K fertilizer is largely specific to the crop and location. Because K is relatively inert in soil, K fertilizers are often applied during field preparation prior to sowing. However, the mobility of K ions within the soil can be influenced by soil type heterogeneity in terms of CEC and texture. Because of its high demand, recommendations for K are dependent on a number of criteria,

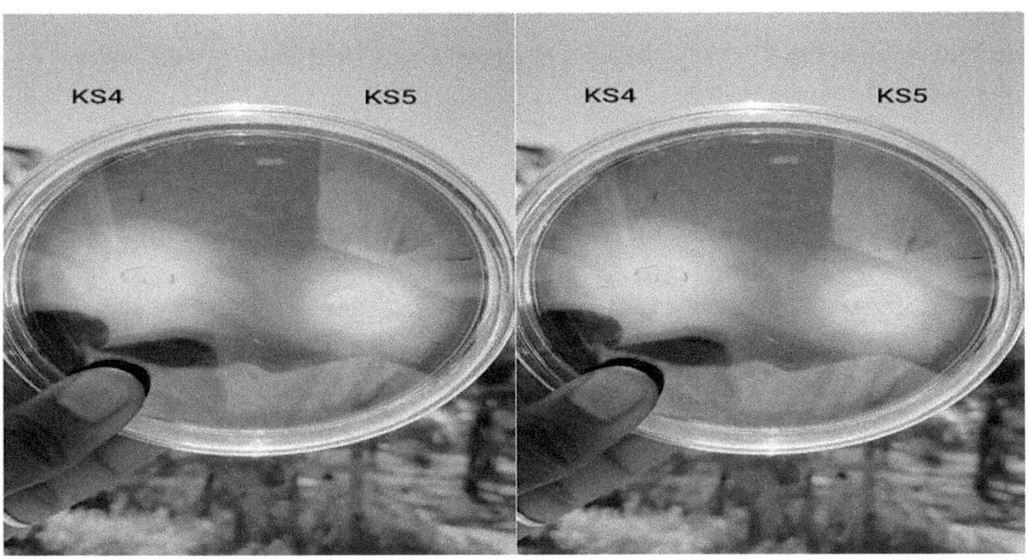

FIGURE 30.2 Potassium solubilization on modified Aleksandrow agar plate supplemented with potassium aluminum silicate. (Laxita and Shruti (2020).)

including crop yield, soil K status, climatic circumstances, management strategies, and the stage and season of plant growth. Although strong K dressing every few years may be more effective, regular mild dressing may be more beneficial in lowering the luxury use of K in particular plants. This may also aid in preventing excessive K loss due to leaching and fixing prior to plant uptake (Brady and Weil, 2012).

K fertilizers are often spread and combined with surface soil. Band placement is only indicated in soils with low accessible K levels or soils with strong K-fixing potential (Subba Rao and Brar, 2009). In certain cases, split application occurs as an alternative to basal application (Mahfuza et al., 2008). Furthermore, cotton and other crops with a high K need benefit from pre-plant soil treatment and/or mid-season side-dressing. Adjusting the time and type of K treatment can enhance K usage efficiency in foliar or fertigation farming systems.

Drip irrigation may improve water and K usage efficiency (Volenec et al., 2021). Due to infrastructure constraints, such systems are best suited for high-value plantation crops such as fruit trees. Fertigation on fruit trees resulted in equal downward movement of K in the soil as broadcasting, while fertigation resulted in greater lateral movement of K than broadcasting (Neilsen et al., 1999). Deep application of K fertilizer (8–10 cm) can be effective in heavy rainfall locations prone to run-off losses, such as the pocket technique, in which fertilizer is placed in or around planting pockets (Aune et al., 2017). In soils with high K-fixing capability, band placement of K fertilizer is more effective than broadcasting.

30.9.4 Use of Decision Support Tools

The International Plant Nutrition Institute (IPNI) and numerous national and international partners created the fertilizer decision assistance tool Nutrient Expert®, which has been successfully utilized to predict site-specific K recommendations for key cereal crops and cropping systems. On-farm studies in several Indian agroecologies revealed that tool-based K recommendations for rice–wheat, rice–rice, rice–maize, and maize–wheat systems, based on the nutrient requirement of a high-yielding cereal crop and soil nutrient providing capability, improved crop yield and K uptake above existing farmers' fertilizer practice (Table 30.6).

TABLE 30.6
Effect of Nutrient Expert® (NE) – Based K Inputs on Grain Yield and K Uptake under Cereal-Based Cropping Systems

		K Use			Grain Yield			K Uptake	
		FFP	NE	Diff. (NE-FFP)	FFP	NE	Yield Grain Over FFP	FFP	NE
Cropping System	Crop	kg K_2O ha^{-1}			Mg ha^{-1}		%	kg K ha^{-1}	
Rice–wheat	Rice	40	67	+27	5.1	5.83	14	104	139
	Wheat	0	64	+64	3.66	4.18	14	89	101
Rice–rice	Rice (m)	60	61	+1	4.28	4.98	16	81	106
	Rice (w)	52	46	+6	4.67	5.38	15	91	124
Maize–wheat	Maize	19	44	+25	2.1	3.5	67	57	78
	Wheat	0	63	+63	2.21	3.51	59	52	81

Source: Singh et al. (2021).
FFP, farmer fertilizer practice; NE, Nutrient Expert recommendations; Rice (m), rice grown during the monsoon season; rice (w), winter season.

TABLE 30.7
Effects of a Single Basal K Application (State Recommendation) and Split K Applications (Nutrient Expert®) on Grain Yield K Partial Factor Productivity (PFP_K) and Grain Produced per Rupee Invested

	Ferilizer K Rate					
	Basal	40 days after Sowing	Total	Grain Yield	PEP_k	Grain Produced per Rupee Invested
K Recommendation		kg K ha^{-1}		Mg ha^{-1}	kg Grain (kg K)$^{-1}$	kg Grain (Rupee Invested^{-1})
State	62	–	62	4.10	66.1	1.68
NE using 6 Mg ha^{-1} yield target	38	25	63	5.59	88.7	2.26
NE using 6 Mg ha^{-1} yield target	40	27	67	6.19	92.4	2.35

Source: Unpublished data from authors Singh et al. (2021).
NE, Nutrient Expert; State, state recommendation.

To establish the K application rate, the Nutrient Expert analyzes yield objectives as well as soil indigenous nutrient providing capacity. It also suggests when to apply the fertilizer to satisfy K needs during the crop's physiological development phases. This integrated K supply to the crops resulted in increased K uptake and K use efficiencies. Recent research in the IGP found that the Nutrient Expert's K recommendations changed with altering yield objectives. With the highest yield objectives (7 t/ha), the yield gain and efficiency metrics (partial factor productivity and economic KUE) were greater (Table 30.7).

The computer software DSSIFER – Decision Support System for Integrated Fertilizer Recommendation – has been developed at Tamil Nadu Agricultural University (Murugappan et al., 2004; Santhi et al., 2010, 2020), and its use for K fertilization, based on Soil Test Crop Response under the Integrated Plant Nutrition System (STCR-IPNS), for various agricultural and horticultural crops in the state of Tamil Nadu, India, resulted in rationalized fertilizer usage and sustained soil health.

30.9.5 Scientific Tools/Interventions for Assessing Soil Fertility

Scientific Tools/Interventions for Assessing Soil Fertility GPS- and GIS-based soil fertility maps generated for 173 districts of the country under the aegis of the All India Coordinated Research Project on Soil Test Crop Response (AICRP-STCR) at ICAR-IISS (Indian Council of Agricultural Research – Indian Institute of Soil Science), Bhopal, provide a pathway to understand the fertility status of different regions and the way to proceed further for soil fertility improvement (http://www.iiss.nic.in/districtmap.html). The GPS/GIS-based maps showing the K status of the Cuddalore District of Tamil Nadu are depicted in Figure 30.3. The validation experiments and frontline demonstrations conducted under AICRP-STCR brought out that the soil test and target-based fertilizer prescription for crops and cropping sequences resulted in higher productivity, efficient fertilizer use, and profitability (Dey and Santhi, 2014). Besides a traditional mineralogical approach, possibilities of K availability indices were also ascertained through mid-infrared (MIR) spectroscopy (Dey et al., 2017b). Decision support tools, *viz.*, the web-based online fertilizer recommendations for different states developed by AICRP-STCR and the National Informatics Centre (NIC), Pune (http.//www.stcr.gov.in), and by AICRP-STCR, NIC and DAC, New Delhi (http:// www.soilhealth.dac.gov.in); besides, the DSSIFER computer software (Santhi et al., 2010, 2020), Nutrient Expert of IPNI have proved to be useful to various stakeholders.

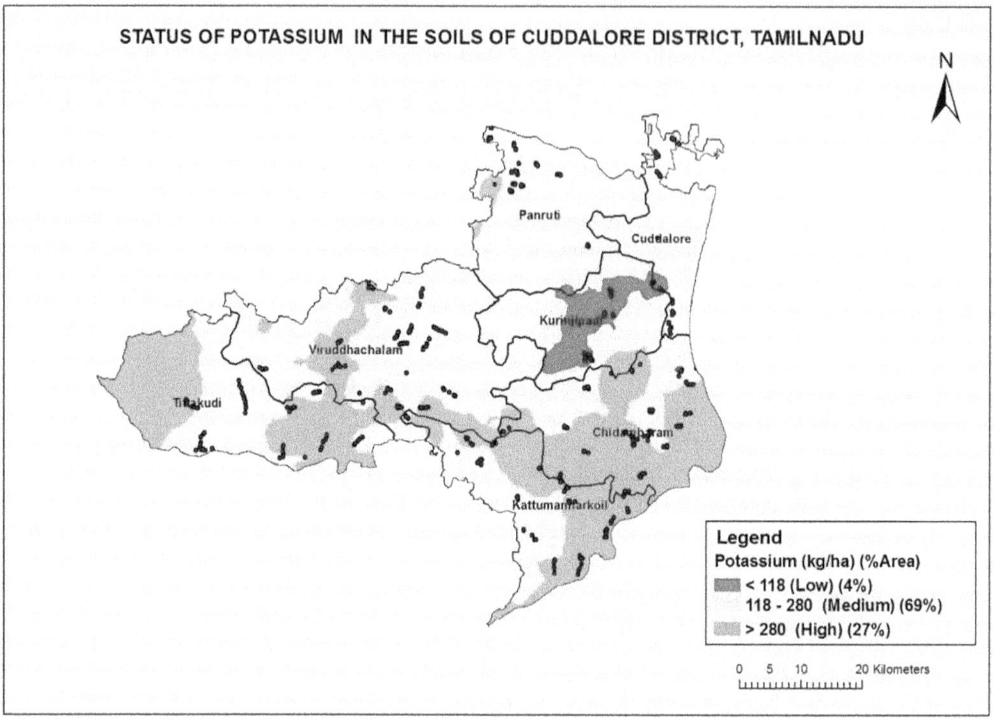

FIGURE 30.3 GPS/GIS-based maps showing the K status of Cuddalore district of Tamil Nadu, India. (Santhi et al. (2013).)

30.10 WAY FORWARD

Research work on K status warrants attention toward continued mining of native soil K in intensive cropping and rainfed areas is a major concern. Research aspects in line with the mineralogical contribution of K, K stratification and addition of K through rainfall, irrigation water and sedimentation warrants in depth investigation and measures for the prevention of leaching loss and solubilization of K from indigenous sources need attention.

REFERENCES

Alfaro, M. A., Jarvis, S. C., and Gregory, P. J. (2004). Factors affecting potassium leaching in different soils. *Soil Use and Management*, 20(2), 182–189.

Anonymous (1998). Potassium application methods. *Better Crops*, 82, 24–27.

Anonymous (2018). *Indian Minerals Year Book 2017*, Part III: Mineral Reviews, 56th ed. Potash, India: Ministry of Mines, Indian Bureau of Mines, Government of India.

Anonymous (2019). *Pocket Book of Agricultural Statistics 2019*. New Delhi, India: Department of Agriculture, Cooperation & Farmers Welfare, Government of India. https://eands.dacnet.nic.in/PDF/Pocket%20Book%202019.pdf, Accessed 12 June 2021.

Arvieu, J. C., and Chaussidon, J. (1964). Etude de la solubilisation acide d'une illite: Extraction du potassium et évolution du résidu. *Annales Agronomiques*, 15(3), 207–229.

Aune, J. B., Coulibaly, A., and Giller, K. E. (2017). Precision farming for increased land and labour productivity in semi-arid West Africa: A review. *Agronomy for Sustainable Development*, 37, 16–32. https://doi.org/10.1007/s13593-017-0424-z.

Basak, B. B., and Biswas, D. R. (2009). Influence of potassium solubilizing microorganism (*Bacillus mucilaginosus*) and waste mica on potassium uptake dynamics by sudan grass (*Sorghum vulgare* Pers.) grown under two Alfisols. *Plant and Soil*, 317(1), 235–255.

Basak, B. B., and Sarkar, B. (2017). Scope of natural sources of potassium in sustainable agriculture. In: A. Rakshit, P. C. Abhilash, H. B. Singh, and S. Ghosh (Eds.), *Adaptive Soil Management: From Theory to Practices* (pp. 247–259). Singapore: Springer.

Bhattacharyya, T., Pal, D. K., Wani, S. P., and Sahrawat, K. L. (2013). Rain-fed agriculture, area and extent. In: G. C. Nelson et al. (Eds.), *Climate Change and Agriculture* (pp. 1–18). New Delhi: Stadium Press (India) Pvt. Ltd.

Bilias, F., and Barbayiannis, N. (2019). Potassium-fixing clay minerals as parameters that define K availability of K-deficient soils assessed with a modified Mitscherlich equation model. *Journal of Soil Science and Plant Nutrition*, 19(4), 830–840.

Bilias, F., and Barbayiannis, N. (2020). Towards a single-point approach for estimating potassium quantity-intensity parameters in soils at different K saturation ratios. *Journal of Soil Science and Plant Nutrition,* 20, 1679–1685. https://doi.org/10.1007/s42729-020-00238-6.

Bilias, F., Tsigili, S., and Barbayiannis, N. (2021). A preliminary evaluation of cation exchange resins as a soil test of potassium avail- ability in soils of northern Greece with different K loadings. *Journal of Soil Science and Plant Nutrition,* 21, 1004–1012. https://doi.org/10.1007/s42729-021-00417-z.

Bolan, N. S. (1976). Study of the distribution of potassium in the soil profiles of the Nilgiris district (Tamil Nadu) and its relationship to various physic-chemical properties. Ph.D. Thesis, Tamil Nadu Agricultural University, Coimbatore.

Bortoluzzi, E. C., Dos Santos, D. R., Santanna, M. A., and Caner, L. (2013). Mineralogy and nutrient desorption of suspended sediments during a storm event. *Journal of Soils and Sediments*, 13(6), 1093–1105.

Brady, N. C., and Weil, R. R. (2012). Soil phosphorus and potassium. In: *The Nature and Properties of Soils* (pp. 594–638). 14th edn. Pearson Education: London.

Brar, M. S., Subba Rao, A., and Sekhon, G. S. (1986). Solution, exchangeable, and nonexchangeable potassium in five soil series from the alluvium soils region of northern India. *Soil Science*, 142, 229–234.

Chuan, L., He, P., Pampolino, M. F., Johnston, Jin, J., Xu, X., Zhao, S., Qiu, S., and Zhou, W., (2013). Establishing a scientific basis for fertilizer recommendations for wheat in China: Yield response and agronomic efficiency. *Field Crops Research*, 140, 1–8.

Das, D., Dwivedi, B. S., Datta, S. P., Datta, S. C., Meena, M. C., Agarwal, B. K., and Jaggi, S. (2019). Potassium supplying capacity of a red soil from eastern India after forty-two years of continuous cropping and fertilization. *Geoderma*, 341, 76–92.

Das, D., Dwivedi, B. S., Datta, S. P., Datta, S. C., Meena, M. C., Dwivedi, A. K., and Jaggi, S. (2021). Long-term differences in nutrient management under intensive cultivation alter potassium supplying ability of soils. *Geoderma*, 393, 114983.

Das, D., Nayak, A. K., Thilagam, V. K., Chatterjee, D., Shahid, M., Tripathi, R., and Biswas, S. S. (2018). Measuring potassium fractions is not sufficient to assess the long-term impact of fertilization and manuring on soil's potassium supplying capacity. *Journal of Soils and Sediments*, 18(5), 1806–1820.

Das, D., Sahoo, J., Raza, M. B., Barman, M., and Das, R. (2022). Ongoing soil potassium depletion under intensive cropping in India and probable mitigation strategies: A review. *Agronomy for Sustainable Development*, 42(1), 1–26.

Das, D., Sarkar, D., Nagarajan, S., and Mitlin, D. (2020). Cobalt phosphide (Co_2 P) encapsulated in nitrogen-rich hollow carbon nanocages with fast rate potassium ion storage. *Chemical Communications*, 56(94), 14889–14892.

Datta, S. C. (2005). Characterization of soil for potential potassium supplying capacity and release of non-exchangeable potassium during cropping through simulation model. *Communications in Soil Science and Plant Analysis*, 36(7–8), 969–992.

Dey, P., and Santhi, R. (2014). Soil test based fertiliser recommendations for different investment capabilities. In: H. L. S. Tandon (Ed.), *Soil Testing for Balanced Fertilisation: Technology, Application, Problem Solutions* (pp. 49–67). Fertiliser Development and Consultation Organisation, New Delhi, India.

Dey, P., Santhi, R., Maragatham, S., and Sellamuthu, K. M. (2017a). Status of phosphorus and potassium in the Indian soils vis-à-vis world soils. *Indian Journal of Fertilisers*, 13(4), 44–59.

Dey, P., Srivastava, S., Sinha, N., Hati, K. M., Chaudhary, R. S., and Patra, A. K. (2017b). Potassium availability indices through traditional mineralogical as well as MIR technology and STCR-based K-fertiliser recommended for sustainable crop productivity. *In Souvenir and Compendium of Papers, 20th Annual Convention.* The Clay Minerals Society of India, p. 76.

Dhillon, S. K., Sidhu, P. S., Dhillon, K. S., and Sharma, Y. F. (1985). Distribution of various potassium forms in some benchmark soils of Northwest India. *Journal of Potassium Research,* 1, 154–165.

Dobermann, A. T. H. F., and Fairhurst, T. H. (2002). Rice straw management. *Better Crops International,* 16(1), 7–11.

Durairaj Muthiah, N. (1986). Investigations on the potassium dynamics in the major soil series of Thanjavur district in relation to response of rice. Ph.D. Thesis, Tamil Nadu Agricultural University, Coimbatore.

Dutta, S., Majumdar, K., Khurana, H. S., Sulewski, G., Govil, V., Satyanarayana, T., and Johnston, A. (2013). Mapping potassium budgets across different states of India. *Better Crops-South Asia*, 7, 28–31.

Ghorban, A. R. (2007). Trying a new potassium uptake model by wheat. *Soil Science Society of America Journal*, 71, 125–132.

Ghosh, A. B., and Hasan, R. (1976). Potassium in soils, crops and fertilizers. *Journal of Indian Society of Soil Science*, 10, 1–5.

Goulding, K., Murrell, T. S., Mikkelsen, R. L., Rosolem, C., Johnston, J., Wang, H., and Alfaro, M. A. (2021). Outputs: Potassium losses from agricultural systems. In: T. S. Murrell, R. L. Mikkelsen, G. Sulewski, R. Norton, and M. L. Thompson (Eds.), *Improving Potassium Recommendations for Agricultural Crops* (p. 75). Springer Nature: London.

Gurav, P. P., Ray, S. K., Choudhari, P. L., Shirale, A. O., Meena, B. P., Biswas, A. K., and Patra, A. K. (2019). Potassium in shrink-swell soils of India. *Current Science*, 117(4), 587–596.

Han, H. S., and Lee, K. D. (2005). Phosphate and potassium solubilizing bacteria effect on mineral uptake, soil availability and growth of eggplant. *Research Journal of Agriculture and Biological Sciences*, 1(2), 176–180.

Hasan, R. (2002). Potassium status of soils in India. *Better Crops International*, 16(2), 3–5.

Helmke, P. A., and Sparks, D. L. (1996). Lithium, sodium, potassium, rubidium, and cesium. In: D. L. Sparks (Ed.), *Methods of Soil Analysis: Part 3 Chemical Methods*(vol. 5, pp. 551–574). SSSA: Madison, WI.

Hosseinpur, A. R., Raisi, T., Kiani, S., and Motaghian, H. R. (2014). Potassium-release characteristics and their correlation with bean (*Phaseolus vulgaris*) plant indices in some calcareous soils. *Communications in Soil Science and Plant Analysis*, 45(6), 726–740.

Katyal, J. C. (2020). Sustainable development of Indian agriculture: Focus natural resources' management. *Indian Journal of Fertilisers*, 16, 198–255.

Laxita, L., and Shruti, S. (2020). Isolation and characterization of potassium solubilizing microorganisms from South Gujarat region and their effects on wheat plant. *Mukt Shabd Journal*, 9, 7483–7496.

Li, T., Wang, H., Zhou, Z., Chen, X., and Zhou, J. (2015). A nano-scale study of the mechanisms of non-exchangeable potassium release from micas. *Applied Clay Science*, 118, 131–137.

Lu, D., Li, C., Sokolwski, E., Magen, H., Chen, X., Wang, H., and Zhou, J. (2017). Crop yield and soil available potassium changes as affected by potassium rate in rice–wheat systems. *Field Crops Research*, 214, 38–44.

Mahfuza, S. N., Islam, M. N., Aktar, S., Amin, M. R., and Islam, M. A. (2008). Influence of split application of potassium on yield and yield contributing characters of rice cultivars-BR11 and Nizersail. *The Agriculturists*, 6(1&2), 84–89.

Majumdar, K., Norton, R. M., Murrell, T. S., García, F., Zingore, S., Prochnow, L. I., and Oberthür, T. (2021). Assessing potassium mass balances in different countries and scales. In: T. S. Murrell, R. L. Mikkelsen, G. Sulewski, R. Norton, and M. L. Thompson (Eds.), *Improving Potassium Recommendations for Agricultural Crops* (p. 283). Springer Nature: London.

Majumdar, K., Sanyal, S. K., Singh, V. K., Dutta, S. K., Satyanarayana, T., and Dwivedi, B. S. (2017). Potassium fertiliser management in Indian agriculture: Current trends and future needs. *Indian Journal of Fertiliser*, 13(5), 20–30.

Menon, R. G., Hammond, L. L., and Sissingh, H. A. (1989). Determination of plant available phosphorus by iron-impregnated filter paper (pi) soil test. *Soil Science Society of America Journal*, 53(1), 110–115. https://doi.org/10.2136/sssaj1989.03615995005300010020x.

Motsara, M. R. (2002). Promoting balanced fertilization in India: Policies and economic issues. *Fertilizer News*, 47, 15–21.

Muralidharudu, Y., Sammi Reddy, K., Mandal, B. N., Subba Rao, A., Singh, K. N., and Sonekar, S. (2011). *GIS Based Soil Fertility Maps of Different States of India*. Indian Institute of Soil Science: Bhopal.

Murugappan, V., Bhaskaran, A., Murugesa Bhoopathi, P., Natesan, R., Santhi, R., and Jagadeeswaran, R. (2004). Decision Support System for Integrated Fertilizer Recommendation (DSSIFER). In: *New Crop Varieties, Farm Implements and Management Technologies*. Directorate of Research, TNAU: Coimbatore, pp. 39–40.

Neilsen, G. H., Neilsen, D., and Peryea, F. (1999). Response of soil and irrigated fruit trees to fertigation or broadcast application of nitrogen, phosphorus, and potassium. *Hort Technology*, 9(3), 393–401.

Nishanth, D., and Biswas, D. R. (2008). Kinetics of phosphorus and potassium release from rock phosphate and waste mica enriched compost and their effect on yield and nutrient uptake by wheat (*Triticum aestivum*). *Bioresource Technology*, 99(9), 3342–3353.

Pal, Y., Wong, M. T. F., and Gilkes, R. J. (1999). The forms of potassium and potassium adsorption in some virgin soils from south-western Australia. *Australian Journal of Soil Research*, 37, 695–709.

Partey, S. T., Quashie-Sam, S. J., Thevathasan, N. V., and Gordon, A. M. (2011). Decomposition and nutrient release patterns of the leaf biomass of the wild sunflower (*Tithonia diversifolia*): A comparative study with four leguminous agroforestry species. *Agroforestry Systems*, 81(2), 123–134.

Ramamoorthy, B., & Bajaj, J. C. (1969). Available nitrogen, phosphorus and potassium status of Indian soils. *Fertiliser News*.

Ramanathan, K. M. (1977). Studies on dynamics of soil potassium. Ph. D. thesis, Tamil Nadu Agriculture University of India.

Raychaudhuri, S. P., and Datta, N. P. (1964). Phosphorus and potassium status of Indian soils. *Agricultural Research Review Series,* 36, 1–44.

Regmi, A., Ladha, J., Pasuquin, E., Pathak, H., Hobbs, P., Shrestha, L., and Duveiller, E. (2002). The role of potassium in sustaining yields in a long-term rice-wheat experiment in the Indo-Gangetic Plains of Nepal. *Biology and Fertility of Soils*, 36(3), 240–247.

Royal Commission on Agriculture. (1929). Report, Government Central Press, Bombay.

Sahoo, S. K., Teixeira, I. F., Naik, A., Heske, J., Cruz, D., Antonietti, M., and Kühne, T. D. (2021). Photocatalytic water splitting reaction catalyzed by ion-exchanged salts of potassium poly (heptazine imide) 2D materials. *The Journal of Physical Chemistry C*, 125(25), 13749–13758.

Santhi, R., Bhaskaran, A., Maragatham, S., Gopalakrishnan, M., Sellamuthu, K. M., Jagadeeswaran, R., Balamurugan, J., Malarvizhi, P., and Dey, P. (2020). Decision Support System for Integrated Fertiliser Recommendation (DSSIFER), Computer software, Department of Soil Science and Agricultural Chemistry, DNRM, TNAU, Coimbatore.

Santhi, R., Bhaskaran, A., Poongothai, S., Maragatham, S., Natesan, R., Sellamuthu, K. M., Duraisami, V. P., Velu, V., Appavu, K., and Murugesa Boopathi, P. (2010). *Decision Support System for Integrated Fertiliser Prescriptions, Computer Software.* Department of Soil Science & Agriculture Chemistry, TNAU: Coimbatore.

Santhi, R., Stalin, P., Arulmozhiselvan, K., Muthumanickam, D., Chitdeshwari, T., Maragatham, S., Elayarajan, M., Sellamuthu, K. M., Natesan, R., Duraisami, V. P., Dey, P., and Subba Rao, A. (2013). GPS and GIS based soil fertility appraisal for selected districts of Tamil Nadu: A monograph, p 172. AICRP-STCR, Department of Soil Science and Agricultural Chemistry, Tamil Nadu Agricultural University, Coimbatore.

Sanyal, S. K. (2014). Nutrient mining in soil-an issue of concern. *Newsletter* 36, Indian Society of Soil Science, New Delhi, pp. 1–3.

Sanyal, A. K., Chattapadhyay, S. K., Pal, T. K., and Karmakar, A. K. (2014). Faunal resources and assessment of the impact of mining activities on fauna of chhotonagpur coalfield areas. *Jharkhand Records of the Zoological Survey of India*, Occasional Paper No. 361, 1–47.

Schaff, B. E., and Skogley, E. O. (1982). Diffusion of potassium, calcium, and magnesium in Bozeman silt loam as influenced by temperature and moisture. *Soil Science Society of America Journal*, 46(3), 521–524.

Sharma, A., Jalali, V. K., and Arora, S. (2010). Non-exchangeable potassium release and its removal in foot-hill soils of North-west Himalayas. *Catena*, 82(2), 112–117.

Shirale, A. O., Meena, B. P., Gurav, P. P., Srivastava, S., Biswas, A. K., Thakur, J. K., and Rao, A. S. (2019). Prospects and challenges in utilization of indigenous rocks and minerals as source of potassium in farming. *Journal of Plant Nutrition*, 42(19), 2682–2701.

Shukla, A. K., Sharma, S. K., Tiwari, R., & Tiwari, K. N. (2005). Nutrient depletion in the rice-wheat cropping system of the Indo-Gangetic plains. *Better Crops*, 89(2), 28–31.

Singh, B., Singh, Y., Imas, P., and Jian-chang, X. (2003). Potassium nutrition of the rice wheat cropping system. *Advances in Agronomy*, 81, 203–259.

Singh, S. K., and Chandran, P. (2015). Soil genesis and classification. In: R. K. Rattan, J. C. Katyal, B. S. Dwivedi, A. K. Sarkar, T. Bhattacharyya, J. C. Tarafdar, and S. S. Kukal (Eds.), *Soil Science: An Introduction* (pp. 57–96). Indian Society of Soil Science, New Delhi.

Singh, V. K., Dwivedi, B. S., Buresh, R. J., Jat, M. L., Majumdar, K., Gangwar, B., … and Singh, S. K. (2013). Potassium fertilization in rice–wheat system across Northern India: Crop performance and soil nutrients. *Agronomy Journal*, 105(2), 471–481.

Singh, V. K., Dwivedi, B. S., Rathore, S. S., Mishra, R. P., Satyanarayana, T., Majumdar, K. (2021). Timing potassium applications to synchronize with plant demand. In: T. S. Murrell, R. L. Mikkelsen, G. Sulewski, R. Norton, and M. L. Thompson (Eds.), *Improving Potassium Recommendations for Agricultural Crops* (p. 363). Springer. https://doi.org/10.1007/978-3-030-59197-7_13.

Singh, V. K., Dwivedi, B. S., Shukla, A. K., Kumar, V., Gangwar, B., Rani, M., and Mishra, R. P. (2015). Status of available sulfur in soils of north-western Indo-Gangetic plain and western Himalayan region and responses of rice and wheat to applied sulfur in farmer's fields. *Agricultural Research*, 4(1), 76–92.

Sirius Minerals (2016). POLY4 brochure. http://www.siriusminerals.com (Accessed on 4 June 2021).

Sparks, D. L. (1987). Potassium dynamics in soils. In: B. A. Stewart (Ed.), *Advances in Soil Science* (pp. 1–63). Springer-Verlag: New York.

Srinivasarao, C., Kundu, S., Ramachandrappa, B. K., Reddy, S., Lal, R., Venkateswarlu, B., Sahrawat, K.L., and Naik, R.P. (2014). Potassium release characteristics, potassium balance, and fingermillet (*Eleusine coracana* G.) yield sustainability in a 27-year long experiment on an *Alfisol* in the semi-arid tropical India. *Plant and Soil*, 374, 315–330.

Srinivasarao, C., Masood, A., Ganeshamurthy, A. N., and Singh, K. K. (2003). Potassium requirements of pulse crops. *Better Crops International*, 17, 8–11.

Srinivasarao, C., and Srinivas, K. (2017). Potassium dynamics and role of non- exchangeable potassium in crop nutrition. *Indian Journal of Fertilizer*, 13, 80–94.

Srinivasarao, C., Veni, G. V., Sharma, L. K., and Raju, B. M. K. (2015). Phosphorus fertility status and strategies to improve its use efficiency in Indian soils. *Indian Journal of Fertilisers*, 11(3), 22–36.

Srinivasarao, C., and Vittal, K. P. R. (2007). Emerging nutrient deficiencies in different soil types under rainfed production systems of India. *Indian Journal of Fertilizer*, 3, 37–46.

Srinivasarao, C., Vittal, K. P. R., Tiwari, K. N., Gajbhiye, P. N., Kundu, S., Pharande, A. L., and Shankar, M. A. (2007). Potassium supplying characteristics of twenty-one soil profiles under diverse rainfed production systems. *Journal-Indian Society of Soil Science*, 55(1), 14.

Stewart, A. B. (1947). Report on soil fertility investigation in India with special reference to manuring. ICAR, New Delhi, pp. 160.

Subba Rao, A., and Brar, M. S. (2009). Potassium. In: N. N. Goswami, R. K. Rattan, G. Dev, G. Narayanasamy, D. K. Das, S. K. Sanyal, D. K. Pal, and D. L. N. Rao (Eds.), *Fundamentals of Soil Science* (2nd ed., pp. 431–432). Indian Society of Soil Science: New Delhi, India.

Subba Rao, A., Brar, M. S., and Sekhon, G. S. (1988). Desorption of potassium from five soil series developed on different parent materials. *Journal of the Indian Society of Soil Science*, 36, 239–245.

Sugumaran, P., and Janarthanam, B. (2007). Solubilization of potassium containing minerals by bacteria and their effect on plant growth. *World Journal of Agricultural Sciences*, 3(3), 350–355.

Supanjani, H. S. H., Jae, S. J., and Kyung, D. L. (2006). Rock phosphate-potassium and rock-solubilising bacteria as alternative, sustainable fertilizers. *Agronomy for Sustainable Development,* 26, 233–240.

Tamhane, R. V., and Subbiah, B. V. (1960). Correlation between soil analysis and response to fertilisers with special reference to rice crop in India. *IRC Newsletter*, 9, 1–10.

Tandon, H. L. S. (2004). *Fertilisers in Indian Agriculture: From 20th to 21st Century.* FDCO: New Delhi.

Tandon, H. L. S., and Sekhon, G. S. (1988). *Potassium Research and Agricultural Production in India.* FDCO: New Delhi.

Tarkalson, D. D., Brown, B., Kok, H., and Bjorneberg, D. L. (2009). Irrigated small-grain residue management effects on soil chemical and physical properties and nutrient cycling. *Soil Science*, 174(6), 303–311.

Tewatia, R. K., Rattan, R. K., Bhende, S., and Kumar, L. (2017). Nutrient use and balances in India with special reference to phosphorus and potassium. *Indian Journal of Fertiliser*, 13, 20–29.

Volenec, J. J., Brouder, S. M., and Murrell, T. S. (2021). Broadening the objectives of future potassium recommendations. In: T. S. Murrell, R. L. Mikkelsen, G. Sulewski, R. Norton, and M. L. Thompson (Eds.), *Improving Potassium Recommendations for Agricultural Crops* (pp. 385–415). Springer Nature: London.

Zhao, S., He, P., Qiu, S., Jia, L., Liu, M., Jin, J., and Johnston, A. M. (2014). Long-term effects of potassium fertilization and straw return on soil potassium levels and crop yields in north-central China. *Field Crops Research*, 169, 116–122.

31 Assessment of the Soil Resources of Sub-Saharan Africa in Relation to Food Security: *Perspectives Past, Present, and Future*

Kenneth Mbene[1], Nelson Mofor Alakeh[2], Godswill Azinwi Asongwe[3], Norbert Nkafu Fomenky[4], Mary Chantal Asong Nkenganang[1], and Aaron Suh Tening[5]

[1]Department of Chemistry, Higher Teacher Training College, University of Yaounde I, Yaounde, Cameroon

[2]Department of Mining and Mineral Engineering, National Higher Polytechnic Institute (NAHPI), University of Bamenda, Bambili, Cameroon

[3]Department of Environmental Sciences, Faculty of Science, University of Buea, Buea, Cameroon

[4]Department of Chemistry, University of Buea, Buea, Cameroon

[5]Deparment of Agronomic and Applied Molecular Sciences, University of Buea, Buea, Cameroon

CONTENTS

DOI: 10.1201/9781003093565-31

31.1 INTRODUCTION

Food security is defined as physical and economic access by all people at all times to sufficient food to meet their dietary requirements for a productive and healthy life. Food security can also be regarded as the availability of food and one's access to it (Upton et al., 2016). A locality is considered food secured when its occupants do not live in hunger and are not likely to experience starvation. Three conditions must thus be satisfied to ensure food security: food must be available through domestic production and imports; food must be accessible or people must have adequate resources to acquire the appropriate foods; and food must be utilized in conjunction with adequate water, sanitation, and health to meet nutritional needs. Often, however, food security is discussed with reference to grains only. This can be misleading, especially for societies (e.g., pastoral societies) that are primarily dependent on sources of food other than grains. In fact, livestock have an important role in achieving food security in sub-Saharan Africa (Herrero et al., 2014).

The availability of food has always been a central preoccupation of mankind. Despite the rapid growth of the global population during the past four decades, farmers have produced sufficient food to allow average per capita food intake to grow gradually. Yet hunger persists and food reserves have fluctuated markedly during this period, sometimes falling to critically low levels. In order to address these concerns, the World Food Conference was organized in 1974, followed by the International Conference on Nutrition in 2016 (Amoroso, 2016). At the Summit, Heads of State reaffirmed the right of everyone to have access to safe and nutritious food, consistent with the right of everyone to be free from hunger. They also committed themselves to reducing the number of undernourished people to half their present level no later than 2035. The achievement of this goal is central to the Strategic Framework of the Food and Agriculture organization (FAO) of the United Nations.

This book chapter takes a new look at the possible causes of hunger and poverty with regard to analysis of farming systems. It recognizes the diversity of the livelihoods of land users and explores various pathways that may improve their living standards within the next 30 years. Rural development ultimately depends on the outcomes of the daily decisions of millions of individual land users. The challenge for governments is to provide incentives that will enable farm households to accelerate agricultural growth and to reduce poverty.

In this chapter, the current extent of rural hunger and poverty in the sub-Saharan African (SSA) countries is highlighted, and the contribution of agricultural growth to poverty alleviation is discussed. Subsequently, the farming-systems concept is introduced, and the ways in which farming systems have affected soil degradation are explained.

31.1.1 Linking Soils and Food Production in Sub-Sahara Africa

Soils are fundamental to supporting plant, animal, and human life on earth (Dixit et al., 2015). For sustainable crop production, it is necessary to maintain soils in situ and in a healthy and fertile state. The production of food is intrinsically linked to soils. The pressure from population increase is the main cause of soil degradation. An estimated 98.8% of the daily calories consumed by humans comes from soil (2,849 kcal per capita) and only 1.2% from aquatic sources (35 kcal per capita) (FAO, 2018; Kopittke et al., 2019). Thus, increased demand for food and fiber results in an increased pressure on soils. Degradation of soil in terms of physical, chemical, and biological properties could adversely influence crop production. When a soil is degraded, this will decrease crop produce, even if all other inputs are optimum. With proper management, it can produce in great quantities. Judicious soil management is proving to be an increasingly difficult and challenging task in many parts of the world and in SSA countries in particular. Although nutritious and adequate food is the focus of ensuring food security and safety, soils also support the production of other essential materials and services, such as fiber for clothing and bedding and timber for fuel and housing. Over most of the earth's land lies a thin layer of soil, which is a complex and variable mixture of minerals, air, water, decaying remains of life, and countless living organisms. The earth's "skin" is not one soil, but many soils-each with its own story. In essence, soil is the thin skin of planet earth that must be protected from damage by soil erosion.

Many constraints to intensive food crop production in tropical Africa are related to tropical soils. Improved technologies are available for different ecological regions. Important technological innovations include manual land clearing, mulch farming, conservation tillage, agroforestry, cover crops, mixed- and relay-cropping, and early sowing for improved and sustained productivity. Irrigation and the use of chemical fertilizers are also important. In fact, the root cause of the perpetual famine in Africa can be traced to the misuse of soil and water resources. Substantial increases in food production are possible, if the proven technologies can be effectively transferred and implemented. Moreover, many of the soils cannot be used for agriculture, because they are too dry, too wet, too cold, too shallow, or too toxic. Follett (2001) estimated that only 20% of earth's soils have no natural limitations in terms of agricultural production and is used for growing crops. The primary cause of soil degradation in sub-Saharan Africa is expansion and intensification of agriculture in efforts to feed its growing population. Effective solutions will support resilient systems and must cut across agricultural, environmental, and socioeconomic objectives. The extent and rate of soil degradation in sub-Saharan Africa is still under debate, as there are no reliable data, just gross estimates (Tully et al., 2015). Nevertheless, certain soils are losing their ability to provide food and essential ecosystem services, and we know that soil fertility depletion is the primary cause. Future research should focus on how soil degradation in sub-Saharan Africa leads to changes in ecosystem services, and how to manage these soils now and in the future.

31.1.2 Current Situation and Trends of Land Degradation in Sub-Saharan Africa

Farmers' perception of land degradation in semi-arid areas varies with the agro-ecology and socio-economic situation of the farming community. However, there are insufficient reports on the topic. Soils are subject to degradation and loss due, principally, to erosion by wind and water and also to physical (e.g., compaction, loss of structure) and chemical (e.g., salinity, toxic contamination) degradation. Tesfahunegn (2019) reported, based on farmers' perceptions, that deforestation (100%), overgrazing (98%), and improper soil management (97%) are the main direct causes of degradation; poverty (100%) and population pressure (98%) are the most commonly identified indirect causes of land degradation. The most frequently used indicators of degradation by farmers are soil depth (100%), soil erosion (95%), and soil fertility (95%). Erosion is part of natural geological processes, but the rate and extent of soil erosion and degradation have increased as a result of human activities. These include overgrazing, excessive cultivation, farming of unsuitable land, and contamination through industrial activity (Tesfahunegn, 2019).

Few, reliable data are available concerning the extent of land degradation in Africa. However, a close observation of land in the sub-Sahara region reveals that land degradation is widespread and serious. The presence of gullies and sand dunes and of degraded forests and grazing lands is obvious, although the effects of sheet erosion and declining soil fertility are less noticeable.

Africa depends on its ability to conserve and manage land resources for wealth. Soil degradation will not only result in decreased food production (Rickson, 2015) but also in droughts, ecological imbalance, and consequent degradation of the quality of life. In Africa, the most conspicuous symptoms of the negative impact of land degradation on food production are stagnating and declining yields and increasing levels of poverty, because a majority of the population depends on farming for income.

Throughout the sub-Saharan region, regardless of the climatic zone, meteorological records show that unpredictability of rains is a common feature. Regions where increasing rainfall was associated with vegetation greening were found in West Africa, Central African Republic, West Cameroon, and the north-eastern part of South Africa, whereas areas with evidence of climatic vegetation degradation were located in Southern Madagascar, Nigeria, Kenya, and the Garden Route region of South Africa (Hoscilo, 2015). Semi-arid and arid areas concentrated in sub-Saharan Africa are particularly prone to extreme rainfall variability and water shortages. The majority of people in sub-Saharan countries (~85%) live in rural areas and are engaged in agricultural activities; thus farming is a principal source of their livelihoods (The World Bank, contamination, pollutants, and nutrient depletion by land users (Braithwaite and Mont, 2009). Barrios et al. (2010) provide evidence that long-term trends in rainfall have affected economic growth rates in sub-Saharan Africa in the past. Recently, long-term drying trends were observed in several African regions, for example, in Nigeria (Oguntunde et al., 2017), the western Sahel (Chappell and Agnew, 2004), tropical North Africa (Kawase et al., 2010), and in the eastern part of southern tropical Africa (Kawase et al., 2010).

Africa also suffers from geologically induced and inherently low soil fertility, as the bedrock consists of mostly granites. The relationship between the parent soils and soil-forming factors are complex, because the land surface has undergone a series of shifts in vegetation and climate. Recent volcanic activity occurred mainly in the eastern and southern parts of the continent, principally between Ethiopia and Lake Victoria. For this reason, most of the soils in Africa are characterized by a low proportion of clay, making them easy to work, but also easy to lose. It is important to monitor rainfall patterns at various spatial and temporal scales, and to understand the consequences of rainfall dynamics on vegetation cover, land conditions, and human populations, which are strongly dependent on natural resources.

Human activities in obtaining food, fiber, fuel, and shelter have altered the soil physicochemical and biological properties. Human settlement has contributed drastically to the degradation of the soil in sub-Saharan regions. Although soil degradation is largely due to anthropogenic activities, and, hence, its pace is governed primarily by the speed at which population pressure mounts, irregular natural events, such as droughts, exacerbate the situation. The 1982–2011 drought, for example, had a dramatic effect on the speed of land degradation and desertification (Vicente-Serrano et al., 2015), and, since then, there has been increased international and regional attention given to the vulnerability of societies in drought-prone tropical environments in sub-Saharan Africa (Rockström, 2003).

Many African countries have already lost a significant quantity of their soils to various forms of degradation. The main causes for land degradation in Africa include, among others, demography growth, conflicts and wars with expanded refugees' settlements, inappropriate soil management, deforestation, shifting cultivation, insecurity in land tenure, variation of climatic conditions, and intrinsic characteristics of fragile soils in diverse agro-ecological zones (Thiombiano and Tourino-Soto, 2007).

An estimated 80% of rural households are involved in farming and contributes about 30% to the Gross Domestic Product (GDP) of Cameroon. However, prolonged dry spells and droughts negatively affect agricultural output and economic development. Soil degradation caused by deforestation is also a serious threat in Africa. Deforestation exposes the soil to high temperatures that break down organic matter, increase evaporation, and make the soils vulnerable to erosion. More serious still is the gradual removal of trees in farms and pastures, which are crucial for protecting productive land from erosion.

In the drier parts of Africa, millions of hectares of grazing land and rangeland are also threatened with degradation. These areas include the arid north, the semi-arid south, the Sudano-Sahelian countries, and the drier parts of the Cameroon, Ethiopia, Kenya, and Nigeria. The 1983–1985 drought and recent droughts killed huge numbers of livestock there, with the result that good breeding stock was lost and the structural balance of herds distorted. Nevertheless, the herds are now recovering, but within 5 or 10 years the trend of increasing overgrazing could be re-established when the next drought reduces livestock numbers again.

Africa's forests and woodlands are also being depleted, threatening one of the continent's most important resources. In Africa, trees play an important role in protecting the environment. They are the principal source of rural energy and provide countless medicinal and industrial products used in both the home and in small-scale industry. They often supply food and feed, are the main source of building materials in the countryside, and, directly, are a source of employment and income for many rural Africans.

Nearly 4 million ha of this resource are now being deforested or degraded annually, largely in humid and sub-humid West Africa. The rate of destruction is alarmingly high in the Cameroon, Côte d'lvoire, and Nigeria (Ogen, 2011). The cause of deforestation is mainly clearing for agriculture, but uncontrolled logging, gathering for fuel wood, fire, and overgrazing are also taking their toll. One of the causes of degradations is that population pressure is forcing farmers to cultivate increasingly marginal land to secure food availability for the increasing population. Even if most of the rain forest is cleared for agriculture, it would not solve the food problems of the Sahel or East Africa. In countries with limited, cultivable land and high population growth rates such as Kenya, Ethiopia, Malawi, Burundi, and Rwanda, fallow periods are no longer sufficient to allow soil fertility to be restored, so that crop yields have fallen. In response, farmers have been forced either to bring increasingly marginal lands into cultivation, or to migrate into tropical forest areas, exacerbating problems of land degradation and deforestation.

According to the review of McGuire (2015) on FAO, IFAD, and WFP meeting whose main goals are: eradication of hunger, food insecurity, and malnutrition; elimination of poverty; economic and social progress; and sustainable management and utilization of natural resources, including land, water, air, climate, and genetic resources, were for the benefit of present and future generations. The FAO in 2015 was working on achieving food security for everyone, with nutrition, diplomacy, technological advances, and environmental sustainability all being considered.

Future prospects must be assessed against a background of diversity. For many countries, a simple extrapolation of current trends to derive estimates of future food availability would be highly inappropriate. Trends that either continues a downward spiral or a rapid improvement in diet are unlikely to be sustained for long. The threat or existence of mass starvation, for example, would provoke corrective action at the national or international level, or both. If the present trend of stagnating or declining productivity found over much of Africa is to be reversed, land degradation will have to be halted. The problems, as well as their underlying causes, vary greatly from country to country and even within individual cities. Consequently, the approaches used, techniques applied, and methods adopted will have to be adapted to the specific needs of different environments.

Drought and conflict have left many countries in sub-Saharan Africa critically dependent on food and emergency aid. The provision of that aid is vital to the region's food security, and for local efforts at post-drought and post-conflict reconstruction. Unfortunately, for several countries, gaps between local needs and donors' pledges remain.

In short, the heart of the problem is that the natural-resource base of Africa is being degraded and destroyed at a rate that will soon make food and agricultural production unsustainable. Poverty, coupled with increasing population pressure, is the biggest single cause of this degradation. The rural inhabitants, which constitute the overwhelming majority of Africa's citizens, destroy their own environment, not out of ignorance, but simply to survive. Peasant farmers are preoccupied with survival, because there is no alternative employment and no better technologies that they can afford. Pastoralists overstock to improve their chances of surviving the next drought. Rural dwellers strip trees and shrubs for fuel wood because they need fuel.

Agriculture, which provides 60% of all employment, is the backbone of most African economies in most countries (Sumberg et al., 2015). Improvement in agricultural productivity has the potential to increase rural incomes and purchasing power for large numbers of people. Thus, more than any other sector, agriculture can uplift people on a mass scale. With greater prosperity, the consequent higher demand for African industrial and other goods would induce conditions that would be a significant source of economic growth. It follows that agricultural productivity has a critical bearing on both national and household incomes. In addition, food imports now constitute a major strain on Africa's balance of payments, accounting for about 15% of total imports.

Sub-Saharan Africa's agricultural trade, by country and type, shows that four countries, Nigeria, Angola, the Democratic Republic of the Congo (DRC), and Somalia, account for most of sub-Saharan Africa's net agricultural imports (Brüntrup, 2006). The rest of the countries in the region are net agricultural exporters. This is good news not only for the present but also for Africa's future economic growth through trade. While the African Development Bank projected that Africa's food imports would reach $110 billion by 2030 (Adesina, 2017), these projections were based on trends during 2000–2010, a period when global food prices rose rapidly, and they do not reflect the more recent 2011–2019 period during which time the value of SSA's food imports has been relatively flat. The fact that, even with rapidly growing demand for food driven by population growth and rising per capita incomes, the value of SSA food imports has not continued to rise over the past decade can be explained by the region's success in expanding food production (Brüntrup, 2006). SSA's top exports are mainly tropical commodities, such as cocoa, coffee, tea, and cotton, while its main food imports are wheat, rice, soybeans, other oilseeds, and frozen meat products.

It is difficult to see how these food gaps could be bridged. Except for some of the more industrialized countries of North Africa, present and potential oil exporters, some mineral producers, and a few net agricultural exporters, it is unlikely that foreign exchange earnings would be sufficient to pay for substantial imports, particularly in the low-income countries that account for half the projected deficit.

A continuation of current trends will place many countries on the edge of survival, even if they were provided with substantial international support. Some countries, however, most of which depend for their incomes on sectors other than agriculture, should be able to preserve and even improve their nutritional standards.

31.1.3 Responses to the Impacts of Land and Environmental Degradation and Desertification

The dominant farming system throughout most of the sub-Saharan countries is shifting cultivation, which is essentially a zero-input form of agriculture because it makes maximum use of the resources available within the system. Sub-Saharan Africa has already demonstrated a competitive advantage in selected cash crops, such as cashews, coffee, processed horticulture, and tea in East Africa and cocoa in West Africa. For some of these crops, such as cocoa, Africa has the lowest cost of production in the world. During cultivation, the soil is usually covered by a mixture of crops that utilize soil nutrients from different depths and provide a varied canopy to shelter the soil surface. At the end of the cropping cycle, the natural system of resting the land restores the nutrients removed by the crops, and the opportunistic pests and diseases are driven out. So long as the fallow period is long enough, the system is stable and sustainable due to the decomposing of organic matter from the cover crops.

Desertification is also a serious problem in the continent where it has been estimated that 319 million ha of Africa are vulnerable to desertification due to sand movement (Ahmady-Birgani et al., 2017). A FAO/UNEP assessment of land degradation in Africa suggests that large areas of countries north of the equator suffer from serious desertification problems. It has been estimated that, globally, as much as two billion hectares of arable and grazing land are affected by moderate to severe degradation, while around 7 million ha of farmland are permanently lost every year (Mackenzie and Mackenzie, 1995). Soil degradation, associated with poor soil fertility management practices, is a major factor underlying poor agricultural productivity in sub-Saharan Africa. About 65% of the agricultural land is degraded, mainly due to low nutrient application, soil erosion, and soil acidification (Mackenzie and Mackenzie, 1995).

The impact of droughts in Africa needs to be seen in a proper global context. Droughts are frequent and severe in many countries of sub-Saharan Africa and have a devastating impact on their peoples and economies. Since the 1960s, rainfall in parts of the Sahel and Southern Africa has been below the norms of the previous 30 years (Sumberg et al., 2015). The overall observation is that most semi-arid lands tend to be naturally plagued by prolonged droughts, followed by periods of plentiful rainfall, thereby presenting agricultural planners with a challenge. The real issue is that a majority of the people who live in these drought-affected areas are not aware of climatic changes, and they are not able to perceive the climatic changes in their environment and do not know what appropriate arrangements they might have to make to manage the climatic resources.

The farming systems of Africa and the techniques used are varied, having evolved to meet the needs of the people in a wide range of environments. The traditional responses to soil degradation vary from one ecological zone to another. In the humid zone, traditional, agricultural practices ensure that soil degradation is kept to a minimum. A mixed-crop canopy protects the soil from physical destruction and maintains organic matter, especially after harvest, because most of the agricultural biomass is abandoned on farms. These traditional farming systems maintain vegetation with properties similar to the native forest. Depending on the type of cropping system in a particular farming zone, it can protect the environment. The systems are maintained by:

- forest or bush, followed by techniques that can allow forest regeneration;
- clearing of limited spaces leaving root stumps and big trees;
- retaining useful trees (such as those with medicinal properties or used for hare wood);
- inter-cropping;
- use of household refuse as fertilizer for crops in the farm or garden.

Some of these traditional practices have had a favorable impact on soils, while some have encouraged soil degradation in Africa. These include the creation of bush fires, the use of trees for fuel wood, overgrazing, and destruction of pasture lands.

Agricultural production in sub-Saharan Africa has, in recent times, remained lower than the rest of the world. Many attribute this to factors inherent to Africa and its people, such as climate, soil quality, slavery, and disease. Agriculture remains an important source of livelihood for the majority of Africans, but the sector is still unproductive. Despite huge agricultural potential, SSA countries have not yet benefited from agriculture and have experienced the highest prevalence of undernourishment worldwide. As a result, countries import increasingly more agricultural products than they export, putting additional strain on scarce foreign exchange reserves. The main challenges of the agricultural sector in sub-Saharan Africa includes gender disparities, dependence on rain-fed agriculture, low use of irrigation, limited public investment, and lack of institutional support. All these factors prevent countries from increasing productivity, adapting to climate change, and promoting agricultural value chains and trade.

Local farmers in the Cameroon form common initiative groups, which are usually required to concentrate on commercial production of the crops given priority by the government and to sell them to the state. Due to the novelty of modern irrigation, and the remoteness of many sites, governments have often had to supplement already costly irrigation works with extensive production, marketing, and social support services. High yields and double cropping are needed to pay for these high investments.

SSA countries have a massive labor force and the highest area of arable, uncultivated land in the world and huge agricultural growth potential (Kanu et al., 2014). But countries have not yet taken advantage of it. Despite the importance of the sector, about one-fourth of the population experiences hunger in sub-Saharan Africa. Out of about 795 million people suffering from chronic undernourishment globally, 220 million live in sub-Saharan Africa. According to Moyo et al. (2015), Africa's use of irrigation and overwhelming dependence on rain-fed agriculture explain the continent's low agricultural productivity. The low main staples of sub-Saharan Africa are unirrigated crops (maize, cassava, millets, sorghum, yams, sweet potatoes, plantains, and rice).

31.1.4 Food Safety Issues

The Global Food Safety Partnership (GFSP) is concerned with what must be done to improve unsafe food in sub-Saharan Africa. Relatively little is being done to reduce foodborne illness among consumers. More must be done to invest in informal markets, food chain infrastructure, and food safety education.

The majority of food bought and consumed in Africa comes from domestic markets, which are often informal and unregulated. These markets are also vital to the African economy. However, the lack of infrastructure in the food industry is contributing to the burden of foodborne illnesses. The absence of safe storage limits the lifetime of produce, increasing the potential that the food is unsafe to consume.

According to a report from the World Health Organization in 2015 concerning the global burden of foodborne diseases, 420,000 deaths worldwide were from unsafe food, with over 600 million foodborne-illness cases. It also identified 31 foodborne hazards, from diarrhea – causing illnesses to infectious agents and chemicals that can cause illness.

Globally, there are differences in the burden of foodborne diseases. The hardest hit areas are Africa, typically sub-Saharan Africa and central Africa. The demand for food and more nutritious food in Africa is rising as the population grows. The largest disease burden of sub-Saharan Africa is non-typhoidal *Salmonella enterica*, an invasive diarrhoeal disease. According to Havelaar et al. (2012), Salmonella is estimated to have the highest mortality of all foodborne hazards, causing about one-third of all deaths. Consumers need to ensure that the food they are eating is safe.

The task of addressing the health impacts of foodborne diseases and food safety in Africa is complex and deeply connected to larger nutritional challenges. Children, the elderly and pregnant women are most at risk from the effects of foodborne disease and are also among the groups most nutritionally at risk. Consuming a more diverse diet is a key way to combat the multiple burdens of malnutrition, including stunting and under-nutrition. But the most nutrient-rich foods, like dairy, meat, and vegetables, are also among the most risky from a food safety perspective.

Solving the health problems of foodborne disease in Africa is only part of the food safety battle. The continent must also be able to catch up with the wide range of economic costs, which range from lost productivity due to illness, to cost of response and prevention, and to trade. The World Bank estimates that in 2016, Nigeria, South Africa, and Angola each had foodborne-disease costs exceeding $US1 billion. Impacts of these costs have further effects, because countries are excluded from markets for high-risk foods, and future trade opportunities are lost. The international markets hold great promise for African financial development. Domestic problems cannot be ignored. At the national level, lack of consumer confidence in the safety of food erodes both the productivity of the domestic market and trust in the government and food safety authorities.

A strategic effort to address these interwoven problems of health and economy, which are both foreign and domestic, will be required to get at the heart of food safety in Africa. The food safety challenges faced by the African continent and globally are daunting, but if partners at all levels are willing to work toward a common goal, lives can be saved and stronger futures in Africa and beyond can be built.

Soil degradation can directly influence the quality of the food crop produced (Oliver and Gregory, 2015). If a soil is deficient in nutrients and if it is not properly managed, it will not only adversely affect crop production, but it will also cause nutrient deficiency in the food produce. Thus, the agricultural products of these areas are a matter of concern with respect to food safety. Meanwhile, concerns remain about the impact of agricultural intensification on the quality of soil, water, food produced, and other environmental resources. To address these challenges, it is essential to understand the links between soil quality, soil degradation, agricultural productivity, and food security.

Improving food safety also means improving local agricultural processes (Oliver and Gregory, 2015). Food often sold at local markets has been produced by the small land holders, so educating them about the use of chemicals and correct food handling will improve food safety. While investing in food infrastructure is important, it is also essential to engage in hygiene and safety education with food producers. Public health in sub-Saharan Africa needs to be prioritized. Increasing awareness of food safety and its importance for producers must be carried out. Consumers and food donors must pay attention to four key food safety topics: disease surveillance, research on hazards, awareness of food safety issues, and informal markets.

31.2 SOIL RESOURCES

31.2.1 Total Arable Lands

The population of sub-Saharan Africa is growing at 2.8% a year, double the rate of South Asia, four times the rate of East Asia and the Pacific, and about 50% higher than that of the Middle East and the North Africa region (Mabiso and Benfica, 2019). The region is relatively well endowed with natural resources. Total land area is 2,455 million ha, of which 173 million ha are under annual cultivation or permanent crops - about one-quarter of the potentially arable area.

Sub-Saharan Africa has six of the world's ten fastest growing economies. North Africa has vast oil and natural gas deposits. The Sahara holds the most strategic nuclear ore, and resources such as gold and copper, among many others, are abundant on the continent.

31.2.2 Oil and Gas

Africa is home to five of the world's top oil-producing countries, with an estimated 57% of Africa's export earnings from hydrocarbons. Algeria, Angola, Cameroon, Chad, Republic of Congo, Egypt, Eritrea, Gabon, Ghana, Kenya, Libya, Nigeria, South Sudan, Sudan, Tunisia, and Mozambique are all rich in oil and gas (Abada and Bouharkat, 2018). Nearly $2tn of investments is expected by 2036. Besides oil and gas, Africa is rich in precious minerals, forests, and diamonds found in Angola, Botswana, the Central African Republic, and the Democratic Republic of Congo.

Natural resources dominate the structure of wealth in Africa. The share of natural capital in the continent's aggregate wealth is the second highest in the world after the countries that form the hydrocarbon-rich Gulf Cooperation Council (GCC). Even resource-poor sub-Saharan Africa has a share of natural capital in total wealth that is higher than the resource-rich countries. While paradoxical at first glance, this finding is the consequence of the fact that the contributions of human and physical capital to total wealth are modest both in resource-rich and resource-poor Africa.

Sub-Saharan Africa is home to nearly half of the world's uncultivated land and possesses about 200 million ha that can be brought into production. Africa uses only 2% of its renewable water resources compared to 5% globally (Abada and Bouharkat, 2018). One measures a country's dependence on natural resources in three ways: the extent to which the country depends on exports of natural resources for foreign exchange, the share of natural resources in their production, and the contribution of resource rents to governmental revenues. The resource dependence of resource-rich countries in sub-Saharan Africa also varies. The Republic of Congo is more dependent than other developing and developed resource-rich countries, including the GCC.

Oil accounts for 43.5% of resource-rich sub-Saharan Africa wealth, substantially less than in resource-rich comparators outside Africa, and land accounts for about 35%. Metals and minerals account for 27% for Zambia, 26% for South Africa, and 14% for Botswana (Abada and Bouharkat, 2018). In resource-poor sub-Saharan Africa countries, land accounts for the largest share of natural capital, as much as 60% in Tanzania.

31.3 SOIL DEGRADATION

Soil degradation associated with poor soil fertility management practices is a major factor underlying poor agricultural productivity in sub - Saharan Africa. About 65% of the agricultural land is degraded, mainly due to low nutrient application, soil erosion, and soil acidification. The primary cause of soil degradation in sub-Saharan Africa is expansion and intensification of agriculture in efforts to feed its growing population. Effective solutions will support resilient systems, and must cut across agricultural, environmental, and socioeconomic objectives.

Land is a limited and non-renewable resource, and experience has shown that, with continuous utilization with the best technologies and skills, its quality is deteriorating. Soil degradation is defined as "a process which lowers the current and/or the potential capability of soil to produce (quantitatively and/or qualitatively) goods or services (FAO et al., 2017). Soil degradation has also been defined as "the decline in soil quality caused through its misuse by humans" (Lal et al., 2016). It is a broad and vague term; however, it refers to a decline in the soil's productivity through adverse changes in nutrient status, SOM, and structural attributes, as well as an increase in concentrations of electrolytes and toxic chemicals. In other words, it refers to a diminution of the soil's current and/or potential capability to produce quantitative or qualitative goods or services. Soil degradation has been defined in many ways, most often referring to the (agro) productive function of the soil (Barrow, 1991). In a general sense, soil degradation could be described as the deterioration of soil quality or, in other words, the partial or entire loss of one or more functions of the soil (Jie, 2002). Hence, it is imperative to not only preserve soil quality but also to adopt measures to ameliorate it while keeping pace with increasing demands from various sectors.

Since soil degradation remains a continuous and gradual process, the amount of soil degradation in sub-Saharan Africa can rely only on gross estimates because there are no reliable data. Nevertheless, most agricultural soils are losing their ability to provide food and essential ecosystem services, and this is attributed to soil degradation. We reviewed data from studies that examined degradation in sub-Saharan Africa at broad scale, and we found degradation depends on multiple indicators. However, different indicators respond differently. The pH and cation exchange capacity tend to decline linearly while soil organic carbon and yields decline non-linearly.

31.4 CAUSES OF FOOD INSECURITY

The traditional usage of the term food security refers to access, availability, and nutritional aspects of food (Wegren et al., 2017). A household is considered food secure when its occupants do not live in hunger or fear of starvation (Mallikarjuna, 2013). Two commonly used definitions of food security come from the World Health Organizatio/Food and Agriculture Organization (WHO, 2003) Food security exists when all people, at all times, have physical and economic access to sufficient, safe, and nutritious food to meet their dietary needs and food preferences for an active and healthy life(WHO, 2003). Food security for a household means access by all members at all times to enough food for an active, healthy life. Food security includes at a minimum of the following: (i) the ready availability of nutritionally adequate and safe foods and (ii) an assured ability to acquire acceptable foods in socially acceptable ways (that is, without resorting to emergency food supplies, scavenging, stealing, or other coping strategies) as reported by Gebhardt et al. (2008).

There are large differences among continents in the prevalence of severe food insecurity. Approximately 27.4% of the population in Africa was classified as severely food insecure in 2016, which is almost four times as high as any other region. Food insecurity is on the rise, specifically in sub-Saharan Africa. Global estimates of undernourishment rose from 777 million in 2015 to 821 million in 2017. Africa has the highest prevalence of undernourishment, estimated in 2016 to be 20% of the population. This is especially alarming in Eastern Africa, where it is suspected that one-third of the population is undernourished (Obayelu, 2019).

31.4.1 Causes of Hunger in Sub-Saharan Africa

Food insecurity and hunger are caused by many factors, often being intertwined with one another. In general, the principal causes of hunger include poverty, conflict, climate and weather, lack of investment in agriculture, and unstable markets (World Food Program, 2018 as reported by Plan, 2018).

31.4.1.1 Poverty

Poverty is a principal cause of hunger in Africa and elsewhere. Individuals living in poverty often cannot afford food of sufficient quality or quantity to live a healthy life. According to the World Bank, in 2013, 42.3% of the population of sub-Saharan Africa lived on $1.90 or less per day, a principal factor of widespread hunger (Sullivan, 2020). Poverty is often a cycle. Children exposed to long-term under-nutrition are often stunted, leading to long-term consequences including decreased labor productivity and income-earning potential (Sullivan, 2020).

31.4.1.2 Conflict

Conflict and violence can have direct and indirect impacts on all levels of the food system, leading to food insecurity and hunger. Conflict often puts constraints on employment and income opportunities, which affects an individual's ability to acquire food. Conflict can also affect exports and imports, which can lead to limited food availability and affordability. Availability of food can also be affected if resources (e.g., land, equipment) used to produce food are destroyed during times of conflict (Sullivan, 2020). The migration of people away from the conflict zones completely prohibits them from being on their farm lands where they solely depend for food and income, and they are bound to face under-nutrition.

In 2017, conflict was the major cause of food insecurity and hunger in 18 countries, affecting about 74 million individuals. Eleven of these countries were in Africa, which totaled about 37 million people. Northern Nigeria, Democratic Republic of Congo, Somalia, and South Sudan accounted for the majority of these individuals (Food Security Information Network, 2018). Since 2013, South Sudan has experienced ongoing conflict, which has caused an increase in food insecurity. In 2017, parts of South Sudan declared famine, and more than 42% of the population faced severe food insecurity (Sullivan, 2020).

31.4.1.3 Environment

Environmental challenges, including erosion, desertification, deforestation, and drought and water shortages, can have detrimental impacts on food security. Land degradation also leads to the diminution of livestock in both quantity and quality so that the living standard of the rural people as a whole is affected by any change in the livestock sector in most African countries (Salami, and Tilakasiri, 2019). Empirical studies discovered that the effects of changes in climate and resulted environment have caused the conflicts, economic shocks and food crises for the millions of people in Africa. Most African countries that experienced a significant rise in undernourishment for 7 years (2011–2017) faced serious economic slowdown or downturn, and there are predictions that there will be a continuous and future increase of migration hotspots around Africa due to the impact of environmental change if actions are not put in place (Salami and Tilakasiri, 2019). Land degradation in Ethiopia also results in enhancing the unemployment rate and out-migration; where there is low agricultural and livestock production (Megerssa and Bekere, 2019). Insufficient land for cultivation leads to a reduction of farm size that creates disguised unemployment. Ethiopia is rankly first among other Africa countries to onlooker the power of land degradation that pushes people out of their home. The conflict between pastoralists occurred as the result of land degradation and continues up to date in Ethiopia. (Megerssa and Bekere, 2019). Due to the fact that more than 85% of the Ethiopian economy is dependent on agriculture, land degradation is seriously affecting the productivity of the economy. In Ethiopia, land degradation is also the causal agent for food insecurity, poverty and low agricultural productivity (Birhanu, 2014). The effect of land degradation on the natural environment is due to the depletion of forests. The impact of increased climatic shocks, due to climate change is felt more by rural poor who highly depend on agriculture (Zidouemba, 2017). However, given the strong relationship between agriculture and the rest of the economy, low-income households in urban areas are also affected. Conversely, the impact of climatic shocks is not as significant for urban non-poor given the level and source of revenue as well as the diversification of their consumption (Zidouemba, 2017). The environmental challenges from drought increased poverty and hunger by reducing agricultural production and people's incomes. Many of these challenges are man-made. Deforestation, for example, is caused by humans seeking new places to live, farm, or obtain firewood. Drought, water shortage, and desertification in Africa all reduce agricultural productivity and, thus, food availability.

31.4.1.4 Population Growth

Africa's population has increased rapidly, from 221 million in 1950 to 1.2 billion in 2018. Africa has the highest population growth rate among world regions; between 2010 and 2015, it grew at a rate of 2.55% per year. It is estimated that more than half of the global population growth between now and 2050 will occur in Africa (WHO, 2019). Rapid population growth can limit increases in per capita income, causing poverty and hunger. Providing nutritious and environmentally sustainable food to all people at all times is one of the greatest challenges currently facing society in general and Africa in particular. This problem is particularly acute in Africa where an estimated one in four people still lack adequate food to sustain an active and healthy life (Hall et al., 2017). Considering the potential impact of future population growth and climate change on food security in Africa, while projecting to 2050, there is an indication that, projected rapid population growth will be the leading cause of food insecurity and widespread undernourishment across Africa (Hall et al., 2017).

31.5 MAGNITUDE OF FOOD INSECURITY IN SUB-SAHARAN AFRICA

31.5.1 Food Security

Food security may have different meanings for different people. In order to achieve food security a country must (i) ensure adequacy of food supplies in terms of quantity, quality, and variety of food; (ii) optimize stability in the flow of supplies, and (iii) secure sustainable access to available supplies by all who need them. A region is food secure when a majority of the people in the geopolitical area have access to food of adequate quantity and quality at all times, while a locality is food secure when a majority of the people in the locality have access to food of adequate quantity and quality at all times (Babatunde and Oyatoye, 2005).

Adequate food availability at the national, regional, and household levels, obtained through markets and other channels, is the cornerstone of nutritional well-being. At the household level, food security implies physical and economic access to foods that are adequate in terms of quantity, nutritional quality, safety, and culturally acceptable to meet each person's needs (Matshe, 2009). Food security is ultimately associated with access to nutritionally adequate food at the household level, i.e. the ability of households or individuals to acquire a nutritionally adequate diet at all times. Household food security depends on an adequate income and assets, including land and other productive resources that are owned. Household food security can be translated into good nutritional status if household members have nutritional security, a condition that combines access to nutritionally adequate and safe food; sufficient knowledge and skills to acquire, prepare, and consume a nutritionally adequate diet, including diets that meet the special needs of young children; and access to health services and a healthy environment to ensure effective biological utilization of foods consumed. Therefore, nutritional well-being depends on a number of interrelated factors, which, besides food security, include health and sanitation, adequate supplies of safe water, parents' education, time to prepare food, and care of vulnerable individuals within the household.

For food security to be ensured at these different levels whether for an individual household, region, or nation three important factors have to be in place: food availability, food accessibility, and food utilization. "Food availability" concerns the physical presence of food, which depends on domestic food production and food importation (FAO et al., 2017). The problem of dwindling food availability has been aggravated because total food production has been constant over the years with a growing socioeconomic challenge. For example, in Nigeria, the value of food imports increased by 53% from N=3.2991 trillion (Naira) in 2008 to N=5.0479 trillion in 2009. Nigeria's import table also shows that 90% of goods imported are for consumption and not for production (Omoh, 2011). Availability is also a function of stock holdings and food marketing, apart from food production. It is one thing for food to be physically present; it is another for it to be accessible, which is dependent on the purchasing power of the people. Food accessibility is not physically or socially constrained, but economically. Food processing and associated activities, in particular, are important factors in the promotion of food access and the production of safe and nutritious foods.

31.5.2 Household Food Security and Nutrition

In sub-Saharan Africa, where ~70% of the population lives in rural areas, crop and animal production, fisheries, and forestry activities are direct sources of food and provide income with which to buy food. Increased and diversified production of food for family consumption or as a source of income is a basic prerequisite for improved household food security. Better home and community food processing, preservation, and storage and access to marketing facilities can also contribute to household food security by alleviating seasonal shortages in food supply and stabilizing market prices. Most of the food crops in sub-Saharan Africa are seasonal crops, and land users need to be made aware of preservation and storage techniques of these perishable crops to ensure a food supply in local markets and in households. Agricultural practices and technology, as well as the establishment of small-scale agro-processing industries, can increase employment and income-generating opportunities and, thus, positively affect household access to food. Food safety and quality, secured through effective food-quality control at all stages of production, processing, and handling, also influence nutritional well-being.

31.5.3 Food Production Trends within Sub-Saharan Africa

The main food crops are rice, wheat, and coarse grains, and they are taken into consideration in this section. Agriculture is the backbone of the economies of most SSA countries. Table 31.1 shows average annual production (in tons) of major food crops in Cameroon, Chad, Kenya, and Nigeria. Among the grains, the fastest growth in output was achieved in rice, but its effect on growth of total cereal output was limited because of the relatively small share of rice in cereal production

For sub-Saharan Africa as a whole, exports of cereals declined and imports expanded rapidly during the last 5 years (from 2015 to 2020) (Table 31.2). In the last 5 years, imports of rice have recorded a marked increase in Cameroon, Chad, Kenya, and Nigeria contrary to the import of wheat, reflecting both food deficits resulting from rapid population growth and land degradation. In the mid-1980s, 20% of the staple food requirements of sub-Saharan Africa were provided by imports. This proportion is predicted to fall by the end of the 2030 because the local population

TABLE 31.1
Average Annual Production (in tons) of Major Food

	2020/2021 Domestic Availability			2020/2021 Utilization (000 tons)			2020/2021 Import Requirements (000 tons)			Per Caput Consumption (kg/year)		
	Wheat	Rice	Coarse Grains	Wheat	Rice	Coarse Grains	Wheat	Rice	Coarse Grains	Wheat	Rice	Coarse Grains
CAMEROON	–	278	3,058	750	768	3,098	750	490	40	27	28	80
CHAD	3	204	2,559	94	274	2,609	92	70	50	6	14	102
KENYA	280	119	4,046	1,880	819	5,446	1,600	700	1,400	28	14	90
NIGERIA	63	5,561	21,373	4,863	7411	21,793	4,800	1,850	420	20	32	79

Source: FAO (2021).

TABLE 31.2
Food Production (tons)

	Production (000 tons)								
	Previous Year's Production			Previous 5 years' Production Average			2019 Production Compared to Average (%)		
	Wheat	Rice	Coarse Grains	Wheat	Rice	Coarse Grains	Wheat	Rice	Coarse Grains
CAMEROON	–	247	2,944	–	200	3,048	–	116	100
CHAD	2	199	2,691	2	180	2,545	121	110	98
KENYA	200	111	3,739	237	85	3,878	118	112	104
NIGERIA	74	5,042	21,430	68	4,627	18,733	93	109	114

Source: FAO (2021).

is gradually getting involved in rice and cereal food farming, as shown by comparing production increase (Table 31.2) and import requirements compared to average percentages

A close comparison of Tables 31.2 and 31.3 shows that, even though production of staple food has increased in the last 5 years, importation of these food stuffs in these countries has not witnessed any great drop and this can be attributed to the fast population growth rate and soil degradation. Population growth is the main factor responsible for increases in food supply and energy requirements. A greater growth in food production will be needed in the future, if dependence on commercial or concessionary imports is to be avoided.

Population data, unlike those of other developing regions, show that growth rates for sub-Saharan Africa are still increasing as of the 2019 census, which showed the following numbers: Nigeria, 201 million; Cameroon, 25.88 million; Chad, 15.95 million; and Kenya, 52.57 million. The 2015 census showed the numbers to be Nigeria, 151.1 million; Cameroon, 23.3 million; Chad, 14.11 million; and Kenya, 47.88 million. These values, combined with other population data, are used to calculate estimates of a country's per caput consumption (kg/year) (Table 31.1). The per caput consumption was not proportional to the population growth with Nigeria (32 kg/year) being the highest while Chad and Kenya (14 kg/year) being the least. It is projected that the overall population growth rate between 2020 and the end of the twenty-first century will be 3% faster than the population growth of the region during the previous decade.

TABLE 31.3
Import Food Tequirements (000 tons)

	Import Requirements (000 tons)								
	Previous Year's Imports			Previous 5 Years' Average Imports			2020 Import Requirements Compared to Average (%)		
	Wheat	Rice	Coarse Grains	Wheat	Rice	Coarse Grains	Wheat	Rice	Coarse Grains
CAMEROON	750	580	40	643	488	49	117	100	81
CHAD	70	75	50	70	57	44	131	123	113
KENYA	1,570	688	1,400	1,440	640	1,188	111	109	118
NIGERIA	4,600	2,670	420	4,540	2,662	406	106	69	103

Source: FAO (2021).

31.5.4 Food Production and Smallholder Farmers

In the past, agriculture has been the major support of the economy in providing raw materials, food, and employment for over 75% of the population. Agriculture is critical to world food security, poverty alleviation, and conservation of natural resources. Farmers are critical for crop production and farm enterprises (Sundrum, 2001). They perform various functions such as clearing, planting, weeding, staking, fertilizer application, harvesting, processing, storage, and marketing. Throughout the world, farmers have been classified as small, medium, and large-scale; however, in Africa, small-scale farming is dominated by peasant farmers living in rural areas, and they have farm holdings of 1–2 ha. They constitute about 70% of the rural areas, which sustain agriculture (Oladele and Fawole, 2007). Over 12 million farmers are scattered in different ecological zones and are engaged in the production of a wide variety of arable crops under traditional subsistence farming. Ninety percent of Nigerian total production comes from small farms. Increasing farmers' productivity and income will require the development of appropriate technology that can be developed through research and then the research can be transferred through efficient extension systems (Abalu, 1988). Promoting productivity among small-scale farmers is, therefore, essential for national growth and food security in Nigeria. Peasant farmers should be encouraged to engage in Integrated Farming Systems (IFS), which are based on integration of livestock production with cropping systems. Through proper management practices, IFS increase soil fertility; minimize insect and disease problems, energy requirements, and risks due to climatic fluctuation; and protect the environment through erosion control (Tirado, 2009).

31.5.5 Supporting Capacity-Building in Food Safety Across Africa

The GFSP is an innovative, public-private initiative dedicated to supporting and promoting global cooperation for building capacity of food safety (Stacey and Garin Fons, 2015). GFSP is uniquely able to assess food safety systems and propose systems-based interventions to address specific food-sector needs and to prioritize hazards and threats.

The report, "Food Safety in Africa: Past Endeavors and Future Directions," considers the quantity and quality of building food safety capacity in sub-Saharan Africa. It provides data, analysis, and recommendations, which organizations working on food safety in sub-Saharan Africa can use to improve the impact of their efforts. The report focuses on investments and activities related to food safety capacity-building that the international donor community can carry out and ways that both donors and African governments can better target and coordinate those investments, with greater cooperation between the public and private sectors.

Current investments by donors for food safety in sub-Saharan Africa largely reflects the concerns of previous decades, and, as a result, substantially focused on access to regional and overseas export markets, with emphasis on national control systems. However, relatively little is being done to reduce foodborne illness among consumers in sub-Saharan Africa. New understanding of the burden and management of foodborne diseases, along with rapid and broad change within sub-Saharan-African societies and agri-food systems, have led to food safety emerging as an important public health and developmental issue. There is need to reconsider strategies of investments by national governments and donor support.

31.6 ORGANIC FOOD PRODUCTION

31.6.1 Organic Farming in Sub-Saharan Africa

In Africa, official statistics contain few data on organic agriculture, even though its products are increasingly available for local and export markets. African consumers' demand is growing, providing an economic opportunity. Today, there is no doubt that organic agriculture is making an important contribution to the agro-ecological transition process underway in Africa. It reduces the adverse environmental and health impacts of agriculture, particularly because it uses no synthetic chemical inputs. It improves the resilience of agricultural systems. If Africans rely on organic farming, there is a high tendency that there will be food shortages in the near future. Under certain conditions, its specific techniques can increase agricultural productivity, even if its yields are on average lower than those in conventional agriculture. Organic systems are labor-intensive and, as such, can be a source of employment for young people and middle-aged women in rural areas.

According to statistics from IFOAM (International Federation of Organic Agriculture Movements) organic agriculture covered 57.8 million ha in 2016, including in-conversion areas, or almost 1.2% of cultivated farmland at the global level. For the African continent, certified organic agriculture covers 1.8 million ha, or only about 0.2% of cultivated land on the continent (Reganold and Wachter, 2016).

The characteristics of organic agriculture make it a mode of production that is potentially suited to African agriculture, which is typically family based with a low level of mechanization, small-scale, labor-intensive, and diversified (a variety of cultivated plants), using local resources and satisfying different needs (feeding the family, generating income). Researchers have highlighted the advantages of organic agriculture in the African context as follows: fewer health risks for producers and consumers, better protection of natural resources, a more resilient environment, increased and enhanced agricultural diversity, higher income for farming families, and more secure outlets. Health risks are a serious concern in Africa, especially because food crops that have been freshly sprayed with fungicides are commonly found in local markets. This is the case for fruit and vegetables consumed in cities, which contain pesticide levels that are often far higher than the maximum residue levels.

Commonly, when agricultural practices are allowed in the hands of local farmers with no training concerning the dangers of pesticides and fungicides, they spray to protect their crops from pests. Consequently, sometimes pesticides are used in excess or for the wrong reasons, including the use of unauthorized or expired pesticides. Finally, at the economic and social level, organic agriculture is labor-intensive (manual weeding, composting), which seems conducive to its development in countries with a large agricultural workforce.

Although Africa is estimated to possess 25% of the world's arable land, and the continent has some of the most fertile land in the world (Jones et al., 2013), it is currently only producing about 10% of global agricultural output. The estimated percentage is a reflection of insufficient food for African consumption. Moreover, most soils are fragile, with low levels of nutrients and concentrations of organic matter. The remaining areas are highly weathered and acidic soils, with high levels of iron and aluminum oxides (Headey and Jayne, 2014), which are known to adsorb plant macronutrients, such as phosphates. Farming in such zones requires careful management, or else the retained minerals will be washed into streams during runoff, causing eutrophication.

31.6.1.1 Organic Farming in Cameroon

Cameroon is well known for its climatic, geographic, and ecological diversity, which enables farmers to grow a large number of crops. The most frequently cultivated organic products are grown for exportation. The production of organic pineapple and papaya in Cameroon is mainly carried out by individual producers and a few groups (Bayiha et al., 2019). The groups are, in general, formed by the exporting producers. The exporting body, which holds the organic certificate, gathers a network of producers that it monitors, controls, and certifies. Common Initiative Groups ("GIC" in French) also exist. There is no international organization yet that works exclusively on promoting the organic sector in Cameroon.

With the support of international aid, the Government of Cameroon set up a professional association named AGROCOM, which was mandated to organize and supervise the whole exporting sector with a view to diversifying Cameroon's agricultural exports (Maldangoï et al., 2003). In particular, it deals with pineapple operators, whether they are under conventional or organic production systems. Papaya is still considered as a crop for domestic consumption. AGROCOM, as a professional organization gathering producers and exporters of the horticultural sectors in Cameroon, has retained a certain number of so-called priority crops (pineapple, onion, potato, and plantain). Only producers of these crops (conventional or organic) can benefit from its support (Maldangoï et al., 2003).

Other products exported as organic are plantain, yam, tubers (sweet potato, potato, macabo, taro), various fruits (corossol, coconut, sagou), various stimulating plants (cocoa, coffee, cola nut, bitter cola, pepper), and legumes (peanuts, soy, beans). Pineapple, mango, avocado, banana, and papaya represent the largest exported quantities.

Pineapple and papaya are principally cultivated by men. The role of women is generally limited to accompanying the men. However, among the labor force there are both women and men. Women often have tasks requiring more attention and care (e.g., crop management, harvest, and storage), while men take care of soil preparation, clearing, felling trees, ploughing, and of all tasks for which more physical strength is required (SHIOYA, 2013).

Cameroon has in its territory all types of inter-tropical climates. Also, Cameroon has good soils (e.g., volcanic, ferralitic) in regions of high, medium, and low altitude. The agriculturally active population of Cameroon accounts for more than 40% of the total population. To this population, mainly rural, must be added professional agricultural farmers searching for new opportunities and endowed with sufficient financial resources (e.g., businessmen, retired high-ranking officials in conversion to agriculture).

The average area in the small-holding production ranges from 1/12 to 1/3 hectare, i.e., 5,000–20,000 plants. Over the last year, some producer-exporters have started to develop organic production on areas of 6–8 ha a year. Regarding papaya, the average areas are in the order of 1 ha, cultivated by small producers and farm workers (Njoya et al., 2017).

Areas under papaya have gone through a dramatic growth since 1995, when difficulties emerged in pineapple trading, which had a negative impact on the financial situation of producers. Papaya is grown all year long (which allows a farmer to obtain cash regularly) with a lower production cost compared to pineapple. In the Moungo region, papaya – mainly the Solo variety – is cultivated on almost 70 ha, 30% of which is grown organically (Njoya et al., 2017).

Constraints of the organizational type are numerous. The grouping of producers in associations is under way, but there is no effective market and price information system. In the local market, organic products (what consumers call "natural products") are bought by all social classes (reportedly because of their good taste); prices are directly related to the relationship between supply and demand, day by day. The strong increase in the prices of synthetic farm inputs (fertilizers, pesticides) and the lack of involvement of public structures have led entire areas to convert to organic farming.

31.6.1.2 Organic Food Production in Nigeria

Nigeria is Africa's largest economy and is a major oil producer with a population of over 162 million people (Nwajiuba and Akinsanmi, 2002). In Nigeria, there are over 30 million ha of farmland under cultivation season to season, falling substantially short of the estimated 78.5 million ha of land that is required for farming to feed Nigeria's growing population.

Organic farming is developing rapidly and practiced in more than 120 countries of the world (Nwachukwu, 2010). In 2007, there were over 26 million ha managed under organic agriculture techniques worldwide also reported that African countries like Uganda Kenya, Tanzania, South Africa, Cameroon, and Ghana have gone far ahead of Nigeria in production of certified organic food.

Majority of small-scale farmers (about 70% of the farming population) practice organic agriculture by default, either because of the prohibitive costs of chemical fertilizers and other agrochemicals or as a habit because according to the farmers, organically farm produce store longer and taste better than inorganic farm products. They produce non-certified organic products, not purely organic according to modern organic farming principles and some of the farmers adopted the use of animal droppings as manure. In an effort to achieve food security for the 162 million people in Nigeria, Nigerian government is advocating massive use of inorganic fertilizers by Nigerian farmers, increasing the distribution of inorganic fertilizers to the farmers. According to AdeOluwa (2010), organic agriculture in an organized manner is still young in Nigeria with few years of experience. Certified organic agriculture has already started in some states (western, southwestern, and few areas in the southeastern states) recording about 3.154 hectares of land under organic farming, of which 50 ha already converted (Willer and Kilcher, 2009). The first certified organic farm in Nigeria is Eurobridge at Odogbolu, Ogun State (Kazeem, 2010). The farm, according to the author, is currently producing five products for local and export markets. Hence, Nigeria can boast of few organically produced food products such as lemon grass tea, under the trade mark of Daradara herbal tea, tumeric *Curcuma longa* for food seasoning; the product is like ginger; Ofada rice (Nigerian indigenous rice), Red hibiscus (Hibiscus sabdariffa) for a local soft drink production (popularly known as Sobo). Other organic farm products include: tropical fruits, herbs, spices, mushrooms, cashew, fruit juices, quail, guinea fowl, chicken, cane rats, snails, ginger, turmeric, and watermelon. Organic farmers in Nigeria rely heavily on crop residues, animal manures, legumes, green manures, organic wastes, and mineral-bearing rocks to feed the soil and supply plant nutrients. The expanding market for organic produce creates opportunities for Nigerians to make decent livelihoods through self-employment in agriculture. Most of them are unable to access this opportunity because they lack skills in organic agriculture and entrepreneurship. Explaining the potential role of organic agriculture in sustainable food security, Kazeem (2010) noted that organic agriculture is something that should be given priority in Nigeria because it is demand-driven as the products are needed in advanced countries. The Nigerian farming system was perceived as non-certified organic agriculture production (Alimi et al., 2006). Presently, most Nigerian farmers do not practice organic agriculture per excellence (Alimi et al., 2006), hence Nigeria is still in the primary or preparatory stage of certified organic farming.

Even though agriculture still remains the largest sector of the Nigerian economy and employs two-thirds of the entire labor force, the production hurdles have significantly stifled the performance of the sector. It is estimated that Nigeria has lost USD 10 billion in annual export opportunity from groundnut, palm oil, cocoa, and cotton due to a continuous decline in the production of those commodities (Alimi et al., 2006). Food (crop) production increases have not kept pace with population growth, resulting in rising food imports and declining levels of national food self-sufficiency. The main factors undermining production include reliance on rain-fed agriculture, smallholder land holding, and low productivity due to poor planting material, low fertilizer application, and a weak agricultural extension system.

Nigeria is the continent's leading consumer of rice; it is one of the largest producers of rice in Africa and simultaneously one of the largest rice importers in the world (Raheem et al., 2021). Rice is an important food security crop and is an essential cash crop, because it is mainly small-scale producers who commonly sell 80% of total production and consume only 20%. Rice generates more income for Nigerian farmers than any other cash crop in the country (Oladimeji and Abdulsalam, 2013). In 2008, Nigeria produced approximately 2 million metric tons of milled rice and imported roughly 3 million metric tons, including an estimated 800,000 metric tons that is suspected to enter the country illegally on an annual basis.

Moreover, the country is the largest producer of cassava in the world, with about 50 million metric tons produced annually from a cultivated area of about 3.7 million ha. Nigeria accounts for cassava production of up to 20% of the world, about 34% of Africa's, and about 46% of West Africa's. The national average yield of cassava is estimated at about 13.63 MT per ha, as against potential yield of up to 40 metric tons per ha. Close to two-thirds (66%) of total production is in the southern part of the country, while about 30% is in the north-central part, and 4% in other parts of the north (Alimi et al., 2006). The crop is predominantly grown by smallholders on small plots for family consumption and for local sale. Large-scale commercial plantations are rare. However, because of the country's massive size and diversity, different regions may face different constraints because of a decentralized approach to designing industrial policies and initiatives that may not be coordinated with the agricultural policies.

31.6.1.3 Organic Food Production in Kenya

Production of organic food in Kenya is still relatively low, although it is increasing at a faster rate. Over 182,000 ha of land are under organic management, which accounts for 0.69% of the total agricultural area in Kenya (CHEBET, 2021). This means that there is need for telling the farming sector about the importance of organic farming. Globally, organic markets are growing especially as organic products are increasingly entering the markets, thereby reducing poverty (CHEBET, 2021). Consumers' awareness of the advantages of organic food over non-organic food has resulted in more and more interest in healthy and sustainably produced food. Expenditures on organic food are rising faster than on non-organic food. These trends are common for both the export market and the local market. A majority of local farmers ends up with farm products in local markets, which has resulted in an increase of organic food crops in the local, organic-farmer markets as well as designated organic sections within super markets. Many consider organic agriculture as an interesting option for smallholder farmers in Africa, because it offers a unique combination of low inputs and environmental conservation and it provides access to premium-price markets.

Some common organic food products from Kenya are moringa and baobab due to the increasing global demand for these superfoods (Evang et al., 2021). Moringa, a natural source of nutrients, grows well in Kenya and in many parts of countries of sub-Sahara African. Moringa farms on the "Green Moringa Gold Coast" offer optimum growth conditions for farming of coconut and macadamia. These crops can be grown on moringa farms due to the fertilization of the soil by the organic residues from moringa (Evang et al., 2021).

Despite the fact that there is a global demand for organic foods, there is still an insufficient supply of organic products in local markets because of a lack of organic farmers in Kenya. It is equally a big challenge to find organically certified growers in Kenya who can meet the demands from their clients. At this moment, the demand is significantly higher than the supply, and this is responsible for the high cost of organic food products in local markets to the extend where local farmers many at times turn to sell more than the amount they consume to overcome poverty at the expense of food insecurity at the level of her household (Weinmann, 2018). Training organic farmers is, therefore, an important investment. Finally, a growing awareness in Kenya for healthy and safe food may generate opportunities in other agricultural sectors. Foods such as moringa can offer a sustainable solution for millions of farmers. Planting crops like moringa in rural areas can improve soil fertility and reverse deforestation. Besides moringa, baobab is a wild fruit now grown and collected by women and youth groups (Weinmann, 2018). Collecting baobab, therefore, creates an income opportunity for people without land or other sources of income.

31.6.1.4 Organic Food Production in Chad

Chad, in Central Africa, has one of the highest levels of hunger in the world –66.2% of its population of 15.5 million lives in severe poverty. It is ranked 187th out of 189 countries in the 2019 Human Development Index (Nogueira et al., 2021).

Conflict and the climate crisis exacerbate hunger and poverty in Chad. The people in Chad are among the most affected by the global climate breakdown, and it also suffers from environmental degradation and rapid desertification. Many depend on farming and livestock for their livelihoods, but the advancement of agricultural practices is challenging as rain patterns are changing and droughts are frequent. These places even more strain on vulnerable families living in the Sahelian belt. Around 40% of children aged below five suffer stunting, according to the World Bank, with low height for their age caused by chronic malnutrition (Feng et al., 2018). Health of pregnant and breastfeeding women is poor, with high maternal mortality rates due to inadequate access to health services. Access to basic education is also limited.

The presence of hundreds of thousands of refugees who have fled conflict in neighboring countries has put additional pressure on Chad's already limited resources. Displaced people and other poor communities in the Lake Chad Basin (in the west-central part of the country) and in the east and south of the country are dependent on humanitarian assistance for survival. According to the 2019 Humanitarian Response Plan (Anderson, 2020), 4.3 million people are in need of humanitarian assistance, of whom only 2 million are targeted with adequate support. The World Food Program (WFP) runs a variety of activities in Chad, with a focus on gradually shifting from emergency relief to strengthening the country's long-term resilience (Anderson, 2020). In the mission to end food insecurity and malnutrition in Chad, the country continues to work alongside the Government and other UN agencies and partners.

Cultivation of maize in Chad needs to be encouraged, despites the growing constraints that impede its emergency. Among several constraints like; continuous cropping, increased population, pressure on arable lands, degradation of land, and low investment in soil fertility, inappropriate production technologies, bad weather and invasive weeds (Ridine et al., 2014), those related to climate cannot be solved by local farmers; however, those related to soil fertility could be improved by providing soils with organic and chemical fertilizers. Averagely 60% of organic matter is used worldwide, but the proportion in use by stake holders in Chad is not significant (Olombo, 2000), despites the high agricultural activities (70%–80%) conducted by Chadian growers.

Persons familiar with the rural regions of Chad know that food is hard to find during lean periods and that the lack of food generally impacts a large number of families between June and September. The depletion of their reserves results in food shortages, particularly for children. With the advancement of the awareness of organic farming in this country, most of their garden plots are not only providing them with food year-round but also with money from the sale of their produce. There is no longer a rush in Goré, a city in the south of Chad, to stock up on grain during the lean period due to the abundant supply of produce from market gardens year-round at affordable prices. Market-garden groups working in the plains surrounding the town have kept the market regularly supplied with tomatoes, cabbage, and maize. Everything is available except the money to buy this produce.

Farmers lack access to services, knowledge, and technology that they need to improve productivity (Ridine et al., 2014). Inadequate access to rural financial services also prevents poor farmers from developing alternative income opportunities or improving productivity. Chad's rural producers are generally engaged in subsistence farming and capture a small part of the value generated by agricultural production, even in areas that are not isolated.

31.7 POOR MANAGEMENT AND LIMITED NUTRIENT SUPPLY

Africa is generally endowed with good soils that have the potential to be highly productive. Yields can be substantially raised from present levels with correct land use and good management, especially of Africa's available water resources. However, this can only be achieved when the local farmers are able to avoid inappropriate land use, poor management, and the lack of inputs, which can quickly lead to decline of productivity and land degradation, a situation that prevails in many parts of sub-Saharan Africa.

Communal and allocated lands are frequently abused, because neither farmers nor grazers feel any responsibility for their conservation. Conservation work becomes practically impossible where holdings are fragmented. For these reasons, campaigns to survey consolidate, and register land are required to establish ownership by individuals (including women), collectives, or corporations. It is estimated that about 4,200 billion m^3 of freshwater flows out of Africa into the ocean every year; 10% of it would increase Africa's food production by 10% (Mounir et al., 2011).

Physical resources, other than land, that must be available include fertilizer, improved seed, tree cuttings, and improved implements. A majority of the land users are the poor and cannot afford these physical resources to boast agricultural productivity. The high cost of these resources is another challenge for a majority of land users in sub-Saharan Africa. The land user needs access to financial resources to obtain these physical resources. Access to physical resources can improve productivity and decrease land degradation, which will reduce hunger and not poverty, even if access to other key factors, like an improved transportation system (roads, bridges, and public transport), is limited. Transport systems have limited the quantity of food crops in local markets. Good transportation facilities also make available an array of communication, marketing, and employment opportunities. Plans for improved food security should consider inclusion of such measures. Land users must realize that sustainable production and productivity are closely linked to both economic and environmental factors.

Without greater investment in infrastructure, proper incentives to farmers, adequate supplies of production inputs, effective marketing, and credit facilities, land will be not be fully utilized and production from cultivated land will remain below its potential. Over-cultivation of farm lands with low levels of inputs will lead to further abuse and degradation.

Educating land users is perhaps the main component of a long-term solution to the many land-husbandry problems in sub-Saharan Africa. Whether it is protecting crops from pests, cultivating crops for soil improvement, or the multitude of other land related practices, the extension worker is undoubtedly a key participant. Training land users, as well as a follow-up and evaluation of extension activities, will provide a guide for improvement. Basic education of people is an important cornerstone in improved resource use.

Through years of misuse and over exploitation, some lands are severely degraded and production is greatly affected. Though these degraded lands have understandably received much attention, it must be realized that, in considerable areas of the continent, the lands are still in good condition and that opportunities for producing food and cash crops are great. One fact that is clear is that the lands of Africa can no longer meet demand unless their farming systems are improved. Overexploitation and misuse must stop, and the optimization of land husbandry must be actively pursued.

Land degradation will remain fundamental and must be critically re-examined. Each country will have to develop its own package of measures for land and water husbandry. Emphasis must be given to the central importance of land conservation; capacity building of humans, institutions, and infrastructure; equity; people's participation; and accountability. Without attention to these factors, no plan or technical solution can guarantee success. The necessity of knowing what one is dealing with is perhaps the most important conclusion policy makers could draw from an analysis of Africa's land degradation. Assisting governments to eliminate hunger and poverty affecting the millions of people in developing countries is one of the major concerns of FAO (2021).

31.8 PERSPECTIVES: PAST, PRESENT, AND FUTURE

Food security has been a problem of concern to humanity from the beginning of time. Kearney (2010) projected that by 2020 there would be an increase in "food consumption per head in most regions," except in sub-Saharan Africa. A meeting held at the Royal Society in England on Land

Resources (Lynch et al., 2015) arrived at a resolution that "if all resources are harnessed and adequate measures taken to minimize soil degradation, sufficient food to feed the population in 2020 can be produced." It is now 2021 and many parts of the world, especially in most parts of sub-Saharan Africa, still face food insecurity. Food security depends on many factors other than soil degradation and soil resources. For example, the rapid population growth in this region is of concern. Governmental policies for land users should be implemented. They must recognize the challenges faced by local farmers that include food preservation, storage, and farm-to-market roads. These are hindrances for sufficient food availability in local markets, and even the rich are not able to afford the food crops in urban areas. Most sub-Saharan countries have the potential to feed their growing populations if water is available, the land is properly managed, inputs are used efficiently, and crop varieties with higher yield potential are developed. Africa is the continent where food shortages are being most acutely felt. Most of the unused land is unexploited, because it is difficult to manage sustainably without easy and economical access to the necessary inputs and without security of tenure. These are the reasons why many farmers are reluctant to invest in land. Thus, if this situation is not changed, the world will not have food security because Africa does not have it.

31.8.1 Past Perspectives

The food shortage has remained persistent in sub-Saharan Africa, where the proliferation of undernourishment is nearly 30% (Popp et al., 2019). A large number of studies have investigated different causes and aspects of the current crisis, or they have evaluated the future of the African agro-food sector based on one or a few particular aspects (e.g. climate change, population dynamics) (Popp et al., 2019).

The food insecurity situation in most regions of Africa is critical, and it is increasing in almost all region of sub-Saharan Africa.

31.8.2 Present Perspectives

Although agricultural production has increased substantially in sub-Saharan Africa over the past decades, it has been due largely to cultivation of more land rather than increases in land productivity (Sanchez, 2002).

Africa's population is expected to increase from 1.01 billion in 2009 to 2 billion in 2050 if current demographic conditions remain constant (UN, 2011). Much of this growth will be concentrated in sub-Saharan Africa, where annual population growth rates are expected to range from between 1.6% to slightly more than 2.4% between 2010 and 2050 (UN, 2011). These growth rates will be unmatched by those observed in other world regions. Within Africa, significant variations in demographic change are expected to occur across geographic regions. In East and West Africa, which have Africa's fastest growth rates, projected growth between 2010 and 2050 is expected to occur at rate of about 2.2% per year, according to the UN's medium-growth-rate assumptions. In absolute terms, annual growth rates in East Africa will be followed by those observed in Central (2%), North (1.07%), and Southern Africa (0.38%) (Thomas and Zuberi, 2012). Although population growth rates are expected to decline in all world regions, including Africa, in the first half of the twenty-first century, SSA countries still will have some of the world's fastest population growth rates during this period. By 2050, for example, eight of the top ten fastest growing countries in the world will be found in the SSA region (Thomas and Zuberi, 2012). Sulemana et al. (2019) reported that sub-Saharan Africa has the world's fastest urbanization rates and its changing urban dynamics are important for understanding future changes in the production, availability, and consumption of food. Despite expected increases in food production in Africa, rapid population growth will lead to increases in the size of its population living in food insecurity.

31.8.3 Future Perspectives

Food security is a complex issue, because it embodies natural and socioeconomic factors. Projections concerning expected future values of these factors are hard to quantify and predict. For example, it is challenging to project the rate of soil degradation and climate change in the next decade since this is not a linear regression.

The increase in growth of the poor will be a major constraint for food consumption in SSA countries. Advanced technology may generate increases in food stocks, but income constraints among the poor will still be a barrier and will negatively affect their ability to buy food in local markets, resulting in increases in the prevalence of hunger. This will remain a problem in Africa, because the majority are poor who settle in rural areas and depend on their farm products for survival. From the socioeconomic point of view, it is still a challenge transporting food products from farms to local markets, which leads to limited food availability in local markets in most SSA countries. This has led to rising food prices, which will further exacerbate the consequences of poverty among Africa's food consumers and will severely affect their prospects for achieving food security.

Addressing sub-Saharan Africa's future food demands will necessitate the efficient exploitation of natural resources to increase the production of basic food staples. For example, increasing agricultural production may require the expansion of crop cultivation to unexploited land areas, which will likely affect forest resources drastically. Water-use patterns will need to be effectively managed, while forest resources will need greater levels of protection. However, because natural resources are finite and a growing population is infinite and depends on these natural resources in greater quantities, there is a high probability that a practical limit will be encountered in the expanded use of these resources in food production.

As cultivated areas expand due to rapid population growth, competing demands will be placed on freshwater resources that will challenge the capacity of Africa's water distribution systems. Climate change will pose additional challenges to the ways in which natural resources are used in Africa's food production systems. Based on their analysis of climatic variability and harvests in the developing countries, Lobell et al. (2008) suggested that major adaptations of Africa's crop production systems are needed to mitigate the negative impacts of these changes on the availability of food staples. Climate change is also expected to decrease the water supply in Africa (Olabanji et al., 2020), and this will accelerate the processes of land degradation within its sub-regions. Specific elements of climate change will also negatively affect levels of food stocks required for achieving food security. Moreover, increases in carbon dioxide and temperature as a result of climate change are likely expected to decrease the region's yields for selected crops, including wheat, while carbon fertilization resulting from climate change is unlikely to produce meaningful increases in food production (Collier et al., 2008). As a result of the fixed nature of land, water, and other natural-resource stocks, sub-Saharan population growth will inevitably reduce the per capita availability of these resources. With the fast population growth in this region that is predicted by 2030, these constraints will require new strategies to manage sustainably the ways in which natural resources are used in food production. Other constraints for natural-resource use, such as declining soil fertility and pollution, if not addressed, will further undermine the fundamental basis of food production. Incorporating new land into agricultural production is unlikely to produce meaningful increases in crop yields, because this is going to affect the forest resources and may speed up climate change. Darkoh (1998) reported that land traditionally used for food production in fragile ecological regions will need protection from drought and desertification that threaten livelihoods and food security. In general, therefore, limited resource stocks, their exploitation in response to population pressures, as well as their sustainable management have far-reaching implications for food security and human development in sub-Saharan Africa.

As the pressure from population increases the demand for food, noticeable changes in the size of crops harvested between 2010 and 2030 will need to occur. Specifically, land areas used for irrigated and rain-fed crop cultivation will require significant expansion to accommodate expected increases in the consumption of food. However, expansion of cultivated areas associated with rapid population growth in Africa will be smaller than land-use changes associated with global population growth, because the cities will expand into cultivated land. Lower population growth rates are expected to result in reductions in the overall size of areas cultivated for crops in the developing world. Studies suggest that the theoretical carrying capacity of Africa's land has been exceeded, resulting in declines in soil fertility and fallow periods (Sayre, 2008). Rosegrant et al. (2009) estimated that more than 70% of the region's arable land is currently undergoing some form of degradation. Nutrient mining continues to affect soil fertility by decreasing the content of phosphorus, nitrogen, and potassium in African soils. In general, if population growth leads to expansion in land areas used for agricultural cultivation, significant challenges will be faced in translating these expansions into increases in crop yield. At current levels of technology, however, such expansions are likely to lead to less than proportional increases in overall crop yields in SSA countries. Therefore, based on these observations and projections, it is difficult to eradicate food insecurity and poverty in this region in the near decades.

31.9 CONCLUSION AND FUTURE PERSPECTIVES

The development of the organic sector has resulted in growth of new crops and the number of operators. Farmers who traditionally farm without fertilizers or chemical products (for financial reasons and due to the lack of commercial opportunities) have been encouraged to convert to organic farming by the market opportunities.

A large portion of the land in sub-Sahara Africa is prone to chemical, biological, and physical degradation, due to both anthropogenic activities and natural processes. Soil degradation has been accelerated by use of artificial means to boast crop production by land users, and, consequently, food production and the quality of the produce have been severely affected. There is an urgent need to check this degradation, and necessary corrective measures should be adopted to reclaim the degraded soils. The respective governments of the affected countries, in consultation with the FAO, should develop strategic plans in this regard. To increase the food production in order to feed the ever-increasing population, the land designated for food crops should be strictly used for farming purposes and not used for other purposes.

REFERENCES

Abada, Z. and Bouharkat, M., 2018. Study of management strategy of energy resources in Algeria. *Energy Reports*, 4, pp. 1–7.

Abalu, G.O.I., 1988. Implementation of a farming systems research strategy: The case of Nigeria, pp. 23–46.

AdeOluwa, O.O., 2010. Organic agriculture and fair trade in West Africa, Food and Agriculture Organization of the United Nations, Office of Knowledge Exchange, Research and Extension, FAO VialedelleTerme di CaracLL, 005153 Rome, Italy.

Adesina, A.A., 2017. The state of African development. *Development Finance Agenda (DEFA)*, 3(2), pp. 4–6.

Ahmady-Birgani, H., McQueen, K.G., Moeinaddini, M., and Naseri, H., 2017. Sand dune encroachment and desertification processes of the Rigboland Sand Sea, Central Iran. *Scientific Reports*, 7(1), pp. 1–10.

Alimi, T., Ajewole, O.C., Olubode-Awosola, O.O., and Idowu, E.O., 2006. Economic rationale of commercial organic fertilizer technology in vegetable production in Osun State of Nigeria. *Journal of Applied Horticulture*, 8(2), pp. 159–164.

Amoroso, L., 2016. The second international conference on nutrition: Implications for hidden hunger. *Hidden Hunger*, 115, pp. 142–152.

Anderson, E., 2020. Beyond toll-free lines and complaints boxes: A study of accountability and participation in humanitarian action in Chad.

Babatunde, R. and Oyatoye, E., 2005, October. Food security and marketing problems in Nigeria: the case of maize marketing in Kwara State. *In International Conference on Research for Development in Agricultural Forestry, Food and Natural Resources Management*. FAO, Rome, Reviewed Strategic Framework.

Barrios, S., Bertinelli, L., and Strobl, E., 2010. Trends in rainfall and economic growth in Africa: A neglected cause of the African growth tragedy. *The Review of Economics and Statistics*, 92(2), pp. 350–366.

Barrow, C.J., 1991. *Land Degradation: Development and Breakdown of Terrestrial Environments.* Cambridge University Press: Cambridge.

Bayiha, G.D.L.P., Temple, L., Mathé, S., and Nesme, T., 2019. Typology and perspective of evolution of organic farming in Cameroon. *HAL.*

Birhanu, A., 2014. Environmental degradation and management in Ethiopian highlands: Review of lessons learned. *International Journal of Environmental Protection and Policy*, 2(1), pp. 24–34.

Braithwaite, J. and Mont, D., 2009. Disability and poverty: A survey of World Bank poverty assessments and implications. *Alter*, 3(3), pp. 219–232.

Brüntrup, M., 2006. Rethinking protection for agricultural markets in Sub-Sahara Africa. *Africa, Regional Co-operation and the World Market, Uppsala: Nordic Institute (Nordiska Afrikainstitutet Discussion Paper 31*, pp. 52–70.

Chappell, A. and Agnew, C.T., 2004. Modelling climate change in West African Sahel rainfall (1931–90) as an artifact of changing station locations. *International Journal of Climatology: A Journal of the Royal Meteorological Society*, 24(5), pp. 547–554.

Chebet, A., 2021. Systematic literature review on organic farming for sustainable agricultural production. *International Journal of Contemporary Research and Review*, 12(03), pp. 20355–20360.

Collier, P., Conway, G., and Venables, T., 2008. Climate change and Africa. *Oxford Review of Economic Policy*, 24(2), pp. 337–353.

Darkoh, M.B.K., 1998. The nature, causes and consequences of desertification in the drylands of Africa. *Land Degradation & Development*, 9(1), pp. 1–20.

Dixit, R., Malaviya, D., Pandiyan, K., Singh, U.B., Sahu, A., Shukla, R., Singh, B.P., Rai, J.P., Sharma, P.K., Lade, H., and Paul, D., 2015. Bioremediation of heavy metals from soil and aquatic environment: An overview of principles and criteria of fundamental processes. *Sustainability*, 7(2), pp. 2189–2212.

Evang, E.C., Habte, T.Y., Owino, W.O., and Krawinkel, M.B., 2021. Can the supplementary consumption of baobab (*Adansonia digitata* L.) fruit pulp improve the hemoglobin levels and iron status of schoolchildren in Kenya? Findings of a randomized controlled intervention trial. *European Journal of Nutrition*, 60(5), pp. 2617–2629.

FAO, 2018. FAO Statistical Data bases Food and Agriculture Organization of the United Nations (FAO), Google Scholar http://apps.fao.org.

FAO, 2021. Cereal supply and demand balances for sub-Saharan African countries: Situation as of February 2021, Rome. https://doi.org/10.4060/cb3674en.

FAO, IFAD, UNICEF, WFP, and WHO, 2017. The State of Food Security and Nutrition in the World 2017. Building resilience for peace and food security. Rome, FAO.

Feng, J., Alam, S., and Eozenou, P.H.V., 2018. Malnutrition gap as a new measure of child malnutrition: A global application. *World Bank Policy Research Working Paper*, 8664.

Follett, R.F., 2001. Soil management concepts and carbon sequestration in cropland soils. *Soil and Tillage Research*, 61(1–2), pp. 77–92.

Gebhardt, S., Lemar, L., Haytowitz, D., Pehrsson, P., Nickle, M., Showell, B., Thomas, R., Exler, J., and Holden, J., 2008. USDA national nutrient database for standard reference, release 21. United States Department of AgricultureAgricultural Research Service.

Hall, C., Dawson, T.P., Macdiarmid, J.I., Matthews, R.B., and Smith, P., 2017. The impact of population growth and climate change on food security in Africa: Looking ahead to 2050. *International Journal of Agricultural Sustainability*, 15(2), pp. 124–135.

Havelaar, A.H., Haagsma, J.A., Mangen, M.J.J., Kemmeren, J.M., Verhoef, L.P., Vijgen, S.M., Wilson, M., Friesema, I.H., Kortbeek, L.M., van Duynhoven, Y.T., and van Pelt, W., 2012. Disease burden of foodborne pathogens in the Netherlands, 2009. *International Journal of Food Microbiology*, 156(3), pp. 231–238.

Headey, D.D. and Jayne, T.S., 2014. Adaptation to land constraints: Is Africa different?. *Food Policy*, 48, pp. 18–33.

Herrero, M., Havlik, P., McIntire, J., Palazzo, A., and Valin, H., 2014. African Livestock futures: Realizing the potential of livestock for food security, poverty reduction and the environment in Sub-Saharan Africa.

Hoscilo, A., Balzter, H., Bartholomé, E., Boschetti, M., Brivio, P.A., Brink, A., Clerici, M., and Pekel, J.F., 2015. A conceptual model for assessing rainfall and vegetation trends in sub-Saharank Africa from satellite data. *International Journal of Climatology*, 35(12), pp. 3582–3592.

Jie, C., Jing-Zhang, C., Man-Zhi, T., and Zi-tong, G., 2002. Soil degradation: A global problem endangering sustainable development. *Journal of Geographical Sciences*, 12(2), pp. 243–252.

Jones, A., Breuning-Madsen, H., Brossard, M., Dampha, A., Deckers, J., Dewitte, O., Gallali, T., Hallett, S., Jones, R., Kilasara, M., and Le Roux, P., 2013. *Soil Atlas of Africa*. European Commission, Publications Office of the European Union: Luxembourg.
Kanu, B.S., Salami, A.O., and Numasawa, K., 2014. Inclusive growth: an imperative for African agriculture. *African Journal of Food, Agriculture, Nutrition and Development*, 14(3), pp. A33–A33.
Kawase, H., Abe, M., Yamada, Y., Takemura, T., Yokohata, T., and Nozawa, T., 2010. Physical mechanism of long-term drying trend over tropical North Africa. *Geophysical Research Letters*, 37(9), L09706.
Kazeem, O., 2010. Organic farming gets a boost in Nigeria. *The Nation Newspaper*, 25 May, 2010.
Kearney, J., 2010. Food consumption trends and drivers. *Philosophical Transactions of the Royal Society B: Biological Sciences*, 365(1554), pp. 2793–2807.
Kopittke, P.M., Menzies, N.W., Wang, P., McKenna, B.A., and Lombi, E., 2019. Soil and the intensification of agriculture for global food security. *Environment International*, 132, p. 105078.
Lal, R., Hansen, D.O., and Uphoff, N., 2016. *Food Security and Environmental Quality in the Developing World*. CRC Press: Boca Raton, FL.
Lobell, D.B., Burke, M.B., Tebaldi, C., Mastrandrea, M.D., Falcon, W.P., and Naylor, R.L., 2008. Prioritizing climate change adaptation needs for food security in 2030. *Science*, 319(5863), pp. 607–610.
Lynch, I., Feitshans, I.L., and Kendall, M., 2015. 'Bio-nano interactions: New tools, insights and impacts': Summary of the Royal Society discussion meeting. *Philosophical Transactions of the Royal Society B: Biological Sciences*, 370(1661), p. 20140162.
Mabiso, A. and Benfica, R.S., 2019. The narrative on rural youth and economic opportunities in Africa: Facts, myths and gaps. *Myths and Gaps* (December 4, 2019).
MacKenzie, J. and MacKenzie, J.M., 1995. *Orientalism: History, Theory and the Arts*. Manchester University Press: Manchester.
Maldangoï, Y., Woin, N., and Cathala, M., 2003. North Cameroon onion producers: Challenges facing a sector looking for an important place. *In Savanes africaines: des espaces en mutation, des acteurs face à de nouveaux défis: Actes du colloque*, Garoua, Cameroun, 27–31 mai 2002 (pp. 6–13), La Librairie du Cirad.
Mallikarjuna, K.G., 2013. Food security and climate change. *International Journal of Research in Applied, Natural and Social Sciences*, 1(1), pp. 45–52.
Matshe, I., 2009. Boosting smallholder production for food security: Some approaches and evidence from studies in sub-Saharan Africa. *Agrekon*, 48(4), pp. 483–511.
McGuire, S., 2015. FAO, IFAD, and WFP. The State of Food Insecurity in the World 2015: Meeting the 2015 International Hunger Targets: Taking Stock of Uneven Progress, Rome: FAO, 2015. *Advances in Nutrition*, 6(5), pp. 623–624.
Megerssa, G.R. and Bekere, Y.B., 2019. Causes, consequences and coping strategies of land degradation: Evidence from Ethiopia. *Journal of Degraded and Mining Lands Management*, 7(1), p. 1953.
Mounir, Z.M., Ma, C.M., and Amadou, I., 2011. Application of water evaluation and planning (WEAP): A model to assess future water demands in the Niger River (in Niger Republic). *Modern Applied Science*, 5(1), p. 38.
Moyo, M., Aremu, A.O., and Van Staden, J., 2015. Medicinal plants: An invaluable, dwindling resource in sub-Saharan Africa. *Journal of Ethnopharmacology*, 174, pp. 595–606.
Njoya, M.A., Nain, C.W., Mahbou, P.Y., Mendi, S.D. and Imele, H., 2017. Physicochemical and sensory properties of pawpaw (*Carica papaya* Linn) flavoured yoghurt. *African Journal of Food, Agriculture, Nutrition and Development*, 17(3), pp. 12311–12324.
Nogueira, R.C., Nigro, M., Veuthey, J., Tigalbaye, C., Bazirutwabo, B., Thior, M.D.F., and Voillat, J., 2021. Can locally produce chlorine improve Water Sanitation & Hygiene (WASH) Indicators in Health Care Facilities (HCF) in Rural Chad? *Health Sciences and Disease*, 22, p. 11.
Nwachukwu, C.A., 2010. Adoption of organic agricultural technologies: Implications for Radio-Farmer-Agricultural Programmes in Imo State, Nigeria. *In 11th World Congress of Scientific and Technical Information and Rural Development*, Montepellier.
Nwajiuba, C. and Akinsanmi, O.A. 2002. Organic manure use among smallholders in the rainforest of southeast Nigeria. In: *Challenges to Organic Farming and Sustainable Land Use in the Tropics and Subtropics*. Deutscher Tropentag, October 9–11, 2002, Witzenhausen. http://www.tropentag.de/2002/abstracts/inks/Akinsanmi_DIZcx5xx.pdf.
Obayelu, A.E., 2019. A preventable tragedy: How close is Africa to end hunger and malnutrition in all its forms by 2030? *African Journal of Food, Agriculture, Nutrition and Development*, 19(3), pp. 1–5.
Ogen, O., 2011. Historicising and globalising the African environmental condition. *Africa Insight*, 41(3), pp. 99–110.

Oguntunde, P.G., Lischeid, G., Abiodun, B.J., and Dietrich, O., 2017. Analysis of long-term dry and wet conditions over Nigeria. *International Journal of Climatology*, 37(9), pp. 3577–3586.

Olabanji, M.F., Ndarana, T., Davis, N., and Archer, E., 2020. Climate change impact on water availability in the olifants catchment (South Africa) with potential adaptation strategies. *Physics and Chemistry of the Earth, Parts A/B/C*, 120, p. 102939.

Oladele, O.I. and Fawole, O.P., 2007. Farmer's perception of the relevance of agriculture technologies in South-Western Nigeria. *Journal of Human Ecology*, 21(3), pp. 191–194.

Oladimeji, Y.U. and Abdulsalam, Z., 2013. Analysis of technical efficiency and its determinants among small scale rice farmers in Patigi local government area of Kwara state, Nigeria. *Journal of Agriculture and Veterinary Science*, 3(3), pp. 34–39.

Oliver, M.A. and Gregory, P.J., 2015. Soil, food security and human health: A review. *European Journal of Soil Science*, 66(2), pp. 257–276.

Olombo, J.R. 2000. Fertilization. Global Report, CTA, Bruxelle, pp. 105–117.

Omoh, G., 2011. For how long will Nigeria continue with this import (syndrome), madness. *Vanguard Newspapers*, 20.

Plan, C.S., 2018. World Food Programme. Access mode: www.wfp.org/operations/lb01-lebanoncountry-strategic-plan-2018-2020.

Popp, J., Oláh, J., Kiss, A., and Lakner, Z., 2019. Food security perspectives in sub-Saharan Africa. *Amfiteatru Economic,* 21(51), pp. 361–376.

Raheem, D., Dayoub, M., Birech, R., and Nakiyemba, A., 2021. The contribution of cereal grains to food security and sustainability in Africa: Potential application of UAV in Ghana, Nigeria, Uganda, and Namibia. *Urban Science*, 5(1), p. 8.

Reganold, J.P. and Wachter, J.M., 2016. Organic agriculture in the twenty-first century. *Nature Plants*, 2(2), pp. 1–8.

Rickson, R.J., Deeks, L.K., Graves, A., Harris, J.A.H., Kibblewhite, M.G., and Sakrabani, R., 2015. Input constraints to food production: The impact of soil degradation. *Food Security*, 7(2), pp. 351–364.

Ridine, W.A.G.U.E., Ngakou, A., Mbaiguinam, M.B.A.I.L.A.O., Namba, F.A.B.I.E.N.N.E., and Anna, P., 2014. Changes in growth and yield attributes of two selected maize varieties as influenced by application of chemical (NPK) and organic (bat's manure) fertilizers in Pala (Chad) grown field. *Pakistan Journal of Botany*, 46(5), pp. 1763–1770.

Rockström, J., 2003. Resilience building and water demand management for drought mitigation. *Physics and Chemistry of the Earth, Parts A/B/C*, 28(20–27), pp. 869–877.

Rosegrant, M.W., Ringler, C., and Zhu, T., 2009. Water for agriculture: Maintaining food security under growing scarcity. *Annual Review of Environment and Resources*, 34, pp. 205–222.

Salami, A.A. and Tilakasiri, S.L., 2019. Impact of climate change on refugees and food shortage crises in Africa. *International Journal of Multidisciplinary Studies*, 6(2), pp. 58–78.

Sanchez, P.A., 2002. Soil fertility and hunger in Africa. *Science*, 295(5562), pp. 2019–2020.

Sayre, N.F., 2008. The genesis, history, and limits of carrying capacity. *Annals of the Association of American Geographers*, 98(1), pp. 120–134.

Shioya, A., 2013. Cassava processing and marketing by rural women in the Central region of Cameroon. *African Study Monographs*, 34(4), pp. 203–219.

Stacey, P. and Garin Fons, T.M.B., 2015. An open ecosystem engagement strategy through the lens of global food safety. *F1000Research*, 4, pp. 129–145.

Sulemana, I., Nketiah-Amponsah, E., Codjoe, E.A., and Andoh, J.A.N., 2019. Urbanization and income inequality in Sub-Saharan Africa. *Sustainable Cities and Society*, 48, p. 101544.

Sullivan, K.L., 2020. The impact of organizational design on poverty development projects. Doctoral dissertation, Northcentral University.

Sumberg, J., Anyidoho, N.A., and Chasukwa, M., 2015. Young people, agriculture, and employment in rural Africa. In: D. Resnick and J. Thurlow (Eds.), *African Youth and the Persistence of Marginalization.* Routledge: London, pp. 129–150.

Sundrum, A., 2001. Organic livestock farming: A critical review. *Livestock Production Science*, 67(3), pp. 207–215.

Tesfahunegn, G.B., 2019. Farmers' perception on land degradation in northern Ethiopia: Implication for developing sustainable land management. *The Social Science Journal*, 56(2), pp. 268–287.

Thiombiano, L. and Tourino-Soto, I., 2007. Status and trends in land degradation in Africa. In: M.V.K. Sivakumar and N. Ndiang'ui (Eds.), *Climate and Land Degradation.* Springer: Berlin, Heidelberg, pp. 39–53.

Thomas, K.J. and Zuberi, T., 2012. Demographic change, the IMPACT model, and food security in Sub-Saharan Africa. *UNDP Regional Bureau for Africa Working Paper.*

Tirado, R., 2009. *Defining Ecological Farming*. University of Exeter, School of Biosciences: Exeter, UK.
Tully, K., Sullivan, C., Weil, R., and Sanchez, P., 2015. The state of soil degradation in Sub-Saharan Africa: Baselines, trajectories, and solutions. *Sustainability*, 7(6), pp. 6523–6552.
Upton, J.B., Cissé, J.D., and Barrett, C.B., 2016. Food security as resilience: Reconciling definition and measurement. *Agricultural Economics,* 47(S1), pp. 135–147.
Vicente-Serrano, S.M., Cabello, D., Tomás-Burguera, M., Martín-Hernández, N., Beguería, S., Azorin-Molina, C., and Kenawy, A.E., 2015. Drought variability and land degradation in semiarid regions: Assessment using remote sensing data and drought indices (1982–2011). *Remote Sensing*, 7(4), pp. 4391–4423.
Wegren, S.K., Nikulin, A.M., and Trotsuk, I., 2017. The Russian variant of food security. *Problems of Post-Communism*, 64(1), pp. 47–62.
Weinmann, S., 2018. Impacts of elephant crop-raiding on subsistence farmers and approaches to reduce human-elephant farming conflict in Sagalla, Kenya.
Willer, H. and L. Kilcher, 2009. *The World of Organic Agriculture: Statistics and Emerging Trends*. IFOAM: Bonn; FiBL, Frick and ITC: Geneva.
World Health Organization (WHO), 2003. Diet, nutrition, and the prevention of chronic diseases: Report of a joint WHO/FAO expert consultation (Vol. 916), World Health Organization.
World Health Organization (WHO), 2019. Trends in maternal mortality 2000 to 2017: Estimates by WHO, UNICEF, UNFPA, World Bank Group and the United Nations Population Division.
Zidouemba, P.R., 2017. Climate variability and food crises in Burkina Faso: A Computable General Equilibrium analysis. *International Journal of Food and Agricultural Economics (IJFAEC)*, 5, pp. 79–95.

32 Effects of Biochar Application on Subtropical Mudstone Soils in Taiwan

Shih-Hao Jien[1] *and Zeng-Yei Hseu*[2]
[1]Department of Soil and Water Conservation, National Pingtung University of Science and Technology, Pingtung, Taiwan
[2]Department of Agricultural Chemistry, National Taiwan University, Taipei, Taiwan

CONTENTS

32.1 INTRODUCTION

32.1.1 Distribution, Characteristic, and Classification of Mudstone Soils in Taiwan

Poorly cemented mudstone covers more than a 1,000 km^2 area in southwestern Taiwan. Mudstone occupied over 1,000 km^2 in Taiwan (Lee et al., 1994) (Figure 32.1). Outcrops of the mudstone formation were only found in southwestern Taiwan and occupied about 3% in the land area of Taiwan. The mudstone formation formed from marine sediments from the littoral zone during the Miocene to Pliocene and Pleistocene Epochs. Sedimentary rocks created during the late Miocene Epoch to the Pleistocene Epoch can be several thousand meters deep. The geological stratigraphy is rather monotonic, consisting mainly of massive mudstones or alternating between mudstone and sandstone.

Soils derived from mudstone sedimentary are also usually saline or alkaline and are widely distributed throughout the world. These soils are found in Australia, Latin America, Africa, and notably in Central Asia (Weil and Brady, 2017). The alkaline characteristics of the mudstone soils are highly adverse to the fertility of these lands. Understanding the process of alkalinization and salinization therefore is also important in the management of productivity and sustainability of agriculture. In high alkali conditions, deficiency of organic matter seems to be another problem, which resulted in destruction of soil structure, and the soils become more vulnerable to erosion and degradation.

With increasing population in Asia, more food demands are required, in order to produce more food, a lot of effort has been made in poor quality lands. For instance, in the irrigated lands of the republic, the 8–10 tons of manure applied per hectare are not nearly sufficient to provide a positive balance of organic matter in the soil (Karajeh et al. 2002). To maintain such a balance, it is

DOI: 10.1201/9781003093565-32

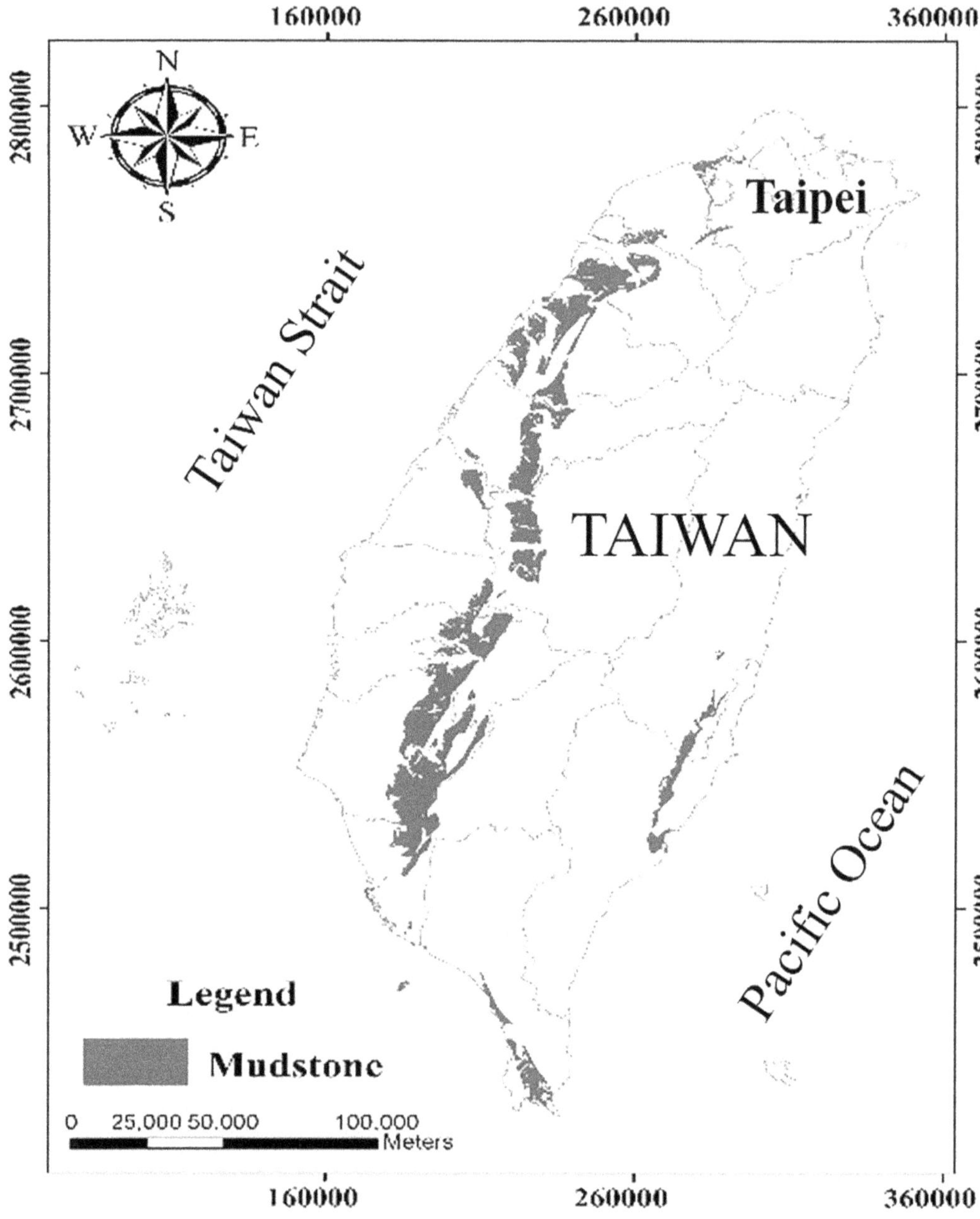

FIGURE 32.1 Distribution of mudstone soils in Taiwan.

necessary to apply at least 15 tons/ha of organic fertilizer. Excessive application of mineral fertilizer to the soil leads to salt accumulation and fertility loss. The process of losing soil fertility will increase and create many problems unless sustainable approaches are carried out.

As an important category of salt-affected soils, the mudstone soils, as well as sodic soils, are characterized by excess levels of sodium ions (Na^+) in the soil solution phase as well as on the cation-exchange complex, exhibiting unique structural problems as a result of certain physical processes (slaking, swelling, and dispersion of clay) and specific conditions (surface crusting and hardsetting). Sodification and salinization affects the water holding capacity, nutrient availability, hydraulic conductivity, water uptake by the plants, root penetration, and seedling germination (Bui, 2017).

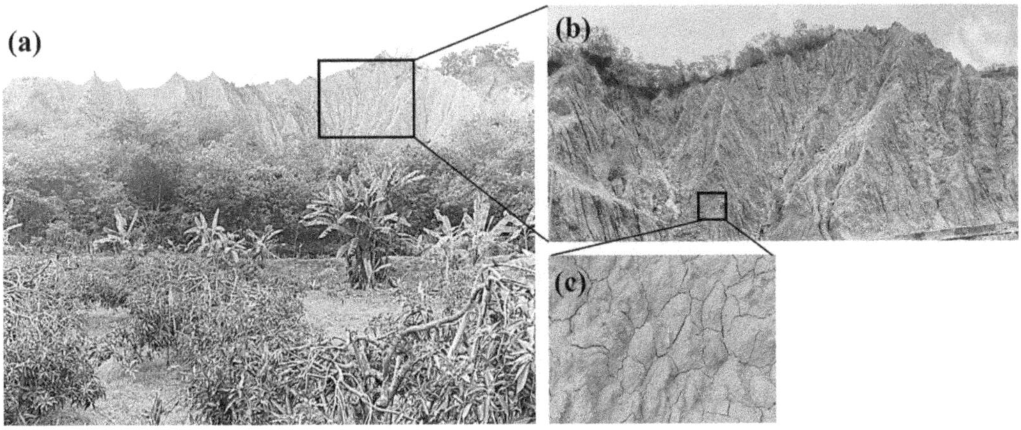

FIGURE 32.2 Landscape and mudstone soils in southern Taiwan. (a) Landscape and some cultivation in toeslope position, (b) mudstone picture, and (c) crust in the surface of mudstone soils.

Hardsetting mudstone soils are considered a degradation soil because the soils are difficult to cultivate until the soil profile is rewetted (Lal, 1990) and exhibit very high erosion potential (Figure 32.2). Mullins et al. (1992) explained that hardsetting soils set to a hard, structureless mass during drying. Chen (1984) reported that approximately 8–10 cm of soil is eroded from the surfaces of mudstone landscapes yearly, corresponding to a soil loss rate of 160 mg/ha per year in Taiwan. Natural organic materials (e.g., green manures) and artificial polymers (e.g., polyacrylamides) have been used to ameliorate degraded soils (Hueso-Gonźalez et al., 2014). However, maintaining the long-term aggregate stability of soil by applying fresh organic residues is difficult because of the rapid degradation of these traditional amendments. In addition, polymer application is costly because polymers are expensive. In Taiwan, techniques such as retaining walls and shotcretes are commonly used to protect soil from erosion in mudstone areas; however, these techniques are ineffective in-situ. Hence, biochar may be a potential amendment that could protect hardsetting soil from rapid degradation in the long term, thus increasing the value of Taiwanese agronomy and engineering.

The soils of mudstone bad land in Taiwan are mainly distributed in the southwestern and eastern regions with about 1,000 km^2. The mudstone soils are characterized by poor cementation, high erodibility, and constrained planting. Less studies were conducted in the genesis and classification of the mudstone soils. The chapter author collected three soil pedons on the summit, backslope, and footslope, respectively, along a mudstone toposequence at Teinliao, Kaohsiung, southwestern Taiwan. The clay contents were more than 40% in the B horizons of the summit and backslope pedons, and they only met the requirement of cambic horizon. Strong and large angular blocky structures were found in all three pedons. The soils with high bulk density were very firm, but some roots and biopores were found. Electrical conductivity (EC) values in all the horizons with high content of exchangeable Na were more than 2 dS/m. Based on morphology and genesis, clear A and B horizons were identified. Therefore, the study profiles are truly soils, not parent materials. The pedons on the summit and backslope are classified as Clayey, mixed, hyperthermic, Typic Haplustepts, and the pedon on the footslope as Loamy, mixed, hyperthermic, Typic Haplustepts (Soil Survey Staff, 2010) (Tables 32.1 and 32.2).

32.1.2 Effects of Biochar Application on Soil Properties and Erosion Potential of Mudstone Soils

The problems of food requirements and land resources have become urgent issues in the recent several decades due to climate changes. Additionally, soil degradation, such as erosion, hardsetting, and desertification, poses a serious problem affecting the productivity of agricultural lands and also

TABLE 32.1
Field Morphological Characteristics of the Three Pedons

Horizon	Depth (cm)	Munsell Color	Texture[a]	Structure[b]	Consistence[c]	Root[d]	Bio-Pore[d]	Boundary[e]
				Summit				
A	0–5	2.5Y 5/3	CL	3f&msbk	vfi, s&p	mvf&f	sf	ds
Bw1	5–17	2.5Y 4/1	C	3m&csbk	vfi, vs&vp	svf, f&m	sf&m	dw
Bw2	17–35	2.5Y 3/1	C	3f&msbk	vfi, s&p	fvf&f	ff	ds
Bw3	35–60	2.5Y 3/1	C	60% 3mabk	vfi, s&p	fvf	fvf	dw
Bw4	60–88	2.5Y 2.5/1	C	90% 3 f&mabk 10% 3f&mp	fi, s&p	fvf	–	dw
Bw5	88–110	2.5Y 3/2	SiC	90% 3 f&mabk	vfi, s&p	fvf	–	dw
Bw5	88–110	2.5Y 3/2	SiC	90% 3 f&mabk 10% 3fp	vfi, s&p	fvf	–	dw
Bw6	>110	2.5Y 3/1	SiC	90% 3mabk 10% 3fp	vfi, s&p	fvf	–	
				Backslope				
A1	0–18	5Y 5/2	SCL	80% 3m&csbk	vfi, s&p	mvf&f	sf	cs
A2	18–25	2.5Y 5/3	SCL	80% 2m&csbk 20% 1fgr	fi, ss&sp	svf, f&m	sf&m	cw
Bw1	25–50	5Y 5/1	SiC	3c&vcabk	vfi, vs&vp	fvf&f	–	ds
Bw2	50–75	N 4/	SiC	3c&vcabk	vfi, vs&vp	fvf	–	dw
Bw3	75–110	N 3/	SiC	90% 3c&vcabk 10% 3m&cp	vfi, vs&vp	fvf	–	dw
Bw4	>110	N 2.5/	SiC	3 c&vcabk	vs&vp	fvf	–	
				Footslope				
A	0–20	5Y 5/2	L	80% 3c&vcabk 20% 2mgr	vfi, s&p	sf&m	svf	ds
Bw1	20–40	5Y 4/2	CL	80% 3vcabk 20% 3mp	vfi, vs&vp	svf, f&m	sf&m	dw
Bw2	40–60	5Y 5/1	CL	80% 3c&vcabk 20% 3mp	vfi, vs&vp	fvf&f	–	dw
Bw3	60–80	5Y 4/2	CL	60% 3c&vcabk 40% 3csbk	vfi, vs&vp	fvf	–	dw

[a] CL, Clay loam; C, Clayey; SiC, Silty clay; SCL, Sandy clay loam; L, Loam.
[b] 3, Strong; 2, Moderate; f, Fine; m, Medium; c, Coarse; vc, Very coarse; abk, Angular blocky; sbk, Subangular blocky; gr, Granular; p, Platy.
[c] vfi, Very firm; fi, Firm; vs, Very sticky; s, Sticky; vp, Very plastic; p, plastic.
[d] f, Few; s, Some; m, Many; vf, Very fine; f, Fine; m, Medium.
[e] d, Diffuse; c, Clear; w, Wavy; s, Smooth.

threatens the world's food security in the future with an increasing global population (Leh et al., 2013; Qadir et al., 2013; Mandal and Sharda, 2013; Ziadat and Taimeh, 2013). The development of new land management techniques to effectively use soil resources and attend land sustainability is necessary for the future, particularly in subtropical/tropical regions. Biochar has attracted attention for remediation of degraded soils recently, which is a byproduct derived from a process in which biofuel is produced using anaerobic pyrolysis at high temperatures (Lehmann and Joseph, 2009). Biochar has been considered a suitable organic amendment for improving the physical properties and

TABLE 32.2
The Selected Chemical Properties of the Three Pedons

Horizon	Depth (cm)	pH H_2O	pH $CaCl_2$	OC[a] (%)	K	Na	Ca	Mg	CEC[b]	BSP[c] (%)	$CaCO_3$	EC[d] (dS/m)
					Exchangeable Bases cmol(+)/kg							
Summit (Clayey, Mixed, Hyperthermic, Typic Haplustepts)												
A	0–5	7.9	7.4	8.4	0.4	ND[e]	4.4	2.8	10	76	3.7	2.2
Bw1	5–17	8.1	7.6	4.9	0.3	ND	5.5	3.4	10	92	3.6	2.1
Bw2	17–35	8.3	7.8	3.0	0.1	0.1	5.7	5.2	13	87	3.2	9.7
Bw3	35–60	8.4	7.8	2.8	0.2	0.3	8.3	5.5	15	95	2.8	2.1
Bw4	60–88	8.5	7.9	2.9	0.2	0.5	8.3	5.8	15	99	2.0	5.2
Bw5	88–110	8.6	7.9	3.1	0.2	0.8	8.3	6.5	17	93	3.2	7.2
Bw6	>110	8.5	8.0	3.0	0.2	2.5	6.9	5.7	16	96	3.2	11.4
Backslope (Clayey, Mixed, Hyperthermic, Typic Haplustepts)												
A1	0–18	8.4	8.0	12.6	0.4	0.2	6.7	2.4	17	57	3.8	6.1
A2	18–25	8.7	8.0	3.5	0.2	0.2	9.3	2.5	16	76	3.7	3.0
Bw1	25–50	9.1	8.6	3.8	0.3	1.5	8.1	8.0	18	69	3.3	2.6
Bw2	50–75	9.1	8.7	3.7	0.4	3.5	6.0	7.5	19	94	2.8	4.2
Bw3	75–110	9.0	8.8	3.7	0.3	4.3	5.2	6.7	18	92	3.0	5.5
Bw4	>110	8.9	8.9	3.6	0.4	4.3	7.1	6.7	19	97	3.1	2.1
Footslope (Loamy, Mixed, Hyperthermic, Typic Haplustepts)												
A	0–20	8.3	8.1	4.4	0.4	1.1	7.7	3.7	13	99	3.7	2.1
Bw1	20–40	8.9	8.7	3.6	0.2	3.0	7.3	5.2	18	87	3.3	4.7
Bw2	40–60	8.9	8.7	3.7	0.2	4.4	7.2	5.0	17	99	2.8	3.8
Bw3	60–80	8.7	8.5	3.0	0.2	2.7	7.0	4.8	17	86	3.2	4.4

[a] Organic carbon.
[b] Cation exchangeable capacity.
[c] Base saturation percentage.
[d] Electrical conductivity.
[e] Not detectable.

maintaining the fertility of the soil, particularly degraded soils in subtropical and tropical regions (Chan et al., 2007, 2008; van Zwieten et al, 2010; Deenik et al., 2011; Jien et al., 2015b). Numerous studies have indicated that biochar effectively improves soil structure (Atkinson et al., 2010; Jien and Wang, 2013), soil aggregate stability (Cerdà, 2000; Kimetu and Lehmann, 2010; Dimoyiannis, 2012; Kocyigit and Demirci, 2012; Herath et al., 2013), and porosity (Liang et al., 2006) because of its high specific surface area (SSA) and inner porous structure. Steiner et al. (2011), Chan et al. (2008), Laird et al. (2010), and Yuan and Xu (2011) have reported that fertilizer-use efficiency and crop production increase following biochar incorporation. Biochar is composed of recalcitrant carbon from microbial degradation and of a charged surface with organic functional groups; therefore, it can increase soil organic carbon (SOC) levels and C stocks (Trompowsky et al., 2005; Tejada and Gonzalez, 2007; Wuddivira et al., 2009; Materechera, 2009). Biochar application could produce agronomic benefits, particularly fertility maintenance, SOC sequestration, and crop production (Jindo et al., 2014; Mukherjee et al., 2014; Ghezzehei et al., 2014) However, a few studies have examined the benefits of biochar application on the hydraulic properties and erosion recalcitrant of soil. In our previous study (Hseu et al., 2014), we recommended biochar application as an innovative management way for the

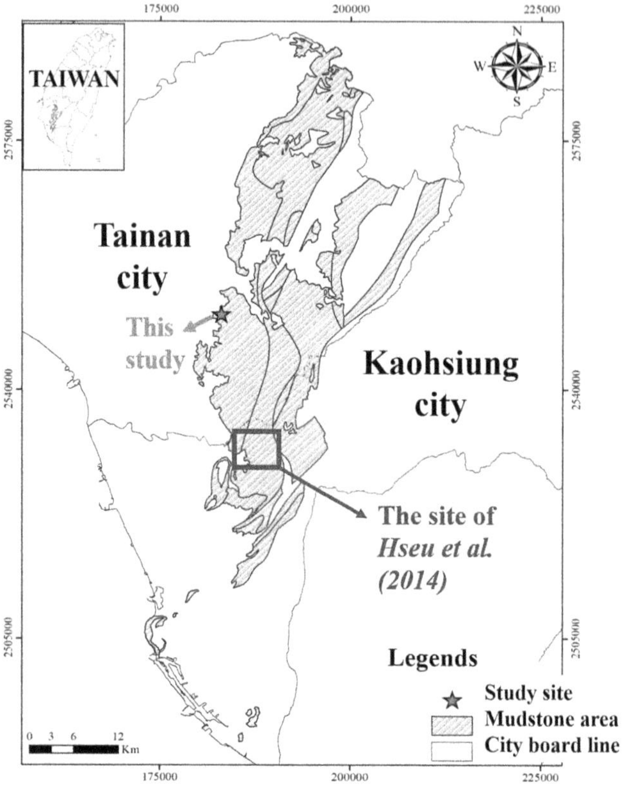

FIGURE 32.3 Location of the study area and the distribution of the mudstone soils in the study area. (This figure is cited from Hseu et al. (2014).)

soils in mudstone areas. In this study, we collected mudstone soil samples from the surface (0–25 cm) in the Teinliao area (E 120°22′58″; N 22°53′02″), a representative mudstone area, Kaohsiung, Taiwan (Figure 32.3). We evaluated the effects of the application of rice husk biochar on changes of physiochemical characteristics and erosion potential of a mudstone slopeland soil. Rice hull biochar pyrolized at 400°C was incorporated into the soil at rates of 2.5%, 5%, and 10% (w/w) and was incubated for 168 days in this study. The results indicated that biochar application reduced the bulk density (Bd) by 12%–25% and the penetration resistance (PR) by 57%–92% after incubation of 6 months, compared with the control. Besides, porosity and aggregate size increased by 16%–22% and by 0.59–0.94 mm, respectively. The results presented that available water contents significantly increased in the amended soils by 18%–89% because of the obvious increase of micropores. The water conductivity of the biochar-amended soils was only found in 10% biochar treatment, which might result from a significant increase of macro-pores and reduction of soil strength (Bd and PR). During a simulated rainfall event of 80 mm/hour, soil loss contents significantly decreased by 35%–90% in the biochar-amended soils. The erosion experiment was conducted by using rainfall simulator, which was equipped with a height of 9.5 m, and the rainfall drop test presented by an average diameter of 2.5 mm for raindrops, and an average terminal velocity of 8.5 m/s for raindrops (Figure 32.4).

32.1.3 Effects of Biochar Application on Non-Point Source Pollution in Mudstone Area

Biochar application could not only improve mudstone soil physical and chemical properties but also mitigate the pollution potential of the reservoir from non-point source pollution (NPS) caused by

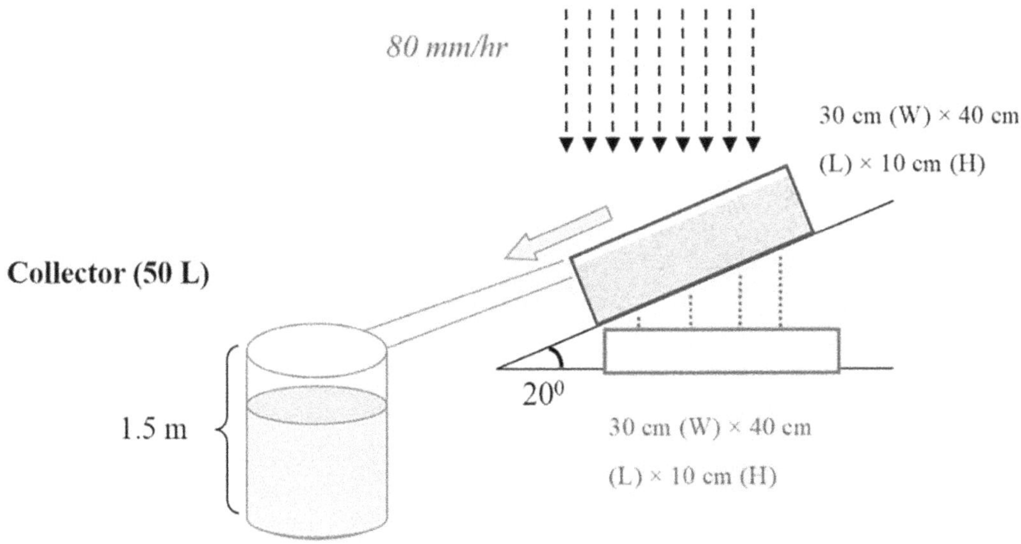

FIGURE 32.4 Schematic of erosion experiment in Hseu et al. (2014).

inappropriate agricultural managements on slopelands. Biochar has been considered as an effective soil amendment which could significantly improve soil fertility and even soil erosion potential. NPS is considered a major factor in eutrophication and sediment deposition in reservoirs, which threaten public health and also impact water resource usage. Mudstone areas in southwestern Taiwan suffer from severe soil erosion and a high degree of NPS. Moreover, the high turbidity and high contents of ammonium, nitrate, and phosphate in rivers due to the high level of soil erosion and intense agricultural activities nearby these areas have led to the overloading of reservoirs in southwestern Taiwan. Biochar can be added to soil to significantly improve its fertility and reduce the potential for erosion. Some nutrients can be incorporated into the micropores of biochar, and thus adsorbed on the surface in order to prevent their loss due to runoff or leaching. This study therefore aims to evaluate the effects of biochar made at different temperatures on the erosion potential and runoff quality in a mudstone area in southwestern Taiwan.

Jien et al. (2015a) have evaluated the improvement efficiency of biochar on erosion potential and runoff quality on a mudstone area in southern Taiwan. The biochar used in this study was produced from rice hulls and supplied by the Industrial Technology Research Institute (ITRI) of Taiwan. Rice hulls are considered agricultural waste, and at least one million tons of this is produced each year in Taiwan. One way to recycle this waste is to use it as a soil additive. The rice hull biochar samples used in this study were produced at pyrolization temperatures of 400°C and 700°C, and denoted as RHB-400 and RHB-700, respectively. After pyrolysis, the biochar was ground to enable it to pass through a 2-mm sieve, ensuring that all the biochar used in the experiments exhibited similar particle sizes. The soil erosion experiment was conducted using simulated rainfall equipment with a height of 9.5 m, a drop diameter of 2.5 mm, and a terminal velocity of 8.5 m/s. The rainfall intensity used in this study was 100 mm/hour.

Jien et al. (2015a) indicated that their results revealed that at least 60% reduction of soil losses and runoff by addition of biochar in the mudstone soil, especially for 10% application rate. In addition, biochar application could decrease the concentrations of NH_4^+-N and total phosphorous (TP) in runoff under the simulated intensive rainfall event. Five percent of biochar application rate could effectively reduce NH_4^+-N contents from 0.19 to 0.70 mg/L and TP from 0.02 to 0.08 mg/L, and corresponded to the regulation of land surface water quality in Taiwan. On the contrary, 10% of application rate increased the contents of NH_4^+-N and TP in runoff, which might be resulted from

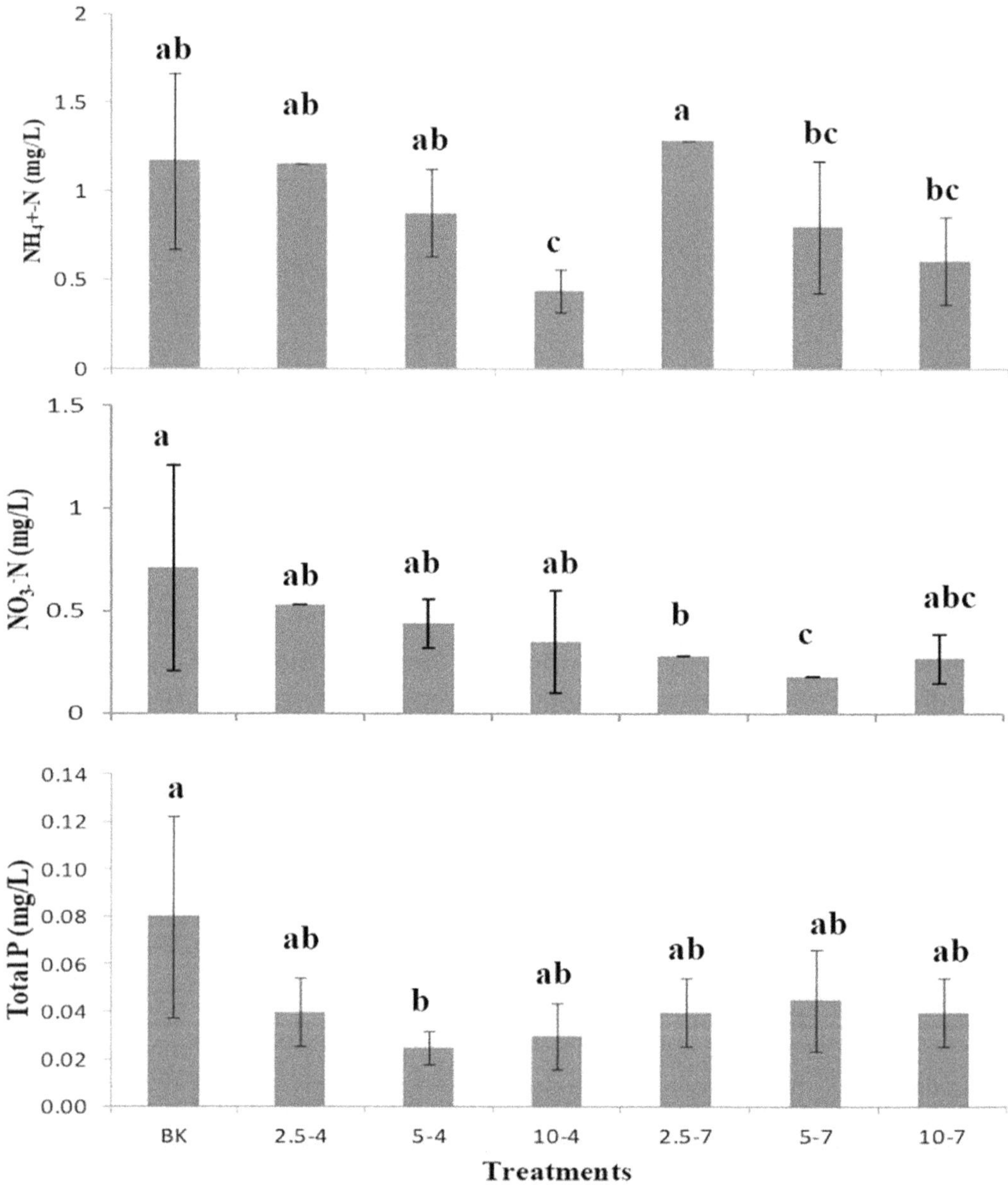

FIGURE 32.5 Contents of NH_4^+-N, NO_3^--N, and total P in runoff in different treatments during simulated erosion process. BK: control; 2.5–4, 5–4, 10–4: 2.5%, 5% and 10% of 400°C rice hull biochar; 2.5–7, 5–7, 10–7: 2.5%, 5% and 10% of 700°C rice hull biochar. (This figure is cited from Jien et al. (2015a).)

the original contents of biochar (Figure 32.5). The present study recommends that 5% is the most suitable application rate of biochar in this mudstone area.

This study examined the recycling of a form of agricultural waste, namely rice hull, into a useful resource to reduce soil erosion and increase the quality of runoff water in a rural soil derived from mudstone. The results revealed that the use of biochar could effectively reduce soil losses by 25%–90%, and reduced the concentrations of NH_4^+-N, NO_3^--N and TP by >50%, 21%–98% and 38%–63%, respectively, in the runoff water from the mudstone area. This innovative practice has potential about concerning reducing eutrophication and sediment deposition in reservoirs in Taiwan (Table 32.3).

TABLE 32.3
Mean Soil Volumetric Moisture Contents Measured at Different Matric Potentials, Macro-Pores and Saturated Water Conductivity of the Control and Biochar-Amended Soils after Incubation (n=3)

Treatments	Rates % (w/w)	Soil Losses g/pot	Soil Losses tons/ha	Reduced Proportion (%)
Control	0	529±3.01 a	36.0	–
	2.5	378±22.7 b	27.0	25
RHB-400	5.0	86.7±9.91 c	6.19	83
	10	47.9±45.6 c	3.42	91
	2.5	327± 29.8 b	23.3	35
RHB-700	5.0	82.9±1.08 c	5.91	84
	10	55.9±26.1 cd	3.98	89

Source: This table is cited from Jien et al. (2015a).
Values followed by the same letter within a column are not significantly different at $p<0.05$ level based on Duncan's test.

32.1.4 Effects of Biochar Application on Vegetation Growing in Mudstone Area

In the southwestern Taiwan, where the soils are marine sedimentary rock (Lee et al., 2013), and was called "Moonscape World" or "Barren lands." The average annual precipitation is about 2,000 mm (Lee et al., 2007), much higher than in other mudstones (barren land) areas (about 645 mm) (Piccarreta et al., 2006). Moreover, in this region, global warming increases the average annual temperature and precipitation, and the intensity of heavy rainfall. Thus, these impacts could lead to more serious hazards in this barren land. Plant cover is known as one of the important factors to affect soil microorganisms (Oh et al., 2012). Under plant canopy in arid soils, such as desert, soil temperature and moisture are stable, and nutrients might be more abundant, resulting in the significantly different in bacterial abundance and composition between shrub and barren soils (Bachar et al., 2012). Microorganisms also have the ability to cope with high levels of salt in arid soils (Mondini et al., 2006). Their responses to the environmental fluctuation could be an important biological indicator in the arid ecosystems, including bad land region.

In mudstone areas in Taiwan, thorny bamboo (*Bambusa stenostachya*) seems to be a dominant species for forestation. As other bamboo plantations, it is one of the important forest resources because of the versatile and wide utilities, including paper and pulp and house construction. In mudstone area, serious soil erosion causes waving landscape. The south-facing slope receives strong sunshine and soil moisture is usually low. Comparatively, the north-facing slope receives weak sunshine and can hold some soil moisture. The high soil moisture in the north-facing slope benefits for the growth of bamboo plantation. Consequently, bamboo covers in the north-facing slope while south-facing slope remains bare in bad land area. As we know, plants are difficult to grow in this degraded soil with high pH, poor fertility, and worse soil physical properties which were indicated by Hseu et al. (2014).

Furthermore, the authors would like to evaluate the effects of biochar application on grass growing and soil erodibility for mudstone soils in southern Taiwan. We constructed three standard erosion plots (22.13 m in length, 4 m in width and 9% in slope gradient) in mudstone areas in southern Taiwan (Figures 32.6 and 32.7). They treated these stand erosion plots with different kinds of biochar treatment (Figure 32.8) in July 2020. The treatments in the three plots included control (without applying any biochar), 2% (w/w) wood biochar (WB) treatment, and 2% (w/w) rice husk biochar (RHB) treatment. The mudstone soil and two different kinds of biochar (WB and RHB)

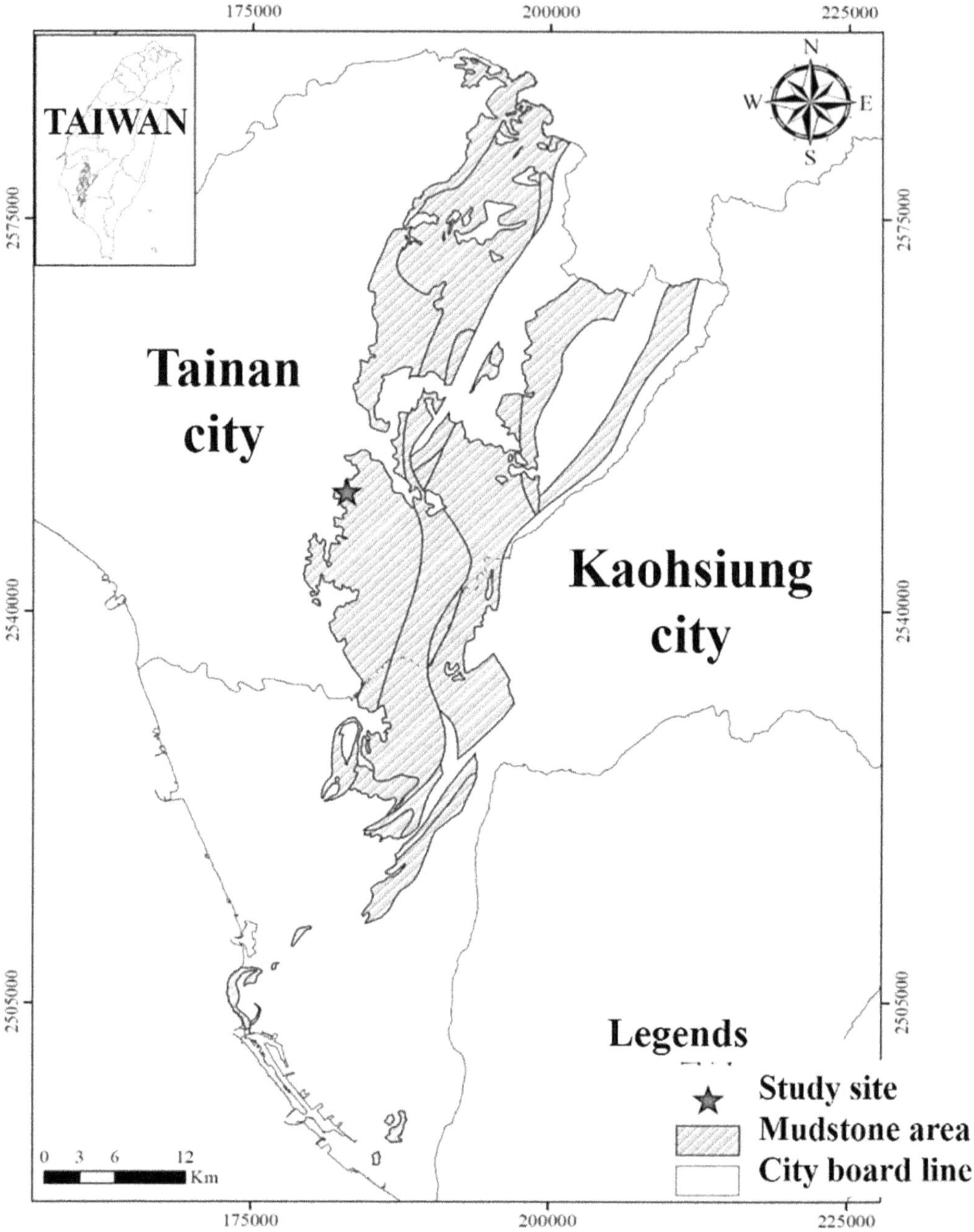

FIGURE 32.6 In-situ study site of mudstone area in southern Taiwan.

were showed in Table 32.4. The two biochar samples used in this study were produced at pyrolization temperatures of about 700°C.

After 120 days incubation, Table 32.5 revealed that biochar presentation could effectively retard the increase of Bd during incubation, as well as EC value, compared with control. For the biochar-amended soils, physical dilution effects might have caused reduced Bd levels, which agreed with Busscaer et al. (2011) who indicated that increasing total organic carbon by the addition of organic amendments in soils could significantly decrease Bd. Furthermore, the decrease

FIGURE 32.7 In-situ constructed standard erosion plots (22.13 m in length, 4 m in width and 9% in slope gradient) in mudstone area.

in Bd of the biochar-amended soils appears to have also been the result of an alteration of soil aggregate sizes, as shown by Tejada and Gonzalez (2007) who amended the following soils by using organic amendments in Spain. Additionally, SOC contents were also obviously increased by biochar application in mudstone soil in initial incubation. Regarding inorganic N and available P, there were no obvious changes among three treatments after biochar application at the beginning and at the end of incubation (Table 32.5). The results indicated that facilitation of vegetation growing in biochar treatment mudstone soils might be attributed to the improvement of physical properties, such as Bd, porosity, and aggregate formation (Lehmann and Rhodon, 2006; Jien and Wang, 2013; Hseu et al., 2014).

FIGURE 32.8 Different kinds of biochar: (a and b) wood biochar, and (c and d) rice husk biochar we used in this study.

TABLE 32.4
Properties of the Study Mudstone Soil and Biochars

Properties	Mudstone Soil	WB (700°C)	RHB (700°C)
Texture	Silty clay loam	–	–
Sand (%)	16	–	–
Silt (%)	54	–	–
Clay (%)	30	–	–
Bulk density (g/cm)	1.52	–	–
pH	7.09	7.79	8.86
EC (dS/m)	3.56	0.15	0.10
SOC (%)	1.23	43.4	47.1
NH_4^+-N (mg/kg)	16.5	36.0	4.50
NO_3^--N (mg/kg)	36.0	49.5	18.0
Bray No. 1-P (mg/kg)	4.19	28.4	25.7
CEC (cmol(+)/kg)	12.8	13.5	14.6
Total N (mg/kg)	648	418	357

Source: This table is cited from Jien et al. (2015a).

BD, bulk density; EC, electrical capacity; SOC, soil organic carbon; CEC, cation-exchange capacity.

TABLE 32.5
Soil Properties before and after Incubation (120 Days)

Treatments	Control		WB		RHB	
Application rate	–		2%		2%	
Sampling date (day)	0	120	0	120	0	120
BD (g/cm)	1.31	1.55	1.20	1.29	1.21	1.32
pH	7.09	7.08	7.19	7.14	7.26	7.20
EC (dS/m)	3.46	3.10	3.57	3.48	3.54	3.46
SOC (%)	1.23	1.20	2.09	1.53	2.17	1.62
NH_4^+-N (mg/kg)	16.5	18.0	20.5	22.5	9.00	15.0
NO_3^--N (mg/kg)	36.0	16.5	42.0	19.5	28.0	15.0
Bray No.1-P (mg/kg)	4.19	2.16	4.12	2.36	4.11	2.45
CEC (cmol(+)/kg)	12.8	12.7	13.0	12.9	13.2	13.0
Total N (mg/kg)	648	472	637	548	614	521

BD, bulk density; EC, electrical capacity; SOC, soil organic carbon; CEC, cation-exchange capacity.

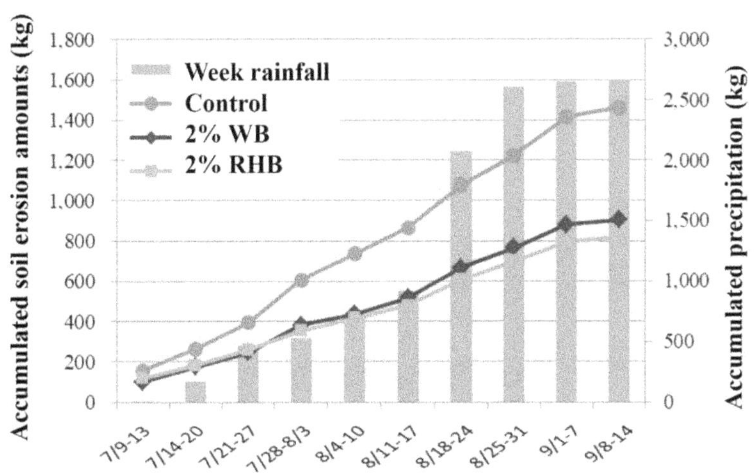

FIGURE 32.9 Accumulation soil erosion amounts and accumulated rainfall data during our study period (July, 2020 to September, 2020).

Furthermore, after 120 days of incubation, our finding presents that the soil loss of the treatment of 2% WB addition was about 1,345 kg, which was significantly lower than the control (2,195 kg) by 38.8% (Figure 32.9). Also, the soil loss of 2% rice hull biochar addition was about 1,206 kg, which was significantly lower than the control treatment by 45.1%. The results implied that biochar addition can greatly improve soil erodibility, namely increasing soil erosion resistance. The authors also evaluated the effects of biochar application on the growing of Bahia grass (*Paspalum notatum Flugge*) in terms of coverage ratio index. The coverage ratio was calculated through K means method and identified by Image J software (National Institutes of Health, USA) (Figure 32.10a and b). Figure 32.10 further indicated that biochar treatments could achieve 80% in coverage ratio much faster than that in control before 50 days of incubation. The results presented that biochar could effectively facilitate vegetation growing in mudstone soils. As a whole, we concluded that biochar application could obviously improve physical properties and organic carbon contents in

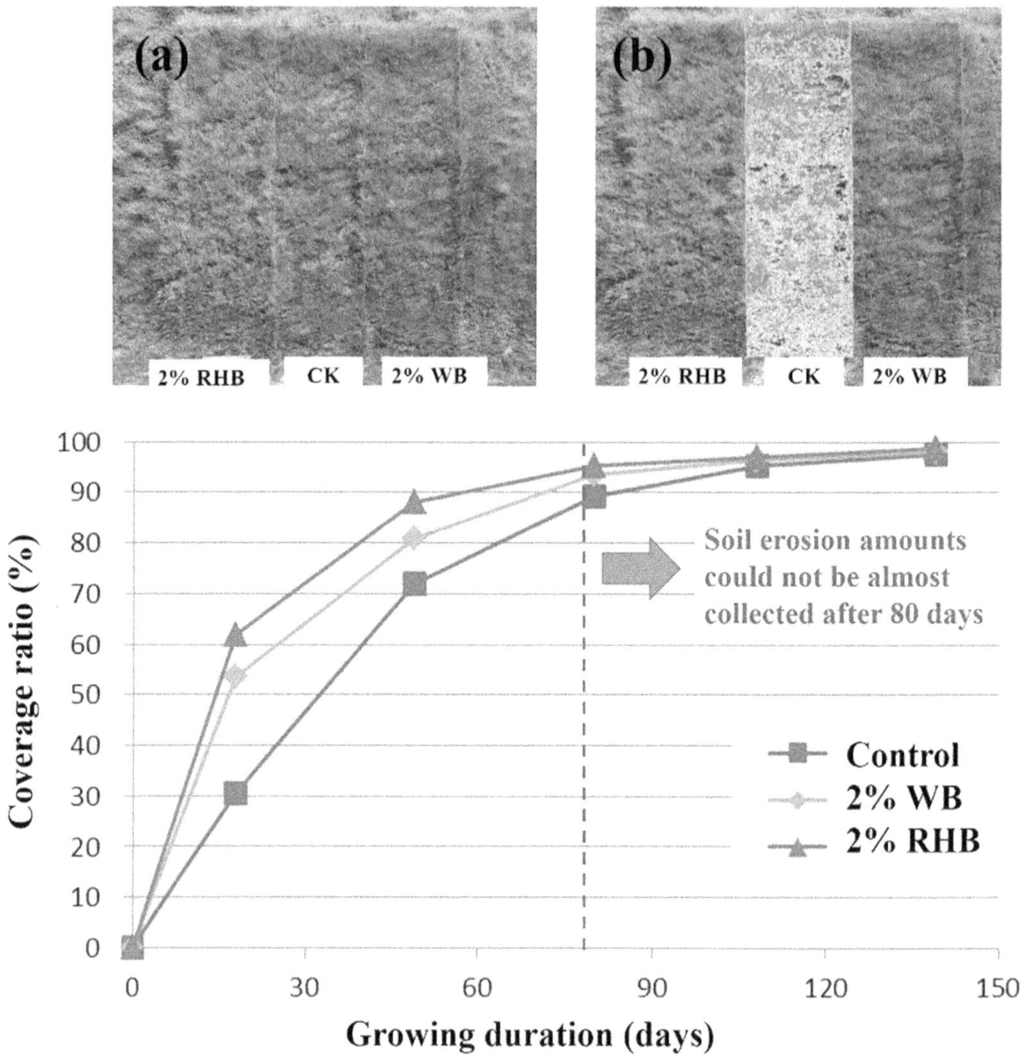

FIGURE 32.10 Calculation of coverage ratio of Bahia grass: (a) photo of grass cover after incubation; (b) schematic of identification of coverage ratio by Image J; (c) accumulation coverage ratio of Bahia grass in different treatments during our study period (July 2020 to September 2020).

the mudstone soil, and which might be the critical reason to facilitate the growth of Bahia grass. Additionally, Xie and Jien (2021) also indicated that yield of Chinese cabbage (*Brassica rapa chinensis*) could increase by 30% at least in treatments of co-application of biochar and compost in a tropical highly weathered soil. Lehmann and Joseph (2009) and Sohi et al. (2009) provided that biochar has been widely shown its beneficial role in increasing crop yield as biochar improves soil structure reportedly due to its effects on bulk density, pore-size distribution, and particle size distribution. With biochar application, beneficial biophysical effects could be expected on the availability of air and water within the root zone and in turn on the germination and survival of plants (Lehmann and Rhodon, 2006; Zhang et al., 2012). Before our study in evaluation of grass growth, biochar has been proposed for the restoration of desertified (Bai et al., 2008) and secondary salinized land (Li et al., 2007; Lehmann and Joseph, 2009), and there has been few field studies on the effects of biochar amendment on salt-affected croplands. Lashari et al. (2013) provided a perspective method of combination of biochar and pyroligneous solution (a byproduct when biochar is produced via pyrolysis of biomass) to improve soil quality and wheat yield of a salt-stressed cropland in China. They indicated that a significant decrease under biochar-pyroligneous solution amendment was observed in soil salinity by 3.6 g/kg, soil pH by 0.3, and in soil bulk density by 0.1 g/cm while increase was seen in SOC and available phosphorous by 2.6 g/kg and by 27 mg/kg, respectively.

32.2 CONCLUSIONS

Agronomy and engineering solutions, including tillage, forest restoration, soil erosion prevention, tunneling, and road construction in mudstone areas, are known as worldwide problems, because of the high salinity and poor physical and chemical properties of these soils. Biochar prepared from the biomass wastes through slow pyrolysis has attracted widespread attention due to its high carbon (C) content, abundant surface functional groups, porous structure, acid-neutralizing capacity, and the ability to retain nutrients. In mudstone soils, we have demonstrated that biochar could not only improve the chemical and biological properties of the soil, including increasing soil pH, CEC, BS, and microbial activity, but also improve the physical properties, such as Bd, K_{sat}, aggregate stability, and erosion resistance. Our previous studies suggested that the biochar application effectively improved poor soil quality in subtropical/tropical degraded soil, and further reduced soil losses under heavy rainfall areas. The problems of food requirement and land resource have become urgent issues in the recent several decades due to climate changes. Additionally, soil degradation, such as erosion, hardsetting, and desertification, poses a serious problem affecting the productivity of agricultural lands and also threatens the world's food security in the future with the increasing global population. The development of new land management techniques to effectively use soil resources and attend land sustainability is necessary for the future, particularly in subtropical/tropical regions.

REFERENCES

Atkinson, C. J., Fitzgerald, J. D., and Hipps, N. A. 2010. Potential mechanisms for achieving agricultural benefits from biochar application to temperate soils: A review. *Plant and Soil* 337:1–18.

Bachar, A., Soares, M., and Gillor, O. 2012. The effect of resource islands on abundance and diversity of bacteria in arid soils. *Soil Microbiology* 63: 694–700.

Bai, Z. G., Dent, D. L., Olsson, L., and Schaepman, M. E. 2008. Proxy global assessment of land degradation. *Soil Use and Management* 24: 223–224.

Bui, E. 2017. Causes of Soil Salinization, Sodification, and Alkalinization. Oxford Research Encyclopedia of Environmental Science. https://doi.org/10.1093/acrefore/9780199389414.013.264.

Cerdà, A. 2000. Aggregate stability against water forces under different climates on agriculture land and scrubland in southern Bolivia. *Soil and Tillage Research* 57: 159–166.

Chan, K. Y., Van Zwieten, L., Meszaros, I., Downie, A., and Joseph, S. 2007. Agronomic values of greenwaste biochar as a soil amendment. *Australian Journal of Soil Research* 45: 629–634.

Chan, K. Y., Van Zwieten, L., Meszaros, I., Downie, A., and Joseph, S. 2008. Using poultry litter biochars as soil amendments. *Australian Journal of Soil Research* 46: 437–444.

Chen, S. T. 1984. A study on the erosion characteristics of the mudstone in southwestern Taiwan (1). The Hazard Mitigation Technology Report No. 73-07, National Science Council, Taipei, Taiwan.

Deenik, J. L., Diarra, A., Uehara, G., Campbell, S., Sumiyoshi, Y., and Antal Jr., M. J. 2011. Charcoal ash and volatile matter effects on soil properties and plant growth in an acid ultisol. *Soil Science* 176: 336–345.

Dimoyiannis, D. 2012. Wet aggregate stability as affected by excess carbonate and other soil properties. *Land Degradation and Development* 23: 450–455.

Ghezzehei, T. A., Sarkhot, D. V., and Berhe, A. A. 2014. Biochar can be used to capture essential nutrients fromdairy wastewater and improve soil physico-chemical properties. *Soil Earth* 5: 953–962.

Herath, H. M. S. K., Camps-Arbestain, M., and Hedley, M. 2013. Effect of biochar on soil physical properties in two contrasting soils: An Alfisol and an Andisol. *Geoderma* 209–210: 188–197.

Hseu, Z.Y., Jien, S.H., Chien, W.H., and Liou, R.C. 2014. Impacts of biochar on physical properties and erosion potential of a mudstone slopeland soil. *The Scientific World Journal* 2014. http://dx.doi.org/10.1155/2014/602197.

Hueso-González, P., Martínez-Murillo, J. F., and Ruiz Sinoga., J. D. 2014. The impact of organic amendments on forest soil properties under Mediterranean climatic conditions. *Land Degradation and Development* 25: 604–612.
Jien, S. H. and Wang, C. S. 2013. Effects of biochar on soil properties and erosion potential in a highly weathered soil. *Catena* 110: 225–233.
Jien, S. H., Chen, J. Y., and Chien, T. S. 2015a. Using rice hull biochar to deal with non-point source pollution in a mudstone area. *Journal of Soil and Groundwater Remediation* 2: 317–327.
Jien, S. H., Wang, Lee C. H., and Lee, T. Y. 2015b. Stabilization of organic matter by biochar application in compost-amended soils with contrasting pH values and textures. *Sustainability* 7: 13317–13333.
Jindo, K., Matsumoto, K., Garćıa Izquierdo, C., Sonoki, T., and Sanchez-Monedero, M. A. 2014. Methodological interference of biochar in the determination of extracellular enzyme activities in composting samples. *Soil Earth* 5: 713–719.
Karajeh, F., Ryan, J. and Studer, C. 2002. *On-farm Soil and Water Management in Central Asia*. ICARDA, Aleppo, Syria.
Kimetu, J. M. and Lehmann, J. 2010. Stability and stabilisation of biochar and green manure in soil with different organic carbon contents. *Australian Journal of Soil Research* 48: 577–585.
Kocyigit, R. and Demirci, S. 2012. Long-term changes of aggregate associated and labile soil organic carbon and nitrogen after conversion from forest to grassland and cropland in northern Turkey. *Land Degradation and Development* 23: 475–482.
Laird, D. A., Fleming, P., Davis, D. D., Horton, R., Wang, B., and Karlen, D. L. 2010. Impact of biochar amendments on the quality of a typical Midwestern agricultural soil. *Geoderma* 158: 443–449.
Lal, R. 1990. *Soil Erosion in the Tropics: Principles and Management*. McGraw Hill, 588 p.
Lashari, M. S., Liu, Y., Li, L., Pan, W., Fu, J., Pan, G., Zheng, J., Zheng, J., Zhang, X., and Yua, X. 2013. Effects of amendment of biochar-manure compost in conjunction with pyroligneous solution on soil quality and wheat yield of a salt-stressed cropland from Central China Great Plain. *Field Crop Research* 144: 113–118.
Lee, D. H., Chen, P. Y., Wu, J. H., Chen, H. L., and Yang, Y. E. 2013. Method of mitigating the surface erosion of a high-gradient mudstone slope in southwest Taiwan. *Bulletin of Engineering Geology and the Environment* 72: 533–545.
Lee, D. H., Jhin, Y. Y., and Tien, K. G. 1994. The basic characteristics of mudstone and the protection methods for mudstone slope. *Sino-Geotechnics* 48: 35–47.
Lee, D. H., Lin, H. M., and Wu, J. H. 2007. The basic properties of mudstone slopes in southwestern Taiwan. *Journal of GeoEngineering* 2: 81–95.
Leh, M., Bajwa, S., and Chaubey, I. 2013. Impact of land use change on erosion risk: An integrated remote sensing, geographic information system and modeling methodology. *Land Degradation and Development* 24: 409–421.
Lehmann, J., and Joseph, S. 2009. *Biochar for Environmental Management*. Earthscan Publishers Ltd., London.
Lehmann, J., and Rhodon, M. 2006. Biochar soil management on highly weathered soils in the humid tropics. In: Uphoff, N. (Ed.), *Biological Approaches to Sustainable Soil Systems*. CRC Press, Boca Raton, FL, pp. 517–540.
Li, X. Y., Wang, Z. M., Song, K. S., Zhang, B., Liu, D. W., and Guo, Z. X. 2007. Assessment for salinized wasteland expansion and land use change using GIS and remote sensing in the west part of Northeast China. *Environ Monit Assess* 131: 421–437.
Liang, B., Lehmann, J., Solomon, D., Kinyangi, J., Julie, M., Grossman, O. B., Skjemstad, J. O., Thies, J. E., Luizão, F. J., Petersen, J., and Neves, E. G. 2006. Black carbon increases cation exchange capacity in soils. *Soil Science Society of America Journal* 70: 1719–1730.
Mandal, D., and Sharda, V. N. 2013. Appraisal of soil erosion risk in the eastern himalayan region of India for soil conservation planning. *Land Degradation and Development* 24: 430–437.
Materechera, S. A. 2009. Aggregation in a surface layer of a hardsettingand crusting soil as influenced by the application of amendments and grass mulch in a South African semi-arid environment. *Soil and Tillage Research* 105: 251–259.
Mondini, C., Cayuela, M. L., Sanchez-Monedero, M. A., Roig, A., and Brookes, P. C. 2006. Soil microbial biomass activation by trace amounts of readily available substrate. *Biology and Fertility of Soils* 42: 542–549.
Mukherjee, A., Lal, R., and Zimmerman, A. R. 2014. Effects of biochar and other amendments on the physical properties and greenhouse gas emissions of an artificially degraded soil. *Science of the Total Environment* 487: 26–36.
Mullins, M. G., Bouquet, A., Williams, L. E. 1992. *Biology of the Grapevines*. Cambridge Univ. Press, 239 pp.

Oh, Y. M., Kim, M., Lee-Cruz, L., Lai-Hoe, A., Go, R., Ainuddin, N., Rahim, R. A., Shukor, N., and Adams, J. M. 2012. Distinctive bacterial communities in the rhizoplane of four tropical tree species. *Microbial Ecology* 64: 1018–1027.

Piccarreta, M., Piccarreta, H., Bentiveng, M. and Capolongo, D. 2006. The influence of physico-chemical material properties on erosion processes in the badlands of Basilicata, Southern Italy. *Geomorphology* 81(3–4): 235–251.

Qadir, M., Noble, A. D., and Chartres, C. 2013. Adapting to climate change by improving water productivity of soils in dry areas. *Land Degradation and Development* 24:12–21.

Sohi, S., Loez-Capel, E., Krull, E., and Bol, R. 2009. Biochar's roles in soil and climate change: A review of research needs. CSIRO Land and Water Science Report 05/09, 64 pp.

Soil Survey Staff. 2010. *Keys to Soil Taxonomy*, 11th ed. USDA-Natural Resources Conservation Service, Washington, DC.

Steiner, C., Melear, N., Harris, K., and Das, K. 2011. Biochar as bulking agent for poultry litter composting. *Carbon Management* 2: 227–230.

Tejada, M., and Gonzalez, J. L. 2007. Influence of organic amendments on soil structure and soil loss under simulated rain. *Soil and Tillage Research* 93:197–205.

Trompowsky, P. M., De Melo Benites, V., Madari, B. E., Pimenta, A. S., Hockaday, W. C., and Hatcher, P. G. 2005. Characterization of humic like substances obtained by chemical oxidation of eucalyptus charcoal. *Organic Geochemistry* 36: 1480–1489.

Van Zwieten, L., Kimber, S., Morris, S., Chan, K. Y., Downie, A., Rust, J., Joseph, S., and Cowie, A. 2010. Effects of biochar from slow pyrolysis of papermill waste on agronomic performance and soil fertility. *Plant and Soil* 327: 235–246.

Weil, R. R. and Brady, N. C. 2017. *The Nature and Properties of Soils.* 15th Edition, Pearson, New York.

Wuddivira, M. N., Stone, R. J., and Ekwue, E. I. 2009. Structural stability of humid tropical soils as influenced by manure incorporation and incubation duration. *Soil Science Society of America Journal* 73:1353–1360.

Xie, Z. H., and Jien, S. H. 2021. Mineralization and nutrient releasing of biochar compound materials in soils with different textures. *Journal of Taiwan Agricultural Engineering* 67(1), 50–59.

Yuan, J. H., and Xu, R. K. 2011. The amelioration effects of low temperature biochar generated from nine crop residues on an acidic Ultisol. *Soil Use and Management* 27: 110–115.

Zhang, A., Liu, Y., Pan, G., Hussain, Q., Li, L., Zheng, J., and Zhang, X. 2012. Effect of biochar amendment on maize yield and greenhouse gas emission from a soil organic carbon poor calcareous loamy soil from Central China Plain. *Plant and Soil* 351: 263–275.

Ziadat, F. M., and Taimeh, A. Y. 2013. Effect of rainfall intensity, slope, land use and antecedent soil moisture on soil erosion in an arid environment. *Land Degradation and Development* 24: 582–590.

33 Cover Crops to Protect Soil during Winter in the Great Plains of the USA

Oliver W. Freeman II[1,2], M.B. Kirkham[1], and Kraig L. Roozeboom[1]

[1]Department of Agronomy, Kansas State University, Manhattan, Kansas
[2]1890 Research and Extension Program, South Carolina State University, Orangeburg, South Carolina

CONTENTS

33.1 INTRODUCTION

Wind erosion during winter fallow periods in semi-arid regions, such as the Great Plains, constrains productivity. Planting cover crops might be one way to hold the soil in place. The Great Plains is a large area in the United States and Canada east of the Rocky Mountains, which act as a natural barrier for precipitation, making it a dry and arid region. The Great Plains contains parts of 10 states: Colorado, Kansas, Montana, Nebraska, New Mexico, North Dakota, Oklahoma, South Dakota, Texas, and Wyoming. However, little work has been done on cover crops in the Great Plains, including Kansas.

A cover crop is a crop, such as a cereal or a legume, grown to protect the soil from erosion and loss of nutrients by leaching. Cover crops grow close to the ground between periods of normal (cash) crop production (Soil Science Society of America, 2008). Their use has been documented in the literature for over 100 years in the United States. Franklin H. King, first professor of agricultural physics in the United States, who was appointed in 1888 at the University of Wisconsin (Tanner and Simonson, 1993), was one of the first to recognize the advantages of cover crops, especially because of their "locking up of the soluble plant foods to prevent them from being lost by soil leaching" (King, 1901, p. 183). Other advantages of cover crops include prevention of water and wind erosion, prevention of surface evaporation, addition of organic matter to the soil, and, if a legume is grown, addition of nitrogen to the soil. However, one disadvantage of cover crops is that they dry the soil (King, 1901, p. 201). Therefore, cover crops traditionally have not been recommended for growth in semi-arid regions, because they use water that may result in limited soil water for the following cash crop (Unger et al., 2010). In the west and southwest of the USA, the use of cover crops is not widely practiced. In the western USA, the adoption of cover crops is less than 5% of the agricultural land (O'Geen et al., 2010).

In a review of cover crops (Clark, 2007), organizations with cover crop experience are given for the northeast, north-central, southern, and western parts of the USA, but no contact is given for the

DOI: 10.1201/9781003093565-33

Great Plains. However, work suggests that cover crops might have a place in agriculture in semi-arid Kansas, especially in the eastern half of the state where rainfall is more plentiful. A gradient in precipitation exists across Kansas. It ranges from 38 to 51 cm (15 to 20 inches) per year in the western part of the state and from 89 to 102 cm (35 to 40 inches) in the eastern part of the state. Mean annual precipitation in the middle part of the state ranges from 64 to 76 cm (25 to 30 inches) per year (Sophocleous, 1998). Even though cover crops do not produce a marketable product, they can potentially benefit rotations by increasing organic matter, maintaining surface residue (which reduces evaporation), reducing nitrate leaching, reducing soil erosion, suppressing weeds, and adding diversity to crop sequences (Roozeboom, 2013). Worldwide, cover crops are being used on small farms to maintain soil cover, increase organic matter, suppress weeds, and add nitrogen (Florentín et al., 2011).

If cover crops can be grown in Kansas, they would provide protection against wind erosion, which is of serious concern. High winds in Kansas cause the erosion. Dodge City, in the western part of Kansas, has the highest wind speeds in the USA (Mary C. Knapp, Kansas State Assistant Climatologist, personal communication, 2015). Its wind speeds are up to 80 km/h (50 mi/h). Blowing dust is a big problem. Interstate Highway 70 (I-70), the main highway across Kansas from the east coast to the west coast of the USA, is closed due to dust storms. The dust storms occur during fallow periods, when the soil is not cropped. Most recently, I-70 was closed on 22 April 2022, when visibility was zero due to dust. The dust storms occur in other states, too. On 15 July 2022, a dust storm caused a 21-vehicle pileup on Interstate 90 in Big Horn County, Montana, resulting in the deaths of six people (Lynch, 2022). Between 2007 and 2011, 125 people died on highways in the USA due to poor visibility from blowing sand, soil, and dirt (Ashley et al., 2015). Farmers in the western part of the Great Plains have to deal with wind on a constant basis. In May 2014, one farmer in western Kansas estimated that the topsoil on his farm, which took generations to create, now sat in a field 316 km (200 mi) away, blown by winds of March and April (Snyder, 2014). Drifting soil is a difficult problem to control. Attempts should be made to keep the soil from spreading and ultimately tie it down with plants. It is important to have plants on the soil to protect it from blowing during the fallow period. Cover crops could serve this role during the winter fallow between summer annual crops.

In central and eastern Kansas in a wheat (*Triticum aestivum* L.) and row crop rotation, the 10–11 months between winter wheat harvest and planting of the next grain crop the following spring provide an opportunity to insert a cover crop (Roozeboom, 2013). Planting a cover crop immediately after wheat harvest can take advantage of the 20–30 cm (8–12 inches) of precipitation usually received in this part of the state from July through September. A number of cover crops have been evaluated in experiments in central and eastern Kansas. Winter non-legume cover crops have included canola (*Brassica napus* L.), barley (*Hordeum vulgare* L.), annual rye (*Secale cereale* L.), oats (*Avena sativa* L.), winter triticale (X *Triticosecale*; *Triticum* x *Secale*), and annual fescue [*Vulpia myuros* (L.) K.C. Gmel.]. Winter legumes studied have included winter pea (*Pisum* sp.) and yellow sweet clover (*Melilotus officinalis* Lam.). Yield of sorghum [*Sorghum bicolor* (L.) Moench] after these cover crops ranged from low (with winter triticale) to medium high (with canola and winter pea) (Roozeboom, 2013).

Cover crops also are being recommended for Kansas because they can increase crop intensity, which reduces evaporation from the soil. A basic principle of efficient crop water use is shifting as much of the total water use, or evapotranspiration, to crop transpiration and away from evaporation (Roozeboom et al., 2012). Rotations that include only winter annuals or only summer annuals typically use water relatively inefficiently. Increasing crop diversity by rotating summer and winter annuals can effectively increase cropping intensity. Research with cover crops in Kansas has revealed that they can increase yield of the following crop. For example, the influence of two cover crops, late-maturing soybeans [*Glycine max* (L.) Merr.] and sunn hemp (*Crotalaria juncea* L.), in a wheat-sorghum rotation was investigated at Hesston, Kansas. Nitrogen was added to the soil at different rates ranging from 0 to 102 kg/ha (0 to 90 lb/acre). When averaged over nitrogen application rates, grain sorghum yielded 560 kg/ha (8.8 bushels per acre) and 1948 kg/ha (4.9 bushels per acre) more when grown with late-maturing soybean and sunn hemp cover crops, respectively, than when grown with no cover crop (Roozeboom et al., 2012).

Annual cool-season grasses are used as cover crops in regions where moisture does not limit their use (Phillips et al., 1996). They reduce soil erosion, add organic matter to the soil, and retain soil moisture during the fall and winter. Usually legumes are preferred as cover crops because of their N_2-fixing ability, but annual cool-season grasses work well as a cover crop in concert with a warm-season crop. Cultivated oat is a cool-season grass, but it is not commonly used as forage (Stubbendieck and Jones, 1996). It can be a cover crop. In Kansas, cool-season grasses include oat, winter wheat, and triticale (a hybrid between wheat and rye).

More information is needed on cover crop growth in the Great Plains. Therefore, six cover crops (three legumes and three small-grain, cool-season grasses) were studied for 3 years (harvested in 2010, 2011, and 2012) at two locations in Kansas: Manhattan in the northeastern part of the state and in Hutchinson in the central part of the state.

33.2 MATERIALS AND METHODS

Field studies were conducted from 2009 to 2012 at Ashland Bottoms Research farm in Manhattan, KS (39°8′39.61″N, 96°37′44.12″W) and the South Central Experiment Field at Partridge, KS (37°96′22.63″N, 98°12′32.37″W). Partridge is near Hutchinson, KS, and so, in the rest of this chapter, reference will be made to Hutchinson instead of Partridge. At Manhattan, the experiment was conducted on a Bismarckgrove Kimo complex. This is a complex of two different soils that cannot be distinguished. The Bismarckgrove series is classified as a fine-silty, mixed, superactive, mesic Fluventic Hapludolls. In the 0–18 cm depth, the series is a silt loam, and in the 18–51 cm depth, the soil is a silty clay loam. The Kimo series is classified as a clayey over loamy, smectic, and mesic Fluvaquentic Hapludolls. In the 0–18 cm depth, the series is a silty clay loam, and in the 18–38 cm depth, the soil is a silty clay. At Hutchinson, the experiment was conducted on a Funmar-Tarver loam (fine-loamy, mixed, superactive, mesic Pachic Argiustolls).

A randomized complete block design with split plots and four replications was used at both locations, which were all in no-till. Six cover crops were grown in the study. Three were legumes: red clover (*Trifolium pratense* L.), Austrian winter pea [*Pisum sativum* var. *arvense* (L.) Poir.], and alfalfa (*Medicago sativa* L.). Three were non-legumes and were grasses: winter wheat (*Triticum aestivum* L.), triticale (X *Triticosecale*; *Triticum* x *Secale*), and winter oats (*Avena sativa* L.).

The cover crops were planted at various rates set forth by Kansas Rural Center Sustainable Agriculture Management Guide and Kansas State University Research and Extension production guides. The planting rates for each cover crop were as follows: red clover: 7,560,000 seeds/ha; Austrian winter pea: 133,633 seeds/ha; alfalfa: 7,426,500 seeds/ha; winter wheat: 1,856,435 seeds/ha; triticale: 1,856,435 seeds/ha; and winter oats: 2,592,000 seeds/ha. Cover crops were planted in 6 m × 12 m plots within each replication. These plots were then split into two 3 m × 12 m plots that were planted and sampled at respective corn and forage sorghum planting times. Even though the cover crops were not in rotation with corn (*Zea mays* L.) and forage sorghum [*Sorghum bicolor* (L.) Moench], they were planted or harvested as if they were. Corn is more cold tolerant than sorghum (Antony et al., 2019). Thus, corn is planted earlier in the spring than sorghum. In Kansas, corn typically is planted in late April about 1 month before sorghum, which is planted in late May. The cover crops were planted after the typical harvest dates of corn or sorghum, and they were harvested before the typical planting dates of corn or sorghum.

At Manhattan for the corn rotation, cover crops were planted on 4 October 2009, 6 October 2010, and 14 October 2011. At Manhattan for the forage sorghum rotation, cover crops were planted on 13 November 2009, 16 November 2010, and 13 November 2011. At Hutchinson for the corn rotation, cover crops were planted on 11 October 2009, 9 October 2010, and 21 October 2011. At Hutchinson for the forage sorghum rotation, cover crops were planted on 21 November 2009, 19 November 2010, and 20 November 2011. At Manhattan for the corn rotation, cover crops were harvested on 25 April 2010, 21 April 2011, and 23 April 2012. At Manhattan for the forage sorghum rotation, cover crops were harvested on 14 May 2010, 12 May 2011, and 16 May 2012. At Hutchinson for the corn rotation, cover crops were harvested on 28 April 2010, 29 April 2011, and 29 April 2012. At Hutchinson for the forage sorghum rotation, cover crops were harvested on 23 May 2010, 15 May 2011, and 18 May 2012.

Cover crops were sampled from a 1 m^2 area from a random location within each plot to determine dry matter. Dry matter was determined by drying the plants at 65°C for 120 hours. After sampling, plots were terminated with 1.5 kg a.i. ha^{-1} glyphosate (isopropylamine salt of N-phosphonomethyl glycine 56) and 0.90 kg a.i. ha^{-1} 2,4-D (2,4-dichlorophenoxy acetic acid).

Table 33.1 gives the rainfall and temperatures during the years of the study in Manhattan and Hutchinson. In 2009, rainfall in Manhattan and Hutchinson was 99 mm and 83 mm above normal, respectively. In 2010, rainfall in Manhattan and Hutchinson was 37 mm below normal and 131 mm above normal, respectively. In 2011, rainfall in Manhattan and Hutchinson was 45 mm and 318 mm below normal, respectively. In 2012, rainfall in Manhattan and Hutchinson was 328 mm and 310 mm below normal, respectively. In 2009, temperatures in Manhattan and Hutchinson were −1.45°C and −0.50°C below normal, respectively. In 2010, temperatures in Manhattan and Hutchinson were −0.39°C below normal and 0.77°C above normal, respectively. In 2011, temperatures in Manhattan and Hutchinson were 0.06°C and 1.22°C above normal, respectively. In 2012, temperatures in Manhattan and Hutchinson were 1.95°C and 2.33°C above normal, respectively.

TABLE 33.1
Total Monthly Precipitation and Average Monthly Temperatures at Manhattan, Kansas, and Hutchinson, Kansas, during the Study (2009, 2010, 2011, and 2012)

Month	Ppt., mm	Temp., °C	Ppt., mm	Temp., °C	Ppt., mm	Temp., °C	Ppt., mm	Temp., °C
	2009		2010		2011		2012	
				Manhattan				
Jan.	1	−2.3	10	−5.3	18	−4.3	1	1.7
Feb.	17	3.1	14	−2.5	23	−1.9	54	2.6
Mar.	76	6.3	71	5.5	30	6.1	69	14.1
Apr.	133	10.8	88	14.2	73	12.7	54	15.3
May	25	17.6	100	16.8	139	17.3	34	21.4
June	215	23.8	195	24.7	132	24.6	105	25.1
July	166	23.0	97	26.4	55	30.1	18	30.0
Aug.	114	22.7	103	26.5	71	26.9	109	24.4
Sept.	52	17.8	89	20.7	35	19.3	72	19.6
Oct.	102	9.1	29	14.9	68	14.6	16	12.6
Nov.	37	8.1	48	6.9	108	6.6	16	7.8
Dec.	51	−4.3	2	-0.7	87	1.4	9	1.1
				Hutchinson				
Jan.	1	−1.0	12	−2.8	5	−2.7	2	2.2
Feb.	6	4.5	26	−1.2	20	−0.8	72	2.8
Mar.	45	6.3	33	6.6	23	7.2	67	13.3
Apr.	151	11.1	47	14.6	10	13.6	33	16
May	99	17.3	138	17.4	48	18.8	51	21.8
June	116	24.6	213	26.6	58	27.3	87	25.6
July	52	24.7	166	27.6	5	32.1	15	30.2
Aug.	105	23.6	121	27.3	84	28.8	87	25.4
Sept.	172	18.7	33	22.7	18	19.4	31	21.1
Oct.	81	9.7	13	15.5	42	14.6	7	12.6
Nov.	15	9.3	95	6.6	74	6.7	5	7.6
Dec.	10	−2.8	2	0.2	65	1.6	3	1.1

As noted, the study had four replications that each had six whole plots of cover crops. Each whole plot was then split into two plots for planting and harvesting dates of corn and forage sorghum. Significance was determined by using the PROC MIXED procedure in SAS (SAS Institute, Inc. 2013). PROC GLIMMIX is a procedure for fitting Generalized Linear Mixed Models. Mean separations were performed if the F-tests for the treatment effects were significant at the $\alpha=0.05$ level. Analyses of variance are given by Freeman (2014).

33.3 RESULTS

Dry matter of the cover crops for 2010, 2011, and 2012 in Manhattan and Hutchinson are shown in Table 33.2. At both Manhattan and Hutchinson, all cover crops planted in 2009 and harvested in 2010 produced dry matter in both the corn and forage sorghum rotations. At Manhattan, triticale

TABLE 33.2
Dry Matter (kg/ha) of Six Cover Crops Grown for Corn and Forage Sorghum Rotations at Manhattan, Kansas, and Hutchinson, Kansas, in 2010, 2011, and 2012

Rotation	Alfalfa	Austrian Winter Pea	Red Clover	Triticale	Winter Oats	Winter Wheat	Mean of All Cover Crops
				Manhattan, 2010			
Corn	1,763ef[a]	865f	2,074ef	1,969ef	1,813ef	2,569de	1,842b
Forage sorghum	7,150c	3,800d	8,050bc	12,488a	8,800b	7,138c	7,904a
Cover crop mean	4,456b	2,332c	5,062b	7,228a	5,306b	4,853b	
				Hutchinson, 2010			
Corn	1,181	1,443	858	1,769	1,611	1,236	1,499b
Forage sorghum	1,421	2,594	1,375	2,995	3,106	2,171	2,542a
Cover crop mean	1,301b	2,698a	1,116b	2,383a	2,359a	2,267a	
				Manhattan, 2011			
Corn	...[b]	1,561	...	654	836	2,118	1,292
Forage sorghum	...	1,251	...	3,303	2,674	2,519	2,437
Cover crop mean		1,406	...	1,978	2,318	1,755	
				Hutchinson, 2011			
Corn	...	696d	...	935d	910d	1,120d	915b
Forage sorghum	...	1,074d	...	3,124b	2,364c	3,888a	2,612a
Cover crop mean	...	855c	...	2,029b	1,637b	2,504a	
				Manhattan, 2012			
Corn	...	...	...	1,670	...	1,119	1,394
Forage sorghum	...	...	...	1,471	...	892	1,160
Cover crop mean	...	...	...	1,581a	...	974b	
				Hutchinson, 2012			
Corn	...	...	...	1,491	1,475	829	1,265b
Forage sorghum	...	...	...	2,313	2,470	1,271	2,018a
Cover crop mean	...	...	...	1,902a	1,973a	1,050b	

[a] Within each set of means, values followed by the same letter are not significantly different ($\alpha=0.05$).
[b] Cover crop did not produce a stand.

produced the greatest amount of dry matter, which was significantly higher than every other cover crop (12,488 kg/ha). Cover crops grown in a rotation with forage sorghum produced the greatest dry matter. This was because the cover crops were terminated later in the sorghum rotation than the corn rotation, which allowed for greater growth.

In 2011, for both Manhattan and Hutchinson, alfalfa and red clover yielded no dry matter because they did not establish a stand, probably as a result of a dry seedbed. In Manhattan in 2012, alfalfa, Austrian winter pea, red clover, and winter oats did not establish a stand, again probably as a result of a dry seed bed. And, probably for the same reason, at Hutchinson in 2012, alfalfa, Austrian winter pea, and red clover did not establish a stand. For both Manhattan and Hutchinson, the only cover crops that produced dry matter in all 3 years were winter wheat and triticale, with triticale producing more dry matter than winter wheat.

33.4 DISCUSSION

The results showed that winter wheat and triticale were the only two cover crops to grow at both locations (Manhattan and Hutchison, Kansas) in all 3 years. All three legumes did not grow at either location in 2012. Austrian winter pea was the only legume that grew at both locations in 2011. In general, triticale produced the greatest amount of dry matter, because it is well adapted to Kansas climatic conditions, which may be a result of its being a hybrid of rye and wheat, which is well adapted to Kansas. Triticale and winter wheat require vernalization. Vernalization is floral induction by low temperatures (Barnes and Beard, 1992). While vernalization is not necessary for vegetative growth, which was our concern in the research with cover crops, it would be interesting to study the association of vernalization with the survival of winter cover crops in Kansas.

The legume cover crops probably did not grow due to a deficiency of moisture for germination. At cover crop planting time in 2010 and 2011, soil moisture was lacking. At Manhattan, in September–October 2010, rainfall was 45 mm below normal, and, in September–October 2011, it was 61 mm below normal. At Hutchinson at cover crop planting time in 2010 and 2011, in Sept-Oct 2011, rainfall was 91 mm below normal, and, in September–October 2011, it was 79 mm below normal. Even though we had no main crop growing, it should be noted that not only can the cover crop remove moisture preventing the main crop from growing, but also the main crop can remove moisture, preventing the cover crop from germinating.

For a winter cover crop to grow in Kansas, it must be both winter hardy and drought tolerant. Laude (1937) in Manhattan, Kansas, found that a winter rye (*Secale cereale* L. 'Dakold') was more resistant to cold than a winter wheat (*Triticum aestivum* L. 'Kanred'). He did not study triticale. The most winter hardy of winter cereals is winter rye, followed by winter triticale and then winter wheat (personal communication, Mark E. Sorrells, Professor, School of Integrative Plant Science, Plant Breeding and Genetics Section, Cornell University, Ithaca, NY, 4 Aug. 2015). According to Sorrells, the reason winter wheat is cold tolerant is because the crown of winter wheat produces oligosaccharides that serve as osmotica, which allow the plant to survive during the winter. An oligosaccharide is a carbohydrate whose molecules are composed of a small number of monosaccharide units. ("Oligo-" is a combining form that means "few.") Oligosaccharides yield monosaccharides on hydrolysis and contain, per molecule, two to ten monosaccharide residues (White et al., 1964, p. 43). Perhaps the osmotica also were important in providing drought tolerance. The 2 years that the legumes did not grow (2011 and 2012) (Table 33.2) were dry years both in Manhattan and Hutchinson. However, the average temperatures for 2011 and 2012 were above normal. It appeared to be the dry conditions rather than cold conditions that caused the legumes to fail.

We found that, of six cover crops studied, only winter wheat and triticale grown in the winter in rotations of corn or forage sorghum survived during a 3-year study and protected soil during the winter. We conclude that cover crops can sustain soil and prevent wind erosion in the winter in the Great Plains. However, they must be adapted to the local climatic conditions. In Kansas, winter cover crops must be able to germinate in dry soil in the fall.

REFERENCES

Antony, R.M., M.B. Kirkham, T.C. Todd, S.R. Bean, J.D. Wilson, P.R. Armstrong, E. Maghirang, and D.L. Brabec. 2019. Low-temperature tolerance of maize and sorghum seedlings grown under the same environmental conditions. *J. Crop Improvement* 33: 287–305.

Ashley, W.S., S. Strader, D.C. Dziubla, and A. Haberlie. 2015. Driving blind. *Bull. Am. Meteorol. Soc.* 96: 775–778.

Barnes. R.F., and J.B. Beard (Editors). 1992. *Glossary of Crop Science Terms*. Crop Science Society of America, Madison, WI. ii+90 pp.

Clark, A. 2007. *Managing Cover Crops Profitably*. 3rd edition. United States Department of Agriculture, Sustainable Agriculture Research and Education. Handbook Series; book 9. Sustainable Agriculture Network, Beltsville, MD. 244 pp.

Florentín, M.A., Peñalva, M., A. Calegari, and R. Derpsch. 2011. Green manure/cover crops and crop rotation in conservation agriculture on small farms. *Integrated Crop Management* Vol. 12–2010. Plant Production and Protection Division, Food and Agriculture Organization of the United Nations, Rome. 97 pp.

Freeman, O.W. II. 2014. Winter cover crops in corn and forage sorghum rotations in the Great Plains. Ph.D. diss. Kansas State Univ., Manhattan, KS. xviii+145 pp.

King, F.H. 1901. *Physics of Agriculture*. Second edition. Published by the Author, Madison, WI. xvi+604 pp.

Laude, H.H. 1937. Cold resistance of winter wheat, rye, barley, and oats in transition from dormancy to active growth. *J. Agric. Res.* 54: 899–917.

Lynch, J. 2022. Dust storm causes a 21-vehicle pileup in Montana, killing 6 people, including 3 children. https://www.cnn.com/2022/07/16/us/hardin-montana-dust-storm-crash/index.html (retrieved 16 Aug. 2022).

O'Geen, A.T., M.J. Singer, and W. Horwath. 2010. Soil and water conservation for California and the desert southwest: Past, present, and future trends, pp. 237–262. In: T.M. Zobeck and W.F. Schillinger (Editors). *Soil and Water Conservation advances in the United States*. SSSA Special Publication 60. Soil Science Society of America, Madison, WI.

Phillips, W.A., S.C. Rao, R.L. Dalrymple, and E.L. Klepper. 1996. Annual cool-season grasses, pp. 781–802. In: L.E. Moser, D.R. Buxton, and M.D. Casler (Editors). *Cool-Season Forage Grasses*. American Society of Agronomy, Crop Science Society of America, and Soil Science Society of America, Madison, WI.

Roozeboom, K., J. Holman, and J. Jennings. 2012. Cropping systems and cover crops, pp. 18–23. In: D. Presley, D. Shoup, J. Holman, and A. Schlegel (Editors). *Efficient Crop Water Use in Kansas*. Publication No. MF3066. Kansas State University Agricultural Experiment Station and Cooperative Extension Service, Manhattan, KS.

Roozeboom, K. 2013. Cover crop opportunities after wheat in central and eastern Kansas. e-Update No. 407, June 7, 2013. Extension Agronomy, Kansas State University, Manhattan, Kansas.

SAS Institute. 2013. Release 9.3 ed. SAS Institute, Cary, North Carolina.

Snyder, M. 2014. Dust Bowl conditions have returned to Kansas, Oklahoma and North Texas. The Economic Collapse Blog. https://readynutrition.com/resources/dust-in-the-wind-dust-bowl-conditions-have-returned-to-kansas-oklahoma-and-north-texas_27052014 (retrieved 16 Aug. 2022).

Soil Science Society of America. 2008. *Glossary of Soil Science Terms*. Soil Science Society of America, Madison, WI. ii+88 pp.

Sophocleous, M. 1998. Water resources of Kansas: A comprehensive outline, pp. 3–59. In: M. Sophocleous (Editor). *Perspectives on Sustainable Development of Water Resources in Kansas*. Bulletin 239. Kansas Geological Survey, Lawrence, KS.

Stubbendieck, J., and T.A. Jones. 1996. Other cool-season grasses, pp. 765–780. In: L.E. Moser, D.R. Buxton, and M.D. Casler (ed.) *Cool-Season Forage Grasses*. American Society of Agronomy, Crop Science Society of America, and Soil Science Society of America, Madison, WI.

Tanner, C.B., and R.W. Simonson. 1993. Franklin Hiram King – Pioneer scientist. *Soil Sci. Soc. Am. J.* 57: 286–292.

Unger, P.W., M.B. Kirkham, and D.C. Nielsen. 2010. Water conservation for agriculture, pp. 1–45. In: T.M. Zobeck and W.F. Schillinger (Editors). *Soil and Water Conservation Advances in the United States*. SSSA Special Publication 60. Soil Science Society of America, Madison, WI.

White, A., P. Handler, and E.L. Smith. 1964. *Principles of Biochemistry*. Third edition. McGraw-Hill Book Co., New York. 1106 pp.

Index

Note: **Bold** page numbers refer to tables and *italic* page numbers refer to figures.

For Product Safety Concerns and Information please contact our EU
representative GPSR@taylorandfrancis.com
Taylor & Francis Verlag GmbH, Kaufingerstraße 24, 80331 München, Germany

www.ingramcontent.com/pod-product-compliance
Ingram Content Group UK Ltd.
Pitfield, Milton Keynes, MK11 3LW, UK
UKHW052051230926
475702UK00025B/1052
* 9 7 8 0 3 6 7 5 5 4 3 9 2 *